BiologyNow CD-ROM Biology Now
Interactive learning!

W9-AER-220

Take charge of your learning with **BiologyNow™**, a powerful, interactive study tool that will help you manage and maximize your study time. This collection of dynamic technology resources will assess your unique study needs, giving you an individualized learning plan that will enhance your understanding and confidence in the course. Designed to maximize your time investment, **BiologyNow** helps you succeed by focusing your study time on the concepts you need to master. And best of all, **BiologyNow** is FREE with every new copy of this text!

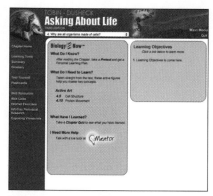

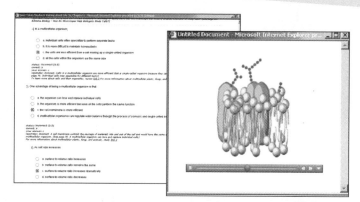

BiologyNow™ directly links with **Asking About Life, Third Edition**. Together, **BiologyNow** and this text enhance each other and provide you with a seamless, integrated learning system.

As you work through the chapters you'll see notes that direct you to the media-enhanced activities on **BiologyNow**. This precise page-by-page integration means you'll spend less time flipping through pages and navigating Web sites for useful exercises. The program's engaging multimedia exercises and examples can make all the difference in your course—after all, it is far easier to understand biology if it's seen in action, and **BiologyNow** enables you to become a part of the action!

You have access to **BiologyNow** by using the free CD-ROM that is packaged with this text. You'll immediately notice the system's easy-to-use, browser-based format. Getting to where you need to go is as easy as a click of the mouse. The **BiologyNow** system is made up of three interrelated parts:

- **How Much Do I Know?**
- **What Do I Need to Learn?**
- **What Have I Learned?**

These three interrelated elements work together, but are distinct enough to allow you the freedom to explore only those assets that meet your personal needs. You can use **BiologyNow** like a traditional Web site, accessing all assets of a particular chapter and exploring on your own.

If you need help at any point, **BiologyNow** includes a convenient link to an exclusive live online tutoring service, **vMentor™**. You will have access, via the Web, to highly qualified tutors with thorough knowledge of this text. When you get stuck on a particular problem or concept, you need only to log on to **vMentor** where you can talk (using your own computer microphone) to **vMentor** tutors who will skillfully guide you through the problem using the whiteboard for illustration.

Explore BiologyNow by using the CD-ROM found in the back of the book!

www.brookscole.com

brookscole.com is the World Wide Web site for
Brooks/Cole and is your direct source to dozens
of online resources.

At *brookscole.com* you can find out about supplements,
demonstration software, and student resources. You can
also send email to many of our authors and preview new
publications and exciting new technologies.

brookscole.com
Changing the way the world learns®

ASKING ABOUT LIFE THIRD EDITION

Allan J. Tobin
University of California, Los Angeles

Jennie Dusheck
Santa Cruz, California

THOMSON

BROOKS/COLE

Australia • Canada • Mexico • Singapore • Spain • United Kingdom • United States

THOMSON

BROOKS/COLE

Executive Editor, Life Sciences: Nedah Rose
Editor in Chief, Michelle Julet
Development Editor: Anne Scanlan-Rohrer
Assistant Editor: Christopher Delgado, Kari Hopperstead
Editorial Assistant: Jennifer Keever, Sarah Lowe
Technology Project Manager: Travis Metz
Marketing Manager: Ann Caven
Marketing Assistant: Sandra Perin
Advertising Project Manager: Linda Yip
Project Manager, Editorial Production: Teri Hyde
Print/Media Buyer: Karen Hunt

Permissions Editor: Joohee Lee
Production Service: Graphic World, Inc.
Text Designer: Jeanne Calabrese
Photo Researcher: Kathleen Olson
Illustrators: Elizabeth Morales, Kimberlee Heldt
Compositor: Graphic World, Inc.
Cover Designer: Larry Didona
Cover Image: Tui de Roy/Minden Pictures
Cover Printer: Transcon-Interglobe
Printer: Transcon-Interglobe

Printed in Canada
1 2 3 4 5 6 7 07 06 05 04 03

For more information about our products, contact us at:
Thomson Learning Academic Resource Center
1-800-423-0563
For permission to use material from this text, contact us by:
Phone: 1-800-730-2214
Fax: 1-800-730-2215
Web: http://www.thomsonrights.com

Library of Congress Control Number:
Student Edition: ISBN 0-534-40653-X
Instructor's Edition: ISBN 0-534-40654-8

Brooks/Cole—Thomson Learning
10 Davis Drive
Belmont, CA 94002
USA

Asia
Thomson Learning
5 Shenton Way #01-01
UIC Building
Singapore 068808

Australia/New Zealand
Thomson Learning
102 Dodds Street
Southbank, Victoria 3006
Australia

Canada
Nelson
1120 Birchmount Road
Toronto, Ontario M1K 5G4
Canada

Europe/Middle East/Africa
Thomson Learning
High Holborn House
50/51 Bedford Row
London WC1R 4LR
United Kingdom

COVER IMAGE, Asking About Tortoises:
These Galápagos Islands tortoises (*Geochelone nigra*) seem to have convened a meeting in the shape of a question mark, and questions are the major theme of *Asking About Life*. What are they asking?

In 1835, when Charles Darwin spent five weeks at the Galápagos Islands, he was amazed by these giant tortoises, which can weigh up to 500 pounds. In his journal, he wrote, "*I frequently got on their backs but I found it very difficult to keep my balance.*" He soon learned that locals could tell which island a tortoise was from by its size and the shape of its shell. Darwin hypothesized that at some point in the past a single species of tortoise had emigrated from South America, more than 500 miles away and then formed separate subspecies on each island.

Recent work on the genetics of Galápagos tortoises has confirmed Darwin's hypothesis. According to this data, *G. nigra* most likely separated from its nearest relative in South America 6 to 10 million years ago and each subspecies most likely evolved on a separate island.

When the Galápagos Islands were first described, more than 200 years before Darwin, explorers said the slopes of the volcanic islands were black with tortoises. Modern biologists estimate that about 250,000 tortoises lived in the islands. But hunting and the introduction of dogs and other domestic animals have reduced the population of tortoises to under 15,000.

Preface

What Is *Asking About Life* About?

Many good conversations start with a question. What do you think of the new restaurant downtown? Is the new sociology professor a good lecturer? Science is one long conversation, like a thousand-year cocktail party. And, as at any party, the topics of conversation often start with questions. Why are plants green? How do cells replicate?

The history of science reveals that the answers to such questions change over time, but the questions themselves, if they are good ones, remain the same. One good question has been, "How do new species form?" Ever since biologists realized that species do form, we have been trying to find out how they do it. But it is a question with many overlapping answers. The answers that seemed correct in 1880 or 1960 continue to be refined, expanded, or even overthrown.

Questions and answers provide a framework. *Asking About Life* consistently emphasizes the importance of questions and the process of finding answers. To remind ourselves and our readers of this emphasis, headings and subheadings are more often questions than statements. In addition, at the end of most subsections we summarize the main points covered. The question heading and the summary statement together act as a reality check for both readers and the authors.

Refining questions and fixing mistakes is the way science works. What scientist hasn't been brought up short by an insightful comment from a colleague? The way questions, or hypotheses, are phrased determines how they can be answered. Sometimes scientists ask the wrong questions. Sometimes they come up with the wrong answers to the right questions. The process is a delightfully social one, in which everyone can play a role.

Real people drive science. Throughout *Asking About Life,* we have emphasized the passionate engagement of individual scientists. Each chapter begins with a story illustrating how an individual biologist or group of biologists pursued a scientific question—often in the face of intense intellectual and social adversity. We tell the story, for example, of how Barry Marshall convinced first himself and then others that bacteria, not "frustrated personalities," cause ulcers; and how Rosalind Franklin struggled in deep social isolation to work out the structure of DNA. Our anecdotes are about real people—their triumphs, their frustrations, their genius, and their persistence. Biology is a story, and, as such, it must be presented as continuously as possible.

In every kind of learning, story and context give shape and meaning to dry facts and ideas. When we are told that someone named Charles Drew invented a way to supply clean blood plasma, a new technology that saved many lives, we may remember this fact briefly. But what if, as happens in Chapter 37, we also learn that Drew was one of only a handful of African-American physicians in the United States in the 1940s? At that time, black physicians did little or no research, but were mostly relegated to quiet country practices in the rural South. Yet Drew ran the "Plasma for Britain" program for the American Red Cross during World War II and invented a way to preserve blood plasma so that it could be shipped across the Atlantic to Great Britain. His work saved the lives of thousands of Londoners injured by Nazi bombs. Midway through the program, however, the American Red Cross decided to refuse blood from any American donor whose skin was black, and Drew resigned in protest. An eminent professor of surgery at Howard University, Drew suffered serious injuries in an auto accident a few years later. Within hours of the accident, he was dead. He bled to death in a small Southern emergency room where neither whole blood nor plasma could save him. Life's too short not to hear the whole story.

Students learn more when they know the whole story. Our philosophy of telling the stories of biology sometimes runs counter to current trends. Many textbooks break up information into modules in which each bit of information is presented dictionary-fashion, as if unrelated to the information in the rest of the book. Important ideas may be marooned on illustration islands. The result is a series of disconnected facts. Some publishers say that such an approach is necessary because today's students belong to the "visual-information generation" and are incapable of sustained reading or synthesis.

We have more faith. We know that, like most people, college students enjoy reading if the reading is interesting and rewarding. Indeed, student reviewers have raved about the clarity of the writing and the engaging stories that reveal scientists as ordinary human beings.

In the evolution section, for example, we mention the intense frustration Charles Darwin experienced while struggling to distinguish among different species of barnacles. Each species seemed to blend into the next. Two chapters later, we show readers that, 200 years later, biologists are still uncertain how to define species. In the context of Darwin's difficulties, the highly politicized debate over whether the red wolf is a species (that deserves legal protection) or a hybrid (that doesn't) takes on a different and deeper meaning. At the same time, the red wolf story brings to life what might otherwise be an abstract discussion of classification.

But even though we stress the continuity of thought, we know that every instructor takes a different approach to teaching biology. Therefore, each unit of the book is understandable on its own terms. Nothing prevents instructors from teaching the units in a different order.

The first edition of *Asking About Life* was well received. The Italian publisher Edizioni Bruno Mondadori selected *Asking About Life* for translation into Italian. And, in 1999, the Text and Academic Authors association conferred on the book an award of excellence for best new textbook in life sciences. But we are most proud of the many instructors and students who have told us they love our book. Thanks to them for their remarkable support.

What Is New in the Third Edition?

In this new edition of *Asking About Life,* we present some new answers and some new approaches. But our starting point, that science is about curiosity, remains the same. We emphasize *how* and *why* scientists ask questions, how they test hypotheses, and how they reach conclusions. As much as possible, rather than merely presenting dry conclusions, we show our readers how science actually works.

Although our philosophy remains the same, we have improved the book in several important ways. We are delighted with the fresh design provided by Robert Hugel. We have made a concerted effort to clarify and simplify explanations and focus on the most important concepts. Through judicious cuts, we've made the third edition substantially shorter than the previous edition.

At the same time, we have added new metaphors and analogies, new lead-in stories, and new art. Some of our metaphors appear in the text alone, while others appear in figures as art or photos. As in the first edition, we compare chromosomes to socks and the functional groups of molecules to the attachments on a Swiss Army knife. But we have also introduced many new metaphors, such as the steel frame of a skyscraper in Figure 35-6, which will remind readers that bone is a structural material. Good visual metaphors help readers remember important concepts.

We have also added new boxed applications on both Health and Business, including, for example, discussions of osteoporosis, diabetes, herbal supplements, and how the discovery of a single steer with mad cow disease affected the beef markets in Canada, the United States, and Japan. Both types of boxes help illustrate the tight connection between biology and everyday life. We hope they appeal to the nonmajor in every reader. Extreme Biology boxes cover the quirky side of biology, from the amazing to the bizarre.

We have streamlined some of the more difficult material on chemistry and greatly improved the art in the animal physiology section, adding, revising, or replacing dozens of figures. Finally, we have once more updated our coverage of ecology and genetics, as well as the classification of the major groups of organisms in the diversity section.

What Kinds of Pedagogy Does Asking About Life Employ?

Asking About Life has a variety of features designed to engage the reader and aid learning. Each chapter begins with a story about a piece of research that draws readers into the subject of the chapter and also introduces one or more important questions or ideas in the area.

At the start of each chapter, readers will also find a list of **Key Questions,** which are the most fundamental questions covered in the chapter. In addition, **question headings** focus the reader's attention on the most significant question to be explored in each major section. Most longer subsections are followed by a **summary statement**—the take-home message. Summary statements provide students with a reality check. If the student doesn't understand the summary statement, that is a cue to study the preceding material more closely.

Visual metaphors rendered as photos or illustrations drive home key points introduced in the text. Figure legends are often written to stand alone, so that a student flipping through the chapter for the first time, glancing at the figures and reading the legends, will learn something. Throughout the text, boldface key terms help students to find important terms and their definitions. At the end of each chapter, all of the boldface terms are used in a highly compressed summary called the **Summary with Key Terms,** organized by the **Key Questions.** The Summary with Key Terms provides students with another opportunity to check their understanding of the chapter. If they encounter terms they don't remember or ideas that seem unfamiliar, they can return to the main text and illustrations. **Key Concepts** further highlight the main ideas.

Following the Summary is a set of **Review and Thought Questions.** Our Thought Questions are especially engaging, frequently bringing ideas from the chapter into the everyday world. Finally, **BiologyNow Resources** at the end of chapter provide references to **Active Figures** and **Post-Test Quizzes** to tie each chapter into the new **BiologyNow CD.**

Supplements

Instructor Supplements:

These supplements are available to qualified adopters. Please consult your local sales representative for details.

Instructor's Manual (0-534-40668-8)

A Guide to Asking About Life for Teachers and TAs, by Donald Cronkite of Hope College, is an instructor's manual for teaching biology using the inquiry-based approach of the textbook. It includes syllabi for teaching, using themes such as biodiversity, cellular and molecular approaches, and social issues; and basic syllabi for one- and two-semester courses. Each chapter offers intriguing demonstrations, alternatives to lecturing, sample handouts, interactive exercises, and group learning exercises. A suggested lecture outline at the beginning of each chapter is annotated to indicate how the different demonstrations, exercises, and activities may be incorporated into your class. Answers to the Review and Discussion Questions from the textbook also appear in the teaching guide.

ExamView® Computerized Testing (0-534-40657-2)
Test Bank (0-534-40667-X)

The *Asking About Life* Test Bank, by Alma Moon Novotny of Rice University, consists of over 3500 questions of assorted type (multiple-choice, matching, fill-in-the-blank, short-answer, and essay questions). The questions are organized by the main chapter headings, and include new items based on the text figures. The third edition test bank now has an additional section with questions that require either integration from more than one area of the text or application of principles presented in the text. We hope that teachers will use these questions, both directly and as a model for writing even better questions of their own. The Computerized Test Bank is available for both Windows and Macintosh platforms. The computer program allows the instructor to sort the questions by chapter, section head, and question type.

Transparency Acetates (0-534-40661-0)

The *Asking About Life* supplements package also includes approximately 200 Overhead Transparencies consisting of new drawings and photos from the book. Many other images from the book are available on the Overhead Transparency set created for the second edition, which can still be ordered using ISBN 0-030-27052-9.

Multimedia Manager (0-534-40660-2)

For the first time ever, the text is accompanied by a powerful **Multimedia Manager**. This full-featured presentation tool includes all of the illustrations and photographs from the text as well as clarifying animations on Microsoft® *PowerPoint®* slides. A unique segmentation feature allows users to select a piece of a figure and blow it up to show more detail. You can even combine a figure with another illustration, or draw on it to highlight ideas during lecture. The Multimedia Manager also includes complete Microsoft® *PowerPoint®* lecture slides—one presentation for each chapter of the book—that you can modify and integrate with the other materials for your unique classroom needs.

CNN® Today Videos: Biology *and* Anatomy & Physiology

(Biology: 0-534-39932-0 • Anatomy & Physiology: 0-534-39930-4)

Launch your lectures with riveting footage from CNN, the world's leading 24-hour global news television network. These exciting videos contain short clips—nearly 300 in all—of high-interest news stories that will spark lively class discussion and generate a deeper understanding of the importance of biological and environmental science in students' day-to-day lives. Includes an accompanying workbook that contains a student worksheet for each video clip.

WebTutor™ on WebCT and Blackboard

(WebCT: 0-534-40664-5 • Blackboard: 0-534-40665-3)

With **WebTutor's** text-specific, preformatted content and total flexibility, you can easily create and manage your own personal Web site. **WebTutor** is a course-management tool for instructors and a study companion for students, and also provides robust communication tools.

Student Supplements:

BiologyNow CD-ROM (0-534-40655-6)

BiologyNow, a new assessment-centered student learning tool for biology and interactive resource, is free with every new copy of the text. **BiologyNow** helps students gauge their particular study needs, then gives them a personalized learning plan that focuses their study time on the concepts and problems that will most enhance their skills and understanding. Through **BiologyNow**, students have access to **vMentor™**, our free, live, online tutoring service that lets students work through problems with highly qualified tutors.

InfoTrac® College Edition

Four months of access to **InfoTrac College Edition** is automatically packaged free with every new copy of this text. This world-class, online university library offers the full text of articles from almost 5,000 scholarly and popular publications—updated daily and going back as much as 22 years. This database of full-length articles (not just abstracts) from thousands of top academic journals and popular sources is ideal for igniting discussions and opening up new worlds of information and research for students.

Book Companion Web Site

http://biology.brookscole.com/AAL3

This free, content-rich Web site includes chapter-specific quizzing, a glossary complete with pronunciations, learning objectives, annotated Web links, **InfoTrac College Edition** access with questions, Internet activities with questions, **Opposing Viewpoint Resource Center** exercises and questions, chapter summaries, and annotated Web links.

Study Guide (0-534-40656-4)

The Study Guide, by Lori Garrett of Danville Area Community College, includes the Key Questions and Key Concepts, an Extended Chapter Outline that gives an overview of the most important topics covered in the chapter, Vocabulary Building exercises, and Chapter Tests that will help students gauge their understanding before an exam. Each Chapter Test has four parts: Multiple Choice, Matching, Short Answer, and Critical Thinking—Using Your Knowledge. All answers are provided, with the exception of the Critical Thinking—Using Your Knowledge questions.

Lab Manual (0-534-40659-9)

Marni Fylling's beautiful, full-color *Laboratory Manual for Asking About Life, Third Edition,* will help students put biology concepts and principles into action. Each of the 15 illustrated labs accompanies a specific chapter (or pair of chapters) from the book and uses readily available materials to give students hands-on, wet-lab experiences. All the labs emphasize basic biological principles. Their interactive design should engage every student's attention.

Who Made This Book?

Jennie Dusheck is a freelance science writer living in Santa Cruz, California. She has been writing for 20 years on topics including biology, geophysics, and astronomy.

Dusheck has written for *Science, Nature, Natural History, Science News, the San Francisco Chronicle,* and other publications. From 1985 to 1993, she was Principal Editor at UC Santa Cruz, where she received several national awards, including a Gold Medal from the National Council for the Advancement and Support of Education. She also coauthored *Life Sciences,* a middle school text in the Holt Science & Technology Series. Her most recent articles—in *Nature* and *Natural History*—concerned the interplay of environment and genome to create phenotype.

Dusheck's undergraduate degree from UC Berkeley is in zoology (1978), as is her master's degree from UC Davis (1983). She also has a certificate in science writing from UC Santa Cruz (1985). Among the subjects of Dusheck's biological research were the effects of light intensity on bird song, social behavior in field mice, competition for food among cattle and three species of deer, and axis formation in *Xenopus laevis.* While working for the Department of Molecular Biology at UC Berkeley, she designed and wrote the protocol for a Space Shuttle experiment that sent live frog embryos into space in the fall of 1992. Her thesis work at UC Davis concerned food and oviposition preferences in skipper butterflies.

Dusheck has taught university classes in science writing, as well as university labs in introductory zoology, embryology, and comparative anatomy. Her current interests—stemming from her early work in ecology and embryology—are in the interplay between ecology and development. She is currently working on another book.

Allan J. Tobin is Director of the Brain Research Institute and the Eleanor Leslie Chair in Neuroscience at UCLA. He has taught introductory biology since 1964, initially as a Teaching Fellow, and then as a faculty member at Harvard and UCLA. A professor of Physiological Science and Neurology, he has taught undergraduate and graduate courses in genetics, cell biology, molecular biology, developmental biology, neuroscience, and neuroengineering—always encouraging students to ask questions and to understand the social nature of scientific inquiry. He is the recipient of a UCLA Faculty Teaching and Service Award.

Tobin's research has similarly crossed disciplinary and departmental boundaries, with the goal of bringing basic science discoveries from the laboratory bench to the patient's bedside. His research topics have included enzymes, red blood cell development, inhibitory signaling in the brain, spinal cord injury, biosensors, juvenile diabetes, and Huntington's disease. Winner of a Javits Neuroscience Investigator Award from the National Institute of Neurological Disorders and Stroke, Tobin has published more than 100 scientific papers and holds more than a dozen patents.

Tobin's undergraduate degree is in literature and biology (MIT, 1963), and his doctoral degree is in biophysics (Harvard, 1969). Tobin did postdoctoral work at the Weizmann Institute of Science, in Israel, and at MIT. He served for nearly 25 years as Scientific Director of the Hereditary Disease Foundation, for which he organized the consortium that identified the gene responsible for Huntington's disease.

Who Else Made This Book?

At the end of this preface are listed all of the reviewers who have helped us improve *Asking About Life.* The quality of this book has depended heavily on their help over the years. But in addition to our hardworking reviewers, we'd also like to thank some of the many other people who helped with this edition of *Asking About Life.*

First, major thanks to our developmental editor, Anne Scanlan-Rohrer, who understood *Asking About Life* from the start and helped us focus a revision that would retain the book's strengths. There was no phase of the project in which she was not involved and highly effective. Her intellectual contributions and moral support were both invaluable. We also thank illustrator Elizabeth Morales for once more helping us make *Asking About Life* the beautiful book it has become; illustrator Kimberlee Heldt for her fresh ideas and boundless enthusiasm for cells, not to mention her suggestions on how to improve the text itself; and photo researcher Kathleen Olson for her ultra-efficient work locating all the right new photos.

Special thanks to science writers Marina Chicurel and Stephen Hart, who made major contributions to this revision of *Asking About Life,* as writers, editors, fact checkers, and friends, and to Harry W. Greene for his friendship, continuing help with the problems of presenting taxonomy, and for his many corrections to the evolution and ecology sections. Thanks also to Kenneth S. Saladin, Darrell J. Moore, Dave Matson, and Ethan Temeles for many important corrections and suggestions; to Scott Gilbert for generously answering our questions; to Alma Novotny for alerting us to mistakes as she revised the test bank; and to freelance developmental editor Anne Scanlan-Rohrer and freelance science writer Bob Holmes for reading the entire book in pages and saving the authors from embarrassing typos, miscalculations, and other mistakes.

Personal thanks to Megan Heller for unflagging moral support and a thousand kinds of help and to Danny Wilkes and Charlie Wilkes for consistently sound advice, perceptive art critiques, help with selecting just the right photographs, not to mention bowls of soup and plates of hot toast.

Thanks to Art Director Robert Hugel, Larry Didona (cover), and Jeanne Calabrese (text) for the beautiful new look for the text and the cover. And thanks to Carol O'Connell and Mike McConnell of Graphic World Inc., not to mention Wadsworth project manager Teri Hyde, all of whom worked like demons to transform a sometimes wobbly manuscript into a beautiful book.

Many thanks also to multimedia expert Travis Metz who managed all things electronic, including the new Biology-Now CD-ROM, as well as assistant editor Kari Hopperstead, who managed the writing and production of all the print ancillaries. Thanks to editorial assistants Rebecca Subity, Jennifer Keever, Sarah Lowe, and Joohee Lee, who kept everything moving smoothly.

Thanks to Wadsworth Executive Editor Nedah Rose and Editor in Chief Michelle Julet, for managing the planning stages and the very few rough spots during production. Thanks also to Pete Marshall and Sean Wakely for taking the time to make sure the project happened.

Finally we once more thank all the reviewers who took the time to read and comment on this manuscript—correcting our errors, asking thought-provoking questions, and suggesting examples, alternative wordings, or new ways of thinking. Although we have met only a few of our reviewers in person, working with them continues to be a rewarding intellectual experience. Both the process of writing this book and the resulting book itself would not have been the same without the reviewers.

Not all our reviewers are listed here. Some preferred to remain anonymous. Some were students who wrote to us on their own, sometimes to ask for clarification. Some were instructors who sent a message by way of a sales representative, and some were casual readers who picked up a copy of *Asking About Life* just to read for fun and wrote to us about their experiences with it. We thank all of you, even if your name is not here. And we welcome other sharp-eyed readers to alert us to possible errors by writing to us at jennie.dusheck@pobox.com. We look forward to hearing from you.

Best wishes,
Jennie Dusheck and Allan J. Tobin

Reviewers for the Third Edition

Tamarah Adair, *Baylor University*

John Aliff, *Georgia Perimeter College*

Sylvester Allred, *Northern Arizona University*

Thomas G. Balgooyen, *San Jose State University*

Kari Benson, *Lynchburg College*

Lesley Blair, *Oregon State University*

Steve Blumenshine, *California State University, Fresno*

Bradley S. Bowden, *Alfred University*

Sara Brenizer, *Shelton State Community College*

George Cline, *Jacksonville State University*

Nancy Cowden, *Lynchburg College*

Prema Dwyer, *LaGuardia Community College*

Robert Evans, *State University of New Jersey Rutgers, Camden*

Richard F. Firenze, *Broome Community College*

Teresa Fulcher, *Pellissippi State Technical Community College*

Lori Garrett, *Danville Area Community College*

Stan Guffey, *University of Tennessee, Knoxville*

Richard Harrison, *Cornell University*

Patricia Hauslein, *Saint Cloud State University*

Chris Haynes, *Shelton State Community College*

Timothy L. Henry, *University of Texas at Arlington*

W. Wyatt Hoback, *University of Nebraska at Kearney*

Laura Jaquish, *Northwestern Michigan College*

Robert M. Kitchin, *University of Wyoming*

Dan E. Krane, *Wright State University*

Mark Lavery, *Oregon State University*

Craig Longtine, *North Hennepin Community College*

Cyndi Maurstad, *Texas A&M University*

Mary Mayhew, *Gainesville College*

Alison M. Mostrom, *University of The Sciences in Philadelphia*

Janine Nelson, *Tulsa Community College*

Vanessa Passler, *Wallace Community College*

David Polcyn, *California State University, San Bernardino*

Scott Porteous, *Fresno City College*

Ann E. Rushing, *Baylor University*

Michael Rutledge, *Middle Tennessee State University*

K. Sata Sathasivan, *University of Texas, Austin*

Heather Smith, *Citrus College*

Linda Smith-Staton, *Pellissippi State Technical Community College*

Robert R. Speed, *Wallace Community College*

Steven R. Strain, *Slippery Rock University*

Eric Strauss, *Boston College*

Emily Willingham, *Texas State University*

Henry H. Ziller, *Southeastern Louisiana University*

Reviewers of Previous Editions

Juan Aninao, *Dominican College of San Rafael*

Edwin A. Arnfield, *Macomb Community College*

Susan Bandoni, *SUNY Geneseo*

Linda W. Barham, *Meridian Community College*

George W. Barlow, *University of California, Berkeley*

Claudia Barreto, *University of Wisconsin, Milwaukee*

Mark S. Blackmore, *Valdosta State University*

Brenda Blackwelder, *Central Piedmont Community College*

Steve Blumenshine, *Arkansas State University*

William Bowen, *Jacksonville State University*

Susan L. Bower, *Pasadena City College*

Bradford Boyer, *Suffolk County Community College*

Mildred Brammer, *Ithaca College*

Sara Brenizer, *Shelton State Community College*

Richard B. Brugam, *Southern Illinois University, Edwardsville*

Arthur L. Buikema, Jr., *Virginia Polytechnic Institute and State University*

Warren Burggren, *University of North Texas*

Naomi Cappuccino, *Carleton University*

Mary Kay Cassani, *Edison Community College, Lee Campus*

Hara Dracon Charlier, *Miami University*

Kerry L. Cheesman, *Capital University*

H. Tak Cheung, *Illinois State University*

Harold Cones, *Christopher Newport University*

Walter Conley, *St. Petersburg Junior College*

Patricia B. Cox, *University of Tennessee*

Charles Creutz, *University of Toledo*

Karen Crombie, *Fresno City College*

Donald Cronkite, *Hope College*

Tom Daniel, *University of Washington*

Forbes Davidson, *Mesa State College*

Jerry D. Davis, *University of Wisconsin, LaCrosse*

David DeGroote, *St. Cloud State University*

Darleen A. DeMason, *University of California, Riverside*

Jean DeSaix, *University of North Carolina, Chapel Hill*

Donald Deters, *Bowling Green State University*

Leah Devlin, *Pennsylvania State University, Abington College*

Matthew Douglas, *Grand Rapids Community College and the University of Kansas*

Ernest F. DuBrul, *University of Toledo*

Peter Ducey, *State University of New York at Cortland*

Patrick Duffie, *Loyola University, Chicago*

Richard P. Elinson, *University of Toronto*

Thomas C. Emmel, *University of Florida*

Carol Erickson, *University of California, Davis*

Kathy McCann Evans, *Reading Area Community College*

Robert Evans, *State University of New Jersey Rutgers, Camden*

Steven H. Everhart, *Campbell University*

Gordon Fain, *University of California, Los Angeles*

Lynn Fancher, *College of DuPage*

Victor Fet, *Marshall University*

Richard Firenze, *Broome Community College*

Cynthia Fitch, *Seattle Pacific University*

James Fitch, *Jones Junior College*

Dietrich Foerstel, *Champlain Regional College & Bishop's University*

Jennifer Fritz, *University of Texas, Austin*

Sally Frost-Mason, *University of Kansas*

Robert Full, *University of California, Berkeley*

Shirley Porteous-Gafford, *Fresno City College*

Lori Garrett, *Danville Area Community College*

Wendy Jean Garrison, *University of Mississippi*

Robert George, *University of Wyoming*

Harry W. Greene, *Cornell University*

Karen F. Greif, *Bryn Mawr College*

Charles J. Grossman, *Xavier University*

Lonnie J. Guralnick, *Western Oregon University*

Ross Hamilton, *Okaloosa-Walton Community College*

Betsy Harris, *Appalachian State University*

Richard Harrison, *Cornell University*

Patricia Hauslein, *Saint Cloud State University*

Chris Haynes, *Shelton State Community College*

Wiley Henderson, *Alabama A&M University*

James A. Hewlett, *Finger Lakes Community College*

Bob Highley, *Bergen Community College*

Nan Ho, *Las Positas College*

Margaret Hollyday, *Bryn Mawr College*

Kathleen L. Hornberger, *Widener University*

Linda Hsu, *Seton Hall University*

Stephen Hudson, *Furman University*

Susan Hughmanick, *University of California, Santa Cruz*

David Inouye, *University of Maryland*

Charles W. Jacobs, *Henry Ford Community College*

John D. Jenkin, *Blinn College, Bryan*

William A. Jensen, *The Ohio State University*

Laura J. Jenski, *Indiana University–Purdue University at Indianapolis*

J. Morris Johnson, *West Oregon State University*

Victoria Johnson, *San Jose State University*

Diane Auer Jones, *Community College of Baltimore County, Catonsville Campus*

Peter Kareiva, *University of Washington*

James Karr, *University of Washington*

Marlene Kayne, *College of New Jersey*

Chris Kellner, *Arkansas Tech University*

Tanseem Khaleel, *Montana State University, Billings*

Robert M. Kitchin, *University of Wyoming*

Ross Koning, *Eastern Connecticut State University*

Dan E. Krane, *Wright State University*

Keith A. Krapf, *John A. Logan College*

James W. Langdon, *University of South Alabama*

Cheryl Laursen, *Eastern Illinois University*

Anton Lawson, *Arizona State College*

Charles Leavell, *Fullerton Community College*

Siu-Lam Lee, *University of Massachusetts, Lowell*

Kathleen Lively, *Marquette University*

Melanie Loo, *California State University, Sacramento*

Thomas Lord, *Indiana University of Pennsylvania*

Jon H. Lowrance, *Lipscomb University*

Linda A. Malmgren, *Franklin Pierce College*

Nilo Marin, *Broward Community College*

Craig E. Martin, *University of Kansas*

Theresa Martin, *College of San Mateo*

Dorrie Matthews, *Sage Jr. College of Albany*

Gary F. McCracken, *University of Tennessee, Knoxville*

Robert J. McDonough, *DeKalb College*

Joseph McGrellis, *Atlantic Community College*

Daniel J. Meinhardt, *St. Olaf College*

John Mertz, *Delaware Valley College*

Timothy D. Metz, *Campbell University*

Debbie Meuler, *Cardinal Stritch College*

Robert Morris, *Widener University*

Alison M. Mostrom, *University of the Sciences in Philadelphia*

Anne-Marie Murray, *Arkansas Tech University*

Ken Nadler, *Michigan State University*

Judy H. Niehaus, *Radford University*

David M. Ogilvie, *The University of Western Ontario*

Bruce Parker, *Utah Valley State College*

Lee R. Parker, *California Polytechnic State University, San Luis Obispo*

Frederick Peabody, *University of South Dakota*

Debra Pearce, *Northern Kentucky University*

Andrew J. Penniman, *Georgia Perimeter College*

Ed Perry, *Faulkner State Community College*

Gary Pettibone, *State College of New York at Buffalo*

Helen K. Pigage, *U.S. Air Force Academy*

Jay Pitocchelli, *Saint Anselm College*

David Polcyn, *California State University, San Bernardino*

Steven M. Pomarico, *Louisiana State University*

Carol Pou, *St. Cloud State University*

Paul A. Rab, *Sinclair Community College*

Eric Rabitoy, *Citrus College*

Paul F. Ramp, *University of Tennessee, Knoxville*

Lynda Randa, *College of DuPage*

David Ribble, *Trinity University*

Laurel Roberts, *University of Pittsburgh*

Franklin Robinson, *Mountain Empire Community College*

Lyndell Robinson, *Lincoln Land Community College*

Frank A. Romano III, *Jacksonville State University*

Earle Rowe, *Walters State Community College*

Donna Rowell, *Holmes Community College*

Christopher S. Sacchi, *Kutztown University*

K. Sata Sathasivan, *University of Texas, Austin*

Andrew Scala, *Dutchess Community College*

Barney Schlinger, *University of California, Los Angeles*

Robert M. Schoch, *Boston University*

Brian W. Schwartz, *Columbus State University*

Shirley Seagle, *Jacksonville State University*

David Seigler, *University of Illinois at Urbana-Champaign*

Angelica P. Seitz, *Auburn University*

Phillip R. Shelp, *Brookhaven College*

John Simpson, *Gadsden State College*

Linda Simpson, *University of North Carolina, Charlotte*

Michael E. Smith, *Valdosta State University*

Cynthia V. Sommer, *University of Wisconsin, Milwaukee*

Fred L. Spangler, *University of Wisconsin, Oshkosh*

Steven Spilatro, *Marietta College*

Herbert Stewart, *Florida Atlantic University*

Cindy Stokes, *Kennesaw State College*

Gail Stratton, *University of Mississippi*

Michael P. Stryker, *University of California, San Francisco*

Robert J. Swanson, *North Hennepin Community College*

Chris Tarp, *Contra Costa College*

Jeffrey Thompson, *California State University, San Bernadino*

Robert Thornton, *University of California, Davis*

Janice Toyoshima, *Bakersfield College*

John Tramontano, *Orange County Community College*

Robert Turgeon, *Cornell University*

Sandra J. Turner, *St. Cloud State University*

Linda Tyson, *Santa Fe Community College*

Michael J. Ulrich, *Elon College*

Kristin Vessay, *Bowling Green University*

Jack Waber, *West Chester University*

Ken Revis-Wagner, *Clemson University*

Timothy S. Wakefield, *Auburn University*

Tom Weeks, *University of Wisconsin, LaCrosse*

J.D. Wilhide, *Arkansas State University*

James A. Winsor, *The Pennsylvania State University, Altoona Campus*

Philip C. Withers, *University of Western Australia*

Daniel Wivagg, *Baylor University*

Edmund B. Wodehouse, *Skyline College*

Tom Worcester, *Mt. Hood Community College*

Anne E. Zayaitz, *Kutztown University*

Henry H. Ziller, *Southeastern Louisiana University*

Contents in Overview

Contents

Contents

How Do Biologists Study Life?

Key Questions

- How do biologists come up with questions and then answer them?

- How are all organisms alike?

- Why are the members of a species alike?

- How do organisms become different?

- How do biologists use statistics to plan and evaluate experiments?

Enough to Give You an Ulcer

In 1984, an obscure Australian physician named Barry Marshall secretly performed a dangerous experiment that would ultimately deprive some of the world's largest corporations of billions of dollars. Marshall's radical idea was born in 1979, when his colleague and friend J. Robin Warren noticed that samples of stomach tissue taken from ulcer patients were often infected with bacteria.

To find one bacterial infection in the stomach would have been strange; to find dozens was bizarre. The human stomach secretes acid so concentrated that few organisms survive it for more than a few minutes, let alone live and reproduce in it. Yet Warren found bacteria flourishing there.

Warren's discovery suggested an alternative to doctors' long-standing belief that ulcers are caused by excess stomach acid. Ulcers, every medical textbook reported, were caused by the oversecretion of stomach acid in people with overanxious, frustrated personalities. Such personality problems were thought to be aggravated by the stressful pace of modern life. But if a bacterium could infect the stomach, Marshall and Warren realized, maybe it could cause ulcers. Intrigued by this idea, Marshall began ordering biopsies for any patient who had stomach problems (Figure 1-1). He found that nearly every patient with ulcers was infected with the same bacterium.

The most common kind of ulcer is a peptic ulcer, an open wound located where the stomach joins the small intestine, at the bottom of the stomach. The word "peptic" comes from the Greek word *peptein*, to digest. Nearly 1 in 10 adults has a peptic ulcer. Some people with ulcers feel no discomfort. But most feel at least mild pain, and many suffer excruciating pain for weeks at a time throughout their adult lives. In rare cases, blood may pour from the wound so freely that the person bleeds to death. The standard treatment for ulcers had always been a bland diet consisting of eggs, milk, creams, custards, and overcooked cereals. In addition, doctors often prescribed tranquilizers, psychotherapy, and, in severe cases, surgery. Mainly, however, doctors prescribed antacids—lots of antacids.

Until recently, prescription antacids were the biggest-selling prescription drugs in the world. In 1992, Americans alone bought $4.4 billion worth. These drugs are remarkably effective at controlling the secretion of stomach acid, but remarkably ineffective at controlling ulcers. Ninety-five percent of ulcer patients have a new ulcer within two years of treatment. That means people who had ulcers took the $100-a-month antacids almost continuously. In a lifetime, an ulcer-sufferer could spend tens of thousands of dollars on antacids.

Yet, if ulcers were caused by a bacterium, as Marshall was suggesting, then a simple two-week course of antibiotics could cure an ulcer forever. Millions of people could be saved from a lifetime of suffering. If Marshall's hunch was right, he had very good news, although not for the companies selling antacids.

In 1983, Marshall decided to present his hypothesis at a scientific conference in Brussels, Belgium. But he had a problem. None of the other researchers had ever heard of him. He was barely out of medical school and he had no credentials at all as a researcher. His idea seemed to have about as much chance of winning recognition as bacteria had of flourishing in the acid environment of the stomach.

Predictably, Marshall's presentation was a disaster. He was unknown; he was young, inexperienced, and overexcited. Worst of all, he had what looked like a screwball idea. "He didn't have the demeanor of a scientist," recalled Martin Blaser, professor of medicine at Vanderbilt University. "He was strutting around the stage. I thought, this guy is nuts." When Marshall's presentation was over, his audience of eminent medical researchers shifted uneasily in their seats, embarrassed. A few laughed. They couldn't believe he was serious. Most bacteria can barely survive a brief passage through the stomach. How could they flourish there for months or years?

In any case, Marshall had no scientific evidence to back up his claim. Maybe, his audience told him, the bacteria had contaminated the stomach samples after the stomach tissues had been removed. Or maybe the bacteria were harmless and unrelated to the ulcers. Or maybe the bacteria were only able to grow in the stomach because the ulcer was there. The audience peppered him with challenging questions, and Marshall realized he couldn't answer any of them.

The only way to settle all these questions was to study the bacterium in an experimental animal. But to do that, Marshall would need to find an animal whose stomach could be infected with the bacterium. After returning to Australia, he began feeding rats the still-unnamed organism. The bacteria all died in the rats' stomachs without having any effect. He fed

Courtesy, Dr. Barry Marshall

Figure 1-1
An independent thinker. Barry Marshall showed that bacteria, not cranky personalities, cause ulcers.

Biology ⓢ Now™ Learn more about Barry Marshall's discovery by clicking on this figure on your BiologyNow CD-ROM.

the stuff to pigs, with the same result. Now he began to wonder, could the bacteria really infect a stomach? Maybe the researchers in Brussels had been right to laugh at him.

Desperate to prove that he was no nut, Marshall planned a highly unusual "experiment." He told no one ahead of time—not the medical ethics board at the hospital, not his wife. They wouldn't have approved, he knew. He began with a stomach exam and biopsy to make sure his stomach was healthy. Then he made himself an "ulcer bug" cocktail containing at least a billion bacteria, grown from bacteria he had isolated from biopsies of ulcer patients. Then, in a few swift gulps, he drank it down. The cocktail was enough, he hoped, to infect his stomach.

Nothing happened. Days passed. Then, eight days later, nausea woke him early and he vomited. For another week he was tired, irritable, and hoarse. He had headaches and foul breath. A second stomach exam and biopsy showed that his stomach was inflamed and swarming with bacteria.

By the third week, Marshall was lucky enough to have recovered completely. In the April 15, 1985, issue of *The Medical Journal of Australia*, Marshall reported the results of his trial. He had not proved that the bacterium could cause ulcers, or even that it could infect the stomach for years at a time. But he had done something that strongly suggested that the bacterium, still unnamed, could infect a healthy human stomach—one that didn't already have an ulcer.

His disastrous debut in Brussels had ensured that other researchers would remember him. They soon began to take an interest in Marshall's idea. Mainly, they were interested in proving him *wrong*. Yet, by the end of the 1980s, the evidence that the bacterium could infect the stomach was unassailable. The bacterium, they found, has a twisted, helical shape and usually lives in the "pylorus," near the bottom of the stomach. *Helicobacter pylori*, as it was finally named in 1989, twisted down into the mucous lining of the stomach and settled there.

By 1993, medical researchers had found that about 80 percent of all ulcers were caused by *Helicobacter pylori*. Nearly all could be swiftly cured with ordinary antibiotics. (The data for these conclusions came not from laboratory experiments but from the painstaking collection of statistics from doctors and health clinics all over the world.)

1

At first, drug companies resisted Marshall's idea. It was true that people treated with $20 worth of antibiotics had a relapse rate of only five or ten percent. It was true that millions of ulcer sufferers would each spend thousands of dollars less on prescription antacids. But, gradually, they began to see a silver lining in the cloud the Australian had created.

The good news, for drug companies, was that infection by *H. pylori* was one of the most common bacterial infections in the world. As many as half of all people worldwide harbor the infection. And studies of war veterans showed that people infected by *H. pylori* had six times the rate of stomach cancer as other people. Here was a potential market for antibiotics consisting of half the population of the world, some 3 billion people.

Drug companies hastily developed diagnostic tests for *H. pylori* and new combinations of antibiotics to treat the infection. Biotechnology companies were trying to develop a vaccine, to be given in childhood, that would protect against *H. pylori*, ulcers, and maybe even stomach cancer, the world's second leading cause of cancer death (after lung cancer).

Early in 1994, the National Institutes of Health declared antibiotics the official treatment for most ulcers. Eleven years after Marshall first presented his hypothesis, doctors could offer their ulcer patients a permanent cure. Even then, change came slowly. As late as 1996, only one-third of doctors in the United States were prescribing antibiotics for ulcer patients. The rest continued to prescribe antacids only. Today, antibiotics are standard treatment for this widespread infection and every family doctor knows what causes ulcers.

In the end, Marshall won full acceptance as a scientist. Ultimately, he moved to the United States and joined the faculty at the University of Virginia Medical School. Marshall succeeded because he possessed many of the attributes of a good scientist. He had curiosity, intelligence, vision, and the dogged determination to pursue an idea—even when his stubbornness made him appear foolish. Perhaps most important, Marshall displayed an unusual independence of thought that allowed him to pursue an idea unimaginable to more dogmatic thinkers.

In this book we will meet many scientists who are as curious, independent, and stubborn as Marshall. Some are impulsive, like Marshall. Others take years to reach their conclusions. Most work long hours for years on end. A very few seem merely to play at science, reaping brilliant discoveries from a few hours' work. We'll see some of them risk their reputations to defend the ideas they believe in. A few even lose their lives. Right or wrong, all are fascinated by questions about what makes living things tick. Good biologists, like other scientists, are people who are intensely engaged with life. And, like all people, they work in intense social and political environments.

In this first chapter of *Asking About Life*, we will examine two important aspects of **biology** [Greek, *bios* = life, *logos* = word, argument], the study of life. We will see how biologists try to answer questions about life, and we will try to define life itself.

Kenneth H. Thomas/Photo Researchers, Inc.

Figure 1-2
Inquiry. Many scientific questions begin when something beautiful or intriguing catches our eye. Here morning dew bedecks a spiderweb. A person with a questioning mind might stop and admire such a web and then wonder: How does dew form? How do spiders build such intricate webs? Why is the web built so that each crosspiece sags under the weight of the dew like a string of pearls?

1.1 How Do Biologists and Other Scientists Ask Questions?

Biology ⊚ Now™ Seeing BiologyNow throughout the text indicates an opportunity for you to test yourself on key concepts, and to explore animations and interactions on our BiologyNow CD-ROM.

From early childhood, everyone asks questions about life: What makes me alive? What goes wrong when I am sick? Where did I come from? How am I like other living things?

The beginning of all scientific inquiry is curiosity. Something in the world captures our attention, and we begin to ask questions. The complex lives of bees may inspire in us pure wonderment. Or perhaps the slow unfolding of a rose arrests our eye. Sometimes, a practical problem commands our attention. Maybe we would like to see a dying friend get well or a hungry village grow enough food to feed its children. Whatever engages our minds—whether beauty, complexity, or misfortune—can be the beginning of scientific inquiry (Figure 1-2).

The physicist Albert Einstein attributed his scientific achievements to his childlike curiosity. Einstein asked the kinds of questions that most of us ask when we are children. But unlike most people, he continued asking questions when he had the intellectual power to answer them.

Many scientists pride themselves on their childlike curiosity. But scientific inquiry is more than asking questions. Science is curiosity that is controlled and channeled. For example, when Barry Marshall asked whether a bacterium could cause ulcers, he knew it was a question that could be answered. Good scientists try to channel their energy into questions that can be answered.

Do Scientists Use the Scientific Method?

In the first part of this chapter, we will discuss how scientists choose questions to ask, and how scientists answer those questions. Because science has been so successful in increasing knowledge, many philosophers and historians have tried to describe how scientists work. One description of the way scientists work is the **scientific method,** a set of formal rules for expressing, testing, and eliminating ideas, or hypotheses. In reality, scientists do not think of themselves as following the rules of this or any other single "method." Rather, they pursue knowledge in a variety of individual and creative ways. Nonetheless, we can give a general description of how most scientists go about learning about the world.

The scientific method refers to a formal set of rules for forming and testing hypotheses. Most scientists use the scientific method only loosely and often unconsciously.

How Do Scientists Begin?

A scientist's first step in understanding something is to focus on a single question or small set of questions. Because science, especially biology, works in tiny steps, scientists rarely answer big, general questions all at once. A scientist does not ask a broad question such as, "What causes ulcers?" and expect to come up with a quick, simple answer to that question.

When we ask how something works, whether it be a toaster or a stomach, we generally want to know how the parts operate. The first two steps are observation and asking a question. Sometimes a scientist observes, then asks a question, then observes some more. Sometimes a scientist just asks a question, then observes. Looking, hearing, smelling, and touching may all contribute to our observations.

In science, the first two steps are to ask a sharply focused question about a process and to observe closely, looking for a pattern that seems to make sense. Sometimes curious observation comes first, sometimes the question comes first.

What Does a Hypothesis Do?

Once we have observed a process, we may have an idea of how it might work. Some scientists may formalize this hunch in a "model," which may be just an abstract idea or a physical working model of a process or structure. Whatever the case, the model or idea must be consistent with what we already know and should suggest things we might not have thought of otherwise. A good model should suggest a **hypothesis**—a possible explanation for the way a process works that allows us to make predictions. Scientists could not work without hypotheses. Hypotheses are as essential to the daily functioning of science as vitamins are to the normal functioning of our bodies. So where do scientists get hypotheses?

In general, if the predictions of a hypothesis are right, we begin to guess that we are on the right track. On the other hand, if the predictions of a hypothesis are wrong, then the hypothesis is likely wrong.

A hypothesis is an informed guess or possible explanation that enables a scientist to make predictions. A hypothesis is a product of logic, previous knowledge, insight, and creativity.

What Is a Testable Hypothesis?

We can quickly generate lots of hypotheses to explain any observation. If we note that ants in the kitchen are crowding around the sink, we might hypothesize: (1) that they are looking for water, (2) that they are looking for food particles, (3) that they are looking for a new nest site, or (4) that they want to clean the dishes for us. Clearly, some hypotheses are better than others. Scientists need ways of deciding which hypotheses to explore.

In general, a hypothesis is valuable only when it is **testable,** meaning that someone can devise an experiment or a set of observations that would disprove the hypothesis if it were incorrect. If we hypothesize that birds evolved from dinosaurs, then we have to be able to go to the fossil record and show that no birds existed before dinosaurs.

Whether a hypothesis is testable does not depend on whether it is correct. The hypothesis that bears eat berries is testable. The hypothesis that bears eat rocks is equally testable. Testable (or *disprovable*) means that we can make a prediction that we can test by experiment.

The hypothesis that extraterrestrial beings influence the movement of ants in your kitchen is untestable—but *not* because it is wrong. We have no way to prove that it is wrong. This hypothesis is untestable—and therefore of no interest to science—because designing an experiment in which the influence of the alleged extraterrestrials is ruled out, for a control, is impossible. We can't prevent their alleged influence because we have no information about what kinds of

Extreme Biology Reductionism and Emergent Properties

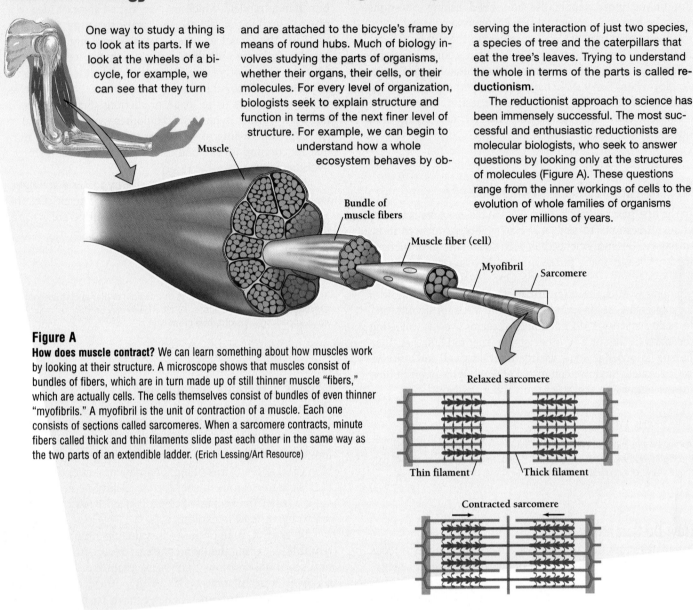

One way to study a thing is to look at its parts. If we look at the wheels of a bicycle, for example, we can see that they turn and are attached to the bicycle's frame by means of round hubs. Much of biology involves studying the parts of organisms, whether their organs, their cells, or their molecules. For every level of organization, biologists seek to explain structure and function in terms of the next finer level of structure. For example, we can begin to understand how a whole ecosystem behaves by observing the interaction of just two species, a species of tree and the caterpillars that eat the tree's leaves. Trying to understand the whole in terms of the parts is called **reductionism.**

The reductionist approach to science has been immensely successful. The most successful and enthusiastic reductionists are molecular biologists, who seek to answer questions by looking only at the structures of molecules (Figure A). These questions range from the inner workings of cells to the evolution of whole families of organisms over millions of years.

Figure A
How does muscle contract? We can learn something about how muscles work by looking at their structure. A microscope shows that muscles consist of bundles of fibers, which are in turn made up of still thinner muscle "fibers," which are actually cells. The cells themselves consist of bundles of even thinner "myofibrils." A myofibril is the unit of contraction of a muscle. Each one consists of sections called sarcomeres. When a sarcomere contracts, minute fibers called thick and thin filaments slide past each other in the same way as the two parts of an extendible ladder. (Erich Lessing/Art Resource)

influence extraterrestrials might exert. We have no reliable observations of extraterrestrials. The same arguments apply to deities.

A hypothesis is valuable only if it is testable.

What Makes a Good Experiment?

An **experiment** is a way of testing a hypothesis or of searching for some unknown effect. An experiment is a procedure that is carried out under "controlled" conditions.

Suppose, for example, that we observe that people who live in countries where cinnamon and other spices are used extensively are shorter than other people. Based on this observation, we hypothesize that cinnamon inhibits growth. We decide to test this hypothesis by measuring cinnamon's effect on the growth of laboratory mice.

We inject baby mice with extract of cinnamon to see if the cinnamon slows their growth. We find that their growth is slow compared with that of mice in other studies. Can we conclude that cinnamon slowed their growth? We cannot. Maybe we were not feeding the mice enough food. Maybe the room they are in is too hot. Maybe chemicals in the extract other than the cinnamon are slowing their growth. The

But reductionism has definite limits. All of the properties of an object are not explainable in terms of the object's parts. To understand how muscles contract, we need to know how animals use them and how they attach to bones. Looking at the individual molecules in a muscle would tell us very little if we had never watched a living animal.

A whole is greater than the sum of its parts. The painting in Figure B illustrates this principle simply. When we examine the painting under high magnification, we see only an abstract pattern of colored dots. It is only when we step back and look at the whole picture at once that we see a river, trees, and flowered hillsides. The landscape is an **emergent property**, a characteristic that arises only at complex levels of organization.

Living organisms are no different. Just as a painting is much more than a collection of brushstrokes, a group of organisms is more than a collection of individuals. A detailed knowledge of the behavior of solitary humans, for example, would never allow us to predict such bizarre group behaviors as war and Tupperware parties. Life is an emergent property of certain combinations of molecules.

Figure B

Art as emergent property A landscape painting is an emergent property—a characteristic that arises only at complex levels of organization. When we look at the individual brushstrokes, we see only an abstract pattern of colored dots. When we step back and look at the whole picture at once, we see a landscape—*La Seine à Herblay,* by Maximilien Luce (Musée d'Orsay, Paris).

outcome of this experiment could depend on many causes besides the cinnamon.

To ensure that the outcome of an experiment depends only on the proposed cause, biologists compare the results of every experiment with the results of a "control" experiment. A **control** is a version of the experiment in which everything is the same except for the one thing being tested. To test our cinnamon hypothesis properly, for example, we need at least two controls. First, we need, in our own lab, a group of mice that receive no injection. These will show us how fast mice normally grow under the conditions in our laboratory (not someone else's). Second, we need a group of mice each of which receive an in-

jection containing everything in the extract except the cinnamon. This second control group will show whether something else in the extract slows the growth of baby mice.

The design of a control is critically important and not always obvious. For example, the trauma of being handled might be enough to slow a mouse's growth. To make sure that is not the case, we might want to include a third control in which researchers merely pretend to give baby mice injections—sticking them with a needle, but not injecting anything. This might sound silly, but we know that both humans and other animals respond in both positive and negative ways to handling. It is good science to test anything that may affect our results. Because an experiment must distinguish causes from irrelevant factors, an ideal experiment allows a researcher to evaluate the effects of one factor at a time.

When Barry Marshall swallowed his mixture of stomach bacteria, he was performing an experiment—but only just barely. It was not a good, controlled experiment. His symptoms—nausea, tiredness, and bad breath—might have been caused by the flu, a sinus infection, or any number of things. Even the inflammation of his stomach shortly after he drank the ulcer bug cocktail might have been a coincidence.

Experiments on a single individual may be strongly suggestive, as Marshall's was. But such experiments do not provide results that mean anything scientifically. To do the experiment properly, Marshall would have had to persuade at least three groups of people to participate. The first group, the "experimental" group, would have gotten the ulcer bug cocktail. The second group would have gotten a different bacterium—one not believed to cause ulcers—to rule out the possibility that any concentrated mixture of bacteria can make someone sick. The third group would have gotten the same mixture except without any bacteria, to rule out the possibility that the cocktail—even without any bacteria—is itself a sickening mixture. For most scientists, such an experiment would have been very difficult to do because of strict regulations governing experiments involving people ("human subjects"). For Marshall, who just wanted a quick idea of whether he might be on the right track, it was far easier to "experiment" on himself.

An experiment must include carefully designed controls.

What Is a Theory?

When different experiments fail to disprove the predictions of a hypothesis, scientists gain confidence in the hypothesis. Scientists often call a group of related hypotheses a **theory**—a system of statements and ideas that explains a group of facts or phenomena. For example, "atomic theory" includes all the ideas connected with the fact that all matter is made of tiny atoms. "Evolutionary theory" includes all the ideas connected with the facts that living organisms have descended from a common ancestor, are related to one another, and have evolved over time.

Scientists use the word "theory" differently from many nonscientists. A nonscientist might use the word theory for any half-baked idea. Joe has a "theory" that if he reads every other paragraph in this book, he'll get an A on the final. To a scientist, that's only a hypothesis, and one that needs testing. It doesn't rate as a formal theory. A scientific theory is a set of interconnected hypotheses that have withstood rigorous experimental testing.

A set of related hypotheses that consistently resist scientists' attempts to disprove them may become recognized as a formal theory.

1.2 How Are All Organisms Alike?

Everywhere on Earth, there is life: from the depths of the Pacific Ocean to the top of Mt. Everest, from the frozen wastes of the Arctic Ocean to the tropical rain forests of the Amazon and the arid Sahara Desert. For each living thing, the same questions arise: How is it put together? How does it work? How did it get here? In answering these questions, biologists have accumulated millions of individual facts (only a tiny fraction of which we will study). Tying these disconnected facts together, however, are a few basic themes. Once we grasp these few generalities, the study of life becomes much easier.

Since biology is the study of life, the first general question we must answer is, What is life? Most of the time we can distinguish the living from the nonliving as easily as we can tell a live bear from a teddy bear. Although agreeing on a definition of life that is both precise and general is harder, we can agree on some common characteristics of all living things. Figure 1-3 lists and illustrates eight features of all living organisms.

In summary, all living things are organized into cells and other parts, perform chemical reactions, obtain energy from their surroundings, change over their lifetimes, respond to their environments, reproduce, and share a common evolutionary history.

How Do Organisms Self-Regulate?

To survive, organisms need to keep their insides separate and different from the outside world. The first step to doing this is to create a barrier. In the case of our bodies, it is our skin. In the case of our cells, it is the cell membrane.

The second step is to actively resist change. The process of resisting certain kinds of change is called **homeostasis** [Greek, *homeo* = same + *stasis* = standing still]. Homeostasis always involves two stages: (1) detecting change and (2) counteracting such change. For example, birds and mammals maintain a remarkably constant internal temperature. A person can be exposed to temperatures as low as −55°F or as high as 140°F and still maintain a body temperature of between 98°F and 99°F.

When our body temperature rises, we perspire, drink water, and look for a cool place to rest. When our body tem-

A. Living things consist of organized parts, as illustrated by this beautiful chambered nautilus. The word "organism" has the same root as "organization."

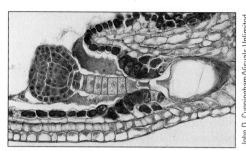

B. All organisms are made of cells, as illustrated by this plant embryo. Just as the atom is a unit of matter, a cell is the unit of life.

C. Organisms perform chemical reactions. Honeybees convert flower nectar into honey.

D. Living organisms obtain energy from their surroundings. Sunlight is the ultimate source of energy for most organisms. Plants obtain energy directly from the sun. Animals obtain energy from plants, or from animals that have eaten plants.

E. Organisms respond to their environments. The plants in this pot are bending toward a sunlit window.

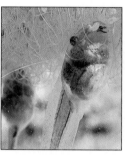

F. Organisms mature over time. A tadpole maintains its organization while developing into a frog.

G. Organisims reproduce. Each organism, including each of these rabbits, comes from another organism or from a pair of organisms.

H. Organisms share a common evolutionary history. All organisms on Earth today are descended from organisms that lived in earlier times. Some have changed dramatically. Others have remained the same. The horseshoe crab of Atlantic beaches shown here is nearly identical to one that lived 200 million years ago.

Figure 1-3
Some characteristics of living organisms.

perature drops, we conserve heat by not perspiring and by reducing blood flow to the skin, hands, and feet. In addition, we increase heat production by moving around, shivering, or increasing our metabolic rate. And, of course, we can also put on warm clothes or turn up the heater. All of these responses help keep our temperature between 98°F and 99°F.

Animals regulate everything from body temperature, thirst, and hunger, to sperm production, while plants regulate their intake of water and carbon. Of course, animals also mature as they age and also tolerate all kinds of other changes, a phenomenon called "tolerance." For example, some tiny animals that live on Antarctic beaches that freeze every time the tide goes out can tolerate freezing and thawing twice a day.

All organisms resist (certain kinds of) change.

What Kinds of Cells Are Organisms Made of?

Whether dandelion or dachshund, *Helicobacter pylori* or bathroom mold, we organisms are all made of cells. A cell is a tiny mass of water, protein, and other molecules, all surrounded by a thin membrane. Indeed, all cells are made from the same few molecular building blocks. But a cell is much more than a bag of protein and water. It can do all of the things listed in Figure 1-3, including reproduce itself.

Cells are wonderfully diverse and specialized. In our own skin, flat "epithelial" cells protect our body, "follicle" cells produce hairs, and nerve cells carry messages. Plant cells are similarly specialized. Tough outer cells protect the plant's body, others transform light energy into sugar, and still others form passages that carry sugar, water, and other materials throughout the plant. There is no easy way to count all the different kinds of cells. Mammals alone have at least 200 different kinds of cells (Figure 1-4).

Nonetheless, we can group cells into just two basic kinds. Our own cells are eukaryotic cells. **Eukaryotic** cells [Greek, *eu* = true + *karyon* = nucleus] contain a **nucleus,** which is a package of genetic material enclosed in a thin membrane. Eukaryotic cells contain about 20 kinds of little membrane-enclosed parts called **organelles** [little organs]. These include the energy-supplying mitochondria and chloroplasts (in plants only). Plants, animals, fungi, and the microscopic protists are all made of eukaryotic cells.

The second kind of cell is the **prokaryotic** cell [Greek, *pro* = before]. Prokaryotic cells contain no nuclei and no membrane-enclosed organelles. They have genetic material, but it is not enclosed in a membrane. All prokaryotic cells are kinds of bacteria and all bacteria are prokaryotic cells. So far, biologists have named only 4,800 species of prokaryotes, compared to more than a million species of eukaryotes.

Yet if the tiny prokaryotes are few in kind, they are great in number. More bacterial cells live in our intestines, for example, than we have cells in our bodies. And it is from the prokaryotes that all eukaryotes—including humans—are descended.

Eukaryotic cells have a nucleus and other organelles, each surrounded by a thin membrane. Prokaryotic cells lack a nucleus or other organelles.

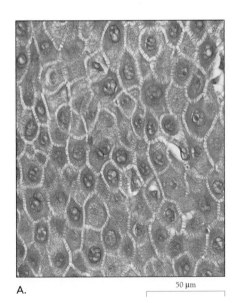 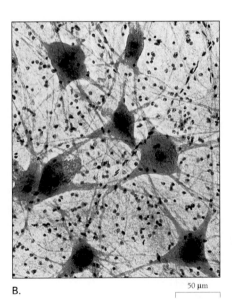 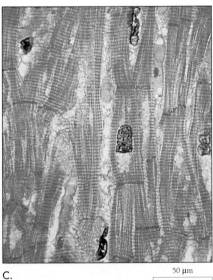

A.   50 μm   B.   50 μm   C.   50 μm

A, G.W. Willis, M.D./Biological Photo Service; B, Manfred Kage/Peter Arnold, Inc.; C, Ed Reschke/Peter Arnold, Inc.

Figure 1-4
Every organism is composed of many kinds of cells. Humans consist of some 200 kinds of cells. Here are just three kinds. A. Human skin cells. B. Nerve cells from the spinal cord. C. Cells of heart muscle.

1.3 How Are Groups of Organisms Different from One Another?

During the last 4.5 billion years, our planet has changed from a lifeless sphere of rock and water to a rich, green world alive with fabulously diverse organisms. This transformation is the result of **evolution**—the process by which species have arisen and changed as they descended from common ancestors.

Organisms come in many sizes, shapes, and colors. They range in size from *Chlamydia*, one of the smallest species of bacteria, to California's coast redwoods, the most massive plants in the world. Organisms range in shape from streamlined dolphins to prickly cacti, and in color from the softest brown moth to the brightest red poppy. The more we look, the more different forms of life we see. Biologists now divide these many organisms into three large categories, called **domains** (Figure 1-5).

The first two domains, **Archaea** and **Eubacteria**, comprise all the prokaryotes, commonly called bacteria. The great majority of these simple, single-celled organisms are free living. Only a handful of bacteria infect humans or other organisms. Each domain is further divided into **kingdoms**.

The third and last domain is **Eukarya**, which consists of four kingdoms. The two biggest kingdoms are the **Plantae** and the **Animalia**, the plants and animals. All plants and animals are multicellular. The approximately 250,000 species of plants range from tiny mosses and ferns to giant redwood and mahogany trees.

Animals are even more diverse, with well over a million known species. Yet, only about five percent of animal species are what many people consider animals—birds, mammals, fish, reptiles, and amphibians. About 200,000 animal species are kinds of sponges, jellyfish, worms, spiders, snails, or sea urchins. But the vast majority of animals, some 800,000 species, are insects. Beetles alone constitute about 30 percent of all known animals, with about 300,000 species.

The third kingdom in the domain Eukarya is the **Fungi**, with just under 70,000 named species. The fungi (plural for "fungus") include both multicellular organisms (such as mushrooms) and single-celled organisms (such as yeast). The fourth kingdom is the **Protista**, most of which are single-celled. These tiny organisms include more than 30,000 species. Protists include different kinds of amebas, such as those that live in pond water; dinoflagellates, some of which cause poisonous red tides in coastal waters; the delicate and lacy foraminifera, and numberless algae, water molds, and slime molds.

Altogether, the named species of organisms within these three domains total about 1.4 million. Biologists believe, however, that most living species have yet to be discovered and named. The world may hold as many as 10 million or even 100 million species. For example, a project to count all the species in Great Smoky Mountains National Park has added more than 100 species new to science and more than 1,000 species no one knew lived in the park. No one knows how many species the world holds. Whatever the current to-

tal, however, it is less than one percent of all the species that have ever lived, since life first appeared on Earth almost 4 billion years ago. The vast majority of species—in all their diversity—have gone extinct over the billions of years since life first appeared on Earth.

Life is enormously diverse. About 1.4 million living species have names, but millions more have yet to be discovered or are extinct.

Why Are the Members of a Species Alike?

Even though organisms are enormously diverse, all share the same chemical building blocks and the same cellular organization. Such common characteristics have persuaded biologists that all organisms are genetically related and are descended from a common ancestor.

Why Do Children Look Like Their Parents?

A visitor to a museum may be surprised to find that a cat living in ancient Egypt, 6,000 years ago, looked just like a house cat from Houston. Species usually change very little from one generation to the next. Just as today's children look much like those born 30 years ago, they also closely resemble children born 3,000 years ago.

Children look like their parents because every individual inherits a set of **genes** [Greek, *gen* = to produce], which carry detailed information about the molecular parts of an organism.

Genes transmit information from generation to generation.

What Are Genes Made of?

As we will discuss in detail later in this book, biologists have learned that organisms store genetic information in long thin molecules called **deoxyribonucleic acid (DNA)**. DNA contains coded instructions for building different kinds of proteins. A cell can read each gene and translate it into a protein. The DNA of humans, for example, contains about 42,000 genes, most of which code for proteins. A set of genes—called a genome—is in many ways like the list of ingredients in a recipe. But the genes are a recipe for building an organism. Instead of calling for "two egg whites," the genome calls for "35,000 molecules of the protein albumin" (the major ingredient in egg white).

Genes are the means by which organisms transmit the information needed to build and maintain new individuals.

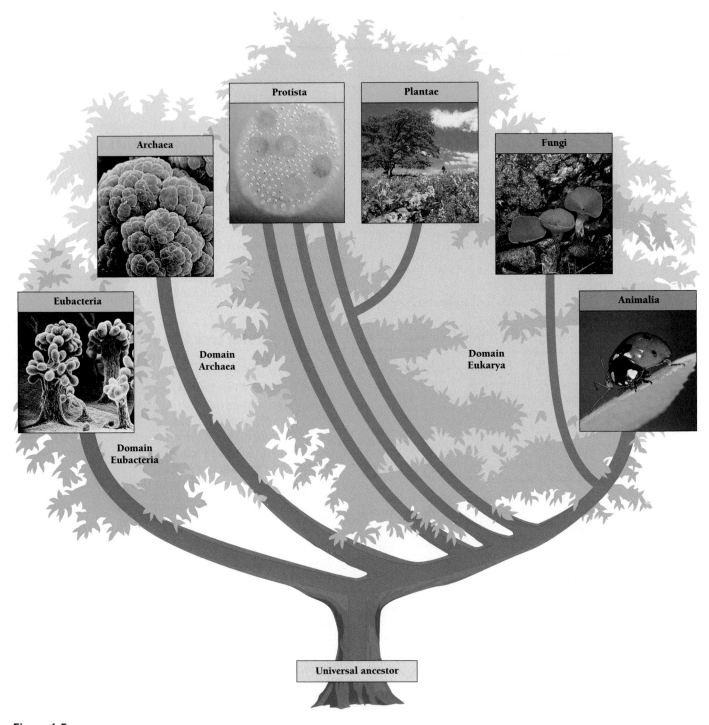

Figure 1-5

Three domains of organisms. At left are the first two domains, two very different kinds of bacteria. The first domain, the Eubacteria, is represented here by aggregating myxobacteria. The second domain, the Archaea, is represented by a colony of methanobacteria, *Methanosarcina*. The third domain, the Eukarya, include four kingdoms, the Protista, Plantae, Fungi, and Animalia. A colony of one-celled organisms called *Volvox* represents the Protista. An oak tree surrounded by wild flowers (arrowleaf balsam root, paintbrush, and lupine) represents the Plantae. The stalked scarlet cup, *Sarcoscypha occidentalis*, represents the Fungi. And the seven-spotted lady beetle represents the Animalia. (*Myxobacteria,* Patricia L. Grilione/ Phototake; Methanosarcina, R. Robinson/Visuals Unlimited; *Volvox,* Biological Photo Service; oak tree, F. Stuart Westmorland/Photo Researchers; stalked scarlet cup, R.M. Meadows/Peter Arnold, Inc.; lady beetle, Runk/Schoenberger from Grant Heilman)

Species Differ in Their Adaptations to Distinct Environments

The members of a species share most of the same genes (and proteins). As a result, all the members of a species share characteristic **adaptations**, special structures or behaviors that fit individuals for life in a particular environment. All adaptations are inherited. Adaptations increase the chance that an individual will live and reproduce. In Figure 1-5, the hard carapace-like upper wings of the spotted lady beetle have evolved to protect the delicate flying wings beneath. Its habit of attacking and eating aphids provides it with a steady supply of nutrients and energy.

Size, form, color, internal structure, and special chemical properties all contribute to an organism's ability to survive in a particular environment and to reproduce more of its own kind. Despite the differences among individuals within a species, all the members of a species are similarly adapted for a particular way of life. Given the opportunity, all cows stand around eating grass, all polar bears hunt seals, all kittens pounce on small objects, and all human children laugh and chase one another.

Life is diverse because different species have different sets of adaptations. The adaptations prompt biologists to ask questions such as, How does a particular structure or behavior help this organism to survive and reproduce? Put more concretely, What good is an elephant's trunk, a mallard's mating dance, or a cactus's thick stems? Of ourselves we might ask, Of what use are a large brain and hairless skin? These are the same general questions that we might ask about the parts of a sewing machine or a car engine: What does it do? How else could it be done?

Because every species has so many adaptations, the same general questions arise again and again. After a while, we might begin to wonder how such a variety of adaptations came to be in the first place. As we mentioned earlier, all organisms appear to share a common origin. Biologists must therefore ask how individual species became so different from one another and why there are so many different species. The question, Why are organisms so diverse? becomes, How did so many different species derive from a single kind of organism?

Every species has a unique set of adaptations.

How Do We Know That the Diversity of Life Resulted from Evolution?

Until the middle of the 19th century, most Europeans believed that the Earth was only about 6,000 years old, and that modern organisms were no different from those present at the beginning of life on Earth. After all, ancient Egyptian

Figure 1-6
Organisms may remain unchanged for many generations. This hippopotamus (giving birth) was carved more than 3,000 years ago, yet it closely resembles hippopotami that live today.

paintings and mummies revealed creatures and plants indistinguishable from those of modern times (Figure 1-6).

However, the discovery of dinosaur and other fossils strongly suggested that organisms very different from the ones we know today once lived on Earth (Figure 1-7). Early in the 19th century, geologists began to recognize that the Earth was far older than 6,000 years. They understood for the first time that the Earth's surface had been constantly shifting and changing over millions of years. Even more exciting, geologists realized that layers of fossil-rich rock contained the history of the Earth and its inhabitants.

By the early 19th century, many scientists understood that some species of plants and animals had disappeared long ago and that new species had appeared to take their places. A few scientists suspected that species of organisms change, or *evolve*, over time from one species into another. At first, no one understood how this could possibly occur. Then, in 1859, the English biologist Charles Darwin published a groundbreaking book titled *On The Origin of Species*, in which he suggested a way that evolution could work.

Darwin's model for how evolution could work came from farming. For thousands of years, plant and animal breeders have bred new varieties of corn, wheat, dogs, or horses. Their method is simple: a breeder selects certain individuals to breed—the most productive wheat plants, the most intelligent dogs, the fastest horses. Slow racehorses and other animals with undesirable characteristics were not allowed to breed.

The power of this strategy is undeniable. Every kind of dog, from Chihuahua to Great Dane, is the result of selection from a line of animals descended from the ordinary wolf. In the same way, broccoli, cabbage, cauliflower and other vegetables have all been bred from a single plant, the wild sea cabbage of the Mediterranean (Figure 1-8). By selecting for plants with big leaves, gardeners developed cab-

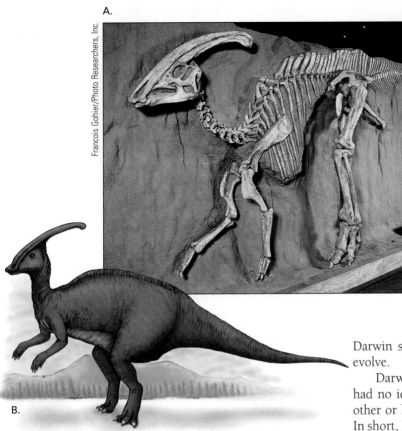

A.

B.

Figure 1-7
Parasaurolophus. The unearthing of fossils in the 18th century suggested that strange organisms had once lived on Earth. Other geologic evidence suggested that the Earth was unimaginably old and that the layers of the Earth's crust contained the history of the Earth and its inhabitants.

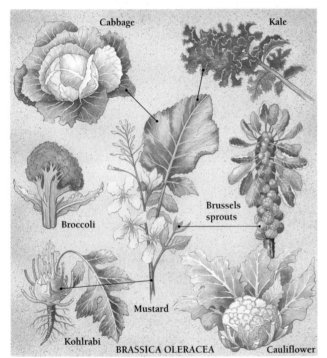

Cabbage

Kale

Broccoli

Brussels sprouts

Mustard

Kohlrabi

BRASSICA OLERACEA

Cauliflower

bage. By selecting for flower buds, gardeners developed broccoli.

Darwin saw that a similar kind of selection, which he called natural selection, could operate in nature. **Natural selection** is the superior reproduction of individuals that have certain inherited traits compared to other individuals who lack those traits.

While artificial selection by human breeders depends on human preferences, natural selection depends on the ever-changing demands of an organism's natural environment. For example, organisms whose adaptations improve their ability to escape predators or to compete for food or sunlight produce more offspring than organisms without these adaptations. To the extent that a trait increases an organism's ability to reproduce, that trait appears more often in the next generation. In this way, Darwin suggested, one organism can gradually change, or evolve.

Darwin understood how evolution might work, but he had no idea what made individuals different from one another or how living organisms pass traits to their offspring. In short, Darwin did not know about genes or DNA. Darwin knew neither why we look like our parents nor why we look slightly different from our parents.

In the 19th century, fossils showed that the kinds of living organisms on Earth have changed over millions of years. Even though Darwin knew nothing about inheritance, he showed that species have evolved from common ancestors and offered a means by which that could happen, which he called natural selection.

We have now seen how biologists define life and how they try to answer questions about living organisms. We have also seen, briefly, how living organisms evolve. In the next few chapters, we will survey the chemistry of life, view the structure of cells, and see how cells obtain energy to fuel life.

Figure 1-8
Artificial selections exploits genetic diversity. European gardeners developed a variety of vegetables from a single species of mustard plant, *Brassica oleracea,* shown at center above. By selecting and breeding mustard plants with large, edible leaves, farmers created cabbage and kale. By breeding the plants with the most flower buds, farmers created broccoli and cauliflower. By breeding the plants with the largest, most edible leaf buds, farmers created Brussels sprouts. Finally, by breeding plants with the largest, most edible stem, farmers created kohlrabi. What other plant parts could be selected for?

Designing experiments is the greatest challenge to a scientist's ingenuity and the greatest source of excitement. Experimental design takes skill and artistry. Like a custom paint job on a car or the cut of a special dress, great experiments are the product of both mastery and inspiration. One of the most famous stories of inspiration in science comes from the German physiologist Otto Loewi.

Loewi was trying to understand how nerves control muscles. He had been studying the heart, a large muscle that rhythmically contracts, giving us our heart beat, and a nerve, called the vagus, that controls how fast the heart beats. When Loewi stimulated a frog's vagus nerve with electricity, the frog's heart rate slowed.

But Loewi didn't know if the nerve slowed the heart through direct contact or by means of some chemical. He suspected that nerves make a chemical that stimulates muscles. If that were true, then a muscle should respond to the right chemical even if the nerve wasn't there. But Loewi had no idea what kind of chemical might do that. And he didn't see how he could possibly test his hypothesis without knowing what chemical it might be.

In fact, 17 years passed before he thought of a way around this problem. Loewi himself later described how the experiment finally came to him:

> The night before Easter Sunday of 1920 I awoke, turned on the light, and jotted down a few notes on a tiny slip of thin paper. Then I fell asleep again. It occurred to me at six o'clock in the morning that during the night I had written down something most important, but I was unable to decipher the scrawl. The next night, at three o'clock the idea returned. It was the design of an experiment to determine whether or not the hypothesis of chemical transmission that I had uttered seventeen years ago was correct. I got up immediately, went to the laboratory, and performed a simple experiment on a frog heart according to the nocturnal design.

In hindsight, Loewi's 3 A.M. experiment seems straightforward, almost routine. Loewi isolated two frog hearts, one with the vagus nerve still attached, the other without the nerve. He stimulated the vagus nerve of the first heart, collected some of the fluid surrounding it, and transferred the fluid to the second heart, the one with no nerve. The heart slowed.

The dramatic result showed that the slowing was a response to some substance produced by the vagus nerve. Loewi called this substance "vagus stuff," which we now know as acetylcholine. Acetylcholine is an important chemical signal that nerves use to communicate. Loewi eventually received a Nobel Prize for his work.

Designing experiments to test a hypothesis requires both creativity and logic.

Why Do Biologists Study Groups of Organisms?

We have talked about experiments in groups of mice or other organisms. Why can't scientists just measure growth rate in one mouse with cinnamon and one mouse without? Why do biologists and other scientists study lots of individuals?

The answer is that every organism is unique. Every fruit fly is unique and every cat is unique (Figure A). So even are identical twins and the individual clones of bacteria or sheep. For example, one pair of identical twin girls differed in height by seven inches as adults. The difference apparently resulted from prenatal trauma in one of the twins. Other identical twins may have cleft palates on opposite sides of their faces, heart defects in one but not the other, and so on. Variation among organisms is the rule. Whether we are talking about humans, mice, worms, oak trees, dandelions, mushrooms, or even one-celled organisms, every individual is unique.

Some of this individuality is due to genetic differences. Some of it is due to the impact of the environment. A *Helicobacter pylori* bacterium on the wrong end of your stomach may not get the same nutrients as a bacterium in the prime position down by the pylorus.

Because each individual organism is unique, each one can react to an experiment differently from every other individual. For example, every mouse in the cinnamon extract experiment could grow at a different rate. Individual mice, like individual humans, grow at different rates. We'd be amazed if every mouse in the experiment grew at exactly the same rate, cinnamon or no cinna-

mon. Because individuals vary, biologists always experiment with groups of organisms rather than single individuals. Then they compare the growth rates of the experimental group with those of the control group.

How many individuals biologists include in an experiment depends on how much individual variation exists. For example, a group of genetically identical mice raised in a laboratory under identical conditions would be expected to show only minimal variation.

Imagine that all the mice are genetically identical, that they all receive the same amount of food and water each day, and that the temperature and humidity in their cages was identical. A biologist wouldn't need many mice to show an effect on growth rate—10 in each group might be enough.

Now imagine a group of wild mice living outdoors. Each mouse is genetically unique. Each one lives a slightly different life from the others. Some get more milk from their mother from the start. Some contract a dis-

Francois Gohier/Photo Researchers, Inc.

Figure A

Why are individuals unique? When organisms reproduce sexually, the mixing of DNA allows the offspring to be both like their parents and unlike their parents. Here, the offspring of a cat look different because each has a unique set of genes inherited from mother and father. Our uniqueness comes not only from our genes, however, but also from our environment. For example, even genetically identical twins can differ dramatically if the environment of the womb favors one over the other.

Continued

ease that the others don't get. The list of possible sources of variation is endless. The result is mice whose growth rates vary by as much as 50 percent. In this case, detecting a 10 percent difference, for example, between the experimental mice and control mice would be difficult. A biologist would need far more than 10 mice to do that. Like wild mice, human populations vary enormously, which is why the best studies of humans often involve thousands of individuals.

The greater the variation, the more individuals we need for an experiment. But we can only find out how many individuals we need by first measuring variation. To measure variation, we have to do a simple version of the experiment. Fortunately, we don't need thousands of animals to measure variation itself. There is a shortcut. Thanks to statistics, it's possible to estimate variation by looking at relatively small numbers of individuals. And, more than that, statistics allows scientists to estimate the difference between two relatively small groups, even when both groups vary enormously.

How Do Biologists Use Statistics?

Biologists often plan experiments and evaluate the results of experiments with **statistics**—the mathematics of collecting and analyzing numerical data. Statistics is a powerful tool with two formal uses. (1) Statistics can help us estimate the actual (or "true" value) of something. (2) Statistics can help us test a hypothesis. Statistics cannot do experiments for us. But it can measure the probability that what we measure is what it seems to be it and not a random effect. The arrangement of data into tables, graphs, and other pictures can also suggest new questions and new hypotheses.

Statistics relies on certain assumptions. But statistics always states its assumptions clearly and then examines those assumptions. If we study the effects of aspirin on heart attack rates in groups of men, we are really interested in the effect of aspirin on *all* men. The men in the experiment are supposed to represent all men. Statisticians call all men the **population** of men. The group selected for the experiment are a **sample** of the population. In statistics, we want to be able to assume that the sample represents the population. In this case, we want to estimate the effect of aspirin on heart attacks in all men by looking at a sample of a few men.

One way of ensuring that the sample represents the population is to take a **random sample.** A sample is random when we ignore any differences among the members of a population while choosing the sample. Suppose we have a population of 100 men, one-third of whom eventually have heart attacks (Figure B). One-quarter of the 100 men agree to volunteer for a study of heart attacks and aspirin.

Figure B (a) shows the whole population. Figure B (b) shows what happens if you choose a sample of 25 men randomly. Figure B (c) shows what happens if you choose a sample based on some difference. In this case, only volunteers were chosen and they do not represent the population. Apparently, the volunteers are more likely to have heart attacks than the general population. Maybe the volunteers were men who volunteered for the study because they knew they were at risk for heart attacks. Compared to the general population, they may include more smokers, more older men, and more sedentary men than the population at large.

A sample can also fail to represent the population because the sample size is too small. Figure B (d) shows two samples of only 10 men each. In neither one does the number of heart attack victims represent the rate in the population. This difference is due to random errors.

If we flip a coin 100 times, we will get equal or almost equal numbers of heads and tails. We might get 52:48 or 47:53, but the ratio will be close to 50:50. On the other hand, if we flip the coin only 10 times, we will probably not get a ratio very close to 50:50. We would have a good chance of getting 2, 3, or 4 heads instead of 5. But chances are we wouldn't get exactly 5 heads and 5 tails. Random effects are big in small samples and tiny in big samples.

In a biological experiment, a sample should represent the whole population of organisms. This can happen only if the sample is chosen randomly and the sample is large enough.

How Do Scientists Estimate a Value?

Biologists are primarily interested in two statistical measures. The first measure is called the mean, which is just the average value of something. The second measure is the **standard deviation,** which is an esti-

mate of the variation in the total population (not just in the sample). If we were studying heart attack rates in men, we would want to know the mean heart attack rate in both the experimental group and in each of the control groups. We would also need to estimate how much the population varies. Both the mean and the standard deviation are essential to thoughtful evaluation of the experimental results.

The **mean** is just what most people call the average. For example, the average, or mean, of the numbers 2, 5, 5, and 8 is 5. Real-life studies have shown that the mean heart attack rate in the population of men over 50 years old who are taking aspirin is lower than that in the population of men over 50 not taking aspirin. Because this is known to be true, millions of older men now take small amounts of aspirin to reduce their chance of getting a heart attack.

How Can Biologists Tell if Two Groups Differ?

It might seem as if all we need to know is whether the mean heart attack rate for aspirin takers is lower than that of the non–aspirin takers. If it is lower, then all men over 50 should take their aspirin. But it turns out that we need to know more.

Imagine that the sample size is too small and the sample mean is different from the population mean. In that case, the mean for the whole population of aspirin takers could actually be the *same* as the mean for all non–aspirin takers. In fact, things could be much worse. The aspirin takers as a whole could have a *higher* rate of heart attack than our sample suggests. In fact, in medical studies of small numbers of people, this often happens. When you read a newspaper article about the latest health findings, always look to see whether the study included thousands of people or just a few dozen. If the article doesn't say, don't take the conclusions too seriously.

Fortunately, statisticians have devised ways of checking for this problem. In order to find out if two means are truly different, we have to estimate the variation in the data. We have to find out, in other words, how much individual variation there is. Statistics measures variation by measuring how far each data point is from the mean. When those distances are averaged we get a measure called the **standard deviation,** which is an estimate of the variation in the total population.

Here's an example. Suppose we want to determine the average length of earthworms we have just dug up from our backyards. We start with only three worms whose lengths are 3 cm, 5 cm, and 7 cm. We say we have three "data points." The mean worm length is simply the average of those measurements, or 5 cm (3 cm + 5 cm + 7 cm = 15 cm, and 15 cm divided by 3 measurements is 5 cm). We can then measure the variation as follows:

$$7 \text{ cm} - 5 \text{ cm} = 2 \text{ cm}$$
$$5 \text{ cm} - 5 \text{ cm} = 0 \text{ cm}$$
$$5 \text{ cm} - 3 \text{ cm} = 2 \text{ cm}$$

To calculate variation, statisticians square each of these differences (to make them all positive numbers), then add them all together, and then divide by the number of data points minus one. That would be $(2^2 + 0^2 + 2^2)/(3 - 1) = 8/2 = 4$. (We leave out the cm here for simplicity.) Then, because they did all that squaring, statisticians bring the number back to reality by taking the square root. (The square root of 4 is 2, of course.) In this case, the variation in the data is the mean difference from the average.

We express this by saying that the length of the earthworms in our sample is 5 cm "plus or minus" 2 cm. The variation here is huge, a full 40 percent of the mean. We can see that our sample of earthworms is highly variable. But does the population vary that much? If we sampled 100 worms, or 1,000, would the mean still be 5 cm × 2 cm? Not necessarily. We may have accidentally picked the very longest and shortest worms. Indeed, the mean length may not be 5 cm at all. It may be something quite different. If the variation is large, we know that our estimates are unreliable. Only with a much larger sample size could we accurately measure the mean of the population.

Imagine, on the other hand, the standard deviation was very small. Suppose the three earthworms were 5.1 cm, 5.0 cm, and 4.9 cm. We would be more confident that our 5 cm sample mean was close to the population mean. And if we had 20 earthworms, whose lengths were so similar, we could be extremely confident that 5 cm represented the population mean.

Calculating variation by hand for lots of data is not something anybody does for fun. We can see intuitively, however, that 2 cm is probably a good estimate of how much vari-

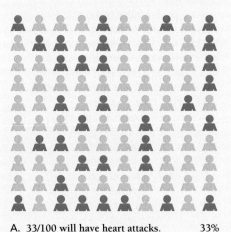

A. 33/100 will have heart attacks.   33%

B. 7/25 will have heart attacks.   28%

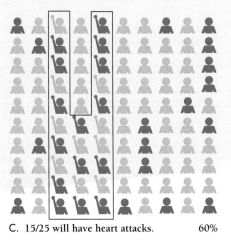

C. 15/25 will have heart attacks.   60%

D. 2/10 and 6/10 will have heart attacks.   20% and 60%

KEY

👤 Heart attack

👤 No heart attack

Figure B

Sampling a population. (a) A population of 100 men includes 33 who have heart attacks at some point. (b) A random sample of 25 men yields 7 who later have heart attacks. This rate, 28 percent, approximately reflects the heart attack rate in the general population, but with some sampling error. (c) 25 men volunteer for a study of heart attacks, but the heart attack rate in the volunteers—15 out of 25, or 60 percent—does not reflect the rate in the whole population. The volunteers are not a random sample. (d) Two small samples of just 10 men each show larger sampling error than in B. In one case, the heart attack rate is 2/10, or 20 percent. In the other case, it is 6/10, or 60 percent. Neither is a good measure of the overall heart attack rate in the general population.

Biology Now™ Learn more about sampling by clicking on this figure on your BiologyNow CD-ROM.

Continued

ation there is in the group of earthworms that measure 3 cm, 5 cm, and 7 cm.

Once we know how much the data varies, we can calculate how big a sample we need to tell if the two population means are truly different. If the variation is large, our sample size may be too small. In that case, we should repeat the experiment using the right sample size. It's possible to calculate the sample size necessary to find out the answer. Alternatively, if the variation is small enough, then we may already have enough data. We can then calculate the likelihood that the two population means are

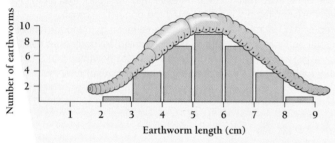

different. Statisticians do this all the time. They might say, "the probability that the aspirin group has a lower heart attack rate than the controls is 95 percent." By this, they mean that there is only a five percent chance that the sample doesn't represent the population. There is only a five percent chance that aspirin takers have the same number of heart attacks as the controls (or even more heart attacks). In other words, they mean that it's very unlikely that aspirin has no effect in all men.

Knowing the standard deviation is essential. But it is accurate only in groups of numbers that are distributed around the mean "normally." A **normal distribution** is one where there are equal numbers of data points on each side of the mean and they are shaped something like a bell (Figure C). Such distributions are surprisingly common in biology. In Chapter 16, Figure 16-5 shows a

graph of student heights. That's a normal distribution.

When data have a normal distribution, about 68 percent of the data will fall within one standard deviation of the mean, and about 95 percent will fall within two standard deviations (Figure D). Once we know that, we can test the hypothesis that the two means are different.

We will stop now. But just remember this. If the numbers in a population are normally distributed (shaped like a bell curve) and if you know the mean and standard deviation of the two samples (experimental and control), then you can calculate the probability that the mean for the experimentals is different from the mean for the controls—in the whole population. This aspect of statistics is a fantastically useful and powerful tool.

In the course of calculating these things, you may learn that the sample sizes were too small because the variation is huge. In this case, you need more data. Or you may learn that the data are not normally distributed, in which case you need more complicated statistics to test your hypothesis. But don't worry. You don't have to calculate anything right now.

Figure C

A normal distribution of worms. Worms can be long or short, but most are medium, as shown in this graph of worm lengths. As many worms are 1 cm shorter than the average (5 cm) as are 1 cm longer than average. This equal distribution of lengths on both sides of the mean is what makes this a "normal distribution." If most of the worms were either very short or very long and few were average, the distribution would not be normal.

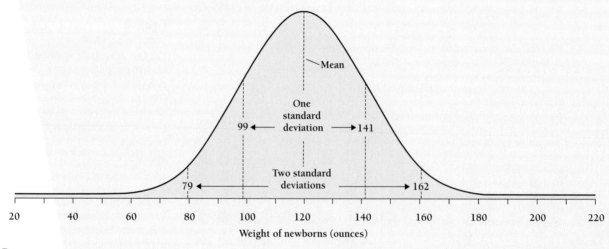

Figure D

What is a standard deviation? This graph shows the weights of newborn babies in ounces. The average, or mean, is 120 ounces (7.5 pounds). One standard deviation is 21 ounces on each side of the mean. So 68 percent of all babies weigh between 99 and 141 ounces. Two standard deviations is 42 ounces to either side of the mean. So 95 percent of all babies weigh between 79 and 162 ounces. Babies born weighing less than 5.5 pounds (88 ounces) are considered premature. They may have problems breathing, staying warm, and digesting, for example. How much did you weigh when you were born?

As you read, keep in mind that in the next 20 years, Earth's rapidly growing human population and dwindling resources will force the world's leaders to make many hard decisions. To participate as citizens in these decisions, we need to do more than simply learn the facts of biology. We need to know how to scrutinize those facts as they come to us from television and radio, the Internet, or a newspaper or magazine. We should know how to ask what biologists and other scientists mean when they describe their findings. To do that we need to know how scientists think, and how they themselves ask questions.

Key Concepts

- Testable hypotheses both answer questions and suggest entirely new questions.
- A good experiment includes one or more controls and a large enough sample to compensate for variation in the data.
- All living things reproduce, evolve, respond to their environment, and consist of a single cell or ordered arrangements of many cells.
- All organisms on Earth have descended from a common ancestor through a process called evolution.

Summary with Key Terms

How do biologists come up with questions and then answer them?

The **scientific method** is a formal set of rules intended to describe the way scientists work, although most working scientists do not stick to a rigid set of rules. Both thinking up questions and finding ways to answer those questions requires inspiration and creativity, as well as hard work, patience, and, often, a methodical approach. In general, a scientist begins by asking a narrow question and then observing. A **hypothesis** is a formal guess at the answer to the question, sometimes accompanied by some sort of model. A good hypothesis allows a scientist to make a **testable** prediction about the way the process or structure will act under specific circumstances. A set of hypotheses that consistently withstand different tests may become accepted as a **theory.** Trying to understand the whole in terms of the parts is called **reductionism.** But the characteristics of a complex system, called **emergent properties,** cannot be predicted using a reductionist approach.

How are all organisms alike?

All organisms are composed of one or more cells, which are themselves alive, and which all come from other cells. All are organized into parts, obtain energy from their surroundings, perform chemical reactions, mature over time, reproduce, share a common evolutionary history, and respond to their environments. Mechanisms that maintain a constant internal environment are a form of **homeostasis.**

Why are all organisms alike?

Over the last 4 billion years, organisms have evolved from a few simple cells into millions of complex multicellular organisms. The millions of species of organisms are organized by biologists into three **domains: Archaea, Eubacteria,** and **Eukarya.** Eukarya comprise just four **kingdoms: Animalia, Plantae, Fungi,** and **Protista.** **Eukaryotic** cells contain a **nucleus** and other **organelles,** each of which is enclosed by a membrane. **Prokaryotic** cells have no membrane-enclosed nucleus.

Why are the members of a species alike?

Genes, which are made of long thin molecules of **deoxyribonucleic acid (DNA),** ensure continuity from generation to generation.

How do organisms become different?

The discovery of fossils suggested to 18th- and 19th-century scientists that organisms had changed, or evolved, over long periods. Charles Darwin assembled compelling evidence for the idea of **evolution** and suggested a mechanism, called **natural selection.** The individual members of a species are unique. Some reproduce more successfully than others. Traits that are inherited in the genes and help a species to thrive and reproduce, called **adaptations,** continue from generation to generation in greater frequency than less useful traits.

How do biologists use statistics to plan and evaluate experiments?

An **experiment** is a way of testing the predictions of a hypothesis or, sometimes, of searching for some unknown effect. An experiment is a procedure that is carried out under controlled conditions. Essential elements of most biology experiments are the **control** experiment and **statistical** analysis. Statistics allows biologists to estimate the true value of something and to test hypotheses.

Meaningful statistics requires a **sample** that represents the whole **population** of organisms. A sample can represent the population only if the sample is a **random** sample and the sample is large enough to overcome random variation. Statisticians can determine if the sample represents the population using the **mean** and **standard deviation.** Statistics is most easily done on data with a **normal distribution,** or bell curve.

Review and Thought Questions

Review Questions

1. Explain the differences between a hypothesis and a theory.
2. Explain why a hypothesis can sometimes be proved wrong, but can never be proved right. Explain how repeated testing of a hypothesis can lead to a theory.
3. Give an example of an untestable hypothesis. Why is it untestable? What would you need to know or do to test it?

4. Suppose you decide to test the hypothesis that your favorite soft drink contains substances that cause cancer. You plan to feed the drink to 100 mice and then count the number of tumors they develop. Describe an appropriate control experiment.

5. Describe eight distinguishing characteristics of living things. Give an example of each one.

6. Define "homeostasis." List two examples of homeostasis.

7. Explain what is meant by natural selection. What is "selected"?

Thought Questions

8. The ants in your kitchen are driving you nuts. Using the "scientific method," how would you find out where they were entering your house? How would you determine why they were entering your house? Think of several alternative hypotheses. Now design a series of experiments whose results will either support or eliminate the various alternative explanations.

9. A smallpox virus is a bit of DNA surrounded by a protein coat. A virus can attach to the wall of a cell and inject the DNA into the cell. Inside the cell, the viral DNA forces the cell to make new viruses. The cell makes so many new viruses that the cell eventually bursts open, releasing thousands of new viruses, which go on to infect other cells. Is the virus alive? Using the definitions of life given at the beginning of this chapter, explain why a virus is alive or why a virus is not alive.

10. From your own experience, describe a moment of inspiration—intellectual or otherwise. How did it feel? How did it change you?

BiologyNow Resources

Biology ⊜Now™

The BiologyNow CD-ROM packaged free with your text uses a learning system that allows you to review your general understanding of a concept. First, you answer a series of diagnostic review questions. Based on your answers, BiologyNow will provide you with a customized learning plan that links you to the text, study guide, animations, Genetics Problem-Solving Guide, and CNN Video for focused study that maximizes your learning. You can also connect to V-Mentor for one-on-one tutoring help from experienced biology teachers.

Web Site

The Web site for this book contains a wealth of helpful study aids, as well as many ideas for further reading and research. Log on to: **http://biology.brookscole.com/AAL3**

- For study and review, **Chapter Outline** gives you an outline of the chapter, **Chapter Summary** allows you to review the chapter's main ideas, and **Glossary** lists concepts and terms for the chapter along with their definitions.

- To test your mastery of important terminology for this chapter, you can use the electronic **Flash Cards**, which may be sorted by definition or by term.

- For testing your knowledge and preparing for in-class examinations, our **Quizzes** pose multiple-choice and/or true-false questions based on each chapter.

- **Hypercontents** takes you to an extensive list of current links to Internet sites with news, research, and images related to individual subjects in the chapter.

- **Internet Exercises** are critical thinking questions that involve research on the Internet with starter URLs provided.

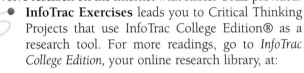 **InfoTrac Exercises** leads you to Critical Thinking Projects that use InfoTrac College Edition® as a research tool. For more readings, go to *InfoTrac College Edition*, your online research library, at:

http://infotrac.thomsonlearning.com

Active Figures

| 1-1: | Barry Marshall's Discovery |
| Box, p. 15, Fig. B(d): | Sampling |

Preparing for an exam? Take a diagnostic test on your BiologyNow CD-ROM.

Online materials relating to this chapter are at:
http://biology.brookscole.com/AAL3

About the Chapter-Opening Image
Most ulcers result from a bacterium called *Helicobacter pylori* that infects the lining of the stomach.

David Spears/Clouds Hill Imaging Ltd./Corbis

PART I
CHEMISTRY AND
CELL BIOLOGY

2

The Chemical Foundations of Life

Key Questions

- What is matter?
- What determines the properties of an atom?
- What kinds of bonds bring atoms together to form molecules?
- What special characteristics of water made it possible for life to evolve?

A Dynamite Discovery

Anyone who has heard of the Nobel prizes knows the name of Alfred Nobel (1833–1896), the 19th-century inventor of dynamite. Much of Nobel's vast fortune came from the use of dynamite—both for the construction of dams, bridges, and canals and for the military destruction of factories, ships, dams, and bridges—as well as his investments in oil wells. A pacifist and lover of literature and science, Nobel regretted that his invention was used as a tool of war. His will created the Nobel Foundation with instructions that his fortune should be used to promote peace and peaceful achievement. Since 1901, Nobel prizes have been given to recognize extraordinary achievements in peace, chemistry, physiology and medicine, physics, literature, and economics. The 1998 Nobel Prize in Physiology and Medicine acknowledged a discovery involving the active ingredient in dynamite, nitroglycerin.

Nitroglycerin is a colorless oily liquid that is so unstable that just dropping it makes it explode. Nobel found a way to stabilize nitroglycerin so that it could be safely transported. But nitroglycerin has an important use besides blowing up bridges. Despite its dangers, physicians have been prescribing nitroglycerine to heart patients for more than 100 years. In small doses, nitroglycerin reduces blood pressure, and it is especially effective in treating the chest pain associated with one form of heart disease (angina). Nobel's own doctor recommended nitroglycerin treatment for his heart condition, but Nobel refused to take it.

In the early 1970s, biologist Ferid Murad, at the University of Virginia, was trying to figure out exactly what nitroglycerin did for the heart. Murad found that nitroglycerin releases the gas nitric oxide, which, he said, causes blood vessels to dilate, which lowers blood pressure. But nitric oxide—abbreviated as NO—is also a major component of smog, a noxious chemical released by automobile engines that causes irritation and burning in the lungs. No one could believe that NO played a major role in our day-to-day physiology. According to Murad's former

Dennis Galante/FPG

professor, "People used to say, 'C'mon, Fred [as they called him]. Why don't you work on something important?' "

About the same time, Robert Furchgott, working at the State of New York Health Science Center at Brooklyn, was struggling to understand the effects of other drugs on blood vessels. Furchgott's work was extremely frustrating, since a chemical might cause vessels to contract at some times and dilate at others. Trying to get to the bottom of these contradictions, Furchgott noticed that drugs that dilated blood vessels worked only when the lining of the blood vessels was intact. He concluded that the drugs were working indirectly, somehow inducing the cells of the blood vessel lining to release an unknown chemical that was in turn triggering a response in the cells of the blood vessel wall. But what was it? In 1986, Furchgott and his colleagues showed that the unknown chemical was nitric oxide.

Physiologists reacted with astonishment and skepticism. Murad, Furchott, and colleagues were claiming that mere gas molecules could carry messages from one cell to another, something no one had ever seen in animals. Furthermore, the skeptics argued, although bacteria were known to produce NO, there was no mechanism by which eukaryotes could possibly make the stuff.

Yet before long, Murad, Furchott, and Louis Ignarro of UCLA showed both how cells produce NO and how it acts—work for which they shared the Nobel prize. Pharmaceutical companies began to develop medicines that increase NO production for heart patients and so stumbled on NO's most dramatic effect in men, the stimulation and maintenance of more-prolonged erections. The anti-impotence drug Viagra acts by increasing the action of NO, thus opening blood vessels and allowing more blood flow to the penis. Because an erection is caused by the penis's engorgement with blood, Viagra and newer drugs similar to it enable many otherwise impotent men to achieve normal erections.

Not surprisingly, too much NO can also be dangerous. For example, in people suffering from extreme bacterial infections, some of the body's immune cells may pump out so much NO that blood pressure drops precipitously, and fatally. And since these same immune cells are known to use NO to attack tumor cells, biologists are using their knowledge of NO to find new anticancer drugs. By 1992, the influential journal *Science* had named NO the "molecule of the year," and the 1998 Nobel prize for NO's discovery once more underscored its importance.

In this chapter, we'll learn what forces hold together simple molecules such as water and NO. In the next chapter, we'll look at more complex molecules.

2.1 What Is Matter?

Anyone whose car has ever run out of gas knows that when the gauge says empty, the gas is gone. But where does it go? Chemistry tells us that when gasoline comes into contact with the oxygen in air, a spark can trigger a chemical reaction in which the gasoline is burned, or "oxidized." Most of the gasoline is transformed into water and carbon dioxide (the same stuff that makes the bubbles in soft drinks). Our bodies perform similar chemical reactions. Like a car's engine, organisms can extract energy from sugars, fats, and proteins, breaking them down into water and carbon dioxide. But unlike cars, we also use these same molecules to build thousands of other chemicals, the stuff of our bodies.

Chemistry is the science dealing with the properties and the transformations of all forms of matter—whether gasoline, air, protons, people, or planets. **Matter** is any substance that occupies space and has mass (what most people understand as weight). Matter can exist in different states or phases—solid, liquid, or gas.

Every bit of matter consists of one or more *pure substances,* called elements, each of which has its own unique properties. **Chemical reactions** transform pure substances into other substances, without changing the total amount of matter. ("Biochemical reactions" are those chemical reactions that occur in organisms.)

Pure substances that cannot be converted into simpler substances in chemical reactions are called **elements.** Hydrogen, helium, oxygen, and carbon are elements. So are silver or gold, the carbon in a pencil, the copper in a wire, and the mercury in a thermometer. Altogether, scientists know of 92 naturally occurring elements. Elements are composed of tiny particles called **atoms**—the smallest units that still have the properties of an element. Atoms themselves are made of smaller "subatomic" particles, including protons, neutrons, and electrons.

Each element has a standard one- or two-letter abbreviation. These usually correspond to the beginning of the element's English name, though sometimes it's the Latin or Greek name. "C" stands for carbon and "O" for oxygen, while "Na" stands for sodium [Latin, *natrium*]. Four elements—carbon (C), oxygen (O), nitrogen (N), and hydrogen (H)—make up more than 99 percent of the matter in organisms. Figure 2-1 shows the main elements of the human body.

Chemical reactions cannot change or destroy elements. Carbon cannot be converted to oxygen. Silver cannot be converted to gold. However, elements can be created or changed by means of **nuclear reactions,** the high-energy transformations of individual atoms that occur in a nuclear reactor or in the core of stars such as our sun.

Although oxygen is an element, we do not breathe oxygen in its atomic form. Instead, we breathe "molecular" oxygen, which is two atoms of oxygen bound together by a "chemical bond." A **molecule** is a stable assembly of two or more atoms.

Most molecules are **compounds**—pure substances forged in chemical reactions from two or more *different* elements. Water, for example, is a compound of the elements hydrogen and oxygen (Figure 2-2). Carbon dioxide is a compound of carbon and oxygen. A compound is unlike the elements that compose it. The properties of water are different from those of either hydrogen alone or oxygen alone,

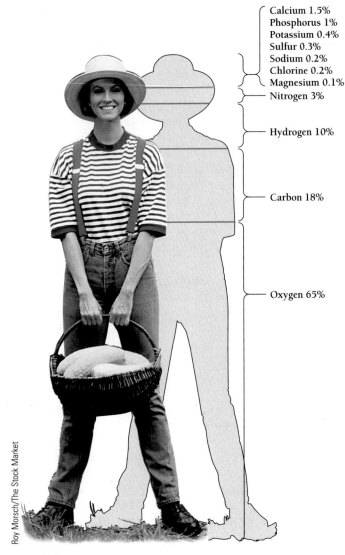

Calcium 1.5%
Phosphorus 1%
Potassium 0.4%
Sulfur 0.3%
Sodium 0.2%
Chlorine 0.2%
Magnesium 0.1%
Nitrogen 3%

Hydrogen 10%

Carbon 18%

Oxygen 65%

Roy Morsch/The Stock Market

Figure 2-1
Hydrogen, carbon and oxygen. Like other organisms, humans are mostly water (hydrogen and oxygen) and different carbon compounds, with a sprinkling of other elements.

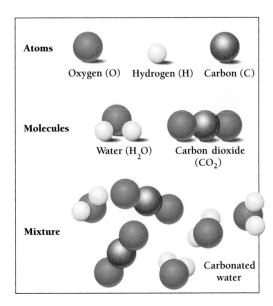

Figure 2-2
Atom, molecule, compound, and mixture. Two atoms of the same element, such as two oxygen atoms, may form a molecule. But most molecules are compounds forged from two or more different elements. Water is a compound of hydrogen and oxygen, and carbon dioxide is a compound of carbon and oxygen. Most substances are mixtures of different compounds. Carbonated water, for example, is a mixture of carbon dioxide and water.

Unlike the components of a compound, which are always present in the same fixed ratio, the components of a mixture may be present in any ratio. A molecule of water always contains twice as much hydrogen as oxygen. But you can mix a little sugar into your cup of coffee, or a lot.

Elements are pure substances that cannot be converted into simpler substances. An atom is the fundamental unit of an element, and a molecule is the fundamental unit of a compound.

and the properties of carbon dioxide are different from those of carbon and oxygen.

A compound is made of a single kind of molecule. Chemists designate the composition of a molecule by writing the number of atoms of each type as subscripts to the right of the symbol for each element. Water is always H_2O, carbon dioxide is CO_2, and ordinary table sugar (sucrose) is $C_{12}H_{22}O_{11}$.

In our day-to-day lives, we rarely encounter either compounds or elements in pure form. Most matter is a **mixture** of different compounds (Figure 2-2). Sand and salt poured together, for example, is a mixture. One component of a mixture does not affect the properties of the other components. If we pour salt into boiling water to make spaghetti, the salt is still salt and the water is still water. If we boil away all of the water, the salt forms crystals at the bottom of the pot, and we can retrieve the same salt we poured into the pot.

What Is Special About the Chemistry of Life?

Until the 16th century, chemistry as we know it did not exist. A few "alchemists" tried to convert lead and other cheap metals into silver or gold by combining different compounds and elements (Figure 2-3). But none of the alchemists of the Middle Ages knew what matter was and why it acted as it did. In their efforts to make gold, however, they learned a lot about the properties and transformations of matter.

Not until the 19th century did scientists begin to understand the chemical nature of matter. And not until recently have they begun to understand the chemistry that underlies all cellular processes—from the workings of a gene to the contraction of a muscle.

One of the great questions of biology was whether living matter was in any way different from nonliving matter.

Figure 2-3

***The Alchemist,* painted by David Teniers the Younger in 1648.** Alchemists of the medieval period and later attempted to turn lead and other cheap metals into gold and silver.

Scala/Art Resource, NY

have a characteristic weight. For example, an atom of carbon weighs 12 times as much as an atom of hydrogen. Chemists and physicists prefer to talk of the "mass" of an atom, rather than its weight. **Mass** is the amount of matter in something. Weight results from the action of gravity on mass. You may weigh 125 pounds on Earth and zero pounds in space, but your mass remains the same no matter where you are. Mass is closely related to weight, but the mass of an object, unlike its weight, is constant.

Are a horse and a plow made of the same basic stuff? In fact, the molecules of living organisms are exactly the same as the molecules of nonliving matter.

From a chemist's point of view, just two aspects of living matter distinguish it from nonliving matter—(1) the elaborate organization of the molecules and (2) the limited number of kinds of molecules that make up living matter. All organisms consist of the same few kinds of small *organic* (carbon-containing) molecules, which we may think of as "building blocks." Just as a good cook can make a dazzling array of cookies, cakes, and pies from just a few ingredients, so organisms can build extraordinary structures from a small number of standard molecules.

The chemistry of living matter is the same as that of nonliving matter.

2.2 What Determines the Properties of an Atom?

By the early 19th century, chemists were able to distinguish compounds (which could be broken down into simpler substances) from elements (which could not). They then began to ask, Does the way in which a substance breaks down reveal something about its basic structure? Are there units that cannot break down further?

To answer these questions, the English chemist John Dalton suggested that matter is actually built from small particles. He proposed that each element is composed of tiny particles called "atoms." The atoms of each element

What Are Atoms Made Of?

Starting with Dalton, chemists discovered that they could measure the relative mass of each element in any compound. Such measurements revealed a fascinating relationship: The mass of every kind of atom is an almost exact multiple of the mass of a hydrogen atom. An atom of helium has four times as much mass as an atom of hydrogen. Carbon has 12 times the mass of hydrogen; nitrogen, 14 times; and oxygen, 16 times (Figure 2-4).

This pattern suggested that the larger atoms were composed of groups of hydrogen atoms. We now know that atoms are not actually made of hydrogen atoms but of subatomic particles, each of which has about the same mass as a hydrogen atom. An atom of carbon is composed of 12 particles, each about the mass of a hydrogen atom, or one "dalton," whereas an atom of oxygen is composed of 16 such particles, or 16 daltons.

Since a molecule consists of a fixed number of atoms, every molecule has a characteristic **molecular weight.** For example, a molecule of water (H_2O), which consists of one oxygen atom and two hydrogen atoms, has a molecular weight of 18.

$$18 = 16 \text{ (oxygen)} + 1 \text{ (hydrogen)} + 1 \text{ (hydrogen)}$$

Atoms are made of subatomic particles similar in size to one hydrogen atom. Every atom has a characteristic mass or molecular weight.

What Is the Internal Structure of an Atom?

Once physicists knew that atoms were made up of hydrogen-sized particles, they began to wonder about those particles. What were the hydrogen-sized particles, and how did they fit together to make an atom? The first important finding

Figure 2-4

A partial Periodic Table of the Elements. Shown here are the first three rows of a periodic table, including the six most important elements to life: hydrogen, carbon, oxygen, nitrogen, phosphorus, and sulfur. The chemical properties of each element depend on its atomic number (the number of protons in the nucleus) and atomic mass (the total number of both protons and neutrons). On the left are highly reactive elements such as hydrogen and sodium. The atoms of these elements contain a single electron in their outer orbital and are "eager" to bond with other atoms. At the far right are unreactive elements such as helium and argon. These "noble gases" remain aloof from other elements and rarely form chemical compounds because their electron orbitals are already full. In between are moderately reactive, or electronegative, elements such as carbon, nitrogen, and oxygen. These last three elements, together with hydrogen, form the basis of all organic molecules.

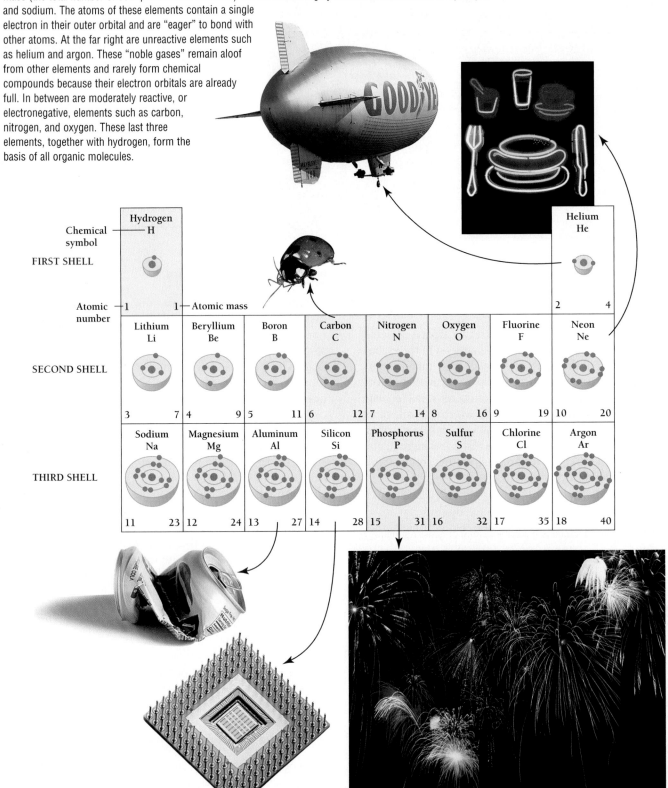

Silicon chip, Alfred Pasieka/Peter Arnold, Inc.; neon sign, Allan Kaye/DRK Photo; blimp, Gianni Tortoli; Photo Researchers, Inc.; aluminum can, Paraskevas Photography; fireworks, AlaskaStock; seven-spotted lady beetle, Runk/Schoenberger from Grant Heilman

about atomic structure was that atoms are almost entirely empty space. In 1910, a New Zealand physicist working in England devised an ingenious way to probe the structure of the atom. Ernest Rutherford's approach was like searching for a stone in a haystack by shooting bullets into the haystack. If a bullet goes straight through the haystack, we know it hasn't hit anything solid. If it comes out at an angle, then it must have bounced off the stone. The "bullets" Rutherford used were small, positively charged alpha (α) particles, produced from the radioactive decay of radium. The "haystack" was a thin piece of gold foil.

Rutherford found that his α particles passed right through most parts of a gold atom, only bouncing off a tiny, central part of each atom. He proposed that the atom is mostly empty space, with almost all of its mass concentrated in a small heavy core, or **nucleus,** at its center (Figure 2-5A).

The diameter of the nucleus is only 1/10,000 that of the whole atom. Imagine a Ping-Pong ball (the nucleus) at the center of two vacant city blocks (the atom), and you will have an idea of the proportions of an atom. The tiny nucleus has a positive charge, which is why it deflected the positively charged α particles. And the positive charge of the nucleus is balanced by the negative charges of even tinier particles, called **electrons,** which have just 1/2,000 the mass of a proton.

Later researchers showed that the nucleus has a positive charge because it is made up of particles called **protons.** The positive charge of a proton exactly balances the negative charge of an electron. The nucleus of an atom also contains **neutrons,** which have no electric charge. A neutron is itself made of a proton and an electron. Protons, neutrons, and electrons and other particles that make up atoms are called "subatomic" particles because they are smaller than an atom.

The mass of an atom depends almost entirely on the total number of protons and neutrons in the nucleus. The atoms of each element always have the same number of protons. Oxygen, for example, always has eight protons. But some forms of an element may have different numbers of neutrons. For example, carbon always has six protons and six electrons. But it comes in three forms, or **isotopes,** each of which has a different number of neutrons and therefore a different mass: "carbon 12" has six neutrons, "carbon 13" has seven, and "carbon 14" has eight. The three isotopes of carbon, sometimes written ^{12}C, ^{13}C, and ^{14}C, have the same chemical properties but different masses.

Although Rutherford envisioned the electrons neatly orbiting the nucleus like tiny planets, and chemists still speak of electron **orbitals,** in fact, electrons actually move around the nucleus randomly, like a cloud of gnats. Like a cloud, an electron orbital has fuzzy boundaries but a definite shape (Figure 2-5A).

The electrons in each orbital have a certain amount of energy. When an electron moves from a high-energy orbital to a low-energy orbital, an atom releases energy, often in the form of heat and light. Conversely, when an atom absorbs energy, one or more electrons may move from a low-energy orbital to a high-energy one. Electrons group together in **shells** surrounding the nucleus. The two tiniest atoms, hydrogen and helium, have just one shell containing one electron (hydrogen) or two electrons (helium). Large atoms have more shells. Each one can hold either 8 or 18 electrons (Figure 2-4).

Atoms are mostly empty space. They consist of a tiny central nucleus orbited by even tinier electrons. The nucleus has a positive charge, balanced by the negative charge of the electrons.

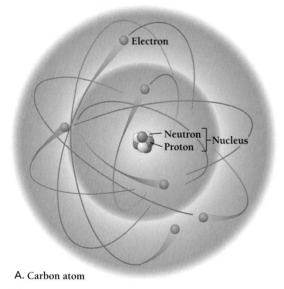

Lowest energy shell—2 electrons maximum

Next shell—8 electrons maximum

Next (!) shell

A. Carbon atom

B. Sodium atom

Figure 2-5

The structure of an atom. A. A carbon atom. By convention, orbitals are shown as flat rings on the page, but in reality, electron orbitals are diffuse, three-dimensional spheres (or other shapes). Each orbital has room for two electrons. Orbitals are grouped into shells, each of which has a different energy and room for different numbers of orbitals (and electrons). B. A sodium atom with three shells. The innermost shell has the lowest energy and room for just two electrons. The next shell out contains four orbitals with a capacity for eight electrons.

Extreme Biology   Why Do Some Atoms Decay?

Some isotopes are very stable, while others are unstable and tend to break down, or *decay*. The nucleus of ordinary carbon, called "carbon-12," consists of six protons and six neutrons. Carbon-12 is quite stable. But another kind of carbon, called carbon-14, is quite *unstable*. Carbon-14 consists of six protons and eight neutrons, and every second about four carbon-14 atoms in a trillion turn into nitrogen atoms. How can that happen?

In carbon-14, one of the neutrons, which is made of a proton and an electron, decomposes into a proton and an extra electron (called a "beta [β] particle), which then speeds away from the nucleus (Figure 2-6A). The new nucleus has lost a neutron and gained a proton, so it now has seven neutrons (having lost one) and seven protons. But an atom with seven protons is not carbon; it is nitrogen. The new nitrogen (nitrogen-14) atom is missing an electron, but it eventually acquires a seventh electron to balance its new positive charge.

Isotopes that release β or other particles are called **radioactive isotopes.** The decay process occurs at a constant rate that is characteristic of each radioactive isotope. The time required for half the atoms of a radioactive isotope to decay is called that isotope's **half-life** (Figure 2-6B). Half-lives for different isotopes range from as little as 0.3 microsecond to as long as 2,000,000,000,000,000 (2×10^{15}) years. The half-life of ^{14}C is about 5,760 years, while that of uranium 235 (^{235}U) is about 704 million years.

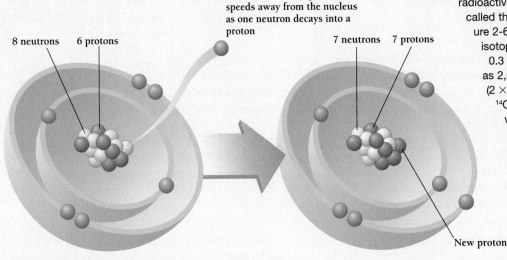

8 neutrons   6 protons

An electron (or "β particle") speeds away from the nucleus as one neutron decays into a proton

7 neutrons   7 protons

New proton

Carbon 14 (^{14}C)

Nitrogen 14 (^{14}N)

A.

Figure A

Half-life: Breakdown of carbon-14. When a neutron decays into a proton, it releases an electron (a β particle). The altered carbon nucleus has one fewer neutron and one more proton, for a total of seven neutrons and seven protons. A nucleus with seven protons is no longer carbon, however, but nitrogen. The new nitrogen atom acquires another electron in its outer shell, and so stays electrically neutral.

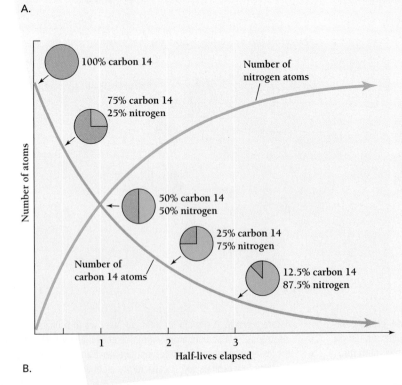

100% carbon 14

75% carbon 14
25% nitrogen

Number of nitrogen atoms

50% carbon 14
50% nitrogen

25% carbon 14
75% nitrogen

Number of carbon 14 atoms

12.5% carbon 14
87.5% nitrogen

Number of atoms

Half-lives elapsed

1   2   3

B.

2.3 What Holds the Atoms of a Molecule Together?

Just six elements compose most of the molecules of life: hydrogen (H), carbon (C), nitrogen (N), oxygen (O), phosphorus (P), and sulfur (S). One way to remember these is by thinking of the nonsense word "SPONCH." The atoms of these six elements can form chemical bonds in a variety of ways to form the building blocks of life.

An atom is most stable when its outermost shell contains eight electrons. The atoms of some elements—such as neon and argon—already contain eight electrons in their outer shells and are chemically unreactive. These are the "noble" elements mentioned in Figure 2-4. But most atoms lack some electrons and will share electrons with some other atom, so that each atom seems to have eight. Sharing electrons is the basis for a chemical **bond** (Figure 2-6A and B).

Biologically, the strongest chemical bonds are "covalent bonds," which hold together all the major molecules of life, including the proteins in a butterfly's wing and the sugars in a ripe peach. (Chemists define bond strength somewhat differently.) In a **covalent bond**, two atoms share one or more electrons. The covalent bonds within a molecule have fixed

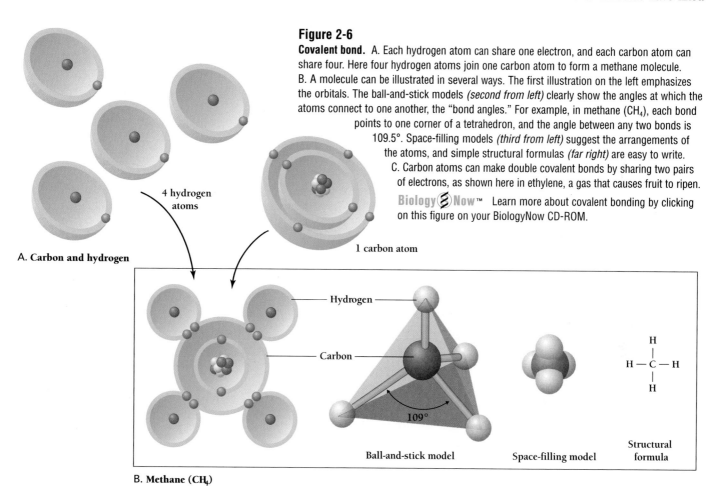

Figure 2-6
Covalent bond. A. Each hydrogen atom can share one electron, and each carbon atom can share four. Here four hydrogen atoms join one carbon atom to form a methane molecule.
B. A molecule can be illustrated in several ways. The first illustration on the left emphasizes the orbitals. The ball-and-stick models *(second from left)* clearly show the angles at which the atoms connect to one another, the "bond angles." For example, in methane (CH_4), each bond points to one corner of a tetrahedron, and the angle between any two bonds is 109.5°. Space-filling models *(third from left)* suggest the arrangements of the atoms, and simple structural formulas *(far right)* are easy to write.
C. Carbon atoms can make double covalent bonds by sharing two pairs of electrons, as shown here in ethylene, a gas that causes fruit to ripen.

Biology ⊗ Now™ Learn more about covalent bonding by clicking on this figure on your BiologyNow CD-ROM.

A. Carbon and hydrogen

4 hydrogen atoms

1 carbon atom

Hydrogen

Carbon

109°

Ball-and-stick model

Space-filling model

Structural formula

B. Methane (CH₄)

Hydrogen

Carbon

Ball-and-stick model

Space-filling model

Structural formula

C. Ethylene (C₂H₄)

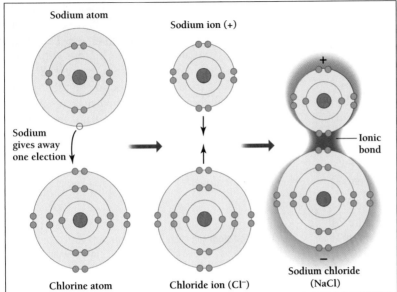

Sodium atom

Sodium ion (+)

Sodium gives away one election

Chlorine atom

Chloride ion (Cl⁻)

Ionic bond

Sodium chloride (NaCl)

A.

Figure 2-7

Ionic bond. A. In table salt, a charged chloride ion and a charged sodium ion are attracted to one another in the same way that a sock is attracted to a shirt just out of the dryer. B. Reaction of sodium and chloride to form salt. C. Salt crystals.

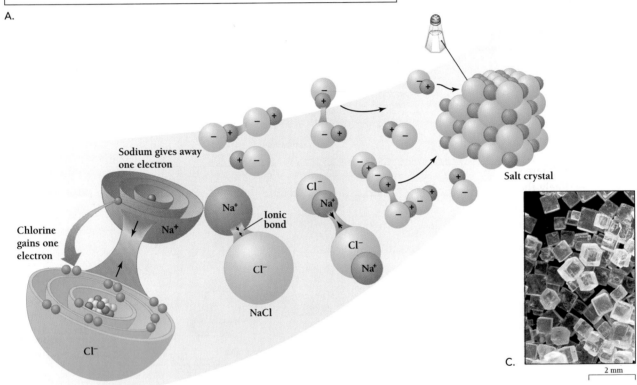

Sodium gives away one electron

Chlorine gains one electron

Na^+

Na^+

Ionic bond

Cl^-

Cl^-

Na^+

Cl^-

Na^+

NaCl

Salt crystal

2 mm

C.

B.

directions and lengths (Figure 2-6B). Because bond lengths and angles are constant, every molecule has a definite size and shape. A water molecule, for example, always has the same boomerang shape, and methane always forms a pyramid.

Another kind of bond, the ionic bond, is subtler. When an atom gains or loses electrons, it acquires an electrical charge (positive or negative), and we call it an **ion**. Positive ions and negative ions are attracted to one another and can form an **ionic bond**. For example, table salt is made of two

kinds of ions—sodium, which has a positive charge, and chloride, which has a negative charge (Figure 2-7A). Together, they form "sodium chloride," or salt.

The sodium ions and the chloride ions in sodium chloride each have eight electrons in their outer shells, so they don't need to share electrons (Figure 2-7B). Instead, the sodium atom gives away an electron and the chlorine atom takes one. The result is positively and negatively charged ions. The positively charged sodium ions and the negatively charged chloride ions (Na^+ and Cl^-) are held together by simple electrical attraction, in ionic bonds. This electrical attraction, or "electrostatic attraction," is the same thing that makes your socks cling to your shirts in the dryer.

Most of the molecules in organisms are made of just six kinds of atoms, usually held together by a covalent bond, in which two atoms share electrons. Because bond lengths and angles are constant, every molecule has a predictable size and shape. An ionic bond is a simple electrical attraction between two oppositely charged ions (Figure 2-7C).

Why Are Some Molecules Polar?

A chlorine atom has a high tendency to gain an electron to form a chloride ion, while a sodium atom has a high tendency to lose an electron to form a sodium ion. The American chemist Linus Pauling realized that the atoms of each element have a characteristic **electronegativity**—the tendency to attract electrons. Oxygen and chlorine, for example, strongly attract electrons; they are two of the most electronegative elements. In contrast, sodium, potassium, and lithium are among the least electronegative elements—they tend to lose electrons. In between are elements such as carbon that have no tendency to lose or to attract electrons. (We can think of electronegativity as a measure of electron greediness.)

The larger the difference in the electronegativities of two atoms, the more likely they are to form an ionic rather than a covalent bond. Sodium and chlorine, for example, have a large difference in electronegativities and so they form ionic bonds. Carbon and nitrogen, on the other hand, have similar, moderate electronegativities, and they form covalent bonds.

Even in a covalent bond, atoms may not share electrons equally. When atoms differ in electronegativity, they do not share electrons equally. Instead, the diffuse clouds of shared electrons tilt toward the more electronegative atoms. In a water molecule, for example, an oxygen atom shares electrons with two hydrogen atoms. But the shared electrons are more concentrated around the oxygen nucleus than around the two hydrogen nuclei. As a result, the oxygen atom has a slight negative charge, and the two hydrogen atoms have a slight positive charge.

Molecules like water that have uneven distributions of electrical charge are said to be **polar,** since they have positive and negative poles in the same way that a magnet has two poles (Figure 2-8). When a polar molecule, such as water, comes close to an ion or to another polar molecule, its negative pole turns toward the other molecule's positive pole.

We mentioned that covalent and ionic bonds are quite strong. Breaking them requires lots of energy. But most of the chemistry of living organisms occurs in water. Inside the watery environment of a cell, different weak interactions bind molecules together. Although individually each interaction is weak, the summation of many weak interactions determines the structure of many complex molecules.

When a bartender mixes an alcoholic drink, such as gin and tonic water, the result is a smooth mixture, uniform

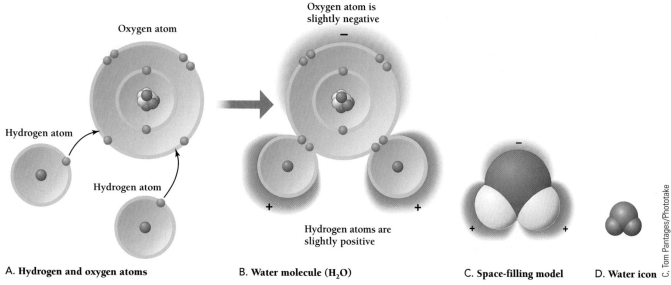

A. Hydrogen and oxygen atoms   B. Water molecule (H_2O)   C. Space-filling model   D. Water icon

C. Tom Pantages/Phototake

Figure 2-8

The formation of water. A. Two atoms of hydrogen and one atom of oxygen share electrons in two covalent bonds to form a molecule of water. B. Because the oxygen atom is more electronegative than the hydrogen atoms, the electrons spend more time hovering around the oxygen end of the water molecule. As a result, the oxygen end has a slightly negative charge, while the hydrogen atoms have a slight positive charge. C. Space-filling model of water molecule. D. Icon for water.

Biology ⋛ Now™   Learn more about the polarity of water by clicking on this figure on your BiologyNow CD-ROM.

throughout. Yet, when a cook mixes oil and vinegar to make salad dressing, the oil runs together in globs and floats to the top of the vinegar. Why is this? The answer is that polar molecules, such as alcohol, mix well with water and are said to be **hydrophilic** [Greek, *hydro* = water + *phili* = love], or water loving. But water molecules attract and surround themselves with other polar molecules, so that nonpolar molecules, such as oil, seem to be repelled by water. We say that nonpolar molecules are **hydrophobic** [Greek, *hydro* = water + *phobos* = fear], or water fearing. In water, hydrophobic molecules are driven together by their mutual tendency to avoid water.

Oils and other hydrophobic molecules cling together when confronted with an aqueous environment.

2.4 How Is Water Especially Well Suited for Its Role in Life?

Life and water are intimately connected. Without water, life could not have evolved on Earth. Wherever life is found, there is water, and wherever liquid water is found, there is life (Figure 2-9). Water makes up more than 70 percent of the material of living organisms themselves and covers more than 75 percent of the Earth's surface. It is the medium in which most cells are constantly bathed and the major component of cells themselves. Although water is common—at least on Earth—its properties are unusual. But water's special attributes make it uniquely fit for its important role in life.

Francis Gohier/ Photo Researchers, Inc.

Figure 2-9
Water is essential to every organism. Killer whales *(Orcinus orca)* spend their entire lives in the ocean.

Why Does Ice Float?

Most solids sink into their corresponding liquids, but water is unusual. As water freezes, the water molecules assume a regular, crystalline arrangement that takes up more room than the disorganized molecules in liquid water. The ice is therefore less dense than water and floats at the surface of water (Table 2-1A). Because ice floats, lakes and ponds freeze from the top down, instead of from the bottom up. The surface ice of a wintry pond, such as that in Figure 2-10, insulates the liquid water below the surface from the freezing air above. As a result, ponds, lakes, and rivers remain liquid at the bottom, allowing fish and other organisms to survive through long winters.

How Does Water Resist Changes in Temperature?

For a given amount of additional heat, water temperature increases much less than in other substances; that is, water has a high **heat capacity.** One consequence of water's heat capacity is that the oceans heat up and cool down very slowly, moderating the climates of nearby land. While the interiors of the world's Northern Hemisphere continents disappear under layers of snow in winter, temperatures in coastal areas remain relatively mild. Conversely, in summer, the continental interiors reach blazing temperatures, while coastal areas remain cool. Every large body of water has a similar influence on the land nearby.

Water not only cools the coasts of continents, it also cools the human body. To keep our body temperature within the narrow range that many biochemical reactions demand, we must rid ourselves of excess heat. Humans do so largely by perspiring. As the water in sweat evaporates, it carries the heat away. Dogs accomplish the same thing by panting (Figure 2-11).

Ice floats on water because ice has a lower density than water. Water's enormous heat capacity stabilizes temperatures in oceans, lakes, coastal areas, and even within individuals.

Why Do Water Molecules Cling to One Another?

An attraction between molecules of the same substance is called **cohesion.** Water and chewing gum are both highly cohesive. Under tension, they stretch but do not break. The cohesion of water molecules also creates high **surface tension**—the tendency of a substance to form a smooth round surface. Because water molecules cohere, a water droplet forms a round pearl and its surface resists pressure. For example, the bowl of a teaspoon balanced on the edge of a teacup will rest on the surface of the tea as though floating (Table 2-1B). Unlike water, a cup of alcohol or olive oil will not support the same spoon. The attraction

Table 2-1
How Does Water Compare with Ethanol and Oil?

	Water	Ethanol	Oil

A. **Density.** In water, the liquid phase is more dense than the solid phase (ice), which is why ice floats. In most substances, including ethanol and olive oil ("oleic acid"), the solid phase is denser than the liquid phase.

B. **Cohesion.** In water, the attraction of the water molecules for one another creates a web of molecules that resists the sinking of the spoon. Neither ethanol nor olive oil is as cohesive as water, and the spoon sinks.

C. **Adhesion.** Water molecules' polarity enables them to form stronger interactions with the molecules in paper, so water rises faster than does ethanol or olive oil.

D. **Ability to dissolve other substances.** Water dissolves table sugar (sucrose) more quickly than does ethanol or olive oils.

E. If you were trapped on a desert island, which of these would you need to live?

(A–D, Charles D. Winters; E, © 1994 Zefa Germany/The Stock Market; Royalty-Free/Corbis; © Camerique/The Picture Cube)

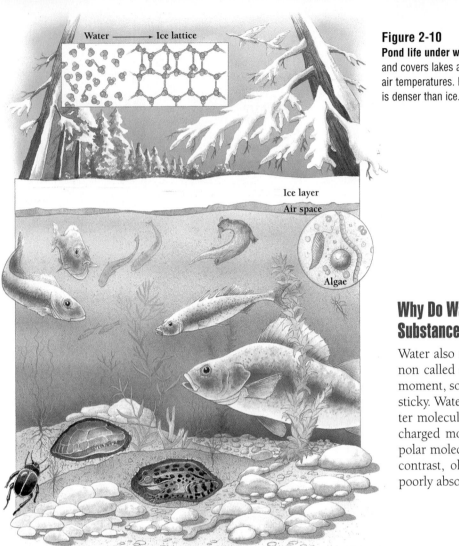

Water ——→ Ice lattice

Ice layer
Air space
Algae

Figure 2-10
Pond life under winter ice. Because water is denser than ice, the ice floats and covers lakes and ponds in the winter, insulating the water from freezing air temperatures. Fish, frogs, and algae survive the winter, all because water is denser than ice.

Why Do Water Molecules Cling to Other Substances?

Water also clings to and "wets" other surfaces, a phenomenon called **adhesion.** If you dip your hand in water for a moment, some of the water will cling to your hand. Water is sticky. Water's adhesiveness results from the tendency of water molecules to form hydrogen bonds with other polar or charged molecules. Paper towels, which are made of long polar molecules called cellulose, absorb water very well. In contrast, olive oil, which is mostly nonpolar, is relatively poorly absorbed by the cellulose in paper towels. That's why

among water molecules is so great that water has the highest surface tension of any liquid except for liquid mercury. The surface of water acts like a thin skin, and some insects, such as the water strider, walk easily over a pond's surface.

Water is cohesive because of special weak bonds among the water molecules. These special bonds are **hydrogen bonds**—a weak attraction between two polar molecules (Figure 2-12). When a hydrogen atom, like that in water, attaches to a highly electronegative atom, such as oxygen, the covalent bond is polar. The hydrogen atom ends up with a slight positive charge (not a full electron's worth of charge) that is attracted to the negative (oxygen) end of other water molecules. The most common hydrogen bonds are those between water molecules, but other hydrogen bonds also play a critical role in holding together the complex structure of DNA and proteins.

Hydrogen bonding makes water molecules cohere, creating surface tension.

Figure 2-11
Why are dogs' noses wet?
Water's enormous heat capacity makes it a useful coolant. As the water in a dog's nose evaporates, it carries heat away, cooling the inside of the nose. On a hot day, overheated blood from the body flows through the cold nose before reaching the brain, so that the dog's brain tissues remain cool.

John Cancalosi/DRK Photo

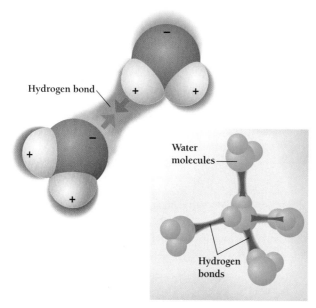

Figure 2-12

Hydrogen bonds between water molecules. Each hydrogen atom on a water molecule is attracted to the oxygen atom of another water molecule.

Biology⊘Now™ Learn more about hydrogen bonds by clicking on this figure on your BiologyNow CD-ROM.

Larry Ulrich/DRK Photo

Figure 2-13

Going up. Because water sticks to itself, trees can raise it hundreds of feet off the ground. Individual California coast redwoods *(Sequoia sempervirens)* may grow to heights of 360 feet, taller than a 25-story building.

cleaning up water is so much easier than cleaning up olive oil (Table 2-1C). Water's adhesiveness allows a close interaction between it and many other biological surfaces.

Because water both clings to itself (cohesion) and clings to other surfaces (adhesion), it readily enters and climbs tiny tubes, in a process called **capillary action.** If we place fine glass tubes (capillaries) into beakers of water, ethanol, and olive oil, the water moves upward a fair distance, the ethanol moves less, and the olive oil not at all. Such capillary action is important in the movement of water through soil, and up the narrow systems of tubes inside of trees and other plants. Water's cohesiveness and adhesiveness help trees raise water from roots deep in the ground to leaves hundreds of feet in the air (Figure 2-13).

Water Is a Powerful Solvent

Most people think of a solvent as the fluid that the dry cleaner uses to clean a wool coat, or maybe the gasoline you use to get grease off the wheels of your car. In fact, a **solvent** is any fluid in which other substances dissolve. Anyone who has poured salt into boiling water or stirred sugar into a cup of hot coffee will have noticed how rapidly salt and sugar dissolve in water. Water is a powerful solvent. It is an especially good solvent for ions, such as those of table salt, and for polar molecules, such as sugar. Salt and sugar, both important components of life, dissolve well in water but poorly

in ethanol and olive oil, which are not polar (Table 2-1D). When sodium chloride dissolves in water, the highly polar water molecules surround the Na^+ and Cl^- ions like devoted fans crowding around a sports star. The water molecules crowding around the ions separate them from one another (Figure 2-14).

In general, polar molecules such as sugar are hydrophilic (water loving) and therefore dissolve well in water. Nonpolar molecules are hydrophobic (water fearing) and tend not to dissolve in water. Oil, for example, is nonpolar and does not mix well with water. On the other hand, it mixes easily with nonpolar solvents such as gasoline.

Some molecules can be hydrophilic and hydrophobic at the same time. Molecules that have both hydrophilic and hydrophobic regions are **amphipathic** [Greek, *amphi* = both + *pathos* = feeling]. When an amphipathic molecule is placed in water, its hydrophobic regions avoid the water, while its hydrophilic regions bond with water. Detergents used to clean dishes and clothes are made of amphipathic molecules that interact with both water and oil. Detergents remove oil by making it soluble in water.

Amphipathic molecules are central to all of biology. The amphipathic "phospholipids" make up the membranes surrounding all cells, from those that form our brains and mus-

Figure 2-14

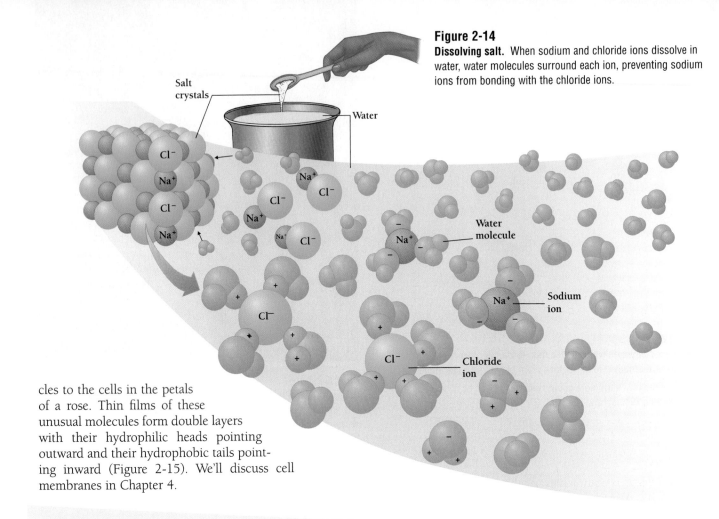

cles to the cells in the petals of a rose. Thin films of these unusual molecules form double layers with their hydrophilic heads pointing outward and their hydrophobic tails pointing inward (Figure 2-15). We'll discuss cell membranes in Chapter 4.

Water's polar properties make it an excellent solvent for ions and polar molecules. Amphipathic molecules called phospholipids form the boundary of every living cell.

How Do Water Molecules Change in Solution?

In a water molecule, the electrons stay closer to the oxygen atom than to the hydrogen atoms. Sometimes the electrons get so close to the oxygen atom that one of the hydrogen atoms has no electron at all and becomes a naked nucleus, or **hydrogen ion** (H^+). A hydrogen ion without an electron is just a proton, and it can jump back and forth between the oxygen atoms of two different water molecules. When that happens, one water molecule can end up with an extra proton and the other water molecule ends up missing a proton. The deprived water molecule is called a **hydroxide ion** (OH^-) and has a negative charge. The water molecule with the extra hydrogen ion (H^+) is called a **hydronium ion** (H_3O^+) and has a positive charge. We can represent this exchange of hydrogen ions by the following chemical equation:

$$H_2O + H_2O \rightarrow OH^- + H_3O^+$$
water + water → hydroxide ion + hydronium ion

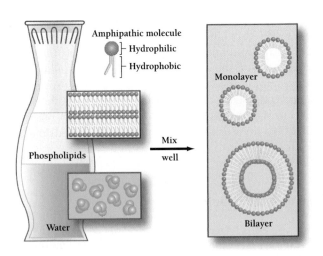

Figure 2-15
Amphipathic molecules form membranes. Amphipathic molecules tend to form layers. In monolayers, the molecules form a sphere with all the hydrophic heads pointing outward. In other cases, two layers of amphipathic molecules form a "bilayer membrane," the inner core of which contains the water-loving tails. Facing outward and inward are the hydrophobic heads. Bilayer membranes form the boundaries for every cell in the body and in the cells of all other organisms.

Extreme Biology The Uses and Dangers of Radioisotopes

Radioactive isotopes are enormously useful. They can be used to determine the age of rocks and fossils; to track individual molecules in viruses, cells, organisms, or ecosystems; and even to identify the origins of products ranging from brand-name beverages to the explosive materials used in bombs.

But radioactive isotopes are also dangerous. As they break down, they release a subatomic particle in the form of high-energy *ionizing radiation.* Like x rays, another form of ionizing radiation, β and other particles can damage and kill cells. When ionizing radiation reaches living tissues, it causes molecules (especially water molecules) to break into *free radicals,* molecules (or atoms) that easily react with and damage DNA, proteins, and other biologically important molecules.

Free radicals can kill cells outright or drastically alter their ability to function. Cells whose DNA has been damaged by radiation often cannot divide. As a result, ionizing radiation most affects tissues where cells divide frequently—those that line the mouth, the stomach, and the intestines, as well as skin cells that produce hair and bone marrow cells that make blood cells. People and other mammals exposed to intense doses of ionizing radiation—whether from nuclear weapons used in World War II, from accidents, or as therapy for cancer—suffer nausea, vomiting, diarrhea, hair loss, and anemia. Because radiation from radioisotopes can prevent cell division, radiation is a standard therapy for many forms of cancer, although, ironically, radiation can also cause cancer.

The Polish-born French chemist Marie Curie did some of the earliest and best work on radioisotopes. Marie Curie and her husband, Pierre, shared the 1903 Nobel prize for Chemistry with H. Becquerel for the discovery of radioactivity. After Pierre's death in 1906, Curie assumed his professorship at the Sorbonne and went on to win a second Nobel prize, in 1911, for the isolation of the element radium. Her subsequent work laid the groundwork for the discovery of the neutron and the synthesis of artificial radioactive elements. This latter work was carried out by Curie's daughter Irène Joliot-Curie and her husband, Frédéric Joliot. In 1934, Marie Curie died of leukemia at

Figure A
Marie Curie and her two daughters, Irène and Eva.

age 67, a result of her lifelong exposure to radiation. The following year, Irène and Frédéric won the Nobel prize for Chemistry. Irène Curie also died of leukemia, in 1956, at age 59, and her husband Frédéric assumed her professorship at the University of Paris.

In the early 20th century, Curie's radium became a major weapon in the war on cancer. No one, especially Marie Curie, wanted to believe that radium could also cause harm. She hid the cataracts that clouded her eyes so people would not think they were caused by her beloved radium, and she could not believe that radium had anything to do with the early deaths of four of her colleagues who also worked with radium. Even today, her desk and laboratory notebooks are still highly radioactive, testimony to the massive amounts of radium that she accumulated for study.

One the ghastliest results of the discovery of radium was the inadvertent poisoning of young women during World War I, whose job was to paint radium onto the faces of watches. The glowing radium helped Allied soldiers to read their watches in the dark and to synchronize attacks. As young women in U.S. factories painted the radium onto each watch, they drew the tips of the brushes to fine points using their tongues and lips, and in so doing ingested radium. Because radium chemically resembles calcium, the women incorporated it into their teeth and bones. As one worker later recalled, after work "we'd get the neighborhood kids to watch Alice glow in a dark closet. They'd get a good laugh out of that." Within a decade all these women began to come down with different kinds of cancers, including leukemia, the failure to make blood cells, and cancers of the jaws and mouths. Gradually, the dangers of radiation became known.

Culver Pictures

Because water molecules that have lost or gained a proton have charges, they very quickly recombine with other hydroxide ions and hydrogen ions to form regular uncharged water molecules. Water is constantly splitting into hydrogen ions and hydroxide ions, and the ions constantly recombine to form water (Figure 2-16).

A molecule that can give up a hydrogen ion is called an **acid**. A molecule that can accept a hydrogen ion from an acid is called a **base**. An acid dissolved in water adds hydrogen ions to the water and raises the concentration of hydrogen ions. Vinegar has a million times more hydrogen ions than hydroxide ions. Stomach acid contains about

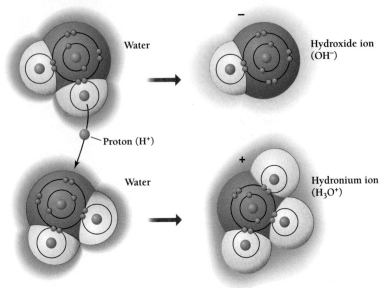

Water

Proton (H⁺)

Water

Hydroxide ion
(OH⁻)

Hydronium ion
(H₃O⁺)

A. A proton jumps from one
water molecule to another.

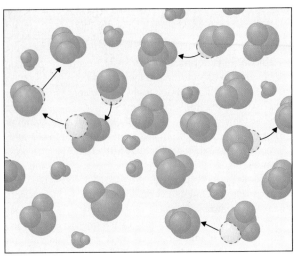

B. Water molecules exchanging protons

Figure 2-16

Water molecules swap protons. A. Hydrogen nuclei, or protons, commonly jump back and forth between water molecules. When a proton jumps from one water molecule to another and stays, the result is a hydronium ion (H_3O^-) and a hydroxide ion (OH^-). B. When the proton from a hydronium ion returns to a hydroxide ion, both ions become ordinary water molecules again. In a glass of water, protons are constantly jumping back and forth, forming and unforming ions.

one trillion times more hydrogen ions than hydroxide ions. The concentration of hydrogen ions—the number of hydrogen ions in a given amount of liquid—is measured as **pH** (Figure 2-17). The pH scale is based on the logarithm (to the base 10) of the concentration of hydrogen ions. We can think of pH as standing for "power of hydrogen." The greater the concentration of hydrogen ions, the lower the pH value. So stomach acid, which has a pH of 1, is a far stronger acid than saliva, which has a pH of 7 (neutral).

Water molecules split into hydrogen ions and hydroxide ions. The exact balance between the two kinds of ions determines the pH (the relative acidity) of the solution.

Why Is pH Important to Organisms?

Changing the pH of a solution affects the properties of other molecules. For example, in humans, the pH of the blood is normally 7.4. If the pH of the blood decreases even slightly—to pH 6.95—the nervous system becomes unresponsive, and coma and death soon follow. Alternatively, if the blood pH increases to just pH 7.7, the nervous system becomes overreactive, and muscle spasms and convulsions begin.

In healthy organisms, blood pH is regulated very closely. But a wide variety of illnesses, injuries, and drugs can cause acid or alkaline blood. One of the most common causes of acid blood, for example, is severe diarrhea. Basic blood can result when a physician gives a patient too much of a diuretic (any drug that causes the body to lose large amounts of water in the urine) or if a person takes too much bicarbonate of soda for an upset stomach.

Small changes in pH do not have to be dangerous. Many organisms use pH to control normal processes. For example, a sperm in the testes remains motionless, saving energy for the time when it is near an egg. Yet something must activate the sperm when the moment is right. In sea urchins, that something is a decrease in pH. The pH of the sea urchin's testes and sperm is about 7.2, very slightly basic. When a sea urchin releases its sperm into seawater, which has a pH of about 8.0, the sperm become more basic (about 7.6), which activates the sperm. Off they swim.

Both inside and outside of cells, pH powerfully influences the chemistry of life.

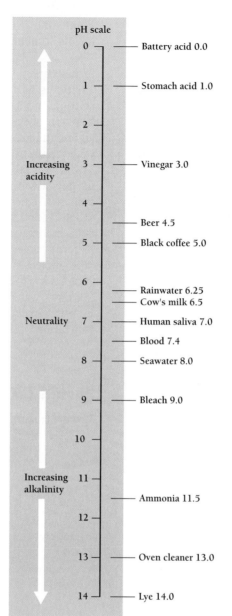

pH scale

0 — Battery acid 0.0

1 — Stomach acid 1.0

2

Increasing acidity   3 — Vinegar 3.0

4

— Beer 4.5

5 — Black coffee 5.0

6 — Rainwater 6.25
— Cow's milk 6.5

Neutrality   7 — Human saliva 7.0

— Blood 7.4

8 — Seawater 8.0

9 — Bleach 9.0

10

Increasing alkalinity   11 — Ammonia 11.5

12

13 — Oven cleaner 13.0

14 — Lye 14.0

Figure 2-17
The pH scale. The pH scale measures acidity and alkalinity. Pure water, with a pH of 7.0, is considered neutral. Anything above 7.0 is alkaline and anything below 7.0 is acid.

Key Concepts

- An atom is the fundamental unit of an element, and a molecule is the fundamental unit of a compound.
- In a covalent bond, atoms share electrons. In an ionic bond, electrical attraction holds atoms together.
- If the atoms in a covalent bond do not share electrons equally, the resulting molecule will be polar.
- Individual molecules may also interact with each other by way of hydrogen bonds or hydrophobic interactions.

Summary with Key Terms

What is matter?

Chemistry is the study of the properties and transformations of matter. All **matter** is composed of **atoms.** An atom is the smallest particle of an element that has the properties of that element. **Elements** are forms of matter that cannot be broken down to simpler substances. **Compounds** can be broken down into elements. Most matter consists of **mixtures** of compounds.

What determines the properties of an atom?

Compounds consist of **molecules**, which are fixed arrangements of two or more atoms, with a characteristic **molecular weight**. Almost all of an atom's **mass**, or weight, is contained in its **nucleus**. The nucleus contains both protons and **neutrons**. (**Nuclear reactions** change the number of protons and neutrons in a nucleus.) Most atoms are electrically neutral; the positive charge of the **protons** in the nucleus is balanced by an equal number of negatively charged **electrons**, which lie outside the nucleus and occupy most of the space of the atom. Electrons are arranged in **orbitals**, each of which can hold a different number of electrons. Groups of orbitals, in turn, form **shells**, which differ in energy. The chemical properties of an atom are largely determined by the number of electrons in the outermost shell.

What kinds of bonds bring atoms together to form molecules?

Chemical reactions change the arrangements of atoms in molecules. In forming chemical bonds, electrons may be rearranged in two distinct ways: (1) by sharing with other atoms to form **covalent bonds**, or (2) by donating or accepting electrons to form **ionic bonds**. Atoms that attract electrons are said to be **electronegative**. Molecules with uneven distributions of electrons are said to be **polar**; those with uniform charge distributions are said to be **nonpolar**. Individual molecules may also interact via **hydrogen bonds** and **hydrophobic interactions**.

What special characteristics of water made it possible for life to evolve?

Water's many special properties include its **cohesiveness**, its higher density as a liquid than as a solid, its high **heat capacity**, its **adhesiveness** (which leads to water's **capillary action**), and its ability to serve as a powerful **solvent**. Water's hydrogen nuclei (protons) can jump back and forth among different water molecules. Many other properties of water result from its polarity. **Hydrophilic** substances readily dissolve in water. Nonpolar, or **hydrophobic**, molecules do not dissolve readily in water. Many biologically important molecules are **amphipathic**, containing both hydrophilic and hydrophobic regions, in particular the molecules that make up cell membranes.

Any molecule that can donate a **hydrogen ion** to another molecule is called an **acid**. And any molecule that can accept a hydrogen ion from an acid is called a **base**. Some of the time, water's jumping hydrogen ions create **hydroxide ions** and **hydronium ions**. The **pH** scale measures the **concentration** of hydroxide ions in a solution.

Review and Thought Questions

Review Questions

1. Define the words "element," "atom," "compound," and "molecule."

2. What is an electron orbital? What is an electron shell?

3. Explain the difference between a covalent bond and an ionic bond. How is each formed? Name a common compound that exhibits each type of bond.

4. Explain how a covalent bond can be polar or nonpolar. Give examples of each type of bond. Name a polar molecule that is of great importance to organisms and their environments.

5. Define "cohesion." How does adhesion differ from cohesion?

6. Give examples of polar and ionic compounds that dissolve easily in water as well as nonpolar ones that do not. Which type of molecule would you use to make waterproof bags?

7. What type of bond is formed between adjacent water molecules? Why do they cohere?

8. What types of bonds tend to produce (1) hydrophilic substances or (2) hydrophobic ones?

9. What is a hydrogen ion? If a hydrogen ion is formed from a water molecule, what other ion is also formed?

Thought Question

10. The tendency for water's hydrogen ions (also called protons) to jump around among different water molecules accounts for water's high heat capacity, a topic not explained in this chapter. Can you try to explain how the jumping hydrogen ions could give water a high heat capacity?

BiologyNow Resources

Biology ⊗ Now™

Active Figures

2-6: Covalent bonding
2-8: Polarity of water
2-12: Hydrogen bonds

Preparing for an exam? Take a diagnostic test on your BiologyNow CD-ROM.

Online materials relating to this chapter are at:

http://biology.brookscole.com/AAL3

About the Chapter-Opening Image

Alfred Nobel invented a way of packaging highly explosive nitroglycerin into dynamite, but he would not take nitroglycerin as a medicine for his heart condition. We now know that nitroglycerin, as a medicine, releases nitric oxide (NO), which dilates blood vessels.

Biological Molecules Great and Small

Key Questions

- How do organisms use chains of carbon atoms decorated with functional groups to build most of the basic building blocks of life?
- What are the main categories of building block molecules?
- What two simple reactions link and unlink these building blocks?
- How are fatty acids, polysaccharides, nucleic acids, and proteins formed from simple building blocks?

Prions: A Slow-Growing Idea

In July of 1972, a medical resident at a San Francisco hospital admitted an elderly woman whose memory was failing. The 30-year-old doctor, Stanley Prusiner, was frustrated to learn that there was nothing he could do to help her. But his interest in her case started him on a career trajectory that would eventually win him a Nobel prize.

The woman had a rare disease of the brain called Creutzfeldt-Jakob disease. No one knew what caused it or how to cure it. Her body's defenses mounted no attack against it, and after only seven weeks, she died. An autopsy showed that her brain was riddled with holes, like a sponge. Intrigued, Prusiner resolved to read everything he could about the mysterious killer. He soon learned that it was one of a group of fatal brain diseases known as transmissible spongiform encephalopathies (TSEs)—literally, diseases that turn the brain into a sponge and are catching ("transmissible").

One TSE, called kuru, killed people in Papua New Guinea whose custom was to honor the dead by eating their brains. Another TSE, called scrapie, damaged the brains of sheep so that they became uncoordinated and also scratched themselves so viciously that they scraped off their own wool.

What caused the different forms of TSEs? One clue was that tissues from an animal with a TSE could be used to infect another animal. From this, biologists concluded that some infectious agent—perhaps a bacterium or a virus—was causing the disease. Most researchers concluded that TSEs were caused by "slow viruses" like the AIDS virus, which take a long time to produce symptoms. But no one had been able to isolate such a virus. And Prusiner soon encountered a radically different hypothesis about TSEs—an idea that slowly took over his life (Figure 3-1).

In 1967, radiobiologist Tikvah Alper had published a scientific paper on scrapie so controversial that scientists around the world buried her in mail. Alper's results implied that scrapie couldn't be caused by a virus or any known organism. Her experiments had shown that tissues infected with scrapie could infect animals even after apparently all DNA and RNA had been destroyed by radiation. Viruses, like bacteria and other organisms, rely on the nucleic acids DNA and RNA to carry genetic information. To destroy their DNA or RNA was to destroy their ability to infect cells.

Alper proposed that whatever was causing scrapie had no genes, or didn't need them. But it was heresy to suggest that a virus or any infectious microbe could exist and reproduce without genes. And it was heresy to suggest that anything else besides a microbe or a virus could infect the body and spread.

Alper knew what her experiments had shown, however, and she was not one to back down. When her paper was initially rejected for publication, she wrote an outraged letter to the editor and persuaded him to change his mind. "Tikvah is remembered by most people for her forceful discussion style, which could be quite alarming," two of Alper's colleagues later wrote. (At 79, she sent a burglar fleeing from her home after tackling him and trying to take his gun away.)

But Alper's funding came from the Agricultural Research Council of Great Britain, which decided that scrapie was unimportant and refused to fund further work, and Alper gamely turned her attention to other interesting projects.

The council's decision proved to be a mistake. In the 1980s, the TSE "mad cow disease" shut down the English beef industry; nearly 2 million beef cattle and other animals had to be slaughtered and burned, at a cost of more than $6 billion. Even so, dozens of people died of mad cow disease, probably from eating tainted beef years earlier (see box).

No one with any interest in TSE was going to forget Alper's discovery. There was, after all, one possible explanation for her results. Shortly after her article was published, another British researcher proposed that proteins could sometimes fold into unusual shapes. The presence of these oddly folded proteins somehow coaxed normal proteins into misfolding in the same way. Could scrapie proteins cause disease by forcing normal proteins in the brains of sheep to refold incorrectly? It was hard to imagine.

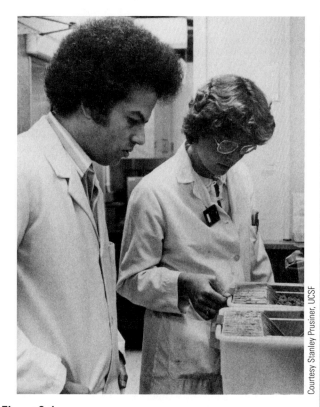

Figure 3-1

Stanley Prusiner. Prusiner's research on the brain disease scrapie persuaded most scientists that the way proteins fold can carry biological information and even cause disease.

Courtesy Stanley Prusiner, UCSF

In 1974, Prusiner had just begun working as an assistant professor at the University of California at San Francisco (UCSF), studying scrapie. Despite Alper's results, Prusiner assumed that a virus caused scrapie. It was just a matter of finding it. That's what everyone said.

But as Prusiner began purifying and testing extracts from scrapie tissues, the evidence gradually forced him to change his mind. The more he repeated Alper's experiments and refined them, the more he came to believe that she was right.

When he damaged the nucleic acids in scrapie extracts, they continued to infect animals. But when he denatured—or unfolded—scrapie's proteins, the extracts lost their ability to infect. These results could only mean that the infectious agent was a protein. But for something to cause an infection, it must make more of itself. And everyone knew that proteins were incapable of reproducing themselves.

Just as Prusiner became completely fascinated by scrapie, however, his research grant was canceled, just as Alper's had been. Almost simultaneously, the university informed him that he would not be given tenure and would have to leave. Prusiner was devastated. He was losing his job, his lab, and the grant he needed to pay for his research. Yet the moral support of a few of his closest colleagues carried him through. Skilled at selling his ideas, Prusiner soon convinced two private foundations to fund his work. And, amazingly, even his tenure decision was reversed. He would be allowed to stay at UCSF.

By 1982, Prusiner still lacked a convincing explanation for scrapie. But he boldly proposed that scrapie was caused by an infectious protein he called a *prion* (pronounced "PREE-on"). Many scientists said his claim was premature, even brazen. Former allies turned against him. One close colleague quit, later telling *Discover* magazine, "There's no point in creating a name for something that we don't know even exists yet. . . . [Prusiner] tends to jump to conclusions, to give less credence to facts that would discount his preferred interpretation of the results." Feeling betrayed, Prusiner refused to talk to the press for a decade after the publication of the *Discover* article. In some circles, he had become a laughingstock.

Business and Biology A Smaller Steak in the Beef Industry

In May of 2003, the discovery of a single case of the TSE "mad-cow disease" virtually shut down the $2.2 billion-a-year Canadian beef industry. Although Great Britain had been hit hard by the disease in the 1980s, this was the first case in Canada (excepting a diseased animal shipped to Canada in 1993).

Because mad-cow disease–infected meat can make humans sick, the United States Food and Drug Administration (FDA) immediately banned the import of beef and live cattle from Canada. The annual flow of 1.7 million cattle and 390,000 metric tons of beef into the United States, about nine percent of the beef consumed in the United States, ground to sudden halt. The ban would cost the Canadian beef industry millions of dollars a day and create shortages elsewhere.

Immediately after the announcement of the ban, stock prices for four of the biggest beef sellers—Tyson foods, McDonald's Corporation, Wendy's International, and Outback Steakhouse—dropped an average of six percent on the New York Stock Exchange, as investors unloaded stocks re-

lated to beef. Japan, South Korea, Mexico, and other major beef importing nations also banned Canadian beef, and the price of Canadian beef plummeted. In the United States, where beef was now in short supply, the price rose.

The immediate mystery facing Canadian investigators was where the eight-year-old steer had contracted the disease. The common name for mad-cow disease is bovine spongiform encephalopathy. Just as in similar diseases in humans and sheep, the symptoms take a long time to develop, so the animal could have contracted the disease anytime in its life. Most researchers believe that the practice of feeding cattle a diet containing animal byproducts allowed mad cow disease to spread in Britain and could elsewhere. To prevent this, current regulations say that cattle feed cannot contain byproducts from cattle, and sheep feed cannot contain byproducts from sheep. But if domestic cattle ate plants only, it is unlikely that they would get prion diseases.

After weeks of investigation, Canadian officials announced that they could not say

where the animal had been raised and could not rule out the possibility that it had come from a 1998 shipment of Angus cattle from the United States. Officials at the U.S. Department of Agriculture hastily countered by saying that there was "no evidence" that anything of the kind had occurred.

In early July, Canadian Prime Minister Jean Chrétien urged President Bush to lift the ban on Canadian beef. Bush obligingly announced that he wanted a "quick end to the ban." But later that month, Senator Tom Harkin of Iowa said that it was too soon to lift the ban on Canadian beef. Japan had refused to buy U.S. beef unless American suppliers could guarantee that it contained no Canadian beef, and the only way to do that was for the United States to maintain the ban on Canadian beef. Harkin's announcement triggered another rise in the price of U.S. beef—1.5 cents a pound, the maximum allowable for one day. In late July, Outback Steakhouse told investors that it would deal with both the beef shortage and high prices by offering its customers smaller steaks.

But like Alper, Prusiner was determined to change the minds of those around him, and to prove his ideas. He and his research team isolated the prion protein, called "PrP."

The next step was to describe PrP's structure. All proteins are made of long chains of small molecules called amino acids. The long chains fold into the complex shapes of functioning proteins. Prusiner and colleagues at other institutions determined the order of the amino acids in the PrP protein, then showed that PrP could fold up in at least two ways. One form was *scrapie PrP*, found in the brains of sick sheep; the other was *cellular PrP*, a protein found in the cells of healthy animals.

The two forms were distinctly different. Where the normal, cellular PrP was twisted into spirals called α "alpha" helices, scrapie PrP was folded into parallel strands called β "beta" pleated sheets (Figure 3-2).

Was it possible that scrapie PrP was forcing the body's own PrP to refold and become scrapie PrP? Prusiner hypothesized that scrapie PrP could turn the α helices into β sheets. Working with short pieces of PrP, they found that β sheets could indeed nudge α helices into becoming β sheets. And, amazingly, other biologists found they could transform full-

length cellular PrP into scrapie PrP by mixing the two versions of PrP together.

These experiments astonished biologists, shattering the established views that genes alone determine the three-dimensional structures of proteins and that genes alone carried biological information. Prusiner and his colleagues showed that proteins too could carry information.

But other scientists continued to reject Prusiner's results. Even Tikvah Alper took on Prusiner. In 1993, Alper began to revisit her early work on scrapie. She argued that the scrapie agent was neither a protein nor a virus, but a rogue piece of membrane from a cell. Just two weeks before her death, in 1995, she was still writing to scientific journals.

But early on October 6, 1997, a boisterous party blossomed in Prusiner's normally quiet lab. After more than two decades of struggling against a resistant scientific establishment, Prusiner had received one of the most satisfying acknowledgments of the importance of his work, a Nobel prize. But even his Nobel prize—for discovering a new class of disease-causing agents and for showing how they could infect an organism—failed to quell the controversy. Some re-

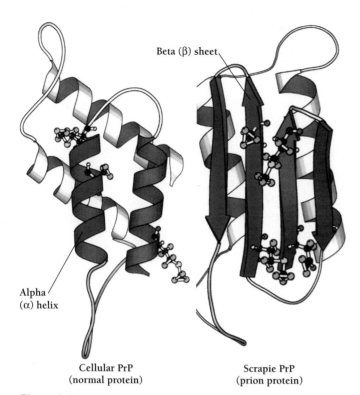

Beta (β) sheet

Alpha (α) helix

Cellular PrP
(normal protein)

Scrapie PrP
(prion protein)

Figure 3-2

An α helix and a β sheet. In the prion protein PrP, the difference between an α helix and a β sheet can spell the difference between life and death. When regions of the protein that are normally α helices (left) fold into β sheets (right), the result may be a fatal disease of the brain. (Courtesy of Stanley Prusiner, UCSF)

searchers continued to insist that TSEs were caused by viruses. More than a few biologists argued that Prusiner's Nobel prize was an outright error.

Five years later, the study of protein folding remains one of the most exciting fields in biology. Understanding how and why proteins fold the way they do is critical to understanding the physiology and genetics of cells and whole organisms. Yet, to understand what causes a protein to fold, we need to understand the chemistry of the smaller molecules that make up proteins. In this chapter, we look at the way cells use small building blocks to build larger molecules and how some of those molecules then fold into three-dimensional shapes.

3.1 How Do Organisms Build Biological Molecules?

If we measure the sizes of all the molecules in a cell, we discover a surprising thing. Cells have many small molecules, with molecular weights less than 300, and many large molecules, with molecular weights greater than 10,000. But cells have very few medium-sized molecules.

How Big Are Biological Molecules?

Biologists call large molecules **macromolecules** [Greek, *macro* = large]. Macromolecules consist of the small molecules linked together in long chains. Specifically, nearly all molecules in living organisms are made of chains of carbon atoms. In fact, living organisms are associated so closely with carbon that all carbon-and-hydrogen-containing compounds are called **organic** molecules (even when they have nothing to do with organisms). **Organic chemistry** is the study of the structures and reactions of carbon-and-hydrogen compounds. **Biochemistry** is the study of the chemistry of living organisms.

The carbon-based building blocks of life are simple and universal. Even though there are a nearly infinite number of carbon-containing molecules, organisms use only a few as basic building blocks. In this chapter, we will discuss the four kinds of small "building block" molecules that alone form almost all biological structures.

The building blocks fall into just four classes: sugars, amino acids, nucleotides, and lipids. Sugars join together to form long-chain macromolecules called polysaccharides, like the starch in bread and potatoes. Amino acids form polypeptide chains, which make up proteins. Nucleotides form long chains called nucleic acids, such as the DNA that our genes are made of. Lipids, which include fats and oils, do not form long chains. However, they do form sheets of thin membrane. Along with other molecules, lipid membranes form the membranes that surround all cells and regulate their interactions with their environment.

All organisms are made from the same four kinds of molecules: sugars, amino acids, lipids, and nucleotides.

Why Are Biological Structures Made from So Few Building Blocks?

Given the enormous diversity of organic molecules, the relatively small number that are used by organisms comes as a surprise. But why should there be so few building blocks? For example, why do almost all organisms use the sugar glucose as their principal energy source? And why do all of the millions of species on this planet use the same 20 amino acids to make proteins?

The most likely explanation for this sameness is that we are all related. We all use the same building blocks because all life on Earth has a common origin. The same molecules are used again and again because organisms inherit from their ancestors the tools to use these few building blocks.

All living organisms are made of the same few small molecules because all organisms are related.

What Kinds of Structures Do Carbon Atoms Form?

Because all the building blocks of life include carbon, we begin by discussing carbon. Carbon atoms often form long chains of atoms. Each carbon bonds to the next by single or double covalent bonds. Each carbon atom can bond with up to four other atoms.

Along the sides of a simple carbon chain, hydrogen atoms fill the remaining slots. Such long chains of carbon

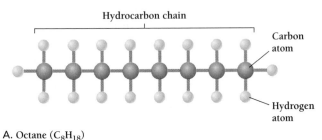

A. Octane (C_8H_{18})

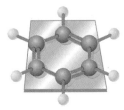

B. Benzene (C_6H_6)

Figure 3-3

Carbon forms chains and rings. A. Chains of carbon atoms form the backbone of many building block molecules, including carbohydrates and fats. Octane, a component of gasoline, has 8 carbons, with 18 hydrogen atoms hanging off the sides. Inside a working automobile engine, the carbon-carbon bonds in octane break and new bonds form with oxygen, releasing stored potential energy. B. Many building block molecules consist of flat rings, whose alternating double bonds make them lie flat. Benzene, shown here, is such a flat ring.

and hydrogen are called **hydrocarbon chains.** One simple hydrocarbon chain is octane, a component of gasoline, which consists of 8 carbon and 18 hydrogen atoms (Figure 3-3A). Most of the building blocks are based on hydrocarbon chains.

Hydrocarbon chains can also form rings (Figure 3-3B). Because carbon can form covalent bonds only at certain angles, most carbon rings have either five or six atoms. A ring sometimes includes other types of atoms, often oxygen or nitrogen. When the atoms of a ring are connected by alternating single and double bonds, the rings lie flat, just like a simple wedding band and just like the benzene ring shown in Figure 3-3B. But when the atoms of a ring are connected by single covalent bonds, the ring can bend into a bumpy structure that looks a little like a chair or a boat (Figure 3-4).

Carbon atoms form long hydrocarbon chains and five-carbon or six-carbon rings.

Functional Groups: How Do the Small Molecules Differ from One Another?

The structure of a molecule determines its biological function. Just as the parts of a car or a bicycle must fit together for the machine to work, the molecules in a cell must also fit together or interact for the cell to work. Every molecule has a distinctive shape and charge distribution. And every molecule interacts with water and other molecules in unique ways.

The special properties of a molecule depend on the arrangement of its atoms. In analyzing the structures of many organic molecules, chemists have repeatedly found standard small groupings of atoms called **functional groups,** which help predict the properties of a molecule.

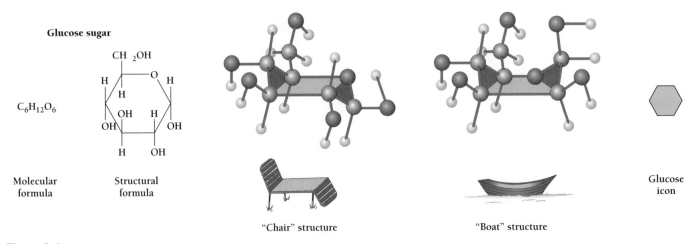

"Chair" structure   "Boat" structure

Figure 3-4

Most carbon rings bend. The simple sugar glucose consists of a ring of carbon atoms (with one oxygen). But glucose's single bonds easily twist around, allowing the molecule to buckle and fold into a "chair" structure *(left)* or a "boat" structure *(right)*. The green hexagon is the icon for glucose.

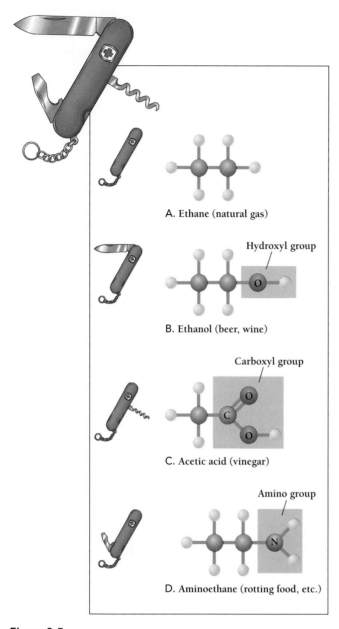

A. Ethane (natural gas)

Hydroxyl group

B. Ethanol (beer, wine)

Carboxyl group

C. Acetic acid (vinegar)

Amino group

D. Aminoethane (rotting food, etc.)

Figure 3-5
Molecular toolkit. Functional groups are like the different attachments on a Swiss Army knife. Functional groups determine the structure and function of a molecule. A. In the nontoxic gas ethane, only hydrogen atoms attach to the two-carbon chain. B. In ethanol, a hydroxyl (OH) group transforms the carbon chain into an alcohol, a toxic liquid. C. In acetic acid, a carboxyl (COOH) group transforms the two-carbon chain into an organic acid. D. In aminoethane, an amino group (NH_2) makes the molecule alkaline, or basic. Like many other amines, aminoethane is strong smelling and toxic.

Functional groups are comparable to the standardized parts of a Swiss Army knife, which may include, for example, a knife, a corkscrew, and a can opener (Figure 3-5). A Swiss Army knife may have only one or two such attachments or it may have many, in a variety of combinations. In the same way, a biological molecule may have a few func-

tional groups or many, in a variety of combinations. The exact combination of functional groups on a biological molecule profoundly affects the way the molecule behaves.

Figure 3-5 shows the structures of four simple molecules: ethane, ethanol, acetic acid, and aminoethane. Each molecule contains two carbon atoms and different numbers of hydrogen atoms. Of these molecules, ethane has no functional groups, while each of the other three molecules has a different functional group. These are a hydroxyl group, a carboxyl group, and an amino group.

Ethane has the simplest structure: two carbon atoms, each with three hydrogens attached. It is an odorless gas that makes up about nine percent of natural gas, the kind we use to heat homes (Figure 3-5A). In high concentrations, ethane has a narcotic effect, but it is not particularly toxic.

Ethanol—the kind of alcohol found in beer, wine, and liquor—looks like ethane except that in place of one hydrogen atom is a **hydroxyl**, or –OH group (Figure 3-5B). That single oxygen atom makes ethanol very different from ethane. Ethanol (or ethyl alcohol) is a sweet-smelling liquid that is comparatively toxic. It is used as an antiseptic to kill both prokaryotic and eukaryotic cells. Ethanol is also used as an industrial solvent, as an additive in gasoline, and, in veterinary medicine, to kill nerve tissue.

Ethanol's most familiar use is in beverages such as beer and liquor. In small amounts, it gives a pleasant high for a short period. In large amounts, ethanol causes nausea, vomiting, mental excitement or depression, loss of coordination, stupor, coma, and even death.

The hydroxyl group in alcohol is an extremely common functional group, which easily forms hydrogen bonds. Because oxygen is more electronegative than hydrogen, electrons in a hydroxyl group spend more time near oxygen than hydrogen. This leaves the hydrogen with a slight positive charge, so it readily forms a hydrogen bond with the slightly negative oxygen atom in a water molecule. As a result, ethanol is extremely hydrophilic and dissolves readily in water. In contrast, ethane, which has no hydroxyl group, does not interact much with water. Ethane bonds are all nonpolar, and ethane molecules do not interact much, even with one another. Most molecules that contain an OH group attached to a carbon atom—including other alcohols and also sugars—have properties similar to ethanol's.

We can take ethane and lop off one carbon with its three hydrogens and substitute a **carboxyl** group (—COOH). The result is acetic acid, the acid that gives vinegar its familiar sour smell and sharp bite (Figure 3-5C). Just as the OH group characterizes alcohols and sugars, the carboxyl group characterizes organic acids.

The carboxyl group is acidic because it gives up its hydrogen easily. Both of the two oxygens in a carboxyl group pull electrons toward them so strongly that the hydrogen atom actually loses the electron that binds it to the carboxyl group. The hydrogen ion breaks away, leaving behind a COO^- ion.

Suppose now that, instead of a hydrogen atom or a hydroxyl or carboxyl group, we substitute an **amino** group, a single atom of nitrogen bonded to two atoms of hydrogen (NH_2). An amino group tends to make molecules basic because it attracts hydrogen ions, forming NH_3^+ ions. Substituting an amino group for one of the hydrogen atoms in ethane creates aminoethane (Figure 3-5D). Like ethanol, aminoethane is a liquid, but it is otherwise very different. Highly alkaline (basic), aminoethane's powerful vapors smell intensely of ammonia and can severely irritate the skin, eyes, and mucous membranes. Aminoethane is 20 times more toxic than ethanol.

Aminoethane belongs to a class of compounds called amines. Amines all have amino functional groups and nearly all smell terrible or at least unpleasant. Two of the most offensive amines are putrescine and cadaverine, which together give memorable odors to rotting fish, aging corpses, urine, semen, and bad breath. Table 3-1 shows the properties and structures of half a dozen of the most important functional groups.

Many more functional groups exist that are not shown. Each functional group is shown attached to an "R." In the amino acids that make up proteins, R has a special meaning, which we'll mention later in the chapter. For now, just think of the R as whatever the functional group attaches to.

A small number of functional groups determine the properties of an array of biological molecules.

How Do Cells Build Complex Molecules?

We are now in a position to understand the properties of many organic molecules. Each building block has a carbon skeleton to which are attached hydrogen atoms or functional groups. These groups determine the properties of the molecule.

Some of a molecule's most important properties depend on how it interacts with water. Recall from Chapter 2 that whether a molecule is hydrophobic, hydrophilic, or both (amphipathic) depends on the polarity of the molecule's bonds. Polar molecules have a positive charge at one end that can form hydrogen bonds with the negative oxygen in a water molecule. This is why polar molecules dissolve easily in water.

As we mentioned earlier, almost all the structures of a cell are built from combinations of the four kinds of small-molecule building blocks listed in Tables 3-2, 3-3, 3-4, and 3-5. These four kinds of building blocks are: (1) **lipids**, fatty or oily nonpolar compounds that tend to dissolve in oils, but not in water; (2) **sugars**, molecules that usually have two hydrogen and one oxygen atom (water) for every atom of carbon; (3) **amino acids**, molecules that contain both amino and carboxyl groups; and (4) **nucleotides**, molecules that each consist of a nitrogen-containing ring, a sugar, and at least one molecule of phosphoric acid.

Of the four kinds of building blocks, three kinds link end to end in long chains, just like a child's plastic "pop beads." Sugars link together to form polysaccharides, nucleotides form nucleic acids, and amino acids form proteins. Only lipids do not form long chains.

In proteins and nucleic acids, the chains are simple, unbranched strings of small molecules. In contrast, polysaccharide chains formed from sugars are sometimes highly branched (Figure 3-6A). Even though proteins and nucleic acids are unbranched, they form much more complex molecules than polysaccharides. Protein chains can coil and fold into complex three-dimensional shapes. As Stanley Prusiner showed with scrapie PrP, how a protein folds is crucial to the way it behaves in a living organism.

Cells build nearly all complex molecules from four types of building blocks.

Table 3-1
Functional Groups

Functional Group	Ball and Stick Model	Structural Formula
Hydroxyl group		R—OH
Carbonyl group		$R-\overset{O}{\overset{\|}{C}}-H$ (or R)
Carboxyl group		$R-C\overset{O}{\underset{OH}{\diagup}}$
Amino group		$R-N\overset{H}{\underset{H}{<}}$
Sulfhydryl group		R—SH
Phosphate group		$R-O-\overset{O}{\overset{\|}{\underset{\|}{\underset{O^-}{P}}}}-O^-$

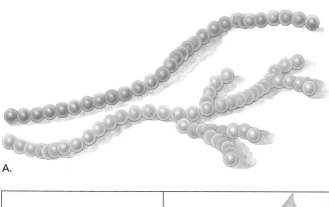

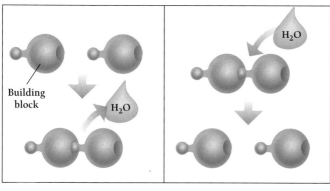

A.

B.

Dehydration   Hydrolysis

Figure 3-6
Making macromolecules. In all living organisms, small molecules link together into long chains that may be proteins, nucleic acids, or polysaccharides. Proteins and nucleic acids are always simple, unbranched chains, while polysaccharides are often branching chains. B. Two simple reactions snap molecular building blocks together or apart. In a "dehydration condensation reaction," building blocks join together after losing the equivalent of one water molecule. In a "hydrolysis reaction," the addition of one molecule of water breaks the bond between each pair of small molecules.

What Links the Building Blocks of Life?

When we digest a slice of bread, we break its carbohydrates into simple sugars and its proteins into amino acids. At the same time, we also break down and rebuild the proteins of our own skin, muscles, and bones. All organisms continually break down macromolecules and reuse the building blocks.

Organisms have to assemble and disassemble macromolecules easily. The bonds that hold macromolecules together must be strong enough so that the macromolecules will not fall apart. But the bonds must not be so strong that organisms can't easily take them apart when they need to. Like children's pop beads and Lego bricks, the building blocks of life easily snap together and easily snap apart.

Amazingly, biological building blocks all snap together in the same way (Figure 3-6B). The building blocks

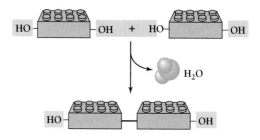

A. Dehydration condensation reaction

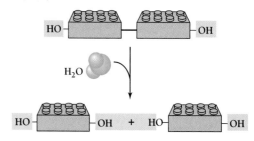

B. Hydrolysis

Figure 3-7
Add water to break bonds, remove water to make them. Shown here are the two simple reactions that link all the basic building block molecules into long chains. A. In the dehydration condensation reaction, two building blocks snap together when a single molecule of water is removed. B. In the hydrolysis reaction, two building blocks snap apart with the addition of a single molecule of water.

of all the major macromolecules join by the same simple chemical reaction. In every case, enzymes (molecules that help make and break chemical bonds) remove two hydrogen atoms and one oxygen atom from between pairs of building blocks, forming a bond. Removing two hydrogens and an oxygen—the equivalent of one molecule of water—is called a **dehydration condensation reaction** (Figure 3-7A), because one water molecule is removed.

To snap apart macromolecules, organisms reverse the dehydration condensation reaction, adding one water molecule to each pair of building blocks. Enzymes detach each small molecule from a macromolecule by adding one molecule of water, a process called **hydrolysis** [Greek, *hydro* = water + *lysis* = breaking] (Figure 3-7B).

Although all the building blocks are joined by similar dehydration condensation reactions, the exact bonds that form are different in each case. For example, sugars form *glycosidic bonds,* while amino acids form *peptide bonds.*

Enzymes link two building blocks by taking away the equivalent of one water molecule. Enzymes add a water molecule to break a single building block from a long chain.

3.2 Lipids Include a Variety of Small Nonpolar Compounds

Look around your kitchen and you'll find lipids. Oils, fats, and candle wax, to name a few, are all lipids. Because lipids contain more chemical energy per gram than other biological molecules, lipids often serve as energy stores.

One characteristic of lipids is that they dissolve poorly in water but dissolve well in nonpolar solvents such as gasoline or olive oil. The reason that lipids do not dissolve in water is that they have few functional groups that are polar. The limited solubility of lipids in water explains why removing a butter or gravy stain often requires dry cleaning, that is, treatment with nonpolar organic solvents such as benzene.

But not all lipids are hydrophobic. Some are amphipathic: they contain a polar functional group as well as nonpolar chains of carbon and hydrogen atoms. An important class of amphipathic lipids is the **fatty acids**—a major component of the membranes that enclose cells.

Fatty acids differ from one another both in the number of carbon atoms they contain and in the numbers of single and double bonds (Table 3-2). The hydrocarbon chain in a fatty acid consists of carbon atoms linked either to one or two other carbon atoms and to hydrogen atoms. When all the bonds between carbon atoms are single bonds, the hydrocarbon chain contains the maximum number of hydrogen atoms and is said to be **saturated** with hydrogen atoms. Stearic acid, found in beef fat, is a highly saturated fatty acid (Figure 3-8A).

When a fatty acid contains one or more double bonds between adjacent carbons, and is therefore missing hydrogen atoms, the fatty acid is said to be **unsaturated**. For example, oleic acid, found in vegetable oils, is unsaturated (Figure 3-8B). A fatty acid with just one double bond is said to be monounsaturated. A fatty acid with many double bonds is said to be **polyunsaturated**.

Fats, waxes, and other lipids that have saturated hydrocarbon chains are solid at room temperature. In contrast, polyunsaturated lipids tend to be liquid at room temperature, and we call them oils. The reason is in the bonds. The single C — C bonds in saturated fats rotate freely, allowing the chains to pack tightly together to form a solid. In contrast, the double C = C bonds in saturated lipids are more rigid and tend to form kinky polycarbon chains that do not pack well. The result is a fluid, or oil. In general, animals contain more fats and plants contain more oils.

Food manufacturers sometimes add hydrogen atoms to polyunsaturated oils in order to make them solid at room temperature. Cookies, chips, and other products made with such "hydrogenated" fats tend to be crisper than if made with oil. Butter would work, but hydrogenated fats are less expensive.

Both naturally and artificially saturated fats seem to increase cholesterol levels and overall risk of heart disease. Nutritionists generally recommend that we avoid saturated fats and stick to diets rich in polyunsaturated and monounsaturated oils, such as those found in nuts, grains, and olive oil. However, this area of research is surprisingly complex. For example, even though stearic acid, found in beef and chocolate, is a saturated fat, it apparently does not increase blood cholesterol levels.

A few years ago, biologists discovered that different kinds of fatty acids affect our health. In particular, people whose diets were very low in "omega-3 fatty acids" suffer more often from a host of diseases, including heart disease and arthritis. What is an omega 3 fatty acid? Technically, it's just a fatty acid with a double bond three steps from the end of the fatty acid.

The first carbon atom in a fatty acid chain is called the "number one" carbon, and the last one is called the "omega" carbon (because omega is the last letter in the Greek alphabet). If the last double bond in the chain is on the third to last carbon, the fatty acid is called an omega-3 fatty acid. But, in the United States, most people's diets are rich in omega-6 fatty acids, whose last double bond is six steps from the last carbon. The body uses both kinds of fatty acids to

LIPIDS

Polar functional group
(carboxyl group)

Hydrocarbon chain of all single bonds

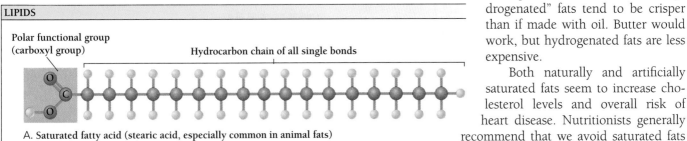

A. Saturated fatty acid (stearic acid, especially common in animal fats)

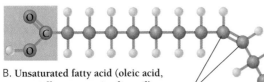

B. Unsaturated fatty acid (oleic acid, especially common in olive oil)

Double bond can accept two or more hydrogen atoms.

Figure 3-8
Saturated and unsaturated. A. Saturated fatty acids, such as the stearic acid in beef fat, have no double bonds. Because saturated fats lack double bonds, they are flexible and pack tightly together into solids such as lard, margarine, and butter. B. Unsaturated fatty acids have one or more double bonds between carbon atoms. Double bonds make the molecules more rigid, so unsaturated fats pack loosely into liquid oils.

Biology Now™ Learn more about lipids by clicking on this figure on your BiologyNow CD-ROM.

Table 3-2
The Building Blocks of Life: Lipids

Like children's Lego bricks, building block molecules easily snap together to form larger molecules. The Lego icon represents the 34 building blocks of life—20 amino acids, 6 lipids, 5 nucleotide bases, 3 simple sugars, and a phosphate group. Shown here are the 6 lipids.

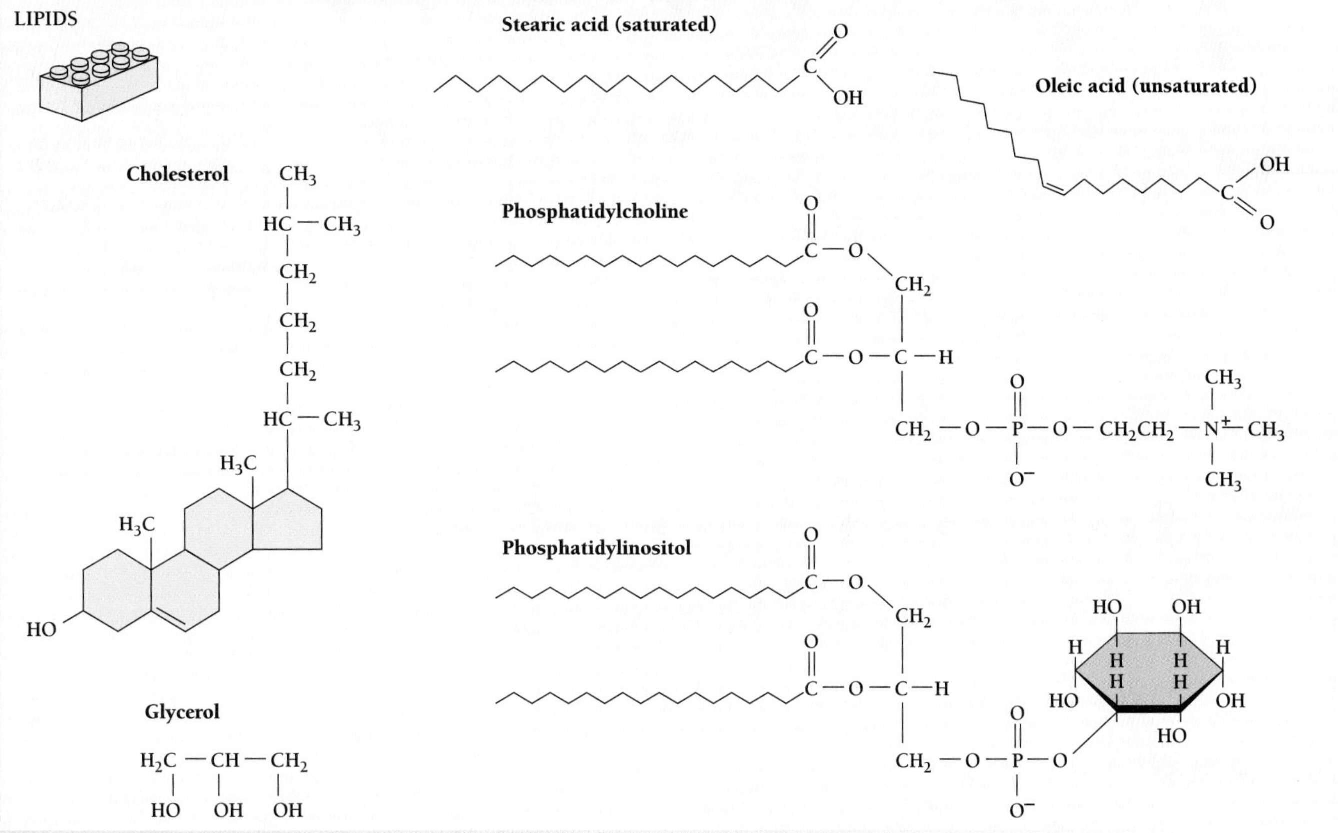

manufacture molecular messengers called eicosanoids. But eicosanoids made from omega-6 fatty acids send different signals than those made from omega-3 fatty acids. Not enough omega-3s seems to stimulate inflammation, in the circulation, the joints, and elsewhere.

Omega-6 fatty acids come to us from terrestrial plants (such as wheat, corn, and rice) or from the animals that eat those plants by way of the food chain. Omega-3 fatty acids are made by algae and tend to accumulate in marine food webs, so fish and other animals that eat a lot of fish (such as seals and polar bears) have high levels of omega-3 fatty acids. Most people associate omega-3 fatty acids with fish oil supplements. But anyone who eats fish or various kinds of edible sea weeds and algae will get omega-3 fatty acids in their diet.

Lipids, which include both fats and oils, are usually oily to the touch, insoluble in water, and high in energy. One class of lipids, the fatty acids, may be "saturated" or "unsaturated."

How Are Fatty Acids Linked Together?

Glycerol is a simple three-carbon molecule with three hydroxyl groups. Each hydroxyl can attach to the carboxyl carbon of a fatty acid by means of a dehydration condensation reaction (Figure 3-9A). A glycerol with all three hydroxyl groups attached to fatty acids, so that the glycerol has three, long hydrocarbon "streamers," is a **triglyceride** (Figure 3-9A). When a triglyceride forms, the hydrophilic character of the carboxyl group is lost. As a result, triglycerides are even less soluble in water than fatty acids. Most of the fats that humans and other animals store, as well as the oils of plants, consist of triglycerides (also called "triacylglycerols").

A glycerol molecule with a phosphate group attached in place of the third hydroxyl group and with just two fatty acid chains instead of three is a **phospholipid** (Figure 3-9B). Phospholipids are highly amphipathic. Each one has a hydrophobic "tail," which consists of the hydrocarbon chains of two fatty acids, and a hydrophilic "head," which contains the charged phosphate group.

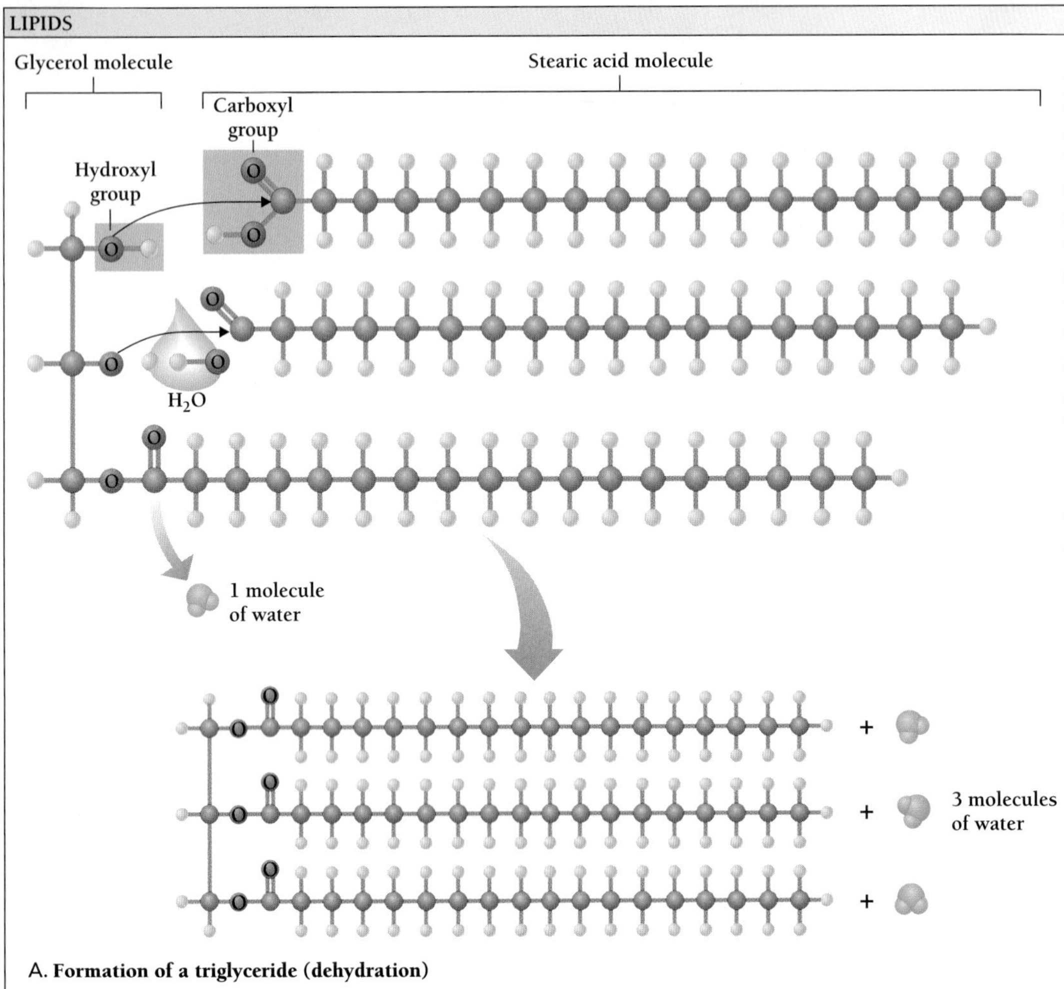

Glycerol molecule

Stearic acid molecule

Figure 3-9
Putting fatty acids to work.
A. A triglyceride forms when three fatty acid chains attach to glycerol. One molecule of water is lost at each attachment. B. A phospholipid forms when just two fatty acids attach to glycerol. C. Phospholipids form membranes in which the hydrophilic heads face out into the aqueous environment and the hydrophobic tails hide together in the middle.

Biology ⊜ Now™   Learn more about lipids by clicking on this figure on your BiologyNow CD-ROM.

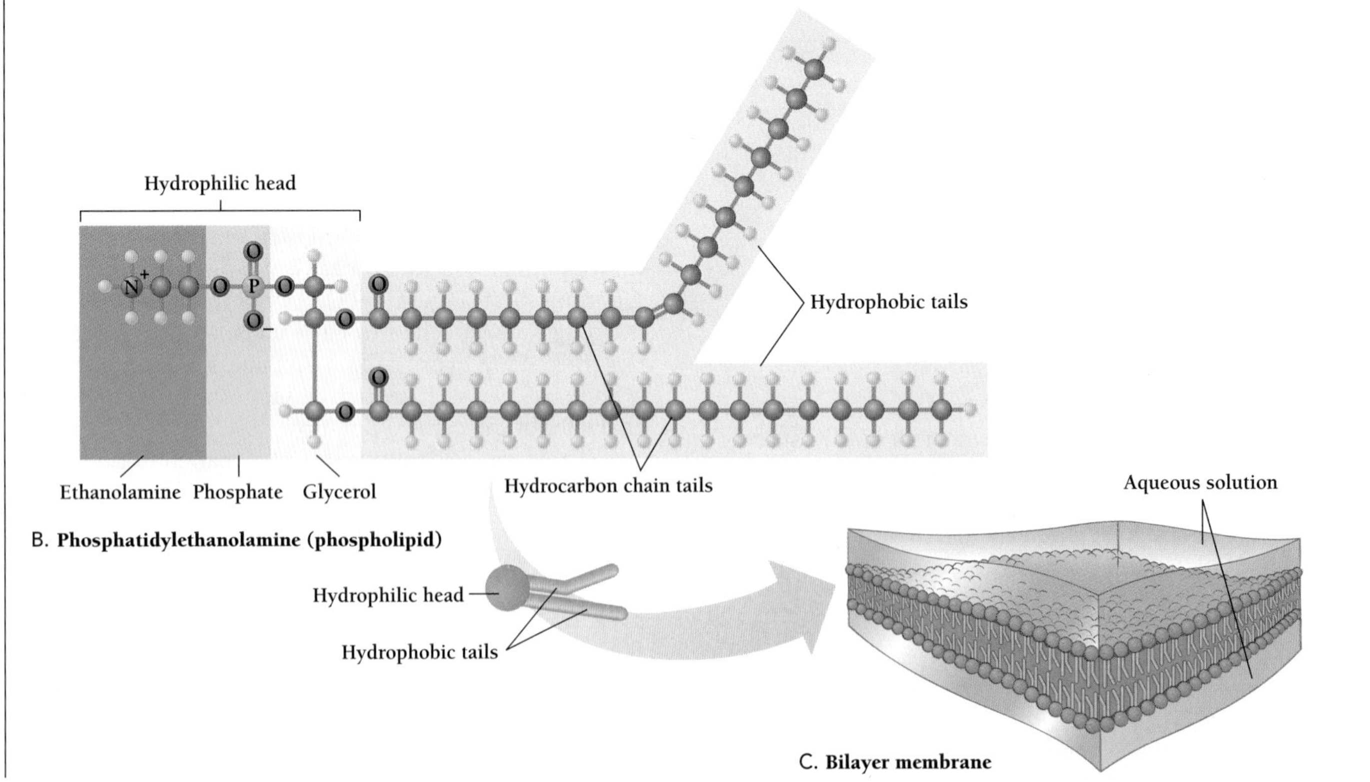

Carboxyl group

Hydroxyl group

H_2O

1 molecule of water

+

+   3 molecules of water

+

A. **Formation of a triglyceride (dehydration)**

Hydrophilic head

Hydrophobic tails

Ethanolamine Phosphate Glycerol

Hydrocarbon chain tails

B. **Phosphatidylethanolamine (phospholipid)**

Hydrophilic head

Hydrophobic tails

Aqueous solution

C. **Bilayer membrane**

A. Cholesterol   B. Progesterone   C. Testosterone   D. Estradiol (estrogen)

Figure 3-10
Cholesterol, a building block of life. A. The basic four-ring structure of cholesterol. B. With the substitution of a few functional groups, cells transform cholesterol into progesterone, a female sex hormone. C. Progesterone can become testosterone, a sex hormone that gives males many of their most striking characteristics. D. A few more changes, and testosterone becomes estradiol, a powerful form of the female hormone estrogen, which likewise endows females with distinctive traits.

In water, the hydrophobic tails of phospholipids clump together to form an oily interior, while the hydrophilic heads point outward into the surrounding water. Both the outer membranes of cells and their interior membranes are composed primarily of phospholipids. All of these membranes have a hydrophobic interior and a hydrophilic exterior (Figure 3-9C).

Another class of lipids are the **steroids**—hydrocarbon chains with four interconnected rings (Figure 3-10). Many of the hormones responsible for sexual development and reproductive functions in animals are steroids. The starting point for the synthesis of all steroids is **cholesterol**—a hydrocarbon with the same pattern of four interconnected rings (see box). Small differences in the attached functional groups can make enormous differences in a steroid's biological properties. For example, cells make the female hormone progesterone from cholesterol by replacing an —OH group with an oxygen atom with a double bond, along with some other changes (Figure 3-10A and B). Cells then make testosterone from the progesterone and then estrogen from the testosterone, as needed (Figure 3-10C and D). All four of these lipids have profoundly different effects in the body, yet differ by only a few functional groups.

Most of the fatty acids in organisms are linked to glycerol to form lipids as diverse as triglycerides, phospholipids, and steroids.

3.3 How Do Sugars Join Together?

We have all been told to avoid eating too much sugar. But while this may be sound advice, avoiding all sugar is neither possible nor desirable. Sugar is the fundamental energy storage molecule of all living organisms.

The smallest sugars are "simple sugars," or **monosaccharides** [Greek, *mono* = one + *saccharine* = sugar), which can have from three to nine carbon atoms. Two of the most important simple sugars are **glucose** and fructose, which have six carbon atoms each (Figure 3-11). Cells link sugar molecules using a dehydration condensation reaction. The result is a **glycosidic bond**, in which one oxygen atom forms a bridge between carbon atoms on two adjacent sugar molecules. Two sugars can be linked to form a **disaccharide** [Greek, *di* = two + *saccharine* = sugar), such as **sucrose**,

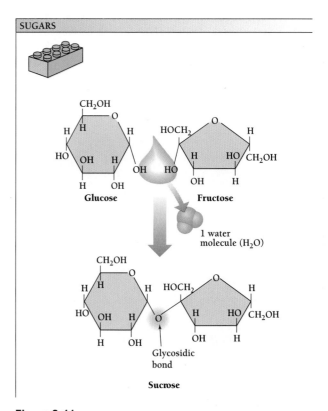

SUGARS

Figure 3-11
Life is sweet. A standard dehydration condensation reaction joins the two simple sugars glucose and fructose, forming the disaccharide sucrose—regular table sugar.

Health and Biology The Biochemistry of Cholesterol

A quick scan of any health magazine would leave most people with the impression that cholesterol is always bad for you. In fact, you can't live without it. The outer membranes of your cells are rich in this lipid, having as much as one cholesterol molecule for every phospholipid molecule. Cholesterol also serves as a chemical precursor to the sex hormones progesterone, testosterone, and estradiol (a form of estrogen). All steroids contain the large, nonpolar ring structure of cholesterol and are fat soluble (hydrophobic), not water soluble.

Cholesterol comes from two sources: the liver and the diet. The liver makes cholesterol as a component of bile, a detergent-like substance secreted into the upper intestine to aid digestion. In addition, most Americans consume about 550 mg of cholesterol a day, all of which comes from butter, eggs, meat, and other animal products.

Some of this cholesterol passes through the digestive tract into the feces, and some gets absorbed into the blood. The blood plasma of a typical college-aged American contains about 180 mg/deciliter (1 deciliter = 100 ml) of cholesterol. Total plasma cholesterol rises slowly with age to peak at about 230 mg/dl in men and 250 mg/dl in women by age 55.

Cholesterol doesn't travel alone in the blood. If it did, its hydrophobic nature would cause it to form big clumps, like oil in a bottle of salad dressing. Instead it travels in lipoprotein complexes, clumps of molecules that are part lipid and part protein that make it soluble in the blood. The most common are called low-density lipoproteins, or LDLs. These are responsible for delivering cholesterol to cells and tissues where it is needed. High-density lipoproteins, HDLs, remove cholesterol from cells and transport it to the liver, where it is secreted into the digestive juices, or "bile."

Sometimes people call LDL cholesterol "bad" and HDL cholesterol "good." This is because if not enough HDLs are present to remove cholesterol, the excess builds up along the walls of arteries, contributing to deposits called "plaques." Plaques cause hardening of the arteries, or atherosclerosis (Figure A). Atherosclerosis reduces blood flow to the organs, including the heart, and damages the arteries. Tiny bits of plaque may break off and lodge in an artery, blocking blood flow to the heart and causing a heart attack.

While limiting cholesterol intake is a good idea, one of the best indicators of the chance of developing heart disease from cholesterol deposits in the arteries is not total cholesterol, but the ratio of bad to good cholesterol. The average American has an LDL-to-HDL ratio of 5:1. Reduce that ratio to 3.5:1, by increasing plasma HDL, and you'll cut your risk of heart disease in half. An even better predictor of heart disease are high levels of IDLs, intermediate-density lipoproteins, which, for now, few doctors measure.

How can people increase HDLs? The following items all tend to raise HDLs: female sex hormones; regular aerobic exercise; monounsaturated fats, such as olive oil (instead of, not in addition to, saturated fats); weight loss; and certain forms of fiber. On the other hand, tobacco smoke lowers plasma HDL levels and also contributes to heart disease in other ways.

But don't assume science has found all the answers. Research in the area of diet and heart disease raises baffling new questions almost daily. For example, while high-fat diets are clearly bad for us, people on extremely low-fat diets have increased numbers of both LDLs and IDLs. And low-fat meals can have other effects. In another study, 12 overweight teenagers were given either a high-fat meal or a low-fat meal, of equal calories. Over the next five hours, the adolescents were allowed to eat whatever they liked. Those who had eaten the low-fat meal ate twice as many calories as those who had eaten the high-fat meal. Apparently, the super-low-fat meal didn't satisfy their hunger.

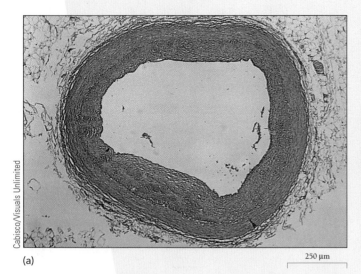
(a)
250 μm

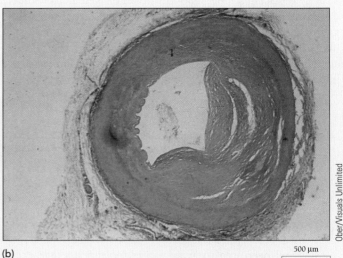
(b)
500 μm

Figure A

An open-and-shut case of atherosclerosis. Atherosclerosis results when plaque deposits narrow and block our arteries. In these cross sections through two arteries, (a) one artery is clear and (b) the other is blocked by plaque.

or table sugar (Figure 3-11). Several sugars can form short chains, called **oligosaccharides** [Greek, *oligos* = few + *saccharine* = sugar]. The longest chains of sugars, called **polysaccharides** [Greek, *poly* = many + *saccharine* = sugar], may include thousands of simple sugars. Polysaccharides include the starch in bread and the cellulose in wood and paper.

Besides glucose, two other important sugars are **ribose** and **deoxyribose**, simple five-carbon sugars that combine with other small molecules to form the building blocks for DNA (deoxyribonucleic acid) and RNA (ribonucleic acid), as shown in Table 3-3.

All sugars and polysaccharides are **carbohydrates,** molecules that contain the equivalent of one water molecule for every carbon atom. Most of the carbons in a carbohydrate have a hydrogen atom on one side and a hydroxyl group on the other, giving big carbohydrates hundreds or even thousands of hydroxyl groups. Because these hydroxyl groups easily form hydrogen bonds with water molecules, most carbohydrates dissolve well in water. Stir some sugar into a cup of hot tea and you'll see.

Table 3-3
The Building Blocks of Life: Sugars

Like children's Lego bricks, building block molecules easily snap together to form larger molecules. The Lego icon represents the 34 building blocks of life—20 amino acids, 6 lipids, 5 nucleotide bases, 3 simple sugars, and a phosphate group. Shown here are the 3 sugars.

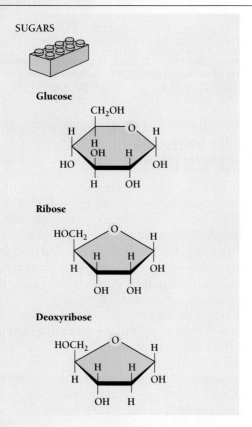

SUGARS

Glucose

Ribose

Deoxyribose

How Do Organisms Use Sugars to Store Energy?

Organisms use monosaccharides and disaccharides to store energy for a few hours, and they use longer polysaccharides, such as starch, to store energy for weeks, months, or years. In the same way, we keep our money in cash if we want to buy something right away, but put it in the bank if we want to save our money for later.

The single most important simple sugar is glucose, what we sometimes call blood sugar. Almost all biochemical reactions are in some way connected to glucose. Synthesizing glucose is a way of storing energy. Breaking down glucose is a way of releasing energy. The energy from glucose drives everything that happens in an organism from cell division to long division. So, linking long chains of glucose molecules is just another way for organisms to store energy.

Animals store energy in a long chain of glucose molecules called **glycogen.** Because glycogen has many short branches, it is easy for enzymes to quickly pick off many glucose molecules at a time. Instead of breaking down the whole molecule to get energy, enzymes harvest glucose from the ends of the branches as needed.

Plants store energy as **starch,** another large polysaccharide made of glucose molecules. Potatoes, pasta, bread, rice, and many other familiar foods contain lots of energy-rich starch, which may be either branched or unbranched. Two of the most common kinds of starch are amylopectin and amylose. Smaller amylose consists of hundreds to thousands of glucose molecules in simple, unbranched chains. Larger amylopectin consists of up to 50,000 glucose molecules in highly branched chains (Figure 3-12).

How Else Do Organisms Use Polysaccharides?

Plants and animals use polysaccharides not only as a way to store energy, but also as a building material. Arthropods such as butterflies and crabs construct skeletons from the polysaccharide **chitin.** Plants build their bodies using **cellulose**—the major component of wood, paper, cotton, and linen.

The structure and properties of polysaccharides depend both on the kinds of sugars they are made of and on the way those sugars are joined together. Cells can use different sugars to make polysaccharides, but many of the most important polysaccharides—including starch, cellulose, and glycogen—consist entirely of glucose. The differences among these important carbohydrates come from the way the glucose molecules are linked.

For us humans, the most important difference between cellulose and starch is that one polysaccharide is digestible and the other is not. But the reason for this is in the chemical bonds. In cellulose, the links between glucose are oriented differently from those in starch. The result is an extremely straight—and strong—chain of glucose molecules that makes an excellent building material for plants. But our digestive enzymes, which work wonderfully on starches, cannot break the bonds between sugars in cellulose, which is why we can't eat hay.

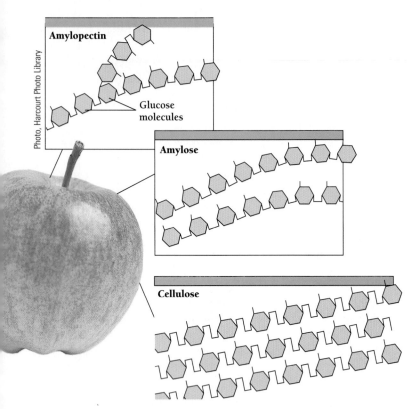

Figure 3-12

An apple contains three important polysaccharides. The starch in most plants mainly consists of two polysaccharides, amylopectin and amylose. In addition, plants like this apple also contain cellulose, an indigestible polysaccharide, and sugars such as fructose. Notice that the bonds in cellulose are different from those in starch; animals do not make the enzymes necessary to break them.

Because most animal cells do not have enzymes that break the links in cellulose, they cannot digest it. The few animals that can digest cellulose—cows and termites, for example—do so only with the help of microorganisms that have the needed enzymes. The animals themselves do not make the enzymes they need. For the rest of us, the cellulose in bran flakes and prunes is just "roughage" or fiber—a water-absorbing material that keeps things moving through the digestive tract. Indeed, cellulose is one of those no-calorie foods that the food industry uses to thicken milkshakes or to make low-calorie breads. Because we cannot digest it, these foods are said to be "low in calories." But if we were cows, these foods would have plenty of calories.

Shorter chains of sugars, called oligosaccharides, are often tacked onto proteins to form **glycoproteins,** which just means "sugar protein." Glycoproteins are commonly found on the outsides of cell membranes. Glycoproteins are signs that help cells to recognize one another and communicate. Inside cells, glycoproteins attached to newly made proteins act as address labels that tell a cell where to ship new proteins.

3.4 How Do Nucleotides Link Together to Form Nucleic Acids?

Nucleotides are the building blocks of RNA (ribonucleic acid) and DNA (deoxyribonucleic acid), the two kinds of molecules that carry genetic information. Each nucleotide building block consists of three parts: a sugar (ribose or deoxyribose), one or more phosphate groups, and a nitrogenous base.

Most cells use just five nitrogenous bases (Figure 3-13A). Each **nitrogenous base** consists of either one or two rings, each of which includes two nitrogens. Attached to these rings are hydrogen atoms and various polar functional groups.

Like all bases, nitrogenous bases accept hydrogen ions. The five bases are divided into two types: the two-ring **purines**—adenine and guanine—and the one-ring **pyrimidines**—cytosine, uracil, and thymine. Uracil and thymine are almost identical, differing by the presence of a methyl group ($-CH_3$) in thymine. Nucleotides serve many important biological roles. First and foremost, they are the building blocks of RNA and DNA.

In addition, nucleotides with one, two, or three phosphate groups play a central role in the production of energy within cells. **ATP (adenosine triphosphate),** which has three phosphates, is the immediate source of energy for all biological processes (Figure 3-14). Energy from the breakdown of glucose is stored in ATP. Cells then release energy from ATP by hydrolyzing its phosphate-phosphate bonds. The resulting energy drives all the chemical reactions of a cell (Figure 3-14).

A nucleotide consists of a sugar, a phosphate, and a nitrogenous base. Nucleotides are essential to life both as the units of DNA and RNA and for energy storage.

How Are the Nucleotides of a Nucleic Acid Linked Together?

Like sugars, nucleotides can connect one to another by a simple dehydration condensation reaction (Figure 3-15). With the removal of one molecule of water, nucleotides form **phosphodiester bonds,** which connect the nucleotides into long, unbranched chains called **nucleic acids.** Two nucleic acids—DNA and RNA—are so important in the storage and transmission of genetic information that this book devotes seven chapters to them.

DNA serves as a library of information for every cell. Along the thin strands of DNA are genes that carry instructions for building every protein a cell will ever need. The genes of all organisms are made of DNA. Some viruses have RNA genes—including the viruses that cause flu and AIDS. But in cells, RNA acts only as an interpreter, carrying information from the DNA to the rest of the cell.

PURINES

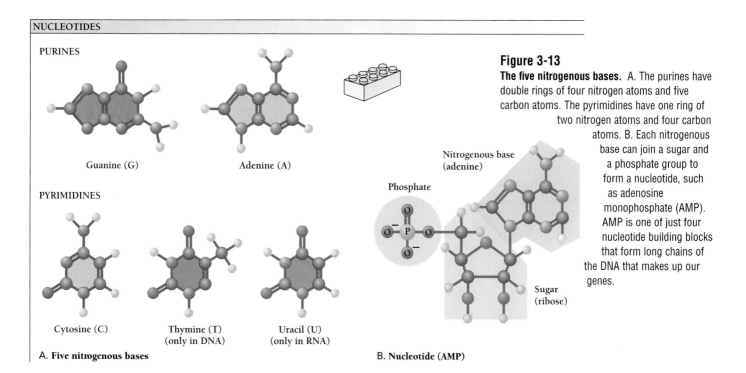

Guanine (G)

Adenine (A)

PYRIMIDINES

Cytosine (C)

Thymine (T)
(only in DNA)

Uracil (U)
(only in RNA)

A. Five nitrogenous bases

Figure 3-13

The five nitrogenous bases. A. The purines have double rings of four nitrogen atoms and five carbon atoms. The pyrimidines have one ring of two nitrogen atoms and four carbon atoms. B. Each nitrogenous base can join a sugar and a phosphate group to form a nucleotide, such as adenosine monophosphate (AMP). AMP is one of just four nucleotide building blocks that form long chains of the DNA that makes up our genes.

Nitrogenous base
(adenine)

Phosphate

Sugar
(ribose)

B. Nucleotide (AMP)

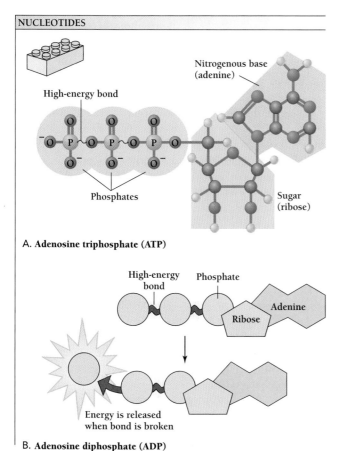

High-energy bond

Nitrogenous base
(adenine)

Phosphates

Sugar
(ribose)

A. Adenosine triphosphate (ATP)

High-energy
bond

Phosphate

Adenine

Ribose

Energy is released
when bond is broken

B. Adenosine diphosphate (ADP)

Figure 3-14

In all organisms, ATP is the common currency of energy. A. The nucleotide ATP consists of the nitrogenous base adenine, the sugar ribose, and three phosphate groups. B. ATP releases energy for use in chemical reactions when one of the phosphate groups breaks off.

In a chain of nucleotides, the sugar and phosphate parts of each nucleotide link together alternately to form a sugar-phosphate "backbone." RNA consists of a single long backbone. DNA usually consists of two parallel backbones connected to each other by thousands of hydrogen bonds between adjacent bases. The result is a molecule that resembles a long ladder, with each rung consisting of a pair of nucleotide bases (Figure 3-16). The whole ladder is twisted along its long axis to form a "double helix."

DNA and RNA each consist of just four kinds of nucleotide bases, but not the same four. Three bases—adenine, guanine, and cytosine—occur in both DNA and RNA. But DNA's fourth base is thymine, and RNA's fourth base is uracil. So there are five bases altogether (Table 3-4). The structures of RNA and DNA are important to the way that they function as information molecules. In Chapters 10 and 11, we discuss the structure and function of DNA and RNA in great detail.

The long chains of nucleotides in the information molecules DNA and RNA are linked—sugar-phosphate-sugar—by phosphodiester bonds.

3.5 How Do We Make and Use Polypeptides?

Proteins perform thousands of jobs. They hold the body together, lend support to bones, transport molecules and ions in and out of cells, and much more. Every organism has

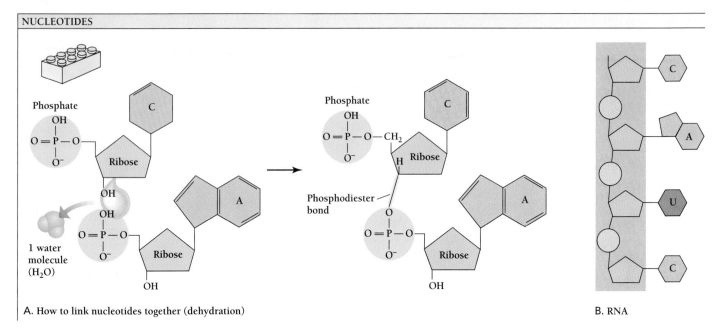

A. How to link nucleotides together (dehydration)

B. RNA

Figure 3-15
Building a library of information. A. All organisms store genetic information in long chains of nucleotides—either DNA or RNA. The nucleotide building blocks join by a dehydration condensation reaction, losing one water molecule. B. RNA consists of a single chain of nucleotides. (DNA consists of two chains.)

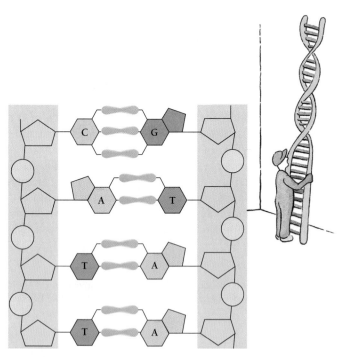

Figure 3-16
The DNA ladder. The DNA of our genes consists of two long chains of nucleotides. The bonds linking the nucleotide bases on the two chains are often compared to the rungs of a ladder. The whole, long double chain is twisted into a helix.

Biology⊛Now™   Learn more about the structure of DNA by clicking on this figure on your BiologyNow CD-ROM.

hundreds to thousands of different proteins. The human body alone contains at least 80,000 different proteins.

Collagen, the most common protein in the human body, gives strength to connective tissues in tendons, ligaments, bones, muscle, and skin. Dissolved in sugar water, collagen gives Jell-O its jiggle. Yet, in the body, collagen is as strong as steel. Elastin makes skin stretchy, while stringy keratin toughens hair, horns, nails, and claws.

Some proteins taxi materials from one place to another. Hemoglobin, for example, binds to oxygen in the lungs and transports it, by way of the blood, to the cells in the rest of the body. Other proteins act as hormones, antibodies, or poisons (rattlesnake venom, for example). Still others transport molecules across the membranes of cells. Some proteins—often glycoproteins with attached oligosaccharides—act as identification tags on the surfaces of cells. Other proteins help cells to communicate with one another. Protein "signaling molecules" carry messages from one cell to another, and protein "receptors" on cell surfaces pick up those signals.

One of the most important roles for proteins is as **enzymes**—biological molecules that speed up reactions between other molecules but are not themselves altered. Enzymes are indispensable to life. Without them, many chemical reactions would occur too slowly to sustain life. Enzymes copy and repair DNA, help cells to extract energy from sugars, digest our food, and much more. Enzymes in the mouth, for example, help break down the polysaccharides in bread into simple sugars. That is why bread gets

Table 3-4
The Building Blocks of Life: Nucleotides

Like children's Lego bricks, building block molecules easily snap together to form larger molecules. The Lego icon represents the 34 building blocks of life—20 amino acids, 6 lipids, 5 nucleotide bases, 3 simple sugars, and a phosphate group. Shown here are the 5 nucleotides.

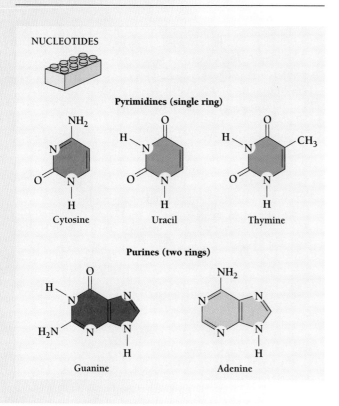

NUCLEOTIDES

Pyrimidines (single ring)

Cytosine   Uracil   Thymine

Purines (two rings)

Guanine   Adenine

sweeter and sweeter as you chew it. We discuss enzymes in greater detail in Chapter 5.

Cells contain thousands of proteins; each one has a different role to play.

How Are the 20 Amino Acids Alike and Different?

Every protein is made of one or more chains of amino acids. Each chain is called a **polypeptide.** A protein may be made of one simple chain or of several different chains. A chain may consist of only a few hundred amino acids or 3,500 or more. Regardless of how long the chains are or how many of them there are, cells build proteins one amino acid at a time.

Each of the 20 amino acids has the same basic structure. Each contains a carboxyl group and an amino group (Figure 3-17). The carboxyl group and the amino group are both attached to the same carbon atom, called the **alpha (α) carbon.**

Each amino acid also has a characteristic side chain, called an R group, which is also attached to the α carbon

atom. *The α carbon, the carboxyl group, and the amino group are exactly the same in 19 of the 20 amino acids.* (The one exception is proline.) But each amino acid has a different R group. Table 3-5 shows the 20 different amino acids cells can use to make proteins.

R groups vary in length and in their functional groups. The R group side chains establish the distinctive properties of each amino acid and the properties of the resulting polypeptide. The 20 amino acids combine into long chains to form a nearly infinite number of different proteins. Most amino acids end up folded inside of massive proteins, just anonymous cogs in a giant machine. But a few amino acids play their own distinct roles in biology. For example, some amino acids are neurotransmitters—chemicals secreted by nerve cells that convey information to cells.

All amino acids have a carboxyl group, an amino group, and an R group, all attached to the α carbon. In each of the amino acids, all the parts are identical except for the R group.

How Do Amino Acids Join Together?

Polypeptides are made of chains of amino acids joined by dehydration condensation reactions, just as polysaccharides and nucleic acids are. The special bond that links the amino acids together into polypeptides is called a **peptide bond.** Two amino acids join when the carboxyl group of one joins to the amino group of the next (Figure 3-18A).

Each amino acid in the string is oriented in the same way as the others—amino, carboxyl, amino, carboxyl— forming an unvarying backbone of repeating atoms. The R groups hang off the sides of the polypeptide backbone. Uninvolved in the peptide bond, the R side chains may interact with one another or with other molecules in their environ-

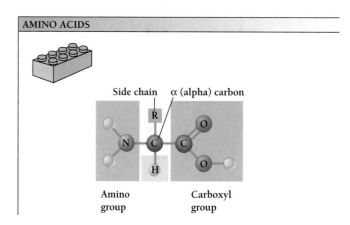

AMINO ACIDS

Side chain   α (alpha) carbon

R

N   C   C   O

O

H

Amino   Carboxyl
group   group

Figure 3-17
What's in an amino acid? Every amino acid includes an amino group, a hydrogen atom, and a carboxyl group, as well as a variable side chain (called R). All four are attached to a single, central carbon atom, called the "α carbon."

Table 3-5
The Building Blocks of Life: Amino Acids

Like children's Lego bricks, building block molecules easily snap together to form larger molecules. The Lego icon represents the 34 building blocks of life—20 amino acids, 6 lipids, 5 nucleotide bases, 3 simple sugars, and a phosphate group. Shown here are the 20 amino acids (another not shown here has recently been added to the list).

AMINO ACIDS

Amino acid

$$H_3N^+ - \overset{\overset{\displaystyle R}{|}}{\underset{\underset{\displaystyle H}{|}}{C}} - COO^-$$

α carbon

Amino acids with basic side chains

Lysine

$$H_3N^+ - \overset{\overset{\displaystyle ^+NH_3}{|}\overset{\displaystyle CH_2}{|}\overset{\displaystyle CH_2}{|}\overset{\displaystyle CH_2}{|}\overset{\displaystyle CH_2}{|}}{\underset{\underset{\displaystyle H}{|}}{C}} - COO^-$$

Arginine

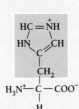

$$H_3N^+ - \overset{\overset{\displaystyle NH_2}{|}\overset{\displaystyle C=NH_2^+}{}\overset{\displaystyle NH}{|}\overset{\displaystyle CH_2}{|}\overset{\displaystyle CH_2}{|}\overset{\displaystyle CH_2}{|}}{\underset{\underset{\displaystyle H}{|}}{C}} - COO^-$$

Histidine

$$HC = \overset{+}{N}H$$
$$HN \quad CH$$
$$\overset{|}{C}$$
$$|$$
$$CH_2$$
$$H_3N^+ - \overset{}{\underset{\underset{\displaystyle H}{|}}{C}} - COO^-$$

Amino acids with acidic side chains

Aspartic acid

$$\overset{\displaystyle COO^-}{\underset{\displaystyle CH_2}{|}}$$
$$H_3N^+ - \overset{}{\underset{\underset{\displaystyle H}{|}}{C}} - COO^-$$

Glutamic acid

$$\overset{\displaystyle COO^-}{\underset{\displaystyle CH_2}{|}\underset{\displaystyle CH_2}{|}}$$
$$H_3N^+ - \overset{}{\underset{\underset{\displaystyle H}{|}}{C}} - COO^-$$

Amino acids with polar uncharged side chains

Asparagine

$$H_2N \searrow C = O$$
$$|$$
$$CH_2$$
$$H_3N^+ - \overset{}{\underset{\underset{\displaystyle H}{|}}{C}} - COO^-$$

Glutamine

$$H_2N \searrow C = O$$
$$|$$
$$CH_2$$
$$|$$
$$CH_2$$
$$H_3N^+ - \overset{}{\underset{\underset{\displaystyle H}{|}}{C}} - COO^-$$

Serine

$$CH_2OH$$
$$H_3N^+ - \overset{}{\underset{\underset{\displaystyle H}{|}}{C}} - COO^-$$

Threonine

$$\overset{\displaystyle CH_3}{\underset{\displaystyle HC-OH}{|}}$$
$$H_3N^+ - \overset{}{\underset{\underset{\displaystyle H}{|}}{C}} - COO^-$$

Tyrosine

$$OH$$
(benzene ring)
$$CH_2$$
$$H_3N^+ - \overset{}{\underset{\underset{\displaystyle H}{|}}{C}} - COO^-$$

Amino acids with nonpolar side chains

Glycine

$$H$$
$$H_3N^+ - \overset{}{\underset{\underset{\displaystyle H}{|}}{C}} - COO^-$$

Alanine

$$CH_3$$
$$H_3N^+ - \overset{}{\underset{\underset{\displaystyle H}{|}}{C}} - COO^-$$

Valine

$$H_3C \searrow CH \diagup CH_3$$
$$H_3N^+ - \overset{}{\underset{\underset{\displaystyle H}{|}}{C}} - COO^-$$

Leucine

$$H_3C \searrow CH \diagup CH_3$$
$$|$$
$$CH_2$$
$$H_3N^+ - \overset{}{\underset{\underset{\displaystyle H}{|}}{C}} - COO^-$$

Isoleucine

$$\overset{\displaystyle CH_3}{\underset{\displaystyle CH_2}{|}}$$
$$H - \overset{}{\underset{\displaystyle }{C}} - CH_3$$
$$H_3N^+ - \overset{}{\underset{\underset{\displaystyle H}{|}}{C}} - COO^-$$

Phenylalanine

(benzene ring)
$$CH_2$$
$$H_3N^+ - \overset{}{\underset{\underset{\displaystyle H}{|}}{C}} - COO^-$$

Methionine

$$\overset{\displaystyle CH_3}{\underset{\displaystyle S}{|}}$$
$$CH_2$$
$$|$$
$$CH_2$$
$$H_3N^+ - \overset{}{\underset{\underset{\displaystyle H}{|}}{C}} - COO^-$$

Cysteine

$$SH$$
$$|$$
$$CH_2$$
$$H_3N^+ - \overset{}{\underset{\underset{\displaystyle H}{|}}{C}} - COO^-$$

Proline

$$\overset{\displaystyle H_2}{\underset{\displaystyle C}{}}$$
$$H_2C \quad CH_2$$
$$H_2N^+ - \overset{}{\underset{\underset{\displaystyle H}{|}}{C}} - COO^-$$

Tryptophan

(indole ring)
$$NH$$
$$C = CH$$
$$|$$
$$CH_2$$
$$H_3N^+ - \overset{}{\underset{\underset{\displaystyle H}{|}}{C}} - COO^-$$

ment. The polypeptide backbone is identical from one end to the other, except for the very beginning and very end.

Peptide bonds link amino acids together into polypeptide chains, in which the R groups hang off the sides.

What Do Proteins Look Like?

Proteins are extraordinarily diverse. For each of the thousands of jobs that proteins perform, there is a unique protein to do it. They can be short or long, as regular in structure as a stack of Lego bricks or as irregular as a wad of crumpled paper.

Despite their diversity, biochemists have found ways to classify proteins. Most are dense, roundish molecules called **globular proteins.** A classic example of a globular protein is hemoglobin, the protein that carries oxygen in the blood. Like many proteins, hemoglobin is gigantic compared to the small molecules of the cells. Hemoglobin is almost 400 times as big as a molecule of glucose and more than 2,000 times as big as each molecule of oxygen that it ferries from the lungs.

In addition to the globular proteins are **fibrous proteins—** long, thin molecules such as the keratin in hair or the collagen in skin and bone. Fibrous proteins provide structural support to the bodies of plants, animals, and other organisms.

Proteins have up to four levels of structure. The **primary structure** of a protein is the sequence of amino acids in each chain (Figure 3-18A). This is like the sequence of letters in a sentence. It is simply the order and kind of amino acids. The **secondary structure** is the coiling, bending, or folding of individual sections of the polypeptide chain, which we discussed at the beginning of this chapter, when we talked about the two forms of the scrapie protein. For example, polypeptides can coil into an **α helix,** which resembles the coil of a phone cord, as in the normal form of the scrapie protein (Figure 3-2A and 3-18B).

Other secondary structures include the β pleated sheet and the collagen helix, found in collagen. The **β pleated sheet** (found in the prion form of the

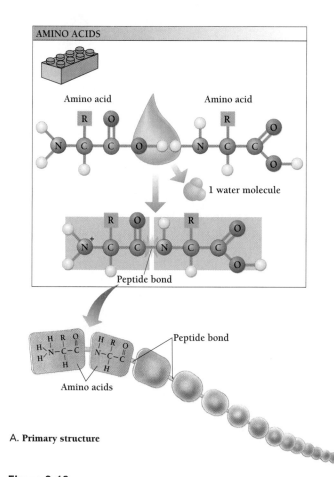

A. Primary structure

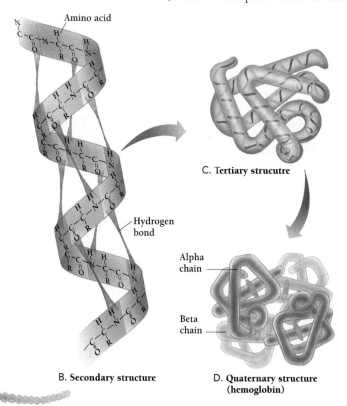

B. Secondary structure

C. Tertiary strucutre

D. Quaternary structure (hemoglobin)

Figure 3-18

Building a protein. The usual dehydration condensation reaction joins pairs of amino acids with a peptide bond, forming long chains called polypeptides. A. The primary structure of a protein is simply the sequence of amino acids in its polypeptide(s). B. The secondary structure is the curling of each polypeptide into a three-dimensional form, such as the α helix shown here. C. Tertiary structure is the folding of the curled polypeptide into a more complex three-dimensional shape. D. Quaternary structure is the fitting together of two or more folded polypeptides, as in the hemoglobin molecule shown here. Hemoglobin is the major oxygen-carrying protein in the blood of many animals.

Biology Now™ Learn more about amino acid structure by clicking on this figure on your BiologyNow CD-ROM.

scrapie protein) is a long chain that folds back and forth against itself—like the coils of a snake or the long lines of people at an airport (Figures 3-2B and 3-19). The folds are held together by sticky hydrogen bonds, like those that hold water molecules together (Chapter 2). Unlike an α helix, which consists of just one polypeptide chain, a **collagen helix** consists of three polypeptide chains wound around each other (Figure 3-20).

Both globular proteins and fibrous proteins contain α helices and β pleated sheets. But fibrous proteins often have many rigid α helices, whereas globular proteins are made of a mixture of many kinds of secondary structures. In globular proteins, the amino acid sequences of the α helices and β pleated sheets are less regular than those of structural proteins. The polypeptide chain of a globular protein may coil into an α helix for a few dozen amino acids, then switch to a β pleated sheet, then back to an α helix. For example, globular proteins contain β structures, with two to five parallel sections of a single chain forming a β sheet.

The **tertiary structure**, or **conformation**, of a protein is the three-dimensional folding of an entire polypeptide chain (Figure 3-18C). If you curl a ribbon into a spiral (or helix) by scraping it with a blade, you have produced a secondary structure.

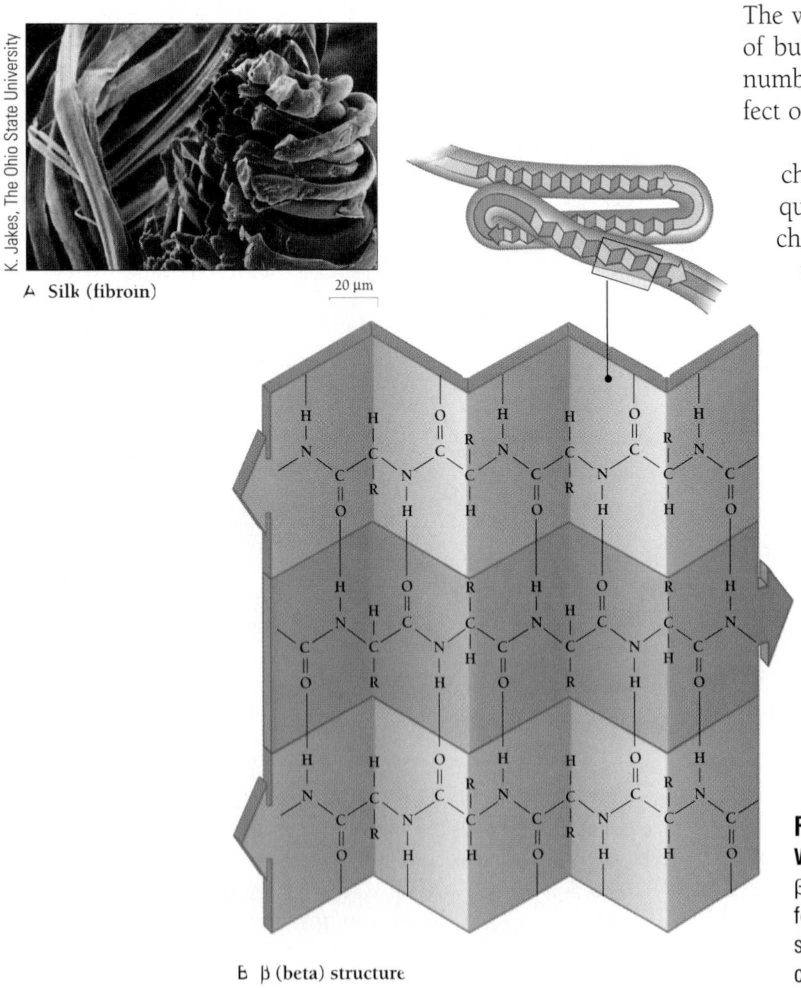

K. Jakes, The Ohio State University

A Silk (fibroin)   20 μm

B β (beta) structure

If you then take the curled ribbon and tie it around a birthday present, you have given the ribbon a tertiary structure. Many proteins also have **quaternary structure**—the fitting together of two or more folded chains (Figure 3-18D). This is analogous to piecing together several different ribbons into an elaborate bow. Just as the ribbons can be either all the same color or a mixture of different colors, proteins can be made of several identical polypeptide chains or of several different ones. Hemoglobin, for example, is made of four polypeptide chains, two of one kind of chain and two of another (Figure 3-18D).

Proteins may be long or short, stringy or globular. One or more polypeptides folded together into a unique shape make a protein.

3.6 How Do Cells Construct Proteins?

The biological role of a protein depends exquisitely on its three-dimensional shape. Because the shape of a protein is so important, the chains that make up proteins must fold in precisely the right way to form biologically active molecules. The way a polypeptide folds depends first on the sequence of building blocks in its chains. By comparison, the exact number and order of sugars in a polysaccharide has little effect on its function.

By making different versions of a polypeptide, biochemists have found that tiny changes in amino acid sequence may affect a polypeptide's behavior. Sometimes changing one amino acid has little effect. For example, a muscle protein from a cow can have a sequence slightly different from that of a human. Yet the difference does not affect the protein's three-dimensional structure or the way it functions in the body. Cow and human proteins differ slightly, but they work the same.

On the other hand, sometimes changing an amino acid can completely alter a protein's structure and destroy its biological activity. The kinds of amino acid changes that make the biggest differences are those that substitute amino acids with very different properties. For example, in the blood protein hemoglobin, the substitution of a hydrophobic amino acid (valine) for a charged one (glutamic acid) produces a defective hemoglobin molecule that crystallizes. The result is the disease sickle cell anemia.

Figure 3-19

What is silk made of? A. The protein that makes up silk consists of β sheets. B. In a β sheet, polypeptide chains can snake back and forth past each other, held fast by hydrogen bonds between adjacent segments. This structure is very strong, in the same way that corrugated cardboard is.

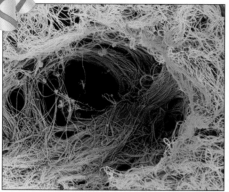

Figure 3-20
Collagen. A. Collagen, the major constituent of skin, tendons, ligaments, cartilage, and bones, is made of three helical polypeptides wound together into a still-larger helix. B. Collagen fibers.

A. Collagen molecule

B. Collagen fibers   5 μm

Professor P.M. Motta and E. Vizza/Science Photo Library/Photo Researchers, Inc.

A small change in the structure of a polypeptide can enormously alter the way it behaves or have no effect at all.

How Does the Folding of a Protein Depend on Interactions Among Its Amino Acids?

Polypeptides are always linear, never branched like some of the polysaccharides. As we've seen, however, polypeptides can coil or fold into helices, sheets, balls, and other shapes. A protein may consist of a single, folded polypeptide chain or of several chains folded together. In proteins with more than one chain, the individual chains are held together tightly, so that, although made of several separate molecules, the protein behaves like a single, huge molecule.

The α helix, the β structure, and the collagen helix all share an important trait. As regular repeating forms, they are strong. Their simple shapes allow them to function as standardized building materials, comparable to the wooden beams and bricks that carpenters use to build a house.

In structural proteins, the α helix, β structure, and collagen helix each result from an extremely regular sequence of amino acids in the polypeptide chain. In the collagen helix, for example, every other amino acid is glycine. This smallest of the amino acids fits snugly in the interior of collagen's triple helix.

In globular proteins, the α helix, β structure, and other regular structures are composed of nonrepeating sequences of amino acids. As a result, globular proteins are more flexible in the way they fold and often form intricate shapes that allow them to play more diverse roles than the fibrous proteins.

Some globular proteins act as hormones, for example, some act as antibodies in the immune system, some act as transport proteins (e.g., hemoglobin), and some act as enzymes.

The tertiary folding of polypeptide chains results from interactions of the side chains on each amino acid. Each side chain has its particular characteristics. Some are polar, some nonpolar. Some can serve as proton donors in hydrogen bonds. Others serve as proton acceptors. Either a hydrogen bond or the attraction between positively and negatively charged side chains can stabilize a particular shape. Conversely, some interactions, such as hydrophobic interactions or the repulsion between two like-charged side chains, can prevent certain shapes.

Alone, any of these interactions contributes only slightly to a given molecular structure. Together, however, the various interactions establish a three-dimensional structure that can be extremely stable in certain chemical environments. Because the internal environment of a cell is relatively constant, each protein usually maintains its characteristic conformation. Subtle changes in the chemical environment of the cell, however, can change the shape of a protein and change the way it behaves.

Structural proteins tend to have regular sequences of amino acids and regular secondary structures, such as α helices and β structures. Globular proteins have nonrepeating amino acid sequences and irregular and unique shapes.

Are the Interactions Among Amino Acids Enough to Determine Protein Structure?

Just seconds after it forms, a polypeptide begins folding into the right shape. A polypeptide chain is as flexible as a snake and could easily take any of thousands of shapes. A short polypeptide just 100 amino acids long has 200 movable angles, not counting the R side chains. Yet, despite all the ways that protein folding could go wrong, these remarkable molecules almost always seem to fold almost inevitably into their "native" functional shape. How do they do it?

The "protein-folding problem" has plagued biochemists for decades. In the 1950s, biochemists wondered if the sequence of amino acids was enough to exactly determine the precise three-dimensional structure of a protein. Might not some outside molecule somehow help direct the formation of a polypeptide? In that case, the shape of a polypeptide would depend not only on its amino acid sequence but also on the presence of some other "instructional" molecule.

In 1957, the biochemist Christian Anfinsen attempted to answer this question with a now classic experiment. He chose to study an enzyme called "ribonuclease," which breaks the bonds in RNA molecules. Anfinsen first exposed ribonuclease to a chemical solution that interfered with hydrogen bonds and hydrophobic interactions. This treatment changed the enzyme's shape and destroyed its ability to function. It could no longer break the bonds in RNA. When Anfinsen removed the ribonuclease from the harsh ("denaturing") solution, however, the ribonuclease spontaneously reverted back to its "native," functional shape. Once again, it could function properly, breaking bonds in RNA.

The enzyme's recovery required help from no other molecule. The amino acid sequence of ribonuclease and the right chemical environment were enough to determine the enzyme's three-dimensional structure and biological activity. Anfinsen's experiment suggested that proteins spontaneously fold into their biologically active forms, a conclusion that satisfied biochemists for many years.

Christian Anfinsen's experiment suggested that in a normal cell environment amino acid sequence alone determines the structure of a protein.

What Factors Determine How Proteins Fold?

For many years, biologists accepted the idea that in a normal cell environment, the amino acid sequence of a polypeptide implied one and only one three-dimensional structure. Yet biochemists cannot themselves predict the structure of a protein simply by knowing the amino acid sequence. What rules governed the folding has puzzled generations of biochemists.

Gradually, molecular biologists discovered that protein folding is more complicated than Anfinsen's experiment suggested. A pure polypeptide in a dilute solution may ultimately adopt a single, stable conformation. But newly made proteins left to themselves usually do not fold properly—whether in the test tubes of protein chemists or in the interiors of cells.

Several obstacles can prevent correct folding. Inside a cell, the concentration of proteins and other molecules is extremely high. At these high protein concentrations, individual peptide chains can easily form bonds with other peptides, rather than with themselves. Molecular biologists call such bonding between separate molecules *promiscuous interactions*. In addition, polypeptide chains are assembled gradually, one amino acid at a time. So one end of a chain is forming among all these other proteins, while the other end does not even exist yet. This is a recipe for promiscuous interactions.

To prevent the parts of a polypeptide chain from linking to the wrong molecule or to the wrong part of itself, a staff of enzymes and special proteins, called **chaperones**, bind to the newly emerging polypeptide and prevent "promiscuous" interactions.

Chaperones not only help other proteins fold into the correct shape, they also help proteins maintain that shape or change shape in order to play a new role.

Proteins' need for chaperones may help explain how scrapie PrP (discussed in the beginning of the chapter) can change the structure of a normal cellular protein.

Amino acid sequence alone was once thought to be sufficient to determine the three-dimensional structure of a protein. But enzymes and chaperone molecules guide the correct folding of a protein in a cell.

In this chapter, we have seen how the structure of a molecule—whether at the level of individual functional groups or at the level of amino acid sequences and tertiary structure—profoundly affects the chemistry of the molecule. The structure of a cell determines its behavior and capabilities just as profoundly, as we will see in the next chapter.

Key Concepts

- Organisms build most structures from just four molecular building blocks: lipids, sugars, amino acids, and nucleotides.
- Functional groups of atoms determine the chemical properties of each building block.
- Dehydration reactions, which remove the equivalent of one molecule of water, join small building blocks into macromolecules.
- A protein folds up according to specific interactions among the functional groups of its amino acid building blocks.

Summary with Key Terms

How do organisms use chains of carbon atoms decorated with functional groups to build most of the basic building blocks of life?

Other than water, nearly all molecules found in living organisms contain one or more carbon atoms. In fact, all **organic** molecules contain carbon. **Organic chemistry** is the study of the structures and reactions of carbon compounds, while **biochemistry** is the study of the chemistry of living organisms.

Most organic molecules are made of **hydrocarbon chains**. But the specific properties of small molecules depend on **functional groups**, such as the **hydroxyl, carboxyl,** and **amino** groups. The properties of **macromolecules** depend on the nature and sequence in which the component building blocks assemble. Some hydrocarbon chains form rings of carbon.

What two simple reactions link and unlink these building blocks?

The small building blocks all assemble into different kinds of long-chain macromolecules, by the same kind of general reaction. This chemical reaction is called a **dehydration condensation reaction** because the equivalent of one molecule of water is eliminated each time two building blocks join together. Conversely, the breaking down of a macromolecule into its constituent building blocks is called **hydrolysis** because the equivalent of one molecule of water is added as each small molecule is removed from the chain.

What are the main categories of building block molecules?

The small building blocks of cells fall into four classes—**sugars, amino acids, nucleotides,** and **lipids.**

How are fatty acids, polysaccharides, nucleic acids, and proteins formed from simple building blocks?

Glycosidic bonds link sugars into macromolecules called **polysaccharides. Peptide bonds** link amino acids together into **polypeptides,** and **phosphodiester bonds** link nucleotides into **nucleic acids.**

Although lipids do not form long chains, some, called **fatty acids,** do contain long carbon chains. Fatty acids may be **saturated** or **unsaturated.** Lipids with more than one double bond are said to be **polyunsaturated.** A glycerol molecule with two fatty acid chains and a phosphoric acid group is a **phospholipid,** an amphipathic molecule that forms cell membranes. A glycerol molecule with three fatty acid chains and a phosphoric acid group is a **triglyceride.** Different kinds of triglycerides make up the fats (solid at room temperature) and oils (liquid at room temperature) of animals and plants. Lipids made of four interconnected rings are called **steroids,** a group of lipids that includes **cholesterol** and the steroid hormones progesterone, estrogen, and testosterone.

Simple sugars, or **monosaccharides,** are the building blocks of **disaccharides, oligosaccharides,** and **polysaccharides.** Sugars, such as **glucose** and **sucrose,** and **polysaccharides,** such as **starch, glycogen, chitin,** and **cellulose,** are all classed as **carbohydrates.** For labels and signals, cells use **glycoproteins,** which are oligosaccharides attached to proteins.

The building blocks of nucleic acids are nucleotides. Each nucleotide consists of a **nitrogenous base,** either a **purine** or a **pyrimidine,** connected to a sugar, either **ribose** or **deoxyribose,** and (by way of the sugar) a phosphate group.

Many proteins, including **enzymes** and transport molecules such as hemoglobin, are dense, roundish molecules called **globular proteins.** But many structural proteins are **fibrous proteins**—long, thin molecules such as the keratin in hair or the collagen in skin and bone.

All polypeptides are made of strings of amino acids. Each amino acid has a carboxyl group, an amino group, and an R group, all attached to the β **carbon.** In each amino acid, all the parts are identical except for the R group. Proteins can have up to four levels of structure: **primary structure,** the sequence of amino acids; **secondary structure,** the coiling, bending, or folding of individual sections of the polypeptide chain into forms such as the α **helix,** the β **pleated sheet,** and the **collagen helix; tertiary structure,** or **conformation,** the three-dimensional folding of an entire polypeptide chain; and **quaternary structure,** the fitting together of two or more folded chains.

The folding of a polypeptide into a precise three-dimensional structure generally requires energy. Some of this energy comes from interactions among the atoms of the polypeptide backbone and those of the amino acid side chains. Additional energy may come from **ATP,** which is used by **chaperones** that help proteins fold.

Review and Thought Questions

Review Questions

1. What evidence persuaded biologists that scrapie could not be caused by a bacterium or virus?
2. Where would you find a large concentration of lipids in a cell? What does the lipid do there?
3. What are the building blocks that make up proteins? How many kinds of these building blocks are there?
4. What three components make up a nucleotide?
5. What small molecule is removed during a dehydration condensation reaction? What is the opposite reaction called? What happens in that case?
6. What is the relationship between polysaccharides, sugars, and carbohydrates? What three roles do carbohydrates play in cells?
7. Define peptide bond, polypeptide, protein, and enzyme.
8. Define nucleic acid. What kinds of nucleic acids are there?
9. Describe the four levels of structure in a protein.
10. Distinguish among the α helix, the β structure, and the collagen helix, all of which occur in fibrous proteins.
11. What is the major component of each of these materials? Use C (carbohydrate), P (protein), and L (lipid) to answer.

   ____ Cotton T-shirt ____ Cooking oil
   ____ Chicken nugget ____ Wooden toothpick
   ____ Apple juice ____ Cheese
   ____ Silk scarf ____ Fat on a piece of steak
   ____ Tortilla ____ Newspaper
   ____ Rice
   ____ Lipstick

Thought Questions

12. How does the shape of a protein determine its properties and therefore its functions?
13. Describe the shapes of a globular protein and a fibrous protein.
14. How are these two different shapes usually put to different purposes in a cell?
15. How can the same building block molecules be used to construct billions of different macromolecules?

BiologyNow Resources

Biology ⑧ Now™

Active Figures

3-8 & 3-9:   Lipids
3-16:   The structure of DNA
3-17:   Amino Acid structure

Preparing for an exam? Take a diagnostic test on your BiologyNow CD-ROM.

Online materials relating to this chapter are at:
http://biology.brookscole.com/AAL3

About the Chapter-Opening Image
This white lab mouse is constructed with building blocks. These represent the building blocks of life.

4 Why Are All Organisms Made of Cells?

Key Questions

- What are cells?

- What are the advantages and disadvantages of being a multicellular organism?

- What organelles and other parts of cells make them work?

- How do membranes act as boundaries and regulate the contents of cells?

Very Little Animalcules

Scott Camazine/Photo Researchers, Inc.

More than 300 years ago, an uneducated Dutch cloth merchant named Antonie van Leeuwenhoek (1632–1723) discovered something so interesting that he received visits from the Czar of Russia, King James II of England, and King Frederick II of Prussia. Like many young men, Leeuwenhoek had a consuming interest that distracted him from his work (Figure 4-1). He liked to make glass lenses by grinding and polishing bits of glass and then mounting the lenses between gold or copper plates. He used these simple, single-lens microscopes to examine anything he encountered.

To Leeuwenhoek's delight, his lenses—some no larger than a pinhead—revealed hundreds of tiny living beings never before seen by human eyes. Leeuwenhoek astonished his friends and acquaintances with descriptions of red blood cells in blood, sperm in semen, and "very little animalcules" in pond water. With his lenses, Leeuwenhoek also demonstrated that fleas, ants, weevils, and other pests do not arise spontaneously from dust or wheat, as was commonly believed, but developed from larvae that hatch from tiny eggs laid by the adults.

In 1673, when Leeuwenhoek was 41, he began sending reports of his discoveries to the recently created Royal Society in London, then one of the few scientific organizations in the world. Over the next 50 years, until Leeuwenhoek's death in 1723, the Society received and published 375 letters from him, the last when the Dutch lens grinder was 90.

Over the years, Leeuwenhoek described how everywhere he looked, he found tiny organisms whose existence few people had even suspected. In 1683, he became the first person to see bacteria. Leeuwenhoek's great contribution was to demonstrate that life is not limited to organisms visible to the naked eye.

In England, the Royal Society welcomed Leeuwenhoek's fascinating observations. The Society had already hired Robert Hooke to be its "curator of instruments." Hooke's job was to demonstrate the microscope and other new technologies for the entertainment of the gentlemen at the Royal Society's meetings. In 1665, Hooke published his book *Micrographia,* a series of illustrations and descriptions of his observations using new and improved microscopes. Hooke showed, for example, the structure of feathers, the stinger of a bee, and the foot of a fly.

Hooke also coined the term "cellulae" [Latin, *cella* = small room] to describe the boxlike cavities (cells) he saw when he examined a slice of cork under a microscope. "Cells," in those days, referred to the small rooms of a monastery, nunnery, or prison.

Hooke later published detailed drawings of cells in other plant tissues, noting that they contained "juices." Leeuwenhoek, Hooke, and the few other microscopists of the day described a whole new world of tiny living organisms, and their research revealed an unsuspected level of structural complexity in larger organisms. These discoveries might well have stimulated the beginning of a new science. Yet, oddly, these discoveries lay dormant for nearly 200 years.

Neither Leeuwenhoek's nor Hooke's studies led directly to any unifying biological principles. Most scientists dismissed Leeuwenhoek's animalcules as little more than amusing natural curiosities. Hooke's "cellulae" were simply bubbly stuff in plants. No one guessed that cells come from other cells by the process of cell division, that cells are the fundamental unit of life, and that all tissues in all organisms are made of cells. No one guessed that some microscopic organisms cause disease. To most people, Leeuwenhoek's microorganisms and Hooke's cells seemed no more important than the specks of dust in a ray of sunshine.

One reason that the important thinkers took neither Hooke nor Leeuwenhoek seriously was the two men's social class. Leeuwenhoek was an uneducated amateur, and Hooke was an employee of the Royal Society, not a full member. Neither man was, in other words, a "gentleman," so neither commanded respect.

Figure 4-1
Antonie van Leeuwenhoek built primitive microscopes so powerful that he saw bacteria 200 years before anyone else.

Science Photo Library/Photo Researchers, Inc.

Equally important, 17th-century scientists and philosophers carried a bias, left over from the Middle Ages, that valued theory over experiments and observations. Hooke's and Leeuwenhoek's microscope work was entirely descriptive. Leeuwenhoek described thousands of objects but, probably because of his lack of education, he did not attempt to present any unifying explanation for why things were as they were.

Most scientists saw the microscope as more of a toy than a scientific instrument. By the end of Leeuwenhoek's life, at the beginning of the 18th century, few people were even making lenses anymore. Those simple microscopes already in existence were playthings for ladies, of whose observations there is little record.

Another important reason that study of the microscopic details of life languished was the poor quality of 17th- and 18th-century microscopes. Leeuwenhoek's handmade lenses were far better than any others in existence. Some of the 400 lenses he left behind could magnify objects 300 times, to show the largest bacteria. And his mysterious techniques—probably sophisticated lighting—which he never revealed, allowed him to see details that neither Hooke nor any of the other prominent microscopists of the time could match. Still, even Leeuwenhoek's microscopes were crude by today's standards (see box). His finely polished lenses distorted both the shapes and the colors of the objects under the microscope. It was not until the late 19th century, 100 years after the beginning of the Industrial Revolution, that technology provided microscopes adequate for the study of the insides of cells.

4.1 Why Are All Organisms Made of Cells?

In the 1820s, better microscopes paved the way for rapid advances in cell biology. Within 10 years, biologists discovered two important parts of cells. They found that both animal and plant cells contain a nucleus, a dark mass that we now know contains the genetic material (DNA). Researchers also finally realized that the "juice" observed by Robert Hooke was a living substance, which they named **protoplasm** [Greek, *proto* = first + Latin, *plasma* = a thing molded or formed].

Extreme Biology   How Do Microscopes Help Biologists Study Cells?

The invention of the microscope fueled all the major developments in cell biology. Every major improvement in the technology for making microscopes led to new a burst of new knowledge. In the 17th century, the development of the first light microscopes made it possible to see cells for the first time. In the 19th century, successive improvements in microscope design revealed that all organisms are made of cells and that their elaborate adaptations allow specific functions. By the mid-20th century, new electron microscopes revealed the rich internal structure of cells.

The lenses of an electron microscope are electromagnets that bend the path of electrons just as a glass lens bends the path of light. Biologists commonly use two kinds of electron microscopes—the transmission electron microscope (TEM) and the scanning electron microscope (SEM). In both microscopes, the electrons form an image on a phosphorescent screen, like that of a television set. Figure A compares a *Paramecium* seen with a light microscope, TEM, and SEM. In TEM, a beam of electrons passes *through* the specimen. In SEM, a narrow beam of electrons moves back and forth across the surface of the specimen.

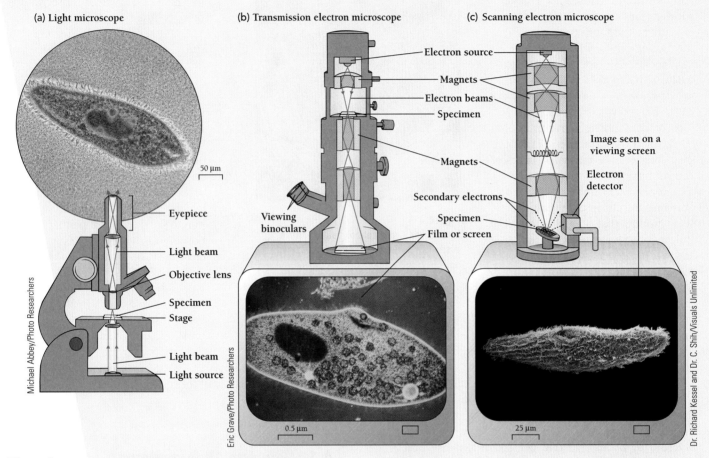

(a) Light microscope

(b) Transmission electron microscope

(c) Scanning electron microscope

Electron source
Magnets
Electron beams
Specimen
Magnets
Secondary electrons
Specimen
Film or screen

Image seen on a viewing screen
Electron detector

Eyepiece
Light beam
Objective lens
Specimen
Stage
Light beam
Light source

Viewing binoculars

50 µm

0.5 µm

25 µm

Michael Abbey/Photo Researchers

Eric Grave/Photo Researchers

Dr. Richard Kessel and Dr. C. Shih/Visuals Unlimited

Figure A

Views of *Paramecium* with different types of microscopes. (a) With a light microscope, we see the nucleus of this one-celled organism and a fringe of hairlike cilia, which these creatures use to propel themselves through water. (b) With a transmission electron microscope, we see a more detailed cross section, but the cilia are nearly invisible. (c) With a scanning electron microscope, we see the cilia protruding from the surface in greater detail than with the light microscope but nothing of the *Paramecium*'s internal anatomy.

All Organisms Are Made of Cells

In 1838, the German botanist Matthias Schleiden proposed that all plants consist of cells—an idea seconded for animal cells in 1839 by German zoologist Theodor Schwann. Cells, they wrote, are the elementary particles of all living organisms. In addition, Schwann and Schleiden argued that all cells are alive—independent of the organisms to which they belong. In other words, a liver cell is alive in its own right even though it is only a small part of a living organism. Remove the cell from the liver and the cell is still alive.

Recall from Chapter 1 that animals and plants are all multicellular, that the Fungi and Protista include both multicellular organisms and single-celled organisms, and that the Archaea and Eubacteria are all single-celled organisms. Noncellular organisms do not exist. With the discoveries of Schleiden and Schwann, biologists recognized the importance of cells. Cells, they saw, were everywhere. But where did cells come from? Where did the new cells that heal a wound come from? Where did the cells that build a child's growing body come from? Where did the burgeoning masses of cells in cancers come from? And where did the one-celled organisms living in a pond come from?

Schleiden and Schwann proposed that cells crystallized spontaneously out of shapeless matter. But ever-improving microscopes allowed more and more scientists to see that individual cells can divide, each giving rise to what biologists call "daughter" cells. In time, biologists accepted the idea that all new cells come from the division of preexisting cells.

In 1858, the widely respected physician and biologist Rudolf Virchow formalized this understanding with the Latin phrase, *omnis cellula e cellula,* meaning "all cells from cells." In a book written for physicians, Virchow summarized his theory about the role that cells play in disease. In doing so, he simultaneously revolutionized both biology and medicine. Virchow argued (1) that cells never arise from noncellular material and (2) that diseases result from changes in specific kinds of cells.

As a result of these arguments, biologists came to accept the principle that cells always come from other cells. Cells healing a wound result when preexisting skin cells begin dividing. A child's growth results from the rapid multiplication of cells. Cancers result from the uncontrolled division of an organism's cells. Single-celled organisms in ponds descend from other single-celled organisms.

Today, we summarize the work of Schleiden, Schwann, and Virchow in the **cell theory,** which states that (1) all organisms are composed of one or more cells; (2) cells, themselves alive, are the basic living unit of organization of all organisms; and (3) all cells come from other cells. We should add qualifications to this. First, as we discussed in Chapter 1, we use the word "theory" to denote a system of statements and ideas that explains a group of related facts or phenomena, *not* to indicate uncertainty. Second, although in today's high-oxygen world all cells come from other cells, when the first cells evolved some 3.8 billion years ago, they may not have come from other cells. We discuss the evolution of the first cells in Chapter 18.

The "cell theory" says that all organisms are composed of one or more cells, that cells are the basic living unit of organization of all organisms, and that all cells come from other cells.

Every Cell Consists of a Boundary, a Cell Body, and a Set of Genes

The cells of organisms from the three domains of life can differ greatly. Even within a single organism, cells come in wildly different types. Nonetheless, all cells have three common features: (1) a boundary that separates the inside of the cell from the rest of the world, (2) a set of genes, and (3) a cell body.

The **plasma membrane** is a cell's boundary, a highly organized and responsive structure. Like the outer wall of a house, with its windows and doors, the plasma membrane not only defines the limits of a cell, but also helps to regulate the cell's internal environment by selectively admitting and excreting specific molecules.

Each cell also contains a set of genes strung out along one or more molecules of DNA. In eukaryotes (animals, plants, fungi, and protists), the DNA is confined within a membrane-enclosed structure called the **nucleus** (Figure 4-2A). The presence of a true nucleus alone defines the eukaryotes and distinguishes them from the prokaryotes (eubacteria and archaebacteria). In the prokaryotes, the DNA occupies a limited region of the cell, called the **nucleoid,** which has no membrane (Figure 4-2B).

Throughout the 19th century and well into the 20th century, biologists continued to think of the protoplasm as a more or less homogeneous jellylike substance. Improved techniques for looking at cells, however, revealed more and more tiny structures in eukaryotic cells, though not in prokaryotic cells. We now know that each of these tiny structures, called **organelles,** performs a specialized task.

Biologists finally discarded the word protoplasm and renamed the cell body—that part of the cell outside the nucleus but inside the membrane—the **cytoplasm** [Greek, *cyto* = a hollow container (a cell) + Latin, *plasma* = a thing molded or formed]. The part of the cytoplasm not contained within membrane-enclosed organelles was named the **cytosol.** Most of a cell's biochemical work occurs within the cytosol. Running through the cytosol is an intricate network of protein fibers, called the **cytoskeleton,** which gives the cell its shape, holds organelles in place, and participates in cell movement.

Every cell is enclosed by a plasma membrane and keeps its DNA in a nucleus (eukaryotes) or nucleoid (prokaryotes). Eukaryotic cells contain other organelles within the cytoplasm.

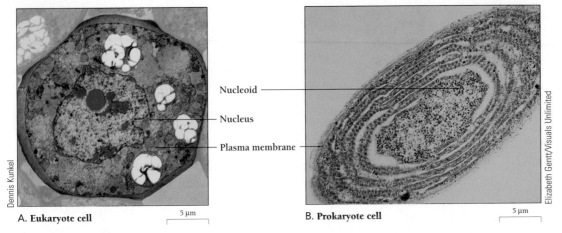

Nucleoid

Nucleus

Plasma membrane

A. Eukaryote cell   5 µm

B. Prokaryote cell   5 µm

Dennis Kunkel

Elizabeth Gentt/Visuals Unlimited

Figure 4-2

Eukaryotes have a nucleus. A. A eukaryotic cell, such as this plant cell (from a voodoo lily), keeps DNA in a nucleus that is enclosed in a plasma membrane like that surrounding the whole cell. Inside the nucleus (green) is an even tinier structure called the nucleolus (purple), where the DNA resides. Also shown are mitochondria (red), which make ATP, and starch (yellow), where energy is stored. These organelles, which also have membranes, characterize the eukaryotes. B. A prokaryotic cell such as this cyanobacterium has no nucleus and no other internal plasma membranes.

How Are Cells Alive?

Cells—from Leeuwenhoek's one-celled "animalcules" to the individual cells of the liver—are the fundamental living units of life. They have all the characteristics of life that we listed in Figure 1–8. Made up of organized parts, cells perform chemical reactions, obtain energy from their surroundings, respond to their environments, change over time, reproduce, and share an evolutionary history. Cell components work together to achieve homeostasis, maintaining an internal cellular environment that is constant enough to carry out the thousands of chemical reactions that allow cells to live and to reproduce.

The ability of cells to obtain energy from their environment depends on a set of chemical reactions, some of which we discuss in Chapters 6 and 7. Almost every cell can capture energy from glucose, usually by oxidizing it to carbon dioxide and water. In eukaryotic cells, many of the energy-producing reactions take place in specialized organelles, while others take place in the cytosol.

Cells change with time, both chemically and mechanically. For example, individual muscle cells shorten as muscles contract. Other cells, too, change their size and shape. To allow for these changes, almost every eukaryotic cell contains a network of fibers, called the cytoskeleton, that supports the cell's components and brings about change in cell shape. The cytoskeleton also serves as train tracks along which cell components can move, pulled by locomotives called molecular motors.

In order to reproduce, every cell must be able to copy its genes for its descendants. All cells have elaborate mechanisms for precisely duplicating DNA.

Cell reproduction also depends on the cell's ability to read and to use the coded information in DNA. In a single-celled organism, cells adjust the readout of their genetic information in response to environmental changes. A bacterial cell growing in a glass of milk, for example, makes different proteins from the same type of cell growing in a glass of sugar water. In a multicelled organism, different types of cells make different sets of proteins, according to their specialized characteristics: a red blood cell, for example, makes lots of hemoglobin, whereas a muscle cell makes lots of contractile proteins.

Once a cell produces the appropriate proteins, however, it must deliver them to the right address within the cell. Cells have elaborate mechanisms for directing molecular traffic to different parts of the cell.

Finally, cells must deal with their own garbage—waste products such as carbon dioxide and ammonia, as well as cell remnants damaged by ordinary wear and tear. Special molecules and mechanisms serve as cellular toilets and garbage disposals.

A cell could conceivably employ a wide variety of solutions to its problems. Different cells might use different strategies to derive energy, to distribute ions, and to organize its organelles. But all cells use the same molecules, share the same biochemical pathways, and contain the same types of organelles. This sharing of solutions reflects the common evolutionary heritage of all cells, which almost certainly descended from a common ancestor that lived about 3.5 billion years ago.

What Are the Advantages of Cellular Organization?

Plants, animals, and other multicellular organisms can be as large as trees and elephants, but free-living single-celled organisms are nearly all small. In fact, with some remarkable exceptions, cells are nearly all about the same size. Most eukaryotic cells range from 10 to 100 micrometers (μm), while prokaryotic cells are smaller, ranging from 0.4 to 5 μm. For comparison, a period on this page is about 100 μm in diameter, the size of a large eukaryotic cell, such as those in an onion skin. Of course, a few dramatically large cells exist, including ostrich eggs, meter-long plant fiber cells, and the several-meter-long nerve cells in the legs of a giraffe.

One reason that cells are nearly all so small is that subdivision into tiny cells offers organisms many advantages. To understand these advantages, we can start by asking a simple question: What problems would you face if you were one very large cell?

A Cell's Need to Regulate Its Internal Environment Limits Its Size

For the biochemical machinery of a cell to function, the cell must maintain a relatively constant internal environment (homeostasis). Otherwise, enzymes and other proteins will not adopt the shapes needed for biological activity. To maintain a constant internal environment, the cell must keep both the concentration of salts and the pH from changing. At the same time, the cell must also be able to take in useful molecules and dispose of waste molecules.

If you were one big cell, you would have to maintain just the right environment for each of the different molecular activities in your body. The highly acidic (low) pH that you use to digest food would have to somehow coexist with the neutral pH needed for protein synthesis.

All the materials that come into a cell or leave a cell must go through the thin membrane that envelops the cell. If you were one big cell, your plasma membrane would have to admit or exclude specific molecules selectively. Somehow you'd have to absorb enough pizza to get you through the day.

Protists, such as the *Didinium* shown in Figure 4-3, perform all these functions and more. In fact, unlike most pro-

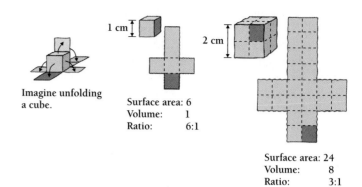

Surface area: 6
Volume: 1
Ratio: 6:1

Surface area: 24
Volume: 8
Ratio: 3:1

Figure 4-4

Surface area. The mathematical relationship between surface area and volume explains some of the properties of cells. In particular, smaller cells have more surface area per volume than larger cells. When we double the linear dimensions of a cube, surface area increases four times, and volume increases eight times.

tists, the *Didinium* even has a sort of mouth with which it can eat other protists. Protists, however, are all small. Accomplishing all that a protist does wouldn't be so easy for a one- or two-hundred-pound cell like yourself. The membrane's ability to admit or to excrete molecules is limited by its size. A big cell certainly has more external membrane than a small one. As a cell grows, however, its volume increases faster than its surface area.

Let's think, for a minute, about the relationship between the surface area and the volume of any object. Suppose we have a tiny cardboard cube, one centimeter on each side. If we double the linear dimensions of the cube, so that each side is 2 cm by 2 cm, the surface area of the cube increases by a factor of 4. But its *volume* increases by a factor of 8 (Figure 4-4).

The larger cube has proportionately less surface area than the smaller cube. We can see this if we look at the **surface-to-volume ratio**—the amount of surface area for each bit of volume. For the small cube, the surface-to-volume ratio is 6:1. For each cm³ of volume, the smaller cube has 6 cm² of surface. But the surface-to-volume ratio for the larger cube is 24:8 (or 3:1). For each cm³ of volume, the larger cube has only 3 cm² of surface.

Because larger objects have smaller surface-to-volume ratios, larger cells have proportionately less membrane with which to regulate their internal space. As a result, larger cells have a harder time obtaining nutrients, getting rid of wastes, and regulating the internal concentrations of ions and molecules. If you were one big cell, molecules of oxygen and food would take decades to get from your membrane to your interior. You could bury yourself in pizza and you'd still go to bed hungry. You simply would not be able to digest the pizza's macromolecules fast enough to survive.

Knowing how a cell's size affects the regulation of its internal environment, we might wonder how eukaryotic cells,

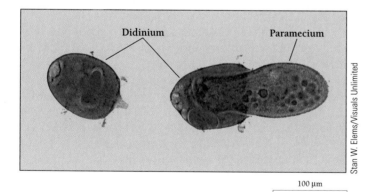

100 μm

Figure 4-3

Even one-celled organisms have to eat. This *Didinium (left)* has a mouth with which it can consume other protists, such as the one-celled *Paramecium* it is ingesting.

which range from 10 to 100 μm, can be so much larger than prokaryotic cells, which range from 0.4 to 5 μm. The answer is that eukaryotic cells have special adaptations to increase their surface areas. For example, many eukaryotic cells have convoluted surface membranes, and almost all have elaborate internal membrane systems, which we will discuss later in this chapter. Indeed, some eukaryotic cells, such as certain white blood cells, have 50 times more membrane inside than outside.

The larger a cell, the smaller is its surface-to-volume ratio. Single-celled organisms must ingest food and expel wastes, just as multi-celled organisms do.

Individual Cells May Specialize for Different Tasks

In multicellular organisms, individual cells often perform separate tasks. For example, red blood cells carry oxygen, while muscle cells contract and move. Cells in the roots of a tree absorb nutrients from the soil, while cells in the leaves harvest the energy of sunlight. Just as a division of labor makes human societies more productive, the specialization of cells allows an organism to function efficiently.

If you were one big cell, organizing your body to perform all its different jobs would be difficult. You wouldn't have a heart or arteries to move oxygen and nutrients around your body. You wouldn't have lungs to extract oxygen from the air. You'd just have 125 pounds (or so) of cytoplasm.

Cellular organization allows organisms to make a division of labor among specialized cells.

A Multicellular Organism Can Lose and Replace Individual Cells

Subdividing the body into cells has another advantage. Cells often live and die independently of the whole organism. In fact, the death of some cells is a normal part of development. The fingers of the hand, for example, develop from paddlelike appendages as a result of the death of cells between the fingers. The lenses of the eyes, the red blood cells that carry oxygen to all the tissues, and the outer layer of the skin all develop because of the death of specific cells. Equally important, the cells of the skin, blood, and intestines continuously replace themselves. The advantage of such replacement is that the life of a multicellular organism can extend far beyond the life of an individual cell.

Cellular organization allows organisms to outlive the cells that compose them.

4.2 What's in a Cell?

Eukaryotic cells contain so many different organelles, each with a special set of tasks, that we can imagine each cell as a tiny walled city (Figure 4-5). Inside the wall of the city are power stations, a central library of genetic information, warehouses for packaging proteins, and much more. Table 4-1 summarizes the characteristics of cells from the six kingdoms of organisms.

When scientists first began using the transmission electron microscope in the 1940s and 1950s, one of the great surprises was the huge amount of internal membrane in eukaryotic cells (see box on p. 66). These membranes divide cells into several kinds of compartments, including the nucleus, cytosol, endoplasmic reticulum, Golgi complex, lysosomes, and mitochondria. In addition, a plant cell contains one more set of compartments—plastids.

Eukaryotic cells also contain a variety of membranous sacs with nothing visible inside. The **vesicles** of plant and animal cells are small, apparently empty, sacs—generally about 100 nanometers (nm) in diameter. Eukaryotic cells (especially plant cells) also contain **vacuoles**—large sacs that may occupy as much as 95 percent of the volume of the cell.

What Role Does the Nucleus Play in the Life of a Cell?

The nucleus usually makes up 5 to 10 percent of the volume of the cell. Inside are the **chromosomes** [Greek, *chroma* = color + *soma* = body], long complexes of DNA and protein. The DNA is the substance of the genes (Figure 4-6).

An electron microscope shows that the nucleus's boundary is a double membrane called the **nuclear envelope.** The two membranes are separated by a space punctuated by **nuclear pores,** channels between the inside of the nucleus and the cytoplasm (Figure 4-7). Nuclear pores control the movement of materials in and out of the nucleus in much the same way that a turnstile controls the movements of people in and out of a subway.

The nucleus serves as the cell's library, and it also provides instructions for the construction of new cells. The nucleus contains the genetic information (coded in DNA) that is passed from one cell to its descendants. When a cell divides, each daughter cell gets a complete copy of the genetic instructions in the nucleus. The DNA in each cell nucleus includes instructions for building every polypeptide the body will ever use. In Part II of this book, we will consider how DNA carries genetic information, how cells use this information, and how they pass it to their daughter cells.

The nucleus of eukaryotic cells contains DNA and protein. The DNA acts as a library of information and a set of instructions for making cell components.

Cell Component	Function	Eubacteria & Archaebacteria	Protist	Fungus	Plant	Animal
Cell wall	Protects and supports cell	Yes	Yes	Yes	Yes	None
Plasma membrane	Regulates communication with other cells and movement of materials in and out of cell	Yes	Yes	Yes	Yes	Yes
Membrane-enclosed nucleus	Isolates DNA from cytoplasm	None	Yes	Yes	Yes	Yes
DNA	Encodes information	Single molecule		Many molecules		
RNA	Constructs proteins from information in DNA	Yes	Yes	Yes	Yes	Yes
Nucleolus	Synthesizes ribosomes	None	Yes	Yes	Yes	Yes
Ribosome	Synthesizes proteins	Yes	Yes	Yes	Yes	Yes
Endoplasmic reticulum	Modifies proteins, makes lipids	None	Yes	Yes	Yes	Yes
Golgi complex	Tags proteins and lipids, packages them for export	None	Yes	Yes	Yes	Yes
Lysosome	Contain digestive enzymes	None	Yes	Yes	Yes	Yes
Mitochondrion	Synthesize ATP	None	Yes	Yes	Yes	Yes
Chloroplast	Photosynthesize	None	Some	None	Yes	None
Other plastids	Store food, pigments	None	Some	None	Yes	None
Central vacuole	Provides turgor pressure; contains water and wastes	None	None	Yes	Yes	None
Cytoskeleton	Shapes, strengthens, and moves	None	Yes	Yes	Yes	Yes
Flagellum, cilium	Moves cell through fluid or fluid past cell surface	None	Yes	Yes	Yes	Yes
Bacterial flagellum	Moves cell through fluid or fluid past cell surface	Yes	None	None	None	None

The Cytosol Is the Cytoplasm That Lies Outside the Organelles

The cytosol makes up about half of a cell's volume. Cell biologists usually isolate the cytosol from the organelles by breaking open cells and then spinning the mixture in a centrifuge. A centrifuge is a machine that spins a set of tubes arranged so that when they are spinning their contents come under a force up to 100,000 times greater than gravity. This force—centrifugal force—is the same force that presses clothes in a washing machine to the outside wall of the tub during the spin cycle. In a centrifuge, a tube spins so fast—2,000 to 80,000 revolutions per minute—that its contents sink to the bottom, with bigger particles and molecules sinking faster than smaller ones. When cell biologists centrifuge cells, the organelles, which are heavier than the cytosol, drop to the bottom of each tube first.

The cytosol contains thousands of different kinds of enzymes responsible for producing building blocks, degrading small molecules to yield energy, and synthesizing proteins. Although the cytosol is aqueous (mostly water), about 20 percent of its weight is protein, giving it the consistency of Jell-O. Embedded in the cytosol are organelles, vesicles, and vacuoles, as well as smaller structures that are not enclosed by membranes. These include granules of energy-rich glycogen, droplets of stored fat, *ribosomes*—tiny, round organelles, about 15 to 30 nm in diameter—and *proteasomes*—barrel-shaped organelles, slightly smaller than ribosomes.

Ribosomes [Latin, *soma* = body + "ribo" because they contain ribonucleic acid (RNA)] are tiny assemblies of RNA and protein, workbenches on which protein molecules are stitched together with peptide bonds. Ribosomes can be bound to membranes or float free in the interior of the cell. But all these proteins eventually become damaged, and a cell must dispose of them somehow. The answer is **proteasomes,** hollow protein complexes that work like garbage disposals to break up old proteins, so the amino acids can be recycled.

The jellylike cytosol, which surrounds the cell's organelles, contains ribosomes, glycogen, fat, thousands of enzymes, and much more.

The Endoplasmic Reticulum Is a Folded Membrane

The **endoplasmic reticulum (ER)** [Greek, *endon* = within + *plasmein* = to mold + Latin, *reticulum* = network] makes proteins and lipids, which the cell exports. The ER

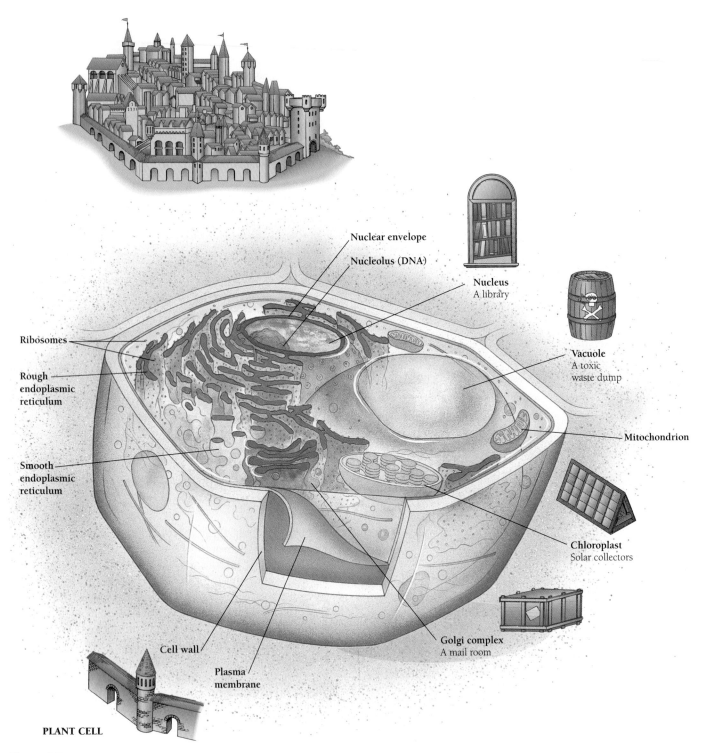

Figure 4-5

A eukaryotic cell is like a walled city. In both plant cells *(left)* and animal cells *(right)*, a plasma membrane regulates the passage of molecules in and out of the cell, just as a city wall regulates the passage of people. Inside both kinds of cells, mitochondria supply power, in the form of ATP instead of electricity. The nucleus is a library, housing most of the information a cell needs to make proteins. The nuclear envelope determines which molecules have access to the information inside and which bits of information (books) may be checked out for use in the cytoplasm. The ribosomes act as workbenches for the production of

proteins. The vacuole serves as a depository for waste materials, much like a toxic waste dump, or for the storage of materials. The Golgi complex is like a mailroom. Here, glycoproteins are packaged and labeled for export or intracellular transport. In addition to all these parts, plant cells also have chloroplasts, which, like solar collectors, capture energy from the sun and turn it into a usable form, sugar.

Biology⊜Now™ Learn more about cell structure by clicking on this figure on your BiologyNow CD-ROM.

Ribosome
A workbench for
making proteins

Flagellum

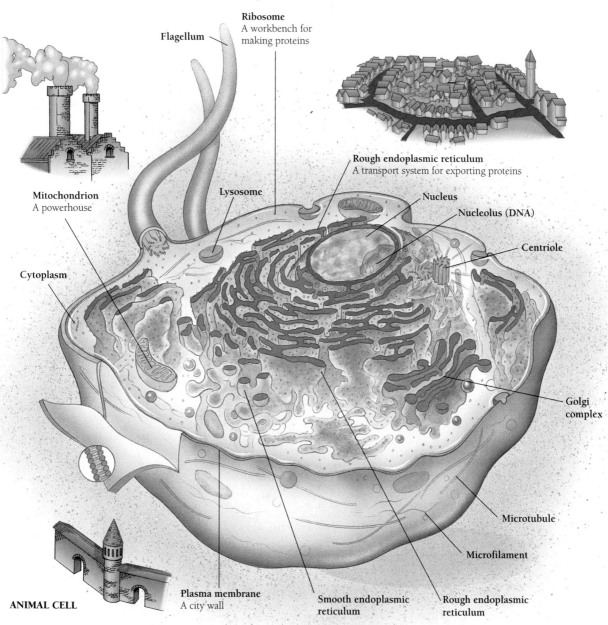

Rough endoplasmic reticulum
A transport system for exporting proteins

Mitochondrion
A powerhouse

Lysosome

Nucleus

Nucleolus (DNA)

Centriole

Cytoplasm

Golgi
complex

Microtubule

Microfilament

Plasma membrane
A city wall

Smooth endoplasmic
reticulum

Rough endoplasmic
reticulum

ANIMAL CELL

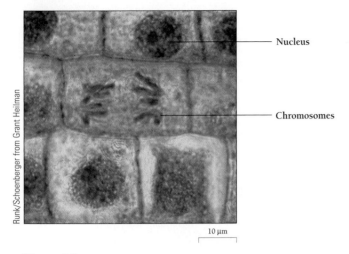

Figure 4-6
Chromosomes. When the chromosomes condense during cell division, they become obvious, but they are always present in the nucleus.

Runk/Schoenberger from Grant Heilman

poses, some specialized cells may have especially large amounts of one or the other. Cells specialized for the production of secreted proteins, such as those in the pancreas, have large amounts of rough ER. Cells specialized for the production of lipids, such as the cells that produce steroid hormones, contain large amounts of smooth ER.

The endoplasmic reticulum (ER), with its mazelike internal lumen, consists of rough ER and smooth ER.

The Golgi Complex Directs the Flow of Newly Made Proteins

The **Golgi complex** functions as both a packaging center and a traffic director. During cell reproduction and maintenance, it helps form structures that stay in the cell, such as lysosomes. It also contributes to specialized cell function by helping prepare materials for export outside the cell.

The Golgi complex, usually found near the nucleus, consists of sets of flattened discs, like stacked dinner plates (Figure 4-9). Each stack is about 1 μm in diameter and typically consists of about six discs. Surrounding the stacks of flat discs are some smaller, spherical vesicles.

Every eukaryotic cell has a Golgi complex, but cells that make lots of "glycoproteins" have more stacks than

Custom Medical Stock Photo, Inc.

R. Kessel-G. Shih/Visuals Unlimited

resembles the marbling in marble cake. When biologists cut a cell into slices, as we would slice a loaf of marble cake, the ER in each slice forms an extensive and convoluted network. Careful study of successive sections shows that the ER is actually a single convoluted sheet of membrane that is continuous with the outer membrane of the nucleus. The ER membrane encloses a network of interconnected cavities and channels called the **lumen** [Latin, *lumen* = light, an opening], which makes up about 15 percent of the cell's volume.

The ER consists of two parts that are distinct in both appearance and function. The **rough ER** is studded with ribosomes on the cytosol side of the membrane and is mostly devoted to modifying recently synthesized proteins, especially those destined for export to organelles or out of the cell (Figure 4-8B). The **smooth ER** has no ribosomes and is mostly devoted to the synthesis and metabolism of lipids (Figure 4-8C). Another important function of the smooth ER is the breakdown of substances such as alcohol.

Both rough and smooth ER are present in most eukaryotic cells. But since the two kinds of ER serve different pur-

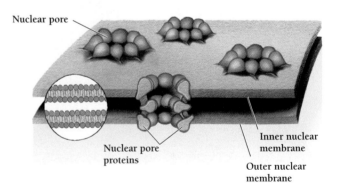

Figure 4-7
Holes in the nuclear envelope. TEM of the nucleus of a mammal cell *(at left)*. The nucleus's double membrane, or envelope, is riddled with tiny holes, called nuclear pores. The nuclear pores help determine which genes are expressed by controlling which molecules can enter and leave the nucleus.

other cells. Glycoproteins are proteins that have attached sugars or carbohydrates. The main protein in egg white (albumin) is a glycoprotein, as are the proteins in mucus (Figure 4-9).

To find out what the Golgi complex did, George Palade and Marilyn Farquhar, at Rockefeller University, labeled newly made glycoproteins with radioactive tracers. Proteins destined for export appeared first in the rough ER and then in the Golgi complex (Figure 4-9). The Golgi complex modifies the glycoproteins, then packages them into membrane-enclosed secretory vesicles. The vesicles eventually fuse with the plasma membrane and discharge their contents to the outside of the cell.

The Golgi complex also manages the flow of proteins to different destinations. The Golgi complex "addresses" glycoproteins by modifying the carbohydrates. Proteins acquire specific "tags" that direct them to specific destinations within the cell—such as lysosomes or for export outside of the cell.

> The Golgi complex packages and labels glycoproteins for destinations both inside and outside the cell.

The Lysosomes Function as Digestion Vats

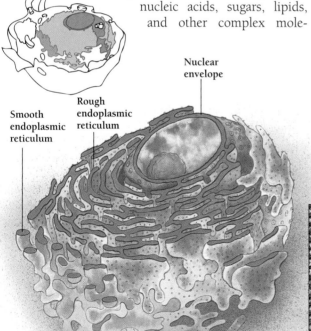

Lysosomes, present in all eukaryotic cells, are small vesicles, enclosed by a single membrane. They vary widely in size, but all contain enzymes that break down proteins, nucleic acids, sugars, lipids, and other complex mole-

cules. The large vacuole of a plant cell contains similar enzymes, and we may consider it to be a giant lysosome.

Lysosomes are particularly numerous in cells that actively perform **phagocytosis** [Greek, *phagein* = to eat + *kytos* = hollow vessel, or cell], the process by which cells take in and consume large particles of solid food. Free-living amebas and many other single-celled eukaryotes feed primarily by phagocytosis (Figure 4-10). Our own white blood cells also regularly engulf and digest bacteria, viruses, and assorted cellular debris. Once a cell has engulfed material within a vesicle, the vesicle fuses with a lysosome, whose enzymes then digest the engulfed material.

The membrane of a lysosome keeps its enzymes from digesting the cytosol's own proteins. If the lysosome membrane breaks down, however, the cell begins to digest itself. Lysosomes first form from membranes derived from the Golgi complex. Their digestive enzymes are made on the ribosomes of the rough ER. They are packaged into lysosomes after passing through the lumen of the ER, where each lysosomal enzyme acquires a special attached sugar. This sugar provides the lysosomal proteins with an address tag, so that the Golgi complex directs them to lysosomes rather than exporting them to the outside of the cell.

> Lysosomes contain enzymes that the cell uses to digest food, wastes, and foreign invaders. Lysosome membranes normally isolate these dangerous enzymes from the rest of the cell.

Mitochondria Convert Energy Into a Usable Form

Mitochondria [singular, mitochondrion; Greek, *mitos* = thread + *chondrion* = a grain] are the organelles most responsible for the cell's ability to obtain energy from nutri-

Smooth endoplasmic reticulum

Rough endoplasmic reticulum

Nuclear envelope

A.

Ribosomes on rough ER

B.

0.5 μm

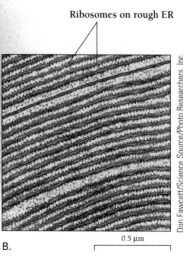

Figure 4-8
Endoplasmic reticulum. A. The smooth ER has no ribosomes and mostly metabolizes lipids and detoxifies substances such as alcohol. B. The rough ER is studded with protein-synthesizing ribosomes, which are what make it rough. Rough ER mostly makes and modifies proteins, especially those destined for export to organelles or out of the cell.

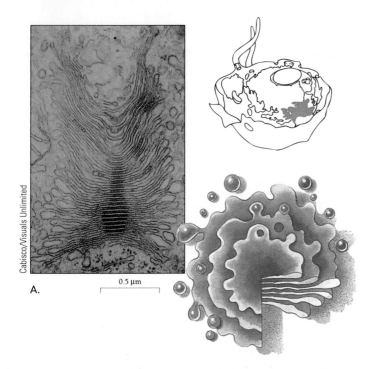

A.

0.5 μm

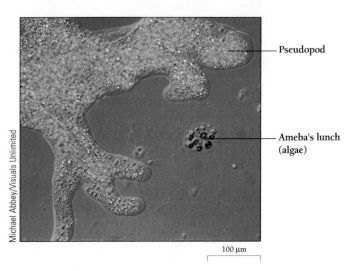

Pseudopod

Ameba's lunch (algae)

100 μm

Figure 4-10

Eat up. An ameba wraps its pseudopods around a bit of algae in the process of phagocytosis. The pseudopods completely enclose a particle, then bring it inside the cell for digestion.

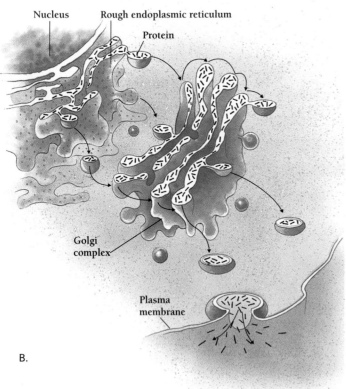

Nucleus   Rough endoplasmic reticulum

Protein

Golgi complex

Plasma membrane

B.

Figure 4-9

The Golgi complex. Newly made proteins pass through the ER to the Golgi complex. A. A Golgi complex or "stack." B. Vesicles on the surface of the Golgi complex shuttle glycoproteins from the rough ER to the Golgi stack, which "labels" the glycoproteins and packages them into vesicles that carry the glycoproteins out through the cell membrane.

ents. These tiny power plants make nearly all of the ATP (adenosine triphosphate) that powers the chemical reactions of the cell. Like a power station that burns coal or oil to make easily used electricity, mitochondria convert the energy in sugars into molecules of ATP, as mentioned in Chapter 3. In Chapter 6, we'll see how the complex membranes of mitochondria help them generate ATP.

Even under the best light microscope, mitochondria, often less than 1 μm in diameter, can barely be seen. However, the transmission electron microscope (TEM) reveals that mitochondria are among the most numerous organelles in most eukaryotic cells (Figure 4-11). Different kinds of cells have different numbers of mitochondria. A liver cell, for example, may contain several thousand mitochondria, which together make up about a quarter of the cell volume. Cells that use particularly large amounts of energy, such as those of the heart, may contain still more.

The transmission electron microscope also reveals the mitochondrion's characteristic structure. Two membranes, 6 to 10 nm apart, separate a mitochondrion's interior from the cytosol. The outer membrane of a mitochondrion is smooth, but the inner membrane forms elaborate folds, called *cristae*. Mitochondria come in a variety of sizes and shapes. Time-lapse photography with a phase contrast microscope reveals that mitochondria may be constantly in motion, changing shape, fusing, and dividing.

Mitochondria are unusual among organelles not only in having double membranes: they also have their own DNA and make some of their own proteins. These characteristics strongly suggest that mitochondria evolved inside

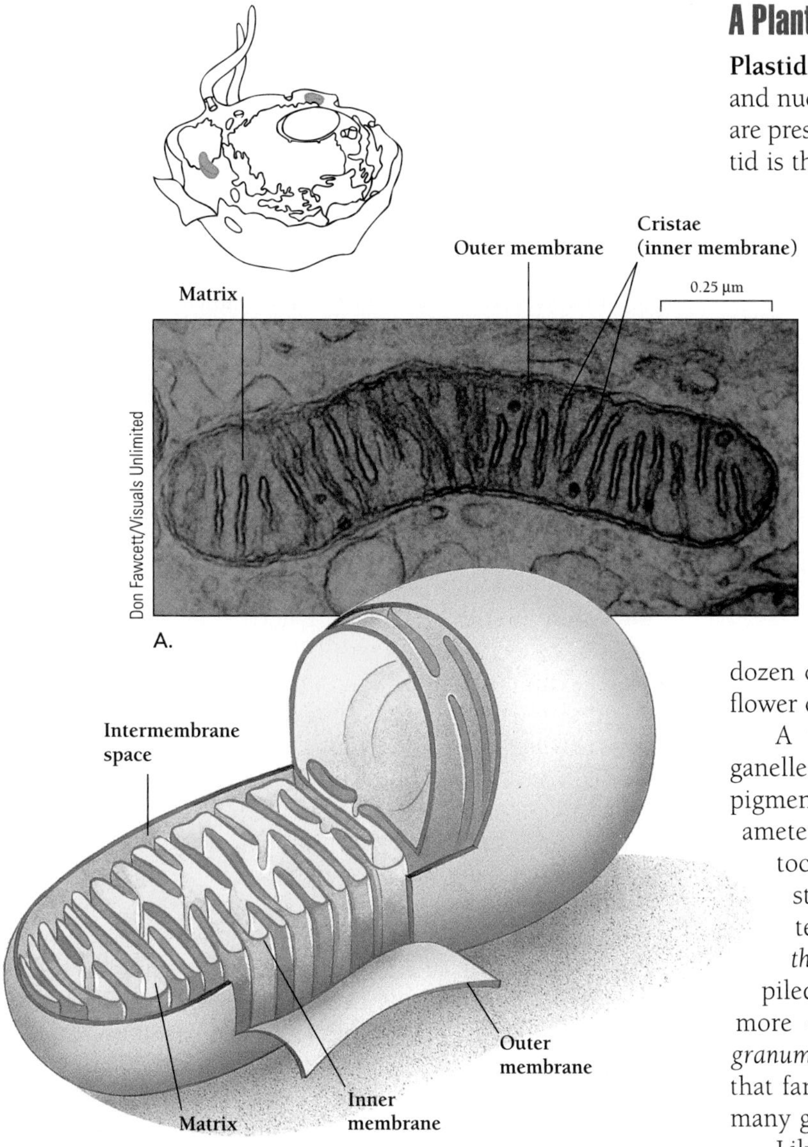

A Plant Cell's Chloroplast Is Just One Kind of Plastid

Plastids are organelles in plant cells that, like mitochondria and nuclei, are surrounded by a double membrane. Plastids are present in nearly all plant cells. The most important plastid is the **chloroplast**, present only in green plant cells and some protists, which uses the energy of sunlight to build energy-rich sugar molecules in the process of **photosynthesis** [Greek, *photo* = light + *syntithenai* = to put together]. Most organisms on Earth therefore ultimately depend on chloroplasts for food. Plants also have other kinds of plastids, which serve as storage depots for various kinds of molecules (Figure 4-12).

During photosynthesis, chloroplasts make sugars. The chloroplasts also serve as temporary storage depots for the newly made sugars (in the form of starch grains). The size, shape, and number of chloroplasts in a cell vary from species to species and from cell type to cell type. A leaf cell in a flowering plant, for example, typically contains several dozen chloroplasts. A stem cell contains only a few, and a flower cell may contain none at all.

A typical chloroplast is a large, round, green organelle. The distinctive shade of green comes from the pigment *chlorophyll*. The chloroplast is 5 to 10 μm in diameter, but only 2 to 4 μm high, a thick disc. Like a mitochondrion, a chloroplast has an intricate internal structure of folded membranes. These lie in flattened, disclike sacs, called **thylakoids** [Greek, *thylakos* = sac + *oides* = like]. The thylakoids are piled like plates (Figure 4-12A). Each stack, of 10 or more discs, is called a **granum** [plural, grana; Latin, *granum* = grain] because the stacks resemble the granaries that farmers use to store grain. Each chloroplast contains many grana.

Like mitochondria, chloroplasts have double membranes and DNA, and they perform protein synthesis. Because of these and other clues, biologists have concluded that chloroplasts also originated as free-living organisms that were captured by early eukaryotic cells and then evolved into chloroplasts.

Two other kinds of plastids are chromoplasts and amyloplasts. **Chromoplasts** contain the pigments that give yellow, orange, and red colors to flowers, fruits, vegetables, and autumn leaves (Figure 4-12B). Chromoplasts arise from chloroplasts through the reshaping of the inner membrane and the breakdown of chlorophyll. **Amyloplasts** store starches and are abundant in roots such as sweet potatoes and turnips, as well as in wheat, rice, and other seeds (Figure 4-12C).

Matrix
Outer membrane
Cristae (inner membrane)
0.25 μm

Don Fawcett/Visuals Unlimited

A.

Intermembrane space
Outer membrane
Inner membrane
Matrix

B.

Figure 4-11

Mitochondria make ATP. The outer membrane of a mitochondrion encloses the whole organelle, and the inner membrane encloses the matrix. Chapter 6 describes how the folded inner membrane makes the energy molecule ATP.

early eukaryotic cells that captured and ate the bacteria but did not digest them. Instead, the bacteria came to live inside the eukaryotic cells and evolved into mitochondria (Chapter 18).

Mitochondria synthesize energy-rich ATP in their mazelike interior.

Chloroplasts turn solar energy into chemical energy in the process of photosynthesis. Other kinds of plastids store pigments or starches.

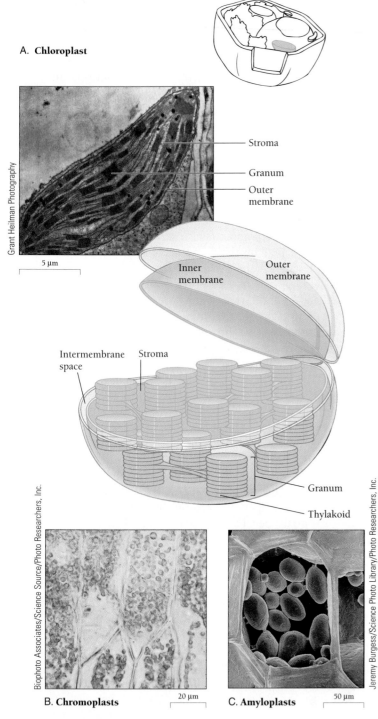

A. Chloroplast

Stroma

Granum

Outer
membrane

5 μm

Outer
membrane

Inner
membrane

Intermembrane
space

Stroma

Granum

Thylakoid

B. Chromoplasts   20 μm

C. Amyloplasts   50 μm

Figure 4-12

Chloroplasts and other plastids. A. A chloroplast uses sunlight to build sugar molecules, which are temporarily stored inside the chloroplast. A chloroplast's green color comes from the pigment chlorophyll, which captures the energy of sunlight during photosynthesis. Like mitochondria, chloroplasts have an outer membrane that encloses the whole organelle and an inner membrane that encloses a matrix. But chloroplasts also have a highly folded third membrane that forms pockets, called thylakoids, arranged in stacks called grana. B. Chromoplasts in the cells of a yellow tulip petal. C. Amyloplasts in raw potato.

What Does the Cytoskeleton Do?

Eukaryotic cells have a complex network of protein filaments, called the cytoskeleton, which is visible with the TEM (Figure 4-13). Three types of filaments make up the cytoskeleton: microtubules, actin filaments, and intermediate filaments (Figure 4-13). Like our own bony skeletons, the cytoskeleton provides structure and support. The name "cytoskeleton" is slightly misleading, however, since the cytoskeleton does more than give mechanical support for the cytoplasm and its component organelles. It also provides the force for most of a cell's movements, including changes in the cell's shape and the transport of materials throughout the cell.

Some of the specialized proteins of the cytoskeleton are closely related to muscle proteins. While muscle proteins are essentially permanent structures, however, the cytoskeleton is constantly breaking down and reforming itself.

Microtubules are hollow cylinders about 25 nm in diameter, which vary in length (Figure 4-13A). Microtubules play a critical role in cell division. When a cell divides, microtubules distribute the DNA and other materials to the two daughter cells.

The assembly of microtubules originates in structures called microtubule organizing centers (MTOCs), shown in Figure 4-14. In the cytoplasm of most eukaryotic cells, the major MTOCs are near the nucleus, in a zone called the **centrosome**—a region that in many cells also contains a distinct organelle called the **centriole.** The MTOCs of the centrosome help cells divide (Chapter 8). The cylinders of microtubules consist of two globular protein molecules called **tubulins.** Pure tubulin in a test tube can form microtubular structures all by itself. But in living cells up to 50 other proteins help form microtubules.

Actin filaments, also called *microfilaments,* are much finer than microtubules—only about 7 nm in diameter (Figure 4-13B). Actin filaments are often anchored to the cell surface and provide the force for movement and shape changes. Actin filaments consist of a single kind of protein, called actin, which is also present in muscle fibers.

In muscle, actin filaments are permanent structures that interact with other proteins to produce the force leading to contraction. In nonmuscle cells, actin filaments are constantly being built and destroyed. Even in the test tube, actin molecules can self-assemble to form a two-stranded helix apparently identical to the actin filaments in intact cells. In cells, they form cross-linked cables that provide mechanical support. They also associate with other proteins to furnish musclelike contractile forces. Other proteins associate with actin filaments to act as molecular winches that pull on actin cables.

Intermediate filaments, usually 8 to 10 nm in diameter, are fibrous proteins such as keratin (Figure 4-13C). Ker-

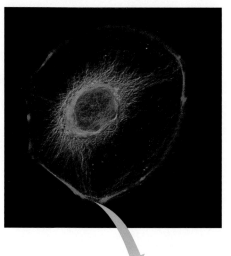

Figure 4-13

The cytoskeleton of a cell. The cytoskeleton is a network of three kinds of protein filaments. A mouse cell *(top)* shows its actin and tubulin cytoskeleton in red and green, with a purple nucleus. A. Microtubules are hollow tubes formed from the protein tubulin that form an internal skeleton for the whole cell *(green)*. B. Actin filaments, made of the protein actin, are often closer to the cell surface *(red)*. C. Intermediate filaments reinforce cells, especially at places of particular stress. (top, David Becker/Photo Researchers; A, Richard Wade; B, Roger Craig; C, Roy Quinlan)

Tubulin subunit

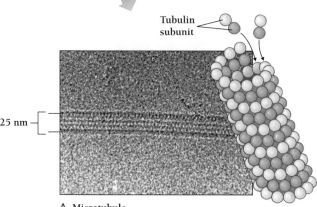

25 nm

A. Microtubule

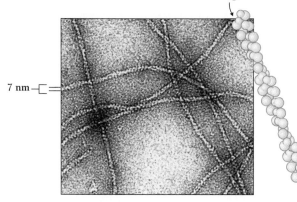

7 nm

B. Actin filaments

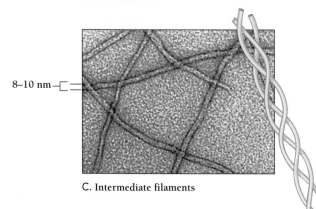

8–10 nm

C. Intermediate filaments

MTOC (fused microtubules)

Microtubule

K.G. Murti/Visuals Unlimited

Figure 4-14

The microtubule organizing center (MTOC). Most MTOCs are near the nucleus in the centrosome, which plays a central role in cell division. Microtubules radiate out from the MTOC.

atin strengthens wool, hair, fingernails, the outer layers of the skin, and other epithelial sheets. All intermediate filaments seem to be common in parts of cells that are subject to mechanical stress, suggesting that these filaments strengthen cells and tissues (see box).

The cytoskeleton is both skeleton and muscle system for the cell. Hollow microtubules provide the cell with a shape and help the cell divide. Filaments supply contractile force, and intermediate filaments provide mechanical support.

Extreme Biology   The 9 + 2 Structure of Flagella

Many prokaryotes, including *Escherichia coli,* have long fibers on their surfaces. Each of these structures is about 10 to 20 nm thick and up to 10 μm long and is called a **flagellum** [Latin, = whip; plural, flagella]. The flagellum's proportions are similar to those of a half-inch rope 40 feet long. Flagella act as propellers, driving bacteria at speeds of tens of micrometers per second. In terms of the bacteria's own dimensions, this would be equivalent to a 15-foot-long car moving at 100 miles per hour.

The flagella of eukaryotes are different from those of prokaryotes. When these structures are long and few in number, they are called **flagella** (Figure A). When they are short and numerous, they are called **cilia** (Figure B). However, both cilia and flagella have the same characteristic cross section. Each consists of a ring of nine sets of microtubules, surrounding two more sets in the center. The microtubule organizing center of a cilium or flagellum also has a nine-membered ring of microtubular structures and is called a basal body. Protists and some cells of animals (including sperm) use flagella or cilia to propel themselves. Cilia on the surfaces of the respiratory and reproductive tracts push fluids along the surface.

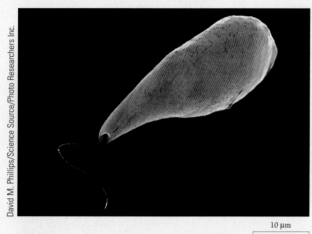
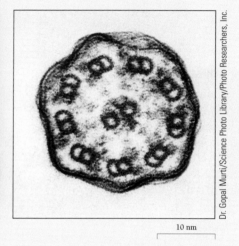

10 μm

10 nm

Figure A
Flagella. The protist *Euglena (left)* has a long flagellum, which it whips around like a propeller. An electron micrograph *(right)* of a cross section of the flagellum shows its 9 + 2 organization much like the cilia in the box on p. 66.

Figure B
Cilia. Cilia of a *Paramecium*.

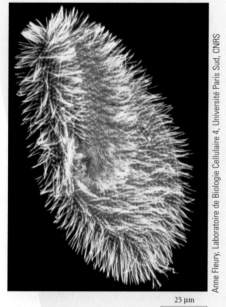

25 μm

4.3 What Do Membranes Do?

We saw earlier in this chapter that cell size is limited by the amount of membrane, which controls the movement of materials in and out of the cell. But membranes do more than just move material. In the past 30 years, cell biologists and biochemists have found four major roles for cell membranes:

1. Membranes are essential boundaries that separate the inside from the outside.
2. Membranes regulate the contents of the spaces they enclose.
3. Membranes serve as a "workbench" for a variety of biochemical reactions, especially those involving the metabolism of lipids and the secretion of proteins.
4. Membranes participate intimately in energy conversions (as we discuss in Chapters 6 and 7).

What Kinds of Molecules Do Membranes Contain?

The plasma membranes of all cells look the same, appearing in cross-section as two dark lines, 7 to 10 nm apart, with a lighter zone between them. The membrane consists, then, of at least two parallel layers separated by a slight space. Our first question concerns the structure of these layers. What molecules make up the layers, and what holds them in the same characteristic structure in all organisms?

The first insight into the molecular structure of membranes came from studies of red blood cells. Mammalian red blood cells are ideally suited for membrane studies: they have no internal membranes, no organelles, and few types of macromolecules other than hemoglobin, the red oxygen-binding protein. Red blood cells do not even have nuclei. Like half-filled water balloons, they can be squashed into many shapes. This flexibility allows them to pass more easily through the tiny vessels of the circulatory system.

Biochemists can obtain essentially pure plasma membrane by breaking open the red blood cells to release the hemoglobin and then using a centrifuge to separate the empty plasma membranes, called red blood cell "ghosts." Biochemical analysis has shown that these membranes consist mostly of lipids and proteins, as well as some carbohydrates. In the 1920s, biochemists calculated that a red cell ghost contained just enough lipid to enclose the cell with a continuous layer two molecules thick. This result suggested that the ghosts might indeed consist of a lipid "bilayer."

The most important lipids in the membrane are phospholipids. Recall from Chapter 3 that phospholipids are highly amphipathic; that is, they have a hydrophilic (water-loving) head and a hydrophobic (water-fearing) tail. When pure phospholipids are placed in water, they can spontaneously organize into sheets that are two layers thick. The oily tails of the lipids point inward toward one another, while their hydrophilic heads point outward. Once biologists understood this, they wondered if a biological membrane was in fact just a phospholipid bilayer. If so, where would the membrane proteins lie? Since biologists knew that proteins studded both the inner and outer surfaces of the cell membrane, they hypothesized that the membrane proteins were scattered on the inner and outer surfaces of the lipid bilayer, in contact with the polar groups of the phospholipid. But this is not quite correct.

Researchers have now determined that proteins not only occupy the two layers of the plasma membrane, but also occupy the space *in between* the two layers. Some membrane proteins are located only on the outside or the inside of the membrane; others, called **transmembrane proteins,** span the entire lipid bilayer (Figure 4-15).

Membrane proteins, like membrane lipids, are amphipathic, with both hydrophilic and hydrophobic regions. The hydrophilic regions interact with the aqueous solutions of the cytosol and the cell exterior, as well as with the hydrophilic heads of the lipids. Proteins' hydrophobic regions interact with the hydrophobic interior of the membrane.

The two lipid layers of the membrane differ slightly from one another (Figure 4-15B). Certain enzymes, for example, can break down the lipids in one layer, but not the other. Another difference is that carbohydrates appear only on the outer face. Some of these carbohydrates are attached to proteins (glycoproteins), and some are attached directly to membrane lipids (glycolipids). The inner and outer layers also differ in the way they interact with membrane proteins. Proteins that span the entire thickness of the membrane, for example, consistently orient themselves in a certain direction (Figure 4-15). We will see later in this chapter that the orientation of a protein is crucial because some membrane proteins work like turnstiles, letting molecules pass through the membrane in only one direction.

The plasma membrane consists of an asymmetric phospholipid bilayer. Some membrane proteins pass through the interior of this bilayer, and others associate with the inner or outer surface. Carbohydrates poke out of the outer surface.

How Does the Structure of a Membrane Establish Its Functional Properties?

The plasma membrane has a consistency similar to that of thin oil, such as you might use to lubricate a sewing machine or a bicycle. Water and certain small molecules can diffuse from one side of the membrane to the other, while the membrane selectively pumps some other molecules from one side to the other. In addition, experiments with chemically marked lipids have shown that (1) individual lipid molecules move freely within the plane of the membrane, and (2) the lipids in one layer do not usually move to the other layer.

Proteins also move freely within each layer of the membrane. To study the movement of proteins, biochemists

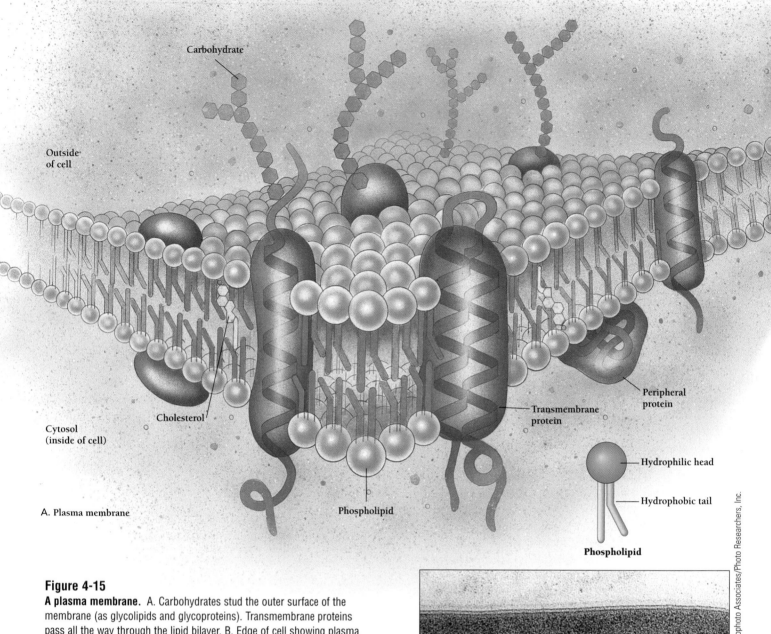

Carbohydrate

Outside
of cell

Cytosol
(inside of cell)

A. Plasma membrane

Cholesterol

Phospholipid

Transmembrane
protein

Peripheral
protein

Hydrophilic head

Hydrophobic tail

Phospholipid

Biophoto Associates/Photo Researchers, Inc.

B. Plasma membrane

100 nm

Figure 4-15

A plasma membrane. A. Carbohydrates stud the outer surface of the membrane (as glycolipids and glycoproteins). Transmembrane proteins pass all the way through the lipid bilayer. B. Edge of cell showing plasma membrane.

Biology ⒺNow™ Learn more about protein movement in the plasma membrane by clicking on this figure on your BiologyNow CD-ROM.

marked membrane proteins with a fluorescent dye. These experiments showed that about half of the proteins in a membrane move freely and about half remain tightly bound in their original position within the membrane.

Lipids and proteins move easily within each layer of the lipid bilayer because the layers are **fluid.** In 1972, the bilayer's obvious fluidity inspired two biochemists at the University of California San Diego, Jonathan Singer and Garth

Nicholson, to propose the **fluid mosaic model,** which has become the basis for understanding all cell membranes. Today we summarize the fluid mosaic model—somewhat modified since Singer and Nicholson first proposed it—as follows:

1. The basic structure of the membrane is a lipid bilayer, with two sheets of phospholipids arranged tail to tail.

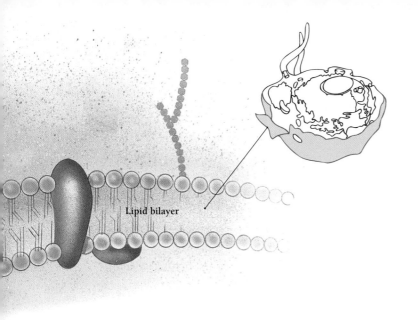

Lipid bilayer

4.4 How Do Membranes Regulate the Spaces They Enclose?

Throw a handful of raisins into a pot of boiling water and the raisins gradually fill with water, swelling until they resemble the grapes from which they were made (Figure 4-16). The raisin skin is a kind of membrane. Water and other small molecules easily pass through the skin, but larger molecules, such as proteins or cellulose, cannot pass through the skin. Large molecules inside the raisin are trapped; those outside are excluded. Such a membrane is said to be **selectively permeable,** meaning that it allows the passage of some molecules (especially of water), but not of others.

Plasma membranes are also selectively permeable. Proteins, for example, cannot directly move through plasma membranes. Nor can many ions, such as sodium ions. The regulation of a cell's internal environment—cellular homeostasis—depends on membrane molecules that actively pull specific small molecules and ions through the membrane.

2. Proteins are dispersed through the membrane like the individual pieces of stone or tile in a mosaic. Proteins contribute to membrane structure and function. Some proteins span the membrane, while others are confined to the inner or outer surface.
3. The membrane is fluid; both protein and lipid molecules move freely within the plane of the membrane.
4. The lipid bilayer serves as a hydrophobic barrier, confining hydrophilic molecules to the inside or the outside of a cell (or organelle).
5. Some membrane proteins help transport specific molecules across the membrane.

Why Does Water Move Across Membranes?

Before we can discuss how membranes regulate a cell's internal environment, however, we must first talk about the general rules that govern the movement of water and dissolved substances. These rules are the same physical laws that govern the movement of water into and out of raisins.

What Is Diffusion?

If you pour a teaspoon of salt into a pot of water, most of the salt will at first drop to the bottom of the pot in a little pile. Gradually, the salt will dissolve into sodium ions and chlo-

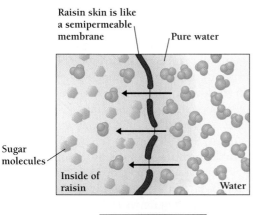

Raisin skin is like a semipermeable membrane

Pure water

Sugar molecules

Inside of raisin

Water

Courtesy Amy Dunleavy

A.

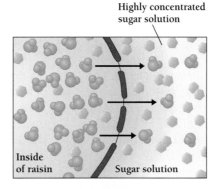

Highly concentrated sugar solution

Inside of raisin

Sugar solution

Courtesy Amy Dunleavy

B.

Figure 4-16
Concentration gradient. A. A raisin skin is a selectively permeable membrane. The high concentration of sugar molecules inside the raisin creates a concentration gradient. Water flows inside. B. In a concentrated sugar solution, the water in the swollen raisin flows outward. Because raisins are so sweet, such a sugar solution would have to be very concentrated.

ride ions. At first, the concentration of ions will be very high near the spot where you first poured the salt. The **concentration** is simply the number of molecules in a given volume. However, as time passes, the salt ions will move about randomly in amongst the water molecules until the ions are distributed evenly throughout the pot of water. Even after the ions are distributed evenly, they continue to move about randomly. Since as many ions move out of a given area as into it, however, the net charge is still zero. The random movement of like molecules or ions from an area of high concentration to an area of low concentration is called **simple diffusion** [Latin, *diffundere* = to pour out]. Stirring and heating both speed this process, but diffusion will happen all by itself if you wait long enough.

Any difference in concentration between two areas is called a **concentration gradient** (Figure 4-16). Concentration gradients are virtually everywhere. Cigarette smoke moving from the smoking section of a restaurant to the non-smoking section creates a concentration gradient from an area of dirtier air to an area of cleaner air. A teaspoon of sugar poured into a cup of coffee creates a concentration gradient—for a while.

Concentration gradients are always temporary: simple diffusion, ordinary mixing, and turbulence all tend to eliminate concentration differences. How fast diffusion occurs depends on a variety of factors, including temperature, the sizes of the molecules, and the steepness of the gradient. At higher temperatures, molecules move faster and so diffuse faster. Small molecules generally diffuse faster than large ones. Finally, the greater the difference in concentration at the two ends of a gradient, the faster diffusion occurs.

In simple diffusion, solutes move from an area of high concentration to an area of low concentration.

What Is Osmosis?

Diffusion occurs across membranes as well as through open solutions. For example, raisins immersed in a pot of plain water fill with water because the raisins have a high concentration of sugar and other molecules. The sugar molecules cannot pass easily through the skin of the raisin and out into the water. However, water molecules move in and out of the raisin easily. From the perspective of a water molecule, then, a raisin is an area of low water concentration. The difference between the sugary inside of the raisin and the watery outside constitutes a concentration gradient, along which the water molecules move.

The movement of the water molecules along this concentration gradient into the raisin is called osmosis. **Osmosis** [Greek, *osmos* = push, thrust] is the movement of water across any selectively permeable membrane in response to a concentration gradient.

As water flows in, it creates a force against the inside surface of the membrane. This **osmotic pressure**, compara-ble to physical pressure, stretches the raisin's skin until it begins to resist. If the skin is weak, it may break. Otherwise, it pushes back, meeting osmotic pressure with mechanical pressure. The physical pressure exerted by the raisin's skin counterbalances the osmotic pressure.

Osmosis is the movement of water in response to a concentration gradient. The mechanical pressure exerted by osmosis is called osmotic pressure.

Which Way Does Water Flow Through a Membrane?

We have seen what happens when a closed membrane (the raisin) with a high concentration of solutes (in this case, various sugars) is dropped into a solution with a low concentration of solutes (plain water). What happens if we take the water-filled raisins and put them into a highly concentrated solution of sugar?

The answer depends on how concentrated the sugar is. Look inside a can of fruit cocktail and you'll see what happens when the concentrations of sugar in a grape and the solution are about the same. The grape neither shrinks nor expands, but retains its normal shape. Put the grape into a highly concentrated sugar solution, however, and you will see it actually shrink as the water molecules inside move out through the skin, following the concentration gradient (Figure 4-16).

When we put a raisin into plain water, the movement of water into the raisin is greater than the movement of water out of the raisin. The water in which the raisin sits is said to be **hypotonic** [Greek, *hypo* = under + *tonos* = tension] compared to the inside of the raisin. In ordinary canned fruit cocktail, the water movements in and out of grapes are balanced perfectly. In that case, the sugary solution in which the grape is sitting is said to be **isotonic** [Greek, *isos* = equal] with the inside of the grape. If we put grapes into too heavy a syrup, however, the concentration of the syrup solution is so high that water moves out of the grapes. In that case, the solution outside the grape is said to be **hypertonic** [Greek, *hyper* = above] compared to the inside of the grape.

Cells, it turns out, behave in the same way. A red blood cell immersed in an isotonic solution, such as blood plasma, or a suitable salt solution, remains intact (Figure 4-17A). But if we immerse the red blood cell in a highly concentrated (hypertonic) solution, the cell shrivels (Figure 4-17B). The selectively permeable membrane allows water to pass freely out of the cell. When enough water flows out so that the new concentration of molecules within the cell is again the same as outside the cell, no more water flows.

Conversely, a red blood cell immersed in a hypotonic solution such as plain water will tend to fill with water and burst. When we decrease the solute concentration outside the cell, water then flows into the cell, causing it to expand (Figure 4-17C). In principle, water will flow in until the total solute concentration is equal on both sides of the mem-

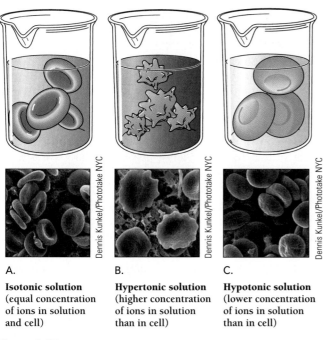

A.
Isotonic solution
(equal concentration
of ions in solution
and cell)

B.
Hypertonic solution
(higher concentration
of ions in solution
than in cell)

C.
Hypotonic solution
(lower concentration
of ions in solution
than in cell)

Figure 4-17
Iso-, hyper-, hypo-. A. A red blood cell in an isotonic salt solution remains intact. B. A red blood cell in a hypertonic salt solution shrivels. C. A red blood cell in a hypotonic salt solution swells with water. In a still more hypotonic solution, such as plain water, the cells in C would fill up and burst open.

brane. In practice, however, there may be too few solute molecules on the outside for this equalization to occur, as, for example, when the cells are suspended in pure water. In this case, the water will continue to flow into the cell until the cell bursts and releases its contents.

When a selectively permeable membrane separates two solutions that differ in their total solute concentrations, water flows in the direction that tends to equalize the solute concentrations on the two sides of the membrane. A solution may be hypotonic, isotonic, or hypertonic, relative to the concentration of solutes on the other side of the membrane.

How Does Turgor Pressure Support Plants?

When plant cells, such as those in the onion skin shown in Figure 4-18, are exposed to a hypotonic solution, water moves into the cell but does not accumulate in the cytoplasm. Instead, the inflowing water passes from the cytoplasm to a vacuole. The expansion of the vacuole pushes the surrounding cytoplasm against the **cell wall**—the rigid boxlike structure that encloses all plant cells. Conversely, when an onion cell is placed in a hypertonic solution, water flows out of the vacuole and the cell shrinks away from the cell wall.

In a hypotonic solution, the flow of water into the cell creates pressure against the cell membrane. The cell wall

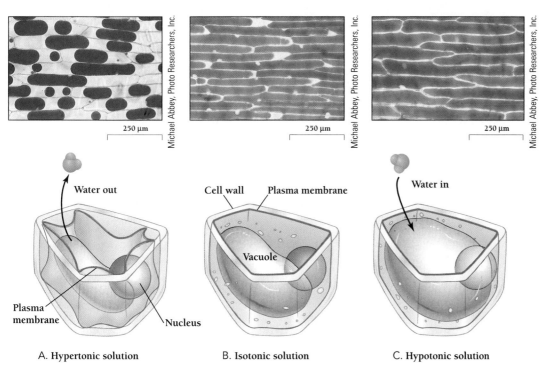

Figure 4-18
Turgor pressure in plant cells. Plant cells like these onion cells behave much like raisins and red blood cells. The water content of the cell's vacuole increases or decreases depending on whether the cell is in a hypertonic, isotonic, or hypotonic solution. A. Hypertonic. The vacuole empties and the cytoplasm pulls away from the cell wall. B. Isotonic. The vacuole fills with water, but without pushing too hard against the cell wall. C. Hypotonic. The water-balloon like vacuole pushes the cell's cytoplasm hard against the cell wall.

pushes back, exerting mechanical pressure that prevents the further flow of water into the cell. (Water actually continues to flow equally in both directions through the plasma membrane, but there is no *net* inflow of water.) The mechanical pressure thus balances the osmotic pressure. The pressure of a plant cell's contents against its cell wall is called **turgor pressure** [Latin, *turgere* = to swell].

Turgor pressure supports nonwoody plants in the same way that water supports the shape of a water balloon. This is why plants (and water balloons) begin to droop, or wilt, when they do not have enough water. The same phenomenon accounts for the way lettuce wilts soon after it is covered with salad dressing. Because the dressing is hypertonic, it draws water out of the cells of the lettuce leaves, causing them to wilt.

The flow of water into a plant cell gives mechanical strength to plant tissue.

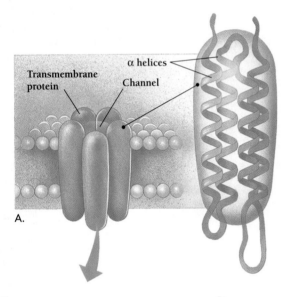

A.

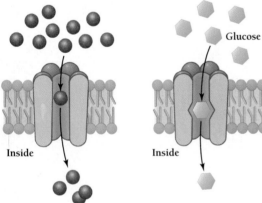

Channel-mediated diffusion   Facilitated diffusion

B. **Passive transport**

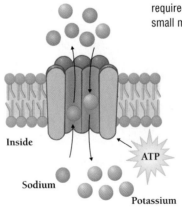

C. **Active transport**
(sodium-potassium pump)

What Determines the Movement of Molecules Through a Cell Membrane?

Cell membranes transport molecules in two main ways, "passively" and "actively." **Passive transport** occurs spontaneously, without the expenditure of energy. Passive transport results in equal concentrations of a molecule on the two sides of a membrane (at least for uncharged molecules). **Active transport** moves molecules against a concentration gradient and requires energy, just as climbing a hill while carrying a pack full of heavy books would. In active transport, cells use the energy of ATP to move ions across the membrane.

Some molecules move passively across membranes much faster than would be expected on the basis of diffusion alone, even though their final concentrations are consistent with passive transport. Such **facilitated diffusion**—an increased rate of passive transport—depends on the action of molecules in the membrane that help, or "facilitate," transport. For example, glucose sugar usually moves across cell membranes by means of facilitated diffusion. A transmembrane protein (called a "glucose transporter") moves glucose across the cell membrane, binding the glucose on one side of the membrane and releasing it on the other (Figure 4-19B). Transport proteins bind only to specific molecules. The glucose transporter, for example, transports only glucose and other similar sugars.

Active transport requires a special "pump" that couples the transport of molecules to an energy source (often ATP). For example, the **sodium-potassium pump** simultaneously transports sodium ions (Na^+) out of cells and potassium ions (K^+) into cells, using the energy of ATP (Figure 4-19C).

Figure 4-19
Passive and active transport. A. Transmembrane proteins help move molecules and ions across the cell membrane. B. In passive transport, no energy is required. In channel-mediated transport, ions or small molecules pass through the transmembrane channels without regulation or help. In facilitated diffusion, molecules move faster than they ordinarily would, thanks to the action of transporter proteins that carry molecules across the membrane. Transporters bind only to certain molecules, such as glucose. C. Active transport requires both energy and a carrier protein. One example is the sodium-potassium pump, which binds to ATP and sodium ions on the *inside* of the plasma membrane and to potassium ions on the *outside* of the membrane. The transporter then carries sodium and potassium in opposite directions using the energy of ATP.

Membranes are important regulators of a cell's contents. They allow water to freely pass in and out of the cell. In some cases, membranes speed the passive transport of specific small molecules such as glucose with transport proteins. In other cases, they harness the energy of ATP or of existing concentration gradients to accumulate or exclude particular molecules or ions.

While we have spoken almost entirely about plasma membrane that encloses the whole cell, the internal membranes of cells are similarly important in regulating the contents of the spaces that they enclose. In Chapters 6 and 7, we will discuss the membranes of mitochondria and chloroplasts.

Molecules pass through plasma membranes by means of diffusion, facilitated diffusion, and active transport. Active transport mechanisms use energy from ATP or from a concentration gradient of an ion or molecule different from the one being transported.

4.5 How Do Membranes Interact with the External Environment?

The plasma membrane of a cell functions like the outer surface of an organism. In multicellular organisms, the cell itself creates its own immediate environment. Most nonanimal cells, for example, surround themselves with a rigid carbohydrate structure, called the **cell wall** (Figure 4-20A). The cell wall can continue to function even after the death of the cell. In fact, many of the vessels that conduct fluids in plants are composed entirely of the walls of dead cells. Animal cells can similarly make their own environment, often a diffuse network of carbohydrates and proteins, called the **extracellular matrix** (Figure 4-20B).

Cells, like organisms, must communicate and must consume and excrete materials. Membranes help accomplish all of these activities and ensure that the cell's contents are best suited to the cell's immediate challenges.

Membrane Fusion Allows the Import and Export of Particles and Bits of Extracellular Fluid

The tendency of membranes to avoid water, together with their fluidity, means that biological membranes can rapidly change shape. In cell division, for example, the plasma membrane of the parent cell divides and reseals itself around each daughter cell.

In **phagocytosis**, cells engulf relatively large particles, such as microorganisms or cell debris. First, part of the plasma membrane extends outward and surrounds the particle (Figure 4-10). The membrane then fuses to form a large vesicle, which travels to the interior of the cell. The vesicle's membrane later fuses with the membrane of a lysosome, thereby exposing the vesicle's contents to the lysosome's digestive enzymes.

In **endocytosis**, the cell takes in tiny amounts of materials in vesicles that arise by the inward folding of the plasma membrane. The pouchlike vesicles formed in endocytosis are tiny, about one-tenth the size of those formed in most phagocytosis.

Cell biologists now recognize two types of endocytosis. In **pinocytosis** [Greek, *pinein* = to drink], the cell takes up any bits of liquid and dissolved molecules. In **receptor-mediated endocytosis,** the cell takes up only specific substances, which it recognizes by means of special proteins on the cell's surface, called receptors. Cholesterol, which travels through the blood in complexes with an amphipathic protein called LDL (low-density lipoprotein), enters cells through receptor-mediated endocytosis (Figure 4-21A; box on p. 52).

In **exocytosis** a cell reverses pinocytosis, accumulating molecules to export in membrane-enclosed vesicles. These vesicles move to the cell surface, fuse with the plasma membrane, and release their contents to the outside (Figure 4-21B).

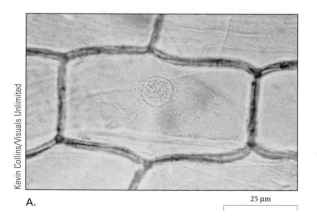

A. *Kevin Collins/Visuals Unlimited*   25 μm
B. *Fred Hossler/Visuals Unlimited*   5 μm

Figure 4-20
Cell wall and extracellular matrix. A. A rigid, external cell wall made of carbohydrate surrounds the plasma membranes of plant and fungus cells, most prokaryotes, and many protist cells. B. An extracellular matrix of carbohydrates and proteins encloses many animal cells.

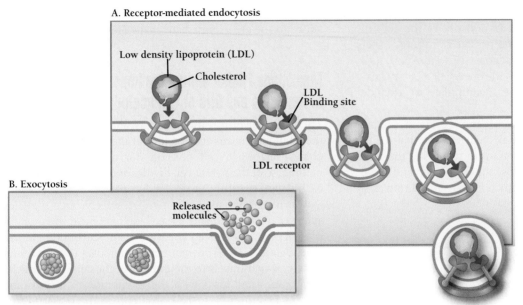

A. Receptor-mediated endocytosis

Low density lipoprotein (LDL)

Cholesterol

LDL Binding site

LDL receptor

B. Exocytosis

Released molecules

Figure 4-21
Endocytosis and exocytosis.
A. Endocytosis plays a central role in the regulation of low-density lipoprotein (LDL) and cholesterol levels in the blood. High levels of LDL and cholesterol are one predictor of heart disease. LDL particles full of cholesterol bind to LDL receptors on the cell membrane. The cell membrane responds by folding around the LDL particle and engulfing and digesting it. B. In exocytosis, the reverse happens: vesicles full of proteins, hormones, or neurotransmitters inside the cell fuse with the cell membrane and diffuse into the outside environment of the cell, a process called secretion.

Cells export many kinds of molecules through exocytosis, from digestive enzymes to hormones and "neurotransmitters," the molecules nerves use to talk to one another.

The regulation of exocytosis is an extremely important part of cellular function. In the nervous system, for example, signaling between cells depends on the precise timing of neurotransmitter release by exocytosis. Rapid increases in intracellular calcium ions stimulate exocytosis, and dozens of proteins are involved in its coordination. Some of these proteins are part of the vesicle's own membrane. Other proteins are part of the plasma membrane, where the vesicle will fuse, while still other proteins form temporary bridges that allow the two membranes to fuse.

Plasma membranes fuse easily and can spontaneously form vesicles. These abilities allow cells to change shape and to perform phagocytosis, endocytosis, and exocytosis.

How Do Cells Communicate in Multicellular Organisms?

Because the individual cells in all multicellular organisms must cooperate, communication among cells is crucial. Cells communicate with one another by means of chemical signals (such as hormones and neurotransmitters) that allow the activities of one cell to influence those of another.

How Do Cells Communicate with Their Immediate Neighbors?

Plant cells are both cemented together and separated from each other by rigid cell walls. Plant cells are nonetheless in intimate contact with one another through fine channels called **plasmodesmata** (singular, plasmodesma) [Greek, *plassein* = to mold + *desmos* = to bond]. Plasmodesmata are narrow channels between cells through which thin strands of cytoplasm stream, allowing molecules to pass directly from cell to cell (Figure 4-22).

Animal cells lack cell walls and can connect and communicate in several different ways. Ions and molecules easily pass through **gap junctions**, in which the membranes of adjacent cells are only 2 to 4 nm apart, instead of the usual 10 to 20 nm (Figure 4-23A). For example, gap junctions connect the cells of the heart, allowing in-

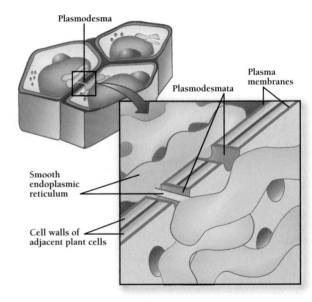

Plasmodesma

Plasma membranes

Plasmodesmata

Smooth endoplasmic reticulum

Cell walls of adjacent plant cells

Figure 4-22
Cell-cell communication (plants). Plasmodesmata are tiny channels between plant cells through which flow thin streams of cytoplasm and sometimes thin strands of smooth endoplasmic reticulum. Plasmodesmata can carry all kinds of molecules from one cell to the next.

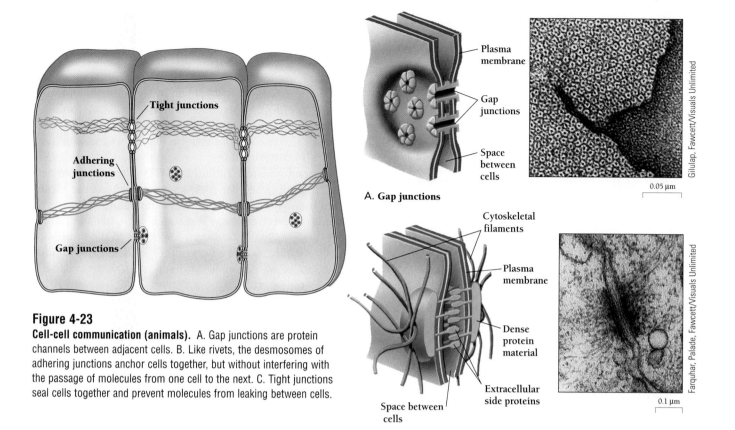

Figure 4-23

Cell-cell communication (animals). A. Gap junctions are protein channels between adjacent cells. B. Like rivets, the desmosomes of adhering junctions anchor cells together, but without interfering with the passage of molecules from one cell to the next. C. Tight junctions seal cells together and prevent molecules from leaking between cells.

A. Gap junctions

B. Adhering junctions (desmosome)

C. Tight junctions

dividual cells to coordinate their contractions each time the heart beats. Other animal cells communicate with **adhering junctions,** which allow the passage of molecules while strengthening the connections between adjacent cells (Figure 4-23B). In the sheets of tissue one might find in skin, for example, adhering junctions not only rivet cells together, but connect the internal cell skeletons of adjacent cells, tightly connecting the individuals cells in a sheet. **Tight junctions** actually fuse the membranes of adjacent cells in a sheet, so that fluid cannot leak between the cells (Figure 4-23C). Such tight junctions are what allow you to chew up a pickle without having the vinegar and spices soak into your own tissues and "pickle" your mouth and jaw.

Plasmodesmata and gap junctions allow materials to pass directly from cell to cell. Adhering junctions strengthen the connections between cells, and tight junctions keep fluids from passing between them.

How Do Cells Communicate with Distant Cells?

In multicellular organisms, cells need to communicate with one another. Whether a nerve cell needs to tell a muscle cell to contract or a leaf cell needs to tell a group of stems cells to change shape and turn the leaf toward the sun, the transmission of information between cells depends on chemical signaling. We've seen briefly that cells can communicate physically through gap junctions and plasmodesmata. But cells often communicate by releasing "signaling molecules" that bind to other cells, triggering a series of specific events. Specifically, signaling molecules bind to other cells at **receptors,** specialized proteins that can lie either on the surface of a cell or within the cell. For example, receptors can bind to and respond to "neurotransmitters," signaling molecules released by nerves, and hormones such as estrogen and testosterone. You can read about cell signaling in Chapters 36, 41, and 42.

Cells respond to chemical signals by means of specialized proteins called receptors.

We have now seen how biologists first recognized the importance of cells, examined the internal structures of cells and their membranes, and seen how cells coordinate the activities of their many internal components. In the next three chapters, we will use some of this knowledge to try to understand the biochemical processes that take place inside of cells.

Key Concepts

- Every cell has a boundary, a cell body, and a set of genes. Different kinds of cells have characteristic organelles and other structures.
- Eukaryotic cells have a nucleus and other organelles enclosed by membranes that help regulate the chemical composition of the spaces inside. Prokaryotic cells have no nucleus and no internal membranes or subcompartments.
- The cytoskeleton contributes to the organization and internal communication of eukaryotic cells.
- Cell membranes establish the boundaries of a cell and its organelles, regulate their contents, serve as a workbench for biochemical reactions, and participate in energy transformations.

Summary with Key Terms

What is a cell?

The **cell theory** states that (1) all organisms are composed of one or more cells; (2) cells, themselves alive, are the basic living units of organization of all organisms; and (3) all cells come from other cells.

What are the advantages and disadvantages of being a multicellular organism?

The subdivision of organisms into cells and of cells into organelles allows for an efficient division of labor and makes it easier to maintain a constant environment within each cell (because of a higher **surface-to-volume ratio**). Multicellular organisms can survive the deaths of individual cells because of cell replacement.

What organelles and other parts of cells make them work?

Cells from all three domains contain three common features: a **plasma membrane** that separates the inside of the cell from the rest of the world, a set of genetic instructions encoded in DNA, and a cell body (or **cytoplasm**). In the cells of all eukaryotes, the genetic material is enclosed in a membrane-enclosed **nucleus.** The nucleus contains the cell's genetic instructions in the form of **chromosomes.** The two membranes of the **nuclear envelope** are fused in many places to form **nuclear pores.**

The Archaea and Eubacteria, or prokaryotes, lack a membrane-enclosed nucleus. They keep their DNA in a region called the **nucleoid.** The cytoplasm of most eukaryotic cells (formerly referred to as the **protoplasm**) is divided into well-defined compartments, or **organelles,** by internal membranes. The jellylike **cytosol,** which surrounds all the cell's organelles, contains **ribosomes, proteasomes,** glycogen, fat, and thousands of enzymes.

Within the cytoplasm are several well-defined kinds of membrane-enclosed organelles: the endoplasmic reticulum (ER), **Golgi complex, lysosomes, mitochondria,** and such **plastids** as **chloroplasts, chromoplasts,** and **amyloplasts.** In addition, eukaryotic cells contain **vesicles** and **vacuoles,** which are also enclosed by membranes but whose structures and functions vary widely.

The **endoplasmic reticulum (ER)** is an extensive network of membranes that encloses a space called the ER **lumen.** One part of the endoplasmic reticulum—the **rough endoplasmic reticulum**—is studded with ribosomes and makes proteins mainly destined for export to the outside or the outer surface of the cell. Another kind of ER—**smooth ER**—contains no ribosomes and appears to function in the synthesis of lipids.

The **Golgi complex** serves as the traffic director in the cell, helping to package some proteins for export and others for inclusion in other membrane-enclosed organelles. Lysosomes are small membrane-enclosed organelles that are responsible for digestion and disposal.

Mitochondria and chloroplasts are the largest organelles outside the nucleus. Each is surrounded by a double membrane, and each contains complicated internal membranes. Mitochondria are the powerhouses of cells, making most of the cell's ATP by breaking down small, organic molecules in the presence of oxygen. Chloroplasts, which contain stacks of **thylakoids,** called **grana,** are present only in photosynthetic cells. They are the sites of **photosynthesis.**

Together with internal membranes, the **cytoskeleton** provides internal organization for eukaryotic cells. It consists of three sets of filaments: **microtubules** (which are made of two kinds of **tubulin**), **actin filaments,** and **intermediate filaments.** In most cells, the cytoskeleton consists of structures that are assembled and disassembled continually. Their assembly seems always to originate in structures called microtubule organizing centers (MTOCs). The major MTOCs are near the nucleus in a zone called the **centrosome,** a region that (in many cells) also contains a distinct organelle called the **centriole.** Some cells have permanent structures, such as **cilia** and **flagella,** which are composed of elements of the cytoskeleton.

How do membranes act as boundaries and regulate the contents of cells?

All cells are surrounded by membranes. Membranes consist mainly of lipids and proteins. The lipids, mostly phospholipids, are arranged in two layers, a "bilayer." Some proteins are located on the external and internal surfaces of the bilayer, in contact with both the hydrophobic interior of the membrane and the hydrophilic cytoplasm or cell exterior. **Transmembrane proteins** pass completely through the membrane. The membrane is **fluid,** both protein and lipid molecules move freely within it. This picture of membrane structure is called the **fluid mosaic model.**

Membranes are **selectively permeable:** they allow some molecules to pass in and out freely and confine others to the inside or the outside. In **simple diffusion**, solutes move from an area of high **concentration** to an area of low concentration. That is, solutes move along a **concentration gradient.** **Osmosis** is the movement of water across any selectively permeable membrane in response to a concentration gradient. A solution may be **hypotonic, isotonic,** or **hypertonic,** relative to the concentration of solutes on the other side of the membrane. The pressure created by the concentration gradient, called **osmotic pressure,** is comparable to mechanical pressure.

In plants, the central vacuole fills with water, which presses against the **cell wall,** creating **turgor pressure**—the principal mechanical support of nonwoody plants. When a plant cell is immersed in a hypertonic solution, water flows out of its vacuole and the cell separates from the cell wall in **plasmolysis.**

Molecules pass through plasma membranes by means of **passive transport** (which requires no energy), **facilitated diffusion,** and **active transport** (which requires energy from ATP). The **sodium-potassium pump** is one active transport mechanism. Certain cells engulf large particles by extending their plasma membranes in a process called **phagocytosis.** Most cells are also capable of **endocytosis** and **exocytosis.** Endocytosis may be nonspecific (**pinocytosis**), or it may depend on specific receptors on the membrane (**receptor-mediated endocytosis**).

The outer surfaces of cells are responsible for their interactions with other cells and with other molecules that bind to **receptors** on the cell surface or within the cytoplasm. Cell walls enclose most prokaryotic and many eukaryotic cells (though not animal cells). Animal cells often are surrounded with **extracellular matrix,** a complex of carbohydrates and proteins.

Multicellular organisms depend on coordination of activities among many cells. Plants and animals have a variety of specialized structures by which cells interact. Some of these structures—**plasmodesmata** in plants and **gap junctions** in animals—allow the free passage of some molecules between adjoining cells. Other specialized junctions between cells—**adhering junctions**—rely on **desmosomes** and provide mechanical strength to sheets (**epithelia**) of **epithelial cells. Tight junctions** weld some cells together and prevent the passage of molecules through them.

Review and Thought Questions

Review Questions

1. State the cell theory. Be sure your statement includes the concepts of what a cell is, where cells occur, where they come from, and what they do.
2. Describe three features that all cells have in common.

3. Explain how the surface-to-volume ratio of a cell is related to its ability to take in resources and to rid itself of wastes. Why are most cells so small? How can a chicken egg, which is one cell, be so big?
4. Sketch and label a mitochondrion and a chloroplast. What is the principal function of each? Where does each obtain fuel?
5. Describe the fluid mosaic model of membrane structure.
6. A plasma membrane functions to keep the material outside a cell separate from the material inside. What is one property of the membrane that allows this?
7. Explain how membrane-enclosed vesicles move materials into, out of, and within cells.
8. What is the role of the cytoskeleton in eukaryotic cells?
9. Compare the sizes of microtubules, actin filaments, and intermediate filaments. Describe one activity or structure in which each participates.

Thought Questions

10. Between conception and death, most of your cells die while new ones are produced by the division of other cells. What makes you really "you"? What is it that is constant about your structure?
11. Explain how the endoplasmic reticulum might increase the effective surface-to-volume ratio for a cell, even though it is located inside the outer perimeter of the cell. Can it really assist in bringing resources in and getting rid of wastes?
12. Chloroplasts are also called the "powerhouses" of cells. What do they produce to earn this description? What types of cells contain chloroplasts?

BiologyNow Resources

Biology ⓔ Now™

Active Figures

4-5: Cell structure
4-15: Protein movement in the plasma membrane

Preparing for an exam? Take a diagnostic test on your BiologyNow CD-ROM.

Online materials relating to this chapter are at:
http://biology.brookscole.com/AAL3

About the Chapter-Opening Image
One-celled organisms, such as this diatom, were among the first studied by early microscopists.

Directions and Rates of Biochemical Processes

Key Questions

- What factors determine which way a reaction will go?
- What factors determine the rate of a chemical reaction?
- How do enzymes work?
- How can cells modify the activity of enzymes?

Ludwig Boltzmann: A Man Left Behind, or Ahead of His Time?

In September 1906, the eminent Austrian physicist Ludwig Boltzmann fell into the last of many deep depressions. Hoping to lift his spirits, he took his wife and daughters on a trip to Duino, Italy, a spectacular resort on the Adriatic Sea.

The mercurial Boltzmann had always been emotional (Figure 5-1). He could be moved to tears of joy by a well-played sonata, the color of the sea, or a piece of poetry. At other times he lapsed into long, grief-stricken silences. He wept at the railroad station whenever he left his wife for any extended period.

Although Boltzmann was unstable, his growing despondency also had roots in the scientific milieu of his time. At 60, the physicist was convinced that much of his life's work had been wasted. For years, he had championed the idea that matter is composed of atoms and that the behavior of atoms accounts for all chemical transformations. Other chemists and physicists discussed atoms extensively, but for them, atoms were just a useful way to think about the formation of chemical compounds. Almost no one thought atoms were real, and no one before Boltzmann thought of them as the key to understanding energy.

The hypothesis that matter was made of tiny particles called atoms was first popularized by the Greek philosopher Democritus in the fifth century B.C., then abandoned for more than two thousand years. Only in the 17th century did "atomism" get serious attention again. But the same renaissance in scientific thought that gave birth to atomism also gave birth to "positivism," the idea that observable facts are the only basis for a true knowledge of the world. Positivism was an idea that would help drive Boltzmann to despair.

Since nobody could see atoms, 19th-century positivists insisted that there was no reason to believe in atoms. Boltzmann's peers, all eminent scientists and heartfelt positivists, insisted

David Muench Photography

that assuming that atoms were real—actually basing scientific theories on their existence—was no better than mysticism. Indeed, one of Boltzmann's stoutest critics, the physicist Ernst Mach, used to harass Boltzmann as Boltzmann lectured on atoms by calling out from the audience, "Have you ever seen one?" Of course, Boltzmann had not. (Indeed, only with the development of atomic force microscopes in the 1980s could scientists actually "see" atoms.)

But Boltzmann felt certain that atoms were real, and for years he had written scathing critiques of even his closest friends' work. In his personal life, Boltzmann was charming, modest, and earnest, but in debates and in the lecture hall, he was a bulldog, relentlessly attacking arguments he believed were wrong and proclaiming his own ideas.

For almost 40 years, Boltzmann worked to show how atoms accounted for the behavior of all forms of matter. His greatest achievements were to find a way to understand the behavior of matter in terms of large collections of molecules and atoms. He showed that the statistical laws of large numbers could be applied to collections of atoms or molecules.

But Boltzmann's work could be correct only if atoms were real. And even after he had spent 40 years arguing for atoms, fewer and fewer people seemed to think that atoms existed. As Boltzmann wrote to a friend, "I am conscious of being only one individual struggling weakly against the stream of time," and later, "I present myself to you therefore as a man left behind."

Boltzmann did not know it, but by 1906, the year he took his family to Italy, the atom was about to take center stage in the world of science. The darkest days for his ideas were over, and bright young scientists were taking a second look at atoms. In 1905, for example, 26-year-old Albert Einstein had published a paper showing that the constant jiggling of particles that scientists noticed when looking down a microscope (called Brownian motion) was most likely due to bombardment by atoms—real, massive objects, not abstract statistical averages. In just a few years, Ernest Rutherford would locate the tiny, central nucleus of the atom. And soon after that, the physicist Niels Bohr would calculate the movements of electrons.

But Boltzmann had given up. On a sunny day in September 1906, while his wife and daughters were swimming, Boltzmann ended his rich and productive life by hanging himself in his hotel room. Had he endured just a few more years of frustration, he would have lived to see the triumph of

Figure 5-1
Ludwig Boltzmann (1844–1906). Boltzmann argued that all matter is made of tiny atoms.

his own ideas and the humiliation of his intellectual enemies.

Indeed, Boltzmann's intellectual victory was so complete that the material in this chapter—written nearly 100 years after his death—rests heavily on his work. The most important applications of Boltzmann's idea were in the field of **thermodynamics** [Greek, *thermo* = heat + *dynamis* = power, force], the study of the relationships among different forms of energy. For example, as a car converts the chemical energy of gasoline into the energy of motion, the car moves forward. Every event, every reaction, every process is accompanied by a "transformation of energy."

The direction a transformation takes depends on the rules of thermodynamics. A scientist can predict the direction of a reaction without knowing exactly how the reaction takes place. In the same way, we know that water always flows downhill, even if we do not know whether it gets there through a pipe, a stream, or a waterfall.

Before Boltzmann, thermodynamics was understood as a set of abstract mathematical laws. These two laws could predict which way a reaction would go. Boltzmann showed that the reason thermodynamics could predict chemical reactions was that the laws actually described the behavior of atoms and molecules. He showed how chemical reactions occur, what heat is, and why some chemical reactions produce heat while others absorb heat. Most important, Boltzmann's work paved the way for the theory of **kinetics** [Greek, *kinetikos* = moving], the study of the rates of reactions. Before Boltzmann, chemists could predict only whether a reaction should take place, not how long it would take. Kinetics, which explains why a reaction takes place, can also predict how fast it will happen.

Understanding a biological or chemical process (or any process, for that matter) requires more than noticing changes in structures. For example, we may observe that most of the gasoline in a car's engine is converted to water and carbon dioxide, with a tremendous release of heat and energy of motion. But we cannot understand what is happening unless we know the answers to two general questions: (1) Which way will the reaction go? and (2) How fast will it go?

We must ask, Why does the conversion of gasoline to water and carbon dioxide occur in one direction only? Why don't the carbon dioxide and water present in the air we breathe combine spontaneously to form gasoline? And why does the conversion from gasoline to carbon dioxide and water occur in fractions of a second? What if converting a tank of gasoline to water and carbon dioxide took several years? In that case, gasoline would be a poor fuel.

We must also ask, Why does the reaction occur as rapidly as it does? What factors determine how fast any reaction goes? Predicting the speed of a process requires that we know something about the process itself. For example, we know that water plunges quickly downhill in a waterfall and meanders slowly in a gentle stream. Thermodynamics predicts only that water will flow downhill. It is kinetics that can tell us how fast the water will flow, which depends on the steepness of its descent. In this chapter, we will see that thermodynamics tells us which way a reaction will go and that kinetics tells how fast.

5.1 What Determines Which Way a Reaction Goes?

Everyone can tell when a movie runs backward: a waterfall goes up instead of down; a rubber ball bounces up a flight of stairs; a smashed apple spontaneously reassembles (Figure 5-2). We all know the direction of time's arrow. Biological processes also have direction: a child matures into an adult; a bud unfolds into a leaf; dead organisms decay.

The key to predicting direction lies in understanding the conversion of energy from one form to another. Thermodynamics provides rules that govern all energy changes, both biological and nonbiological, and establishes the direction of time's arrow.

Photograph by Dr. Harold E. Edgerton, Harold and Esther Edgerton Foundation, ©1999. Courtesy Palm Press, Inc.

Figure 5-2
We all know the direction of time's arrow. If we saw two photos of an apple taken before and after a bullet passed through it, we would know instantly which photo was taken first.

How May Work Be Converted to Kinetic or Potential Energy?

Most of us think of work as any process that requires energy. There is, however, a scientific definition of work: **work** is the movement of an object against a force. For example, lifting a sack of concrete against the force of gravity is work. So is winding the spring in a child's toy (Figure 5-3).

Work, in this sense, can be stored for later use, the way energy is stored in a battery. The wound-up spring in a toy beetle can propel the toy by operating its legs. When the spring is wound, we say that the work of winding the spring is stored as **potential energy.** When the spring unwinds and the beetle leaps forward, we say that the toy has **kinetic energy**—the energy of moving objects. A boulder at the edge of a cliff has potential energy. The same boulder falling to the bottom of the cliff has kinetic energy. A moving car, a running athlete, and a contracting muscle all have kinetic energy. Fast-moving molecules also have kinetic energy, which we call heat. Potential energy also exists in many forms, including gravitational, electrical, and chemical.

Rearranging atoms in a chemical reaction is a form of work. Recall from Chapter 3 that the energy-rich molecule ATP consists of a molecule of adenosine linked to three negatively charged phosphate groups (Figure 5-3). Making a molecule of ATP requires the squeezing together of three negatively charged phosphate groups. This squeezing is work against an electrical force. Once the atoms are squeezed together into a molecule, the work is stored as potential energy. Thus, ATP has potential energy. This energy can later be released to do work. If, for example, the third phosphate group is removed, stored energy is released. That energy can drive another chemical reaction, which can make a muscle move, heat the body, or help us understand a sentence.

Work, the movement of an object against a force, can be stored as potential energy. The construction of a molecule of ATP stores potential energy that can be released later as kinetic energy.

How Does Thermodynamics Predict the Direction of a Reaction?

Scientists and engineers first began to understand the relationship among different forms of energy in the late 18th century. The major stimulus for the development of the rules of thermodynamics was the increasing use of the steam engine, which converted the potential energy of coal into kinetic energy capable of powering a locomotive or a factory. As we will see, the same rules that apply to steam engines also apply to apples and humans.

One of the first and most important theories of the new field of thermodynamics was that, in any process, energy is neither created nor destroyed—it only changes form. The **First Law of Thermodynamics** states that the total amount of energy in any process stays constant (Figure 5-4A). Energy may switch from potential energy to kinetic energy and back again, but it is neither created nor destroyed. Any heat generated during a process is merely another form of kinetic energy. When Ludwig Boltzmann was only 22, he showed that the movements of atoms perfectly obeyed and explained the First Law of Thermodynamics. If atoms and molecules actually exist, he argued, then heat is just their rapid movement. Molecules and atoms—though invisible to us—have kinetic energy in the same way that a ball moving through the air has.

The second major theory of thermodynamics—the **Second Law of Thermodynamics**—said that in any process the energy available for work decreases (Figure 5-4B). The Second Law explains why we can tell when a film is running backward—why time has a direction (Figure 5-2). At 27, Boltzmann worked out a mathematical description of the behavior of large groups of atoms. He showed that the properties of these groups accounted for the Second Law, as well the first.

Everyone accepted the laws of thermodynamics because they accurately predicted the behavior of chemical reactions. But, oddly, Boltzmann's peers mostly ignored his ideas about atoms, which explained *why* the laws worked so well. They all assumed he was a genius, but few understood or wanted to understand what he was talking about. In the rest of this book, we will see this pattern again and again. Interesting and important ideas often meet rejection when they are first proposed.

Today, scientists have come to accept that the two laws of thermodynamics together predict the direction of a reaction. For example, we know that if we do a pushup, our muscles contract and perform work. The energy for this work comes from the chemical energy in ATP. In the muscle cells, ATP molecules give up part of their stored energy by splitting off one of three phosphate groups. The remaining

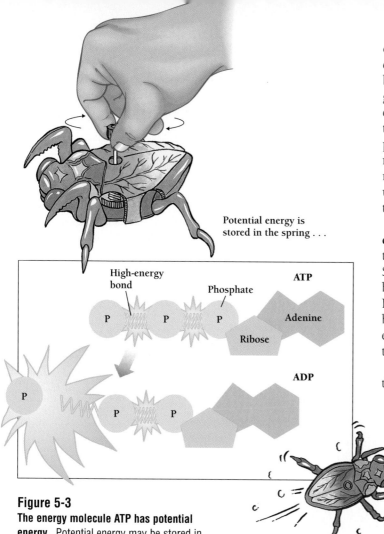

Figure 5-3

The energy molecule ATP has potential energy. Potential energy may be stored in the spring of a windup toy or in the high-energy bonds of ATP. Tightening a spring takes work. Releasing the spring discharges the potential energy stored in the spring. Likewise, creating a high-energy phosphate bond takes work. Breaking the bond discharges the potential energy stored in the bond.

Potential energy is stored in the spring . . .

. . . and is released as kinetic energy.

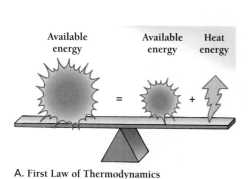

A. First Law of Thermodynamics

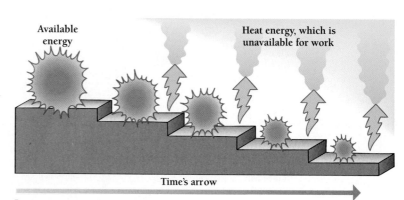

Available energy

Heat energy, which is unavailable for work

Time's arrow

B. Second Law of Thermodynamics

Figure 5-4

The two laws of thermodynamics. A. First Law: Energy may change from one form to another, but it is never lost. B. Second Law: The energy available to do work decreases as energy is lost as heat.

molecule has two phosphate groups and is then called adenosine *di*phosphate, or ADP [Greek, *di* = two]. ADP (with two phosphates) still has potential energy, but less than ATP (with three phosphates).

But, according to the Second Law, muscle contraction can convert only a part of the energy obtained from ATP into useful work. The rest of the energy from splitting ATP becomes heat—the random motion of molecules. Even though some energy is lost as heat, however, no energy is actually destroyed. All of the energy released from splitting ATP either contracts the muscle or leaves as heat.

All forms of energy can be measured in **calories**—by definition the amount of energy needed to raise the temperature of 1 gram of water by 1°C. We measure the energy in food in Calories (with a capital C). One Calorie is 1,000 calories, also called a "kilocalorie" or a kcal. The daily diet of a typical college student is 2,000 to 3,000 kcals.

Every time energy changes form—whether from chemical to electrical or from electrical to light energy—a little more of the energy turns into heat (Figure 5-5). Over time, more and more energy turns into heat, and less and less energy becomes available for work (Figure 5-4B). Heat loss is a natural consequence of all energy transformations. If we ask which way a reaction will run, the answer is that it will usually run in the direction that releases heat.

Energy changes form but is neither created nor destroyed. In any process, the energy available to do work decreases as energy is lost in the form of heat. The loss of heat establishes the direction of the process. Without an input of energy, the process cannot reverse itself.

How Do Changes in Free Energy Predict the Direction of a Reaction?

To predict the direction of a chemical reaction, we need a measure of how much energy is available. **Free energy** is the energy in a system available for doing work. Free energy is abbreviated *G,* after the American chemist J. Willard Gibbs, who, in the 1870s, single-handedly worked out the theory of chemical thermodynamics.

Every chemical has a certain amount of free energy. And every chemical reaction involves a "change in free energy." When two molecules (or **reactants**) react to form one or more new molecules (or **products**), the change in free energy is equal to the free energy of the products minus the free energy of the reactants. The change in free energy is abbreviated **Δ*G*.** (Δ, pronounced "delta," is the mathematical

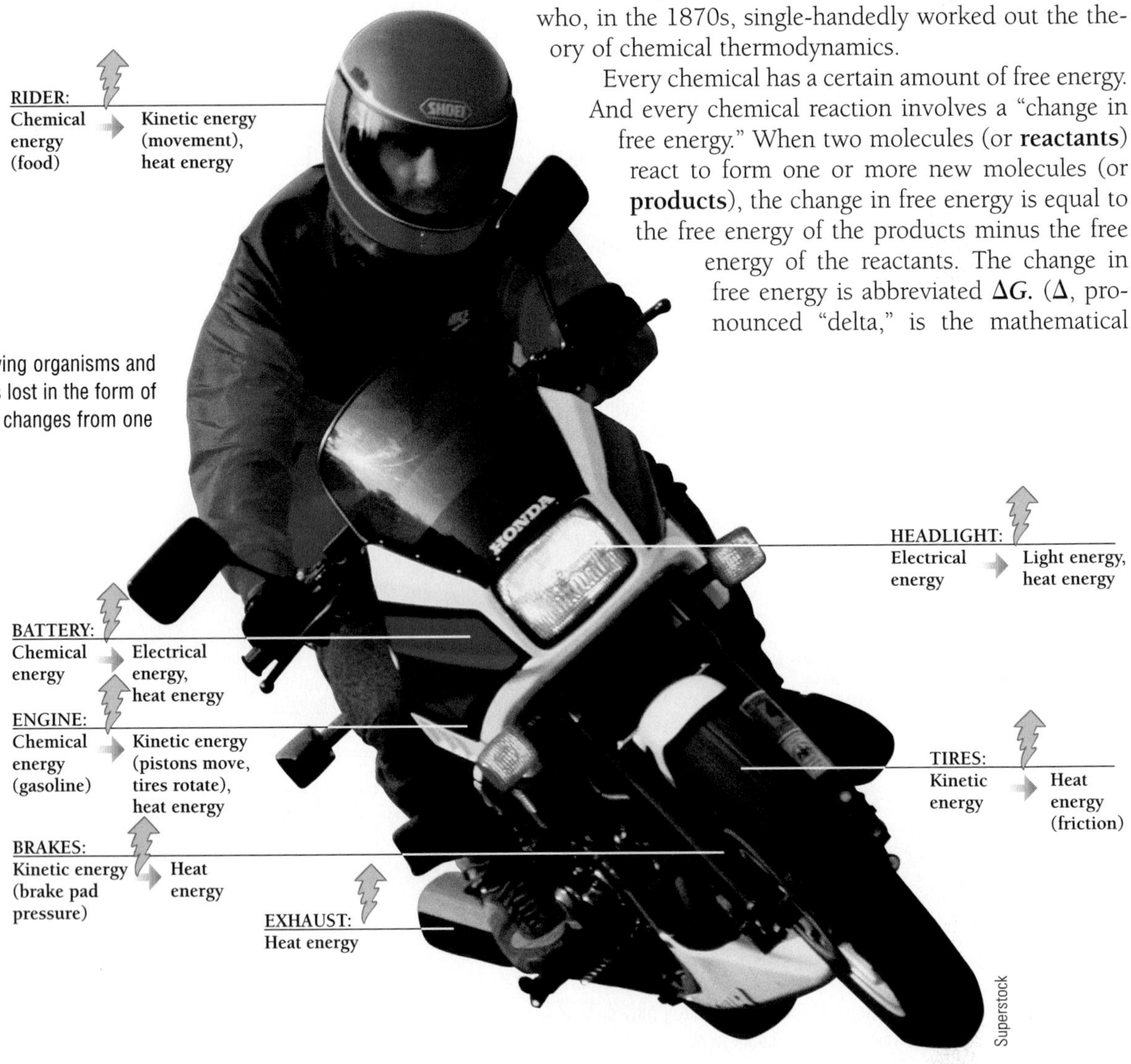

Figure 5-5

A hot bike. In both living organisms and in machines, energy is lost in the form of heat each time energy changes from one form to another.

RIDER:
Chemical energy (food) → Kinetic energy (movement), heat energy

HEADLIGHT:
Electrical energy → Light energy, heat energy

BATTERY:
Chemical energy → Electrical energy, heat energy

ENGINE:
Chemical energy (gasoline) → Kinetic energy (pistons move, tires rotate), heat energy

BRAKES:
Kinetic energy (brake pad pressure) → Heat energy

EXHAUST:
Heat energy

TIRES:
Kinetic energy → Heat energy (friction)

Superstock

symbol for a change in a quantity.) Chemists abbreviate this statement by writing

$$\Delta G = G_{\text{products}} - G_{\text{reactants}}$$

This equation might look intimidating, but it is no different from saying that the amount of weight you have gained or lost in a week is equal to the amount you weigh today minus the amount you weighed last week:

$$\Delta \text{pounds} = \text{pounds}_{\text{today}} - \text{pounds}_{\text{last week}}$$

Here, Δpounds ("delta") equals the change in your weight, the amount of weight you lost or gained. In fact, without adding energy to your body's "system" by eating, you will inevitably lose weight, and Δpounds will be negative. That's because your body's activities are energy-releasing processes.

All processes that release energy are said to be **exergonic.** The breaking of ATP into ADP and phosphate, for example, is exergonic. Likewise, a coconut falling from a tree releases energy as it falls. Stand beneath it, and you will know for sure that it is an exergonic process.

On the other hand, carrying a coconut up a tree is work and therefore an **endergonic** process, one that requires an input of energy (Figure 5-6). In the same way, forcing another phosphate onto ADP to make ATP is an endergonic, or energy-consuming, process.

In all exergonic processes, free energy decreases, so ΔG is negative. As a result, exergonic processes, such as ATP changing to ADP, occur spontaneously (given enough time). In endergonic processes, free energy increases and ΔG is positive. Endergonic reactions will not occur spontaneously, but require energy from some other source.

A change in free energy depends only on the beginning and end of a reaction. The change does not depend on anything that happens in between or on how long the process takes. A ball on a staircase might fall down the whole stair-case in one big bounce or in several smaller bounces. The ball might even sit unmoving on the top step for years on end until it is jostled by an earthquake or just a curious child.

Thermodynamics does not predict how long it will take for a ball to fall down the stairs or for ATP to break down into ADP and phosphate. All that thermodynamics tells us is that, if we wait long enough, the ball will roll down the stairs—not up—and that ATP will break down into ADP and phosphate.

If the change in free energy is negative, a reaction is said to be exergonic and the reaction can occur spontaneously.

What Is the Source of the Free Energy Released or Consumed During a Reaction?

When two isolated atoms form a chemical bond, energy is released. The tighter the bond, the more energy is released. During a chemical reaction, some of the bonds in the reactants break, while other bonds form to create the products. For example, consider the breakdown of ATP to ADP:

$$\text{ATP} + \text{H}_2\text{O} \rightarrow \text{ADP} + \text{phosphate}$$
$$\text{(Reactants)} \quad \rightarrow \quad \text{(Products)}$$

In this reaction, the oxygen atom in the water molecule separates from one of its hydrogens and replaces the oxygen atom in ATP's third phosphate group (Figure 5-7). Altogether, two bonds break—an O—H bond (in water) and a P—O bond (in ATP). In addition, one new bond forms—a P—O bond (in phosphate). The new P—O bond (in the phosphate) is more stable than the old one. In general, free energy decreases when unstable bonds break and stable ones form. And, overall, the free energy of the products

ADP + P + Energy → ATP   **ATP → ADP + P + Energy**

Endergonic   Exergonic

Figure 5-6

Work and play. Carrying a coconut up a tree is an endergonic (energy-consuming) process that scientists call "work." Dropping a coconut is an exergonic (energy-releasing) process. In the same way, adding a phosphate group to ADP consumes energy and removing the same phosphate group releases energy. It's work to make ATP.

REACTANTS: Water molecule and ATP molecule

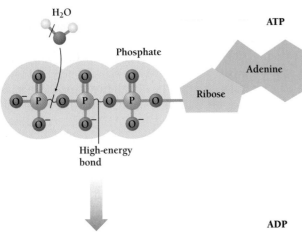

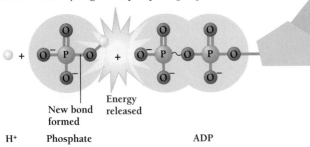

PRODUCTS: Hydrogen ion, phosphate group, ADP molecule

Figure 5-7

The most important exergonic reaction in biology. Splitting a high-energy phosphate bond in ATP releases energy that can be used in thousands of other reactions. A water molecule breaks into a hydroxide ion, which combines with the outermost phosphate group, leaving ADP and one hydrogen ion (H^+), or proton.

(ADP and phosphate) is less than the free energy of the reactants (ATP and water).

The ATP → ADP reaction favors the breaking of less stable bonds and the forming of more stable bonds. Biochemists call unstable chemical bonds that give up energy easily **high-energy bonds.** They are sometimes represented as a squiggle (~). The bonds that hold the second and third phosphates of ATP are both high-energy bonds.

Chemists can predict the direction of a chemical reaction by measuring the change in free energy for the making and breaking of chemical bonds. For example, under standard conditions and concentrations, the change in free energy (ΔG) for the breakdown of ATP to ADP and phosphate is -7.3 kcal/mole. (A mole is a fixed, huge number of molecules—6×10^{23}.) Because ΔG is negative, we can tell that free energy decreases and the reaction is exergonic. It will occur spontaneously.

The bonds between the atoms of a molecule have characteristic energies that change in the course of a reaction. Rearrangements of these chemical bonds may consume or release energy.

How Can One Process Provide the Energy for Another?

We saw earlier that an energy-consuming, or endergonic, reaction needs a source of energy. That energy often comes from an exergonic reaction. The most common source of energy for biological processes is the breakdown of ATP. When ATP breaks down into ADP and phosphate, it releases energy that becomes available to drive other chemical reactions, such as the linking of glucose molecules into a polysaccharide.

When an exergonic reaction drives an endergonic reaction, we say that the two processes are **coupled.** The chimpanzees and coconuts in Figure 5-8 are a coupled system. The falling coconut releases energy that lifts the small chimpanzee. A coupled reaction must be exergonic overall. That is, the exergonic reaction and the endergonic reaction must together release energy. If the picture in Figure 5-8 showed an elephant at the end of the log, we could guess that the coconut could not lift it. In the same way, the energy from splitting a single ATP molecule can drive only small reactions. But most reactions in biology occur in small steps.

How do cells couple endergonic and exergonic reactions? The answer is **"enzymes,"** large molecules, usually proteins, that speed up chemical reactions.

Enzymes do not affect the change in free energy of a reaction. That is, enzymes do not change whether a reaction will happen or not, just how fast. You can slide down a hill faster if it's covered with ice than if it's just dirt. But no matter how slippery a hill is, you can't slide up the hill. Enzymes

Figure 5-8

A coupled reaction. When the adult chimpanzee drops the coconut, the baby chimp is thrown up into the air. The falling coconut is *coupled* to the flying chimp.

are like the ice on the hill: they speed up a spontaneous reaction but cannot reverse its direction.

Enzymes can also couple endergonic and exergonic reactions. In muscle contraction, for example, an enzyme couples the breakdown of ATP to the contraction of tiny muscle filaments.

Enzymes couple exergonic reactions, such as the splitting of ATP, to endergonic reactions.

How Do Concentration (and Entropy) Affect Equilibrium?

The Second Law of Thermodynamics—that the energy available to perform work always decreases—also implies that uncoordinated motion is more probable than coordinated motion. Our common experience with "time's arrow" is that it is more likely that the pieces of a burst balloon will fly apart than it is that they will spontaneously reassemble. It is more likely that the books in your room will become disorganized than that they will remain neatly shelved in alphabetical order. The Second Law is a formal statement that the amount of disorder in the universe always increases.

Physicists have defined a formal measure of disorder, called **entropy.** Entropy has a high value when objects are disordered or distributed at random and a low value when they are ordered. Surprisingly, because entropy tends to increase, it can be used to do work (Figure 5-9).

Any process that converts an orderly arrangement to a less orderly one can perform work. For example, when we burn wood (converting well-ordered molecules of cellulose to disordered molecules of carbon dioxide and water), we can use the energy from this exergonic reaction to warm a house, to grill a hamburger, or to power a steam engine.

On the other hand, any process that converts a disorderly arrangement to an orderly one consumes energy. An oak tree uses energy from the sun and carbon dioxide from the air to build cellulose in its massive trunk and branches. Likewise, a cell uses energy to maintain concentrations of molecules inside the cell that are different from those outside the cell. But cells must perform work to keep their internal environments constant. When no energy supply is available to perform this work, the differences between the inside and the outside of the cell disappear, and the cell dies.

The movement of molecules from a more concentrated to a less concentrated state constitutes an increase in entropy, an increase that can be harnessed to do work. The free energy—the energy available to perform work—depends on the concentration of the molecules (Figure 5-9). The greater the difference in concentration between two sides of a cell membrane or other barrier, the more work can be done. In summary, whether a reaction occurs depends not only on the energy in the individual bonds but also on the concentrations of both chemical reactants and products.

Imagine a chemical reaction in which reactant A is converted to a product B. If the reaction is exergonic (energy releasing), then the reaction will proceed, and product B will begin to accumulate. As the concentration of B increases, however, so does B's free energy (the energy available to do work). Gradually, the difference in energy between the reactants and the products decreases. When the free energy of B equals the free energy of A (and $\Delta G = 0$), the reaction is finished. It is no longer "downhill." The products and reactants have reached **equilibrium.**

Thermodynamics predicts where the equilibrium lies, but gives no information about how long a reaction will take to get to equilibrium. In the next section, we will find out how the rules of kinetics predict the rates of chemical reactions.

The equilibrium of a reaction depends both on the free energy in the bonds between the atoms of each molecule and on the concentration of the reactants and products.

5.2 What Determines the Rate of a Chemical Reaction?

In the 19th century, archaeologists opened the tomb of an Egyptian pharaoh and found, among other artifacts, a perfectly preserved breakfast laid out for him. If the pharaoh

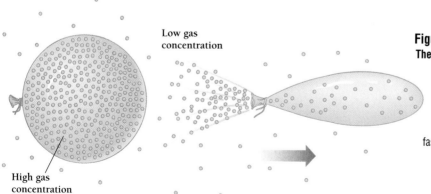

Low gas concentration

High gas concentration

Figure 5-9

The more you blow it up, the faster it flies. Any difference in the concentration of molecules creates potential energy. The greater the concentration of gas molecules inside a balloon, the faster the balloon flies when we release it. More precisely, the greater the difference between the concentration inside the balloon and outside, the faster it flies.

had lived long enough to eat this meal, the food would have been oxidized—converted to carbon dioxide and water—in a matter of hours. Yet, even after sitting for several thousand years in an oxygen-filled room, the pharaoh's breakfast was chemically unchanged. What kept the pharaoh's breakfast from being oxidized? In other words, How can cells initiate and speed up chemical reactions that run slowly or not at all outside the body?

We have seen that thermodynamics predicts both whether a reaction will occur and the direction of a reaction, independent of the precise mechanism. But to find out how fast a reaction occurs, we must understand its mechanism. We need to know what actually happens—what "path" a process takes from start to finish. Predicting the rate of a chemical reaction requires an understanding of how molecules behave.

How Does Molecular Motion Help Explain Reaction Rates?

Molecules and atoms are constantly in motion. When a substance absorbs heat, its temperature increases and the atoms and molecules bounce and jostle about more rapidly. This simple fact is the basis for kinetic theory, or kinetics. Kinetics allows us to understand the rates of chemical reactions and how temperature and other environmental factors influence these rates.

For a moment, picture molecules as tiny frogs constantly jumping about at random (Figure 5-10A). Some of these frogs (molecules) are in an area called "Level A," and some are in an adjacent, but lower area, called "Level B." Some frogs will naturally jump down to Level B. But frogs will have a harder time

A. High activation energy barrier

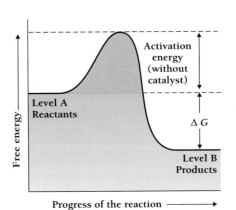

B. Low activation energy barrier

Figure 5-10

Leaping frogs and bouncing molecules. The random bouncing of molecules resembles random leaping in a group of frogs. A. Even if all the frogs begin at Level A, more will end up at Level B because it is easy to jump down but harder to jump back up. B. With a decreased activation energy barrier, frogs can more easily jump between the two levels, and it will take them less time to reach equilibrium.

Biology⊗Now™ Learn more about enzymes and activation energy by clicking on this figure on your BiologyNow CD-ROM.

Figure 5-11

Hot frogs, cold frogs. The equilibrium point of a reaction varies with temperature. A. At a low temperature, few, if any, frogs have the activation energy to jump over the barrier and reach Level B, so all the frogs remain at Level A. B. At a higher temperature, a few frogs have enough energy to leap the barrier. C. At an even higher temperature, all the frogs are jumping back and forth, but more end up down at B than up at A.

jumping back up to A. Because more frogs can jump from A to B than can jump back up, frogs will tend to accumulate at B. Only the frogs (molecules) with the highest kinetic energy will be able to jump back to Level A. (Of course, real frogs do not hop about at random, but in molecules, the kinetic energy of "hopping" is exactly the same as heat energy.)

A. Low temperature

Kinetic theory, which is based on the random movements of atoms and molecules, helps predict the rates of chemical reactions.

What Stops a Chemical Reaction?

In the 1870s, Ludwig Boltzmann proposed that molecules are constantly in motion and that their behavior explains many of the properties of matter. Boltzmann's ideas inspired the concept that heat is the kinetic energy of moving molecules. The faster molecules move, the higher their temperature. However, temperature measures only the average energy of the molecules. Individual molecules have a range of kinetic energies.

In our frog model, we would say that some frogs jump higher than others. At high temperatures, for example, most frogs can jump the barrier (Figure 5-11C). Only a few will be stuck at Level B. After the frogs have jumped for a while, they reach equilibrium—meaning that just as many frogs are jumping from A to B as from B to A. At equilibrium, the number of frogs at each energy level remains constant.

B. Increased temperature

A chemical reaction stops when it reaches equilibrium. Equilibrium depends in part on temperature.

What Starts a Chemical Reaction?

A ball sitting at the top of a flight of stairs will not bounce down the stairs until something starts it moving. Just so, every process—no matter how much it is favored thermodynamically—needs to be started in the right direction. In chemical reactions, we must ask, Where does this push come from? The answer is that the energy for the reactions between molecules comes from the random movements of molecules.

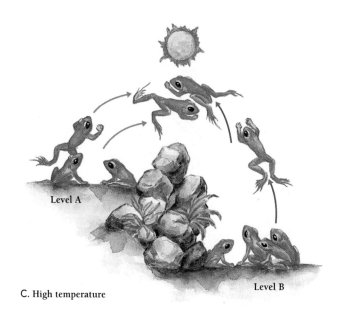
C. High temperature

To understand the factors that start chemical reactions, let us return to the jumping-frog model. The difficulty in starting a chemical reaction is comparable to the height of the barrier between the frogs. To get from Level A to Level B, the frogs first must jump over the barrier (Figure 5-10). The molecules or frogs must have a minimum energy to do this. The minimum energy needed for a process to occur is called the **activation energy.**

The number of the molecules in a space that are able to react depends first on the temperature and second on the height of the activation energy barrier (Figure 5-10). The higher the temperature, the greater the number of molecules with enough energy to jump the barrier. The lower the activation energy, the greater the number of molecules that can jump the barrier and the more rapidly equilibrium will be achieved.

Laboratory and industrial chemists can speed up a specific chemical reaction by using a **catalyst**—a substance that lowers the activation energy of a reaction but is not consumed or changed in the reaction. A catalyst, in essence, lowers the barrier between the two groups of frogs (Figure 5-10B).

An enzyme is a biological catalyst. Almost all enzymes are proteins. Enzymes allow organisms to lower the activation energy for thousands of specific chemical transformations and so shorten the time required to attain equilibrium. Hastening reactions is essential to life. We all need to be able to digest our breakfast in something less than 5,000 years.

Kinetic theory helps us understand why some thermodynamically favored reactions take place while others do not. Spontaneous reactions will actually occur only when most of the molecules have the needed activation energy. The carbohydrates in a piece of bread can be oxidized either by increasing the temperature (who hasn't burned toast?) or by exposing them to enzymes such as those in the human digestive tract.

The presence of an active enzyme allows a cell to use a molecule such as ATP in specific ways. Depending on how particular enzymes couple the splitting of ATP to other reactions, the result may be the contraction of muscle, the pumping of ions into or out of cells, or the synthesis of other molecules. Cells control their chemistry by regulating the production and activity of individual enzymes.

The initiation of a chemical reaction depends on the temperature of the molecules (their kinetic energy) and on the activation energy of the reaction. In organisms, enzymes can hasten a chemical reaction by lowering the activation energy.

5.3 How Do Enzymes Work?

In the last section, we saw that enzymes speed up biological reactions by lowering activation energies. In this section, we will see how enzymes accomplish this feat.

How Does an Enzyme Bind to a Reactant?

Enzymes speed up chemical reactions by binding to the reacting molecules, which are called **substrates.** Enzymes have enormous *catalytic power*. The right enzyme can make a chemical reaction go a million times faster than it would without the enzyme. For example, at 0°C, the freezing point of water, a single molecule of an enzyme called catalase can break down 5 million molecules of hydrogen peroxide each minute.

Enzymes are also highly *specific*. Each enzyme usually catalyzes only a single chemical reaction. What makes enzymes specific is their shape and also subtle interactions with the substrate. Most enzymes are large, globular proteins with irregular shapes. Every enzyme has an **active site**—a groove or "cleft" on its surface with which it binds to the small substrate molecule (Figure 5-12A). The binding of the enzyme to the substrate lowers the activation energy for a particular chemical reaction. Different kinds of enzymes may be able to bind to the same substrate, but in most cases *each enzyme lowers the activation energy for only a single kind of reaction.*

The enzyme temporarily binds to the substrate usually by weak noncovalent bonds—including hydrogen bonds, ionic bonds, and hydrophobic interactions. Sometimes, as the two molecules bind, the shape of the enzyme changes to fit them (Figure 5-12A).

The binding of the enzyme and the substrate is reversible, and the reactant or product molecules are quickly released back into the solution. Some enzymes are capable of binding, altering, and releasing individual substrate molecules more than 600,000 times a second.

Even so, the interactions between the enzyme and the substrate occur one molecule at a time. So although the rate of any biochemical reaction depends on temperature, as well as the concentration of the reactants, it also depends on how many molecules of enzyme are present.

The active site on an enzyme binds to a substrate usually by means of weak noncovalent bonds. Each enzyme lowers the activation energy for one reaction.

How Does an Enzyme Lower the Activation Energy of a Chemical Reaction?

The formation of noncovalent bonds between enzyme and substrate lowers the activation energy for a reaction. In all cases, the enzyme itself remains unchanged after the completion of the reaction. So, if the enzyme is unchanged, how does it change the activation energy of its substrate? The answer is that the enzyme briefly changes the shape of the substrate.

First, enzymes bring substrates together in a position that favors a reaction (Figure 5-12B). Without the enzyme, substrate molecules in solution bump into each other at

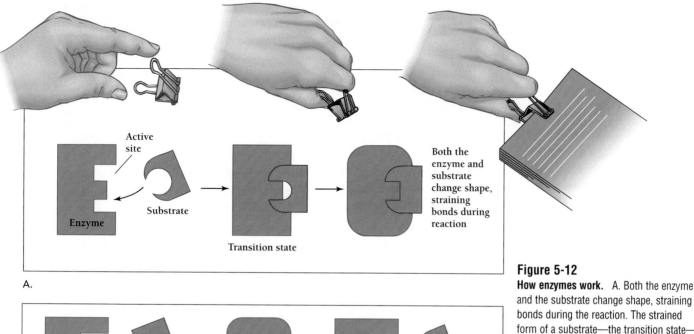

Active
site

Substrate

Enzyme

Transition state

Both the enzyme and substrate change shape, straining bonds during reaction

A.

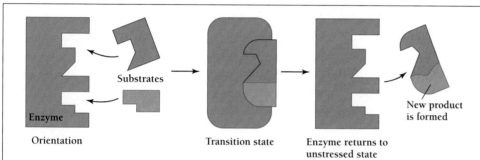

Substrates

Enzyme

Orientation

Transition state

Enzyme returns to unstressed state

New product is formed

B.

Figure 5-12

How enzymes work. A. Both the enzyme and the substrate change shape, straining bonds during the reaction. The strained form of a substrate—the transition state—exists for as little as a billionth of a second. B. An enzyme can catalyze a reaction between two other molecules by bringing both substrates close together in the proper orientation and also by straining the bonds of the substrate molecules.

Biology ⓔ Now™ Learn more about enzyme function by clicking on this figure on your BiologyNow CD-ROM.

random, but only a few of these collisions lead to a reaction. Most molecules do not react. We say the reaction is occurring very slowly.

Second, when an enzyme binds to a substrate molecule, the enzyme may strain and distort the covalent bonds within the substrate. Bonds under such strain can break more easily. The distorted form of a substrate is called a **transition state**—a molecule intermediate in form between the starting reactant and the final product. The transition state exists only for a brief moment—as little as a billionth of a second.

If we use our fingers to flex a spring clip just before clipping together a sheaf of papers, we are performing a role similar to that of an enzyme (Figure 5-12A). Our hand is the enzyme, the spring clip is one of the substrate molecules, and the sheaf of papers is a second substrate. Our hand briefly alters the shape of the substrate (spring clip) so that it can bond with the second substrate (sheaf of papers). We have created a transition state. At the end of the "reaction," our hand is unchanged, just as an enzyme is unchanged.

An enzyme lowers the activation energy barrier. In some cases, enzymes may contribute directly to the chemical reaction—for example, by lending or temporarily accepting an atom or an ion. This change strains the substrate mole-

cule, bringing it to its transition state. Again, the net effect is to lower the activation energy barrier to the reaction.

Enzymes lower the activation energy for a reaction by aligning two or more substrate molecules, by straining the covalent bonds within the substrates, and even by participating directly in the reaction.

How Do Environmental Conditions Affect the Rates of Enzymatic Reactions?

Enzymes work most efficiently at certain temperatures, pH values, and salt concentrations. In mammals, for example, most enzymes work best at body temperature. In our previous discussion of kinetics, we saw that all chemical reactions occur more rapidly at higher temperatures because more molecules possess enough kinetic energy to get over the activation energy barrier.

But increasing the temperature of an enzymatic reaction increases the rate only to a point. Once the temperature reaches a certain point, the bonds that maintain the enzyme's structure begin to loosen and break. When an enzyme loses its shape and activity, the enzyme is **denatured.** When an organism's enzymes denature, it dies. We have all seen denatured proteins in cooked egg white, for example. Because heat denatures enzymes and other proteins, most organisms cannot survive temperatures much higher than 40°C. A few unusual organisms live at higher temperatures. Bacteria that live in hot springs, for example, have evolved enzymes that function at high temperatures.

Besides temperature, pH (or acidity) also matters. Most enzymes work best in only a narrow range of pH values. The pH of the environment can, for example, alter the shape of an enzyme and add or remove hydrogen ions from the enzyme, changing the way the enzyme binds to the substrate. Most cellular enzymes work best at a pH of between 6 and 8. (Recall that pure water is neutral, with a pH of 7.) However, pepsin, the protein-digesting enzyme secreted by the stomach, works best at a pH of about 2. The fuller the stomach becomes, the more the stomach acid is diluted and the higher the pH rises, though usually no higher than about 3.5. If the pH of the stomach contents were to rise above 5, the pepsin would no longer digest our food. This is one reason it's a good idea not to eat too much at once.

Many enzymes work by acting as acids or bases. That is, they donate or accept hydrogen ions to a reaction. For example, when the amino acid histidine is part of the active site of certain enzymes, it can act as a base or an acid. If the pH of the environment is about 6.5 (slightly acid), histidine is an acid. At a pH of about 7.5 (slightly alkaline), however, histidine loses a hydrogen ion and becomes a base. An enzyme with histidine at its active site can donate a hydrogen ion and help break a particular chemical bond, but only if the pH is right. By regulating pH closely, cells control what enzymes do.

Most enzymes, however, do not work in highly acid environments. This is why vinegar is such an effective preservative. The acetic acid in vinegar disables any enzymes present either in food itself or in any microorganisms that slip into the food. Without their enzymes, bacteria are unable to break down the nutrients in food. In vinegar, most reactions take place so slowly that pickled cucumbers, onions, peppers, herring, and cabbage can remain almost unchanged for decades or longer.

The rates of enzymatic reactions increase with temperature, but only up to a point. Most enzymes work in a narrow range of pH values.

5.4 How Does a Cell or Organism Regulate Its Own Metabolism?

Organisms and cells are enormously flexible in the way they process, or metabolize, food. They function in lean as well as in fat times. They can obtain energy when food is available or break down their own stores when it is not. They can

make building blocks not available in their diet, then stop making a particular building block when it becomes available. An organism with a diet rich in protein, for example, need not waste hard-won ATP making its own amino acids.

Organisms do all this by regulating the amounts and activities of enzymes. They both control the total flow of energy and focus that energy on specific tasks.

Why Do Enzymatic Reactions Often Occur in Small Steps?

When organisms and cells transform one molecule into another, they frequently do so by means of a series of small steps or subreactions, each one mediated by a different enzyme. For example, if a chemical reacts to form product B, which reacts to form C, which reacts to form D,

$$A \rightarrow B \rightarrow C \rightarrow D$$

each step is catalyzed by a different enzyme. In humans, separate reactions number in the thousands. For example, 10 different reaction steps and 10 different enzymes are necessary in the initial extraction of energy from the sugar glucose. Almost all organisms—including all bacteria, all eukaryotes, and most archaebacteria—use the same 10 enzymes to extract energy from this sugar (Chapter 6).

The network of biochemical reactions in organisms is complex for at least four reasons. First, some of the chemical transformations that cells perform are quite complex and can only be accomplished in many steps. Second, vast numbers of reactions are endergonic. For these to proceed, each one must be coupled to an exergonic reaction, such as the breakdown of ATP to ADP. Coupled endergonic reactions are especially common when cells synthesize large molecules from small ones, a process called **anabolism.**

Third, many exergonic transformations produce more energy all at once than the cell can spend. If you wanted to buy a soft drink, you couldn't use a $100 bill in most cities. A cell "makes change" by separating a highly exergonic reaction into steps, so that the cell can capture smaller denominations of energy. A series of smaller reactions allow the cell to more easily store the energy for later use and also minimize the buildup of heat. Stepwise transformations of this kind are especially common in the breakdown of food molecules, called **catabolism.**

Fourth, in some stepwise reactions, intermediate products are the starting materials for the synthesis of essential building blocks. For example, some of the intermediate steps in the breakdown of sugars to carbon dioxide and water are precursors for amino acids, the building blocks from which proteins are made.

Many biological reactions occur in series of small steps. Stepwise reactions allow cells to carry out complex transformations, to couple reactions, to capture smaller denominations of energy, and to generate useful intermediate products.

How Do Organisms Regulate Enzyme Function?

The huge variety of molecules in a cell can undergo countless thermodynamically favored chemical reactions. The only reactions that happen rapidly enough to matter, however, are those that are catalyzed by enzymes. These include all the reactions that provide cells with the energy they need for life. Not only do enzymes serve as catalysts for specific reactions, they also regulate the rates of reactions, often in response to environmental changes.

For example, organisms need to regulate the quantities of building blocks they make. For an organism to make unnecessary building blocks would waste energy. Organisms that conserve energy by regulating the activity of enzymes therefore have an advantage over organisms that waste energy by building molecules they do not need. An example of such regulation is the synthesis of the amino acid isoleucine. When isoleucine is not available, cells make it from another amino acid, threonine. Making isoleucine from threonine requires five enzymatic reactions. But this succession of steps is energetically expensive, so the pathway from threonine to isoleucine shuts down when enough isoleucine accumulates. Isoleucine, the end product of the pathway, specifically inhibits the first step of this pathway (Figure 5-13).

One way cells regulate enzyme activity is through inhibitors that decrease the activity of enzymes.

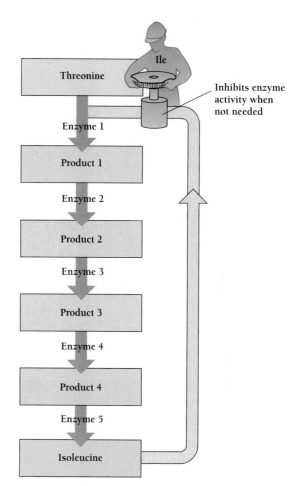

Figure 5-13
How can cells regulate enzymes? If a cell contains little or no isoleucine, a sequence of five enzymatic reactions synthesizes isoleucine from threonine. However, when isoleucine molecules accumulate, they inhibit the enzyme that catalyzes the first of the five reactions, shutting down the isoleucine pathway.

How Do Inhibitors Influence Enzymes?

Other molecules besides end products can change the activity of enzymes—sometimes inhibiting enzyme action, sometimes increasing enzyme action. Those that inhibit enzymes are called **inhibitors.**

Because an enzyme recognizes the shape of its substrate, enzymes can be "fooled" into binding to molecules that are similar in size and shape. Molecules that resemble the substrate can block an enzyme's active site and prevent the enzyme from working (Figure 5-14). Such a molecule is called a **steric inhibitor** [Greek, *stereos* = solid, three-dimensional] because it inhibits the work of the enzyme by means of its physical shape.

In carbon monoxide poisoning, for example, carbon monoxide (produced by burning gasoline or other fuel) competes with oxygen for a site on the hemoglobin molecule in our blood. Carbon monoxide binds so tightly to the hemoglobin that the hemoglobin can no longer carry oxygen to the brain and body. Unlike carbon monoxide, however, most steric inhibitors are reversible. They detach from the enzyme, so that a given enzyme sometimes binds to the substrate and sometimes to the inhibitor.

A steric inhibitor can be overcome. If the concentration of substrate is much greater than the concentration of a steric inhibitor, the enzyme will mostly bind to the substrate, not the inhibitor. The substrate and the inhibitor are competing for access to the enzyme. So the more concentrated the substrate, the more effective it will be at binding enzyme. And the more concentrated the steric inhibitor, the more effective it will be at binding enzyme. Because of such competition, a steric inhibitor is also called a "competitive inhibitor."

Many inhibitors are "noncompetitive." Such molecules do not compete directly with the substrate. Lead and other heavy metals, for example, work by blocking specific groups outside the active site, sometimes changing the overall shape and chemistry of an enzyme. An inhibitor that binds to an enzyme at a place other than the active site is called an **allosteric inhibitor** [Greek, *allos* = other, different]. For example, an allosteric inhibitor blocks an enzyme by binding to a site on the enzyme other than the active site.

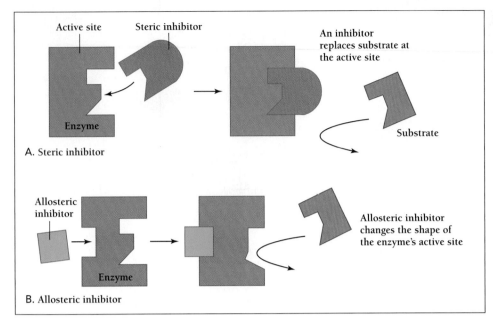

Figure 5-14
What can prevent an enzyme from catalyzing a reaction? A. A steric inhibitor is a molecule that binds to an enzyme's active site, preventing the substrate from binding. A steric inhibitor "competes" with the substrate for the active site. B. An allosteric inhibitor does not compete with the substrate but binds to another site on the enzyme in such a way that the active site no longer functions.

Some noncompetitive inhibitors act only temporarily. Others act irreversibly by forming a permanent bond with the enzyme and altering its shape. The antibiotic penicillin, for example, irreversibly damages an enzyme that bacteria use to build their cell walls.

Many noncompetitive inhibitors are poisons, like lead. Other inhibitors are essential parts of cells. Cells use allosteric inhibitors, for example, to regulate their own chemical processes. Most enzymes have evolved so that their most active shape is also their most stable shape. When an allosteric inhibitor binds to an enzyme, however, it changes the enzyme's shape to a less active form. The allosteric inhibitor acts as a switch, changing the enzyme from its active shape to its less active shape.

Inhibitor molecules decrease the activity of enzymes—either by competitively binding to the active site or by altering the shape of the enzyme.

Key Concepts

- The Second Law of Thermodynamics predicts the direction of any process because energy becomes less available for work as a reaction or process occurs.
- The rate of a chemical reaction depends on the random movements of molecules.
- Enzymes speed up biochemical reactions by lowering their activation energies.
- Cells and organisms regulate their own metabolism by changing the shapes and activities of enzymes.

Summary with Key Terms

What factors determine which way a reaction will go?

All processes—biological and nonbiological—are accompanied by changes in the form of energy, which are subject to the laws of **thermodynamics**. The **First Law of Thermodynamics** states that the total amount of energy does not change. The **Second Law of Thermodynamics** states that the total amount of energy in a closed system becomes less and less available for **work**; more and more of it is converted to heat. In short, **entropy** increases.

The laws of thermodynamics allow scientists to predict whether a reaction will occur spontaneously. The making and breaking of chemical bonds between atoms leads to changes in available energy. A reaction will occur spontaneously if the energy available to do work—the **free energy** (**G**)—decreases. A reaction is **exergonic** if free energy (**ΔG**) decreases and **endergonic** if free energy increases.

Organisms can accomplish "uphill reactions" by **coupling** endergonic processes to exergonic reactions with protein **catalysts** called **enzymes**. In many cases, the energy that drives endergonic reactions comes from the **high-energy bonds** of ATP.

During chemical reactions and other processes, **kinetic energy** may be converted to **potential energy**. The free-energy change in a chemical reaction depends on the concentrations of **reactants** and **products**, as well as on the number of chemical bonds that are made and broken. Reactions tend to go in the direction of increased disorder, or entropy, in which the concentrations of molecules become more uniform.

What factors determine the rate of a chemical reaction?

To predict how long a process will take to occur, we need to know how it occurs. The study of **kinetics** allows scientists to predict the rates of chemical reactions from the properties of molecules. The basis of kinetics is that molecules are constantly in motion, with the average energy of a group of molecules determined by the temperature, measured in **calories.**

At chemical **equilibrium,** the direction of a reaction is balanced by the opposite reaction. Enzymes and other catalysts speed up specific chemical reactions but do not affect the free-energy changes of chemical reactions. They increase the speed at which equilibrium is approached, but they do not affect the character of the equilibrium.

How do enzymes work?

Enzymes hasten chemical reactions by lowering the **activation energy** of the reaction. Enzymes accomplish this by binding to the reacting molecules by means of noncovalent interactions. An enzyme may join two or more **substrate** molecules. Sometimes enzymes strain the bonds of a single substrate molecule, forcing it into a **transition state.** Some enzymes actually participate in the chemical reaction by temporarily giving or receiving atoms or ions. The activity of an enzyme is highly sensitive to environmental conditions that can affect its activity. Temperature, pH, and the presence of other molecules may **denature** an enzyme.

How can cells modify the activity of enzymes?

Many chemical transformations performed by organisms in **anabolism** and **catabolism** consist of numerous small reactions. Such small reactions more easily allow the cell to capture the energy for later use. Cells inhibit the activity of many enzymes with **steric inhibitors** and **allosteric inhibitors.** Steric inhibitors block the enzyme's **active site,** while allosteric inhibitors change the enzyme's shape and activity by binding to a nonactive site.

Review and Thought Questions

Review Questions

1. State the First and Second Laws of Thermodynamics and explain their relationships to an example of work.
2. What is free energy? What is its relationship to the Second Law of Thermodynamics? In what direction must it change in all reactions (or processes of work)? Compare this to the meaning of "entropy."
3. Define endergonic and exergonic in terms of the free-energy change associated with reactions. Which reaction involving ATP is endergonic? Exergonic? Which reaction occurs spontaneously?
4. Which part of downhill skiing is endergonic (work), and which part is exergonic?
5. What is the activation energy for a reaction? How might it influence the speed with which a reaction proceeds?

6. Enzymes are biological catalysts. How do they perform the task of speeding up biological reactions?
7. What is meant by the "active site" of an enzyme? How is it related to the substrate of that enzyme?
8. Describe four benefits that cells derive from organizing their reactions into pathways.
9. Decide if these reactions are exergonic or endergonic. Mark "X" for exergonic and "N" for endergonic.
   ____ Growing hair
   ____ Digesting pizza
   ____ Baking soda and acetic acid (vinegar) reacting to give off fizzy carbon dioxide that makes a cake rise
   ____ Liver cells produce fibrinogen, a blood clotting protein
   ____ The adrenal glands produce cortisol derived from cholesterol
   ____ ATP splits during muscle contraction

Thought Questions

10. Where does the energy come from when you perform the work of riding your bicycle? Did *all* of the energy in the food you used for this task actually help turn the wheels of your bicycle? Explain using the First and Second Laws of Thermodynamics.
11. Explain why unraked leaves might form a neat pile in a specific spot in your backyard. Does this mean that the Second Law of Thermodynamics is not working? What is the source of energy for this organization of the leaves?
12. Why do you think that some exergonic reactions will *not* go at all in the absence of a catalyst? If the change in free energy is no obstruction, what is?

BiologyNow Resources

Biology ⑤ Now™

Active Figures

5-10: Enzymes and activation energy
5-12: Enzyme function

Preparing for an exam? Take a diagnostic test on your BiologyNow CD-ROM.

Online materials relating to this chapter are at:

http://biology.brookscole.com/AAL3

About the Chapter-Opening Image

A car rusting in a field is a good example of increasing entropy. The Second Law of Thermodynamics says that entropy (disorder) always increases.

How Do Organisms Supply Themselves with Energy?

Key Questions

- How do organisms supply themselves with energy?
- How do organisms extract energy from glucose?
- How is the energy in glucose used to make ATP?

Paraskevas Photography

Louis Pasteur and Vitalism

In 1835, the French scientist Charles Cagniard de la Tour observed that the yeast in fermenting beer consisted of tiny living cellular organisms. Cagniard de la Tour noted that these tiny organisms multiplied by budding, and he suggested that they were somehow linked to the process of alcoholic fermentation—the conversion of sugar to ethanol and carbon dioxide. In the same year, two other scientific papers described similar observations. One was by Theodor Schwann, the German botanist whose work helped establish the cell theory described in Chapter 4.

Schwann also showed that if grape juice were boiled to kill the yeast, the juice would not ferment to wine unless yeast cells were added back to the juice. Schwann correctly concluded that yeast plays an essential role in fermentation. He then showed that fermentation began at the same time that the yeast cells first appeared, progressed with their multiplication, and stopped as soon as the yeast cells stopped multiplying.

Amazingly, the scientific community rejected these important discoveries out of hand, owing to a handful of powerful personalities and the scientific and philosophical temper of the time. The most influential scientists insisted that yeast was not a living organism at all, but a substance, and probably an unimportant one.

This mistaken idea was held by no less than the father of modern chemistry, Antoine Lavoisier (1743–1794). Lavoisier had already argued that fermentation could be understood as a simple chemical reaction. Fermentation, he and others had written, consisted of a reaction in which the sugar glucose broke down into ethanol and carbon dioxide, a reaction we can now write:

$$C_6H_{12}O_6 \rightarrow 2\ C_2H_5OH +\ \ 2\ CO_2$$

glucose $\rightarrow$ ethanol + carbon dioxide

This nicely balanced formulation seemed so simple and so right that chemists, riding high on their tremendous successes (such as Lavoisier's own discovery of oxygen), could not imagine why yeast would be necessary. Even if yeast did turn out to be necessary, where could it possibly go in the equation? They accepted that yeast was present during fermentation. But there was no question of yeast actually taking part in the reaction. At most, they said, dead and dying yeast cells might release some substance that could speed up the chemical reaction. Nowadays, we would call such a substance a "catalyst."

Nineteenth-century chemists firmly believed that all biological processes were chemical in nature. To believe otherwise, to insist on some mysterious role for living organisms that was not purely chemical in nature, was condemned as **vitalism**—the belief that living systems have powers beyond those of nonliving systems.

The idea that a simple chemical reaction needed tiny living organisms to occur seemed absurd. In 1839, two of the most prominent early chemists, Friedrich Wöhler and Justus von Liebig, composed and published a lampoon designed to humiliate Schwann and the other yeast researchers. Wöhler and Liebig pretended to give a scientific description of the anatomy and behavior of yeast cells. A yeast cell, they said, looks exactly like a miniature still (the apparatus used to distill whiskey). Yeast cells suck in sugar and excrete ethanol and carbon dioxide, Wöhler and Liebig joked, writing, "Although teeth and eyes are not to be seen, one can distinguish a stomach, intestine, the anus (a rose-pink spot), and the organs of urine secretion. . . . The urine bladder in the full condition is shaped like a champagne bottle."

The lampoon must have worked, for Schwann retired from the debate, and Liebig and Wöhler's mistaken opinions prevailed unquestioned for 20 years. However, in 1854, young Louis Pasteur, a bright new star in the scientific sky, was appointed professor of chemistry at Lille, France.

Lille was a major center for the manufacture of ethanol, through the fermentation of beet juice, and it was not long before one of Pasteur's students appealed to him to help solve problems at the family distillery. Pasteur immediately became intrigued by the process of fermentation. Just as quickly, he realized that fermentation depended entirely on the presence of yeast.

Pasteur was ambitious, and he must have known that proving the eminent Liebig wrong would be a magnificent coup. In September 1855, Pasteur began a series of studies in fermentation. As his wife, Marie Pasteur, wrote to her father-in-law, "Louis is up to his neck in beet juice. He spends all his days in the distillery. He has probably told you that he teaches only one lecture a week; this leaves him much free time which, I assure you, he uses and abuses."[1]

By 1860, Pasteur had succeeded in growing the yeast in a broth seeded with tiny numbers of yeast cells. The ethanol produced was proportional to the multiplication of the yeast. Normally, vast numbers of yeast cells are present in fermenting solutions. And Liebig had long insisted that some chemical released by the decomposing bodies of yeast cells initiated fermentation. But Pasteur's yeast cells were multiplying, not dying, and the scientific community soon rallied to Pasteur's side. Liebig stubbornly rejected Pasteur's results to the end of his life. Fermentation, he said, was a chemical process, and vitalism was not science.

Ironically, although fermentation normally depends on cells, Liebig was largely right. In fact, fermentation *is* a chemical process—though far more complicated than that envisioned by early chemists. And fermentation can take place outside of living cells—provided the right enzymes are available.

In 1897, Hans and Eduard Büchner decided to make an extract of yeast, called "zymase," to be used as a patent medicine, what we would nowadays call a supplement. However, the extract kept rotting, so the brothers added sugar to slow bacterial growth, the same reason that fruit canners add sugar to canned peaches and berries. In high concentration, sugar is an excellent preservative. To their amazement, the yeast extract began bubbling furiously, as carbon dioxide gas burst from the surface of the extract. Hans Büchner immediately recognized that the sugar was fermenting in the absence of living yeast cells. The Büchners' zymase contained essential enzymes that catalyzed the alcoholic fermentation reaction—just as Liebig had predicted.

Nonetheless, Pasteur also was right. Fermentation normally does not occur outside of living organisms. It is a highly specialized process by which microorganisms extract energy from sugar. And although biologists now all agree with Liebig's idea that life will ultimately be understood in terms of chemistry, organisms are more than mere bags of enzymes. Their organization and complex structures guide and control the chemical reactions that contribute to life. In this chapter, we will see how cells of all kinds extract energy from biological molecules.

6.1 How Do Organisms Supply Themselves with Energy?

All organisms need energy, and the ultimate source of energy for most organisms is sunlight. Plants convert light energy from the sun into chemical energy through photosynthesis. Organisms that make their own food, such as plants, or obtain energy and synthesize organic molecules from inorganic material are called **autotrophs** [Greek, *auto* = self, same + *trophe* = to nourish]. The vast majority of autotrophs, including most plants and many algae and bacteria, photosynthesize. Nonetheless, not all autotrophs photosynthesize. Autotrophs include a few bacteria (called *chemotrophs*) that obtain energy by oxidizing inorganic substances such as sulfur and ammonia.

[1]From Dubos, Rene, Louis Pasteur, Free Lance of Science, Little, Brown, 1950.

Animals, fungi, and other organisms that obtain chemical energy from other organisms are called **heterotrophs** [Greek, *hetero* = other + *trophe* = to nourish]. Both heterotrophs and autotrophs obtain energy by breaking down organic molecules in the process called cellular respiration. The chemical reactions of cellular respiration are part of a complex network of biochemical conversions that are collectively called **metabolism** [Greek, *meta* = after + *ballein* = to throw]. The difference between heterotrophs and autotrophs is that autotrophs make these organic molecules themselves, while heterotrophs take them from others—either from autotrophs or other heterotrophs. Ultimately, nearly all the energy that powers living organisms comes from sunlight.

What Is the Common Currency of Energy for Organisms?

If you spend all your time in the next 24 hours studying biology, you will use the energy from about 10^{25} molecules of ATP (adenosine triphosphate)—some 90 pounds (40 kg) of the stuff. If you were to run a marathon, you would use the same amount in just two hours. Almost every time one of your cells performs an energy-consuming reaction, it gets the energy it needs by splitting a phosphate group off of ATP (Chapter 3). ATP is the universal currency of biological energy. Fortunately, however, cells use the same ATP and ADP molecules over and over, recharging ADP to make ATP, then draining the energy from ATP once more. So you don't really have to carry around a 90-pound ATP/ADP "battery."

All cells use ATP. Plants produce ATP during photosynthesis, as we will see in Chapter 7. But both plants and animals can produce ATP in another way—by capturing energy from the breakdown of carbohydrates, lipids, and proteins. The main way that organisms break down organic molecules is **cellular respiration**, the oxygen-dependent process by which cells extract energy from food molecules. Animals, fungi, and other heterotrophs obtain almost all of their energy through cellular respiration. Plants also use cellular respiration for ATP. Cellular respiration perfectly illustrates how eukaryotes all share the same metabolic pathways.

Cellular respiration is a chemical process inside of cells that is different from ordinary respiration, or breathing. Breathing—inhaling and exhaling—is nonetheless closely related to cellular respiration. Breathing supplies our cells with the oxygen needed for cellular respiration and disposes of the carbon dioxide that is the waste product of cellular respiration.

Most eukaryotic cells (and many prokaryotic cells) are **aerobic,** that is, they depend on oxygen for life. All aerobic organisms, from earthworms to arctic terns, produce the bulk of their ATP by means of cellular respiration (Figure 6-1).

Figure 6-1
When it comes to sustained aerobic activity, the arctic tern, *Sterna paradisaea,* is the champion. This arctic tern nests each summer near the Arctic Circle in North America, then migrates across the Atlantic Ocean to Europe, then south to South Africa, then across the South Atlantic to Antarctica, a distance of 18,000 km (11,000 miles). In the spring, the bird flies all the way back around the world to the Arctic Circle to nest once more.

Many prokaryotes and a few eukaryotes can live **anaerobically,** without oxygen. The microbe *Clostridium botulinum,* which grows in contaminated canned meats and vegetables and causes botulism (food poisoning), grows anaerobically (Figure 6-2A). Even our own cells, which are mainly aerobic, can produce energy anaerobically for short periods in a process similar to alcoholic fermentation. Some organisms, such as the red-eared turtle shown in Figure 6-2B, can sustain themselves without oxygen for days at a time.

In all organisms, the common currency of energy is ATP.

How Do Organisms Extract Energy from Macromolecules?

Cellular respiration must begin with small molecules such as the simple sugar glucose. Cells cannot extract energy directly from carbohydrates, proteins, or fats. In all animals, macromolecules must first undergo **digestion,** the process of splitting macromolecules into simple building blocks—amino acids, simple sugars, or nucleotides that can enter cells. These small energy-rich molecules move inside of cells, and mitochondria transfer their energy into the chemical bonds of ATP.

Organisms digest macromolecules into their component building block, which can enter cells and the biochemical pathways of cellular respiration.

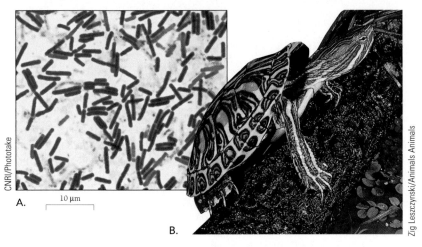

A.

10 μm

B.

Figure 6-2

Life without air. A. The obligate anaerobe *Clostridium botulinum,* which causes a serious form of bacterial food poisoning, cannot reproduce in the presence of oxygen. It is known, however, for its ability to multiply inside of sealed canned goods. B. Among vertebrates, the red-eared turtle, *Chrysemys scripta elegans,* is unusual in its ability to live without oxygen. It can stay under water for two weeks at a time, relying on glycolysis for energy production.

What Are the Steps of Cellular Respiration?

The most important small molecule from which cells extract energy is the simple, six-carbon sugar glucose. Cells extract energy from glucose by breaking the bonds between each of its six carbon atoms and transferring the energy from those bonds to the high-energy bonds of ATP. During cellular respiration, glucose is completely broken down to carbon dioxide and water. The overall reaction is:

$$C_6H_{12}O_6 + 6\,O_2 \rightarrow 6\,CO_2 + 6\,H_2O + ATP$$

glucose + oxygen → carbon dioxide + water + ATP

The complete breakdown of glucose to carbon dioxide and water and the packing away of the energy into ATP occurs in four steps. In eukaryotes, the first step is **glycolysis,** which takes place in the cytosol. In glycolysis, a six-carbon glucose is broken in half, forming two molecules of pyruvate, each with three carbon atoms (Figure 6-3).

In the second step, the three-carbon pyruvate molecules each lose a carbon atom, which is ultimately released as carbon dioxide. The remaining two-carbon molecules go into the third step, the **citric acid cycle,** where the last remaining carbon-carbon bonds are broken and the single carbon atoms combine with oxygen to form more carbon dioxide. The second and third steps take place inside the matrix of the mitochondria.

The fourth step of cellular respiration is **oxidative phosphorylation,** when the energy released from the breaking of carbon-carbon and carbon-hydrogen bonds in glucose is transferred to phosphorus-oxygen bonds in ATP. The fourth step is not actually the last in time, since much of it occurs simultaneously with the other three steps. As we will see later, much of the action of the fourth step occurs during the process of "electron transport" in the molecules of the inner membrane of the mitochondria. Of the four steps, oxidative phosphorylation makes the most ATP by far.

> Cellular respiration occurs in four steps, one of which takes place in the cytosol of eukaryotic cells, and three of which take place in mitochondria.

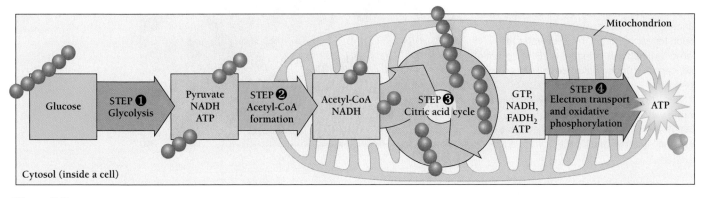

Figure 6-3

Cellular respiration. The complete breakdown of glucose occurs in four steps. (1) In glycolysis, six-carbon glucose is broken in half, forming two molecules of pyruvate, each with three carbon atoms. Glycolysis produces two ATPs from ADP and two NADHs from NAD⁺. (2) In the second step, each pyruvate molecule loses a carbon atom (as carbon dioxide) to form an acetyl-CoA molecule and one NADH. (3) The citric acid cycle converts two more carbon atoms into carbon dioxide. The cycle generates two high-energy phosphate bonds (in GTP) and uses the eight electrons of each acetyl-CoA molecule to generate six NADHs and two FADH₂s. (4) Oxidative phosphorylation, which occurs during electron transport in the mitochondria's inner membrane, moves energy from GTP, NADH, and FADH₂ to ATP.

Extreme Biology   Sprints, Dives, and Marathons

Respiration yields at least 15 times more ATP than glycolysis. It's no wonder that animals—especially warm-blooded ones—depend on respiration for most of their energy. However, because oxygen is the ultimate electron acceptor in respiration, available energy depends on a steady supply of oxygen. For vertebrates, continuous energy production requires efficient lungs (to take in oxygen-rich air) and hearts (to pump oxygen-rich blood from the lungs to the muscles).

Training can improve the performance of the systems that deliver oxygen, but humans and other animals often need to spend energy more quickly than our oxygen supply will allow. We all need to sprint occasionally, whether we are a pedestrian dashing to avoid an inattentive driver, an outfielder swooping down on a wayward baseball, a cheetah chasing a gazelle, or a gazelle eluding a cheetah. Several energy reservoirs allow such bursts of energy: (1) muscles have stores of ATP and other compounds with similar high-energy phosphate bonds, enough to allow several minutes of work without any new ATP production; (2) muscles can use blood glucose and stored glycogen to produce ATP by anaerobic glycolysis, producing lactate as the end product.

The length of time that an organism can depend on anaerobic glycolysis for ATP production is limited both by training and by genetic capacity. Humans can manage only a few minutes. But some animals, such as whales, seals, and other diving animals, can function by means of glycolysis alone for much longer periods. The freshwater red-eared turtle shown in Figure 6-2B, for example, can stay under water for as long as two weeks.

Anaerobic glycolysis provides animals with a way of borrowing energy. But the loan is only short term. After anaerobic exertion, all animals must get rid of the accumulated lactate and restore their supplies of glucose and glycogen. These processes require oxygen. In most animals, the "oxygen debt" must be repaid almost immediately—through heavy breathing and increased circulation. Anyone who has run after a bus or away from an angry sibling has experienced the recovery from an oxygen debt. Animals extract energy from the accumulated lactate by converting it to pyruvate, which can be used in aerobic respiration. In some vertebrates, some of the lactate travels via the blood to the liver, where it can be recycled into glucose. The liver then resupplies the glucose reservoir of the blood.

Anaerobic glycolysis allows untrained schoolchildren to sprint up to 17 miles per hour (27 km/hr) in a 100-yard (or 100-meter) dash. But the necessity for oxygen prevents even champion athletes from averaging more than about 11.5 miles per hour in a 26-mile marathon (Figure A). Nonmammalian vertebrates (fish, amphibians, and reptiles), which live their lives at a slower metabolic pace than we do, may function without air for long periods.

Instead of using oxygen to recycle the accumulated lactate, fish, amphibians, and reptiles may excrete the lactate in their urine. This represents a huge waste of energy, but it allows these animals to feed and function in environments in which they could not otherwise live.

David Madison

Figure A
1995 World women's marathon. The speed that a marathoner can attain is limited by the need for oxygen.

6.2 How Do Cells Extract Energy from Glucose?

The first of the four steps in the metabolism of glucose is **glycolysis** [Greek, *glykys* = sweet (referring to sugar) + *lyein* = to loosen], a set of 10 chemical reactions (or steps). Glycolysis converts glucose, a sugar with six carbon atoms, into two identical smaller molecules, called **pyruvate,** with three carbon atoms each. The 10 reactions also convert some of the energy of glucose into ATP, and another high-energy molecule (called NADH). All 10 reactions take place in the cytosol, outside the mitochondria.

The most remarkable fact about glycolysis is that *all* organisms do it in exactly the same way. This universality suggests that the common ancestors of all present-day organisms performed glycolysis in the same way that we do. The enzymes that catalyze the 10 reactions of glycolysis must have evolved at least 3.8 billion years ago, at the very dawn of life on Earth. Its evolutionary stability supports the idea that it has been a key process for all life on Earth.

What Is Oxidation?

Most of the energy-producing processes in cells involve the transfer of electrons from one molecule to another. A molecule that accepts an electron is called an electron acceptor, or **oxidizing agent.** A molecule that donates an electron is called an electron donor, or **reducing agent.** Oxidation and reduction reactions always go hand in hand. The oxidation of one atom or molecule provides the electrons for the reduction of another atom or molecule.

In cellular respiration, electrons from the break-up of glucose combine with oxygen and hydrogen ions to form water. The six carbon atoms in glucose donate electrons, and oxygen atoms ultimately accept those electrons. When glucose loses its electrons and its bonds break, we say the glucose has been "oxidized." Oxygen is the most electron-hungry atom in the environment, and organisms have evolved mechanisms for managing oxygen's enormous hunger for electrons.

When wood burns, the sugar molecules that make up the cellulose in wood are also oxidized to carbon dioxide and water. Burning wood (or gasoline) is pretty much the same oxidation reaction as cells perform with glucose. The chemical similarity between glycolysis and actual burning of wood or gasoline is one reason people talk informally about "burning" food calories. The most important difference between the way glucose burns (oxidizes) in a cell and the way wood burns is that burning wood is an uncontrolled reaction in which a huge amount of heat energy is released all at once. Cells do it differently, breaking the reaction into small steps, which allows them to control the destinations of the electrons and capture tiny units of energy to be packed away into ATP.

Besides oxygen, the two most important electron acceptors are **NAD$^+$** (nicotinamide adenine dinucleotide) and **FAD** (flavin adenine dinucleotide), which take electrons from glucose and transfer them to oxygen in steps. Each NAD$^+$ can accept two electrons and a hydrogen ion and become NADH. When a molecule gains an electron, chemists say it is reduced because it is more negative than before, so NADH is the "reduced form" of NAD. Each FAD can also accept two electrons and two hydrogen ions to form **FADH$_2$,** the reduced form. As we'll see, most of the ATP made during cell respiration comes from energy that was originally captured as NADH.

Step 1: Glycolysis

As glycolysis breaks glucose into two molecules of pyruvate, it converts two molecules of NAD$^+$ to NADH and two molecules of ADP to ATP. Glycolysis captures this energy in 10 chemical reactions, which fall into three parts (Figure 6-4):

Part 1. Splitting glucose's six carbons into two three-carbon molecules and phosphorylating them (adding phosphates), which actually consumes ATP energy.

Part 2. Removing phosphates and electrons from these three-carbon molecules and adding the electrons to NAD$^+$ to make NADH (generating two molecules of ATP).

Part 3. Removing more electrons and phosphates from the three-carbon molecules and adding them to ADP molecules to make ATP molecules.

Part 1 requires five reactions, each catalyzed by a different enzyme. Together, these reactions transfer two phosphate groups from two ATP molecules to glucose and splits glucose in half. The energy from splitting ATP goes to make a three-carbon molecule called glyceraldehyde-3-phosphate (G3P).

In part 2, NAD$^+$ accepts two electrons from each molecule of G3P, which each accept a second phosphate group. This second phosphate then joins an ADP molecule to form the first two ATP molecules of glycolysis (Figure 6-4).

The reaction steps of part 2 require both phosphate ions and NAD$^+$. If either is missing, the reaction simply cannot occur. Phosphate and NAD$^+$ are essential: cell survival requires a constant supply. Phosphate is usually not a problem, since it is present at fairly high concentrations in most cells. It is continuously released as ATP is split. In aerobic organisms, cells make new NAD$^+$ by transferring NADH's electrons to oxygen. Without oxygen to accept electrons, however, cells make new NAD$^+$ in other ways, one of which is fermentation (see box on p. 116).

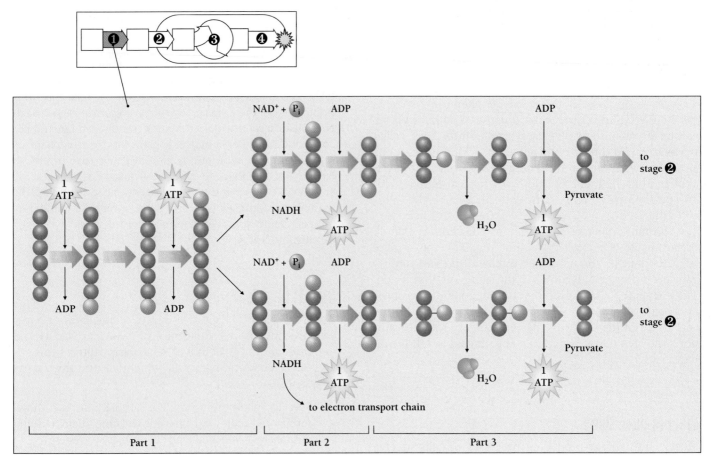

STEP 1: GLYCOLYSIS

Figure 6-4
Step 1: the first step of metabolism is glycolysis. Glycolysis is a set of 10 chemical reactions that break each six-carbon glucose into two molecules of three-carbon pyruvate. The 10 reactions—nine of which are shown here—also convert some of the energy of glucose into high-energy ATP and NADH. All these reactions take place in the cytosol, outside the mitochondria.

Biology⊗Now™   Learn more about glycolysis by clicking on this figure on your BiologyNow CD-ROM.

Part 3 transfers a phosphate to ADP, leaving pyruvate. For each starting glucose molecule, then, the glycolysis of one glucose molecule yields two molecules of ATP and two molecules of NADH.

Glycolysis breaks glucose into two molecules of pyruvate, generating two molecules of ATP and two molecules of NADH.

Step 2: Converting Pyruvate to Acetyl-CoA

The pyruvate molecules still contain lots of energy, some of which will be extracted in Step 2 of cellular respiration (Figure 6-4). Step 1 (glycolysis) occurs in the cytosol of the cell,

but all the subsequent steps of cell respiration take place in the mitochondria. The pyruvate produced by glycolysis enters the mitochondria where it is oxidized to a two-carbon compound called "acetyl-CoA" (Figure 6-5).

First, pyruvate loses a carbon and two oxygens in the form of carbon dioxide, the first waste carbon dioxide produced by cell respiration. The remaining two-carbon fragment loses two electrons to NAD^+ (forming higher-energy NADH), leaving two-carbon acetate (the same as the acetic acid of vinegar). Finally, enzymes link a molecule called **coenzyme A (CoA)** to the acetate, forming acetyl-CoA. For this reaction to occur, oxygen must be present to accept electrons from the NADH (and so regenerate NAD^+).

Acetyl-CoA has only two principal fates: (1) it can enter the third step of energy metabolism and help make

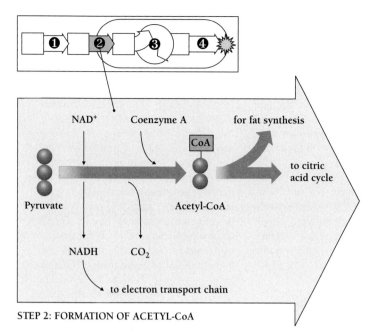

STEP 2: FORMATION OF ACETYL-CoA

Figure 6-5
Step 2: pyruvate to acetyl-CoA. After glycolysis, pyruvate enters the mitochondria for the second step of energy extraction. In the mitochondria, pyruvate is oxidized to form acetyl-CoA. Acetyl-CoA can either enter the third step of energy metabolism and generate more ATP or it can be used to synthesize body fat.

more ATP, as discussed next, or (2) it can be used to form fatty acids or cholesterol. When a cell already has plenty of ATP, it uses acetyl-CoA to make energy-rich fats and oils, allowing the body to store energy as fat for later use. The redirection of acetyl-CoA is the main reason that we accumulate fat when we consume too many energy-rich molecules. It doesn't matter to our cells whether the acetyl-CoA was derived from the starches in spaghetti or the stearic acid in a steak.

The pyruvate produced by glycolysis enters a cell's mitochondria, where it is converted to acetyl-CoA. Acetyl-CoA can generate still more ATP. Oxygen must be present for this reaction to occur.

How Does a Cell Generate ATP from Acetyl-CoA?

Acetyl-CoA still contains most of the energy of the original glucose molecule. It is Step 3 that captures most of the high-energy electrons in the form of NADH. Oxidative phosphorylation (Step 4) then transfers the energy in NADH to ATP.

In most organisms, acetyl-CoA enters into a set of reactions that converts its carbon atoms into carbon dioxide. To-gether, the reactions of Step 3 are called the **citric acid cycle,** or the **Krebs cycle,** after Hans Krebs, who began working out the details of the cycle in the 1930s.

In eukaryotes, the citric acid cycle takes place in the matrix of the mitochondria. Briefly, the two-carbon compound acetyl-CoA combines with a four-carbon molecule (oxaloacetate, or OAA) to form a six-carbon molecule (citric acid), which then loses two carbons as carbon dioxide, leaving four-carbon OAA available to combine with the next molecule of acetyl-CoA (Figure 6-6).

Altogether, the citric acid cycle removes eight electrons from the two carbons of acetyl-CoA. Three NAD^+ molecules accept six of these electrons, and one molecule of FAD accepts the other two. The citric acid cycle also forms one high-energy phosphate bond. Unlike the other reactions that we have discussed, however, the phosphate acceptor in the citric acid cycle is guanosine diphosphate (GDP), rather than ADP. The resulting guanosine triphosphate (GTP) carries the same amount of energy as ATP.

Because oxygen is the final acceptor of all these electrons, oxygen's presence in the cell is absolutely essential for the citric acid cycle to proceed. In Step 4, oxygen accepts electrons from NADH and $FADH_2$, turning them back into NAD^+ and FAD. Their regeneration permits both the citric acid cycle (Step 3) and the oxidation of pyruvate (Step 2) to continue.

Besides making energy molecules, the citric acid cycle also serves as the major source of building blocks for many amino acids and other small molecules. The flow of acetyl-CoA into the cycle depends both on a cell's need for energy and for amino acids.

The reactions of the citric acid cycle fall into three parts: (1) the formation of the six-carbon citrate (citric acid) and then isocitrate; (2) the conversion of isocitrate into a four-carbon compound called succinate, which is linked to CoA; and (3) the production of another molecule of OAA, which starts the cycle all over again. With each turn of the cycle, citrate loses a total of eight electrons to the electron acceptors such as NAD^+ and FAD.

The citric acid cycle transfers eight electrons from acetyl-CoA to form NADH and FADH2.

The Citric Acid Cycle Intersects Other Biochemical Pathways

The citric acid cycle lies at the center of the complex network of metabolic reactions. Metabolism is divided into two classes of pathways: **catabolism** [Greek, *cata* = down + *ballein* = to throw]—the breakdown of complex molecules, such as those in food; and **anabolism** [Greek, *ana* = up + *ballein* = to throw]—the synthesis of complex molecules (Figure 6-7). In general, catabolism pro-

Extreme Biology   How Can Glycolysis Continue Without Oxygen?

To sustain glycolysis, a cell must regenerate more NAD+ from NADH. Just how it does this depends on the species of organism and on whether oxygen is present. In the presence of oxygen, most cells can oxidize NADH to NAD+ by the reactions of oxidative phosphorylation. In the absence of oxygen, however, cells must regenerate NAD+ by using NADH to reduce pyruvate. Their manner of doing so depends on what enzymes they have.

Many microorganisms live by **fermentation**—the anaerobic extraction of energy from organic compounds (Figure A). Yeast cells, for example, ferment the sugars in beer, wine, and bread, converting glucose into carbon dioxide and ethanol. The carbon dioxide makes beer bubble and bread rise. The ethanol gives beer its punch and freshly baked bread its sweet aroma. To perform this feat, yeasts employ all the reactions of glycolysis that we have discussed. In addition, they use the accumulated NADH to reduce pyruvate to form ethanol. In the process, NADH is oxidized back to NAD+. Yeast is a *facultative anaerobe*—an organism that can live either anaerobically, by fermentation, or aerobically, by oxidative phosphorylation. Other microorganisms, such as *Clostridium botulinum,* are *obligate anaerobes,* meaning that they can grow only in the absence of oxygen (Figure 6-2A).

Animals, including humans, often suffer a temporary lack of oxygen in some of their cells. For example, during sprinting or other vigorous exercise, muscle cells need more oxygen (to oxidize NADH) than the blood and lungs can deliver (Figure A). In this case, and for short periods only, the cells can regenerate NAD+ in the absence of oxygen. Instead of producing carbon dioxide and ethanol (as a yeast cell does), however, muscle cells make lactic acid. During a vigorous sprint, lactic acid accumulates in the muscles, sometimes causing soreness. Once oxygen becomes available again, however, cells convert most of the lactic acid back to pyruvate. Heart muscle is especially efficient at extracting energy from lactic acid by first converting it to pyruvate.

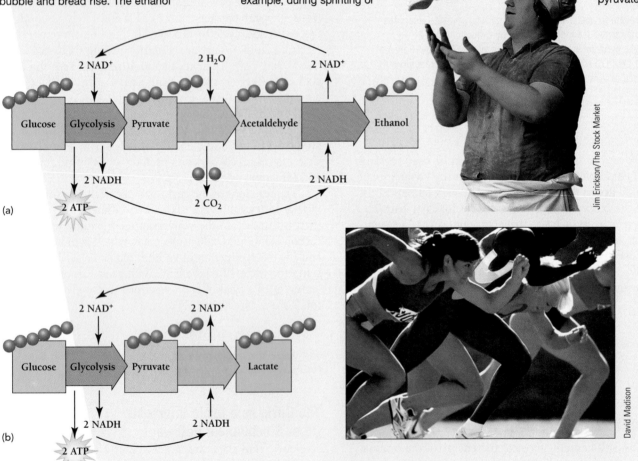

Figure A

Two kinds of fermentation.   In each case, cells regenerate NAD+ for further glycolysis. (a) When yeast cells grow in limited oxygen (as in a fermentation vat or a bread dough), pyruvate forms acetaldehyde and ethanol. (b) In muscle cells that have exceeded their oxygen supply, pyruvate forms lactic acid (lactate).

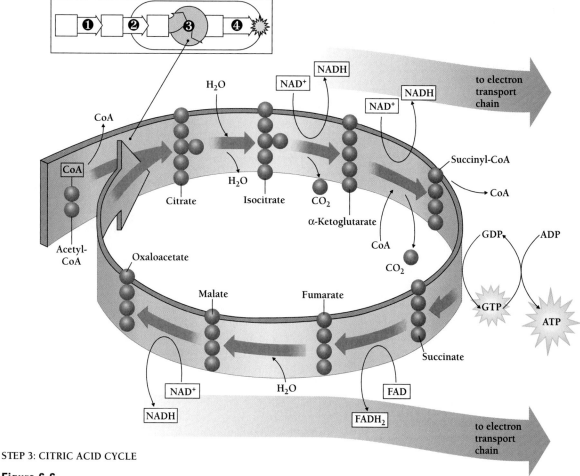

STEP 3: CITRIC ACID CYCLE

Figure 6-6

Step 3: the citric acid cycle. During the citric acid cycle, a single molecule of acetyl-CoA generates three NADH, one $FADH_2$, one GTP, and two carbon dioxide (CO_2) molecules.

Biology ⊘ Now™   Learn more about the citric acid cycle by clicking on this figure on your BiologyNow CD-ROM.

duces energy and usually involves oxidation—the removal of electrons and the production of NADH. Anabolism, on the other hand, uses up electrons, which renews NAD^+.

Figure 6-8 shows that different molecules enter the cellular respiration pathway at different points. Carbohydrates in such foods as bread are broken down into simple sugars that usually take the same path as glucose. Fats and oils are broken down into glycerol, which may enter glycolysis, and into fatty acids, which enter the pathway as acetyl-CoA. Proteins are broken down into component amino acids, which are then converted to pyruvate, acetyl-CoA, or other intermediates of the cellular respiration pathway.

Cells regulate the rates at which pathways or sets of pathways operate. High levels of ATP favor anabolism, and low levels favor catabolism. When ATP levels are high but the products of anabolism are abundant, the energy in food is stored for future use—as carbohydrates, proteins, or fats. Despite the importance of glucose and other carbohydrates in energy metabolism, the preferred form of energy storage in animals is fat.

The end product of the catabolism of fats is acetyl-CoA. Fats are more reduced compounds than sugars; that is, for each carbon atom, fats provide more electrons for oxidative phosphorylation, the most productive step of cellular respiration. Fats, therefore, contain more energy—

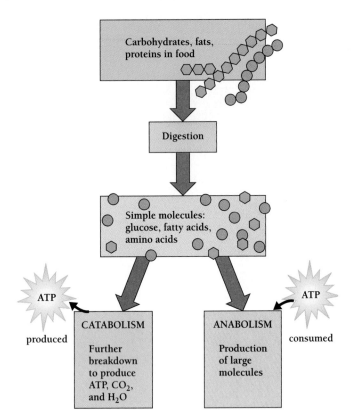

Figure 6-7

Metabolism consists of two pathways: catabolism and anabolism. Catabolism produces energy through the oxidation of molecules. Anabolism usually consumes energy in the process of building large molecules.

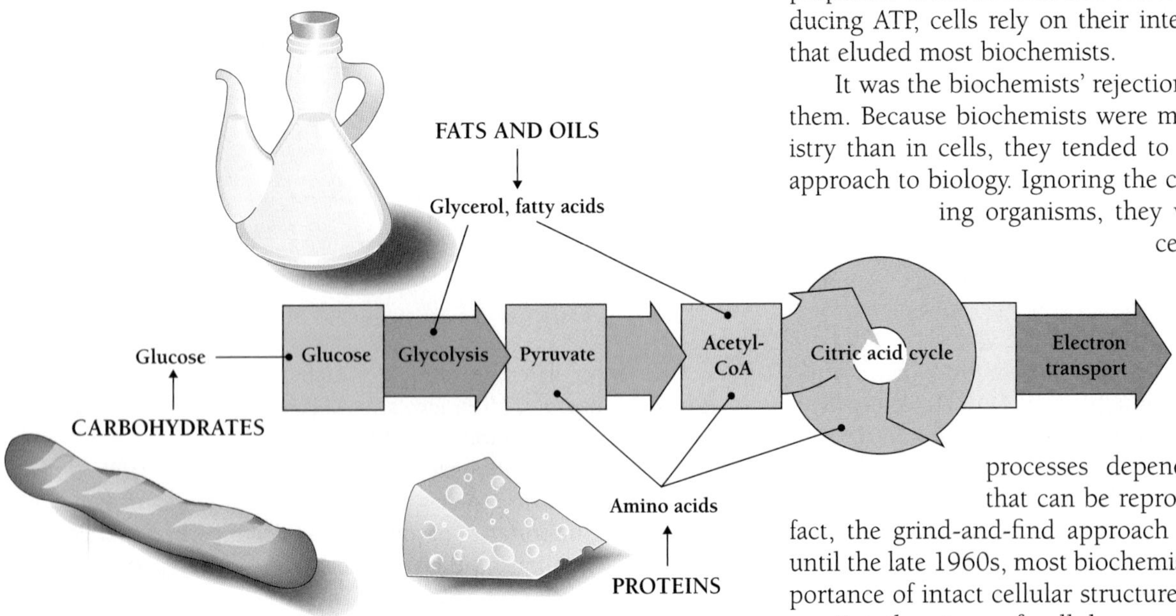

about 9.4 kcal per gram of fat, compared to 4.1 kcal per gram of carbohydrate (glycogen) and 5.6 kcal per gram of protein.

The citric acid cycle lies at the intersection of the catabolic and anabolic pathways. High levels of ATP favor anabolism, and low levels favor catabolism. When ATP levels are high, the most efficient way to store energy is in the form of fat.

6.3 How Is Glucose Energy Finally Converted to ATP?

A major goal of biologists and biochemists has been to learn how cells convert the energy of sunlight or food into the high-energy bonds of ATP. Until the early 1960s, biochemists believed that they would soon understand how organisms make ATP simply by studying chemical reactions catalyzed by enzymes in test tubes. Decades of ingenious research provided detailed maps of complex molecular transformations, many of which made or used ATP.

Yet researchers gradually realized that they were missing something fundamental. Most ATP in living cells was made in some other way. But how? Surprisingly, the first clue came not from the properties of enzyme reactions, but from the properties of membranes and subcellular organelles. In producing ATP, cells rely on their internal structure in a way that eluded most biochemists.

It was the biochemists' rejection of vitalism that misled them. Because biochemists were more interested in chemistry than in cells, they tended to take a "grind-and-find" approach to biology. Ignoring the complex structure of living organisms, they would simply grind up cells to find out what chemicals are in them. This approach has worked very well. Decades of research have shown that the majority of cellular processes depend on simple reactions that can be reproduced in a test tube. In fact, the grind-and-find approach was so successful that, until the late 1960s, most biochemists were blind to the importance of intact cellular structures.

But the steps of cellular respiration that release the most energy occur in functioning, intact mitochondria, not in test tubes full of enzymes. Cellular respiration depends so completely on mitochondria that biologists can instantly recognize the specific cells most involved in cellu-

Figure 6-8

How do other food molecules enter metabolism? Fat, carbohydrates, and proteins enter the cellular respiration pathway at different points. The busiest crossroad is at acetyl-CoA, which then enters the citric acid cycle. All fatty acids and many amino acids are converted into acetyl-CoA.

FATS AND OILS

Glycerol, fatty acids

Glucose → Glucose → Glycolysis → Pyruvate → Acetyl-CoA → Citric acid cycle → Electron transport

CARBOHYDRATES

Amino acids

PROTEINS

Health and Biology The Metabolism of Alcohol

For thousands of years (at least), ethanol has provided humans (and birds that eat fermented berries) with a popular means of changing mood and behavior. But too much ethanol can permanently damage cells in the liver, nerves, brain, and heart. The metabolic pathways described in this chapter help us to understand how this happens.

Alcohol accumulates in the blood because it moves from the intestines into the blood more rapidly than the body can eliminate it. The blood carries the alcohol to cells throughout the body. Nerve and brain cells are particularly vulnerable to damage by alcohol. Heavy alcohol use irreparably damages heart muscle after about 10 years in men, sooner in women.

The most common organic disease among heavy drinkers is liver failure. In humans, liver cells are practically the only cells that can metabolize ethanol.

In these cells, NAD^+ oxidizes ethanol, first to acetaldehyde and then to acetic acid, generating lots of NADH. The NADH donates its electrons to the electron transport chain to make ATP. But the citric acid cycle is out of the loop. The combination of the shutdown of the citric acid cycle and active oxidative phosphorylation results in strange-looking mitochondria.

Meanwhile, carbohydrates that would ordinarily enter the citric acid cycle are instead converted to fat. After a few years, the liver cells fill with fat and to cease to function. The dead cells create scarring in the liver—a condition called *cirrhosis*. Once the liver cells lose their ability to metabolize alcohol and other toxins, damage to the rest of the body accelerates.

---

lar respiration simply by noting which cells have the most mitochondria.

Biochemists' "grind-and-find" approach to cell biology prevented them from guessing the role of cellular organization in cell respiration.

Step 4: Electron Transport and Oxidative Phosphorylation

The process of cellular respiration transfers 24 electrons from glucose to oxygen. NAD^+ accepts 20 electrons and FAD accepts the other four. Altogether, about 90 percent of the energy stored in the chemical bonds of glucose goes to form the high-energy compounds NADH and $FADH_2$.

NADH and $FADH_2$, which have each received two electrons, ultimately pass their electrons to an oxygen atom, releasing energy that goes to make high-energy phosphate bonds in ATP. (The reduced oxygen atom, together with two hydrogen ions, forms a molecule of water, or H_2O).

The electrons from NADH and $FADH_2$ make their way to oxygen by way of a series of other molecules, called **electron carriers.** The pathway of the electrons (from one carrier to another) is called the **electron transport chain.** Biochemists compare the electron transport chain, which carries electrons to oxygen, to the "bucket brigade" that old-time firefighters used to get water to a fire (Figure 6-9D). Each electron carrier passes its electrons to the next carrier in the line. A reduced carrier becomes oxidized as it gives up its electrons, and the next carrier becomes reduced as it receives electrons. In eukaryotes, the transfer occurs in the mitochondrial membrane. In prokaryotes, these reactions occur in association with the cell membrane.

In the 1920s, biochemists discovered that some of the "firefighters" in the electron transport chain were pigment proteins called **cytochromes** [Greek, *kytos* = cell + *chroma* = color], whose color changes as they accept or donate electrons. The cytochrome electron carriers accept electrons from NADH, then pass these electrons along through a series of electron carries to oxygen.

Cellular respiration transfers electrons from glucose to either NAD^+ or FAD, forming NADH and $FADH_2$, which then release their electrons to oxygen by way of the electron carriers in the electron transport chain.

How Do Cells Harvest the Energy of Electron Transport?

Although oxygen's acceptance of electrons is a downhill (energy-releasing) reaction, making high-energy phosphate bonds to form ATP from ADP is definitely an uphill (energy-consuming) process. It is the downhill flow of electrons that drives the uphill production of high-energy bonds in ATP, a process called "oxidative phosphorylation." Oxidative phosphorylation and the electron transport chain together make up Step 4 of cell respiration.

How do cells use the energy released by the flow of electrons in the electron transport chain to make ATP? Once biochemists understood the electron transport chain, they began looking for some intermediate molecules that make ATP. They spent decades looking for these mysterious "intermediates," but with no luck. They found that an electron transport chain in a test tube could pass electrons, but it would generate no ATP. Intact mitochondria (isolated from cells) could transport electron transport *and* make ATP, but as soon as

Figure 6-9

Step 4: the electron transport chain is like a bucket brigade. A. A single mitochondrion. B. The electron transport chain is embedded in the inner membrane of the mitochondrion. The electron transport chain pumps protons out of the matrix, so the matrix has a lower concentration of protons than the intermembrane space. C. The electron transport chain consists of a series of protein complexes that pass electrons from one to another. As the electrons move along the chain, energy becomes available to pump protons out of the matrix, which creates an electrochemical gradient. D. The electron transport chain passes electrons in the same way that the members of an old-fashioned fire brigade pass buckets of water. Instead of passing buckets of water from a well to a fire, however, electron carriers pass electrons that ultimately come from glucose to NADH and $FADH_2$ and finally to oxygen.

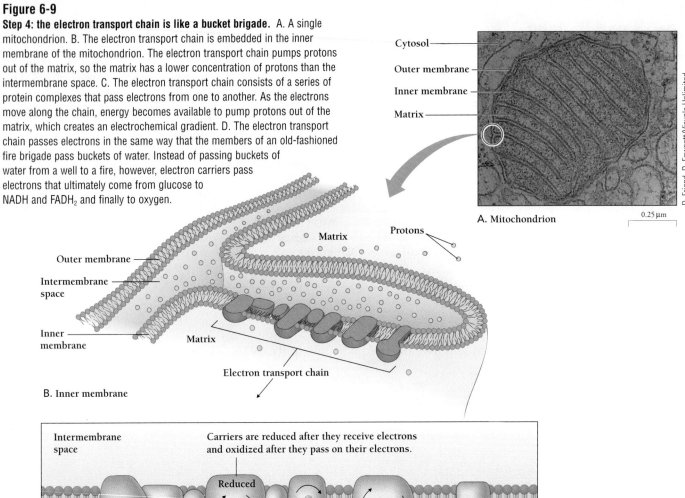

Cytosol
Outer membrane
Inner membrane
Matrix

A. Mitochondrion

0.25 μm

D. Friend, D. Fawcett/Visuals Unlimited

Outer membrane
Intermembrane space
Inner membrane
Matrix
Matrix
Protons
Electron transport chain

B. Inner membrane

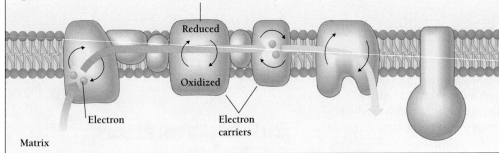

Intermembrane space

Carriers are reduced after they receive electrons and oxidized after they pass on their electrons.

Reduced

Oxidized

Electron

Electron carriers

Matrix

C. Electron transport chain

Glucose

Oxygen

D. Bucket brigade

biochemists broke up the mitochondria to see how they worked, the little organelles refused to make anymore ATP.

The biochemists were beside themselves with frustration. These experiments showed that ATP synthesis depended on the structure of the mitochondria. We know that a telephone works only if it is put together properly; a pile of loose wires and connectors in a cardboard box won't help you talk to your friends. In the same way, a lot of loose molecules in a test tube won't necessarily do what they do inside a cell or organelle. This might seem obvious, but at one time biochemists hoped that cell processes would all be simple chemical reactions that could be reproduced in a test tube.

The puzzle was solved in the early 1960s by the English biochemist Peter Mitchell. Mitchell had been studying the way bacteria pump hydrogen ions, also called protons, across their membranes. The inside of a bacterium has a higher pH (fewer protons) than its outside environment, a difference called a **proton gradient.**

Mitchell showed that bacteria could use the energy in this proton gradient to transport other molecules and ions across the membrane and into the cell. In 1961, he suggested that mitochondria might be able to use a proton gradient to make ATP. His idea was thoroughly unconventional, but Mitchell was an independent thinker who also did all his research at home in his country manor house.

Mitchell proposed a two-part hypothesis. He suggested first that the flow of electrons through the electron transport chain created a proton gradient across the mitochondrial membrane. Second, he argued, some machinery in the membrane captured the potential energy in the proton gradient by using the energy to make ATP. Mitchell called his proposed mechanism for making ATP—the harnessing of the energy stored in the chemical gradient created by the electron transport chain—**chemiosmosis** [Greek, *osmos* = to push]. His hypothesis implied that the intermediate molecules biochemists had been searching for did not exist.

Mitchell, his collaborator, Jennifer Moyle, and others later showed that chemiosmosis could also explain how chloroplasts capture light energy. Mitchell's radical hypothesis, called the chemiosmotic hypothesis, became the basis for understanding the harvesting of energy in both cell respiration and photosynthesis.

The English biochemist Peter Mitchell proposed that mitochondria convert energy from the electron transport chain into ATP by harnessing the energy in a proton gradient.

How Do Mitochondria Generate a Proton Gradient?

Recall from Chapter 4 that a mitochondrion has both an inner and outer membrane. Between the two membranes is a space called the intermembrane space. And inside the inner membrane is another space, the **mitochondrial matrix** (Figure 6-9A).

The mitochondrial matrix makes up about two-thirds of the volume of a typical mitochondrion, and it is where most of a cell's NADH and $FADH_2$ is made. Embedded in the inner membrane is the electron transport chain. As the electron carriers (the firefighters) pass electrons from one to another, they cause protons to move across the inner membrane from the matrix to the intermembrane space, creating a proton gradient across the inner membrane (Figure 6-9B).

The electron transport chain embedded in the inner membrane of the mitochondrion pumps protons out of the matrix and generates a proton gradient.

What Pumps Protons Out of the Mitochondrial Matrix?

By taking apart the inner membrane, biochemists found four protein complexes (Figure 6-10). As protons leave the matrix, its pH increases to about pH 8. But the protons that arrive in the intermembrane space easily pass through the outer mitochondrial membrane into the cytosol, so the intermembrane space has the same neutral pH as the bulk of the cytosol, about pH 7.

Because protons have a positive charge, their movement out of the matrix leaves behind an excess of ions with negative charges, which creates an electrical gradient as well as a pH gradient. The proton pump generates an **electrochemical gradient,** a double gradient composed of a *chemical* gradient (the difference in proton concentration) and an *electrical* gradient (the difference in charge).

The electron transport chain pumps protons out of the matrix, generating an electrochemical gradient.

How Does the Flow of Protons Back into the Matrix Cause the Synthesis of ATP?

The electrical and chemical gradients both make protons tend to return to the matrix—from an area of high concentration to an area of low concentration and from an area of net positive charge to one with a net negative charge. But once protons leave the matrix, there is only one way back in, through special channels in the inner membrane (Figure 6-10). As the protons flow through these channels back into the matrix, a protein complex called **ATP synthase** uses the flow of protons to drive the synthesis of ATP. ATP synthase works remarkably like a turbine in a hydroelectric dam on a river, which converts energy from the channeled flow of water to electrical energy. ATP synthase converts the energy in a channeled flow of protons into the chemical energy of ATP (Figure 6-11).

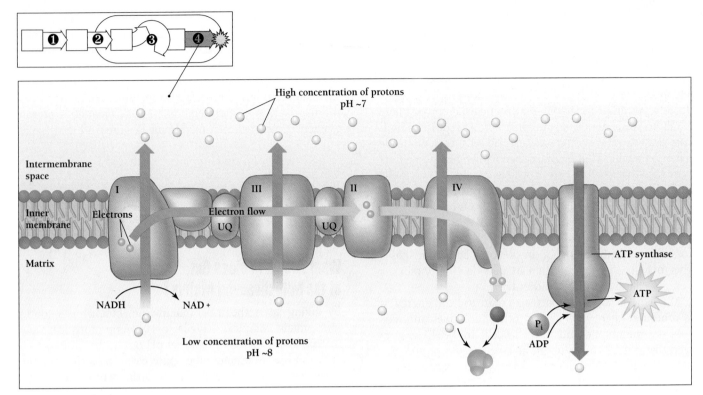

High concentration of protons
pH ~7

Intermembrane space

Inner membrane

Matrix

I

Electrons

Electron flow

UQ

III

UQ

II

IV

ATP synthase

ATP

NADH   NAD +

P_i

ADP

Low concentration of protons
pH ~8

STEP 4: ELECTRON TRANSPORT CHAIN

Figure 6-10

More details of the electron transport chain. Each electron carrier is alternately reduced and oxidized as it gains and loses electrons. As the electrons pass, some of their energy goes to pump protons, or hydrogen ions, out of the matrix. The electrons passing through the chain ultimately combine with oxygen (and two protons) to form water.

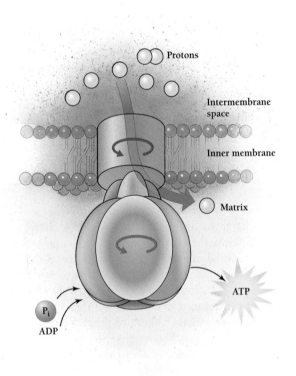

Protons

Intermembrane space

Inner membrane

Matrix

ATP

P_i

ADP

A.

B.

Figure 6-11

ATP synthase. A. A proton gradient across the inner mitochondrial membrane causes ATP synthase to spin and make ATP. Protons flow through the cylinder embedded in the inner membrane, causing the blades sticking out of the membrane to rotate and generate ATP. B. Hydroelectric turbine generators. Water flowing under high pressure from behind a dam makes these turbines spin, generating electricity. Just as a turbine's spinning blades generate electricity, ATP synthase's spinning blades generate ATP.

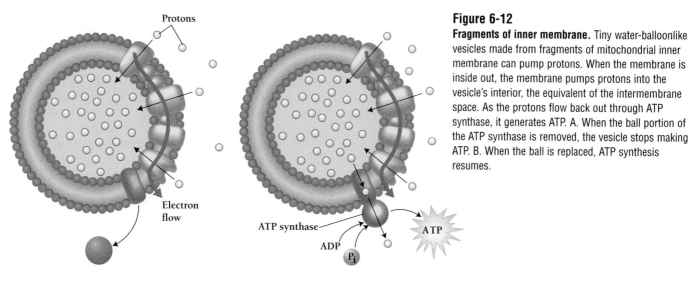

Protons

Electron flow

A. ATP synthesis halts.

ATP synthase

ADP

P$_i$

ATP

B. ATP synthesis resumes.

Figure 6-12

Fragments of inner membrane. Tiny water-balloonlike vesicles made from fragments of mitochondrial inner membrane can pump protons. When the membrane is inside out, the membrane pumps protons into the vesicle's interior, the equivalent of the intermembrane space. As the protons flow back out through ATP synthase, it generates ATP. A. When the ball portion of the ATP synthase is removed, the vesicle stops making ATP. B. When the ball is replaced, ATP synthesis resumes.

ATP synthase looks something like a miniature lollipop sticking out of the inner membrane into the matrix (Figure 6-11). Biochemists have found that each ATP synthase complex contains two major components: (1) a hollow "stick," which runs through the inner membrane, and (2) a round top that sticks into the matrix. The protons flow through the stick into the matrix. As the protons come through, the round top actually spins like a turbine and makes ATP.

Removing the ball portion of ATP synthase prevents it from making ATP, even though protons continue to flow through the stick portion. Replacing the spinning turbine allows it to resume making ATP (Figure 6-12). In Chapter 7, we will see that proton gradients not only produce ATP in mitochondria but also produce ATP in chloroplasts during photosynthesis.

> As protons flow through ATP synthase back into the matrix, ATP synthase uses the energy of the flowing protons to make ATP.

Key Concepts

- Cellular respiration converts the chemical energy of food molecules into the chemical energy of ATP.
- Almost all cells can use glycolysis to derive energy, but cellular respiration provides much more energy.
- ATP production from cellular respiration depends on oxidative phosphorylation, which couples electron transport to the production of high-energy phosphate bonds in ATP in the mitochondria.

Summary with Key Terms

How do organisms supply themselves with energy?

Plants and other photosynthesizing organisms that produce their own food from sunlight are called **autotrophs**. Organisms that get their energy from other organisms are **heterotrophs**. Heterotrophs prepare for cellular respiration by capturing macromolecules and digesting them into their component building blocks.

Almost all eukaryotic cells and many prokaryotic cells are **aerobic**, producing most of their ATP by **cellular respiration**. All the chemical reactions of cell respiration are called **metabolism**. Most energy-producing reactions in cells transfer electrons from the **reducing agent** to the **oxidizing agent**.

How do organisms extract energy from glucose?

Cellular respiration extracts 24 electrons from the breakdown of glucose into carbon dioxide and water and transfer them to oxygen. For 20 of these electrons, the first acceptor is **NAD$^+$**, which can accept two electrons and a hydrogen atom to give its reduced form **NADH.** The other four electrons of glucose are transferred to **FAD**, whose reduced form is **FADH$_2$**. These reduced **coenzymes**—the major energy storage molecules in energy metabolism—ultimately transfer their electrons to oxygen in a series of steps. In eukaryotes, electron transport occurs within the inner mitochondrial membrane. In prokaryotes, these reactions occur in association with the plasma membrane.

Glycolysis is the first step in the metabolism of glucose. Glycolysis is a set of 10 chemical reactions that breaks glucose into two molecules of **pyruvate.** These 10 reactions also convert some of the energy of glucose into two high-energy ATP bonds.

Microorganisms obtain energy from **anaerobic** glycolysis, or **fermentation.** Yeast cells, for example, can avoid the need for oxygen by recycling NADH back to NAD by reducing pyruvate to ethanol.

In eukaryotic cells, energy metabolism (other than glycolysis) takes place in the mitochondria. Pyruvate enters the mitochondria, where cellular respiration proceeds. Pyruvate then undergoes an oxidation reaction, in which NAD^+ serves as the electron acceptor. In acetyl-CoA, acetate is linked, via a high-energy bond, to a molecule called **coenzyme A (CoA).**

The **citric acid cycle,** or **Krebs cycle,** converts the carbon atoms of acetyl-CoA to carbon dioxide. The citric acid cycle is a nearly universal set of reactions. The citric acid cycle adds the four-carbon molecule oxaloacetate (OAA) to two-carbon acetyl-CoA to form a six-carbon molecule, from which two carbons are removed in succession. OAA then reenters the cycle once more.

The citric acid cycle, which not only provides energy but also serves as the major route for the synthesis of many small molecules, lies at the center of the complex network of biochemical conversions that collectively are called **metabolism.** Metabolism is divided into two classes of pathways: **catabolism** and **anabolism.**

How is the energy in glucose used to make ATP?

Oxidative phosphorylation couples the oxidation of NADH and $FADH_2$ to the production of high-energy phosphate bonds in ATP. The electrons from NADH and $FADH_2$ reach oxygen,

their ultimate acceptor, through the **electron transport chain** (or respiratory chain), which consists of a series of **electron carriers.** Five of the electron carriers are **cytochromes.**

Modern biologists reject **vitalism,** the belief that living systems have powers beyond those of nonliving systems. But the production of ATP by the **mitochondria** presents an excellent example of how the chemical transformations of living organisms depend on cellular structures.

Oxidative phosphorylation operates by a mechanism (called **chemiosmosis**) summarized in the chemiosmotic hypothesis: (1) the flow of electrons through the electron transport chain pumps protons (hydrogen ions) out of the **mitochondrial matrix,** establishing an **electrochemical gradient** across the mitochondrial membrane; and (2) molecular machinery in the membrane captures the energy, as protons flow back into the mitochondrial matrix.

The orientation of the electron carrier complexes establishes the direction of the **proton gradient.** As the protons flow down the electrochemical gradient through channels in the inner membrane, their free energy decreases. The protein complex ATP synthase uses the free energy from the flow of protons to make ATP.

Review and Thought Questions

Review Questions

1. Write the overall reaction for breaking down glucose.
2. Name the process in which most of the energy from glucose is converted to ATP. What role does oxygen

play in this process? What gas is produced as a waste product? Where in a cell do these reactions occur?

3. Why is the electron transport pathway divided into so many small reactions?

4. Across what eukaryotic membrane is a gradient established during cellular respiration? What is unequally distributed across these membranes? What caused this unequal distribution?

5. Describe and name the three steps that aerobic organisms use to extract energy from glucose.

6. How many carbon atoms does a glucose molecule have? Name the end product molecule that is produced during glycolysis. How many carbon atoms does it have?

7. How many ATPs are produced in the citric acid cycle? How does this number compare with that produced in glycolysis?

8. Define metabolism, anabolism, and catabolism, and give examples of catabolic and anabolic pathways. Which type of metabolism is most favored when a cell (a) has abundant ATP? (b) contains little ATP?

Thought Questions

9. Why might you lose weight (a lot of weight) if your mitochondria suddenly lost the ability to couple electron transport to the production of ATP?

10. When you exercise heavily, you consume oxygen faster than you can take it in through your lungs and circulatory system, so your cells have to rely at least partly on anaerobic processes to extract energy from food molecules. Which of the three steps in energy extraction can your cells not perform under these conditions? How much ATP can you form from food molecules under these conditions?

BiologyNow Resources

Biology ⓔ Now™

Active Figures
6-4:   Glycolysis
6-6:   The citric acid cycle

Preparing for an exam? Take a diagnostic test on your BiologyNow CD-ROM.

Online materials relating to this chapter are at:
http://biology.brookscole.com/AAL3

About the Chapter-Opening Image
Louis Pasteur proved that the fermentation of wine and beer depends on the presence of microorganisms (yeasts). Twentieth-century biochemists figured out just what the yeasts were doing.

Photosynthesis: How Do Organisms Get Energy from the Sun?

Key Questions

- How do plants use solar energy to make sugar?

- How do plants cope with too little water or carbon dioxide?

The Chemical Evangelist

Early in the 17th century, a Belgian alchemist named Johann Baptista van Helmont (1579–1646) performed a simple and dramatic experiment that scientists have been admiring for nearly 400 years. Although van Helmont dared not publish his results, his studies laid a firm foundation for the modern understanding of **photosynthesis** [Greek, *photos* = light + *syntithenai* = to put together]—the process that plants and other organisms use to transform the energy of light into the chemical bonds of sugars.

Van Helmont was a man who, even in our time, would be considered highly unusual. He was deeply religious, nearly fanatical in his beliefs. Yet he was a born skeptic, always questioning the beliefs of those around him. It was an attitude that would bring upon him the unremitting fury of the Spanish Inquisition.

As a teenager, van Helmont found his education at the Catholic University of Louvain so disappointing that he refused the title "Master of Arts" because he said he had learned nothing substantial or true. On leaving the university, he then refused a well-paid position as a clergyman because, he said, he opposed "living on the sins of the people." For a while, he tried living the austere life of a Christian Stoic, until the hardship made him ill.

Perhaps illness inspired the young van Helmont, for he then began to study medicine, with particular emphasis on herbal remedies. Again he was disappointed, however. The herbalists, he complained, had discovered nothing new about medicine since ancient Greek times, and argued endlessly over details that had nothing to do with making people well. When van Helmont was 20 years old, he accepted a medical degree from Louvain. Yet he never practiced medicine, believing that he had learned nothing that would actually help anyone who was sick.

His new interest was alchemy, a result, he later wrote, of a call from God. (Some called him the "chemical evangelist.") He had frequently claimed to detest money, but at 28 he married a woman so wealthy that he was freed from ever having to work for a living again. For 15 years, he happily pursued his own researches into alchemy, religious philosophy, medicine, and natural science.

He found, for example, that humans digest food with acid and bile. He recognized gas as a form of matter, defining it as the volatile part of a substance, distinct from the mixture of gases we call air. He recognized that the lung is an organ for gas exchange. By weighing reactants and products, he showed that matter is conserved during chemical reactions.

Van Helmont wrote profusely of his discoveries and investigations, never hesitating to criticize the thinking of others. He especially singled out physicians—for their ignorance and superstition—and the Jesuits, a powerful order of Catholics—for what he viewed as their ignorance and arrogance.

Van Helmont was smart enough not to publish his offensive tirades against physicians and Jesuits. But, eventually, one of his many enemies published one of the worst. The faculty at the Louvain medical school in 1623 denounced the pamphlet as "monstrous." And two years later, the Spanish Inquisition declared that 27 statements in van Helmont's manuscript were almost certainly heretical. Between 1624 and 1642, the Spanish Inquisition imprisoned and interrogated van Helmont, then locked him up in a convent, and finally confined him to house arrest for many years. During all of this time, van Helmont published nothing. Yet he continued to work and, equally important, to write.

Much of van Helmont's work was directed at undermining the teachings of the Jesuits. The Jesuits based their teachings on the writings of Aristotle, the most influential of the Greek philosophers. Aristotle had argued that plants derive their sustenance from the soil. A plant's roots, Aristotle said, extract materials from soil in much the same way that an animal's stomach extracts materials from its food. This view seemed logical: farmers had long known that even the richest soils eventually lose the ability to support luxuriant plant growth.

But van Helmont knew Aristotle was wrong. Anyone could see that a potted plant didn't use much soil; it used water. Besides, van Helmont believed that all natural bodies come from water. To knock Aristotle from his pedestal and bolster his own theory, van Helmont performed a now famous experiment. He planted a five-pound willow sapling in 200 pounds of soil, which he had carefully dried in a furnace, weighed, and then placed in a large pot (Figure 7-1). He covered the pot with an iron plate to keep dust from falling into it and watered the growing tree as needed. He did not weigh the leaves that fell from the tree each autumn. After five years, he again weighed both the tree and the soil. The willow tree now weighed 169 pounds, while the soil weighed only 2 ounces less than the starting 200 pounds. Van Helmont concluded, "164 pounds of wood, bark, and roots arose out of water alone."

Van Helmont's experiment was excellent, but his interpretation was partly mistaken. His experiment did prove that Aristotle (and the Jesuits) were wrong: the tree's mass could not have come from the soil. However, van Helmont himself was also wrong in assuming that his ex-

Figure 7-1

Plants are made of water and air. Van Helmont's willow tree experiment showed that the material that makes up plants does not come from the soil. After five years, the willow tree gained 164 pounds, but the soil lost only a few ounces.

periment showed that the tree came entirely from water. It was ironic that van Helmont, the discoverer of gas, never guessed that something in the air might also have contributed to the growth of his willow tree. Only years after he died did other scientists discover the role of gas in plant growth.

7.1 How Do We Know How Plants Obtain Carbon and Oxygen?

The first suggestion that the atmosphere contributed to the stuff of plants came not from chemists or alchemists but from microscopists. With the new microscopes developed in the late 17th century, an Englishman and an Italian almost simultaneously discovered minute pores in the leaves of plants, called **stomata** [Greek, *stoma* = mouth]. Stomata

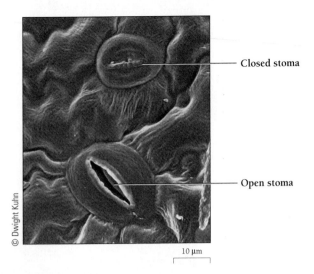

Figure 7-2
Stomata from the underside of a leaf. Tiny pores in the surfaces of leaves called stomata open if a plant needs to collect carbon dixoide and close if a plant needs to conserve water. The discovery of stomata in the surfaces of leaves suggested that plants need gases from the air as well as water from the ground. Scientists concluded that "vegetation is planted in the air as well as in the earth."

10 μm

appeared to allow air to pass to the interior of the leaves (Figure 7-2). The activities of plants, they speculated, must somehow depend on interactions with air. By the mid-18th century, scientists had concluded that "vegetation is planted in the air as well as in the earth."

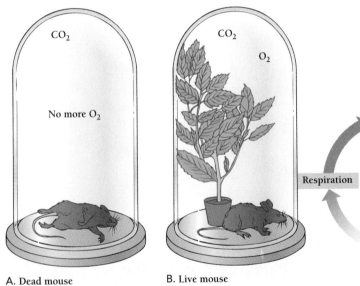

A. Dead mouse   B. Live mouse

CO₂
Sunlight
CO₂ and H₂O
Respiration
Photosynthesis
Glucose O₂

Figure 7-3
What did Priestley's experiment show? Plants put something into air that mice need. A. The mouse in the jar uses up the oxygen and fills the jar with carbon dioxide, which suffocates the mouse. B. A photosynthesizing plant placed in the jar takes up the carbon dioxide released by the mouse and replaces it with oxygen. The mouse is fine.

The Granger Collection, New York

Figure 7-4
Joseph Priestley. Even as modern chemistry took its first breath, its greatest exponents—Joseph Priestley and Antoine Lavoisier—were caught up in the turmoil surrounding the American and French revolutions. In 1791, Priestley's sympathies for the French Revolution inspired an angry mob to destroy his home, library, and laboratory. Within three years, he fled to the United States. In the same year, on the other side of the English Channel, Lavoisier, who had served King Louis XVI, was guillotined.

What Do Plants and Air Do for Each Other?

The next big clue to the interaction between plants and air came from a series of experiments in which a mouse or a candle was placed in an airtight container. After some time, the mouse would die from lack of oxygen or the candle would go out for the same reason. This was because the mouse, or the burning candle, consumed the oxygen in the container.

In the 1770s, the English clergyman Joseph Priestley performed an experiment that showed that a sprig of mint could restore the air in the jar and keep a mouse alive. Priestley showed that plants put back what living animals (or burning candles) take out (Figure 7-3). That "something" that plants release is the gas oxygen. Around the same time, the French chemist Antoine Lavoisier independently discovered oxygen, so credit for the discovery usually goes to both men (Figure 7-4).

Like animals, plant cells respire, consuming oxygen and releasing carbon dioxide (Figure 7-5A). In light, however, plants can put back the oxygen and remove something very different—carbon dioxide. When plant cells are photosynthesizing, using the energy of sunlight to make sugar, they

Figure 7-5
Photosynthesis. A. Like animals, plant cells respire, using oxygen to break sugars into carbon dioxide and water. But unlike animals, plants can reverse this reaction. B. In light, plants also photosynthesize, using solar energy to make sugar from carbon dioxide and water.

Biology ⦿ Now™ Learn more about photosynthesis and the forms of plant metabolism by clicking on this figure on your BiologyNow CD-ROM.

remove carbon dioxide from the air and release oxygen as a waste product (Figure 7-5B).

Today we know that during photosynthesis plants take in water and carbon dioxide with which they synthesize the simple sugar glucose. To a large extent, then, plants are made of water and air. However, as we saw in Chapter 3, proteins and most other biological molecules also include atoms of nitrogen, phosphorus, and other elements. These come, in small quantities, from the soil. Indeed, the 2 oz of soil that van Helmont measured most likely included nutrients essential to the growth of the willow tree.

In this chapter, we will ask how cells accomplish photosynthesis. How does photosynthesis capture light energy? How does it store the captured energy? What molecules and subcellular structures participate?

During photosynthesis, plants take carbon dioxide from the air and water from the soil and release oxygen back into the air.

Where Do the Atoms Go in Photosynthesis?

Now that we know the raw materials of photosynthesis, we can summarize the overall chemical transformation in the following equation:

$$6\ CO_2\ +\ 6\ H_2O \rightarrow C_6H_{12}O_6\ +\ 6\ O_2$$
carbon dioxide + water → glucose + oxygen

Notice that this is the exact reverse of the overall equation for respiration given in Chapter 6 on page 111. Another way of writing the equation for photosynthesis is:

$$CO_2\ +\ H_2O \rightarrow C(H_2O)\ +\ O_2$$
carbon dioxide + water → carbohydrate + oxygen

Here, $C(H_2O)$ is the general formula for a carbohydrate, with each carbon atom associated with the equivalent of one molecule of water.

To early 20th-century chemists, this equation suggested a hypothesis for how photosynthesis might occur: light splits the carbon dioxide into oxygen and carbon. Then oxygen is expelled into the atmosphere, and the carbon combines with water to form carbohydrate. Although this idea is appealing, it is wrong.

The first realization that this view of photosynthesis was wrong came from studies of photosynthetic bacteria. In the early 1930s, a graduate student at Stanford University named Cornelius van Niel was examining photosynthesis in bacteria that used hydrogen sulfide (H_2S) instead of water (H_2O). Instead of producing oxygen as a by-product, these bacteria produce sulfur (Figure 7-6).

$$CO_2\ +\ 2\ H_2S \rightarrow C(H_2O)\ +\ 2\ S\ +\ H_2O$$
carbon dioxide + hydrogen sulfide → carbohydrate + sulfur + water

Van Niel reasoned that these bacteria probably used the same biochemical pathways as other photosynthesizing organisms. Water (H_2O) and hydrogen sulfide (H_2S) probably served the same biochemical role. Yet, the sulfur produced

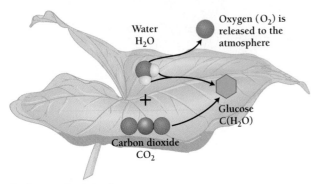

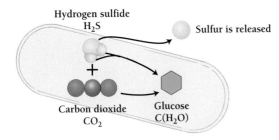

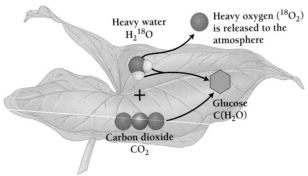

A. Photosynthesis in plants

B. Photosynthesis in bacteria

C. Photosynthesis in plants using heavy water

Figure 7-6
Where do the oxygen atoms come from? A. The oxygen released by plants during photosynthesis comes from water, not from carbon dioxide. We know this because: B. First, the sulfur released by certain bacteria during photosynthesis comes from hydrogen sulfide, which is chemically analogous to water. C. Second, in plants supplied with "heavy water," the heavy oxygen released during photosynthesis clearly comes from the water.

by photosynthetic bacteria clearly comes from splitting hydrogen sulfide. It stood to reason, then, that the oxygen produced by plants might come from splitting water, not carbon dioxide.

However, van Niel had no evidence to support this hypothesis. No one knew which oxygen atoms came from which molecules. More than 10 years passed before scientists found a way to distinguish between the oxygen molecules in water and those in carbon dioxide. The key was the isotope ^{18}O, an oxygen atom that is slightly heavier than usual. Biochemists provided plants with normal carbon

dioxide and so-called heavy water, $H_2^{18}O$. The plants split the heavy water and produced heavy oxygen, but the carbohydrates were normal.

$$CO_2 + 2 H_2^{18}O \rightarrow C(H_2O) + H_2O + {}^{18}O_2$$
carbon dioxide + water → carbohydrate + oxygen

This experiment showed conclusively that plants split water, not CO_2 (Figure 7-6C).

The oxygen released by plants comes from water, not from carbon dioxide.

Biochemists now had a clear understanding of what raw materials plants used to synthesize glucose. But two difficult questions still remained: How do plants channel light energy into this chemical reaction? And how do plants use that chemical energy to synthesize sugar?

Photosynthesis consists of two parts. In the first part, energy from sunlight powers the addition of a phosphate group to ADP to form ATP. Adding a phosphate group is called "phosphorylation." Because light comes in packets of energy called "photons," using light energy to phosphorylate some-

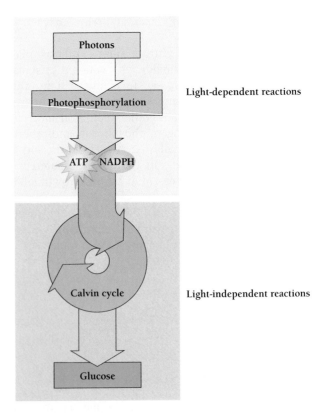

Figure 7-7
The two parts of photosynthesis. The light-dependent reactions, called photophosphorylation, generate energy-transfer molecules such as ATP. The light-independent reactions use the energy packaged in the light-dependent reactions to make glucose.

thing is called **photophosphorylation.** Sometimes plants also use light energy to reduce NADP⁺ to NADPH, another energy-storage molecule. Both kinds of reactions are called photophosphorylation, or **light-dependent reactions.**

The second part of photosynthesis uses the energy from the light-dependent reactions (ATP and NADPH) to make glucose. Because this second set of reactions uses no photons, they are called **light-independent reactions.** In the rest of this chapter, we will review both the light-dependent reactions and the light-independent reactions (Figure 7-7).

7.2 How Do Plants Collect Energy From the Sun?

Julius Mayer, one of the discoverers of the First Law of Thermodynamics, elegantly summarized the business of photosynthesis: "Nature has put itself the problem of how to catch in flight light streaming to the earth and to store the most elusive of all powers in rigid form."

What Is Light?

In the morning, sunlight streams in your window, shining on the curtains, the furniture, and the floor. All of these objects absorb sunlight and transform it into heat. In photosynthesis, plant cells absorb light and transform it into chemical energy. To understand how plants use the energy in light, we need to understand what light is.

Light is the energy of particles, called **photons,** that travel at the speed of light (186,000 miles per second). Like other moving particles, photons travel in straight lines. They can hit a surface and bounce off at an angle (Figure 7-8). Light can also vary in intensity. The brightness of the light

Figure 7-8
Light is made of particles. Particles of light called photons move in straight lines and bounce off surfaces in the same way as marbles or billiard balls.

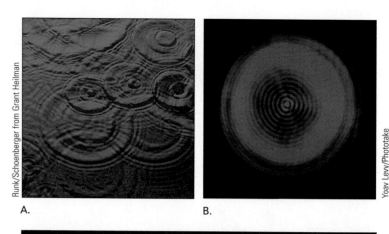

A.   B.

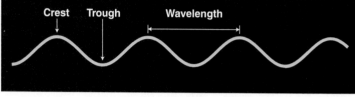

C.

Figure 7-9
Light is a wave. A. The circular ripples created by raindrops on a pond make a pattern of troughs and crests. B. Light also forms waves with crests and troughs. C. Wavelength is the distance from the crest of one wave to the next. Light with a long distance from one crest to the next (long wavelength) is red. Light with a short wavelength is blue.

shining on the pages of this book corresponds to the number of photons hitting the page each second. If you turn on a bright light, you will increase the total number of photons hitting the page.

Photons all travel at the same speed, but they nevertheless have different amounts of energy. How much energy depends on their wavelength, for light behaves like a wave as well as a particle. Light has troughs and crests that interfere with one another, just like the ripples in a rain puddle (Figure 7-9A and B). (The crest is the highest point of a wave and the trough is the lowest point.)

One of the most important properties of light is its **wavelength**—the distance between successive crests (Figure 7-9C). Light energy with a long wavelength has long troughs and crests; light with a short wavelength has short troughs and crests. The shortest wavelengths are measured in nanometers (nm). A nanometer is a billionth of a meter. Wavelengths vary in length from as short as millionths of nanometers, which we call gamma rays, to meters or even kilometers, which we call radio waves.

Most light, including sunlight and the light from most lightbulbs, is a mixture of many wavelengths (Figure 7-10). But our eyes and brains allow us to see only those wavelengths between 390 nm and 760 nm—what we call **visible light** (Figure 7-11). Within this narrow range, we can distinguish all the different wavelengths of the rainbow. The

A. Sunlight is a mixture of many wavelengths

Light

Iris absorbs all colors except purple.

Tiger lily absorbs all colors except orange.

B.

shortest wavelengths of light appear to us as blue or purple, and the longest wavelengths appear red. For example, we see 400-nm light as violet, 500-nm light as blue-green, and 600-nm light as orange-red.

A photon is a package of energy. Unlike an electron or a proton, a photon has no mass. It is pure energy. The amount of energy in a photon depends on its wavelength. Photons of shorter wavelengths (more bluish) have more energy than photons of longer wavelengths (more reddish). Blue light, for example, has about twice as much energy per photon as red light. The energy, wavelength, and color of light are all measures of the same thing. (But light intensity, how bright the light seems to us, is different. Light intensity is a measure of numbers of photons, not the energy, or wavelength, of the individual photons.)

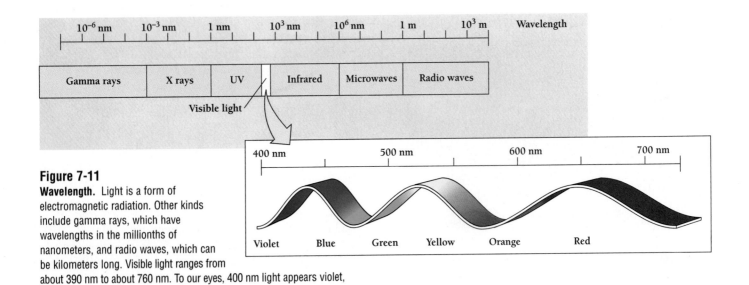

Figure 7-11
Wavelength. Light is a form of electromagnetic radiation. Other kinds include gamma rays, which have wavelengths in the millionths of nanometers, and radio waves, which can be kilometers long. Visible light ranges from about 390 nm to about 760 nm. To our eyes, 400 nm light appears violet, 500 nm is blue-green, and 600 nm is orange-red.

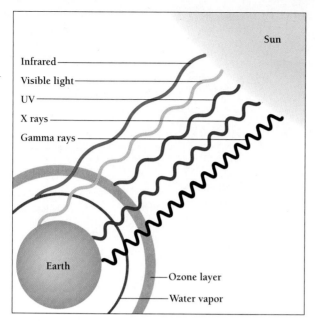

Figure 7-12

Our protective atmosphere. Because the Earth's atmosphere absorbs infrared light and reflects ultraviolet light, most sunlight that reaches the Earth's surface is visible light. Organisms have evolved biochemical pathways that allow us to use this light energy to build complex molecules.

Plants photosynthesize by taking advantage of the way light interacts with matter. Our own vision is the result of visible light exciting electrons in special molecules in our eyes. Visible light can also alter the chemicals in photographic film to create photographs or induce an electrical signal in a video camera or the photoelectric cell of a solar collector. Any atom or molecule can absorb a photon of light if the energy of the photon matches the energy needed to boost an electron to a higher energy level.

The wavelength of a photon is a measure of its energy. Short-wavelength photons have high energy, and long-wavelength photons have low energy. Photons with the right amount of energy (the right wavelength) can

increase the free energy of molecules by exciting their electrons.

In photosynthesis, light excites electrons in special molecules, increasing their free energy. The light-dependent and light-independent reactions capture some of this energy and convert it into high-energy chemical bonds.

How Can We Tell Which Molecules Help with Photosynthesis?

Much of the light from the sun does not reach the Earth's surface (Figure 7-12). High-energy ultraviolet ("beyond violet") light is absorbed by ozone in the upper atmosphere, and low-energy infrared ("below red") is absorbed by water vapor and carbon dioxide in the atmosphere. Not surprisingly, living organisms have evolved molecules that take full advantage of the energy in visible light—the light that best penetrates our atmosphere. The energy contained in a visible photon stimulates the chemical reactions of both photosynthesis (in plants) and vision (in our own eyes).

Different molecules absorb different wavelengths of light. For example, a red flower absorbs almost all the wavelengths in the visible range. The only ones the flower does not absorb are those that we see as red, which the flower's petals reflect into our eyes and elsewhere. Likewise, the green pigment in plants absorbs every color but green.

When any molecule absorbs a photon, the extra energy kicks an electron to a higher-energy orbital. The high-energy electron can return to the lower orbital by losing energy as another photon (called fluorescence), or as heat. Alternatively, the high-energy electron may move to another molecule, leaving the first molecule in an oxidized state. We will see that such a transfer of electrons drives photosynthesis.

The particular set of wavelengths that a molecule absorbs is called its **absorption spectrum.** A molecule's absorption spectrum tells us how much energy the molecule needs to move an electron from one orbital to another (Figure 7-13).

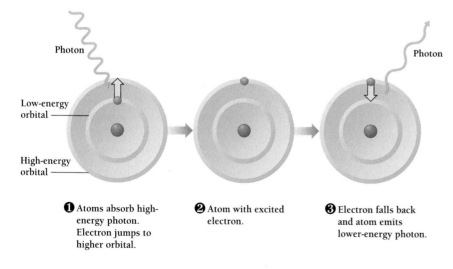

❶ Atoms absorb high-energy photon. Electron jumps to higher orbital.

❷ Atom with excited electron.

❸ Electron falls back and atom emits lower-energy photon.

Figure 7-13

Exciting electrons. ❶ We say an atom *absorbs* light when a photon excites an electron to a higher-energy orbital. ❷ The period of time when the atom is excited may be very brief. ❸ This temporary state of excitement ends when the electron falls back to a lower-energy orbital, releasing the energy as either light or as heat. Because a pigment molecule absorbs photons of only a certain range of wavelengths (or "spectrum"), biochemists say that every pigment has an "absorption spectrum."

Most of the molecules in an ordinary cell absorb ultraviolet light only. This is because the energy required to move an electron to a higher-energy orbital is usually quite high. But some molecules, called **pigments,** can absorb low-energy light from the visible spectrum. Any molecule that absorbs visible light is a pigment.

In plants, the most important molecule of photosynthesis is the green pigment **chlorophyll,** which absorbs every color but green and yellow. Plants use two kinds of chlorophyll—chlorophyll *a* and chlorophyll *b*, each with a slightly different absorption spectrum. A solution of pure chlorophyll absorbs photons with red and blue wavelengths, while photons with green and yellow wavelengths pass through or are reflected. These unabsorbed photons excite pigment molecules in our eyes, and we perceive a green color.

Every light-dependent process has an **action spectrum**—how well different wavelengths make the process happen. For example, in plants we can measure how well different wavelengths promote photosynthesis. Figure 7-14 shows that the rate of photosynthesis increases under blue and orange-red light and decreases under green and yellow light. The action

Figure 7-15
Autumn leaves along the Kancamagus Highway, in New Hampshire.
Some of the beautiful colors of autumn leaves result from the carotenoid pigments that help plants photosynthesize sugar. In the fall, trees and shrubs pull chlorophyll from their leaves. Left behind are the red and organge carotenoid pigments.

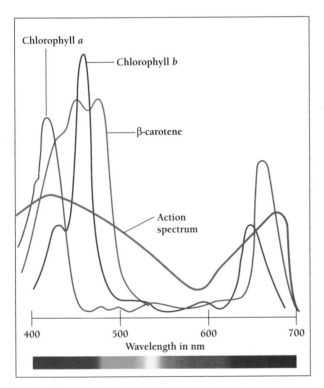

Figure 7-14
Absorption and action. Plants absorb light using two kinds of chlorophyll and carotenoids such as β-carotene. The set of wavelengths that a molecule absorbs is called its absorption spectrum. For example, chlorophylls absorb blue and red light but reflect green light. The range of wavelengths that promote photosynthesis is called the action spectrum. The action spectrum of photosynthesis roughly follows the combined absorption spectra of the chlorophylls and β-carotene.

spectrum of photosynthesis roughly follows the absorption spectrum of chlorophyll.

But, surprisingly, the action spectrum of photosynthesis does not correspond exactly to the absorption spectrum of chlorophyll *a*. Chlorophyll *a* reflects rather than absorbs blue-green light (500 nm). Yet blue-green light does promote photosynthesis, because plants have pigments that can take advantage of other wavelengths. These pigments include both chlorophyll *b* and the yellow and orange **carotenoids,** such as beta (β)-carotene (Figure 7-15). Carotenoids, such as β-carotene, absorb light that chlorophyll doesn't, then transfer the energy to chlorophyll.

In fact, chlorophyll molecules also transfer energy to one another. Most of the pigment molecules in a leaf are not directly involved in photosynthesis. Instead, hundreds of chlorophyll molecules, often associated with a few carotenoids, form a weblike **antenna complex** that traps light energy. The pigment molecules of the antenna complex transfer energy among themselves and ultimately send excited electrons to a **photochemical reaction center,** where a small number of chlorophyll molecules actually convert the captured light energy to chemical energy (Figure 7-17).

The action spectrum of photosynthesis corresponds to the absorption spectrum of chlorophyll and other pigments. Chlorophyll and other pigments in the antenna complex trap light energy and transfer it to the photochemical reaction center.

Extreme Biology   How Do Herbicides Kill Plants?

When we think of farmers and gardeners, we imagine them nursing tender seedlings into healthy plants. But growing plants also involves killing other plants, weeds that compete with whatever a farmer or gardener is trying to grow. A weed is simply a plant that is unwelcome for any reason. The easiest way to kill a weed, or any plant, is to spray it with herbicide, a chemical that kills plants. Farmers began using modern herbicides long before anyone understood how these chemicals hurt plants. In many cases, however, research has revealed precisely the metabolic processes in which herbicides act.

Some herbicides selectively kill different types of plants. For example, 2,4-D—a synthetic compound that mimics a plant hormone—kills C_3 plants more effectively than it kills C_4 plants. Many weeds are C_3 plants, while corn, a major U.S. crop, is a C_4 plant. Thus, farmers can apply 2,4-D to cornfields, killing the weeds but not the corn.

Most herbicides kill all plants indiscriminately. Such herbicides act by interfering with photosynthesis or other basic processes. Two herbicides, atrazine and diuron, do so by interfering directly with electron transport in photosystem II. (Atrazine also interferes with development in frogs by boosting levels of an enzyme that turns testosterone in estrogen.) Both compounds prevent the transfer of electrons from excited P_{680} to one of the cytochromes. The site of interference is a thylakoid protein, called QB, that helps transfer electrons. Atrazine and diuron bind to QB and prevent electron transfer.

Corn farmers have used atrazine and related compounds heavily, since corn contains enzymes that make it resistant to atrazine. In contrast, diuron is as toxic to corn as it is to weeds. So farmers can only use it when there is no risk of killing their crop. In recent decades, however, many species of weeds have evolved tolerance to atrazine. In other words, these weeds thrive despite being sprayed with massive amounts of atrazine.

Researchers have been able to identify the exact cause of this resistance. Atrazine-resistant weeds have an altered QB protein. In nonresistant weeds, QB becomes disabled when it binds to atrazine. But a single change in one of QB's 300 amino acids can prevent atrazine from binding, rendering the weed safe from atrazine.

Another widely used herbicide that inhibits photosynthesis is paraquat. Paraquat and related compounds accept electrons from photosystem I and transfer them directly to oxygen to produce *free radicals,* highly reactive ions that destroy delicate cellular machinery. These radicals rapidly attack thylakoid membranes, destroying their ability to sustain photophosphorylation and leading to plant death. Because paraquat serves as an electron shuttle—taking electrons from photosystem I and transferring them to oxygen—it continues to do damage for as long as photosystem I is operating. Unfortunately, paraquat is also toxic to humans and must be used with great caution. In humans and other animals, paraquat can potentially damage the digestive tract, the kidneys, and the lungs—thereby causing diarrhea, vomiting, and sometimes death.

Some herbicides act on other basic metabolic pathways that are unique to plants. The widely used herbicide glyphosate (Roundup), for example, interferes with the synthesis of aromatic amino acids (phenylalanine, tyrosine, and tryptophan). Unlike paraquat, however, glyphosate is not particularly toxic to animals. In animals, the synthetic pathways for the aromatic amino acids are completely different from those in plants.

A number of herbicide-producing companies are now using genetic engineering techniques to produce herbicide-resistant crop plants. For example, some crops now carry the atrazine-resistant QB protein. Chemical companies argue that such herbicide-resistant crops allow farmers to destroy weeds selectively by applying a particular herbicide. As a result, farmers do not have to plow their fields to get rid of weeds, and less topsoil is lost to erosion. However, most environmentalists doubt the wisdom of this strategy, since it will ultimately increase herbicide use.

Hundreds of herbicides are used everywhere today: in fields where fruits and vegetables are grown; on crops we feed to animals, themselves bound for our dinner table; along roadsides; and on our lawns and gardens. We find herbicides useful, indeed, hard to live without. Some break down soon after they are applied and are nearly harmless to the environment in the long run. Most do not harm humans. A few are hazardous in high doses, such as those that farmworkers might be exposed to. Some herbicides remain in the environment for months or years, continuing to poison plants and, in some cases, animals. All plants regularly exposed to herbicides ultimately evolve resistance, forcing researchers to develop new toxins.

What Happens in the Reaction Centers?

Plant physiologists have discovered that some of the wavelengths that are readily absorbed by plant pigments do not promote much photosynthesis. In fact, red light—light with a wavelength longer than 680 nm—promotes very little photosynthesis (Figure 7-14). On the other hand, light that is even redder—700 nm or longer—will promote photosynthesis if shorter-wavelength light (680 nm) also illuminates the plant. When 680-nm (red) and 700-nm (deep red) light both shine on a plant at the same time, photosynthesis is more efficient than with 680-nm light alone.

Photosynthesis works better with light of two wavelengths than with either alone because photosynthesis consists of two separate sets of reactions. One set of reactions, called **photosystem I**, absorbs light with wavelengths of about 700 nm (deep red). The other set, called **photosystem II**, absorbs light with wavelengths of about 680 nm (red light).

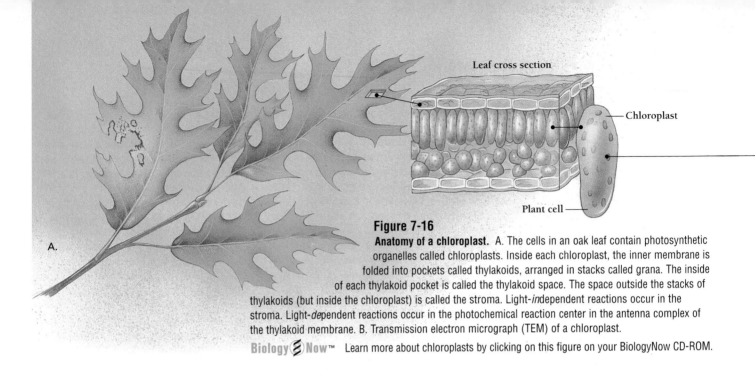

Figure 7-16

Anatomy of a chloroplast. A. The cells in an oak leaf contain photosynthetic organelles called chloroplasts. Inside each chloroplast, the inner membrane is folded into pockets called thylakoids, arranged in stacks called grana. The inside of each thylakoid pocket is called the thylakoid space. The space outside the stacks of thylakoids (but inside the chloroplast) is called the stroma. Light-*in*dependent reactions occur in the stroma. Light-*de*pendent reactions occur in the photochemical reaction center in the antenna complex of the thylakoid membrane. B. Transmission electron micrograph (TEM) of a chloroplast.

Biology 🄬 Now™   Learn more about chloroplasts by clicking on this figure on your BiologyNow CD-ROM.

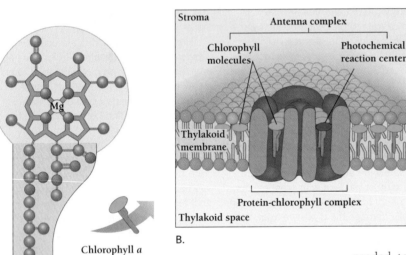

Figure 7-17

The heart of photosynthesis. A. Plants use two kinds of chlorophyll, chlorophyll *a* (shown here) and chlorophyll *b*. B. The photochemical reaction centers are made up of clusters of chlorophyll molecules embedded in the thylakoid membrane. Light excites electrons in the chlorophyll molecules, stimulating electron flow similar to that in the electron transport chain of mitochondria.

carotenoids, specific proteins, and other molecules and ions. But the two kinds of photosystems are separate entities and they perform different biochemical reactions (Figure 7-17). In photosystem I, the chlorophyll absorbs available light best at 700 nm and is called P_{700} ("P" stands for "pigment"). In photosystem II, the chlorophyll absorbs available light best at 680 nm and is called P_{680}.

The excited electrons of photosystem I can take either of two paths. The first path, called "cyclic photophosphorylation," produces ATP directly. The second, called "noncyclic photophosphorylation," does not make ATP but NADPH, which provides electrons needed to make sugars. Noncyclic photophosphorylation occurs only when photosystem I receives electrons from photosystem II. The excited electrons of photosystem II all take one path—to the electron-hungry P_{700} molecules in photosystem I.

The two separate photosystems of plants each absorb a different wavelength of light most efficiently. Photosystem II supplies electrons to P_{700} molecules in photosystem I.

What Path Do the Excited Electrons Take in Cyclic Photophosphorylation?

When P_{700} absorbs a photon from the antenna complex, its free energy increases, and it becomes a powerful electron donor. An excited P_{700} readily gives up a high-energy elec-

Where are these photosystems? Recall from Chapter 4 that inside a plant cell's chloroplasts are stacks of tiny saclike "thylakoids" (Figure 7-16). Each thylakoid is a flattened, hollow disc of thylakoid membrane. Both photosystems lie in the thylakoid membrane, and both photosystems consist of chlorophyll molecules together with

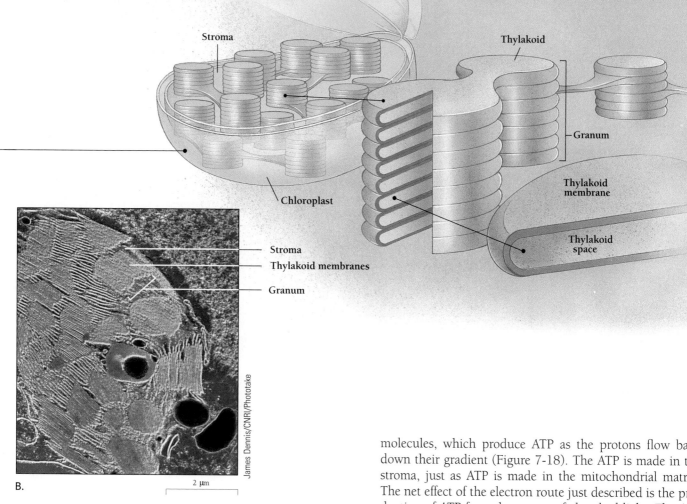

Stroma

Thylakoid

Granum

Thylakoid membrane

Thylakoid space

Chloroplast

Stroma

Thylakoid membranes

Granum

James Dennis/CNRI/Phototake

B.

2 μm

tron to one of two electron transport chains (Figure 7-18). When P_{700} loses its excited electron, however, it also loses most of the free energy it gained from the photon. Because P_{700} has now lost an electron, it is "oxidized."

One of the two paths an electron can follow in photosystem I leads the excited electron back to oxidized P_{700}. This cycle restores P_{700} to its original state and makes it ready to absorb another photon, a process called **cyclic photophosphorylation** (Figure 7-18).

The electron acceptor in cyclic photophosphorylation is a cytochrome complex that passes the electron on to another acceptor and back to oxidized P_{700}, completing the cycle. As in mitochondria, the passage of electrons through the cytochrome complex results in a proton gradient across a membrane. Protons move from a space just outside the thylakoids, called the **stroma,** into the thylakoid space (Figure 7-19). The stroma corresponds to the mitochondrial matrix. In mitochondria, the electron transport chain pumps protons out of the matrix and into the intermembrane space. In chloroplasts, a similar electron transport chain pumps protons out of the stroma and into the thylakoid space.

Like the inner mitochondrial membrane, the thylakoid membrane contains spinning, turbinelike ATP synthase

molecules, which produce ATP as the protons flow back down their gradient (Figure 7-18). The ATP is made in the stroma, just as ATP is made in the mitochondrial matrix. The net effect of the electron route just described is the production of ATP from the energy of absorbed light. The ATP made in the stroma is later used to synthesize glucose.

In cyclic photophosphorylation, excited electrons from P_{700} pass through an electron transport chain and back to P_{700}. This cyclic passage of electrons creates a proton gradient across a membrane in the chloroplast capable of generating ATP.

What Path Do the Excited Electrons Take in Noncyclic Electron Flow?

Photosystem I also includes an alternate, noncyclic electron path, which does not produce ATP (Figure 7-19). Instead, the energy of the excited electron of P_{700} is used to make NADPH, a derivative of NADH. NADPH is the electron-carrying coenzyme that chloroplasts use (called $NADP^+$ when lacking two electrons). Recall from Chapter 6 that the similar molecules NADH and NAD^+ are used in pathways such as glycolysis and respiration (catabolism). But NADPH and $NADP^+$ are different. They are used to build high-energy molecules such as glucose—in chloroplasts, in photosynthetic bacteria, and even in nonphotosynthetic organisms.

The noncyclic route of photosystem I supplies most of the reducing power for glucose synthesis. But after two P_{700}

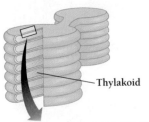

—Thylakoid

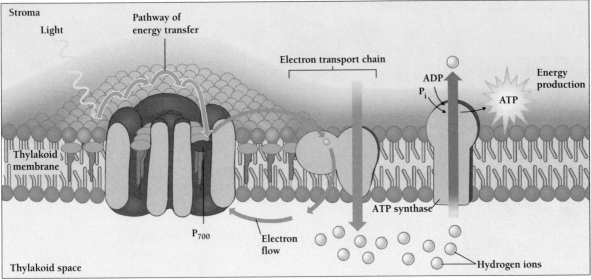

PHOTOSYSTEM I: Cyclic photophosphorylation

Figure 7-18

Cyclic photophosphorylation in photosystem I. Excited electrons from P_{700} pass through an electron transport chain *(blue)* and back to P_{700}. This cycle of electrons creates a proton gradient across the thylakoid membrane. ATP synthase uses the flow of protons back across the membrane to generate ATP.

Figure 7-19

Noncyclic photophosphorylation in photosystem I. An excited electron from P_{700} goes to help form NADPH. P_{700} then is missing an electron, giving it a positive charge.

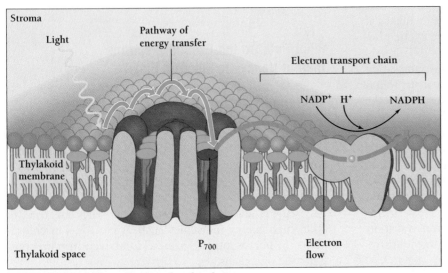

PHOTOSYSTEM I: Noncyclic photophosphorylation

molecules have given up two excited electrons to form NADPH, each P_{700} needs another electron before it can again serve as an electron donor. That is, some other molecule with an available electron must reduce P_{700} before it can function again.

In the noncyclic route of photosystem I, the excited electron from P_{700} goes to help form NADPH. P_{700} is then missing one of its electrons.

How Does Photosystem II Provide Electrons to P_{700}?

The chlorophyll molecules of photosystem II absorbs available light best at 680 nm and are called P_{680}. Photon-excited P_{680} is an even more powerful electron donor than photon-excited P_{700} (Figure 7-20). Electrons move through photosystem II's electron transport chain from excited P_{680} to oxidized P_{700} (photosystem I). By regenerating reduced P_{700}, photosystem II supplies electrons to photosystem I.

Photosystem II also produces ATP as protons flow down the gradient generated by its electron transport chain. This ATP production is called **noncyclic photophosphorylation,** to distinguish it from the ATP production in the cyclic path of photosystem I.

But we have the same question all over again: if an electron from excited P_{680} is transferred to oxidized P_{700}, how is the reduced form of P_{680} regenerated? How does it get an electron back again? The answer is that the electrons for regenerating P_{680} come from water. Recall from the beginning of this chapter that, using heavy oxygen isotopes, biochemists were able to show that plants split water molecules into oxygen and hydrogen.

In all plants, photosystem II strips electrons from water molecules, separating the hydrogens from the oxygen molecule. The waste product of this reaction is oxygen gas, which we gladly breathe. If plants and other organisms around the world ever ceased to photosynthesize, not only would we have no food to eat, but we animals would also suffocate from lack of oxygen. Indeed, until bacteria evolved the capacity to split water and release oxygen, about 3 billion years ago, the Earth's atmosphere contained virtually no oxygen. No animals existed; nor could they have existed. Only when photosynthesis evolved did oxygen begin to accumulate in our atmosphere and allow the evolution of animals, fungi, and other heterotrophic organisms. Today 50 to 70 percent of all of the Earth's oxygen comes from photosynthetic marine algae.

In noncyclic photophosphorylation, the breakdown of water by photosystem II supplies electrons to P_{680}, which in turn supplies electrons to P_{700}. The flow of electrons also creates a proton gradient that generates ATP. Photosynthesis supplied the oxygen that allowed the evolution of all animals.

How Do the Light-Dependent Reactions Generate NADPH and ATP?

Figure 7-20 shows how photosystems I and II together make up the light-dependent reactions of photosynthesis. The absorption of light by the two photosystems causes (1) the splitting of water to oxygen and hydrogen ions, (2) the production of ATP, and (3) the production of NADPH. Oxygen is a waste product released into the atmosphere through the stomata. ATP and NADPH mostly remain in the chloroplast, where they contribute energy to the synthesis of sugar—the light-independent reactions of photosynthesis.

The mid-20th century biochemist Albert Szent-Györgyi characterized this efficient process as "a little electric current driven by the sunshine." Each photon of light excites a single electron. Reducing a molecule of $NADP^+$ to NADPH, however, requires two electrons. To move these two excited electrons along the electron path to $NADP^+$, photosystems I and II must each absorb two photons. The production of each molecule of NADPH requires four photons of light.

As electrons flow from photosystem II to photosystem I, they supply the energy to pump protons across the thylakoid membrane. These protons then flow back through ATP synthase, as in cell respiration, driving the production of ATP.

The energy stored in the chemical bonds of NADPH and ATP is used to make glucose, a more stable form in which to store energy. Plants then convert glucose to sucrose, starch, or some other carbohydrate. Starch forms solid deposits in chloroplasts. In contrast, sucrose, or table sugar, remains in solution and can move throughout the plant. Just as animals transport sugar from cell to cell in the form of glucose, plants transport sugar from cell to cell in the form of sucrose.

The process by which plants transform the energy in ATP and NADPH into more stable forms such as glucose is the subject of the next section. The synthesis of sugar is the second part of photosynthesis.

The light-dependent reactions of photosynthesis split water into oxygen and hydrogen ions and generate a flow of electrons that produces ATP and NADPH. These molecules supply the energy to synthesize sugar.

7.3 How Do Plants Make Glucose?

Glucose is both plants' most important way to store energy and a basic building block for making larger molecules. We've seen that the hydrogen for glucose comes from water, and the carbon and oxygen come from carbon dioxide in the atmosphere.

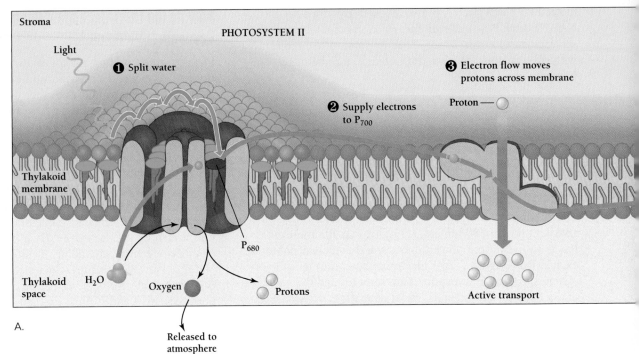

Stroma

PHOTOSYSTEM II

Light

❶ Split water

❸ Electron flow moves protons across membrane

❷ Supply electrons to P_{700}

Proton

Thylakoid membrane

P_{680}

Thylakoid space

H_2O

Oxygen

Protons

Active transport

A.

Released to atmosphere

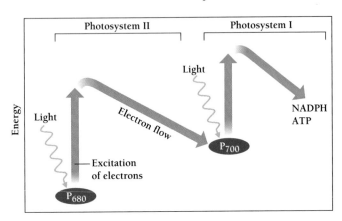

Photosystem II

Photosystem I

Energy

Light

Electron flow

Light

Excitation of electrons

P_{680}

P_{700}

NADPH ATP

B. Z scheme of photosynthesis

Figure 7-20

Photosystems I and II. A. ❶ Photosystem II's P_{680} molecule acquires electrons from splitting water, then ❷ supplies these electrons to P_{700} by way of an electron transport chain. ❸ The flow of electrons generates a proton gradient. ❹ The proton gradient causes protons to flow back across the thylakoid membrane, and ATP synthase uses the energy from that passive flow to generate ATP. The breakdown of water into oxygen and hydrogen by photosystem II is the ultimate source of the oxygen we breathe. The evolution of photosystem II underlies the evolution of animals and other heterotrophs. B. The "Z" scheme of photosynthesis summarizes the flow of electrons in photosystems I and II.

The overall reaction of photosynthesis is the reverse of respiration (though the individual steps are not). So we can guess, from what we learned in Chapter 6, that the synthesis of glucose from carbon dioxide and water must consume a lot of energy:

$$6\ CO_2\ +\ 6\ H_2O \rightarrow C_6H_{12}O_6\ +\ 6\ O_2$$
carbon dioxide + water → glucose + oxygen

Recall that the individual reactions in the breakdown of glucose divide the huge downhill process into many small steps whose energy can be captured. In the same way, the individual reactions of glucose synthesis allow the energy of ATP and NADPH to be converted to glucose a little bit at a time. The energy "hill" is climbed in a set of small steps.

We know, from the last section, that photosystems I and II split the H_2O in this equation and provide electrons. But

what is the process by which the single carbon atoms in carbon dioxide are combined into six-carbon glucose?

How Did Melvin Calvin Work Out the Calvin Cycle?

The discovery of the radioactive isotope ^{14}C ("carbon 14") provided a way to answer this question. In 1940, Martin Kamen and Samuel Ruben, then working at the University of California, Berkeley, discovered the isotope ^{14}C and immediately showed that $^{14}CO_2$ (carbon dioxide made from ^{14}C) could be used to study photosynthesis. In 1945, Melvin Calvin, another researcher at Berkeley, began the crucial experiments.

Calvin injected a small amount of $^{14}CO_2$ into a flask of algae that were actively growing in light. Quickly, sometimes within five seconds, he killed the algae by adding alcohol.

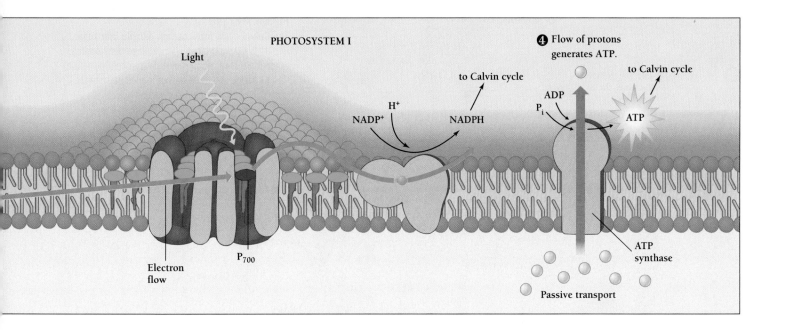

Then he worked out what compounds the algae had formed using the radioactive carbon atoms. To do this, Calvin used a technique called paper chromatography, in which the compounds made by the algae were dotted on a piece of paper and then treated with solvents. As the solvents moved across the paper, the various compounds were carried along, too. But the compounds moved across the paper at different rates, separating from one another like runners in a foot race.

To detect these invisible compounds, Calvin pressed the paper against a photographic film. Wherever ^{14}C was present, the radioactive compounds turned the film black. Calvin later wrote that "The [film] ordinarily does not print out the names of these compounds, unfortunately, and our principal chore for the succeeding ten years was to properly label those blackened areas on the film."

The first detectable product made by the algae, Calvin concluded, was a three-carbon compound. When he collected his algae within five seconds after the injection of $^{14}CO_2$, all the radioactivity was in a single spot. At successively later times, the pattern of spots became much more complicated. It took years for Calvin and his colleagues to discover the order in which the compounds formed. Eventually, however, his team worked out a set of reactions that converts the carbon of carbon dioxide into glucose. The reactions form a cycle called the **Calvin cycle**, after its principal discoverer (Figure 7-21).

Melvin Calvin used paper chromatography to obtain an ordered list of the chemical products of photosynthesis. Calvin was then able to suggest a reaction sequence that explained the order in which the various chemicals appeared in time. This reaction sequence came to be known as the Calvin cycle.

What Happens to Carbon Dioxide in Photosynthesis?

Carbon enters the Calvin cycle as carbon dioxide and exits as sugar. The transformation of carbon dioxide into sugar occurs in three parts:

> **Part 1. Capturing the carbon.** In the first part of the Calvin cycle, carbon dioxide combines with a five-carbon compound called ribulose bisphosphate. The resulting six-carbon molecule immediately breaks down into two molecules of phosphoglycerate.
>
> **Part 2. Making sugar.** In the second part of the Calvin cycle, each of the two molecules of phosphoglycerate is converted into a true three-carbon sugar. These two sugars may be converted to either glucose or other compounds, or they move into part 3.
>
> **Part 3. Making new ribulose bisphosphate.** In the third part of the Calvin cycle, the three-carbon sugars made in part 2 are used to regenerate ribulose bisphosphate, which, recall, captures carbon dioxide.

What Is the First Thing That Happens to Carbon Dioxide?

The first reaction of the Calvin cycle is the addition of carbon dioxide to a five-carbon compound called ribulose bisphosphate, to form a six-carbon molecule (Figure 7-21). This six-carbon molecule is unstable, however, and immediately breaks into two molecules of three-carbon phosphoglycerate.

The enzyme that catalyzes this reaction is ribulose bisphosphate carboxylase, but it has a friendlier name—**Rubisco.** Rubisco has the world on its shoulders. This remarkable enzyme is almost solely responsible for recycling

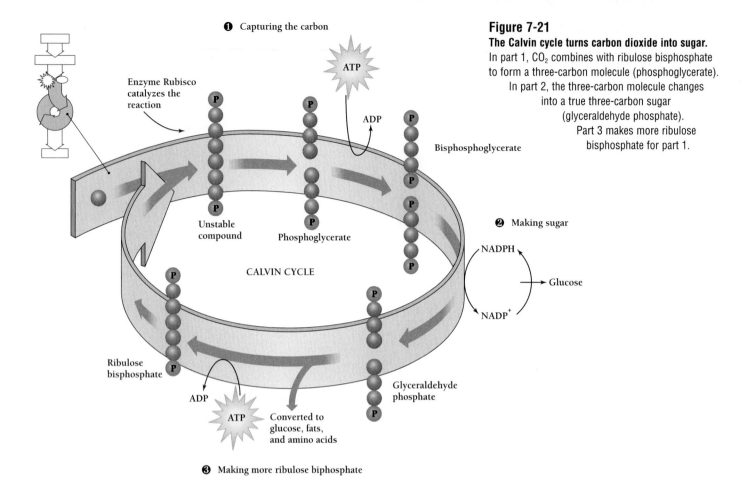

❶ Capturing the carbon

Enzyme Rubisco
catalyzes the
reaction

ATP

ADP

Bisphosphoglycerate

Unstable
compound

Phosphoglycerate

CALVIN CYCLE

❷ Making sugar

NADPH

Glucose

NADP⁺

Ribulose
bisphosphate

ADP

ATP

Converted to
glucose, fats,
and amino acids

Glyceraldehyde
phosphate

❸ Making more ribulose biphosphate

Figure 7-21
The Calvin cycle turns carbon dioxide into sugar.
In part 1, CO_2 combines with ribulose bisphosphate
to form a three-carbon molecule (phosphoglycerate).
In part 2, the three-carbon molecule changes
into a true three-carbon sugar
(glyceraldehyde phosphate).
Part 3 makes more ribulose
bisphosphate for part 1.

carbon dioxide from the atmosphere into plants and, ulti-mately, for supplying carbon to all living things.

But Rubisco is no superenzyme. Most enzymes can pro-cess 1,000 molecules per second, and some champions even manage 600,000 per second. But Rubisco is so slow that it barely deserves to be called an enzyme, managing only three molecules per second. To compensate for this enzyme's slug-gishness, plants typically produce huge amounts of Rubisco. In some plants, Rubisco makes up more than 50 percent of the total protein in the chloroplasts. Rubisco is probably the most abundant protein on the planet.

In the first reaction of the Calvin cycle, carbon dioxide combines with ribulose bisphosphate with the aid of the enzyme Rubisco. The result-ing six-carbon molecule immediately breaks down into two molecules of three-carbon phosphoglycerate.

What Happens to the Two Molecules of Three-Carbon Phosphoglycerate?

The second part of the Calvin cycle involves two steps that together convert phosphoglycerate into a three-carbon sugar called glyceraldehyde phosphate. Glyceraldehyde phos-phate is a true carbohydrate.

These two reactions require one molecule of ATP and one molecule of NADPH for each molecule of phosphoglyc-erate. Each molecule of carbon dioxide entering the cycle re-sults in two molecules of phosphoglycerate, so each turn of the cycle requires two molecules of ATP and two molecules of NADPH.

Some of the glyceraldehyde phosphate enters the cyto-plasm, where it is converted to glucose, fats, or amino acids. And some of it remains in the chloroplast, where it regenerates ribulose bisphosphate, the task of part 3 of the Calvin cycle.

Each of the two molecules of phosphoglycerate is converted into a three-carbon sugar. This sugar is either converted to glucose, fats, or amino acids (in the cytoplasm), or it is used to regenerate the ribulose bisphosphate consumed in part 1 of the Calvin cycle.

How Is Ribulose Bisphosphate Regenerated?

The molecule that first combines with carbon dioxide is ribulose bisphosphate. But the resulting six-carbon mole-cule immediately splits into two molecules of three-carbon phosphoglycerate, and the ribulose bisphosphate is gone. How can a plant cell make more of this important molecule?

The short answer is that a cell uses some of the glyceraldehyde phosphate to make more ribulose bisphosphate. This process is the messy part of the Calvin cycle and involves seven different enzymatic steps. In the last of the seven steps of part 3, ATP contributes a second phosphate group to form more ribulose bisphosphate.

The important molecule ribulose bisphosphate is regenerated in seven steps from glyceraldehyde phosphate.

7.4 How Do Plants Cope with Too Little Water or Carbon Dioxide?

The reactions of photosynthesis do not take place in a test tube, but in living, growing plants, subject to storms, drought, and attacks by plant-eating animals. Grasses, small plants, shrubs, and trees all live in complex and demanding ecosystems where many plants die without reproducing. Many factors determine whether a plant survives and even thrives. One of these factors is how well the plant uses sunlight to synthesize sugars.

We have seen that plants use the three-carbon sugar glyceraldehyde phosphate to make glucose, fats, and amino acids. These, in turn, are used to make the carbohydrates and proteins that make up the bulk of the body of the plant. The construction of leaves, stems, trunks, and flowers creates a massive demand for glyceraldehyde phosphate from the Calvin cycle.

If a plant lacks any of the raw materials of photosynthesis—water, carbon dioxide, or photons—meeting this demand becomes impossible. As a result, plants have evolved a variety of adaptations for increasing their productivity, the efficiency with which they collect and use these three raw materials.

The factors that influence plant productivity are not only important to plants but important to humans as well. All the world's major crops—including rice, wheat, and corn—depend on plant breeders' efforts to increase the productivity of those crops. The world food supply and the lives of billions of people depend on these super-productive crops.

What Factors Limit Productivity?

The three major factors that affect photosynthesis are the availability of water, carbon dioxide, and photons of a useful wavelength. We have already seen how the wavelength of light might influence the efficiency of photosynthesis. Plants cannot use light that is the wrong wavelength. Only light in the action spectrum of photosynthesis is of any use at all. Equally important is the *intensity* of light, the number of photons striking a leaf surface per day. Plants cannot produce as much sugar in the shade as they can in full sun.

Many factors influence how much light a plant gets. These include, for example, the length of the day, the length of the growing season, and the amount of cloud cover. In a tropical rain forest, 15 to 20 layers of vegetation may absorb and reflect more than 95 percent of the light before it reaches plants at ground level.

Too little water can also limit photosynthesis. Plants pull water from the soil and transport it through stems and trunks to their leaves. Some of this water is used in photosynthesis, but most passes out of the stomata in the leaves and evaporates into the surrounding air, pulling along with it water and minerals from the roots.

When the weather is hot and dry and water is in short supply, plants shut their stomata to avoid losing water. But a photosynthesizing plant needs carbon dioxide with which to make sugar and also a way to get rid of waste oxygen. If the stomata are closed, waste oxygen accumulates inside the plant, and carbon dioxide levels plummet. With the stomata closed, a plant can't photosynthesize any better than we could eat with our mouths shut. For desert plants, which must keep their stomata closed as much as possible to avoid drying out, stocking up enough carbon dioxide for photosynthesis is a problem.

How Do Plants Prevent Photorespiration?

Plummeting carbon dioxide levels not only slow photosynthesis, they also initiate a wasteful chemical pathway called **photorespiration,** in which plants squander much of their hard-won carbohydrate. The problem occurs because Rubisco requires a lot of carbon dioxide to work well. Normally, in the first step of the Calvin cycle, Rubisco converts ribulose bisphosphate into two molecules of phosphoglycerate. However, when carbon dioxide levels are low, Rubisco converts ribulose bisphosphate into one molecule each of phosphoglycerate and another, almost useless, two-carbon molecule. The result is that during photorespiration plants waste as much as half of the carbohydrate they produce. Photorespiration, unlike cellular respiration, produces no ATP or NADH. Its function, if any, is unknown.

There is only one cure for photorespiration: lots of carbon dioxide. Some plants have evolved a special biochemical pathway that allows them to maintain high levels of carbon dioxide inside their cells, even when their stomata are closed most of the time.

In these plants, the first reaction of carbon dioxide does not produce the usual three-carbon molecule phosphoglycerate. Instead, they make a four-carbon compound called oxaloacetate, or OAA (which is also part of the citric acid cycle, described in Chapter 6). The cells then convert OAA to another four-carbon compound, which moves into special cells that make carbohydrates. There it breaks down to release carbon dioxide, which immediately enters the Calvin cycle.

Plants that use the three-carbon phosphoglycerate, which includes most plants, are called **C₃ plants.** Plants that use the four-carbon OAA pathway are called **C₄ plants.** C₄ plants, such as corn and sugar cane, have a distinctive anatomy and grow well in hot, dry climates. The C₄ pathway and anatomy are adaptations that allow plants to minimize water loss by keeping the stomata closed, while still providing enough carbon dioxide to supply photosynthesis and prevent photorespiration. C₄ plants waste far less sugar than C₃ plants, but C₄ plants need more ATP (and therefore more light) than C₃ plants.

In hot, dry climates with plenty of light, C₄ plants predominate. In temperate zones with less light and more water, C₃ plants have the advantage. In Kentucky, for example, the humid summers allow plants to leave their stomata open much of the time, and C₃ plants, such as Kentucky bluegrass, thrive. But Kentuckians who move to California are in for a rude surprise. When they plant beautiful Kentucky bluegrass lawns, ugly yellow-green crabgrass gradually takes over. Crabgrass is a C₄ plant that grows better in the bright but dry California summers.

C₄ crops are especially efficient at capturing sunlight, converting as much as eight percent of the solar energy falling on a field to carbohydrates. As oil and coal reserves decline, some countries have begun to fuel their cars and trucks with alcohol produced from corn and sugar cane.

Desert-adapted plants, such as cacti, use a modified C₄ pathway called crassulacean acid metabolism, or **CAM,** a pathway that conserves even more water. CAM plants open their stomata to collect CO_2 at night only, when the air is cool and water loss is minimal. They then store the CO_2 in a four-carbon molecule until the next day, when photosynthesis can proceed. But collecting and storing CO_2 at night requires enormous amounts of ATP. This extra ATP comes from the breakdown of carbohydrates made during the day. The enormous energy cost of CAM plants' nighttime respiration keeps their growth rates extremely low.

C₄ plants minimize water loss and photorespiration and maximize photosynthesis by storing high carbon. C₄ plants are more efficient than C₃ plants, but they need more light than C₃ plants.

In these first few chapters of *Asking About Life,* we have examined some of the basic chemical processes that enable organisms to live. In the next few chapters, we will see how cells multiply and grow. Then we will see how organisms recombine their genetic material in novel arrangements in the process of sexual reproduction. We will learn how the structure of DNA was discovered and find out how genes are expressed. Finally, we will explore the ways that molecular biologists have learned to control the expression of genes and how this dizzying array of new knowledge applies to humans.

Key Concepts

- Photosynthesis is the process by which plants use energy from the sun to build sugars from carbon dioxide and water.
- Specialized photosynthetic machinery absorbs light energy, transforms it into the chemical energy of NADPH and ATP, and then stores it in the chemical bonds of sugar.
- Photosynthesis in plants includes two light-dependent reactions—photosystem I and photosystem II—as well as a light-independent reaction.

Summary with Key Terms

How do plants use solar energy to make sugar?

In **photosynthesis,** green plants and other organisms use the energy of **photons** to synthesize carbohydrates from carbon dioxide and water. In the **light-dependent reactions (photophosphorylation),** photosynthesis splits water into oxygen and hydrogen atoms, generating oxygen and high-energy electrons. In the **light-independent reactions,** the electrons from the light-independent reactions help link the carbon atoms in carbon dioxide into three-carbon sugars.

Each photon carries a fixed amount of energy, which corresponds to its **wavelength. Visible light** ranges from about 400 nm to about 750 nm. Plants contain **pigments** that absorb light, including the **chlorophylls** and the **carotenoids.** Every molecule has an **absorption spectrum,** and every light-dependent process has an **action spectrum.**

Antenna complexes, consisting of large numbers of chlorophyll and carotenoid molecules, absorb photons and transfer them to one of two kinds of **photochemical reaction centers,** located in the thylakoid membrane of the chloroplast. Each kind of reaction center participates in a separate photosystem.

In **photosystem I,** the main pigment is P_{700}. In **photosystem II,** the main pigment is P_{680}. When P_{700} or P_{680} absorbs a photon, the pigment's energy increases, which excites its electrons. The excited electrons transfer to special electron acceptors.

The excited electrons of photosystem I can take either of two paths. In **cyclic photophosphorylation,** excited electrons from P_{700} pass through an electron transport chain and back to P_{700}. This cyclic passage of electrons creates a proton gradient between the **stroma** and the thylakoid space in the chloroplast that generates ATP. In **noncyclic photophosphorylation,** the excited electrons of photosystem I produce NADPH. But the noncyclic path leaves P_{700} short an electron. Only electrons from photosystem II can make up this deficit.

The breakdown of water by photosystem II supplies electrons to P_{680}, which in turn supplies electrons to P_{700}.

The flow of electrons from P_{680} to P_{700} also creates a proton gradient that generates ATP. The breakdown of water by photosynthesis supplied the oxygen that allowed the evolution of all animals.

The **Calvin cycle** is the heart of the light-independent reactions of photosynthesis. The Calvin cycle uses the energy of ATP and NADPH (both generated by the light-dependent reactions) to make sugar out of carbon dioxide. In the Calvin cycle, the enzyme **Rubisco** synthesizes three-carbon sugars from ordinary carbon dioxide. These small sugars can then be converted to glucose, fats, amino acids, or ribulose bisphosphate.

How do plants cope with too little water or carbon dioxide?

The efficiency of photosynthesis depends on the amount of light, water, and carbon dioxide. In hot, dry weather, plants close their **stomata** to limit water loss. With the stomata closed, however, waste oxygen builds up, and carbon dioxide levels plummet. Under such conditions, **C_3 plants** may begin **photorespiration,** an unhealthy biochemical pathway that wastes about half of all carbon dioxide. **C_4 plants** and **CAM** plants avoid photorespiration by maintaining high levels of carbon dioxide inside the cells that make glucose.

Review and Thought Questions

Review Questions

1. What is photosynthesis? What material does it synthesize? Where in a cell does this process occur?

2. Write the overall reaction for photosynthesis. How do the reactants get into the leaf and into the cells? How does oxygen get out of the leaf cells and back into the air outside the plant?

3. Which parts of the electromagnetic spectrum do plants use in photosynthesis? Which parts of the spectrum are reflected or absorbed by the Earth's atmosphere?

4. Describe the cyclic photophosphorylation associated with photosystem I. What is the relationship between the absorption of light and the excitation of electrons? How does the transfer of electrons lead to the synthesis of ATP?

5. Describe the reactions that make up the Calvin cycle. What enzyme captures carbon dioxide from the atmosphere? What are the products of this reaction? How is the energy from light introduced in the Calvin cycle? Why must the cycle produce more ribulose bisphosphate?

6. Under what environmental conditions might a plant close its stomata? Compare the rates at which plants can carry out photosynthesis when their stomata are open and when they are closed. Why should the opening of stomata affect the rate?

7. Explain how photorespiration causes C_3 plants to waste some of their photosynthetic efforts under conditions of low carbon dioxide concentration.

8. Describe Priestley's experiment that showed that a plant could restore the air in a bell jar. Make a sketch of a bell jar containing only a mouse, and explain why the mouse dies. Then sketch a bell jar containing both a mouse and a plant, and show with arrows how oxygen is produced and consumed.

9. What advantages do C_4 plants have over C_3 plants? Why don't C_4 plants take over everywhere? What limits them?

10. Put the letter "R" in the blank if the event is part of cell respiration. Use the letter "P" for photosynthesis. Use the letter "B" for both.
    ____ Sugar is product
    ____ Uses oxygen
    ____ Requires light
    ____ Uses electron transport chain
    ____ Releases carbon dioxide
    ____ Produces a proton gradient
    ____ Releases oxygen
    ____ Uses ATP synthase to make ATP

Thought Questions

11. What prevents the roots of a plant from carrying out photosynthesis? By what metabolic process do you suppose roots extract energy from their available resources?

12. Describe in a few sentences the concept of how light energy is used to drive the reaction of glucose synthesis. (Be careful to give the outline of the story, so that a nonbiologist could understand what you mean. Don't just list a lot of details.)

13. Why is the process of photosynthesis important to feeding our entire planet? Can you name any living things that do not depend on plants (directly or indirectly)?

BiologyNow Resources

Biology ⑤ Now™

Active Figures

7-5: Photosynthesis and the forms of plant metabolism
7-16: Chloroplasts

Preparing for an exam? Take a diagnostic test on your BiologyNow CD-ROM.

Online materials relating to this chapter are at:

http://biology.brookscole.com/AAL3

About the Chapter-Opening Image

A large plant in a small pot of soil emphasizes the idea that plants make their bodies primarily from air and water, not from soil.

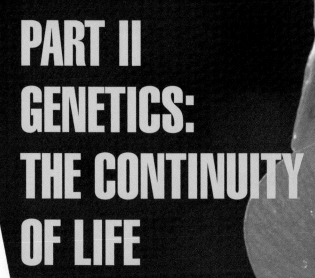

PART II
GENETICS:
THE CONTINUITY
OF LIFE

Animals Animals/Earth Scenes

147

Cell Reproduction

Key Questions

- How is cell division critical to the lives of single-celled and multicelled organisms?
- How do eukaryote cells divide during mitosis and cytokinesis?
- How does the DNA fit into the nucleus without becoming tangled?
- How do cells regulate the cell cycle?

The Legacy of Henrietta Lacks

Early in 1951, an anxious 31-year-old woman checked into the medical clinic at Johns Hopkins University in Baltimore (Figure 8-1). Her anxiety was confirmed when her doctor found a one-inch purple growth on her cervix, the neck-shaped entrance to the uterus. He told her he would have to do a biopsy, and he cut out a small piece of the growth and sent it to a pathologist, a medical specialist who can tell, among other things, whether a human cell is cancerous or not. The report came back positive, indicating cervical cancer. Henrietta Lacks's growth was dangerously malignant; its cells were dividing rapidly, and without treatment the tumor would kill her.

The news came as a blow. She and her husband had two children to raise. She was too young, she thought, for so serious a disease, one that does not usually afflict women until their 40s or 50s. Within days, the doctors at Johns Hopkins began treating her with high doses of radiation, in what was to be a futile effort to save her life. In the autumn of 1951, Henrietta Lacks died. Ironically, her tumor cells survived and live today in laboratories all over the world.

The day of her biopsy, Lacks's doctor took a piece of her tumor to a group of researchers who were struggling to grow human cells in glass dishes containing a mixture of nutrients and blood products. They had tried adding different nutrients and changing the pH of the mixture. But no one had yet succeeded; most human cells died almost immediately without dividing more than once or twice. Still, the researchers hoped that any day they would find the right cells or the right conditions for growing human cells in the laboratory.

When Henrietta Lacks's cells arrived in the lab, they were different from anything anyone had ever seen before. They grew without restraint, dividing and redividing every 24 hours. The same virulence that would finally kill Henrietta Lacks made these cells unstoppable in the laboratory. Before Lacks's life was over, her cells took up an independent existence, thanks to the work of medical researchers.

Although Henrietta Lacks's cells, named HeLa cells, were the first human cells to be successfully cultured, hundreds of others soon followed. New culturing techniques allowed researchers to grow cells from other tumors, and, eventu-

Amy Dunleavy

ally, normal tissues. Unlike HeLa cells, most of these were difficult to culture. Even with the best culture techniques today, human cells die quickly in culture. And beginning a culture of human cells requires care and attention.

Still, in the 1950s and 1960s, growing human cell cultures was exciting work. For the first time, scientists could study human biology without experimenting on humans themselves. They could study the way human cells divide, the way human cells move, the way cancer cells differ from normal cells. The possibilities seemed endless. Every week a lab somewhere developed a new kind of human cell to culture. These new cell lines were rapidly distributed to other labs—some down the hall, others on the other side of the globe.

In the early 1970s, odd things began to happen. The National Institutes of Health (NIH), in Washington, D.C., sent five kinds of human cells, which had come from five different research labs in the Soviet Union, to a lab in Berkeley. There, an alert researcher named Walter Nelson-Rees discovered that the five supposedly different cell lines were all HeLa cells.

Soon HeLa cells began to turn up everywhere. Nelson-Rees began checking every cell line he knew of. He published lists of cell cultures that were not what they were supposed to be. By 1981, he had listed 90 HeLa cell lines, each one originally thought to be a unique cell line. Researchers studying the biology of breast cells, both normal and cancerous, were horrified to find that their years of research had not been done on breast cells—as they had thought—but on Henrietta Lacks's cervical cells. The story was the same for research on prostate cells, kidney cells, liver cells, and more.

The repercussions were enormous. Years of research by scores of researchers had become meaningless. Much of the biology of cancer had to be rethought. For years researchers had believed that all cancer cells have the same nutritional needs and all cancer cells have abnormal chromosomes, the so-called mark of cancer. Then, as one cell line after another proved to be not prostate cells, not kidney cells, not this and not that, but HeLa cells, biologists came to realize that, in fact, cancer cells are not all alike. The only cancer cells that were all alike were HeLa cells. The abnormal chromosomes were typical of HeLa cells, but not of all cancer cells.

But how had it happened? The answer was that Henrietta Lacks's cancer cells were unbelievably vigorous. If so much as a single one of her cells fell by accident onto a culture dish, her cells could take over, no matter what other cells

Figure 8-1

Henrietta Lacks. Young Henrietta Lacks died of cancer in 1951, but biologists have kept her cancer cells alive ever since. In the 1950s, her cells, called HeLa cells, were used to develop a vaccine against polio. Today, genetic engineers insert genes for valuable human proteins into HeLa cells, which then synthesize the proteins.

Science Photo Library/Photo Researchers, Inc.

were living in the dish. One group of researchers found that HeLa cells could float through the air inside the tiny bits of spray created when a rubber stopper was pulled from a bottle containing HeLa cells. The floating cells could then drop down into nearby culture dishes. Within a week, a culture of prostate cells was taken over by cervical cells. It was a scenario that scientists had never imagined.

Today, cell biologists have learned to be more cautious. They work with cell cultures whose ancestry they've checked. They use techniques designed to prevent cross contamination of cultures.

But HeLa cells are still used, and Henrietta Lacks's cells live on. All over the world her cells continue to divide relentlessly day after day. In the 1950s, her cells were used as hosts for growing poliovirus, enabling researchers to develop the vaccine against polio. Today, genetic engineers insert genes for valuable human proteins into HeLa cells, which then synthesize the proteins. Henrietta Lacks died before she could finish raising her own two children, yet her cells have been instrumental in basic research and have helped save the lives of millions of children all over the world.

8.1 Cell Division

Cells are the fundamental living units of all organisms. An organism may consist of a single cell or many cells. Bacteria such as *Lactobacillus* and yeasts such as *Candida albicans* are single-celled organisms made of just one cell per individual. But most organisms are made of thousands or millions of cells. An adult human, for example, consists of about 10 trillion (10^{13}) cells. But no matter how many cells an organism is made of, all the cells originate from a single cell.

How Do Cells Divide?

In the five minutes or so that it takes you to read this page, your body will produce a billion (10^9) new cells. Most of these will replace battered and dying cells of your skin, intestines, and blood. Each newly formed cell contains both a copy of the genetic information you inherited from your parents and the molecular machinery and materials needed to interpret that information—including membranes, organelles, macromolecules, and small molecules.

Health and Biology Cervical Cancer and the Pap Smear

In the 1930s, more women died of cervical cancer in the United States than of any other form of cancer. Today, the rate of cervical cancer is less than one-tenth that of breast cancer and declining rapidly, thanks to a diagnostic test invented by George N. Papanicolaou in 1943.

In the past 50 years, the widespread use of the Pap test, or Pap smear (named after its inventor), has allowed medical professionals to detect and remove precancerous lesions of the cervix before they become cancerous, saving thousands of lives. Even when a lesion has advanced to an early form of cancer, the five-year survival rate (the standard for all cancers) is 91 percent. (For women who are not checked and who develop late stage cervical cancer, five-year survival is less than 15 percent.) The Pap smear consists of a simple office procedure in which a doctor or nurse scrapes a few living cells from the surface of the cervix. The scraping may hurt very slightly, but the life-saving test is well worth the second or two of discomfort. Once the cells have been collected from the cervix, they are sent to a laboratory pathologist for analysis.

In most forms of cervical cancer, cells undergo changes in appearance years before they become cancerous. Such "precancerous" cells have distinctive characteristics that a good pathologist can easily recognize (Figure A). The cervical cells are usually classified into five grades: normal, atypical, low-grade precancerous lesion, high-grade precancerous lesion, and invasive cancer. In the great majority of cases, all the cells collected will be normal. Occasionally, some of the cells will be "atypical," usually because of some temporary infection of the cervix or vagina. This is nothing to worry about, but a second test may be necessary.

Any precancerous growths that are found are removed from the cervix, often using cryosurgery (freezing) or lasers. After treatment, a woman is normally cancer free and able to bear children if she wishes. If the Pap smear reveals cancerous cells, as opposed to precancerous ones, more tissue needs to be removed, depending on how far the cancer has spread.

Cervical cancer strikes women of all ages and ethnic groups, but Asian-American, Native American women, Hispanic women, and poor women in general run a higher risk. For example, African-American women are twice as likely as European-American women to die of the disease. Researchers attribute the difference to lack of good medical care, especially regular Pap tests. Fortunately, increasing numbers of African-American women are getting Pap tests, and the difference in survival is therefore decreasing. Worldwide, cervical cancer kills more women than all but two other cancer types. But in the United States, it ranks 18th, primarily because of widespread use of the Pap smear. Because most cervical cancer develops slowly, it is quite easy to diagnose the disease while it is still in one of the precancerous stages. Public health experts recommend that all women have annual Pap smears starting at age 18 or at the onset of sexual activity, whichever occurs first. After three or more negative tests, a woman is safe in having an exam every other year. Women who are at risk for cervical cancer should continue to have the exam every year.

The main risk factor for cervical cancer is infection with human papilloma virus (HPV), similar to the virus that causes genital warts. In 93 percent of all cases of both cervical cancer and precancerous lesions, HPV DNA is present. Indeed, HPV incorporates two genes into the genomes of previously healthy cervical cells. These two genes encode proteins that cause the cells to form tumors. In Chapter 12, we will learn more about how viral DNA infects cells and how tumor viruses cause these cells to form tumors. Of more than 100 types of HPV, about 30 are spread through sexual contact and about 20 appear to cause cervical cancer. Researchers are actively working on a vaccine against HPV.

Because HPV is a sexually transmitted disease (STD), the best ways to avoid infection—and therefore the frontline defense against cervical cancer—are the same as the recommendations for other STDs: delay the first sexual experience, always use a condom, keep the lifetime number of sex partners low. Finally, as is true for many cancers, smoking increases the risk of cervical cancer.

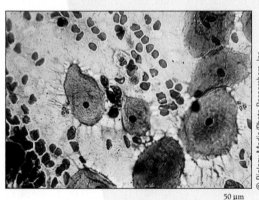

50 μm

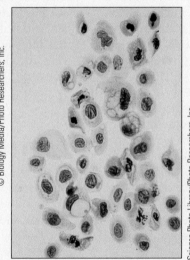

100 μm

100 μm

Figure A
Normal cervical cells *(left)* **are distinctly different from precancerous** *(middle)* **and cancerous** *(right)* **cervical cells.** Note the sperm-shaped cells that characterize invasive cancer cells.

Cell reproduction is so common that it is easy to forget how extraordinary it is. Our wonder at this copying process may be heightened by our everyday experience with non-biological copying—the recording of music, the filming and copying of movies and television programs, and the printing of books. In each case, complicated machinery carries out a complex but ordered process. But cell reproduction is even more remarkable than these other processes, for the cell is both the thing being copied and its own copying machine.

When a single-celled organism divides, the process creates a new individual. But when a cell in a multicelled organism divides, the process can allow growth or the replacement of damaged cells. A dandelion root can regenerate its leaves and flowers after they are cut away by a lawnmower. In the same way, children quickly replace the skin cells that they scrape and tear off in their inevitable bangs and falls. In the case of a fertilized egg cell, however, cell division leads to the formation of a wonderfully complex organism such as a fly, a frog, or a human. In the simple division of a cell into two new cells, a process called **mitosis** [Greek, *mitos* = thread], the cell copies its DNA then divides in two.

In plants, animals, and other sexually reproducing organisms, another very different kind of cell division forms the cells that become sperm and eggs. This second process, called *meiosis,* is more complex than mitosis, and we introduce it in Chapter 9.

Simple cell division, called mitosis, allows growth and cell replacement in multicelled organisms and creates new individuals in single-celled organisms. Before a cell divides, it copies its DNA.

What Did Early Biologists Discover About Chromosomes?

In the 19th century, the cell theory established that all organisms are made of cells and that all cells come from other cells. But the cell theory suggested a problem. How does a cell know how to make another cell just like itself? How does the cell pass on the knowledge of what kind of cell it is? How, for example, does a cell from Koko the gorilla know how to make another Koko cell, a cell unlike that found in any other gorilla and unlike that found in other primates such as your classmates?

Today we know that the answer is in the **chromosomes** [Greek, *chroma* = color + *soma* = body]—the threads of DNA and protein that carry the genes. Yet, 19th-century microscopists saw chromosomes and examined them in great detail without recognizing that they were the means by which organisms inherit traits. Early biologists held the key to the problem of inheritance without knowing it. In the late 1800s, improvements in microscopes for the first time allowed microscopists to see the tiny structures inside of cells.

Other advances in microscopy came with the use of natural dyes that had been recently developed to dye cloth. The two most useful dyes, still used to dye cells today, were purple-and-blue hematoxylin (which comes from the logwood tree of Central America) and the synthetic red dye eosin. Thanks to these dyes, microscopists of the 1870s were able to see and describe the chromosomes of animal and plant cells. In fact, during one four-year period, researchers published more than 200 papers describing the complex dance the chromosomes perform during cell division.

Staining cells showed that the cell's nucleus is ordinarily full of a fuzzy material now known to be chromosomes that have unwound into invisibly thin threads—long molecules of DNA (Figure 8-2). By the 1880s, biologists had discovered that during **cell division,** when a single parent cell gives rise to two **daughter cells,** the nucleus also divides. During this division, the chromosomes became obviously visible as darkly stained bars and Xs. The early microscopists watched in fascination as these purple objects divided up, a complete set going to each new cell. What it meant they had no idea.

The German physician and microscopist Walther Flemming pieced together a detailed description of the process by which chromosomes go to the two new cells during cell division. It was Flemming, whose description of mitosis remains accurate today, who gave a name to this elaborate division of what we now know is the genetic material.

First, Flemming noted, a fuzzy-looking material he called chromatin appeared to condense into visible X-shaped bars (chromosomes). He also noticed the duplication of an object he called an **aster** [Latin, *aster* = star]—the center of a finer set of threads just outside the nucleus. Dur-

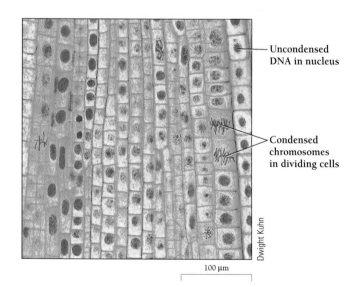

Dwight Kuhn

100 µm

Figure 8-2
Onion cells reproduce by dividing in two. In cells stained with a purple dye that binds to DNA, we can see the DNA in the cell's nucleus. In nondividing cells, the DNA appears as a diffuse fuzzy material. In dividing cells, the DNA condenses into distinct visible chromosomes.

ing mitosis, the two resulting asters migrate to opposite ends of the cell. While the asters are migrating, the X-shaped bars (the chromosomes) line up in a row in the middle of the cell, and then each one splits. Each part migrates toward one of the asters. At the end, the chromosomes seemed to disappear, and the nuclei became fuzzy again.

Late 19th-century microscopy revealed all the major details of cell division in eukaryotes without explaining the role of the nucleus or chromosomes in the life of the cell.

How Did Biologists Discover the Function of the Chromosomes?

But what was the point of this careful division? For the moment, no one knew. The nucleus of the cell had been recognized as a standard component of cells as early as the 1830s, but its function was elusive. During cell division, the nuclear membrane breaks up, and the nucleus seems to disappear. Because the nucleus disappears, biologists did not at first consider that it might play an important role in cell division. Many biologists assumed that the cell constructed a new nucleus after each cell division.

The first hint that the nucleus might be important came with studies of *fertilization,* the process by which a sperm and egg fuse to form a *zygote.* In the 1870s, several microscopists noticed that a sperm cell is little more than a nucleus with a tail to propel it. Further studies showed that a zygote (a fertilized egg) forms from exactly one sperm and one egg and that the sperm nucleus and the egg nucleus fuse to form a single nucleus. It was getting harder to avoid the realization that the nucleus must be important.

But the nucleus's peculiar role in fertilization did not imply to 19th-century biologists that the nucleus contained the hereditary material. First, they were not necessarily looking for a hereditary material. In fact, many biologists of the time thought that the process of fertilization did not transmit a material substance, but merely triggered the development of the embryo. They had no way of knowing that the material in the nucleus carried information.

Biologists assumed that the individual chromosomes in a cell were all alike, no more different from one another than are individual pretzels. Although they might look different from one another, dividing them would be similar to dividing a bag of pretzels between two people. It wouldn't matter who got which pretzel, as long as each person got half.

Flemming and others had demonstrated the presence of individual chromosomes. For example, we now know that every human cell has the same set of 23 pairs of chromosomes. Each member of a pair has a distinctive and unique shape and size. Flemming saw the uniqueness of chromosomes in other organisms and described how each chromo-

some splits down its length, one-half "predestined for one new cell and [one-half] for the other." If we divided a bag of pretzels in half, however, we would not cut each pretzel down its whole length. We would just give each person half the pretzels in the bag.

It was the great embryologist Wilhelm Roux who first argued that the complex way the X-shaped chromosomes split during mitosis strongly suggested that the chromosomes were not uniform. Why, he asked, would the cell use such a complex process to divide the nucleus, dividing each X neatly in half, if simply cutting the nucleus in two would work? Each chromosome, said Roux, must contain some unique material, making it crucial that each daughter cell receive exactly half of each chromosome.

In 1889, the German biologist August Weismann insisted that the nucleus contained the hereditary material, which he called the *germ plasm.* "Heredity," Weismann wrote, "is brought about by the transference from one generation to another, of a substance with a definite . . . molecular constitution."

But Weismann was known for his radical ideas, and most biologists of the time dismissed Weismann's claim. In fact, as recently as the 1930s, a few respected biologists still rejected the idea that the genes were on the chromosomes. Today we know that each daughter cell carries genetic information inherited from the parent cell and that the genetic material determines how the organism interacts with its environment. In all cells, the genetic material is DNA and it is carried on the chromosomes.

Despite detailed descriptions of the division and distribution of chromosomes to daughter cells during mitosis, 19th-century biologists did not realize that chromosomes carry hereditary information.

8.2 How Does a Dividing Cell Ensure That Each Daughter Cell Receives an Exact Copy of the Parent Cell's DNA?

One of the main tasks of cell reproduction is to **replicate,** or copy, the cell's DNA (Chapter 10). After the DNA replicates, the cell's machinery distributes one copy to each of the two daughter cells. In most bacteria, all the genes are on a single molecule of DNA and equal distribution is simple. The cells of eukaryotes have more than one chromosome and more than one DNA molecule, so distributing the DNA to two new cells is more complex. During cell division, the cell must also split the rest of the cell's contents between the two daughter cells. Cell reproduction, then, accomplishes three tasks: the replication of DNA, the equal distribution of the DNA to the two daughter cells, and the division of the rest of the cell material into two daughter cells. We can describe the process of cell reproduction in terms of three M's—*materials* (the small molecules that carry energy or serve as building blocks), *machinery* (the organelles and macromolecular structures needed to carry on cellular processes, including their DNA), and *memory* (the information, contained in the nuclear DNA). Before a cell divides, it usually doubles its materials, machinery, and memory. In most cases, each daughter cell then receives an equal share of each.

Each of the three M's is essential to cell reproduction. DNA, for example, contains the information needed to produce the next generation of cells and organisms. Without the memory encoded in DNA, the cell's materials and machinery are useless.

On the other hand, without the cell's materials and machinery, the DNA has no environment in which to express itself.

The precise copying of each DNA molecule employs some of the cell's most elaborate molecular mechanisms, which we discuss Chapter 10. In this chapter, we see how dividing cells coordinate the duplication and distribution of DNA and cytoplasm.

How Do Prokaryotic Cells Divide?

Cell division in prokaryotes is fairly straightforward. Prokaryotes divide by **binary fission,** in which a cell pinches in two, distributing its materials and molecular machinery more or less evenly to the two daughter cells. But what about the cell's memory? In most prokaryotes, genetic information is stored in a single molecule of DNA, which usually forms a closed loop (Figure 8-3). (This loop is sometimes referred to as a "circle" of DNA, even though it looks like a circle only in a diagram.) Before a prokaryotic cell divides, it duplicates its DNA molecule. Each of the two resulting daughter DNA molecules then attaches to the cell membrane in such a way that one ends up in each daughter cell.

Bacteria divide by binary fission in a process that does not involve mitosis.

How Do Eukaryotic Cells Divide?

Eukaryotic cells have many separate DNA molecules and many internal membranes and compartments, which makes the process of cell reproduction far more complicated than in prokaryotes. Just the existence of a nucleus poses a problem: DNA cannot possibly migrate to daughter cells as long as it is surrounded by the nuclear membrane. Furthermore, a eukaryotic cell can contain more than 1,000 times as much DNA as a prokaryotic cell.

We can observe the doubling and distribution of DNA if we stain dividing cells with a dye that binds to DNA. Figure 8-2 shows individual cells from an onion root in

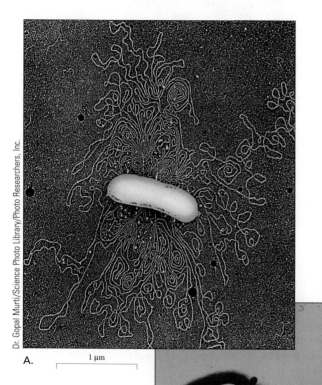

A.   1 μm

B.   1 μm

Figure 8-3
Prokaryote cells divide by pinching in two.
A. In *Escherichia coli* and most other prokaryotes, genetic information is stored in a loop of DNA. Here, circular DNA is spilling out from a broken *E. coli* bacterium. B. Dividing bacteria.

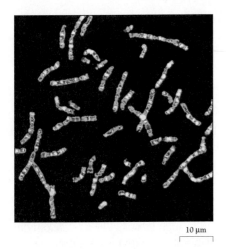

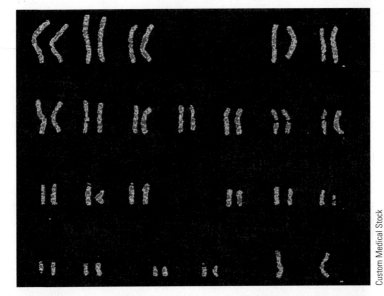

10 μm

A.

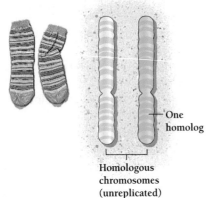

One
homolog

Homologous
chromosomes
(unreplicated)

B.

Figure 8-4
Chromosomes come in pairs, like socks. A. If we take a photograph of a set of chromosomes from a dividing cell, then arrange the images of the 46 individual chromosome in pairs of decreasing size, we create a tidy picture called a karyotype. In this micrograph, we can see chromosome bands, which can provide landmarks for locating genes. In each chromosome, the DNA stains in characteristic ways to form a recognizable pattern of bands. Using a karyotype, biologists can detect abnormal numbers of chromosomes and abnormal chromosomes (from abnormal banding patterns). B. Like pairs of socks, each member of a pair of chromosomes is similar, or "homologous," to its mate, but not identical.

various stages of division. We can see that, in some cells, the material of the nucleus has become organized into chromosomes. The cells of each species contain a characteristic number of chromosomes: onions have 16, humans have 46, and the fruit fly *Drosophila melanogaster* has 8. (Don't conclude, however, that 46 chromosomes make you smarter than a fruit fly or an onion. Your baked potato had 48 chromosomes in each cell before it went into the oven.)

Nearly every cell in the human body has copies of the same 46 chromosomes. (One exception is the red blood cell; as a red cell develops, it sheds its nucleus and ends up with no chromosomes at all. Another exception are the eggs and sperm, which have only one copy of each chromosome, for a total of 23.) The chromosomes of most sexually reproducing eukaryotes come in pairs, like socks (Figure 8-4). Humans, for example, have 23 pairs of chromosomes, whereas fruit flies have 4 pairs. Each member of the pair is

Figure 8-5

The cell cycle consists of G₁, S, G₂, mitosis, and cytokinesis. After the first growth phase (G₁), the DNA doubles (S). The cell then increases materials (G₂) in preparation for the separation of identical chromatids (mitosis) and the separation of the two daughter cells (cytokinesis).

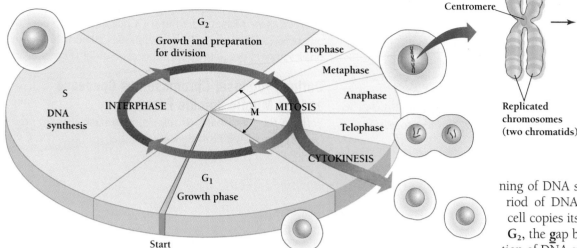

similar but not identical to its mate. Pairs of matching chromosomes are called **homologous chromosomes** [Latin, *homo* = same], and each member of the pair is called a **homolog.** (We will discuss the significance of these pairs in Chapter 9.)

Everything that happens to a cell from the time that it first forms until it divides is part of the **cell cycle**—the orderly sequence of events that accomplish cell reproduction. The time from one division to the next is one cycle. Biologists distinguish three major stages in the cell cycle:

1. **Mitosis,** the division of the nucleus.
2. **Cytokinesis,** the division of the cytoplasm and formation of two separate plasma membranes.
3. **Interphase,** the time when the DNA replicates and the cells grow.

The amount of time that a single cell cycle takes depends on the organism and on its circumstances. Most plant and animal cells are capable of dividing in about a day, with mitosis and cytokinesis occupying one or two hours. But every cell is different.

For example, a neuron spends most of its life in interphase, while rapidly dividing cancer cells spend very little time in interphase. The replication of DNA during interphase is invisible, and nothing appears to be happening in the cell. Only when biologists recognized the importance of DNA replication did they realize that what happens during interphase is just as important as what happens during mitosis and cytokinesis.

The cell cycle is divided into several phases (Figure 8-5). The most visually dramatic phase is **M,** for **m**itosis, which includes both mitosis and cytokinesis (the actual division of the cell). Interphase includes three parts: **G₁,** the **g**ap, or growth phase, between the completion of M and the begin-

ning of DNA synthesis; **S,** the period of DNA **s**ynthesis, when a cell copies its chromosomes; and **G₂,** the **g**ap between the completion of DNA synthesis and the beginning of M (of the next cell cycle). During G₁, a cell accumulates the materials and machinery needed for the DNA synthesis that will occur during S. In G₂, the cell prepares for mitosis and cytokinesis, assembling the molecular machinery needed to sort the chromosomes and divide the cell.

With each new phase of the cell cycle, the characteristics of the chromosomes change. At the beginning of mitosis, when the chromosomes first become visible, they have already duplicated. Each copy is called a **chromatid,** and the two chromatids are joined at a stretch of DNA called the **centromere.** Each chromatid is a single molecule of DNA, along with various proteins. The two chromatids (a single replicated chromosome) are copies with exactly the same genetic information. During mitosis, the chromatids separate and go to opposite poles of the cell.

During cytokinesis each chromosome gradually unfolds and becomes invisible in a microscope. The result is the fuzzy-looking DNA visible in stained, nondividing cells. Chromosomes remain unchanged throughout G₁. During the S phase, each chromosome—still in its stringy, invisible form—duplicates. Duplication takes 8 to 10 hours in most eukaryotic cells. During G₂, the replicated chromosomes remain invisible; only at the beginning of mitosis do the replicated chromosomes again condense and become visible. During G₁, the cell doubles nearly all of its materials and machinery (although not its DNA). If the cell runs out of nutrients, the cell cycle stops in G₁.

The cell cycle consists of four parts, G₁, S, G₂, and the M phase, which includes mitosis and cytokinesis. During G₁, a cell doubles all of its material. During S, it replicates its DNA. During mitosis, chromosomes are distributed to the daughter cells. During cytokinesis, the cell divides into two daughter cells.

8.3 How Does Mitosis Distribute One Copy of Each Chromosome to Each Daughter Cell?

The major task of mitosis is to separate pairs of chromatids and then to ensure that each daughter cell inherits a full complement of chromosomes. Remarkably, all eukaryotic cells accomplish mitosis with the same chromosomal ballet that Flemming first described more than 100 years ago. Modern staining and optical techniques, however, allow biologists to watch the whole process in much more detail.

Mitosis Is a Continuous Process, but Biologists Distinguish Four Phases

Cell biologists divide mitosis into four phases. These are prophase, metaphase, anaphase, and telophase. The boundaries between these phases are arbitrary, since mitosis is a continuous process. Nonetheless, the events described in these four phases always occur in exactly the same order (Figure 8-6).

During Prophase, Chromosomes Condense and the Mitotic Spindle Forms

Early in the first phase of mitosis, **prophase** [Greek, *pro* = before], replicated chromosomes condense and become vis-

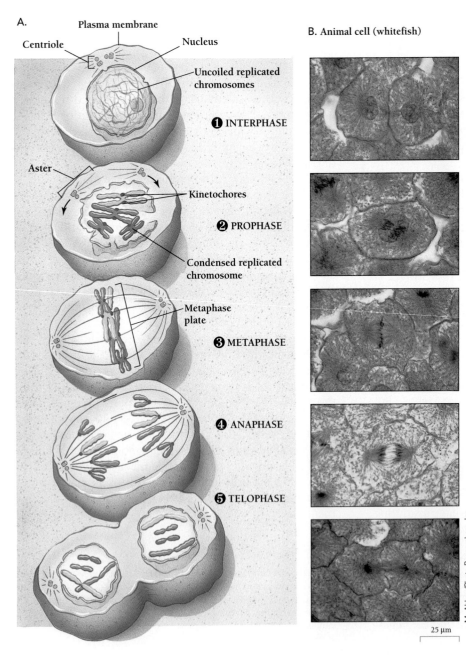

A.

Centriole · Plasma membrane · Nucleus · Uncoiled replicated chromosomes · **❶ INTERPHASE**

Aster · Kinetochores · **❷ PROPHASE** · Condensed replicated chromosome

Metaphase plate · **❸ METAPHASE**

❹ ANAPHASE

❺ TELOPHASE

B. Animal cell (whitefish)

M. Abbey/Photo Researchers, Inc.

25 μm

Figure 8-6

Mitosis. The stages of mitosis appear clearly in microscopic images of cells whose chromosomes have been stained. ❶ Interphase. The uncoiled chromosomes replicate. They remain invisible in the microscope but give the stained nucleus a fuzzy look. ❷ Prophase. The chromosomes condense and the nuclear membrane breaks up. ❸ Metaphase. The chromosomes line up along a central plane of the cell and the asters become visible. ❹ Anaphase. Spindle fibers pull the chromatids to opposite ends of the cell. ❺ Telophase. Two new nuclear membranes appear, the asters disintegrate, and the chromosomes uncoil, again becoming invisible.

Biology⊗Now™ Learn more about mitosis by clicking on this figure on your BiologyNow CD-ROM.

ible, each appearing as two chromatids, joined together at the centromere. This process takes 10 to 15 minutes in mammalian cells growing in culture dishes.

At the same time, a new structure, called a **mitotic spindle,** develops outside the nucleus. The mitotic spindle, which is shaped like an American football, consists of prominent bands of **microtubules**—hollow tubes made of the protein tubulin. Microtubules form the machinery that moves the chromosomes apart. The microtubules of the mitotic spindle are identical in molecular structure to those found in cilia, flagella, and the cellular cytoskeleton (Chapter 4).

In animal cells (and in some other eukaryotes as well), each pole of the mitotic spindle contains a pair of small cylinders, called **centrioles,** that lie at right angles to one another. Microtubules originate from each pole of the spindle. Some run toward the equator, others extend outward from each centriole. Together, they form a star shape, or aster, around the centriole, just as Flemming first observed. The cells of most plants have no centrioles or asters.

By late prophase, the spindle has started to elongate, and the two poles (with or without asters) begin to move apart. Researchers give the transition between prophase and metaphase its own name, "prometaphase." Now the nuclear membrane disappears and the chromosomes attach to the mitotic spindle.

In mammalian cells, the chromosomes take 10 to 20 minutes to attach to the mitotic spindle. Each chromatid develops a **kinetochore,** a specialized disc-shaped structure to which attach the microtubules of the mitotic spindle (Figure 8-6). The two kinetochores make up a centromere. After the spindle apparatus has formed, the attached chromosomes are dragged along the spindle toward the cell's equator.

During Metaphase, Chromosomes Align

By the end of **metaphase,** the second and longest stage of mitosis, the chromosomes have lined up between the two poles of the spindle, where they form a disc, called the **metaphase plate.** For about an hour, all the replicated chromosomes lie in a single plane at the equator (at right angles to the spindle fibers). At this stage, each pair of replicated chromosomes has a distinctive appearance that distinguishes it from all the others. The differences in size and in the position of the centromere on each chromosome allow researchers to lay the chromosomes out neatly in rows, a pattern called a karyotype (Figure 8-4).

During Anaphase, Chromatids Separate

The most dramatic stage of mitosis is **anaphase,** the time of chromatid separation. The microtubules attached to the centromeres begin to shorten, splitting the centromere of each replicated chromosome. Then, all at once, the two separated chromosomes begin to move away from the metaphase plate as the spindle microtubules continue to shorten. For 5 to 10 minutes, all the chromosomes slowly move toward the two poles of the spindle, pulled by the centromere, with the rest of the chromosome dragging behind like a small dog being dragged by its leash. Even as the shortening microtubules drag the passive chromosomes toward the spindle poles, the poles themselves are also moving farther apart, often doubling the distance between them.

During Telophase, the Mitotic Apparatus Disappears

During **telophase** [Greek, *telos* = end], the last phase of mitosis, the mitotic apparatus breaks down. The chromosomes unwind, and the distinctive bar or string shapes are no longer visible with a microscope. The nucleus reappears, with new nuclear membranes enclosing the two new sets of daughter chromosomes. Each daughter nucleus normally has the same number of chromosomes as the parent nucleus, but now each chromosome consists of a single chromatid. Each chromosome will replicate during the next S phase. After the completion of telophase, the cell usually completes cytokinesis, which we discuss later.

What Propels the Chromosomes During Mitosis?

We have described three sets of movements whose molecular mechanisms we would like to understand: (1) the movement of the chromosomes toward the equator to form the metaphase plate, (2) the movement of the separated chromatids toward the poles, and (3) the movement of the spindle poles away from each other during prophase and again during anaphase. The chromosomes themselves do not play active roles in their own movement. Rather, the major players are the microtubules. These seem to move the chromosomes around by lengthening and shortening, as necessary.

Experiments have demonstrated the importance of microtubules in mitosis. In one experiment, a researcher used a laser to cut the kinetochore microtubules that attach one of the two chromatids to one pole at metaphase. The loose chromatid followed its twin chromatid to the pole opposite the one to which it normally would have gone. This experiment showed that the kinetochore microtubules normally pull the two chromatids in opposite directions. Cell biologists still do not fully understand the mechanism of mitosis, however, and the identities and roles of the components of the mitotic machinery are the subject of intense research.

During mitosis, microtubules move the chromosomes by attaching to the chromatids and growing or shortening.

8.4 How Does a Cell Fit All Its DNA into a Nucleus?

The DNA in a cell carries information that the cell uses to construct cell machinery and to carry out cell processes. When a cell divides, the cell must duplicate this information and pass it on to its two daughter cells.

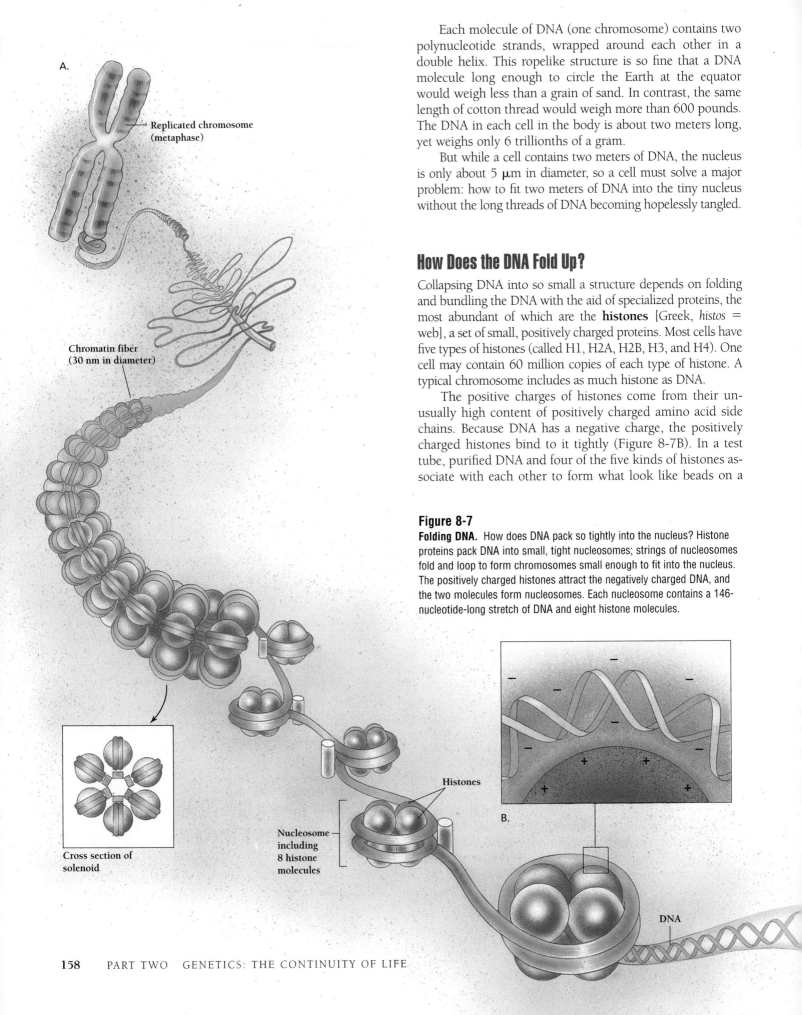

A.

Replicated chromosome
(metaphase)

Chromatin fiber
(30 nm in diameter)

Cross section of
solenoid

Nucleosome
including
8 histone
molecules

Histones

B.

DNA

Each molecule of DNA (one chromosome) contains two polynucleotide strands, wrapped around each other in a double helix. This ropelike structure is so fine that a DNA molecule long enough to circle the Earth at the equator would weigh less than a grain of sand. In contrast, the same length of cotton thread would weigh more than 600 pounds. The DNA in each cell in the body is about two meters long, yet weighs only 6 trillionths of a gram.

But while a cell contains two meters of DNA, the nucleus is only about 5 μm in diameter, so a cell must solve a major problem: how to fit two meters of DNA into the tiny nucleus without the long threads of DNA becoming hopelessly tangled.

How Does the DNA Fold Up?

Collapsing DNA into so small a structure depends on folding and bundling the DNA with the aid of specialized proteins, the most abundant of which are the **histones** [Greek, *histos* = web], a set of small, positively charged proteins. Most cells have five types of histones (called H1, H2A, H2B, H3, and H4). One cell may contain 60 million copies of each type of histone. A typical chromosome includes as much histone as DNA.

The positive charges of histones come from their unusually high content of positively charged amino acid side chains. Because DNA has a negative charge, the positively charged histones bind to it tightly (Figure 8-7B). In a test tube, purified DNA and four of the five kinds of histones associate with each other to form what look like beads on a

Figure 8-7

Folding DNA. How does DNA pack so tightly into the nucleus? Histone proteins pack DNA into small, tight nucleosomes; strings of nucleosomes fold and loop to form chromosomes small enough to fit into the nucleus. The positively charged histones attract the negatively charged DNA, and the two molecules form nucleosomes. Each nucleosome contains a 146-nucleotide-long stretch of DNA and eight histone molecules.

string (Figure 8-7A). One histone bead, along with its two wraps of DNA and a length of DNA linking to the next bead, is called a **nucleosome.** The formation of nucleosomes is the first stage in the packing of DNA into a nucleus.

Histone proteins from all eukaryotes are remarkably similar. For example, the histone H4 of cows differs by only two amino acids out of 102 from that of peas. Since the ancestors of these organisms diverged more than a billion years ago, the extraordinary similarity in their histones suggests that the structure of histone H4 has remained almost unchanged for a billion years or more. This method of packing DNA must therefore have evolved early in evolutionary history.

Proteins called histones participate in packing DNA into small, tight bundles called nucleosomes.

How Do DNA, Histones, and Other Proteins Form Such Compact Structures?

Histone H1 appears to help coil and fold groups of nucleosomes to form a fiber that is about 30 nm in diameter. As a result of this packing, the DNA of a typical human chromosome could be condensed about five times. But this is still not small enough for it to fit into the nucleus. We still do not know exactly how DNA is made as compact as it is. In some specialized cells, microscopists can see loops extending from chromosomes (Figure 8-8). Many biologists now think that all cells contain such loops. Throughout interphase, parts of chromosomes unfold and then refold, allowing proteins involved in gene expression access to the DNA.

During cell division, the DNA becomes even more condensed. The looped fiber itself forms loops. The specific pattern of DNA looping, coiling, and folding in each replicated chromosome causes **chromosome banding,** a striped pattern visible in a light microscope because dyes darken each band a little differently. Biologists can detect abnormal chromosomes by comparing their banding pattern with that of a normal chromosome. It is possible to determine the location of individual genes on specific chromosomes, using chromosome bands as landmarks (Figure 8-4A).

Strings of nucleosomes fold and loop to form a highly compact arrangement that fits into the nucleus. No one knows exactly how this occurs.

8.5 How Does a Cell Divide Its Cytoplasm?

So far we have discussed how a eukaryotic cell distributes the two copies of its genetic information. But we still need to describe cytokinesis—the process by which a dividing cell partitions its cytoplasm.

Cytokinesis and mitosis are separate processes, and they depend on different molecular machinery. In some organisms mitosis can occur without cytokinesis, leading to two or more nuclei within a single cell. Most of the time, however, mitosis and cytokinesis go together. Cytokinesis usually begins during anaphase and finishes shortly after the end of telophase.

During cytokinesis, the cell divides perpendicularly to the long axis of the mitotic spindle. In early anaphase, divid-

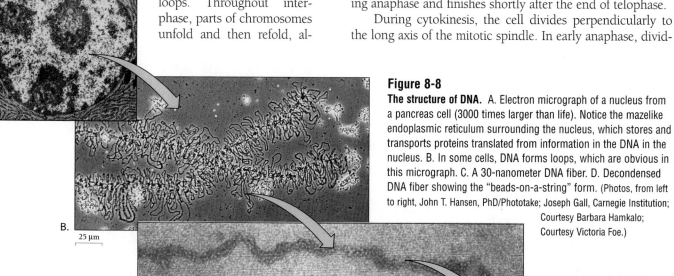

A.

B.

25 μm

C.

D.

Figure 8-8

The structure of DNA. A. Electron micrograph of a nucleus from a pancreas cell (3000 times larger than life). Notice the mazelike endoplasmic reticulum surrounding the nucleus, which stores and transports proteins translated from information in the DNA in the nucleus. B. In some cells, DNA forms loops, which are obvious in this micrograph. C. A 30-nanometer DNA fiber. D. Decondensed DNA fiber showing the "beads-on-a-string" form. (Photos, from left to right, John T. Hansen, PhD/Phototake; Joseph Gall, Carnegie Institution; Courtesy Barbara Hamkalo; Courtesy Victoria Foe.)

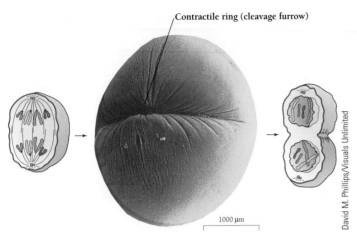

Contractile ring (cleavage furrow)

1000 μm

David M. Phillips/Visuals Unlimited

Figure 8-9
A dividing cell. During cytokinesis, actin filaments inside the cell form a contractile ring, which forms a "cleavage furrow" that pinches the cell in two.

two daughter cells. Cell division begins during telophase as the cell constructs a small, flattened disc in the space between the two daughter nuclei. The disc of new cell wall surrounded by cell membrane, called the **cell plate**, grows until its edges fuse with the cell membrane of the parent cell, and so separates the two daughter cells (Figure 8-10). Later, the cell produces cellulose fibers to strengthen this new cell wall.

During cytokinesis in animal cells, actin filaments form a contractile ring that pinches the cytoplasm in two. In plant cells, telophase includes the building of new cell membranes and cell walls between the two daughter cells.

ing animal cells form a **contractile ring**, a belt of actin filaments just inside the cell membrane of a dividing cell. Like a belt being cinched tight, the contractile ring pulls a ring of cell membrane inward, forming a deepening groove called the **cleavage furrow**, that divides the cell in two (Figure 8-9).

As telophase proceeds, the two daughter cells become almost, but not quite, completely separated. A thin connection between the daughter nuclei, called the midbody, persists until the end of mitosis. The midbody is packed with microtubules from the spindle apparatus. Finally, at the end of telophase, the microtubules disassemble, and the midbody breaks, leaving the daughter cells completely separated.

As the two daughter cells separate, they must enlarge their cell membranes (because two cells have more surface area than one). The additional membrane material comes from a supply of extra membrane made by the parent cell during interphase.

A plant cell, which has a rigid cell wall outside the cell membrane, cannot simply pinch its cytoplasm in two the way an animal cell does. A dividing plant cell must build two new sections of cell membrane and a new cell wall between the

8.6 How Does a Cell Regulate Passage Through the Cell Cycle?

In this chapter, we have seen how normal cells accomplish the equal distribution of the chromosomes by mitosis and the division of the cytoplasm by cytokinesis. But how do cells know when to begin the cell cycle? How do they coordinate its separate parts? And how do they know when *not* to start the cycle?

If the cell cycle repeated over and over without stopping, single-celled organisms would soon run out of space and food. For example, after just four days of undisturbed growth, a single yeast cell could have divided 48 times and produced about 10 million billion descendants, as many cells as are in 100 people.

In healthy multicellular organisms, cells normally subordinate such growth and division to the needs of the body's tissues and organs. Cells divide rapidly during periods of growth and more slowly (or not at all) in mature organisms. The great exception to this rule is the uncontrolled repetition of the cell cycle characteristic of cancer cells, such as those of Henrietta Lacks.

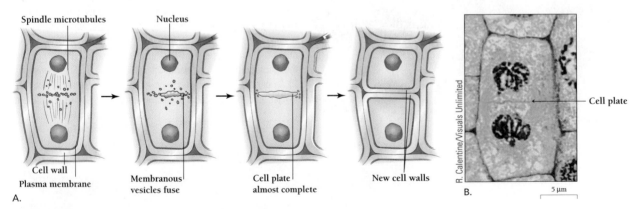

Spindle microtubules   Nucleus

Cell wall
Plasma membrane

Membranous
vesicles fuse

Cell plate
almost complete

New cell walls

A.

B.   5 μm

R. Calentine/Visuals Unlimited

Cell plate

Figure 8-10
New cell wall. In plants, a new cell wall forms between daughter cells. The cell wall begins during telophase as a small, flattened disc, called the early cell plate, then grows into a full-sized cell wall, complete with interconnecting plasmodesmata.

How Do Normal Cells Determine When to Stop Dividing?

What normally prevents such runaway cell division? Two general mechanisms appear to operate—cell senescence (aging) and growth control. **Cell senescence** limits the number of times a normal cell can divide: the more times a cell has divided (at least under conditions of laboratory culture), the more likely it is that the cell will not divide. An average cell taken from a newborn baby, for example, will divide about 50 times in a standard culture medium. But cells taken from an 80-year-old stop cycling after about 30 divisions. Despite the general occurrence of cell senescence, however, even a mature mammal contains cells that are capable of unlimited cell divisions in the presence of the right "growth factors."

A second set of regulatory mechanisms—those involved in **growth control**—prevent uncontrolled division by allowing the cell cycle to proceed under some conditions and to stop under others. As an example of growth control, think about what happens when you cut yourself. The skin cells on the edges of the cut begin to divide and fill the space left by the wound. The cells divide until the two edges again touch. When the cells on each side of the cut meet, and the wound is healed, cell division ceases. In some people, however, the cells don't stop properly and an overgrowth of scar-like tissue, called a *keloid*, may result.

In a culture dish, cells regulate their division in much the same way (Figure 8-11). Cells attach to the bottom of a plastic dish and divide until they form a single, continuous layer that occupies the whole surface. Then they stop dividing. If we make a "wound" in this monolayer by scraping away a swath of cells, the cells on the margins move into the open space and begin to divide. Division stops when the space is filled.

We can summarize the growth control of normal cells in culture with the following rules: (1) cells stop dividing during G_1 when they run out of free space on which to spread—that is, when neighboring cells all touch each other; and (2) cells that have proceeded beyond G_1 begin to divide when contact with their neighbors ceases. This kind of growth control is called **contact inhibition** of cell division.

Cancer cells provide an excellent example of what happens when contact inhibition fails. Cancer cells do not exhibit growth control in cell culture (Figure 8-11). Instead of forming a continuous monolayer, they pile up on top of one another and continue to divide.

Growth control depends on the actions of several types of proteins. Some of these proteins are closely related to proteins specified by genes, called *oncogenes*, present in tumor-causing viruses (Chapter 12). Among these growth-regulating proteins are (1) growth factors, which stimulate cell division or promote cell survival, depending on the type of cell; (2) proteins that bind to growth factors and contribute to the sequence of events that they trigger; and (3) proteins that regulate gene expression. A fourth type of protein, called "tumor-suppressor" proteins, regulates the passage of cells through the cell cycle.

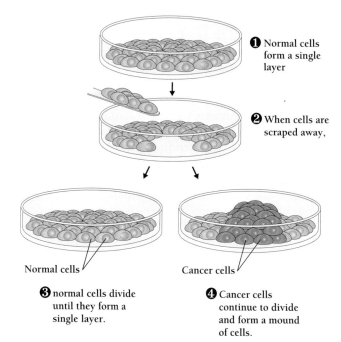

❶ Normal cells form a single layer

❷ When cells are scraped away,

Normal cells

❸ normal cells divide until they form a single layer.

Cancer cells

❹ Cancer cells continue to divide and form a mound of cells.

Figure 8-11

❶ In cell cultures, normal cells divide until they form a single, continuous layer of cells, then stop. ❷ If we scrape a "wound" in this monolayer of normal cells, cells from the edge move into the open space and divide until the wound is filled. ❸ Normal cells stop dividing when they come into contact with other cells. ❹ Cancer cells do not self regulate in this way. Instead, they continue to divide even after they come into contact with other cells, piling up on top of one another.

Biology ⊗ Now™   Learn more about growth of cells by clicking on this figure on your BiologyNow CD-ROM.

Changes in just one kind of tumor-suppressor protein, called *p53*, occur in more than 50 percent of human cancers.

Growth control mechanisms such as contact inhibition prevent normal cells from growing out of control the way cancer cells do.

How Do Normal Cells Determine When It Is Time to Divide?

Most cells divide only after they have first doubled their volume. If they didn't, the daughter cells would be too small. Somehow cells sense when they have reached a critical size and have enough nutrients to supply two daughter cells.

We do not know how eukaryotic cells accomplish this sensing, but once they do, they become irreversibly committed to cell division. Once a cell proceeds beyond a certain point in G_1, the cell proceeds through the rest of the cycle, including mitosis and cytokinesis. This "point of no return" is called **Start**. Arrival at Start depends heavily on the environment of the cell—including the availability of nutrients and signals from other cells. Once the cell has gone through Start, however, the rest of the cycle proceeds, independent of the extracellular environment.

One important exception to the general rule that cells double their mass before dividing are egg cells, which grow to enormous sizes without dividing. For example, the ostrich egg is a single cell. Another important exception is the early embryos of many animal species. Embryos derived from large eggs begin life after fertilization with a series of rapid cleavage divisions, which divide the embryo into many cells without increasing its mass. A frog zygote, for example, is an enormous single cell about one millimeter in diameter. Cleavage divisions rapidly divide it into much smaller cells that are more nearly the size of cells in the adult.

> The cells of mature organisms divide only when they have doubled their mass, unless they are egg cells or the cells of an early embryo.

What Triggers the Main Events of Mitosis?

Cell reproduction requires the coordination of growth, DNA synthesis, mitosis, and cytokinesis, and all eukaryotes, from yeasts to humans, use almost the same mechanisms to coordinate these processes.

Researchers often study cell cycles by creating **synchronous cell populations,** groups of cells that are all at the same stage of the cell cycle. One source of synchronous cell populations are the rapidly dividing cells of embryos from marine invertebrates; a researcher can initiate cell division in thousands of embryos at once just by mixing eggs and sperm together. Another way to make synchronous populations is with a chemical inhibitor that stops the cell cycle just before S. When the inhibitor is removed, all the cells begin S at the same time and continue in lock step through the rest of the cell cycle.

Studies of synchronized cells have led to the discovery of specific proteins, called **cyclins,** whose concentrations change through the cell cycle. Each type of cyclin accumulates during a particular phase of the cell cycle and stimulates the passage of the cell into the next phase of the cycle.

Each type of cyclin stimulates a specific enzyme, called a cyclin-dependent kinase, which alters proteins of the cell (by adding phosphates from ATP). Among the targets of these enzymes are components of the mitotic spindle. These protein modifications are the "ticks" of the cell cycle clock, propelling the cell to enter the next phase. As the cell moves into the next phase of the cycle, the cell's protein-digesting machines (the proteasomes) digest each type of cyclin, causing the oscillations that give the cyclins their names.

> Various events regulate a cell's passage through the cell cycle, including the completion of DNA synthesis and fluctuations in cyclins.

In this chapter we have seen how eukaryotic cells with two sets of chromosomes divide their genetic material to create two daughter cells, each with a full complement of chromosomes. Eukaryotes, such as ourselves, employ such mitotic division to create genetically identical copies of cells. In the next chapter, we will see how eukaryotic cells divide a cell with two sets of chromosomes to create gametes (for example sperm and eggs)—cells with just one set of chromosomes. This process, called meiosis, creates cells that are genetically different from the parent cells and also genetically different from one another. Along the way, we will see that sex and meiosis are two sides of the same coin.

Key Concepts

- Cell reproduction in eukaryotes involves three distinct but interconnected processes: the replication of DNA in the nucleus, its equal distribution to each of two daughter cells, and the division of the cytoplasm.
- During mitosis, microtubules and other subcellular components bring about chromosome movements that distribute one copy of each chromosome to each daughter cell.
- Eukaryotic cells, ranging from yeast to human cells, use similar molecular mechanisms to regulate passage through the cell cycle. Failure of these mechanisms can result in cancer.

Summary with Key Terms

How is cell division critical to the lives of single-celled and multicelled organisms?

Cells reproduce themselves in a process called **cell division,** in which a parent cell gives rise to two **daughter cells,** each with the same genetic information as the parent cell. Dividing cells replicate DNA and distribute it during an orderly sequence of events called the **cell cycle.**

How do eukaryote cells divide during mitosis and cytokinesis?

Prokaryotes divide by a simple process called **binary fission,** which distributes one copy of the parent cell's DNA to each daughter cell. Cell reproduction is more complicated in eukaryotic cells, which have many chromosomes that are enclosed in a nucleus. Eukaryotes employ a spindle apparatus, but prokaryotes do not. In dividing cells, **chromosomes** condense and become visible. In nondividing cells, chromosomes unwind and become invisible (but they are still there). Chromosomes come in nonidentical pairs called **homologous chromosomes.** Each member of a pair is a **homolog.** When the DNA in a cell has replicated, each replicated chromosome consists of two identical **chromatids.**

Cell biologists divide the cell cycle into four phases: M, mitosis and cytokinesis, G_1, S, and G_2. The process of evenly distributing chromosomes to daughter cells is called **mitosis.** The process of dividing a cell's cytoplasm is called **cy-**

tokinesis. During G_1, a cell accumulates the materials and machinery necessary to **replicate** its DNA; during **S**, it synthesizes DNA, making two copies of its DNA; during G_2, it prepares for mitosis and cytokinesis. G_1, S, and G_2 together are called **interphase**. The point of no return is **Start**.

Just before a cell begins mitosis, each replicated chromosome consists of two identical chromatids. Mitosis is a continuous process, but biologists distinguish four major phases: prophase, metaphase, anaphase, and telophase. During **prophase**, chromosomes condense and the **mitotic spindle** forms. As the **centrioles** of animal cells move apart, **microtubules** radiating outward from the centrioles form **asters**. Between prophase and metaphase, the nuclear membrane disappears and the chromosomes attach to the spindle fibers. During **metaphase**, the chromosomes align in a disc called a **metaphase plate**, then bind to the **kinetochores** that form on the **centromere** of each chromatid. During **anaphase**, the two chromatids that make up each replicated chromosome separate. During **telophase**, the mitotic apparatus disappears.

During interphase, the cell duplicates its single centriole. During prophase, a mitotic spindle assembles, with microtubules emerging from each centriole to form a mitotic spindle. Microtubules connect to each chromatid in a structure called a kinetochore. The mitotic apparatus is responsible for chromosome movements during mitosis.

During cytokinesis in animals cells, a **contractile ring** drags a circle of membrane inward to form a **cleavage furrow**, which deepens and separates the two daughter cells. Cytokinesis in plant cells requires a special mechanism for building a new cell wall. During telophase, a small membrane-covered disc grows to become a **cell plate**.

How does the DNA fit into the nucleus without becoming tangled?

The DNA of eukaryotic cells is highly folded. Protein molecules called **histones** bind to DNA and form particles that resemble beads on a string. One bead and the DNA linking it to the next bead make up one **nucleosome**. DNA, histones, and other proteins fold to form still more compact structures. The orderly packing of the DNA differs from chromosome to chromosome, so that each metaphase chromosome has a characteristic **chromosome banding** pattern, which dyes can reveal.

How do cells regulate the cell cycle?

Cells actively regulate passage through the cell cycle. Regulated cell division in multicellular organisms depends on **cell senescence** and **growth control**. Cell senescence limits the number of times a cell can divide. Growth control regulates cell reproduction according to external conditions. For example, many cells exhibit **contact inhibition** of cell division, in which cells divide only when they are not in contact with their neighbors. Growth control may also involve growth factors—extracellular proteins that stimulate cell division. Cancer cells fail to show normal growth control. Studies of **synchronous cell populations** show that changes in specific proteins, called **cyclins**, control a cell's passage through the cell cycle.

Review and Thought Questions

Review Questions

1. Why must a cell double its materials, machinery, and memory before it can divide?
2. Why did the discovery that DNA was the genetic material change the way that cell biologists regarded interphase?
3. How does the amount of DNA in a cell change during the cell cycle?
4. Describe and draw the events of the four stages of mitosis. When does the actual division of the replicated chromosomes occur?
5. Using diagrams, compare the events in prophase to those in telophase. Why do you think they are so nearly opposite?
6. Compare the role of the centriole in mitosis with that of a kinetochore.
7. What factors regulate a cell's ability to proceed to the next round of cell division?
8. Describe the role of histones in the structure of eukaryotic chromosomes. How do histones help DNA fit into a nucleus?
9. How does cytokinesis in a plant cell differ from that in an animal cell?

Thought Questions

10. How would preventing DNA synthesis affect mitosis?
11. What is the relationship between the materials accumulated in G_1 and G_2 and the processes that follow?
12. How might cancer be associated with improper regulation of the cell cycle?

BiologyNow Resources

Biology ⊗ Now™

Active Figures

8-6: Mitosis
8-11: Growth of cells

Preparing for an exam? Take a diagnostic test on your BiologyNow CD-ROM.

Online materials relating to this chapter are at:
http://biology.brookscole.com/AAL3

About the Chapter-Opening Image
Socks come in matched pairs, as do chromosomes. Just as a sock can be damaged or wear out, so chromosomes can change, or mutate. Just as two matched socks can different slightly, the two chromosomes of a pair can differ.

From Meiosis to Mendel

Key Questions

- How do sexually reproducing organisms use meiosis to keep the same number of chromosomes from generation to generation?

- How do the mechanics of meiosis explain Mendel's rules of segregation and independent assortment?

- What is the value of recombination and genetic variation?

- Why are some diseases "sex-linked"?

Why Is the Yellow Dog Yellow?

In the spring of 1902, a young graduate student at Columbia University, in New York City, approached his professor with barely suppressed excitement and declared that he had solved the problem of heredity. The student, Walter S. Sutton (1877–1916), was a six-foot-tall, 215-pound Kansas farm boy who towered over everyone around him (Figure 9-1). Gleefully, he announced, "I know why the yellow dog is yellow." At just 25, Sutton had discovered that tiny bars in cells, called chromosomes, must carry the hereditary material.

Chromosomes, realized Sutton, were the mechanism by which organisms inherit traits. They were thus the key to the then-infant science of genetics. But disappointingly, Sutton's professor only nodded indulgently. Neither he nor anyone else took Sutton seriously.

It was disappointing, especially because Sutton's professor was E. B. Wilson (1856–1939), the most distinguished cell biologist in the United States. But when summer came, Wilson invited Sutton to the seaside at Beaufort, North Carolina—to study marine animals, to sail, to swim, and to talk. For the usually preoccupied Wilson, it was also a slow time when he would really listen to young Sutton. "It was only then," Wilson later wrote, "that I first saw the full sweep and the fundamental significance of his discovery."

Sutton had come to biology partly by chance, and he hadn't yet decided if he would stay. In 1896, he had begun studying engineering at the University of Kansas. When he returned home for the summer in 1897, he took with him a case of typhoid fever that sickened his entire family. When it was over, Walter Sutton's beloved 17-year-old brother John was dead. In his grief, Sutton decided to abandon engineering for medicine. That fall, he returned to the University of Kansas at Lawrence and began to study biology.

Before long, Sutton became friends with a young professor of zoology named Clarence McClung. On vacations home, Sutton collected giant grasshoppers from the wheat fields of his family's farm to send to McClung

(Figure 9-2). The grasshopper's enormous cells were perfect for studying the fine structure of a cell.

McClung was especially interested in the fact that some of the grasshopper sperm cells seemed to have an extra chromosome. At the time, no one knew what chromosomes were, but McClung hypothesized that this "accessory chromosome" determined sex. In essence he was correct. Today, we know that in many animals (including mammals, birds, and insects), males and females differ in one chromosome. In humans and fruit flies, for example, females have two X chromosomes and males have one X chromosome and one Y chromosome. The X and Y chromosomes are the "sex chromosomes." But at the end of the 19th century, no one knew that, and McClung's novel idea that a chromosome could give an individual grasshopper a specific trait—its sex—suggested to Sutton that the other chromosomes might do the same thing for other traits.

In April 1900, Sutton published a paper describing in great detail the sex cells of the lubber grasshopper, including 41 drawings and 10 photographs. A year later, Sutton received his master's degree based on his work on cell division in the sperm cells of the grasshopper. In the fall of 1901, encouraged by McClung, he enrolled at Columbia University to work with the eminent cell biologist E. B. Wilson.

Within six months, Sutton was hanging on Wilson's sleeve trying to persuade him that he had solved the mystery of heredity. That fall, the famous British biologist William Bateson came to Columbia to talk about the new science of genetics. Two years before, Bateson told his audience, three botanists had rediscovered the laws of inheritance first described and published by the Austrian monk Gregor Mendel in 1866. These laws laid a foundation for all future work on inheritance. The years between 1900 and 1910 would be a breathtaking period in biology, and Bateson's inspired talk about inheritance electrified his American audience.

Bateson banished any doubts Sutton may have had about the importance of his idea. Everything Bateson said made perfect sense in the light of Sutton's knowledge of chro-

Figure 9-1

Walter Sutton. At age 25, Sutton discovered the location of the genes in a cell. Genes, he said, had to be in the nucleus on the chromosomes.

University of Kansas Archives

mosomes. Within one month, Sutton had published a brief paper describing his idea that the chromosomes are the physical basis for heredity. Four months later, he had published a full-length paper describing the theoretical implications of his idea.

Research in the 100 years since has confirmed nearly every detail of Sutton's hypothesis. His ideas are so well accepted today that it is hard to imagine why the more experienced biologists in his time did not notice the role that chromosomes play in inheritance. Yet, as is so often the case in science, not only did older biologists not see what now seems obvious, but some continued to resist Sutton's idea for nearly 40 years.

Maybe it was partly the cool response of his elders. Maybe Sutton thought he could do more immediate good in the world as a doctor. In any event, after doing work so important that it would forever influence biology, Sutton dropped out of graduate school, abandoning biology forever. Two years passed before, pressured by his father, he returned to Columbia, where he resumed his long-delayed medical training. He was as outstanding a medical student as he had been a biologist, and in 1907 he received his medical degree. Returning to Kansas in 1909, he practiced as a surgeon for the rest of his short life. Still a bachelor, he died of a burst appendix at 39.

Nevertheless, his research in biology would change the world. At 25, he had discovered a fundamental fact that had eluded scientists for decades. Sutton revealed the basic mechanism by which organisms inherit traits—why, as he put it, "the yellow dog is yellow." Sutton's work is the foundation for all of modern genetics and molecular biology. A hundred years later, we even know the precise nature of the two genes responsible for the yellow, chocolate, and black colors of Labrador retrievers. In the next section, we will see why it took a young biologist such as Sutton to recognize what now seems so obvious. In the rest of this chapter, and in succeeding chapters, we will try to find out exactly what the chromosome does.

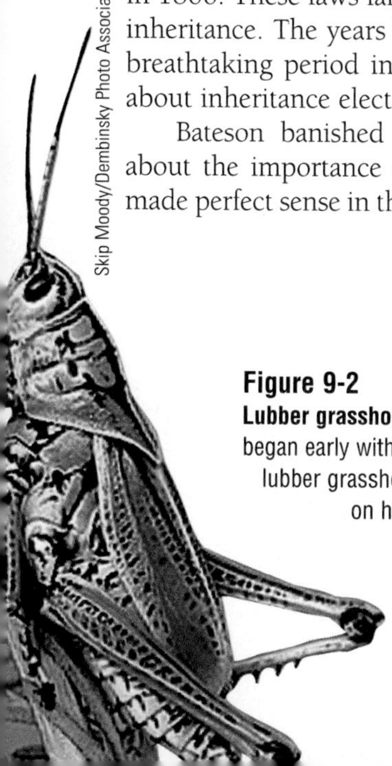

9.1 Why Was the Chromosomal Theory of Inheritance So Hard to Accept?

Historically, biologists had no trouble posing theories to explain how organisms stay the same between generations. Before the 17th century, a favorite theory was that miniature versions of future organisms were already "preformed" in

Skip Moody/Dembinsky Photo Associates

Figure 9-2

Lubber grasshopper. Walter Sutton's interest in biology began early with his research on the chromosomes of the lubber grasshopper, *Brachystola magna,* which he collected on his parents' farm in Kansas.

sperm or eggs, so that the properties of all future generations were already determined (Chapter 45). Many philosophers believed that the egg contributed the materials for an individual and that the sperm contributed some life force that determined the form or shape of the individual. Philosophers compared the creative action of the sperm to the divine creation of the universe from formless matter.

Blending Inheritance: A Wrong Turn

In contrast, plant breeders had long known that when two varieties of plants are crossed, both parent plants contribute equally to the form of the next generation. No matter which kind of plant provided the sperm (pollen) and which kind provided the egg, the resulting cross was the same. In the face of such evidence, most biologists reasoned that the different types of offspring somehow resulted from the "blending" of the genes of the parents. We are not surprised, for example, when a black mare and a white stallion produce a gray foal or when a red snapdragon crossed with a white snapdragon produces a plant with pink flowers (Figure 9-3). The color of the offspring appears to be the result of blending, as if two cans of paint have been poured together and mixed.

The **blending** model of inheritance appeared to work for traits that varied smoothly, such as height and skin color in humans. But plenty of evidence did not fit the blending model. Pink flowers can give rise to red or white offspring, and brown-eyed parents often have blue-eyed children. If traits blend, how could unblended characteristics reappear in later generations?

The blending model also created a serious problem in understanding the mechanism of evolution (Chapter 15). If traits blended, variation would eventually disappear and natural selection would have no inherited variation on which to act. For 19th-century biologists, the problem of blending inheritance was so intriguing that, in time, nearly every important biologist of the second half of the 19th century had a different hypothesis about how inheritance worked.

Most 19th-century biologists mistakenly accepted blending inheritance.

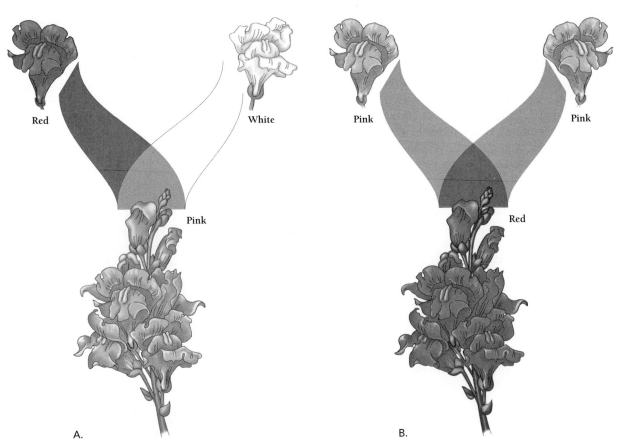

Figure 9-3
Blending inheritance? The offspring of red and white snapdragons are pink. A. Blending inheritance seemed reasonable to early biologists. But organisms inherit color and other traits by way of discrete units called genes. B. Blending inheritance could not explain how a pink flower could have offspring with red flowers.

Why Did Biologists Doubt That Chromosomes Could Carry Information?

Recall from the last chapter that a wealth of clues suggested that the chromosomes carry the hereditary material. Walther Flemming had shown that during mitosis, the chromosomes appeared to divide down the middle, so that half of each chromosome went to each cell. Wilhelm Roux had argued that this complex way of dividing the chromosomes at mitosis suggested that each chromosome was unique.

But the idea that the chromosomes were responsible for heredity was still no more than a possibility, and most 19th-century biologists did not take the idea seriously. One exception was the German biologist August Weismann. Weismann hypothesized that organisms inherit traits by means of some information-carrying molecule in the nucleus of the cell. Moreover, he argued, since the hereditary material from the mother and father fuse at fertilization, the egg and sperm must each contain only half the normal amount of hereditary material. Otherwise, said Weismann, the hereditary material would double every generation.

Within three years, cell biologists confirmed Weismann's prediction. First, they showed that sperm and eggs contain only half the usual number of chromosomes. Second, they observed the special process of **meiosis,** by which cells produce daughter cells (eggs and sperm) with only half the normal number of chromosomes. By the late 19th century, most cell biologists accepted that chromosomes were central to the process of both mitosis and meiosis and probably somehow essential for the normal development of an embryo into an adult. But they did not know what the chromosomes did.

In 1900, however, an intellectual earthquake rocked the scientific landscape. In that year, three scientists independently rediscovered the work of an obscure but brilliant scientist named Gregor Mendel. In 1866, Mendel had discovered several important principles of heredity while studying inheritance in pea plants. In particular, Mendel had shown that every organism has two sets of what we now call genes.

The rediscovery of Mendel's work gave birth, in just a few months, to an entire new field of science—genetics. Biologists began to suspect that the behavior of chromosomes during the creation of sperm and egg cells (meiosis) might explain Mendel's principles. Chromosomes suddenly claimed the intense attention of a great many biologists.

But it was one thing to suspect that chromosomes *might* carry genes, quite another to show that they *did*. Two important problems stood in the way. First, when biologists looked at cells under a microscope, the chromosomes simply vanished between cell divisions. It seemed possible, even likely, that cells create new chromosomes after each cell division. In short, biologists had no way of knowing that chromosomes passed intact from generation to generation. And if the chromosomes didn't pass intact from generation to generation, biologists reasoned, they could not carry information from parent cell to daughter cell.

Second, biologists had no evidence that the individual chromosomes in a cell were different from one another. The chromosomes seemed to come in a variety of shapes, but they all behaved in the same way, and they all seemed to be made of the same material. Like pretzels, they might come in slightly different shapes, but they were all the same stuff. There was nothing to suggest that each one might be carrying unique information.

Weismann's argument that each chromosome always maintains its individuality and integrity was only a guess. Young Walter Sutton, the Columbia University graduate student, was the first person to state explicitly that Mendel's genes were on the chromosomes and to give good reasons why this must be so.

Nineteenth-century biologists doubted that the chromosomes were the genetic material because chromosomes disappeared following cell division and because biologists had no reason to think that each chromosome carried unique information.

9.2 How Do Organisms Pass Genetic Information to Their Offspring?

Genetics, the study of inheritance, today faces the same contradictory challenges it did 100 years ago. Genetics must explain both how like begets like—why children resemble their parents—and also the origin and maintenance of variation—why children differ from their parents. In this chapter, we will discuss **transmission genetics,** the study of how variation is passed from one generation to the next. We will see that the behavior of chromosomes during meiosis and fertilization does much to explain the inheritance of variation.

Distinct from transmission genetics is **molecular genetics,** the study of how DNA carries genetic instructions and how cells carry out these instructions. Molecular genetics helps explain how a single cell divides and eventually develops into liver cells, skin cells, nerve cells, and more. Molecular genetics also explains how the DNA in these different kinds of cells helps direct day-to-day operations in the cell. We discuss molecular genetics in Chapters 10 through 14. In this chapter, we begin our discussion of transmission genetics by briefly defining how the zygote (the fused egg and sperm) uses the genetic information it receives from its parents to become a unique individual. No two organisms are identical, and the unique collection of traits that define an individual is called the "phenotype."

How Is Phenotype Related to Genotype?

Everything about an organism, from its size and chemistry to its behavior, constitutes its **phenotype** [Greek, *phenein* = to show]. Phenotype encompasses both physical and behavioral characteristics. In humans, eye color, height, and skin color are obvious *phenotypic traits,* the individual aspects of phenotype. In contrast, the **genotype** is the particular collection of genes of a cell or organism. Biologists use the word "genotype" to refer both to all the genes in an organism, its **genome,** or, alternatively, to the subset of genes (or even one gene) that influences a particular trait, such as eye color. A genome, the set of all the genes of an individual, is a relatively passive entity within the cell. Like a cookbook, a genome provides lists of ingredients (proteins) and directions for how much of each protein to use.

Phenotype results from the interaction of the genes with each other and with their environment. A **gene** is a region of the DNA that does one or two things: it specifies a particular protein and it regulates the "expression" of other genes. A gene is *expressed* when a cell makes a protein encoded by the gene. How much protein is made, and when, is *regulation.* Some genes are not expressed at all; others are expressed constantly in virtually every cell in the body. Each gene responds to signals within a cell that tell the cell when to make more and when to make less of the gene's particular protein.

An organism's phenotype reflects both its genotype (what genes it has) and its environment. In the case of plants and animals, the phenotype reflects the environment in which an individual has developed. Organisms can express different phenotypes in response to different environments, a phenomenon called **phenotypic plasticity.** For example, the tadpoles of spadefoot toads in the Sonoran desert grow and develop differently depending on how fast their small ponds dry up. In a deep pool, the tadpoles grow up in a leisurely fashion, eating green algae and brine shrimp. But in shallow pools, tadpoles are at risk of dying if the pond dries up before they have turned into toads. Tadpoles in rapidly drying ponds assume a fiercer, more-predatory phenotype, developing large mouths and powerful jaws, which they use to eat anything they can, including their own sisters and brothers. The carnivorous tadpoles grow up fast, and they end up being smaller toads than if they had developed in a deeper pool, but they are alive. All of the adaptations that the carnivorous toads display result from differences in gene expression in response to the tadpoles' particular environment.

Environments continue to affect adult organisms. For example, women living in groups ovulate in synchrony with one another. Each woman releases molecules called pheromones that speed up and slow down the menstrual cycles of other women in the group until everyone in is cycling synchronously. The pheromones influence the expression of genes that help control when an egg is released, the cycling of hormones, the production of a uterine lining, and dozens of other traits that change in women over the course of each four-week menstrual cycle.

"Environment" includes many effects that most people don't associate with the word. For example, an egg cell contains proteins made by the mother that influence the development of an embryo. But which proteins and how much of each depends on the external environment of the mother. In zebra finches, for example, red stripes painted on the father bird's legs induce the mother bird to load up her eggs with an extra dose of the hormone testosterone. The extra testosterone influences the expression of genes and makes the baby birds more aggressive. Testosterone directly affects the expression of genes throughout the body, but its own production is influenced by the social environment of the individual. In humans and other mammals, hormones and nutrients in a mother's blood similarly affect the development of an individual.

Gene expression is something that is happening throughout the body every minute of an individual organism's life. As we'll see in Chapter 11, the bacteria in our intestines express different genes depending on what we have for breakfast. In men, winning a chess game, watching the home team win a basketball game, and picking up a paycheck can all increase the production of testosterone. Losing a game or enduring a chewing out from the boss can reduce testosterone levels. And testosterone influences the expression of possibly hundreds of genes. We can think of the DNA as a keyboard on which our environment sends signals to the cells of the body. Each gene is a separate key, which may be pressed down (expressed) or not.

Studies of genetically identical twins illustrate that both genes and environment shape individual differences during development. Identical twins occur when a zygote or early embryo divides in two. Identical twins thus have identical genetic information. Yet, they can have different phenotypes because of environmental influences (Figure 9-4). (Fraternal twins, which develop from two eggs fertilized by two sperm, are no more related than ordinary sisters or brothers.) Most of the differences between identical twins result from differences in their environments, in the womb and outside it as they grow to adulthood.

The conjoined ("Siamese") twins Eng and Chang Bunker, born in 1811 in Siam (now Thailand), toured the United States as a circus act with P. T. Barnum. They were almost certainly identical twins who shared all the same genes, but they had notoriously different personalities. Chang was a gloomy alcoholic, while Eng was a cheerful teetotaler (even though they shared the same liver) and an expert poker player. But while Chang and Eng's genes may have been identical, their environments were not. Eng's environment included Chang, and Chang's included Eng. Bit by bit, in reaction to one another and to the way others' treated them, the two individuals grew up to be different from one another.

Phenotype includes all of the physical characteristics of an organism. The word "genotype" is used to mean either all the genes an individual has—the genome—or, alternatively, a particular gene or group of genes that produce a trait such as blue eyes.

Figure 9-4

More than genes. Even identical twins, with exactly the same genes, often look and behave differently. Identical twins occur when a single-celled embryo divides into two separate cells, which then develop into two separate people with the same set of alleles. As the two individuals develop in the womb, however, subtle differences in environment—such as a different blood supply—can lead to distinct differences in phenotype. The 8-year-old twins shown here are genetically identical. Yet the facial bones of one twin are more robust, those of the other twin, more delicate. Prenatal differences in the womb account for the differences in their faces.

The Same Laws of Inheritance Apply to All Sexually Reproducing Organisms

All sexually reproducing eukaryotes—from the most primitive algae to the most complex primates—follow the genetic rules that we describe in this chapter. The notable exceptions to these rules include the inheritance of genes in the DNA of mitochondria and chloroplasts (Chapter 12) and, of course, bacteria, which do not reproduce sexually. Four important points underlie the universality of genetic inheritance in eukaryotes:

1. All cellular organisms use DNA as the genetic material.
2. The DNA of all eukaryotic organisms is organized into chromosomes.
3. Almost all chromosomes exist in pairs at some time during a sexual life cycle.
4. These pairs of chromosomes behave in the same ways during meiosis and at fertilization in all eukaryotes.

is to divide in two through mitosis (Figure 9-5A). Even multicelled eukaryotes can reproduce through mitosis, by splitting off a single cell or a group of cells, which then develops into a whole individual. Plants, for example, often reproduce "vegetatively," by sprouting a new individual from a root or branch (Figure 9-5B). Orchardists can create hundreds of genetically identical trees by taking cuttings of branches from a single individual. Indeed, this method has been in use for thousands of years and is called "cloning" [Greek, *klon* = twig].

But all reproduction through mitosis, called **asexual reproduction**, produces offspring with genes from just one parent. All the offspring have the same genes, those of their one parent. We say the offspring that result from asexual reproduction constitute a **clone**, a set of genetically identical individuals.

(Biologists have even learned to clone some animals, including frogs, mice, pigs, sheep, cows, and mules, by inserting nuclei into egg cells whose own nuclei have been re-

9.3 How Do Sexually Reproducing Organisms Keep the Same Number of Chromosomes from Generation to Generation?

In Chapter 8, we saw how cells distribute two identical sets of chromosomes to their daughter cells. In this chapter, we will see how sexually reproducing organisms provide single sets of chromosomes to their sperm and egg cells. Later, when a sperm and egg fuse, the resulting individual has a full, double set of chromosomes. This is the essence of sexual reproduction.

For single-celled eukaryotes, such as the common freshwater protozoan *Paramecium*, the simplest way to reproduce

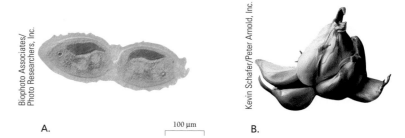

A.　　　　　100 μm　　　　B.

Figure 9-5

Asexual reproduction. A. The single-celled protozoan such as this *Paramecium* can reproduce by simply doubling its DNA and dividing into two cells, through mitosis. B. Many plants also reproduce "vegetatively," making genetically identical "clones" through mitosis. Shown here is a head of garlic. Each clove can take root and form a separate individual that is genetically identical to the parent plant.

moved. Cloning farm animals may be useful in agriculture, though the process results in more dead and deformed embryos than healthy ones. But, despite extravagant claims to the contrary, nuclear transfer and cloning are neither suitable nor necessary for human reproduction. Cloning is discussed further in Chapters 13 and 44.)

In contrast, **sexual reproduction** produces offspring that inherit genetic information from two parents. Sexual reproduction combines two sets of genes in new ways, so that, in every generation, a new set of unique individuals is created.

Sexual reproduction creates enormous diversity. But sexual reproduction creates a problem: How can each new organism receive chromosomes from both parents and still maintain the same total number of chromosomes?

The answer depends on the fact that chromosomes, like socks, come in pairs. Figure 9-6 shows the 46 chromosomes

of a human being. Notice that for each chromosome with a characteristic size and banding pattern there is another that looks just like it. So the 46 chromosomes in a human cell are actually 23 *pairs* of chromosomes. Cells that contain paired sets of chromosomes are said to be **diploid** [Greek, *di* = double + *ploion* = vessel]. Pairs of matching chromosomes are called **homologous chromosomes,** and each member of the pair is called a **homolog.** Like the individual socks in a pair, homologous chromosomes are basically alike, but they are not perfect copies. One may have a slight defect at one end; the other may have a different color in one spot near the middle.

Organisms that reproduce sexually ensure that their offspring have the right number of chromosomes by giving offspring only one of the chromosomes from each pair. In humans, for example, each parent contributes 23 chromosomes. This is analogous to receiving for your birthday 23 unmatched socks from your mother and 23 unmatched socks from your father. You put them together and discover that, miraculously, you have 23 matching pairs.

An organism that reproduces sexually passes the half-set of chromosomes to its offspring in specialized reproductive cells called **gametes** [Greek, *gamos* = marriage]. Each gamete is **haploid** [Greek, *haploos* = single], meaning that it carries a single set of chromosomes—in humans, 23. (Sometimes the word "haploid" is confusing, since it sounds like "half." But haploid means "single," not "half.")

Nearly all sexually reproducing organisms produce two kinds of gametes. One kind, called the **egg,** or **ovum** [Latin, = egg; plural, **ova**], is large and usually cannot move itself. The other, called the **sperm,** or **spermatozoon** [Greek, *sperma* = seed + *zoos* + living], is small and able to move under its own power. Males and females of other species may be completely unlike humans and other mammals. But no matter how small or strange an organism seems to us, if it produces sperm, we say it is male. Likewise, anything that produces eggs is female.

Gametes are the genetic link between generations. But where do these haploid cells come from? The answer is that haploid gametes arise from diploid cells by **meiosis,** a process that allots one haploid set of chromosomes to each of four daughter cells. Gametes and the special cells from which they arise are called **germ cells** or the **germ line** (Figure 9-7A). Gametes can come only from germ cells. In adult

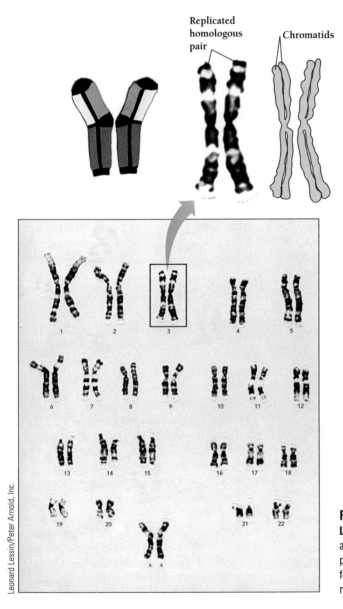

Replicated homologous pair

Chromatids

Figure 9-6
Like socks, chromosomes come in matched ("homologous") pairs. Shown here are the 23 pairs of (replicated) chromosomes of a human. They are laid out like pairs of socks in an arrangement called a "karyotype." Is this individual male or female? How can you tell? Each member of a homologous pair has already replicated and consists of two identical "chromatids."

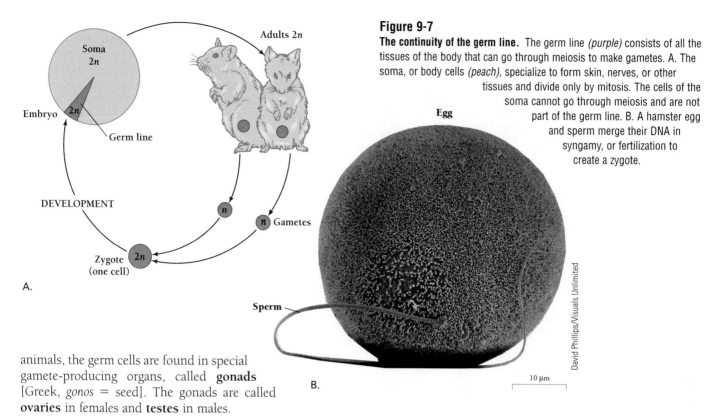

A.

B.

David Phillips/Visuals Unlimited

Figure 9-7
The continuity of the germ line. The germ line *(purple)* consists of all the tissues of the body that can go through meiosis to make gametes. A. The soma, or body cells *(peach),* specialize to form skin, nerves, or other tissues and divide only by mitosis. The cells of the soma cannot go through meiosis and are not part of the germ line. B. A hamster egg and sperm merge their DNA in syngamy, or fertilization to create a zygote.

Egg

Sperm

10 μm

animals, the germ cells are found in special gamete-producing organs, called **gonads** [Greek, *gonos* = seed]. The gonads are called **ovaries** in females and **testes** in males.

All of the rest of the cells in the body of a multicelled organism are called **somatic cells** [Greek, *soma* = body]. Mature somatic cells may divide by means of mitosis, but not by means of meiosis. In normal sexual reproduction, genetic instructions pass exclusively through the germ line.

The germ cells of sexually reproducing organisms undergo meiosis to produce haploid gametes, which fuse during fertilization to form a diploid individual.

The First Cell of the New Generation Has Two Sets of Chromosomes

Every new generation begins with **syngamy,** or **fertilization,** the union of the two haploid gametes (one from each parent) to form a single diploid cell called a **zygote** (Figure 9-7B). In humans, syngamy and fertilization are often called "conception." Many biologists prefer the term "syngamy" to "fertilization," because "fertilization" wrongly suggests that the sperm makes the egg fertile. In fact, the egg and sperm are both equally fertile before they fuse to form the zygote. Nonetheless, in this book we will use the older and still commonly used term, "fertilization."

Fertilization depends heavily on adaptations of the gametes. For example, the spermatozoon's small size,

whiplike tail, and many mitochondria specialize it for rapid movement. The egg cell, in turn, is a virtual supermarket of useful molecules and organelles that are essential for the early development and growth of the embryonic organism. In addition, the egg has special adaptations for moving the sperm nucleus and its own nucleus together, for fusing the two haploid sets of chromosomes into one diploid set, and for dividing rapidly after fertilization.

How Do the Egg and Sperm Find One Another?

Organisms bring the egg and sperm together in many ways. Some aquatic animals produce millions of eggs and sperm, which they "broadcast" into the water. Only a tiny fraction of these eggs and sperm meet and fuse, however. Most of the rest are filtered from the water by filter-feeding animals such as clams and sponges. Many plants broadcast their sperm (but not their eggs) in the form of grains of pollen blown by the wind. Other kinds of plants entice insects, bats, birds, and other animals to carry pollen from plant to plant. Females of many species of fishes and amphibians lay eggs externally while a male sheds his sperm over them.

In many animals, fertilization is "internal," meaning that it takes place inside the female's reproductive system and therefore requires copulation. Mammals, birds, reptiles, insects, and many other animals that practice internal

A.

B.

C.

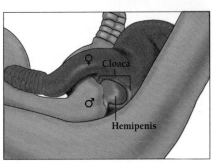

B.

C.

Figure 9-8
Adaptations to internal fertilization. Internal fertilization, or copulation, requires special behavioral and physical adaptations. A. Here a pair of black-tailed rattlesnakes (*Crotalus molossus*) copulate, lying together under a juniper tree in Arizona with their tails crossed (*center of photo A*). The male is at the left, and the female at the right. In males, the most striking physical adaptation to internal fertilization is the penis. Male snakes and lizards have two penises, called "hemipenes," but use only one at a time. In A, the male rattlesnake's left hemipenis is inserted into the female's bulging cloaca. B. A close-up shows the female's cloaca and the base of the male's purple-red left hemipenis. C. The two hemipenes of a male rattlesnake under anesthesia. Sperm travel into the female along a groove (barely showing on the left hemipenis). The spines most likely keep the two snakes from separating before copulation is completed; it's not unusual for the female to begin crawling away, literally dragging the male with her.

fertilization often rely on elaborate behavioral and structural adaptations to bring the egg and sperm together (Figure 9-8).

Despite the many different ways that species achieve fertilization, it always serves the same purpose: to combine two single (haploid) sets of chromosomes into one double (diploid) set of chromosomes (Figure 9-9). Meiosis, in turn, always serves to produce the haploid gametes. All sexually reproducing eukaryotes accomplish meiosis in essentially the same way, in an elaborate chromosomal ballet that both resembles and differs from mitosis.

In all sexually reproducing eukaryotes, meiosis creates haploid gametes, which combine to form diploid individuals at fertilization, or syngamy. In animals, fertilization may be either external or internal.

9.4 How Does Meiosis Distribute Chromosomes to the Gametes?

Like mitosis, meiosis is a continuous process. Biologists divide meiosis into separate stages for convenience only. Meiosis consists of two cell divisions, called **meiosis I** and **meiosis II**. These divisions occur only in cells of the germ line during the production of gametes. Biologists divide each of these two meiotic divisions into four phases (Figure 9-10). These are prophase, metaphase, anaphase, and telophase.

In general outline, each phase of meiosis resembles the corresponding stage of mitosis.

- During each **prophase**, replicated chromosomes condense, and the nuclear membrane breaks down.
- During each **metaphase**, the chromosomes move to the equator of the spindle apparatus.

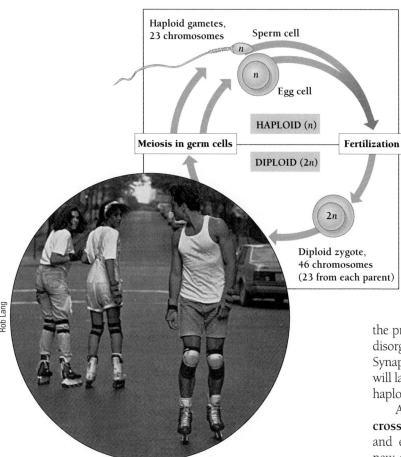

Haploid gametes,
23 chromosomes

Sperm cell

n

n

Egg cell

HAPLOID (*n*)

Meiosis in germ cells ——— **Fertilization**

DIPLOID (2*n*)

2*n*

Diploid zygote,
46 chromosomes
(23 from each parent)

Rob Lang

Figure 9-9
Meiosis and fertilization. In mature organisms, meiosis produces haploid gametes, which have half-sets of chromosomes. At fertilization, the two haploid sets of chromosomes, from the egg and the sperm, combine into one diploid set of chromosomes in a zygote.

in one essential way: after the chromosomes begin to condense, the homologous pairs come together in the process of **synapsis** [Greek, = union], in which homologous chromosomes align exactly—at some places, even gene for gene and nucleotide base for nucleotide base.

During prophase I, each doubled chromosome itself consists of two **chromatids,** so that the four chromatids form a group (visible in a microscope), called a "tetrad" (Figure 9-10). Synapsis resembles the process of pairing socks before putting them away. From a disorganized pile of 46 socks, we can put together 23 pairs. Synapsis is the central event of meiosis. It is the process that will later allow homologous chromosomes to separate into two haploid sets.

Another important event that occurs during synapsis is **crossing over,** in which homologous chromosomes break and exchange equivalent pieces. Crossing over results in new combinations of genes from the two parents. Crossing over is visible under a microscope as a crosslike configuration called a **chiasma** [Greek, = cross; plural, **chiasmata**] (Figure 9-11). At each chiasma, homologous chromosomes may exchange chromatid segments. Crossing over is an important way that sexually reproducing organisms recombine genes and increase variation. In humans, each pair of homologous chromosomes exchanges an average of two or three segments. We will discuss such recombination in detail later.

Prophase I is the longest stage of meiosis, taking up more than 90 percent of the total time. One reason that prophase I takes so long is that during this time, the egg cell builds up materials and machinery to be passed on to the gametes. In the females of some species, meiosis may seem to stop for months or even years during prophase I. In frogs, for example, prophase I may last for several years. In humans, meiosis begins before birth, pauses in prophase I, and does not resume until after puberty. Starting at around age 12, one or two eggs complete meiosis each month. Since humans may continue to ovulate into their 50s, some individual cells may not resume meiosis for more than 50 years.

- During each **anaphase,** spindle fibers move the chromosomes toward the poles.
- During each **telophase,** the nuclear membrane reforms.

Despite the resemblance to the stages of mitosis, the end result of meiosis is crucially different. In mitosis, each and every division produces two diploid daughter cells. In meiosis, by contrast, DNA replication occurs before the first division (meiosis I), but *not* before the second division (meiosis II). As a result, the two divisions of meiosis produce four haploid daughter cells.

In mitosis, a cell replicates its DNA once and divides once, resulting in two diploid daughter cells. In meiosis, a cell replicates its DNA once but divides twice, resulting in four haploid daughter cells.

Prophase I: Homologous Chromosomes Form Pairs

During interphase, before prophase I, the cell copies, or replicates, the chromosomes. As in mitotic prophase, the replicated chromosomes condense, the nucleoli and the nuclear membrane disappear, and the spindle apparatus begins to form. But meiotic prophase differs from mitotic prophase

During prophase I, a cell accumulates materials for its daughter cells and matches up homologous pairs of chromosomes, a process called synapsis. In addition, homologous chromosomes cross over.

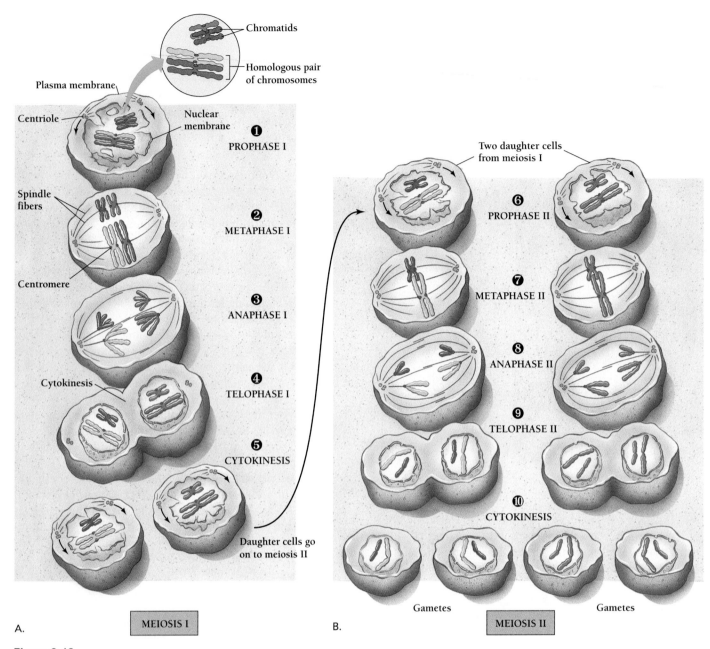

Chromatids

Homologous pair
of chromosomes

Plasma membrane

Centriole

Nuclear
membrane

❶ PROPHASE I

Spindle
fibers

❷ METAPHASE I

Centromere

❸ ANAPHASE I

Cytokinesis

❹ TELOPHASE I

❺ CYTOKINESIS

Daughter cells go
on to meiosis II

A. MEIOSIS I

Two daughter cells
from meiosis I

❻ PROPHASE II

❼ METAPHASE II

❽ ANAPHASE II

❾ TELOPHASE II

❿ CYTOKINESIS

Gametes   Gametes

B. MEIOSIS II

Figure 9-10

Meiosis. During meiosis, a cell replicates its DNA once but divides twice, producing four haploid daughter cells. Of the eight (homologous replicated) chromosomes in the original cell (top, left), each of the four resulting daughter cells (bottom, right) receives two chromosomes—one from each tetrad. In meiosis I, one chromosome from each homologous pair goes to each daughter cell. Meiosis II resembles mitosis, but in a haploid cell. A. ❶ During prophase I, replicated chromosomes condense and pair up and the nuclear membrane breaks down. ❷ During metaphase I, the replicated chromosomes move to the equator of the spindle apparatus. ❸ During anaphase I, spindle fibers move the chromosomes to the poles. ❹ During telophase I, the cell divides in two and the nuclear membrane reforms. ❺The two resulting daughter cells, which contain different homologous chromosomes, may then enter meiosis II. B. ❻ During prophase II, replicated chromosomes condense and the nuclear membrane breaks down in each daughter cell. ❼ During metaphase II, the chromosomes move to the equator of the spindle apparatus. ❽ During anaphase II, spindle fibers move the chromosomes to the poles. ❾ During telophase II, each cell divides in two and the nuclear membrane reforms. ❿ The two resulting daughter cells complete meiosis II.

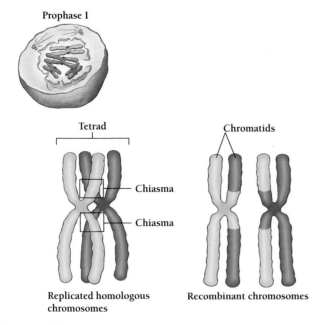

Figure 9-11
Crossing over. At the very beginning of meiosis, during prophase I, homologous chromatids pair up in synapsis. While they are paired up, the chromatids may break at places called chiasma and exchange equivalent pieces. The result is new combinations of genes on each chromatid.

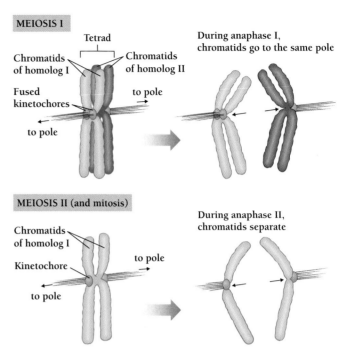

Figure 9-12
Separation of the chromatids. During meiosis I, homologous pairs separate but replicated chromosomes remain intact. During meiosis II, chromatids separate and go to opposite poles (as in mitosis).

During the Rest of Meiosis I, One Chromosome from Each Homologous Pair Goes to Each Daughter Cell

After prophase I comes metaphase I, in which spindle fibers pull each pair of chromosomes to the equator. Then, during anaphase I, something dramatic happens that is very different from what occurs in mitosis: whole replicated chromosomes remain intact and are pulled to the same pole (Figure 9-12). (In contrast, in mitosis, the two chromatids of each chromosomes separate and go to *opposite* poles.) During metaphase I, the homologous chromosomes go their separate ways, while the chromatids of each replicated chromosome stay together. Telophase I and cytokinesis rapidly follow. The daughter cells of meiosis I move right into meiosis II without any interphase and without any further DNA synthesis. The chromosomes may even remain partly condensed.

An important question is how the replicated chromosome pairs sort out during metaphase I. If all the chromosomes from the mother went to one pole, and all the chromosomes from the father went to the other, meiosis would recreate the same combinations of chromosome generation after generation. Instead, the replicated chromosomes sort randomly. One pair of homologs aligns one way, another pair aligns another way. Biologists say that each pair aligns "independently" of all the other pairs.

Because chromosome pairs sort themselves independently, meiosis is a major source of genetic diversity. Sutton showed that just two pairs of chromosomes can form 4 (2^2) kinds of gametes (Figure 9-13). Similarly, three pairs of chromosomes can form 8 (2^3) kinds of gametes. A single human germ cell, with 23 pairs of chromosomes, can therefore form 2^{23} (about 8 million) kinds of gametes, each one unique. Together, a woman and a man could produce more than 64 trillion genetically unique offspring (8 million kinds of eggs × 8 million kinds of sperm). In addition, crossing over increases the actual diversity of possible gametes and zygotes far beyond 64 trillion. All this diversity comes from (1) crossing over and (2) the random alignment and subsequent separation of replicated chromosomes in metaphase I. Given this random recombination, it is no wonder that all sexually reproducing species (including our own) are so wonderfully diverse.

During anaphase I of meiosis, chromatids go to the same pole in the cell, but homologous chromosomes go to opposite poles. Telophase I and cytokinesis I resemble their counterparts in mitosis.

Meiosis II Distributes Chromatids to Daughter Cells

Meiosis II looks just like mitosis would look in a haploid cell. At the end of meiosis I, the cell is haploid. After it divides again in meiosis II, it is still haploid. Meiosis II simply separates the chromatids of replicated chromosomes.

Prophase II (unlike prophase I) is brief, since the chromosomes are still mostly condensed from meiosis I. If a

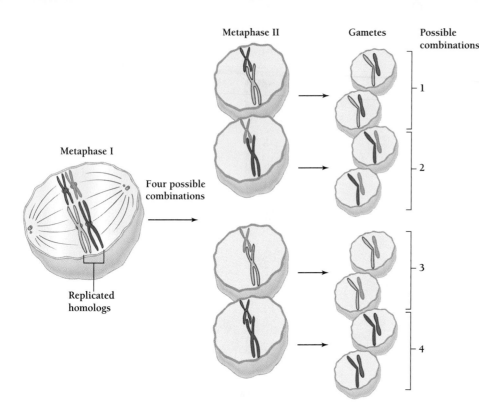

Metaphase II Gametes Possible combinations

Metaphase I

Four possible combinations

Replicated homologs

1

2

3

4

Figure 9-13
Just two pairs of chromosomes can form four kinds of gametes. If during prophase I the chromosomes arrange themselves as shown in the first (left hand) cell in this figure, the resulting gametes have chromosome combinations 1 and 2. If, in other germ cells, the chromosomes randomly swap places, the resulting gametes have combinations 3 and 4.

nuclear membrane has reappeared during telophase I, it breaks down during prophase II. The replicated chromosomes attach to newly assembled spindle fibers. Then, during metaphase II, the chromosomes line up across the equator of the spindle apparatus, with each centromere connected (by kinetochores) to spindle fibers from both poles. During anaphase II, the replicated chromosomes split, and one chromatid moves to each pole. In telophase II, the nuclear membrane forms once more. The daughter cells each have a complete haploid set of unreplicated chromosomes.

Meiosis II resembles mitosis except that the cells at all stages are haploid rather than diploid.

Disjunction and Nondisjunction

The separation of homologous chromosomes in anaphase I and of chromatids in anaphase II is called **disjunction.** Occasionally, separation fails to proceed normally, an event called **nondisjunction.** This results in a gamete having too many or too few chromosomes. In humans, for example, a sperm or egg might have 22 or 24 chromosomes instead of 23. After fertilization, a zygote would likewise have the wrong number of chromosomes—45 or 47.

In humans, one of the most common results of nondisjunction is **Down syndrome,** a disorder that leads to mental retardation and the abnormal development of the face,

heart, and other parts of the body (Figure 9-14A). Down syndrome is almost always associated with a chromosomal abnormality called **trisomy 21** [Greek, *tri* = three + *soma* = body], the presence of three, rather than two, copies of chromosome 21 (Figure 9-14B).

Trisomy 21 usually results from nondisjunction during meiosis I. If, during anaphase I, the two homologous chromosomes number 21 fail to separate, then two of the four possible gametes produced during meiosis will each have two copies instead of one, and the other two will have no copies of chromosome 21 (Figure 9-15). When these gametes unite with normal gametes at fertilization, they produce abnormal zygotes—some with three copies of chromosome 21, and some with only one copy (see box on page 178).

If, during anaphase I, homologous chromosomes fail to separate, the gametes will have too many or too few chromosomes.

9.5 Why Sex?

We saw earlier that many organisms can reproduce either sexually or asexually. For example, plants can reproduce sexually, as when pollen finds its way to a flower, or asexually, as when the roots of an aspen tree sprout new trees. Even some animals, such as the many-armed hydra, can reproduce asexually, by budding.

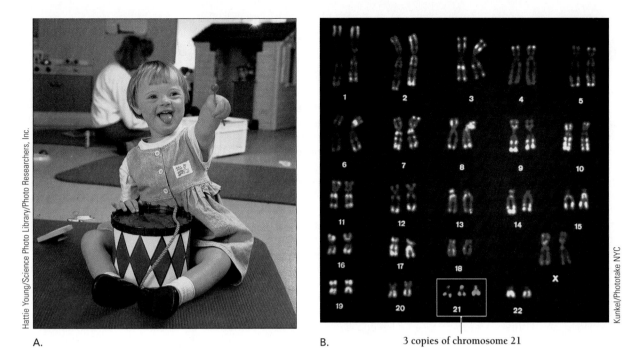

A.

B.

3 copies of chromosome 21

Hattie Young/Science Photo Library/Photo Researchers, Inc.

Kunkel/Phototake NYC

Figure 9-14

Down syndrome. A. Children with Down syndrome have distinctive characteristics, including mental retardation, heart defects, a flat facial profile, excess skin on the neck, double-jointedness, and an unusual crease in the palm of each hand. B. This karyotype of a person with Down syndrome shows the unusual extra chromosome number 21, called "trisomy 21." The replicated nature of each chromosome is quite obvious in this photo.

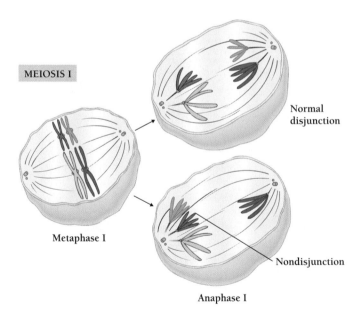

MEIOSIS I

Metaphase I

Normal disjunction

Nondisjunction

Anaphase I

Figure 9-15

How does trisomy occur? In nondisjunction, a pair of homologous chromosomes fails to separate during anaphase I. The resulting gametes have either two copies of a chromosome or none. Later, when the gamete combines with another, normal gamete, the resulting zygote has one homolog or three, but not two.

Asexual reproduction is the quickest and easiest way to reproduce. Sexual reproduction is enormously expensive. To reproduce sexually, organisms must make hundreds, thousands, or even millions of gametes to ensure that a few meet for fertilization. A female codfish produces some 9 million eggs per year, of which only a small fraction are fertilized. A typical male human may produce 20 to 30 billion sperm over the course of the weeks or months it takes him to successfully fertilize a single egg. Even a female human is born with 2 million egg cells (oocytes). In addition, humans and other animals invest huge amounts of energy into finding, courting, and mating with other individuals. Many plants construct elaborate and showy flowers stocked with sweet nectar, adaptations for attracting insects and other animals that carry pollen from one plant to another. What makes sexual reproduction worth a great expense?

One answer is genetic variation in one's offspring. We have seen that, because of crossing over and because of the independent assortment of chromosomes during metaphase I, a single pair of humans can (theoretically) produce a seemingly infinite number of unique zygotes. In contrast, asexual reproduction produces offspring that are genetically identical to the parent.

Among human embryos with Down syndrome, or trisomy 21, about 80 percent die before birth in **spontaneous abortions,** natural abortions that occur because the embryo develops abnormally and cannot live. Trisomy 21 embryos that survive until birth may live for many years. For comparison, about 70 percent of all human embryos normally survive to birth. Although the death rate among embryos with three copies of chromosome 21 is high, the death rate among those with only one copy is even higher—nearly 100 percent.

About one baby in 700 has Down syndrome. Although the extra chromosome usually comes from the mother, about one-fifth of the time, the extra chromosome 21 comes from the father. Older women are much more likely than younger women to make eggs with an extra chromosome 21. As a rule, physicians test the embryos of women older than 35 years for Down syndrome, but there is nothing magic about age 35. The likelihood of giving birth to a child with Down syndrome increases gradually (Table A).

Cell biologists believe that the chance of nondisjunction (at least for chromosome 21) increases with maternal age because of the pattern of meiosis in humans. The cells that form human eggs begin meiosis while the mother is herself still an embryo in the womb. But meiosis stops in the middle of prophase I, well before birth, and does not resume until after the egg is released from the ovaries, anywhere from 13 to 55 years later. Nondisjunction of chromosome 21

may be the result of interrupting the complicated process of meiosis for so long a time. Although trisomy is most common in chromosome 21, it also occurs in other chromosomes.

Table A
Risk of Trisomy 21

Age of mother (years)	Risk of giving birth to child with trisomy 21
Under 25	1/1,400
Under 30	1/1,000
At age 35	1/350
Over 40	1/100

What Good Is Genetic Variation?

In an unchanging environment, a single, perfectly adapted genotype might persist unchanged for millions of years. Most environments, however, change constantly. A polar bear with short hair might survive fine in a period of global warming. But as temperatures drop, its siblings with longer fur do better. Parents that produce a diversity of offspring are more likely to have some of them survive and reproduce than parents that produce offspring that are all identical. Indeed, evolutionary biologists generally find that the less predictable an environment is, the more likely that the organisms living in it will have evolved methods for increasing diversity.

Some species actually change their mode of reproduction depending on how predictable their environment is. For example, a female water flea reproduces asexually when food and space are plentiful, giving birth to huge numbers of genetically identical offspring, but reproduces sexually when food is limited (Figure 9-16). The resulting diversity increases the chance that at least some individuals will survive.

Many biologists think that the need for genetic variation alone is not a good enough reason for sex. After all, many organisms, including all bacteria, reproduce asexually. Bacteria have survived quite well for billions of years. In Chapter 43, we discuss one alternative hypothesis for why so

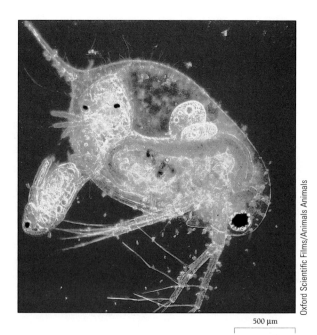

500 µm

Oxford Scientific Films/Animals Animals

Figure 9-16
Where's Dad? When food is abundant, the water flea, *Daphnia,* reproduces asexually. Each female is born pregnant with more female water fleas, each one a clone of its mother. A drop in water temperature or a decrease in food supply results in the birth of males and sexual reproduction.

many eukaryotes reproduce sexually. The reasons for sex remain controversial.

Sexual reproduction is energy expensive, and biologists wonder why organisms bother with it. One reason may be that sex increases genetic diversity, and genetic diversity increases the chances that some individuals will survive, no matter how the environment changes.

What Are Sex Chromosomes?

In mammals, birds, and some other organisms, either the male or the female may have a pair of unmatched chromosomes. In mammals (including humans), for example, the female has two homologous "X" chromosomes, while the male has one "X" chromosome and one "Y" chromosome (Figure 9-17). In birds, the males have two matched sex chromosomes ("ZZ"), and the females have one "W" chromosome and one "Z." In many insects, including the lubber grasshoppers that Sutton collected, the female has two homologous X chromosomes and the male has only one X chromosome. The **X** and **Y** chromosomes (as well as the W and Z chromosomes) are called **sex chromosomes.** The rest of the chromosomes are called the **autosomes.** A human cell has 22 pairs of autosomes and 1 pair of sex chromosomes.

Protists, fungi, plants, and many animals lack sex chromosomes. In most organisms, males and females both have complete sets of homologous (matching) chromosomes. In these organisms, what sex an individual becomes depends on different factors, including, for example, plant hormones in plants or temperature during development in some animals.

In humans and other mammals, the X chromosome in a male is the same as the X chromosome in a female. The Y chromosome, however, is much smaller and bears far fewer genes than the X chromosome. In both females and males, the two sex chromosomes (XX in females and XY in males) pair and segregate during meiosis in the same way as homologous pairs of autosomes.

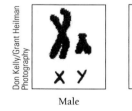

Figure 9-17
Sex chromosomes in humans. In certain organisms, one sex has a pair of unmatched sex chromosomes. For example, a female human has 23 pairs of matched chromosomes, including two X chromosomes, which are called sex chromosomes. A male human has 22 pairs of matched chromosomes, plus two sex chromosomes called X and Y.

How can X and Y pair off during synapsis if they differ so much in size and form? The answer is that they share a region that is homologous. Despite their striking differences in size and form, then, they still behave as homologs during meiosis.

Sex chromosomes are involved in sex determination. In mammals, the female has two homologous X chromosomes and the male has one X chromosome and one Y chromosome, which are not homologous.

9.6 How Do the Number and Movements of Chromosomes Explain the Inheritance of Genes?

If a cell has two copies of each chromosome, then the cell must also have two copies of every gene. This is key to understanding genetics in diploid organisms. The copy of a gene on one chromosome may differ slightly from its counterpart on the other chromosome. Alternate versions of the same gene are called **alleles.** For example, in humans, both chromosomes in a pair may have a certain gene that influences eye color, but one chromosome might carry an allele for brown eyes, while the other chromosome carries an allele for blue eyes.

Most genes are simply instructions for the production of a particular protein. Alleles are variations in those instructions. Just as your aunt June makes chocolate brownies with extra chocolate and your cousin Ralph makes them with a teaspoon of vanilla, different alleles specify slightly different proteins. Sometimes the differences are subtle and the resulting protein is the same either way; sometimes an allele specifies a protein so different from normal that the protein cannot function at all—as if your roommate had put a half cup of salt into the brownies instead of sugar.

For a given gene, an individual can carry two identical alleles or two different alleles. In snapdragon flowers, for example, one allele of a particular gene specifies white flowers and the other allele specifies red flowers. If a plant has two red flower alleles, its flowers are red. If the snapdragon has two white-flower alleles, its flowers are white. On the other hand, if the snapdragon has one red-flower allele and one white-flower allele, the flowers are pink (Figure 9-3). If an individual has two copies of the same allele, we say it is **homozygous** for that allele. If an individual has two different alleles of a gene, we say it is **heterozygous.**

Biologists name the alleles of a gene with letters or short abbreviations. In snapdragons, for example, we can call the red-flower allele R_1 and the white-flower allele R_2. We can imagine that the R_1 allele allows the plant to make a red protein pigment, while the R_2 allele allows the plant to make no functioning pigment at all. The genotype of the homozygous red snapdragon would be "R_1R_1," and that of the homozygous white snapdragon would be "R_2R_2." The genotype of the pink heterozygote would be "R_1R_2."

Plant breeders knew about homozygous plants for hundreds of years before the first geneticists, calling plants that were homozygous for flower color "true breeding." In **true-breeding** plants, all the offspring of the red-flowered variety, for example, produce red flowers only, generation after generation. A line of plants (or other organisms) can be true breeding for other traits besides flower color. In all true-breeding varieties, both chromosomes carry the same allele for a certain trait. (We say they are homozygous.)

Snapdragons normally "self-fertilize." Each snapdragon flower produces both pollen (which produces sperm) and egg cells. The sperm from the pollen fertilizes eggs within the same flower. So, on a red snapdragon, gametes carrying red-flower alleles normally pair up with gametes carrying red-flower alleles. The same is true for white-flowered snapdragons.

A plant breeder can cross a true-breeding red-flowered plant with a true-breeding white-flowered plant only by preventing self-fertilization. To cross the two homozygous plants, the breeder must prevent each plant from self-pollinating. This is done by snipping off the pollen-carrying structures and then artificially pollinating the flower with pollen from another plant. This method of breeding two genetically distinct organisms is called **cross breeding** (or **crossing**). The offspring of such a cross are called **hybrids.** In snapdragons, crossing a homozygous red snapdragon with a homozygous white snapdragon results in a heterozygous hybrid that has pink flowers (Figure 9-3).

Geneticists call the original parents in such a cross the **parental,** or **P,** generation, and the offspring the "first filial" [Latin, *filia* = daughter, *filius* = son], or **F1,** generation. The **F2** generation are the offspring of the F1 generation. In other words, the F2 generation are the grandchildren of the P generation. In the snapdragon cross, then, the parents are R_1R_1 and R_2R_2 homozygotes, and the F1 generation are R_1R_2 heterozygous hybrids. The genotypes of the heterozygous hybrids are all R_1R_2, and their phenotypes are pink flowers.

Eukaryotes that have two sets of chromosomes (diploid) have two copies of each gene. An individual may have two identical alleles, or two different alleles.

What Are the Genotypes and Phenotypes of the F2 Generation?

Both a plant breeder and a geneticist would want to know what color flowers result from crossing two hybrid snapdragons. We can easily figure this out before we even cross the plants. We know that meiosis produces haploid gametes

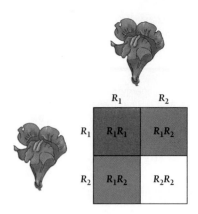

Figure 9-18
Inheritance of flower color in snapdragons. A cross between two R_1R_2 heterozygotes *(pink)*, shown here in a Punnett square, produces three genotypes: snapdragons homozygous for R_1 *(red)*, snapdragons homozygous for R_2 *(white)*, and R_1R_2 heterozygous snapdragons *(pink)*.

with just one of each chromosome. The F1 hybrids have one R_1 allele and one R_2 allele, each on a separate chromosome of a homologous pair. So half of all the gametes get the R_1 allele, and half get the R_2 allele. During fertilization, an R_1 can pair up either with another R_1 or with an R_2, and an R_2 can pair up with either another R_2 or with an R_1. So the F2 generation can have R_1R_1 (red flowers), R_2R_2 (white flowers), or R_1R_2 (pink flowers). But how many of each kind will we get?

A convenient way to find out is to use a checkerboard scheme invented by an early-20th-century British geneticist named Reginald Punnett (Figure 9-18). In a **Punnett square,** we show each kind of gamete made by one parent along the top of the square and each kind of gamete made by the other parent down the left side of the square. Within the central squares, we write the genotypes that would be produced by each combination. The R_1R_2 genotype is the same as the R_2R_1 genotype. The flowers are pink no matter which allele comes from which parent. But we end up with twice as many pink flowers as white or red.

The Punnett square shows that the three possible genotypes of the F2 generation of our snapdragon experiment should be in the ratio 1:2:1. If we actually cross the flowers, we can directly count the number of red, white, and pink flowers. Because the pink phenotype of the heterozygote is different from the phenotypes of either homozygote (red or white), we can directly count the ratios of the genotypes in the F2 generation. Experiments of this kind always produce a ratio fairly close to the expected 1:2:1 (if the number of individuals counted is large enough).

If we cross two R_1R_2 heterozygotes, one-quarter of the offspring are homozygous for R_1, one-quarter are homozygous for gene R_2, and half are heterozygous R_1R_2.

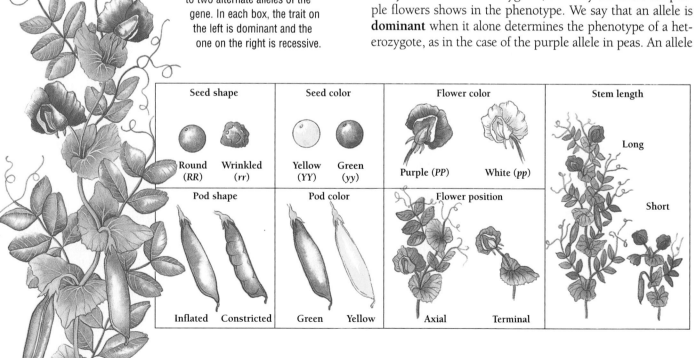

Figure 9-19
Gregor Mendel in his monastery garden.

9.7 How Did Gregor Mendel Demonstrate the Principles of Genetics?

Today, the inheritance of flower color in snapdragons makes sense to us because we know about chromosomes and how they assort during meiosis. But 19th-century biologists knew little about chromosomes and nothing about genes. For them, inheritance was a mystery. Our knowledge of the behavior of chromosomes and genes gives us a great advantage.

The first person to penetrate the mystery of inheritance was Gregor Mendel (Figure 9-19). In his monastery's garden, in what is now the Czech Republic, Mendel studied seven different pairs of traits in pea plants (Figure 9-20): round versus wrinkled seeds, yellow versus green seeds, purple versus white flowers, inflated versus constricted pods, green versus yellow pods, axial versus terminal flowers, and tall versus dwarf stems. Each of these pairs of traits showed the same pattern of inheritance. Mendel chose pairs of traits in which each variant was distinct (in other words, tall could not be confused with short).

In each case, Mendel showed that the parental stocks were true breeding. For example, self-pollination of plants with round seeds always produced offspring with round seeds. Mendel then cross-pollinated flowers with different traits, a purple-flowered pea with a white-flowered pea, for example (Figure 9-21).

But F1 hybrids may not have an intermediate phenotype like the pink flowers of hybrid snapdragons. Instead, a heterozygous hybrid can have the same phenotype as one of the parents. For example, when Mendel crossed true-breeding, purple-flowered peas with true-breeding, white-flowered peas, all the hybrids had purple flowers (Figure 9-22A). The F1 hybrids are all heterozygotes, but only the allele for purple flowers shows in the phenotype. We say that an allele is **dominant** when it alone determines the phenotype of a heterozygote, as in the case of the purple allele in peas. An allele

Figure 9-20
Mendel's seven traits in pea plants.
Mendel chose seven traits, each with two forms that corresponded to two alternate alleles of the gene. In each box, the trait on the left is dominant and the one on the right is recessive.

Seed shape		Seed color		Flower color		Stem length
Round (*RR*)	Wrinkled (*rr*)	Yellow (*YY*)	Green (*yy*)	Purple (*PP*)	White (*pp*)	Long
Pod shape		**Pod color**		**Flower position**		Short
Inflated	Constricted	Green	Yellow	Axial	Terminal	

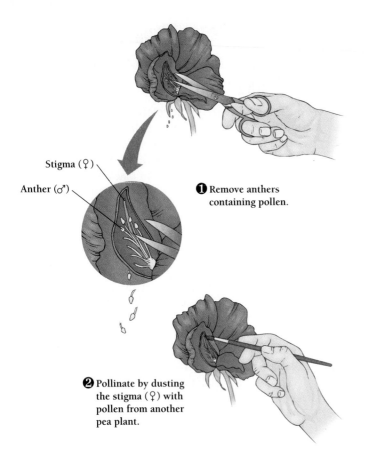

Stigma (♀)

Anther (♂)

❶ Remove anthers containing pollen.

❷ Pollinate by dusting the stigma (♀) with pollen from another pea plant.

Figure 9-21

Cross-pollinating peas. Pea flowers carry both male and female parts, which allows them to self-pollinate. But Mendel's studies of inheritance required him to prevent the peas from self-pollinating so he could cross different lines. He snipped the male, pollen-producing anthers from each pea flower and then sprinkled the female stigma with pollen from another flower.

is **recessive** when it contributes nothing to the phenotype of a heterozygote, as is the case of the white allele of peas. (Dominant alleles begin with a capital letter, and those of recessive alleles with a lower case letter.)

Today we know that a dominant allele typically specifies a functional protein, while a recessive allele usually specifies a nonfunctional protein. Recessive alleles that are harmless, such as the one that causes type O blood, may be extremely common. Others, such as the allele for the genetic disease Tay-Sachs, which kills children at three or four years of age, are quite rare. Lethal recessives such as Tay-Sachs are rare because people who are homozygous for this allele never pass the gene on to any offspring.

When two alleles each contribute to the phenotype, as in the case of the R_1 and R_2 alleles of snapdragons (which together gave a color different from either homozygote), the alleles are said to lack dominance. In the snapdragons, for example, both alleles are expressed and we say they each have **incomplete dominance** or "codominance."

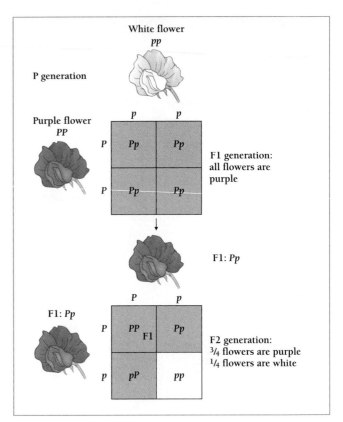

A. HYBRID CROSS: Pp × Pp

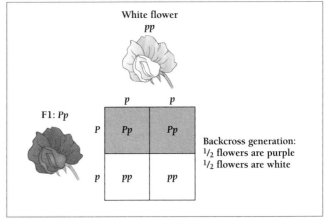

B. BACKCROSS: Pp × pp

Figure 9-22

A backcross. Shown here in a Punnett square is a backcross between purple- and white-flowered peas. A. First, a homozygous white flower is crossed with a homozygous purple flower to produce purple heterozygotes. Crossing these heterozygous offspring (the F1 generation), produces a mix of purple homozygotes and heterozygotes and white homozygotes. B. In the backcross, a white homozygote is crossed with a purple heterozygote. Half the offspring are purple heterozygotes and half are white homozygotes. (By a convention begun by Mendel, we begin the names of dominant alleles with a capital letter and those of recessive alleles with a small letter. But unlike Mendel, modern geneticists often also use the letter of the recessive trait to name the gene.)

Biology ⓔ Now™   Learn more about crosses by clicking on this figure on your BiologyNow CD-ROM.

In each of Mendel's seven crosses, all of the F1 hybrids had the same phenotype. Round pea stocks crossed with wrinkled pea stocks gave F1 hybrids with round peas, tall pea stocks crossed with short pea stocks gave tall F1 hybrids, and so on. We say that the purple flower allele is dominant to the white flower allele, the round allele is dominant to the wrinkled allele, and the tall allele is dominant to the short allele.

Mendel began his experiments with true-breeding (homozygous) pea plant stocks. He chose pairs of traits in which the two alternate forms were distinct. In retrospect, we can see that each pair of traits was determined by just two alleles, and each allele in a pair was either dominant or recessive.

What Did Mendel's F1 Crosses Show?

Mendel allowed the F1 hybrids to self-pollinate, then counted the F2 offspring. In every case, the dominant allele phenotype made up about three-fourths of the offspring, and the recessive allele phenotype made up the remaining one-fourth of the offspring.

This is the same result we got with the snapdragons, except that in Mendel's peas heterozygotes (which make up half of the offspring) look the same as the dominant homozygotes (which make up a quarter of the offspring). This was because, in each of Mendel's crosses, one allele was dominant and one was recessive.

In the case of the round versus wrinkled cross, for example, $\frac{1}{4}$ were homozygous for the round allele, $\frac{1}{4}$ were homozygous for the wrinkled allele, and $\frac{1}{2}$ were heterozygous. Since the round allele is dominant, the heterozygotes had the same phenotype as the homozygote for the round allele. So $\frac{3}{4}$ ($\frac{1}{4} + \frac{1}{2}$) were round, and $\frac{1}{4}$ were wrinkled.

The Principle of Segregation

Mendel had to make sense of his results without the knowledge of genes and chromosomes that we now have. Amazingly, he was able to do it by keeping detailed records and studying the ratios of phenotypes in his breeding experiments. Previous biologists had studied inheritance in peas and kept records consisting of "some" and "many" and "most." Mendel, who had training in mathematics, counted and wrote down specific numbers, the key to his discovery of the ratios.

Mendel's first conclusion was the **principle of segregation**: *Each sexually reproducing organism has two genes for each characteristic; these two genes segregate (or separate) during the production of gametes.* (Mendel didn't use the word "gene," which is a modern term.)

To test the principle of segregation, Mendel performed a different kind of cross, called a **backcross**. Instead of allowing the F1 heterozygotes to self-pollinate, he crossed them with the homozygous recessive parental stock—those that contained only the recessive allele (Figure 9-22B). By crossing the purple F1 plants to a white homozygous recessive, Mendel could find out how many of the F1 plants had recessive alleles and how many didn't. A cross with a homozygous recessive that is supposed to reveal the genotype of the other parent is called a "test cross."

In Mendel's test cross, the parents looked just like those in the original parental cross—a purple-flowered plant with a white-flowered plant (Figure 9-22A). But Mendel had discovered that the purple-flowered F1 hybrids could produce two kinds of gametes, one with the purple allele, and the other with the white allele (whereas the original purple parents produced only gametes with purple alleles). The plants with white flowers, on the other hand, would only produce gametes with the white allele. He correctly predicted that half the offspring would have purple flowers and half, white flowers.

Mendel's hybrid crosses and backcrosses produced offspring with ratios of phenotypes that suggested the principle of segregation; each sexually reproducing organism has two genes for each characteristic, which segregate during the production of gametes.

The Principle of Independent Assortment

Mendel next wanted to know if different traits were inherited together. In other words, could traits be "linked"? To find out, he chose two true-breeding stocks, one that made round, yellow peas and one that made wrinkled, green peas. From previous crosses, he knew that round was dominant to wrinkled and yellow was dominant to green. As he expected, all the F1 peas were yellow and round. He then determined how these traits were inherited in the F2 generation. We can call the allele specifying round peas *R,* the one for wrinkled peas *r,* the allele specifying yellow peas *Y,* and the one for green peas *y* (Figure 9-23). The genotypes of the parents are therefore called *RRYY* and *rryy.*

Homozygous parents can produce only a single kind of gamete: *RRYY* plants produce only *RY* gametes, and *rryy* plants produce only *ry* gametes. The genotype of the F1 plants therefore must be *RrYy.*

But what happens in the F2 generation? Mendel allowed the F1 to self-pollinate and then counted 556 peas in the F2. Of these, 315 were yellow and round (showing the action of the dominant alleles), and only 32 were green and wrinkled (Figure 9-23). The rest represented new phenotypes, unlike either of the parental stocks: 101 were yellow and wrinkled, and 108 were green and round. The two crosses had produced new combinations of genes, illustrating again how sexual reproduction provides diversity in a population. But how did Mendel make sense of the numbers?

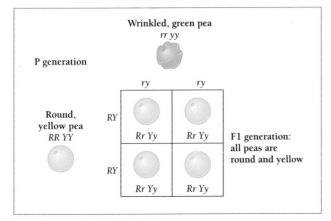

P generation

Wrinkled, green pea
rr yy

Round,
yellow pea
RR YY

	ry	*ry*
RY	*Rr Yy*	*Rr Yy*
RY	*Rr Yy*	*Rr Yy*

F1 generation:
all peas are
round and yellow

A.

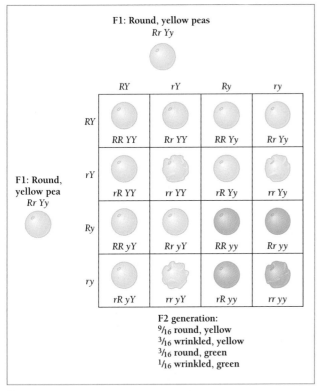

F1: Round, yellow peas
Rr Yy

F1: Round,
yellow pea
Rr Yy

	RY	*rY*	*Ry*	*ry*
RY	*RR YY*	*Rr YY*	*RR Yy*	*Rr Yy*
rY	*rR YY*	*rr YY*	*rR Yy*	*rr Yy*
Ry	*RR yY*	*Rr yY*	*RR yy*	*Rr yy*
ry	*rR yY*	*rr yY*	*rR yy*	*rr yy*

F2 generation:
$^9/_{16}$ round, yellow
$^3/_{16}$ wrinkled, yellow
$^3/_{16}$ round, green
$^1/_{16}$ wrinkled, green

B.

Figure 9-23

Testing two traits at once. In this two-factor cross—shown here in two Punnett squares—Mendel crossed round, yellow peas with wrinkled, green peas. A. In the first (F1) generation, all the peas were round, yellow heterozygotes. B. When he crossed the F1 generation, he got an assortment of different peas. Notice that three-quarters (12 out of 16) of the peas are yellow and three quarters are also round (not wrinkled), but not the same set of 12 for the two traits.

First, Mendel noted that the traits associated with each gene were in a ratio of about 3:1, just as in his experiments with only one trait. In the F2 generation there were three times as many yellow as green peas (416:140 = 3:1) and three times as many round as wrinkled (423:133 = 3.2:1).

Mendel immediately saw that the two traits were behaving independently; whether a pea was yellow or green had no effect on whether it was wrinkled or round.

Mendel could explain the data if he assumed that each F1 plant produces four kinds of gametes: *ry, Ry, rY,* and *RY.* The Punnett square of Figure 9-23B predicts the distribution of genotypes and phenotypes of the F2 generation of this cross. It shows that 9 out of the 16 combinations of gametes produce plants with round, yellow peas, but only one combination produces plants with wrinkled, green peas. The overall ratio of phenotypes predicted by this analysis is 9:3:3:1.

The correspondence of Mendel's experiments to this prediction is called the **principle of independent assortment.** This principle states that each pair of genes is distributed independently from every other pair during the formation of the gametes. The independent assortment of homologous chromosomes during meiosis causes the independent assortment of genes that are on different chromosomes.

Mendel tested his understanding of the genotypes of the gametes produced by the (heterozygous) F1 hybrids by doing another test cross with plants that produce green, wrinkled peas (Figure 9-24). Because both *r* (for wrinkled peas) and *y* (for green peas) are recessive, the genotype of the plants used in the backcross must be *rryy.* The gametes produced by these plants all must be *ry.*

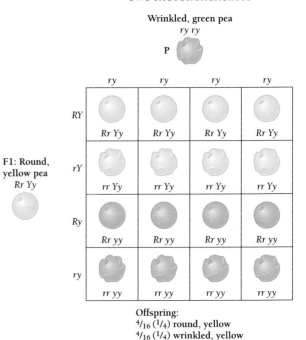

Wrinkled, green pea
ry ry

P

F1: Round,
yellow pea
Rr Yy

	ry	*ry*	*ry*	*ry*
RY	*Rr Yy*	*Rr Yy*	*Rr Yy*	*Rr Yy*
rY	*rr Yy*	*rr Yy*	*rr Yy*	*rr Yy*
Ry	*Rr yy*	*Rr yy*	*Rr yy*	*Rr yy*
ry	*rr yy*	*rr yy*	*rr yy*	*rr yy*

Offspring:
$^4/_{16}$ ($^1/_4$) round, yellow
$^4/_{16}$ ($^1/_4$) wrinkled, yellow
$^4/_{16}$ ($^1/_4$) round, green
$^4/_{16}$ ($^1/_4$) wrinkled, green

Figure 9-24

Independent assortment. By backcrossing the heterozygous F1 peas with the homozygous recessives (green, wrinkled), Mendel was able to show that these genes assorted independently.

Biology ⑧ Now™ Learn more about independent assortment by clicking on this figure on your BiologyNow CD-ROM.

If the genotype of the F1 plants is really *RrYy* and the chromosomes bearing the two genes assort independently, then the four kinds of gametes (*RY, rY, Ry,* and *ry*) should be made in equal numbers (Figure 9-24). When these gametes form zygotes with *ry* gametes, we expect equal numbers of the four phenotypes, in contrast to the 9:3:3:1 ratio expected in the case of self-fertilization of the F1 plants. Again, Mendel obtained the expected ratios, confirming his understanding of the process.

Mendel's principle of independent assortment states that each gene in a pair is distributed independently during the formation of the gametes.

9.8 What Was the Evidence for the Chromosomal Theory of Inheritance?

The rediscovery of Mendel's laws in 1900 hit British biologist William Bateson like a bolt of lightning. Within hours of reading the news, Bateson began devoting himself to spreading the word. As we saw earlier in this chapter, Bateson's talk at Columbia inspired Walter Sutton to quickly publish his own theory of inheritance.

In 1902, Sutton and the German biologist Theodor Boveri independently published papers that showed that the behavior of chromosomes during meiosis perfectly explained Mendel's principles of segregation and independent assortment. Both biologists concluded that the genes had to be on the chromosomes.

What Were Sutton's Arguments?

Sutton's argument consisted of six separate points that together formed an almost fortresslike argument for what came to be known as the **Sutton-Boveri chromosomal theory of inheritance.** Sutton's first three points merely confirm the ideas of others; the second three are his own new ideas.

First, said Sutton, chromosomes come in pairs, one of which comes from the mother and one from the father. Second, synapsis is the pairing of homologous maternal and paternal chromosomes. Because the chromosomes of the lubber grasshopper had distinctive sizes and shapes, Sutton was able to see that the chromosomes occurred in pairs and that they separate during meiosis. Third, the individual chromosomes retain their identity throughout the cell cycle. Although they may become invisible under a light microscope, they do not actually dissolve or otherwise go away.

Fourth, said Sutton, the chromosomes contain Mendel's genes. Fifth, meiosis creates new combinations of genes in

Table 9-1
Walter Sutton's Six Rules of Inheritance

1. Chromosomes come in pairs. One chromosome comes from the mother and one from the father. We now call these "homologous" pairs.

2. Synapsis is the pairing of homologous maternal and paternal chromosomes. These pairs separate into different daughter cells during meiosis.

3. The chromosomes are not broken down and created anew during each cell cycle.

4. The chromosomes carry genes.

5. Meiosis creates new combinations of genes in each generation. During meiosis, either homolog may end up in either new cell, regardless of the way all the other homologous chromosomes divide. This accounts for Mendel's principle of segregation.

6. Each chromosome carries a different set of genes, and all of the genes on one chromosome are inherited together. That different chromosomes are inherited independently accounts for Mendel's principle of independent assortment

each generation. During meiosis, when a diploid cell is reduced to four haploid gametes, each homolog in a pair orients randomly with respect to the two poles of the dividing cell. As a result, either homolog may end up in either new cell. This, said Sutton, accounted for Mendel's principle of independent assortment. Sixth, argued Sutton, each chromosome carries a different set of genes, and all of the genes on one chromosome are inherited together. That different chromosomes are inherited independently accounts for Mendel's principle of independent assortment.

We now see Sutton's insights as the basis for understanding the chromosomal basis of the laws of inheritance. Table 9-1 summarizes the rules of inheritance that we can take both from Sutton's work and from later work. Because of Sutton's synthesis and because of vast amounts of subsequent work in genetics, we are in a better position to understand genetics than were the 19th-century biologists who first discovered the physical basis of heredity.

How Did Thomas Hunt Morgan Bolster the Theory of Chromosomal Inheritance?

With the publication of Sutton and Boveri's papers, the field of genetics took off. One of the most important contributors to the new field was Thomas Hunt Morgan, who began breeding tiny fruit flies called *Drosophila melanogaster* in his laboratory at Columbia University, in New York City, in 1909. Fruit flies have many advantages for genetic research. Among them are: (1) they breed rapidly, growing from eggs to sexually mature adults in less than two weeks (Figure 9-25); (2) they are prolific, with each female laying hundreds of eggs; (3) they are

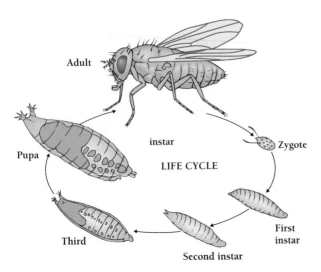

Figure 9-25

Growing up fast. Fruit flies *(Drosophila melanogaster)* mature from egg to adult in just two weeks, making them an ideal organism for studies of inheritance.

small (about 3 mm long) and easy to keep; (4) they have dozens of easily identifiable traits; and (5) they have only four pairs of chromosomes.

Morgan first collected his flies by leaving a piece of ripe pineapple on his windowsill. He kept the flies captive in small milk bottles "borrowed" from his home milk delivery-man. Before long he was raising thousands of the tiny flies,

and his laboratory, called "the fly room" by other biologists, became the center of genetic research in the United States for more than 25 years. Since then, thousands of geneticists and hundreds of thousands of undergraduate biology students have studied inheritance in fruit flies, demonstrating over and over that traits such as eye color, body color, and wing shape are all inherited according to Mendel's laws.

One of Morgan's first discoveries concerned a gene for eye color, which, he showed, lies on the X chromosome. Most wild fruit flies have red eyes. One day, however, Calvin Bridges, an undergraduate hired to wash the hundreds of dirty milk bottles in Morgan's laboratory, noticed a single fly with white eyes (Figure 9-26). Morgan crossed the white-eyed male with several normal, red-eyed females. All the F1 offspring had red eyes. He then crossed the F1 males and females and obtained an overall ratio of 3:1 of red-eyed flies to white-eyed flies. White eyes seemed to be a recessive trait. But not one of the females had white eyes. Half of the F2 males had red eyes, but all of the females had red eyes. What was going on?

Drosophila have only four pairs of chromosomes (Figure 9-27A). Males have three pairs of autosomes plus one pair of unmatched X and Y sex chromosomes, while females have three pairs of autosomes plus one pair of X chromosomes. As in humans, some of the same genes are on X as on Y. But the Y chromosome is smaller than the X chromosome and missing many of the genes that are on the X chromosome. As a result, where females have two copies of each

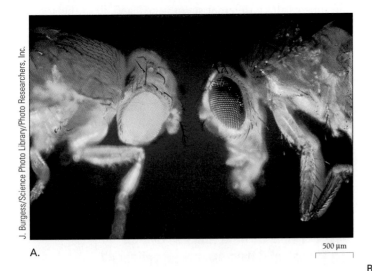

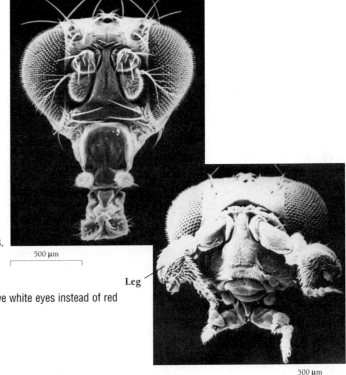

Figure 9-26

Mutant flies. Geneticists have isolated scores of mutant fruit flies. A. Some have white eyes instead of red eyes. B. Some have legs growing where their antennae should be.

gene on the X chromosome, males have only one of some genes.

Morgan hypothesized that one gene determined eye color and it was on the X chromosome. In Figure 9-27B, Punnett squares explain Morgan's experiment, showing the dominant red allele as X^W (because it is on the X chromosome) and the recessive white allele as X^w (because it too is on the X chromosome). The *W* stands for *white,* the name of the eye-color gene. Geneticists frequently name a gene for what happens when it isn't functioning. So, for example, *wingless* is a gene involved in the formation of wings. Likewise, flies with a normal *white* (*W*) allele have red eyes, but white eyes when *W* is mutated.

The males and females in the F1 generation have different genotypes. The females are all heterozygous, with the white-eye allele (*w*) on one X chromosome and the dominant red-eye allele (*W*) on the other X chromosome. So all the females have red eyes. But the males have the red-eye allele on their single X chromosome and no eye-color allele at all on their Y chromosome. So all the males have red eyes, too, which they got from their mothers.

Morgan went on to confirm his conclusions, both by crossing the hybrid F1 flies with each other and by performing a backcross, mating F1 females with a white-eyed male. His results were exactly as he predicted, establishing the location of *white* eye-color gene on the X chromosome. The *white* gene is one of almost 1,000 genes in Drosophila that are sex linked. In flies and other organisms with unmatched sex chromosomes, **sex-linked genes** have different patterns of inheritance in males and females because they lie on chromosomes that also determine the sex of the offspring. In mammals and flies, for example, the X chromosome contains many more genes than the Y chromosome, and sex linkage therefore nearly always refers to genes on the X chromosome. The many demonstrations of sex linkage—from Morgan on—provided early convincing support for Sutton's proposals, a decade earlier, that genes were indeed on chromosomes.

Thomas Hunt Morgan's experiments with sex-linked genes confirmed Sutton's theory that the genes lie on the chromosomes.

Do Genes on the Same Chromosome Assort Independently?

The principle of independent assortment depends on the independent assortment of chromosomes, but it cannot hold for genes that are close together on the same chromosome. Sutton had recognized this problem in 1903, but Morgan and his students were the first to design experiments around this question. In so doing, they further established the chromosomal basis of inheritance.

Morgan crossed two types of flies. One kind had wings of a more or less standard length and red eyes—characteristics

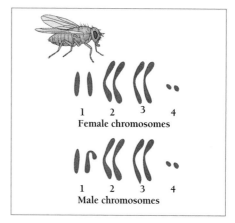

A.

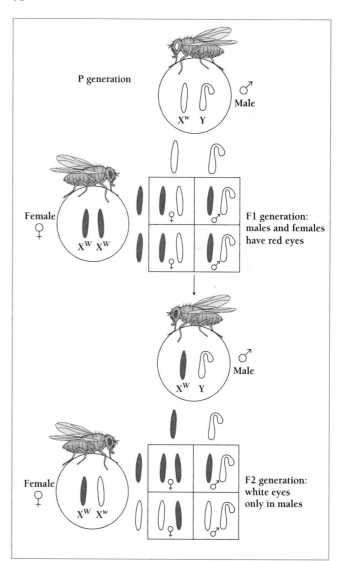

B.

Figure 9-27

Sex-linked genes. A. Fruit flies have four pairs of chromosomes. As in humans, the male has an unmatched set of sex chromosomes. B. The white gene is on the sex chromosome, and we can trace its descent through two crosses by looking at the way the sex chromosomes assort.

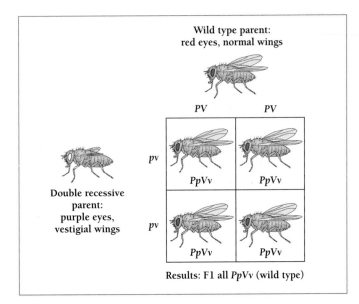

Wild type parent:
red eyes, normal wings

	PV	PV
pv	PpVv	PpVv
pv	PpVv	PpVv

Double recessive
parent:
purple eyes,
vestigial wings

Results: F1 all *PpVv* (wild type)

A. ORIGINAL CROSS

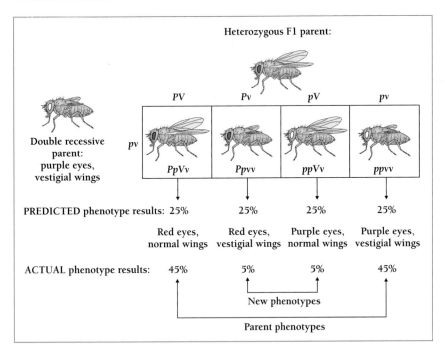

Heterozygous F1 parent:

	PV	Pv	pV	pv
pv	PpVv	Ppvv	ppVv	ppvv

Double recessive
parent:
purple eyes,
vestigial wings

PREDICTED phenotype results: 25%   25%   25%   25%

Red eyes,   Red eyes,   Purple eyes,   Purple eyes,
normal wings   vestigial wings normal wings   vestigial wings

ACTUAL phenotype results: 45%   5%   5%   45%

New phenotypes

Parent phenotypes

B. BACK CROSS

Figure 9-28

Independent assortment? No, genetic linkage. *P* and *V* indicate the dominant wild-type alleles, red eyes and normal wings. Small *p* and *v* indicate the recessive alleles, purple eyes and vestigial wings (which are small and crumpled). A. In a cross between a red-eyed parent with normal wings and a purple-eyed parent with vestigial wings, all the offspring are heterozygotes with red eyes and normal wings ("wild type"). B. If the two genes assorted independently, a backcross between the heterozygous and homozygous flies should produce a ratio of 1:1:1:1. Instead, nearly all the flies had either red eyes and wings or purple eyes and vestigial wings. The two genes do not assort independently because they are on the same chromosome.

P and *V* were dominant. Since *p* and *v* were recessive, the genotype of the purple vestigial flies was *ppvv*.

To find out whether the *p* and *v* genes would assort independently, like Mendel's pea genes, Morgan performed a backcross. He mated the heterozygous F1 flies with purple vestigial flies. If the two genes assorted independently, this cross should have yielded four possible phenotypes in equal proportions. Morgan expected: ¼ purple eyes, vestigial wings; ¼ purple eyes, normal wings; ¼ red eyes, vestigial wings; and ¼ red eyes, normal wings (Figure 9-28B).

Instead, almost 90 percent of the flies had the same phenotypes as the two original parents—either purple eyes and vestigial wings or red eyes and normal wings. The remaining 10 percent of the flies were divided into flies with purple eyes and normal wings and flies with red eyes and vestigial wings.

Morgan concluded that the two genes did not assort independently. The *v* and *p* alleles almost always stayed together, and the *V* and *P* alleles almost always stayed together. Morgan guessed that these alleles stayed together because they were on the same chromosome and, more generally, that genes assort independently only when they are on separate chromosomes.

The tendency of genes on the same chromosome to stay together is called **genetic linkage**. Morgan and his students found that all of the fly alleles they studied fell into just four distinct **linkage groups**, sets of genes that do not assort independently. The four linkage groups, Morgan realized, corresponded to the four pairs of chromosomes in Drosophila. One linkage group consisted of sex-linked traits, indicating

usually found in wild populations. These flies were called *wild type*. The second kind of fly had short, stubby "vestigial" wings and purple eyes (Figure 9-28). Morgan had previously shown that purple-eyed flies with vestigial wings bred true, and he concluded that the purple, vestigial phenotype results from the homozygous condition of two genes. Morgan named the purple allele *pr* and the vestigial allele *vg*. For simplicity we call them *p* and *v* and the corresponding wild-type alleles *P* and *V*.

Purple vestigial flies crossed with true-breeding, wild-type flies produced normal (wild-type) F1 flies (Figure 9-28A). This showed that *p* and *v* were recessive alleles and that

that it included all of the genes on the X chromosome. The fact that genes lie on chromosomes explains not only Mendel's laws but also the existence of linkage groups.

When Morgan repeated Mendel's studies of independent assortment, he found that genes on the same chromosome do not assort independently. This discovery further bolstered Sutton's chromosomal theory of inheritance.

Why Did Some Flies Have Normal Wings but Purple Eyes?

Genetic linkage explains why 90 percent of the flies in the cross shown in Figure 9-28B resemble one or the other parental phenotype. But why do some flies in the backcross have purple eyes and normal wings?

The answer is that homologous chromosomes frequently exchange material with each other when they pair up during meiosis I (Figure 9-11). After crossing over, a new chromosome contains some genes from each of the two homologs. These new combinations of genes then stay together through the rest of meiosis. In this way, the chromosomes of the resulting gametes can differ dramatically from the chromosomes of the parents.

The physical basis of crossing over was first demonstrated in 1931 by graduate student Harriet Creighton and her mentor, geneticist Barbara McClintock, at Cornell University. McClintock, an expert on the genetics of corn, had noticed a pair of homologous chromosomes that differed in appearance. One homolog had a knob at one end and a long tail at the other, while the other homolog was relatively plain (Figure 9-29). McClintock suggested that Creighton find out whether homologous chromosomes literally traded parts by comparing the inheritance of alleles on the odd pair of chromosomes with their appearance after meiosis. Creighton, not realizing that the work was important, dawdled, and McClintock had to press her to finish the work. In time, young Creighton was able to show that crossing over wasn't just an abstract concept but that the chromosomes actually swapped long sections, so that the new homologs had either a knob but no tail or a tail but no knob (Figure 9-29).

In 1931, Thomas Hunt Morgan came to visit Cornell. When he heard what Creighton had done, he insisted that she write up her results and publish them immediately. Although he didn't mention it, Morgan knew that another famous geneticist was finishing up the same work in Drosophila and would soon publish. Morgan sat down on the spot and wrote a letter to the editor of a scientific journal, telling him to expect Creighton and McClintock's paper in two weeks. Because Drosophila reproduce quickly, each experiment lasts just a couple of weeks, while the same experiment with corn takes a whole summer. As Morgan later

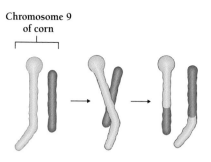

Chromosome 9
of corn

Figure 9-29
Crossing over made easy. In corn, one homolog of chromosome 9 has a knob at one end and a tail at the other. When homologous chromotids cross over, one gets the knob and one gets the tail. In 1931, Harriet Creighton and Barbara McClintock easily showed that what looked like crossing over in chromosome 9 corresponded to the pattern of genetic linkage of genes on chromosome 9.

admitted, "I thought it was about time to give corn a chance to beat Drosophila!"

Crossing over recombines alleles in new ways. Harriet Creighton and Barbara McClintock showed that crossing over is a physical exchange of material between homologous chromosomes.

Can All the Genes on a Chromosome Cross Over and Recombine?

Once Sutton and Boveri had persuaded biologists that genes were on chromosomes, the door was open for a new kind of question: Just *where* on a chromosome did a particular gene lie? How could they map the 1,000 genes on the X chromosome of a fruit fly?

The answer came from Alfred Sturtevant, another undergraduate in Morgan's laboratory. Sturtevant discovered that the farther apart two genes are on a chromosome, the more likely they will be separated because of crossing over. Sturtevant expressed the distance between two genes as the likelihood that they would recombine separately. He defined a **map unit** as the distance between two genes that would produce one percent recombinant gametes (and 99 percent parental gametes). He then constructed a **genetic map** (or **linkage map**) that summarized the distances between many different genes. Geneticists refer to the position of a gene on a chromosome as its **locus** [Latin, = place; plural, **loci**]. The distance between the *v* locus and the *p* locus, for example, is about 10 map units.

The chance that any two genes on a chromosome will recombine is proportional to the physical distance between them.

In this chapter, we have seen how cells divide to create haploid gametes by means of meiosis and how the twin processes of syngamy and meiosis complement each other. We have also seen that the behavior of the chromosomes during meiosis explains the patterns in which genes are inherited. In the next chapter, we learn what a gene does and what it is made of. We will also see that the structure of DNA explains how it replicates during the S phase of the cell cycle.

Key Concepts

- Because genes lie on chromosomes, the inheritance of genes parallels the inheritance of chromosomes.
- Sexually reproducing organisms have *pairs* of homologous chromosomes, which meiosis distributes—one chromosome from each pair—to each gamete (egg or sperm).
- At fertilization, two gametes unite to form a zygote, which has paired chromosomes. Each gamete contributes one chromosome to each pair.
- The phenotype of an individual results from the two-way interaction between its genotype and the environment.

Summary with Key Terms

Genetics is divided into **transmission genetics,** the study of how variation is passed from one generation to the next, and **molecular genetics,** the study of how DNA carries genetic instructions and how cells carry out these instructions. For many years, biologists thought that the characteristics of offspring represented a **blending** of the characteristics of their parents. Most genes are present in two copies, one from each parent, and—most of the time—these genes do not change as they pass from generation to generation. The **Sutton-Boveri chromosomal theory of inheritance** states that the units of heredity, now called genes, are on the chromosomes.

Phenotype encompasses all physical and behavioral characteristics of an organism. Organisms can express different phenotypes in response to different environments, a phenomenon called *phenotypic plasticity*. **Genotype**—the genetic constitution of a cell or organism—is a collection of genes. The word "genotype" can refer to the entire **genome** or to a subset of genes. A **gene** is a region of DNA that specifies the amino acid sequence of a polypeptide and its pattern of expression.

How do sexually reproducing organisms use meiosis to keep the same number of chromosomes from generation to generation?

Asexual reproduction, through mitosis, produces offspring with genes from just one parent, a **clone. Sexual reproduction** produces offspring that inherit genetic information from two parents. This genetic information is contained in two sets of chromosomes, one set from each parent.

The somatic cells of an organism all have two copies of each chromosome and are said to be **diploid.** The **sperm,** or

spermatozoon, and **ovum (egg),** or **gametes,** each carry only one copy of each chromosome and are said to be **haploid.** Gametes arise from the **germ line** or **germ cells** in the **gonads—ovaries** in females and **testes** in males.

Male and female gametes come together at fertilization (or syngamy) to form a **zygote. Fertilization** (or **syngamy**) restores the diploid chromosome number. The genetic information in the zygote, however, is a mixture of information from the parents. Adult organisms produce new haploid gametes by means of **meiosis,** a process that allots one haploid set of chromosomes to each of four daughter cells.

What is the value of recombination and genetic variation?

Because the distribution of chromosomes occurs randomly, meiosis generates combinations of genes that are different from those present in the parents and thus accounts for much of the genetic diversity of individuals within a species.

Meiosis consists of two cell divisions (**meiosis I** and **meiosis II**), each of which consists of **prophase, metaphase, anaphase,** and **telophase.** During prophase I, homologous chromosomes form precisely aligned pairs in a process called **synapsis.** During the rest of meiosis I, one replicated chromosome from each pair (with two **chromatids**) goes to each daughter cell. Meiosis II separates the chromatids of each replicated chromosome and distributes the chromatids to daughter cells.

Each member of a matching pair of chromosomes, or **homologous chromosomes,** is called a **homolog.** Homologous chromosomes pair and **cross over** during synapsis, forming **chiasmata.** Harriet Creighton and Barbara McClintock showed that crossing over is a physical exchange of material between homologous chromosomes, confirming the chromosomal theory of inheritance. **Disjunction,** the separation of homologous chromosomes sometimes fails. **Nondisjunction** leads to the production of gametes with an abnormal number of chromosomes. One example is **Down syndrome,** or **trisomy 21.** Most embryos with the wrong number of chromosomes die long before birth in **spontaneous abortions.**

Alternative versions of the same gene are called **alleles.** An organism with two copies of the same allele is **homozygous,** one with two different alleles is **heterozygous,** and one with only a single allele is hemizygous.

An allele is **dominant** when it alone determines the phenotype of a heterozygote and **recessive** when it contributes nothing to the phenotype of a heterozygote. In **incomplete dominance,** or **codominance,** both alleles are expressed.

How do the mechanics of meiosis explain Mendel's rules of segregation and independent assortment?

Mendel made use of **true-breeding** varieties of peas to show his **principles of segregation** and **independent assortment.** Breeding two genetically distinct organisms is called **cross breeding** (or **crossing**), and the offspring of such a cross are called **hybrids.** To test the principle of segregation, Mendel performed a new kind of cross, called a **backcross.** The original parents in a cross are called the **parental,** or **P,** generation, and the offspring are the "first filial," or **F1** generation, **F2** generation, and so on. All kinds of crosses can be represented in a **Punnett square,** with each kind of gamete produced by one parent along the top of the square and each kind of gamete produced by the other parent along the left side of the square.

The position of a gene on a chromosome is its **locus.** The tendency of genes on the same chromosome to stay together is called **genetic linkage.** Geneticists can construct a **genetic map** (or **linkage map**), which summarizes the distances between genes in **map units. Linkage groups** are sets of genes that lie on a single chromosome and therefore do not assort independently. Compelling evidence that genes are on chromosomes came from the analysis of genetic linkage in Drosophila. Genetic maps of Drosophila and other organisms identified the location of the genes on the chromosomes and explained deviations from the principle of independent assortment.

Why are some diseases "sex-linked"?

In mammals and fruit flies, males have two unmatched **sex chromosomes,** called **X** and **Y,** as well as homologous **autosomes.** Females have the same autosomes plus a pair of matched X chromosomes. The X and Y chromosomes have homologous regions and pair during synapsis. Because men have only one X chromosome, they are susceptible to **sex-linked** genetic defects such as hemophilia and color-blindness.

Review and Thought Questions

Review Questions

1. How are homologous chromosomes similar to each other? How are they different?
2. Describe and draw the steps of meiosis I and explain what they accomplish. Then do the same for meiosis II. Which part of meiosis actually accomplishes the reduction in chromosome number?
3. Explain how crossing over of chromatids in prophase I leads to genetic recombination, so that the gametes produced from one parent are virtually all unique.

4. Describe the main ways in which meiosis differs from mitosis.
5. Define the principles of independent assortment and segregation. What aspects of meiosis explain these principles?
6. Explain the difference between autosomes and sex chromosomes. How many of each do you have?
7. What causes Down syndrome? Why should the frequency of eggs with the wrong number of chromosomes increase as a mother ages?

Thought Questions

8. In what way is fertilization the opposite of meiosis? Why is it important for a complete sexual life cycle to contain both processes?
9. If you found a new species in which the sexes look different, what criteria would you use to decide which to call "female" and which to call "male"?
10. Textbooks make a point of contrasting the genotype and the phenotype as mutually exclusive. If the phenotype is every measurable aspect of an organism, including its structure, chemical makeup, and behavior, could an organism's genotype be considered a part of the phenotype? Why or why not?

BiologyNow Resources

Biology ⓔ Now™

Active Figures
9-22:   Crosses
9-24:   Independent assortment

Preparing for an exam? Take a diagnostic test on your BiologyNow CD-ROM.

Online materials relating to this chapter are at:
http://biology.brookscole.com/AAL3

About the Chapter-Opening Image
When young Walter Sutton solved the mystery of heredity in 1902, he cried, "I know why the yellow dog is yellow!"

The Structure, Replication, and Repair of DNA

Key Questions

- How did biologists discover what genes were made of and what they did?

- What is the structure of DNA?

- How does DNA's structure allow it to act as a template for its own replications?

- What is a mutation and why are mutations important?

Rosalind Franklin and the Double Helix

On February 28, 1953, James D. Watson and Francis Crick discovered the structure of DNA, the hereditary material of the genes. Their breathtaking discovery laid a foundation for all of the most spectacular advances in biology until the present time. It was "the secret of life," as they announced at the time.

The two men were helped in this momentous accomplishment by conversations with many other scientists. Yet, a young scientist named Rosalind Franklin, working in near-total isolation, almost beat them to the punch (Figure 10-1).

In 1951, Jim Watson was a 23-year-old former radio Quiz Kid, cherished by his elders as an up-and-coming genius. He had just received his Ph.D., studying the genetics of bacteria and the viruses that infect them, and his professors had arranged for him to go to Copenhagen to study the chemistry of nucleic acids. But the chemistry of nucleic acids bored him, and he began looking about for something else to work on.

While on a trip to Naples, Italy, Watson found that something—a question so fundamental that its answer seemed guaranteed to bring a Nobel prize. At a lecture by an English physicist turned biochemist named Maurice Wilkins, Watson saw a slide of an x-ray diffraction pattern from what Wilkins said was a crystalline form of DNA. Wilkins argued that his photos, taken with the help of graduate student Raymond Gosling, were the first step in deducing the structure of DNA. The molecular structure of DNA would, of course, shed light on how genes work.

Watson was transfixed. He understood that if DNA could form crystals, it must therefore have a regular and relatively simple shape. Then, within just a few days, he read a sensational paper by famed chemist Linus Pauling on the alpha-helical structures within proteins. Watson longed to do something equally spectacular. With DNA and helices swirling in his head, Watson persuaded his former professors back in the United States to get him out of Copenhagen and into a lab in England where he could study DNA.

In a few months, Watson arrived at the Cavendish Laboratories in Cambridge, in the same building as Max Perutz and John Kendrew, pioneers in the study of macromolecules using x-ray diffraction techniques. Also in the lab was Francis Crick, a 31-year-old graduate student working on his Ph.D. project. Working under Perutz, Crick was studying the x-ray diffraction of polypeptides and proteins, especially the blood protein hemoglobin.

Crick immediately liked Watson, and the two began to do an enormous amount of talking, so much so, in fact, that Perutz and Kendrew finally banished the two younger scientists to a room of their own. Watson knew nothing about x-ray crystallography except that he needed it to solve the structure of DNA. He nagged Crick relentlessly to help him work on DNA.

Crick liked to talk about DNA, but he wasn't about to start working on its structure. With a wife and daughter to support, he had to finish his Ph.D. and get a job. Besides, etiquette forbade Crick from working on the DNA problem. Crick and Wilkins were old friends going back to their joint work for the British Admiralty during World

Figure 10-1
Rosalind Franklin on a walking tour of France in about 1950. Franklin produced most of the data needed to discover the structure of DNA.

War II. Just as important, the Cavendish and the King's College Laboratory in London, where Wilkins worked, were sister facilities—and the DNA problem formally belonged to King's. At King's College, Wilkins and Rosalind Franklin were already working on the structure of DNA, and they hadn't asked for help.

Still, it didn't hurt to talk. The first step to solving the three-dimensional structure of DNA was to study its density, size, and other properties using x-ray crystallography. Because DNA is a long, thin molecule, it would be impossible to obtain true DNA crystals until the 1960s, when researchers learned to produce crystals from short pieces of DNA synthesized in the laboratory (Figure 10-2). The DNA Wilkins had acquired was not crystals, but fibers, in which hundreds of millions of DNA molecules lay parallel to one another.

In 1951, the best x-ray diffraction photos of DNA fibers were those taken by Rosalind Franklin at the King's Laboratory. In November 1951, Watson took the train down to London to visit Maurice Wilkins and to hear Franklin give a lecture on the structure of DNA.

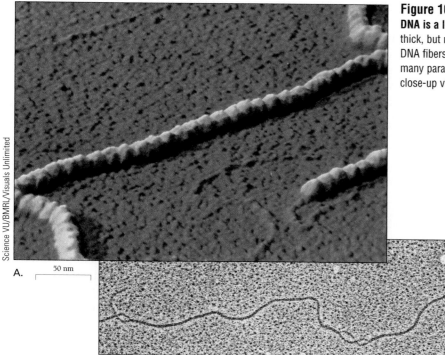

Figure 10-2

DNA is a long, thin molecule. A double strand of DNA is only 2 nm thick, but many DNA molecules reach a length of many centimeters. DNA fibers such as those that Rosalind Franklin studied consist of many parallel molecules. A. Atomic force micrograph showing a close-up view of DNA. B. Electron micrograph of a strand of DNA.

A. 50 nm

B. 1 μm

Franklin had arrived at King's at the beginning of 1951. At 31, she was a talented and recognized authority in industrial physical chemistry, and she had just completed four years of work on the structure of coals at a lab in Paris. Her work laid the foundation for modern industrial carbon-fiber technology. Today, carbon-fiber technology is used in making strong, lightweight bicycle frames and sailboats, for example. For her successful analysis of coals, Franklin learned and improved on known x-ray diffraction techniques, developed new mathematical techniques for interpreting the resulting photos, and constructed scale models like those that Linus Pauling had used in his work. She returned to her native England eager to apply her skills to an important biological molecule.

John Randall, the head of King's Laboratory, invited Franklin to set up and head an x-ray crystallography lab, with all the newest, most powerful equipment available. Once the lab was set up, her assigned task was to work out the structure of DNA. Randall assigned a graduate student to assist Franklin. It was Raymond Gosling, the same young man who had taken pictures for Wilkins.

What Randall did not explain to Franklin was that Maurice Wilkins, just down the hall, had already begun work on DNA and considered the DNA problem his own. In fact, even in his letter offering her the job, Randall had never mentioned Wilkins. Worse, at her first meeting with Randall and Gosling to discuss the project, Wilkins was not present. With that peculiar oversight, Randall set the stage for one of the most disastrous personality clashes in 20th-century science.

Wilkins was equally in the dark about Franklin. He viewed her more as a technician than a scientist, and he appears to have believed that she was hired specifically to do his x-ray crystallography work, in which he had no expertise. He handed over his best DNA fibers, hoping she would produce for him x-ray diffraction images that would reveal their structure. But Franklin displayed none of the deference that Wilkins expected from a female assistant. Instead, she treated Wilkins as a colleague, vigorously arguing with his ideas about DNA in the spirited style she had picked up in Paris. It was her way of trying to make friends. Offended by what he perceived as combativeness, he struggled to put her in her place. Franklin found him unaccountably touchy; he was given to turning away in the middle of conversations he didn't like.

Wilkins spitefully attacked her behind her back, undermining her relationships with other scientists at the lab. This was all too easy, for Franklin was never around when the other scientists gathered to talk at lunchtime or teatime: women were not allowed in the King's dining room.

Over the next six months, Franklin set up the x-ray crystallography lab and then, with Gosling's assistance, began working out the structure of DNA. Though she liked the work, the social atmosphere at Kings was poisonous, unlike anything she had experienced in other labs.

Franklin was used to making warm friendships with her colleagues. But between Wilkins's hostility and her banishment from the dining room, this seemed impossible. She longed for her happy days in Paris. Meanwhile, unable to do the x-ray crystallography work himself, Wilkins resentfully moved on to other projects. Randall, their boss, might have reconciled the two researchers, but he did nothing.

10.1 What Is the Structure of DNA?

When Franklin began her work in 1951, the chemical makeup of DNA—but not its three-dimensional structure—had been known for about 30 years. Everyone knew that DNA consisted of long chains of nucleotides linked together into polynucleotides. Biochemists knew that each nucleotide consisted of a sugar, a phosphate, and a base, and that the sugar and phosphate units were linked together alternately (sugar-phosphate-sugar-phosphate, etc.) into a sugar-phosphate "backbone" (Figure 10-3A).

They also knew that from the sugar-phosphate backbone hung four kinds of nucleotide bases. Two of the four bases are **pyrimidines,** which consist of rings made up of four carbon atoms and two nitrogen atoms. The other two bases are **purines,** which contain a double ring of six carbon atoms and four nitrogen atoms (Figure 10-3B). The purine bases are **adenine** and **guanine,** and the pyrimidine bases are **cytosine** and **thymine.**

Each nucleotide has a long, correct name: deoxyadenosine monophosphate, or dAMP, for adenine; deoxyguanosine monophosphate, or dGMP, for guanine; deoxycytidine monophosphate, or dCMP, for cytosine; and thymidine monophosphate, or TMP, for thymine. But biologists rarely use these names. In fact, they generally refer to each nucleotide by a single letter abbreviation: A, G, C, or T. This convention not only saves space, it also emphasizes that the nucleotides are indeed the four letters in the DNA alphabet.

What was *not* known by anyone at the time that Franklin set to work was that a molecule of DNA consists of exactly two polynucleotide chains that are twisted together into a long helix, like a ladder twisted about its long axis (Figure 10-4). The two uprights (or legs) of the ladder are the two sugar-phosphate chains. Each rung of the ladder consists of a pair of bases (A always goes with T, and G always goes with C). Each twist of the ladder contains 10 rungs, which are 0.34 nanometer (nm) apart. A single molecule of DNA may be either short or long. In humans, a molecule of DNA may be very long—up to 9 centimeters long—and may consist of hundreds of millions of nucleotide pairs. The width of the ladder, however, is the same from one end to the other—2 nm.

Importantly, each sugar-phosphate chain has a direction, and the two chains—the uprights of the ladder—run

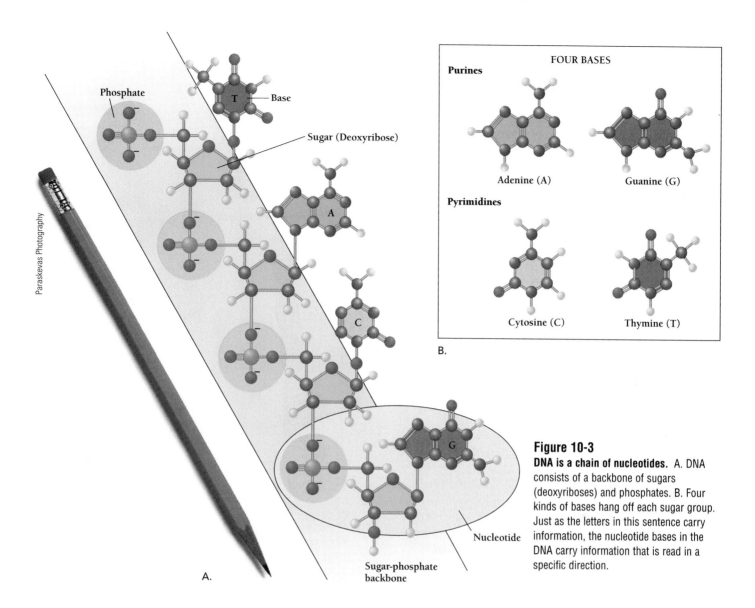

Phosphate

Base

Sugar (Deoxyribose)

FOUR BASES

Purines

Adenine (A)

Guanine (G)

Pyrimidines

Cytosine (C)

Thymine (T)

B.

Nucleotide

Sugar-phosphate
backbone

A.

Figure 10-3
DNA is a chain of nucleotides. A. DNA consists of a backbone of sugars (deoxyriboses) and phosphates. B. Four kinds of bases hang off each sugar group. Just as the letters in this sentence carry information, the nucleotide bases in the DNA carry information that is read in a specific direction.

Paraskevas Photography

in opposite directions (Figure 10-5). In each sugar-phosphate backbone, the third carbon of one sugar is attached by a phosphate group to the fifth carbon of the next sugar. The third carbon is called the 3′ carbon, pronounced "three-prime carbon," while the fifth carbon is called the 5′ carbon, or "five-prime carbon." One end of each polynucleotide has a free 5′ carbon, called the **5′ end**, while the other end has a free 3′ carbon, called the **3′ end.** If we draw an arrow from the 5′ end to the 3′ end of each strand in Figure 10-5, one arrow points up, the other, down.

Each nucleotide in DNA consists of a sugar, a phosphate, and a base. The sugar and phosphate units are linked together alternately (sugar-phosphate-sugar-phosphate, etc.) into a sugar-phosphate "backbone." The two parallel backbones (running in opposite directions) resemble a ladder twisted around its long axis, with each rung consisting of a pair of bases (A-T or G-C).

How Much Did Franklin Discover About the Structure of DNA?

By the time Franklin gave her talk in November 1951, her five months of work had already revealed some essential facts about the structure of DNA. She believed that DNA was probably a big helix with two, three, or four parallel chains, with the phosphates wound around the outside of the helix, like the uprights of a ladder. She had accurately measured the density of DNA and the number of water molecules associated with each nucleotide, and discovered that DNA fibers have two slightly different structures, depending on whether they are wet or dry. Finally, she had identified the "symmetry" of DNA. Crystallographers classify crystals into 230 types according to their symmetry. Which of these symmetries DNA had might seem like an obscure point, but it would be crucial to Watson and Crick's solution to the structure of DNA. DNA's symmetry was the clue that the two strands run in opposite directions.

Only the crystallographers working at the Cavendish, particularly Crick, could fully appreciate the meaning of the particular symmetry Franklin had found. If Crick had heard Franklin speak, he would have understood that Franklin's parallel chains were symmetrical in a particular way. Just as the two ends of a sharpened pencil look different, a chain that ran in one direction only would look different when rotated 180°. In contrast, two pencils running parallel but in opposite directions look the same at both ends, and a pair of symmetrical chains running opposite to each look the same when rotated. The symmetry Franklin had discovered implied that DNA consisted of two chains, one running up, the other running down. The chains could have been four (two up and two down), but Franklin's density measurements ruled that out.

Franklin lacked the extensive training of a full-fledged crystallographer, and unlike Crick, had never worked with a large biological molecule, so she couldn't yet deduce the shape of the DNA molecule. Nonetheless, she was on her way.

Watson, who knew almost nothing about crystallography, watched Franklin talk, but he understood little of what she said and took no notes. After the talk, he visited Wilkins, who said that Franklin didn't know what she was doing. Back at Cambridge, Watson could tell Crick almost nothing about Franklin's talk. Crick was annoyed, but it would be 14 months before he learned how much he had missed. None of the three men took Franklin seriously, as Crick himself later conceded.

What Watson did bring back was one wrong fact, the amount of water in the molecule, and the mistaken impression that neither Franklin nor Wilkins was capable of solving the structure of DNA. Clearly, the two at King's were not working together, and Wilkins seemed to be going nowhere by himself. Watson does not appear to have considered that Franklin might solve the structure. It seemed to him ridiculous to allow Franklin to bumble away, not solving this important problem, when he and Crick could be deciphering DNA themselves.

Back in Cambridge, Watson finally persuaded Crick to help him attack the problem. Within days, the two produced a model of DNA consisting of three intertwined helices, sugar-phosphate backbones on the inside, and far too little water. They called Wilkins and

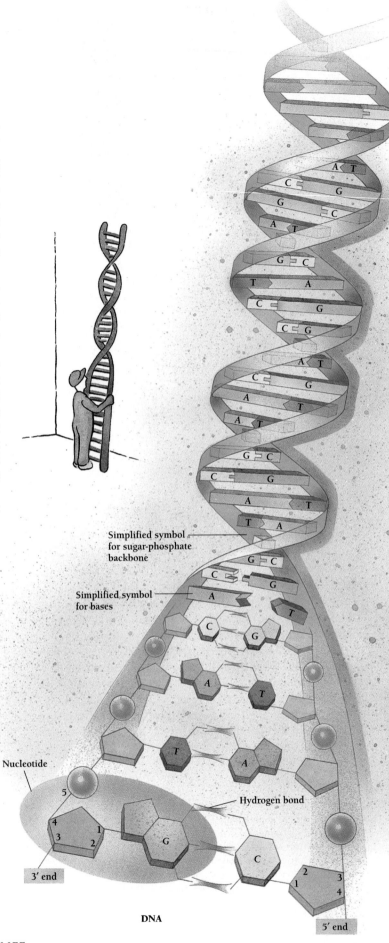

Simplified symbol for sugar-phosphate backbone

Simplified symbol for bases

Nucleotide

Hydrogen bond

3' end

5

4

3

2

1

G

C

2

1

3

4

DNA

5' end

Figure 10-4

DNA is like a twisted ladder. The two sugar-phosphate backbones run in opposite directions. The bases hanging off the sugars connect in the middle, like the rungs of a ladder.

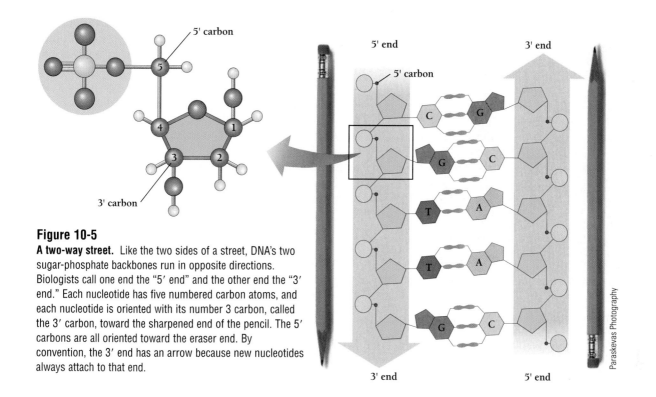

Figure 10-5

A two-way street. Like the two sides of a street, DNA's two sugar-phosphate backbones run in opposite directions. Biologists call one end the "5′ end" and the other end the "3′ end." Each nucleotide has five numbered carbon atoms, and each nucleotide is oriented with its number 3 carbon, called the 3′ carbon, toward the sharpened end of the pencil. The 5′ carbons are all oriented toward the eraser end. By convention, the 3′ end has an arrow because new nucleotides always attach to that end.

Franklin to come up to Cambridge the next day and look over their model. Franklin glanced at the model, knew at once that it was completely wrong, and said so. Crick was humiliated, for he knew instantly that she was right.

Watson had missed everything important that she had said. Worse, Watson and Crick had worked on a problem that belonged to King's, and they had been wrong. The head of the Cavendish Lab, Sir William Lawrence Bragg, sternly reminded them that the DNA problem was not theirs to work on. Reluctantly, they agreed to stop trying to model its structure, but they continued to talk about DNA. They couldn't help it. They loved to talk (Figure 10-6).

For another year, Franklin studied DNA in near solitude, talking only to Gosling. For a while she thought that maybe the structure couldn't be a helix after all. The dry form of the DNA fiber gave x-ray diffraction images that seemed to suggest DNA was not a helix. There was, in fact, no certain evidence for a DNA helix early on. Pauling had argued that a repeating chain of any kind, whether a molecule or a telephone cord, will tend to twist about its axis into a helix. A helix seemed likely. But the pictures of the dry form did not lend support to a helix.

Franklin played with other possibilities, including repeating figure eights. In the summer she took an excellent picture of wet DNA fibers, showing a pattern that fairly shouted "helix." She labeled it photo number 51. She liked it a lot, and spent several days thinking about it (Figure 10-7). She had been working on the dry form for so long, however, that she soon returned to that. If she had had a collaborator, someone to argue with, she might have given more thought to photo number 51. But the atmo-

Figure 10-6
Francis Crick and James D. Watson share a cup of tea in the early 1950s.

sphere at King's was as bad as ever. She was already making plans to go to another lab.

Back at the Cavendish, Watson frittered away his time, playing tennis, going to parties, and fretting vaguely about DNA. He was, in his own words, "totally underemployed." Then, in the fall of 1952, two young researchers arrived at the Cavendish—Linus Pauling's son Peter and one of Pauling's graduate students. In mid-December, Peter announced that his father had discovered the structure of DNA. It was more than Watson could stand. In January, he begged Peter to get his father to send the manuscript describing his structure for DNA.

Franklin, who had also heard the news, wrote to Pauling's lab with the same request. Meanwhile, she began calculating

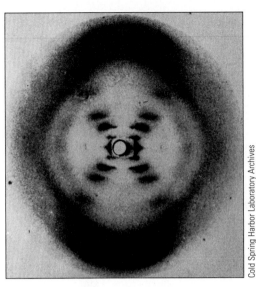

Figure 10-7

DNA is a double helix. Shown here is the famous x-ray diffraction "photo 51" of DNA taken by Rosalind Franklin. The clear "X" pattern across the center indicated that DNA was a double helix.

bond lengths and angles for the different building blocks of DNA. Her months of work had yielded the main dimensions of the molecule. She suspected that DNA was a double helix. She knew for certain that it was 2 nm in diameter and that if it was a helix, each complete turn of the helix would be 3.4 nm long. Now, with these and other hard-won clues at hand, she quietly prepared to build a model of DNA, the same approach that had worked so well for Pauling in deducing the α helix of proteins.

On January 30, 1953, Watson went to King's College and stopped to visit Wilkins. Finding Wilkins busy, Watson dropped in on Franklin to taunt her with his copy of Pauling's manuscript. When she looked at the manuscript she could see as well as Watson that Pauling's three-chain helix was hopelessly wrong.

She also discovered that Pauling and his coauthor, Robert Corey, had snubbed her. The previous spring, Corey had come to King's and asked to see her pictures, and she had graciously shown him her work. Wilkins and Gosling had taken earlier pictures of DNA also, but these were vastly inferior to Franklin's. Now Pauling and Corey had not only ignored her letter requesting their manuscript, they had also cited Wilkins's inferior x-ray diffraction pictures of DNA—which they had never seen—and not her own, which they had. Watson later recalled that he then began to goad her again, implying that she was incompetent to interpret her own x-ray diffraction photos. Although few people believe him, Watson claimed that the petite Franklin was so angry that she chased him from her lab.

According to Watson, he then found Wilkins, and the two men took turns complaining about Franklin. Wilkins reminded Watson of Franklin's ideas on DNA, the same ideas she had presented 14 months earlier, then showed

Watson her striking image of the wet form of DNA, number 51. Crick and another scientist had already worked out a mathematical understanding of the way that a helix would interact with x rays. That work, which Crick had published and which Franklin was fully aware of, had shown that a helix should generate a cross-shaped pattern. Now, with that knowledge and without Franklin's scores of other pictures to confuse him, Watson leapt to the conclusion that DNA must be a helix. It was obvious to everyone who had seen it— Franklin, Wilkins, and Watson—that this photo was of a helix. Nonetheless, Wilkins told Watson that Franklin stubbornly insisted that DNA was *not* a helix. What she had really said, however, was that although a helix was possible and even likely, there was still some doubt, since her dry form pictures didn't support a helix.

Watson could only conclude that Franklin really was incompetent. He was frustrated by the British laboratory etiquette that prevented him from working on DNA himself. He would build a model of the wet form of DNA, whose picture looked so simple. Instinctively, he decided to try two backbones. He had a feeling that for DNA to duplicate itself most easily, it should have two chains. Four chains seemed needlessly complex, and three chains seemed illogical. He was still convinced, however, that the sugar-phosphate backbones must be on the inside of the helix, with the bases hanging off the outside, despite Franklin's assertion that the sugar-phosphate backbones were on the outside. Watson rushed back to the Cavendish and at last succeeded in persuading Bragg to let him work on DNA. Feverishly, Watson began working to build a model. Crick, still writing his Ph.D. thesis, watched from across the room with amusement and growing interest.

Watson made little progress until, on February 5, 1953, Crick finally persuaded him to try putting the two sugar-phosphate backbones on the outside, as Franklin had suggested. For the first time, Watson had the beginnings of a model. The following weekend, Wilkins came up to Cambridge and reluctantly gave Watson and Crick permission to work on the DNA problem. Wilkins promised Crick in a letter to "tell you all I can remember & scribble down from Rosie."

No one asked Franklin's permission. Unaware that Wilkins and Bragg had now given away her research problem, she restarted her work on a structure for the easier, wet form of DNA. She was trying to decide if it was definitely two chains, or if there was any way it could be just one chain.

For the first time, Crick felt free to work on DNA. The best source of information about DNA was Franklin, but Crick had no more intention of talking to her than had Watson or Wilkins. Crick soon learned, however, that Franklin and Gosling had summarized their unpublished work in an annual report on research at King's. Max Perutz had a copy of the report, and when Crick asked for it, Perutz handed it over.

Now Crick knew almost everything that Franklin and Gosling had so laboriously worked out. For the first time, he began to take the whole enterprise seriously. When he read Franklin's report describing DNA—its density, its water content, and, critically, its symmetry—it gradually dawned on Crick that DNA had to have two chains and that they had to run in opposite directions.

Franklin and Gosling also reported that the distance along the wet form of the molecule through one complete turn was 3.4 nm. Previous work had shown that the distance between the bases was 0.34 nm, so it was obvious that there were probably 10 bases per 360° turn. Wilkins and Gosling had earlier shown that the diameter of DNA was about 2 nm, and Franklin and Gosling had confirmed this. Things were beginning to come together.

Crick tried to persuade Watson to build a model with the chains running in opposite directions. But Watson couldn't figure out how to make it come out right. So, while Watson was off playing tennis, Crick abandoned his Ph.D. dissertation for an afternoon and took over the model. He built a more open double helix than Watson's, with sugar-phosphate chains running in opposite directions as required by the symmetry, and 10 bases per turn of the helix. The building blocks, with their bond angles and lengths, all fit nicely. Each complete turn was 3.4 nm, as Franklin had measured, and the distance from one base to the next was 0.34 nm. The only problem now was to find a way to fit the bases together in the middle.

It wasn't immediately obvious how this could be done. The four bases were of very different sizes and shapes. Some pair combinations would be wider than 2 nm and would make the molecule bulge and strain. Other combinations would have been too narrow and would have left a gap in the middle or pinched the two backbones in. According to Franklin's report, however, the helix had a constant diameter and neither bulged nor pinched in, so Crick and Watson knew they didn't yet have the right structure.

Watson began reading all the biochemistry of bases that he could find. If DNA were to carry genetic information, it should be able to accommodate any sequence of nucleotides, just as each line of print on this page must be able to accommodate any sequence of letters. Two weeks after he first put the sugar-phosphate backbones on the outside, Watson guessed that like bases paired (guanine with guanine and so on). But within days, Jerry Donohue, the young chemist from Pauling's lab, completely demolished Watson's hypothesis, explaining that the molecular forms that Watson had used, from standard textbooks, were all wrong. Besides, Crick now remembered, Watson's like-with-like base pairing did not square with Chargaff's rules.

What Were Chargaff's Rules?

In the 1920s and 1930s, chemists believed that DNA contained almost equal amounts of each base. Then, in the 1940s, the Austrian-American biochemist Erwin Chargaff, at Columbia University, developed a new method for analyzing the base composition of DNA. Chargaff found that the proportions of the bases could vary widely from species to

species. But the amount of A always equaled the amount of T, and the amount of G always equaled the amount of C. Neither Chargaff nor anyone else at the time understood the significance of his findings, summarized in what came to be called **Chargaff's rules.**

In the spring of 1952, nine months before Watson and Crick guessed the correct structure of DNA, 46-year-old Chargaff had come to Cambridge. Watson and Crick were introduced to Chargaff, and the older scientist took an immediate dislike to the two young men. "It struck me as a typical British intellectual atmosphere, little work and lots of talk," Chargaff later complained. Crick knew almost nothing about Chargaff's work, but the older man brusquely set Crick straight.

At the time, Crick saw that Chargaff's rules might imply complementary base pairing. That is, A always pairs with T, and G always pairs with C. Such a relationship could, he reasoned, provide a means for DNA to copy, or replicate, itself. Nine months later, however, Crick had forgotten his own insight.

By Saturday, February 28, Watson had constructed cardboard cutouts representing the shapes of the four bases. He fiddled with them, fitting them together this way and that. He struggled to come up with pair combinations that would have the same symmetry and diameter, and still allow any order of bases on the backbone. Finally, he tried pairing A with T and G with C. The fit was unbelievably good. An A-T pair was the same width as a G-C pair (Figure 10-8).

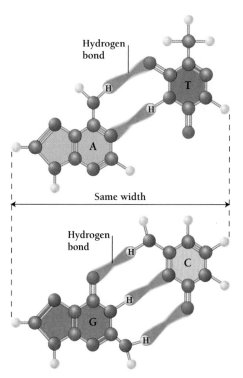

Figure 10-8

Base pairing. In DNA the bases connect in the middle of the twisted ladder by means of hydrogen bonds. The nucleotides A and T can form two stable bonds with each other, while G and C can form three. When matched in this way, the width of a G-C pair is the same as the width of an A-T pair, as predicted by Franklin's x-ray diffraction data.

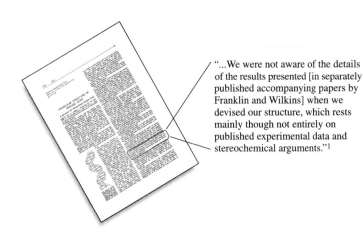

"...We were not aware of the details of the results presented [in separately published accompanying papers by Franklin and Wilkins] when we devised our structure, which rests mainly though not entirely on published experimental data and stereochemical arguments."[1]

Donohue quickly assured Watson that the two sets of bases could in fact fit together that way. The arrangement of atoms in the bases is such that A is perfectly arranged to make two hydrogen bonds with T, while G can form three hydrogen bonds with C. The chemistry worked, the model conformed to Franklin's data, and for the first time, Watson and Crick thought they understood the entire structure. It would be another week before they had the materials to build a complete model, but their work was done. Euphoric, they announced to everyone they met that they had "discovered the secret of life."

On Saturday, March 7, Watson and Crick eagerly took model bases that had been made for them in the Cavendish machine shop and built an accurate model of the DNA double helix. On the same day, Wilkins, unaware of the model, wrote a note to Crick, including the following:

> I think you will be interested to know that our dark lady leaves us next week and much of the . . . data is already in our hands. I . . . have started up a general offensive on Nature's secret strongholds on all fronts: models, a theoretical chemistry and interpretation of data, crystalline and comparative. At last the decks are clear and we can put all hands to the pumps!
>
> It won't be long now.
>
> > Regards to all,
> > Yours ever,
> > M.

Franklin Comes Within Two Steps of the Correct Model

But the race was over. Wilkins little knew how close Franklin had come to solving the problem on her own, or how late he was in restarting his own work. Nearly two weeks earlier, on February 23, Franklin had begun remeasuring photo number 51. By the end of the next day, she had correctly concluded that DNA was a double helix, with a diameter of 2 nm. She was now certain that the sugar-phosphate backbones were on the outside. She guessed that the backbones had some peculiar relationship to each other, but what? She hadn't yet realized that the two backbones ran in opposite di-

rections. She knew about Chargaff's rules, and she had begun playing with the problem of how the bases would fit together in the middle. She was two short steps from the answer.

But Franklin's work on DNA was at an end. Randall had told her she could leave King's for another lab if she wanted, but she could not take the DNA problem with her. Accordingly, she left King's and, with Gosling, wrote up the results of their 18 months of work on the structure of DNA. Their work was not complete, but it was substantial. The potent combination of Franklin's experimental work, Crick's theoretical work, and Watson's infectious enthusiasm had provided everything needed to solve the problem. Yet, amazingly, when Watson and Crick wrote up their results for the prestigious British journal *Nature,* Franklin was the one person whose work was not formally cited. This deficiency was apparently no oversight. According to writer Horace Judson, Wilkins asked Watson and Crick to delete a sentence from their manuscript that said, "It is known that there is much unpublished experimental material." Instead, Watson and Crick explicitly asserted that they were unaware of Franklin's work, writing that "We were not aware of the details of the results presented [in separately published accompanying papers by Franklin and Wilkins] when we devised our structure, which rests mainly though not entirely on published experimental data and stereochemical arguments."[1]

In fact, Watson and Crick's model rested heavily on the unpublished data of Franklin and Gosling, the details of which Watson and Crick knew as well as anyone.

Not until March 18, when Wilkins first received Watson and Crick's manuscript, did Franklin learn of their achievement. Eagerly, she took the train up to Cambridge to see their model. Watson was astonished at how quickly and graciously she agreed that the model was correct. He had been sure that she would find some grounds on which to criticize it, for he little appreciated how much she knew and understood. She, for her part, had no reason not to be gracious, for she little suspected how much Watson and Crick had relied on her own work. The elegant beauty of the structure delighted her. She was filled with admiration for Crick's genius.

Franklin was glad, in any case, to be hard at work in her new lab. She was now studying the structure of tobacco mosaic virus. Over the next five years, Franklin directed a highly productive research team. Later, she collaborated with researchers at Berkeley and Yale in the United States, and at Tübingen in Germany. When she discovered she had cancer, in 1956, she continued to work cheerfully and productively until the very end, so devoted was she to the research she loved. She died in 1958, only 37 years old, without ever learning of her own crucial contribution to the discovery of the structure of DNA.

When, a couple of years later, it appeared that Watson and Crick might get a Nobel prize for their work, Sir

[1]From Watson, J.D., and Crick, F.H.C., "Molecular Structure of Nucleic Acids: A Structure for Deoxyribose Nucleic Acid," *Nature,* April 25, 1953, vol. 171, pages 737–738.

Lawrence Bragg, who had headed the Cavendish, made sure that Wilkins received the prize with them. Recounting the story to writer Horace Judson, Bragg said that "when it came to the Nobel prize, I put every ounce of weight I could behind Wilkins getting it along with them. It was just frightfully bad luck, really." Wilkins was lucky enough, however, to receive in 1963 a third share of a Nobel prize for the discovery of the structure of DNA.

In 1968, 10 years after Franklin's death, James Watson wrote a personal account of the discovery of the structure of DNA. In *The Double Helix,* Watson for the first time made clear how essential Franklin's experimental results had been to their discovery. He also falsely depicted her as Wilkins's rebellious and shrewish assistant, who seemed to deserve any bit of ill treatment that came her way. Watson's book was attacked and reviled by nearly everyone who had been present during the discovery. It has never been considered an accurate account. Yet, as a work of literature and as a window into the mind of a successful molecular biologist, *The Double Helix* is a classic.

Modern Science Is a Social Endeavor

The discovery of DNA's structure is the best-documented story in the history of science. Watson, Crick, and others have all written extensively and passionately about both the science and the scientists. It is clear that Franklin's failure to discover the structure of DNA before Watson and Crick was a direct result of her social isolation. Watson himself said in 1993, "[Franklin] would be famous for having found DNA if she'd just talked to Francis [Crick] for an hour."

Watson and Crick had each other to talk to. They had Max Perutz, John Kendrew, Jerry Donohue, Peter Pauling, and even, grudgingly, Erwin Chargaff. All told, at least a half dozen scientists besides Rosalind Franklin herself provided essential clues that guided Watson and Crick toward the correct answer. Franklin, isolated by her feud with Wilkins, as well as her exclusion from the King's dining room, had no one to talk to but Gosling, who had little useful knowledge.

Watson and Crick's discovery of the structure of DNA was a borrowed victory that rested heavily on Franklin's discoveries. Their main contribution to biology was their early recognition that DNA's structure explains how it works. In fact, Watson and Crick's insight into the implications of base pairing stands as one of the most influential advances in the history of biology. It has provided the sole basis for the twin fields of molecular genetics and genetic engineering, and transformed every other area of biology.

What was so important about their model? We've seen that the structure was like a twisted ladder with the uprights running in opposite directions. Each rung of the ladder was a pair of bases held together by hydrogen bonds. Watson and Crick's rules of complementary base pairing meant that the sequence of one strand exactly specified the sequence of the second strand. That is, one strand could give directions for building the other. This explained how a chromosome

could replicate itself during cell division. Further, it was clear that the bases could be arranged in a nearly infinite variety of different sequences. That meant that the sequence of bases in DNA, like the sequence of letters on a page, could contain coded information.

In the rest of this chapter, we will examine the structure of DNA in more detail. Along the way, we will find out how biologists found out that the genetic material was DNA, what a gene is, and how DNA replicates and repairs itself. In the next chapter, we will see how cells actually use the information encoded in the DNA.

10.2   What Is a Gene?

Defining a gene remains a problem not only to students taking biology examinations but also to scientists searching for biological molecules. To Mendel in 1865, a gene was an abstraction—some form of information that determined the traits of individuals. To Sutton and Morgan in the early years of the 20th century, a gene was both information and a concrete physical structure that was part of a chromosome.

But biologists still did not know what genes actually were or how cells used and copied the information that genes contain. That knowledge came only when geneticists tried to understand genes in chemical terms. In the rest of this chapter, we will see how researchers established (1) what genes do, (2) what they are made of, and (3) how they replicate. We will learn that a gene spells out the sequence of amino acids in a polypeptide, that a gene is made of DNA, and that the structure of DNA explains its ability to replicate and repair itself.

In this and later chapters we will also see that genes are not only physical entities or places on the chromosomes but pure information. In this sense, genes are abstractions, as Mendel conceived them. Just as information about biology can be communicated as printed pictures or text or as spoken words, genetic information can also be relayed in many forms. The physical expression of that information is separate from the information itself (Figure 10-9).

How Do Genes Affect Biochemical Processes?

The first hint of the biochemical importance of genes came in 1902 from the work of an English physician, Archibald Garrod. Among Garrod's patients at the Hospital for Sick Children, in London, was a baby whose diapers developed black urine stains. The baby's disease, alkaptonuria, had first been described in 1649 and named for dark "alkapton bodies" that formed after the urine was exposed to air. The precursor of the alkapton bodies was identified as homogentisic acid (HA), which oxidized to form the alkapton bodies.

Garrod, working just two years after the rediscovery of Mendel's paper, suspected that alkaptonuria might be an in-

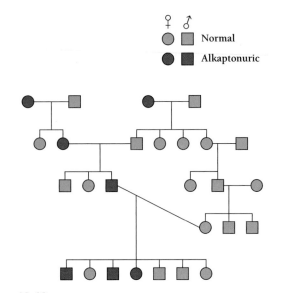

Figure 10-10
How do people inherit alkaptonuria? Part of a pedigree from a Lebanese family with alkaptonuria. The marriage of a man to his younger cousin—shown midway down the pedigree—increases the incidence of the homozygous recessive alkaptonuria allele. In general, marriages among family members increases homozygosity.

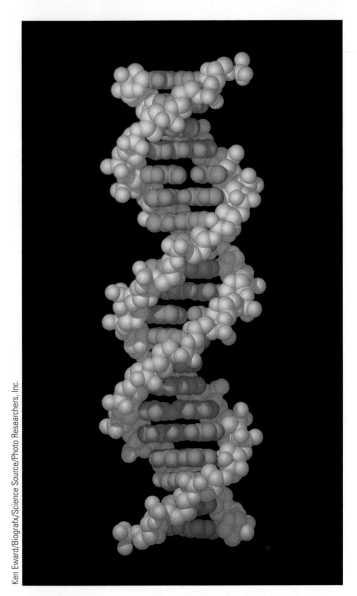

Figure 10-9
Computer-generated model of DNA. The structure of DNA is now so well known that most of us have seen the familiar double helix somewhere.

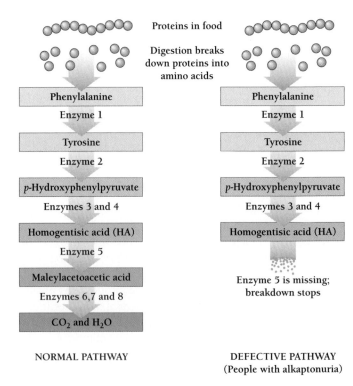

Figure 10-11
What causes alkaptonuria? Cells can break down phenylalanine into homogentisic acid (HA) and, with a few more reactions, carbon dioxide and water. But if a crucial enzyme is missing, HA accumulates in the body and is excreted in the urine. Once outside the body, oxygen in the air turns HA in the urine to dark "alkapton" bodies. Although the dark urine would scare any parent, alkaptonuria is a relatively harmless genetic anomaly.

herited disease because the baby's parents were first cousins (Figure 10-10). After studying the way the disease was inherited, he realized that alkaptonuria behaved as if it were caused by a recessive allele, much like the white flowers of Mendel's peas. Each parent had contributed the same recessive allele to the baby, who therefore had two copies of the same allele. We say the baby was homozygous for that allele.

Like the amino acid phenylalanine, HA contains a benzene ring, so Garrod hypothesized that HA was an intermediate in the break down of phenylalanine. Most people have an enzyme called *HA oxidase*, which helps break down HA (Figure 10-11). Garrod realized that the absence of HA oxidase might cause HA to accumulate and turn to alkapton bodies in the urine. Furthermore, the recessive allele whose inheritance he had charted could explain the missing enzyme. If an alkaptonuria patient had two copies of a defective allele of the gene for HA oxidase, he or she would not be able to make the enzyme.

Archibald Garrod's study of alkaptonuria suggested that genes might work by specifying enzymes.

One Gene—One Enzyme

Garrod's insight that genes work by specifying enzymes at first seemed to apply to only a few rare diseases. In alkaptonuria and albinism, for example, the body's failure to make a single enzyme explained an obvious abnormality. But do all genes specify enzymes?

In 1941, two Stanford University geneticists, George Beadle and Edward Tatum, began looking for other genetic defects in metabolism that might be caused by faulty enzymes. Beadle and Tatum chose to study a simple fungus, the pink bread mold *Neurospora*. They induced mutations in *Neurospora* with x rays and then looked for biochemical defects. The biochemical defects were easy to spot because *Neurospora* is a haploid organism—it has only a single copy of each gene. When that single copy is altered, the change immediately shows up in the phenotype. (In diploid organisms such as humans, a normal allele on the homologous chromosome can compensate for a defective recessive allele.)

Wild-type *Neurospora* can grow on an extremely simple food medium (called a minimal medium) that contains glucose sugar, ammonia, and a few salts. But Beadle and Tatum identified a large number of mutant *Neurospora* molds that could not grow on such a minimal medium. These mutants required various compounds that the molds could not make themselves. One mutant strain, for instance, would grow only if the minimal medium was enriched with vitamin B_6. Another mutant strain needed the amino acid arginine (Figure 10-12).

Beadle and Tatum mapped the chromosomal location of each of their mutants by studying genetic linkage, as described in the last chapter. The mutants that needed arginine, for example, fell into three groups. Beadle and Tatum established that the three groups corresponded to mutations in three separate genes, on three separate chromosomes. Each gene coded for a separate enzyme needed to make arginine.

One group of mutants (*arg*-1) would grow when arginine was added and also in the presence of two related compounds, ornithine and citrulline. The second group of mutants (*arg*-2) would grow in the presence of arginine or citrulline, but not ornithine alone. The third group (*arg*-3) required arginine; neither citrulline nor ornithine would do.

Beadle and Tatum realized they could explain their results with the following model of arginine synthesis:

enzyme 1   enzyme 2   enzyme 3
precursor $\longrightarrow$ ornithine $\longrightarrow$ citrulline $\longrightarrow$ arginine

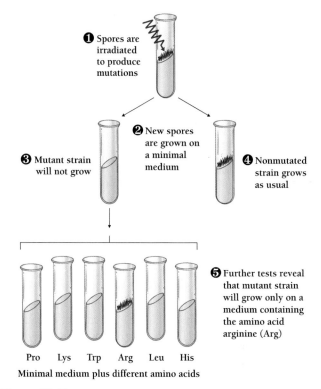

Figure 10-12
Mutant *arg* genes in the bread mold *Neurospora crassa*. Irradiation causes various mutations in bread mold spores. One set of mutant strains of *Neurospora* could not make the amino acid arginine in the normal way and needed either arginine or other amino acid supplements to survive.

Each step in the synthesis of arginine, they reasoned, is catalyzed by a different enzyme. The *arg*-1 mutations, they argued, must lie in the gene for enzyme 1. The absence of enzyme 1, then, could be overcome by ornithine, citrulline, or arginine. Similarly, defects in the gene for enzyme 2 can be overcome by adding citrulline or arginine, but not ornithine, since the mold cannot process it. Finally, defects in the gene for enzyme 3 can only be overcome by the addition of arginine itself. Beadle and Tatum's work suggested that every gene somehow affects the function of a single enzyme. Beadle and Tatum captured this concept in a single phrase: "**One gene—one enzyme.**"

Beadle and Tatum showed that one gene could specify one enzyme.

One Gene—One Polypeptide

Recall from Chapter 3 that all enzymes are proteins, but cells contain thousands of proteins that are not enzymes. Do genes contain information for making enzymes only? Are enzymes the only link between genotype and phenotype? Or can genes code for other proteins besides enzymes?

The answers to these questions came with the unraveling of the molecular basis of another genetic disease, sickle cell

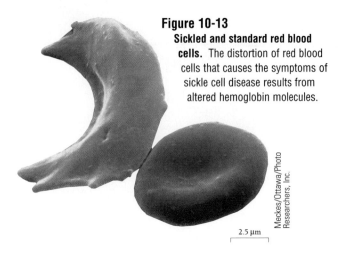

Figure 10-13
Sickled and standard red blood cells. The distortion of red blood cells that causes the symptoms of sickle cell disease results from altered hemoglobin molecules.

Meckes/Ottawa/Photo Researchers, Inc.

2.5 µm

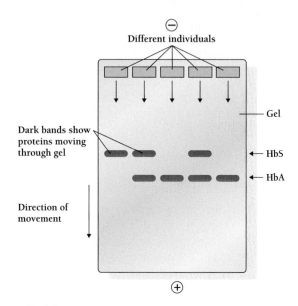

Different individuals

Dark bands show proteins moving through gel

Gel

← HbS
← HbA

Direction of movement

Figure 10-14
Electrophoresis. Biologists can separate macromolecules according to charge, shape, and size. In electrophoresis, an electric current pulls DNA, RNA, or proteins through a gel. Molecules move through the gel at different rates depending on their size and charge. Smaller ones move faster than larger ones, and those with a greater negative charge move faster than those that are less negative. Sickle cell hemoglobin, HbS, is less negative than normal hemoglobin, HbA. So HbS moves more slowly through the gel than HbA.

anemia. The name "sickle cell" comes from the crescent or sickle shape of the red blood cells of people with this disease (Figure 10-13). The distorted blood cells clump and block the small blood vessels of the circulation, causing excruciating pain, anemia, heart trouble, and brain damage. The body's defense mechanisms eliminate the defective cells as rapidly as possible, but this causes severe anemia. People with sickle cell disease almost always die young. Sickle cell disease, like alkaptonuria, results from a recessive allele. Unlike alkaptonuria, however, sickle cell anemia is both serious and common—1 in 13 African Americans carry the allele, and 1 in 625 have two copies of the allele and have the disease.

In 1949, the biochemist Linus Pauling and his collaborators showed that the red, oxygen-binding blood protein hemoglobin of sickle cell patients was different from that of healthy people. He called normal hemoglobin protein HbA and sickle cell hemoglobin HbS. Pauling and his colleagues used a technique called electrophoresis that reveals even tiny differences among molecules to show that the healthy parents of sickle cell patients have a mixture of half HbS and half HbA (Figure 10-14). But the children who have sickle cell anemia lack HbA entirely. A simple explanation of the disease emerged:

1. A dominant allele specifies the normal protein, HbA.
2. A recessive allele specifies the abnormal protein, HbS.
3. Heterozygotes (such as the parents) have alleles for both HbA and HbS.
4. Heterozygotes, who make both HbA and HbS, are healthy.
5. Homozygotes, who have two copies of the HbS allele and can only make HbS, are sick.
6. In the absence of the normal protein, HbA, HbS causes sickle cell disease.

In sickle cell disease, differences in the gene had caused differences in the hemoglobin protein, not in an enzyme. Pauling's work suggested that genes controlled other proteins besides enzymes. Pauling extended "one gene—one enzyme" to "one gene—one protein."

Exactly how does an abnormal gene change a protein? To answer this question, Vernon Ingram, a young bio-

chemist working at Cambridge University in 1956, used new methods of protein chemistry to analyze the structural differences between HbS and HbA.

Normal hemoglobin is made of four polypeptides folded together—two of one kind, α-globin, and two of another, β-globin. Ingram showed that the two α-globins in sickle cell hemoglobin were normal, but the β-globins contained a tiny "point" mutation. In the β chain's entire sequence of 146 amino acids, a single amino acid change produced an altered protein that could cause a deadly disease.

In β-globin, amino acid number 6, which is normally glutamic acid, has been changed to valine. This single amino acid substitution alters the way hemoglobin molecules pack into the crowded red cell. When oxygen concentration drops—due to heavy exercise, for example—the HbS forms clumps that distort the cell.

Ingram found that the sickle cell gene produced abnormalities only in the β-globin polypeptide, never in α-globin. Biologists concluded that each gene is responsible for the production not of an entire protein, but of a single polypeptide chain. "One gene—one protein" became **"one gene—one polypeptide."** But even this rule has important exceptions. As we'll see later on, some genes specify no polypeptide at all and some genes specify several different polypeptides. And in the immune system's protein antibodies, several genes determine the sequence of a single polypeptide.

Generally, however, we can say that genes influence phenotype by specifying polypeptides. Exactly how does a gene code for a polypeptide? To answer this question, we must look more closely at what a gene is.

Genes carry information necessary for the synthesis of polypeptides.

10.3 How Did Biologists Learn What Genes Are Made of?

Research on an amazing variety of organisms—including humans, peas, flies, and bread mold—has contributed to our understanding of what genes do. To find out what genes are made of, biologists turned to less complex, smaller organisms: bacteria and the viruses that infect them.

How Could Dead Bacteria Change the Genetic Properties of Other Bacteria?

Before the development of antibiotics during World War II, many people died of bacterial pneumonia. Efforts to develop a vaccine against pneumonia led, by chance, to the discovery of the nature of genes. In 1928, Frederick Griffith, an English bacteriologist, was trying to stimulate the immune defenses of mice against the pneumonia bacterium *Streptococcus pneumoniae,* also called pneumococcus. Griffith worked with two strains of pneumococcus. One caused pneumonia when injected into the mice, and one was harmless. When Griffith grew the two strains in dishes of gelatin, the harmless pneumococci formed rough (R) colonies, while the disease-causing strain formed smooth (S), glistening colonies. Today, we know that the virulent S bacteria form smooth colonies because they produce a polysaccharide capsule that protects them from attack by the body's immune cells. The R bacteria are mutants—they lack an enzyme needed to produce the polysaccharide capsule, which is why they cannot cause disease.

Griffith tried to produce immunity to pneumonia in mice by injecting them with different combinations of live and heat-killed pneumococci. As he expected, the mice injected with live S strain bacteria got pneumonia and died, while the mice injected with either live R strain bacteria or with heat-killed S bacteria survived. Griffith was amazed, however, by the results of injecting a mouse with a mixture of heat-killed S strain bacteria and live R strain bacteria (Figure 10-15). The mice died! Equally astonishing, the blood of the sick mice contained live S bacteria. Something in the dead S bacteria had *transformed* the harmless R bacteria into virulent S bacteria.

Frederick Griffith discovered that dead virulent bacteria could transform harmless bacteria into virulent ones.

What Was the Transforming Factor?

Other scientists soon found that they could convert R strains into S strains in a test tube as well as in a mouse. Extracts of S strains, dead or alive, could give R strains the ability to make the protective polysaccharide coat and become virulent. Miraculously, once the R strains were transformed into S strains, their descendants inherited the properties of S strains. How could the R bacteria "inherit" traits from individuals they were not descended from? What was transforming them into S bacteria?

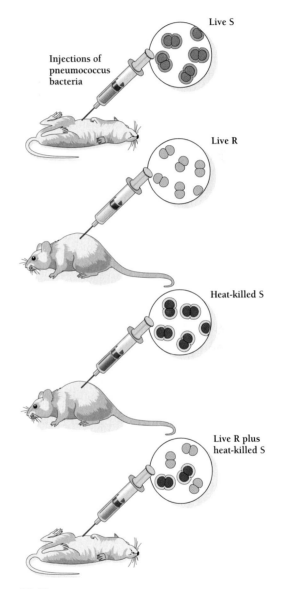

Figure 10-15
Transforming bacteria. In 1928, Frederick Griffith showed that harmless bacteria could be "transformed" into deadly ones. Virulent S (smooth) *Streptococcus pneumoniae* bacteria kill mice, while harmless R (rough) bacteria and heat-killed S bacteria do not. Yet, amazingly, a mixture of harmless R bacteria and dead S bacteria results in a deadly infection. Oswald Avery showed that DNA from the virulent S bacterium can transform harmless bacteria into deadly ones.

It looked as if whatever was transforming the bacteria was some sort of genetic material. In the early 1930s, Oswald Avery at the Rockefeller Institute in New York set out to determine what that material was. With his coworkers Colin MacLeod and Maclyn McCarty, Avery carefully isolated different kinds of molecules from heat-killed S bacteria and tested to see which ones transformed the R bacteria.

At the time, almost everyone expected that the transforming factor would turn out to be a protein, since proteins were already known to perform so many other cellular functions. (Chromosomes were known to be made of both DNA and protein, so it was reasonable to hypothesize that protein carried the genetic information.)

Avery and his colleagues worked for almost 10 years, isolating substance after substance from heat-killed S bacteria and then testing each substance to see if it could transform R bacteria. They treated each isolated substance with enzymes that would break down proteins into amino acids and nucleic acids into nucleotides. Finally, in 1944, they were forced to conclude that the transforming factor was DNA. First, only DNA from the S bacteria transformed the R bacteria. Second, the extract's ability to transform was destroyed only by an enzyme (DNAase) known to break down DNA. Enzymes that broke down proteins did not affect the transforming factor, which suggested that proteins were not important to transformation.

Avery's work strongly suggested that genes are made of DNA. Yet despite the care with which he and his colleagues had carried out their work, most biologists rejected their conclusion. Biologists were skeptical for two reasons. First, if DNA were the genetic material, the DNA of bacteria should differ from the DNA of butterflies and bananas. Yet, all DNA seemed to be chemically identical. Second, most biochemists believed that DNA was a "tetranucleotide," the same sequence of four nucleotides repeated over and over. If that were true, DNA could not carry any information. So, as late as 1944, most biologists continued to believe that DNA was a simple repeating molecule, no more capable of carrying information than a string of identical beads. Biologists thought DNA was a structural component of the chromosomes, no more.

By 1944, Oswald Avery and his colleagues had shown that bacteria are transformed by DNA, not proteins. Yet most biologists rejected the idea that DNA could be the genetic material.

What Is a Bacteriophage?

In 1915, long before Griffith's work on pneumococcus, a young French microbiologist named Felix d'Herelle discovered that bacteria themselves could be killed by a tiny infectious agent called a bacteriophage [Greek, *bakterion* = little rod + *phagein* + to eat], or simply **phage.**

D'Herelle first became aware of phages while trying to stop a plague of locusts that was then sweeping the Yucatan,

in Mexico. The young biologist noticed that some of the locusts were dying from some disease that gave them diarrhea. He collected bacteria from the diarrhea of dying locusts and cultured it. Then, as hordes of the locusts advanced across the Mexican countryside, d'Herelle dusted the vegetation before them with the bacterium. As the locusts ate, they sickened and died. D'Herelle's plan worked magnificently.

D'Herelle was puzzled, however, by the appearance of clear spots in his cultures, places where the bacteria themselves seemed to have died. What was killing his bacteria? At first, he had no idea. The answer did not come until a moment of inspiration some five years later. Back in France, d'Herelle was studying an epidemic of dysentery in a squadron of French cavalry stationed near Paris. The soldiers had Shiga dysentery, and d'Herelle had been culturing the Shiga bacteria when, once more, the phages made their presence known.

> The next morning . . . I experienced one of those rare moments of intense emotion which reward the research worker for all his pains: at the first glance I saw that the broth culture, which the night before had been very turbid [cloudy], was perfectly clear: all the bacteria had vanished, they had dissolved away like sugar in water . . . in a flash I . . . understood: what caused my clear spots was in fact an invisible microbe, . . . a virus parasitic on bacteria. Another thought came to me also: "If this is true, the same thing has probably occurred during the night in the sick man, who yesterday was in a serious condition. In his intestine, as in my test tube, the dysentery bacilli will have dissolved away under the action of [the virus]. He should now be cured."

And, indeed, d'Herelle dashed to the hospital to find the sick man greatly recovered.

D'Herelle found that phages were small enough to pass through the filters he used to remove bacteria. Also, these "filterable agents" could reproduce themselves within bacteria and, therefore, seemed to be alive.

Biologists do not consider phages alive, however, because they are not made of cells. Phages are **viruses,** assemblies of protein and either DNA or RNA molecules. Viruses are not cells, and they cannot reproduce outside of a host cell. A single virus can, however, infect a cell and take over the cell's molecular machinery to make more viruses. Besides infecting bacteria, viruses infect plant cells, animal cells, and other kinds of eukaryotic cells.

Phages are a kind of virus that infect bacteria. Viruses are assemblies of protein and DNA (or RNA) capable of forcing host cells to make more viruses.

What Convinced Biologists That Genes Are Made of DNA?

In the 1940s, Max Delbruck and Salvador Luria began to study the genetics of the phages that infect the common intestinal bacterium *E. coli* (Figure 10-16A). A bacterium such as *E. coli* infected with a single phage makes about 200 new

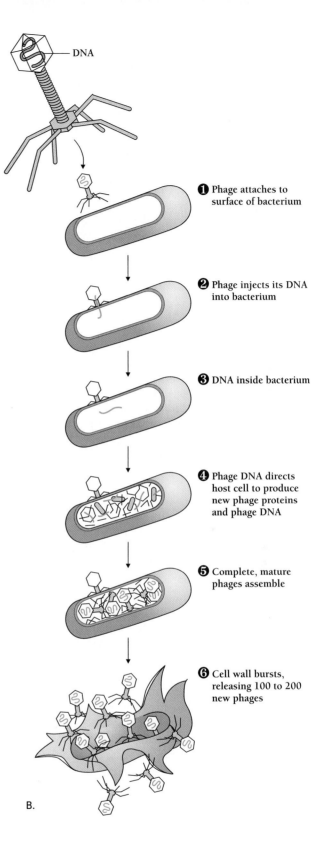

❶ Phage attaches to surface of bacterium

❷ Phage injects its DNA into bacterium

❸ DNA inside bacterium

❹ Phage DNA directs host cell to produce new phage proteins and phage DNA

❺ Complete, mature phages assemble

❻ Cell wall bursts, releasing 100 to 200 new phages

B.

A.

Figure 10-16
The life of a phage. A. Max Delbruck and Salvador Luria examining a petri dish containing bacteria and phages. B. Life cycle of a bacteriophage.

phages, then breaks open after 20 minutes and releases them (Figure 10-16B). Each new phage can infect another bacterium and kill it. If the bacteria are growing on a solid culture medium, the phages create a clear spot, or "plaque," in the "lawn" of bacteria.

Delbruck, Luria, and others catalogued many inherited variants of the phage. Some produced clear plaques, and others produced cloudy plaques. Some infected one strain of *E. coli* but not another. The different kinds of phages passed their traits to their descendants by means of specific alleles, just as organisms do. Yet, the researchers knew phages consisted only of complexes of protein surrounding a molecule of nucleic acid, usually DNA. The genes of phages had to be in either the protein or the DNA.

Avery had already shown that DNA is the hereditary material in pneumococcus bacteria. In 1952, Alfred Hershey and Martha Chase performed an experiment specifically designed to discover if DNA was also the hereditary material in phages (Figure 10-17A). Hershey and Chase knew from microscope studies that during infection only part of a phage enters the host cell. They reasoned that the material that actually enters the cell must carry the phage's genetic information.

To find out whether the genes were DNA or protein, Hershey and Chase used radioisotopes to label the DNA and

protein. The work on the atomic bomb during World War II had greatly increased the availability of radioisotopes, so Hershey and Chase grew bacteria on food containing ^{32}P-("phosphorus 32") labeled phosphate and ^{35}S-("sulfur 35") labeled sulfate. Because amino acids (which make up pro-

A.

Figure 10-17

What convinced biologists that the genetic material was DNA?
A. Martha Chase and Albert Hershey showed that when a phage infects a bacterium, it is the viral DNA that enters and infects the bacterium, not protein. B. In the Hershey-Chase experiment, radioactive DNA from a phage shows up inside the infected bacterium. Radioactive protein stays outside. This 1952 experiment finally convinced biologists that DNA, not protein, was the genetic material. (Later work showed that a few viruses use RNA as their genetic material.)

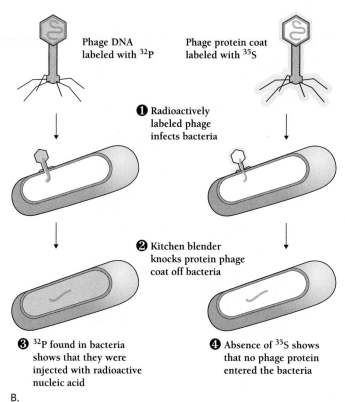

Phage DNA labeled with ^{32}P

Phage protein coat labeled with ^{35}S

❶ Radioactively labeled phage infects bacteria

❷ Kitchen blender knocks protein phage coat off bacteria

❸ ^{32}P found in bacteria shows that they were injected with radioactive nucleic acid

❹ Absence of ^{35}S shows that no phage protein entered the bacteria

B.

teins) contain sulfur but no phosphorus and because nucleic acids have phosphorus but no sulfur, the bacteria had ^{35}S incorporated only in their proteins and ^{32}P only in their DNA. When phages infected the radioactive bacteria, newly made phages picked up the radioactive labels, so their DNA contained ^{32}P and their protein contained ^{35}S.

Hershey and Chase infected clean, nonradioactive bacteria with the radioactive phages, waited just long enough for the infection to begin and then interrupted the process by dumping the bacteria into a kitchen blender. The agitation in the blender knocked the empty phage particles off the outside of the bacteria and separated these particles from the bacteria. Hershey and Chase then examined the infected bacteria to see what part of the phage had entered the bacteria (Figure 10-17B). It was the DNA. The ^{32}P-labeled phage DNA was inside of the bacteria, and the ^{35}S-labeled phage protein was outside. Hershey and Chase also showed that phage DNA alone could force cells to make new phages. Hershey and Chase therefore concluded that the genes of the phage were made of DNA, not protein.

Hershey and Chase's 1952 experiment made a huge splash, immediately drawing the attention of a large group of influential biologists in a way that Avery's paper of eight years earlier had not. Hershey and Chase's experiment was more persuasive than Avery's for a couple of reasons. When Avery's paper came out in 1944, most molecular biologists still believed that DNA was a simple repeating structural

molecule. By 1950, however, Erwin Chargaff had shown that all DNA was not the same—that different species had different proportions of bases. This implied that DNA *could* carry information. And just months later, Watson and Crick's 1953 paper showed *how* DNA could carry information (and also replicate). Chargaff's work, Hershey and Chase's experiment, and Watson and Crick's model together removed all doubt. After 1953, virtually all biologists accepted that DNA was the genetic material.

Hershey and Chase's experiments with radioactively labeled phages showed that when a phage infects a bacterium, it is the DNA that enters and infects the bacterium, not the protein. This experiment, together with Chargaff's data on nucleotides and Watson and Crick's model of DNA, convinced biologists that DNA was the genetic material.

10.4 How Does One DNA Strand Direct the Synthesis of Another?

Recall that DNA consists of strings of nucleotides held together by phosphate groups linking the third carbon of the sugar of one nucleotide and the fifth carbon of the sugar of the next nucleotide. The third carbon is called the **3′** ("three prime") **carbon,** and the fifth carbon is called the **5′** ("five prime") **carbon.** Each polynucleotide chain has a direction with a 5′ end and a 3′ end. Hanging off the sugar-phosphate backbone of DNA are four kinds of nitrogen-containing bases. The bases are two **purines** (**adenine** and **guanine,** abbreviated A and G) and two **pyrimidines** (**cytosine** and **thymine,** abbreviated C and T). C always pairs with G, and A always pairs with T. Inside a cell, DNA consists of two polynucleotide strands twisted around each other in a double helix. These two strands of DNA run in opposite directions. Because each base pairs with only one other base, each strand of DNA can act as a template for the manufacture of another, "complementary" strand.

Watson and Crick's complementary base-pairing rule established the basis of DNA's ability to replicate itself. In the same way that a photographer can make a positive print from a negative, or a negative from a positive print, each strand of DNA has the capacity to direct the synthesis of the other. This complementarity is the basis for understanding both how a chromosome replicates itself during cell division and how genes actually work. The structure of DNA explains both how cells copy their DNA and how genes specify proteins (Chapter 11).

How Do Dividing Cells Copy the DNA?

Each of the four bases in DNA can pair with only one other base—A with T and G with C. The rules of base pairing in DNA mean that the sequence of one polynucleotide strand specifies the sequence of the other strand. The two strands of DNA are complementary, meaning that they fit together to make a unified whole—the double helix.

The basis for DNA replication is that the sequence of bases in one chain specifies the sequence of bases in the other. That is, the two strands of the DNA double helix contain the same information in two different forms. Picture a molecule of DNA unzipping to form two complementary single strands. If the bases and the proper cellular machinery are available, each strand can act as a template for the manufacture of a new complementary strand. When each strand has created its one complement, the result is two new double-stranded DNA molecules, each a replica of the parent DNA.

According to this model, DNA replication is **semiconservative.** That is, half of each parent molecule is present in each daughter molecule (Figure 10-18). Each new DNA molecule, then, is really half old and half new.

DNA replication is semiconservative: when DNA replicates, each half of the double helix acquires a new mate.

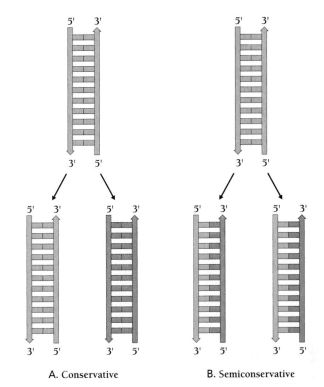

Figure 10-18
Semiconservative replication. A. At one time biologists hypothesized that DNA replication was conservative. Both halves of the old molecule stayed together, they thought incorrectly, while a whole new molecule was created. B. In fact, when DNA replicates, each half of the old double helix acquires a new mate.

Our discussion has so far focused on how DNA passes information to daughter DNA molecules. But what is the mechanism? How do cells actually produce new DNA? As in the case of other biochemical reactions, DNA replication depends on enzymes. The full process of DNA replication requires at least a dozen enzymes. Even now, more than 45 years after molecular biologists first synthesized DNA in a test tube, they still do not know all the details of DNA replication. The most basic steps of the process, however, are well understood.

How Does a Cell Begin Assembling a New Strand of DNA?

The assembly of DNA involves linking many small molecules (nucleotides) into a long chain, a process called **polymerization** [Greek, *polys* = many + *meros* = part]. The enzyme that strings together the nucleotides is called **DNA polymerase.** In each strand, the nucleotides must polymerize in a fixed order, guided by the order of bases in the complementary strand (Figure 10-19).

The Watson-Crick model for DNA replication suggests how each strand of DNA serves as a **template,** or guide, for

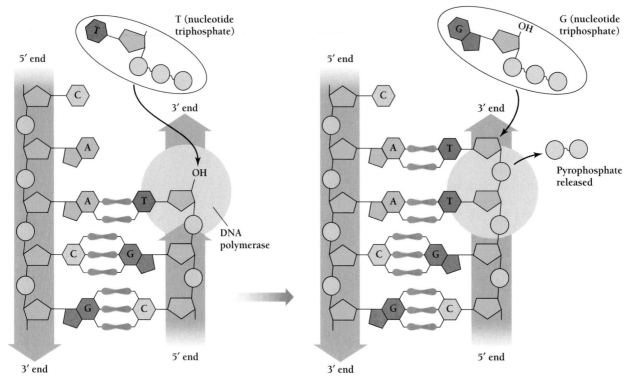

Figure 10-19
DNA polymerization. The enzyme DNA polymerase makes a copy of a strand of DNA by adding nucleotides to the 3′ end (*arrow*) of an adjacent but incomplete polynucleotide chain. A always pairs with T, and G always pairs with C.

the assembly of a complementary sequence. DNA polymerase adds nucleotides to the 3′ end, so that the polynucleotide grows in a single direction—from its 5′ end toward its 3′ end (Figure 10-20). The particular nucleotide added at each step depends upon the base sequence of the template strand. Each added nucleotide forms hydrogen bonds with the next base on the template strand. So each growing strand extends in a single direction, starting with the sequence closest to the 5′ end and extending from its 3′ end.

Once a free 3′ end exists, it is easy to see how the process continues. But DNA polymerase can only add nucleotides to a free 3′ end, so how does the replication process begin? The answer is that DNA replication begins from an RNA **primer.** Using DNA as a template, an RNA polymerase makes short stretches (fewer than 10 nucleotides) of RNA. Once the RNA primer is in place, DNA polymerase begins adds a nucleotide its 3′ end and begins making a new strand of DNA.

An RNA polymerase makes an RNA primer provides a 3′ end for DNA polymerase. DNA polymerase then links together the nucleotides that assemble opposite the template into a new strand of DNA.

How Does a Cell Copy Both Strands of DNA Simultaneously?

Researchers can actually see cellular DNA in the process of replication with the electron microscope. The DNA in a dividing tissue culture cell looks like a long, thin molecule punctuated by bubbles. As DNA replicates, these bubbles expand. Beyond each bubble are the two strands of parental DNA. Within each bubble are the four strands of replicated DNA—two parental strands and two newly made strands.

Each bubble consists of two **replication forks,** Y-shaped regions of DNA where the two strands of the helix have come unzipped (Figure 10-21A). At the replication fork, each strand begins to direct the assembly of a new complementary strand. As replication proceeds, the forks move away from each other. In bacteria, the replication forks move away from each other at about 500 nucleotides per second, and, in mammals, at about 50 nucleotides per second.

DNA replication occurs in replication forks, Y-shaped regions of DNA where the two strands of the helix have come apart.

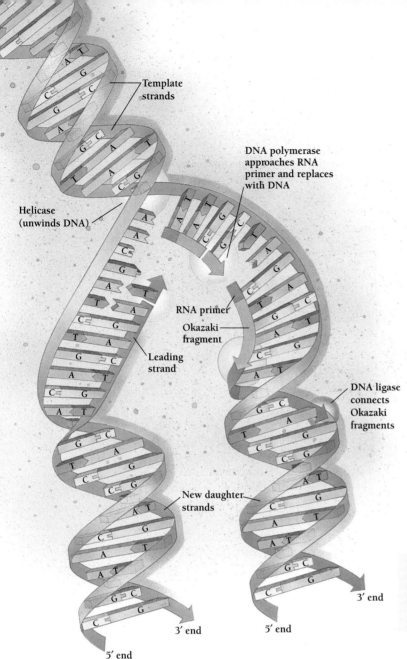

Template
strands

DNA polymerase
approaches RNA
primer and replaces
with DNA

Helicase
(unwinds DNA)

RNA primer

Okazaki
fragment

Leading
strand

DNA ligase
connects
Okazaki
fragments

New daughter
strands

3′ end

3′ end   5′ end

5′ end

Figure 10-20
How do enzymes replicate DNA? The blue strand of DNA acts as a template for the creation of a new complementary strand of DNA (rust). The enzyme *helicase* unwinds the strands. On the left, DNA polymerase adds nucleotides in a continuous string, working from the 3′ end of the template to the 5′ end. On the right (*center*), DNA polymerase works in the opposite direction, starting with RNA *primers* (also rust) to create Okazaki fragments, which are later linked together by *DNA ligase*.

Biology ⊜ Now™   Learn more about replication by clicking on this figure on your BiologyNow CD-ROM.

The answer is that the second strand is produced in bits and pieces. After the moving fork has exposed 100 to 1,000 nucleotides of template for the second, or "lagging," strand, a short piece of complementary RNA serves as a primer for DNA polymerase, which attaches nucleotides to its 3′ end. This whole process makes DNA fragments, called **Okazaki fragments**, after their discoverer Reiji Okazaki. Each Okazaki fragment contains an RNA primer connected to a few hundred nucleotides (Figure 10-21B). DNA polymerase then removes the RNA primer, and an enzyme, called **DNA ligase**, stitches the fragments together.

In one strand of a replication fork, nucleotides add directly to the 3′ end of an RNA primer. The other strand is produced in short fragments that are joined together by DNA ligase.

How Does DNA Synthesis Begin?

DNA always begins replication at a special site called an **origin of replication**. *E. coli*, the organism whose DNA synthesis has been most studied, has a single piece of circular DNA with a single origin of replication. Starting from this single site, two replication forks move in opposite directions away from the origin of replication, and DNA synthesis stops when the fork that is moving clockwise meets the counterclockwise fork (Figure 10-22).

Several proteins participate in DNA replication. An enzyme called *helicase* uses energy from ATP to untwist the helix, while another enzyme called DNA gyrase prevents the accumulation of kinks in the stems of the replication forks. Still another protein binds to the unwound DNA and keeps the jaws of the replication fork open.

How Does DNA Replicate in the 3′ to 5′ Direction?

As a replication fork moves along both strands of DNA, it moves from the 5′ end to the 3′ end along one strand, but from the 3′ end to the 5′ end on the other strand. We have seen how simply and continuously the first strand is replicated. But the second strand cannot be replicated in this way because nucleotides must be added to its 5′ end, a task that DNA polymerase cannot perform. How is this strand made?

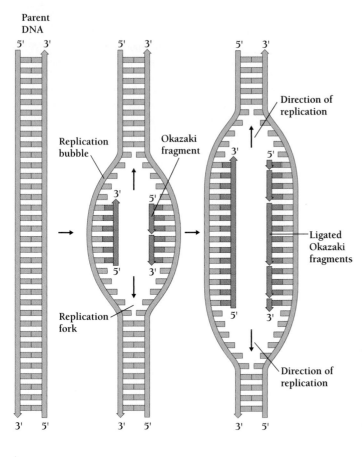

Parent
DNA

5'   3'

Replication
bubble

Okazaki
fragment

3'

5'

3'

5'

3'

5'

3'

Replication
fork

3'   5'

3'   5'

Direction of
replication

5'   3'

5'

3'

Ligated
Okazaki
fragments

5'

3'

Direction of
replication

3'   5'

A.

David Hogness and Henry Kriegstein, Stanford University School of Medicine, *Proceedings of the National Academy of Science*, January 1974, vol. 71, no. 1, pp. 135–139

Figure 10-21
Replicating DNA. A. A simplified view of a DNA replication bubble. Replication begins in the middle of the bubble and moves away from the middle in two directions at once. B. TEM of a replication bubble, with a replication fork at each end.

Biology ⬧ Now™ Learn more about replication forks by clicking on this figure on your BiologyNow CD-ROM.

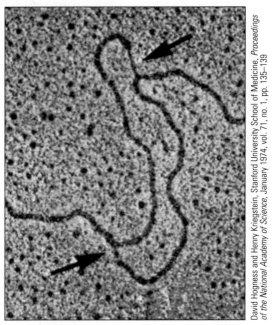

B.

In eukaryotic cells, DNA replication is more complex. Each cell contains thousands of origins of replication. Groups of 20 to 50 origins, called **replication units,** form replication forks at the same time. Within a replication unit, the forks move in opposite directions from each origin until they encounter a fork from an adjacent origin.

DNA synthesis always begins at special sequences in the DNA called origins of replication. Proteins bind to these origins of replication and open and untwist the double helix.

Figure 10-22
Replication of bacterial DNA. Replication begins at the origin of replication (*arrows*), then works in both directions around the circle.

10.5 What Is the Ultimate Source of Genetic Diversity?

Every living organism shares the same chemistry of life. Your DNA is made of the same stuff and follows the same rules as the DNA of the roses nodding in your neighbor's yard. Yet you are different from a rose. If, for example, you want lunch, you must get up and make yourself a sandwich, while the rose simply waits passively for the sun to emerge from behind a cloud. What makes us so different from a rose?

Extreme Biology How to Sequence DNA

One of the most important questions a biologist can ask about a gene or any stretch of DNA is, What is its sequence of nucleotides? With that information, medical researchers can begin to understand genetic diseases such as sickle cell anemia or alkaptonuria. The most widely used method for DNA sequencing was developed in England in the 1960s by Fred Sanger. Sanger's method depends on DNA polymerase and five "rules" of DNA synthesis:

1. DNA synthesis is semiconservative.
2. A new DNA strand grows in a single direction, adding one nucleotide at a time to the 3′ end.
3. The exact sequence of nucleotides in the new strand depends on the sequence in the original "template" strand.
4. One enzyme, DNA polymerase, is responsible for copying the new complementary strand from the old template strand.
5. DNA polymerase requires a primer polynucleotide.

Finally, sequencing DNA also requires that a researcher be able to separate DNA fragments of different lengths.

Sanger began by heating the DNA until it separated into two single strands. He then added DNA polymerase and plenty of A, C, G, and T. The DNA polymerase automatically set to work extending the primer until it produced a complete copy of the template strand.

Sanger then devised a way to stop the polymerase reaction at different places. He could stop the reaction at A, C, G, or T by replacing some of each base with a modified base (lacking a 3′-hydroxyl group) that stopped further polymerization. For example, a growing strand with a modified A stopped DNA polymerase at A. The result was a set of fragments of different lengths, each ending in A. Sanger did the same for C, G, and T.

The next problem was to put all these fragments in order, shortest to longest. Fortunately, electrophoresis does just that. If we run a small electric current through a flat sheet of gel containing the fragments, the shortest fragments move quickly across the gel, while the longest ones move slowly, threading their way through the complex structure of the gel. Using electrophoresis, Sanger could put the fragments in order, shortest to longest—even ones that differ in length by no more than one nucleotide (Figure A). Each fragment showed up on the gel as a short, dark band. Each column of bands represents the different fragments ending in A, G, C, or T. From these four columns of bands, researchers can work out the complete sequence of any newly made DNA.

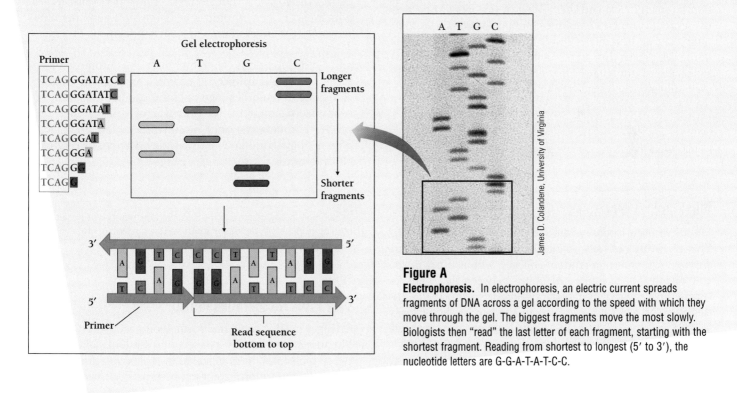

Figure A

Electrophoresis. In electrophoresis, an electric current spreads fragments of DNA across a gel according to the speed with which they move through the gel. The biggest fragments move the most slowly. Biologists then "read" the last letter of each fragment, starting with the shortest fragment. Reading from shortest to longest (5′ to 3′), the nucleotide letters are G-G-A-T-A-T-C-C.

Each species has its own characteristic DNA. But life most likely began on Earth just once. If that is the case, then at one time all organisms were of one kind and all of their genetic material was the same. How, over billions of years of evolution, did we organisms become so different from one another? The short answer is **mutations,** permanent changes in the sequence of DNA. The different alleles of a gene are created by changes in the sequence of the original gene. The original sequence itself is a mutation of some other gene. All genetic differences ultimately depend on the accumulation of mutations in the genome.

Although the evolution of organisms depends on the accumulation of changes in the DNA, the vast majority of mutations are either harmful or meaningless. If we opened up the back of a television set and switched two wires at random, the chances that this would improve the picture or the general performance of the television are almost zero. The same is true of mutations in large, complex genomes, such as those of butterflies or mammals. Most gene mutations are a disaster for the phenotype. Still, every now and then, a new mutation turns out to be useful. Under the right circumstances, natural selection increases the mutation's representation in a population of organisms.

We have already seen in Chapter 9 how the recombination (crossing over) and the mixing of gametes during sexual reproduction create unique combinations of alleles. But, ultimately, all new alleles result from mutations. Our focus in the last part of this chapter will be on the origin and consequences of mutations: How do mutations arise? How do they influence the phenotype of the organism?

Mutations are the ultimate source of all genetic diversity.

What Causes Mutations?

Mutations occur spontaneously as random events in the cell. The structure of any molecule—including DNA—can change as a result of a random collision with another molecule or ion. For example, the bonds between the A and G (purine) bases in DNA and the backbone sugar are slightly unstable. As a result, each cell in the human body loses some 5,000 to 10,000 A and G nucleotides every day. (Of course, most of this damage is repaired, as we'll see later.)

Occasionally, mutations occur because DNA polymerase places the wrong nucleotide in a sequence during the synthesis (S) phase of the cell cycle (Chapter 8). In addition, certain agents, called **mutagens,** increase the rate of mutation. Ultraviolet light, high-energy radiation such as x rays, and many chemicals can all act as mutagens. Each mutagen has a characteristic spectrum of effects. Ultraviolet light, for example, may link together adjacent Ts in the same DNA strand. X rays are particularly effective producers of chromosomal mutations, causing DNA to break and rejoin in strange new arrangements. Chemical mutagens, such as alcohol and dioxin, for example, also increase mutation rates.

Mutations occur spontaneously. But they occur more frequently when DNA is exposed to ultraviolet light, x rays, chemicals, or other mutagens.

What Kinds of Mutations Are There?

We can categorize mutations on the basis of the sizes of the changes in DNA. Geneticists often distinguish between **point mutations,** those which change one or several nucleotide pairs, and **chromosomal mutations,** those which change relatively large regions of chromosomes. A point mutation may be (1) a **base substitution**—the replacement of one base (nucleotide) by another, (2) an **insertion**—the addition of one or more nucleotides, or (3) a **deletion**—the removal of one or more nucleotides.

Chromosomal mutations may affect large regions of chromosomes or even whole chromosomes. Chromosomal mutations include (1) **deficiencies**—deletions that are larger than a few nucleotides, (2) **translocations**—in which part of one chromosome is moved to another chromosome, (3) **inversions**—in which a segment of a chromosome is flipped over 180° from its normal orientation, and (4) **duplications**—in which part of a chromosome appears twice. **Aneuploidy,** abnormal chromosome number, is regarded as another type of chromosomal mutation. For example, **trisomy** is the presence of three copies of a certain chromosome, and **monosomy** is the presence of only a single copy in a cell that normally has two of each chromosomes.

A mutation is any change in the sequence or number of nucleotides, ranging from small changes of one or a few nucleotides to chromosomal inversions or aneuploidy.

How Often Do Mutations Occur?

Changes in DNA sequence occur spontaneously all the time. Mutations can initially result from random collisions between molecules, from mistakes in replication (for example, from a failure to form a correct base pair), or from changes in the structure of a nucleotide.

The **mutation rate** is a measure of how often mutations occur. The mutation rate depends both on how often a sequence mutates and on how efficiently cells repair these mutations. Some sequences of DNA, called **mutation hot spots,** are more likely to mutate than others. In bacteria, for example, some sequences mutate 25 times more rapidly than others. Even the highest rates of mutation, however, are

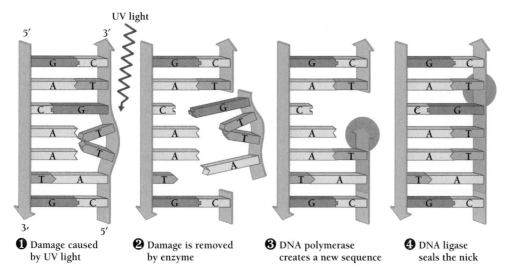

Figure 10-23
Healthy cells are quick to repair damaged DNA. Ultraviolet light can damage DNA by linking together adjacent Ts in the same strand. Repair enzymes excise the abnormal nucleotides and restore the original sequence, then DNA polymerase and DNA ligase repair the gap as in normal replication.

UV light

5′ 3′

3′ 5′

❶ Damage caused by UV light

❷ Damage is removed by enzyme

❸ DNA polymerase creates a new sequence

❹ DNA ligase seals the nick

far lower than the error levels to which we are accustomed in everyday life. In the days before computers and word processors, for example, an outstanding typist could type this whole chapter without going back to correct anything and with only one mistake. Yet, the average rate of mutation in bacteria during ordinary DNA replication is 1,000 times more accurate than the best human typist.

Mutation rates are usually low, and they vary from gene to gene.

10.6 How Do Cells Handle Mistakes in the Nucleotide Sequence of DNA?

Because DNA mutates, cells must be able to repair damage to DNA if they are to maintain the continuity of life. This is especially important for the germ line cells that pass information to the next generation by way of eggs and sperm. But mutations in somatic (body) cells are dangerous as well. In skin cancer, for example, a cell with a defective gene for growth control multiplies so fast that immature daughter cells take over the area. More than half of all forms of cancer arise from cells with one or more mistakes in just a few genes. The normal forms of these genes specify proteins called "tumor suppressors."

All cells have sophisticated mechanisms for repairing damaged DNA and correcting mistakes in replication. These mechanisms depend on the redundancy of information in DNA's structure and on special enzymes that recognize common mistakes.

To correct a mistake, an enzyme detects something wrong in one strand of the DNA and removes the error. For example, cytosine may be accidentally converted to uracil, a base that is ordinarily not present in DNA. A special enzyme

can recognize the unusual base and remove it. In another example, two T nucleotides may bond to one another instead of to their respective A nucleotides (Figure 10-23).

Once correction enzymes remove the offending nucleotides, DNA polymerase goes to work. It copies the information in the intact second strand and creates a new stretch of DNA that is properly matched. DNA ligase then seals the gap, and the original sequence is restored. Over 50 enzymes "proofread" DNA for errors such as unusual bases, gaps, and bulges.

Errors still can slip through this elaborate system. For example, the genes for error-correction enzymes may themselves be damaged. In that case, large numbers of mutations can accumulate (Figure 10-24). If these uncorrected muta-

Figure 10-24
Sunburn damages DNA and leads to skin cancer. The ultraviolet light in sunlight can damage DNA repair mechanisms. In the absence of DNA repair, many genes become damaged, including those that control the cell cycle. In early rounds of sunburn, most cells with damaged DNA self-destruct, and the skin peels. One theory of skin cancer holds that a few cells with damaged DNA remain but do not divide and multiply because they are surrounded by healthy cells. During subsequent sunburns, however, more cells with damaged DNA self-destruct, clearing the field for cancerous cells to begin growing.

tions occur in genes that help control cell division, the cells become cancerous.

Cells have sophisticated mechanisms for repairing damage and for correcting mistakes made during DNA replication.

In this chapter, we have seen that DNA has a special structure that enables it to replicate. We have also seen how we know that genes are made of DNA and that genes are physical regions in the DNA. In the next chapter, we will see how DNA's structure allows the "expression" of genes as polypeptides and how cells control the expression of genes—so that a given polypeptide is synthesized at the right time and in the right amount. We will also further explore how mutations in the structure of DNA change the information in genes.

Key Concepts

- DNA consists of two polynucleotide strands, which wind together to form a double helix.
- The two strands of DNA contain complementary versions of the same information. During DNA replication, each strand directs the synthesis of a complementary strand.
- In all organisms, genes are made of DNA, and each allele of a gene usually specifies the structure of a single polypeptide.

Summary with Key Terms

How did biologists discover what genes were made of and what they did?

Biologists first hypothesized that genes specified the structures of enzymes. When an organism cannot make an enzyme, it cannot carry out the biochemical conversions catalyzed by the enzyme. In alkaptonuria, for example, the absence of a specific enzyme causes the accumulation of a compound that turns the urine black. Studies of similar interruptions in biochemical pathways in the bread mold *Neurospora* in the 1940s suggested that every gene specifies the structure of a single enzyme. Researchers Beadle and Tatum coined the phrase **One gene—one enzyme.** But within just a few years, other research showed that genes could specify the structures of proteins other than enzymes. Genes could specify single polypeptides only, however, not necessarily whole proteins. A more accurate statement of gene function is, **One gene—one polypeptide.** Different alleles of a gene produce different versions of the same polypeptide.

The idea that genes are made of DNA came from work on bacteria and bacterial **viruses,** called bacteriophages, or simply **phages.** A virulent strain of bacteria can transfer the genetic instructions for virulence to a nonvirulent strain by means of DNA.

What is the structure of DNA?

DNA consists of strings of nucleotides held together by bonds between the third (3′) carbon of the sugar of one nucleotide and the fifth (5′) carbon of the sugar of the next nucleotide. Each polynucleotide chain has a direction with a 5′ end and a 3′ end. Hanging off the sugar-phosphate backbone of DNA are four kinds of nitrogen-containing bases. The bases are two **purines** (**adenine** and **guanine,** abbreviated A and G) and two **pyrimidines** (**cytosine** and **thymine,** abbreviated C and T). C always pairs with G, and A always pairs with T. According to **Chargaff's rules,** the amount of C always equals the amount of G, and the amount of A equals the amount of T in the DNA of a given species. Inside a cell, DNA consists of two polynucleotide strands twisted around each other into a 2 nm-wide double helix.

How does DNA's structure allow it to act as a template for its own replications?

These two strands of DNA run in opposite directions. Because each base pairs with only one other base, each strand of DNA can act as a template for the manufacture of another, "complementary" strand. Each strand is a template for the other, and each strand can direct the synthesis of a complementary strand during DNA replication.

DNA replication is **semiconservative:** When DNA replicates, each strand of the double helix acquires a new mate. The enzyme **DNA polymerase** assembles nucleotides into a sequence specified by one strand of DNA, the **template** strand, a process called **polymerization.** DNA polymerase can add nucleotides only to a preexisting RNA **primer** made by an RNA polymerase. DNA polymerase can add nucleotides only to the 3′ end of the primer and continues to add nucleotides to the 3′ end of the growing polynucleotide.

DNA replicates in Y-shaped regions of the DNA called **replication forks.** Nucleotides add directly to the 3′ end of one strand in the fork, called the leading strand. The other strand is made discontinuously in **Okazaki fragments** of 100 to 1,000 nucleotides. Each fragment starts with a short RNA primer. **DNA ligase** links the fragments into a continuous strand. Special sequences in DNA serve as **origins of replication,** where DNA synthesis always begins. The sequences bind to proteins that open and untwist the double helix. Groups of 20 to 50 origins are called **replication units.**

What is a mutation and why are mutations important?

DNA molecules suffer inevitable damage in the form of mutations. A **mutation** is any change in the sequence or number of nucleotides. **Point mutations** include multiple **base substitutions, insertions,** and **deletions. Chromosomal mutations** include **deficiencies, translocations, inversions,** and **duplications,** as well as **aneuploidies** such as **trisomy** and **monosomy.** Mutations are the ultimate source of all genetic diversity, which is the basis for the evolution of all species on Earth. **Mutation rates** are usually low. But **mutation hot spots** mutate more often than other parts of the genome. Mutations can occur spontaneously, but ultraviolet light, x rays, certain chemicals, and other **mutagens** can increase the mutation rate. Cells have sophisticated mechanisms to repair damage to DNA and to correct mistakes made during DNA replication. One result of uncorrected mistakes is cancer.

Review and Thought Questions

Review Questions

1. What facts about the structure of DNA did Franklin contribute? What did Watson and Crick contribute?
2. How does Watson and Crick's theory of base pairing explain how DNA replicates itself?
3. Do genes specify enzymes, proteins, or polypeptides? Why did biologists worry about this distinction?
4. What does it mean for a bacterial cell to be *transformed*? What does the transforming?
5. How is semiconservative replication different from conservative replication?
6. Describe the main steps of DNA replication.
7. What serves as a template for the replication of a single strand of DNA?
8. What enzyme polymerizes nucleotides into a new strand of DNA?
9. Why does the cell need an RNA primer for DNA replication to begin?
10. Where on the DNA does DNA polymerase begin replication?

11. What is the ultimate source of all genetic variation?
12. Why do cells need mechanisms for repairing DNA?

Thought Questions

13. D'Herelle and others thought that phages might provide a way to kill disease-causing bacteria—an idea that was especially exciting in the days before the discovery of penicillin and other antibiotics. Yet, until recently, no one found a way to use phages to destroy bacteria that have already infected the body. What problems might prevent medical researchers from making phages kill bacteria in the body?
14. Why does the dentist place a lead shield over your lap before x-raying your teeth? What might happen if the dentist did not take this precaution?

BiologyNow Resources

Biology⑤Now™

Active Figures
10-20: Replication
10-21: Replication forks

Preparing for an exam? Take a diagnostic test on your BiologyNow CD-ROM.

Online materials relating to this chapter are at:
http://biology.brookscole.com/AAL3

About the Chapter-Opening Image
Like these two parallel pencils, the two strands of a molecule of DNA are parallel but run in opposite directions.

How Are Genes Expressed?

Key Questions

- How did molecular biologists decode the genetic code?

- What is the central dogma?

- How does RNA polymerase transcribe DNA into RNA?

- How do ribosomes and tRNAs translate mRNA into polypeptides?

- How do cells regulate the expression of individual genes?

How Was the Genetic Code Discovered?

As soon as Watson and Crick discovered the structure of DNA, scientists all over the world wanted to become involved in the obvious next phase of the work, solving one of life's oldest secrets—the genetic code. Especially attracted were mathematicians and physicists, who believed they could crack the genetic code through logic alone. Yet, after years of effort, the code was ultimately cracked, not by the theoreticians, but by the rigorous experimental work of two young and unknown molecular biologists.

The first chink in the armor of the coding problem came in 1955, when Marianne Grunberg-Manago, working at New York University, discovered an enzyme that would link nucleotides together into long polymers. Grunberg-Manago took a year to isolate the enzyme. Once she had the enzyme, however, she began making artificial RNA polymers. She could, for example, link together long chains of A (adenine) nucleotides, which she called "poly A," or long chains of C (cytosine) nucleotides, called "poly C." In addition, she could make poly U, poly G, poly AU, and other combinations.

For the moment, however, no one knew what combination of nucleotides could actually be translated by a cell into a polypeptide. It was six more years before the next big break came. Then, in 1960, 31-year-old Johann Heinrich Matthei arrived at the National Institutes of Health (NIH) near Washington, D.C. Matthei had come from Germany looking for an interesting scientific project. He was especially interested in working on protein synthesis. At NIH, he got the names of three researchers who were trying to synthesize proteins artificially outside of a cell. Thirty-three-year-old Marshall Nirenberg stood out from the other two (Figure 11-1). His interests and instincts matched Matthei's own, and in November 1960, Matthei and Nirenberg embarked on a productive, if sometimes strained, collaboration.

Paraskevas Photography

Nirenberg's goal was to synthesize the polypeptides that make up proteins in a test tube. To do this, he planned to add ATP (for energy) and free amino acids to a mixture of ribosomes, nucleic acids, and enzymes extracted from cells. Other researchers had used this technique and had shown that the amino acids were incorporated into polypeptides. No one, however, knew what substances in the cell extracts were making the polypeptides, and no one knew what kinds of polypeptides were being synthesized.

The obvious next step was to try to make a particular polypeptide by putting genetic information into the test tube. Several labs in the United States were hotly pursuing this idea. Before Matthei arrived, Nirenberg had tried to synthesize penicillinase, the enzyme that antibiotic-resistant bacteria use to break down the antibiotic penicillin. Nirenberg added bacterial DNA and RNA to his test-tube system. In experiment after experiment, however, he got no penicillinase. But without knowing how the code worked, no one would ever be able to synthesize any polypeptide, let alone a particular one such as penicillinase. Nirenberg needed to ask a simpler question.

After Matthei arrived, the two researchers asked a basic question: What kinds of RNA stimulate the synthesis of polypeptides? To answer this question, they took an extremely methodical approach. Before trying to synthesize any protein, they would make sure that their test-tube system could respond to RNA, any RNA. The two spent months perfecting their system. In the end, they could detect even minute amounts of protein synthesis. Their test-tube system was now highly sensitive to added RNA.

Nirenberg and Matthei knew that protein synthesis required ribosomes and that ribosomes contained RNA. Like most other scientists of the period, they believed that RNA must carry a sort of message from the DNA in the nucleus to the ribosomes in the cytoplasm, where protein synthesis normally happened. But although Nirenberg and Matthei's test tubes were packed with ribosomal RNA, they were seeing very little protein synthesis. Instinctively, they felt that, besides ribosomal RNA, there must be another kind of RNA—a kind that carried information that could specify a polypeptide and would make protein in their test tubes. But where was it? What was it?

They drew up a list of some 200 kinds of RNA to test. One of the first possibilities was RNA from tobacco mosaic virus. (In some viruses, RNA, rather than DNA, carries the genetic information.) The results were spectacular. Nirenberg

Figure 11-1
Marshall Nirenberg. Nirenberg and Johann Matthei cracked the genetic code in 1960.

and Matthei's test-tube system incorporated huge amounts of amino acids into some mystery protein.

Nirenberg instantly recognized a golden opportunity. Another molecular biologist, Heinz Fraenkel-Conrat, at the University of California, Berkeley, had recently worked out the sequence of the tobacco mosaic virus protein. Nirenberg called Fraenkel-Conrat, forged an instant collaboration, and in May 1961 left to spend a month in California to try to synthesize tobacco mosaic virus proteins. If he succeeded, he would gain scientific celebrity.

Matthei stayed behind at NIH, patiently testing the other possibilities on the list of 200 kinds of RNA. High on the list were the lengths of artificial RNA made using the methods invented by Marianne Grunberg-Manago. A week after Nirenberg left, Matthei set up the test tubes, the different enzymes, the ribosomes, the ATP, and the 16 amino acids he had on hand. Then to the different test tubes he added poly U, poly A, and poly AU. He knew that if he got any synthesis at all, poly U should make a polypeptide consisting entirely of one kind of amino acid. Poly A should too, but it should be a different amino acid.

The results again were dramatic. Poly A and poly AU did almost nothing, but poly U stimulated 12 times as much protein synthesis as when no RNA was added. The only question now was, What polypeptide had been made? In other words, which of the amino acids had been polymerized? Matthei attacked the problem aggressively. For five days, he worked almost round the clock, testing one amino acid after another.

On Friday night, Matthei stayed up all night again. Early on Saturday morning, bleary eyed but elated, Matthei had his answer. Poly U coded for a polypeptide made exclusively of the amino acid phenylalanine (Figure 11-2). He now knew that U nucleotides coded for phenylalanine, but how many nucleotides it took—whether one, two, three, or even more nucleotides—was another question. The number of nucleotides required to specify one amino acid was still unknown. Nonetheless, as Matthei stood in the lab, he realized that he alone in all the world knew the first word of the genetic code. He told Nirenberg's boss at NIH, but Matthei did not call Berkeley.

Nirenberg did not hear the news until he returned from California nearly three weeks later. His experiments at Berkeley had been inconclusive. The protein that tobacco mosaic virus RNA seemed to be synthesizing could not be identified. He must have heard Matthei's news with a mix-

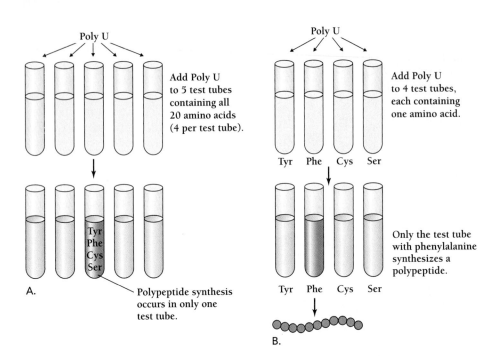

Figure 11-2

A simplified view of Nirenberg and Matthei's experiments. A. Matthei added four different radioactively labeled amino acids to each of five test tubes containing enzymes, ribosomes, and ATP, as well as poly AU. One test tube synthesized large amounts of an unknown polypeptide. B. To find out which amino acid had been joined into polypeptides, Matthei added just one amino acid to each of four more test tubes. The test tube that synthesized protein was the one that contained the amino acid phenylalanine. The synthesized polypeptide consisted entirely of phenylalanine.

Add Poly U to 5 test tubes containing all 20 amino acids (4 per test tube).

Add Poly U to 4 test tubes, each containing one amino acid.

Tyr Phe Cys Ser

Tyr Phe Cys Ser

Polypeptide synthesis occurs in only one test tube.

A.

Only the test tube with phenylalanine synthesizes a polypeptide.

Tyr Phe Cys Ser

B.

ture of emotions—delight at Matthei's success, disappointment that he had not been there to share in it.

Everyone at NIH soon knew what Matthei and Nirenberg had done. But the rest of the world either had not heard or did not believe it. Nirenberg was quiet, soft spoken. As one well-placed molecular biologist later said apologetically in reference to Nirenberg, "Either a person . . . is someone who's in the club and you know him, or else his results are unlikely to be correct, because he [isn't] in the club." Nirenberg was definitely not in the club. And, as far as the fast-moving world of molecular biology was concerned, Matthei might as well have not existed.

Together Nirenberg and Matthei wrote up the results of all their work so far and sent it to a scientific journal. A week later, NIH sent Nirenberg to Moscow for the Fifth International Congress of Biochemistry. In a small, nearly empty room, Nirenberg presented the results of his and Matthei's months of work. Of the handful of scientists present, only molecular biologist Matthew Meselson really listened.

As Meselson later recalled, "I was bowled over by the results, and I went and chased down Francis [Crick], and told him that he must have a private talk with the man." Crick talked to Nirenberg, then arranged for him to give his talk once more in front of an audience of hundreds of biochemists and molecular biologists. Nirenberg's scientific fortune was made. He might not belong to the club, but he had been heard by nearly everyone who did.

While Nirenberg was in Moscow, Matthei nailed down still another "word" in the genetic code. Poly C, he found, made a polypeptide composed entirely of the amino acid proline.

Nirenberg returned to NIH in triumph. Yet within days of his announcement in Moscow, another research lab began using Nirenberg and Matthei's refined technique to crack the rest of the genetic code. Although initially taken aback by the aggressiveness of this move, Nirenberg soon threw himself into the competition, determined to be first.

In his excitement, Nirenberg began making all the decisions, and Matthei began to feel like a fifth wheel. His friendship and collaboration with Nirenberg soured, and 18 months after they began to work together, Matthei was back in Germany working by himself once more.

Nirenberg never missed a step. He began working with another collaborator, Philip Leder, and the two soon showed that a length of RNA just three nucleotides long could specify an amino acid. They were able to define a **codon** as a group of three nucleotides that specifies a single amino acid within a polypeptide.

Then another researcher stepped into the picture. Organic chemist Gobind Khorana found a way to link nucleotides together in any order he wanted. In one experiment, he put together long chains of poly AC, which contained the sequence ACACACACAC and so on. Khorana found that poly AC RNA stimulated the production of a polypeptide consisting of two alternating amino acids—threonine and histidine.

Khorana showed that ACA specified threonine and CAC specified histidine. Thus ACACACACAC coded for threonine-histidine-threonine-histidine. From a series of such experiments, Khorana and his colleagues established a dictionary of codons. It took five years, but by 1966, Nirenberg, Khorana, and others had written a complete dictionary for the genetic code (Figure 11-3). There were 64 different codons. Sydney Brenner and Francis Crick showed that 3 of the 64 possible codons coded for no amino acids. They were dubbed **nonsense,** or **stop,** codons whose role was to signal the end of a polypeptide chain, just like the period at the end of a sentence.

First letter (5′ end)	Second letter				Third letter (3′ end)
	U	**C**	**A**	**G**	
U	UUU Phe UUC UUA Leu UUG	UCU UCC Ser UCA UCG	UAU Tyr UAC UAA Stop UAG Stop	UGU Cys UGC UGA Stop UGG Trp	U C A G
C	CUU CUC Leu CUA CUG	CCU CCC Pro CCA CCG	CAU His CAC CAA Gln CAG	CGU CGC Arg CGA CGG	U C A G
A	AUU AUC Ile AUA AUG Met	ACU ACC Thr ACA ACG	AAU Asn AAC AAA Lys AAG	AGU Ser AGC AGA Arg AGG	U C A G
G	GUU GUC Val GUA GUG	GCU GCC Ala GCA GCG	GAU Asp GAC GAA Glu GAG	GGU GGC Gly GGA GGG	U C A G

Nucleotide base   Codon   Amino acid

A.

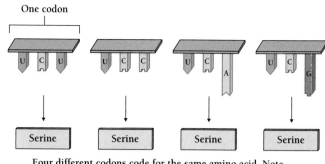

Four different codons code for the same amino acid. Note difference is in last base.

Figure 11-3

The genetic code. A. This table shows which amino acid corresponds to each RNA triplet. All the DNA Ts are replaced by Us, since this is RNA. B. Four different codons that all say "serine." Some amino acids have more than one codon. In such cases, the first two bases are usually the same, but the third base is different.

In 1968, Nirenberg and Khorana won a Nobel prize for their elucidation of the genetic code. In the same year, James Watson's best-selling book, *The Double Helix,* was published. In the world of biology, the first era in the history of molecular biology had come to an end.

The genetic code was deciphered by means of careful laboratory experiments.

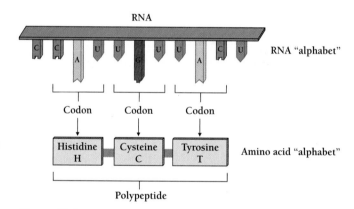

Figure 11-4

Three nucleotides code for one amino acid. The DNA/RNA language is written with an alphabet of nucleotides, while the polypeptide language is written with an alphabet of amino acids.

What Else Do We Know About the Genetic Code?

The genetic code serves to translate the language of nucleic acids (DNA and RNA) into the language of polypeptides. The DNA/RNA language is written with an alphabet of just 4 nucleotide "letters," while the polypeptide alphabet is written with a language of 20 amino acid letters. Although the DNA and protein languages are different, the **genetic code** clearly specifies which nucleotide sequences correspond to which amino acid sequences. Each set of three nucleotides, called a codon, specifies an amino acid (Figure 11-4).

Since different combinations of any three nucleotides can code for 64 different amino acids, and there are only 20 amino acids, what do the extra 44 codons do? The answer is

that most of these extra codons also code for the same 20 amino acids. The genetic code has several ways of specifying each amino acid. Because several codons may have the same meaning, we say that the genetic code is "redundant."

The genetic code is also said to be **universal,** because all eukaryote organisms use the same code, for example, when globin RNA from rabbit red blood cells is added to wheat germ cells, the wheat cells make rabbit globin, demonstrat-

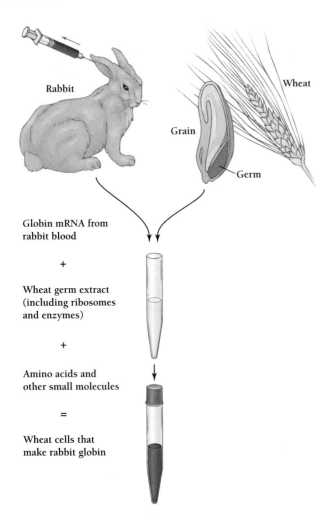

Figure 11-5
All organisms use the same genetic code. The protein-making machinery of a wheat plant can synthesize blood proteins normally found only in rabbits.

ing that wheat cells and rabbit cells use the same genetic code to translate RNA into protein (Figure 11-5).

The genetic code is not completely universal. Mitochondria—which have their own DNA, separate from that in the nucleus—use a slightly different code from the one in Figure 11-3. And yeast mitochondria use a different genetic code from human mitochondria.

We have seen how molecular biologists solved the genetic code, which helps explain how genetic information flows from DNA to RNA to proteins. But we must also ask about the mechanism by which this information flows. What happens, step by step?

The genetic code translates the language of DNA and RNA into the language of polypeptides. Three nucleotides—a codon—specify one amino acid. The genetic code is redundant and nearly universal, and 3 of the 64 codons are nonsense, or stop, codons.

11.1 Genetic Information Flows from DNA to RNA to Polypeptides

By the mid-1950s, biologists knew that genes were sequences of nucleotides that specify the sequences of amino acids in polypeptides. The next question was, How does a sequence of nucleotides specify a sequence of amino acids? Molecular biologists such as Nirenberg and Khorana confirmed what Watson and others had long suspected: RNA is always an intermediate in the production of proteins.

Francis Crick summarized cellular information flow in a statement he called the **central dogma** of molecular biology: "DNA specifies RNA, which specifies proteins." According to Crick's central dogma, information can flow *only* from DNA to RNA to proteins (Figure 11-6). It cannot flow from proteins to RNA or DNA. Crick put this idea another way, writing, "Once 'information' has passed into protein, *it cannot get out again*." Proteins cannot change the information in the genes.

Crick used the word "dogma" ironically because "dogma" means an opinion, belief, or a tenet of religious faith. It is not the word to use to describe scientific facts or well-supported theories. At the time Crick formulated the central dogma, mo-

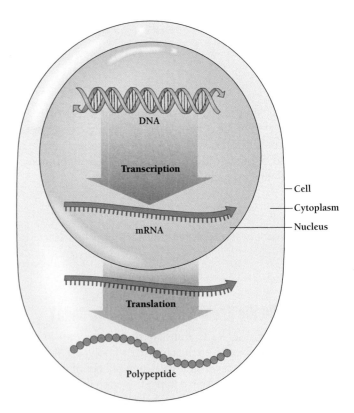

Figure 11-6
The central dogma. DNA specifies RNA, which specifies protein. DNA in the nucleus is *transcribed* into RNA. Outside the nucleus, in the cytoplasm, ribosomes *translate* RNA into polypeptides. In prokaryotes, which lack a nucleus, both transcription and translation occur in the cytoplasm.

Biology⑧Now™ Learn more about transcription by clicking on this figure on your BiologyNow CD-ROM.

lecular biologists had no good evidence supporting the idea. Crick and others just thought it must be true. Crick's joke was lost on most other researchers, however, and the "central dogma" has now been presented in textbooks *as dogma* for decades. As we will see in the next chapter, Crick had good reason for his twinge of doubt. The central dogma is mostly true, but not completely. Some viruses—such as the AIDS virus—carry information in RNA, which infected cells can incorporate into their own DNA. In addition, information in RNA and proteins strongly influences the expression of the genes.

Nevertheless, the idea that DNA specifies RNA, which specifies proteins, is a good starting point for understanding how genes are expressed. DNA transmits its information to RNA through a process called **transcription,** in which the information in the DNA is rewritten, or "transcribed," to RNA. Cells then use the information in RNA to make polypeptides. Using RNA to carry messages from the DNA is like working from a photocopy of an important document while keeping the original in a safe place. We can think of DNA as the original, safely stored in the nucleus, and the RNA as the photocopy.

The RNA copy of the information from DNA is composed of **messenger RNA (mRNA).** Molecules of mRNA have the same information as the DNA original. RNA, like DNA, is a polynucleotide. It differs from DNA in just three ways:

1. The backbone sugar is ribose instead of deoxyribose.
2. Uracil (U) replaces thymine (T) as one of the pyrimidine bases.
3. RNA is usually single stranded, whereas DNA is usually double stranded.

Despite these differences, a message in the RNA alphabet (A, G, C, and U) is just the same as in the DNA alphabet (A, G, C, and T), but with U replacing T.

The conversion of the message carried by the RNA into strings of amino acids—actual polypeptides—is called **translation** (Figure 11-6). Translation takes the mRNA text

and translates it into the language of amino acids. This is the essence of the central dogma.

$$\text{DNA} \xrightarrow{\text{Transcription}} \text{RNA} \xrightarrow{\text{Translation}} \text{Polypeptide}$$

In eukaryotes, transcription occurs in the nucleus and translation occurs in the cytoplasm (outside the nucleus). Messenger RNA passes through the nuclear membrane out to the cytoplasm. There, ribosomes, acting like molecular sewing machines, stitch together amino acids into polypeptides.

The message in the DNA is transcribed into mRNA, which is translated into polypeptides. The central dogma says that information flows from DNA to RNA to protein only. Information in protein is not supposed to be able to influence the information in the DNA.

11.2 RNA Polymerase Transcribes DNA into RNA

When a cell makes a polypeptide specified by a particular gene, we say that the cell is "expressing" that gene. Gene expression begins when the enzyme **RNA polymerase** transcribes the DNA gene into mRNA (Figure 11-7). Actually, there are several kinds of RNA polymerases, but since they all make RNA in the same way, we will discuss them as if they were a single enzyme.

How Does RNA Polymerase Begin Transcription?

In the last chapter, we saw that DNA polymerase can build DNA in two directions (extending the 3′ end of each DNA strand) and that the enzyme begins by adding nucleotides to

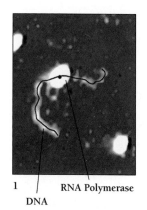 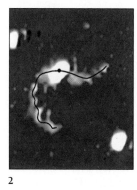 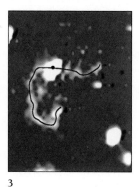 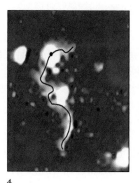 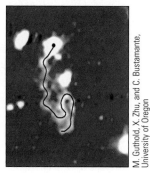

1   RNA Polymerase   2   3   4   5

DNA

Figure 11-7

Transcription. In this sequence of micrographs, RNA polymerase—the large, round blob—transcribes a strand of DNA into mRNA (which is not visible). RNA polymerase moves from the middle of the strand (1) to the end (5) in less than five minutes.

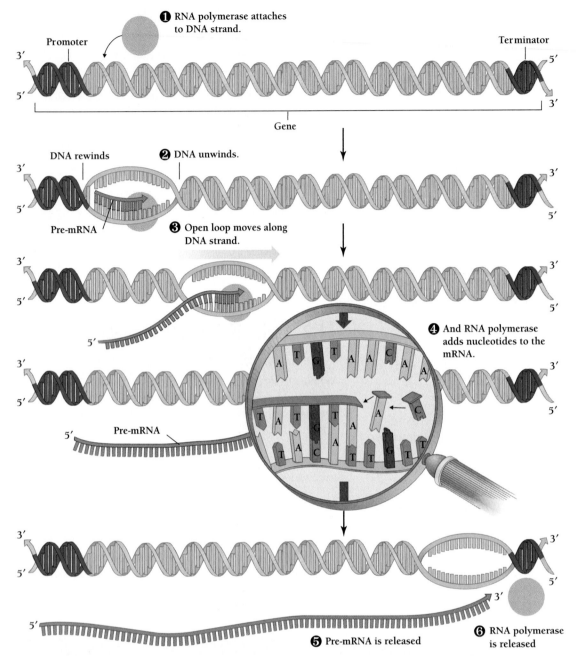

Figure 11-8

How does RNA polymerase begin transcription? Starting at the DNA promoter, RNA polymerase transcribes DNA into mRNA until it reaches the terminator.

Biology⊛Now™ Learn more about the role of RNA polymerase in transcription by clicking on this figure on your BiologyNow CD-ROM.

① RNA polymerase attaches to DNA strand.

Promoter

Terminator

Gene

DNA rewinds

② DNA unwinds.

Pre-mRNA

③ Open loop moves along DNA strand.

④ And RNA polymerase adds nucleotides to the mRNA.

Pre-mRNA

⑤ Pre-mRNA is released

⑥ RNA polymerase is released

an existing polynucleotide called a primer. In contrast, RNA polymerase transcribes DNA into RNA from just one of DNA's two strands and does not require a preexisting RNA primer (Figure 11-8). The start signal for mRNA synthesis is a special sequence of DNA called a **promoter.** RNA polymerase (working with other proteins) recognizes the promoter site, unwinds and separates the two strands of DNA near the promoter site, and then attaches to the promoter

and begins to "read" the sequence of bases in the gene. Wherever the template strand has a G, RNA polymerase will put a C; wherever the template has an A, it puts a U.

In bacteria, the promoter consists of two characteristic sequences, each about six nucleotides long, with about 25 other nucleotides between them. RNA polymerase binds to the promoter and begins to make an RNA molecule that is complementary to the template strand.

Extreme Biology   Which Strand of the DNA Is Transcribed?

At the beginning of this chapter, we saw that protein synthesis in a test tube helped solve the genetic code. Such test-tube protein synthesis also allowed researchers to discover how RNA polymerase transcribes the DNA.

Just as DNA polymerase assembles new molecules of DNA one nucleotide at a time, RNA polymerase assembles RNA by adding one nucleotide at a time. However, DNA polymerase copies both strands of the DNA double helix, while RNA polymerase copies only one of the two strands of DNA. The strand that is copied is called the *template strand*. Although the two strands are chemically the same, the information in the template strand is completely different from that in the nontemplate strand of DNA. The template strand, also called the *antisense strand,* is a reverse copy of its mate, the *sense strand.* If this sentence were the sense strand, we might think the template strand would look something like this: :siht

ekil gnihtemos kool dluow dnarts etalpmet eht kniht thgim ew ,dnarts esnes eht erew ecnetnes siht fI

Actually, however, the template strand is not only a reverse copy of the sense strand. It is also composed of the Watson-Crick base pairs, as if we had substituted different letters into our reversed sentence. For example, if the sense strand of DNA read CATTAG, then the antisense strand, read forward by RNA polymerase, would be CTAATG.

sense strand

→

CATTAG
GTAATC

antisense strand

←

The most important result of this reversal and substitution is that the antisense strand cannot code for a protein. The antisense strand is a chaotic sequence of codons that includes so many stop codons that any

strange polypeptides made following its instructions could not be more than a few amino acids long. The sense strand, in contrast, has stop codons only at the end of each gene.

The antisense strand is useful, however, because the mRNA created from the antisense strand is not nonsense. The mRNA is an RNA version of the sense strand, the reverse of a reverse. The antisense strand at left, for example, would be transcribed as CAUUAG, since wherever DNA has a T, the mRNA has a U.

RNA polymerase copies the antisense DNA strand by holding in place an RNA nucleoside triphosphate base, which pairs with the next nucleotide in the template.

The mRNA strand grows from its 5′ end to its 3′ end, as it copies the template DNA strand from its 3′ end to its 5′ end. The new mRNA molecule has the same sequence of bases as the sense strand of DNA, except that wherever DNA has a T the RNA has a U.

In eukaryotes, the production of RNA involves an extra step. Eukaryotic RNA polymerases do not themselves recognize the promoter (unlike prokaryote RNA polymerases). Instead, they depend on the assistance of other proteins, called **activators,** or **transcription factors,** which lead the RNA polymerase and its accompanying proteins to the promoter. One set of activators, for example, guides RNA polymerase to a DNA sequence called a "TATA box," since it usually contains the sequence TATA. Other activators guide RNA polymerase to other specific DNA sequences.

Both eukaryotic and prokaryotic RNA polymerases stop transcription at special sequences, called **terminators,** which stop the enzyme from transcribing any more of the DNA.

In eukaryotes, proteins called activators guide RNA polymerase to the promoter sequence of gene. There, RNA polymerase begins transcribing the gene.

How Does the Cell Alter the mRNA Transcribed from the DNA?

In prokaryotes, the mRNA binds to the ribosomes and begins directing protein synthesis as soon as the mRNA is made. Sometimes translation begins even as the DNA is being transcribed. In eukaryotes, however, transcription and translation occur separately. Transcription occurs in the nucleus, and translation occurs—minutes, hours, or even days later—in the cytoplasm.

Transcription and translation in eukaryotes are separated by much more than just the nuclear membrane. Just after the mRNA is transcribed from the DNA, in the nucleus, it is called pre-mRNA. While still in the nucleus, pre-mRNA undergoes a number of modifications that transform it into a molecule of "mature" mRNA that can be translated into a polypeptide. These modifications include: (1) capping the 5′ end of the RNA with a molecule of GTP (a relative of ATP), (2) adding a "tail" of 150 to 200 As to the 3′ end of most mRNAs, and (3) removing large pieces of the RNA and splicing the remaining pieces together (Figure 11-9A). In addition, using a process called RNA editing (recently discovered), cells can add, remove, or change a few selected nucleotides in some regions of the mRNA.

The cutting and splicing of pre-mRNAs was a surprising discovery, since it demonstrated that genes are interrupted by nucleotide sequences that do not code for proteins. These interruptions in pre-mRNA (as well as in the corresponding DNA sequences) are called **introns,** for "intervening" sequences. The RNA sequences that directly code for proteins are spliced together to form the mature mRNA molecule.

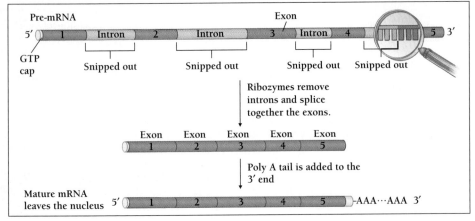

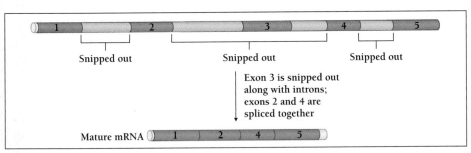

Figure 11-9

RNA splicing. A. In this example, introns are snipped from the mRNA, the five exons are spliced together, and each end of the spliced mRNA is punctuated. Like a capital letter, the GTP cap indicates the beginning of the gene sentence. Like a period, the poly A tail indicates the end. B. The same pre-mRNA can be snipped and spliced in different ways to produce mRNAs that code for different polypeptides.

These sequences (and the corresponding sequences in DNA) are called **exons,** for "expressed" sequences. Sometimes introns are responsible for their own excision, acting in a way like enzymes. When RNA displays such an "enzymatic" behavior it is called a *ribozyme*. Ribozymes might have been the first biomolecules with catalytic activity, perhaps before the first cells appeared.

In many eukaryotic genes, most of the sequence consists of introns, with exons making up only a tiny fraction of the total length. A single pre-mRNA molecule may contain as many as 70 introns, each one ranging in size from 80 to 10,000 nucleotides. Yet all the exons in a single mRNA are, together, rarely more than 3,000 nucleotides long.

Most of the DNA in a gene consists of introns, which do not code for any part of a functioning polypeptide. In fact, more than 98 percent of human DNA does not directly specify polypeptide sequences, and a large part of this apparently useless DNA is introns. No one knows what introns do. Some scientists believe that the DNA in the introns is basically "junk DNA" that has accumulated in the genome in the same way that old keys and odd parts of things accumulate in the back of a kitchen drawer. Others believe that all DNA must be in the genome for a purpose, even if that purpose is not yet apparent.

Pre-mRNA includes long stretches of noncoding DNA sequences, called introns, that must be removed. The remaining exons are spliced together to form mature mRNA.

One Gene—Heaven Only Knows How Many Polypeptides

We have now seen that most eukaryotic genes consist of exons interspersed with introns. Molecular machinery within the nucleus cuts and splices the pre-mRNA to produce mature mRNAs without introns.

As it turns out, transcription can be far more complicated than this simple picture suggests. In some cases, eukaryotic cells can actually transcribe many different mRNAs from the same gene. For example, the mRNA transcript may start or end at different sites on the same gene. Or a single pre-mRNA may be broken up and spliced together in several different ways, so that one mRNA's exon becomes another mRNA's intron (Figure 11-9B). As a result, a single gene may encode several polypeptides. What is more, the cell may express all of these various polypeptides at once.

Geneticists spent decades arriving at our definition from Chapter 10, "one gene—one polypeptide." But studies of gene expression beginning in the late 1970s have conclusively shown that one gene can actually encode several polypeptide chains, occasionally with very different functions. Most molecular biologists now say that a gene is a DNA sequence that is transcribed as a single unit and encodes either a single polypeptide or a set of related polypeptides.

A single gene can encode several different polypeptides.

11.3 How Does a Cell Translate mRNA into a Polypeptide?

The synthesis of proteins requires many kinds of molecular tools. Among them are three distinct kinds of RNA: (1) messenger RNA (mRNA) carries genetic information from DNA that specifies the amino acid sequence of a polypeptide; (2) transfer RNAs (tRNAs) carry each amino acid to a codon in the mRNA; (3) ribosomal RNA (rRNA), together with proteins, forms ribosomes, which serve as the site for polypeptide synthesis. In addition, an array of enzymes and other proteins assist in the synthesis and folding of new proteins.

What Do Ribosomes Do?

Ribosomes are like subcellular sewing machines that link amino acids into polypeptides. All cells contain ribosomes—a single bacterial cell contains about 20,000 ribosomes. Each ribosome consists of a small and a large subunit (Figure 11-10). Each subunit contains both RNA and proteins. The *E. coli* ribosome, for example, consists of three molecules of RNA, called **ribosomal RNAs, or rRNAs,** and 54 kinds of protein molecules. Eukaryotic ribosomes are even more complex, with four molecules of rRNA and some 82 kinds of protein.

Ribosomal RNA is very different from messenger RNA. The sequence and length of messenger RNA varies, depending on the kind of polypeptide it encodes. But the RNA that makes up the ribosome is always the same, no matter what kind of protein the ribosome is producing. In fact, all ribosomes of all organisms on Earth are remarkably alike. Such similarity suggests that the exact shape of the ribosome is important to how it works. Any individual with even minor changes in the standard ribosome design would probably be unable to make proteins and would die immediately.

Ribosomes' main task is to assemble amino acids into polypeptides in the order specified by mRNA. In all ribosomes, the rRNA acts as a "ribozyme," helping to catalyze the joining of amino acids into polypeptides. The proteins of a ribosome are structural units that help keep the ribozymal RNA in the right orientation to do its job.

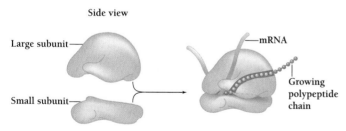

Figure 11-10

How do ribosomes work? The two parts of a ribosome fit together around the mRNA molecule and the growing polypeptide chain.

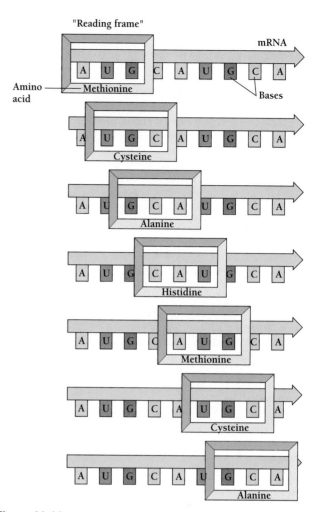

Figure 11-11

The reading frame. The exact base at which the ribosome begins translating the mRNA determines how it interprets all of the three-base codons from there onward.

Ribosomes must recognize where to start and where to stop reading the mRNA for the production of each polypeptide (Figure 11-11). For example, consider the translation of the mRNA sequence AUGCAUGCA. If a ribosome started reading at the first nucleotide, it would assemble the peptide methionine-histidine-alanine. If, however, the ribosome started with the second nucleotide, it would assemble cysteine-methionine. If the ribosomes started with the third nucleotide, it would assemble alanine-cysteine. The grouping of nucleotide triplets for translation is called the **reading frame.**

Because starting at the right place is crucial to reading mRNA correctly, mRNAs must contain signals that say "start" and "stop." Start and stop signals are like punctuation that tells where a sentence begins and ends. The smaller ribosomal subunit finds the right place to start.

Each ribosome contains two grooves. On the small subunit is a groove for reading the mRNA. On the large subunit is a groove for the growing polypeptide chain. Protein synthesis requires the participation of both subunits. The small subunit is responsible for starting synthesis at the right place,

Extreme Biology How Do the Ribosomes Read Different Kinds of Mutations?

A base substitution in a gene may result in a *missense mutation,* which changes the codon for one amino acid into the codon for another amino acid. The resulting polypeptide may be either nonfunctional or altered in the way it functions. For example, some altered polypeptides result in *conditional mutations,* in which the resulting protein may be able to function under some conditions, but not under others. In one *Drosophila* mutant, for example, the fly functions normally at 20°C but becomes paralyzed at 30°C. This is due to a protein that regulates ion transport across nerve cell membranes, which fails at 30°C. Such temperature sensitivity results when an amino acid substitution reduces a protein's stability at higher temperatures, so that it denatures, or changes shape, more readily than the normal protein.

A base substitution can also result in a *nonsense mutation,* which changes the codon for an amino acid into a nonsense or stop codon, leading to a shortened polypeptide (Figure A). Such a shortened polypeptide almost certainly will not function at all in its former role.

Mutations that have no effect on phenotype are called *silent mutations.* Some are simply single-base substitutions that result in a codon that codes for the same amino acid (because of the redundancy of the genetic code). In that case, the polypeptide is unchanged. Most silent mutations, however, lie in DNA sequences that do not actually code for a polypeptide. Other supposedly silent mutations may merely produce changes in the phenotype so subtle that researchers cannot tell the difference.

Insertions and deletions in structural genes produce *frameshift mutations,* which alter the groupings of nucleotides into codons (Figure B). The addition or subtraction of a single base can wreak havoc in the ribosomes' reading of mRNA downstream from the mutation. The result is often a string of missense changes, followed by a stop codon.

Sometimes point mutations can reverse themselves in a process called *reversion.*

Reversion mutations, sometimes called *suppressor mutations,* consist of a second mutation that reverses the effects of the first mutation. A reversion might be an exact restoration of the original nucleotide sequence. Alternatively, the DNA might have a nucleotide sequence that is not identical to the original but that specifies the same amino acid—a silent mutation.

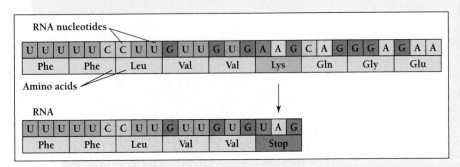

Figure A
Nonsense mutation. A mutation changes an AAG codon (lysine) into UAG (stop) in RNA. This nonsense, or stop, codon shortens the polypeptide.

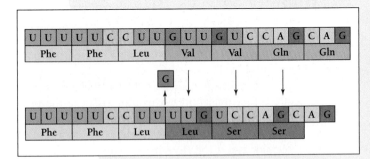

Figure B
Frameshift mutation. A deleted nucleotide causes a "frameshift," in which each codon after the deletion reads differently. A frameshift mutation causes dramatic random changes in the resulting polypeptide.

---

while the large subunit contains the machinery for making the peptide bonds that hold together the polypeptide.

Ribosomes, which are made of rRNA and proteins, link amino acids together into polypeptides. The small ribosomal subunit reads the mRNA, and the large ribosomal subunit makes the peptide bonds between each pair of amino acids in the growing chain.

How Do tRNAs Serve as Adapters Between mRNAs and Amino Acids?

Once molecular biologists began to suspect that RNA dictated the sequences of amino acids in polypeptides, they tried to imagine how this could occur. An early idea was that each codon in an mRNA molecule bound directly to a specific amino acid. But no one could find any such binding.

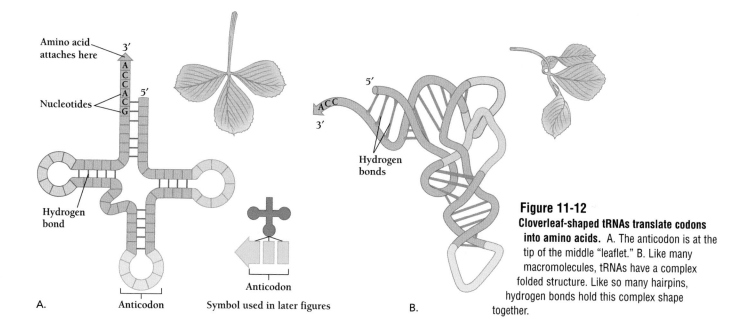

Figure 11-12

Cloverleaf-shaped tRNAs translate codons into amino acids. A. The anticodon is at the tip of the middle "leaflet." B. Like many macromolecules, tRNAs have a complex folded structure. Like so many hairpins, hydrogen bonds hold this complex shape together.

Labels in figure: Amino acid attaches here; 3'; Nucleotides; 5'; Hydrogen bond; Anticodon; Anticodon; Symbol used in later figures; A.; 5'; 3'; Hydrogen bonds; B.

Besides, it was hard to imagine how a sequence of just three nucleotides could distinguish among similarly shaped amino acids.

The actual explanation, proposed by Francis Crick, is that the amino acids are first linked to special adapter molecules, called **transfer RNAs,** or **tRNAs.** The mRNA codons bind to the tRNA, not to the amino acid itself. Each tRNA consists of 75 to 85 nucleotides and folds to form a secondary structure shaped like a cloverleaf (Figure 11-12). The "stems" of the tRNAs are double strands of RNA held together by base pairing (with hydrogen bonds). The "leaves" are loops of RNA with unpaired bases. X-ray diffraction studies of tRNA's tertiary structure show that the whole cloverleaf bends into a shape resembling the letter "L."

One stem of each tRNA attaches to the amino acid corresponding to that codon (Figure 11-12). The opposite side of each tRNA molecule binds to a specific codon in the mRNA. In one of the loops is a sequence of three nucleotides, called the **anticodon,** that forms base pairs with the matching codon sequence in the mRNA. Each tRNA molecule has only one kind of anticodon and can pick up only one kind of amino acid.

We may wonder how many tRNAs a cell needs—one for each amino acid or one for each of the 64 possible codons? A cell would have at least one tRNA for each of the 20 amino acids. Yet we might expect 61—one for every codon except the three codons that usually do not specify any amino acid. In fact, different species have different numbers of tRNAs. But every species has between 20 and 61 kinds of tRNA. Organisms can have fewer than 61 tRNAs because some tRNA anticodons recognize more than one codon.

How do we know that mRNA recognizes the tRNA rather than the amino acid itself? To find out, researchers "tricked" the mRNA. They took the tRNA for cysteine, at-

tached the cysteine, then chemically converted the attached cysteine into alanine so that the tRNA for cysteine actually carried an alanine. Finally, researchers added this incorrectly loaded tRNA to a cell extract that was synthesizing protein. Wherever the mRNA specified cysteine, the tRNA inserted alanine instead. This result showed that the mRNA recognizes the tRNA, not the amino acid itself.

For each tRNA to load the correct amino acid, it must recognize the unique shape and charge distribution of each amino acid. Special "charging enzymes" bind each amino acid to its corresponding amino acid, then bind them together with high-energy bonds. Each charging enzyme has two binding sites: one site holds a particular tRNA, and the other site holds the corresponding amino acid.

Transfer RNA (tRNA) helps assemble amino acids into polypeptides by binding first to an amino acid and then to the correct codon in the mRNA. Charging enzymes help tRNAs hook up with the correct amino acid.

How Do Ribosomes Begin Translation?

We have seen how each molecule of mRNA contains signals that tell the ribosomes where to start and where to stop. We have also seen that starting translation at the right place is crucial, since a mistake of only a single nucleotide would lead to a complete misreading of the information.

The first step of translation is **initiation,** which begins with the attachment of the ribosome's small subunit to the mRNA (Figure 11-13). The actual initiation site, or **start site,** in mRNA is always the codon AUG, which specifies methionine. This means that *all* recently translated polypep-

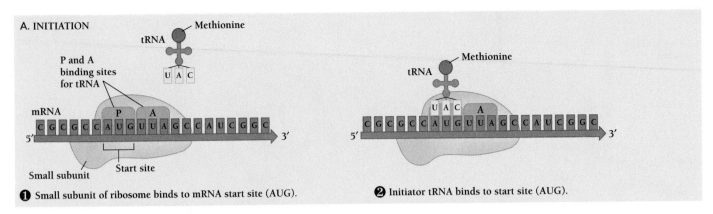

A. INITIATION

❶ Small subunit of ribosome binds to mRNA start site (AUG).

❷ Initiator tRNA binds to start site (AUG).

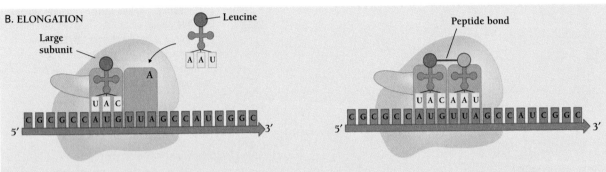

B. ELONGATION

❸ During elongation, another charged tRNA arrives, and the large subunit of ribosome binds the two amino acids together.

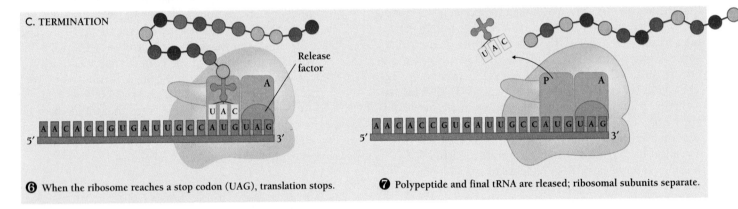

C. TERMINATION

❻ When the ribosome reaches a stop codon (UAG), translation stops.

❼ Polypeptide and final tRNA are released; ribosomal subunits separate.

Figure 11-13

Translation. Translation occurs in three stages—initiation, elongation, and termination. ❶ During initiation, the small subunit of the ribosome binds to the start site, AUG, and ❷ moves the amino acid methionine into place. ❸ During elongation, another charged tRNA arrives and the large subunit of the ribosome holds the two charged tRNAs side by side, in the P and A sites, and binds the two amino acids together. ❹ The ribosome and its attached polypeptide chain then move forward, and ❺ another charged tRNA moves into place. ❻ During termination, the ribosome reaches a stop codon and a protein called release factor binds to the stop codon. ❼ An enzyme cuts the newly made polypeptide chain from the last tRNA, and the ribosomal subunits then separate from each other.

tides begin with methionine. In most cases, however, a special enzyme later clips away the methionine. In prokaryotes, a sequence of about six nucleotides, just ahead of the start site AUG, distinguishes an initiation site from an ordinary methionine. In eukaryotes, each ribosome simply attaches to the 5′ end of the mRNA and moves along until it encounters the first AUG, where translation then begins.

In the first step of translation—initiation—the ribosome attaches to the start site—the codon AUG. In prokaryotes, a sequence of six nucleotides distinguishes the start AUG from the AUG codon for methionine. In eukaryotes, the start AUG is simply the first one from the 5′ end of the mRNA.

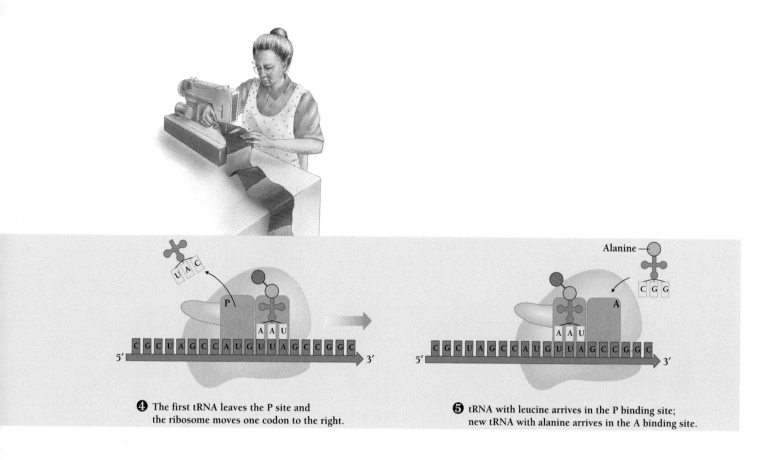

④ The first tRNA leaves the P site and the ribosome moves one codon to the right.

⑤ tRNA with leucine arrives in the P binding site; new tRNA with alanine arrives in the A binding site.

How Do Ribosomes Make Peptide Bonds?

The process of polypeptide-chain **elongation** consists of three steps (Figure 11-13): (1) putting the next amino acid into position, (2) forming a peptide bond between the growing polypeptide and the next amino acid, and (3) moving the ribosome to the next codon. Every amino acid is joined to its neighbor in exactly the same way.

Elongation, unlike initiation, requires the participation of the large ribosomal subunit (which binds to the initiator tRNA) as well as the mRNA and the small ribosomal subunit. The large subunit contains two special pockets, called the P site (for *p*eptide) and the A site (for *a*mino acid). The P and A sites bind two tRNAs to adjacent codons in the mRNA. The binding of the two charged tRNAs to the large subunit is the first step of elongation.

After the P and A sites are filled, the ribosome links the attached amino acids to form a peptide bond. This linkage is the second step of elongation. The tRNA in the P site then falls away from the ribosome. Left behind is the tRNA in the A site and its attached amino acid, which is now linked to the entire polypeptide chain. In the third step, the ribosome moves forward so that the tRNA in the A site with its attached polypeptide chain ends up in the P site. The repositioning of the tRNA and the attached polypeptide to the P site completes the three steps of the elongation cycle. The cycle now repeats over and over until all the codons in the mRNA are read.

During elongation, amino acids are joined one after the other in exactly the same way. A tRNA first moves an amino acid into position next to the last amino acid. Once two amino acids are in place, the ribosome joins them together with a peptide bond. The ribosome then moves forward on the mRNA, repositioning the tRNA and its attached polypeptide chain.

How Does Polypeptide Synthesis Stop?

The appearance of a stop codon in the mRNA signals the end of a polypeptide chain, a process called **termination.** When the ribosome reaches one of the three stop codons (UAA, UAG, or UGA), there will be no corresponding tRNA. Instead a protein called a **release factor** binds to the stop codon (Figure 11-13). The enzyme that makes peptide bonds during elongation now cuts the polypeptide from the last tRNA, releasing the completed polypeptide into the cytoplasm. The ribosomal subunits then separate from each

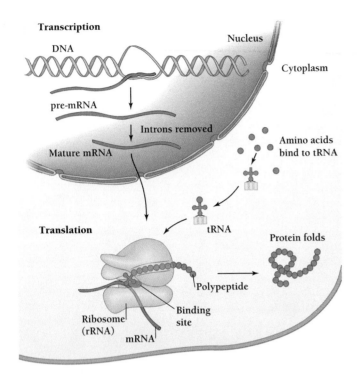

Transcription

DNA

Nucleus

Cytoplasm

pre-mRNA

Introns removed

Mature mRNA

Amino acids bind to tRNA

Translation

tRNA

Protein folds

Polypeptide

Ribosome (rRNA)

Binding site

mRNA

Figure 11-14
Transcription and translation in a eukaryotic cell.

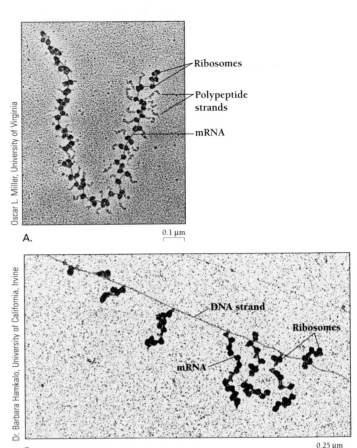

Ribosomes

Polypeptide strands

mRNA

Oscar L. Miller, University of Virginia

A.   0.1 μm

DNA strand

Ribosomes

mRNA

Dr. Barbara Hamkalo, University of California, Irvine

B.   0.25 μm

Figure 11-15
A large polysome. A. Individual ribosomes move along a strand of mRNA *(red)*, translating as they go. The ribosomes at the upper left are just beginning translation and the polypeptide strands are not yet visible. The ribosomes on the right and in the middle have produced visible polypeptide strands. B. In prokaryotes, which lack a nuclear membrane, transcription and translation can occur simultaneously. In this electron micrograph, the DNA *(blue)* runs diagonally from the upper left to the lower right. Transcription began at the upper left end of the DNA, so the longest mRNAs (with the most ribosomes) are at the lower right. Clusters of up to 13 ribosomes read individual strands of mRNA, even as the mRNA strands are themselves being transcribed from the DNA.

other and become available for another round of initiation and elongation.

All of the steps of protein synthesis, including transcription and translation, are summarized in Figure 11-14. After a polypeptide chain is assembled, it must fold into a functioning protein. We can also, again, summarize the central dogma as follows:

$$\text{DNA} \xrightarrow[\text{(RNA polymerase)}]{\text{Transcription}} \text{RNA} \xrightarrow[\text{(Ribosomes)}]{\text{Translation}} \text{Polypeptide}$$

When the ribosome reaches a stop codon, a protein called release factor binds to the stop codon and an enzyme cuts the newly made polypeptide chain from the last tRNA.

Several Ribosomes Can Read a Single Strand of mRNA Simultaneously

As soon as a ribosome has moved far enough away from the AUG start site, the start site is free for another ribosome to attach and begin translation. As a result, several ribosomes, spaced as little as 80 nucleotides apart, may be attached to a single prokaryotic mRNA at one time, each one overseeing the production of identical polypeptides. This complex of mRNA and ribosomes is called a **polysome,** short for polyribosome (Figure 11-15A). In this way, the cell can synthesize many copies of a polypeptide at once, greatly speeding the mass production of proteins. We will see later in this chapter, for example, that an *E. coli* bacterium can begin producing enzymes to break down milk sugar (lactose) as soon as the bacterium detects the milk sugar. In prokaryotes, which lack a nuclear membrane, transcription and translation can occur simultaneously (Figure 11-15B). In eukaryotes, transcription occurs inside the nucleus and translation occurs outside the nucleus.

11.4 How Do Cells Regulate Gene Expression?

We have seen now how cells express the information in the DNA. But cells contain thousands of genes. Some of these are expressed as polypeptides all the time, some only occasionally, and many are never expressed. Prokaryotes transcribe almost all of their genes into RNA. But eukaryotes regulate gene expression more closely. Cells, especially eukaryotic cells, do not make all of their polypeptides at once for the same reason that we do not leave all of the water taps in our homes flowing at once. Just as we turn on hot or cold water according to our needs, a cell produces different proteins according to its needs. Further, just as a water tap can be turned on full blast or at a trickle, a cell can turn out thousands of copies of a polypeptide, or only two or three copies.

Eukaryotes, most of which are multicellular, face an additional problem. The dramatic cell specialization that characterizes most multicellular organisms requires the differential expression of genes in cells that have identical genomes. A human, for example, consists of more than 200 distinguishable types of cells. Nearly every one of these cells contains the same set of about 40,000 genes. Yet each kind of cell expresses only a fraction of its 40,000 genes.

A typical eukaryotic cell transcribes only about 20 percent of its genes into RNA. In a human skin cell, for example, this 20 percent would include information for basic life-support proteins that all human cells need—such as membrane proteins, ribosomal proteins, and metabolic enzymes—as well as information for specialized proteins, such as the stretchy collagen that makes skin so tough. In a skin cell, the 80 percent of genes *not* transcribed would include, for example, the genes for α-and β-globin, various muscle proteins, and a host of digestive enzymes.

Cells can regulate the expression of genes at many levels. The most efficient and usual way is called **transcriptional control,** in which cells increase or decrease the amount of mRNA transcribed from the DNA. Every cell in the body has the gene for the globin polypeptides, for example, yet only the developing red blood cells transcribe them. In every other kind of cell in the body, the globin genes sit unused.

Cells can regulate the expression of genes in other ways. In **posttranscriptional control,** a cell transcribes the DNA as usual into mRNA, but modifies the mRNA before it reaches the cytoplasm by altering the pattern of splicing exons. In **translational control,** a cell varies the rate at which it translates different mRNAs, making more or less as needed. Finally, in **posttranslational control,** a cell modifies the structure of a protein. Even after a protein has been made, the cell can regulate protein activity.

Cells regulate the expression of genes in many ways, but transcriptional control is the most common way.

How Do Prokaryotes Regulate Gene Expression?

Bacteria provided the first well-understood examples of gene regulation. Most of the time, bacteria regulate gene expression by controlling transcription. Like other organisms, bacteria must respond to their environment. They have evolved sophisticated and sensitive molecular mechanisms that ensure the production of enzymes when they need them.

Negative Regulation in Prokaryotes

The best-studied example of regulated gene expression in prokaryotes controls the use of lactose (the main sugar in milk) by *E. coli,* a common bacterium in the intestines of humans and other mammals. This elegant system enables bacteria to produce costly enzymes only when they are needed. As long as glucose is available in the intestines, *E. coli* will not digest the milk sugar lactose. When glucose is absent, however, *and* the host organism (that's you) drinks milk, then *E. coli* begins synthesizing three proteins that allow it to digest lactose. One of these is the enzyme β-galactosidase (which splits lactose into the sugars galactose and glucose).

The three genes encoding these three proteins lie next to each other on the DNA of *E. coli* and are part of the lactose operon, or **lac operon.** An operon is a stretch of DNA that includes several genes under coordinated control. *E. coli* transcribes the DNA of the *lac* operon into an mRNA that encodes all three proteins.

In the absence of lactose, each bacterium contains only about three molecules of β-galactosidase. However, when glucose is absent and lactose becomes available, the lactose stimulates, or **induces,** the expression of the *lac* operon, and the bacterium synthesizes about 3,000 molecules of β-galactosidase. Lactose is called the "inducer" because it induces gene expression.

Two genes regulate the expression of the *lac* operon. One gene, which lies a short distance from the gene for β-galactosidase, is the gene that encodes the **lac repressor,** a protein that prevents the expression of the *lac* operon (Figure 11-16A). The other gene, a sequence of 21 nucleotides called the **operator,** lies immediately next to the gene for β-galactosidase and is part of the operon. The operator, we now know, is not actually a separate gene because it does not encode a protein (or RNA). It is a regulatory site to which the *lac* repressor can bind.

The operator overlaps with the operon's *promoter,* which, recall, is the DNA sequence that signals the start of a gene. When the *lac* repressor binds to the operator, it prevents RNA polymerase from binding to the promoter and transcribing the three enzyme genes in the operon (Figure 11-16B). The *lac* repressor was the first example of a protein that regulates gene expression. The operator was the first example of a regulatory site.

How do the repressor and operator allow lactose to control gene expression in the *lac* operon? The key to understanding the induction of the *lac* operon is the realization that the *lac* repressor has two binding sites: one site binds to the operator (a DNA sequence), and the other binds to lactose (a sugar). But the repressor cannot bind to both at the same time. When the repressor binds to lactose, it can no longer bind to the operator and prevent transcription (Figure 11-16C). RNA polymerase can then bind to the *lac* operon's promoter and transcribe the RNA that specifies the three proteins needed to digest lactose. The *lac* repressor is an example of a **negative regulator,** a molecule that inhibits transcription.

When lactose binds to the *lac* repressor (a protein), the *lac* operon's promoter becomes unblocked, and RNA polymerase can transcribe the three genes that encode proteins needed to digest lactose.

Positive Regulation in Prokaryotes

Bacteria also use **positive regulators,** or activators, molecules that increase the rate of transcription. One such activator is the **catabolite activator protein,** or **CAP** (Figure 11-17). Like the *lac* repressor, CAP has two binding sites. One site binds to a particular sequence in DNA, and the other binds cyclic AMP, a nucleotide derived from ATP. Cyclic AMP is a *signaling molecule,* a molecule that conveys signals from one part of the cell to another. In bacteria, cyclic AMP is a "hunger" signal: high concentrations of cyclic AMP tell the cell that glucose is unavailable and that the cell needs to get energy from another source, such as lactose.

In the absence of glucose, CAP binds to the *lac* promoter and stimulates the transcription of the *lac* operon. When the cell has lots of glucose and little cyclic AMP, however, the *lac* operon is not active even in the presence of lactose.

The combination of negative control (by the *lac* repressor) and positive control (by the CAP activator) allows bacteria to respond quickly to changes in their

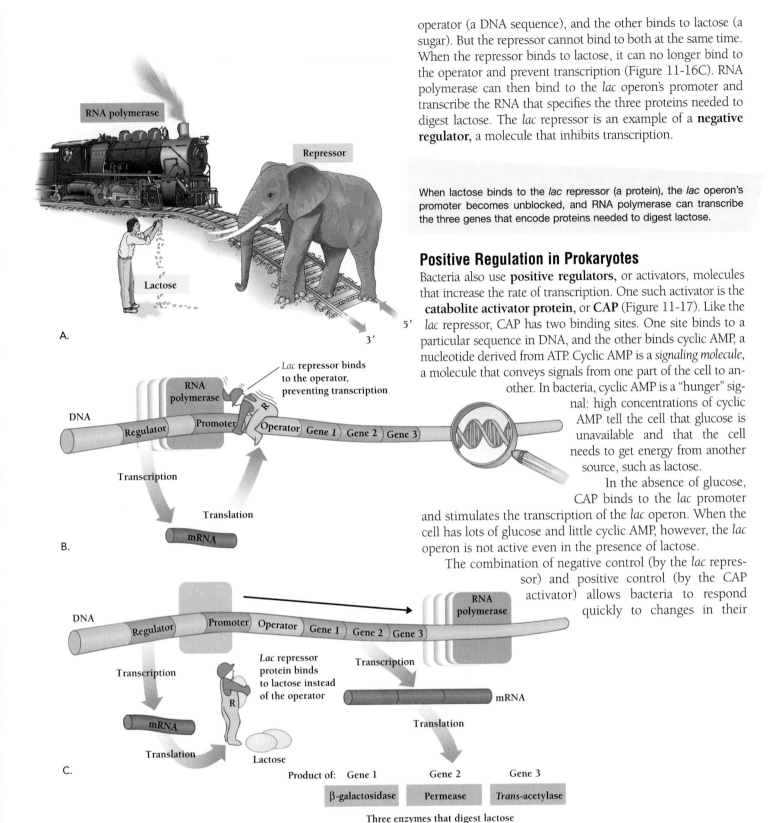

Figure 11-16

Negative feedback regulation in the *lac* operon. A. Peanuts entice an elephant off the tracks so that a locomotive can pass. In the same way, lactose binds to the repressor so that RNA polymerase can transcribe the DNA. B. The repressor sits on the operator and blocks RNA polymerase. C. When lactose is present, the repressor binds to lactose instead of the operator, and the RNA polymerase transcribes the mRNA for three enzymes that break down lactose.

environment. When lactose is present but glucose is not, *E. coli* quickly begin to make the enzyme β-galactosidase. Similarly, when glucose becomes available or when lactose is exhausted, *E. coli* almost immediately stops production of the enzyme. This system saves the bacteria from expending energy making polypeptides they do not need.

CAP activates transcription not only of the *lac* operon but also of several other genes. All of these genes code for enzymes needed to obtain energy from molecules other than glucose. CAP provides an example of how a single activator can affect the transcription of genes that are physically separated but functionally related.

The activator protein CAP increases the transcription rate for certain mRNAs.

How Do Eukaryotes Regulate Gene Expression?

Until the late 1980s, molecular biologists knew far more about regulation in prokaryotes than in eukaryotes. More recent work, however, has resulted in an explosion of information about gene regulation in eukaryotic cells. We now know that eukaryotes control gene expression at many more levels than prokaryotes. Nonetheless, as in prokaryotes, most regulation occurs at the level of transcription.

The major differences between regulation in prokaryotes and eukaryotes result from the presence, in eukaryotes, of a nuclear membrane and of a much larger amount of DNA. The presence of the nuclear membrane has several important consequences. One is that only proteins in the nucleus can contribute to regulation. Another consequence of the nuclear membrane is that the pre-mRNA can be modified before it moves into the cytoplasm. Posttranscriptional

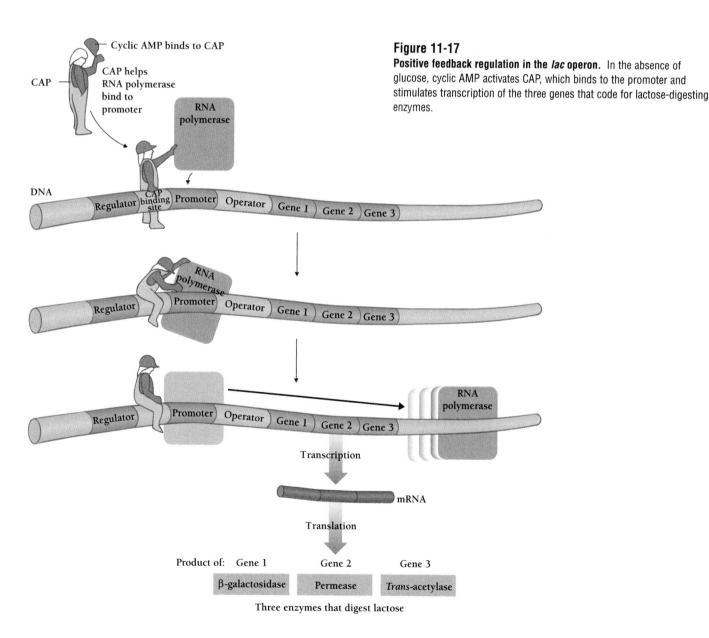

Figure 11-17
Positive feedback regulation in the *lac* operon. In the absence of glucose, cyclic AMP activates CAP, which binds to the promoter and stimulates transcription of the three genes that code for lactose-digesting enzymes.

regulation may occur in any of the steps involved in modifying the pre-mRNA.

Eukaryotes also have much more DNA than prokaryotes. Each human cell, for example, contains about 1,000 times as much DNA as is in an *E. coli* bacterium. Recall from Chapter 10 that to fit the DNA of a eukaryotic cell into its nucleus, the cell packages the DNA into tightly folded nucleosomes in which proteins called *histones* bind tightly to DNA. The nucleosomes are part of more elaborate chromatin structures, which limit transcription by restricting access to RNA polymerase. In general, the highly packaged DNA of eukaryotes is much less available for transcription than that of prokaryotes. As a result, eukaryotic transcription requires an elaborate set of activators. In eukaryotes, these activators turn on specific genes in specific cell types at specific times. In mammals, for example, genes that promote milk production in the cells of mammary glands turn on during pregnancy and after the mother gives birth, but not at other times of life and not in other parts of the body.

Eukaryotes also have many intracellular compartments and elaborate mechanisms for the distribution of newly made proteins. Sending the right proteins to the nucleus, for example, is especially important for building the chromatin and regulating gene expression. Varying the delivery of these regulatory proteins is another way by which cells regulate gene expression.

Eukaryotes can regulate gene expression in more ways than prokaryotes because of the presence of the nuclear membrane, because of the way the DNA is folded and packaged, and because of the many compartments in eukaryotic cells.

How Complex Is the Problem of Gene Regulation in Eukaryotes?

Just as in prokaryotes, the best-studied cases of transcriptional regulation in eukaryotes depend on the interaction of proteins with regulatory sites on the DNA. Unlike prokaryotic DNA, however, most eukaryotic DNA is largely "repressed" because of its tight association with histones. Most eukaryotic regulation therefore depends on specific activators of transcription.

Regulatory proteins may affect transcription in two ways. The first class of regulators change the rate of transcription, like the *lac* repressor and the CAP protein. The second class of regulators (sometimes called "chromatin-remodeling machines") produce permanent changes in chromatin that make a particular gene more or less accessible to the transcriptional machinery. Such a change in chromatin structure can persist long after the regulator itself has disappeared.

Many of the proteins that contribute to transcriptional regulation bind directly to specific sequences in DNA.

Among these sequences are (1) *promoters,* which specify the starting point and direction of transcription; (2) *response elements,* also called *enhancers* because they usually stimulate the transcription of neighboring DNA without themselves serving as promoters; (3) *silencers,* which help inhibit transcription; and (4) *termination signals.*

Some regulatory proteins bind not to DNA but to other proteins, forming large assemblies of proteins. Such a protein assembly can interact with activator proteins around a set of enhancer elements in a specific gene.

The regulation of the 40,000 or so genes in the human genome is an enormously complex problem. Biologists estimate that mammals have at least 5,000 transcriptional activators and a corresponding number of response elements (silencers and enhancers). Each transcriptional activator may interact with more than one response element, and each response element can respond to more than one activator. The number of possible interactions at the transcription level alone is huge.

Indeed, organisms with similar DNA can look and act quite differently because of changes in the expression of transcription factors. For example, human DNA is 99.8 percent identical to chimpanzee DNA, yet our behaviors and reasoning abilities are strikingly different. Some evidence suggests that much of the difference in our behavior can be accounted for by differences in the way certain transcription factors are expressed in the brains of humans and chimps.

In eukaryotes, the huge numbers of regulatory proteins and their corresponding binding sites in DNA allow nearly unlimited varieties of interactions. The problem of gene expression is enormously complex.

A Single Transcription Factor May Affect Many Genes

The hormone testosterone, which works by binding to a transcriptional activator, shows the diverse effects of a single regulator. In human and other mammalian embryos, testosterone is one of a group of hormones called androgens that stimulate the formation of the male genitalia (penis, scrotum, and the internal ducts that carry sperm). Later, at puberty, androgens stimulate the development of male secondary sexual characteristics—male distribution of fat, muscle, and body hair; deepening of the voice; and other patterns of growth and behavior.

We know that androgens accomplish all this by binding to an **androgen receptor.** This androgen receptor binds both to testosterone and to response elements in the DNA, thereby influencing the expression of genes related to the secondary sexual characteristics.

In a condition called **testicular feminization,** a chromosomal (XY) male produces testosterone but develops into a female. This situation results from a mutation in the gene that specifies the androgen receptor. Because every cell in the body has a nonfunctioning androgen receptor, no matter how much testosterone is in the body, the individual cells are incapable of detecting it and responding to it.

So, despite the presence of the Y chromosome and plenty of testosterone, a person with this mutation is insensitive to androgen (testosterone) and develops into a woman, with breasts and feminine external genitalia. However, she has testes and no ovaries, and either no uterus at all or only a rudimentary one. Testicular feminization shows how the actions of testosterone and its receptor increase the rate of transcription of a host of important genes.

> The absence of a single transcription factor—the androgen receptor, for example—can greatly alter gene expression and the resulting phenotype.

In this chapter, we have seen how the coded information in the gene flows from DNA to RNA to polypeptides and how cells regulate that flow. In the next chapter, we will see that genetic information (in the form of individual genes) is not confined to the DNA in the nucleus but can move about within the cell or from cell to cell. The consequences of this genetic mobility are enormous.

Key Concepts

- Protein synthesis in test tubes, using cell extracts, revealed how information flows from messenger RNA (mRNA) to polypeptide.
- DNA and RNA carry information in a 4-letter alphabet, while proteins carry information in a 20-letter alphabet.
- Translation of messenger RNA requires correct initiation, elongation, and termination of polypeptide synthesis.
- In both prokaryotes and eukaryotes, many proteins cooperate to accomplish and influence the expression of a single gene.

Summary with Key Terms

How do molecular biologists decode the genetic code?

DNA and RNA carry information in the sequences of nucleotides. Protein synthesis translates this information into a sequence of amino acids. The **genetic code** is the **universal** dictionary that specifies the relationship of the nucleotide sequence to the amino acid sequence. A **codon** is a group of three nucleotides that specifies a single amino acid. There are 64 codons. Of these, 61 specify the 20 amino acids and 3—**nonsense**, or **stop**, **codons**—signal the end of a polypeptide.

What is the central dogma?

According to Francis Crick's **central dogma**: "DNA specifies RNA, which specifies proteins." Cells **transcribe** information from DNA into RNA and **translate** information from **messenger RNA (mRNA)** into protein.

How does RNA polymerase transcribe DNA into RNA?

RNA polymerase copies information from one strand of DNA into RNA. The starting point for transcription is called a **promoter**. In eukaryotes, **transcription factors** lead the RNA polymerase to the right promoter. Eukaryotic and prokaryotic RNA polymerases stop transcription at **terminators**.

In prokaryotes, RNA polymerase produces mature RNAs directly. In eukaryotes, cells modify the initially transcribed pre-mRNAs into functional mRNAs. Such modifications include cutting out **introns** and splicing together the remaining **exons** to form a patchwork mRNA.

How do ribosomes and tRNAs translate mRNA into polypeptides?

Ribosomes, made of **ribosomal RNAs (rRNAs)** and proteins, are the molecular factories that assemble amino acids into polypeptides. Several ribosomes may associate with a single mRNA molecule to form a **polysome**. Before ribosomes can arrange the sequence of amino acids according to the information encoded in mRNA, however, the amino acids are first linked to adapter molecules called **transfer RNAs (tRNAs)**. Each tRNA contains an **anticodon**, a specific nucleotide sequence that forms base pairs with a codon in mRNA. The pairing of specific tRNAs with their appropriate amino acids is the task of enzymes that recognize one amino acid and the corresponding tRNA and link them together.

The translation of information from mRNA into the sequence of amino acids in a polypeptide occurs in three distinct steps, called initiation, elongation, and termination. **Initiation** positions the mRNA in the correct **reading frame**. Initiation always starts with AUG, which is both a signal for methionine and the **start site**. **Elongation** consists of three steps: (1) putting the next amino acid in place, (2) forming a peptide bond, and (3) moving the ribosome to read the next codon. **Termination** occurs when the ribosome encounters a stop codon to which a protein called a **release factor** has bound.

Most molecular biologists now say that a gene is a DNA sequence that is transcribed as a single unit and encodes either a single polypeptide or a set of related polypeptides.

How do cells regulate the expression of individual genes?

Cells regulate the expression of genes in many ways, including **posttranscriptional control**, **translational control**, and **posttranslational control**. But **transcriptional control** is the most efficient and the most common way.

Two types of transcriptional control in prokaryotes are positive regulation and negative regulation. The *lac* operon is an example of **negative regulation.** When lactose binds to the *lac* **repressor** (a protein), the *lac* operon's promoter becomes unblocked, and RNA polymerase can transcribe the three genes that encode enzymes for digesting lactose. We say lactose **induces** the expression of the *lac* operon. When lactose is absent, the *lac* repressor binds to a regulatory site called the **operator** and prevents transcription.

CAP is an example of an **activator,** or **positive regulator.** CAP increases the transcription rate for certain mRNAs, provided that other regulatory mechanisms will allow these mRNAs to be transcribed at all.

Eukaryotes can regulate gene expression in more ways than prokaryotes because of the presence of the nuclear membrane, because of the way the DNA is folded and packaged, and because of the many compartments in eukaryotic cells. In eukaryotes, the huge numbers of transcription factors and regulatory sites allow nearly unlimited varieties of interactions. The problem of gene expression is enormously complex. In **testicular feminization,** the absence of a single transcription factor—the **androgen receptor,** for example—can greatly alter gene expression and the resulting phenotype.

Review and Thought Questions

Review Questions

1. What is the central dogma?
2. What is the difference between the template, or antisense, strand of DNA and its mate, the sense strand? Which one is transcribed into RNA? What would happen if the sense strand were transcribed?
3. What kinds of proteins help RNA polymerase recognize the right promoter? What is a promoter?
4. How do tRNAs mediate between the mRNA and the amino acids? What does "t" stand for?
5. What is the difference between positive regulation and negative regulation of gene expression? Give an example of each.
6. Why do bacteria need to regulate the expression of their genes?
7. What is a regulatory site? Give an example of one.
8. Why do eukaryotes need to regulate the expression of their genes? Why do multicellular organisms need to regulate gene expression?

9. As described in this chapter, there are three kinds of point mutations: base substitutions, insertions, and deletions. Considering the redundancy of the genetic code and the notion of "reading frame," which of these mutations would usually have the least biological effect?

10. The expression of some genes can be regulated after their transcription, by means of an "antisense" RNA synthesized from a different gene. The "antisense" RNA binds to the "sense" mRNA, preventing its binding to the ribosome, hence inhibiting the synthesis of the protein. Can you imagine a way to use this mechanism to fight cancer?

11. Most of the cells in our body carry exactly the same genetic information, and differences between highly specialized cells, such as neurons, muscle cells, or liver cells, lie in the selective expression of different sets of genes. Yet at least three kinds of cells carry a different genetic "baggage." Which ones are they? Why?

BiologyNow Resources

Biology ⓔ Now™

Active Figures

11-6: Transcription
11-8: The role of RNA polymerase in transcription

Preparing for an exam? Take a diagnostic test on your BiologyNow CD-ROM.

Online materials relating to this chapter are at:

http://biology.brookscole.com/AAL3

About the Chapter-Opening Image

The sewing machine is a metaphor for the ribosome, a tiny molecular machine that stitches amino acids together into polypeptides.

12

Viruses, Jumping Genes and Other Unusual Genes

Key Questions

- How are viruses, plasmids, and transposons alike?

- What is a virus?

- How do viruses and other mobile genes replicate?

- How can unconventional genetic systems evolve?

Jumping Genes and Indian Corn

In the 1940s, a geneticist named Barbara McClintock made a discovery that would eventually revolutionize molecular biology, a science that did not even exist yet (Figure 12-1). Like Gregor Mendel, McClintock was so far ahead of her time that her research made no impact on the world of biology for 30 years.

McClintock began her career in 1918 at Cornell University, where she was first an undergraduate student, then a graduate student, and finally a researcher. In the 1920s and 1930s, geneticists there noticed that the normal purple color of corn kernels switched to pale yellow or white as a result of a mutation in a pigment gene. Multicolored "Indian" corn is normal corn: the pale corn we buy at the store has been bred to prevent the expression of these purple pigment genes. Oddly, however, the mutation that turns purple kernels yellow or white can reverse itself, or partly reverse itself, from generation to generation.

Each kernel of corn on a cob contains a separate, unique individual—the result of a separate pairing of sperm and egg, a separate pairing of maternal and paternal chromosomes. As a result, some kernels on a cob might have the mutation and some might have it reversed: some kernels are white and some are purple. Still others may be streaked or spotted—the result of partial expression of the pigment gene. When McClintock was at Cornell, no one understood how these "unstable mutations" could flip back and forth so easily.

In 1941, McClintock moved her own research on corn to Cold Spring Harbor Laboratory on Long Island, New York. She was especially interested in broken chromosomes. She had noticed a corn plant in which all of the chromosomes always broke in the same place—an oddity, since chromosomes usually break randomly. McClintock called the place where the chromosome always broke the *Ds* gene. *Ds* was short for "dissociator," the gene that took apart (dissociated) the chromosome. She soon discovered that the *Ds* gene could move

Gene Kelly: The Kobal Collection

Runk/Schoenberger from Grant Heilman

around on the chromosome, or "transpose," from generation to generation. Then came an even more startling revelation. McClintock noticed that the same corn plants in which *Ds* had moved also acquired the unstable pigment mutation.

That connection suggested to McClintock that the unstable pigment mutation might be caused by the movement of the *Ds* gene into the middle of the pigment gene. Taking another step, she saw that if the *Ds* gene jumped back out of the pigment gene, the pigment gene would function again, coloring the kernel purple. This would explain how the kernels seem to gain and lose the mutation, switching back and forth between a normal phenotype and an abnormal one, from generation to generation. McClintock's hypothesis not only explained how unstable mutations worked, but introduced a completely new idea to the science of genetics: the idea that genes could move. She coined the word **transposon** to describe these genes that jumped around the genome. Others would come to call them "jumping genes."

In time, McClintock found more complex examples of transposons. She also realized that just as the *Ds* gene could influence the expression of the pigment gene, regulatory genes could regulate the expression of structural genes. McClintock was the first person to recognize that one gene could influence the expression of another. McClintock was also the first person to discover promoter and suppressor genes. But, unfortunately, she discovered them in corn, an organism far more complex than the bacteria in which Jacob and Monod would later discover the *lac* operon. The system of regulation that McClintock discovered was far more difficult—both to understand and to explain—than the *lac* operon, and most geneticists did not understand her discoveries.

Despite other geneticists' failure to appreciate her work, McClintock was a highly respected geneticist. In 1931, she and Harriet Creighton had demonstrated that chromosomes cross over, exchanging genes during meiosis (Chapter 9). The two women had provided conclusive evidence for the theory of chromosomal inheritance. Yet, in 1951, when McClintock presented her work on transposons to a meeting of geneticists specifically interested in mutations, her talk "met with stony silence," as McClintock's biographer Evelyn Fox Keller wrote. "With one or two exceptions, no one understood. Afterward, there was mumbling—even some snickering—and outright complaints. . . . It was impossible to understand." McClintock knew her work was complex, but she had not anticipated such rejection. "It was such a surprise that I couldn't commu-

Figure 12-1
Barbara McClintock. McClintock in the early 1920s, at her parents' home in Brooklyn, New York.

Cold Spring Harbor Laboratory Archives. Permission of Marjorie M. Bhavnani

nicate," she told Keller. "It was a surprise that I was being ridiculed, or being told I was really mad."

The idea that genes can move from one part of the genome to another is not, in fact, extraordinarily difficult to understand. But it was extraordinarily difficult for geneticists to accept. McClintock's colleagues just assumed that—except for crossing over and other changes that occur during meiosis—the genome is a rigid, stable structure. McClintock might as well have told them that the streets they drove every day moved from place to place. She was viewed as something of a mad genius. Her explanations were extraordinarily complicated, and her writing lacked the spare elegance of Jacob and Monod's discussions of the *lac* operon. For 30 years, most geneticists did not even try to understand McClintock's work.

Not until the early 1970s, when bacterial geneticists began to discover the same things that McClintock had discovered in corn in the 1940s and 1950s, did biologists begin to grasp the importance of her work. They realized, finally, that McClintock's transposons, or jumping genes, are just one example of a whole class of **mobile genes**—genes that move around. Finally, biologists began to see how broadly McClintock's ideas applied to all sorts of mobile genes, not just jumping genes and not just in corn, but to mobile genes in all living organisms. In the 1980s, younger molecular biologists began to use mobile genes as tools for the study of how genomes work. And, as we will see in the next chapter, mobile genes now play a pivotal role in many genetic engineering techniques. In 1983, when McClintock was 81, her great contributions to biology were finally recognized with a Nobel prize.

Since the beginning of the 20th century, geneticists have carefully studied both the information in a gene and the gene's place in the chromosome. While the information in a gene is important, it is a gene's location that determines how the gene will be inherited. We have already seen, for example, that genes that are far apart on the chromosome are more likely to recombine than genes that lie next to one another. Barbara McClintock's work showed that individual genes on the chromosome can actually move from place to place, with profound consequences.

In this chapter we will examine an odd collection of **unconventional genetic systems.** Unconventional genes include both mobile genes and any genes not in the regular genome of a cell—that is, genes not in either the nucleus (in eukaryotes) or the nucleoid (in prokaryotes). The latter include the genomes of mitochondria and chloroplasts, which pass on their DNA, organelle to organelle, separately from the DNA in the nucleus.

The replication and expression of all genes, unconventional or conventional, occurs only in cells. The simplest mobile genes are jumping genes, or transposons, which insert themselves into the host cell's DNA and replicate as part of a host cell's DNA. In contrast, **plasmids** can exist as simple loops of DNA separate from the host cell's DNA—usually in prokaryote or yeast cells—that pass from cell to cell. Sometimes plasmids integrate into the host cell's own DNA. Another group of mobile gene are the viruses, familiar to us as the cause of colds, flu, and other diseases. Viruses usually force host cells to make lots of viruses, but many viruses—including the one that causes AIDS—insert genes directly into the genomes of their host cells.

The study of unconventional genes has fundamentally changed biology. It has given biologists new insights into diseases ranging from the common cold to cancer and AIDS, helped explain how antibiotic resistance spreads among bacteria, suggested new views about the origin of eukaryotic cells, and fueled the extraordinary growth of the modern biotechnology industry.

12.1 What Is a Virus?

A **virus** is an assemblage of nucleic acid (DNA or RNA) and proteins. Occasionally, viruses also include other components, such as lipids or carbohydrates. Viruses have many properties of life: they are highly organized, they reproduce, they respond to their environments, and they evolve. But viruses can neither obtain energy from their surroundings nor form the building blocks needed for the synthesis of their components. Because a virus cannot perform such metabolic reactions, it can only replicate by hijacking the biochemical machinery of a host cell. For this reason, biologists say that a virus is not alive.

Most viruses infect bacteria, plants, and animals without causing disease, and some even introduce useful genes into the genomes of other organisms. In fact, in the next chapter, we will see how molecular biologists have used this attribute of viruses to create new strains of bacteria, plants, and animals.

But most of the first studies of viruses involved diseases. In the 19th century, the word "virus" [Latin = slimy liquid or poison] came to mean any material associated with death and disease. By the late 19th century, Louis Pasteur and others had shown that bacteria and other small cells cause many infectious diseases and also spoil food. Pasteur's colleague Charles Chamberland developed a way of removing even the smallest cells from liquids by forcing them through a fine filter (Figure 12-2A). But biologists soon discovered that some infectious agents managed to get through the bacterial filters. These mysterious entities were called "filterable viruses." One by one, the bacteria specifically responsible for many diseases were identified and named. The only infectious agents left unnamed were the mysterious viruses that could not be filtered out and studied.

In the 1890s, just a few years after Chamberland's filter became available, researchers demonstrated that a virus caused a disease of tobacco leaves (Figure 12-2B). Because the diseased leaves have a patchy, or mosaic, appearance, the disease was called *tobacco mosaic disease,* and the virus that causes it is now called *tobacco mosaic virus.* Other microbiologists soon showed that viruses also caused hoof-and-mouth disease in cattle and yellow fever in humans. In 1915, the English bacteriologist Frederick Twort discovered that viruses can even infect bacteria. Twort called his bacteria-infecting virus a **bacteriophage** [Greek, *phagein* = to eat]. Discoveries such as these demonstrated that disease-causing agents could be smaller than the tiny (0.5-μm) pores in a Chamberland bacterial filter. But no one knew what viruses actually were. Most people thought that they were just cells too small to be stopped by a filter or seen through a microscope.

In 1935, however, the biochemist Wendel Stanley stunned other researchers by making crystals of tobacco mosaic virus (Figure 12-3). A crystal forms only when identical molecules arrange

A. Porcelain filter removes bacteria from sour milk.

B. Tobacco mosaic virus passes through filter.

Figure 12-2
Viruses are much smaller than bacteria. A. Fine filters remove the bacteria that cause milk to sour.
B. Fine filters do not remove the virus that causes tobacco mosaic disease.

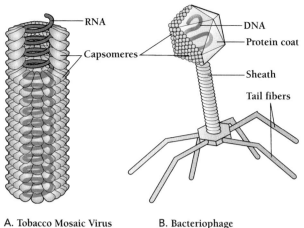

A. Tobacco Mosaic Virus   B. Bacteriophage

Figure 12-3
Tobacco mosaic virus. Proteins form a capsule *(yellow)* around the virus's DNA *(blue)*.

themselves in a regular lattice (as in the case of the familiar salt and sugar crystals on the breakfast table). How could a living cell form the precisely ordered arrays of a crystal? No one could imagine a crystal of bacteria or of mice.

Stanley's discovery seemed to imply that a virus was just a large molecule and that viral diseases were similar to chemical poisoning. But a poisonous substance would be diluted and broken down if it were passed from individual to individual, while tobacco mosaic virus could be passed indefinitely from plant to plant. Clearly, tobacco mosaic virus was replicating itself—something ordinary molecules do not do.

We now understand that viruses are relatively large, regular molecular complexes, some of which can form crystals. By analyzing such crystals with x-ray diffraction, researchers have been able to determine the structure of some viruses in great detail. Most viruses consist of a nucleic acid

Health and Biology Can Viruses Make You Well?

Imagine one day going to the doctor with a case of strep throat and getting a spoonful of virus to cure it. As more and more bacterial infections become resistant to antibiotic treatments, that day may not be so far off. In some parts of the world, bacteriophages, the viruses that infect bacteria, are already in use to combat disease, and a handful of companies in the United States are actively pursuing "phage therapy."

Many bacteria that cause infections that were once easily cured have become resistant to all but a few antibiotics. The list includes some kinds of pneumonia, ear infections, strep throat, urinary tract infections, cholera, and tuberculosis. The heavy use of antibiotics in our food supply has created hundreds of antibiotic resistant strains of bacteria. And the now-standard use of antibiotics during and after surgery has made hospital recovery rooms into breeding grounds for antibiotic-resistant strains of bacteria. In recent years, some hospitals have reported outbreaks of hospital-acquired pneumonia that are resistant to every known antibiotic.

Bacteriophages are natural parasites of bacteria. Once a phage invades a bacterial cell, it reproduces rapidly, and within an hour the cell bursts open to release millions of phage progeny, which then hunt down more bacteria. And phages are some of the most

abundant life forms on Earth. A single drop of seawater contains millions of phages, and soils and our bodies are thick with thousands of different strains of phages.

Phages have several advantages over antibiotics. Besides the fact that so many bacteria quickly evolve resistance to antibiotics, antibiotics are also toxic and cause side effects such as intestinal disorders and hearing problems. Some people are violently allergic to antibiotics. Phages produce no side effects and tend not to be allergenic. And, since they multiply themselves in the process of slaughtering the bacteria, forgetting to take a dose or stopping treatment too early is never a problem.

In addition, bacteria are less able to develop resistance to phages than to antibiotics. If the bacteria do develop resistance, researchers can develop a new strain of phage that will overcome the bacterium's resistance in a matter of days or weeks. For comparison, it takes drug companies years or decades to come up with new antibiotics. One U.S. company has developed a phage that kills a strain of bacterium that is resistant to even the most powerful antibiotics. And a Canadian company is offering phages that can be used on raw vegetables to kill the bacteria that cause food poisoning.

But phages are no panacea, so antibiotics will continue to be useful. Each

phage strain is so specific to a particular type of bacterium that doctors have to know exactly which bacterium is causing an infection. That could mean a delay of several days while the bacterium is identified. Also, some harmful bacteria have no known phage predators.

In the 1930s, researchers in many parts of the world attempted to treat infections with phages, and many hoped that phages would be the ultimate treatment for bacterial infections. In Sinclair Lewis's popular novel *Arrowsmith,* written in 1924, the hero tried to combat a devastating epidemic with a bacteriophage treatment. In the West, the method was virtually abandoned with the advent of antibiotics in the 1940s.

But in parts of the former Soviet Union, doctors and researchers kept the technique alive—for example, using phages to treat burn infections and food poisoning. Some hospitals used phages to disinfect operating rooms and surgical instruments, as well as to reduce the spread of antibiotic-resistant pneumonia and strep throat.

In the last decade, phage therapy has reemerged in the West. As of 2002, the U.S. government's Food and Drug Administration was still uncertain how to regulate this new therapy, but the potential uses for phage therapy are enormous.

core (either DNA or RNA) that carries the virus's genes surrounded by a protein coat.

> A virus is a small complex of nucleic acid and protein capable of infecting a cell.

What Is the Form of a Virus?

Only with the development of the electron microscope in the 1940s did scientists see what viruses looked like. Viruses are large particles, much smaller than cells but larger than single protein molecules. The biggest viruses are the pox viruses such as smallpox virus, which range from 250 to 400 nm in diameter. The smallest ones, including, for example, poliovirus, are only 20 to 30 nm in diameter.

Viruses come in many shapes, but most have either of just two forms: a long rodlike helix, such as that of tobacco mosaic virus (Figure 12-4A); or a compact, nearly spherical particle, such as that of poliovirus (Figure 12-4B). Poliovirus is not actually spherical. It is a 20-sided object, called an **icosahedron** [Greek, *eikosi* = twenty + *hedra* = base], a little like a soccer ball. Other viruses have more complex forms. The bacteriophage T4 is among the most complex of viruses, with an icosahedral head and an intricate tail structure (Figure 12-4C). T4 looks and works like a tiny medical syringe, injecting its DNA into a bacterial host.

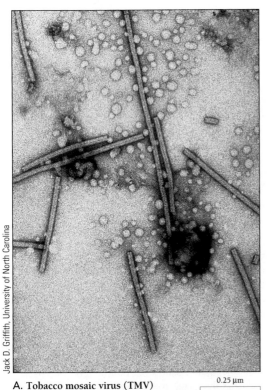

Jack D. Griffith, University of North Carolina

A. Tobacco mosaic virus (TMV) 0.25 μm

Andrew O. Jackson, University of California, Berkeley

B. Tomato bushy stunt virus 50 nm

Dr. Thomas Broker/Phototake NYC

C. T4 bacteriophage 0.1 μm

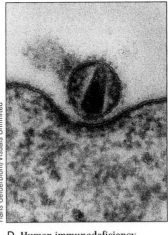

Hans Gelderblom/Visuals Unlimited

D. Human immunodeficiency virus (HIV) 0.1 μm

Figure 12-4
Viruses come in many sizes and shapes. A. Tobacco mosaic virus. B. Like the poliovirus, tomato bushy stunt virus is a 20-sided icosahedron. C. T4 bacteriophage. D. HIV, which causes AIDS.

Biology Now™ Learn more about the HIV virus by clicking on this figure on your BiologyNow CD-ROM.

A virus's genes may be either DNA or RNA. (In contrast, a cell's genes are always DNA.) Just as in humans and other living organisms, viral genes contain the information for most, if not all, of the virus's proteins. Some of a virus's genes direct the synthesis of proteins and nucleic acids that make up the virus, while other genes specify proteins that regulate the host cell's DNA. Such regulatory genes force the host cell to make new virus particles.

Surrounding the viral genetic material is a **protein coat.** A few viruses also direct the host cell to make a viral **membrane envelope** out of the host cell's own plasma membrane. Enveloping itself in the host's own membrane is a trick that helps an infecting virus conceal its identity. Many viruses that infect animals, such as influenza virus and human immunodeficiency virus (HIV, the AIDS virus) use this trick.

Viruses have regular shapes, such as an icosahedron or a helix. They consist of a core of genetic material—DNA or RNA—enclosed in a protein coat. Sometimes the viral protein coat is further by enclosed a membrane envelope.

How Do Viruses Make More Viruses?

Although viruses can make more viruses in many different ways, their life histories always include the following stages: ❶ the virus attaches to the host cell; ❷ the viral nucleic acid enters the cell; ❸ the cell synthesizes proteins specified by the virus's genes; ❹ the cell replicates the virus's DNA or RNA; ❺ the new viral protein and DNA or RNA assembles into new viruses; and ❻ the new viruses are released from the cell (Figure 12-5).

Each stage in a virus's multiplication cycle depends on the interaction between the host cell and the virus. A given kind of virus can infect only certain kinds of cells. We say that viruses are adapted to specific hosts. Most cold viruses and the HIV virus, for example, are specific to humans and sometimes other primates. But our cats and dogs don't catch colds from us and we don't usually catch their illnesses either. In this section, we will briefly discuss four of the six stages of viral infection. After that, we will discuss the replication of the viral DNA or RNA in greater detail.

How Does a Virus Attach to a Cell?

The attachment of the virus to a cell depends on an interaction between a viral protein and receptor molecules on the host cell's membrane (Figure 12-6). Specific proteins on the surface of the virus bind to molecules on the surface of the host cell. The attachment site may be on all of the host's cells or just on a special set of cells. Human cold and

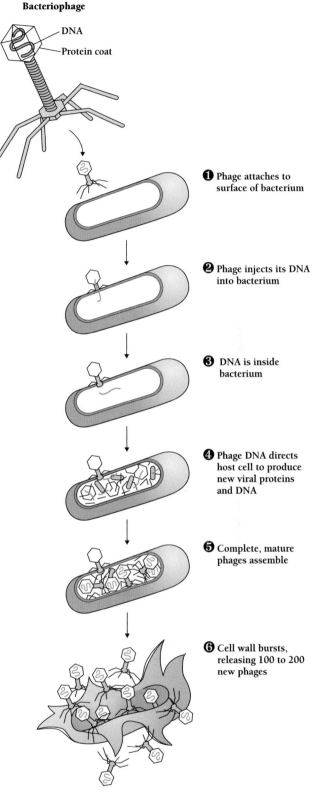

Bacteriophage
DNA
Protein coat

❶ Phage attaches to surface of bacterium

❷ Phage injects its DNA into bacterium

❸ DNA is inside bacterium

❹ Phage DNA directs host cell to produce new viral proteins and DNA

❺ Complete, mature phages assemble

❻ Cell wall bursts, releasing 100 to 200 new phages

Figure 12-5
The life of a bacteriophage. A phage attaches to the surface of its bacterial host and injects its DNA inside. The host cell transcribes and translates the phage genes to make more phages, which eventually burst from and kill the host cell.

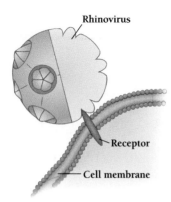

Rhinovirus

Receptor

Cell membrane

Figure 12-6
Catching on. Viruses attach to the surfaces of cells by latching onto cell surface receptors.

flu viruses, for example, specifically infect cells of the respiratory tract. Others—such as chickenpox and other types of herpes viruses—infect skin and nerve cells. Poliovirus initially attacks cells in the digestive tract, then moves into the bloodstream, where it can gain access to all the cells in the body. In one percent of polio cases, poliovirus attaches to and destroys nerve cells, causing paralysis or even death. In comparison, HIV primarily infects cells of the immune system and the brain. This specificity of attachment is what makes viruses able to infect one species of organism but not another, why, in short, we so rarely catch diseases from our pets.

Viruses attach to specific protein receptors on the surfaces of cells, which is the basis for the species and tissue specificity of viral infection.

How Does the Viral DNA or RNA Enter the Host Cell?

The mechanism by which the viral nucleic acid enters the host cell depends on the kind of virus and the kind of host cell. The bacteriophage T4, for example, contains an enzyme that digests away part of a bacterial cell wall, allowing the phage to inject its DNA into its host. Other viruses, such as the flu virus, are taken into cells by endocytosis, the process by which the cell's membrane enfolds particles. Still other viruses enclose themselves in bits of the cell's own membrane, which the virus particles capture as they leave an infected cell. The virus's stolen membrane envelope then fuses with another host cell's membrane and the virus easily enters (Figure 12-7).

Viruses enter cells by cutting a hole in the cell membrane, by inducing the cell to engulf the virus particle, or by fusing with the cell's membrane.

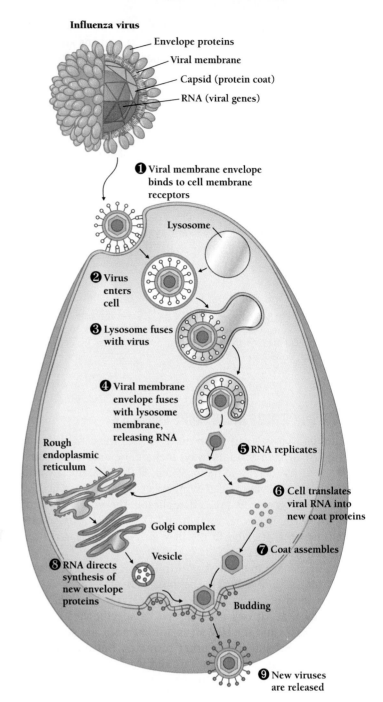

Influenza virus

Envelope proteins
Viral membrane
Capsid (protein coat)
RNA (viral genes)

❶ Viral membrane envelope binds to cell membrane receptors

Lysosome

❷ Virus enters cell

❸ Lysosome fuses with virus

❹ Viral membrane envelope fuses with lysosome membrane, releasing RNA

❺ RNA replicates

Rough endoplasmic reticulum

❻ Cell translates viral RNA into new coat proteins

Golgi complex

Vesicle

❼ Coat assembles

❽ RNA directs synthesis of new envelope proteins

Budding

❾ New viruses are released

Figure 12-7
Like the Trojan horse. An influenza virus tricks a cell into taking up the virus. The virus attaches to cell surface receptors, which initiate endocytosis of the virus.

How Are Virus Particles Manufactured and Released?

Once inside the host cell, the virus's genes direct the synthesis of viral proteins in a variety of ways. Some viruses direct the host cell's protein-synthesizing machinery to make an enzyme that destroys the host's own DNA. Other viruses rely on powerful promoters and other regulatory genes to ensure rapid replication, transcription, and translation. In most cases, viral protein completely dominates protein synthesis.

For a bacteriophage, the whole cycle, from attachment and entry to assembly of new virus particles, may take only 20 minutes and may produce 100 or more new bacteriophages (Figure 12-5). After these new viruses are produced, a viral enzyme causes the host cell membrane to break open in a process called **lysis** [Greek, *lysis* = loosening]. The ruptured cell then releases the newly made viruses, which are ready to infect more host cells. Viruses that destroy their host cells in this way are called **lytic viruses.**

Not all viruses cause cell lysis, however. Killing the host—whether the host is a single cell or a large multicellular organism—is not the only strategy by which a virus can successfully replicate. After all, a host that a virus kills may be the last one it encounters for a while, so killing the host could spell the end of the line for a virus. Some viruses have a multiplication cycle similar to that of a lytic virus, but they exit the cell without destroying the host. Other viruses stay inside the cell for long periods, as we will see.

Viruses force the host cell to make viral proteins—sometimes by destroying the host's DNA, sometimes through subtler forms of gene regulation. Many viruses can escape from their hosts by breaking open the cell membrane.

12.2 How Do Mobile Genes Replicate Their Genes?

"A hen is only an egg's way of making another egg," wrote the English novelist and essayist Samuel Butler. Just so, a virus is only a gene's way of making another gene. This statement might sound extreme, but the idea behind it helps us to understand the strategies that viruses and other mobile genes use to replicate.

A virus is the most elaborate of the mobile genes, but there are simpler mobile genes that consist only of DNA or RNA. Like viruses, however, these stripped-down mobile genes induce the host cell to replicate them.

Mobile genes—including viruses, plasmids, and transposons—have two reproductive strategies. In **autonomous replication,** the genes are separate from the DNA of the host cell and can replicate even when the host DNA is not replicating. In **integrated replication,** the genes become integrated into the host's DNA and are replicated along with

those of the host. Integrated genes can replicate *only* when the host DNA replicates. Some mobile genes can switch back and forth from autonomous to integrated replication depending on circumstances.

In the next few sections, we will discuss plasmids, which are mobile genes that usually replicate autonomously only; transposons, which are mobile genes that replicate only when integrated into the host's DNA or into a plasmid; and **episomes,** which are viruses and plasmids that switch back and forth between integrated and autonomous replication.

In autonomous replication, the genes of a virus, plasmid, or transposon are separate from the DNA of the host cell. In integrated replication, the genes become integrated into the host's DNA. Episomes are viruses or plasmids that can switch back and forth between autonomous and integrated replication.

How Do Plasmids Replicate Autonomously?

A plasmid is a small piece of circular DNA or RNA in a bacterial or yeast cell that includes from 3 to 300 genes. Most plasmids do not integrate themselves into the host cell's DNA but exist as independent entities inside cells. Usually, plasmids replicate only when the DNA of their hosts does. However, many plasmids can replicate even when host DNA synthesis has stopped. Such plasmids are said to replicate autonomously. A plasmid can do this because it contains an origin of replication, the DNA sequence where DNA replication begins. As a result of this autonomous replication, a cell may contain up to 1,000 copies of each plasmid. And a cell may carry many different kinds of plasmids.

Plasmids, unlike viruses, do not produce protein coats that would allow them to move easily from one cell to another. However, plasmids have another way to get around. Their bacterial hosts sometimes engage in a sexlike process called **conjugation,** during which two bacteria temporarily attach to one another and exchange genetic material. During conjugation, some plasmids can move from one bacterium to another (Figure 12-8).

Plasmids often contain genes that bring their hosts some advantage. For example, some plasmids—called **resistance factors,** or **R factors**—contain genes for enzymes that make the cell resistant to antibiotics. Other plasmids contain genes for resistance to heavy metal poisoning, for the metabolism of unusual compounds (such as those found in oil spills), or for the production of toxins that kill other bacteria. In addition, molecular biologists have synthesized thousands of **recombinant plasmids,** plasmids that contain combinations of genes never found in nature. We will dis-

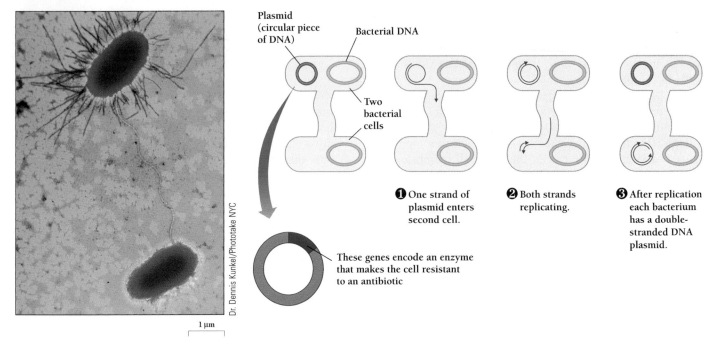

❶ One strand of plasmid enters second cell.

❷ Both strands replicating.

❸ After replication each bacterium has a double-stranded DNA plasmid.

These genes encode an enzyme that makes the cell resistant to an antibiotic

Dr. Dennis Kunkel/Phototake NYC

1 μm

Figure 12-8

Sex in bacteria. During a brief encounter called conjugation, bacteria can exchange genes carried either on the bacterial DNA or on plasmids. If the bacteria transfer plasmids that carry genes for antibiotic resistance, the genes for antibiotic resistance can easily spread among different strains and even different species of bacteria.

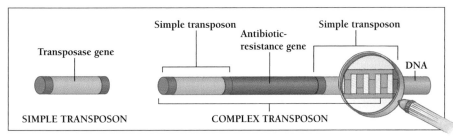

SIMPLE TRANSPOSON

COMPLEX TRANSPOSON

Figure 12-9

Transposons. A simple transposon carries the gene for transposase and little else. A complex transposon is two simple transposons with some other genes carried between them.

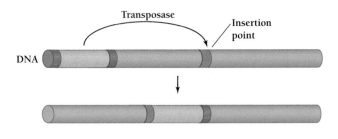

cuss the production and uses of such recombinant DNA in Chapter 13.

A plasmid is a circular piece of DNA or RNA capable of autonomous replication that carries genes, such as those for antibiotic resistance, from one bacterial cell to another.

Many Mobile Genes Can Replicate Only When Integrated into a Host Cell's DNA

Unlike a plasmid, a transposon is a mobile DNA sequence that can replicate only when it is part of the host cell's DNA. A transposon, or jumping gene, is a DNA sequence that moves, either by moving somewhere else on the chromosome or by copying itself and repeatedly inserting itself elsewhere. A transposon always includes a gene for an enzyme called **transposase**, which inserts the transposon into new places (Figure 12-9). While plasmids are found mostly in bacteria and in some yeast cells, transposons are common in all cells, including Barbara McClintock's corn.

Transposons in eukaryotes usually reproduce while integrated into the host DNA, but they are surprisingly mobile. In eukaryotic organisms from corn to humans, transposons can jump from place to place within the genome, duplicating themselves, disrupting other genes, deleting and repeatedly duplicating whole stretches of DNA, breaking chromosomes, and sometimes even attaching part of one chromosome to another. The mapping of eukaryotic genomes has revealed that virtually all genomes are littered with DNA sequences that appear to be relics of ancient transpositions of jumping genes.

A transposon may be copied many times and each copy separately moved from place to place in the host's genome. For example, the genome of the bacterium *E. coli* includes at least six different transposons, several of which have been copied 5 to 10 times.

A **simple transposon** specifies transposase and nothing else. As far as biologists know, a simple transposon's only function is to make more simple transposons. But when two simple transposons lie near each other on the host genome, they can jump together, carrying with them any genes that lie between them. Such combinations are called **complex transposons**. Some complex transposons carry genes for antibiotic resistance (Figure 12-9). Because complex transposons can move genes into plasmids, the transposons can help plasmids accumulate genes for resistance to several antibiotics (such as tetracycline and penicillin). The plasmids can then carry this resistance from one bacterium to another, spreading antibiotic resistance even to different species of bacteria.

As we will see in Chapter 20, transposons' ability to move genes ultimately explains how antibiotic resistance spreads so rapidly among bacteria. For example, some strains of the bacterium that causes tuberculosis (TB) are resistant to every known antibiotic. Patients with multiple-antibiotic-resistant TB are kept isolated to prevent them from passing on this dangerous form of TB, but there is no specific medical treatment that will help them get well.

A simple transposon codes for nothing but transposase, which is all the transposon needs to insert itself into a new site in the host's DNA. Complex transposons carry other genes as well.

Tumor Viruses Also Insert Themselves into Chromosomal DNA

We saw earlier that many viruses replicate autonomously, lyse the cell, and escape to infect more cells. But viruses that cause tumors are a special case. Rather than multiplying as rapidly as possible and killing their host cells in a lytic cycle, tumor viruses multiply only when their eukaryotic hosts do (Figure 12-10).

Once a tumor virus infects a cell, its genes usually become integrated into the DNA of the host cell. The virus makes the cell express the viral genes, and the resulting viral proteins force the cell to divide rapidly, which causes a tumor to form. With each round of cell division, the tumor grows

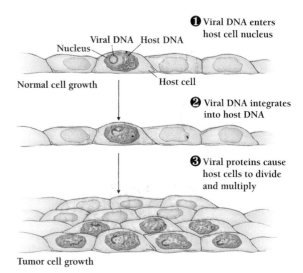

❶ Viral DNA enters host cell nucleus

Viral DNA   Host DNA

Nucleus

Normal cell growth   Host cell

❷ Viral DNA integrates into host DNA

❸ Viral proteins cause host cells to divide and multiply

Tumor cell growth

Figure 12-10
The life of a tumor virus. How do DNA tumor viruses cause tumors? DNA tumor viruses integrate their DNA into the DNA of eukaryotic cells, then disrupt the cell cycle, forcing infected cells into excessive cell division, resulting in the growth of the tumor.

larger and a new generation of viral genes is created. The viral genes whose proteins interfere with the host cell's ability to control cell division are called **oncogenes** [Greek, *onkos* = mass, swelling], a word that means "genes that cause tumors."

Tumor viruses integrate their DNA into that of a eukaryotic host, then force the host to replicate both viral and genomic DNA more often than normal, resulting in too many cells (a tumor) and many copies of the viral DNA.

Episomes Can Replicate Both Autonomously and Through Integration

Episomes are viruses or plasmids that switch between autonomous replication and integrated replication. Episomes can reproduce either autonomously or else as part of the host's chromosome. Episomes were first found in bacteria, but, as we will see, they also exist in eukaryotic cells.

The bacteriophage lambda (λ), for example, is an episome with two alternative reproductive cycles in its *E. coli* hosts. In the **lytic cycle**, λ's DNA is separate from that of its host, and λ lyses its host. But sometimes λ integrates its DNA into *E. coli*'s DNA. The integrated form is called a **prophage** [Greek, *pro* = before]. The prophage and the lytic cycle together make up the viral **lysogenic cycle** (Figure 12-11).

When λ is in the prophage stage, the host cell does not make phage proteins because λ's DNA directs the synthesis of a powerful repressor that prevents the transcription of any other λ genes. Ultraviolet light and other environmental insults that threaten to destroy the host cell shut down the repressor, and the phage then enters a lytic cycle, destroys the host cell, and new λ bacteriophages escape the dying cell.

A virus that is not actively lysing host cells cannot infect other cells. However, the virus's genes are copied along with the cell's own DNA each time the cell divides. As a result, all of the daughter cells of an infected cell are also infected with the virus, and an infected eukaryotic organism cannot rid itself of the virus.

In humans, herpes viruses are well known for remaining in this "latent" form for long periods of time. But if the host cells are exposed to ultraviolet light or certain DNA-damaging chemicals, the viral DNA then escapes and begins

a lytic cycle (Figure 12-11). This may explain why sunlight can activate herpes viruses in the cells of the mouth and lips. When the virus becomes active, it enters a lytic cycle, lysing cells and causing a painful cold sore.

Episomes are viruses or plasmids that switch back and forth between integrated and autonomous replication as the occasion demands.

How Do RNA Viruses Reproduce and Express Their Genes?

All plasmids and transposons are made of DNA, but many viruses use RNA genes, including those that cause polio, influenza, AIDS, and many animal tumors. RNA viruses replicate by using special enzymes encoded in the virus's genome. Some viruses carry their genes on a strand of RNA that works just like mRNA (messenger RNA) and actually codes for various viral proteins. Such RNA is translated directly into polypeptides by the host cell as soon as the virus enters the cell. Such viruses are called "plus-strand" RNA viruses (Figure 12-12A). Other RNA viruses, called minus-strand viruses, use an "antisense" RNA strand. Such viruses contain an enzyme that copies the viral RNA into a comple-

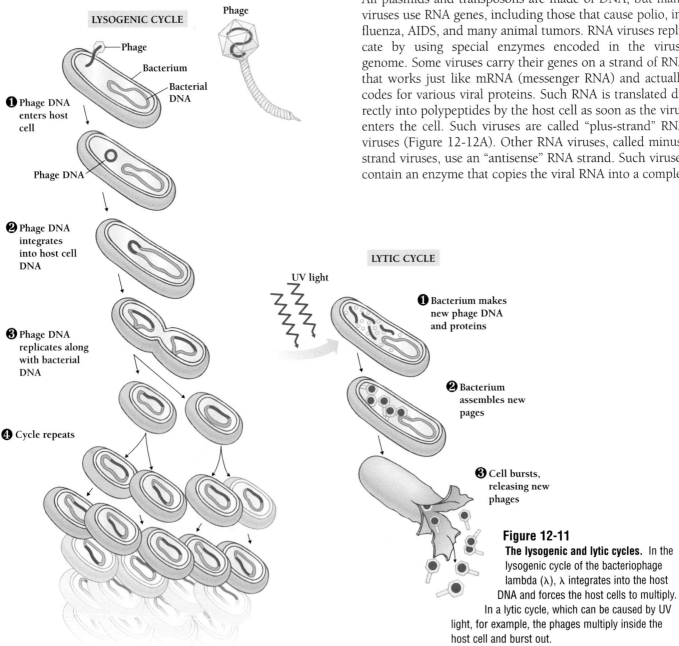

LYSOGENIC CYCLE

Phage

Phage

Bacterium

Bacterial DNA

❶ Phage DNA enters host cell

Phage DNA

❷ Phage DNA integrates into host cell DNA

❸ Phage DNA replicates along with bacterial DNA

❹ Cycle repeats

UV light

LYTIC CYCLE

❶ Bacterium makes new phage DNA and proteins

❷ Bacterium assembles new pages

❸ Cell bursts, releasing new phages

Figure 12-11

The lysogenic and lytic cycles. In the lysogenic cycle of the bacteriophage lambda (λ), λ integrates into the host DNA and forces the host cells to multiply. In a lytic cycle, which can be caused by UV light, for example, the phages multiply inside the host cell and burst out.

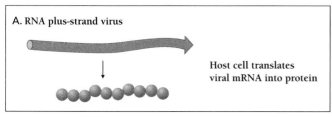

A. RNA plus-strand virus

Host cell translates viral mRNA into protein

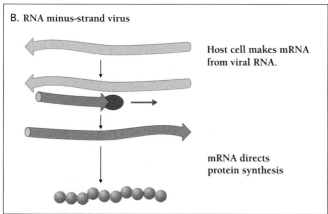

B. RNA minus-strand virus

Host cell makes mRNA from viral RNA.

mRNA directs protein synthesis

Figure 12-12
How RNA viruses express their genes. A. An RNA plus-strand virus acts just like mRNA, and the host cell translates its genes into polypeptides. B. An RNA minus-strand virus uses an "antisense" RNA strand, and then makes a complementary strand, which is read as mRNA.

mentary (sense) strand, which functions as mRNA and codes for proteins (Figure 12-12B).

Plus-strand viral RNA works like mRNA. Minus-strand RNA is antisense RNA, which is transcribed into mRNA.

RNA Tumor Viruses Present Special Problems

By the mid-1960s, virologists had discovered many of the ways that RNA viruses could reproduce. But the reproductive cycle of RNA tumor viruses, such as Rous sarcoma virus, mystified them. DNA tumor viruses inserted their genes directly into the DNA of their hosts, permanently changing their genetic makeup. But where could an RNA tumor virus leave a permanent copy of its genes? Certainly single-stranded RNA could not insert directly into chromosomal DNA.

The answer was first suggested in the 1960s by Howard Temin, a virologist at the University of Wisconsin. Perhaps, Temin argued, the RNA is recopied into DNA, and the DNA is then inserted into the host cell's genome. Many scientists found this suggestion heretical, since it violated Francis Crick's central dogma: "DNA specifies RNA, which specifies proteins."

In 1970, however, Temin and David Baltimore, a molecular biologist then at MIT, independently discovered that

RNA tumor viruses contain an enzyme, called **reverse transcriptase**, which indeed copies RNA into DNA. The enzyme received its name because the direction in which information flows is backward from that found in normal transcription (where information goes from DNA to RNA). As we will see in the next chapter, reverse transcriptase has become an essential tool for molecular biologists. Viruses that use reverse transcriptase in this way are called **retroviruses**. The discovery of reverse transcriptase supported Temin's hypothesis.

Subsequent work showed that Temin was exactly right. In most viruses, only the virus's nucleic acid actually enters the host. In retroviruses, however, the reverse transcriptase enzyme is actually packaged *with* the viral RNA, so the enzyme (not just the gene for the enzyme) enters the cell along with the virus's RNA (Figure 12-13). The reverse transcriptase then copies the RNA first into single-stranded and then into double-stranded DNA. The double-stranded DNA then integrates into the host cell's DNA, so that the virus's genes are replicated along with those of the host. Like DNA-containing tumor viruses, many (but not all) retroviruses contain an oncogene, whose expression leads to tumor formation. Retroviruses are known to cause cancer in birds, mice, monkeys, and humans. HIV is a retrovirus that infects cells of the human immune system but does not contain an oncogene.

Retroviruses use reverse transcriptase to transfer information from RNA to DNA, in the opposite direction to that specified by the central dogma.

12.3 How Do Mobile Genes Evolve?

Viruses and other mobile genes live so closely with cells that we might not realize at first that they are influenced by evolutionary forces independent of those that influence their hosts. Even though mobile genes are not alive, they are still independent entities whose replication is subject to natural selection—the mechanism by which Darwin explained the evolution of organisms.

Although Viruses Are Not Alive, They Are Still Subject to Natural Selection

Even when a virus doesn't kill its host cell, it uses energy-expensive materials from the host cell, impairing the cell's ability to grow and reproduce. As a result, hosts that evolve ways to resist viral infection are more likely to survive and reproduce. This is simply natural selection. Cells have therefore evolved general mechanisms for resisting most kinds

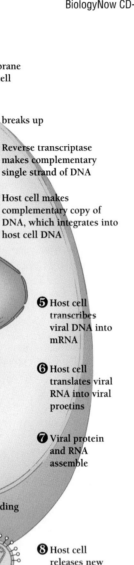

Retrovirus

- Envelope protein
- Lipid bilayer
- Protein coat
- Viral RNA
- Reverse transcriptase

Figure 12-13

How retroviruses reproduce. A. HIV (which causes AIDS) and other retroviruses reproduce by copying their RNA genes into double-stranded DNA, which integrates into the host cell's DNA. The host cell then transcribes and translates the viral genes, making more viruses. B. HIV viruses *(blue)* bind to the surface of a human immune cell—a T cell. HIV kills people by killing T cells, so the body cannot mount a defense against infections.

Biology⊗Now™   Learn more about viral integration by clicking on this figure on your BiologyNow CD-ROM.

❶ Retrovirus membrane fuses with host cell membrane

❷ Viral coat breaks up

RNA

❸ Reverse transcriptase makes complementary single strand of DNA

Host cell

DNA

❹ Host cell makes complementary copy of DNA, which integrates into host cell DNA

Nucleus

❺ Host cell transcribes viral DNA into mRNA

Viral RNA

❻ Host cell translates viral RNA into viral proetins

❼ Viral protein and RNA assemble

Budding

❽ Host cell releases new retroviruses

A.

B.

of viruses and also resistance mechanisms aimed at specific viruses. Many strains of *E. coli,* for example, guard against viruses by producing **restriction enzymes,** enzymes that cut foreign DNA into little pieces. Most restriction enzymes do not cut the DNA at random, but at special "target sites"— particular sequences of nucleotides that occur commonly in all genomes. Although a restriction enzyme can protect *E. coli*

(or another bacterial strain) from infection by a bacterio-phage, such an enzyme would be self-destructive if it also digested *E. coli's* own DNA. To prevent such molecular suicide, *E. coli* modifies its DNA by adding methyl groups that protect the bacterial DNA from digestion. Since the discovery of restriction enzymes, geneticists have put them to work in genetic engineering, as we will see in the next chapter.

At the same time, viruses that can overcome a cell's defenses themselves reproduce and persist. So, as hosts evolve mechanisms for resisting a virus, the virus, in turn, evolves mechanisms for getting past the cell's defenses. The kinds of viruses that are able to infect *E. coli,* for example, all have DNA that is missing the sequence targeted by *E. coli* restriction enzymes. Because the viral DNA lacks this target sequence, the cell's restriction enzymes cannot destroy the viral DNA.

Viral multiplication cycles—from infection to release— have evolved to become highly efficient. Cycles are generally rapid, with lots of new viruses produced in each infected cell. Viral regulatory genes are so powerful that the host's transcription and translation machinery recognizes them in preference to the host's own genes.

One way that viruses speed up their cycle times is to keep their genetic material to a minimum. Unlike eukaryotic genomes, which are loaded with "junk DNA," every gene in a virus matters. One kind of cold virus, for example, specifies several different proteins from a single gene merely by arranging to have the mRNA spliced in different ways.

How May Viruses Have Evolved?

Some scientists believe that viruses are descended from free-living cells that became parasites on other cells. According to this view, viruses gradually lost most of their cellular components, letting the host cells do all the work.

An alternative hypothesis is that viruses evolved from plasmids, which evolved from transposons. Knowing what we now know about mobile genes, we can imagine that mobile genes could have first appeared as simple transposons that could move from place to place in the genome. Transposons that also included genes that benefited the host cell—such as genes that allow the host to break down poisons or antibiotics—would have allowed the host cells that had them to reproduce faster than cells without them. As a result, transposons with beneficial genes would have replicated much faster than those without beneficial genes, an example of natural selection. Later, beneficial genes could have acquired the ability to replicate independently as plasmids do. Finally, some autonomously reproducing plasmids could have acquired genes for protein coats that allowed

them to move efficiently from cell to cell. A protein-wrapped plasmid would look just like a virus.

Over evolutionary time, viruses could have evolved either from free-living cells or from transposons.

Mitochondria and Chloroplasts Have Their Own DNA

Chloroplasts and mitochondria contain their own DNA. Since all eukaryotes have mitochondria and many have chloroplasts as well, almost every eukaryotic cell therefore carries at least one genetic system independent of the nucleus. In a mammalian cell, each mitochondrion contains 5 to 10 identical DNA molecules, each about 16,000 nucleotides long. Since a single cell often has hundreds of mitochondria, up to one percent of the total DNA in the cell may consist of mitochondrial DNA. Each piece of mitochondrial DNA is circular, meaning that it forms a continuous loop, like the DNA of bacteria and many viruses (Figure 12-14A).

Plant mitochondria also contain circular DNA molecules, but usually of several different kinds. The total length and the total information content of mitochondrial DNA in plants can be more than 20 times greater than in animal mitochondria. Chloroplasts contain DNA as well (Figure 12-13B). Each species has a single kind of chloroplast DNA. The chloroplast DNA is circular and usually about 150,000 nucleotides long. In higher plants, chloroplast DNA may constitute up to 15 percent of the total DNA in a cell.

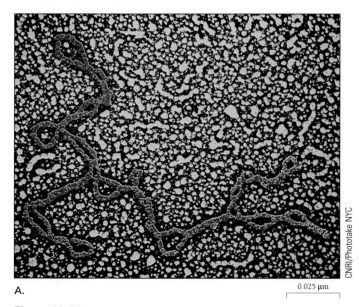

A.   0.025 μm

CNRI/Phototake NYC

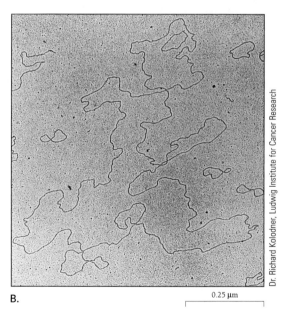

B.   0.25 μm

Dr. Richard Kolodner, Ludwig Institute for Cancer Research

Figure 12-14
Circular DNA. A. DNA from a mitochondrion. B. DNA from a chloroplast. Chloroplast DNA is 10 times longer than DNA from the mitochondria of animals. The small circles of DNA shown here are from viruses.

Health and Biology   How Can Researchers Design Drugs That Will Combat AIDS?

The greatest epidemic in recent history is acquired immune deficiency syndrome (AIDS). AIDS is caused by one of several subtypes of human immunodeficiency virus (HIV). HIV infects specialized white blood cells, called T lymphocytes, that normally help find and destroy invading pathogens. Over several years, the virus reduces the number and effectiveness of these crucial immune cells, making the body susceptible to infections and cancers that a healthy immune system normally fends off. These secondary illnesses, rather than the HIV virus itself, are what kill most people infected with AIDS.

Because HIV grows and replicates in the body for years before producing any symptoms, a person can test positive for HIV antibodies, indicating the presence of the virus, long before the onset of any illness. During this early stage of infection, bodily fluids such as blood and semen harbor active viruses that can transmit the infection to a new host. Most people contract the virus by sharing contaminated hypodermic needles or by having unprotected sex with an infected partner.

Like all retroviruses, HIV carries its genetic information in a single RNA strand, which is accompanied by a molecule of reverse transcriptase. Once inside the cell, the reverse transcriptase generates a DNA strand copied from the RNA template. A second round of synthesis produces a double-stranded DNA copy of the viral RNA, which inserts itself into a chromosome in the host cell's nucleus. Without reverse transcriptase, HIV could get into the cell, but it would never be able to replicate.

Since human cells do not depend on reverse transcriptase, the virus's enzyme makes an excellent target for anti-AIDS drugs. One type of antiviral therapy deceives the enzyme with synthetic molecules that resemble ordinary nucleotides, the raw materials for DNA synthesis. Reverse transcriptase incorporates these nucleotide impostors into the growing strand of viral DNA. But since the false nucleotides are subtly different from real ones, replication

stops. AZT and 3TC, two drugs now commonly used to slow viral replication in patients with AIDS, are examples of false nucleotides. Human DNA polymerase is more picky than reverse transcriptase in selecting nucleotides and rarely incorporates AZT or 3TC. As a result, human cells exposed to the drugs continue to grow and divide normally. Unfortunately, HIV mutates rapidly, and strains that no longer use the false nucleotides quickly emerge. These resistant strains thrive in patients treated with these drugs, rendering the drugs useless.

Another class of HIV drugs works by interfering with a different viral enzyme—one that cuts a viral protein. After viral DNA integrates into the host's chromosome, host proteins transcribe the viral genes. Rather than producing a separate mRNA for each viral protein, however, the messages for several proteins form a single, long mRNA molecule. The resulting mRNA gets translated into a long polypeptide that must be cut into separate proteins before they will function. One of the proteins in the viral polypeptide chain is the enzyme that does the cutting. This enzyme, called a *protease,* begins cutting even while it is still part of the "polyprotein." Without the protease, the polyprotein remains unclipped and inactive, and the virus cannot go on to infect more cells.

In an attempt to stop HIV by knocking out the protease, researchers devised tiny molecules that nestle into the active site of the protease enzyme and prevent it from doing its job. The first of these protease inhibitors, called *saquinavir,* was released in December 1995—with dramatic results. A drug cocktail containing saquinavir, AZT, and 3TC can reduce the amount of virus in a patient's body to undetectable levels. For the first time since the discovery of HIV, this "triple therapy" has transformed AIDS from a death sentence to a potentially survivable disease. Since 1995, five additional protease inhibitors have proved useful for treatment, and others are now being tested.

But triple therapy has its problems. The cost of these AIDS drugs, particularly pro-

tease inhibitors, is still extremely high, placing an economic barrier between many sufferers of AIDS and good treatment. In the United States, not all insurance covers AIDS therapies, and in Africa, where most of the world's AIDS cases occur, the price is out of reach of almost everyone who has the disease. Also, triple therapy doesn't work well for everyone, and it will be several years before researchers know whether the drug cocktail can permanently clear the virus out of the body. Indeed, researchers worry that HIV may evolve resistance to protease inhibitors, as it has to other drugs. In that case, it could still return in full force, insensitive once more to all drug treatments.

To help stave off that possibility, still another class of drugs may close the door on the virus, preventing it from getting into T cells in the first place. To enter T cells, an HIV virus must bind to a pair of proteins on the cell surface, one of which is a chemokine receptor, called CCR5. Chemokines are chemical signals that induce inflammation at a site of injury or infection. Researchers noticed in 1995 that chemokines block HIV from infecting cultured cells. Not long after, more work revealed that the CCR5 receptor helps HIV enter cells and that the chemokine was interfering with that process. In addition, people with a deficit in CCR5 have no obvious health problems but are resistant to some HIV subtypes. Drug companies have tried to develop a drug that binds to CCR5 without setting off the inflammatory pathway. One such drug—enfuvirtide (Fuzeon)—received FDA approval in March 2003.

The battle against the AIDS epidemic in the United States may finally be succeeding, even as millions of people are dying elsewhere, particularly in Africa. The epidemic is far from over. As has been true of many other infectious diseases, the AIDS virus will continue to evolve, and we can still hope that researchers will be at least one step ahead. So far, however, the virus is winning.

The reproduction of mitochondria and chloroplasts depends on cellular machinery that is specified both by the DNA in the nucleus and by the DNA in the organelle itself. The replication and expression of mitochondrial or chloroplast DNA, for example, requires about 90 proteins—far more than the organelle's DNA can usually encode. These proteins are specified by the DNA in the nucleus and made on cytoplasmic ribosomes. The DNA of mitochondria and chloroplasts nonetheless contains important information. For example, both mitochondria and chloroplasts contain the proton-pumping complex ATP synthase that harnesses the energy of proton gradients to make ATP. Each complex contains nine different polypeptides, only some of which are specified by the DNA in the nucleus. The other genes are in the mitochondria or chloroplasts.

Mitochondria and chloroplasts contain complete systems for transcription and translation, including not only DNA but also ribosomes, transfer RNAs (tRNAs), and an RNA polymerase similar to that of bacteria. Some components of the machinery of protein synthesis are made in the cytoplasm, and other components are made within the organelles.

Like prokaryotic DNA, the DNA of most mitochondria and chloroplasts is circular. The ribosomes of mitochondria and chloroplasts are also more like bacterial ribosomes than like the ribosomes in the cytoplasm of eukaryotic cells. These and many other similarities between the energy organelles and prokaryotes support the **endosymbiotic theory** [Greek, *endon* = within + *syn* = together + *bios* = life]. This theory holds that present-day mitochondria and chloroplasts are descended from prokaryotes that came to live inside ancient eukaryotic cells. In support of this view, biologists point to many modern examples of cells that host endosymbionts. Recent studies of DNA sequences have revealed that *all* mitochondrial DNA shares many common features. Researchers now suspect that all mitochondria have evolved from a single type of mitochondrion in an ancient eukaryote.

The genetic code used in mitochondria is slightly different from that used in prokaryotes or in eukaryotic nuclei, with four codon assignments that are different. If mitochondria are descended from free-living prokaryotes, their different genetic code suggests an origin so ancient that it predates the establishment of a universal genetic code. In Chapter 18, we will discuss the origin of mitochondria and chloroplasts in more detail.

Mitochondria and chloroplasts, which have their own DNA separate from the nuclear DNA, evolved when a prokaryote ancestor came to live in symbiosis with another cell. The genetic systems of mitochondria and chloroplasts differ from those in both eukaryotic nuclear DNA and in prokaryotes, suggesting that these organelles have an ancient and separate origin.

In this chapter, we have seen that not all genes reside in the chromosomes, nor do all genes pass neatly from generation to generation by means of meiotic divisions. Genes exist in mitochondria and chloroplasts, suggesting that these organelles were once independent organisms. Even more intriguing, however, genes also exist in mobile forms, such as viruses, plasmids, and jumping genes. In the next chapter, we will see that mobile genes are the power behind genetic engineering, a new field that may have profound effects on society in the next century.

Key Concepts

- Unconventional genetic systems include all genes that are not a stable part of a cell's nuclear genome.
- A virus is an assembly of protein and nucleic acid that can replicate within a living cell.
- The genomes of both prokaryotes and eukaryotes contain DNA that originally derived from mobile genes.
- Energy organelles—mitochondria and chloroplasts—contain independent genetic systems, which probably evolved from ancient associations of prokaryotes.

Summary with Key Terms

How are viruses, plasmids, and transposons alike?

Barbara McClintock showed that some genes can move from place to place. She showed that jumping genes, or transposons, cause the curious dappled colors of Indian corn. **Unconventional genetic systems** are DNA or RNA sequences that can replicate and function apart from chromosomal DNA—including the DNA of mitochondria and chloroplasts. **Mobile genes**, which include **viruses, plasmids**, and **transposons**, can move from cell to cell or from place to place within a chromosome. Unconventional genes can replicate only by using the cellular machinery of cells, which is specified by chromosomal DNA.

What is a virus?

Viruses, including **bacteriophages**, are assemblages of nucleic acid and proteins that can reproduce within a living cell, by forcing the machinery of their host cells to produce more viruses. Whether shaped like an **icosahedron** or a long rod-shaped helix, viruses consist of an inner core of nucleic acid surrounded by a **protein coat** or **membrane envelope**. Viruses were first discovered as disease-causing agents that could pass through filters fine enough to remove bacteria. Viruses reproduce, respond to their environments, and evolve, but they cannot obtain energy from their surroundings or form the building blocks needed for the synthesis of their components and so are not considered to be living organisms.

How do viruses and other mobile genes replicate?

The growth of viruses in laboratory cell cultures has allowed researchers to study their multiplication cycles. These cycles consist of six steps: (1) attachment to the host cell, (2) entry of viral nucleic acid into the cell, (3) synthesis of viral proteins, (4) replication of viral nucleic acid, (5) assembly of new virus particles, and (6) release of new viruses.

Many viruses kill their hosts in the process of replicating themselves. These viruses, which break open the host membrane in a process called **lysis**, are called **lytic viruses.** Sometimes, viral genes integrate into the chromosomal DNA of their hosts (a form called a **prophage** or provirus). When the host cell replicates, it replicates the viral genes as well, a process called **integrated replication.** Other viruses make new viruses through **autonomous replication.** Actual virus particles are produced only when something induces the provirus to enter a lytic cycle. In a **lysogenic cycle,** a virus or bacteriophage switches integrated and autonomous replication (a **lytic cycle**).

RNA serves as the genetic material in some viruses. In plus-strand viruses, the viral RNA directs polypeptide synthesis. In minus-strand viruses, an enzyme copies the viral RNA into a complementary strand that directs polypeptide synthesis. In AIDS and other **retroviruses, reverse transcriptase** copies viral RNA into DNA, which then integrates into the host cell's chromosomal DNA.

Mobile genes include plasmids, which can replicate only autonomously; transposons, which can replicate only in integrated form; and **episomes**, which can replicate either autonomously or in integrated form. **Simple transposons** carry nothing but the gene for **transposase**, while **complex transposons** carry genes that benefit their bacterial hosts such as those for antibiotic resistance. Plasmids artificially constructed in the lab are called **recombinant plasmids.**

Plasmids called **resistance factors,** or **R factors,** contain genes for enzymes that make the cell resistant to antibiotics. Bacteria engaging in **conjugation** exchange genetic material, including both bacterial DNA and plasmids. In eukaryotic cells, some transposons, or jumping genes, move within the chromosomal DNA, disrupting the genome. Bacteria defend themselves against bacteriophages using **restriction enzymes,** which cut DNA at special "target sites." **Oncogenes** specify proteins that help regulate cell division. The oncogenes of tumor viruses interfere with a host cell's ability to regulate cell division.

How can unconventional genetic systems evolve?

Although viruses are not alive, they evolve through natural selection. Over evolutionary time, viruses could have evolved either from free-living cells or from transposons. According to the **endosymbiotic theory,** the circular form of the DNA of chloroplasts and mitochondria, as well as the particular genes they contain, suggest that they evolved from ancient prokaryotic cells that came to live inside eukaryotic cells.

Review and Thought Questions

Review Questions
1. What characteristics of a virus make it nonliving?
2. How can a virus or other nonliving genetic entity evolve through natural selection? Give two examples.

3. How does a virus attach to a cell and transfer its genetic material into the cell?

4. What advantages does integrated replication have over autonomous replication?

5. What is a retrovirus? How does it differ from other RNA viruses?

6. What is the difference between a plasmid and a transposon?

7. What evidence suggests that mitochondria and chloroplasts were once prokaryotes that lived independently of other cells?

Thought Questions

8. If you take antibiotics frequently, you will select for *E. coli* that carry genes for antibiotic resistance. Suppose that you contracted tuberculosis (TB) from your roommate. Could plasmids then transmit antibiotic resistance from the *E. coli* in your intestines to the TB bacteria? Why or why not?

9. Many biologists suspect that transposons move genes among different species of eukaryotes. For example, some evidence suggests that within the past 100 years, a parasitic mite has carried transposons called P elements from one species of fruit fly to another. What evolutionary consequences might result from such transfers?

BiologyNow Resources

Biology(Ɛ)Now™

Active Figures

12-4D: The HIV virus

12-13: Viral integration

Preparing for an exam? Take a diagnostic test on your BiologyNow CD-ROM.

Online materials relating to this chapter are at:

http://biology.brookscole.com/AAL3

About the Chapter-Opening Image

Gene Kelly (1912–1996), dancing star of Broadway shows and Hollywood films. A photo of this "jumping Gene" next to a photo of Barbara McClintock's experimental organism, Indian corn. The colored kernels of the corn result from the movement of jumping genes, or transposons.

13

Genetic Engineering and Recombinant DNA

Key Questions

- What does it mean to "genetically engineer" an organism?

- What steps must biologists take to insert a gene into the genome of another organism?

- How are transgenic plants and animals useful to biologists and others?

The Maverick

Early in the morning of October 13, 1993, reporters and a photographer showed up at the door of 48-year-old unemployed surfer and molecular biologist Kary Mullis (1944–). The Nobel prize winners for 1993 had been announced in the middle of the night, and all the winners were being interviewed. The reporters gathered around Mullis's beachfront apartment in southern California and caught him on his way out for a morning of surfing (Figure 13-1). They asked him, Did he know he had won? Was he happy? Surprised? How did he feel when the phone call came from Stockholm, Sweden? Mullis knew—he had already gotten his call in the small hours of the morning. Naturally, he was happy. "I realized I did something significant, the kind of thing you get the Nobel prize for," he told reporters, "but never in my wildest dreams did I ever think I'd win."

Paraskevas Photography

He wasn't the only one surprised. Explained one colleague to *Esquire* magazine, "He doesn't fit the normal mold of a Nobel-prize winner. . . . He's a wild man. There's a certain amount of politicking involved, and I was not sure his personality and his lifestyle would have allowed him to win." Mullis is indeed a maverick. Eventually, he offends nearly everyone, and he seems to enjoy doing it. At one scientific conference, he began laughing when someone else was speaking. When the speaker chided Mullis for laughing at a serious matter, he responded, "I'm not laughing at that. I'm laughing at you." Mullis is a far cry from the serious, dedicated researcher who still goes to the lab seven days a week long after retirement. In fact, he hasn't worked in a lab in years.

Mullis, whose personality is equal parts boyish charm and astonishing boorishness, grew up in South Carolina, where, like many a southern boy, he learned about life by roaming the woods and building rockets from drugstore supplies. At Georgia Tech, he earned a degree and got married for the first time.

In 1966, he arrived at the University of California, Berkeley, with a young wife and baby daughter, to work on a Ph.D. in biochemistry. Berkeley was all he had hoped it would be, and then some. "Six years in the biochemistry department didn't change my mind about DNA," he told UC Berkeley's alumni magazine, "but six years of Berkeley changed my mind about almost everything else." Besides psychedelics such as LSD, Mullis also discovered free love and divorce. "It wasn't a good place to be married."

By 1981, he was on his third marriage, with two young sons. For a while he worked as the night manager at the Buttercup Bakery in Berkeley. When he took a job as a chemist for one of the earliest biotechnology companies, Cetus Corporation, it was as a low-level technician. At Cetus he fell in love with another biochemist and, at 36, ended his third marriage.

One Friday night in the spring of 1983, the two biochemists headed for Mullis's weekend cabin in the coastal mountains of northern California. Late that night, as Mullis

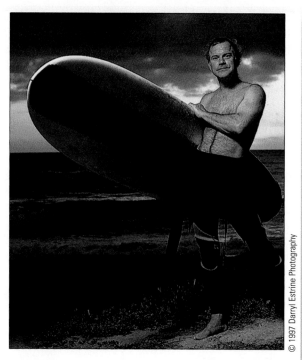

Figure 13-1
Kary Mullis, inventor of PCR.

© 1997 Darryl Estrine Photography

drove along a twisting mountain road through vineyards and redwoods, he filled the silence by thinking about DNA. Specifically, he was interested in making his job at Cetus—building short pieces of synthetic DNA—a little easier.

As the winding road unfolded before him, he thought about the way the two strands of DNA untwist and separate so that they can be copied. He designed experiments in his head, wondering how he might single out one particular nucleotide sequence in the DNA and make copies of it. He saw ways his imaginary experiments might go wrong, and twisting his thoughts this way and that, fixed them. His idea seemed foolproof—he couldn't find anything wrong with it. In theory, he should be able to take a test tube filled with thousands of genes, single out a stretch of nucleotides, and make millions of copies within a day (Figure 13-2). For a molecular biologist, it was undreamed-of power.

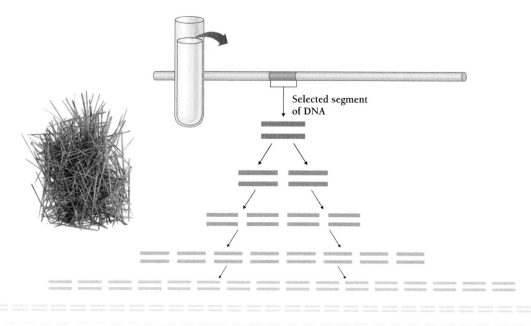

Selected segment of DNA

Figure 13-2
PCR makes more genes fast. When DNA is heated, the two strands uncoil. They are then cooled and replicated. A PCR machine repeats the cycle of heating, cooling, replicating, and then heating again, making millions or billions of copies of the DNA sequence.

He realized with a shock that his idea was almost good enough to win him a Nobel prize. There had to be something wrong with his idea, he thought. It was too obvious, too slick, too easy. The idea was this: Mullis could make an oligonucleotide, a short piece of DNA, that binds to a complementary sequence in a far longer strand of DNA (Figure 13-3). Once the oligonucleotide was bound to the DNA, it could serve as a primer for DNA polymerase. The DNA polymerase would copy only the DNA immediately "downstream" (in the 3′ direction) from the bound oligonucleotide. The newly made DNA could then be used as a template to make still another strand. The result would be a chain reaction: one copy of a specific DNA sequence yielding 2 copies, then 4, then 8, then 16, and so on.

Back at Cetus, Mullis's boss told him to take all the time he needed to develop his idea. Colleagues pitched in and helped him make the idea work. By mid-December, Mullis had worked out all the bugs in a new technique called the **polymerase chain reaction (PCR)**. PCR specifically and repetitively copies any segment of DNA between any two defined nucleotide sequences. Today, specialized machines about the size of a toaster perform this process automatically, so that a single copy of a chosen sequence can be multiplied into millions or billions of copies in just hours.

Finding a short sequence of nucleotides in a cell's DNA is like searching for a needle in a haystack. PCR finds the right sequence, then copies it repeatedly until, after 20 to 30 rounds of replication, the original haystack is dwarfed by a haystack of needles. Not only does PCR help find particular sequences, it also makes them in usable amounts. For molecular biologists to study a particular nucleotide sequence, they need more than one copy of a sequence; they

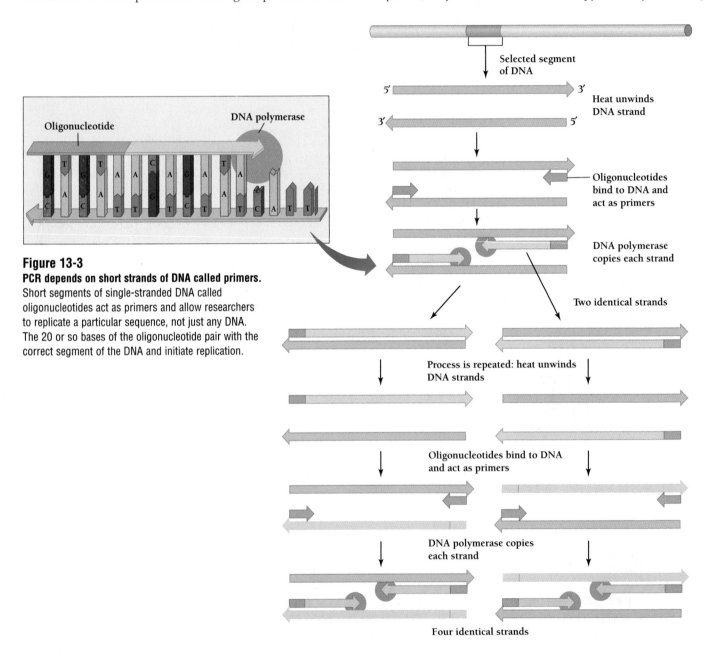

Figure 13-3
PCR depends on short strands of DNA called primers.
Short segments of single-stranded DNA called oligonucleotides act as primers and allow researchers to replicate a particular sequence, not just any DNA. The 20 or so bases of the oligonucleotide pair with the correct segment of the DNA and initiate replication.

need thousands of identical copies. Mullis's idea was a singularly valuable technique that would allow thousands of other molecular biologists to make important discoveries.

The uses for PCR seem limitless. PCR enables crime experts to use DNA profiling on mere specks of blood and other tissues to identify who has been present at the scene of a crime. PCR allows researchers to pick out trace amounts of DNA in blood from HIV and other viruses, as well as pick up early signs of cancer and genetic defects in human eggs and embryos. PCR allows archaeologists and paleontologists to study minute amounts of DNA from ancient animals and plants—from the 130-year-old tissues of Abraham Lincoln to an 18-million-year-old magnolia leaf preserved in a peat bog.

Mullis's discovery might have launched his career, even made him happy. But it did not. Almost as soon as Mullis had perfected the technique, other researchers at Cetus began using the technique to do experiments. Mullis thought his name would be on those papers—as a coauthor—but his colleagues didn't agree. A simple acknowledgment of his invention near the end of the paper seemed like enough to them. Tempers flared, and Mullis was asked to leave if he couldn't get along with his colleagues. In the end, Mullis left Cetus, taking with him a $10,000 bonus for his invention. It was little enough. Cetus made millions of dollars from his invention and ultimately sold the patent rights to PCR for a mind-boggling $300 million. But even that was a bargain. Today, PCR technology is a $1-billion-a-year industry.

Mullis has mostly abandoned conventional science. Friends and colleagues suggest that his talents are being wasted. Molecular biology, however, is as much a social endeavor as an intellectual one, and Mullis's intellectual talents are of little value in an environment where getting along with people is paramount. Doing science, especially doing science that others notice, means fitting in, belonging to one or another "club." A few researchers—loners like Barbara McClintock—do not entirely mind if no one appreciates their discoveries. They enjoy their work so much that they continue in a virtual social vacuum.

Mullis will be starved for neither money nor attention. He eventually won $450,000 from the prestigious Japan Prize and a $412,500 Nobel prize for his invention of PCR. And although he may not have a desk at a university or his name on a door at a biotechnology company, his Nobel prize has provided him with many opportunities for fame and fortune.

13.1 What Is Genetic Engineering?

Ten thousand years ago, humans invented agriculture and began growing grains such as wheat and rice and breeding plants and animals. Early breeding consisted of saving seeds for the next year's crop from the plants that produced the most seeds, the largest seeds, or the best-tasting seeds. Sheep with the thickest wool were bred together. Beer brewers and cheese makers learned to select the right strains of

Figure 13-4
Prehistoric corn. A. Cob from a cave in Mexico. Humans have been breeding bigger and bigger corn cobs for thousands of years. Some biologists believe that early corn was bred specifically to be popped. B. A recently popped prehistoric corn kernel *(left)* compares favorably with the modern version *(right).*

Science VU/Visuals Unlimited

A.

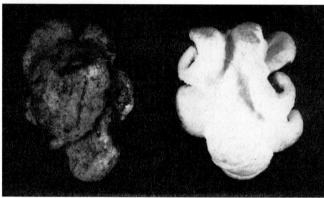

B.

Courtesy C.P. Mangelsdorf

microorganisms by starting each batch with the right "starter." Some of the earliest farmers developed corn about 7,000 years ago (Figure 13-4).

Beer brewing, cheese making, and agriculture are all early examples of **biotechnology**—the use of living organisms for practical purposes. Modern 21st-century biotechnology depends as much on the selection of organisms with the right traits as the first biotechnology did 10,000 years ago. Today, genetic engineers can alter some genetic traits in a variety of organisms. The biggest difference between ancient methods and modern ones is the precision and rapidity with which organisms can be altered. Genetic engineers now happily move genes among all kinds of organisms, including humans, mice, tomatoes, yeasts, and bacteria. But while traditional selection techniques allow breeders to select for almost any trait, modern genetic engineering is limited to traits whose genetics and biochemistry biologists understand.

Biotechnology first arose 10,000 years ago when humans began selecting and breeding useful plants, animals, fungi, and microorganisms.

Genetic Engineering Makes Use of Both Natural and Induced Variation

In 1929, a microbiologist named Alexander Fleming discovered the first antibiotic, penicillin, when he noticed that a mold was growing on and killing his cultures of bacteria. Within 10 years, other biologists had shown that an extract of Fleming's mold could protect mice from an otherwise fatal bacterial infection.

But the mold made very little of the antibiotic we now know as penicillin. To treat just one adult for a bacterial infection, a doctor needs about 7.5 grams of penicillin (the weight of two and a half animal crackers). To make enough penicillin to treat just one sick person would require that drug manufacturers grow about 225 million bathtubs full of mold. Penicillin was so rare that the precious drug was sometimes extracted from the urine of patients and given to them once again in subsequent doses. It was clear that to make penicillin in the amounts needed to control human disease, scientists needed to find a way to make much more penicillin.

Microbiologists began by searching through soil samples from all over the world for the one that made the most penicillin, ultimately choosing one from a U.S. Department of Agriculture Laboratory in Peoria, Illinois. Microbiologists then increased the genetic variation in the Peoria mold by subjecting it to mutagens such as x rays, ultraviolet light, and DNA-damaging chemicals. The changes in the DNA were random and most were useless. But a few increased the production of penicillin. After each round of mutagenesis, researchers chose the strains with the highest yields of penicillin and subjected them to more mutagens. By the end of this process of inducing mutations and selection, researchers had created a mold that produced nearly 100 times as much penicillin as the original Peoria strain. To treat one sick person, they would now need only 2 million bathtubs of mold. It was a small but hard-won step forward.

Biotechnology selects useful traits from a range of variation. This variation can be natural or induced by means of mutagens.

Knowing Biochemical Pathways Helps Molecular Biologists Design Useful Organisms

In the 1940s, researchers had no idea which mutation in the mutated Peoria strain made it produce so much more penicillin than other strains. But today molecular biologists know so much more about biochemical pathways that they can sometimes predict and even design mutations that will produce a desired trait. The genetic engineering of tomatoes provides one example.

In order for tomato growers to get tomatoes safely to the supermarket, they must pick the tomatoes at just the right time—not red, but not too green either. If the tomatoes are picked when they are soft, red, and ripe, the fruits will be crushed to a pulp by handling and the weight of other tomatoes. To avoid this, growers pick tomatoes when they are still green and hard. Ideally, growers pick tomatoes when they have a slight reddish blush and "flavor components," chemicals that make them taste good once they ripen. Just before the tomatoes are ready to be delivered to a store, they are treated with the plant hormone ethylene gas, which triggers their ripening and the production of their own ethylene (Chapter 33). If the tomatoes have been picked at the right time, they ripen into sweet, flavorful red tomatoes.

But the problem for growers is that the key flavor components are completely invisible and growers can't necessarily check every individual tomato for just the right amount of reddish blush. It's more tempting to pick the fruits when they are too green than too red. (Nobody can sell a ton of overripe, rotting tomatoes.) As a result, about 40 percent of all tomatoes are picked when too green. These are the tomatoes we know too well, the ones that are red but taste like wet cardboard.

Genetic engineers solved this problem by deliberately damaging a gene that controls the production of natural ethylene in the tomato. Such tomatoes can be left on the vine until they turn a pinkish orange, which signals that they have developed their flavor components. These tomatoes will never soften (or turn red and sweet) on their own, because they cannot make ethylene. Only after the tomato grower has harvested these hard tomatoes and the tomato distributor has treated them with ethylene will they ripen. Since growers can safely wait until the tomatoes have plenty of color (and flavor components), growers won't pick them too soon.

Modern molecular biologists can design useful organisms by inserting or destroying genes that code for proteins involved in specific biochemical pathways.

13.2 What Is Recombinant DNA and How Is It Used?

In the 1950s, biologists first learned that genes are nothing more than chains of nucleotides. By the 1970s, they had learned to cut and paste together pieces of DNA from different organisms. And by the 1980s, biologists could replace individual nucleotides in a gene. Earlier work had already revealed how viruses and plasmids shuttle genes around the genomes of bacteria and other host cells. Molecular biologists easily induced these tiny molecular parasites to go to work in the lab, reproducing pieces of DNA that had been patched together in a test tube.

How Do Molecular Biologists Use Recombinant DNA?

Recombinant DNA is any DNA molecule consisting of two or more DNA segments that are not found together in nature. For example, researchers can add a gene for antibiotic resistance to a plasmid, then infect a cell with the plasmid, so that the infected cell carries a gene for antibiotic resistance. Because molecular biologists can now design and make recombinant DNA that fulfills particular requirements, the resulting recombinant DNA molecules are sometimes called "designer genes."

Recombinant DNA technology is powerful: it can turn plants, animals, and microorganisms into chemical factories that churn out vast quantities of proteins and other substances that these organisms would never make on their own; and it helps biologists study how genes work, the molecular basis of genetic diseases, and why people differ in their responses to medicines (Figure 13-5). Recombinant DNA promises purer vaccines, new drugs, and genetic tests for hereditary diseases. But recombinant DNA also presents moral and ethical questions that must ultimately be answered by nonscientists.

Recombinant DNA techniques join pieces of DNA in combinations that do not occur in nature.

How Do Restriction Enzymes Cut Up a Genome?

The development of recombinant DNA technology depended on the union of many different lines of research, including work on restriction enzymes in bacteria, DNA repli-

cation and repair, and the replication of viruses and plasmids. DNAase is an enzyme that cuts the links between the nucleotides in DNA. DNAase usually cuts DNA into short fragments at random. In a test tube filled with DNA, each of the many copies of a genome will be cut differently. The resulting fragments are all cut into different sizes, and the pieces from one genome overlap those in another. In such a mess, there is little hope of finding a specific sequence. To isolate a specific piece of DNA, a researcher must instead use **restriction enzymes,** special DNAases that cut DNA only at particular sequences (Figure 13-6).

Recall from Chapter 12 that bacteria use restriction enzymes to defend themselves against viruses and other mobile genes. When a bit of alien DNA enters a bacterium, the restriction enzymes cut it to bits. Bacteria have lots of different restriction enzymes, each of which cuts a different nucleotide sequence. The short sequences of nucleotides recognized and cut by a given restriction enzyme are called a **restriction site.**

A restriction enzyme will randomly cut all the DNA in a test tube at a restriction site. That sequence of nucleotides may occur thousands of times, but the restriction enzyme cuts only some of them and not others. The result is a matching set of **restriction fragments**—pieces of DNA that all begin and end with a certain restriction site—of all different lengths. By comparing the lengths of restriction fragments, researchers can create a **restriction map,** which maps the restriction sites on a length of DNA.

Some restriction enzymes cut each of the two strands of DNA in the same place, so that the resulting fragment of DNA has "blunt" ends (Figure 13-6A). Other restriction en-

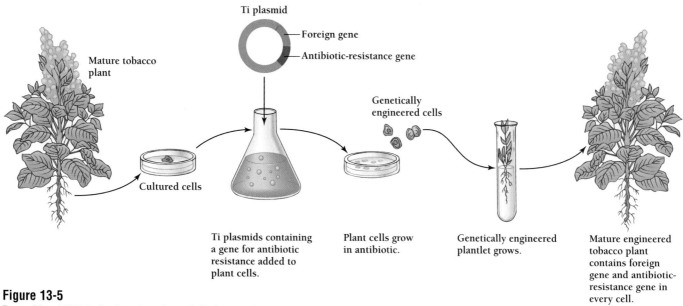

Figure 13-5
Recombinant DNA technology in action. Cells from a tobacco plant are grown in culture, then infected with a Ti (tumor inducing) plasmid carrying a gene for antibiotic resistance and other genes. The growing cells are treated with antibiotic, and only the cells expressing the new genes survive. These antibiotic-resistant cells can then be grown into mature plants, which will bear seeds carrying the new genes.

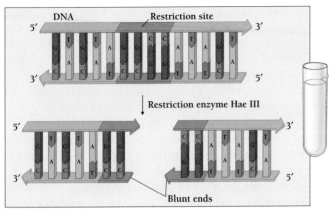

Figure 13-6

Cutting up DNA. A. DNAases cut DNA at random. Each copy of DNA is cut differently. B. Restriction enzymes cut each copy of DNA at the same places. Matching fragments have the same length and sequence. Restriction enzymes can cut the DNA with blunt ends or sticky ends.

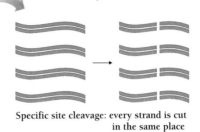

Specific site cleavage: every strand is cut in the same place

A.

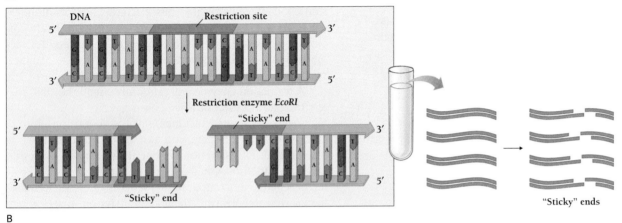

B

zymes make staggered cuts that leave "sticky" single-stranded ends that easily join with other such sticky ends (Figure 13-6B). For example, a strain of the bacterium *E. coli* makes a restriction enzyme, called *EcoRI*, that makes a staggered cut wherever the enzyme finds the sequence GAATTC

(Figure 13-6B). Genetic engineers can easily join restriction fragments together—even linking fragments of DNA from humans, mice, and bacteria (Figure 13-7A).

Restriction enzymes cut DNA molecules at specific sequences. Restriction enzymes allow the preparation of DNA fragments with defined lengths and sequences.

One of the first products that early biotechnologists wanted to make was human insulin. Many people who have diabetes must give themselves daily injections of the hormone insulin, which helps regulate blood sugar level. But

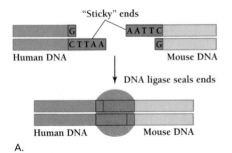

A.

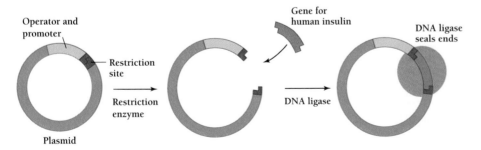

B.

Figure 13-7

Genetic engineering. A. Sticky ends enable researchers to join pieces of unrelated DNA. DNA from a mouse and human can be joined, as long as they have been cut at the same sequence. B. Restriction enzymes cut a gap in a plasmid, and DNA ligase adds the cDNA gene for human insulin and seals it in place.

until the advent of recombinant DNA, insulin was available only from the pancreases of pigs and cattle slaughtered for their meat. In the 1980s, the demand for insulin began to outpace production from slaughterhouses. So molecular biologists looked for a way to persuade immense colonies of bacteria growing in vats to make pure human insulin.

To do this, they need to join fragments of human DNA with bacterial DNA. Recall from Chapter 10 that when cells replicate DNA, the enzyme DNA ligase links together the separate pieces of DNA into one continuous strand. Bioengineers realized they could use DNA ligase to link restriction fragments together. To make insulin, they used DNA ligase to insert the gene for human insulin next to the operator and promoter of the *lac* operon in the bacterium *E. coli* (Chapter 11). The operator and promoter regulated the expression of the human insulin gene in the bacterium, just as if it were a bacterial gene (Figure 13-7B). Inside the bacterium, the recombinant DNA—a combination of bacterial and human DNA—expressed the gene whenever the bacteria were exposed to the milk sugar lactose. Ordinarily, lactose would signal the bacteria to make enzymes that digest lactose. But bacteria carrying the recombinant DNA responded to lactose by making human insulin instead.

Biologists make recombinant DNA by linking together different kinds of DNA with the enzyme DNA ligase.

How Do Molecular Biologists Express Recombinant DNA in Bacteria and Other Hosts?

Once scientists taught themselves to make recombinant DNA using DNA ligase and restriction fragments, they needed to find a way to make bacteria or other host cells express recombinant genes as usable proteins. The next problem was to get a gene into a cell. Recall from Chapter 12 that some natural plasmids carry genes for antibiotic resistance. Using restriction enzymes and DNA ligase, researchers can build simple plasmids that consist of only two parts: (1) an origin of replication—the DNA sequence needed to start DNA synthesis and (2) one or more genes for antibiotic resistance. The antibiotic resistance acts as a marker, a way to distinguish and select the bacteria that carried the genetically engineered plasmid. With the help of DNA ligase, researchers can take any gene they want and hook it to a plasmid. A host bacterium then copies the plasmid DNA as if it were its own. Finally, researchers grow the bacteria in a medium containing an antibiotic, eliminating any bacteria without the resistance gene.

Microbiologists have long used the word **vector** [Latin, = carrier, rider] for any organism or virus that carries disease from organism to organism. Molecular biologists use the word "vector" to describe anything that carries genes from organism to organism. Plasmids are one important kind of vector for recombinant DNA, but researchers also use genetically modified phages, animal and plant viruses, transposable elements, and even fragments of chromosomes.

Vectors are DNA sequences from viruses, plasmids, and other mobile genes that can carry recombinant DNA into cells.

How Can Bacteria Be Induced to Express Eukaryotic Genes?

It is easy to get a bacterium to express a gene from a bacterium or bacteriophage. But bacteria often cannot express useful genes from humans, plants, and other eukaryotes. Recall that a eukaryotic gene is interrupted by long sequences of introns. But bacteria themselves do not have introns in their DNA and so they lack the cellular machinery to remove introns from the messenger RNA transcribed from a human gene. Because the bacteria cannot remove the introns, they cannot make mRNA that can actually be translated into a protein. How can researchers acquire a clean, intronless copy of a eukaryotic gene?

The solution came from retroviruses, the viruses that use RNA as their nucleic acid, yet insert themselves into the DNA genomes of eukaryotes. Recall from Chapter 12 that retroviruses use the enzyme **reverse transcriptase** to copy RNA into DNA. Reverse transcriptase copies almost any RNA into DNA. Researchers who want a gene that bacteria can translate into protein obtain the mature mRNA for that gene, which has already had all its introns removed, from the cells of a eukaryote. For example, researchers could gather mRNA for human insulin from the pancreatic cells that normally make insulin. Reverse transcriptase then copies that mRNA back into DNA. The result is a DNA molecule, called **complementary DNA,** or **cDNA,** that is complementary to the mRNA from which it was copied. The cDNA "gene" is a version of the gene, but without the introns (Figure 13-8). Once the cDNA version of the gene is made, another enzyme copies the single-stranded cDNA into double-stranded cDNA, which can then be joined to a vector and replicated.

Actual genes from eukaryotes cannot be expressed in bacteria. Using reverse transcriptase, molecular biologists make a DNA copy (called cDNA) of a eukaryote's mRNA. The cDNA can then be inserted into the DNA of a vector.

How Do Researchers Make Multiple Copies of Recombinant DNA?

Once the cDNA is made, researchers need to make many copies of it, so it can be inserted into lots of cells. At one time, replicating the cDNA required that plasmids and bacteria make thousands of copies of the gene. Since Kary Mullis's invention of PCR, however, researchers have been able to make millions, or even billions, of copies of cDNA se-

Figure 13-8

How do researchers make human genes without introns? In a human cell, the gene for human insulin is transcribed into pre-mRNA. The cell edits out the introns to form mature mRNA. Researchers then isolate the mature insulin mRNA and copy it into DNA, which can be joined to a plasmid vector and expressed by a bacterium.

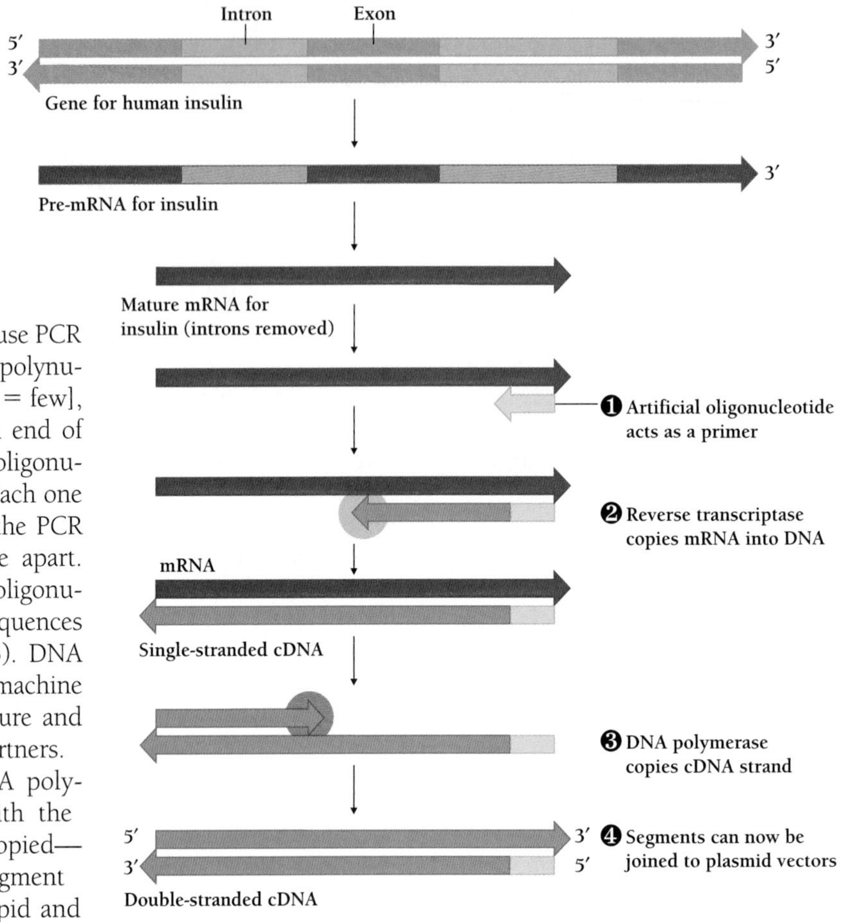

Intron   Exon

Gene for human insulin

Pre-mRNA for insulin

Mature mRNA for insulin (introns removed)

❶ Artificial oligonucleotide acts as a primer

❷ Reverse transcriptase copies mRNA into DNA

mRNA

Single-stranded cDNA

❸ DNA polymerase copies cDNA strand

❹ Segments can now be joined to plasmid vectors

Double-stranded cDNA

quences in a test tube. A researcher who wants to use PCR to copy a gene begins by making two short polynucleotides, called **oligonucleotides** [Greek, *oligos* = few], that correspond to nucleotide sequences at each end of the gene to be copied. These two synthetic oligonucleotides are usually about 20 nucleotides long. Each one serves as a primer for DNA polymerase. When the PCR machine heats the DNA, the two strands come apart. When the machine cools the DNA, the two oligonucleotide primers bind to their complementary sequences on the DNA—one on each strand (Figure 13-3). DNA polymerase then copies each strand until a PCR machine stops the reaction by again raising the temperature and causing the new strands to separate from their partners.

Lowering the temperature then allows DNA polymerase to go back to work, again starting with the oligonucleotide primers. The only region that is copied—again and again with further cycles—is the segment flanked by the two primers. The result is the rapid and exclusive multiplication of one specific sequence of DNA.

PCR allows researchers to make thousands or millions of copies of individual genes.

Can Host Cells Be Induced to Express Polypeptides in a Usable Form?

Despite the many ingenious tricks molecular biologists use to get bacteria to make eukaryotic proteins, recombinant *E. coli* bacteria still cannot actually make insulin that will function properly in the human body. Bacteria simply cannot express all eukaryotic genes. Recall from Chapter 11 that many polypeptides require extensive modification after they are translated from the mRNA. Bacterial cells lack the equipment to perform these modifications. All the same, molecular biologists have been able to manufacture hundreds of functional proteins using recombinant DNA molecules in bacteria. How do they do it? The answer is that because eukaryotic cells *can* modify proteins after translation, the solution is to insert genetically modified genes into eukaryotic cells and get them to express the genes.

One useful animal is a moth that is susceptible to infection by a virus called baculovirus. Once the moth is infected by the virus, the moth's cells devote almost all their protein synthesis to the production of the virus's coat protein. By sub-

stituting another gene for the coat protein gene, molecular biologists can get the moth cells to make any protein they like.

Other useful cells that can be genetically modified and then grown in immense vats include yeast cells, human kidney cells, and hamster ovary cells. Among the commercially important proteins made by eukaryotic cells from recombinant DNA are insulin (used to treat diabetes), tissue plasminogen activator (TPA, used to treat heart attacks and strokes), and erythropoietin (used to stimulate red blood cell production, especially after kidney transplants). Another approach is to insert a gene into the fertilized egg of an animal or plant. When the animal or plant grows up, its cells express the gene. We'll discuss such "transgenic" organisms later in this chapter.

Genetic engineers can use recombinant DNA technology to make an extraordinary number of products. These products include purer vaccines, ingredients for processed foods, and enzymes needed to make valuable small molecules or to destroy pollutants, and human insulin and growth hormone (Figure 13-9).

Some of the proteins made using recombinant DNA are so rare and so hard to isolate that little was known about them before until recombinant DNA technology made it possible to make large enough amounts to study. *Growth fac-*

Figure 13-9
Effects of human growth hormone (hGH). hGH cures certain forms of dwarfism. Formerly hGH had to be isolated from human cadavers, and recipients of the growth hormone too often contracted hepatitis and other viral diseases. Today, hGH is produced by genetically engineering bacteria and is free of viruses that infect humans.

tors, for example, are rare proteins that keep specific kinds of animal cells alive, stimulate cell division, or trigger particular types of cell specialization. *Cytokines* are other rare proteins that coordinate the immune response. Many laboratories are now investigating the possibility that growth factors may be able to prevent nerve degeneration in diseases such as Alzheimer's disease, Parkinson's disease, amyotrophic lateral sclerosis (ALS), and Huntington's disease.

Bacteria and various eukaryotic cells can produce large amounts of specific proteins that would otherwise be available only in small amounts. Genetically engineered bacteria and eukaryotic cells can make useful proteins, including enzymes, vaccines, drugs, and human proteins.

13.3 How Do Biologists Study Genes?

Molecular biologists have developed dozens of clever techniques for working with genes. Two of the most useful are (1) the creation of "gene libraries" for different organisms, from which researchers can select the genes they want to work with and (2) "DNA arrays" that reveal patterns of expression in thousands of genes at once.

How Do Biologists Find the Right DNA in a Gene Library?

We have seen that the genome of any organism—whether bacterium, plant, or animal—can be cut into pieces with restriction enzymes. The collection of thousands or mil-

lions of restriction fragments from a single genome is called a **gene library.** These fragments can be combined with plasmid vectors and introduced into populations of bacteria, which can then replicate large numbers of each fragment in the library. If the library contains at least one representative of every sequence in a genome, it is called a "complete genomic library." Alternatively, a recombinant DNA library may be constructed from mRNA alone (using reverse transcriptase), in which case it is called a "cDNA library." In either case, a gene library is useful only if researchers can find the gene they want. Until the invention of PCR, isolating the right gene was like trying to find a needle in a haystack.

Molecular biologists have developed techniques for finding and cataloging the tens of thousands of sequences in a gene library. Just as a magnet can help find a needle even in a haystack, molecular biologists can use two kinds of molecular "magnets" to find particular pieces of DNA. One kind of magnet is a DNA **probe,** a short bit of single-stranded DNA whose sequence is complementary to a small part of the sequence in the gene researchers are looking for, much like the DNA primers used in PCR. The other magnets are **antibodies,** proteins from the immune system that recognize and bind to other proteins. As we will see, antibodies detect not the recombinant DNA itself but rather the proteins specified by the DNA.

A DNA probe, also called a **hybridization probe,** can hybridize by base pairing with DNA in the library to form **hybrid DNA**—DNA whose two complementary strands come from different sources (Figure 13-10A). Such a probe, dropped into a test tube filled with thousands of restriction fragments, can locate and pair with the one complementary fragment as easily as a bloodhound can track down a lost child. If the probe is radioactively labeled, then once it binds to its complement, biologists can easily locate the probe and its complementary sequence of DNA. The technique of hybridizing a labeled probe to find the right piece of recombinant DNA is called molecular hybridization.

Researchers who know what protein they are looking for can also use antibodies to find the gene for that protein (Figure 13-10B). Antibodies are proteins with surfaces that recognize the shapes of foreign molecules, thereby giving animals an important defense against infection. When an animal is exposed to a foreign protein, it responds by making antibodies that bind specifically to that protein. Researchers start with a complete cDNA library laid out in a half-million colonies of bacteria, each colony carrying and expressing a different cDNA "gene" from the human genome. Over a few days, the 500,000 colonies can be treated with labeled antibody, which binds to and identifies the colony that is synthesizing the protein and therefore carries the gene the researchers are looking for.

Two kinds of molecular probes allow biologists to locate specific genes, hybridization probes, and antibodies.

Figure 13-10

Two techniques for locating a gene. A. A hybridization probe locates a specific DNA sequence. B. Antibodies locate the protein product of the same sequence.

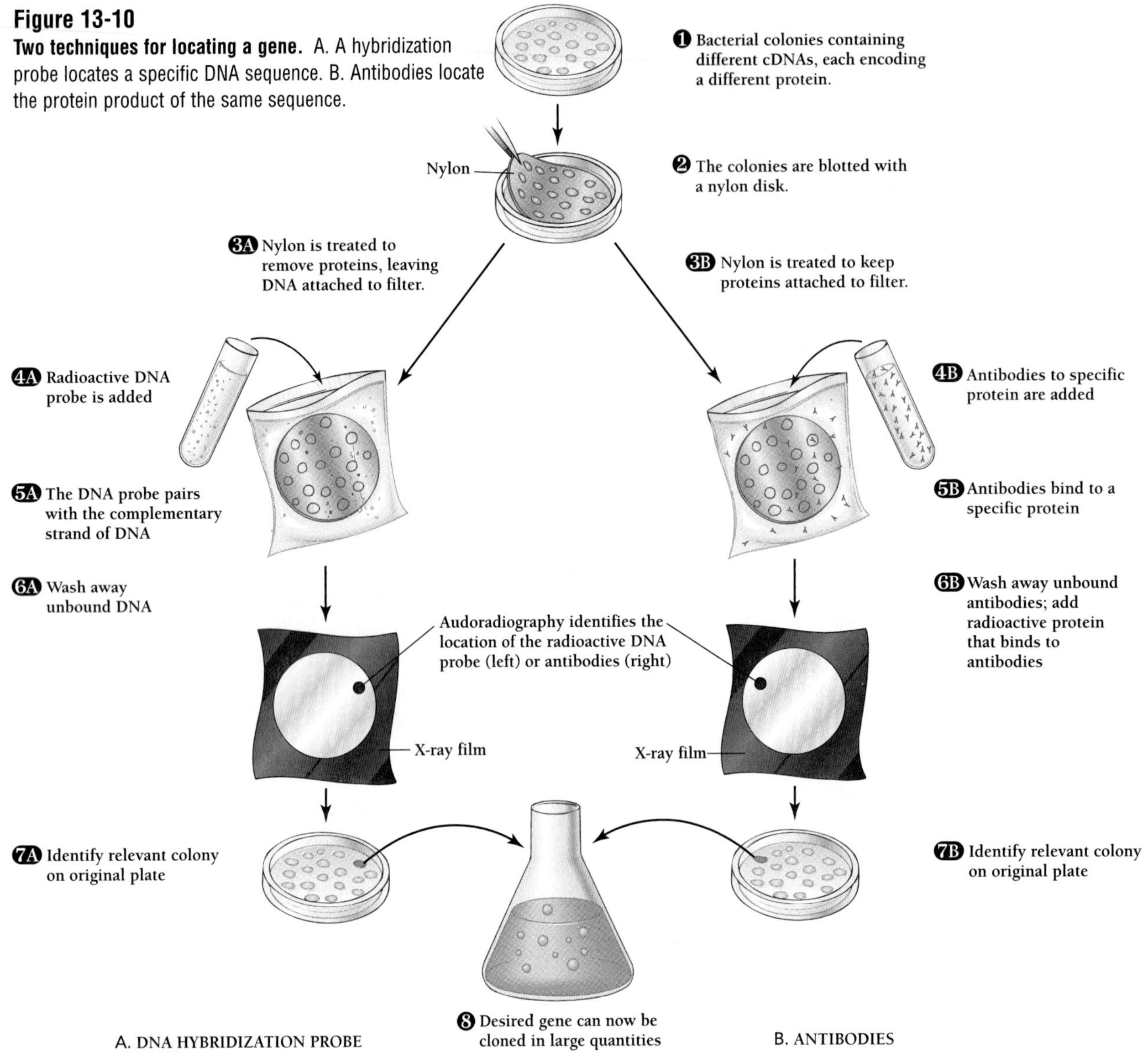

❶ Bacterial colonies containing different cDNAs, each encoding a different protein.

❷ The colonies are blotted with a nylon disk.

Nylon

3A Nylon is treated to remove proteins, leaving DNA attached to filter.

3B Nylon is treated to keep proteins attached to filter.

4A Radioactive DNA probe is added

4B Antibodies to specific protein are added

5A The DNA probe pairs with the complementary strand of DNA

5B Antibodies bind to a specific protein

6A Wash away unbound DNA

Autoradiography identifies the location of the radioactive DNA probe (left) or antibodies (right)

6B Wash away unbound antibodies; add radioactive protein that binds to antibodies

X-ray film

X-ray film

7A Identify relevant colony on original plate

7B Identify relevant colony on original plate

❽ Desired gene can now be cloned in large quantities

A. DNA HYBRIDIZATION PROBE

B. ANTIBODIES

How Can Biologists Study Thousands of Genes at Once?

A DNA probe for a specific sequence can hybridize to the DNA of the corresponding gene. But the same probe can also hybridize to the mRNA transcribed from that gene. Researchers can therefore use a cDNA probe to measure the amount of a particular mRNA in a tissue. The amount of mRNA indicates how much the tissue is expressing that particular gene. Not surprisingly, different cells express different genes. Muscle cells contain lots of mRNA for the muscle protein myosin and skin cells contain lots of mRNA for the skin protein collagen. But researchers could also look at the expression of much less common proteins, growth factors, and other regulatory genes. For the first time, they were able to study how the expression of small amounts of one gene deeply affects the expression of dozens of other genes.

Until recently, researchers had to test gene expression in tissues one gene (one mRNA) at a time (see box, page 269). But in the last few years, a powerful new technology has allowed them to study the expression of thousands of genes at a time. In organisms with small genotypes, such as the roundworm *C. elegans,* it is now possible to measure the simultaneous expression of every polypeptide the organism makes. Even in humans, researchers can now look at the expression—at the level of mRNA—of *every* gene in a particular cell type. For example, it's possible to compare the gene expression profile of thousands of genes in healthy breast tissue with that in cancerous breast tissue and discover which genes are turned on or off in the unhealthy tissue compared to normal breast tissue.

To measure so many kinds of mRNA, researchers use thin, dime-sized wafers, containing thousands of tiny DNA spots in a fixed arrangement. Each DNA spot is put in place by a robot

Extreme Biology Southern, Northern, and Western Blotting

Stop by a molecular biologist's lab bench today and chances are you will find a plastic box filled with gel and wired up to run an electric current through the slab of clear gel. Using a technique called gel electrophoresis, researchers can separate large molecules on the basis of their size. Gel electrophoresis is one of the most useful techniques available for studying DNA, RNA, and proteins.

In a gel containing sequences of DNA, the current draws the negatively charged DNA through the gel—a porous matrix that slows down the larger molecules as they snake and bump their way toward the positive electrode. A special stain reveals bands of DNA molecules, which group at different positions in the gel according to their size. The smaller fragments have moved farther, and the larger fragments remain closer to the rectangular well in the gel where they were first "loaded."

Sometimes, however, there are so many bands that they blur together and it can be hard to tell one fragment from another. For such complex samples, molecular biologist Ed Southern in 1975 developed a technique that lets the experimenter choose which fragments to look at. Usually, only a particular part of the genome is of interest—perhaps a specific gene or a polymorphic region used for DNA profiling. In Southern blotting and hybridization, the DNA is separated by means of electrophoresis (Figure A). Then, for more convenient handling, the DNA is transferred from the thick, watery gel to special paper. The paper, placed between the gel and a stack of paper towels, wicks the DNA and all the liquid out of the gel. The result is a Southern blot.

Next, the researcher applies a radioactive DNA probe that base pairs, or hybridizes, with the region of interest. (One or more of the phosphorus atoms in the DNA probe is radioactive.) Since the DNA is normally already double stranded, heat is applied to melt the strands apart, allowing the probe to find its target. When the temperature is again lowered, some of the single-stranded probe molecules hybridize with DNA fragments in the blot.

The result of this Southern hybridization is a sheet of paper the size of the original gel with radioactive bands of DNA wherever there were fragments of the gene of interest. The Southern blot is then exposed to a sheet of film, which reveals only the bands of DNA lit up by the probe. The DNA sequences of interest light up, while irrelevant bits of DNA remain invisible.

Several years after Southern developed his blotting technique for DNA, scientists began studying complex samples of RNA with a similar procedure and dubbed the technique northern blotting. Other scientists blotted proteins out of protein gels and called it western blotting. There is no such thing as an eastern blot, but the other three points of the compass come up in laboratory discussions every day.

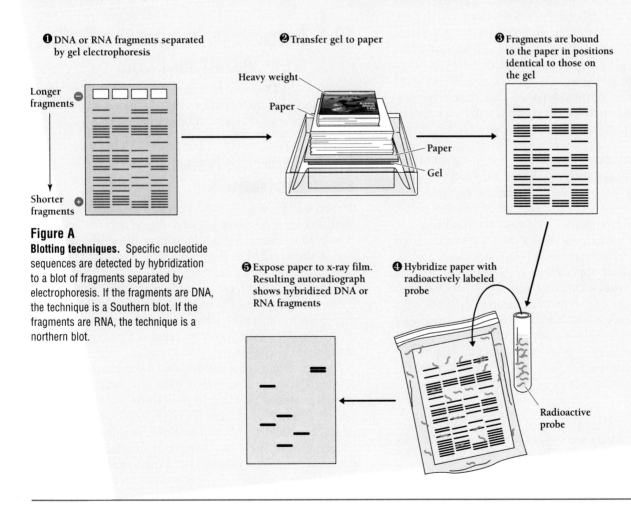

❶ DNA or RNA fragments separated by gel electrophoresis

Longer fragments

Shorter fragments

❷ Transfer gel to paper

Heavy weight

Paper

Paper

Gel

❸ Fragments are bound to the paper in positions identical to those on the gel

❺ Expose paper to x-ray film. Resulting autoradiograph shows hybridized DNA or RNA fragments

❹ Hybridize paper with radioactively labeled probe

Radioactive probe

Figure A

Blotting techniques. Specific nucleotide sequences are detected by hybridization to a blot of fragments separated by electrophoresis. If the fragments are DNA, the technique is a Southern blot. If the fragments are RNA, the technique is a northern blot.

or (in some cases) by a lithographic process derived from integrated circuit technology. The whole arrangement of spotted DNA is called a **DNA array** or chip. Each spot in the array contains some of the DNA from just one gene, enough to indicate its unique sequence of nucleotides. Although a DNA array may contain 6,000 or even 10,000 spots, each one corresponds to a particular gene in the genome. (Researchers now think that mammals express 80,000 to 100,000 different mRNAs.)

To actually measure gene expression, researchers isolate mRNA from a tissue, then copy each mRNA into a length of cDNA that is tagged with a fluorescent marker. These thousands of kinds of fluorescent cDNAs are then applied to the DNA array, and each bit of cDNA hybridizes to the DNA spot for its particular gene. In doing this, all the cDNA becomes arranged, as if ushers at a concert had directed each one to the correct seat in a stadium or concert hall.

Researchers can then examine how much cDNA has ended up at each spot. Very bright fluorescent spots that are "lit up" indicate lots of cDNA. Lots of cDNA means the tissue had lots of mRNA for that particular gene. By looking at the pattern of brightness for thousands of genes at once, researchers can perform a variety of analyses, looking for patterns of gene expression in different kinds of tissues (Figure 13-11). They can compare healthy and unhealthy tissues. They can compare the same tissue at different stages of development. They can compare similar tissues that seem to do slightly different things, like the different parts of the brain.

The simplest and crudest way to look at the patterns revealed by DNA arrays is to look for just a handful of the brightest spots, find out what genes those are, and then try to figure out why they are lighting up. But an array contains far more information than that, and looking at just a few spots is like focusing on only the 10 brightest pixels on your computer screen and ignoring the rest of the image.

More sophisticated approaches use a statistical approach called "cluster analysis" to look for groups of genes that light up together or go dark together, for example. Such clusters of bright spots (which may not be together on the array) might indicate genes that are all regulated together. The analysis of patterns of biological information, using computers and statistics, is called "bioinformatics."

One of the first applications of DNA arrays was the study of cancer tissues. Different kinds of tumors have different patterns of gene expression, and researchers hope that they can predict which tumors will grow rapidly or spread throughout the body and which will be easier to treat. Such differences in gene expression will help physicians of the future predict how best to treat a particular tumor.

DNA arrays allow researchers to study patterns of expression in thousands of genes at the same time. The analysis of the data from DNA arrays is an important and rapidly moving area of biotechnology and one form of bioinformatics.

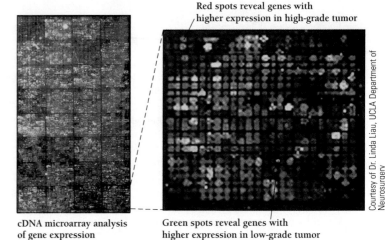

Red spots reveal genes with higher expression in high-grade tumor

cDNA microarray analysis of gene expression

Green spots reveal genes with higher expression in low-grade tumor

Courtesy of Dr. Linda Liau, UCLA Department of Neurosurgery

Figure 13-11

DNA array. DNA arrays allow the simultaneous monitoring of the expression of thousands of individual genes. The array on the left shows the expression of 18,000 different genes. In the blow up at right, differences in expression (as measured by mRNA) are coded by green and red spots. In this array comparing two kinds of tumors, the faintly red spots indicate individual genes whose expression is higher in a high-grade, malignant brain tumor than in a low-grade, less malignant brain tumor. Faint green spots indicate genes whose expression is higher in the low-grade tumor.

13.4 Recombinant DNA Can Genetically Alter Animals and Plants

Recombinant DNA can be inserted into the cells of whole plants and animals. Organisms that carry recombinant DNA in their genomes are called **transgenic organisms.**

How Do Researchers Produce a Transgenic Mammal?

The only way for a gene to be present in all the cells of an animal is for researchers to put the gene into the animal at the very beginning of its development before the fertilized egg (a single cell) has begun dividing into lots of smaller cells. Once a gene has been inserted into the genome of the fertilized egg, all the cells that develop from that egg will contain the engineered DNA. One major difficulty in creating transgenic animals has been inserting the gene in a place in the genome where it will be expressed properly and not disrupt other genes and kill the animal.

The first transgenic animals were mice created a quarter century ago. The mice were given the gene for human growth hormone (hGH), which made them grow twice the size of normal mice (Figure 13-12A). Today, molecular biologists have created thousands of different mice, as well as transgenic pigs, goats, and sheep (Figure 13-12B). The first genetically modified primate was a rhesus monkey carrying a gene from a jellyfish.

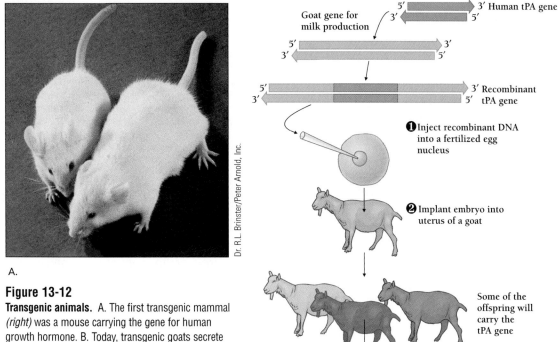

A.

Figure 13-12

Transgenic animals. A. The first transgenic mammal *(right)* was a mouse carrying the gene for human growth hormone. B. Today, transgenic goats secrete valuable proteins in their milk. The proteins are isolated from the milk for use as pharmaceuticals. This goat carries the gene for tissue plasminogen activator (tPA), a protein that activates the anticlotting system and is used to save the lives of heart attack patients.

Biology ⑧ Now™ Learn more about transgenic animals by clicking on this figure on your BiologyNow CD-ROM.

Goat gene for milk production

5′ ———————————————— 3′ Human tPA gene
3′ ———————————————— 5′

5′ ———————————————————— 3′
3′ ———————————————————— 5′

5′ ———————————————————— 3′ Recombinant
3′ ———————————————————— 5′ tPA gene

❶ Inject recombinant DNA into a fertilized egg nucleus

❷ Implant embryo into uterus of a goat

Some of the offspring will carry the tPA gene

❸ Milk is collected from lactating transgenic goats

❹ tPA is extracted from milk

B.

To produce a transgenic mammal, researchers must surgically remove eggs from the ovaries in the abdomen of a female mammal such as a mouse. Only a few eggs can be gathered in this way. These few eggs are put into a dish and fertilized with sperm from a male. The resulting zygotes are then injected with the engineered gene. Only about one transgenic animal is born for every 100 injected zygotes. And since only a few eggs are available from each female, creating transgenic animals is a labor-intensive and therefore expensive process. On the other hand, once a healthy transgenic animal grows up, it can pass its engineered genes to its offspring through its own eggs or sperm.

Why is creating a transgenic animal so difficult? First, few of the zygotes actually incorporate the recombinant DNA. And even if a gene inserts into a chromosome, it does so randomly and may end up in the middle of an important gene. Randomly disrupting key genes creates mutant animals that are sickly or die young, a major reason why it would be unethical to make transgenic humans.

Replacing an animal's genes with a recombinant version, instead of merely adding a gene randomly, is easy in prokaryotes and yeast but difficult in animal cells, although it has been done. Much easier to do is to create genetic

knockouts—animals (usually mice) in which a particular gene has been completely inactivated (Figure 13-13).

Knockouts are a useful way of creating research animals with specific genetic defects, similar to the ones that cause genetic diseases in humans. In the past 20 years, researchers have produced thousands of kinds of transgenic mice, many of them knockouts. These mice help researchers learn what causes certain diseases and what kinds of treatments work.

Knockout mice are often used to find out what a particular gene does. But the results are often a surprise. Figure 13-13, for example, shows a mouse in which researchers knocked out a gene specifying a growth factor they believed would influence early embryonic development. The researchers expected the knockout mice to stop developing and die long before birth. Instead, the mother mouse gave birth to healthy baby mice with hair so long that they looked a bit like Persian cats. The researchers had to go back and rethink their understanding of what this growth factor does.

Surprises like hairy mice aren't the only drawback to the knockout approach to understanding what genes do. Quite often, all of the individual knockouts die so early in development that little can be learned about what the gene did. On the other hand, other knockouts show no effects at all

Figure 13-13

Knockout mouse. Researcher Gail Martin decided to knock out a growth factor gene, called FGF5, believed to play a role in the very early formation of an embryo. Martin and her colleagues expected that none of the homozygous embryos would finish developing. Instead, the knockouts all developed normally, but with hair nearly 50 percent longer than normal. Although FGF5 is expressed in muscle cells, brain cells, and motor neurons, all of these tissues were normal.

from the loss of a gene. It seems that our mammalian genomes are flexible and redundant enough to compensate for the loss of certain genes. For example, when researchers created knockout mice unable to make the hormone oxytocin, which is produced by mammals during labor and commonly given to women to induce labor, the knockout mice delivered their pups with no difficulty. It was hard for medical researchers to believe that oxytocin wasn't essential to delivering pups, but there is no other explanation. Knockouts are full of surprises.

But interpreting those surprises is problematic. Knocking out genes to see what they do is like trying to find out what the different objects in a house do by removing them one at a time. If we remove the washing machine, the laundry no longer gets washed. But the same thing occurs if we remove the laundry detergent, or the clothes, or the electrical outlet into which the washing machine is normally plugged. Each of them is essential to getting the clothes done, but no single one of them is solely responsible for clean clothes. Molecular biologists sometimes announce that they have found a "master gene" for a trait because damaging the gene results in animals that lack wings or some other dramatic malformation. But the "master" gene could equally well be described as the "weak link" in the intricate chain of events and materials that make up a phenotype. Teasing apart that chain of events has become a central focus of modern research.

Currently, molecular biologists are studying hundreds of different and overlapping "cascades" of gene regulation. In such a cascade, a hormone or other signaling molecule, usually influenced by some aspect of its cellular environment, triggers the expression or suppression of one or more other genes. The turning on or off of these genes, in turn, affects the expression of another set of genes. Gene products (or their products) can circle back in loops or split up and turn on one gene while shutting another one off. The permutations and complexities are endless.

Transgenic mammals are made by inserting an engineered gene into a zygote (fertilized egg), the eggs for which are surgically harvested from female mammals. Only about 1 embryo in 100 will grow up to be a healthy transgenic animal, so the procedure is unsafe for humans. Transgenic animals are useful for studying diseases, protein function, and the complex cascades of gene regulation that make animals normal and healthy.

Genetically Engineering Plants Is Easy and Common

Genetically engineering crop plants is a routine and lucrative industry. Plants are easier to engineer than animals, and the market for plants with the right traits is enormous. For example, if an engineered tomato like the one mentioned earlier in this chapter became popular among growers, its inventors could gain a virtual monopoly in the huge tomato market. More than two-thirds of all genetically modified crops on Earth are grown in the United States, mostly in the form of soybeans, corn, and cotton. In 2001, more than two-thirds of all soybeans and cotton grown in the United States were genetically modified (GM), with the numbers going steadily up. About one-quarter of all corn was genetically modified, as well. In some states, virtually all of the main crops are genetically modified. For example, in 2001, 92 percent of all cotton grown in Louisiana was genetically modified.

One way that researchers carry genes into plant cells involves a species of soil bacterium called *Agrobacterium tumefaciens*. *A. tumefaciens* carries a plasmid that causes plants to grow tumors. The plasmid, called the **Ti plasmid** (for Tumor inducing), inserts itself into the genome of a plant cell and triggers rapid cell division and therefore a tumor. The Ti plasmid can be used to carry engineered genes into a plant's genome. As in other forms of genetic engineering, the recombinant DNA usually includes an antibiotic marker that allows researchers to select the cells that have taken up the recombinant DNA.

In an even more direct approach, called the **gene gun**, tiny particles of gold or tungsten are coated with fragments of recombinant DNA, then fired into a young plant with a miniature "shotgun" powered by gunpowder or compressed gas. The young plant is then chopped up and its cells grown on a culture containing an antibiotic. The antibiotic kills all of the cells except those that have taken up the recombinant DNA introduced with the gene gun.

The kinds of genes introduced tend to make the crops resistant to herbicides, insects, or diseases. For example, most GM corn carries the gene for a protein that is toxic to insects, called Bt toxin. All the cells in a corn plant make the toxin, which kills any insects that feed on the corn. In 1999, farmers began to plant soybeans engineered to resist a fun-

Extreme Biology   Can Biologists Clone Dinosaurs?

Scary as *Tyrannosaurus rex* must have been, few people would pass up a chance to see the great flesh-eating dinosaur alive. Is it possible that DNA technology will someday bring dinosaurs back from extinction to populate theme parks everywhere? Can we scrape together bits of prehistoric DNA from fossils, stuff them in a big egg, and watch a baby brontosaurus poke through the shell?

The answer is no. Researchers cloned frog embryos in the early 1950s, and cloning adult animals became reality in 1997 with the birth of Dolly, a lamb whose genetic makeup was taken from the mammary tissue of an adult ewe. But the creation of a living organism from scraps of ancient DNA is an entirely different problem.

The first challenge for would-be dinosaur cloners is to find the genetic material of a race of animals that died out 65 million years ago. Little more remains of the ancient beasts than fossilized bones, in which minerals have replaced the organic material, essentially turning the bone into rock. Fossils therefore contain little or no DNA. Blood from ancient mosquitoes preserved in amber may be a more likely source of Jurassic genes. But tapping into that resource raises other questions: What animal provided the insect's last meal? Did mosquitoes even bite dinosaurs? And if they feasted on sev-

eral different species, which DNA in the insect's belly came from whom?

Even if scientists found a reliable source of dino DNA, a complete dinosaur genome would be nearly impossible to recover. For one thing, many scientists believe that DNA is not likely to survive intact for so long. One study has shown that DNA in a water solution will break down almost entirely into individual nucleotides after about 50,000 years. Still, under the right conditions, DNA can last longer. In 1990, geneticists successfully extracted DNA from a 17-million-year-old fossilized magnolia leaf dug up in Idaho. Two years later, researchers used PCR to amplify short gene fragments from a termite and a stingless bee, both encased in amber for 30 million years.

Attempts to recover dinosaur genes have been less successful. In 1993, scientists in Montana found blood cells in a *T. rex* bone that had partly fossilized, thus protecting the central cavity from the elements. Unfortunately, the only sequences isolated from the sample later turned out to be human DNA contaminating the specimen. Other labs have had similar contamination problems in trying to squeeze dino DNA from ancient bones. Nevertheless, some researchers remain confident that PCR will eventually produce authentic dinosaur gene fragments. But even then,

those few hundred bases will be a far cry from the millions and possibly billions of bases of sequence that once resided in dinosaur chromosomes.

The lack of an appropriate egg is yet another nail in *T. rex*'s coffin. Even in the unlikely event that scientists discovered a cache of perfectly preserved dinosaur chromosomes, they would need to place them in an egg from some suitable modern animal, such as a bird or snake. But it's unlikely that eggs from animals that have evolved for millions of years longer than dinosaurs would provide a suitable environment for dinosaur DNA.

If dinosaurs lie on the evolutionary pathway to modern birds, as many biologists theorize, then the ostrich egg might not be a bad one to start off with. But an egg is more than just a protective coating for DNA. Eggs contain information about how cell division should proceed, or how the embryo will be oriented (which side is up, for example). It seems improbable that an ostrich egg, or any modern egg, would resemble a dinosaur egg closely enough to support the growth of a baby dinosaur, even with the proper genetic instructions.

Cloning living animals is technologically feasible, though not so far in humans. But for better or worse, the great dinosaurs are certainly gone for good.

---

gus (Phyophthora) that causes root rot. Large numbers of seeds made by the Monsanto Corporation and marketed as "Roundup-Ready" carry a gene that allows the resulting plants to resist the herbicide glyphosate (Roundup), also made by Monsanto. These engineered crops allow farmers to spray an entire field with herbicide and kill the weeds but not the crops (Figure 13-14).

One engineered crop, "high-oil" corn, makes twice as much oil as conventional corn. This crop reduces the cost of making cooking oils and feeding livestock. Engineered soybeans make more of the amino acids lysine and methionine, which are in short supply in regular grains, and canola and sunflowers come with more of the fatty acids useful in cooking or in making soap. Genetically modified plants can also make valuable human proteins. For example, tobacco plants that make human antibodies can be grown in fields with a yield of hundreds of pounds per acre. Plant-made antibodies seem to be indistinguishable from the ones made by the mammal they came from. Even GM tomato plants can do

more than make good pizza. One tomato makes human serum albumin, a protein used to treat burn victims.

Using either the Ti plasmid or the gene gun, molecular biologists can genetically engineer plants that can, for example, synthesize animal or plant proteins, resist herbicides, or resist infection by plant viruses.

What Are the Risks of Recombinant DNA in Agriculture?

The incredible power of recombinant DNA technology lies partly in how precisely and quickly changes can be made and partly in the ability of molecular biologists to transfer genes from one organism to another. The long-term consequences of moving genes from one organism to another are entirely unknown. Genetically engineered plants can, for ex-

Figure 13-14

Transgenic cotton. Both the normal cotton plant *(left)* and the transgenic cotton plant *(right)* have been sprayed with a potent herbicide. Only the transgenic plant flourishes, however, for it alone possesses a gene for herbicide resistance.

Biology ⑧ Now™ Learn more about transgenic plants by clicking on this figure on your BiologyNow CD-ROM.

Calgene

ample, exchange their new genes with wild cousins, either through conventional interbreeding or by way of viruses or other mobile genes. A gene for herbicide resistance could conceivably make its way from a GM crop to a population of weeds and other wild plants. Whether engineered plants and animals might interact with their environments in unforeseen, potentially destructive ways is unknown.

The United States biotech industry is enthusiastic about GM crops and most Americans accept them, but they are controversial in other parts of the world. Popular opinion in Europe, especially, is strongly opposed to the use of genetically modified food crops. People who object to GM foods specifically object to one or more of the following: dispersing the genes and genetic material of viruses and bacteria that are known to cause diseases in plants; dispersing genes for antibiotic resistance in the food supply; using gene promoters that can function in all plants, green algae, yeasts, and bacteria.

They also object to the tendency of engineered genes and their promoters to jump to the genome of other organisms. For example, engineered genes can jump to soil fungi and bacteria or even to bacteria inside the animals that eat the plants. In one study, pollen from GM rapeseed that was resistant to a herbicide was fed to bee larvae. Bacteria and yeast from the gut of the bee larvae carried genes for herbicide resistance.

A different objection to GM foods is more political and economic than biological. Because genetically engineered seeds are patented, farmers are not supposed to save seed from one year for planting the following year. Seed companies have been known to sue farmers who replanted with (patented) seeds from their own crop. Many critics are concerned that the increasing use of patented seeds will make farmers in developing countries economically dependent on just a few corporations.

Critics of genetically modified crops are concerned with the spread of genes that may be ecologically harmful as well as with the increasing dependence of farmers on a few seed producers.

We do not yet know all the limitations of the new technology. We must each evaluate the risks, benefits, and moral implications of genetic engineering. In the next chapter, the last chapter of this section on genetics, we will explore some of these issues in more detail.

Key Concepts

- Genetic engineering started with the selective breeding of organisms with high levels of natural or induced variation.
- Recombinant DNA technology allows biologists to create genes that do not exist in nature.
- Using recombinant DNA, molecular biologists can engineer both prokaryotic and eukaryotic cells to make specific proteins in large amounts.
- Biologists can also engineer plants and animals to produce new proteins and to have new biological properties.

Summary with Key Terms

What does it mean to "genetically engineer" an organism?

Biotechnology is thousands of years old. Recombinant DNA technology is new. A molecule of **recombinant DNA** consists of two or more DNA segments that are not found together in nature. For example, by combining DNA pieces from humans and *E. coli*, researchers made a recombinant DNA molecule with the regulatory region of the *lac* operon and the gene for insulin. Bacteria that contain this DNA make insulin in response to the presence of lactose.

Making recombinant DNA depends on the use of enzymes that can cut, link, and modify DNA. **Restriction enzymes** cut DNA at **restriction sites,** where specific sequences occur, allowing molecular biologists to isolate individual DNA fragments, called **restriction fragments. DNA ligase,** an enzyme used in DNA replication, links pieces of DNA from different sources. A **restriction map** shows the positions of the restriction sites in a piece of DNA.

DNA molecules derived from viruses and plasmids can serve as **vectors** for the propagation of pieces of recombinant DNA within host cells. Some of the most widely used vectors derive from R factors, plasmids that carry antibiotic resistance genes. Other vectors include bacteriophages, as well as plant and animal viruses.

Polymerase chain reaction (PCR) is now a basic tool for finding and multiplying, or **cloning**, fragments of DNA. **Oligonucleotides**, short pieces of single-stranded DNA, act as primers for DNA polymerase. Alternating heating and cooling do much of the rest.

What steps must biologists take to insert a gene into the genome of another organism?

The starting materials for recombinant DNA may come from the genomes of viruses or organisms or from DNA copies of mRNA. Most plant and animal genes contain introns, which would interfere with their expression in bacteria. Recombinant DNA that codes for plant and animal proteins are derived from **complementary DNA (cDNA)**, copied from mRNA by the **reverse transcriptase** of a retrovirus.

Collections of recombinant DNA, called **gene libraries**, may contain thousands or millions of different DNA sequences. Finding a desired DNA sequence usually depends on the use of a **probe**. A **hybridization probe**—often a piece of DNA made in the laboratory—bonds with one strand of the DNA to form **hybrid DNA**. By labeling the probe with radioactive nucleotides or a chemical marker, the experimenter can identify recombinant DNA. **Antibodies**, which have surfaces that recognize the shapes of specific molecules, can help identify bacteria that are producing a particular protein.

How are transgenic plants and animals useful to biologists and others?

Recombinant DNA can reprogram organisms to make new products. Products such as insulin and **growth factor** that are made by genetically engineered bacteria are already commercially important as pharmaceuticals. Recombinant DNA can be used to alter animals and plants genetically. Plants, animals, and other organisms whose cells carry a gene added by scientists are said to be **transgenic.** Transgenic animals can only be made by injecting genes into a fertilized egg, or zygote. Molecular biologists can also damage a specific gene to make a **knockout** organism.

Making transgenic plants is easier than making transgenic animals because experimenters can easily add or replace genes in single cells in culture. Genes are introduced either by means of a tumor-inducing (or **Ti**) **plasmid** or by means of a **gene gun.** The use of GM crops is controversial, especially outside of the United States.

Review and Thought Questions

Review Questions

1. How does modern genetic engineering using recombinant DNA differ from breeding techniques practiced over the last 10,000 years? How are these two kinds of genetic engineering the same?

2. How can knowing a biochemical pathway help molecular engineers design useful organisms?
3. What is recombinant DNA?
4. What two enzymes are basic to assembling recombinant DNA? What do the two enzymes do, and what is their role in nature?
5. Why do molecular biologists need multiple copies of a gene?
6. How can molecular biologists copy a gene?
7. Why do molecular biologists want to engineer cells to express large amounts of the gene product, or protein? How do they accomplish this?
8. What prevents bacteria from expressing a gene cut directly from a human genome? How do molecular biologists get around this problem?
9. How do molecular biologists find the right DNA sequence in a recombinant DNA library?
10. Describe the steps of PCR.
11. What kinds of genes are added to transgenic crop plants?
12. How are transgenic mice useful to biologists?

Thought Questions

13. What is your opinion about creating transgenic humans? Suppose that you knew that you carried a gene that would predispose your children to a mental illness. Would you opt to conceive by means of *in vitro* fertilization so that your zygotes could be screened and the gene replaced? Suppose large numbers of other people chose not to use this service. Some of their children would continue to develop severe mental illness, eventually becoming rather expensive wards of the state. Would you feel that everyone should be required to screen his or her zygotes? What if you knew that the families of people carrying this gene tend to be more intelligent and more creative than average? Would that make a difference in your opinion? Why or why not?

14. Would you mind eating a genetically engineered tomato? Why or why not?

BiologyNow Resources

Biology ⦿ Now™

Active Figures

13-12: Transgenic animals
13-14: Transgenic plants

Preparing for an exam? Take a diagnostic test on your BiologyNow CD-ROM.

Online materials relating to this chapter are at:

http://biology.brookscole.com/AAL3

About the Chapter-Opening Image

Finding a particular DNA sequence is like finding a needle in a haystack. PCR simplifies the process by making many copies of the needle.

Human Genetics

14

Key Questions

- How different is one human being from another?

- How do environment and genome interact to create a phenotype?

- What was the Human Genome Project?

- Can we prevent or cure genetic diseases?

To Map the Genome or Hunt the Huntington's Gene?

Burly and bombastic, the professor jumped to his feet and paced the small conference room, loudly lecturing the other 12 scientists still seated at the table. It was October 1979, and the 13 scientists had met at the National Institutes of Health near Washington, D.C., to discuss the impact of molecular biology on inherited diseases. The discussion had come to a critical point of disagreement, and David Botstein, a brilliant and temperamental molecular biologist from the Massachusetts Institute of Technology (MIT), could no longer contain himself (Figure 14-1).

Beginning a search for the gene for a single disease now, argued Botstein, was both premature and a waste of time. New technology would soon revolutionize human genetics, he said, allowing scientists to map and identify every human gene, including those that caused genetic diseases. Once a general map was established, it would be child's play to find the molecular defects that cause Huntington's disease, cystic fibrosis, muscular dystrophy, and any of the other 3,000 inherited diseases that affect humans.

Seated at the table was David Housman, another molecular biologist from MIT. Housman watched Botstein impassively, waiting for a pause in the lecture. Then, speaking quietly, Housman made his own argument. Yes, eventually, all would be known, he conceded. But wasn't it better to start right now with a single disease? And why not begin with Huntington's disease, whose peculiar genetics made it a little easier than other genetic diseases? Researchers didn't need to wait for a complete genetic map to find the gene for Huntington's disease. There was no reason not to jump ahead, applying knowledge and techniques that were already available.

Botstein scoffed. It was short-sighted to jump into problems that might seem dramatic or even personally pressing but were not yet scientifically "ripe." Better for scientists to wait for basic knowledge.

But although the more passionate Botstein would turn out to be right, Housman prevailed at the 1979 meeting. The conference had been sponsored and paid for by a small foundation called the Hereditary Disease Foundation (HDF), dedicated to promoting research on Huntington's disease.

The HDF had been founded 10 years earlier by Milton Wexler, a prominent Beverly Hills psychoanalyst whose ex-wife Leonore had been diagnosed with Huntington's disease. Their two daughters, Alice and Nancy, were at risk for developing Huntington's, and the father was determined that researchers should find the gene and perhaps a cure. When Milton Wexler founded the HDF in 1969, his two daughters were still young, and he knew that if either of them became ill it would not be for about 30 years. It had to be enough time to help them (Figure 14-2).

Huntington's disease is a devastating neurological disease that causes jerky movements, mood swings, personality changes, and premature death. Huntington's usually begins in middle age with a characteristic unceasing motion of the arms and legs. Over a decade or two, people with Huntington's disease gradually lose control of their movement, memory, and mood, wasting away and finally ceasing to move at all.

Huntington's is an unusual genetic disease. Most genetic diseases are caused by recessive alleles, and people become ill only if they get two copies of the disease allele, one from each parent. But Huntington's disease is different: the allele that causes disease is dominant and people only have to receive a single copy of the allele to become ill. The reason Huntington's is different has to do with when it make people ill and with the way natural selection works. Most genetic diseases cause illness and death early in life, and people who have both copies of the disease allele die without leaving any offspring. The disease allele is maintained in the gene pool by "carriers," people with a single copy of the disease allele. In such diseases, a dominant allele would never be passed on at all and would be rapidly eliminated from the gene pool.

Because Huntington's doesn't make people ill until late in life, most people with the allele for Huntington's can lead normal lives until sometime after their 40s or 50s. Crucially, they

Figure 14-1
David Botstein. Irascible but right.

can have children, half of whom (on average) will carry the dominant disease allele and become sick themselves. Virtually everyone who inherits the Huntington's disease allele eventually develops the symptoms. The allele's high level of penetrance made its inheritance unusually clear.

In 1969, Milton Wexler had founded the HDF and begun engaging young scientists to brainstorm new approaches to understanding Huntington's disease. He attracted these biologists both by inviting them to parties that included some of the most famous artists and entertainers in the Los Angeles area and by encouraging these young biologists to think freely and creatively. Few young biologists could resist these brushes with Hollywood glamour, and most scientists who were invited came. Many were swept up in the excitement of finding a cure for Huntington's. On the other side, the Hollywood entertainers valued both the cause and the opportunity to talk to working biologists. The entertainers, some of whom were Milton Wexler's patients, also donated money to his foundation.

The purpose of the workshops was to bring together small groups of biologists who did not ordinarily meet—researchers in fields ranging from genetics to neurology. These informal workshops were intellectual incubators for new ideas not only about Huntington's disease but about the application of new techniques to the study of all inherited diseases, especially disorders of the brain. Among the young scientists stimulated by these workshops was one of this book's authors, Allan Tobin. In 1979, Wexler appointed Tobin Scientific Director of the Hereditary Disease Foundation, a position he held until 2002. It was Tobin who orga-

Figure 14-2
The fight against Huntington's disease. The famous singer Woody Guthrie *(left)* and the little known biology teacher Leonore Wexler *(far right)* both died of Huntington's disease. In both cases, their familes used considerable influence to initiate and support the search for the Huntington's gene. The portrait of the Wexler family shows Wexler's husband Milton and his two daughters Nancy and Alice Wexler.

Culver Pictures

Courtesy of Alice Wexler

nized the 1979 workshop on molecular genetics, with help from his former Cambridge neighbor, David Housman. One of the other researchers at the table was Nancy Wexler.

Nancy Wexler, the younger daughter of Milton Wexler, herself a researcher who has devoted her professional life to the search for the Huntington's gene, helped provide the raw data that led to the gene's discovery. A professor of neuropsychology at Columbia University, Wexler documented the inheritance of Huntington's disease within an extended family that lives on the shores of Lake Maracaibo in Venezuela. Within this population of about 10,000 people, several hundred have Huntington's disease at any given time, and hundreds more carry the gene.

By tracking the simultaneous inheritance of Huntington's disease and DNA markers within this family, James Gusella, Nancy Wexler, and their colleagues were able to determine the approximate chromosomal location of the Huntington's disease gene. They did this in 1983, 13 years before the rough-draft map of the human genome that Botstein so eagerly anticipated. But the accomplishment depended directly on the methods advocated by Botstein. The Huntington's allele was the first disease gene ever found with that method.

It took another 10 years to identify the actual gene for Huntington's disease. By then (1993), recombinant DNA techniques had already led to the precise identification of many other disease-causing alleles, including that for cystic fibrosis, the most common genetic disease among Caucasians. But the Hereditary Disease Foundation's gene search—undertaken despite Botstein's fierce arguments for patience—dramatically demonstrated the power of molecular genetics to help understand human disease.

But Botstein's fears that mapping the Huntington's gene would be difficult had turned out to be well founded. Despite the efforts of an international consortium of dozens of researchers—organized by Allan Tobin and Nancy Wexler for the Hereditary Disease Foundation—the Huntington's gene was not sequenced for 14 years after the 1979 meeting. The early mapping and sequencing of individual disease genes moved excruciatingly slowly.

To find a particular gene among the 40,000 or so genes on the human chromosomes, geneticists of the 1970s and earlier had to look for markers, obvious phenotypic characteristics, such as eye color, that are always inherited with the defective allele and are therefore near the defective gene. Geneticists had constructed linkage maps in fruit flies by carefully crossing and backcrossing flies with obvious phenotypes such as white eyes. But geneticists cannot set up matings of humans and instead must look at complex pedigrees that aren't always complete or even perfectly accurate. A few markers in humans, such as those for blood proteins, work well, but none of these were inherited with the Huntington's allele.

The early hunt for the Huntington's allele went extraordinarily slowly. Only when basic research by Botstein and others provided new, faster techniques did the mapping of Huntington's and other disease alleles begin to proceed quickly. Specifically, the picture changed when molecular biologists began to use recombinant DNA techniques to copy and sequence specific genes. They then discovered numerous small differences in the DNA of individuals.

Even though two people may make exactly the same hemoglobin proteins, their hemoglobin genes are likely to differ from one another. These "silent" differences lie in DNA sequences that are not transcribed and translated into protein—for example in introns or flanking sequences that aren't expressed in the mature mRNA. By the late 1970s, researchers understood that (except for identical twins) every person is genetically unique.

The technology that had so excited David Botstein in 1979 was a method for identifying the differences among individuals, which he called restriction fragment length polymorphisms, or RFLPs (pronounced "RIFFlips"). Within a few years, differences in DNA sequences would allow geneticists to produce a map of the human genome and also give police and military laboratories a tool for matching blood and tissue samples with a previously unimagined accuracy. This tool, commonly called DNA fingerprinting, has changed forever the prosecution of criminal suspects, and made it possible to identify the dead in crimes, disasters, and battles. By 1980, Botstein, his collaborators, and his competitors had established the feasibility of making a map of the human genome based on differences in DNA sequence. By 1994, researchers had succeeded in making a relatively complete RFLP map of the human genome, allowing researchers to locate genes within about 1 million base pairs of DNA. And by 2003, a consortium of labs around the world had finished sequencing the entire human genome.

Until the late 1980s, enthusiasm for making a map of the human genome was not universal, even among molecular biologists. The undertaking was staggeringly expensive, and the research required was repetitive and dull. But the successful hunt for the Huntington's disease helped provide momentum for the Human Genome Project.

Because of the efforts of Milton Wexler's HDF, dozens of researchers focused on applying known techniques to the specific problem of Huntington's disease. Choosing a single disease on which to focus their efforts gave biologists a sense of mission and a single, clear-cut goal that was emotionally appealing. They could imagine that their day-to-day research might one day save the charming Wexler daughters and other sufferers of Huntington's disease. In addition, the HDF's intimate workshops promoted fruitful scientific friendships and collaborations. And finally, James Gusella's location of the Huntington's gene in 1983 provided the "proof of principle" that may well have convinced both scientists and scientific administrators that mapping the human genome was feasible. But in the rest of this chapter, we will see that it was the basic research carried out by Botstein and others that invested the science of molecular genetics with the promise it now holds.

The genetics of humans is no different from the genetics of other organisms. For this reason, many biologists object to the very idea of having a chapter on human genetics. The *study* of human genetics, however, does differ in one very significant way from the study of the genetics of fruit flies, pea plants, or corn. Humans cannot be crossed at the whim of geneticists to provide the complex crosses and backcrosses that classical geneticists traditionally rely on.

Another difference between the genetics of humans and that of other organisms is that our own traits are, to us at least, infinitely more interesting and subtle. In addition, the degree to which a trait such as intelligence is the product of genetic influences or environmental influences rouses people to intense debate. In this chapter we will try to present enough information for readers to make intelligent judgments about what they read and hear. We will discuss genetic diversity in humans, the complex relationship between genotype and phenotype, and both the power and limitations of applying knowledge about human genetics.

14.1 Genetic Diversity in Human Beings

On a Saturday evening in the winter of 1983, five-year-old Alice Brown was abducted just steps from her own front door. She and her mother, Sarah Brown, were about to sit down to dinner with a neighbor in their 200-unit apartment complex in northern California when Alice's three-year-old brother, Brad, begged for his boots. Alice took Brad home to get his boots. It was just two doors down, but when Brad returned moments later he was alone. Alice's mother never saw her daughter alive again.[1]

When Alice's body was found a week later, police discovered that she had been raped and then suffocated. Sarah said that Alice would never have gone with any man except one: a neighbor in the same apartment complex whom Sarah had dated briefly. Police considered Harry Gordon, then 31, a prime suspect, but they had no concrete evidence to link him to the case. They had to have "probable cause."

All of the evidence was circumstantial. He'd given contradictory stories about his whereabouts on the night of the murder. Police had arrested him in 1981 for exposing himself to a little girl and again, in 1984, on suspicion of molesting a third girl. He was eventually convicted of a misdemeanor for exposing himself while drunk. Gordon's ex-wife said he was a likely suspect, and police pursued dozens of leads. But Gordon denied any involvement in Alice's death, and nothing gave the police the evidence they needed to arrest him.

Sarah and Brad moved away and changed their names. Thirteen years passed. Then, in the spring of 1996, when Alice would have been 18 years old, police finally arrested

[1]We have changed the names of those involved.

Gordon for the rape and murder of Alice Brown. It turned out that police had had the concrete evidence they needed all along. Subtle changes in California law and advances in molecular biology had finally made it possible for prosecutors to use that evidence to link Gordon to the crime and to know that they could probably convict him. (The same evidence had cleared a previous suspect.)

Just as every individual has a unique set of fingerprints, so every individual also has a unique DNA sequence. The set of unique restriction fragments that can be obtained from someone's DNA is called a "DNA fingerprint" or "DNA profile." The DNA for a DNA fingerprint can be extracted from any cell that has a nucleus. And thanks to the invention of PCR, researchers can analyze DNA from just a few cells. Such cells can be found in bloodstains or tears, in saliva on an envelope, and in skin cells under fingernails or in hair. In rape cases, the most important source of DNA is usually semen. When police found Alice's body in 1983, they had collected and stored samples of semen from her body and clothes. Whoever had killed her had left evidence that would someday be incriminating. But in 1983 DNA fingerprinting had not yet been invented, and there was no way to connect the semen samples to any particular man. Yet investigators carefully stored the sample, hoping it might someday be useful (Figure 14-3).

By the early 1990s, prosecutors knew that the DNA in the semen could tell them who had murdered Alice Brown. But a legal problem prevented them from acting. Unlike other states, California had not yet issued a ruling on the admissibility of DNA fingerprinting as evidence. As a result, lower courts had issued conflicting rulings, and a conviction based on a DNA fingerprint stood a good chance of being overturned on appeal. In 1992, a northern California judge had ruled that DNA fingerprinting was not as reliable as a real fingerprint. If prosecutors tried to convict Gordon and the court rejected the DNA fingerprinting, prosecutors knew they would have lost their only chance to convict Gordon, and at great cost. Just arguing the admissibility of DNA profiling in court could cost $50,000—the price of obtaining testimony from expert witnesses. The prosecutors handling the Alice Brown case were biding their time, but they had not forgotten their prime suspect, Harry Gordon, by any means.

How Reliable Is DNA Fingerprinting?

The haploid human genome contains some 3 billion base pairs of DNA. Yet only two to four percent of this DNA codes for polypeptides. The other 97 percent consists of noncoding DNA, including introns that are removed from mRNA before translation. As far as biologists can tell, noncoding DNA has no central role in forming the phenotype, the actual physical body, which means the noncoding DNA can mutate without affecting an individual's ability to survive and reproduce. As a result, noncoding DNA accumu-

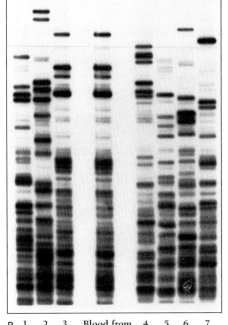

Cellmark Diagnostics

Figure 14-3

DNA fingerprinting. A. A gel of six samples of DNA treated so that it glows under ultraviolet light. B. DNA fingerprint of a criminal *(center)* compared to the DNA fingerprints of seven suspects *(numbered at bottom)*. The criminal's pattern of DNA fragments matches that of the third suspect from the left *(number 3)*.

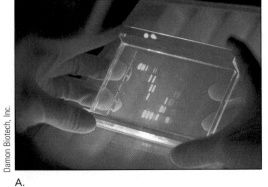

Damon Biotech, Inc.

A.

B. 1   2   3   Blood from   4   5   6   7
              crime scene

lates large numbers of mutations and contains far more variation than coding DNA. Thanks to this highly variable noncoding DNA, every person's DNA sequence is unique (not counting identical twins).

There are two ways to do a DNA fingerprint. The first DNA fingerprints were constructed using RFLP fragments. But RFLP analysis requires a large amount of DNA and takes several days to do. Once PCR was invented, researchers could do a profile with mere traces of DNA and in only eight hours. PCR profiling is now more common for most purposes.

Recall from Chapter 13 that restriction enzymes, sometimes called "chemical scissors," cut the extremely long DNA molecule into fragments. Each restriction enzyme cuts a different pattern of bases in the DNA. Because the exact sequence of bases is unique in each individual, the particular sequence that a restriction enzyme cuts occurs in different places in different people. And because the pattern occurs in different places, the length of the cut fragment differs from person to person. Geneticists call fragments that vary in length "polymorphic" (many forms) and so the fragments are called *restriction fragment length polymorphisms*, or RFLPs.

In the next step, electrophoresis, the RFLP fragments are exposed to an electric current that moves them through a jellylike substance, called a "gel." The DNA fragments are negatively charged, so they move through the gel toward the positive electrode. The different-length fragments travel through the gel at different rates. The shorter fragments travel through the gel rapidly, while the longer fragments get a bit tangled in the gel and move more slowly. After a while, the various fragments are spread out across the gel like runners in a race (see box on page 284). When the fragments

are marked (by DNA probes in a Southern blot), they produce a pattern of dark bands.

In very rare cases, two people might have the same pattern of fragment lengths. Therefore, all DNA samples are tested with more than one enzyme. The probability that two people will have the same pattern for two or more restriction enzymes is very small because the probabilities are multiplied together. For example, if the probability of two people having the same pattern for one enzyme were 1 in 1,000, then the probability for two enzymes would be 1/1,000 × 1/1,000, or 1 in a million.

If enough RFLPs are examined, every genome will show a unique pattern of dark bands. In practice, laboratories that analyze DNA samples examine only a small number of DNA fragments, those known to vary a lot from person to person. Researchers then calculate the probability that someone else in the general population has that exact set of base changes. Researchers can come up with a probability of uniqueness ranging from 1 in 500 to 1 in several billion.

Two factors affect this number. The first is the number of different markers examined. The greater the number of markers surveyed, the more specific the DNA profile will be, and the more other people in the world can be excluded from consideration. The second factor is how "the general population" is defined. As we will see later, different populations are more or less likely to have certain genotypes. The chances of someone having the same DNA profile as Harry Gordon might be one number for the population of San Francisco Bay Area and another number for the population of all of California. In any case, geneticists do not necessarily know what those probabilities are for different populations. They have estimates for large populations that have been studied, such as those of the United States, Argentina, or Spain, but not for individual cities such as San Francisco or Los Angeles.

For forensic fingerprinting, the RFLP technique is considered more accurate than PCR because it analyzes sequence differences on much longer strands of DNA. But RFLPs cannot be used in situations where only a few, short strands of DNA are present, such as a crime scene where a small bloodstain is the only evidence. PCR is so fast and can be done with such small amounts of DNA that it is virtually a miracle for forensics specialists.

The major drawback of DNA profiling with PCR is that because investigators rely on so few DNA molecules, it's very easy to contaminate a sample with stray bits of DNA from some other source. Because any carelessness could result in contamination, technicians must collect and process the sample with great care, both at crime scenes and back in the lab. Later, at the trial, lawyers closely examine the way the DNA evidence (like other evidence) was collected, labeled, stored, retrieved, and tested, looking for any mistakes the technicians may have committed.

Why Did Police Finally Arrest Gordon?

In the fall of 1995, prosecutors in the Harry Gordon case began to sense that DNA fingerprinting was coming to be accepted in California. Confident that if they took Gordon to court their evidence would be accepted, the police obtained a court-ordered search warrant that allowed them to compel Gordon to give a blood sample. They sent the blood sample and the sample of the semen from Alice Brown's body to a laboratory in Maryland. There, technicians used the older RFLP approach to determine that the two samples matched. In six different highly variable RFLPs, the pattern of bands was the same.

These six RFLPs each vary so much from person to person that it was highly unlikely that a perfect match between the recovered semen DNA and Gordon's DNA could have arisen by chance. The testing laboratory estimated that the DNA recovered from Alice's body could be matched to only 1 in 5.7 billion male Caucasians, some 10 times the number of male Caucasians in the world. Prosecutors finally had the tangible evidence for which they had waited so long. In the spring of 1996, they arrested Gordon and charged him with Alice's murder—13 years after her death. Many people hoped that his arrest would bring to an end a string of similar unsolved murders in the area.

How Can DNA Fingerprints Be "Wrong"?

The use of DNA-based evidence has been much debated in U.S. courts and in the pages of scientific journals. Even if the probability of someone else in the world having Gordon's DNA profile is virtually zero, his defense attorneys can argue that the evidence is inconclusive. Defending attorneys can argue (1) that the DNA samples could have been mixed up, so that the perfect match came from duplicate samples of Gordon's own DNA; (2) that the semen samples are so old that the evidence is unreliable; (3) that another person be-

sides Gordon could have the same RFLP pattern; and (4) that the RFLP analysis was itself faulty.

Questions about the proper handling of evidence apply to all physical evidence (hairs, threads, or fingerprints, for example). A prosecutor must demonstrate—beyond any reasonable doubt—that the police have properly collected the biological samples, preserved them appropriately, and not mixed them up.

In some cases, defense attorneys have argued that the tissue samples from the site of a murder have been contaminated by DNA from other sources. Because PCR-based DNA profiling is so sensitive, even a few cells from a different person could change the results. But unless the contaminating cells came from the suspect, they would not match the suspect's DNA. In some situations, DNA from blood or other tissues may prove that someone was present, but not when or under what circumstances. Such evidence makes up only a part of a prosecutor's arsenal of evidence.

Improper storage can prevent a proper identification but would not convict an innocent person. If over time the DNA were to break down as a result of exposure to heat, sunlight, or chemicals, the changes in base sequence would be random and extremely unlikely to exactly match Gordon's own DNA.

Can two people have the same RFLPs by chance? Geneticists are unanimous in the conclusion that everyone (other than identical twins) has a unique genome. If a laboratory sequenced all the DNA in the blood or semen from two individuals, they would certainly find differences. But labs do not sequence the whole genome, only a sample of it, concentrating on sequences that are highly variable within the population. Prosecutors can calculate the chance that each of these DNA variants would appear in a large population. In general, however, no one knows the frequency of each variant in a local population. For the RFLPs used for DNA profiling, however, the probability of obtaining six identical DNA patterns by chance is typically 1 in 10 billion. And, of course, so far the Earth has only 6 billion people.

Finally, there is the question of faulty laboratory analysis. Early DNA analyses were sometimes very poor, but certification of laboratories and technicians has greatly reduced purely technical errors. Still, DNA evidence, like other evidence, must be scrutinized for artifacts and misinterpretations.

DNA profiling has changed the way that the police investigate rapes and murders by providing a powerful new kind of physical evidence. DNA analysis has put many criminals behind bars, but it has also set free people who were wrongly convicted. For example, DNA evidence freed a 32-year-old Baltimore man after nine years on death row.

DNA profiling has also allowed the identification and return of kidnapped children. One of the most egregious crimes of the Argentine military dictatorship (1975–1983) was the abduction and killing of some 15,000 people, called "the disappeared." Among these "disappeared" were pregnant women, whose babies were delivered and given to military families and their friends. After the fall of the dictatorship, a group of women whose daughters or granddaughters had "disappeared" contacted Professor Mary Claire King, a human geneticist then at the University of California Berkeley. King was using RFLP analysis to track a gene variant responsible for breast cancer. By applying her profiling methods to the DNA of the stolen children and their surviving relatives, she was able to help reunite more than 50 children with their families.

DNA fingerprinting identifies individuals on the basis of variations in DNA sequences.

What Are Human "Races"?

We all know that the human species varies in appearance and in physiology geographically. People of Indian descent, for example, are recognizably different from those of Chinese descent. We can look at the shape of the nose, the eyes, and the ears, for example, and we see differences (Figure 14-4). We can look at the distribution of individual alleles and see differences. Fifty-one percent of Nigerians have type O blood, compared to only 30 percent of Japanese. Twenty percent of Russians have type B blood, while the Amerindians of Lima, Peru, have no detectable levels of type B blood at all.

Nineteenth-century anthropologists struggled to classify human groups into a few major races. Some systems identified only 12 races, while other systems listed 30 or more. One problem was that no matter how anthropologists classified humans, there always seemed to be tribes or nations that would not fit into any known group. The Basques, who live in the Pyrenees mountains between France and Spain,

A. B.

Figure 14-5
Classifying humans into a few discrete races has been unsuccessful. Named races frequently include markedly distinct peoples. A. A Bushman from Namibia. B. A rubber plantation foreman from the Ivory Coast.

for example, appear European. Yet their language and culture are unlike any other in the world. Similarly, the Bushmen are unique among African groups in both appearance and physiology (Figure 14-5).

A more serious problem with the grouping of humans into races is that most groups do *not* stand out from those around them; they blend. Because groups of humans inevitably mix, through migration, warfare, and trade, human "races" are never pure. Both the Japanese and British, for example, take pride in the purity of their island races. Yet the Japanese are a graded mixture of Korean and Ainu north islanders (a people of possibly European descent). This mix shows up in the distribution of blood types from one end of Japan to the other (Figure 14-6).

The British are even more of a melting pot than the Japanese. The Bronze Age Beaker Folk mixed with the Indo-European Celts in the first thousand years B.C. In the next thousand years, the Angles, the Saxons, the Jutes, and the Picts arrived, followed by the Vikings and their descendants, the French Norman invaders. In the 19th and 20th centuries, Africans, Indians, Pakistanis, and others have lent spice to this already heady mix.

If the Japanese and the British, seemingly isolated by geography, are well mixed, most mainland groups represent a true continuum of traits. In fact, studies of the distributions of different alleles have convinced researchers that human races—in the biological sense—do not exist. This is because, in humans, genetic variation within populations is greater than that between nations or races. According to recent studies by geneticist Marcus Feldman, at Stanford University, of all human genetic variation, 93 to 95 percent is among the individuals within a country or a continent. Less than five percent of all genetic difference corresponds to differences in skin color and the other usual indicators of "race."

A. B.

Figure 14-4
People from different geographic regions vary in such traits as height, blood type, skin color, and facial features. A. An Indian woman. B. An ethnic minority woman from China.

and hair color reflect variations in a single gene. That gene specifies the structure of a protein, called MC1R. MC1R affects our response to a hormone that regulates the balance of different types of skin pigments (melanins). Caucasians who tan well, for example, have the same form of MC1R that is present in Africans. Research suggests that variations in skin color represent adaptations to different amounts of sunlight (see the box on p. 284). A mere handful of variants are probably responsible for the most obvious traits that have historically been used to distinguish human "races." As geneticists Mary Claire King and Kelly Owens state, "The myth of major genetic differences across 'races' is…worth dismissing with genetic evidence."

Geneticists and anthropologists have also begun to combine DNA-based differences among current populations with fossil evidence to study early human history. The data suggest that modern humans originated in Africa. By 100,000 years ago, human skeletons were anatomically identical to those of modern humans. But it was not until 50,000 years ago that substantial numbers migrated out of Africa through the Arabian Peninsula into Asia.

Modern humans appear to have reached Europe only about 40,000 years ago. By analyzing genetic variations among European populations, geneticists have concluded that the modern human colonization of Europe occurred in five distinct waves. The first two of these were from the Middle East, first westward across the Mediterranean Sea to Greece, Italy, and Spain, and then northwest to northern Germany and Britain. Subsequent migrations gave rise to other distinctive populations—first, Finns, Lapps, and other northern European peoples; then, Celtic populations of Britain, Ireland, and Brittany; and, later, the Basques of Spain and France.

Human races blur into one another, so that defining races is impossible. Almost all human variation is present in an individual population.

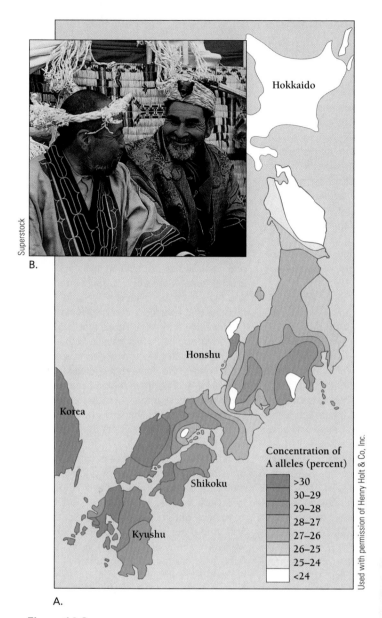

Superstock

B.

A.

Hokkaido

Honshu

Korea

Shikoku

Kyushu

Concentration of A alleles (percent)

>30
30–29
29–28
28–27
27–26
26–25
25–24
<24

Used with permission of Henry Holt & Co., Inc.

Figure 14-6
Variation in blood group A in modern Japan. A. Even the Japanese, a relatively homogeneous island race, show variation that demonstrates intermarriage with other groups. Populations closest to mainland Korea have the greatest concentration of A alleles. Populations closest to Hokkaido have the lowest concentrations of A alleles. B. In Hokkaido lives a remnant population of Ainu, sometimes called the "hairy Ainu." The Ainu, who were once a white minority in Japan, are considered primitive by other Japanese. The Ainu had white skin, abundant body hair, and round eyes. Descendants of Causasoid peoples from northern Asia, the Ainu once lived on all four of Japan's major islands. Today they live only in the north. Intermarriage and cultural assimilation by the Japanese have made the traditional Ainu virtually extinct.

In practice, we focus on the small numbers of traits (e.g., skin color and eye shape) that vary geographically, and we use just those few traits to identify groups of people who differ from us more in their cultural practices than in their genetic constitutions. For example, many differences in skin

14.2 How Do Environment and Genome Interact to Create a Phenotype?

Genes provide the continuity between the generations, but they do not by themselves define individuals. Every individual's phenotype results from complex interactions between genes and environment. The environment includes both the physical and chemical environment of the developing organism and the ongoing environment of the mature organism. In humans, for example, environmental influences include the quality of the cytoplasm in the egg, hormonal and nutritional influences in the womb, upbringing, and the physical and social atmosphere of the home or workplace.

Health and Biology   The Evolution of Human Skin Color

People have long known that skin color in humans goes from dark to light, along a gradient from the equator to the north. And over the last hundred years anthropologists have struggled to classify everyone into discrete races according to the exact shade of their skin. But the distribution of genes for pigment doesn't correlate that well with other genes. What it does correlate with is sunlight, of course.

Specifically, research in the 1970s established that skin color correlates well with ultraviolet (UV) radiation from the sun. UV is known to damage skin and cause skin cancer, and some scientists suggested that skin cancer selected for people with darker skin. But skin cancer virtually never affects humans in their reproductive years, so it was unlikely that UV was selecting for dark skin by way of skin cancer. So why would people in the tropics need dark skin and why would people in the far north need light skin?

In the 1990s, biologist Nina Jablonski gathered together threads of information from several unrelated sources, including remote sensing data from NASA satellites, and wove them into a coherent answer. Biologists have known since the 1970s that UV radiation in sunlight can break down the B vitamin folate as it flows through blood vessels in the skin. But that small fact seemed unimportant until researchers realized that a primary predictor for a major birth defect called spina bifida was low levels of folate in pregnant women.

The U.S. government responded by asking cereal and bread makers to fortify their products with folate and the rate of spina bifida in the United States immediately plummeted. It was one of the minor triumphs of public health policy. Jablonski began to wonder if exposure to sunlight could cause spina bifida. Experiments with humans were out of the question, but she quickly located a report by an Argentine pediatrician, three of whose patients' babies died of this developmental defect. All three women seemed healthy but had spent time in tanning parlors in the early weeks of their pregnancies. Tanning parlors, of course, sell UV radiation.

Intrigued by this potent bit of anecdotal evidence, Jablonski hypothesized that UV destroys folate, leading to birth defects. In that case, melanin, the main dark pigment in human skin, would protect against spina bifida (and related birth defects) by preventing UV radiation from penetrating the skin to the folate in the blood. This explained why dark skin could be adaptive. But what about light skin? Why didn't everyone have dark skin?

Once again, the answer seems to lie exactly where a good biologist would expect it, in the reproductive biology of humans. Nowadays people take milk fortified with vitamin D for granted, not always aware of why the vitamin D is there. The reason we fortify our milk is that vitamin D is normally synthesized in the blood only when we are exposed to UV radiation in sunlight. Vitamin D makes it possible for the intestines to absorb calcium, which is required for the normal development of the skeleton.

But not all sunlight contains enough UV radiation to synthesize all the vitamin D we need. During a Boston winter, for example, the sun isn't above the horizon enough to provide most people with enough vitamin D. Until mid-March, people living there need to get vitamin D from their food. (The rare foods that provide plenty of vitamin D are different kinds of fish, which northern humans tend to eat a lot of.)

Jablonski and colleagues looked at large data sets of UV radiation patterns on Earth and classified solar radiation into three "vitamin D zones": the tropics, the subtropics, and the temperate regions. Populations of humans that evolved in the tropics have permanently dark skin. Those that evolved in the subtropics have light skin in the winter, but tan easily when exposed to sun in the summer. And those that evolved in the temperate and northern regions have permanently pale skin that is ready to allow the synthesis of vitamin D at any opportunity, but offers no defense against the destruction of folate.

If Jablonski turns out to be right, many of us are just on the edge of getting too much UV or too little. If our skin is too light for where we live, we may develop a folate deficiency and have a baby with spina bifida. And if our skin is too dark for where we live, we may not get enough vitamin D (and therefore calcium). Fortunately, we can drink vitamin D–fortified milk on folate-fortified cereals (and avoid tanning parlors).

Studies of other animals illustrate the influence of environment on gene expression. For example, some tadpoles develop different phenotypes depending on whether predators are present. Butterflies may develop wings of one pattern when the weather is cold and completely different colors when the weather is warm. In crocodiles, extreme hot or cold temperatures make eggs develop into females, while moderate temperatures favor the development of males. The list of such environmentally induced variants is endless. In every case, something in the environment throws a genetic switch that regulates a whole cascade of gene expression that can determine anything about an animal or plant, including color, sex, physiology, and behavior.

One of the most dramatic examples of how environment affects phenotype comes from the tiny Japanese goby fish *Trimma okinawae*. In these little fish, social environment decides whether an individual is male or female (Figure 14-7). The fish can change sex repeatedly from female to male and back again. Schools of female fish breed with a single male. If the male leaves, one of the females becomes male and takes his place. On the other hand, if a large breeding male shows up, juvenile males in the school turn into females and remain to breed with the new male. These fish can change the structure of their gonads, genitals, and brain, as well as their behavior, in as little as four days. The anatomy and behavior of a new "male" or "female" is identical to that of fish that have

A.

B.

Figure 14-7
The phenotype of an animal can change with its social environment.
A. The Japanese goby fish *Trimma okinawae* is one of dozens of fish that can change sex in response to a change in its social environment. If a school of females loses a breeding male, the dominant female changes into a male and begins breeding. Over a period of a few days, a female's ovaries convert into testes, and "her" coloring changes. Just days after the fish's transformation, he begins courting females. His anatomy, coloring, and behavior are indistinguishable from those of any other male's. B. Many other species of fish also change sex. In this school of scalefinned anthias, the magenta-colored male of the day is at center, surrounded by orange females.

always been one sex. In short, the same genotype can impart alternate phenotypes whose expression depends on environment. In this section, we will explore the complex relationship between genotype and phenotype.

Genes Are Expressed to Different Degrees

The possession of a particular allele usually does not by itself predict what an individual's phenotype will be. Genes vary enormously in their **expressivity**—the range of phenotypes associated with a given genotype. Two individuals

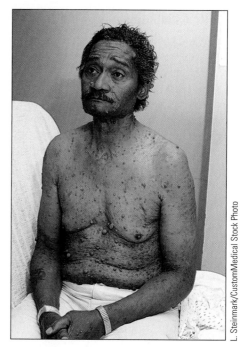

Figure 14-8
Neurofibromatosis and gene expressivity. Neurofibromatosis, whose distinctive symptoms include lumpy "neurofibromas" and brown spots, is caused by a dominant allele. But the allele is expressed to different degrees, so that some people with the allele have dramatic symptoms such as blindness, while others with the allele have no symptoms at all.

with the same allele, for example, may express it very differently. The human disease neurofibromatosis, for example, is caused by a dominant allele whose expressivity varies so much that in some people it causes blindness or disfiguring growths, whereas in others the only symptom is a coffee-colored spot (Figure 14-8). Still others with the same neurofibromatosis allele have no symptoms at all. Huntington's disease also varies in how strongly it is expressed and when. Although most people show no symptoms until middle-age or after, a few children may exhibit severe symptoms as early as two years of age.

Genes also vary in their **penetrance,** the likelihood that an individual with a dominant allele will show the phenotype usually associated with that genotype. A dominant allele that has complete penetrance, such as the allele for purple flowers in Mendel's peas, will always be expressed.

A dominant allele that has incomplete penetrance, however, may not be expressed, depending on its physical or genetic environment. In flowers, this would simply mean that only a certain proportion (say, 10 percent or 75 percent) of individuals with a given genotype would show the phenotype. But penetrance can be a slippery concept. The allele for Huntington's disease, for example, is said to be completely penetrant. Theoretically, those who have the allele always get the disease. Because the allele may be expressed late in life, however, some people with the allele die of other causes before they experience any symptoms.

How Can Environment Regulate Gene Expression?

For most genes today, the regulatory mechanisms that make a gene vary in expressivity and penetrance are unknown. Both the physical environment (temperature and nutrition, for example) and interactions with other genes play crucial roles. Increasingly, however, biologists are working out the exact mechanisms by which the environment communicates with the genome by means of molecular switches.

One remarkable example comes from studies of ants by young biologist Ehab Abouheif. Abouheif, working at Duke University with his advisor, showed that switches can develop anywhere in a network of genes that direct the formation of a trait. Ants in a nest have different "castes," including soldiers and workers (which have no wings), and reproductive males and queens (which do have wings). All of the female castes have the same genes, so whether they develop into one type of ant or another depends on their environment.

Abouheif and his advisor began by showing that the same network of six genes controls wing formation in the ant *Pheidole morrisi* as in the fruit fly *Drosophila*. Gene A encodes a protein that turns on gene B, which encodes a protein that turns on other genes. In ants with wings (the reproductive males and queens), the expression of these six genes is nearly the same as in flies. But in the soldier and worker castes, which do not make wings, things are different. In soldiers, five of the six genes are normally expressed, but the last one in the cascade of gene expression is not expressed, and so the soldiers develop no wings. In the smaller worker ants, wing formation is interrupted even earlier.

The workers and soldiers have no special mutations in their genes. They are genetically identical to the winged queens. So what developmental switches keep them from becoming queens? The fate of a female ant embryo, explains Abouheif, depends on two switches. At the first switch, females that experience the correct light and temperature experience a burst of juvenile hormone that sets them on the path to becoming queens. If not, they become sterile workers. At the second switch, the right diet triggers another pulse of juvenile hormone, which causes the ants to become soldiers. Otherwise, they become the smaller workers.

Abouheif looked at worker castes in three other species of ants and discovered that in each case, wing formation was interrupted at a different place. Even though the network of genes for wing formation is stable—the same genes conserved over some 300 million years—ants can turn off wing formation anywhere in a network of genes. Making wings is conservative, but not making them is a flexible process.

Because alleles vary in their expressivity and penetrance, the genotype does not define the phenotype. In all organisms, molecular switches provide a means for the environment to control which genes are expressed.

A Single Gene Can Affect Many Traits

When a baby is born with the genetic disease cystic fibrosis (CF), the baby's doctors know that the baby is destined for a life with chronic respiratory infections. Cystic fibrosis sufferers tend to have unusually salty perspiration and thick, sticky mucus that clogs the lungs and leads to repeated bouts of pneumonia, bronchitis, and other respiratory infections (Figure 14-9). Many have blocked pancreatic ducts, which prevents the normal secretion of digestive enzymes, and males are frequently sterile because the tube that carries sperm from the testes is blocked. Cystic fibrosis is caused by a mutation in a single gene. How can one defective gene cause so many different symptoms?

People with cystic fibrosis are unable to make a polypeptide called cystic fibrosis transmembrane conductance regulator, or CFTR. The normal alleles are dominant and the disease-causing alleles are recessive. The CFTR polypeptide is a cell membrane protein that pumps chloride ions out of epithelial cells, which are the cells that line the inner and outer surfaces of cavities. Epithelial cells can be found in every kind of hollow duct—in the intestines, the lungs, the nose, and the mouth. Skin cells are also epithelial cells. The altered version of the CFTR polypeptide functions poorly, and epithelial cell fluids do not go where they are supposed

Figure 14-9
A child who has cystic fibrosis. Children who have cystic fibrosis must be thumped on the chest several times a day to loosen accumulated mucus in the lungs.

Biology ⒺNow™ Learn more about cystic fibrosis by clicking on this figure on your BiologyNow CD-ROM.

to. Thus, this single mutation can have wide-ranging effects on the phenotype.

The allele that causes cystic fibrosis affects many aspects of phenotype. The capacity of one gene to have diverse effects is called **pleiotropy** [Greek, *pleios* = more + *trope* = turning]. In fact, most biologists now agree that all genes are pleiotropic.

A small mutation in a single gene, such as the one for the CFTR polypeptide, can affect many aspects of phenotype, a phenomenon called pleiotropy. All genes are pleiotropic.

A Single Trait Can Be Influenced by Many Genes

Phenotype depends on the interactions of many genes, as well as the effects of environment. Most traits are **polygenic**, meaning that they are influenced by more than one gene. In mice, for example, fur color depends on the action (and interaction) of at least five genes: one gene determines the distribution of pigment within individual hairs, a second gene determines the color of the pigment, a third gene allows or prevents the production of the pigment molecules, a fourth gene determines the intensity of the pigmentation, and a fifth gene controls the presence or absence of spots (Figure 14-10). The gene that regulates the intensity of pigmentation is called a modifier gene because it regulates the action or "expression" of another, separate gene.

Many genes are now thought to be backup systems for other genes. That is, the genome has several ways of accomplishing the same thing. Knocking out seemingly important genes may have no effect on phenotype. The long-haired mice mentioned in Chapter 13 are an example of such redundancy. In 1993, Gail Martin, at the University of California, San Francisco, knocked out a growth factor gene (*FGF5*) thinking that the mice would have developmental defects and probably die in their mothers' womb. The researchers were frankly disappointed to see healthy mice born, but astonished when all the young mice grew hair that was nearly 50 percent longer than normal (Figure 13-12). *FGF5* normally is expressed in early embryo tissues, in muscle cells, in the hippocampus of the brain, and in motor neurons. Yet, only the hair follicles were obviously affected by the absence of the gene. Every other aspect of phenotype appeared normal. Martin and other developmental biologists think that nerve, muscle, and brain function survive knocking out the *FGF5* gene because other genes provide backup systems.

In humans, as in mice and other organisms, the vast majority of traits are polygenic and probably redundant as well. Studying polygenic traits is difficult, since with each added set of alleles, the number of possible genotypes increases geometrically. If groups of genes are difficult to study in research animals or plants, they have been nearly impossible to study

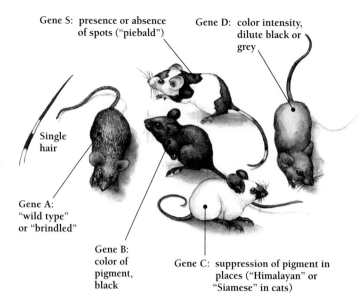

Figure 14-10

Several interacting genes determine coat color in mice. Gene A determines where on the hair the pigments go. In wild mice, a band of yellow in the middle of the hair creates an "agouti" or "mousy" coat color. One allele confers a narrow yellow band, which gives a dark coat. Another allele confers a broader yellow band, which results in a lighter coat. In mice with a coat of a solid color, the yellow band is not present at all. The normal allele of gene B encodes a black pigment. Gene C suppresses the expression of pigment in the warmest parts of the body. Only the cooler ears, nose, tail, and paws are dark. Gene D regulates how much pigment each hair receives. Gene S determines the presence or absence of pigment in patches.

in humans. The techniques of DNA array analysis discussed in Chapter 13 may help, however. For now, almost nothing is known about polygenic traits in humans, and that is most traits. For example, even after years of study of the heritability of IQ, researchers' best estimates say only that IQ scores (similar to SAT scores) are between 30 and 70 percent genetic, which isn't a very precise estimate.

The expression of each trait is the result of many genes acting together.

A Single Gene May Have Multiple Alleles

Most of the time we have spoken as if each gene can have only two alleles—a dominant allele and a recessive allele. But a gene may exist in many alternate forms, each of which contains a distinct sequence of nucleotides that specifies a distinct polypeptide. Some of the polypeptides cannot function at all, while others may function differently.

One familiar example of a gene with multiple alleles is the one responsible for the ABO blood types. These blood types are vitally important for blood transfusions. For ex-

Table 14-1
Human ABO Blood Groups

Blood genotype	Red blood type	Plasma cell surface	Antibodies
$I^A I^A$	A	A glycoprotein	anti-B
$I^A i$	A	A glycoprotein	anti-B
$I^B I^B$	B	B glycoprotein	anti-A
$I^B i$	B	B glycoprotein	anti-A
$I^A I^B$	AB	A and B glycoproteins	neither
ii	O	neither	anti-A and anti-B

ample, people with type O blood have antibodies against type A and B blood and cannot accept transfusions of these blood types. The gene for blood type has three alleles, called I^A, I^B, and i, which specify different versions of a surface marker on red blood cells. (Although three alleles exist in most human populations, each individual has only two of the three alleles, one on each homologous chromosome.) Alleles I^A and I^B each specify an enzyme that helps to synthesize a glycoprotein marker on the surface of red blood cells. These two glycoproteins are called, respectively, glycoproteins A and B. Allele i specifies no enzyme and so neither cell surface glycoprotein is synthesized. An individual human may have one of six genotypes (Table 14-1): $I^A I^A$, $I^B I^B$, $I^A I^B$, $I^A i$, $I^B i$, or ii. Because alleles I^A and I^B are dominant to i, individuals with $I^A I^A$ or $I^A i$ genotypes have type A blood, individuals with $I^B I^B$ or $I^B i$ have type B blood, and individuals with ii genes lack these surface glycoproteins and are said to have type O blood. Alleles I^A and I^B are codominant, meaning that in individuals with both alleles, both alleles are fully expressed. $I^A I^B$ individuals have type AB blood.

A single gene may have many functioning alleles. While some amino acid changes result in proteins that do not function at all, other amino acid changes alter the activity of the resulting protein. When both alleles are expressed, they are said to be codominant.

14.3 The Human Genome Project

If each of the 6 billion human beings on Earth is genetically unique, we can conclude that human beings harbor enormous genetic diversity. At the same time, we must remember that we are still about 99 percent identical. For this reason, it makes sense to most people to study "the human genome" as if it were a definable entity.

What Is the Human Genome Project?

In 1988, the National Institutes of Health (NIH), the federal agency that oversees and funds research in the health sciences, launched the largest and most expensive single project in the history of biology. The Human Genome Project was a 15-year, $3 billion project jointly funded by the Department of Energy and by NIH's National Human Genome Research Institute (NHGRI). The project, completed in April 2003, provided four kinds of data about the human genome:

1. Genetic linkage maps of some 40,000 genes, along with DNA-based markers, so that geneticists could determine the chromosomal position of any gene.
2. A physical map of each human chromosome, made by cutting chromosomes into restriction fragments, then identifying unique landmark sequences.
3. The sequence of all 3 billion base pairs on one set of human chromosomes and a database with all that information.
4. The genome sequences of several other species, including E. coli and several other prokaryotes, brewer's yeast, the nematode worm Caenorhabditis elegans, the fruit fly Drosophila melanogaster, the mouse Mus musculis, and the plant Arabidopsis thaliana.

The Human Genome Project sequenced the entire human genome, as well as the genomes of several other organisms, and constructed genetic maps and physical maps of each human chromosome.

How Do Researchers Locate Genes for Specific Diseases?

The Human Genome Project provided both an electronic database of the genome's sequence information and a gene library of actual gene fragments stored in numbered test tubes. Researchers can express many of these genes to find out what sorts of proteins they make. But for the most part, molecular biologists do not know what these proteins do or even in what kinds of cells they are expressed. So, although the contents of the gene library are considered enormously valuable, they are still something of a mystery.

One approach to finding a particular gene for a particular disease is to look for genetic markers that seem to travel with a disease from generation to generation, a technique called **pedigree analysis,** the same approach used by James Gusella in looking for the Huntington's gene (Figure 14-11). Researchers collect blood samples from individuals in families that carry a given disease, then look for landmark sequences that are consistently associated with the disease. In the search for the Huntington's disease allele, Gusella began looking for a particular RFLP sequence that traveled with the Huntington's disease allele in some families. The human

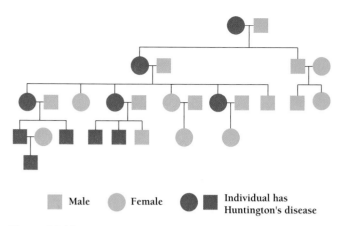

Male   Female   Individual has Huntington's disease

Figure 14-11
Pedigree for a family with Huntington's disease. The marriages and offspring of some unaffected individuals have been omitted.

Biology ⊘ Now™ Learn more about pedigrees by clicking on this figure on your BiologyNow CD-ROM

genome is enormous, and no one knew even on what chromosome the Huntington's gene might lie. But miraculously, Gusella immediately happened on a polymorphic marker that was always inherited with Huntington's disease (in certain families). This RFLP sequence was on the short arm of chromosome 4.

Pedigree analysis allows researchers to find a genetic marker associated with a particular trait or disease that shows approximately where on the genome a gene lies.

Of What Use Is a Sequenced Gene?

Once researchers know the sequence of a disease allele, they are in a good position to begin to understand how the gene does its damage. Huntington's disease provides a good example of this. In 1993, researchers found that the gene involved in Huntington's disease codes for a huge protein called Huntingtin, which is expressed in every cell in the body. The disease-causing form of Huntingtin codes for more than 37 glutamines in a row in one place, whereas the normal form usually has fewer than 34 at that same place. On average, the more glutamine repeats, the sooner a person will become sick, so the glutamine tract itself is somehow involved in the disease.

If researchers could understand how this expanded glutamine tract can lead to cell destruction, then they may be able to devise a treatment for this disease. Researchers have transferred the disease-causing allele of the gene *huntingtin* to other cells and organisms, including transgenic mice with the disease-causing allele. The characteristics of these genetically engineered cells and mice have led researchers to unexpected ideas about the causes of cell death not only in Huntington's disease but also in other diseases in which

nerve cells progressively die or fail to function properly. For example, transgenic mice that make the disease form of the Huntingtin protein develop normally but begin to show abnormal behavior within the first month of life. When researchers examined the brains of the sick mice, they found abnormal deposits of protein in the nucleus of many brain cells. When other researchers looked in the brains of people who had died from Huntington's disease, they found similar deposits. In each case, the protein deposits contain the protein specified by the disease-causing gene.

Many researchers have hypothesized that these protein deposits are the direct cause of cell death. Other researchers, however, have suggested that the deposits are a cell's way of getting rid of a potentially toxic protein. The best way to decide between these alternatives would be to find a chemical that can prevent the deposits without otherwise damaging the cell.

By using robots that can add individual compounds to hundreds of cell cultures at the same time, researchers hope to screen large sets of chemical compounds. This approach, called *high-throughput screening,* takes advantage of the huge "libraries" of compounds that have been assembled by pharmaceutical companies and others. Compounds that are effective in such cell-based assays could then be used to treat transgenic mice to see if they prevent protein deposits and abnormal behavior. If one or more compounds are successful in mice, researchers could then begin to plan clinical trials on people who have Huntington's disease.

Once a gene has been sequenced, geneticists can also test individuals to see if they have a normal allele or a disease-causing allele of the gene. For example, someone who is at risk for Huntington's disease can take a test and learn whether they carry the allele or not. If they do not carry it, they can stop worrying about becoming ill from Huntington's. If they do carry the allele, they may decide not to have children and to live their life a little differently. Many people prefer not to know, including Nancy Wexler and her sister Alice.

Another example comes from breast cancer research. Two genes, called *BRCA1* and *BRCA2,* each of which has several alleles, increase a woman's chance of developing breast cancer. Most alleles of these genes are moderately harmless, but some alleles can give a woman a staggering 85 percent chance of developing breast cancer. There are hundreds of different mutations, each, apparently, with a different risk for breast cancer.

If a woman has the most dangerous alleles, it's important for her to know about it. But most women do not have these mutations. More important, the vast majority of women who get breast cancer do not have any abnormal allele of either *BRCA1* or *BRCA2.* For that reason, it only makes sense to test women who come from families where breast cancer is strongly inherited and common. Even for that minority of women, however, the test is of limited value. Molecular biologists' knowledge of the sequences of *BRCA1* and *BRCA2* does not enable them to either prevent or cure breast cancer. The test may allow a healthy woman who has one of the dangerous alleles to choose to have a "preventative mastectomy," removing the breasts before breast cancer begins. And she may choose not to have children, so she won't pass on the gene to future generations.

But one significant downside to genetic tests is that once such tests reveal that a person has a gene that predisposes him or her to a particular disease, insurance companies will not want to provide health insurance to that person for that disease. Most geneticists argue that no genetic information should be made available to insurance companies. But an insurance company has a powerful economic incentive to overturn or sidestep any rules limiting their access to genetic information. For the average person, a genetic test could, right now, do more harm than good.

The ability to transfer a disease-causing gene to cells and transgenic mice may allow researchers to develop new therapies. Genetic tests give people the information to make decisions with long-term consequences, such as whether to have children. But genetic tests can also increase a person's risk of losing a job or health insurance.

Table 14-2
Top 14 Causes of Death in the United States in 2000

	Number	Percent of Total
All causes	2,403,351	100.0
1. Heart disease	710,760	29.6
2. Cancer	553,091	23.0
3. Stroke	167,661	7.0
4. Lung diseases	122,009	5.1
5. Accidents	97,900	4.1
6. Diabetes	69,301	2.9
7. Flu and pneumonia	65,313	2.7
8. Alzheimer's disease	49,558	2.1
9. Kidney diseases	37,251	1.5
10. Blood poisoning	31,224	1.3
11. Suicide	29,350	1.2
12. Cirrhosis of the liver and other liver diseases	26,552	1.1
13. High blood pressure	18,073	0.8
14. Homicide	16,765	0.7
All other	408,340	17.0

14.4 Preventing Genetic Disease

Only about three percent of all human diseases are caused by defects in a single gene. The most common genetic disease in the United States is cystic fibrosis, which affects about 1 person in 10,000. None of these single-gene diseases—several of which we have discussed in this chapter—are the ones that kill most people. For comparison, heart disease and stroke together killed 36.5 percent of the 2.4 million people who died in the United States in the year 2000 (Table 14-2). And another 23 percent died of cancer. People fall prey to one disease or another—lung cancer versus stroke, for example—because of lifelong behaviors such as smoking and because a constellation of different genes slightly predisposes them to one weakness or another. Every person carries at least 5 to 10 genes that could make them seriously ill under certain circumstances. As geneticist Michael Kaback puts it, "We are all mutants. Everybody is genetically defective."

Genetic Counseling: How Can Parents Decide?

The first kind of genetic testing is simply to test couples who are considering having a baby to see if they carry alleles for genetic diseases, such as sickle cell anemia, Tay-Sachs disease, cystic fibrosis, or Huntington's disease. Geneticists can now test for more than 100 different genetic defects. Usually, however, geneticists test only for the alleles that are relatively common or for those that the parents seem most likely to carry. A genetic counselor can calculate the likelihood

that a child will have a certain disease and offer information that will help the couple decide whether to risk having children or not.

Tay-Sachs, for example, is a genetic disease that results from a single defective gene for an enzyme normally synthesized by nerve cells. In healthy children, this enzyme breaks down lipids. But in children homozygous for the Tay-Sachs allele, lipids accumulate in the brain, causing mental retardation, blindness, and failure to develop control of the muscles. All Tay-Sachs children die in early childhood, usually by age four years. Tay-Sachs has a simple pattern of inheritance because all children homozygous for the single Tay-Sachs allele become ill. Heterozygous individuals, who have one defective allele and one normal allele, are called "carriers." Carriers can be identified by their low levels of the lipid-digesting enzyme, but they are otherwise healthy.

The Tay-Sachs allele is most common among eastern European Jews and their descendants, about three percent of whom are heterozygote carriers. If both parents are of eastern European Jewish descent, a genetic counselor will likely recommend that they be tested for the allele. If only one parent is Jewish, however, the counselor will probably not recommend the test, because a baby can become ill only if he or she receives the allele from both parents. This population has taken such care to avoid having Tay-Sachs children that Tay-Sachs is now quite rare. In America, the disease may now be more common in the population at large than among those descended from eastern European Jews. This decrease in the prevalence of Tay-Sachs is one of the great successes of genetic testing.

Genetic counselors can tell couples whether they carry certain alleles and the likelihood that their offspring will express certain traits.

Prenatal Testing: Can Parents Tell Before?

Once a woman becomes pregnant, physicians can obtain cells from the embryo and search for genetic defects. Besides looking for chromosomal abnormalities by examining the chromosomes, genetics clinics can now detect genetic diseases in the cells of an early embryo. Two techniques exist for testing the developing embryo (less than eight weeks after conception) or fetus (more than eight weeks). But no techniques now exist for curing the disease, and parents must then decide whether to continue with the pregnancy or to abort the embryo or fetus.

Chorionic Villus Sampling

Chorionic villus sampling (CVS) uses cells present in the placenta—the tissue that carries nutrients and oxygen from the mother's blood to the embryo and wastes from the embryo back to the mother (Figure 14-12). CVS may be done as early as 6 to 12 weeks after conception, when the embryo or fetus is still relatively undeveloped. Because the results of the test become available so much sooner, a decision to abort is more acceptable for many people. Likewise, news that the embryo is healthy means a shorter period of worry for the parents. The risk of the procedure itself is low, but not zero. About 0.5 percent of mothers who have CVS lose their babies through miscarriage due to CVS.

Chorionic villus sampling allows geneticists to test for genetic defects early in a pregnancy.

Amniocentesis

An older technique designed to look at the fetal chromosomes can only be used at about 15 weeks after conception. Geneticists usually obtain the fetal chromosomes by taking a sample of fetal cells from the **amniotic fluid** [Greek, *amnion* = membrane around a fetus], the watery fluid that surrounds the fetus. This procedure, called **amniocentesis** [Greek, *centes* = puncture], is usually recommended to women over 35 years old and to women with a family history of chromosomal abnormalities or other diseases (Figure 14-11B). The most common chromosomal abnormality found in humans is trisomy 21, or Down syndrome. But amniocentesis can detect more than 100 genetic abnormalities besides Down syndrome.

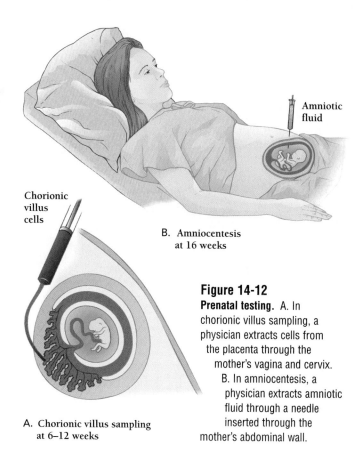

B. Amniocentesis at 16 weeks

A. Chorionic villus sampling at 6–12 weeks

Figure 14-12
Prenatal testing. A. In chorionic villus sampling, a physician extracts cells from the placenta through the mother's vagina and cervix. B. In amniocentesis, a physician extracts amniotic fluid through a needle inserted through the mother's abdominal wall.

At 15 or 16 weeks after conception, a physician inserts a needle through the mother's abdominal wall, just below her navel, and into her uterus. The needle is used to remove from the womb amniotic fluid that contains cells from the fetus. This procedure is virtually painless and as safe as chorionic villus sampling. To prepare the chromosomes for examination takes three to four weeks, during which a geneticist grows the fetal cells in culture. During this time, the fetus continues to mature and approaches the age when it becomes viable outside of the womb, which is about 20 weeks after conception. (A baby is full term at 30 to 38 weeks after conception. Physicians add two weeks to these times, counting from the date of the last menstrual period.) For many parents, this period of anxious waiting is difficult, as they think about what they will decide if the fetus turns out to have Down syndrome or some other genetic defect.

Amniocentesis allows a geneticist to look for chromosomal or other genetic abnormalities in a four-month-old fetus.

Making a Decision

In the next section, we see that researchers are working to develop techniques for actually curing and treating genetic diseases. For now, however, prenatal testing gives only a few

Figure 14-13
A child who has Down syndrome *(at right).*

mones that induce her to ovulate several eggs at once. A surgeon then removes these freshly ovulated eggs from the mother's abdomen and takes them to a laboratory. Freshly ovulated eggs have barely completed meiosis, and they are still attached to a sister cell. Because the two daughter cells do not divide evenly, however, the egg cell is very large, while the other cell, called the *polar body,* is very tiny (Figure 44-3).

The polar body is normally sloughed off, but geneticists have found a use for it. Because homologous chromosomes separate during meiosis, only one of the two cells gets the defective gene. If researchers find that the polar body has the defective gene, then they know that the egg has the normal copy. The normal eggs can then be combined with sperm and implanted in the mother's uterus.

Medical researchers have invented ways to select human eggs that are free of certain genetic defects, but this technology is cumbersome and expensive.

options. It's important to remember that the vast majority of babies are born perfectly healthy, with no health problems of any consequence. For most people, genetic testing ends with a sigh of relief. For parents whose test reveals a genetic defect, the choices are limited. For many people, abortion is unacceptable under any circumstances. For them, prenatal testing gives them time to plan for the care of their baby.

For most people in the United States, abortion is acceptable under certain circumstances. Each family's circumstances are unique, and what is acceptable varies from individual to individual and from situation to situation. For example, many people find it more acceptable to abort an embryo with Tay-Sachs than one with Down syndrome. A child with Tay-Sachs always dies young and cruelly. The effects of Down syndrome, however, vary enormously, and many of its symptoms can be treated. A child with Down syndrome can experience a happy childhood and may live well into adulthood (Figure 14-13).

Another factor may be the age of the embryo or fetus. Parents who could accept aborting a six-week embryo, which is only about 1.3 centimeters long and only partially formed, might hesitate if the much slower procedure of amniocentesis put them in the position of making the same decision for a kicking 20-week fetus. No matter what decision parents make, it is not made lightly.

Test-Tube Babies from Selected Gametes

Recently, medical researchers have come up with a way for parents to avoid having a baby with a specific genetic defect without confronting the question of abortion. Very simply, a doctor takes unfertilized eggs from a mother who is heterozygous for a particular genetic defect and selects the eggs that lack the allele in question.

The techniques for doing this are rather complex. In one example, the woman must first be treated with hor-

Gene Therapy: Can We Fix It Later?

Gene therapy is a developing technology whose purpose is to deliver functioning genes into diseased or damaged cells. Over the last decade, researchers have attempted to treat more than a dozen different genetic diseases, as well as AIDS, heart disease, and cancer. They have used dozens of different methods. So far, however, not one variety of gene therapy has resulted in a complete cure of even one person. Neither the genotype nor the phenotype has been "fixed."

In the best cases, researchers can provide a partial, temporary treatment that must be renewed every few months, often in combination with conventional drug therapy (Figure 14-14). And in the worst cases, gene therapy experiments can be extremely dangerous. In 2002, gene therapy experiments in France induced leukemia in two previously cancer-free individuals. The U.S. Food and Drug Administration responded by calling a temporary halt to two dozen gene therapy trials, but when the trials were allowed to proceed three months later, one scientific advisor said, "We really have to come to grips with what steps need to be taken to try to prevent this from happening again." It was hardly the first crisis in gene therapy. In 1995, trials had been temporarily halted also. And in 1999, a series of experiments with gene therapy ended with the death of a healthy 18-year-old named Jesse Gelsinger.

Gelsinger had a very mild version of a serious genetic disease, but volunteered for the experiments in order to help others less fortunate. Although several previous volunteers had experienced serious immune reactions to the treatments, the experiments continued, and four days after Gelsinger received an injection of a common cold virus en-

Courtesy Yan De Silva

Figure 14-14
Still the most successful example of gene therapy. In 1990, researchers at the National Institutes of Health treated Ashanthi DeSilva for the genetic disease SCID (severe combined immunodeficiency disease), which almost completely disables the immune system. Researchers removed immune cells from the four-year-old DeSilva, added a functioning allele for the enzyme adenosine deaminase, then injected the genetically engineered cells back into her body. With repeated treatments, DeSilva has been transformed from a chronically sick child with a short expected life span to an evidently healthy child.

gineered to carry a copy of the allele some of his cells needed, he died of a massive immune reaction.

Despite these frightening setbacks, medical researchers continue trying to make gene therapy work, as surgeons did in the early days of heart transplants and kidney transplants 30 years ago. All human gene therapy so far has focused on treating the somatic cells (the cells of the body) rather than trying to fix a fertilized egg, or zygote. Genetically engineering zygotes, called "germ-cell therapy," would be ideal because it would theoretically fix all the cells in the body and the "cure" would be passed on to any children. But germ-cell therapy would also require that researchers apply to humans the techniques now used to make transgenic animals. Making transgenic humans is theoretically possible, but ethical, legal, and scientific limitations have so far prevented experiments on human zygotes.

The two major challenges in gene therapy research are, first, to find vectors that will carry specific genes into human cells and put them in a safe part of the genome and, second, to introduce those genes and vectors into cells that can pass the genes to descendant cells. Viruses make effective vectors, because they have evolved to deliver genes into cells. The

second challenge is more difficult because most cells of the body have "differentiated" into skin cells, brain cells, kidney cells, or muscle cells. Even if a genetic fix is delivered into these cells, they cannot pass it onto other cells, because differentiated cells are dead ends. New kidney or skin cells come not from old kidney and skin cells but from the much rarer "stem cells," which have not yet become specialized ("differentiated").

Vector researchers have focused much of their energy on designing vectors that are both effective and safe. Vectors can be unsafe because they cause disease, because they stimulate an immune response, or because they insert genes randomly and affect nearby genes. For example, an inserted gene could turn off an adjacent tumor suppressor gene and trigger cancerous growth. Most current vectors modify disease-causing vectors such as the viruses that cause colds, cancers, and AIDS so that they do not cause disease but still deliver the genes they carry into cells. Early trials with cold viruses failed when ordinary immune responses destroyed the painstakingly engineered viruses before they could deliver their load of DNA. Current research with adenoviruses (a kind of cold virus) is focused on making a genetically stripped-down model that will neither give the patient a cold nor set off an immune response. Some researchers are also trying to avoid viruses altogether by using tiny lipid globules ("liposomes") that fuse with cell membranes and deliver their load of DNA directly into cells. Another virus-free approach is to use small proteins (one of which came from the AIDS virus) that allow naked DNA to pass through a cell membrane and into a cell.

A new area of research that may have important implications for treating diseases is in "interfering RNA." Researchers have discovered in the last decade that cells are loaded with short lengths of RNA, called "micro RNA." This RNA binds to messenger RNA (mRNA) so that the ribosomes cannot translate the mRNA into polypeptides. Cells use microRNAs to shut off the expression of certain genes. Researchers can do the same, shutting off particular genes without touching the genome itself.

Another RNA called "interfering RNA" also silences genes, but in a different way. Interfering RNA is double stranded, like the genomes of certain viruses. When a cell detects double-stranded RNA, an enzyme called "dicer" chops the RNA into short lengths of 21 to 23 bases, which themselves attract an assortment of enzymes. These enzymes then seek out the corresponding mRNA throughout the cell and destroy it, "silencing" the corresponding gene from which the mRNA was transcribed. The interfering RNA acts like a wartime censor, destroying certain messages from the genome.

The cells of mammals and other organisms respond to long lengths of double-stranded RNA by destroying both the RNA and themselves, a ploy that biologists believe is a defense against viruses. (It is better for an infected cell to die than to pass the virus to its neighbors.) But short lengths of double-stranded RNA injected into cells shut off the expression of the corresponding gene without triggering the self-destruct mechanism. Interfering RNA is a much easier way to "knock out" a gene than altering the genome itself and is allowing researchers to systematically shut off hundreds of genes in organisms such as fruit flies and roundworms to see what the genes do.

Both academic researchers and a small horde of start-up biotech companies are also hoping to use interfering RNA to treat diseases. For example, researchers engineered mouse stem cells that make a double-stranded RNA version of an important gene in the genome of HIV, the virus that causes AIDS. These cells later mature into T cells (the cells that the virus infects). When HIV infects these cells, the cells inactivate the viral gene and slow the infection. However, other, nonengineered cells in the body would remain vulnerable to infection.

Still another approach is to use interfering RNA to knock out the T cell gene that codes for the cell surface receptor that HIV latches onto during infection. If interfering RNA could be distributed to all the T cells in the body, they would not make the receptor protein and HIV could not infect any of them. But, so far, as with other forms of gene therapy, actually delivering the interfering RNA to all these individual cells is not possible.

Much of current gene therapy research ignores genetic disease altogether. Researchers want to use biotechnology to deliver genes whose products can help mobilize the immune system to combat cancer or AIDS. For example, one treatment promotes the growth of new blood vessels in conjunction with bypass heart surgery.

The second major problem facing gene therapy researchers is getting engineered genes into stem cells, the cells that can develop into many alternative cell types, instead of ordinary, already-differentiated cells. The stem cells in bone marrow, for example, give rise to both red and white blood cells. Other stem cells give rise to muscle and even to brain cells. If researchers could deliver genes to stem cells in the body, those cells would provide a reservoir of healthy replacement cells for many types of cells. Another advantage of stem cells is that researchers can grow stem cells in cultures in the laboratory, so delivering the genes to these cells is easier than trying to get them into the millions of differentiated cells in the body. A population of engineered stem cells could then be implanted in the specific tissues affected by a genetic disease, theoretically providing a permanent treatment.

Although many researchers are excited about the prospects of stem cell therapies, the potential use of human stem cells raises ethical problems, since most stem cell populations come from aborted embryos. People opposed to abortion are concerned that researchers and physicians will directly or indirectly encourage abortions in order to harvest stem cells. You can read more about the stem cell controversy in Chapter 45.

Despite expenditures of more than $400 million per year and more than 100 clinical trials in humans, human gene therapy is still in its infancy. Several critical reports have accused gene therapy researchers of "overselling" gene therapy, of understating its dangers, and of neglecting basic questions about gene regulation, vectors, stem cell function, and other areas.

Until researchers are able to insert genes into dividing cells, especially into stem cells, gene therapy will be only a temporary cure that can last only as long as the engineered cells live. Such genetic medicines will need to be repeated every few weeks or months. And an individual with a genetic disease will always harbor the defective allele in the germ cells and can continue to pass it on to offspring. For now, gene replacement therapy remains an area of active basic research, not of practical results.

> Using new gene therapy techniques, researchers can cause human cells to express a missing or otherwise desired gene, but expression is limited, so far, to the differentiated cells that originally take up the engineered DNA. As those cells gradually die off, the treatments must be repeated.

In this chapter on genetics, we have used the knowledge gained from previous chapters to examine genetic diversity in humans, the complex relationship between genotype and phenotype, the Human Genome Project, and the promise and limitations of applied human genetics. Now that we understand inheritance, we are ready to tackle evolution, the most important idea that biology has ever produced.

Key Concepts

- Even though humans are more than 99 percent genetically identical to one another, all humans, except identical twins, are genetically unique.
- Phenotype results from a complex interplay between genotype and environment.
- Many aspects of human genetics, including, for example, diagnostic testing and gene therapy, raise ethical, legal, and social issues.

Summary with Key Terms

How different is one human being from another?

DNA fingerprinting, also called DNA profiling, identifies variations in DNA derived from blood or other tissues. Every individual who is not an identical twin has a unique genotype. Human races blur into one another, so that defining consistent races is essentially impossible. The differences among the members of one race are greater than the average differences between races.

How do environment and genome interact to create a phenotype?

A small mutation in a single gene, such as the one that causes most cases of cystic fibrosis, can affect many aspects of phenotype. But the relationship between genotype and phenotype is generally loose. One reason is that alleles vary in their **expressivity** and **penetrance**. In addition, each gene affects many aspects of phenotype—a phenomenon called **pleiotropy.** The expression of each trait is **polygenic,** the result of many genes acting together. Finally, a single gene may have many functioning alleles. Because of all this complexity, the expression of a gene depends on its physical environment and its genetic environment (what other genes are being expressed).

What was the Human Genome Project?

The Human Genome Project sequenced the entire human genome, as well as the genomes of several other organisms. For a particular trait or disease, **pedigree analysis** allows researchers to find a genetic marker that shows approximately where on the genome a gene lies.

Can we prevent or cure genetic diseases?

Genetic counselors can tell couples both whether they carry certain alleles and the likelihood that their offspring will express certain traits. In **amniocentesis,** a geneticist samples a four-month-old fetus's **amniotic fluid** and looks for chromosomal or other genetic abnormalities. **Chorionic villus sampling (CVS)** allows geneticists to test for the same genetic defects as amniocentesis, but earlier in pregnancy.

Using new **gene therapy** techniques, researchers can induce cells to express a gene that the cells were previously unable to express. Expression is presently limited, however, to relatively few molecules of the desired protein. Because the cells cannot pass the gene on to new daughter cells in the body, treatments must be repeated. Also, the treated patient cannot pass on the "good" allele acquired in this way to any offspring.

Review and Thought Questions

Review Questions

1. What factors ensure that the relationship between genotype and phenotype is a loose one? What is pleiotropy?

2. What are the limitations and strengths of DNA profiling?
3. What did the Human Genome Project accomplish?
4. How do researchers locate genes for specific diseases such as Huntington's disease or cystic fibrosis?
5. How is prenatal testing different from genetic counseling?
6. If a prenatal test reveals a genetic defect in a fetus, what choices do the parents have?
7. Give an example of how gene therapy could cure a genetic disease.
8. How can the environment influence the development of an organism?

Thought Questions

9. If you and your spouse both carried the gene for cystic fibrosis, would you get prenatal testing? If you decided to have a test, would you prefer amniocentesis or chorionic villus sampling? Why?
10. How would your current knowledge about research on gene therapy for cystic fibrosis influence your decision?
11. If you knew you had an allele that slightly predisposed you to not get heart disease, would that influence your eating and exercise habits? If you had an allele that slightly predisposed you to get liver cancer, would you care if your insurance company and your employer knew?

BiologyNow Resources

Biology⊗Now™

Active Figures
14-9:   Cystic fibrosis
14-11:   Pedigrees

Preparing for an exam? Take a diagnostic test on your BiologyNow CD-ROM.

Online materials relating to this chapter are at:
http://biology.brookscole.com/AAL3

About the Chapter-Opening Image
Mapping the human genome is symbolized by this woman, wrapped in a map of the world.

PART III
EVOLUTION

Frans Lanting/Minden Pictures

297

15

What Is the Evidence for Evolution?

Key Questions

- What is biological evolution?

- Why was Darwin's argument for evolution more persuasive than those of earlier biologists?

- What is the evidence for evolution?

- How does natural selection drive evolution?

Daniel J. Cox/Natural Selection

The Scopes Trial

It was a hot July morning in Dayton, Tennessee, when high school teacher John T. Scopes went on trial for teaching evolution in the public schools. In the stuffy courtroom, spectators mopped their brows and swatted at flies. A dozen journalists crowded together on top of a table at the back of the room, trying to get a better view. Midway through the interminable testimony the table collapsed, dumping the men on the floor. Cheers and boos from the rowdy audience outside came through the open windows and drowned out the judge's calls for order. Also outside were a score of evangelists. Their trumpeting voices could be heard throughout days of testimony.

The town of Dayton was bedecked as if for a carnival, with banners and posters of monkeys. Newly constructed stands selling hot dogs, lemonade, and sandwiches lined the sidewalks, and a circus man ran a sideshow with two chimpanzees. The animals had been brought to town to "testify for the prosecution," the man said. Two hundred twenty-five newspapermen had arrived from New York, Baltimore, Philadelphia, and points beyond. It was 1925, before television, and these newsmen would put the tiny southern town of Dayton on the map forever.

The law prohibiting the teaching of evolution—passed by the Tennessee legislature in a moment of boredom—was never intended to be enforced. It nonetheless attracted the attention of the American Civil Liberties Union (ACLU), which believed the law violated the First Amendment right to free speech. Eager to test the law in a federal court, the ACLU advertised in the *Chattanooga News* for a teacher willing to be arrested. Dayton boosters jumped at the chance to liven up their sleepy town with some national attention.

John Scopes agreed to be arrested. An affable high school physics teacher and athletic coach, Scopes had never actually taught evolution, but he'd once substituted for a biology teacher who was ill and assigned the textbook pages that covered evolution. For this crime he was arrested.

The trial—loudly and forcefully argued by two nationally famous lawyers, William Jennings Bryan and Clarence Darrow, and covered by every major newspaper in the country—was, nevertheless, a rather dull affair. The attorneys representing each side wanted the same outcome—Scopes's conviction. The ACLU wanted a clean conviction so they could appeal the case in a higher court. Their goal was to have the antievolution law overturned.

But it was not to be. In an ironic twist, the ACLU won the battle it was trying to lose and so lost the war. Scopes was initially convicted, but then acquitted on a technicality. The judge technically violated the law when he mistakenly levied Scopes's $100 fine himself instead of letting the jury do it. Because Scopes was acquitted, the ACLU never got to appeal the case, and the Tennessee antievolution law stayed on the books until the U.S. Supreme Court declared such laws unconstitutional in 1968.

No one was again arrested for teaching evolution. But American textbook publishers—ever shy of controversy—nevertheless took the hint, reducing or even eliminating discussions of evolution in most high school textbooks. By 1942, fewer than half of all high school science teachers covered evolution at all. Only with the Soviet Union's 1957 launching of the first satellite, Sputnik, did the United States begin emphasizing better science education and textbooks. But the increasing appearance of evolution in textbooks sparked a new reaction among antievolutionists, who realized that their fight against science was far from won.

In the 1970s and 1980s, antievolutionists in Arkansas, Tennessee, and Louisiana passed identical bills calling for "equal time" for teaching evolution and **creationism,** the biblical myth that the universe was created by the Judeo-Christian God in six days. But a court ruled that the "equal-time" bill was unconstitutional on the grounds that it violated the separation of church and state. (The United States Constitution outlaws teaching religion in public schools because not everyone has the same, or any, religion; and the United States was founded on the idea of religious freedom.) Creationists argued that the account given in the Bible's book of Genesis was the basis for "creation science." But creation "science" is not science. In 1982, Federal Judge William R. Overton overturned the Arkansas law, writing, " . . . the evidence is overwhelming that both the purpose and effect of Act 590 is the advancement of religion in the public schools." About creation science, he said, "While anybody is free to approach

a scientific inquiry in any fashion they choose, they cannot properly describe the method used as scientific if they start with a conclusion and refuse to change it regardless of the evidence. . . ."

On the basis of Overton's decision, another judge struck down the Louisiana law as well. And in June 1987, the United States Supreme Court affirmed this decision in a 7 to 2 vote, effectively ending a dispute that had begun 128 years before.

Creationists of the 1990s have found another way to fight evolution, however. Rather than try to persuade public schools to teach the biblical version of creation, they now simply ask schools not to teach evolution. But it is unconstitutional to outlaw the teaching of ordinary science. So some school boards simply say students cannot be tested on any origin science, including the origin and age of the universe (about 14 billion years), the origin and age of the Earth (4.6 billion years), or the origin of species (kinds) of organisms.

In December 1999, for example, the Kansas State Board of Education adopted a curriculum stipulating that Kansas schoolchildren not be tested on evolution in any biology or other science class. Students would be expected to learn about natural selection (Darwin's mechanism for evolution) but not that, over millions of years, primitive mammals evolved into modern monkeys, bats, lions, and bears. Students would be "protected" from knowing that all organisms on Earth are related. American scientists expressed bewilderment. And the presidents of Kansas's six public universities wrote a letter stating that the new standards would "set Kansas back a century" and drive away science teachers. "The argument that teaching evolution will destroy a student's faith in God," the university presidents argued, "is no more true today than it was during the Scopes trial in 1925."

Efforts to prevent biology teachers from teaching evolution continue. In April 2003, for example, a Tennessee school board rejected three new biology textbooks because they presented evolution but not religious creationism.

Modern debate about evolution began in 1859, when Charles Darwin published his arguments for the evolution of species in *On the Origin of Species by Means of Natural Selection*. In that book, Darwin rejected the idea that every kind of organism had been individually created. The book disturbed a great many people. Three of Darwin's arguments offended most often: (1) that the Earth was extraordinarily old, (2) that one species could change into another, and (3) that humans and apes were related (Figure 15-1).

In this chapter we will discuss the evidence that has persuaded biologists that organisms are all descended from a common ancestor. At the end of the chapter, we will introduce the mechanism by which most evolution probably occurs—natural selection—which Charles Darwin first proposed about 150 years ago.

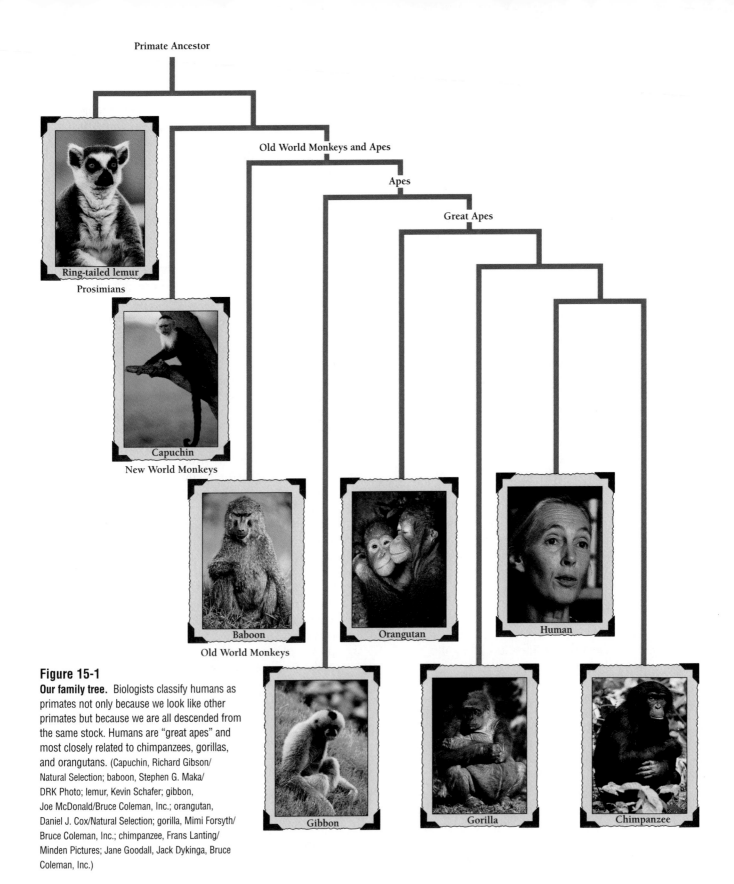

Primate Ancestor

Old World Monkeys and Apes

Apes

Great Apes

Ring-tailed lemur
Prosimians

Capuchin
New World Monkeys

Baboon
Old World Monkeys

Orangutan

Human

Gibbon

Gorilla

Chimpanzee

Figure 15-1

Our family tree. Biologists classify humans as primates not only because we look like other primates but because we are all descended from the same stock. Humans are "great apes" and most closely related to chimpanzees, gorillas, and orangutans. (Capuchin, Richard Gibson/Natural Selection; baboon, Stephen G. Maka/DRK Photo; lemur, Kevin Schafer; gibbon, Joe McDonald/Bruce Coleman, Inc.; orangutan, Daniel J. Cox/Natural Selection; gorilla, Mimi Forsyth/Bruce Coleman, Inc.; chimpanzee, Frans Lanting/Minden Pictures; Jane Goodall, Jack Dykinga, Bruce Coleman, Inc.)

15.1 What Is Evolution?

Biological evolution is the process that has taken life on Earth from perhaps a few mere flecks of RNA to the breathtaking variety of more than a million species. (A **species** is a group of organisms that only breed with others in the same group.) What is popularly understood to be Darwin's theory of evolution is really a group of several ideas that can be roughly divided into two parts.

The first part is **evolution** itself, the idea that species change over time. Darwin called evolution **"descent with modification."** A concise definition of evolution says it is heritable change in organisms over time. But the theory of evolution, along with extensive molecular and fossil evidence, also suggests that all the different kinds of life on Earth are related and have descended from a few simple organisms that came into being billions of years ago. Just as the members of a large family that includes hundreds of cousins can trace their family history back to a few great-great-grandparents, all species on Earth can trace their lineages back to common origins. Dogs and foxes truly are cousins, of a sort, as are humans and orangutans.

The second part of Darwin's theory is his specific mechanism for evolution, which he called natural selection. For Darwin, natural selection was the heart of his theory. It was a mechanism that set his theory of evolution apart from similar theories about evolution. Briefly, **natural selection** is the differential reproductive success of individuals with particular, genetically influenced phenotypes.

Of course, the word "genetically" is not one Darwin would have used. Genes had not yet been discovered, and Darwin was, in fact, frustrated by his ignorance of how inheritance worked. As we saw in the genetics section of this book, every individual has unique combinations of genetically influenced traits. Some of these combinations are **adaptive,** which means that they give the individual an advantage in the struggle for survival and reproduction and so increase its chances of perpetuating these useful combinations of genes in future generations. Darwin's studies of the natural history of organisms had given him a good general understanding of inheritance even if he couldn't know the specifics of how chromosomes and genes actually make it all work.

For example, in the Galápagos Islands, off Equador, the average beak size of a species of finches increases or decreases according to changes in weather and the sizes of the seeds the birds eat. In some conditions, plants that make big seeds flourish, and the big-beaked finches flourish. In other conditions, plants that make small seeds flourish, and the small-beaked finches flourish. The average beak size changes, or evolves, over periods of just a few years, depending on the weather. When many large, hard seeds are present, a large beak is useful, or "adaptive." When many small, soft seeds are present, a small beak becomes more adaptive.

Darwin compared natural selection to **artificial selection,** the process by which animal breeders and farmers cre-

ate new lines of dogs, pigeons, corn, and other organisms. In artificial selection, human beings select the most desirable individuals for breeding, generation after generation. In natural selection, the individuals best adapted to their environment tend to leave the most offspring.

At the end of this chapter, we will discuss Darwin's theory of natural selection. In later chapters, we will encounter other mechanisms for evolution. Some evolutionary change can occur through random processes. We will also discuss the rate at which evolution occurs. But in this chapter, we will focus on the evidence for evolution itself.

Darwin's theory of evolution, which he called "descent with modification," depended on four ideas. First, he recognized that the world is ever changing and very old (rather than constant and recently created). Second, he recognized that species change. (As we'll see, neither of these ideas was original with Darwin.) Third, Darwin argued that species are composed of populations of individuals, not identical organisms. Every individual is unique. And fourth, he argued that every species and group of species is descended from a common ancestral species. Ultimately, Darwin said, all organisms on Earth—including humans—may derive from a single origin of life.

These ideas conflicted with the biblical account of creation and with a major idea in Western philosophy, called **essentialism.** According to the Greek philosopher Plato (427–347 B.C.), individuals, whether chairs, daisies, or people, are only distorted shadows of an ideal, or essential, form. Only the ideal form was real. The variations that make every daisy unique, said Plato, are distortions and flaws.

Because of essentialism, most scientists in Darwin's time viewed species in terms of their ideal forms. Like so many Lego bricks, every member of a species was supposed to be identical to every other. Species were supposed to be perfect, unchanging, sharply defined "natural kinds." In contrast, Darwin viewed every individual as an individual, not a distortion of some ideal form. As evolutionary biologist Ernst Mayr has written, "It was Darwin's genius to see that this uniqueness of each individual is not limited to the human species but is equally true for every sexually reproducing species of animal and plant."

Darwin asserted (1) that the world is ever changing and very old; (2) that species are made up of individuals; (3) that species change; and (4) that all organisms are related.

Evolution Before Darwin

Although Darwin was the first to persuade the world that plants and animals had evolved, he was hardly the first to hypothesize that organisms evolve. His own grandfather, Erasmus Darwin (1731–1802), argued in 1794 that all species originated from the same primitive "filament." However, Erasmus Darwin propounded many ideas that seemed outlandish in the 18th century, and no scientist took his ideas on evolution seriously. In fact, in the senior Darwin's time, to be accused of "Darwinizing"—engaging in wild speculation—was an acute embarrassment.

How Old Is the Idea of Evolution?

Evolution as an idea actually predates even Plato. In the 6th century B.C., 100 years before Plato, the Greek philosopher Anaximander argued that the world was not created abruptly as most ancient religions said. He believed, instead, that animals evolved and that humans, like all other vertebrate animals, were descended from fishes. Although we now know that Anaximander was mainly correct, it was 2,500 years before his ideas found scientific support and general acceptance.

In large part, no one took evolution seriously because Christian beliefs incorporated Plato's essentialism. And essentialism, along with creationism, established a nearly unbreachable intellectual barrier to the idea of the evolution of species. Another serious barrier to the idea of evolution was the notion—accepted by nearly everyone in 17th- and 18th-century Europe—that the Earth had been created in 4004 B.C. This estimate had been worked out in 1650 by Archbishop James Ussher, who added together the life spans of all the patriarchs named in the Bible, beginning with Adam. Most Christians believed that the Earth was less than 6,000 years old.

A.

Tim Flach/Getty Images

By the end of the 18th century, however, the biblical account of creation and the notion that species do not change were undermined by three separate lines of evidence. First was the discovery of hundreds of thousands of species not mentioned in the biblical story of creation. Explorers returning to Europe from Africa, the Americas, and the Pacific brought back plants and animals that no European had ever seen or heard of before (Figure 15-2A). Equally disturbing were the great number of fossil plants and animals turning up in Europe itself. These fossils were stone remnants of species as alien as any from the far side of the world (Figure 15-2B). Yet, they appeared to have lived in Europe in some previous era. But when? Finally, by the middle of the 19th century, geologists found compelling evidence that the Earth was far older than Ussher's estimated 6,000 years.

The enormous difficulty of reconciling these new facts with the story of creation in the Judeo-Christian Bible is apparent in the thinking of five of the greatest scientists of the 18th and early 19th centuries. All of them discovered evidence for evolution, yet because of creationism and essentialism, all of them interpreted the evidence in nonevolutionary terms.

How Did Linnaeus's Ideas Lay the Groundwork for Evolution?

The first of the five scientists is Carolus Linnaeus (1707–1778), a Swedish physician whose most familiar contribution to biology was the two-part Latin names—genus and species—that biologists assign to all species. The scientific name for a human being, for example, is *Homo sapiens*. *Homo* refers to the **genus** to which we be-

B.

Kevin Schafer/Peter Arnold, Inc.

Figure 15-2

Another world. A. Western grey kangaroo and joey. In the eighteenth century, European explorers and naturalists brought back strange animals and plants no one had ever heard of, let alone seen. These organisms were not mentioned in the Bible. B. Even at home in Europe, paleontologists dug up fossil animals as strange as this fossil pterosaur, a flying reptile of the Jurassic Period.

long, and *sapiens* (Latin, = "wise") refers to our particular **species.** Linnaeus named more than 4,000 plants and animals, and biologists since Linnaeus have used his system to name some 1.4 million more species. Biologists estimate that 10 to 30 million more remain to be discovered and named. We discuss Linnaeus in more detail at the beginning of Chapter 22.

Even more important than Linnaeus's system of naming organisms was his system of grouping, or classifying, them. For example, the genus *Homo* belongs to the **family** Hominidae. And the family Hominidae belongs to the **order** Primates, one of 16 orders in the **class** Mammalia. Thus, we group species into genera, genera into families, families into orders, and orders into classes (Chapter 19). Each of these groupings is a **taxon** [Greek, *taxis* = arrangement; plural, **taxa**]. The science of classifying organisms is called **taxonomy.**

Linnaeus firmly believed that each species was individually created by God, yet he arranged species into groups that reflected their similarities to one another. He could see, for example, that all the species of mice were more closely related to rats than to horses, and his formal system for recognizing such natural relations was fundamental to Darwin's arguments 100 years later.

Linnaeus avoided asking *why* organisms fall so naturally into families, piously writing, *"Deus creavit, Linnaeus disposuit"*—God creates, Linnaeus arranges. But one of his contemporaries, the great French biologist George-Louis Leclerc, or Count Buffon (1707–1788), couldn't leave the question alone.

Linnaeus introduced a modern system of classifying organisms.

Why Was Count Buffon Afraid to Discuss Evolution?

Buffon was an accomplished mathematician as well as biologist. He was also a science popularizer, loved by the public but resented by fellow scientists. In his early manhood, Buffon boldly stated that the Earth must be far older than 6,000 years. But when church elders threatened him with excommunication for this heresy, he quickly took back his arguments.

In later years, he carefully recanted each heresy as he committed it. In a 44-volume natural history, in which he interspersed descriptions of plants and animals with discussion of natural philosophy, Buffon noted the remarkable similarity of the horse and the ass, or donkey. He wrote, in 1753, "But if we once admit that there are families of plants and animals, so that the ass may be of the family of the horse, and that the one may only differ from the other through degeneration from a common ancestor, we might be driven to admit that the ape is of the family of man." And, as if that were not bad enough, he added, " . . . then there is no further limit to be set to the power of nature, and we

should not be wrong in supposing that with sufficient time she could have evolved all other organized forms from one primordial type."

To placate the church, however, he quickly contradicted this insight: "It is certain through Revelation that all animals have shared equally in the grace of creation and that each emerged from the hands of the creator as it appears today."

Buffon's theory of evolution resembled Darwin's, but without natural selection. Fearing excommunication, however, he denied that he believed his own arguments.

What Made Hutton Think that the Earth Was Older than 6,000 Years?

Beginning in the mid-18th century, the Industrial Revolution gave rise to an insatiable demand for coal and metals that turned geology into a practical and hugely popular science. For the first time, scientists began to seriously explore the surface of the Earth. They learned that soil, rock, and mountains were eroded by wind, frost, and running water. In some places, whole mountain ranges seemed to have eroded completely away. But could all this have happened, geologists asked, in just 6,000 years?

In 1788, the Scottish geologist James Hutton (1726–1797) suggested that it could not. The Earth, he said, was eternal. The perpetual erosion of mountains was repaired by the uplift of sediments and the expulsion of magma from the Earth's depths. At the time, most geologists doubted continents and mountains could rise from the Earth's surface. But Hutton insisted that the Earth was not a static, passive body but a "beautiful machine," powered by the heat of its interior.

He proposed that endless cycles of erosion, sedimentation, and uplift molded the Earth's surface. The Earth, he said, might be immeasurably old. In his now famous conclusion, he wrote, "The result, therefore, of our present inquiry is, that we find no vestige of a beginning—no prospect of an end."

It was heresy. But if Hutton was a thorn in the side of theologians, Baron Georges Cuvier (1769–1832) was their savior. In 1812, the great French anatomist and paleontologist found an ingenious way to reconcile geology and Genesis—a way to reconcile the great age of the Earth with theological doctrine.

Catastrophism: How Did Cuvier Try to Reconcile Genesis and Geology?

Cuvier was a gifted anatomist and the first to study fossils systematically. He invented both **comparative anatomy,** the detailed comparison of the anatomies of different species, and **paleontology** [Greek, *palaios* = ancient + *on* = being + *logos* = discourse], the study of fossils. **Fossils** [Latin, *fossilis* = dug up] are the remains or imprints of past life. The most impressive fossils are the intact skeletons of whole an-

A.

B.

Figure 15-3

Catastrophism and uniformitarianism. A. Noah's flood. The deluge as portrayed in the first edition of the Luther Bible, published in 1534. B. The Grand Canyon. Eighteenth-century geologist James Hutton argued that the surface of Earth was constantly changing and very old, not new and not static. Hutton wrote, "Thus . . . from the top of the mountain to the shore of the sea . . . everything is in a state of change; the rock and solid strata slowly dissolving, breaking and decomposing, for the purpose of becoming soil; . . . without those operations which wear and waste the solid land, the surface of the earth would become sterile."

imals, such as 70-million-year-old dinosaurs. Such spectacular finds are rare, however, compared to the hundreds of thousands of small fossilized shells and tiny bone fragments paleontologists have unearthed. Fossils include preserved bones and teeth, petrified trees, bits of fossilized dung, footprints, impressions of the surfaces of ferns, as well as ants and other insects preserved in hardened tree resin (amber).

Fossils had puzzled people for thousands of years. Fossils look like plants and animals or parts of plants and animals. But they are made of stone, and they rarely look like *familiar* plants and animals. Some people thought fossils were God's first experiments with creation or maybe the remains of plants and animals that had literally missed the boat. Failing to board Noah's ark, they had perished in the great flood described in the Bible (Figure 15-3A).

But Cuvier knew that all the different organisms represented by fossils could not have died in the same flood. The different kinds of fossil organisms had lived and died out at different times in the Earth's history. By 1760, geologists had already shown that rocks could be arranged in a series of layers, or strata, according to their age. The oldest layers were usually deepest (Figure 15-3B). The newest layers were on top.

William "Strata" Smith, a geologist and canal engineer, compared these layers to "slices of bread and butter on a breakfast plate." Smith showed that different fossils appeared and disappeared from layer to layer. Older fossils appeared in the deepest layers, then disappeared and were replaced by new fossils—on up to the topmost layers.

Starting with the assumption that species could not evolve, Cuvier concluded that the fossils in the oldest layers had nothing to do with modern life. To account for the different kinds of life that appear in the fossil layers, Cuvier hypothesized that the organisms from the different layers had perished in a series of catastrophes. Noah's flood was merely the most recent and dramatic catastrophe. (The previous catastrophes were not mentioned in Genesis, Cuvier argued, because God didn't think Moses needed to know about them.)

After each catastrophe, argued Cuvier, new species appeared in another creation. **Catastrophism,** as his theory came to be called, allowed for the great age of the Earth and the fossil record without implying that Genesis was wrong. This idea was appealing to a great many pious Christians, and catastrophism remained popular for decades. In fact, the 1981 Arkansas equal-time bill called for a revival of this very theory.

But, although paleontologists have identified several mass extinctions, the evidence does not support catastrophism. First, although Cuvier had originally proposed only four or five catastrophes before Noah's flood, fossil hunters continued to find so many distinct layers of fossils that more and more catastrophes had to be invented to account for them all. The Swiss-American biologist Louis Agassiz (1807–1873) maintained, for example, that the fossil record revealed 50 to 80 separate catastrophes, each one requiring a separate creation.

And while some organisms appeared once in the fossil record and then disappeared forever, others kept reappearing virtually unchanged in "creation" after "creation" (Figure 15-4). Lastly, archaeologists found human bones and other artifacts that suggested that humans predated not only the biblical flood but the hypothetical creation of the world in 4004 B.C. Catastrophism didn't seem to fit the facts.

Figure 15-4

The continuity of life. Many groups of organisms have changed hardly at all over millions of years. The modern horseshoe crab *(left)* almost perfectly resembles its fossilized Jurassic ancestor *(right)*, which lived nearly 200 million years ago.

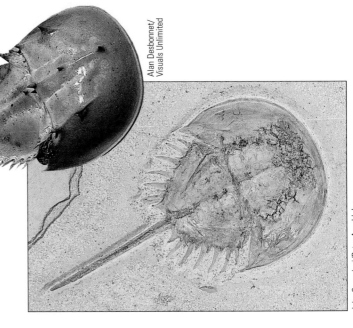

Alan Desbonnet/ Visuals Unlimited

John Cancalosi/Peter Arnold, Inc.

Uniformitarianism: Lyell Rejects Catastrophism

Catastrophism received its most damaging blow in 1830, when the Scottish geologist Charles Lyell (1797–1875) published the first volume of his three-volume *Principles of Geology.* Lyell rejected catastrophism. Instead, expanding on Hutton's ideas, Lyell argued that the slow processes that now mold the Earth's surface—erosion, sedimentation, and upheaval—are the same ones that have always molded it. Geologic change, he said, is slow, gradual, and steady—not catastrophic. Given enough time, mountains could erode down to mere hillocks, seabeds rise to great heights, and rivers cut deep canyons (Figure 15-3B). Gradual processes alone could create the world we know.

Lyell, who had been trained as a lawyer, provided huge amounts of evidence in his book to back up this hypothesis, later named **uniformitarianism.** Darwin was so influenced by Lyell that he later wrote, "I have always thought that the great merit of the *Principles* was that it altered the whole tone of one's mind, and therefore that, when seeing a thing never seen by Lyell one yet saw it partially through his eyes." But even though Lyell's *Principles* provided fertile soil for Darwin's ideas, Lyell himself believed that species could not evolve. It was Darwin's own book *The Origin of Species,* published almost 30 years after the *Principles,* that changed Lyell's mind.

The Industrial Revolution's demand for fossil fuels created the science of geology. The fossil record discovered by geologists showed the Earth was extremely old.

Why Was Lamarck Ignored?

Even together, the fossil record and the great age of the Earth were not enough to convince most scientists that species evolve. Among reputable scientists, evolution was at best a dubious and controversial hypothesis. Count Buffon and Erasmus Darwin had proposed that species evolve, but neither man was taken seriously by fellow scientists. In 1809, the great French biologist Jean-Baptiste Lamarck (1744–1829) again proposed the idea that species evolve over time.

Lamarck came from an impoverished noble family and served as a tutor to Count Buffon's son. He studied medicine and botany in Paris and published two works on botany before being appointed professor of zoology at the Jardin des Plantes in 1793. A renowned taxonomist, Lamarck named hundreds of species, wrote more than a dozen volumes on fossil "invertebrates" (a word he invented), and argued that plants and animals are all related. To express this union, he invented the word "biology."

In his studies of fossils, Lamarck saw a clear line of descent—from older fossils to more recent fossils to modern species. *The oldest fossils looked the least like modern organisms. Later groups of fossils included more and more familiar organisms.*

Lamarck hypothesized that the accumulated changes were adaptive—that is, the changes made organisms better able to survive under new conditions. A giraffe, for example, reaching ever higher in the trees for leaves to eat, would constantly stretch its neck and thus acquire a permanently longer neck. The giraffe's offspring would then inherit longer necks. Lamarck hypothesized that adaptive characteristics acquired by the parent could be inherited by the offspring.

In Lamarck's day, most scientists accepted his mechanism for evolution, the theory of the **inheritance of acquired characteristics.** But they vehemently rejected evolution itself, and he was vilified by the entire scientific community. The fiercest opposition came from Cuvier. As both an eminent scientist and a French cabinet minister, Cuvier's opinion carried enormous weight, and Cuvier criticized Lamarck relentlessly. In 1832, three years after Lamarck had died—poor and almost forgotten—Cuvier delivered an "elegy" to Lamarck at the prestigious French Academy, in which he once more dismissed not only Lamarck's theory of evolution but also his reputation as a biologist. Cuvier succeeded in blackening Lamarck's name for 170 years and only now are biologists beginning to acknowledge his significant contributions to biology.

With this cloud over Lamarck and his ideas, it was not surprising that Darwin did everything he could to distance himself from Lamarck (as well as his own grandfather Erasmus) and to anticipate every possible objection to his own ideas.

How Did Charles Darwin Convince Others That Species Evolve?

By the middle of the 19th century, natural science had become enormously popular with the lay public. Every sector of English society was obsessed with the exotic and the unusual in nature. Even as explorers brought back exotic animals and plants from Africa, Australia, and Asia, every middle-class English family acquired an aquarium, a fern case, a butterfly cabinet, or a shell collection. Mothers routinely taught their children the names of ferns and fungi. Aristocrats turned their estates into parks for exotic animals such as elands, beavers, and kangaroos. Books on natural history were so popular that one—*Common Objects of the Country*—sold 100,000 copies in a week, a total that even today would earn the book a place high on *The New York Times* best-sellers list.

Darwin's Education

Charles Robert Darwin (1809–1882) came of age in the middle of this national obsession and was himself entirely caught up by it. An upper-class English boy whose father was a doctor, Charles spent his boyhood tramping through the woods and fields of Shrewsbury, hunting, fishing, and collecting insects. As a young man, his interests narrowed to hunting, and his father despaired of Charles ever making anything of himself. Darwin senior once wrote to his son, "You care for nothing but shooting, dogs and rat-catching, and you will be a disgrace to yourself and all your family."

Darwin's utter lack of ambition forced his father to pick a profession for him, and in 1825 Robert Darwin sent young Charles to Edinburgh University to study medicine. In the course of his studies, Darwin attended lectures on geology and zoology and thought they were the most boring thing he'd ever had to endure in his life. His only interest, outside of shooting birds, was the meetings of a natural history club. As for medicine, if he had ever considered becoming a doctor, he changed his mind after watching two horrifying surgical operations done before the invention of anesthesia, one on a child. His resistance to studying medicine, or anything else, only hardened when he realized that his wealthy father would probably support him for the rest of his life, whatever he chose to do (or not to do).

Dr. Darwin then decided that Charles should study for the clergy and in 1827 sent him to Cambridge University. Fortunately for modern biology, the clergy was then a haven for naturalists, and 19-year-old Charles Darwin soon shook off his perpetual boredom and rediscovered his boyhood passion for nature.

In three years, he managed to complete his B.A., and when he graduated from Cambridge in 1831, one of his professors recommended him for a position on board the H.M.S. *Beagle*. The *Beagle* was a survey ship about to embark on a trip around the world, its five-year mission to map the coast of South America. Darwin's duties were to collect specimens of the plants, animals, and rocks from this unexplored new world. In December 1831, just before the *Beagle* sailed, a professor friend sent Darwin the first volume of Lyell's *Principles of Geology*. His course was set.

During the long trip across the Atlantic, Darwin immersed himself in Lyell's arguments about the slow processes that formed the Earth. He thought of nothing but geology for months. Lyell formalized two important ideas: first, that the Earth is very old, and second, that geologic processes occur very slowly and gradually. Darwin's theory of evolution, as he would later conceive it, likewise stated that life is very old (millions of years) and change (evolutionary, in this case) has been slow and gradual.

As Darwin explored South America over the next five years, he began to suspect that all animals and plants were descended from a common ancestor. He wondered if Count Buffon, Lamarck, and even his own grandfather had been right about evolution. He took notes; he collected specimens. He noticed what animals ate, where they lived, what kind of soil this plant or that seemed to prefer.

Almost as soon as he returned to England in 1836, Darwin began to classify his specimens and to organize his notes. But it was six more years before he "allowed himself the satisfaction" of even outlining his theory. In the meantime he read the work of the second great intellectual influence in his life after Lyell—Thomas Malthus.

Malthus: Too Many Children

In 1798, the English economist and clergyman Thomas Robert Malthus (1766–1834) wrote a highly controversial paper, *An Essay on the Principle of Population*. Malthus argued that human populations tend to increase exponentially, while food supplies and other resources increase only arithmetically, if at all. For example, if a couple had three children, and each of their children had three children, and the nine grandchildren each had three children, the family farm could not possibly support the 27 great-grandchildren. The family may be able to increase the amount of food they can produce on their farm, but they cannot possibly triple production every generation. Poverty, famine, overcrowding, disease, and war are all the inevitable consequences of this excessive population growth, argued Malthus. He predicted that without some check on population growth, human populations would face a continuing, ferocious struggle for existence in the face of limited resources.

When Darwin read Malthus in 1838, he saw for the first time how evolution could work. Malthus's arguments applied to all species, reasoned Darwin, not just humans. Many more individuals of each species are born than can possibly survive, and, consequently, there is a constant struggle for existence.

"It follows," Darwin later wrote, "that any being, if it vary however slightly in any manner profitable to itself, under the complex and sometimes varying conditions of life, will have a better chance of surviving, and thus be *naturally selected.*" His theory of evolution through natural selection was beginning to take shape.

Wallace Scoops Darwin

But Darwin delayed publishing. Month after month, year after year, he added new facts to bolster his theory. Lyell, who was by then a friend, still didn't accept evolution, but he urged Darwin to publish his idea before someone else thought of the idea and published first. Then, in 1858, Alfred Russel Wallace (1823–1913), a land surveyor and naturalist working in the Malay Archipelago (in what is now Indonesia), independently hit upon natural selection as a mechanism for evolution.

Wallace had written to Darwin the year before asking for information about selective breeding. Darwin wrote back with the information and added tactfully that he had been working on similar ideas for some 20 years (Figure 15-5). Again Wallace wrote, this time outlining his theory of evolution—but without natural selection. Darwin responded politely, but hinted that he knew more. Wallace was obviously hot on his heels, yet Darwin did nothing.

Then, a few months later, in 1858, Wallace wrote once more. In the interim, he had read Malthus and experienced the same flash of insight Darwin had. Eagerly, Wallace wrote to Darwin, outlining his theory of evolution by natural selection. Darwin, utterly crushed, was ready to give all of the credit to Wallace. In a letter to Lyell, he wrote:

> I never saw a more striking coincidence; if Wallace had my [manuscript] sketch written out in 1842, he could not have made a better abstract! Even his terms now stand as heads of my chapters. Please return me the [manuscript], which he does not say he wishes me to publish, but I shall, of course, at once write and offer to send it to any journal. So all my originality, whatever it may amount to, will be smashed.

Then worse misfortune struck. Scarlet fever swept Darwin's family, killing his 19-month-old son Charles in a matter of days. Grief stricken, Darwin threw up his hands and turned everything over to Lyell, who gladly rescued Darwin. Lyell presented both Wallace's paper and an extract from Darwin's unpublished outline to the Linnaean Society, giving credit to both men. The following year, 1859, Darwin finally published his theory in a book, *On the Origin of Species by Means of Natural Selection.*

The Origin, as biologists call it affectionately, is the single most influential scientific book ever written. When it appeared in 1859, all 1,250 copies of the first printing sold out in a single day. Its arguments persuaded even those, such as Lyell, who had previously rejected the idea that species evolve. What was so convincing?

A.

B.

Bridgeman Art Library

The Granger Collection, New York

Figure 15-5

Darwin and Wallace. A. Charles Darwin and Emma Wedgwood Darwin. Even in his most productive years, Charles Darwin was sickly and worked only for short periods. B. Alfred Russel Wallace's ideas about evolution generally matched Darwin's. However, although Wallace accepted that humans had evolved from ancestral primates, he did not think that our capacity for thought and spiritual passion could have evolved in the same way.

Era	Period	Epoch	Millions of Years Ago	Major Events
CENOZOIC	Quaternary	Holocene	* 0.01	Humans
CENOZOIC	Quaternary	Pleistocene	1.65	New mountains
CENOZOIC	Tertiary	Pliocene	5.3	Early humans
CENOZOIC	Tertiary	Miocene	23.7	Apes
CENOZOIC	Tertiary	Oligocene	36.6	Monkeys
CENOZOIC	Tertiary	Eocene	57.8	Grasslands
CENOZOIC	Tertiary	Paleocene	66.4	Early primates
MESOZOIC	Cretaceous		* 66.4	Flowering plants, insects, and reptiles diversify
MESOZOIC	Jurassic		144	First birds
MESOZOIC	Triassic		208	First mammals
PALEOZOIC	Permian		* 245	First conifer trees
PALEOZOIC	Carboniferous		286	First reptiles and forests
PALEOZOIC	Carboniferous		320	Age of amphibians
PALEOZOIC	Devonian		360	First insects and amphibians
PALEOZOIC	Silurian		* 408 / 438	First jawed fishes
PALEOZOIC	Ordovician		438	First land plants
PALEOZOIC	Cambrian		* 505	First fishes
PALEOZOIC			545	
PRECAMBRIAN			900	First invertebrate animals
PRECAMBRIAN			3500	Photosynthesis in bacteria
PRECAMBRIAN			4600	Origin of Earth

* Mass extinctions

Zion National Park

Grand Canyon National Park

15.2 What Is the Evidence for Evolution?

Evolution is heritable change in species over time, which Darwin called "descent with modification." But evolution is also "speciation," the division of one species into two or more. Such "diversification" is the means by which a few kinds of simple cells living 3.8 billion years ago gave rise to the millions of complex organisms that live on Earth today. *The Origin's* great strength was Darwin's gluttony for evidence. Darwin provided evidence for evolution from

Figure 15-6

The history of life. A composite of the Grand Canyon and a similar canyon in Zion National Park showing the geologic eras and periods as discrete layers of rock. The Grand Canyon includes layers from the Precambrian, at the bottom, all the way up to the Permian, near the rim of the canyon. Nearby Zion National Park includes Triassic and Jurassic layers. At both canyons, the most recent layers, from the Cretaceous onward, have all been washed away. But these layers showing life's history can be found intact in other places. If you wanted to look for bones from the first amphibians would you go to the Grand Canyon or to Zion?

Figure 15-7 (part 1)

Distinctive fossils from the geologic past from the Precambrian to the Permian extinction. (Photos from left to right: Dickinsonia, Biological Photo Service; fossil jellyfish, William E. Ferguson; trilobites, James L. Amos/Photo Researchers, Inc.; Devonian fish *Dinichthys terrelli*, James L. Amos/Photo Researchers, Inc.; calamites, Ted Clutter/Photo Researchers, Inc.; fern fossil, J.C. Carton/Bruce Coleman, Inc.)

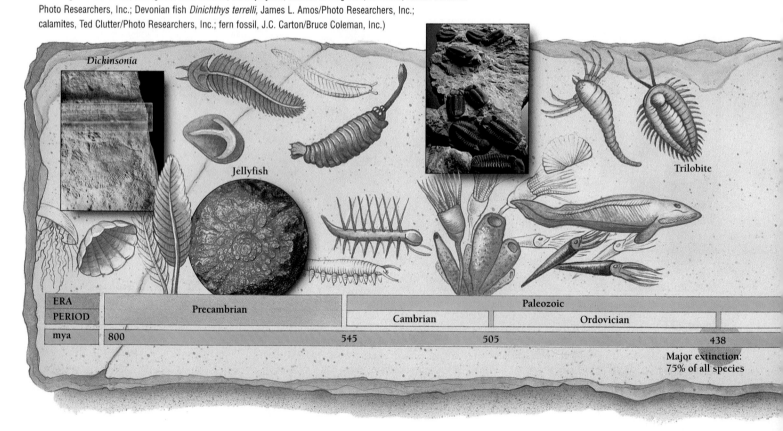

Dickinsonia

Jellyfish

Trilobite

ERA	Precambrian		Paleozoic		
PERIOD		Cambrian	Ordovician		
mya	800	545	505		438

Major extinction: 75% of all species

the fossil record, from the distribution of plants and animals (biogeography), from comparative anatomy, from taxonomy (the classification of organisms), from embryology, and from domestic breeding. Since Darwin's time, other biologists have further buttressed his arguments with evidence from other fields, as well, including genetics, population ecology, and animal behavior. In the rest of this chapter we will briefly examine some of the evidence that persuaded Darwin and all modern organismal biologists.

The Fossil Record Tells a Story of Evolution

The sequence of organisms in the fossil record tells a story with an obvious meaning. At the bottom, where sediments are oldest, lie the primitive prokaryotes and eukaryotes (Figures 15-6). Above them come simple multicellular organisms, then trilobites and other invertebrates, and then the first vertebrates, the fishes. Many more fish follow. Then, in rapid succession, amphibians and reptiles arrive. Newer, and higher in the sedimentary strata, are the dinosaurs, crocodiles, birds, and mammals.

Two facts stand out. First, fossils are distributed consistently. Rocks of the same age contain approximately the same groups of organisms. For example, the 500-million-year-old rocks of the "Cambrian Period" usually contain trilobites, but never dinosaurs or horses. More recent rocks contain no trilobites. Such a pattern is consistent with the

idea that once organisms go extinct, they stay extinct. Extinct species do not suddenly reappear millions of years later as a new creation.

Second, the order in which organisms are laid down in the fossil record suggests a sequence of evolution that is independently confirmed by several other fields of biology. Recent fossils look most like modern organisms, while the oldest fossils generally look least like modern organisms (Figure 15-7). In the layers of rock, obvious patterns of change suggest evolution. The reptiles first appear, followed by the mammal-like reptiles, followed by true mammals. Such patterns are consistent with the idea that mammals evolved from mammal-like reptiles, which evolved from earlier reptiles. Similar patterns appear over and over again in the fossil record in thousands of lineages of animals and plants.

Rocks of different ages tell a story of change. Each layer, or stratum, of rock comes from a different chapter of Earth's history. The particular order in which new species appear in these chapters irrefutably illustrates that organisms have changed over time, or evolved.

How Do Fossils Form?

An animal or plant becomes a fossil only if it dies under the right conditions. The vast majority of dead plants and animals are consumed by other organisms and leave no trace of their existence. Dead things leave a trace only if their bodies

Dinichthys terrelli

Cooksonia

Horse-tails

Cycad

Paleozoic			
Silurian	Devonian	Carboniferous	Permian
408	360	286	245

Major extinction: 70% of all species

Major extinction: 90% of all species

or imprints are rapidly buried in a bog or at the bottom of a sea or lake. There, protected from scavengers, erosion, and decay, organisms may in time become more deeply buried as successive layers of mud and sand settle over them.

Over time, pressure from all these layers of sediment turns the deepest layers of mud and sand into "sedimentary rock." After millions of years, geologic forces raise the rock up into new mountains. Rivers flow down the sides of the new mountains, eroding canyons through the layers of old sediment, revealing the ancient—and to us, amazing—fossils.

Because organisms fossilize most often when buried in sediment at the bottom of a sea or lake, most fossils are marine or freshwater organisms. Terrestrial plants and animals are usually fossilized only if their bodies happen to fall into a lake or sea. Furthermore, only the hard parts of animals—shells, teeth, and bones—and the woody parts of plants are likely to survive long enough to form fossils. The probability is remote that a soft, terrestrial organism such as a slug will be fossilized.

The chanciness of fossilization means that the fossil record is incomplete. For example, most insects are too small and delicate to be preserved, and so the fossil record for insects is poor. However, where insects are preserved, they often appear in great numbers. Large numbers of insects may appear in one layer, disappear—leaving a gap—then reappear millions of years later. Even vertebrates, some

of the best preserved of all organisms (because of their hard bones), show a discontinuous record. A mammal may disappear from the record, then reappear, slightly altered, ten thousand years later. (These discontinuities reflect the chanciness of the fossilization process and do not imply that the same species has evolved twice.) Despite the fossil record's limitations, however, it clearly shows that life has evolved according to the sequence summarized in the next section.

What Does the Fossil Record Show?

In the 18th and 19th centuries, scientists recognized that the fossils in the top layers of sediment were the most recent, while the fossils in the deepest layers were the oldest. For the first time, scientists were able to read the layers of rock as if they were the pages of a history book. They discovered a startling history of the last 570 million years. Twentieth-century paleontologists have now extended this history back to the appearance of the first cells on Earth nearly 4 billion years ago.

Geologists divided this history into chapters. Each chapter, or **period**, is 30 to 75 million years long and contains distinctive forms of life in its fossil record (Figure 15-6). Periods are grouped into four long **eras**, of from 65 million to several billion years each. The most recent periods are subdivided into relatively short **epochs**. The same distinctive layers appear in rock formations around the world.

Figure 15-7 (part 2)
Distinctive fossils from the geologic past from the Triassic until today. (Photos from left to right: *Archaeopteryx* fossil, James L. Amos/Photo Researchers, Inc.; fossil dinosaur eggs *(Protoceratops andrewsi)*, Ken Lucas/Visuals Unlimited; ichthyosaur fossil, E.R. Degginger/Bruce Coleman, Inc.; fossil flower, William E. Ferguson; plesiosaur fossil, E.R. Degginger/Earth Scenes; orangutan and baby, Tim Davis/Photo Researchers, Inc.)

Nineteenth-century geologists could not give exact ages to each layer. However, they could tell the order in which the layers had been deposited because the same order of layers appeared wherever geologists looked. So they could estimate the ages of the various periods and eras recorded by these fossil layers remarkably well. The modern development of radioactive dating techniques, in the 1940s and 1950s, allowed paleontologists to assign actual dates to evolutionary events, which confirmed many of the old estimates.

The **Precambrian Era** comprises most of the history of the Earth, beginning with the planet's formation 4.6 billion years ago. Before the discovery of the first Precambrian fossils in the 1950s, paleontologists thought that no life existed for the first three and half billion years of Earth's history. But the discovery of ancient bacteria and algae proved otherwise and paleontologists now subdivide the Precambrian into six eras. The first fossils, some as old as 3.8 billion years, resemble modern bacteria. These ancient fossils vividly illustrate the breathtaking antiquity of life on Earth. Fossils of the first eukaryotes (cells with membrane-enclosed nuclei) appear in rocks about 2 billion years old. And new evidence—in the form of decayed cholesterol molecules in ancient rocks in northwestern Australia—suggests that eukaryotes may have appeared as early as 2.7 billion years ago. In Chapter 18, we discuss the formation of the Earth and how life might have originated.

The Cambrian Explosion

The first multicelled organisms appear in the fossil record near the end of the Precambrian, about 800 million years ago. And the first multicelled animals—which resemble jel-

lyfish, worms, and tiny shelled creatures—appear about 640 million years ago. But it is in the Cambrian Period [Latin, *Cambria* = Wales], at the beginning of the **Paleozoic Era** [Greek, *palaios* = ancient + *zoos* = life], that multicelled life first appeared in all its staggering diversity. By 543 million years ago, at the time of the "Cambrian explosion" of life, the Earth was home to trilobites, jellyfishes, corals, segmented worms, fungi, algae, and organisms unlike anything alive today (Figure 15-7, part 1). Indeed, by the end of the Cambrian, 11 of the 35 phyla that live today had already evolved, including the first chordates, ancestors of all vertebrates. We discuss both the Cambrian explosion and vertebrate animals in more detail in Chapter 23. Geologic studies suggest this burst of evolution may have occurred in as little as 5 million years. At the end of the Cambrian a mass extinction wiped out whole classes and families of organisms. But the first vertebrates—the armor-plated fishes—survived.

During the next two periods of the Paleozoic Era—the Ordovician and the Silurian—the first fossil plants, great numbers of starfishes, nautiluslike animals, and fishes with bony skeletons appeared (Figure 15-6). During the Devonian Period, the oceans receded, leaving vast new tracts of dry land. On land, fungi, mosses, and other bryophytes evolved, as well as the first "vascular" plants, those with a system of tubes for transporting water and nutrients. Arthropods such as scorpions, centipedes, millipedes, and springtails appeared as well. In the seas and lakes, so many kinds of fishes evolved that the Devonian is called the Age of Fishes.

During the Carboniferous, the seas rose and fell repeatedly, and the first amphibia evolved. They were the first ver-

Flowers

Early mammals

Toxodon

Mesozoic		Cenozoic	
Cretaceous		Tertiary	Quaternary →

66.4

1.65

Major extinction:
75% of all species

Extreme Biology Radioactive Dating

The fossil record provides a useful chronicle of the sequence of organisms that have lived on Earth. Fossils also provide an accurate way to determine the relative ages of different layers. Still, even if scientists know the relative ages of every single layer, they still want to know the **absolute age**—the age in years—of each of the layers. Scientists had no way of estimating the absolute ages of rocks until the discovery of radioactivity by Marie Sklodowska Curie and Pierre Curie in 1898.

As we discussed in Chapter 2, radioactive atoms, such as uranium, break down into other materials by giving off alpha, beta, and gamma rays. The breakdown, or **decay**, occurs at a steady rate that is unaffected by temperature, pressure, or other environmental variables. Every form of an atom, called an **isotope**, breaks down into a series of other isotopes. Half of the isotope carbon-14, for example, decays to nitrogen-14 every 5,600 years. We say that the **half-life** of carbon-14 is 5,600 years (Figure A).

By comparing the relative proportions of different isotopes in a rock sample, scientists can tell the age of the rock. Different radioactive isotopes have different half-lives. Carbon-14 is useful for dating objects that are only a few thousand years old. In contrast, uranium-238 has a half-life of about 5 billion years, and rubidium-87 has a half-life of about 50 billion years. These isotopes are useful for dating very old rocks such as those that originated at the same time as the solar system (about 5 billion years ago).

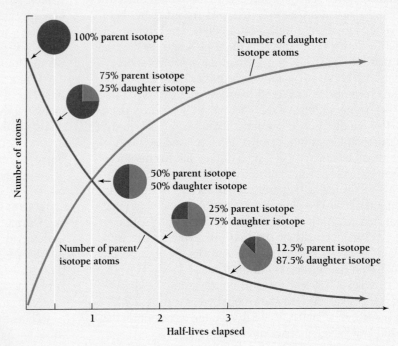

Figure A

Radioactive decay. The half-life of a radioactive element is the time it takes half the atoms to decay to another element. Scientists can use this steady, predictable decay to measure the ages of rocks, fossils, and other objects.

---

tebrates to live permanently on land. Sometimes called the Age of Amphibia, the Carboniferous and Permian Periods were also the age of the first great forests. The first reptiles appeared, as well as flying insects, such as cockroaches and dragonflies, some with wingspans of 75 cm (2.5 ft). Swamps filled with ferns and horsetails, and cone-bearing trees covered huge areas of the Earth. These fossilized swamps persist today as the coal deposits from which we get coal, oil, and natural gas.

The extraordinary explosion of plants and animals that occurred during the Carboniferous ended with another mass extinction in the Permian. The continents drifted together to form a single supercontinent, called Pangaea; the seas dropped to their lowest level ever; and the climate became extremely dry (see box above). About 90 percent of all marine species went extinct over a period of 5 to 8 million years.

The **Mesozoic Era** [Greek, *mesos* = middle + *zoos* = life], often called the Age of Reptiles, began about 248 million years ago and ended 65 million years ago with the ex-

tinction of the dinosaurs. The Mesozoic comprised three periods—the Triassic, Jurassic, and Cretaceous. During the Triassic, more kinds of reptiles lived than ever before or since. The appearance of the reptiles depended on the evolution of an egg that could be laid on dry land. The amniotic egg is responsible for the success of the reptiles and their descendants, the birds and mammals.

Soon after the reptiles appeared, they split into three groups: the turtles, the first mammals (small, shrewlike animals that did not diversify much, but, luckily for us, hung on for 200 million years until the dinosaurs were gone); and all the other reptiles—including the lizards, snakes, dinosaurs, and, eventually, the birds. In the seas, long-necked plesiosaurs and ichthyosaurs fed on Mesozoic fish. In the air, pterosaurs flapped across the sky, some with wingspans of as much as 10 meters (30 feet), others as small as sparrows. Late in the Triassic, the first dinosaurs appeared, diversified throughout the Jurassic, then disappeared in another mass extinction at the end of the Cretaceous (Fig-

Extreme Biology Continental Drift

Studies in biogeography reveal that species resemble nearby species more than species that live far away. However, similar organisms can occur in places that are remote from one another. For example, rheas, ostriches, and emus, all large flightless birds, occur in South America, Africa, and Australia, respectively. We may wonder if these birds are related to one another and, if so, how they came to be so widely dispersed.

The fossil record shows many examples of similar changes in distribution, with closely related species in the southern continents. The distinctive fossil plant *Glossopteris,* for example, appears in similar rock formations from the late Paleozoic Era in both Brazil and South Africa. How could organisms that are so closely related live so far from one another?

In 1915, the German geologist Alfred Wegener, observing the jigsaw-puzzle fit of the African and South American coastlines, proposed the theory of "continental drift." Wegener argued that Africa and South America, as well as Australia, Antarctica, and India, were once a single supercontinent, which broke up into separate continents that drifted apart. Geologists rejected Wegener's theory, however, and biologists and paleontologists continued to puzzle over the similarities between the species on the southern continents.

Then, in the early 1960s, geophysicists discovered that the Earth's surface consists of large blocks, or **plates,** which move with respect to one another at the rate of a few centimeters a year. The discovery that both the continents and the sea floor were moving about on the surface of the Earth immediately revived Wegener's theory.

Geologists now understand that this movement comes from the continuous formation of new crust under the ocean. As lava emerges from below the surface, the new rock pushes apart the plates on either side. Powered by this pressure, plates can collide, scrape past one another, or dive beneath one another, pushing up mountains and precipitating earthquakes and volcanic activity. The Himalayas, for example, rise

a little higher each year as the Indian-Australian plate crashes—ever so slowly—into the Eurasian plate. Similarly, the Pacific plate, beneath the Pacific Ocean, scrapes northward past the North American plate, creating California's earthquake-prone San Andreas Fault.

Geologists have been able to puzzle out many of the past movements of the continents by studying the magnetic orientation of rocks. This history explains many features of the fossil record and the distribution of modern species as well. About 250 to 200 million years ago, all of the continents were united into a supercontinent called Pangaea. Pangaea was surrounded by one huge ocean, Panthalassa (Figure A). In time, Pangaea began to break apart, allowing independent evolution to occur on the two resulting continents—*Laurasia* (Eurasia, Greenland, and North America) and *Gondwanaland* (India, Africa, South America, Madagascar, Antarctica, and Australia). The historical distributions of hundreds of species follow the movements of the continent. For example, fossilized pollen from the southern beech shows that this tree lived in Gondwanaland, in what is now South America, Antarctica, and Australia. Today the southern beech tree is extinct in Antarctica, which is now too cold for any trees, but it still lives in southern South America and parts of Australia.

By the beginning of the Cretaceous, 135 million years ago, Gondwanaland had also begun to break into the separate continents. By about 65 million years ago, South America had split from Africa and Australia had separated from Antarctica, after which Australian species evolved in isolation. The land connection between North and South America formed only about 3 million years ago.

Any change in the distribution of an organism that is due to populations being divided—whether by continental drift, rising sea levels, or other changes in geography—is called "vicariance." Vicariance is distinct from changes in distribution that are due to organisms dispersing, for example from a mainland population to an island.

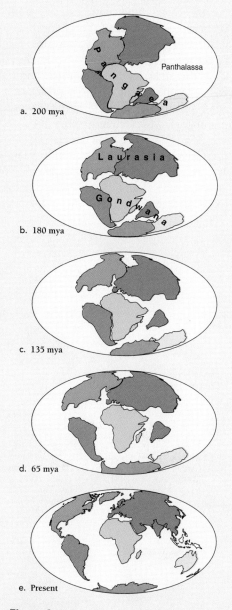

a. 200 mya

b. 180 mya

c. 135 mya

d. 65 mya

e. Present

Figure A

The history of the continents. (a) Alfred Wegener's view of the Earth as it existed 200 million years ago. At that time, there was one immense ocean and one continent (Pangaea). (b) About 20 million years later, the supercontinent had begun to split into northern Laurasia and southern Gondwanaland. (c) Separation continued, and (d) looked this way about 65 million years ago. (e) Further widening of the Atlantic Ocean and the northward migration of India brings the Earth to its present state.

Biology ⒺNow™ Learn more about continental drift by clicking on this figure on your BiologyNow CD-ROM.

ure 15-7). The disappearance of the dinosaurs left an open playing field for insects, flowering plants, birds, mammals, and many other reptiles.

The shortest era of geologic time, the **Cenozoic Era** [Greek, *cainos* = recent], extends from about 65 million years ago to the present. The Cenozoic has seen the diversification of the insects and flowering plants, the first grasslands, together with grassland animals, and the evolution of all of the other modern mammals and birds. The Cenozoic has seen the origin of most primates, including the first apes; and the coevolution of the flowering plants and their pollinators, which include birds, bats, and insects.

Because so little time has elapsed since the beginning of the Cenozoic—a mere 65 million years—fossils from this era are common and well preserved. Fossils of mammals are especially abundant, and paleontologists have been able to piece together nearly continuous evolutionary histories

for many modern species, including those of humans and horses.

How Does Taxonomy Suggest the Family Tree of Life?

The taxonomist Carolus Linnaeus did not intend for his classification of organisms to be interpreted literally as a family tree, yet it provided Darwin with persuasive evidence for evolution. We now understand that the careful anatomical comparisons that Linnaeus and his successors made usually reflect actual lines of descent. We can group species into hierarchies for the same reason that we can group individual humans into families and other hierarchies of relatedness. All organisms are related (Chapter 19).

In Figure 15-8, we see that all the dogs, wolves, coyotes, and other species in the genus *Canis* look alike, just as mem-

Figure 15-8
Strangers and cousins. No dog could be mistaken for a bear, a cat, a hyena, or any of the other noncanid families in the order Carnivora. All the members of the dog family (Canidae) are recognizably different from bears, raccoons, cats, and other carnivores. Yet the different genera of dogs are recognizably doggy. In their natural forms, all the canids look similarly doglike. At the bottom, notice how artificial selection by humans has exaggerated certain traits and deformities to such an extent that the physical variation expressed by just five breeds of dog surpasses that shown by all the other species in the genus.

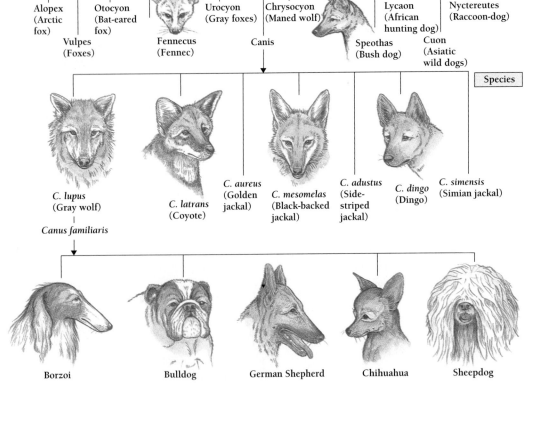

bers of the same family should. They are like siblings. The members of related genera—*Canis* and the foxes *Vulpes* and *Fennecus*, for example—are also similar, but not as much. They are like cousins. All the dogs, foxes, and wolves belong to the same family, the Canidae, for the members of the dog family are more closely related to one another than they are to the bears, cats, and other families in the order Carnivora.

How Does Comparative Anatomy Show How Organisms Are Related?

Species descended from a common ancestor may evolve in quite different directions and yet retain many of the same characteristics. For example, consider the forelimbs of humans, horses, bats, moles, and whales (Figure 15-9). Each of the different forelimbs possesses the same bones arranged in the same way. But each animal uses its forelimb differently. Humans write, horses stride, bats fly, moles dig, and whales swim. Even though these various bones serve different func-

tions today, they all arose from the same structures in a common ancestor. Similar structures, despite differences in function, imply common ancestry and, therefore, evolution.

Similar structures in two or more species are called **homologous structures** if the structures are similar because they evolved from the same ancestral structure. Homologous structures don't have to perform different functions. For example, the bird, the bat, and the pterosaur evolved different arrangements of the same bones to perform the same function—flying. All these wings evolved separately from homologous structures, and each in its own way allows flight.

And some homologous structures are hidden unless you know what you are looking for. For example, the four tiny bones in the mammalian middle ear are homologous with much larger bones from the jaws of our ancestors (Figure 15-10). Evolution is opportunistic. Just as a handyman can use a bit of wire here, a bit of glue there, to fix a bicycle or a light fixture, so evolution may employ a humerus bone here, a radius bone there, to allow one organism to gallop, another to fly.

Not all similar structures are homologous. The wings of birds and of insects, for example, evolved from unrelated structures, yet they serve the same function and have a similar shape. All wings must be light and have broad surfaces (Figure 15-11). Similar structures that have independently evolved to serve the same function are an example of convergent evolution, which is one kind of "homoplasy." Homoplasy is the opposite of homology.

Paleontologists can trace changes in homologous structures through the fossil record. For example, the first recognizable horse lived about 50 million years ago. The size of a large dog, *Hyracotherium* had four toes on its forefeet and three on its hind feet (Figure 15-12). Its teeth were too weak for it to chew the tough grasses that modern horses live on, so paleontologists think that these ancestral horses lived on leaves, which are softer and more easily digested.

Subsequent horse fossils show a progressive lengthening of the legs, strengthening of the teeth, and reduction in the number of toes. In the foreleg of a modern horse, the lower

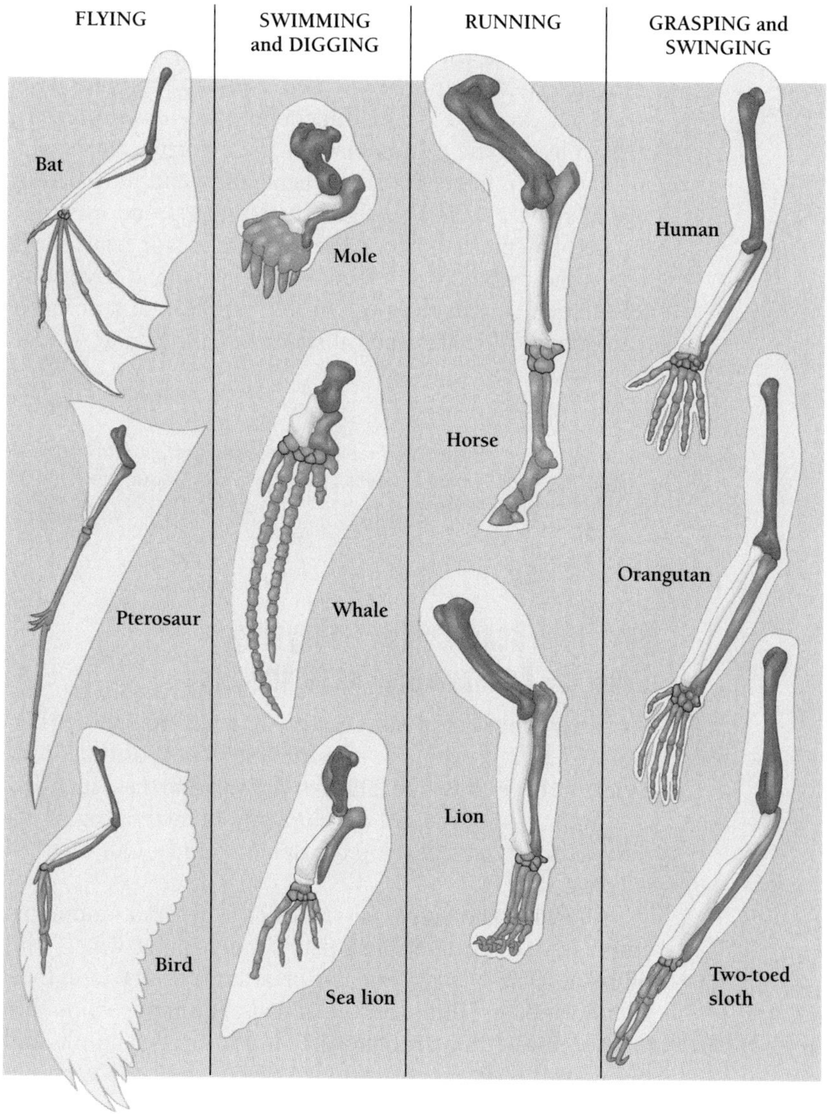

FLYING   SWIMMING and DIGGING   RUNNING   GRASPING and SWINGING

Bat

Mole

Human

Pterosaur

Horse

Whale

Orangutan

Bird

Lion

Sea lion

Two-toed sloth

Figure 15-9
Homology. The limbs of bats, moles, horses, and humans contain the same set of bones, modified for flying, digging, running, and grasping, respectively. This limb homology extends to all vertebrates, including birds, reptiles, and amphibians.

Lobe-finned fishes

Amphibians and reptiles

Mammal-like reptiles

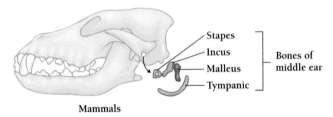
Stapes
Incus   } Bones of
Malleus   middle ear
Tympanic

Mammals

Figure 15-10
Evolution is opportunistic. The four tiny bones that make up the mammalian middle ear evolved from the jawbones of early reptiles. The reptilian jawbones evolved, in turn, from the branchial arches (structures that carry gills) of ancient fishes.

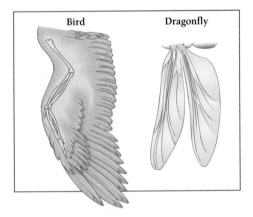
Bird   Dragonfly

Figure 15-11
Convergence in the wings of bird and dragonfly. Not all similarities in structure are due to homology. Birds and insects have evolved wings separately. The two kinds of wings look similar because flying exerts similar selective pressures on all flying organisms. When unrelated organisms independently evolve similar adaptations, biologists call it convergent evolution (a form of "homoplasy").

bones (radius and ulna) are fused and long. The last segment of a horse's leg is actually a long middle finger, the only one remaining of the five digits found in primitive ancestral mammals. These successive changes in the homologous bones give the modern horse longer legs and extraordinary fleetness. The many fossils that grade from the tiny, horse-like animal Hyracotherium to the modern horse are called **intermediate forms.**

Some homologous structures have no apparent use. The modern horse has a single, useless side toe, called a splint. The splint is an example of a **vestigial structure** [Latin, *vestigium* = footprint], a part of an organism with little or no function but which had a function in an ancestral species. A vestigial structure is a remnant. Our own tailbone, invisible without an x-ray image, is a vestige of another way of life. Similarly, some snakes and whales have vestigial pelvic and leg bones, left over from ancestors that walked (Figure 15-13). Fossil whales show substantial hind limbs, teeth like those of land animals, and nostrils near the tips of their noses (instead of blowholes on top). All these clues suggest that whales evolved from land mammals. Vestigial organs, like other homologous structures, are indicators of biological history and thus offer evidence for evolution.

Even genes can be vestigial. Molecular biologists call vestigial genes "pseudogenes" because they look like other genes but do not code for functional proteins or play any other role in gene expression. Yet these corrupt genes persist, passed from generation to generation and inherited by related species. Like a vestigial structure, a pseudogene is a vestige of a common ancestry and a marker for relatedness. Just as the members of a family of humans may have some defect such as a missing tooth, the members of a group of species may all share vestigial structures and genes.

Homologous structures, intermediate forms, and vestigial structures all suggest an evolutionary process whereby ancient structures are redesigned or abandoned.

How Does Comparative Embryology Show the Common Origins of Animals?

The early embryos of vertebrates are amazingly alike (Figure 15-14). For example, all vertebrate embryos, including humans, have tails and gill-like *branchial arches*. In fishes and amphibians, the tail develops into a swimming structure. Other groups develop the tail differently. Our own tail is vestigial.

Likewise, the branchial arches of fishes and amphibian embryos develop into gills for breathing under water. Amphibians that change into air-breathing adults lose their gills. In reptiles, birds, and mammals, the branchial arches develop into other structures in the adult ears, mouth, and respiratory tract. These include the bones of the ears and

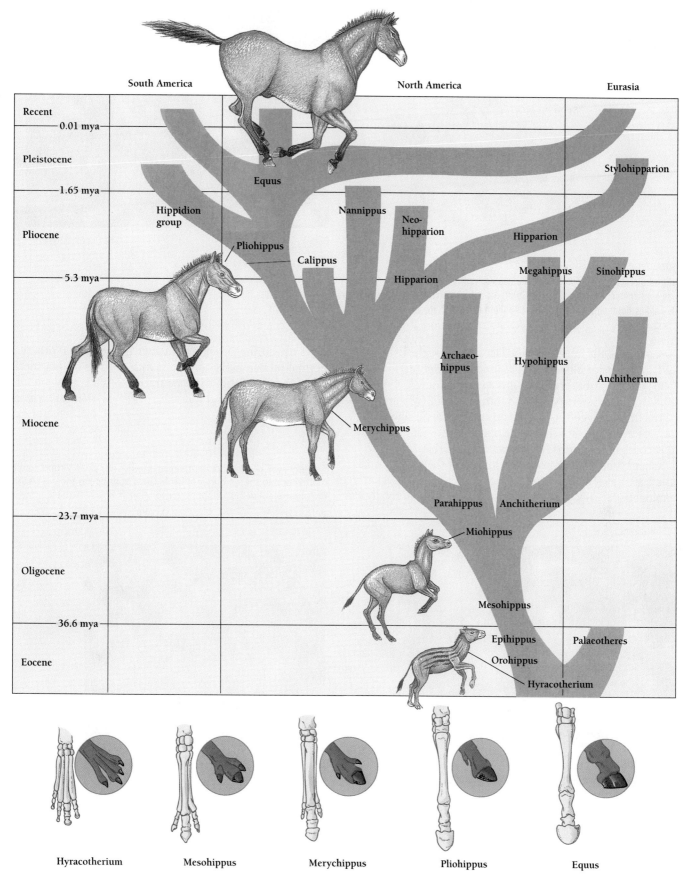

Figure 15-12

The evolution of horses. The fossil record for horses shows many intermediate forms, from the tiny Eocene horse that had four toes to the large modern horse that has one toe on each foot.

Biology❻Now™ Learn more about the evolution of the horse by clicking on this figure on your BiologyNow CD-ROM.

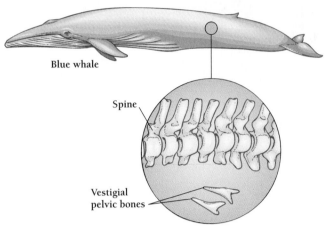

A.

B.

Figure 15-13

Vestigial structures. A. The blue whale has vestigial pelvic bones, remnants from the whale's ancestors' life as land animals with legs. B. The South African python's short claw is useless for walking (although male snakes fight with them and also use them to scratch the backs of females during courtship).

throat, some of the muscles of the face and neck, the thymus gland, and a major blood vessel of the lungs (Figure 15-15).

In embryos, the branchial groove or opening between the branchial arches may open to form a gill slit (where water exits from the gills). These are obvious in sharks and tadpoles, for example. Amazingly, the opening between the first and second arches forms, in humans, a hole that forms the Eustachian tube leading from the outside of the ear inward to the eardrum. The outer hole (where we clean our ears) is homologous with a gill slit. The other gill slits normally close up during development, but in rare cases, otherwise normal human babies are born with gill slits in their necks. All of these structures can be traced to the embryonic pharyngeal pouches and provide evidence that we are descended from fishes.

The embryos of developing organisms frequently pass through stages that resemble the embryos of organisms from which they evolved, a fact consistent with the theory of evolution.

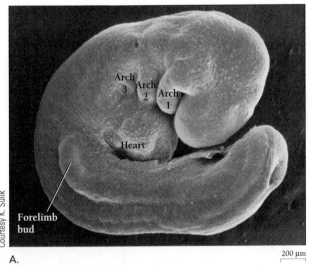

A.

200 μm

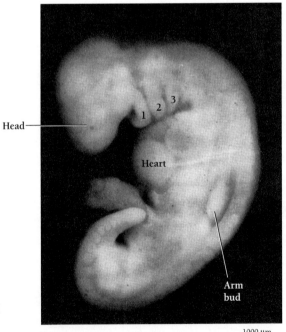

B.

1000 μm

Figure 15-14

Early embryos of a mouse and a human. These two mammals are at about the same stage of development. A. A mouse. B. This one-month-old human embryo is about four millimeters long, smaller than the word "bud" on the photo. The early embryos of vertebrates are remarkably alike, all displaying branchial arches, paddlelike limbs, and obvious tails.

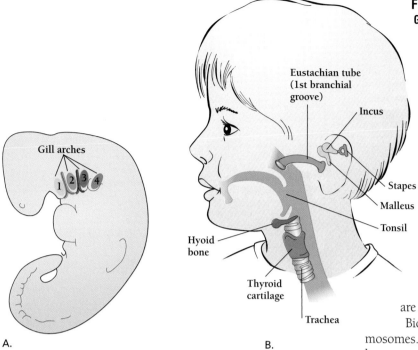

Figure 15-15

Gills in adult humans? Not quite. In all adult humans, the opening into the outer ear and the eustachian tube derive from the branchial groove, an undeveloped gill slit *(blue)*. In addition, the bones of the inner ear derive from the first and second branchial arches of the gills, as shown in A and B. Bones in the throat derive from the 3rd and 4th branchial arches *(purple and orange)*. Rarely, an opening between the throat and the neck survives development, a remnant of the second primitive gill slit, or branchial groove *(red)*.

Gill arches — 1 2 3 4

A.

Eustachian tube (1st branchial groove)

Incus

Stapes

Malleus

Tonsil

Hyoid bone

Thyroid cartilage

Trachea

B.

How Can Comparative Molecular Biology Show How Organisms Are Related?

Just as the morphology and embryology of organisms provide clues about their relatedness, so too does their molecular makeup. An organism's genes and the products of those genes, proteins, are a clear record of its lineage.

All cells rely on the same molecular machinery. All use DNA to carry the genetic information; RNA, ribosomes, and the same genetic code to translate that information into proteins; the same 20 amino acids to build proteins; and the energy molecule ATP to carry energy. The universal homology of this system suggests that it is a common heritage, passed down from some ancient common ancestor.

Certain cell proteins are nearly as universal. One of the earliest proteins that evolutionary biologists analyzed was cytochrome *c*, a protein in the electron transport chain of the mitochondria (Chapter 6). Cytochrome *c* is a single polypeptide chain of 104 amino acids, whose exact sequence is known in dozens of species. In all of these species, 35 of the 104 amino acids are always identical. The remaining 69 amino acids can vary from species to species. Similar species have similar cytochrome *c* sequences, while very different species have more changes. For example, humans and chimpanzees have exactly the same sequence of amino acids in cytochrome *c*, but humans and tuna differ by about 29.

Nowadays, biologists more often look directly at gene sequences that code for proteins than at the proteins themselves. The more closely related two species are, the greater the similarity between their gene sequences. In conjunction with anatomical data, biologists can use DNA sequences from various organisms to construct "phylogenetic trees," which, like a family tree, illustrate how different species or groups of species are related. We discuss phylogeny in Chapter 19.

Biologists can also look at the organization of chromosomes. Such comparisons show that less-related species have more chromosomal rearrangements. A cat genome, for example, differs from a human genome by about 13 rearrangements. These rearrangements occur rarely, and geneticists estimate that in mammals, one rearrangement occurs about every 10 million years.

The evolutionary history suggested by amino acid sequences usually agrees rather closely with the history implied by the fossil record. For example, the cytochrome *c* proteins of mammals and reptiles are far more different from one another than the cytochrome *c* of horses and donkeys. This suggests that mammals and reptiles diverged from each other long before horses and donkeys did. And, indeed, the fossil record shows that the first mammal-like reptiles emerged at least 180 million years ago, while the first horse-like fossils appeared only about 36 million years ago. This is logical: groups of mammals, such as horses, can't diverge from each other if mammals haven't evolved yet. The fossil record, anatomical comparisons, and molecular comparisons are internally logical and independently confirm that modern organisms have evolved from older ones.

Comparative molecular biology has also revealed that genes tend to change, or mutate, at a constant rate within certain groups of organisms. Molecular changes seem to accumulate in each type of protein, or in the corresponding piece of DNA, steadily, like the ticks of a *molecular clock*. Molecular clocks help researchers estimate when two species diverged. By comparing many genes, biologists can construct reasonably reliable phylogenetic trees.

But molecular clocks are imperfect, for individual genes evolve at different rates. And the same gene may evolve faster in one lineage than in another—faster in rodents, say, than in primates. Because genes evolve at different rates in different lineages, molecular clocks have limited value for measuring actual evolutionary time. Nonetheless, molecular clocks can tell us which groups diverged first, second, or third, if not exactly when.

The molecules of closely related organisms are more similar than the molecules of distantly related organisms. Comparative molecular biology confirms and elaborates on the story told by comparative anatomy and the fossil record.

How Does Biogeography Explain Why Organisms Evolved Where They Have?

When Darwin wrote *The Origin of Species,* most Europeans believed that every species had been individually created—a rhinoceros for this continent, a horse for that continent. If there were no tigers in North America, it was because God had not put any there. But Darwin saw that the distribution of many groups of species made no sense unless they had arisen from common ancestors.

From his studies of the distributions of species, called **biogeography,** Darwin concluded that each species has a definite region of origin, becomes distributed by migrating, and evolves descendant species in the regions to which it migrates. Darwin's view is so well accepted that nowadays if we ask someone why are there no tigers in North America, we will be told it's because tigers have no easy way to get from India and China across the Pacific Ocean. We all take it for granted that species move from place to place. But 19th-century biologists did not.

One of Darwin's strongest arguments for evolution was the close resemblance of island species to those on nearby continents. A 19th-century biologist would have expected that the closest relative of a finch from the Galápagos Islands, off the coast of South America, would be found in a similar group of islands, matched for climate and geology. But Darwin found that the closest relatives of island species were not on similar islands halfway around the world but right next door, usually on the nearest continent.

Other evidence that species come from other parts of the world comes from ecological studies of species richness. In general, the farther an island is from the nearest mainland source of species, the fewer species it has. On the other hand, islands that are close to an area with lots of species tend to have more species (Figure 15-16).

Biologists can also trace the history of which species lived where. Newer groups of organisms may appear on a particular continent and then disperse to other regions of the world. Older groups of species may have originated in areas that were later divided by rising ocean levels or by continental drift.

Species all over the world are most closely related to those that live nearby. Darwin's theory of common descent accounts for this pattern.

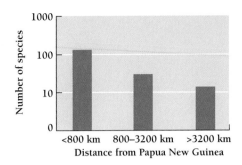

Figure 15-16
Few and far between. The more isolated a small island, the fewer species live there. Isolated islands are harder for new species to get to.

15.3 Natural Selection: Darwin's Mechanism for Evolution

Darwin made a strong case for the theory of evolution. And the evidence that scientists have accumulated since then has strengthened this theory. Indeed, the biologist Theodosius Dobzhansky summed up biologists' thinking best, writing, "Nothing in biology makes sense except in the light of evolution."

We know that organisms have changed over time, and we know that modern organisms are descended from ancient ones. But how does evolution happen? What is natural selection, the mechanism that Darwin proposed for evolution? Darwin hypothesized that natural selection, acting gradually over immense spans of time, created not only new species but new genera, new families, and all the higher taxa. Why did he think that was possible?

Once Darwin conceived the idea of evolution by natural selection, he developed an obsessive interest in the work of gardeners, farmers, and stockbreeders. It was perfectly apparent to him that many of the different breeds of dogs and pigeons, for example, would, if found in nature, be identified as separate species.

How Is Artificial Selection a Model for Natural Selection?

Breeders derive such varied forms from a single species by carefully selecting and breeding only individuals that show certain desired traits most strongly. Under this selective regimen, each generation will show the selected trait more clearly than the last.

Dramatic examples of such intense selection come from all kinds of farm animals and crops. Most amazing are the differences among domestic dogs (Figure 15-8). The English bulldog, for example, is the result of selection for a developmental deformity in which the upper jaw fails to develop at the same rate as the lower jaw. The lesson here is that nearly all species possess tremendous untapped genetic variation.

Figure 15-17
Artificial selection in pigeons. Charles Darwin collected different breeds of pigeons to study the variation possible within a single species. Shown here are a pair of baldheads, a carrier cock, two white fantails, two white pouters, and several other varieties.

The Illustrated London News Picture Library

Darwin's particular interest was the work of Victorian pigeon fanciers. He collected many varieties of pigeons, studied the differences in their skeletons, and bombarded fanciers with questions (Figure 15-17). Ultimately, he concluded that all breeds of pigeons were descended from a single species—the European rock pigeon. Yet, the more than 20 different breeds were so different, Darwin argued, that a zoologist who discovered them in the wild would certainly classify them into a dozen or more species and quite possibly into half a dozen genera, as well.

Clearly, artificial selection by breeders could transform the characteristics of a breed. Success required only (1) that individual organisms vary in their characteristics, (2) that these variations be heritable, and (3) that the breeder consistently select certain traits in each generation.

Artificial selection does not result in the production of new species. At least new breeds are not recognized as species, no matter how distinctive they are. A mammologist from Mars wouldn't hesitate to classify a bulldog as a species separate from a Chihuahua or a German shepherd. Morphologically they are distinct. But we know their history and that they can interbreed, so we classify them all as one species. Artificial selection demonstrates how much genetic variation exists in populations of organisms and how much change can occur in just a few years. More importantly, however, artificial selection hints at the much greater degree of change that might occur under the constant pressure of natural selection over millions of years.

Artificial selection provides a model for natural selection, a possible mechanism for evolution.

How Can We Summarize Darwin's Argument for Natural Selection?

How can nature make selections in the same way as human breeders, generation after generation? In Malthus, Darwin found the answer. Malthus, recall, had argued that the production of food and other resources can never keep up with the enormous increases that human populations are capable of, that early death is the fate of a large proportion of individuals born in every generation.

Darwin saw that Malthus's arguments applied to populations of any species. Populations of organisms, although capable of enormous increases, do not continue to grow exponentially, but instead are limited by resources, predation, and disease. Natural selection must occur, Darwin argued, because nature cannot sustain unlimited exponential growth. To illustrate this point, he considered the descendants of a single pair of elephants, among the slowest breeders of all organisms. Even elephants, however, have the potential to breed exponentially. Over the course of 750 years, two elephants could have some 19 million descendants. Yet, historically, the elephant population has been fairly stable. So it is likely that two elephants in the year 1000 had only two descendants alive in the year 1750. The question Darwin posed was, Which two?

According to Darwin's theory, the survivors would be the descendants of whichever elephants were best adapted to the environment over the course of those 750 years. In artificial selection, the decisions of the breeder determine which individuals produce the next generation. In nature, the individuals that manage to survive and reproduce produce the next generation.

We can easily see how this argument might apply to the evolution of horses, for example. Individuals with long legs and teeth capable of chewing the tough grasses so common on the plains apparently survived and reproduced, whereas those with short legs and small teeth died out.

Darwin argued that if large changes can occur in domestic plants and animals over just a few generations, then even larger changes can certainly occur over the millions of years available for evolution. The rate at which changes accumulate depends both on the intensity of selection and on the extent of inherited variation. For example, if only a few young horses from each generation survive, because most cannot escape a fast predator or cannot digest grass well enough, selection is strong and advantageous traits would accumulate rapidly. However, if most of the young horses from each generation survive and reproduce, then selection is weak and change occurs slowly or not at all.

The rate of change also depends on the extent of genetic variation within the population. To take an extreme case, if all the individual horses of a generation were genetically identical, then the next generation of horses could not differ from their parents. But, in fact, individuals vary enormously.

Natural Selection Analyzed

Modern evolutionary biologists have broken down natural selection into a series of basic facts and inferences.

1. Fact: **Superfecundity.** All species of organisms are so fertile that their populations would increase exponentially if all offspring survived.

2. Fact: **Not all offspring reproduce.** Populations of organisms do not attain these staggering proportions because many offspring succumb to disease, predation, and other fates before they reproduce. Those that do reach reproductive age find that the world does not contain enough food, territory, and other resources for all of them to reproduce.

   Inference: **Struggle for existence.** Whenever extinction does not occur, some individuals will escape death and disease and avail themselves of the limited resources more successfully than others. In the "struggle for existence," as Darwin called it, some individuals will survive to reproduce, while others will not.

3. Fact: **Individual variation.** Among sexually reproducing species, the individuals of a population differ. Most individuals are unique.

4. Fact: **Heredity.** Many of the differences between individuals are heritable—that is, individual differences can be passed on to offspring. Darwin knew nothing about genetics, but he could see, for example, that the short legs of the dachshund were inherited from its parents.

   Inference: **Traits may be passed on differentially.** Darwin argued that in the struggle for existence, individuals whose traits best suited them to their environment survive and reproduce more successfully than those less well suited to the environment. Traits that are both advantageous and heritable are more likely to be passed on to the next generation. For example, a hummingbird that lives in an area where only one kind of flower provides enough nectar to support reproduction may develop traits, such as a longer bill, that best allow the bird to quickly get all the nectar out of a particular species of flower.

   Inference: **Adaptive traits accumulate within a population.** As a result of natural selection, various traits become more or less common in a population, and the overall character of the individuals within that population changes. For example, if a long bill is both heritable and adaptive, then, over time, more and more hummingbirds will have the longer bill. In this example, the genetic traits that allow a hummingbird to get more nectar faster are *adaptive* traits. They are traits that allow it to reproduce more successfully than birds with shorter bills.

Biologists accept the reality of both evolution and natural selection. The evidence that natural selection operates in natural populations is now irrefutable, yet some modern biologists question whether natural selection is the *primary* means by which new species and higher taxa such as genera and families are created. Over the years, biologists have proposed a host of alternatives to natural selection. A few biologists have argued that mechanisms other than natural selection may be responsible for macroevolution, the evolution of species, genera, and higher taxa of organisms. In the next chapter, we will examine natural selection more closely and also consider alternative mechanisms for evolution.

Key Concepts

- Evolution is heritable change in populations of organisms, which Darwin called "descent with modification."
- Darwin's theory of evolution described evolution itself, natural selection (a mechanism for how evolution occurs), and the time scales over which evolution occurs.
- The evidence for evolution includes a clear history of evolution in the form of fossils in the ordered layers of the Earth's crust; biogeography; obvious similarities among groups of related organisms; and similarities in structure of all parts of organisms from limbs and organs to individual molecules.
- Natural selection is the differential reproductive success of individuals with particular, genetically influenced phenotypes.

Summary with Key Terms

What is biological evolution?

When Darwin published his book *On the Origin of Species* in 1859, he changed the way we view the world. The theory of **evolution,** or **"descent with modification,"** contradicted **creationism,** the Bible's version of the story of creation, the philosophy of **essentialism,** and the Christian doctrine that the Earth was only 6,000 years old.

By the middle of the 19th century, Hutton's theory of gradual geologic change and Lyell's **uniformitarianism** had persuaded the scientific community that the Earth was very old, and that change occurs slowly and gradually, not catastrophically. Examination of the fossil record by Cuvier and others suggested a very long history of life on Earth—a history far longer than the 6,000 years implied by the Bible. In a theory that came to be called **catastrophism,** Cuvier suggested that fossil species were organisms that had perished in periodic catastrophes, each followed by a creation.

Linnaeus had created a system of naming organisms and a hierarchy that resembled a family tree. Biologists such as Buffon and Lamarck understood that species can change over time, and Lamarck formally introduced the idea of evolution. But Lamarck thought that species evolved in response to "felt needs" and proposed that evolution depended on the **inheritance of acquired characteristics.** Many European scientists accepted the inheritance of acquired characteristics, but rejected evolution.

Why was Darwin's argument for evolution more persuasive than those of earlier biologists?

In 1859, Charles Darwin published the influential *On the Origin of Species,* which not only showed that evolution has occurred but also described a mechanism by which evolution could work, **natural selection.** Natural selection is the differential reproduction of individuals with traits that are both advantageous and heritable. Darwin argued that natural selection worked like **artificial selection.** In artificial selection, the breeder selects the varieties that will produce the next generation. In natural selection, the individuals who best survive and reproduce are those that will produce the next generation. Natural selection, like artificial selection, favors some forms over others.

What is the evidence for evolution?

Artificial selection suggested a mechanism for evolution, but other lines of evidence suggested evolution itself. One of the most persuasive was the fossil record. **Fossils** are the remains or imprints of past life. Most come from layers of sedimentary rocks, which form from the compression of settling mud, sand, and debris. The order of the layers (or strata) of sedimentary rock establishes the sequences in which ancient life evolved. The simplest organisms lie at the bottom of the record and through **intermediate forms** become both increasingly complex and increasingly like modern organisms, a pattern that suggests evolution. **Paleontologists,** scientists who study fossils, usually divide the history of life into four long **eras**—the **Precambrian,** the **Paleozoic,** the **Mesozoic,** and the **Cenozoic** eras. Each era in turn consists of several shorter divisions called **periods,** and, more recently, **epochs.**

Another line of evidence for evolution comes from **biogeography.** Species tend to resemble neighboring species in different habitats more than they resemble species that are in similar habitats but far away. This pattern of geographical distribution suggests that species evolved from one another.

Linnaeus's **taxonomy,** a hierarchical grouping of organisms by **species, genus, family, order, class,** and other **taxa,** suggests patterns of relatedness among different species, not separate creations. The relationships implied by taxonomy—so-called **phylogenies**—are derived from studies in **comparative anatomy.** Species that closely resemble one another are considered to be more closely related than species that do not resemble one another. **Homologous structures,** such as limb bones, provide additional evidence for evolution. **Vestigial structures,** such as the horse's splint, indicate evolutionary history.

Embryology provides another highly persuasive line of evidence for evolution. Embryos repeat the development of their ancestors to a limited but nonetheless remarkable degree. The early embryos of fish, amphibians, reptiles, birds, and mammals all have the tails and pharyngeal pouches. Comparative molecular biology confirms the lines of descent suggested by both comparative anatomy and the fossil record.

How does natural selection drive evolution?

Natural selection, Darwin's mechanism for evolution, rests on several facts and inferences. All species are **superfecund,** capable of producing far more offspring than the world can support. Not all of these offspring survive long enough to reproduce. In each generation, the offspring of sexually reproducing organisms vary. That is, each individual is unique, and this

uniqueness has a distinct heritable component. Those individuals that survive to reproduce determine which traits will predominate in the next generation. Such individuals can be said to be naturally selected.

Review and Thought Questions

Review Questions

1. What three ideas kept most scientists from seriously considering the idea of evolution until the 19th century?
2. Which of these three ideas did Cuvier's theory of catastrophism address? Describe the theory of catastrophism.
3. Describe the theory of uniformitarianism. In what two ways did it influence Darwin?
4. Who wrote the essay that inspired both Darwin and Wallace to invent the theory of natural selection? What did the essay say?
5. Explain why the fossil record is not complete. Describe two possible interpretations for the gaps in the fossil record.
6. How did Darwin's observations on the distribution of species suggest that species have evolved?
7. If you were a bat, which bones (that you have as a human) would you use for wings?
8. If you were a fish, instead of a human, how would you use the hole that in your human form lets sound into your ear?
9. Describe the molecular clock. What is its use?
10. Why is the eastern coastline of South America a mirror image of the western coastline of Africa?
11. Darwin's theory of natural selection can be divided into four facts and three inferences. What are they?

Thought Questions

12. The world population is currently about 6 billion. If every couple in the world had 20 surviving children, what would the world population be in two generations?
13. If the individuals of a species did not vary but were all identical genetically, what would be the consequences of natural selection?

BiologyNow Resources

Biology ⊗ Now™

Active Figures

Box 15-2, Figure A: Continental drift
15-12:   Evolution of the horse

Preparing for an exam? Take a diagnostic test on your BiologyNow CD-ROM.

Online materials relating to this chapter are at:

http://biology.brookscole.com/AAL3

About the Chapter-Opening Image

The orangutan represents our close relations with other primates, an idea that disturbs many nonbiologists.

16

Microevolution: How Do Populations Evolve?

Key Questions

- What is the basis of the variation on which natural selection operates?
- How did the modern synthesis reconcile genetics and natural history?
- What factors cause the allele frequencies of natural populations to change?

Frans Lanting/Minden Pictures

A Fatal Disagreement

On January 26, 1943, an internationally famous and much beloved plant geneticist died of starvation in a Soviet prison hospital. He must have wondered at the irony. Nikolai Vavilov (1887–1943), director of the prestigious Institute of Plant Breeding in Leningrad, had devoted his life to the study and improvement of wheat, corn, and other cereal crops, in order to eliminate hunger in the vast, new Union of Soviet Socialist Republics (USSR). His "crime" had been to become involved in a long-running dispute about how evolution works. He contradicted the party line and paid for it with his life.

A cheerful, good-natured scholar, Vavilov traveled all over the world in the 1920s collecting seeds from different kinds of crops. The world's expert on the biogeography of wheat, he was a friend to eminent biologists in every part of the world. In the United States, he picked up the expression "Keep smiling" and liked to use it on anxious Soviet colleagues.

Vavilov's own troubles began in the mid-1920s when he befriended a hard-working and charismatic young scientist named Trofim Lysenko (1898–1976). Lysenko, the son of a peasant, worked from sunup until sundown, talking to his experimental plants as if they were people. He continually imagined new ways of "training" crops to yield more grain or grow in different climates. Lysenko passionately wanted to improve the lot of the Soviet people. However, he had little education in biology and disdained what he did not know. He especially disdained genetics, which he considered to be "harmful nonsense." Even his friends in those days joked, "Lysenko is sure that it is possible to produce a camel from a cotton seed."

Vavilov tolerated the young worker's ignorance, and, although Lysenko was not well received by other biologists, Vavilov began inviting him to scholarly meetings, hoping to win him over to genetics. Vavilov had reason to promote the young man. The Soviet government was pressuring Soviet scientists to abandon basic science and come up with "practical" results. Moreover, the government preferred uneducated workers of peasant stock to educated bourgeois elitists. Peasant scientists were especially hard to come by, and Vavilov leaped at the chance to bring along one who also worked entirely in the pursuit of practical results.

In the early 1930s, Lysenko joined forces with Isai Prezent, a philosophy student. Prezent could see that Lysenko needed a theoretical framework to make his shoot-from-the-hip approach to biology seem more scholarly. Prezent knew no more about genetics than did Lysenko, but he had studied Lamarck's theory of the inheritance of acquired characteristics, finally in its death throes in the West. The theory perfectly supported Lysenko's approach to plant breeding, stating as it did that traits developed during the life of an organism in response to the environment—long roots grown during a dry season, for example—could be passed on to the offspring. By then, modern genetics had supplanted Lamarckism, and few biologists in the world believed that acquired characteristics could be inherited. Yet, surprisingly, no one had actually disproved Lamarckism.

For this reason, and because Lysenko's peasant background and pragmatic philosophy were above criticism, government officials granted him more and more privileges and opportunities to carry out his research. Lysenko promised spectacular results with his techniques, claiming, for example, that he would double or triple the wheat crop by "training" wheat seedlings to grow in the far North. Grain would be superabundant, Lysenko promised. Soviet bureaucrats responded by meeting his every demand, and newspaper reporters fell all over themselves to interview the young genius.

No one seemed to notice that Lysenko's techniques consistently failed to increase harvest yields. When one idea failed, he came up with another. Each one's promise eclipsed the failures of the last. In the late 1930s, Lysenko was put in charge of all of Soviet agriculture. The whole nation became his "experimental garden," and every peasant farmer became a scientist. Once, Lysenko ordered tens of thousands of collective farms to test a theory about cross-pollinating different varieties of wheat. He requisitioned 800,000 pairs of tweezers for transferring pollen, one pair for each of the peasant farmers who would implement his "experiment."

Joseph Stalin, the totalitarian premier of the Soviet Union, was entranced by the enormous scale of Lysenko's work. Imagine! Eight hundred thousand tweezers! But Vavilov and his fellow biologists were horrified. They knew that cross-pollinating most of the wheat crop in the Soviet Union would eliminate many of the country's best varieties of wheat. And it did.

Despite Lysenko's failures, Lysenko and Prezent became more outspoken than ever in their disdain for genetics. Under their influence, Stalin agreed that genetics would no longer be taught at the universities, medical schools, or high schools. Increasingly, seminars at Vavilov's Institute of Plant Breeding, a world-class research center for genetics, deteriorated into confrontations between geneticists and Lysenko's student supporters. Lysenko came to regard Vavilov's persistent advocacy of genetics as a personal affront. In an era when millions of Soviet citizens were executed for far less than irritating a friend of Stalin's, Vavilov's defense of genetics was courageous but fatal.

In the summer of 1940, Vavilov was arrested for "wrecking" Soviet agriculture and sentenced to be shot. That sentence was later commuted to life in prison. But in the two and a half years he spent in prison, he was never moved to a work camp. Instead, along with thousands of other Soviet intellectuals, he died a slow and humiliating death in prison. In his last year, starving and wasted by dysentery, Vavilov and two other imprisoned professors passed the time by giving a series of lectures on history, biology, and forestry. Vavilov delivered more than a hundred hours of lectures. But his efforts to bring a bit of civilization to the dirty, crowded cells could not stave off the hungry prisoners who stole his bread and the slow death that finally took him.

Vavilov knew that by continually contradicting Lysenko he had signed his own death warrant. Yet he never renounced genetics. Why did the truth not prevail? If Lysenko was wrong in his belief in Lamarckian inheritance, why couldn't Vavilov and his fellow geneticists prove that inheritance was genetic?

The short answer to this question is that although geneticists of the 1920s and 1930s knew that genes were somehow passed from generation to generation, they could not prove beyond doubt that genes were real physical objects rather than just theoretical constructs. Nor could they prove Lamarck and Lysenko wrong in their belief that acquired characteristics somehow influenced the genes.

The proof that genes have a physical reality in DNA did not come until the 1950s—too late for Vavilov. Later work demonstrated that information goes from the genes to the body, not the other way around. In general, an individual's behavior—whether reaching for higher leaves, lifting weights, or tanning in the sun—cannot alter the genes it receives from its own parents and passes to its offspring. Nonetheless, actual disproof of Lamarck's theory of acquired characteristics has remained surprisingly difficult to pin down.

Lysenko's influence continued only a few years after Stalin died. In the early 1960s, disastrous crop failures forced the Soviet Union to buy wheat from the United States and Canada, a humiliation that irreparably weakened then Soviet premier Nikita Khrushchev's political power. In 1964, Soviet biologists rejected Lysenko's ideas and resumed the research in genetics that had been suspended for 30 years. Lysenko himself, however, retained his academic titles and his freedom until his death in 1976.

16.1 How Are Variants Created and Maintained?

We saw in Chapter 15 that evolution as an idea has a long and venerable history, dating back to the ancient Greeks. Yet even when biologists finally accepted that evolution must have occurred, they couldn't agree on how it could work. Darwin had proposed natural selection. But without an understanding of inheritance, Darwin's mechanism was incomplete. We know now that Darwin's theory of natural selection was, in fact, largely correct. But it took biologists more than 70 years of patient research and bitter dispute to appreciate Darwin's ingenious mechanism for evolution—natural selection.

In this chapter we will see how biologists gradually came to understand how organisms inherit traits. We will study variation in populations, the basis for Darwin's theory of natural selection, the sources of variation, and the means by which that variation is passed from generation to generation. We will examine, as well, some of the theories that have been proposed to account for the evolution of populations.

Why Did Charles Darwin Accept Blending Inheritance?

Darwin wrote the *Origin of Species* primarily to show that evolution had occurred and to discredit the idea that species were specially and individually created by God. His theory of natural selection was a separate idea—advanced to account for how evolution might have happened. Darwin thoroughly understood selection—the differential survival and reproduction of genetic variants. But he did not know where the variation came from. The gene was unknown in 1859.

Most scientists, including Darwin, believed that inheritance was "blended"—much as yellow and blue watercolors blend to make green (Figure 16-1). Biologists could see that when two races or species were crossed, the resulting offspring generally appeared to possess a smooth blending of the characteristics of each of the parents. For example, a cross between a horse and a zebra produces an intermediate animal, called a "zebroid," not simply a horse with stripes or a zebra without stripes (Figure 16-2). Many 19th-century scientists assumed that the hereditary material, whatever it was, must blend smoothly also.

Darwin suspected that inheritance did not smoothly blend all traits, but he had no evidence for

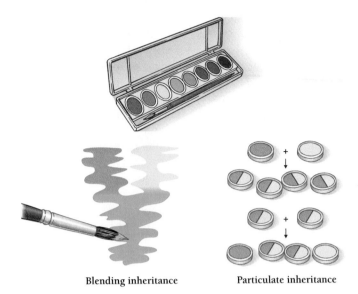

Blending inheritance   Particulate inheritance

Figure 16-1

Blending inheritance or particulate inheritance? Nineteenth-century biologists believed that the hereditary material from each parent blended together in the offspring much as watercolors blend. Darwin suspected the truth: each parent's genetic contribution remains separate.

the idea. Just a few years after Darwin's publication of the *Origin of Species*, Gregor Mendel published his groundbreaking research on peas that showed that traits could be inherited separately and not blend from generation to generation (Chapter 9). For evolutionary theory, Mendel's most important conclusion was that the individual units of inheritance,

Figure 16-2

Particulate inheritance. The characteristics of a horse and zebra remain distinct in this cross between a horse and a zebra, called a zebroid. How are the zebroid's stripes not quite like those of a zebra?

which we now call genes, remained intact and did not blend with one another. But hardly anyone read or understood Mendel's paper and it had no impact scientifically when it came out in 1866. Meanwhile, Darwin's public acceptance of blending inheritance left him open to a criticism for which he had no answer and led him into a second fundamental misunderstanding about the nature of inheritance.

Why Was the Idea of Blending Inheritance a Problem?

In 1867, Fleeming Jenkin, an enterprising Scottish engineer, announced that natural selection could never work in sexually reproducing organisms. Favorable variants, Jenkin argued, would not be able to pass on their advantages to the next generation. Instead, interbreeding would quickly "dilute" any new traits. For example, when a fast cheetah bred with a slow one, Jenkin would have argued, the offspring would be only as fast as the average of the two parents. Eventually, any increased speed from an unusually fast cat would disappear altogether. How, he asked, could natural selection operate on variants that faded away as soon as they appeared? Assuming blending inheritance, he had a point.

But we now know that inheritance is particulate, not blending. Sexual reproduction does not necessarily dilute traits, but instead creates an assortment of different traits. In the cheetah example, the offspring would not all be exactly halfway intermediate in speed between the faster parent and the slower parent. Each young cheetah would be different, ranging from fast to slow, and each one's ability to reproduce would be correspondingly different.

Jenkin had no idea about real plants and animals; he was mistaken in nearly everything he said. Darwin could easily have demolished the engineer's argument on a variety of grounds. Yet Darwin was so uncertain about the nature of inheritance that, rather than counter the particulars of Jenkin's arguments, or question the basic assumption behind it—blending inheritance—Darwin made a second mistake.

He stopped defending natural selection and argued for the inheritance of acquired characteristics, an idea that had been universally accepted for hundreds of years.

Darwin knew that individuals are unique. He knew that natural populations contain enormous reservoirs of variability on which natural selection can act. If variability was constantly blending away, where were new variants coming from? The **inheritance of acquired characteristics** seemed to provide the only way for new variants to be continuously generated and passed on. According to the theory:

1. Changes in the environment create changes in the needs of organisms.
2. Changes in needs occasion changes in behavior to satisfy these needs.

3. Changes in behavior result in the increased use and development of certain parts, or the use and development of new parts.
4. These changes are passed on to the offspring.

Fleeming Jenkin proposed that blending inheritance and evolution by natural selection were incompatible. Darwin's acceptance of blending inheritance, as opposed to particulate inheritance, forced him to accept the theory of acquired characteristics.

16.2 How Did Biologists Come to Reject the Theory of Acquired Characteristics?

Of all the world's biologists, German biologists were the most enthusiastic about Darwinism. Among them was a man named August Weismann (1834–1914), who became one of the first biologists to unequivocally reject the inheritance of acquired characteristics and the first to suggest a coherent theory of molecular genetics.

Weismann realized—as Darwin apparently had not—that the controversy about natural selection could not be resolved until inheritance was fully understood. Like other biologists of the time, Weismann at first accepted the inheritance of acquired characteristics. But in the 1870s he began testing the theory in a series of experiments. By the time he was 49 years old he had completely changed his mind, and, in 1883, he published a paper rejecting Lamarckism entirely.

It was a radical proposal, and most biologists rejected Weismann's arguments. Only Wallace, Darwin's codiscoverer of evolution through natural selection, publicly supported Weismann. As Weismann saw it, no one had any idea how the inheritance of acquired characteristics could work, and it was *unnecessary* for natural selection to work. Besides, Weismann had a better idea—another mechanism for inheritance. In 1882, a German cell biologist had discovered the cell's nucleus and noticed the way the chromosomes divide at mitosis. Weismann theorized that the genetic material was all in the nucleus. His theory, he said, was "founded upon the idea that heredity is brought about by the transmission from one generation to another of a substance with a definite chemical, and above all, molecular constitution."

Weismann went on to argue that the hereditary material, once it is formed, is essentially isolated from the rest of the body. Embryologists had already noted that in vertebrate animals the tissue that eventually forms the ova or testes—the **germ plasm**—separates from the rest of the body early in development. In humans, for example, these cells separate early in the fourth week of development, when the embryo is about half the size of a chocolate chip.

Weismann concluded that the germ plasm is separate from the rest of the body from the very beginning, and nothing in the development or environment of the body normally influences the germ cells or the hereditary material (which we now know is DNA). The genetic material, he said, is passed from generation to generation unchanged. Because the hereditary material was passed from generation to generation intact and unaltered by environment, Weismann argued, the inheritance of acquired characteristics must be impossible.

Modern biology has mainly confirmed Weismann's conclusion. A parent who works outside all day and develops a deep tan does not pass this trait to his or her children. Tanning is not known to influence the DNA in the ova or sperm.

August Weismann proposed that the hereditary material was in the nucleus and that it was molecular. He further argued that because the hereditary material was not susceptible to environmental influence, the inheritance of acquired characteristics was impossible.

How Did Weismann Explain Variation in Individuals?

Weismann's theory of inheritance ruled out Lamarckism. However, he still needed to explain the origin of the variation on which natural selection acts. Remarkably, he had an answer that was partly correct—sexual reproduction.

By 1883, cell biologists had observed that maternal and paternal chromosomes do not fuse during fertilization, but merely restore the two sets of chromosomes normally found in most cells. From this observation, Weismann leapt to the conclusion that the effect of sexual reproduction was not to blend parental characters (as blending inheritance implied) but to recombine separate "hereditary tendencies."

Weismann rightly concluded that sexual reproduction, far from blending parental traits, brings together old traits into new combinations. Even though the field of genetics did not yet exist, Weismann had mostly solved the problem of heredity. His views remained a minority opinion throughout the last part of the 19th century and into the beginning of the 20th century. But he and Wallace kept alive the idea of natural selection. Without these two scientists, the idea might have been forgotten. Not until the 1940s did biologists appreciate Weismann's enormous contribution to evolutionary theory.

In the meantime, one major piece of the heredity puzzle remained to be discovered. Sexual reproduction might mix up old traits in new ways, but where did genuinely new traits come from?

Weismann showed how sexual reproduction, by creating new combinations of traits, could act as a source of new variants for natural selection.

How Did the Rediscovery of Mendel's Work Transform Biology?

In 1900, three biologists working in three different countries independently rediscovered Mendel's laws. In the Netherlands, Hugo Marie de Vries (1848–1935) published a paper showing that if he crossed even numbers of hairy and smooth primroses he got 3:1 ratios of hairy and smooth plants in the second generation. From this he concluded that the two traits—smooth and hairy—were controlled by a single pair of what we now call genes. Each parent, he explained, passes on only one of each of its pairs.

As soon as de Vries's paper came out, he immediately received copies of papers from two younger biologists—Karl Franz Joseph Correns (1864–1935), in Germany, and Erich Tschermak (1871–1962), in Austria—each claiming that they too had discovered the 3:1 ratios, each offering the same genetic explanation as de Vries, and each pointing out to him that Gregor Mendel had anticipated them all by 35 years.

Historians believe that both Correns and Tschermak probably read Mendel's paper before they had their respective "flashes of insight." De Vries himself received a copy of Mendel's paper from a fourth biologist sometime before he published his first paper. Whether there were actually any "independent discoveries" besides Mendel's we may never know. One thing is clear: Mendel's ideas made a splash.

In 1900, the British biologist William Bateson (1861–1926) was traveling to London to deliver a paper to the British Horticultural Society. As the train trundled through the English countryside, Bateson passed the time reading. When he got to de Vries's paper, he was thunderstruck. In London, he cast aside his prepared notes and proclaimed the death of Darwinism (by which he meant gradual evolution) and "the birth of a new science" (later called "genetics").

Bateson, de Vries, and others, who came to be known as the Mendelians, argued that evolution occurred not gradually, through natural selection, but in leaps and bounds by means of genetic mutations. These earliest geneticists firmly rejected the idea that evolution occurred through the slow selection of variants in a population. The Mendelians rejected natural selection and any suggestion that evolution occurred among groups of individuals. They believed that evolution occurred suddenly, from one generation to the next, when genetic mutations transformed a single individual into a radically new type.

The Mendelians contended that every evolutionary change, large or small, was the result of a separate individual mutation, and that natural selection, individual variation, and sexual reproduction played no important role in evolution. Normal continuous individual variation, such as adult height in humans—the very foundation for Darwin's theory of natural selection—was not even heritable, argued the Mendelians.

Naturalists of the era were horrified. They knew, from their close observations of wild populations of plants and animals, that most variation was continuous and that continuous variation was indeed heritable. But they, in turn, overreacted. The naturalists claimed that although the rules of Mendelian inheritance might apply to discontinuously inherited traits such as smooth or hairy plants or blue or brown eyes, such traits were unusual and irrelevant to evolution. Continuous variation, the stuff of evolution, they argued, was not governed by the rules of Mendelian inheritance. Rejecting all of genetics, they clung to a variety of other theories, including Lamarckism. The polarized debate between the Mendelians and the naturalists raged for nearly 40 years.

> The rediscovery of Mendelian genetics precipitated an intense debate. Early geneticists believed that evolution occurred from one generation to the next as a result of radical mutations in individuals. Biologists who studied natural populations of organisms insisted that Mendelian genetics did not apply to the kind of variation that was the basis of evolution. Both sides were wrong.

How Did the Modern Synthesis Reconcile Genetics and Natural History?

In the late 1930s, a remarkable marriage of genetics and evolutionary theory occurred, which, for the first time, elucidated the genetic basis of variation and natural selection. This marriage, called the **modern synthesis,** did much to end the international feud between geneticists and naturalists. The modern synthesis fully embraced Weismann's theories and added to them all that was known about genetics. Much of what we will consider in the rest of this chapter comes from a field of genetics that contributed substantially to the modern synthesis—population genetics. Population genetics combined Mendelian genetics with the recognition that evolution can occur at the population level.

Population genetics describes in precise mathematical terms the processes by which organisms generate new traits and pass them on. These processes are sometimes called **microevolution.** Population geneticists define microevolution narrowly as *changes in the frequencies of alleles of genes in a population.* For example, consider a wildflower population that includes some individuals with a gene for red flowers and other individuals with a gene for white flowers. If the relative proportions of red and white flowers change over the course of several generations, the population of flowers is said to evolve. Microevolution is usually discussed separately from **macroevolution**—the processes by which species and higher groupings (taxa) of organisms originate, change, and go extinct, a topic we will discuss in the next chapter. The appearance of new groups of organisms occurs over longer spans of time than microevolution, which can occur over years or, in bacteria and viruses, just days.

Population genetics explains variation at the population level and provides a basis for natural selection and other evolutionary forces. In the next section, we will discuss what genetic variation is, where it comes from, and how it spreads and persists in populations of organisms.

16.3 What Is the Source of the Variation That Fuels Evolution?

Until the 1930s, few biologists believed that genes were real entities. No one had ever seen a gene, and there was no other evidence for their existence. Geneticists carefully stated that genes were purely theoretical constructs. Gradually, however, geneticists began to notice more and more cases where one genetic trait, such as eye color, was consistently linked to some other genetic trait, such as size. It began to look as if the genes themselves were physically linked together when they were passed from parent to offspring.

As we saw in Chapter 10, genes are very real, consisting of strings of nucleotides in the DNA. Each set of three nucleotides codes for one amino acid. Strings of amino acids, constructed on the ribosomes, fold into proteins.

Recall also that most sexually reproducing organisms have two of each chromosome. The two chromosomes in a pair are said to be **homologous.** One homologous chromosome possesses the same genes in the same positions as its homologue (unless they are unmatched sex chromosomes). Eggs and sperm have only one of each chromosome. When the egg and sperm fuse into a zygote, the two sets of chromosomes come together. Thus each parent supplies one of the chromosomes in each pair of chromosomes.

Genetic variation originates when a gene mutates. The different forms of a single gene are called **alleles.** If each chromosome in a pair has the same allele at a particular gene position, the organism is said to be **homozygous** for that gene. If each chromosome has a different allele at a particular gene position, the organism is said to be **heterozygous** for that gene (Chapter 9).

What Is Genetic Variation?

All genetic variability begins with mutations. Some mutations change only one or a few nucleotides in a gene and, thus, alter only a single protein. Chromosomal mutations, which change the number of chromosomes or the number or arrangement of genes on a chromosome, can affect a large number of proteins at once. Most chromosomal mutations

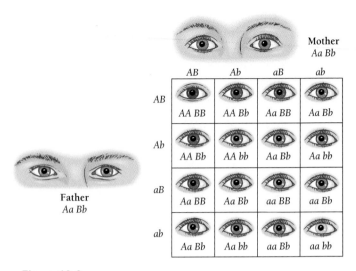

Figure 16-3

Polymorphism. In many animal species, every individual has eyes of approximately the same color. Human eye color varies from pale blue to nearly black. Most of this variation derives from just two alleles of two genes. The four alleles can combine in nine ways.

are lethal or extremely damaging in animals, though less often in plants. But chromosomal rearrangements can, on occasion, benefit the organism. For example, the duplication of a chromosome segment, if harmless, can be passed on. In time, mutations in the new copy of the gene can allow it to take on new functions. As a result, the species has more genes than it did formerly.

In a population at large, a given gene may have just one form, or allele. In that case, every member of the population would have to be homozygous for that gene. On the

other hand, a given gene may have two or even hundreds of alleles. If a gene had several alleles, members of the population could be either homozygous or heterozygous in one of several different ways. For example, members of a population possessing a gene with three alleles *A*, *a*, and *a′* could be homozygous in three ways—*AA*, *aa*, or *a′a′*—or heterozygous in three ways—*Aa*, *Aa′*, or *aa′*. If a population has two or more alleles of a given gene, it is said to be **polymorphic** [Greek, *poly* = many + *morph* = form]. If the members of a population come in two or more forms, the population is also said to be polymorphic, even though many genes may be involved. Humans, for example, are polymorphic for eye color (Figure 16-3).

> The original source of variation is gene mutations or chromosomal mutations. A population is polymorphic for a given gene if that gene has more than one allele.

Are Traits Determined by Single Genes?

The Human Genome Project focused the public's attention on single genes that, when mutated, cause certain genetic diseases. A single-gene trait with just two alleles (one on each chromosome) creates a pattern of variation like that in the left-hand panel of Figure 16-4. But most genes work in concert with other genes and under the influence of the environment. The more genes that contribute to a trait, the more continuously the trait can vary (Figure 16-4).

In a genetics lab, most research deals with alleles that specify discrete characters such as the presence or absence of a specific enzyme. In the natural world, however, most traits that matter, such as flower color or claw length, are

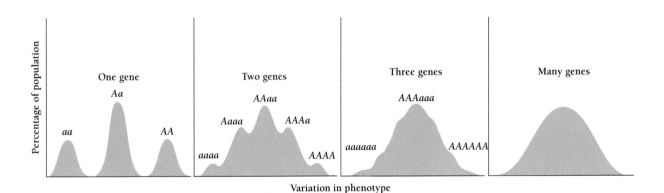

Figure 16-4

Most traits are polygenic. The more genes that influence a trait, the more continuously the trait will vary in a population. If only one gene with two alleles is involved, then no more than three basic types will appear. Two genes with two alleles each can result in five phenotypes (as in eye color). However, even those individuals with the same genotype differ from one another because of environmental influences. Most phenotypic traits are influenced by many genes and display a smooth bell-shaped curve like that at right.

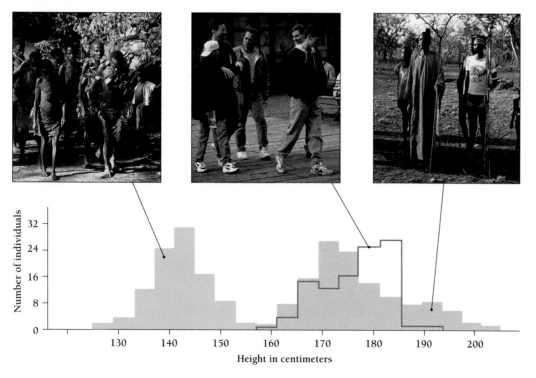

Figure 16-5
Height is a polygenic trait. Because adult height in humans is controlled by many genes, the distribution of heights in adults is a smooth curve, as illustrated by this photo of students. The students are arranged according to height, with the shortest on the left and the tallest on the right. Most students are of about average height, with very few at the extremes.

Photo, courtesy of Joiner Associates, Inc.

Number of individuals

Height in inches

polygenic, influenced by many genes. Everything about individual organisms—height and weight, bone structure, ability to withstand stress or infection, hormone levels, personality, and intelligence—is influenced by multitudes of genes. The constellation of traits we call personality, for example, is probably influenced by thousands of genes, as well as by environment. Height, weight, and skin color vary smoothly and continuously within a population, often forming a bell curve (Figure 16-5).

In Chapter 1, we discussed bell curves in more detail. But two aspects of such curves deserve special attention—the average value, or mean, and the breadth of the curve.

The breadth of the curve is a measure of the variability of the trait. We all know that the average value of a trait may vary from population to population. African Pygmies are, on average, shorter than people living in Boston (Figure 16-6). But populations may also vary in the variability of a trait. The population of African Dinkas pictured in Figure 16-6, for example, shows a greater range of heights than that of either the Pygmies or the Bostonians. An adult Dinka can be taller than almost any Bostonian, or as short as the tallest Pygmies. Genetic variation, the fuel of evolution, varies from population to population. We should not forget, however, that we cannot tell from the graph if the Dinkas' great vari-

Figure 16-6
Even variation varies. Heights for male African Pygmies, male American Bostonians, and male African Dinkas, who are known for their tallness. Nearly all Pygmies are shorter than all Bostonians and Dinkas. Dinkas are, on average, taller than Bostonians, but they have a greater range of heights, so that many Dinkas are actually shorter than many Bostonians. Which group varies most? (Pygmies, Klaus Payson/Peter Arnold, Inc.; Bostonians, Frank Siteman/Stock, Boston; Dinkas, F. Jackson/Bruce Coleman, Inc.)

Number of individuals

Height in centimeters

ability is all due to genetic variation or also to environmental effects such as diet.

Most traits are influenced by many genes.

What Determines the Genetic Variability of a Population?

Populations that contain many alleles vary more than populations that have few alleles per gene. Highly variable populations can evolve more rapidly than more uniform populations because natural selection has more to select from. Three factors determine genetic variability: (1) how fast mutations accumulate in the DNA, (2) how fast such mutations spread through a population, and (3) how fast selection eliminates mutations from a population, which we discuss later in this chapter.

How Often Do Mutations Occur?

Mutations in the DNA cause random changes in protein structure and, thus, most mutations damage the organism. Occasionally, however, random changes may cause no harm or may even give an advantage to the organism that bears them. Such rare advantageous mutations are the fuel of evolution. Without a constant supply, evolution would stop. Instead, mistakes and mutations persist and occasionally provide every species with adaptive variants. For example, on average, each new human baby carries one or two new mutations.

Mutations are not the result of directed molecular surgery but arise constantly as the result of random processes. Because they occur constantly and randomly in all genes, geneticists have been able to estimate the probability that a given nucleotide will mutate in each generation. The allele for eyelessness in the fruit fly *Drosophila melanogaster* appears as a mutation in 6 of every 100,000 fly gametes. Because mutation occurs constantly, and because the total number of genes is great (about 40,000 for humans, for example), the total number of variations within a species can be enormous.

How Do Mutations Spread Through a Population?

The rate at which a population evolves depends on how rapidly new mutations spread through the population. In fact, although evolution could not continue without new mutations, the spread of those mutations through a population, rather than their original occurrence, is the most important factor in increasing variability in a population. The main way for a particular mutation to increase in frequency is through sexual reproduction, which mixes different mutations into endless new combinations. Mutations that prevent an individual from reproducing are immediately elimi-

Figure 16-7
Every individual is unique. This tray of meadowlarks from a zoological collection illustrates how every individual in a population is unique. Notice the variation in the shape of the black band around each bird's throat, as well as the variation in birds' overall size and even bill shape. These are just the differences that are obvious on the outside of a dead animal. A biologist looking for differences in physiology and behavior could find even more variation. The ordinary variation shown here that has a genetic basis is the material on which natural selection can operate to propel evolutionary change.

nated, of course. But mutations that confer any advantage at all may persist.

Genetic variation originates when genes or chromosomes mutate. Genetic variation spreads when sexual reproduction and recombination mix these changes into endless new combinations.

Genetic Variation

Since variability is the raw material for selection, one of the first questions population geneticists ask is, How much variation is there in single genes within a population? Figure 16-7 shows a tray of meadowlark specimens. These birds were killed as part of a scientific collection. Notice how each bird is unique in its particular combination of traits, including size, coloration, and bill length. All organisms are similarly diverse and part of the basis for this variation is genetic.

To examine the extent of variation of single genes, we can look at the proteins that are the products of genes or at the genes (DNA) themselves. In populations of the fruit fly, for example, about 53 percent of genes that code for proteins are polymorphic. In a typical natural population, between 33 and 50 percent of enzymes have more than one form.

Another way of expressing the level of genetic variation in a population is to estimate the likelihood that any individual will be heterozygous for a given gene. Any two fruit flies, for example, are likely to differ at about 25 percent of their gene loci. In most natural populations, the average individual has two different alleles (one on each chromosome) at 4 to 15 percent of its genes. Humans are average; about 7 percent of an individual's genes are heterozygous, or come in more than one form. For comparison, the heterozygosity of most invertebrates is about 13 percent, and plants, about 17 percent.

Most natural populations vary enormously.

Is All Genetic Variation Subject to Natural Selection?

The only variants important for natural selection are those that contribute to changes in the whole individual in ways that affect the individual's ability to reproduce. The expressed traits of a whole individual organism are called the **phenotype** [Greek, *phainein* = show]. The phenotype, which includes the appearance, the physiology, and the biochemical makeup, is a result of both the genetic makeup of the individual—its **genotype**—and the effects of the environment on the development of the individual.

Selection acts on the phenotype. For example, an individual homozygous for some lethal, recessive trait, such as cystic fibrosis, expresses the trait and dies. The gene has expressed itself in the phenotype, and selection has occurred. Another individual, heterozygous for the trait, will have only one copy of the deadly allele and will have a phenotype that is perfectly healthy. In the heterozygote, the allele is not expressed in the phenotype, and so selection against the allele cannot operate even though the allele is present in the genotype. Most lethal mutations are recessive and are usually not subject to selection unless they are homozygous.

With the advent of high-speed DNA sequencing machines and the ability to look at thousands of genes at once, evolutionary biologists have begun to accumulate more information on the extent of variation in DNA. On average, about 1 nucleotide in every 500 is different in the two homologous chromosomes. Most of these differences, however, lie in nucleotides that do not code for proteins or otherwise regulate the expression of genes. Many of these mutations contribute to high variability at the DNA level but do not alter the structure of a protein or change the way the

organism functions. Such mutations do not affect phenotype and are therefore not subject to natural selection.

Mutations that do not affect the phenotype are not subject to selection.

Why Is Genetic Variation Essential to Evolution?

If mutations are the fuel of evolution and recombination is its motor, then the driver—the guiding force—of evolution is natural selection. Because some variants reproduce more successfully than others, the mutations that best equip their possessors to survive and reproduce will increase in frequency in each generation.

When biologists study genetic variation within a species, they almost always find lots. In some exceptional cases, however, genetic variation is limited. For example, most Thoroughbred horses are direct descendants of a few dozen horses that lived in the late 18th century. Although breeders have allowed some outbreeding—breeding with non-Thoroughbreds—the relative lack of genetic variability has led to stagnation. Despite the best efforts of highly skilled breeders, the winning times for the Kentucky Derby have not changed significantly in 50 years. The lack of change in winning Derby times could be considered an example of microevolution that has come to a standstill. Later in this chapter we will discuss some examples of natural populations that lack genetic diversity.

Evolution by natural selection depends on the continued existence of variation. Yet selection tends to reduce variation by eliminating organisms less able to survive and reproduce. We might expect, then, that species would evolve so that all individuals will have the same "best" genotype. If this were to happen, evolution of that species could not continue. (However, environments change constantly, and so do selection pressures. What is "selected" changes over time.)

In the absence of genetic variation, evolution would cease.

In the rest of this chapter, we will see how populations maintain variability, even as their gene frequencies change during evolution. Such variability allows populations to adapt to continuing changes in their environment.

16.4 Are Natural Populations at Genetic Equilibrium?

In 1908, G. Hardy, an English mathematician, and G. Weinberg, a German physician, independently reconsidered the idea that sexual reproduction eliminates variation in light of the fresh knowledge that heredity is nonblending. Hardy and Weinberg each reached the same conclusion, namely that: *Sexual reproduction does not change the frequencies of alleles within a population.*

To see what the **Hardy-Weinberg Principle** means, we need to define some terms. We speak of all the genes of all the individuals in a population as the **gene pool.** And we can calculate the *frequency* of a given allele within the gene pool by dividing the number of times the allele is present in a population by the number of times it could be present if every individual had the allele on both chromosomes. *The allele frequency is the number of times the allele is present in the population divided by the total number of chromosomes on which the gene appears.*

To understand the Hardy-Weinberg Principle, let us consider a gene that has two alleles, *A* and *a*. For example, the color of mussels on the north coast of California depends on a single gene with two alleles. One allele, *A*, gives the mussels a blue color and is *dominant*. The other allele, *a*, makes the mussels brown and is *recessive*. That is, whether the mussel is *AA* or *Aa*, it will be blue. Only when the mussel is *aa* will it be brown.

The fraction of sperm and ova in the population that carries the *A* allele we call "*p*," and the fraction that carries the *a* allele we call "*q*." The fraction of the blue allele (*A*) in a particular mussel population is $p = 0.6$, while the fraction of the brown allele (*a*) is $q = 0.4$. Since there are only two alleles for this gene, $p + q = 1$. Geneticists call *p* and *q* allele "frequencies." Thus the *A* allele has a frequency of 0.6 and the *a* allele has a frequency of 0.4.

But the frequencies of the alleles are not the same as the frequencies of the genotypes. Suppose the mussels release their eggs and sperm into the seawater. We know that 60 percent will carry the *A* allele and 40 percent will carry the *a* allele. Now the sperm and eggs begin joining up at random and fusing into zygotes with genotypes *AA, Aa,* and *aa*. What are the frequencies of those genotypes? We can calculate the frequencies of the three genotypes among the offspring, or F1 generation, using either a Punnett square, as described in Chapter 9, or simple math.

If the frequency with which a sperm or egg carries the *A* allele is 0.6, then the frequency of an *AA* homozygote will be the probability that an *A* sperm will fertilize an *A* egg. That probability is $p \times p = p^2 = 0.36$. Similarly, the frequency of an *aa* homozygote is $q^2 = 0.16$. The frequency of a heterozygote is the probability that an *A* sperm will fertilize an *a* egg plus the probability that an *a* sperm will fertilize an *A* egg, that is:

$$(pq) + (qp) = 2pq = 2(0.4)(0.6) = 0.48$$

The frequencies of *AA, aa,* and *Aa* are 0.36, 0.16, and 0.48, respectively. Notice that the sum of the frequencies is $p^2 + 2pq + q^2 = (p + q)^2 = 1^2 = 1$. Likewise, $0.36 + 0.48 + 0.16 = 1$. In other words, we have accounted for 100 percent of the population.

From the genotype frequencies we can derive the allele frequencies in the F2 generation and see if they are the same as in the previous generation, in short, $p = 0.6$ and $q = 0.4$.

First, let's assume that we have 100 individuals, each with 2 alleles, for a total of 200 alleles in the gene pool. What fraction of these 200 alleles are the *A* allele? Each of the *AA* homozygotes carries two copies of *A*, so they contribute $0.36 \times 200 = 72$ *A* alleles. In addition, each of the *Aa* heterozygotes, which make up 0.48 of the population contribute $0.48/2 \times 200 = 48$ *A* alleles (since only half of their alleles are *A* alleles). So $72 + 48$ *A* alleles is 120 *A* alleles (out of 200). In Table 16-1, we derive the frequency for the *a* allele in the same way. So despite sexual reproduction, the allele frequencies do not change. In successive generations, the frequencies for the genotypes *AA, Aa,* and *aa*—p^2, $2pq$, and q^2, respectively—would always remain the same.

So far, we have considered only two alleles for each gene, but we know that mutations are possible at every nucleotide of DNA. The number of possible alleles for each gene is large indeed. We could easily accommodate more alleles in our discussion.

We might, for example, designate the frequencies of three alleles (*A, a,* and *a'*) as *p, q,* and *r,* so that $p + q + r = 1$. We would find that the Hardy-Weinberg Principle remains the same. The frequencies of the individual alleles in a population do not change as a result of sexual reproduction.

Table 16-1
The Essence of Hardy–Weinberg

The frequency of the *A* allele remains *p* and the frequency of the *a* allele remains *q*. Generally, then, sexual reproduction does not change allele frequencies. The total number of *A* alleles is $2p^2 + 2pq$. Because there are two chromosomes, the *frequency* of the *A* alleles is this number divided by two, or $(2p^2 + 2pq)/2 + p^2 = pq + p(p + q)$. Remember, however, that $p + q = 1$, so $p(p + q) = p$. So if a population starts with gene frequencies of $p = 0.6$ and $q = 0.4$, the next generation will have the same frequencies.

Genotype	*AA*	*Aa*	*aa*
Frequency	p^2	$2pq$	q^2
	0.36	0.48	0.16
Number of individuals with that genotype in the population	36	48	16
Total alleles	72A	40A 48a	32a

A alleles = 72 + 48 = 120

a alleles = 32 + 48 = 80

Number of individuals is 100.

Total alleles is 200.

So *A* alleles have a frequency of 120/200 = 0.6

And *a* alleles have a frequency of 80/200 = 0.4

Biology⊗Now™ Learn more about the Hardy-Weinberg equation by clicking on this figure on your BiologyNow CD-ROM.

The frequencies of each allele therefore stay constant in successive generations, and sexual reproduction does not lead to the dilution of variation, as Fleeming Jenkin suggested. Equally important, genotype frequencies stay at the same **Hardy-Weinberg equilibrium,** a stable distribution of genotype frequencies maintained by a population from generation to generation.

Such an equilibrium occurs, however, only when the population meets all of the following five conditions:

1. Random mating
2. Large population size
3. No mutations
4. No gene flow (interbreeding among populations)
5. No selection

In reality, these conditions are probably never met in natural populations. Few organisms select mates totally randomly. Zero mutation rates are unheard of. If a population is large and healthy, it's likely to be interbreeding with other populations, which will likely have different allele frequencies. And while selection pressures on a particular gene may be nil, the likelihood of zero selection on every gene in a species genome is unlikely. In real life, gene frequencies change and evolution occurs. So natural populations are probably never at genetic equilibrium. However, the Hardy-Weinberg principle showed that sexual reproduction does not dilute variation. By disproving Fleeming Jenkin's argument, Hardy-Weinberg provided support for Darwin's theory of natural selection.

The Hardy-Weinberg Principle shows that sexual reproduction does not lead to the dilution of variation but maintains a stable distribution of genotype frequencies from generation to generation.

16.5 What Causes Allele Frequencies to Change?

Many factors can cause changes in the frequencies of alleles or genotypes. These include nonrandom mating, small population size, mutation, interbreeding between populations (gene flow), and natural selection. But only natural selection *consistently* leads to adaptive changes in allele frequencies.

How Can New Mutations Cause Changes in Allele Frequencies?

Random mutations can potentially generate every possible change in the amino acid sequence of each protein during the lifetime of a species. Mutations provide the raw material of evolution. Without mutations, evolution could only reshuffle the

same deck over and over. As mentioned earlier, the rate of mutation is high enough that each new human baby carries one or two new mutations. But mutations have very little immediate effect on allele frequencies (Figure 16-8A). One or two mutations out 3 billion nucleotides isn't going to change overall allele frequencies in an entire population of thousands of people, each with 3 billion nucleotides. Furthermore, most of the mutations that do occur are in "junk" DNA and most likely do not to affect the phenotype. The number of new mutations in a single generation is minuscule compared to the entire gene pool and hardly affects allele frequencies or phenotypes.

Mutation barely alters the overall frequency of any given allele.

Genetic Drift: How Does Chance Alter Gene Frequencies?

Flip a coin 1,000 times and your chance of getting all heads is virtually zero. You can reasonably expect that the number of heads will be close to 500, the "expected" number. However, if you flip the coin only four times, your chance of getting all heads is rather high (1 in 16), and you're not likely to get heads exactly half of the time. In fact, your chance of getting two heads is less than 40 percent. Such chance fluctuations, which affect small samples much more than large ones, are known as "sampling errors." In small populations of organisms, sampling errors will likely cause allele frequencies to change randomly from generation to generation (Figure 16-8B).

For example, in a small population of mussels, the actual proportion of eggs and sperm with the A allele may differ widely from 0.6. The smaller the population, the greater the likelihood that the frequency of various alleles will deviate from what they are "supposed" to be. The effects of chance in small populations can lead to **genetic drift**— changes in gene frequency due to random events (rather than to selection, mutation, or immigration).

A Population Bottleneck Is a Decrease in Genetic Diversity

In small populations, genetic drift can have pronounced effects. For example, in the early 19th century, large populations of elephant seals flourished along the California coast from Point Reyes, just north of San Francisco Bay, south to Baja California. By 1890, hunters had almost eliminated the population (Figure 16-9). Best estimates are that the population was reduced to 20 to 100 individuals. Under federal protection, the elephant seal population increased to 100,000 animals, but all of these animals have descended from the same few individuals. When, as in this case, a large population is reduced to a few individuals by some disaster,

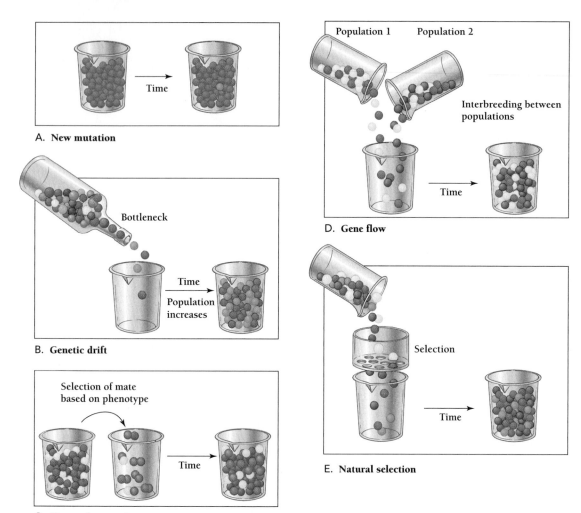

A. New mutation

B. Genetic drift

C. Nonrandom mating

D. Gene flow

E. Natural selection

Figure 16-8

What causes changes in gene frequency? In these simplified diagrams, colored marbles represent individuals with different genotypes. Beakers represent populations. A. A new mutation has little effect on overall allele frequency. B. Random events can cause single alleles to vanish. When a population is reduced to a small number of individuals, in a "bottleneck," large numbers of alleles may be lost. C. Nonrandom mating (preferring one mate over another) can lead to the accumulation of particular alleles in the population. D. Gene flow. E. Natural selection.

Biology ⓔ Now™ Learn more about Hardy-Weinberg equilibrium and gene frequency changes by clicking on this figure on your BiologyNow CD-ROM.

population biologists say the population has passed through a **bottleneck** (Figure 16-8B). Because so many alleles were eliminated by hunters, the new population harbors a tiny fraction of the original gene pool. The elephant seals are said to be "inbred" and to lack genetic diversity.

The cheetah, of Africa, has some of the lowest genetic diversity found in any animal, with an average "heterozygosity" of only 0.07 percent (compared to about 7 percent for humans). That is, only 0.07 percent of genes per individual come in two forms. The other 99.93 percent of genes are homozygous; the two chromosomes carry copies of the same allele. Cheetahs most likely passed through a bottleneck sometime in their past.

A population bottleneck produces a population whose gene pool is a tiny sample of the original. In the new inbred population, the frequency of a particular allele is likely to be quite different from that in the original population. Many alleles will not be represented at all. Since the allele frequencies have changed, the population can be said to have evolved. But the change may make the individuals in the population less well adapted to their environment, the opposite of what natural selection does.

In humans, population bottlenecks are thought to be responsible for the high incidence of genetic diseases in certain human populations. For example, population bottlenecks are responsible for the high frequency of otherwise rare dis-

Frank S. Balthis Photography

Figure 16-9

A classic bottleneck. In the 19th century, heavy hunting reduced the number of elephant seals to 20 to 100 individuals. These individuals could not possibly carry all the alleles that the original large population of thousands of animals carried. We can think of a bottleneck as the extinction of alleles. The elephant seals have been legally protected since 1884 and are now flourishing along the Pacific Coast of the United States. But because the animals passed through a bottleneck, they lack genetic diversity. We say they are highly "inbred." Genetically, they are all very similar to one another.

ease alleles among Afrikaaners of South Africa (descended from about 30 Dutch families in the 17th century), the Old Order Amish in the United States (descended from only a few individuals in the 18th century), and Swedes in Lapland (descended from three families in the 18th century).

How Can Genetic Drift Change Allele Frequencies?

Genetic drift can play an important role in the spread of new mutations. Even in the absence of selection, a new mutation will either disappear completely or spread through the population over many generations. The smaller the population, the less time it takes for either to happen. If a new mutation appears in the gene pool, it will at first exist in only a few individuals and will have a significant chance of being eliminated. For example, imagine a population of 25 elephants, 3 of which have a new allele for hairy ears. A volcano that erupts every 10,000 years happens to eliminate the three hairy-eared elephants. Their elimination is not selection, since smooth ears do not protect elephants from volcanoes. It's just chance. Yet, the result is the same as if natural selection had eliminated the hairy-eared elephants—a population of smooth-eared elephants.

Even in large populations, genetic drift will tend to cause alleles to go to either 0 percent or 100 percent. It would happen slowly, but it would happen eventually, leading, ultimately, to a population that was completely homozygous. In such a population, evolution would cease. New mutations "restock" genetic variation, and the constant mixing of such mutations, within populations and between populations, is key to evolution.

Does Nonrandom Mating Change Allele Frequencies?

The Hardy–Weinberg Principle assumes that individuals within a population mate randomly. But birds, mammals and many other animals choose their mates on the basis of the mate's genotype (as reflected in the phenotype). In peacocks, for example, the female bird, the hen, prefers a male with a large and colorful tail to a drabber male. Over time, the choices of hens cause alleles that promote fancy tails to increase in the population. The nonrandom mate choices of both males and females can profoundly affect allele frequencies. Mate selection is a form of natural selection that is usually called "sexual selection." We discuss sexual selection again later in this chapter and in Chapter 29.

Individuals may also choose mates on the basis of similarities to the individual's own genotype. Humans mate preferentially with those of similar height and skin color. When individuals choose mates from among their own relatives, children are more likely to carry two identical alleles (be "homozygous"). For example, Ellis–van Creveld syndrome is a genetic defect frequent among the Amish that has resulted from 200 years of inbreeding. Recessive genes may have negative effects that are expressed only in children that are homozygous for that allele. In sexual species, inbred individuals are usually less healthy than individuals with greater heterozygosity. So most species have adaptations that promote outbreeding, so that individuals choose mates somewhat unlike themselves.

In small, isolated populations, genetic drift can cause significant changes in allele frequencies. In large populations, drift operates so slowly that its importance is small compared to the effects of mutation, gene flow, and selection.

Inbreeding within small populations can increase the proportion of homozygotes without changing overall allele frequencies. But mate selection, which is a form of natural selection, alters allele frequencies

Figure 16-10

Gene flow. Humans travel singly or in groups, spreading their customs and genes. In November 1996, in the wake of a devastating civil war, hundreds of thousands of Rwandan refugees began walking home after living in Zaire for months. Without doubt some of them returned to Rwanda carrying alleles from Zaire.

How Does Breeding Between Populations Change Allele Frequencies?

Hardy-Weinberg equilibrium requires that a population not interbreed with other populations. But few populations are completely isolated from all others of the same species. Animals immigrate to nearby populations, taking new alleles with them. Plant pollen, carried into a population by wind or insects, introduces new alleles. In either case, the resulting **gene flow** causes changes in allele frequencies. In our own species, we have many obvious examples of gene flow as people move to new countries. Despite a general tendency to find a mate within one's own group, gene flow among different subpopulations is almost universal (Figure 16-10).

Gene flow has two major effects on allele frequencies. Within a given local population, gene flow from outside populations increases genetic variation. The same gene flow makes adjacent populations more like one another.

Gene flow increases variation within a local population but makes adjacent populations more alike.

How Does Selection Decrease the Frequency of Certain Traits?

A group inept
Might better opt
To be adept
And adopt
Ways more apt
To wit, adapt.
—John M. Burns

A large, isolated population whose members mated randomly and whose mutation rate was low would maintain constant allele frequencies—except for one thing: **natural selection.** Because natural selection reduces the frequency of some alleles and increases the frequency of others, it is a powerful way to change gene frequencies (Figure 16-11).

A simple example occurs when an individual carries two copies of a lethal allele, such as that for Tay-Sachs disease. Children with Tay-Sachs die before they are four or five years old. Having Tay-Sachs therefore eliminates the possibility of reproducing. Although the allele can survive in heterozygotes, homozygotes *never* pass the allele to offspring. As a result, the allele survives only at low frequencies.

Lethal recessive alleles such as Tay-Sachs do not disappear entirely because these alleles "hide" from natural selec-

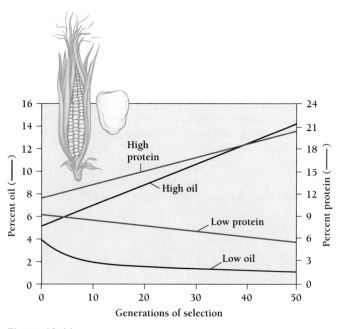

Figure 16-11

Selection is a powerful way to change gene frequencies. In four long-running experiments, researchers selected for corn with more protein or less protein and for corn with more oil or less oil. After 50 generations, researchers continued to produce plants with greater and greater amounts of protein and oil. However, in the experiments selecting for less protein or less oil, the amounts of oil and protein reach a minimum and microevolution stops. Apparently, plants need a minimum amount of oil and protein to survive.

tion in the form of heterozygotes, who are perfectly healthy. (In contrast, natural selection eliminates dominant alleles that kill before the age of reproduction.)

How long does it take to eliminate one allele? To estimate the rate at which selection eliminates recessive lethals, imagine that all *aa* homozygotes die before they reproduce (as in Tay-Sachs). Recall that, according to the Hardy-Weinberg Principle, the frequencies of the *AA, Aa,* and *aa* genotypes are p^2, $2pq$, and q^2, respectively. In each generation q^2 individuals will die without reproducing. When the lethal *a* allele is present at relatively high frequencies, many individuals are eliminated in each generation. When the allele is present at low frequencies, however, selection eliminates very few individuals. For example, if the frequency (q) of *a* is 0.5, then 25 percent of the population will die young. But if $q = 0.01$, then only 1 infant in 10,000 will die from having the *aa* genotype, and only $2(0.01)(0.99) = 2$ percent of the population will carry the gene. Thus, while selection can dramatically change allele frequencies, even the most lethal recessive alleles can still persist indefinitely as heterozygotes.

Nevertheless, even a small difference in the probability of reproduction can lead, over time, to the effective elimination of a deleterious allele. Suppose, for example, that individuals with the homozygous *aa* genotype are slightly less well adapted than individuals with either *AA* or *Aa* genotypes. Say that each *aa* individual leaves only 999 progeny for every 1,000 of the *AA* or *Aa* individuals. Even if we start with a gene frequency for the *A* allele of only 0.00001, we can predict a steady increase in the frequency of the *A* allele, until after 23,400 generations its frequency becomes 0.99. For a fruit fly, this could occur in less than 1,000 years. For species with generation times as long as those of humans, it would take about half a million years—certainly a long time, but only a tick in the history of life. Even subtle selection pressures thus have large effects on allele frequencies.

Natural selection cannot, by itself, eliminate most alleles, but natural selection is, nevertheless, a powerful way to change allele frequencies.

How Does Selection Increase the Frequency of Certain Traits?

When selection eliminates or reduces the incidence of one trait, it increases the frequency of others. One example is the development of resistance to toxins by natural populations of "pests." Most people are now aware that populations of bacteria, parasites, and pests have developed "resistance" as a result of humans waging chemical warfare against them.

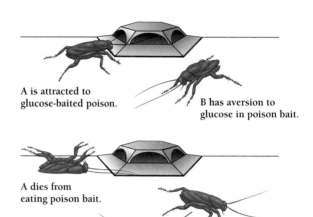

Two apparently identical cockroaches touch/taste glucose-baited poison with their antennae.

A is attracted to glucose-baited poison.

B has aversion to glucose in poison bait.

A dies from eating poison bait.

B survives to reproduce generations with useful aversion to glucose.

Figure 16-12
Natural selection in action. Because cockroach traps and poisons are commonly baited with the sugar glucose, intense selection pressure favors cockroaches that dislike the flavor of glucose. Researchers have discovered populations of German cockroaches from Florida to California that no longer take such bait because selection has allowed only the glucose-haters to survive.

For example, in many parts of the United States, household cockroaches have even evolved an aversion to the glucose (sugar) used to bait roach traps (Figure 16-12). And, after World War II, insects developed resistance to the pesticide DDT. In the 1940s, countries around the world began spraying massive amounts of DDT to kill disease-carrying insects such as flies and mosquitoes. At first, the heavy doses of DDT killed all or most insects. But each time the DDT was applied, a few individual flies would survive. These survivors happened to carry alleles that made them resistant to DDT, and they founded a new population of resistant flies. In time, DDT would no longer kill certain insects in some areas. Because DDT is toxic to many forms of life, its use is now banned in the United States. In some parts of the world, it is still widely used agriculturally, as well as to control malaria mosquitoes. Unfortunately, the regions most affected by malaria can least afford to buy DDT.

Of more pressing concern is the evolution of pathogenic bacteria, more and more of which carry genes for resistance to antibiotics. It is now all too common to find patients in big-city hospitals infected with bacteria that have genes for

Figure 16-13
Directional selection. The English peppered moth *Biston betularia* comes in two forms, light and dark. The light ones, whose colors blend well with the pale lichens found on trees, formerly predominated throughout England. The dark form predominates in areas where industrial pollution has blackened the landscape.

resistance to all known antibiotics. In this case, the antibiotics themselves select for resistant bacteria, killing off all that are not resistant. We discuss this in more detail in Chapter 20.

A classic example of selection in nature concerns the peppered moth (*Biston betularia*). Because of a long British tradition of collecting moths and butterflies, biologists in Great Britain have been able to trace a change in the moth population over more than 100 years. In the early 19th century, all the peppered moths collected in Britain were pale and mottled. Collectors netted the pale moths as they fluttered through the air. Occasionally, a collector found a dark variant of the peppered moth. These were rare and therefore highly prized (Figure 16-13).

The Industrial Revolution dramatically changed this scene, however. In the 19th century, thousands of coal furnaces filled the air with dark soot, which blackened the landscape. The light peppered moth became rare, and collectors found more and more of the dark variants. While lighter moths continued to prevail in unpolluted rural areas, the frequency of dark moths around factory cities increased to about 98 percent.

Looking at the historical data (from old collections of moths and butterflies), the British evolutionary biologist J.B.S. Haldane calculated that, in industrial areas, the darker variant was twice as likely to survive and reproduce as its lighter counterpart. A reasonable explanation for this was that, against the now dark background of trees and rocks, the dark variant had a selective advantage because it was less likely to be seen and eaten by birds. A fine theory, but the naturalists of the time said that they had never seen a bird eating a peppered moth. Indeed, no one had actually seen a

peppered moth resting on a tree or rock. The moths were always caught while flying. The question remained: Was the change in frequency of dark moths the result of selection by birds?

To answer this question, H.B.D. Kettlewell, a biologist at Oxford University, undertook experiments that now provide among the best evidence concerning the operation of natural selection. In the 1950s, Kettlewell captured several hundred dark and light moths and marked them with little spots of paint. He released some of the dark and light marked moths in a smoky industrial area near Birmingham and the rest in an unpolluted woodland near Dorset. He then recaptured as many moths as he could and calculated the percentage of recovered dark and light moths (Figure 16-13). Near blackened Birmingham, the fraction of dark moths recovered was twice that of the light. Near rural Dorset the situation was reversed; the percentage of light survivors was twice that of the dark. This experiment supported the idea that the change in frequency was related to the color of the tree trunks.

But were birds the agents of selection? Kettlewell showed that the moths really were being eaten by birds and that the color of the moth does, indeed, change the chance of being eaten. After placing moths on tree trunks, Kettlewell recorded the menu decisions of the birds in the area with hidden cameras. As expected, the birds preferentially chose the dark moths on the lighter trees of Dorset and the light moths on the darker trees near Birmingham.

Kettlewell's work seemed to demonstrate both the operation and the actual agents of natural selection. The darkening of moths and butterflies to give better protective coloration in sooty environments is now called "industrial melanism." It has occurred in dozens of species not only in Britain, but also near coal-burning industries in the United States and Germany. Beginning in 1956, clean-air laws returned the forests to their normal greens and browns. Gradually, the fraction of dark moths decreased. In Michigan, black peppered moths increased throughout the 1950s, and then, after the United States instituted its first clean-air laws in 1965, the white form came to predominate.

But there is a flaw in this story. At one site where the moths were studied, the pale moths began to return long before the lichens began coming back. And, despite intensive searching, no researcher has ever found a peppered moth resting on a lichen-covered tree (unless the researcher put it there). Some researchers now postulate that the moths roost in the topmost branches of trees and that selection occurs there.

Selection increases the frequencies of certain traits. Kettlewell showed that in a pollution-blackened environment, dark forms of the peppered moth have a selective advantage over lighter forms. Researchers do not yet know whether selection by birds accounts for such changes in trait frequency.

Extreme Biology Does Selection Act on Preexisting Variants?

Many 19th-century biologists believed, as Lamarck did, that variation arose in response to the needs of the organism. Thus even after mutations and genetics were understood, an important question arose: Do mutations arise spontaneously without reference to the needs of the organism or selective pressure, or do they arise in response to selective pressure? In short, are the variants that selection acts on preexisting ones, or not?

In the early 1950s, the American geneticists Joshua and Esther Lederberg decided to test the idea that variants preexist. They used bacteria, because they grow so fast, and an elegant technique they had devised, called replica plating. Replica plating allowed them to study the descendants of thousands of individual cells in different environments.

After growing bacteria in a nutritious broth, the Lederbergs spread the bacteria onto a petri dish containing the Jell-O-like medium agar. Each single bacterium on the agar divided to produce a visible colony composed of thousands of cells. The Lederbergs then touched the surface of the agar to a piece of velvet. The velvet picked up a sample of each colony. By touching the velvet to another agar plate, they made a replica of the original plate. Every colony of the replica had the same position as the "parent" colony on the original plate. The corresponding colonies all derived from the same cell in the liquid broth and so had the same genotypes.

The Lederbergs used their replica plating technique to study antibiotic-resistant variants of the human gut bacterium *E. coli*. In plates containing the antibiotic streptomycin, most of the bacteria died, but a few survived and grew into colonies. All of the cells in each of the colonies themselves produced streptomycin-resistant progeny, showing that the resistance to the drug is a genetic property.

But were individuals with genetic resistance present in the original population, or did they arise only as a result of exposure to the drug? Replica plating provided an easy way to answer this question. The Lederbergs simply examined the replicate colonies that had not yet been exposed to streptomycin. Because they knew the identity of each colony from its position on the plate, they could pick out the ones that had been able to flourish on streptomycin-spiked agar. Here were colonies of the same genotype that had not yet been exposed to streptomycin. Were they resistant to the drug, too? The answer was yes. The Lederbergs showed that the variation in resistance preexisted in the original bacterial population, even though the bacteria had no "need" for such resistance. This result supported Charles Darwin's theory that selection operates on preexisting variants.

16.6 In What Ways Can Natural Selection Change Populations?

We think of natural selection as pushing populations in some adaptive direction. And, indeed, in many populations that is true. For Galápagos finches, for example, selection enlarges the beaks of birds that have only big seeds to choose from. But selection can have other effects as well. Evolutionary biologists name several kinds. Here we briefly discuss directional selection, stabilizing selection, disruptive selection, coevolution, and sexual selection.

How Does Directional Selection Influence Traits?

Directional selection shifts the frequency of one or more traits in a particular direction (Figure 16-14). Industrial melanism is a classic example of directional selection. Selection for resistance, whether to a pesticide or an antibiotic, is also a form of directional selection. Likewise, agricultural breeders have long practiced directional artificial selection to produce such things as chickens that lay more eggs and cows that give more milk.

In nature, directional selection is typical of a changing environment and is a likely explanation for the directional changes that occur during the evolution of new species. The fossil record shows, for example, that the descendants of early humans and horses have become progressively larger. These changes presumably reflect the action of various selective pressures. In the case of horses, paleontologists suggest that the major selective force was the presence of predators. As horses became larger, they could run faster and were better able to avoid being caught and eaten.

It is important to understand, however, that selective pressures can change over time. Traits that are adaptive in one decade or millennium may not be selected for in the next. Thus directional selection can push and pull an organism in many different directions over time.

How Does Stabilizing Selection Influence Traits?

Stabilizing selection tends to act against extremes in the phenotype, so that the average is favored (Figure 16-14). For example, variability in wasps drops dramatically during the winter, when their mortality is high. An average wasp stands a better chance of surviving a harsh winter than those with unusual phenotypes, which die in great numbers.

Stabilizing selection is characteristic of stable environments because particular traits are probably already optimized (by directional selection) for that particular environment. In that case, selection tends to favor the average rather than extreme variants. Stabilizing selection is one explanation for the almost unchanging morphological characteristics

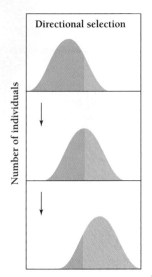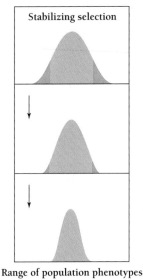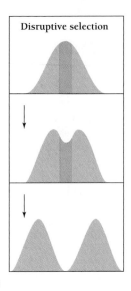

| Directional selection | Stabilizing selection | Disruptive selection |

Number of individuals

Range of population phenotypes

Figure 16-14
Directional, stabilizing, and disruptive selection. In each column, selection eliminates organisms in the orange range of trait distribution and thereby promotes those in the grey regions. In directional selection *(left)*, selection pushes the phenotype in a particular direction. In stabilizing selection *(center)*, selection eliminates extremes (such as a giraffe with too long a neck or too short a neck). In disruptive selection *(right)*, selection does the opposite, selecting for individuals with extreme but not intermediate traits.

of such "living fossils" as the horseshoe crab, the coelacanth, and the ginkgo tree (Figure 16-15). If the environment for such organisms were sufficiently stable, the same phenotype could be adaptive for hundreds of millions of years.

How Does Disruptive Selection Influence Traits?

Disruptive selection is the opposite of stabilizing selection; it increases the frequency of extreme types in a population, at the expense of intermediate forms (Figure 16-14). Like directional selection, it is a likely cause of the evolution of

new species. A nice example of disruptive selection is the African swallowtail butterfly *(Papilio dardanus)*, which comes in three distinct variants, each of which lives in a different part of the butterfly's range (Figure 16-16). All three variants appear unappetizing to their predators because of their resemblance to another particularly foul-tasting butterfly. However, each variant resembles a different foul-tasting butterfly that lives in the same area. Predators in each area avoid the three extreme forms, but they eat any intermediate variants because they resemble none of the three foul-tasting butterflies.

Coevolution

Selection comes from both the physical environment (light levels, weather, and geography, for example) and the biological environment (other organisms). Because species mutu-

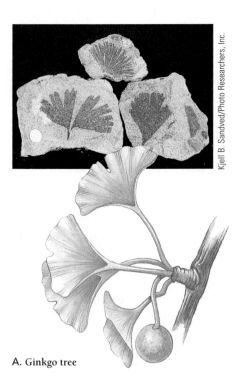

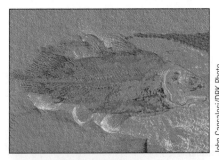

A. Ginkgo tree

B. Coelocanth

Figure 16-15
They haven't changed a bit. Stabilizing selection probably accounts for organisms that hardly change over millions of years of evolution. Shown here are the ginkgo tree (modern and 250 million years old) and the coelacanth (modern and 350 million years old).

Figure 16-16
Disruptive selection. Female African swallowtail butterflies, *Papilio dardanus (top row)*, taste good to birds. The butterflies avoid bird predators by mimicking one of three foul-tasting species of butterflies that also live in the area *(bottom row)*. Intermediate forms of the African swallowtail are snapped up by birds, and only the swallowtails that most closely resemble the bad-tasting species survive and reproduce.

Hope Entomology Collection, Oxford/Biological Photo Service

ally affect one another's environments, many species **coevolve** with one or more biological partners. For example, cat fleas are adapted to feeding on cats and human fleas are adapted to feeding on humans. In one recent example, researchers showed coevolution in carib hummingbirds and the flowers they pollinate. On the island of St. Lucia in the Caribbean, male carib hummingbirds specialize on the nectar of tropical Heliconia flowers. Because the caribs are the only pollinator of these flowers, the flowers' pollen only goes to other flowers that these birds visit. The male birds defend patches of flowers that have the most nectar and therefore pollinate and promote the reproduction of the flowers with the most nectar. The males also have short straight bills that are well adapted to getting the nectar from the short, straight flowers.

The males are so aggressive about defending their patches of flowers that they won't even let female hummingbirds drink from the flowers. The females therefore feed from another species of Heliconia. The female birds are smaller, need less nectar, and have longer curved bills that fit this other species of flower. Finally, the flowers the females visit produce less nectar than the males' flowers. The males have coevolved with one species and the females have coevolved with the other species (Figure 16-17).

How Does Sexual Selection Influence Traits?

Physical or behavioral differences between the sexes are extremely common. Among birds, for example, the male is often showy, with brilliant plumage, a long tail or crest, and a lively song, while the female is drab and inconspicuous. In mammals, the two sexes more often resemble one another, but the male tends to be larger and more aggressive. Many male mammals have horns, antlers, or other rather theatrical weapons. These are used in displays or fights with other males in competitions for territory or access to females. Differences in the anatomy of males and females are called **sexual dimorphism.** Some (but not all) sexual dimorphism results from **sexual selection,** the differential ability of

Female hummingbird   **Male hummingbird**

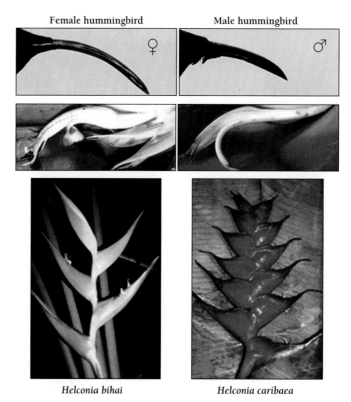

Helconia bihai   *Heliconia caribaea*

Figure 16-17
Coevolution. These two species of Heliconia flowers have coevolved with the hummingbirds that pollinate them. On an island in the Caribbean, the female carib hummingbird gets nectar from (and pollinates) *Heliconia bihai*, while the male carib hummingbird gets nectar from (and pollinates) *Heliconia caribaea*. Notice that coevolution has given the birds' bills and the corresponding flowers similar shapes. (Temeles, E. J., and W. J. Kress. 2003. Adaptation in a plant-hummingbird association. *Science* 300:630–633.)

individuals with different genotypes to acquire mates and to reproduce (Chapter 29).

Biologists generally recognize two kinds of sexual selection, both of which revolve around sexual reproduction. These are **female choice** (or, rarely, **male choice**) and **male competition.** In male competition, a male competes with

Figure 16-18
Male competition. Male gemsbok antelopes sparring. Male competition leads to selection for traits that increase the reproductive success of males but are of no value to females. In most animals that have horns, for example, only the males have elaborate horns.

Figure 16-19
Female choice, in humans. Among the Wodaabe people of Niger, the men dress up in elaborate costumes and makeup for the Geerewal dance, where women observe and judge, selecting the most beautiful men. In America and in many other cultures, it is women who compete in beauty contests and men who do the choosing.

other males for territory or access to females (Figure 16-18). The male is more likely to leave numerous descendants if he can win the contest. Anything that gives him an edge in a contest with other males, whether larger overall size, larger horns, or greater aggressiveness, will give him a selective advantage.

The second variety of sexual selection, female choice, specifically involves courtship behavior. Courtship allows the female to assess the male for health and other traits that may be valuable to her offspring (Figure 16-19). In most cases, all females breed but produce a limited number of offspring. Males may or may not breed. But, at least potentially, a male can breed with several females and monopolize their eggs, while other males fail to breed at all. In short, males' reproductive success is more like the lottery. Because of this imbalance, males tend to be less selective than females. In general, females try to mix their own genes with the best genes they can find, sometimes with just one male, often with several. As a result, most species have evolved complex rituals that enable the female to decide if a prospective mate is genetically "good enough" and choose accordingly.

For example, males of outbred and inbred populations of the fruit fly *Drosophila subobscura* differ in their ability to court female flies. To gain a female's favor, a male must perform a series of complex maneuvers, including tapping the female on the head with his front legs and moving around to approach her head-on while extending his proboscis, the tonguelike structure with which insects feed and taste. The female quickly sidesteps back and forth, forcing the male to

keep adjusting his position. This she does, apparently, to prolong the courtship. The longer the courtship, the more easily she can assess his vigor and skill. Outbred males, which show far greater athletic ability in performing these fly dances, are more successful in their courtship. They have a distinct selective advantage over inbred males.

How Can Natural Selection Maintain Genetic Variation?

Given the power of natural selection to change allele frequencies, we may well wonder why significant polymorphism persists at all. The maintenance of genetic heterogeneity in a population is not particularly surprising if different alleles of the same gene are equally useful. For example, within the range of a single interbreeding population, local environments may select for different alleles. The climate of a mountainside may select for short trees that are less likely to be toppled by winter winds, while a milder valley climate may favor larger trees that produce more seeds. Or one allele may be better adapted at one time of year than another. Among the members of one species of ladybug, for example, black ladybird beetles are more frequent in the fall, while red ladybird beetles predominate in the spring. Such a situation creates a balance of different alleles in a population, called a **balanced polymorphism.** The most famous case of a balanced polymorphism in nonhumans is, of course, the current balance between dark and light peppered moths.

Populations may maintain high frequencies of alleles that are deleterious or even lethal when homozygous. In such cases, polymorphism results from the selective superiority of the heterozygote. Sickle cell anemia is an excellent example of a case in which heterozygote superiority creates a balance of different alleles.

As we discussed in Chapter 10, people with two copies of the "β^S-globin" ("beta S globin") allele suffer from sickle cell anemia. Recall that hemoglobin consists of four polypeptide chains, two βs ("betas") and two αs ("alphas"). In people of African and Middle Eastern descent, the normal allele for the β chain is sometimes replaced by an allele known as β^S. In people with two copies of the β^S allele, low oxygen levels cause the abnormal hemoglobin to precipitate, leading to distortion and stiffening of the red blood cells. The sickled cells are so stiff that they can block capillaries and so fragile that they break open in the strong currents of larger arteries. In the United States, victims of sickle cell disease rarely live beyond age 40, dying of infections, kidney failure, or the blockage of blood vessels in vital areas such as the lungs. Most suffer periodic bouts of pain, increasing weakness, and constant infections.

In Central and West Africa, where health care is worse, homozygotes often die in their teens and 20s and rarely reproduce. Nevertheless, the β^S allele persists in the population, for it has strong selective value in β^A/β^S heterozygotes. Heterozygotes, whose blood contains a mix of normal and abnormal hemoglobins, are not anemic, yet they are resistant to infection by the parasite that causes the deadly disease malaria. The heterozygotes therefore have a distinct selective advantage over people who are homozygous for the "normal" copies of these genes, who are susceptible to malaria.

In some regions of Central and West Africa, the frequency of the β^S allele is about 20 percent (Figure 16-20). Using the Hardy-Weinberg Principle, we can calculate that the frequency of heterozygotes ($2pq$) is then $2(0.2)(0.8) = 0.32$, or 32 percent. The expected frequency of homozygotes would be $(0.2)^2 = 0.04$, or 4 percent. The homozygotes usually die before reproducing, and these deaths represent the "cost" of maintaining the β^S allele in the population. There is a trade-off between the deaths caused by the homozygous condition and the lives saved by the heterozygote condition.

In the United States, where malaria is nearly nonexistent, the β^S allele probably gives people no selective advan-

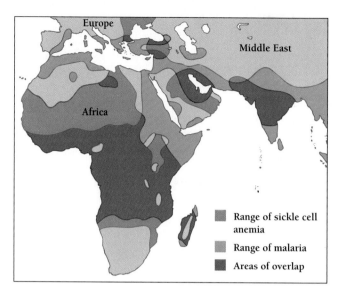

Figure 16-20
A balanced polymorphism. The distribution of the sickle cell allele (β^S) closely parallels that of malaria. Blue shows areas of southern Europe, Africa, and the Middle East where malaria is or was once common. Red illustrates areas where sickle cell anemia occurs. Considerable overlap *(purple)* suggests selection for the sickle cell allele (β^S) in areas with malaria.

tage, either in the homozygote or the heterozygote individual. In the United States, the frequency of homozygotes among African Americans is far less than the frequency in Central and West Africa.

The reduction in the frequency of the β^S allele is due not only to its selective disadvantage but also to the effects of gene flow. On average, one-third of the genes of contemporary African Americans come from their European ancestry, so the frequency of the β^S allele would be reduced by dilution as well as by selection.

A balanced polymorphism may result from heterozygote superiority or from different selective pressures in time and space within the same population.

In this chapter we have seen where variation comes from, how it is inherited, and some of the ways that allele frequencies can change. Changes in allele frequency in a population, called microevolution, can occur through natural selection or, in small, isolated populations, through genetic drift. However, explanations for the larger patterns of evolution—the multiplication of species and higher taxa, as well as patterns of extinction, for example—remain controversial. In the next chapter we will consider why it is so hard to define a species, how new species form, and why species multiply and go extinct.

Key Concepts

- Genetic variation is the fuel that drives evolution.
- Mutations in genes and chromosomes are the ultimate source of all genetic variation.
- Natural selection can bring about evolution by changing gene frequencies in a population.
- Natural selection is only one of several forces that can change gene frequencies in a population. Other forces are mating patterns, genetic drift, mutation, and breeding between populations.

Summary with Key Terms

Natural selection was once controversial because heredity was not understood. Darwin believed in **blending inheritance,** which led him to wrongly embrace the theory of **the inheritance of acquired characteristics.** Weismann correctly guessed that inheritance is particulate, that the genetic material is in the **germ plasm** and is unaffected by the life of the parent, and that sexual reproduction is a means of creating new variants by mixing and recombining old traits.

How did the modern synthesis reconcile genetics and natural history?

The **modern synthesis** wedded natural selection and genetics into an essentially modern theory of evolution. Population geneticists attempt to understand evolution in terms of the changes in the frequency of alleles within populations and species. **Microevolution** is shifts in gene frequencies in a population. **Macroevolution,** which we will study in the next chapter, deals with the processes by which species and higher groupings of organisms originate, change, and go extinct.

What is the basis of the variation on which natural selection operates?

Genetic variation is the fuel of evolution. Gene and chromosome mutations are the ultimate source of all genetic variation. Genetic recombination, enhanced by sexual reproduction, spreads new **alleles** through a population by recombining different alleles into new genotypes.

Most organisms have two sets of **homologous** chromosomes. When the same allele is on each chromosome in a pair, the organism is said to be **homozygous** for that gene. When a different allele is on each chromosome in a pair, the organism is said to be **heterozygous** for that gene. Traits determined by two or more genes are said to be **polygenic.** Organisms displaying two or more alternative forms are said to be **polymorphic.**

The **Hardy-Weinberg Principle** shows that under certain conditions, a single generation of random mating establishes a distribution of genotype frequencies called the **Hardy-Weinberg equilibrium** ($p^2 + 2pq + q^2$) that will not change in future generations. The Hardy-Weinberg Principle disproves Fleeming Jenkin's assertion that sexual reproduc-

tion by itself would change the genetic characteristics of a population.

What factors cause the allele frequencies of natural populations to change?

The Hardy-Weinberg Principle applies only if the following conditions are met: random mating, large population size, no mutations, no gene flow between populations, and no selection. **Natural selection** is only one of several forces that can change the frequency of alleles within a population. Others are nonrandom mating (including nonrandom mating, **inbreeding,** and **outbreeding**), **genetic drift, mutation,** and **gene flow.** A **population bottleneck** occurs when a small population founds a larger one. Chance affects gene frequencies in small populations more than in larger ones. A **gene pool** is all the alleles of all the individuals in a population.

Natural selection can bring about evolution by changing allele frequencies in a population. Natural selection can lead to changes in allele frequencies even when individuals with different genotypes have only slight differences in reproductive success. Natural selection acts on the **phenotype** directly, and the **genotype** only indirectly. Although natural selection reduces the frequency of deleterious alleles, deleterious recessive alleles—even when they are lethal in homozygotes—persist at low frequencies for many generations. Natural selection can also act to decrease the genetic variability in a population (**stabilizing selection**), to change the quantitative characteristics of a population in a particular direction (**directional selection**), to increase the frequency of extreme types in a population (**disruptive selection**), or to cause **sexual dimorphism (sexual selection).** Species may **coevolve** with other species. Biologists believe that sexual selection results most often from **female choice** and **male competition.** Selection for heterozygotes and simultaneous selection for different alleles maintains genetic variation in a population (**balanced polymorphism**).

Review and Thought Questions

Review Questions

1. Darwin understood how natural selection works in the presence of a population with many variant types. What does that variation consist of? How does the variation arise? How does it spread through the population?
2. How much genetic variation do human populations exhibit?
3. Do the eye colors exhibited by your classmates fall neatly into the five colors shown in Figure 16-4? If not, why might they be different?
4. Why doesn't sexual reproduction, by itself, change allele frequencies in a population?
5. Why do the frequencies of the alleles in a gene pool add up to 1?
6. What conditions must be met for a population to be at a Hardy-Weinberg equilibrium?
7. How does chance lead to changes in allele frequencies?

8. Why are certain genetic diseases more common in certain human groups than in others?

9. Give an example of gene flow in a human population and in another animal population.

10. Why would a dominant lethal allele die out in one generation, while a recessive lethal might persist for hundreds of generations?

11. Why is the lethal recessive allele for sickle cell anemia so common in some human populations?

12. Why did Darwin think that sexual reproduction must eliminate variation in a population? What was his mistake?

Thought Questions

13. Why doesn't natural selection eliminate variation in a population?

14. What facts tended to suggest that the theory of the inheritance of acquired characteristics was wrong? What facts were needed to prove that it was wrong?

15. Is it possible that the apparent changes in the proportions of light and dark peppered moths are due to the moths being invisible to collectors rather than to birds? Discuss your reasoning.

BiologyNow Resources

Biology⑤Now™

Active Figures

16-8:   Hardy-Weinberg equilibrium and gene frequency
Table 16-1: Hardy-Weinberg equation

Preparing for an exam? Take a diagnostic test on your BiologyNow CD-ROM.

Online materials relating to this chapter are at:

http://biology.brookscole.com/AAL3

About the Chapter-Opening Image

The cheetah has some of the lowest genetic variability known in a wild organism. Biologists believe that cheetahs passed through a bottleneck within the past few thousand years, in which the entire population was reduced to a few individuals.

17

Macroevolution: How Do Species Evolve?

Key Questions

- What is a species?
- How do species form?
- How do species multiply over time?
- How fast can evolution occur?
- How do scientists think humans evolved?

Tui de Roy/Minden Pictures

What Darwin Saw at the Galápagos Islands

When Charles Darwin arrived at the Galápagos Islands in the summer of 1835, he found a group of dreary volcanic islands consisting of hundreds of craters, the largest some 3,000 or 4,000 feet high. Five main islands dominated the little group of 14 islands, 600 miles off the coast of Ecuador. Despite their position on the equator, the Galápagos were not the lush, tropical islands typical of other parts of the Pacific, but barren desert islands populated by a small number of cacti, reptiles, and small, brown birds. It was the dry season.

"Nothing could be less inviting than the first appearance," Darwin wrote in his diary. "A broken field of black basaltic lava, thrown into the most rugged waves, and crossed by great fissures, is everywhere covered by stunted, sun burnt brushwood, which shows little signs of life. The dry and parched surface, being heated by the noonday sun, gave to the air a close and sultry feeling, like that from a stove: we fancied even that the bushes smelt unpleasantly." The most striking animals were the reptiles, large marine iguanas (Figure 17-1), and immense Galápagos tortoises (*Geochelone elephantopus*), some of them nearly six feet tall and weighing 500 pounds (*left*).

Darwin spent a month in the islands, exploring and collecting rocks, plants, and animals. He discovered that almost all of the terrestrial animals and at least half of the plants were endemic species, that is, found nowhere else in the world. Yet, all of the Galápagan species resembled species found on the nearby South American mainland—all, that is, but a European rat, probably brought by ship to the islands.

The island-bound plants and animals were related to those on the mainland. But how were they related, and why were they different? Darwin's observation that so many Galápagan species were unique to the islands did not at first strike him as meaningful. Even when the vice-governor of the islands casually mentioned that tortoises from different islands were so distinctive that he could easily tell which island a particular tortoise had come from, Darwin paid no attention. Then, just days before the H.M.S. *Beagle* was to set sail for Tahiti, the vice-governor's remark struck Darwin with sudden force.

Figure 17-1

The voyage of the *Beagle*. In 1831, Charles Darwin left England on board the British surveying ship H.M.S. *Beagle* for a five-year trip around the world. Darwin and the *Beagle* spent nearly four years in South America. Darwin was struck by the fact that each island had its own kind of tortoise. Darwin eventually realized that on each island the tortoises had evolved separately from those on the other islands. (Darwin, The Bridgeman Art Library; H.M.S. *Beagle,* The Granger Collection, New York; marine iguana, Michio Hoshino/Minden Pictures)

Maybe, Darwin thought, each island supported a different set of organisms. Frantically, he went back through his specimens to see if different species came from different islands. But he had labeled most of them "Galápagos." He had hopelessly mixed up the different kinds of tortoises and the 14 kinds of finches. There was no way to tell which islands they had come from.

But one group of birds, the mocking thrushes, "thoroughly aroused" his attention. Of three species, one was unique to Charles Island, one to Albemarle, and one to James and Chatham Islands. Darwin's plant collection, more carefully labeled, was a gold mine. For example, of 71 species found on James Island, 38 were endemic to the Galápagos and 30 of those were endemic to James Island alone. Every island had the same story to tell.

"One is astonished," he wrote in his diary, "at the amount of creative force, if such an expression may be used, displayed on these small, barren, and rocky islands;

and still more so, at its diverse yet analogous action on points so near each other."

Two years after his visit to the Galápagos, Darwin began a notebook on the "transmutation of Species." It would be years before he realized the full implications of all that he had seen in the Galápagos. It was no accident, he ultimately realized, that Galápagos species were so closely related to each other and to those on the South American mainland. Each endemic species had evolved on one of the islands, having descended from migrants from other islands or the mainland.

Nowadays, every child knows that different sorts of animals and plants populate each of the different continents. Africa has its lions and camels, South America its jaguars and llamas. Lions and jaguars, although descended from the same ancestral cat, have evolved independently for millions of years. The same is true of camels and llamas.

In Darwin's day, however, the presence of similar animals in very different habitats, or of completely different assemblages of animals in similar habitats, would have been surprising. According to the 19th-century theological concept of special creation, similar habitats should support identical species. Every dark forest should have its bear and every clear stream its trout. And, since each species was thought to have been created by the Judeo-Christian God less than 6,000 years ago, species were not supposed to be related to one another. Most biologists, like Linnaeus, could see that animals and plants group naturally into rough hierarchies that resemble family trees. Yet they carefully ignored the suggestion of genealogy suggested by these hierarchies.

But Darwin—heir of a distinguished and wealthy family—knew the importance of genealogy. He realized that 5 million years ago, the volcanic Galápagos Islands had burst steaming and erupting from the sea, devoid of life. Gradually, plants and animals began to colonize the islands. Seabirds roosted on the barren rocks, leaving seeds that had stuck to their wings or passed through their digestive tracts. Some of these seeds took root and managed to survive. Ocean currents brought more plant seeds and an occasional reptile. Winds also brought seeds, as well as finches and sparrows and other land birds. Some plants and animals survived and reproduced; others did not. Some moved from island to island, spreading and interbreeding with other populations; others remained confined to one island. These slowly evolved away from their cousins on the other islands and the parent population on the mainland. In this way, one species became many. The formation of new species is called **speciation.**

In this chapter we will study **macroevolution**, the origin and multiplication of species and higher groups of organisms. We will see what a species is, and explore how it comes into being. We will ask what the patterns in the fossil record tell us about the rate at which evolution can occur. And we will discover why, after over 150 years, evolutionary biologists are still intrigued by the questions raised by Darwin after his visit to the Galápagos Islands. Near the end of the chapter, we will briefly discuss the evolution of humans.

17.1 What Is a Species?

In the next section of this book, we will explore the enormous biological diversity of life on Earth. For now, we will merely note that biologists have so far identified about 1.7 million species of living organisms—give or take a hundred thousand. Newly discovered species pour into the museums monthly, so this number is probably less than one-tenth of the total number of species currently living. The actual number could be anywhere from 10 million to 30 million. Of the known species, more than half are insects (Figure 17-2). Nearly half of all insects—290,000 species—are beetles, some 20 percent of all species of organisms combined. The biologist J.B.S. Haldane once ironically remarked that God "must have had an inordinate fondness for beetles."

Why Is It So Hard to Define a Species?

The average person, faced with groups of familiar organisms such as mammals or birds, easily recognizes one species from another. Even when presented with a set of unfamiliar mammals, the average person can generally distinguish most species. In 1927, for example, a young biologist named Ernst Mayr, who later became a leading figure in evolutionary biology, led an expedition into the Arafak Mountains of Papua New Guinea to catalogue new species of animals. After he had identified 138 species of birds, he was surprised to discover that the local Papua natives themselves recognized 137. Two

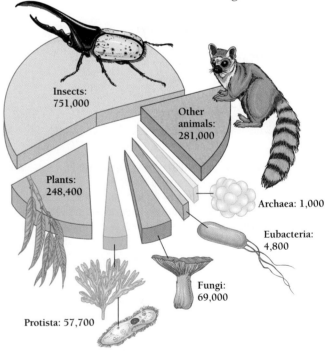

Insects: 751,000

Other animals: 281,000

Plants: 248,400

Archaea: 1,000

Eubacteria: 4,800

Fungi: 69,000

Protista: 57,700

Figure 17-2
The diversity of life. Shown here are some older estimates of the number of species in the major groups of organisms. At least 1.7 million species now inhabit the Earth.

of Mayr's species were so similar that the Papuans considered them a single species.

Defining a species is awkward. When biologists believed that organisms were all variants of "ideal types," as 19th-century essentialists believed, the problem was much easier. The definition of a species was a description of the physical features of the ideal type, a sometimes arbitrarily chosen "type specimen." Type specimens are still essential for naming a new species and for identifying other individuals of the same species that are collected later. When a new species is found, a single individual is chosen as the type specimen, or "holotype," and preserved in a research museum. Researchers keep a single type specimen for every named species of organism.

But we now realize that every individual in a species is unique—ideal types do not exist. Even an individual that happened to be completely average in every respect would not represent its species better than any other individual in the population (Figure 16-7). Research museums therefore store thousands of versions of each species so that biologists can study variation.

The definition of species must take account of the enormous variation among individuals and recognize that two species can be distinct even if they are very much alike. Some species of frogs, for example, are distinguishable only by their calls. Adult forms of the malaria-carrying mosquito *Anopheles maculipennis* look just like several other species that do not carry malaria. Appearance alone doesn't define a species.

Darwin found himself exasperated by the problem of classifying species. Writing to his friend J. D. Hooker in 1853, Darwin noted, "After describing a set of forms as distinct species; tearing up my [manuscript], & making them one species; tearing that up & making them separate, & then making them one again (which has happened to me) I have gnashed my teeth, cursed species, & asked what sin I committed to be so punished."

Darwin concluded that since species gradually evolve one from another, the term "species" is a convenient but arbitrary designation that has no biological meaning. He found this a great "relief," for he no longer felt he had to worry about whether organisms were truly separate species or not.

How Do Biologists Define Species?

Darwin may have been relieved, but biologists need to define and name species in order to discuss them and understand their evolution. Formally, species are evolutionarily independent populations. But identifying such groups in natural environments and in the fossil record is anything but straightforward. One definition is the **biological species concept,** which states: *"Species are groups of actually or potentially interbreeding populations, which are reproductively isolated from other such groups."*

Many animals seem to conform both to the biological species concept and to our sense that species should look different from one another. Bullfrogs (*Rana catesbeiana*) and

wood frogs (*Rana sylvatica*), for example, look different from each other and do not interbreed (Figure 17-3). They are members of different species.

But the biological species concept is hard to apply to species, such as lions and tigers, that can interbreed but don't (Figure 17-4). Nor does the biological species concept apply to asexual organisms that do not ordinarily breed, including bacteria, certain fungi, and even some plants. *E. coli,* the bacterium that lives in our gut, mostly reproduces asexually by dividing. It rarely exchanged genetic material with other *E. coli.* Taxonomists classify such bacteria anyway, giving them a genus and species name, according to the way they look and act.

Finally, for the vast majority of living species, no one has any idea who interbreeds with whom, how often, or where. Although DNA fingerprinting and similar techniques have now given biologists the tools to find out, most of the million or so named species were named on the basis of their appearance.

Organisms classified according to their form are *morphological species*. The morphological species concept is useful for naming and classifying asexual organisms, fossils, and species whose breeding habits are unknown. Since there is no way anyone can know which groups of fossils interbred in the past, fossils are classified only by their bones, shells, and other

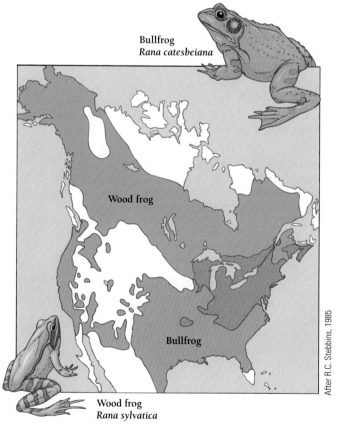

After R.C. Stebbins, 1985

Figure 17-3

Separate species. The wood frog *Rana sylvatica* and the bullfrog *Rana catesbeiana* look different and occupy different, though overlapping, habitats. They probably do not interbreed; they are true species.

Figure 17-4
Separate species? In captivity, a lion and tiger can mate and produce a creature called a "liger." But in the wild, these two cats do not meet, let alone mate. Would they mate if their habitats overlapped? No one knows.

Gerard Lacz/Animals Animals/Earth Scenes

Central California
(*M. m. heermani*)

Desert form
(*M. m. saltonis*)

San Francisco Bay
(*M. m. samuelis*)

Alaskan form
(*M. m. inexpectata*)

Figure 17-5
Subspecies. The song sparrow *(Melospiza melodia)* includes 34 known subspecies. Here are just four. Each one has slightly different size, coloring, bill shape, or song.

hard parts. These same organisms might be judged differently if they were still alive. This point will be important to keep in mind when we discuss the evolution of humans at the end of this chapter.

A newer way of classifying and naming species is the *phylogenetic species concept,* which says that a species is the smallest group in a family tree of populations that has only one lineage. For example, Figure 1-5, in Chapter 1, shows that several separate lineages contribute to the kingdom Protista. Technically, because the protists did not evolve from a single lineage, they aren't comparable to the fungi or the animals, which are more proper groupings. The disadvantage of the phylogenetic species concept is that it requires a careful analysis of the traits of large groups of related species in order to establish a family tree, and well-accepted family trees are still unusual. We'll discuss phylogenetic classification in more detail in Chapter 19.

According to the biological species concept, species are groups of actually or potentially interbreeding populations that are reproductively isolated from other such groups. For most organisms, the biological species concept is hard to apply. More practical definitions include the morphological species and phylogenetic species concepts, but these have drawbacks, too.

What About Other Taxonomic Categories?

Many species have **races** or **subspecies,** morphologically distinct subpopulations, which can interbreed. For example, along the western coast of North America, the song sparrow has been divided into as many as 34 subspecies (Figure 17-5). If you look hard enough, almost any sub-

population can be distinguished from others on the basis of some cluster of characteristics. The people living in Los Angeles might differ from those living in New York in having, on average, big toes that are two millimeters shorter. But we would not necessarily conclude that the two populations were separate subspecies. Many evolutionary biologists therefore consider the concept of race or subspecies to be just a normal manifestation of natural variation, which may not always have any real biological meaning. Still, even though most races or subspecies interbreed easily, a few do not and may be in the process of separating into species.

In Chapter 19, we discuss how species and higher taxa are named, and we'll see why higher taxonomic categories are even harder to define than species. The hierarchical system of classification that taxonomists use to organize biological species into higher groupings generally represents degrees of anatomical and molecular difference and, ideally, evolutionary distance as well. A higher taxon such as genus or family is usually defined by a few characteristics that consistently distinguish its members from the species in other taxa. Mammals, for example, are distinguished by hair, two sets of teeth, and mammary glands. If we find an animal with these characteristics, we can be sure it is a mammal.

However, the division of a group of species into separate genera, families, or higher taxa can be very subjective. For example, in 1947, a botanist decided that a recently discovered daisylike plant whose flowers had no petals was so novel that it should be put into its own genus. Only later did botanists discover that the plant was a simple variant of a well-known species that normally had petals. The "new genus" was not even a new species. In many animal species, the male is much smaller than the female, and taxonomists have sometimes unwittingly classified the two sexes of such species into unrelated groups.

Species exist as biological entities independent of human definitions. Higher taxa, as well as subspecies and races, are somewhat subjective groupings.

17.2 How Do Species Form?

We have seen that for two populations to become separate species, they must become **reproductively isolated**, unable to interbreed. Two populations may happen to become reproductively isolated and then evolve apart later. For example, in the group of flowers called *Delphinium*, or larkspur, most species have blue or purple flowers. Bees visit blue larkspur, carrying pollen from flower to flower. But one species of larkspur has a mutation that gives it bright red flowers—a color bees cannot see (Figure 17-6). Fortunately, hummingbirds prefer red flowers, and they pollinate the red larkspurs. The bees and hummingbirds do not intrude on one another's flowers, and the blue and red larkspurs are reproductively isolated. In this case, a single mutation has suddenly reproductively isolated the flowers.

More often, however, the process by which one species becomes two begins when one population is geographically isolated from the rest of the species. If that happens, the physically isolated population may evolve separately. Once gene flow between the two populations ceases, the allele frequencies in each population can begin to change independently of the other. If the genetic changes in the isolated population happen to make it impossible for individuals to breed with members of the parent species, the population becomes a separate species.

Figure 17-6

Reproductive isolation. Sometime in the past, the scarlet larkspur, shown here, became isolated from the rest of its species. A mutation produced individuals with red flowers instead of the usual blue. But red flowers tend to attract hummingbirds instead of the bees and other insects that pollinate the blue flowers. Because the hummingbirds do not carry pollen between the red and blue flowers, the scarlet larkspur has become reproductively isolated from its blue relatives.

Biology Now™   Learn more about another type of isolation, temporal isolation, by clicking on this figure on your BiologyNow CD-ROM.

What Barriers Reproductively Isolate Populations and Species?

Any barrier to interbreeding, called a "reproductive barrier," will minimize or eliminate interbreeding between the two groups and keep them from losing their separate adaptations. Many remarkable features of organisms maintain reproductive isolation, including the colors and structures of flowers and the elaborate courtship rituals of animals.

Barriers fall into two broad categories: (1) **prezygotic barriers,** which prevent the fusion of the sperm and egg to form a zygote, and (2) **postzygotic barriers,** in which the zygote is either certain to die or sterile. The red and blue larkspurs are an example of a prezygotic barrier, since the

flowers' different colors prevent cross-fertilization. Prezygotic barriers also include differences in the ways that species live (**ecological isolation**), in the times at which they reproduce (**temporal isolation**), in their mating behaviors (**behavioral isolation**), in the complementarity of male and female reproductive organs (**mechanical isolation**), and in the compatibility of their gametes (**gametic isolation**).

Postzygotic barriers operate when fertilization has already occurred, but the resulting zygote either dies or fails to reproduce. An individual that results from cross-fertilization between two different species, or, sometimes, between two distinct populations, is called a hybrid. For example, a cross between a horse and a donkey results in a mule, a hybrid that is nearly always sterile. The mule is an example of **hybrid sterility.** However, hybrids between related species often die during development or early in life because of chromosomal and genetic incompatibilities, a barrier referred to as **hybrid inviability.** In some cases, the hybrids of two species may be viable and fertile, but *their* offspring are weak or sterile (**hybrid breakdown**).

In plants, changes in chromosome number can simultaneously create new variants and cause their immediate reproductive isolation. Plants that contain more than two complete sets of chromosomes are said to be **polyploid.** Polyploidy has played an important role in the evolution of grasses, ferns, and other vascular plants (Figure 17-7). If homologous chromosomes fail to separate during meiosis, then the resulting sperm and eggs will have twice as many chromosomes as normal, and fertilization may produce a zygote with four sets of chromosomes. Tetraploid plants again make sperm and eggs with $2n$ chromosomes. These gametes cannot form fertile zygotes with gametes of the parental stock. Like the red larkspur mentioned earlier, tetraploid plants become immediately isolated from the main population.

> Populations become reproductively isolated either by prezygotic or by postzygotic barriers.

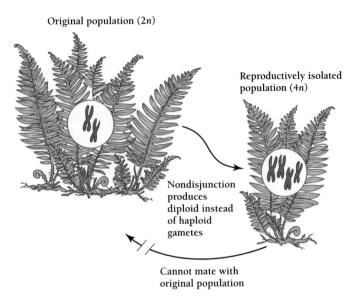

Original population (2n)

Reproductively isolated population (4n)

Nondisjunction produces diploid instead of haploid gametes

Cannot mate with original population

Figure 17-7

Reproductive isolation by polyploidy. An error during meiosis produces diploid gametes, which form tetraploid offspring. The offspring can breed only with one another, however, because if the diploid gametes fuse with normal haploid gametes, they form triploid zygotes, which cannot survive.

How Does Geographic Isolation Initiate Speciation?

In most cases, populations must be geographically isolated from the parent population before other barriers to reproduction arise. Oceans, mountain ranges, and rivers, for example, are all capable of sharply reducing or eliminating gene flow from a parent population. Once two populations are isolated from one another, different environments can exert different selective pressures and set each population on its own evolutionary course (Figure 17-8). For example, one side of a mountain range might have twice the rainfall of the other, so that different plants would grow on each side. A population of field mice on one side of the mountain range, separated from the parent population, might have to live on very different grasses, seeds, and other foods.

Selection, mutation, and genetic drift can all cause gene frequencies (and phenotypic traits) in the isolated population to become different from those in the parent population (Chapter 16). In time, the members of the isolated population may acquire a distinct ecology and behavior that prevents interbreeding with the parent population even if the two should again come into contact. The population has become a separate species. Species formation by geographic isolation is called **allopatric speciation** [Greek, *allos* = other + *patra* = country].

> Once two populations are geographically isolated, selection, mutation, and genetic drift can allow the two populations to differentiate and form two distinct species, a process called allopatric speciation.

A Ring of Species

In some species, gene flow between adjacent populations prevents full speciation while populations that are farther apart begin to separate. For example, over evolutionary time, the ensatina salamander has spread down the two sides of California's Great Central Valley (Figure 17-9). Increasing geographic distance correlates with both increasing genetic differences and increasing differences in appearance. The populations on either side of the valley are separated from one another and mostly do not interbreed. One exception is the population of yellow-eyed salamanders on the coast, which can move up a river system to the mountains in the east, where they hybridize with the Sierra Nevada salamanders. In the south, the Monterey salamander and the yellow-blotched salamander, which do not interbreed, are different enough to be considered different species, yet they are linked by a continuous chain of subspecies that can and do interbreed.

How Can Species Form Without Geographic Isolation?

The formation of separate species without geographic isolation is called **sympatric speciation** [Greek, *sym* = together + *patra* = country]. For a single population to sponta-

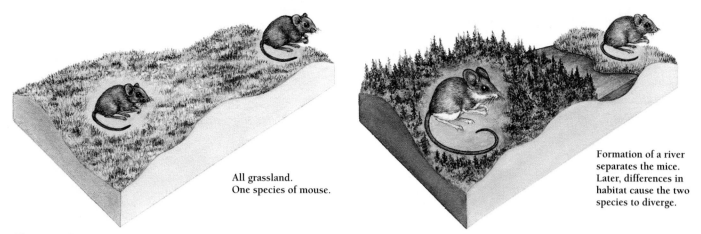

All grassland.
One species of mouse.

Formation of a river separates the mice. Later, differences in habitat cause the two species to diverge.

Figure 17-8

Allopatric speciation. When a single population of mice becomes divided—by a river, for example—the two subpopulations evolve separately and can form separate species.

Biology ⑤ Now™ Learn more about allopatric speciation by clicking on this figure on your BiologyNow CD-ROM.

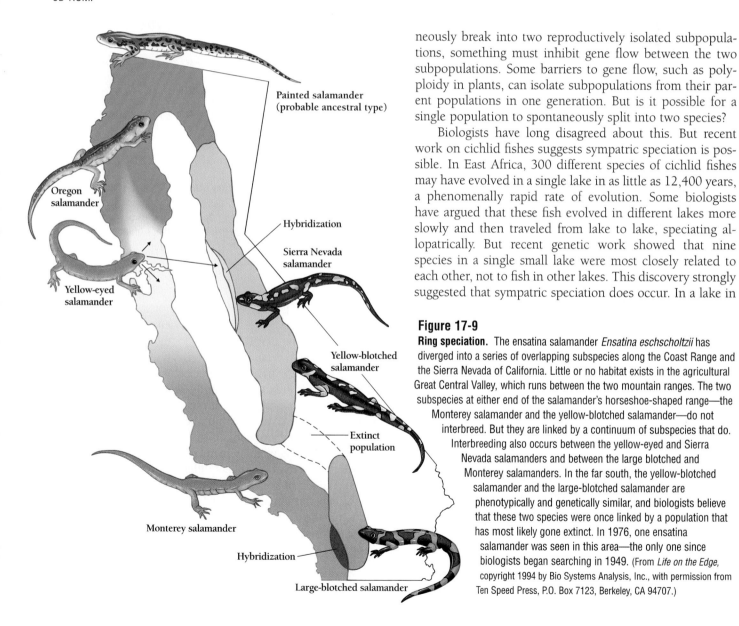

Painted salamander (probable ancestral type)

Oregon salamander

Hybridization

Sierra Nevada salamander

Yellow-eyed salamander

Yellow-blotched salamander

Extinct population

Monterey salamander

Hybridization

Large-blotched salamander

neously break into two reproductively isolated subpopulations, something must inhibit gene flow between the two subpopulations. Some barriers to gene flow, such as polyploidy in plants, can isolate subpopulations from their parent populations in one generation. But is it possible for a single population to spontaneously split into two species?

Biologists have long disagreed about this. But recent work on cichlid fishes suggests sympatric speciation is possible. In East Africa, 300 different species of cichlid fishes may have evolved in a single lake in as little as 12,400 years, a phenomenally rapid rate of evolution. Some biologists have argued that these fish evolved in different lakes more slowly and then traveled from lake to lake, speciating allopatrically. But recent genetic work showed that nine species in a single small lake were most closely related to each other, not to fish in other lakes. This discovery strongly suggested that sympatric speciation does occur. In a lake in

Figure 17-9

Ring speciation. The ensatina salamander *Ensatina eschscholtzii* has diverged into a series of overlapping subspecies along the Coast Range and the Sierra Nevada of California. Little or no habitat exists in the agricultural Great Central Valley, which runs between the two mountain ranges. The two subspecies at either end of the salamander's horseshoe-shaped range—the Monterey salamander and the yellow-blotched salamander—do not interbreed. But they are linked by a continuum of subspecies that do. Interbreeding also occurs between the yellow-eyed and Sierra Nevada salamanders and between the large blotched and Monterey salamanders. In the far south, the yellow-blotched salamander and the large-blotched salamander are phenotypically and genetically similar, and biologists believe that these two species were once linked by a population that has most likely gone extinct. In 1976, one ensatina salamander was seen in this area—the only one since biologists began searching in 1949. (From *Life on the Edge*, copyright 1994 by Bio Systems Analysis, Inc., with permission from Ten Speed Press, P.O. Box 7123, Berkeley, CA 94707.)

Nicaragua, biologists found two varieties of the same species of fish, whose genetics suggested that they were in the process of splitting away from one another. How can this happen? One explanation is "assortative mating." Imagine a population of blue and gold fish, and that each gold fish prefers to mate with another gold fish, while each blue fish prefers a blue one. If the fishes enforce this preference, the gold and blue fishes will eventually evolve apart.

Biologists have debated for many years whether sympatric speciation can occur, but recent studies of cichlid fishes suggest that it can.

17.3 How Do Species Multiply?

The separation of one species of organism into two (or more) species is also called **divergent evolution.** In some sense, evolution tends to be divergent. From a few simple forms, millions of species have evolved. Species are constantly splitting into two, three, or more new species. These separate species diverge even more to form separate families and orders.

Species often diverge and multiply rapidly when a new place to live becomes available. Biologists call such an event an **adaptive radiation,** the generation of many new species with widely varying adaptations. For example, nearly all the major groups of modern mammals—the carnivores, whales, bats, rodents, and primates, for example—evolved during a 10- to 15-million-year period in the Cenozoic. DNA sequence analysis suggests that these diverse groups of mammals split from one another almost simultaneously.

How Does Adaptive Radiation Occur on Chains of Islands?

Adaptive radiation is the key to understanding why there are so many species. We will first discuss examples of adaptive radiation in chains of islands and then try to understand how diversity increases worldwide.

Some of the most dramatic cases of adaptive radiation occur on chains of islands (or archipelagos), such as the Galápagos. We have seen that the formation of new species generally requires either reproductive isolation (so that descendant populations cannot interbreed with one another) or one or both of two other conditions. Even if two populations are not reproductively isolated, strong selection pressure (so that different descendant species acquire different adaptations) and small population size (so that genetic drift can operate) can allow two populations to split into two species.

Archipelagos sometimes provide all of these conditions. Geographic isolation from the mainland ensures reproductive isolation from mainland species. Of course, islands don't have to be surrounded by water to encourage specia-

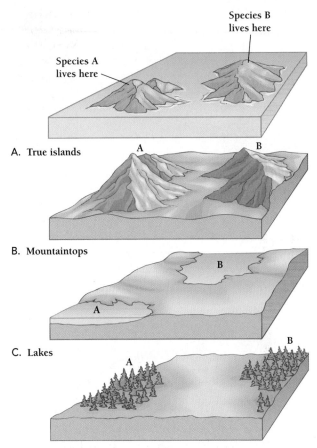

A. True islands

B. Mountaintops

C. Lakes

D. Variation in vegetation

Figure 17-10
Geographic isolation comes in many shapes. Although real islands surrounded by oceans have often caused allopatric speciation, other kinds of "islands" also facilitate speciation. To a fish, a series of lakes or ponds is a series of islands. To a mouse, the individual tops of mountains or plateaus are islands. A. True islands. B. Mountaintops. C. Lakes, rivers, and oceans. D. Variation in vegetation.

tion. A series of isolated lakes, mountaintops, or even forests are, for many species, equivalent to islands (Figure 17-10).

Whether the isolation is caused by water or by a surrounding desert, relatively few individuals arrive at each island. Plants may arrive as seeds carried by wind or birds. Animals arrive only if they can fly, swim, or float on a piece of driftwood or other natural raft. In short, these populations tend to be so small in number that random mutations alone cause gene frequencies in the island populations to differ from those in their mainland parent populations. In addition, the environments of different islands differ enough to exert different selective pressures.

When Darwin visited the Galápagos Islands in the Pacific and the Cape Verde Islands in the Atlantic, he noted three surprising facts:

1. The species in each chain of islands resembled those on the closest mainland rather than those of other islands. Despite the similar climate and terrain of the Cape Verde Islands

and the Galápagos Islands, the birds on the Cape Verde Islands were similar to those of Africa and those of the Galápagos to those of South America. This suggested that the islands were colonized by migrants from the mainland.

2. The archipelagos contained a much smaller number of species than the mainland. The islands' biological poverty supported Darwin's idea that only a small number of ancestral organisms had colonized the islands.

3. The individual islands in a chain had distinct species. Tortoises from the drier islands had longer necks and peaked shells that allowed them to reach the cactus and tree branches, while those from wetter islands had shorter necks, better adapted to feeding on ground vegetation.

Darwin recognized that the distribution of species in chains of islands provides a record of the arrival of immigrants from the mainland. For example, of the hundreds of bird species in South America, only eight succeeded in reaching and colonizing the Galápagos. Yet, one of those eight gave rise to no fewer than 14 species of finches. The Galápagos finches offer the most famous example of adaptive radiation (Figure 17-11).

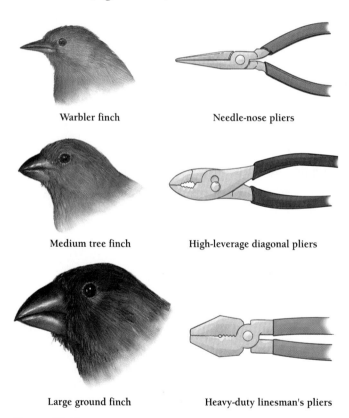

Warbler finch   **Needle-nose pliers**

Medium tree finch   **High-leverage diagonal pliers**

Large ground finch   **Heavy-duty linesman's pliers**

Figure 17-11

Adaptive radiation. Less than 5 million years ago, the Galápagos Islands began erupting from the floor of the Pacific Ocean. Sometime after the islands had been colonized by plants, a single species of finch colonized the islands. This finch then radiated into 14 separate species, which taxonomists have grouped into four genera. Here are just three, representing three genera. Each finch is specialized for a different way of life. The finch with the largest beak, for example, breaks open large, hard seeds, impossible for the other two finches to crack.

Figure 17-12

Vampire finch. Galápagos finches seem to have evolved to take advantage of any opportunity. This sharp-billed finch *(Geospiza difficilis)*, sometimes called the vampire finch, feeds on the blood of larger birds such as this masked booby.

The Galápagos finches are all distinct from any on the continent of South America. Yet they were enough alike to suggest to Darwin that they all stemmed from a common ancestor. Contemplating the finches, Darwin wrote his first statement on adaptive radiation: "One might really fancy that . . . one species had been taken and modified for different ends" (Figure 17-12).

The probable mechanism of adaptive radiation on islands such as the Galápagos is relatively easy to understand. Over the course of the 5 million years since the islands first appeared—and possibly as recently as a half-million years ago—a few finches (at least one male and one female) managed to fly the 1,000 kilometers (600 miles) from the South American mainland—no easy task, since finches are poor long-distance flyers. The founders were probably ground birds, with short bills useful for crushing seeds. Descendants of these birds managed to colonize the 13 islands in the Galápagos chain, as well as Cocos Island, about 1,000 kilometers away. However, because of the finches' limited flying abilities, flights among the islands must have been rare. So the finches on each island mated only with other finches from the same island. In short, the finches on each island became reproductively isolated.

Because each island differs slightly in geology, terrain, vegetation, and insect life, the finch population on each island experienced different selective pressures. Acted upon by both natural selection and genetic drift, the populations came to differ in many ways: size and form of beaks, nesting patterns, songs, plumage, and courtship behavior. Separate species formed through allopatric speciation. Over time, many of the new species managed to colonize other islands, so that many islands now sustain several distinct species of Galápagos finches.

Detailed studies of the finches have shown that pairs of species that share an island are more different from one another—in beak size and other measures—than are species that live on different islands. In short, where several finch species live together, individual species tend to evolve away from each other and become more specialized. Like all organisms on Earth, these birds will evolve or go extinct under the tug and pull of natural selection. Only in these finches and a few other organisms, however, have researchers documented the effects of natural selection so clearly.

Yet, it is the absence of more specialized birds in the Galápagos that has allowed the finches to radiate. For example, because the Galápagos have no true woodpeckers, one species of Galápagos finch has evolved into a sort of imitation woodpecker (Figure 17-13). The Galápagos wood-

pecker finch does not have the special tongue and beak and other adaptations of a true woodpecker. Yet the woodpecker finch manages to feed on the insects that live in tree bark by using a cactus spine or a twig clamped in its beak to dig the insects from the bark.

Similarly, on the island chain of Hawaii, which also lacks true woodpeckers, a species of honeycreeper digs insects from tree bark by chiseling the bark with its lower jaw and picking out insects with its curved upper bill. Neither the Hawaiian honeycreeper nor the Galápagos finch seems as well adapted as a true woodpecker, but each manages to flourish in the "woodpecker" niche in the absence of direct competition from a true mainland woodpecker.

Plants also undergo adaptive radiation. The sunflower family (*Asteraceae*) provides two especially dramatic examples. In the Hawaiian Islands, 28 species of the most spectacular plants in the islands have all derived from a single small and unremarkable California tarweed. The amazing Hawaiian tarweeds, some of which are shown in Figure 17-14, include the Haleakala silversword, which grows within the cone of a volcano on the island of Maui, as well as the yuccalike iliau, which grows among the scrub vegetation of Kauai. Most of these plants, like their relatives in California, grow in dry areas, but a few have adapted to wet forests and bogs.

Adaptive radiation is the evolutionary divergence of a single line of organisms.

North American hairy woodpecker

Galápagos woodpecker finch

Hawaiian honeycreeper (akiapolaau)

Figure 17-13
Opportunity knocks. In the absence of mainland woodpeckers, other kinds of island birds can evolve adaptations for taking grubs from tree bark.

How Do Species Diversify on Continents?

The multiplication of species is not confined to chains of islands. On continents, two ways that species diversity increases are **dispersal,** the spread of a taxon through a large area, and **vicariance,** the fragmentation of a widely dispersed species or group of species.

Birds and bats and even insects are, of course, very good at dispersing all over the world. So are plants, whose tough seeds hitch rides on animals and winds. Ants, horses, and other terrestrial organisms may not be able to cross oceans, but they can travel enormous distances by simply walking a little farther, generation after generation (Figure 17-15).

In vicariance, a widely dispersed population can become fragmented either through the extinction of populations in between or through the creation of geographic barriers. As we saw in Chapter 15, tectonic plates move around, alternately drifting apart and bumping into one another over millions of years. Such movement not only separates and joins whole continents, it also raises mountain ranges and lifts old sea beds that are then cut by deep canyons. Once populations are isolated by such geologic changes, they can speciate allopatrically.

The fossil record makes sense in terms of the movements of the continents (see Chapter 15). For example, Aus-

Figure 17-14
Adaptive radiation in Hawaii. In the isolation of the Hawaiian Islands, an unremarkable plant, represented in California by tarweeds in the genus *Madia,* has radiated into 28 species. A. The California tarweed *Madia madiodes.* B. Hawaiian iliau *Wilkesia gymnoxiphium.* C. The Haleakala silversword *Argyroxiphium sandwicense.*

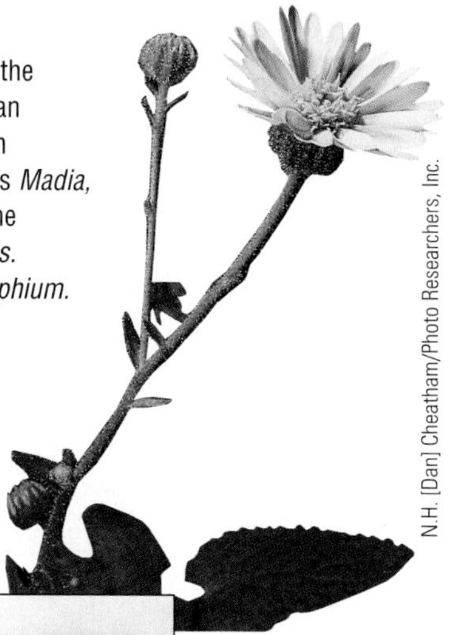

N.H. [Dan] Cheatham/Photo Researchers, Inc.

A.

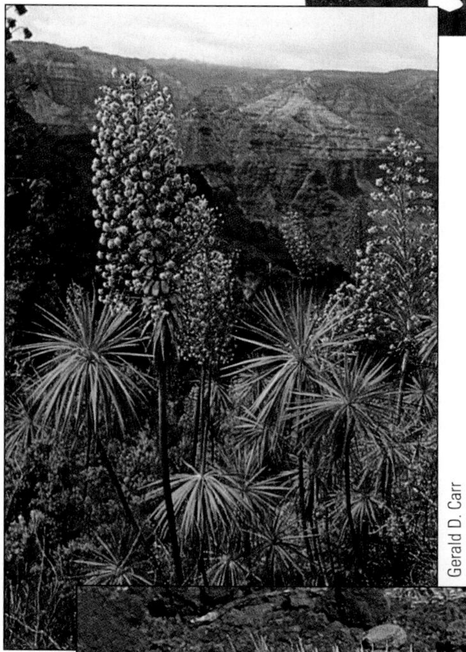

Gerald D. Carr

B.

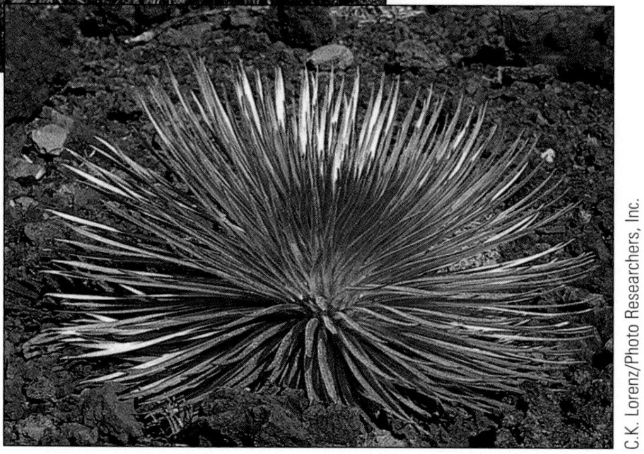

C.K. Lorenz/Photo Researchers, Inc.

C.

North America
Armadillo
South America
Llama
Anteater
Horse
Mastodon

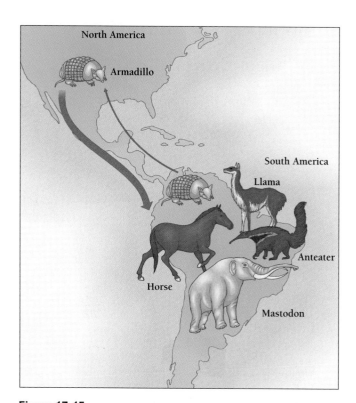

Figure 17-15
Land bridge. When North and South America were separated by an ocean, separate groups of organisms evolved (in allopatry) on each continent. In North America, many groups also came from Asia and even as far away as Africa. When the ocean level dropped, the Central American land bridge emerged, enabling horses, mastodons (elephants), and camels (llamas) to emigrate from North America to South America and anteaters and armadillos to emigrate north.

tralia, which became separated from all the other continents about 65 million years ago, has a unique lineage of mammals. Nearly all mammals in the world are either "placental" mammals like ourselves or "marsupial" mammals, like kangaroos. Placental mammal mothers nurture their young in the uterus with the aid of a specialized tissue, called the *placenta* [Latin, *placenta* = flat cake], which passes nutrients, oxygen, and wastes between the mother and the developing embryo (Chapter 45). Marsupial mammal mothers give birth to tiny fetuses that are barely developed, which the mother carries in a pouch until the young are able to fend for themselves. A handful of marsupials live in North and South America, but on every continent except Australia the vast majority of mammals are "placental" mammals—mammals such as humans and mice that give birth to more-developed young.

The two kinds of mammals diverged from each other in the early Cretaceous, and the marsupials happened to colonize Australia first, just before it split off from the rest of the continents. For 65 million years, the marsupial mammals evolved in isolation from the rest of the mammals. With the exception of bats, which evolved elsewhere and later flew to Australia, and various species of rats, which probably rafted across the ocean from island to island, all mammals that are native to Australia are marsupials. And marsupials are not all koala bears and kangaroos. Over the millennia, marsupial mammals evolved to do everything that placental mammals do. Australia evolved marsupial wolves, marsupial cats,

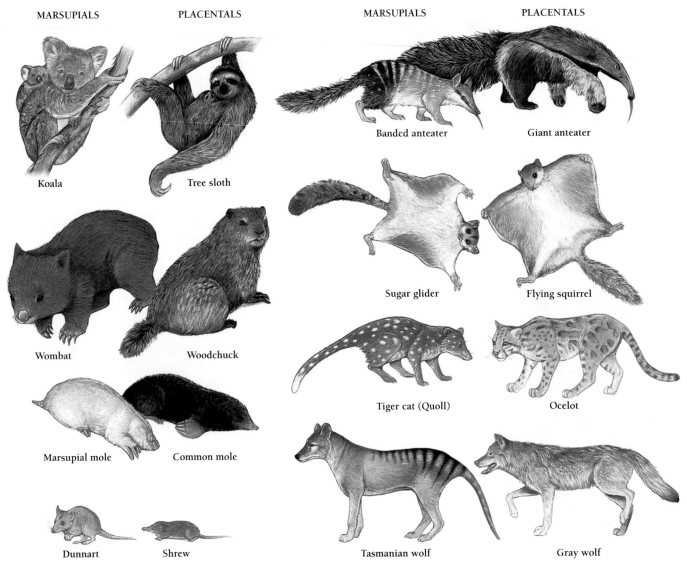

| MARSUPIALS | PLACENTALS | MARSUPIALS | PLACENTALS |

Koala

Tree sloth

Banded anteater

Giant anteater

Wombat

Woodchuck

Sugar glider

Flying squirrel

Marsupial mole

Common mole

Tiger cat (Quoll)

Ocelot

Dunnart

Shrew

Tasmanian wolf

Gray wolf

Figure 17-16

Adaptive radiations in parallel. The marsupial mammals of Australia belong to an ancient and very distinct lineage of mammals. Yet, marsupial mammals *(left in each pair)* and placental mammals *(right)* have separately evolved many of the same body types. The sugar glider and the flying squirrel, for example, both have long, bushy tails and webbed limbs—and for the same reason. Both animals quickly cover great distances by sailing from the tops of tall trees down to the ground.

marsupial squirrels, and marsupial moles, to name a few (Figure 17-16).

The marsupial mammals of Australia are a spectacular example of adaptive radiation. The marsupial radiation illustrates the opportunism of evolution and helps explain the enormous diversity we see around us. But the marsupial radiation also parallels a similar radiation among placental mammals. Together, these two examples provide one of the best examples of another important idea in evolution, called "convergent evolution." **Convergent evolution** is the independent development of similar features in separate lineages of organisms.

Recall from Chapter 15, the convergent evolution of wings in groups as unrelated as birds, bats, and insects. Figure 17-16 shows that time after time, marsupial and placental mammals evolved the same sets of traits to do the same kinds of things. The marsupial "tiger cats" are nocturnal hunters that sometimes kill chickens. The now extinct marsupial "wolf" was an aggressive carnivore that looked and behaved something like its placental counterpart, trotting "relentlessly" after its prey for days at a time until the prey was exhausted. The marsupial wolf had large jaws, "canine" teeth, and shearing premolars that remarkably resemble those of the placental wolves of North America and Eurasia.

Species evolve convergently because the living world has only a limited number of roles for organisms to play. Plants photosynthesize, and animals either eat plants or they eat each other (or both). As we discuss in Chapter 27, every habitat has certain ecological "niches," which different organisms can evolve to fill. A niche is approximately what an organism does for a living. Dogs and wolves run down their prey. Cats pounce. In Australia, all the niches were open to marsupials, and natural selection slowly shaped animals that filled each niche. In the rest of the world, placental mammals filled most of those same niches.

Continental diversity increases through both dispersal and vicariance. The evolution of Australian marsupials illustrates both adaptive radiation and convergent evolution.

17.4 The Cambrian Explosion

Up until the Cambrian Period, almost 600 million years ago, life on Earth seems to have been limited to a few soft, one-celled organisms, such as algae and bacteria. Until quite late in the Precambrian, the only complex organisms that paleontologists find are a few jellyfish-like creatures.

Then, at the beginning of the Cambrian, life bloomed. In a burst of diversification unmatched in the history of the world, all of the modern animal phyla that have fossilizable skeletons appeared, though not necessarily in their current forms. For example, the arthropods, which today include the insects, spiders, and crustaceans, were represented by the trilobites, crustaceans, and other marine animals. Also present in the Cambrian were the brachiopods (lamp shells), mollusks (snails, clams, and squid), sponges, echinoderms (starfish and sea urchins), and cnidarians (jellyfish and corals), not to mention the first chordates (today including amphibians and mammals).

Taxonomists have based the classification of animal phyla on basic body plans—the two-layered radial symmetry of the jellyfish; the three-layered bilateral symmetry of flatworms; the three-layered bilateral symmetry, plus notochord, of chordates, such as ourselves, to name a few. Although enormous evolutionary change has occurred since the early Cambrian and the level of diversity is arguably as great today as it ever was, almost no new body plans have evolved in the half-billion years since these basic forms first appeared.

Evolutionary biologists refer to this one-time evolutionary radiation as the **Cambrian explosion.** Paleontologists remain divided over the exact time span during which this explosion occurred. Some evidence suggests that the major animal phyla may have diverged up to 900 million years ago, hundreds of millions of years before the Cambrian. Other evidence suggests that all the phyla may have ap-

peared in a brief period of time at the very end of the Precambrian. Either way, evolutionary biologists are left with a major puzzle. What has prevented major new body plans from evolving since the Cambrian?

One argument is that evolution has somehow slowed since then, that changes simply do not occur as rapidly as they did in the Precambrian and the Cambrian. But this argument is flawed; modern populations are capable of evolving quite rapidly. For example, both bacteria and insects evolve resistance to antibiotics and pesticides within a few years or even months. Vertebrates, such as fish and birds, are also capable of rapid changes. Research on the finches in the Galápagos has shown that natural selection can change their average beak size, according to what size seeds are available, from season to season. All of the Galápagos finches probably evolved from a single species within the last half-million years, a short period in geologic time.

Another argument is that once the major body plans formed, **developmental constraints**—the rules of embryological development that determine the general form of an organism—prevented new body plans from evolving. Many biologists believe that in most organisms, development has become so finely tuned since the Cambrian explosion that any dramatic change results in organisms too defective to flourish. The idea is that modern organisms are so specialized that an organism with an entirely new form could not compete with modern organisms.

But the ancient body plans are not necessarily optimal designs that could never be improved on. Indeed, very likely they are just accidents of history, like the homologous vertebrate limbs illustrated in Chapter 15.

Animal body plans, such as the particular arrangement of bones in the limbs of horses, whales, and humans, may simply be a case of organisms "making do" with what they have. The first airplanes were made of wood, not because this is a better material than light alloy metals but because that was the best material available at the time. Modern airplanes can be completely redesigned, with new shapes and new materials. But organisms tend to reuse the same parts and materials, reshaping them for new purposes.

Evolution at the species level occurs as rapidly as ever, but the evolution of new phyla seems to have ceased. The Cambrian explosion may have been a one-time phenomenon.

17.5 Extinction and the Rate of Evolution

Evolutionary biologists estimate that more than 99 percent of all the species that have ever lived are now extinct and that the vast majority of these went extinct millions of years ago. But why species go extinct is, in most instances, something of a mystery. By looking at species that have gone extinct in historical times, we can say that some species go extinct because of competition from other species, some because their habitat is drastically changed, and some because of a new predator. Human beings, for example, have driven a variety of species to extinction—the American passenger pigeon, for example—through heavy hunting. But in most extinctions, probably no single factor determines whether a species will go extinct. Some species survive such onslaughts, while others do not.

The Mass Extinctions

The fossil record shows that, in the long run, species go extinct at a more or less regular rate. However, superimposed on this ongoing extinction rate are five intriguing mass extinctions, including the death of the dinosaurs (Figure 17-17). Near the end of five geologic periods—the Ordovician, the Devonian, the Permian, the Triassic, and the Cretaceous—the number of extinctions increased by as much as 2.5 times the background extinction rate.

Why these mass extinctions occurred is one of the most controversial questions in paleontology. Some researchers deny that the mass extinctions are qualitatively different from the background extinction rate. They argue that these mass extinctions are just random fluctuations in the usual loss of species over time. Other researchers argue that the five extinctions are the consequence of major worldwide catastrophes. The exact nature of the catastrophes is hotly debated.

The most controversial mass extinction is the one at the end of the Cretaceous. Sixty-five million years ago, all the dinosaurs and half of all species of plants and animals went extinct. Every conceivable catastrophe has been suggested to explain this extinction. But the majority of researchers now seem to believe that this most recent extinction occurred when a giant asteroid hit the Earth with enough force to blast a 180-km crater and plunge the world into almost total darkness for months.

We are now in the midst of a sixth mass extinction. The full extent of this extinction is somewhat controversial. No one knows precisely how many families of organisms have gone extinct, nor how many will go extinct before this extinction event is over. But the cause of this mass extinction is not at all controversial. It is entirely the result of the expansion of the human population into virtually every part of the world (Chapters 26 and 28).

After each major extinction in the past, diversity increased dramatically (Figure 17-17). For example, the mammals radiated spectacularly after the Cretaceous extinction of the dinosaurs and other reptiles 65 million years ago. We might guess that the only circumstances under which really

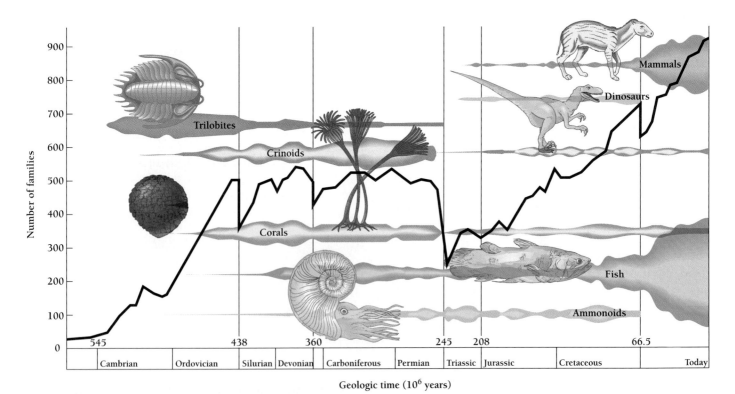

Figure 17-17
Mass extinctions and increasing diversity. This selection of animals shows that numbers of species and families tend to increase over time. Each mass extinction is followed by an enormous adaptive radiation. Recovery from a mass extinction like the one the Earth is experiencing today, however, takes 5 million years or longer. Many groups, such as the dinosaurs and the trilobites, never recover.

Extreme Biology Did It Come from Outer Space?

In 1979, geologist Walter Alvarez discovered a thin layer of mineral that told him that the Age of Dinosaurs ended with a bang, not a whimper. Alvarez was studying rocks in Gubbio, Italy, at what geologists call the K/T boundary. Here, rocks laid down at the end of the Cretaceous (K), 65 million years ago, are overlaid by rocks from the Tertiary (T).

At Gubbio, the boundary is clearly marked by a 2- to 3-cm layer of clay (Figure A). To find out how many years of Earth's history was represented by that inch-deep clay layer, Alvarez measured its iridium content. The metal iridium is rare on the Earth's surface, but a constant hail of it from space lightly dusts the Earth's surface. Like the sand in an hourglass, the buildup of iridium can be used to determine how much time has passed.

Alvarez expected to find less than 1 part per billion (ppb) of iridium, and, indeed, both the Cretaceous rock below the clay and the Tertiary rock above had the expected amount—1 ppb or less. To his surprise, however, he found a whopping 10 ppb in the clay layer itself. What had caused the spike in iridium in the clay?

Collaborating with his father, Nobel prize–winning physicist Luis Alvarez, and geochemist colleagues Frank Asaro and Helen Michel, Alvarez found similar iridium spikes at two other K/T sites—in Denmark and New Zealand. In 1980, the team published a groundbreaking paper, "Extraterrestrial Cause for the Cretaceous-Tertiary Extinction," in the journal *Science*. In it, the researchers argued that the iridium spike, or *iridium anomaly,* as it is often called, was the result of a gigantic 10-km asteroid that slammed into the Earth, exploded, and drove to extinction all of the dinosaurs and many other forms of terrestrial and marine life.

Could this really have happened? Paleontologists initially scoffed dismissively. But the Alvarez team argued back, and the controversy spread, so that soon even nonscientists began taking sides.

The excitement stimulated other scientists to investigate, and to date they have found iridium anomalies at more than 100 different K/T boundary sites, both on land and in ocean sediments. Even more intriguing is evidence that the asteroid may have fallen into the Gulf of Mexico.

If an asteroid 10 km across smashed into the Earth, the pressure generated would liquefy the meteorite and the rock it smashed into, spewing molten meteorite and rock into the air—just as water droplets fly into the air when a boulder is heaved into a pond. As the molten meteorite droplets cooled, they would solidify into smooth round pebbles called *microtektites* and fall back to Earth. Such microtektites, as well as *shocked quartz—* deformed pieces of the mineral quartz—have been found in association with known meteor sites, such as Meteor Crater in northern Arizona.

But the most dramatic evidence would be a huge hole in the ground. Did it exist? Until the late 1980s, no one could find the 150- to 300-km crater that scientists predicted such an asteroid would make. If the crater was deep in the ocean and now filled with sediment, it would not be easy to find.

Off the Gulf Coast of Texas, however, researchers found that, just at the K/T boundary, huge boulders had been moved around as if by a tidal wave. Then another researcher reported a subterranean bowl-shaped structure, 180 km across, located in the northwestern corner of the Yucatan Peninsula near the town of Chicxulub (pronounced CHICKS-uh-loob). In 1991, other researchers reported that samples from the Chicxulub bowl contained shocked quartz. The layer of melted rock underneath the huge bowl has now been dated at 65 million years old—the age of the Cretaceous-Tertiary (K/T) boundary.

Few people now doubt that the structure at Chicxulub is the crater from an enormous asteroid that slammed into Earth 65 million years ago. That this impact caused global catastrophe is also indisputable. The impact would have incinerated plants and animals across the North American continent and spewed debris into the atmosphere, blotting out the sun worldwide for months or even years.

The only debate is over the exact consequences of that global catastrophe.

Some scientists still insist that the asteroid alone did not cause the mass extinction at the end of the Cretaceous. For one thing, whole lineages of organisms disappeared long before the asteroid hit. This fact suggests that there was a more gradual increase in extinctions during the late Cretaceous, not a sudden catastrophic reduction in species.

But why else might they have gone extinct? The best alternative is massive volcanic activity in India and the Pacific Rim that released enormous amounts of volatile gases and dust into the atmosphere. Such gases could have altered the global climate, caused acid rain, and damaged the ozone layer, and caused a large-scale climatic catastrophe.

For now, paleontologists still disagree about whether the late Cretaceous extinctions were rapid or gradual. If the extinctions were sudden, the asteroid is the most likely culprit. If the extinctions were more gradual, the asteroid may have acted merely as a *coup de grâce.*

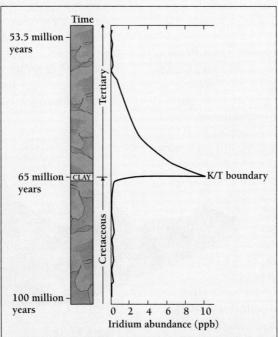

Figure A

An asteroid impact? At the end of the Cretaceous, the amount of iridium present in deposits increases dramatically from 1 part per billion (ppb) to 10 ppb, then gradually decreases again over a period of about 10 million years. Iridium is a metal rare on Earth but common in asteroids and meteors.

new organisms might flourish are where a mass extinction has eliminated much of the competition.

All species go extinct eventually. The rate of extinctions since the Cambrian has been relatively constant except for five mass extinctions. Evolutionary biologists have yet to agree on what caused these mass extinctions.

How Fast Can Species Evolve?

In 1859, Charles Darwin was nearly alone in his belief that evolution was a gradual process. He held that not only individual species but also genera, orders, and other higher taxa all arose through the slow transformation of species. His closest friends tried to persuade him that evolution might very well occur in jumps, or **saltations.** On the eve of the publication of the *Origin of Species,* in 1859, Darwin's friend Thomas H. Huxley wrote to him, "You have loaded yourself with an unnecessary difficulty in adopting *natura non facit saltum* [nature makes no jumps] so unreservedly."

The argument over the exact pace of evolution was one that would continue for 150 years. Even today, biologists argue about how fast evolution can happen and also how fast it normally happens. This conflict has sometimes led people to say that biologists do not agree about whether species evolved, which is not the case. Biologists frequently argue about how fast evolution happens and also in what ways it can occur, but not whether it occurs.

For Darwin, the conflict was between what he saw everywhere around him in nature and what the fossil record showed. As we noted in Chapter 15, gaps in the fossil record pose a problem in understanding the evolution of species. To demonstrate that a species evolved gradually, we would like to see a fossil representing each small step in an evolutionary line, much like those in the fossil records for horses and humans. But for most organisms in the fossil record, intermediate forms are disappointingly unusual.

The lack of intermediate forms may merely represent the incomplete preservation of ancient organisms, as Darwin argued. Just as most individual human beings are not named in the pages of recorded history, so most species are not preserved in the layers of rock that make up the fossil record. However, another interpretation for the gaps in the fossil record is that species change suddenly, evolving in short bursts, or saltations. The intermediate forms are not preserved, either because they never existed or because they existed only momentarily in geologic time (less than 10,000 years).

The gap-filled fossil record convinced many 19th-century biologists that evolution must occur by fits and starts, a view known as **saltationism.** After the rediscovery of Mendel's work in 1900, Mendelians argued strenuously that species arose abruptly from one generation to the next because of so-called macromutations—mutations that caused large changes in individual organisms. By the 1940s, however, Darwin's gradualism was finally in vogue and saltationism was out. By the 1960s, most biologists subscribed to **gradualism**—the view that most species evolved gradually and continuously in direct response to selective pressures from the environment.

A few renegades held out for saltationism. The most famous of these was the German-American biologist Richard Goldschmidt (1878–1958). He argued that even though macromutations, such as chromosome rearrangements, usually result in wildly deformed organisms, some of these monstrosities have potential if they happen to arise in the right environment. He called these potential species **hopeful monsters.** "A fish undergoing a mutation which made for a distortion of the skull carrying both eyes to one side of the body is a monster," he argued. "The same mutant in a much compressed form of fish living near the bottom of the sea produced a hopeful monster, as it enabled the species to take to the life upon the sandy bottom of the ocean, as exemplified by the flounders" (Figure 17-18). Goldschmidt's term "hopeful monsters" was both memorable and unfortunate. Evolutionary biologist Ernst Mayr's equally memorable term of derision for Goldschmidt's idea—"hopeless monsters"—was quickly picked up by other biologists.

Nevertheless, by 1950, the fossil record, which Darwin had found so imperfect, had improved little. A hundred years of research still revealed few cases with all the intermediate stages in the evolution of a lineage. When new species appeared in the fossil record, they did so suddenly, the intermediate stages leaving no trace. And many groups hardly seemed to change at all over long periods of geologic time. For exam-

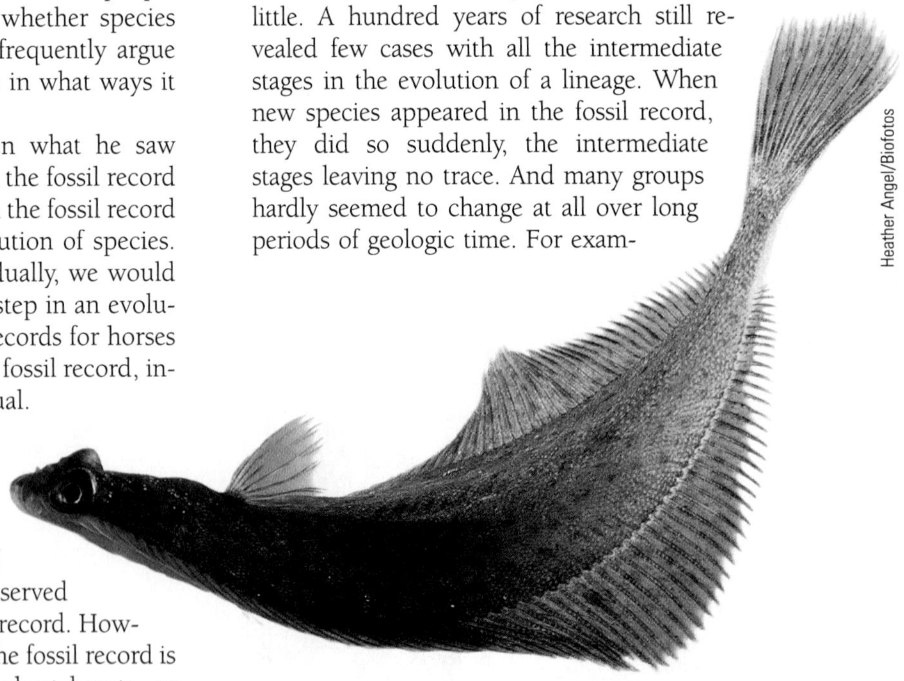

Heather Angel/Biofotos

Figure 17-18

Hopeful monster? The flounder lies on the bottom of the sea on its side. To compensate for this position, it has evolved so that its eyes are both on one side of its body (the top side). The flounder begins its life like any other fish, with one eye on each side of its head. As the embryonic fish develops, however, one eye migrates over the top of the head to join the other eye. As an adult, the fish lies on its side in the sand and the mud at the bottom of the ocean and looks up.

Extreme Biology   Modern Tropical Rain Forests: Mass Extinction or Not?

The fossil record shows five natural mass extinctions during the past 600 million years. Many biologists believe that a sixth, human-induced mass extinction is now under way.

The most species-rich regions of the world are tropical environments, such as the Amazon Basin. For any group of organisms—whether birds, insects, trees, or fungi—the number of species is greatest in the tropics. Although tropical rain forests cover less than 7 percent of the world's land area, one-half to two-thirds of the Earth's 10 million to 100 million species inhabit these rich forests.

Few of these species are known to science—only 1.7 million of the world's species have been named. A disproportionate number of these are European and North American species, where most trained taxonomists have lived and worked. Researchers are struggling to keep ahead of the chainsaws and forest fires that daily destroy the habitats of thousands of tropical forest species, half of which are insects. Tropical rain forests are dense with species. A 1-hectare (2.5-acre) plot of Brazilian forest may contain 200 species of trees, whereas the same size plot in Canada would support fewer than 10. And each species of tree may support 5 to 10 species of insects that specialize only on that one species of tree. Eliminating that tree species causes all the insect species to vanish also. Such cascades of extinctions are not limited to plant-insect interactions. When peccaries (a piglike mammal) were eliminated in one Brazilian forest, so were three species of frogs. All the frogs, it seemed, lived in the mud wallows the peccaries created.

As of today, fewer than one-half of the original tropical forests remain. And logging, agriculture, cattle ranching, and firewood harvesting continue to degrade and destroy tropical forests (Figure A). In some cases, regions of burning forests are so vast that

Earth Scenes/© 1996 Dr. Nigel Smith

Figure A
Slash-and-burn agriculture. In Brazil, farmers burn small areas of tropical rain forest. They farm for a while and then move on to burn another plot of land as young trees grow up in the last plot. As human population size increases, however, rain forests are being burned faster than they are growing back.

the fires are visible from space. Harvard biologist E. O. Wilson estimates that if tropical forests continue to be destroyed at current rates, fewer than 10 percent of the original forests will remain by the year 2050, and between 2.5 and 25 million species will go extinct.

How does this loss of species compare with the five prehistoric mass extinctions? By examining the fossil record, scientists estimate that the normal, or background, extinction rate may be as high as one extinction per year. In a mass extinction, the number of extinctions increases to at least 2.5 times the background rate. Since 1600, more than 1,000 species of plants and ani-

mals have gone extinct—yielding a minimum extinction rate of more than 2.5 species per year (Figure B).

However, there were probably at least 10 times as many extinctions during that time period. Recall that scientists have, at best, named only about one-tenth of the 10 million to 100 million species thought to exist. Just as the named species are a small fraction of the total species, the tally of documented extinctions includes a similarly small fraction of the total number of extinctions. For the 400 years since 1600, the minimum extinction rate may well approach 25 species per year. In addition, as the remaining tropical forests grow smaller, species are lost at a faster rate. The current loss of tropical forest species is comparable in magnitude to the prehistoric mass extinctions.

Although evolution will continue, the world will not recover its former biological diversity in our lifetime—or in our great-great-great-grandchildren's lifetime.

If we can look to the fossil record for an answer, it may be 5 million years before the Earth once more supports the diversity of life that existed when Columbus sailed to the New World.

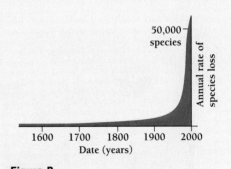

50,000 species

Annual rate of species loss

1600   1700   1800   1900   2000
Date (years)

Figure B
The rate of species loss has increased precipitously in the last 100 years.

---

ple, garfishes and sturgeons show only minimal morphological change over 80 million years of "evolutionary change." Horseshoe crabs have barely changed at all in 230 million years. Such stability over long periods is called **stasis**.

Clearly, some kinds of organisms have not had to change. But the periods of apparent stasis in the fossil record may not be what they seem. Ironically, paleontologists often ignore intermediate forms. Faced with a series of fossils, paleontologists tend to classify individual fossils in a layer as one species or another, not as "something in between." Like Darwin before them, they may have trouble making distinctions between two organisms, but they have to call them something. And classifiers normally say that two fossils are either the same species or two different species.

An intensive study of 15,000 trilobite fossils from Wales provided a good example of what happens when a scientist looks instead for the in-between species. The trilobites, a massively successful group of arthropods that flourished through the Permian, showed so many intermediate forms that classifying individual fossils into species became an exercise in arbitrariness.

Nonetheless, even if we accept that species may not actually evolve quite as abruptly as paleontology suggests, we can still try to explain why speciation should seem sudden at all. Species' abrupt appearances suggest two alternatives. In one scenario, new species appear almost instantaneously, as, for example, in a single generation through saltations. In the other scenario, new species evolve quite rapidly—by geologic standards—over just millions or even mere thousands of years. In either case, the intermediate forms would not appear in the fossil record. The Mendelians and Richard Goldschmidt had hypothesized that new species appeared almost instantaneously.

But the second idea—rapid, but not instantaneous, speciation—is entirely consistent with what population genetics had suggested about speciation in small, isolated populations. The bottleneck effect, allopatric speciation, and even sympatric speciation are all consistent with the idea of rapid speciation in small, isolated populations. Ernst Mayr himself had suggested this idea in 1942. In the case of rapid speciation, intermediate forms would exist in such small numbers and for such a short time geologically that the chances of their being preserved in the fossil record would be almost nil.

Until the 1970s, biologists paid little attention to the connection between rapid speciation and the gaps in the fossil record. Most biologists were interested in microevolution and population studies, not macroevolution. And paleontologists, interested primarily in macroevolution, wondered how biologists could go on believing in gradualism when there were so few examples of it in the fossil record.

In the 1970s, two paleontologists—Niles Eldredge at the University of California, Berkeley, and Stephen J. Gould at Harvard—proposed that the fossil record is exactly what we would expect it to be if the formation of new species was very rapid compared to the accumulation of changes within a species.

This view, called **punctuated equilibrium,** proposes two main ideas. First, *species change very little most of the time* (stasis). Second, *most anatomical or other evolutionary change in individual species occurs during a geologically brief period at the time of speciation.* In this view, speciation events punctuate an otherwise stable equilibrium, and the most important events in evolution are those that lead to reproductive isolation. Discontinuities (even catastrophes) are paramount, and gradual changes within species are mere fine-tuning.

But punctuated equilibrium does not contradict the findings of population geneticists or Darwin's view that evolutionary change occurs gradually. The brief periods during which populations can probably speciate—5,000 to 50,000 years—are invisible in the fossil record (and therefore "instantaneous"), but these brief periods in evolutionary history provide ample time for *gradual* change to occur.

How fast can species form? At the extreme end, plants can speciate in one generation through polyploidy. Several species of fish have formed in a lake near Lake Victoria, in Africa, in just 4,000 years. And Lake Victoria itself gave rise to 300 species of fish in only 200,000 years. One group of researchers has even suggested that the lake had dried out 12,000 years ago, which would mean that the fish evolved since then, a blink of the eye in geologic time.

Punctuated equilibrium and gradualism are extreme ends of a continuum of possible scenarios (Figure 17-19). The two ideas are not truly alternative theories that can be tested against one another. It's just a question of how much

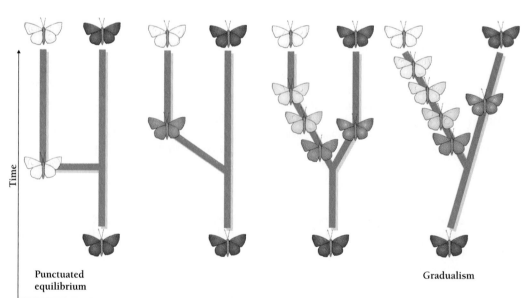

Punctuated
equilibrium

Gradualism

Figure 17-19
Is evolution gradual or abrupt?
Evolutionary changes in traits can occur over relatively brief periods of time (a few hundred to a few thousand years) or over longer periods of time. Different lineages may evolve at different rates. At left, the yellow butterfly appears relatively abruptly. The other phylogenies suggest increasingly gradual speciation. Punctuated equilibrium and gradualism are simply the ends of a continuum of possible evolutionary rates.

change accumulates how fast. No biologist believes that *all* evolutionary change occurs only gradually over millions of years. And few biologists would argue that all speciation occurs suddenly in just a few generations. Real species can most likely evolve rapidly or slowly, depending on circumstances such as how much genetic variation they have and what their environment is like. Whether *most* species have formed in relatively brief time spans, of 5,000 to 10,000 years, or in longer time spans, say 75,000 to 100,000 years, is a question we may never answer. The fossil record is mute on this point.

The gaps in the fossil record between an ancestral species and its descendant species represent periods of rapid evolution that are invisible in the fossil record. However, such rapid evolution—on the order of 5,000 to 10,000 years—is consistent with Darwin's ideas about gradual evolution.

What Is Phylogeny?

Phylogeny is the history of descent of groups of organisms from their common ancestors. Phylogeny includes the order in which different groups branch off and the time in history when branching occurred. Phylogeny is very like what we humans called "genealogy," except that phylogeny applies to all living organisms instead of just one human family.

Biologists classify species according to their phylogeny, how they are related. Groups of related species are put into

a genus, groups of related genera are put into families, and so on. (In the same way, we may classify our cousins by which side of the family they're from.)

But species don't always leave a clear record of who's descended from whom, which makes biologists' job difficult. When the fossil record is poor, dozens of large groups of species may appear to emerge all at once. And DNA studies suggest that rapid speciation leaves an equally muddy record in the genes.

Yet, we know that these species did evolve and that there is one true phylogeny. We just don't always know what it is. And sometimes we know what it is, but it's not what we used to think. As we'll see in Chapter 19, biologists now know that crocodiles are more closely related to birds than to lizards. Despite this knowledge, however, birds and reptiles are two separate classes. And the crocodiles are tucked in with the lizards. Even though we still classify crocodiles this way, we know better.

17.6 How Did Humans Evolve?

When we visit the zoo, we are often drawn to the animals that most resemble ourselves—the monkeys and apes. Taxonomists say that apes, monkeys, and lemurs all belong to the same order of mammals. This group, called the **primates,** are remarkably unspecialized. Like the earliest mammals, we have five digits on each hand or foot and unspecialized teeth. All primates lack, for example, the sharp cutting teeth and broad grinding surfaces that characterize the teeth of horses, rabbits, and other herbivores. We also lack the long canines and shearing teeth of carnivores such as dogs.

Nonetheless, despite our lack of specialization, we primates have a set of common characteristics that together distinguish us from other mammals. These include large brains and binocular vision, flexible shoulder joints, and hands

A.

B.

Tom & Pat Leeson/Photo Researchers, Inc.

Glenn Vanstrum/Animals Animals

Figure 17-20
The two major classes of anthropoid monkeys and apes. A. The monkeys of South and Central America such as this howler monkey have nostrils that point sideways. B. Old World monkeys and apes, from which both humans and this African green monkey have descended, appeared later and have nostrils that point downward.

and feet with five grasping digits, a grasping thumb, and flat fingernails rather than claws.

Most of the 166 species of living primates are tree dwellers. Our flexible shoulder joints and grasping hands and feet enable us to climb trees and swing from one branch to another. As a group, primates are basically herbivores that live on leaves and fruits. Nonetheless, many primates are opportunistic omnivores who will occasionally eat grains, grubs, and even flesh. Most 21st-century humans eat grains and flesh regularly.

How Do We Classify the Apes?

Primates are one of the oldest orders of mammals, dating back some 55 million years. Early primates had smaller bodies and longer snouts than modern monkeys and apes. The Anthropoid primates—all the monkeys and apes—consist of two large groups whose first representative appeared about 42 million years ago, in China. The monkeys of South and Central America first appeared about 26 million years ago and have flat noses with widely separated nostrils and often prehensile (grasping) tails (Figure 17-20A).

A newer lineage of both monkeys and apes appeared in East Africa 20 to 17 million years ago and spread throughout Africa and Asia. These newer monkeys and apes have downward-pointing nostrils and no prehensile tail (Figure 17-20B). Taxonomists group them into the "Old World" monkeys and the apes, or **hominoids** [Latin, *homo* = human]. Hominoids include gibbons, orangutans, gorillas, chimpanzees, and humans. All have large brains and long arms, and they tend to walk at least partially erect (Figure 17-21B).

Exactly how all the different apes are related to each other and to humans is a subject of intense interest (Figure 17-22). Gibbons apparently diverged early in the hominoid lineage, and are sometimes called "lesser apes" (Figure 17-21A). But the Great Apes—the orangutans, gorillas, chimpanzees, and humans—are all relatively closely related (figure 17-21B). For example, a comparison of a 10,000-nucleotide DNA sequence from the X chromosomes of a human and a common chimpanzee showed that nearly 99 percent of nucleotides were identical in the two species.

If it were not for humans' tendency to distance ourselves from other animals, we would be in the same family as the other apes, the gorillas, chimpanzees, and orangutans, which together are called the **Pongidae.** But, instead, humans have given themselves a separate family, the **Hominidae,** informally called the hominids. Fossil and molecular data suggest that the common ancestors of orangutans diverged from those of other pongids and hominids about 12 million years ago, and the hominids probably diverged from the chimpanzees and bonobos 5 to 7 million years ago. The oldest fossil hominids are about 6 million years. Anthropologists are still debating whether a recently unearthed 7-million-year-old skull is a hominid or just another ape.

The species name for modern humans is, of course, *Homo sapiens*. But although *H. sapiens* is the only hominid alive today, it is not the only hominid. We count as relatives a num-

A.   B.

Gerard Lacz/Peter Arnold, Inc.

M. Long/Visuals Unlimited

Figure 17-21
Two hominoids. The hominoids include the gibbons (lesser apes) and the more evolved orangutans, gorillas, chimpanzees, bonobos and humans (the great apes). A. White-handed gibbon. B. Bonobo and her baby. Notice how she stands upright.

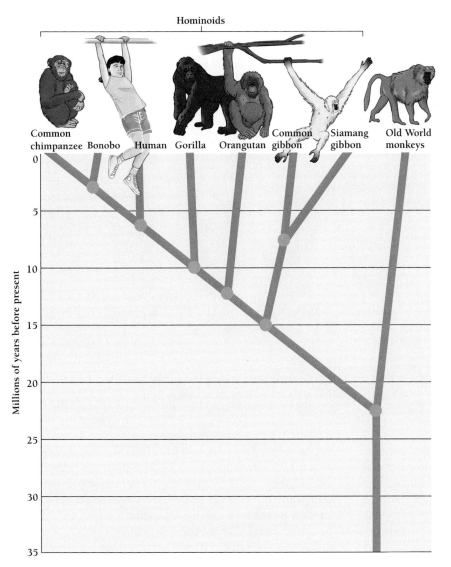

Hominoids

Common chimpanzee · Bonobo · Human · Gorilla · Orangutan · Common gibbon · Siamang gibbon · Old World monkeys

Millions of years before present

Figure 17-22
Phylogeny of the hominoids. By comparing matching sequences of DNA from different species, biologists can estimate the time in years since two species diverged.

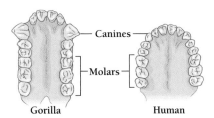

Canines
Molars
Gorilla
Human

Figure 17-23
Round jaws. Several characteristics distinguish hom*inids* from other hom*inoids*. Hominids tend to have more rounded jaws than chimps and gorillas. In addition, because hominids walk upright, we have arched rather than straight spines.

ber of different species of hominids. Two characteristics distinguish hominids from other apes: (1) hominids are **bipedal,** that is, they consistently walk on two rather than four feet, and (2) hominids have rounded rather than rectangular jaws (Figure 17-23).

All hominid species except ours are now extinct, but anthropologists have discovered the buried bones of more than a dozen species of hominids throughout Africa, Asia, and Europe. One of the oldest hominids is *Australopithecus anamensis,* which had a small brain, a squarish jaw, and long apelike arms. But it walked upright.

How are we different from *Australopithecus?* Later hominids such as ourselves have large brain cases. In addition, we are adapted for an upright posture and for distance running, much like dogs and horses (Figure 17-24). We modern humans combine extremely sparse body hair (compared to other modern apes) with an extravagant growth of hair on the head and, in the case of males, on the face as well. When illustrating hominids, artists must decide whether to draw our ancestors naked like us, or hairy like pongids. Most artists compromise and draw our ancestors as semihairy, but with short hair on the head. In reality, the degree to which other species of *Homo* shared our strange distribution of hair will probably never be known, as hair is rarely fossilized.

All the apes together are called hominoids. Humans and their extinct relatives are hominids, while chimpanzees, gorillas, and orangutans are pongids.

Why Did *Australopithecus* Stand Up?

The question of how walking hominids evolved has haunted anthropologists for more than a century. Some researchers have argued that the evolution of an upright stance and a hairless body was an adaptation for remaining cool on a hot African savanna millions of years ago. The idea is that a person (or ape) standing up exposes less surface area to the sun. Other researchers have suggested that the first hominids stood up because they could run faster that way—either from predators, or, alternatively, in pursuit of prey. Still oth-

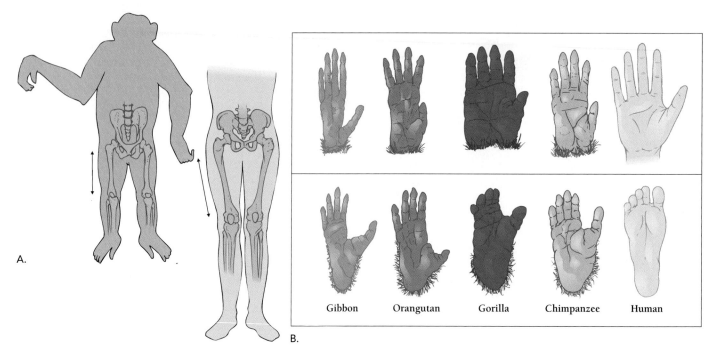

Figure 17-24

Tools for walking. A. Humans and ancestral hominids evolved the long lower limbs and reduced toes that characterize dogs, cheetahs, horses, and other animals that walk or run. Our legs attach to our hips differently from those of chimps, making us relatively "knock-kneed." B. Our toes are short and our big toe is almost useless for grasping. Our hands have evolved less and more closely resemble those of other hominoids.

ers have argued that hominids stood up so that they could carry and wield weapons and other tools. Some now argue that hominids became walking apes when they began scavenging over large territories. Most of these ideas lack supporting evidence. The only certainty is that about 4 million years ago, some apes evolved an upright stance and the long legs of a long-distance walker.

Nearly all of the hypotheses about why hominids evolved an upright stance depend on the idea that hominids left the forest to live in an open grassland, or savanna. Researchers have theorized for many years that the dry climate caused African forests to turn into savanna. Our herbivorous, tree-dwelling forebears then abandoned both the remaining forest and their vegetarian habits and began hunting other animals out on the savanna, or so goes this widely accepted story. Out on the savanna, hominids evolved an erect posture, stone tools, large brains, and hairless skin. This story has been repeated so many times and with such assurance that most nonscientists are under the impression that it is an accepted scientific fact.

Unfortunately, it is just a story. Studies of the dust in ocean sediments have shown only a single *dramatic* change in climate in Africa in the last 5 million years. This dry period came about 2.8 million years ago, more than a million years too late to account for the evolution of the first known bipedal hominid, 4.2 million years ago.

The idea that standing up and losing the fur were adaptations to the heat of the savanna likewise has little support, as most other mammals in Africa walk and run on four legs and have fur. The ostrich, which is bipedal and runs well, has a broad, horizontal, black back, well suited to absorbing heat. It seems unlikely that the ostrich's bipedal gait is an adaptation for staying cool. Excluding naked mole rats, which live underground, the few African mammals that lack fur are immense herbivores—elephants, rhinoceroses, and hippopotamuses, for example.

Did the first hominids stand up in order to carry something? Possibly, but that something was emphatically not stone weapons. The earliest stone tools, which experts say were most likely used to chop and mash vegetables (not to kill zebras), do not appear in the fossil record until about 2.5 million years ago, long after our ancestors stood up. Absent stone tools, it's difficult to imagine our australopithecine ancestors fashioning any other tools or objects important enough to confer a selective advantage on an upright posture. Some possibilities are wooden tools or woven baskets, which usually rot away and so do not appear in the paleontological record.

Another object an ape or australopithecine might have found worth carrying around is a baby. The babies of other monkeys and apes cling to their mother's fur, however. Human babies are born relatively undeveloped compared to other primate infants and are unable to cling. Biologists hypothesize that this is because our heads are so big that if babies waited any longer in the womb, they could not pass through the mother's pelvic bones. But our large heads apparently evolved *after* our upright stance. So researchers would have to hypothesize that hominids lost their fur be-

Figure 17-25

Australopithecus afarensis. One of the earliest australopithecines, *A. afarensis* is represented both in the nearly complete skeleton of "Lucy"—discovered in Ethiopia in the 1970s—and in other, more fragmented, fossils from Tanzania. Evidence that this australopithecine was bipedal comes not only from the pelvis but from the extraordinary footprints found by Mary Leakey and her colleagues. A. Reconstruction of the *A. afarensis* fossil "Lucy." B. Footprints of *A. afarensis*.

A.

B.

fore they evolved an upright posture, an idea for which there is no evidence.

With luck, researchers may someday agree about what selective pressures caused hominids to evolve an upright stance. For the moment, we can only say with certainty that they did. From the earliest australopithecines to the later ones, the skeleton increasingly became specialized for walking and running while upright.

The traits that most set humans apart from other apes are our large brains, bipedalism, and curious distribution of fur. No hypothesis that is consistent with the anthropological data has accounted for the evolution of any of these traits, let alone all of them.

Who Were the Earliest Known Hominids?

The australopithecines fall into three to six distinct species, which lived at slightly different but overlapping times. All of them come from Africa, south of the Sahara desert. In fact, all hominids older than about 2 million years come from Africa, and most anthropologists agree that the first homi-nids evolved in Africa. The different species of australopithecines appear to have lived between 4.2 and 2.5 million years ago. For example, *A. afarensis* has turned up in Chad and in East Africa, where a nearly complete skeleton was unearthed (Figure 17-25). This specimen of *A. afarensis*, named "Lucy" by her discoverer, was small, only about 1.1 meters (3 feet 7 inches) tall, and weighed less than 30 kg (about 60 pounds). Lucy's adult body was the size of a 9-year-old *Homo sapiens*, but her brain was far smaller. Brain size for most normal adult humans ranges from 1,000 to 1,600 cubic centimeters (cm^3) (although the full range is 900 to 2,000 cm^3). In contrast, the brain of *A. afarensis* was about 400 cm^3, about the size of a chimpanzee's brain.

Other australopithecine species, taller and with larger brains, appeared during the next 2 million years. But the first fossils to be recognized as members of the same genus as ourselves—*Homo*—date from 2 to 3 million years ago. These hominids all used crude stone tools, which is one measure that sets them apart from the australopithecines. Members of the genus *Homo* include *H. ergaster, H. habilis, H. erectus, H. heidelbergensis,* and *H. sapiens*.

Homo habilis [Latin, *habilis* = able], the "handy man," used stone tools and also apparently used them to butcher large animals, which *H. habilis* either hunted or scavenged.

Extreme Biology What Is the Nature of Anthropological Evidence?

The kinds of evidence that molecular biologists use when studying living organisms and their relationships to one another are different from the evidence that paleontologists and anthropologists must use. The essence of most scientific evidence is reproducibility, and studies of living organisms tend to yield reproducible results, provided the studies are done carefully. For example, if a molecular biologist finds that chimpanzees share a certain proportion of the same alleles with humans on Tuesday in Los Angeles, a different researcher, using the same material and methods, should be able to get the same result on Friday in Boston.

Anthropologists, however, have limited material with which to work, and much of their evidence is not reproducible. For example, the number of fossils of early apes

and humans is minuscule. An entire species of ape may be represented by a single jawbone or a few teeth.

One of the most complete skeletons of an early hominid is "Lucy," a 3.2-million-year-old specimen of *Australopithecus afarensis.* Researchers strongly suspect that Lucy was female. Based on that assumption, they have hypothesized what a male of the same species would look like. They have further assumed that the male would be larger by a certain percentage than Lucy, and some textbooks report the height and weight of this theoretical male as if it were a fact. But other researchers have argued that there is no certainty that Lucy is female. Maybe "she" was a male. Further, without hundreds of complete skeletons of the same species, there is no way to know if Lucy was in any way typical of males *or* fe-

males. Lucy may have been unusually tall or unusually short.

Nonetheless, the scarcity of evidence in no way detracts from the fact that a long line of species has existed and that those species are morphologically intermediate between modern apes and modern humans. We may never know either the true average dimensions of skulls and limbs of each species or whether males and females were mostly the same size or different sizes. Nor may we ever know with certainty how each ancient hominid species is related to any other. But we do know that we are closely related genetically to modern chimpanzees and other apes. And we do know that the 3.2-million-year-old fossil Lucy walked upright, just as we do. There may be only one Lucy, but Lucy lived.

H. habilis stood about 1.7 m (5 ft) tall and had a 590- to 700-cm³ brain, considerably larger than that of *A. afarensis.* Researchers vigorously dispute which of the various australopithecines and hominids are our true ancestors (Figure 17-26).

The early hominids, the australopithecines, were short and small brained compared to modern humans and other members of the genus *Homo.*

Where Did We Come From?

Nearly 2 million years ago, hominids left Africa and migrated to Asia. Fossils of *Homo erectus* have been found not only in Africa, but also from Java, in eastern Asia, to northern China, so it's reasonable to think that the *H. erectus* in Asia migrated out of Africa. *Homo erectus* was by far the longest-lived species of *Homo,* with a total species life span of 1.7 million years, nearly 17 times as long as our own species has lasted so far. *Homo erectus* skeletons are larger than those of *H. habilis,* and the brain case is also proportionately larger, ranging to more than 1,000 cubic centimeters (cm³) in the earliest specimens. In Java, where *H. erectus* lived until just 100,000 years ago, the brain cases are as large as 1,300 cm³, about average for a modern human (1,000 to 1,600 cm³).

Some anthropologists and paleontologists have suggested that *H. erectus* migrated from Africa, starting about

1.5 million years ago, with subsequent waves of migration evolving separately into varieties, or races, that differ from one another. Some of these, they argue, evolved into *H. sapiens.*

Others argue that *H. erectus* living in different parts of the world were replaced first by *Homo heidelbergensis* and later by *H. sapiens,* each of which migrated separately from Africa. In this view, *H. erectus* was a side branch in the evolution of humans, and we are descended from another line of *Homo* that left Africa after *H. erectus.* In that case, the most likely candidate ancestor is *H. heidelbergensis,* found in Ethiopia and dated at about 130,000 years old. Several skulls that seem transitional between *H. heidelbergensis* and *H. sapiens* have been found in Ethiopia.

A last alternative is that *H. sapiens* evolved in Africa 200,000 to 100,000 years ago, separately and at the same time as *H. heidelbergensis,* then emigrated to Asia and Europe. The earliest known representatives of *H. sapiens* were found at two sites in Israel and are dated at about 90,000 years old.

The early *H. sapiens* are commonly divided into two kinds, the **Cro-Magnons** and the **Neanderthals.** The Neanderthals, who looked much like us, but with a heavier build, evolved between 200,000 and 100,000 years ago and occupied parts of Europe until just 35,000 years ago. Their thick arms and legs and heavy brows have provided cartoonists with the classic "caveman" image. Because of differences in Neanderthal throat bones, some researchers doubt that they spoke as well as we do, if at all. But Neanderthals probably had a fair amount

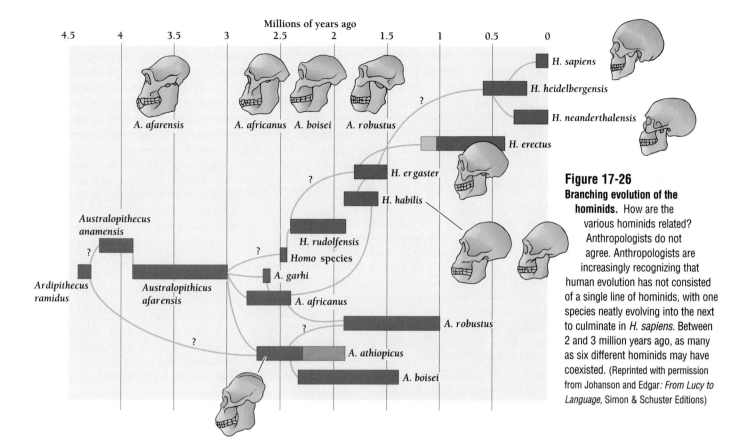

Figure 17-26
Branching evolution of the hominids. How are the various hominids related? Anthropologists do not agree. Anthropologists are increasingly recognizing that human evolution has not consisted of a single line of hominids, with one species neatly evolving into the next to culminate in *H. sapiens.* Between 2 and 3 million years ago, as many as six different hominids may have coexisted. (Reprinted with permission from Johanson and Edgar: *From Lucy to Language,* Simon & Schuster Editions)

of mental horsepower. Their brains were as large or larger than those of the Cro-Magnons, and some Neanderthals may have played musical instruments (Figure 17-27).

Researchers do not know why the Neanderthals became extinct or if even if they went extinct. They may have interbred with Cro-Magnons. At one site, in Israel, the Cro-Magnons and Neanderthals appear to have lived within a few miles of each other at the same time. If we knew that they interbred, we could apply the biological species concept and conclude that they were one species. But no one knows, and researchers do not agree whether the Neanderthals were a separate species *(H. neanderthalensis),* a subspecies, or just a now-extinct race of *H. sapiens.* Recent analysis of DNA extracted from a Neanderthal bone suggested that they were genetically distinct from modern humans. But we do not know if they were equally different from our Cro-Magnon forebears.

Even if we ultimately conclude that the Neanderthals were *H. sapiens,* we probably had other *Homo* cousins in the early days of our existence. Both *H. heidelbergensis* and *H. erectus* lived up to 100,000 years ago. What happened to them? No one knows, but anthropologists strongly suspect that *H. sapiens* exterminated them.

Anthropologists do not agree whether *Homo sapiens* evolved in Africa. In the last 2 million years, several species of *Homo* appear to have lived on Earth together.

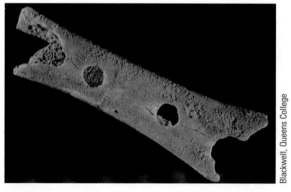

Blackwell, Queens College

Figure 17-27
Neanderthal flute? In 1996, a team of researchers discovered what appeared to be a bone flute in a 43,000-year-old Neanderthal cave in Slovenia, in eastern Europe. Apparently fashioned from the thigh bone of a young bear, the hollow bone has four holes spaced as if for a diatonic scale (do, re, mi . . .). Similar flutes have been found in newer, Cro-Magnon caves, and researchers around the world reacted with delight to the idea that Neanderthals might have played music. Some researchers, however, flatly reject the idea. They argue that the holes were made by wolves' canine teeth and that the spacing is purely accidental.

In the first two chapters of Part III, we reviewed the evidence for evolution and studied the ways that changes in gene frequency allow evolution to occur. In this chapter we discussed how species change and multiply and examined some of the larger patterns of evolution that are evident in the fossil record. In the next chapter, the last in Part III, we will explore a topic that is related, but in many ways quite different—the question of how and where life first originated. The evolution of life is, like gravity, a scientific fact. The origin of life, however, is truly mysterious. We know that life appeared on Earth quickly, within a few million years of the cooling of the Earth's crust. But we can only surmise how it got there.

Key Concepts

- A species is a population of organisms whose members can interbreed to produce viable offspring but that usually do not interbreed with members of other such populations.
- A population may form a separate species if it becomes reproductively isolated for a long enough period of time.
- Adaptive radiation is the generation of diverse new species or other taxa from one or more ancestral species.
- Species probably evolve at different rates at different times.

Summary with Key Terms

What is a species?

Macroevolution is the origin and multiplication of species. According to the commonly accepted **biological species concept**, a biological species is a group of organisms that actually or potentially interbreed to produce viable offspring but cannot interbreed with members of other such populations. Populations that do not interbreed with other populations are said to be **reproductively isolated**. **Races** or **subspecies** are subpopulations of a species that are distinct yet not reproductively isolated. In other words, races can interbreed with other races of the same species.

How do species form?

Populations become reproductively isolated from one another through two kinds of barriers to gene flow: **prezygotic barriers**, which prevent fertilization, and **postzygotic barriers**, which either make the zygote inviable or make the resulting adult sterile. Prezygotic barriers include **ecological, temporal, behavioral, mechanical,** and **gametic isolation**. Once fertilization occurs, postzygotic barriers can come into play. These include chromosomal and genetic incompatibilities that cause the hybrid zygote to die young (**hybrid inviability**); or, if the individual reaches maturity, to produce no gametes (**hybrid sterility**); or, in some cases, if the hybrid produces offspring, to have *its* offspring be weak or sterile (**hybrid breakdown**).

Two other postzygotic barriers, peculiar mostly to plants, reproductively isolate hybrid offspring from the parent population in just one generation. These include the multiplication of chromosomes, or **polyploidy**.

How do species multiply over time?

Geography frequently isolates populations from one another well before other isolating mechanisms develop. Populations can become isolated through **dispersal** and through **vicariance**. When reproductively isolated populations diverge in morphology or behavior, they are said to **speciate**. Geographic speciation follows several patterns. In **allopatric speciation**, geographic isolation halts nearly all gene flow. Genetic drift or different selective pressures complete the speciation process. In **sympatric speciation**, populations that occupy the same geographic area become reproductively isolated through barriers to gene flow—temporal, ecological, or other.

Phylogeny is the history of descent of lineages of organisms from a common ancestor. **Divergent evolution** is the separation of a lineage into two or more species or lineages. **Convergent evolution** is a separate idea, referring to the acquisition of similar characters in unrelated lineages. An important kind of divergent evolution is **adaptive radiation**, the generation of diverse new species from a single ancestral species. Because no new body plans evolved after the **Cambrian explosion**, some biologists have suggested that **developmental constraints** limit evolutionary pathways.

How fast can evolution occur?

Despite the gaps in the fossil record, Darwin believed in **gradualism**, the view that evolution was a gradual process. Early biologists rejected Darwin's gradualism in favor of **saltationism**, the view that evolution occurred in sudden leaps, or **saltations**. The last well-known saltationist suggested that genetic monstrosities, which he called **hopeful monsters,** began new evolutionary lineages.

Although gradualism gained favor from the 1930s through the 1960s, the fossil record seemed to show that many species remained unchanged for millions of years (**stasis**) and then abruptly gave rise to new species. A full record of intermediate stages was usually lacking. Thus, evolution, according to the fossil record, was neither gradual nor continuous. It seemed to happen in fits and starts.

In the 1970s, two paleontologists, Niles Eldredge and Stephen J. Gould, argued that the formation of new species is rapid compared to the rate of accumulation of changes within a species. This view, called **punctuated equilibrium,** rests on two main ideas. The first is that species change very little most of the time. The second is that most anatomical or other evolutionary change in individual species occurs during a geologically brief period at the time of speciation.

How do scientists think humans evolved?

Anatomical and molecular evidence suggests that humans are **primates** most closely related to chimpanzees, gorillas, and other pongids. The **Pongidae** are a family of **hominoids.** Humans and their ancestors all belong in the family **Hominidae.**

Hominids are **bipedal** and have rounded jaws. Anthropologists call the earliest hominids australopithecines. Later hominids also have large brains, arched backs, and reduced toes. No one knows exactly why hominids evolved a bipedal gait. Anthropologists also do not agree whether *H. sapiens* evolved in Africa, as did all or most other species of *Homo,* or if *H. sapiens* evolved from an earlier ancestor that evolved in Africa but then dispersed to Europe or Asia. In the last 2 million years, several species of *Homo* appear to have been living on Earth at the same time. The longest-lived species was *H. erectus,* which lived in parts of Africa and Asia for nearly 2 million years and became extinct only about 100,000 years ago. Early *H. sapiens* are divided into the heavy-boned **Neanderthals** and the more delicate **Cro-Magnons.** Modern humans more closely resemble the Cro-Magnons.

Review and Thought Questions

Review Questions

1. Define the word "species"; then explain why that definition is tricky. What is a subspecies? Why is a subspecies not a separate species?
2. What conditions are required for speciation?
3. Distinguish between prezygotic and postzygotic barriers to gene flow. Give several examples of each. Is polyploidy a prezygotic barrier or a postzygotic barrier?
4. Distinguish between sympatric and allopatric speciation.
5. What clues from the Galápagos Islands suggested to Darwin the idea of adaptive radiation?
6. Describe the Cambrian explosion. What happened 600 to 900 million years ago that has not happened since? Why hasn't it happened again?
7. Name a postzygotic isolating barrier in plants that is consistent with the saltationist view of evolution.

8. Define gradualism and punctuated equilibrium. What characteristics of the fossil record favor the theory of punctuated equilibrium? Explain why these two ideas are or are not in conflict with each other.
9. What are some weaknesses in the argument for stasis?

Thought Questions

10. Consider an oceanic archipelago consisting of only 10 islands. Describe a pattern of colonizations whereby 20 or more species could arise from a colonization by a single mainland species.
11. The current mass extinction will likely result in an adaptive radiation over several million years. What organisms do you think will survive and what niches will they evolve to fill 5 million years from now?

BiologyNow Resources

Biology ⑧ Now™

Active Figures

17-6: Temporal isolation
17-8: Allopatric speciation

Preparing for an exam? Take a diagnostic test on your BiologyNow CD-ROM.

Online materials relating to this chapter are at:

http://biology.brookscole.com/AAL3

About the Chapter-Opening Image

The Galápagos tortoises are a classic example of the adaptive radiations that lead to speciation, the evolution of new species. The tortoise shown here is a giant tortoise (*Geochelone elephantopus vandenburghi*).

How Did the First Organisms Evolve?

Key Questions

- Can living organisms arise from inanimate material?
- When did the first life appear on Earth?
- How could living cells developed from inanimate material?

Can Organisms Arise Spontaneously from Nonliving Matter?

Until the last few hundred years, most people believed that new life arose spontaneously and daily from nonliving materials. Anyone could see that flies and maggots arose from rotting meat, frogs from swamps, mice from moist grain. Nearly everyone believed that "spontaneous generation" of life was not only possible but common among the simple—so-called "lower"—plants and animals. Some people even argued that, under carefully controlled conditions, it might be possible to generate a man from a corpse.

Of course, every farmer knew that wheat and many other plants came from seeds, and that the cows and sheep were born of their mothers. Here and there, a few skeptics also insisted that insects and other creatures come from the eggs or sperm of parent organisms. But not until 1668 did anyone put spontaneous generation to the test.

The skeptic was Francesco Redi (1626–1697), an Italian physician, poet, and naturalist, and one of the first modern experimental biologists. Heavily influenced by Galileo's belief that the natural world could be understood best through the senses, Redi strove to perform controlled experiments to test his ideas. Among these was a series of experiments challenging the idea of spontaneous generation. In his own account of these experiments, Redi first states his hypotheses:

> I shall express my belief that the Earth, after having brought forth the first plants and animals at the beginning by order of the . . . Creator, has never since produced any kinds of plants or animals . . . and everything which we know in past or present times that she has produced came solely from the true seeds of the plants and animals themselves, which thus . . . preserve their species. And, although it be a matter of daily observation that infinite numbers of worms are produced in dead bodies and decayed plants, I feel . . . inclined to believe that . . . the putrefied matter in which they are found has no other office than that of serving as a place . . . where animals deposit their eggs at the breeding season, and in which they also find nourishment; otherwise, I assert that nothing is ever generated therein.

James Cotier/Getty Images

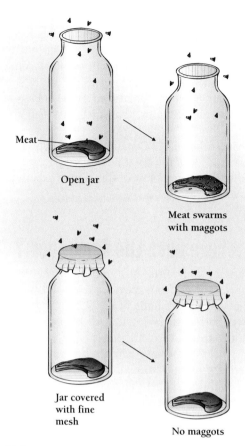

Figure 18-1
Francesco Redi's experiment. Do flies arise spontaneously from rotten meat? Or do they develop from eggs laid on the meat by adult flies?

Redi's 1668 experiments convinced most scientists that the common conception of spontaneous generation was incorrect. However, within 10 years, van Leeuwenhoek used his new microscope to find little "animalcules" (microorganisms) everywhere. Most people felt that even if maggots and flies could not arise by spontaneous generation, very likely these microorganisms could.

It was not until a hundred years later that anyone showed that they could not. In the 18th century, Lazzaro Spallanzani (1729–1799), an Italian physicist, microscopist, and physiologist, demonstrated that no organisms grew in a rich broth if it were first heated and allowed to cool in a stoppered flask (Figure 18-2). Other experimenters less careful than Spallanzani, however, contaminated their broth and so obtained the opposite result. Moreover, some scientists argued that, by heating the broth and air, Spallanzani may have destroyed some substance needed for spontaneous generation.

Redi then describes how, after preliminary studies of the development of different kinds of maggots into various species of flies, he put several kinds of fresh meat into two sets of jars. One set of jars was open, the other closed. Those that were open all developed maggots and, later, flies. Those that were closed developed no maggots, though, Redi notes, "Outside on the paper cover there was now and then a deposit, or a maggot that eagerly sought some crevice by which to enter and obtain nourishment."

Finally, Redi wanted to prove that the absence of air was not keeping the maggots from developing on the meat. He placed fresh meat into large vases that were either open or covered with very fine mesh. To prevent maggots from crawling through the holes in the netting, he put the covered vases inside a large net-covered frame and carefully removed any maggots that managed to penetrate the outer netting (Figure 18-1). The results were the same. "I never saw any worms in the meat, though many were to be seen moving about on the net-covered frame." The meat in the open vases was, of course, riddled with maggots, which, Redi observed, eventually turned into various kinds of flies. This experiment showed that maggots did not appear spontaneously.

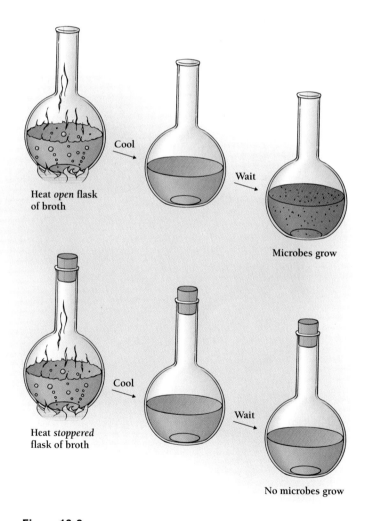

Figure 18-2
Lazzaro Spallanzani's experiment. Do microorganisms arise spontaneously in a rich broth?

For another hundred years, these objections stood. Finally, in 1860, the controversy over spontaneous generation became so intense that the prestigious French Academy of Sciences offered a prize to anyone who could resolve the matter. In 1864, Louis Pasteur, the founder of microbiology, designed a simple but irrefutable experiment that won the prize.

Pasteur repeated Spallanzani's experiment, but with one essential difference. Instead of stoppering the flask to prevent organisms or spores from falling into the broth, Pasteur used a flask with a long neck like that of a swan (Figure 18-3). Air could pass freely in and out of such a flask, but microbes could not get through the long, curved neck. Only when Pasteur broke the neck of the flask was the broth "seeded" by bacteria and fungi falling from the air. Some of Pasteur's open flasks—still sterile after more than 130 years—are on display in Paris.

"All life from life," Pasteur emphatically concluded. "Never," he said, "will the doctrine of spontaneous generation recover from the mortal blow of this simple experiment."

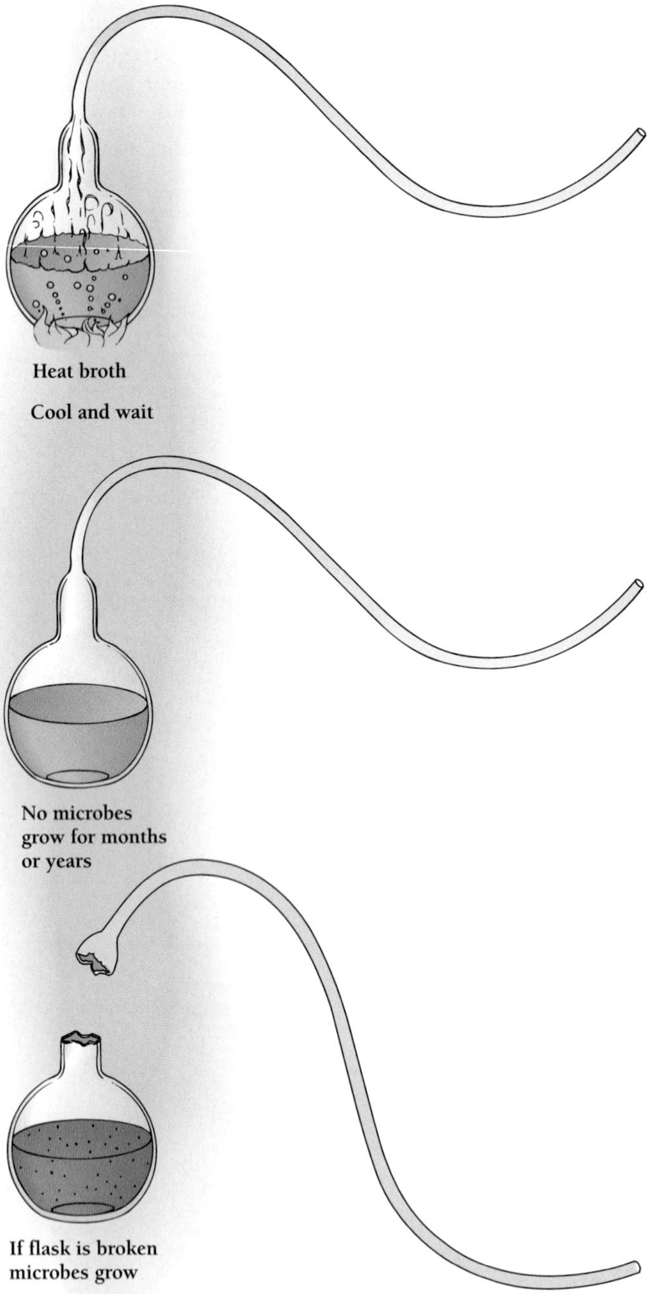

Heat broth

Cool and wait

No microbes grow for months or years

If flask is broken microbes grow

Figure 18-3
Louis Pasteur's experiment. Do microorganisms arise spontaneously in a rich broth provided that air is available?

18.1 Where Does Life Come From?

Since Darwin, biologists have known that all modern species have evolved from previous ones. But we must ask, From what did the millions of species on Earth evolve? If Pasteur was correct in saying that all life comes from other life, we must account for the presence of evolved life on a planet that itself came into being a finite number of years ago. Where did life first come from?

Every culture on Earth—ancient, primitive, or modern—has its own answer to this question. Many invoke an intelligent, creative force. However appealing many of these myths may be, they do not constitute scientific explanations, and they tell us little more than that "life happened."

Scientists cannot say with certainty how life arose on Earth, either. However, they can speculate—with a modern understanding of astronomy, planetary science, geophysics, and biology—about what might reasonably have happened. Scientists see essentially two alternative answers to the question, Where did life on Earth come from? Either life on Earth arose spontaneously from nonliving organic molecules in the primitive environment or else it arrived from outer space by means of a meteor, comet, or asteroid.

In this chapter, we will see why neither of these ideas is quite as outlandish as it might at first seem. The bulk of our discussion will be devoted to exploring how life might have originated on Earth. As one prominent biologist put it, this discussion resembles a mystery novel with too many suspects.

Evolutionary biologists have inferred part of the story from the shared characteristics of modern organisms. But much of the story must be guesswork. Because the origin of life is not an ongoing, testable process, as evolution is, biologists can only surmise how life might have arisen. The point is to show that life could have arisen by some series of steps, each of which is reasonable in terms of our current understanding of the laws of physics, chemistry, and biology. By showing that they can tell such a reasonable story, evolutionary biologists have attempted to refute the idea that there is an uncrossable gulf between living and nonliving matter. We will also see why, despite all this, Pasteur was after all correct: spontaneous generation could not happen today.

When and Where Did Life First Appear?

Before 1950, no one had seen a fossil older than about 570 million years (the beginning of the Cambrian Era). Older fossils went unnoticed, it turns out, because they are extremely rare, usually quite small, and found in a kind of rock different from that of later fossils.

Cambrian and later fossils are usually hard bones and shells preserved in shales, sandstones, limestones, and other sedimentary rocks. But most Precambrian organisms had few or no hard parts, and their soft remains usually dissolved or decayed. Their soft, tiny bodies persisted only when caught in an extremely fine-grained rock called chert. Such rocks are now generally far below the Earth's surface, covered by later deposits. Even cherts that have returned to the surface may be hopelessly deformed by heat and pressure. In only a few places on Earth are the cherts that bear Precambrian fossils both accessible and intact.

But for scientists interested in Precambrian life, finding fossil-bearing cherts was only half the battle. In order to see the fossils, paleontologists had to slice the chert into thin sections, polish them until they were transparent, and then examine the sections under a microscope.

The oldest and most primitive fossils known resemble modern prokaryotes—simple rods, spheres, and filaments with no nucleus (Figure 18-4). These ancient fossils, embedded in cherts from southern Africa and western Australia, are about 3.5 billion years old. More recently, researchers have dated fossil organisms that are even older, some 3.8 billion years.

Another important group of ancient fossils are the stromatolites, banded columns of limestone or chert, as old as 3 billion years (Figure 18-5). For many years, scientists did not realize that stromatolites were the remains of living organisms. This failure was because they consist of an unusually orderly arrangement of sediments, with no trace of the original organisms. But living stromatolites, discovered in a few remote intertidal areas, solved the mystery. Stromatolites are formed when huge, mounded colonies of cyanobacteria and other microorganisms trap fine sediment. The sediment settles in thin layers within each mound. The cyanobacteria

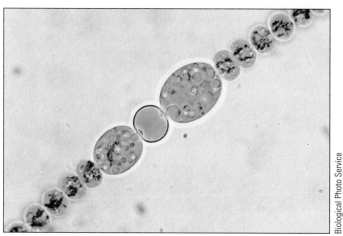

A.

Biological Photo Service

20 μm

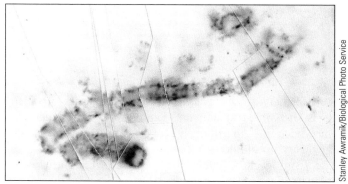

Stanley Awramik/Biological Photo Service

10 μm

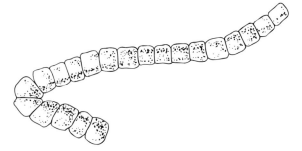

B.

Figure 18-4
Filamentous cyanobacteria. Some filamentous fossils closely resemble their modern counterparts. A. Modern filamentous cyanobacterium. B. A 3.5-billion-year-old filamentous cyanobacterium from the Warrawoona rocks of western Australia.

John Reader/Science Photo Library/Photo Researchers, Inc.

Figure 18-5
Living stromatolites like these once filled vast shallow seas. These cushions of photosynthetic bacteria at Shark Bay, western Australia, resemble 3.5-billion-year-old fossil stromatolites.

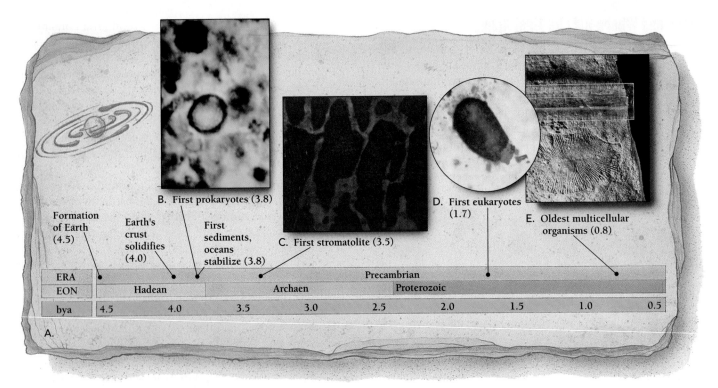

Figure 18-6

A calendar of Earth's early history. Life appeared on Earth almost as soon as conditions would allow it. The first stromatolites (photosynthetic cyanobacteria) appeared 3.5 billion years ago. Eukaryotes appeared nearly 2 billion years after the first prokaryotes. (B, Science VU-USM/Visuals Unlimited; C, William E. Ferguson; D, Andrew H. Knoll, Knoll and Calder, *Palaeontology* 26:467, 1983; E, Biological Photo Service)

Biology(Ⓢ)Now™   Learn more about the conditions for life in Earth's early history by clicking on this figure on your BiologyNow CD-ROM.

in modern stromatolites are photosynthetic. Paleontologists assume that fossil stromatolites were also photosynthetic. In other words, photosynthesis probably evolved at least 3 billion years ago (Figure 18-6). Later in this chapter, we will see how the photosynthetic abilities of early cyanobacteria profoundly changed the world, allowing the evolution of all higher organisms.

As we have noted, there are two general explanations for the relatively sudden appearance of prokaryotic organisms soon after the Earth had cooled: (1) life arose spontaneously from nonliving organic molecules in the primitive environment, or (2) life arrived from elsewhere in the universe. It is almost impossible to imagine how we could test either of these explanations. We can, however, examine the evidence in favor of or against each hypothesis. The second hypothesis at first appears to be the simpler explanation for the appearance of life. According to this view, the present diversity of life all derives from a colonization of Earth followed by a huge adaptive radiation. An appealing feature of this hypothesis is that the universe is thought to have originated at least 12 to 16 billion years ago, 8 to 12 billion years before the cooling of the Earth. This far longer history would give life more time to develop. Furthermore, an extraterrestrial origin of life explains some otherwise unexpected chemical properties of organisms. It would explain, for example, the requirement for molybdenum—an extremely rare element on this planet—in key reactions of photosynthesis and nitrogen fixation.

Until recently, no evidence existed for life on the other planets in the solar system. In 1996, researchers discovered what they thought were signs of life in a meteorite composed of 3.6-billion-year-old Martian rock, apparently dislodged from the Martian surface by an asteroid collision about 15 million years ago. Most scientists now doubt that the rock contained life. Even if Mars did once harbor living organisms, there is no evidence that life on Earth was seeded by Martian organisms. Earth may have seeded Mars. Or life may have arisen independently on each planet. Another possibility is that life arrived on both planets from some source outside the solar system.

For life to colonize Earth, it would have to have come from at least as far away as the nearest star, some 4 light years away—38 trillion km (23 trillion miles). Living organisms would have had to survive extremely violent events, for nothing else would have propelled them into interstellar space. There they would have had to survive intense radiation for untold millions of years. Recent research has sug-

Extreme Biology   Is There Life Elsewhere in the Universe?

In 1959, a young astronomer assigned to the Greenbank Observatory in rural West Virginia began using the big observatory radio telescope to listen for signals from extraterrestrial beings. Frank Drake and his colleagues tried to keep it a secret, but word got out quickly. Newspaper reporters besieged the observatory with calls, but Drake could not honestly report any definite communications from outer space.

Yet, for nearly 40 years, Drake hasn't stopped listening (Figure A). In 1974, he used Cornell University's Arecibo radio telescope, a giant, 300-m telescope in Puerto Rico, to send the first deliberate message from Earth. The message described Earth's position in the Milky Way galaxy, what humans look like, and how many of us there were.

Today, Drake and others are still listening for signs of intelligent life in the universe. This project, called the Search for Extraterrestrial Intelligence (SETI) asks, Is anyone out there? (SETI's work was depicted melodramatically in the Hollywood film *Contact*.) Biologists tend to be conservative about such things. Many believe that the emergence of life on Earth is so remarkable that it would be astonishing if it occurred elsewhere in the universe. Indeed, the biologist and writer Jared Diamond has argued that the development of the radio on Earth was extremely unlikely. (He adds that, given the way humans treat less advanced civilizations and less intelligent creatures on our own planet, we will be fortunate indeed if we are alone in the universe.) Biologists are also eager to distance themselves from people who claim encounters with intelligent extraterrestrials on Earth.

Astronomers tend to take a different view. They argue that to insist that life could have arisen only on Earth is arrogant and excessively geocentric, or Earth-centered.

Figure A
Are we alone? Probably not. Frank Drake, president of the SETI Institute and professor emeritus of astronomy at the University of California, Santa Cruz, first began searching for intelligent extraterrestrials in 1959.

Of course, they dismiss flying saucers—so-called UFOs (unidentified flying objects). Even the closest stars are too far away for spaceships to get here, let alone home again. And if intelligent life exists, the chances are that it is far away, indeed. Any signals that SETI picks up will be thousands or millions of years old.

But we are almost certainly not alone. Drake has suggested that the number of civilizations in the galaxy can be calculated using the following equation, called the Drake Equation:

$$N = Rf_p\, n_e\, f_l\, f_i\, f_c\, f_L$$

In English, this equation says, the number of technological (detectable) civilizations in the galaxy (N) is equal to the rate (R) of star formation, times the fraction of stars with planets (f_p), times the number of planets ecologically suitable for life (n_e), times the fraction of those planets where life actually evolves (f_i), times the fraction that develops intelligent life (f_i), times the fraction that communicates (f_c), times the fraction of the planets' life occupied by the communicating civilization (f_L).

Clearly, the answer depends on what values are plugged into this equation. However, many of these values can be estimated with some confidence. For example, astronomers estimate that the Milky Way contains about 400 billion stars. Conservative estimates for most of the rest of the variables give predictions such as 4 billion planets with some form of life and 4,000 existing technological civilizations in the Milky Way. And that is in our galaxy only. From astronomers' perspective, the evolution of life was practically inevitable.

But if extraterrestrial life is out there, we haven't found it yet. Soil samples from other planets have revealed no life, and SETI has so far turned up nothing concrete. The only tantalizing hints so far are some complex organic compounds from a meteorite believed to have come from Mars.

Organic molecules seem to abound in the universe. Astronomers have detected the presence of a number of carbon- and nitrogen-containing compounds in interstellar gas clouds. These compounds include those that we think were present in the Earth's primitive atmosphere—hydrogen, hydrogen cyanide, carbon monoxide, and ammonia.

Soil samples brought back from the moon contained several amino acids. Meteors, such as the one that fell in Murchison, Australia, in 1969 (Figure 18-10), contain the amino acids found in organisms as well as other amino acids and more complex molecules. The fragments from the Murchison meteorite contained 1 or 2 percent organic material. If such an endowment is typical of comets, asteroids, and meteors in the rest of the galaxy, it is hard to imagine how life could not have evolved elsewhere.

gested, however, that bacteria might survive the intense radiation if they were encased in ice.

If such a trip were possible, we would still want to understand how (and where) those organisms first arose. The problem of the origin of life would be virtually insurmountable, since we would not know where in the universe such organisms might have come from.

For now, the colonization-from-space hypothesis is only remotely likely. Perhaps more important, the hypothesis is so inaccessible to inquiry that most scientists interested in such things have focused their attention on the possible origin of life on Earth. Most biologists now believe that life must have developed from nonliving molecules on the early Earth's surface. They have accepted the challenge of showing that the development of organisms from nonliving matter—called **prebiotic evolution**—is possible.

strate that the steps in the presumed process can actually occur. To discuss these questions, however, we must first understand the conditions on Earth shortly after it formed.

Most biologists believe that life originated on Earth. No one knows enough about conditions in the rest of the universe to guess how extraterrestrial life might have evolved or safely arrived on Earth.

The conditions available for the origin of life differ from those for spontaneous generation in three dramatic ways—the long period of time available, the absence of life, and the absence of oxygen.

18.2 Prebiotic Evolution: How Could Complex Molecules Evolve?

By the 1930s biochemists realized that all organisms are made from the same building blocks—the sugars, lipids, amino acids, and nucleotides that make up the "biochemical alphabets" described in Chapter 3. Two influential scientists—Alexander Ivanovich Oparin, a Russian biochemist, and J.B.S. Haldane, the English evolutionary biologist—proposed that these building blocks had been formed in the primitive environment. The original synthesis of these materials, they argued, was **prebiotic** and therefore **abiotic**—before life and therefore without life.

Important differences separate spontaneous generation from theories of prebiotic evolution. First, while spontaneous generation had to occur nearly instantly, the origin of life may have taken as long as 300 million years to unfold. Second, the conditions during which life must have first appeared were dramatically different from those that exist today. One of the most important differences was that the atmosphere contained virtually no free oxygen. Had oxygen been present in any quantity, it would certainly have oxidized the starting materials for life. A second important difference was the absence of life itself. Today, life could not arise on Earth because a plethora of hungry organisms find and consume organic molecules.

Darwin himself grasped this distinction, writing to a friend in 1871:

> It has often been said that all of the conditions for the first production of a living organism are now present which could ever have been present. But if (and oh! what a big if!) we could conceive in some warm little pond, with all sorts of ammonia and phosphoric salts, light, heat, electricity, etc., present, that a protein compound was chemically formed ready to undergo still more complex changes, at the present day such matter would be instantly devoured or absorbed, which would not have been the case before living creatures were formed.

Was Darwin right? In the absence of devouring organisms and destructive oxygen, can the building blocks of life manage to assemble themselves into living organisms? To decide, we must ask whether we can imagine a reasonable scenario for the "spontaneous" production of the building blocks of life and whether laboratory experiments demon-

What Was the Earth Like When It Was Young?

The Earth formed about 4.6 billion years ago, when a spinning cloud of rocks, dust, and gas formed the sun and its planets. Astronomers believe that the spinning cloud gradually spread out and flattened as it rotated, like a lump of spinning pizza dough (Figure 18-7). The bulk of the material, at the center, condensed into an enormous compact mass that became the sun. The rest of the spinning cloud gradually formed the planets, comets, meteors, and other objects that make up the solar system.

The debris orbiting closest to the protosun [Greek, *protos* = primitive] moved fastest, and there the friction of

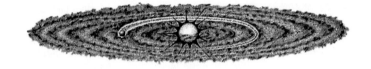

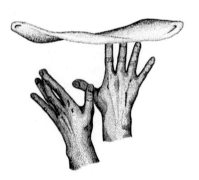

Figure 18-7

Formation of the solar system. Like a spinning lump of pizza dough, a cloud of dust and rocks flattened and then condensed into a central mass that became the sun. Material further out condensed into rotating balls of rock and gas that we now know as the planets. (Source: Kristin Levier, Science Notes, University of California, Santa Cruz)

constant collisions generated the most heat. Close to this hot center, pieces of silicon, metals, metal oxides, and other materials that remain solid even at high temperatures fell together, drawn to one another by their own gravities. These hot lumps of solid material gradually formed the inner protoplanets—Mercury, Venus, Earth, and Mars. Water, methane, and other compounds that solidify only at low temperatures formed the cold, outer protoplanets, such as Jupiter, far from the center of the spinning disk. The huge, outer planets attracted and held cold atmospheres of hydrogen, helium, and other gases. But the tiny, inner planets had gravities too weak to hold the hot gases, which escaped into space.

The early protosun was much cooler than today's sun. But Earth was far hotter. Gravity compacted the Earth, and huge meteors, many of them hundreds of miles in diameter, bombarded the planet. (The impact of one of these collisions was so great that it dislodged a huge chunk of Earth, which became the moon.) The compaction and the bombardment kept the molten Earth from solidifying for another half a billion years. The heaviest elements sank to the center of the Earth, while the lightest ones floated to the top. Iron and nickel sank to the core, the lightest silicates (granite, for example) rose to the surface, while denser silicates of iron and magnesium floated in between.

After 500 to 600 million years, the surface cooled enough to form a thin crust. Dense gases escaped from the interior through cracks and volcanoes in the new crust to form a primitive atmosphere. Among these gases was water vapor, immense quantities of which then condensed into clouds. Torrential rains pummeled the still hot Earth for millions of years, creating the oceans and eroding the Earth's rocky surface. Although the oceans formed relatively early, constant bombardment by meteors, large and small, may have repeatedly vaporized the oceans until about 3.8 billion years ago. About that time, the first sediments and the first life appeared.

The Evolution of the Atmosphere

The exact composition of the gases in the primitive atmosphere has been a matter of much debate. The modern atmosphere is 78 percent nitrogen, 21 percent oxygen, and about 1 percent argon, with varying amounts of water vapor close to the Earth. Another 0.1 percent of the atmosphere consists of traces of carbon dioxide, carbon monoxide, sulfur oxides, nitric oxides, and other gases.

Scientists long assumed that the early atmosphere contained hydrogen (H_2), ammonia (NH_3), methane (CH_4), and water (H_2O), with traces of carbon dioxide (CO_2) and hydrogen sulfide (H_2S). Some research now suggests that the early atmosphere contained much more CO_2 and no hydrogen. But about one important gas there has been little argument. The early atmosphere had almost no oxygen. The strongest evidence for this is the near-total absence of oxi-

dized iron (rust) in the earliest sediments. Had oxygen been present in any quantity, it would certainly have oxidized the starting materials for life. Oxygen, we will see later in this chapter, came much later, probably as a by-product of photosynthesis.

The absence of oxygen had other important consequences. Chief among these was the absence of an ozone layer. The ozone (O_3) layer, about 40 km (25 miles) up, today shields the Earth from intense ultraviolet (UV) light from the sun. Modern organisms need that protection because UV light, extremely good at making and breaking chemical bonds, interferes with normal life functions.

However, 4 billion years ago, any chemical synthesis would have required considerable energy. Recall from Chapter 5 that synthetic reactions are almost all thermodynamically uphill. So UV radiation from the sun would have helped to promote novel chemical reactions. Heat from the hot interior of the Earth provided more energy. But the most dramatic energy source was the lightning accompanying the storms that filled the oceans.

The Earth's crust solidified about 4 billion years ago, but the oceans were not stable until about 3.8 billion years ago. Since the first fossil organisms are about 3.8 billion years old, we can conclude that life on Earth originated relatively quickly—within a 300-million-year window.

Is Prebiotic Synthesis Plausible?

By the early 1950s, scientists had some idea of what the early Earth was like: they believed it had the raw materials for organic synthesis—water, methane, ammonia, and hydrogen—and it had sources of energy—UV radiation, heat, and lightning. But scientists did not know whether the conditions of the Earth's early history would allow the abiotic synthesis of biologically important molecules.

Then, in 1953, a 23-year-old graduate student working in the laboratory of Harold Urey at the University of Chicago, undertook to simulate the primitive conditions of the Earth. Urey's student, Stanley Miller, simulated the early Earth inside a simple glass apparatus built from the standard equipment found in a sophomore organic chemistry laboratory (Figure 18-8). The simulated Earth consisted of water in a boiler (the ocean) and a chamber fitted with electrodes (the atmosphere, with lightning). In one experiment, Miller mixed together an atmosphere of hydrogen (H_2), methane (CH_4), and ammonia (NH_3). Miller let the gases circulate past the flashing electrodes for a week and then analyzed the contents.

He found that more than 10 percent of the carbon from the methane was in other organic molecules. These included at least five amino acids, as well as compounds that are precursors of other amino acids and of the bases of nucleic acids. Among these precursors was hydrogen cyanide

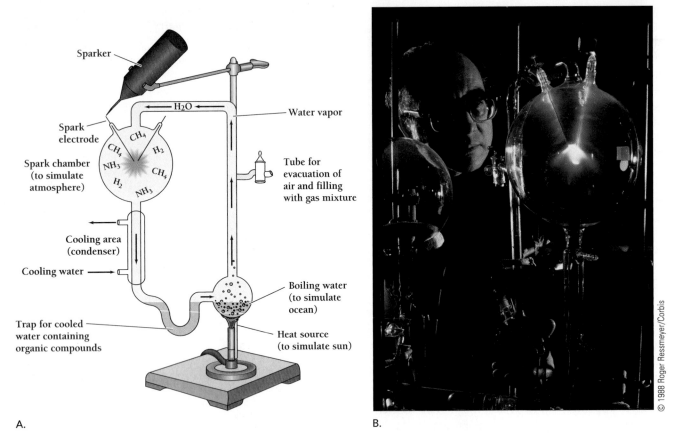

A.

B.

© 1988 Roger Ressmeyer/Corbis

Figure 18-8

Stanley Miller's experiment. Can conditions similar to those on Earth 4 billion years ago give rise to the building blocks of life? A. Miller treated a mixture of water and gases with flashes of artificial "lightning." Among the complex molecules produced after one week were five amino acids. B. Miller and his apparatus.

Biology❊Now™ Learn more about the Miller experiment by clicking on this figure on your BiologyNow CD-ROM.

(HCN), five molecules of which can directly form adenine (Figure 18-9). The mixture was unexpectedly rich in some (but not all) of the same basic building blocks of life, which we listed in Chapter 3.

Miller and others repeated this basic experiment, with different atmospheres and different sources of energy. As long as free oxygen (O_2) was absent, the same building blocks appeared. Other sources of energy also work—most notably UV light, the principal energy source available on the Earth's surface before the Earth developed its protective ozone layer.

As a result of these experiments, it seemed likely that—within a few million years—the primitive oceans loaded up with organic molecules capable of becoming the building blocks of life. This primordial soup may have contained as much as 1 percent organic molecules in some places, about the same as a watery chicken broth. Remarkably, the amino acids most easily produced in these abiotic experiments are the ones most common in proteins today. And the most easily synthesized base—adenine—is biologically the most important of all the nitrogenous bases.

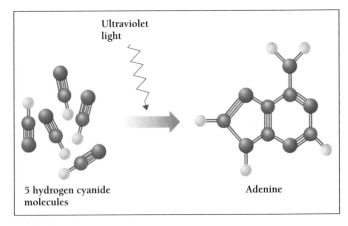

5 hydrogen cyanide molecules

Ultraviolet light

Adenine

Figure 18-9

First steps in the formation of life? Miller's apparatus also produced hydrogen cyanide. In the presence of ultraviolet light, five molecules of hydrogen cyanide form adenine, the nitrogenous base in ATP.

A distinct problem with this scenario, however, is that recent research by geoscientists suggests that the early atmosphere contained neither methane (CH_4) nor ammonia (NH_3). Heavy doses of UV light from the sun would have destroyed such hydrogen-based molecules, and even free hydrogen is so light that it would have escaped into space. Geoscientists say that the major components of the atmosphere were more likely carbon dioxide and nitrogen, with water and traces of carbon monoxide and free hydrogen, the same gases that are ejected from modern volcanoes.

Such an atmosphere would not have allowed the kind of amino acid synthesis suggested by Miller's experiment. Formaldehyde (H_2CO), essential to the formation of sugars, would have formed easily enough. But a likely chemical pathway for the synthesis of hydrogen cyanide, the building block for adenine and other amino acids, has yet to be found.

Miller, now a professor of chemistry at the University of California, San Diego, defends his early study, saying that smoke and clouds might have shielded the fragile methane and ammonia molecules from UV light. And several Japanese researchers, supporting Miller, argue that solar particles and cosmic rays would have countered the effect of UV light by releasing free hydrogen from water molecules and so promoting the synthesis of methane and ammonia.

Other researchers have suggested another possible source of the first building blocks—the seas surrounding thermal vents in the ocean. These hot springs lie deep in the oceans, at the crests of underwater ridges. Seawater circulates through the vents, where its temperature rises to 350°C and then abruptly plunges to 2°C as it mixes with cold ocean waters. Modern-day vents are the sites of lots of underwater life, sustained by unusual metabolic reactions that do not depend on solar energy.

Some biologists suggest that the abundant hydrogen sulfide and iron sulfide in the superheated water could have provided the perfect environment for the synthesis of primordial building blocks. In this scenario, the building blocks accumulated on the surfaces of underwater rock instead of dissolving in the ocean.

Prebiotic synthesis of biological building blocks may have taken place in the open ocean or on underwater rocks.

Was Prebiotic Synthesis Necessary?

Some astronomers have suggested that the debate over the exact composition of the early atmosphere may be moot. They believe that another source of simple organic molecules is possible. Researchers have discovered that asteroids, meteors, and comets are rich in complex organic compounds created during the formation of the solar system (Figure 18-10). Further, based on the calculated number and size of comets and meteors likely to have entered the Earth's atmosphere, astronomers estimate that as much as 10 million kg of complex organic molecules might have showered the Earth each year. At the end of a billion years, they calculate, the Earth could have accumulated as much as a million billion (10^{15}) kg of organic molecules, more than the total mass of all life on Earth today.

The first organic molecules could have come from asteroids, meteors, or comets.

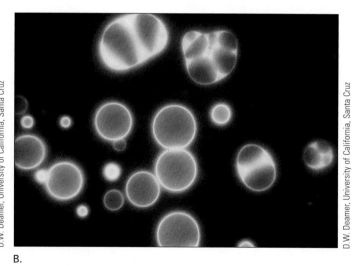

A.

B.

Figure 18-10

Remains of extraterrestrial life? A. This fragment of a meteorite hurtled through the atmosphere and landed in Murchison, Australia, in 1969. Scattered through the rock are particles of organic compounds. B. When extracted, these compounds self-assembled into vesicles. The yellow-green color is the fluorescence of complex organic compounds called polycyclic aromatic hydrocarbons.

How Can Biological Building Blocks Assemble Outside of Cells?

If we assume that the Earth's early environment either allowed the production of amino acids and other building blocks or received them passively from space, we must still account for the assembly of those building blocks into other amino acids, nucleic acids, ATP, and other macromolecules. Can this process be simulated in the laboratory?

In the Miller atmosphere, adenine easily forms from hydrogen cyanide (Figure 18-9). The other nucleic acid bases require more complex reactions, so adenine could have been the first nitrogenous base to form. If so, it would not be surprising for its activated product, ATP, to have become the universal energy currency of life.

ATP itself may have formed rather easily. Laboratory experiments show that mixtures of adenine, ribose, and phosphate can form ATP when exposed to UV light. Other triphosphates may have arisen in the same way.

Even prebiotic replication of nucleic acids is not at all far-fetched. Biochemists have discovered that nucleotides can combine in the test tube to form short RNAs. These RNAs, moreover, form double-stranded molecules that follow the Watson-Crick rules for base pairing, which means they can replicate, as DNA does.

Simulating the polymerization of amino acids into polypeptides is more of a challenge. Many researchers, however, think that ATP in the primordial broth may have somehow contributed to the synthesis of activated amino acids. Solutions of such activated amino acids can spontaneously give polypeptide chains up to 50 amino acids long. Polypeptides also form when researchers heat dry mixtures of amino acids. Sidney Fox, a biochemist at Southern Illinois University, has suggested that such polypeptides could have formed on volcanic cinder cones and washed into the sea.

The earliest polypeptides almost certainly would not have had specific sequences of amino acids. In fact, Fox has dubbed the early polypeptides **proteinoids,** to distinguish them from the true proteins, which are the products of living organisms and whose sequences are defined by the genes in DNA.

The abiotic synthesis of RNA, ATP, and even polypeptides seems plausible.

How Could Prebiotic Assemblages Reproduce Themselves?

Most biologists think that the building blocks of life became available for prebiotic evolution up to 4 billion years ago. But the materials of life are not the same as life. Two important steps would have to have occurred—the evolution of reproduction and the evolution of metabolism. The evolution of reproductive ability depended on establishing a highly specific relationship between the self-replicating nucleic acids and the sequence of amino acids. This meant the development of both the genetic code and a mechanism of translation.

Before this could happen, however, polypeptides and other molecules probably began to come together into organized (and concentrated) associations. As we will see, such primitive cells, or **protobionts,** could concentrate organic molecules and maintain an internal environment different from that of their surroundings. They could, in short, begin to develop the first suggestion of metabolism.

Life required the evolution of reproduction and metabolism.

How Can Prebiotic "Cells" Form?

Polypeptides, nucleic acids, and polysaccharides in solution will spontaneously concentrate themselves into discrete tiny, balloonlike droplets, which the Russian biochemist Oparin called **coacervates.** Coacervates have a discrete inside and outside separated by a membranelike boundary.

If enzymes are present in the solution, these will be bound in the coacervates, which then act like little cells. They absorb molecules from the solution and release the products of the reactions catalyzed by the enzymes. For example, Oparin showed that coacervates containing the enzyme phosphorylase would take up glucose-1-phosphate from the surrounding medium, convert it to starch, and concentrate it inside the droplet (Figure 18-11).

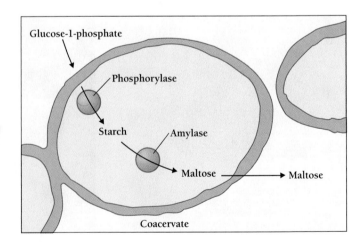

Figure 18-11
Prebiotic metabolism? Macromolecules in solution can form droplets, which Alexander Oparin called "coacervates." In one experiment, coacervates that contained the enzymes phosphorylase and amylase could convert glucose-1-phosphate to starch and then to maltose. These coacervates resemble extremely simple cells.

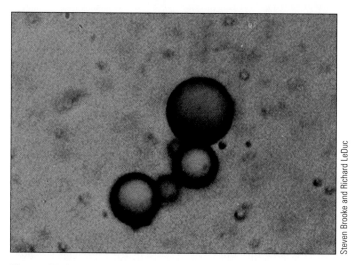

Figure 18-12
Proteinoid microspheres. These nonliving polymer bubbles can reproduce themselves by growing and budding.

Other mixtures can also give rise to isolated, cell-like structures. For example, Sidney Fox found that the proteinoids formed from dry amino acids would form tiny spheres. These structures, called **proteinoid microspheres,** could absorb more polymers from solution, forming buds or even a new generation of spheres (Figure 18-12). Microspheres dropped in water are differentially permeable. They swell or shrink according to the concentration of salts in the water. Some will even store energy, in the form of a differential electric charge across the membrane. Such microspheres can release a charge in much the same way a nerve cell does. A mixture of phospholipids and proteins can form yet another kind of enclosed structure, called a **liposome.** Some liposome membranes have two layers, much like those in modern cells.

None of these self-organizing structures are cells. Yet they look so much like cells that experienced biologists have occasionally mistaken them for bacteria and even tried to classify them. And protobionts show some of the characteristics of life, including growth, metabolism, and responsiveness to their environment. The existence of protobionts shows that cell-like structures could have formed spontaneously in the primordial soup.

Further, we can envision the evolution of primitive metabolic pathways. As coacervates used up a given compound in the environment, those coacervates able to make the compound from another component of the soup would persist. Once the precursor was exhausted, further growth of protobionts would depend on the ability to produce the needed product from still another precursor. According to this scheme, a number of pathways would evolve—one step at a time (Figure 18-13).

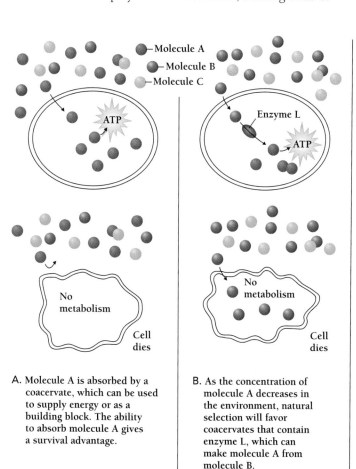

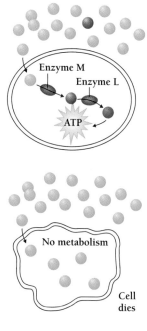

Figure 18-13
How might metabolic pathways have evolved? Here a primitive cell begins to perform simple metabolism. In a process of natural selection, cells containing the right enzymes *(top)* would persist, while those lacking these enzymes *(bottom)* would die out.

A. Molecule A is absorbed by a coacervate, which can be used to supply energy or as a building block. The ability to absorb molecule A gives a survival advantage.

B. As the concentration of molecule A decreases in the environment, natural selection will favor coacervates that contain enzyme L, which can make molecule A from molecule B.

C. As the concentrations of molecules A and B decrease, natural selection will favor coacervates that contain enzyme M, which can make molecule B from molecule C, and enzyme L, which can make molecule A from molecule B.

18.3 Early Biotic Evolution: How Did the First Genetic System Evolve?

Even with the evolution of primitive metabolic pathways, protobiont "life" would have been severely limited. Despite protobionts' ability to grow and to bud, they could not reproduce. Even if they had accumulated catalysts for specific chemical reactions, protobionts could not make more of these specific catalysts. As unique molecular catalysts were passed on to increasing numbers of "offspring" protobionts, the catalysts became increasingly diluted. To invest the new protobionts with equal concentrations of functional molecules required the invention of a way of "teaching" the next generation to make its own catalysts. Life would need a way of storing, replicating, and translating the information for the building of enzyme catalysts—in short, a genetic system.

Could RNA Have Served as the Original Genetic Material?

A modern cell stores its genetic information as DNA, transcribes the information into RNA, and then translates the message into specific polypeptides. When the cell divides into two cells, it gives a copy of the DNA to each new cell, and so passes on the recipe for each polypeptide.

The DNA-to-RNA-to-polypeptide scheme is too complicated to have arisen all at once in the first cells. So biologists have tried to guess which of these three kinds of molecules could have first stored information, directed the synthesis of proteins, and also replicated itself. But DNA cannot form without catalytic proteins, and proteins cannot form without DNA.

That leaves RNA. In modern cells, RNA plays a central role in translation, supporting the idea that RNA could have been the primitive genetic material. Still, if RNA acted as the original genetic material, some sort of mechanism had to exist for aligning amino acids along sequenced strands of genetic RNA. Such a mechanism would have to allow RNA to store information, direct the synthesis of proteins, and also replicate itself. Considerable evidence now supports the idea that RNA can do this.

First, scientists have found that some RNA can spontaneously create itself and replicate itself in a test tube. Short strands of nonordered abiotic RNA can self-assemble from individual ribonucleotides. And, in the presence of an RNA template, ordered sequences of 5 to 10 nucleotides will polymerize according to the standard base-pairing rules.

In the early 1980s, molecular biologists Thomas Cech and Sidney Altman showed that certain kinds of RNA could act as enzymes, snipping off a bit at an end or cutting out a bit and splicing the remaining pieces together. It was a dramatic discovery that earned the two researchers a Nobel prize in 1989 and suggested that perhaps RNA could replicate itself without the help of protein enzymes (polymerases).

Indeed, in 1989, Jennifer A. Doudna and Jack Szostak created an artificial RNA enzyme, or **ribozyme**, that could make new RNA molecules. However, although the molecule can splice preassembled strands of RNA along a template, it cannot add individual bases to a growing chain. These discoveries suggested to many researchers that the first cells evolved in an **RNA world,** in which RNA molecules functioned both as enzymes and as carriers of genetic information. The hypothesized RNA world is not without problems, however. RNA is notoriously unstable, and it is unlikely that RNA enzymes could have survived outside of a living cell. Many researchers therefore think that the first enzymes were neither RNA nor protein, but a different type of molecule, perhaps a nucleic acid with a sugar other than ribose.

Cells based on such a system allowed the evolution of RNA-based cells. RNA-based cells in turn allowed the evolution of cells that used proteins as enzymes and DNA as the genetic material. Reverse transcriptase, so important in many tumor viruses and in HIV, may have played an important role in this last process (Chapter 12).

How Did Translation Evolve?

We know that RNA is capable of storing information, and it might, perhaps, be able to replicate itself. But the hardest question to answer is how modern translation—of RNA into protein—could have evolved. How a correlation between nucleotide sequences and amino acid sequences could have arisen remains a mystery.

Since the genetic code plays such a pivotal role in translation, we may look to it for clues about the evolution of translation. The first clue is the fact that all organisms use the same triplet genetic code. This suggests that the system originated with life itself. Second, most of the information for each codon is in the first two nucleotides; for 7 of the 20 amino acids, the third "letter" is irrelevant. This does not mean that the code was originally a doublet, however. If the code had started out as a doublet, then all previously evolved genes would have become useless when the code became a triplet. Instead, the present characteristics of the code were almost certainly fixed after its initial success.

As molecular biologist Francis Crick put it, the code is a "frozen accident." Information in mRNA was probably read three bases at a time from the beginning. But initially, the first two letters of each codon contained all the meaning.

Because the code is universal among both prokaryotes and eukaryotes, it must have completed its refinement before the time of the common origin of all modern organisms.

How Did Ribosomes Evolve?

Modern ribosomes contain more than 50 different proteins as well as ribosomal RNAs (rRNAs). Since each protein has a defined amino acid sequence, ribosomal proteins could not have been part of the original translational machinery. It seems more likely that the primitive translational machinery depended primarily on RNA. Indeed, the mRNAs of many prokaryotic and eukaryotic bacteria can form specific base pairs with part of the smaller rRNA.

The anatomy of rRNAs also supports the idea that rRNAs directly participated in protein synthesis. For example, researchers have determined the nucleotide sequences of the smaller rRNAs of many widely divergent organisms, ranging from bacteria to vertebrates, as well as of ribosomes from mitochondria and chloroplasts. For each rRNA, researchers predicted the pattern of stems and loops that could be formed. While the sequences of the various rRNAs vary widely, the patterns of stems and loops are much the same. Such conservation suggests an essential function for the pattern of stems and loops, probably as a catalyst for directed protein synthesis.

How Did tRNAs Evolve?

All modern cells attach amino acids to transfer RNAs (tRNAs) before assembling a polypeptide. Transfer RNAs play a critical role in translation, allowing unlike molecules—nucleotides and amino acids—to be connected. All modern tRNAs have similar stem-and-loop structures, and all attach to an amino acid at their 3′ ends. In addition, all tRNAs contain an anticodon with common surrounding elements. Thus, like the genetic code, tRNAs probably diverged from a single tRNA early in the history of life.

How Did the First Cells Use Genetic Information?

The first true cell would have had to have the following properties:

1. A boundary, such as that in artificial protobionts
2. Enzymes for extracting energy from chemicals in the environment
3. Energy storage in ATP
4. RNA that specified the structure of specific enzymes
5. RNA replication, using the Watson-Crick rules of base pairing
6. Specification of each amino acid by a triplet codon, with most of the information conveyed by one or two bases
7. Primitive tRNAs to connect amino acids to nucleotides

The first enzymes probably arose by the random assembly of amino acids into short polypeptide segments. Such primitive enzymes were probably neither as active nor as stable as modern enzymes. But if a polypeptide segment could form the active site of a catalyst, then its possessor would have enormous advantages over its competitors.

Similarly, the first genetically useful RNAs probably had only short stretches of information. Once the first genetic system had evolved, however, natural selection would begin

to work. Natural selection would favor cells with increased capacity to reproduce, as well as cells better able to extract energy from chemicals in the environment. Reproduction, protein synthesis, and metabolism must therefore have evolved together. The most successful organisms would be those that best coupled their genetic and metabolic systems—those best able to pass on useful catalysts to their descendants.

18.4 How Did Modern Cells Evolve?

The first fossils that are clearly eukaryotic are about 1.7 billion years old. More than 2 billion years elapsed between the first appearance of life 3.8 billion years ago and the first eukaryotes. What, we may ask, happened during that period to allow more complex forms to evolve? What biochemical pathways did the earliest organisms develop, and, equally important, how did these developments fundamentally change the Earth's environment?

How Did Early Life Affect the Environment?

Most biologists think that the first cells were **heterotrophs** [Greek, *heteros* = other + *trophe* = nourishment], organisms that obtain energy only by degrading existing organic molecules. Human beings and other modern animals are all heterotrophs, too. We obtain both energy and building blocks, such as amino acids, by eating other organisms. We can extract energy in the process of respiration by oxidizing preexisting organic molecules from other organisms.

The earliest organisms probably obtained both energy and building blocks from preexisting organic molecules in a rich primordial soup. Like all heterotrophs, they consumed what was in the soup but contributed no energy. But unlike animals, they could break down molecules only by means of anaerobic glycolysis, or fermentation. They were incapable of true respiration, since there was no oxygen. And they were incapable of obtaining energy from the sun by photosynthesis or from inorganic molecules, as plants and other modern **autotrophs** [Greek, *auto* = self + *trophe* = nourishment] do.

Of the three ways to extract energy from molecules—photosynthesis, respiration, and fermentation—only fermentation is common to all bacteria and eukaryotes. It seems likely, then, that fermentation is the oldest metabolic pathway and that it had already evolved in the earliest cells—at least those whose descendants have survived.

The suspected sources of nourishment for early heterotrophs—the organic molecules in the primordial soup—were in limited supply. Early organisms would have gradually exhausted them of their nutritive value, since no autotrophs existed to renew them. Early organisms probably forestalled the exhaustion of the Earth's first resources by evolving elaborate chemical pathways for exploiting more and more of the molecules in the soup. Eventually, though, all the original molecules, and all the energy in them, would have been exhausted. One solution was to begin extracting energy from inorganic chemicals by oxidizing them and to use the energy to build new organic molecules from carbon dioxide. The present-day *Sulfolobus* bacterium, for example, gets its energy by converting sulfur to sulfur dioxide.

The Evolution of Photosynthesis

The most abundant energy source was not inorganic molecules, however, but sunlight. We may guess at how photosynthesis gradually developed, 3.0 to 3.5 billion years ago, by examining the variety of ways in which modern organisms harness solar energy.

The first step in the direct harnessing of solar energy by organisms was probably the development of a light-driven proton pump, such as that in the modern purple bacterium. This bacterium pumps protons across its cell membrane, which allows the bacterium to store energy in ATP, as well as actively transport other substances across the membrane.

More advanced modern bacteria use light to generate reducing power (hydrogen atoms or electrons) to synthesize carbohydrates from carbon dioxide. Some photosynthetic bacteria obtain the needed electrons from such donors as hydrogen gas or hydrogen sulfide.

The largest potential source of electrons, however, is water. Water is difficult to break down, but it is the electron source for nearly all modern photosynthetic organisms. The first ancient organism to synthesize organic molecules from sunlight, water, and carbon dioxide won for its descendants a freedom from want that has lasted more than 3 billion years. Cells that could photosynthesize would no longer have to compete for energy-bearing organic molecules. They could simply make their own from the abundant water, carbon dioxide, and sunlight.

The First Global Toxic Waste Problem

However, breaking down water created a serious global toxic waste problem. The very steps needed to separate the hydrogen atoms from water also released huge quantities of molecular oxygen. Oxygen is a powerful oxidant that easily rips apart many molecules. Sensitive organic molecules would have been destroyed by oxygen. Many of the original heterotrophs, which depended on organic molecules, probably starved, their food supply destroyed by the wastes of their own autotrophic cousins.

The development of photosynthesis drastically altered the Earth's early environment. Major contributors to this change were the cyanobacteria making up the stromatolites, mentioned at the beginning of this chapter. The wide distribution of stromatolite fossils suggests that they covered vast areas of the world's shallow waters from about 3.8 billion years ago until the end of the Precambrian. In the absence of modern grazing sea animals, photosynthetic cyanobacteria flourished nearly unchecked for millions upon millions of years. For perhaps as long as 2.5 billion years, they pumped oxygen into the oceans and the atmosphere. The cyanobacteria began to decline about a billion years ago. However, single-celled algae began to appear throughout the open oceans, their photosynthetic activity further boosting the oxygen content of the seas and the atmosphere.

The availability of oxygen allowed the evolution of a new class of heterotrophic organisms that could use oxygen to break down organic molecules. The complete oxidation of organic compounds to water and carbon supplies more energy than glycolysis alone. The advantage of exploiting oxygen was so great that today anaerobic organisms persist in only a few isolated environments.

The vast flats of cyanobacteria and oceans of algae must have presented another undeniable opportunity. Some of the new organisms that had developed the ability to breathe oxygen now developed the ability to eat the very organisms that had created this caustic waste. The first autotrophs thus provided the basis of a food chain that still survives today. Most animals that live in the ocean eat smaller animals that eat planktonic animals that eat single-celled algae.

How Did Mitochondria and Chloroplasts Arise?

Almost all eukaryotes—protists, fungi, plants, and animals—have mitochondria, the organelles that produce most of a cell's energy. Only plants and some protists, however, have chloroplasts. Biologists conclude from this that the first eukaryotes probably had mitochondria but no chloroplasts.

Because mitochondria depend on oxygen to generate useful energy, they could have evolved only as the photosynthetic prokaryotes filled the world with oxygen. Mitochondria are probably extremely ancient, however. Like chloroplasts, mitochondria have their own DNA. But their DNA uses a genetic code that differs from the universal code used by the nuclear DNA of all modern eukaryotes and even the DNA of most prokaryotes. In fact, the mitochondria of different organisms use slightly different codes (Figure 18-14).

Mitochondria are probably older than chloroplasts.

Are Mitochondria and Chloroplasts Endosymbionts?

In both size and internal organization, chloroplasts resemble cyanobacteria (Figure 18-15). These similarities have long suggested that chloroplasts derived from free-living organisms. Indeed, there are many examples today of photosynthetic organisms living within animals or protists. Such a close association of two organisms, one of which lives inside the other, is called **endosymbiosis** [Greek, *endon* = within + *syn* = together + *bios* = life]. Endosymbiosis is mutually

Codon	UGA	AUA	CUA	AGA/AGG
Universal code	STOP	Ile	Leu	Arg
Human mitochondria	Trp	Met	Leu	STOP
Drosophila mitochondria	Trp	Met	Leu	Ser
Yeast mitochondria	Trp	Met	Thr	Arg
Plant mitochondria	STOP	Ile	Leu	Arg

(Left side label spanning the mitochondrial rows: **Mitochondrial codes**)

Figure 18-14

Genetic code in mitochondria. Mitochondrial DNA encodes polypeptides using a slightly different code than that used by nuclear DNA. The top codons translate to stop, isoleucine, leucine, and arginine, respectively, in the universal code. The mitochondria of humans, fruit flies, and yeast each translate one or more of these codons differently. In contrast, plant mitochondria use the universal code. What does that suggest about plant mitochondria?

beneficial, since it provides an internal source of food for the host and mobility and protection for the inner organism.

Most biologists now favor the view that both chloroplasts and mitochondria are the endosymbiotic descendants of free-living prokaryotes (Figure 18-16). One of the most forceful proponents of this view has been Lynn Margulis, an evolutionary biologist at the University of Massachusetts.

Some biologists still consider the endosymbiosis theory too speculative. Many feel that chloroplasts may well have originated as free-living cyanobacteria but that the case for mitochondria is much weaker. Recent studies of mitochondrial DNA, however, suggest that all mitochondria have evolved from a single kind of ancient bacteria.

Those who favor the endosymbiotic theory still disagree about the origin of eukaryotic nuclei, though a strong candidate has recently emerged. Most prokaryotes lack the proteins histone and actin. But the archaebacterium *Thermoplasma acidophilum*, a heat-loving prokaryote, does make histones and actin. The ancestors of this organism, which lacks a cell wall, are a popular best guess for the origin of the nucleus.

Did Eukaryotes Arise Through the Proliferation of Internal Membranes?

Opponents of the endosymbiotic theory argue that eukaryotes arose instead by the proliferation of internal membranes to form new cellular compartments. Lysosomes and the endoplasmic reticulum, for example, form from cellular membranes—not from free-living forms. Perhaps mitochondria and chloroplasts formed when membranes surrounded pieces of mobile DNA (plasmids or transposons) that detached from nuclear genes.

The endosymbiotic theory produces a more complicated evolutionary tree than the traditional view. With this complexity, however, comes a consistency with current views of prebiotic evolution. For the moment, at least, most biologists seem to find endosymbiosis the most convincing general explanation for the origin of eukaryotic cells.

Mitochondria and chloroplasts, which resemble prokaryotes, probably began their association with eukaryotic cells as endosymbionts.

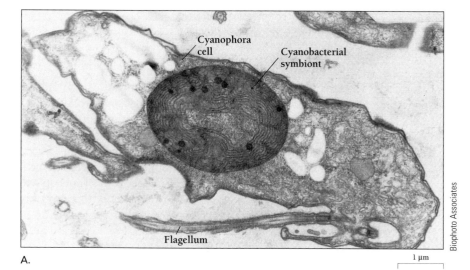

1 μm

Figure 18-15

Endosymbiosis. A. Living inside the photosynthetic protist *Cyanophora* is a photosynthetic cyanobacterium. B. A chloroplast, which resembles a cyanobacterium.

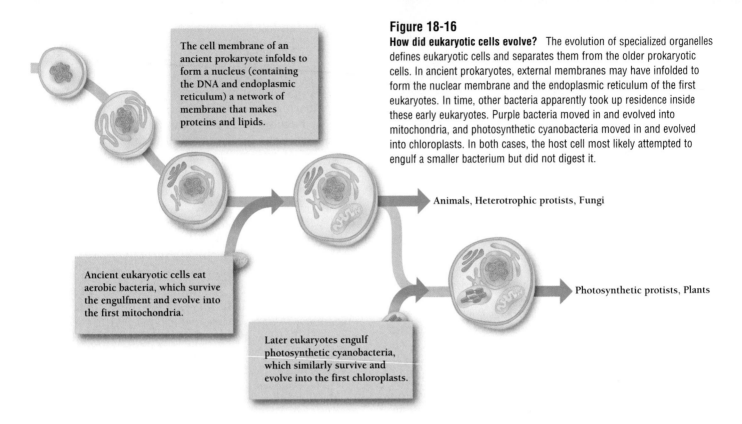

The cell membrane of an ancient prokaryote infolds to form a nucleus (containing the DNA and endoplasmic reticulum) a network of membrane that makes proteins and lipids.

Figure 18-16

How did eukaryotic cells evolve? The evolution of specialized organelles defines eukaryotic cells and separates them from the older prokaryotic cells. In ancient prokaryotes, external membranes may have infolded to form the nuclear membrane and the endoplasmic reticulum of the first eukaryotes. In time, other bacteria apparently took up residence inside these early eukaryotes. Purple bacteria moved in and evolved into mitochondria, and photosynthetic cyanobacteria moved in and evolved into chloroplasts. In both cases, the host cell most likely attempted to engulf a smaller bacterium but did not digest it.

Ancient eukaryotic cells eat aerobic bacteria, which survive the engulfment and evolve into the first mitochondria.

Later eukaryotes engulf photosynthetic cyanobacteria, which similarly survive and evolve into the first chloroplasts.

Animals, Heterotrophic protists, Fungi

Photosynthetic protists, Plants

In this chapter, we have explored scientists' hypotheses about how life might have evolved on Earth billions of years ago. Our discussion of such long-ago events has been, necessarily, speculative. In the next section of *Asking About Life,* we return to the present, with a review of the great diversity of modern life. We will examine how biologists classify organisms and then survey the three domains of life.

Key Concepts

- Life appeared just a few hundred million years after the Earth's crust hardened.
- Scientists can construct a reasonable scenario for the evolution of the first biological molecules and for the generation of abiotic cell-like structures.
- The first cells required a means of coupling metabolism and protein synthesis to genetic instructions. These cells may have used RNA as both genetic instructions and enzymes.
- Eukaryotic cells probably arose through a symbiotic association of prokaryotic cells.

Summary with Key Terms

Can living organisms arise from inanimate material?

By the middle of the 19th century, the work of Redi, Spallanzani, and Pasteur had convinced scientists that life could not arise spontaneously. Yet all life must have some ultimate origin. While some scientists have argued that life

may have arrived from elsewhere in the universe, most researchers assume that life evolved from nonliving, or **abiotic,** organic molecules here on Earth.

The theory of the origin of life on Earth is different from the theory of spontaneous generation. Life could not evolve on Earth today because oxygen would oxidize the components and other organisms would eat the components before enough time elapsed for anything to happen. Ideas about the origin of life rest on the demonstration that plausible **prebiotic** chemical reactions could have produced molecules and organized structures like those in modern organisms.

When did the first life appear on Earth?

After the Earth formed some 4.6 billion years ago and its crust had hardened half a billion years later, torrential rains began creating the first sediments. Precambrian rocks contain fossils as old as 3.8 billion years, only 300 million years after the formation of the oldest known sedimentary rocks. The Earth's early atmosphere contained no free oxygen. Consequently, no ozone layer protected the atmosphere from ultraviolet (UV) light. UV light and lightning probably contributed the intense energy needed for the formation of the first organic molecules.

How could living cells develop from inanimate material?

The early atmosphere may have contained hydrogen, ammonia, and methane, as well as other gases and water. In the laboratory, such a mixture, exposed to electrical discharges, produces many organic molecules, including amino acids and precursors of nucleotides. However, recent research by geoscientists suggests that the early atmosphere was more likely composed of carbon dioxide and nitrogen. Such an atmosphere could have easily given rise to sugars, but to amino acids only with difficulty.

Biologists assume that the primitive oceans became loaded with organic molecules. These organic molecules can form polymers, called **proteinoids.** Mixtures of different kinds of molecules can assemble into organized cell-like structures, called **protobionts.** Polypeptides, nucleic acids, and polysaccharides form **coacervates.** Proteinoids will form tiny spheres called **proteinoid microspheres.** A mixture of phospholipids and proteins form **liposomes,** which have a bilayered membrane similar to that of a true cell.

Although protobionts resemble cells and are capable of cell-like behavior and **prebiotic evolution,** they are not alive because they cannot reproduce themselves. Living cells must be able to store, translate, and replicate genetic instructions. By comparing the characteristics of protein synthesis in contemporary organisms, scientists have inferred that the first organisms used RNA as the genetic material, with an early version of the triplet genetic code. RNA can store information and replicate itself. It can also snip and splice itself like an enzyme. RNA catalysts, called **ribozymes,** can catalyze the synthesis of tRNA, rRNA, and mRNA, the major components of translation. Most evolutionary biologists think that the first cells functioned in an **RNA world,** using RNA both for genes and enzymes.

The first organisms were probably **heterotrophs.** Photosynthetic and other forms of **autotrophic** organisms probably evolved after early life had depleted the primitive oceans of energy-rich organic molecules. The photosynthetic organisms pumped the oceans and the atmosphere full of deadly free oxygen.

The presence of oxygen allowed the evolution of a new class of heterotrophs that could respire. The presence of autotrophs also allowed the development of herbivores and the beginnings of a modern food chain.

Eukaryotes evolved long after prokaryotes, probably after the appearance of oxygen. Most biologists believe that mitochondria and chloroplasts derived from previously free-living prokaryotes that became **endosymbionts.**

Review and Thought Questions

Review Questions

1. Give two reasons why it took paleontologists so long to find Precambrian fossils.

2. Describe the two possible alternative compositions of the Earth's primitive atmosphere, including the disagreements scientists have about them. List two major sources of prebiotic organic molecules.

3. Describe Stanley Miller's landmark experiment to test the idea of prebiotic synthesis. What is the "primordial soup"?

4. What is a coacervate made of, how is it formed, and what can it do? What is a proteinoid microsphere made of, how is it formed, and what can it do? What is a liposome made of, how is it formed, and what can it do?

5. What would a protobiont need to become a living cell?

6. Why do scientists think that RNA was the first genetic material?

7. In what way did the early heterotrophs change the environment? What were some of the consequences of this change?

Thought Questions

8. Can natural selection operate on nonliving structures such as coacervates? Explain.

9. Pasteur's sterile broths contain no organisms that could consume any life that might arise spontaneously. Why have these flasks not given rise to new life in 130 years?

BiologyNow Resources

Biology ❸ *Now*™

Active Figures

18-6:   Conditions for life in Earth's early history
18-8:   The Miller Experiments

Preparing for an exam? Take a diagnostic test on your BiologyNow CD-ROM.

Online materials relating to this chapter are at:

http://biology.brookscole.com/AAL3

About the Chapter-Opening Image

Until Francesco Redi proved otherwise in the mid-17th century, most people believed that houseflies arose spontaneously.

PART IV
DIVERSITY

Frans Lanting/Minden Pictures

Classification:
What's in a Name?

Key Questions

- Why is it so difficult to identify species and classify them?

- How do biologists classify organisms?

Is the Red Wolf a Species?

In the 18th century, when European naturalists first began mapping and naming the animals and plants of North America, the destruction of the eastern forests had already driven countless species from the landscape. One of the first to go was the grey wolf, which had, before the time of Columbus, occupied nearly the whole of North America. By the 1970s, the grey wolf (*Canis lupus*) had been driven out of virtually all of the lower 48 states, its range reduced to Alaska, Canada, and tiny bits of northern Michigan and Minnesota (Figure 19-1). In the grey wolf's wake were left a rapidly shrinking population of red wolves (*Canis rufus*), which had once lived throughout the American South, from Florida to Texas; and a rapidly expanding population of coyotes (*Canis latrans*), which gradually spread from the Rockies and the Great Plains—north, east, and west—into the once-forested areas vacated by the grey and red wolves.

In 1967, the federal government designated the red wolf an endangered species, and biologists working for the U.S. Fish and Wildlife Service began surveying the South for red wolves. The red wolf, they discovered, was all but extinct. By 1972, the last few red wolves clung to the southern edge of North America, where Texas and Louisiana drop into the Gulf of Mexico. These last wolves were actively breeding with coyotes.

Government biologists realized with a shock that the only way to save the red wolf from complete extinction was to capture all the remaining wolves and breed them in captivity. From 1974 to 1980, government biologists trapped more than 400 animals. In 1980, in a marsh edging an industrial area of Galveston, Texas, the last wild red wolves in America walked into traps. Of the 400 captured animals, only 43 actually looked like wolves, however, and, when bred, only 14 produced descendants that looked like wolves rather than coyotes.

Under the watchful eyes of biologists, this group of 14 wolves has produced hundreds of offspring. Many of these captive-bred wolves have been released into places such as the Great Smoky Mountains of Tennessee and the Alligator River National Wildlife Refuge in North Carolina. The rest of the captive wolves continue to produce offspring

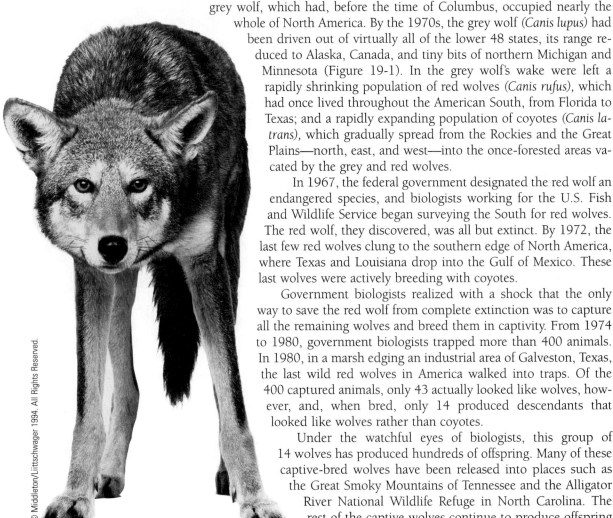

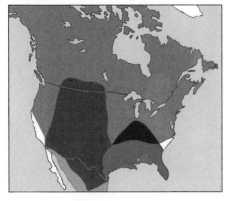

Historical

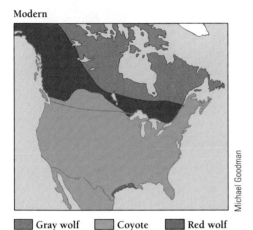

Modern

■ Gray wolf   ■ Coyote   ■ Red wolf

Michael Goodman

Figure 19-1

The disappearance of the grey wolf. As settlers moved across North America in the 1900s, they used guns, traps, and poison to nearly eliminate the grey wolf from the southern part of its range, shown in grey in the top map. Historically *(top)*, the grey wolf and the coyote used to overlap in the central United States, the wolves preying on bison and other large prey and the coyotes preying on smaller animals. But even though the coyote has been subjected to organized massacres like those that drove the wolf from the United States, the coyote has flourished. And, as shown on the lower map, today the coyote occupies nearly all of the wolf's former range.

to be released in years to come. The Red Wolf Recovery Program, as this project is now called, is the most successful captive-breeding program in existence. By the early 1990s, the red wolf had been brought back from the very brink of extinction. Although the wolves introduced into the Great Smoky Mountains survived poorly and also began breeding with local coyotes, the red wolves in Alligator River are doing much better. There, 100 wolves roam free on a half a million acres of public and private land. About two-thirds of the wolves have radio collars so that researchers can closely monitor their activities.

Not everyone initially applauded the program's success. Rural residents worried about the safety of their children and pets. Others questioned the price of saving the red wolf.

The Red Wolf Recovery Program costs about $1 million a year. Yet the program's most vocal opponents were not taxpayers' groups, but sheep and cattle ranchers. For them, the reintroduction of wolves to areas of the country now occupied by cattle and sheep might realize the romantic notions of city-dwelling environmentalists, but at the cost of valuable livestock.

Into this heated conflict stepped David Mech, one of the world's leading wolf experts. In 1989, at an Atlanta meeting of experts on wolf biology, Mech challenged his fellow researchers to tell him how they could justify spending so much money rescuing the red wolf when it might not even be a species.

Since the 18th century, biologists had argued over the status of the red wolf. In 1791, naturalists designated the red wolf a subspecies of the grey wolf. But later biologists decided that the red wolf was a separate species. In 1979, U.S. Fish and Wildlife Service biologist Ronald Nowak carefully compared the skulls of grey wolves, red wolves, and coyotes and noticed that the size and shape of the red wolf skull fell midway between that of the coyote and the grey wolf (Figure 19-2). Nowak's interpretation of the fossil record further suggested to him that intermediate skulls like that of the red wolf skull first appeared in North America more than a million years ago, well before the first wolves or coyotes. Nowak concluded that the red wolf was not only a unique species but also the ancient ancestor of both the grey wolf and the coyote.

Additional paleontological evidence suggested that the grey wolves of Europe and Asia originated in North America. Therefore, argued Nowak, the red wolf must be a direct descendant of the first wolf. The rescue of the red wolf from the brink of extinction, then, could be seen as the salvation of a southern species of nearly royal lineage. It was a compelling idea, one that persisted almost unchallenged for 10 years, throughout the early years of the Red Wolf Recovery Program.

But David Mech had a different theory about red wolves. In a 1970 book, Mech had proposed that the red wolf was neither species nor subspecies but a hybrid produced by interbreeding between the grey wolf and the coyote.

Mech's hypothesis accounted for Nowak's data showing that the size and shape of the red wolf skull was midway between that of the grey wolf and the coyote. Furthermore, northern grey wolves were known to interbreed with coyotes in some areas, and the offspring had skull measurements similar to those of red wolves. Nowak's skull measurements could be interpreted in two ways, Mech argued: either the red wolf species had split over evolutionary time into three species, the red wolf, the grey wolf, and the coyote; or red wolves were hybrids of grey wolves and coyotes (Figure 19-3).

In 1989, two University of California biologists, Robert Wayne (of UCLA) and Susan Jenks (of UC Berkeley), approached the U.S. Fish and Wildlife Service and offered to settle the matter once and for all. Like Nowak, Wayne was

Figure 19-2

In between. The red wolf *(center)* is intermediate in size, shape, and behavior between the grey wolf *(left)* and the coyote *(right).*

Biology ⓔ Now™   Learn about the connection between wolves and dogs by clicking on this figure on your BiologyNow CD-ROM.

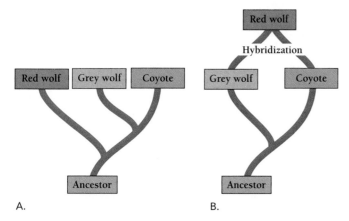

Figure 19-3
Alternate hypotheses. Biologists used to believe that the grey wolf and the coyote were descended from ancestors of the modern red wolf. But DNA evidence suggests that the red wolf is a hybrid of the grey wolf and the coyote.

an expert on the morphology and taxonomy of wolves and other canids (members of the dog family). But unlike Nowak, Wayne had added molecular genetics to his repertoire of research techniques. Wayne and Jenks offered to search the red wolf genome for some unique stretch of DNA that neither the coyote nor the grey wolf had, something that would clearly identify the red wolf as a separate species.

The government agreed to fund the study, and the two biologists began examining DNA from red wolves, grey wolves, and coyotes. Wayne and Jenks began with mito-

chondrial DNA because, in general, it accumulates mutations more rapidly than DNA from the nucleus. This rapid evolution makes mitochondrial DNA particularly useful for comparing groups of closely related organisms. The base sequence of the mitochondrial DNA of grey wolves (*Canis lupus*) and coyotes (*Canis latrans*), for example, differs by about 4 percent.

The mitochondrial DNA of the grey wolf, Wayne and Jenks found, included sequences that the coyote did not have. Similarly, the DNA of the coyote included sequences that the grey wolf did not have. The grey wolf and the coyote were clearly separate species (even though they can occasionally interbreed). But, to Wayne and Jenks's surprise, the mitochondrial DNA of the red wolf (*Canis rufus*) included *no sequences unique to red wolves.* Red wolf mitochondrial DNA only included sequences identical to those of grey wolves and sequences identical to those of coyotes. The two biologists tentatively and somewhat reluctantly concluded that the red wolf was most likely a hybrid of the grey wolf and the coyote.

Nowak and the other biologists at the U.S. Fish and Wildlife Service could not believe what they were being told. Maybe, argued the government biologists, Wayne and Jenks had simply missed the DNA sequences that distinguished the red wolf. Maybe they had not looked at enough DNA.

To put to rest any lingering doubts, Wayne and other colleagues turned to special repetitive regions of the DNA in the nucleus, called microsatellites. Wayne checked 10 microsatellite regions in blood samples taken from hundreds of living wolves and from 16 pre-1930s skins stored in a vast collection of furs at the Smithsonian Institution, in Washington, D.C.

The results were the same. Neither the samples of blood from living red wolves nor the samples from the skins of pre-1930s red wolves showed any unique sequences. By 1994, Wayne had found no evidence that the red wolf had ever been reproductively isolated from either grey wolves or coyotes. The red wolf had to be a hybrid of the grey wolf and the coyote.

Wayne's genetic data proved to be an embarrassment to the U.S. Fish and Wildlife Service, which had poured millions of dollars into the reintroduction program in the belief that the red wolf was a unique and endangered species. Yet, the agency had acted in good faith. Until Wayne and his colleagues finished their research, the U.S. Fish and Wildlife Service had no way of knowing that the red wolf was not a species.

Now the government agency was faced with a terrible dilemma. Wayne's results threatened to discredit the wolf recovery program, strip the red wolf of its endangered status, and further undermine the increasingly battered public image of the federal Endangered Species Act. The act had been under attack from its very inception by real-estate developers, ranchers, miners, and other industries that find species preservation inconvenient and expensive.

For now, the red wolf hybrid has not lost protection under the Endangered Species Act, but its hybrid status certainly weakened its claim to protection. The American Sheep Industry Association, for example, formally petitioned the U.S. Department of the Interior to take the animal off the endangered list. The Department of the Interior turned down this petition in 1992, arguing that "limited genetic exchange" between groups does not undermine species status. In short, the agency did not acknowledge that the red wolf was probably not a species.

To protect the red wolf, the U.S. Fish and Wildlife Service began pressuring Wayne to avoid the word "hybrid" in his research papers and to substitute the term "intergrade species" and other similar phrases. By 1994, it was clear, however, that the red wolf was not a species engaging in "limited genetic exchange" with other species.

In 1995, the U.S. Department of the Interior issued a legal opinion that said that hybrids would be protected under the Endangered Species Act if *morphological* evidence showed that the hybrids were similar to the endangered "pure" form. In essence, if they looked like red wolves, they would be protected. But the genetic data did not support the idea that a "pure" form of the red wolf had ever existed, certainly not in the last 100 years. In issuing this opinion, the agency excluded all the genetic evidence regarding the red wolf's species status. The only question was whether the red wolf looked different from the coyote and the grey wolf. It did, and, therefore, until such time as the government acknowledges the genetic data, the red wolf will be considered a species.

From the point of view of a taxonomist, the agency's decision might seem illogical. But from a political point of view, the decision made sense. The Endangered Species Act is the keystone law for environmental protection. Anything that weakens it weakens all environmental protection in the United States. Furthermore, the red wolf is enormously popular. One study by researchers at Cornell University found that nearly 80 percent of people surveyed in eight states surrounding the red wolf reintroduction area favored the wolves. Researchers estimated that tourists attracted by the wolves were bringing $37 million per year to eastern North Carolina and about $132 million per year to the national park area. In addition, the researchers estimated that people in the eight states surveyed were willing to pay about $3 million per year for the Red Wolf Recovery Program, and nationally, the public was willing to pay $18 million per year, figures that far exceed the actual cost of the program.

Some private groups continue to fight against the wolves. In the late 1990s, a group of private landowners took the U.S. Department of the Interior (which includes the U.S. Fish and Wildlife Service) to court, arguing that the agency could not legally restrict or limit the killing of red wolves on private property. This legal case went all the way to the Supreme Court, which in 2001 decided against the private property owners. The Supreme Court said that the Endangered Species Act gives Fish and Wildlife the right to regulate the killing of endangered species whether they are on public lands or private lands.

In any case, biologists themselves continue to disagree about both the status of the red wolf and the political meaning of that status. In a 1998 issue of the journal *Conservation Biology,* Nowak and another biologist wrote, "Studies support the biological hypothesis that the red wolf is systematically valid. . . . Hybridization has been advanced by individuals and groups who oppose reintroduction of the red wolf as a conservation strategy," while in the same issue, Wayne and his colleagues insisted, "Enough scientific evidence exists to suggest that the red wolf is a hybrid of the grey wolf and the coyote. . . . Conservation efforts directed solely at the red wolf may therefore compromise preservation efforts for other wolf species."

Despite these disagreements, programs to reintroduce both grey wolves and red wolves back into the wild continue to be funded and the courts continue to defend the practice. A major factor in the survival of these animals— whether one species or two—is the stubborn determination of the wildlife biologists who breed and release the wolves. The biologists don't give up easily.

19.1 Why Is It Difficult to Name and Classify Organisms?

The story of the red wolf illustrates that deciding how to classify organisms into tidy groups is no easy task. Surprisingly, recognizing species is usually the *least* controversial task facing a **taxonomist,** a biologist who classifies organisms. In contrast, determining whether a species should be subdivided into subspecies (and, if so, how many subspecies) nearly always raises an argument. Likewise, grouping species into higher taxa—such as families, orders, or classes—immediately stirs up a host of controversies. In this chapter, we discuss how biologists classify organisms. Along the way, we will see why classification is nearly always a controversial topic.

Why Do We Name Things?

"What's in a name?" asks Shakespeare's Juliet, "That which we call a rose by any other name would smell as sweet."

Why do we name things? By naming the rose, we distinguish it from other flowers—both in our own minds and in the minds of others with whom we speak. The name allows us to talk with others about the rose's significance—romantic, evolutionary, or otherwise. To the extent that everyone agrees on a name, we can each talk of roses with confidence that we will be understood. Naming things also allows us to communicate what we know about something, whether the something is an individual object or organism or a class of objects or organisms. We can distinguish poisonous plants from palatable ones, and dangerous animals from docile ones.

Further, by naming things, we force ourselves to make distinctions that we might not otherwise have noticed. In thinking about a rose, most people picture a red flower, yet we would not confuse a red rose with a red poppy or red carnation. Nor would we mistake a yellow rose for a buttercup or a dandelion. Naming an object helps us to know it.

Another reason for naming an object is to classify it, or assign it to a group. A couple might name their children David and Nina simply to distinguish them from others, for example. But parents also give their children a last name to identify the children as part of a particular group, in this case a family.

More and more biologists are taking an interest in naming and classifying all of the organisms that live on Earth. Questions about which organisms are protected by the Endangered Species Act are one reason for this increased interest. In addition, now that the human genome as well as other genomes have been completely sequenced, biologists want to know what all the different genes do. One way to find out is to compare the sequences of similar genes in different groups of organisms. When biologists see that humans share a gene with plants, that says something different about the gene than a gene we share only with other mammals. A gene we share with plants must be very old and more likely to be involved in some fundamental processes in the cell.

Naming objects allows us to both distinguish one object from another and assign objects to groups of similar objects. Understanding the evolutionary relationships among organisms has practical value.

What Did Linnaeus's System of Classification Accomplish?

To classify objects, we usually establish different categories, or levels, of classification and place these categories in a **hierarchy,** an arrangement in which larger groups contain smaller groups, which contain still smaller groups. We can illustrate a hierarchy of classification with a familiar example. Your address probably falls into each of the following categories: apartment number, street number, street, zip code, city, county, state, country, continent, planet, and so forth. Notice that the categories form a hierarchy, with each category often (but not necessarily) including more than one example of the lower category (Figure 19-4A). Each state, for example, includes many counties, and most counties include several cities and towns.

Recall from Chapter 15 that biologists use a hierarchical taxonomy established by the 18th-century Swedish physician and biologist Carolus Linnaeus. Before Linnaeus's work, biologists had already used a category called a **genus** [plural, **genera;** Greek, *genos* = to be born], a group of species that share many morphological characteristics. Linnaeus and his later followers established higher categories, so that today we traditionally recognize the following hierarchy of classification: species are grouped into genera, genera into families, families into orders, orders into classes, classes into phyla, and phyla into kingdoms. Some people remember these categories—in the reverse order—using the mnemonic, "Kings Play Chess On Fine-Grained Sand." But you can invent your own mnemonic. Biologists have further arranged the kingdoms into three domains: the Eubacteria (the true bacteria), the Archaea (the ancient bacteria), and the Eukarya (all organisms whose cells contain a nucleus). Figure 19-4B shows the hierarchical classification of several species, including our own. We are members of the **domain** Eukarya, the **kingdom** Animalia, the **phylum** Chordata, the **class** Mammalia, the **order** Primates, the **family** Hominidae, the genus *Homo,* and the species *Homo sapiens.*

Biologists use the word **taxon** [plural, taxa; Greek, *taxis* = to put in order] to mean a group of organisms at any level of the classification hierarchy. So we may speak of all the primates as one taxon and all the bats as another. At the same time, we may speak of all the individuals of a species such as *Homo sapiens* as another taxon.

Linnaeus not only introduced a system of classification but also standardized the naming of organisms. The official name of a species is a **binomial,** a two-part name that contains both the genus (*Homo,* for example) and species (*sapiens,* for example). It is not correct to leave out the genus name or the species name when referring to a species. The fruit fly (or, more properly, the vinegar fly) most often used in genetics experiments is *Drosophila melanogaster.* We may, after writing the name out fully once, abbreviate the genus to its initial letter, as in *D. melanogaster.* But scientists and laypersons alike often refer to *Drosophila melanogaster* as simply Drosophila or *Drosophila.* Both are informal and technically incorrect. North America alone has nearly 200 different

Figure 19-4

Hierarchical classification. A. In an address, the information in each level is included within the levels above. B. A domain can include one to several kingdoms, each of which can include several phyla. Each phylum can include several classes, and so on.

Your address	
Galaxy	Milky Way
Star	Sun
Planet	Earth
Continent	North America
Country	United States
State	California
City	San Francisco
Street	Haight St.
Number	555

A.

TAXON	Human	Barn owl	Box turtle	Maize
Domain	Eukarya	Eukarya	Eukarya	Eukarya
Kingdom	Animalia	Animalia	Animalia	Plantae
Phylum	Chordata	Chordata	Chordata	Anthophyta
Class	Mammalia	Aves	Reptilia	Monocotyledones
Order	Primates	Strigiformes	Chelonia	Commelinales
Family	Hominidae	Tytonidae	Emydidae	Poaceae
Genus	Homo	Tyto	Terrapene	Zea
Species	*Homo sapiens*	*Tyto alba*	*Terrapene carolina*	*Zea mays*

B.

species that all belong to the genus *Drosophila*. We cannot call all of them *Drosophila* and know what we are talking about.

Binomial names are always Latin or Greek (or at least they are meant to be). Until the 20th century, Latin was regarded as the "universal language" by educated Europeans. The use of Latin names eliminates uncertainty in translation from one language to another (as the Latin name need never be translated). Latin also prevents confusion that results when people in different regions give the same common name to different species. For example, the English robin (*Erithacus rubecula*) bears only passing resemblance and relation to the American robin (*Turdus migratorius*) we know in the United States (Figure 19-5).

Linnaeus standardized both the naming of organisms (with a Latin binomial) and the classification of organisms (into a strict hierarchy).

Figure 19-5

Why we use scientific names.
Although unrelated, both the English robin and its American namesake are called "robins." Scientific names are intended to minimize such confusion.
A. English robin, *Erithecus rubicula*.
B. American robin, *Turdus migratorius*.

A.

B.

Who Names a Species?

The first person to describe a new species assigns it its name. Sometimes biologists discover that two people have described and named the same species. The name given first is almost always considered the correct name and has "priority." This rule is a way of ensuring that species names are stable, that they stay the same from one decade to the next. That way, biologists always know exactly which species is meant.

Still, sometimes this is too bad. One hundred years ago, the Yale paleontologist Charles Marsh assigned the charming name eohippus ("dawn horse") to the 55-million-year-old ancestor of the modern horse. Marsh did not know that the tiny fossil horse had already been described and named by someone who thought it was a hyrax (an animal that looks like a guinea pig, only bigger). Eohippus, which is both poetic and accurate, lost out to *Hyracotherium,* a mouthful that means hyraxlike mammal.

Taxonomists use their naming privileges for different purposes. Some name species after colleagues they admire. Some try to describe the species: *Rosa alba,* for example, is a white rose [Latin, *alba* = white, as in albion or albumen), while *Rosa chinensis* is a Chinese rose. Other biologists just want to have fun: the names for some genera of stink bugs (family Pentatomidae), for example, end with the suffix -chisme, pronounced "kiss me."

Species are named by the first person to recognize the species as new.

How Do We Recognize Species?

Long ago, the Greek philosopher Plato described how we make ordinary classifications. Plato argued that the individual objects of the world are all variations of universal "types," forms that share an ideal underlying plan (Chapter 15). A chair, for example, always has a seat large enough to support a sitter and, usually, has four legs and a back.

Aristotle extended Plato's concept of types to organisms. In this view, called the **typological species concept,** each species consists of individuals that are variants of a fixed underlying plan. The key word here is "fixed." A typological species cannot evolve and become something different. Aristotle would have argued that a robin and a cat each have a fixed underlying plan.

Eighteenth- and 19th-century biologists applied the philosophy of types in the following way. When they first recognized a new species, they killed and collected several examples of the organism. One individual was then selected as the "type specimen" for each species and kept in a museum for reference. All other organisms believed to belong to that species were then compared to the type specimen to verify that they were the same. Because of Charles Darwin,

however, biologists have recognized that the typological species concept has two important limitations:

1. All species change over time.
2. The individual members of a species vary.

Modern biologists have mostly defined **biological species** as groups of actually or potentially interbreeding populations that are reproductively isolated from other such groups (Chapter 17). For example, the cardinals in a Virginia suburb can easily breed with one another, as well as with other cardinals in nearby Maryland. But they cannot breed with mockingbirds. In this view, called the **biological species concept,** individuals are members of a species only in relation to other species. In the same way, "brother" is not a category that can be applied to a single individual. Someone is a brother only in relation to someone else.

To decide whether a group of organisms constitutes a separate species, most biologists like to use the biological species concept. For most organisms, however, the biological species concept is impossible to apply. Certainly we can obtain no information about breeding populations for fossil organisms or for organisms that reproduce asexually, including great numbers of plants, fungi, protists, and bacteria.

Even living, well-studied, sexually reproducing organisms pose problems. No one was certain, for example, how frequently grey wolves and coyotes interbred until biologists tested DNA in blood samples drawn from hundreds of wild wolves and coyotes. Such efforts are expensive and therefore unusual. Many sexually reproducing plants and animals readily form hybrids with groups traditionally classified as separate species.

We can see two objections, then, to the biological species concept: (1) the frequent lack of information about whether two groups can interbreed; and (2) the fact that many recognized species can and do interbreed.

Biologists cope with the practical problems presented by the biological species concept by recognizing that a biological species is usually a reproductive unit, a genetic unit, and an ecological unit. A species is a **reproductive unit** in the sense that the species is the group of potential mates. Among organisms that do not reproduce sexually, that definition of a species does not apply. A species—whether sexually reproducing or not—is also a **genetic unit,** because a species can be defined as a gene pool that is usually isolated from other species by reproductive barriers. A reproductive barrier, discussed in Chapter 16, is anything that prevents two organisms from mating, including differences in behavior, differences in the timing of reproduction, and physical or physiological incompatibility.

Finally, a species is an **ecological unit:** within a given habitat, a species interacts in a predictable fashion with other species in the environment. Grey wolves, for example, hunt in packs of 7 to 20 if large animals such as deer, moose, and caribou are available. In contrast, coyotes do not hunt in packs and prefer to prey on smaller animals such as mice and hares. Because of their size and behavior, coyotes interact with their environment differently from wolves. Biologists now know, for example, that under certain circumstances wolves and coyotes interbreed. They also know that the two species generally remain distinct, both genetically and ecologically.

Even these rules are not enough to help classify all organisms. As we will see in the next chapter, bacteria frequently exchange genes with other, unrelated species. Thus bacteria are not generally reproductive units or even genetic units, but only ecological ones, and the way that bacteria interact with their environment can change radically with the sudden acquisition of a single gene.

In reality, biologists usually classify species by the way they look, relying on morphology—the form or visual appearance—of organisms for the provisional definition of a species. As we saw at the beginning of this chapter, biologists name even living, sexually reproducing species such as the red wolf on the basis of their morphology, simply because it is easier.

The morphologically defined species usually corresponds to the biologically defined species. The grey wolf and the coyote, for example, look different and behave differently. In large part, they occupy different habitats. In short, they look and act like true species. Genetic data confirm that the two species have distinct sequences of DNA, showing that they have been reproductively isolated from each other. That these two species can interbreed extensively does not make them the same species. In general, the morphologically defined species corresponds to the biologically defined species because reproductive isolation tends to lead to the evolution of divergent morphological characters. Despite most biologists' noisy objections to typological species, then, they frequently begin defining a species on a typological basis.

Biologists use both biological and typological concepts to name and classify species. There is no single definition of a species that is always both true and practical.

19.2 How Do Biologists Group Species?

Although biologists sometimes argue about whether particular groups deserve the species label, all modern biologists consider species to be real entities, not abstract or arbitrary classifications. Because species are natural entities, recognizing what constitutes a species is obvious most of the time. Biologists disagree surprisingly rarely on the boundaries that separate individual species. Yet, grouping species into larger groups such as genera, families, orders, phyla, kingdoms, and now domains has been a source of endless disagreement.

Naming and classifying species is the main business of **systematics**, the study of the diversity of organisms and of the relationships among them. In contrast, **taxonomy** is the theoretical study of classification, including the study of the principles, procedures, and rules of classification. These two words have different meanings, but, for simplicity, we use the word "taxonomy" only in this book.

Biologists would like to classify organisms according to *relatedness*, but they must do so from measurements of *similarity*. In classifying any kinds of objects—organisms, books, or buildings—we naturally group things according to their similarities. We can understand the problems this presents if we imagine attending a big family reunion of 200 people and trying to decide who is related to whom. A group of redheads might appear to be related. We might find, however, that one of them is simply married to someone else in the family and is not related to anyone else at the reunion. Similarity by itself does not always prove relatedness. And deciding which similarities are the important ones is often hard to decide.

For example, in Papua New Guinea, the giant tropical island just north of Australia, biologists have compared their own classification of plants and animals with that used by the tribespeople who live there. The biologists and the tribespeople agree nearly exactly on the number and the identity of species. They disagree, however, on the higher classifications. The biologists group organisms according to their biological similarities—lumping all the birds together in the class Aves, for example. The tribespeople classify according to use—lumping together all the species that are edible, all that are poisonous, or the ones that provide materials for clothing. Both systems of classification serve legitimate purposes. The accompanying box discusses still another taxonomy of animals.

The 18th-century taxonomist Carolus Linnaeus divided all organisms into two kingdoms, the plants and the animals. As biologists learned more about life on Earth, however, they became dissatisfied with Linnaeus's groupings. The fungi might not fly like birds or swim like shrimp, but mushrooms were clearly not photosynthetic organisms like plants either. In 1969, the ecologist Robert H. Whittaker proposed a five-kingdom classification that most biologists accepted. Whittaker put all the prokaryotes into the kingdom Monera, then divided the eukaryotes into four kingdoms: the protists, fungi, plants, and animals. In recent years, biologists have all but abandoned the five-kingdom classification, as more and more evidence suggests a more complex phylogeny, something like that in Figure 19-6.

And while Linnaeus classified just 10,000 species, biologists have now named more than 1.7 million living species. Experts estimate the total number of living species is 4 to 10 million. Classifying all these organisms has become a monumental task. At current rates of research, all the world's traditional taxonomists can only describe and classify about 250 species a year. At that rate, a complete phylogeny of life is centuries away. In recent years, biologists have begun looking for shortcuts to produce a rough-

Extreme Biology Foucault's Chinese Encyclopedia

"This book first arose out of a passage in Borges, out of the laughter that shattered, as I read the passage, all the familiar landmarks of my thought—our thought, the thought that bears the stamp of our age and our geography—breaking up all the ordered surfaces and all the planes with which we are accustomed to tame the wild profusion of existing things, and continuing long afterwards to disturb and threaten with collapse our age-old distinction between the Same and the Other.

"This passage quotes a 'certain Chinese encyclopaedia' in which it is written that 'animals are divided into: (a) belonging to the Emperor, (b) embalmed, (c) tame, (d) sucking pigs, (e) sirens, (f) fabulous, (g) stray dogs, (h) included in the present classification, (i) frenzied, (j) innumerable, (k) drawn with a very fine camelhair brush, (l) et cetera, (m) having just broken the water pitcher, (n) that from a long way off look like flies.'

"In the wonderment of this taxonomy, the thing we apprehend in one great leap, the thing that, by means of the fable, is demonstrated as the exotic charm of another system of thought, is the limitation of our own, the stark impossibility of thinking that."

—From the introduction to *The Order of Things, An Archaeology of the Human Sciences,* by Michel Foucault, Vintage Books, New York, 1970.

Figure A
A Chinese hanging scroll depicting hens and roosters in a barnyard.

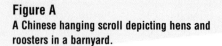

Figure 19-6
The Tree of Life. Here is a simplified tree of life showing the three domains (Eubacteria, Archaea, and Eukarya) and the main lineages of eukaryotes. Notice the four separate lineages of protists. The protists are not a single lineage like the fungi or the animals, but a hugely diverse group of organisms with many separate lineages.

draft phylogeny. For example, University of Guelph biologist Paul Hebert dreams of a device that will take an insect leg, grind it up, and instantly classify it according the sequence of a single gene in the mitochondria called cytochrome oxidase. The idea is controversial and his device does not yet exist, but the idea of a simple way to roughly classify more species faster has captured the imaginations of taxonomists, evolutionary biologists, and ecologists.

Taxonomy seems to invite disagreement, and working biologists rely on three separate methods for classifying organisms. The first is classical evolutionary taxonomy, in use for the last 150 years. In the last few decades, biologists proposed two other ways to classify organisms: phenetics, which proposed a mechanical step-by-step method for classifying organisms, and cladistics, which ambitiously proposed to classify organisms according to their evolutionary history.

Classical Evolutionary Taxonomy

Classification by means of classical evolutionary taxonomy relies explicitly on the judgment and experience of the taxonomist. In fact, in one definition of a species, biologists like to say, "A species is what a good taxonomist says it is. And a good taxonomist is one whose work has stood the test of time." A good taxonomist examines the similarities and differences in as many traits as possible, but giving more weight to some traits than others.

The traits that biologist most commonly used to classify living or well-fossilized animals used to be morphological ones—including external form, internal anatomy, tissue types, and chromosomal morphology. Mammals, for example, are classified into three broad groups according to their skeletons. (Biologists tend to focus on bones because these are the parts of mammals and other vertebrates preserved in the fossil record.) Adult "monotremes"—including the duck-billed platypus and the spiny anteaters—have no teeth and lack certain bones of the skull that other mammals have. They also lay eggs instead of giving birth to live young, and their limbs stick out to the sides like those of lizards (and the most primitive fossil ancestors of lizards and mammals). Because of these and other traits, monotremes are considered primitive mammals, separate from "true mammals" like ourselves.

Classical taxonomists also rely on embryological evidence that implies common ancestry. For example, the early embryos of all mammals, including humans, include a stage with gill-like pharyngeal slits and pouches (Figure 15-15). This fact, with others, suggests that we are descended from aquatic organisms that needed gills to extract oxygen from water. Finally, taxonomists may also classify on the basis of behavioral traits and, increasingly, on the basis of molecular characteristics, either protein sequences or DNA sequences.

Not all similarities between organisms mean that they are related. Bats and birds have both evolved wings, for example, but bats are no more related to birds than are humans. We say that the wings of birds and bats are "homoplasious." **Homoplasy** [Greek, *homos* = same + *plasis* = mold, form] is the possession by two species of a similar trait that is not derived from a common ancestor. ("Homoplasy" has replaced the older word "analogy," which requires that two similar traits perform the same function. In an obscure organism about which biologists don't know much, it is sometimes difficult to decide if two similar-looking structures actually perform the same function, or if they just look as if they do.) Cases where two species share similar structures that have evolved to serve the same role, such as the flippers of penguins and seals, for example, are examples of "convergent evolution," discussed in more detail in Chapter 17.

A second form of homoplasy is a **reversal,** where a trait that has evolved into one thing reverts to its original primitive form. For example, dolphins have evolved somewhat fishlike bodies, similar to those of the fish they evolved from millions of years ago. But such reversals in body structures are unusual and also obvious to someone trying to classify an organism. In contrast, a reversal in the base sequence of a gene is not unusual, and there's no way to know if such a reversal has occurred.

In contrast to homoplasy, similarities that *are* inherited from a common ancestor are said to be **homologous.** The leg of a horse and the leg of a human are homologous. They share similar bones and muscles because they evolved from

a common ancestor. Supporting evidence comes from embryology: in humans and horses, the bones and muscles of the leg develop from the same embryological tissues.

When classifying organisms, a classical taxonomist must take into account which characteristics are likely to be homologous and which homoplasious, giving weight to homologous characters and disregarding homoplasious ones. Although the wings of bats and birds seem obviously different to us, homoplasious characters in closely related species are sometimes hard to distinguish from homologous ones.

Homologous characters indicate relatedness, while homoplasious characters indicate similar adaptations but not necessarily relatedness. Taxonomists try hard to distinguish homologous characters from homoplasious ones.

How Does Phenetics Classify Organisms?

Phenetics is a method of classification that avoids subjective choices by making none. **Phenetics** classifies species using *all* observable characteristics. The word "phenetics" comes from the same Greek root as "phenotype" [Greek, *phainein* = to show]. The goal of phenetics is to list as many traits as possible for each organism and then to use computers to sort organisms according to their degree of similarity. Pheneticists argue that the resulting groupings are objective measures of the similarities among organisms. Phenetic classification does not depend on an organism's ancestry.

A phenetic taxonomist describes the phenotype of an organism with a series of yes-or-no questions. Each of these questions defines a *unit character*—a characteristic of an organism that cannot be split into finer descriptions. For example, some unit characters for an animal are the answers to questions such as: (1) Does it have five digits on its front limb? (2) Does it have four digits on its front limb? (3) Does it have a backbone? The pheneticist records the answers to these questions in a binary (1 or 0) code: 1 for yes, 0 for no.

In phenetics, each unit character is as important as any other. That is, each of hundreds or thousands of characters is considered to be equally important. Whether an animal has spotted fur would be considered as important as whether it has a backbone.

The drawback to phenetics is that while it accurately groups organisms according to their similarities and differences, it doesn't necessarily group them according to lines of descent. Because phenetics views all traits as equivalent, it doesn't distinguish between homologous and homoplasious traits. As mentioned above, unrelated species often evolve similar adaptations in convergent evolution (Chapter 17). For example, marsupial moles share many traits with placental moles, but not because they are related (Figure 17-14). Both penguins (which are birds) and seals (which are mammals) have independently evolved flippers and round, streamlined

Extreme Biology Are Guinea Pigs Rodents?

One group of animals whose classification came under close scrutiny recently is the guinea pigs and their relatives. One of the defining characteristics of the order Rodentia is a single pair of razor-sharp incisors with which they can gnaw through the husks of nuts and seeds. In fact, the word "rodent" means "to gnaw." Because guinea pigs possess these distinctive gnawing incisors, as well as other adaptations for gnawing, taxonomists have traditionally

Figure A
Guinea pigs. Rodents or not?

classified them in one of three suborders of rodents.

The three suborders—the squirrels, the mice and rats, and the guinea pigs and their relatives—are classified according to the anatomy of their gnawing muscles. The principal jaw muscle is the masseter muscle, which closes the jaws and also pulls the lower jaw forward, a movement critical to gnawing. In mice and rats, the two sections of the masseter—the lateral masseter and the deep masseter—both attach well forward on the jaw, and these animals' ability to gnaw is unparalleled. In squirrels, only the lateral masseter extends forward. In guinea pigs, only the deep masseter extends forward. Consequently, neither squirrels nor guinea pigs gnaw as well as mice and rats.

The suborder to which the guinea pigs belong, called the caviomorphs, differs from other rodents in numerous ways. Caviomorphs are more diverse in their morphology and habits than either the squirrels or rats and mice. They include porcupines, chinchillas, various guinea pigs (called cavys), capybara, pacas, and agoutis. The capybara, for example, is a large, social herbivore. Individuals may reach 150 pounds. Caviomorphs all have proportionately larger heads than other rodents and they reproduce differently (Figure A). While a rat may bear litters of 11 every three weeks, a guinea pig bears 2 or 3 young every few months.

Like mice and rats, guinea pigs are frequent subjects of scientific experiment. As a result, quite a bit is known about guinea pig physiology and genetics. In 1991, researchers suggested that the protein sequences of guinea pigs are sufficiently different from those of rats and mice to justify giving the guinea pigs and their relatives a separate order.

A series of studies in the early 1990s on both protein and nucleotide sequences gave contradictory answers, however, and the guinea pigs remained classified as rodents. Then, in 1997, a group of Italian and Swedish researchers carefully analyzed differences in the mitochondrial DNA (mtDNA) of 16 animals whose complete mtDNA sequences were known.

The results were dramatic. Three different methods of classification all put the guinea pigs in a separate taxon from the mice and rats. Indeed, the guinea pigs appeared to be more closely related to rabbits and primates than to rodents. Although the five researchers called for further research—"the phylogeny will be clearer when a larger number of species can be considered in phylogenetic analysis"—they titled their research report in the journal *Nature* decisively: "The guinea pig is not a rodent." Despite their certainty, however, the question remains open.

---

bodies that are ideal for swimming. Penguins' and seals' many common traits say nothing about their relatedness.

What most biologists really want to know is how species are related to one another. At that big family reunion, we want to know which individuals are siblings and which ones are cousins. And sometimes cousins look more alike than siblings, so grouping only by similarity wouldn't necessarily give us the right answers. Phenetics would not reveal whether the red wolf was a hybrid of a grey wolf and a coyote or a common ancestor of both. Phenetics can say that red wolves are intermediate in morphology but not *why* they are intermediate.

Pheneticists classify species and higher groupings by assigning equal value to all characters, whether homologous or homoplasious.

How Do Advocates of Cladistics Classify Organisms?

The goal of **cladistics** [Greek, *clados* = branch] is to group organisms according to their evolutionary history, or **phylogeny** [Greek, *phylon* = race or tribe + *geneia* = birth or origin]. Cladistics starts with the assumption that there is one true family tree and that, if we can discover it, we can produce a natural grouping or taxonomy of organisms (Figure 19-7). In contrast to phenetics, which classifies according to all traits, cladistics constructs family trees using only certain traits. These certain traits are called **shared derived characters**. For example, all mammals share fur and mammary glands, both of which are shared characters. In the context of other vertebrates—amphibians and reptiles—fur is a shared derived character of mammals. But context is everything. In primates, fur is a **primitive character**, one that is shared with all other mammals.

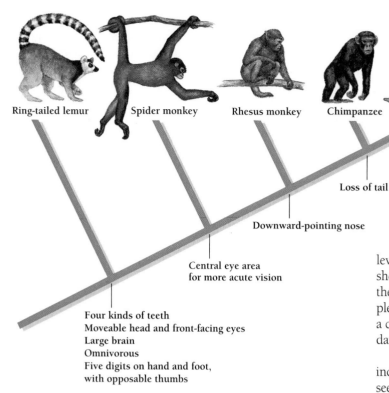

Ring-tailed lemur   Spider monkey   Rhesus monkey   Chimpanzee   Human

Loss of opposable
thumb on foot

Loss of tail

Downward-pointing nose

Central eye area
for more acute vision

Four kinds of teeth
Moveable head and front-facing eyes
Large brain
Omnivorous
Five digits on hand and foot,
with opposable thumbs

Figure 19-7

A cladogram for five primates. At each node, some "shared derived characters" are listed. Rhesus monkeys, chimpanzees and humans all have downward-pointing nostrils. But only chimpanzees and humans are without tails. A "primitive character" for apes such as humans, chimpanzees, and gorillas (not shown) is front-facing eyes. All primates have front-facing eyes.

Biology ⓔNow™ Learn more about the process of classification of primates by clicking on this figure on your BiologyNow CD-ROM.

Cladists use shared derived characters to construct a family tree called a **cladogram.** For example, the cladogram in Figure 19-7 shows the relationship among five primates: a ring-tailed lemur, a spider monkey, a rhesus monkey, a chimpanzee, and a human. A cladogram is a hypothesis suggesting a set of relationships. It's an educated guess that can be tested, just like any other hypothesis. If it withstands lots of tests, it may come to be accepted.

Constructing the cladogram requires that we note characters in which these species differ. Four of the five primates in Figure 19-7 (all except the lemur) possess a shared derived character—a specialized area in the center of the eye's retina that allows more acute vision. Three species (rhesus monkey, chimp, and human) have nostrils that point downward. Humans and chimps differ from the other three species in that they lack tails.

The cladogram in Figure 19-7 is a classification hierarchy that represents a series of hypotheses about the evolutionary history of the primates. Biologists can test these hypotheses by checking to see if other features of these five

animals show a pattern of primitive and derived characters consistent with this phylogeny. Biologists can also check to see when each of these animals appears in the fossil record relative to the others. For example, if chimpanzees appeared in the fossil record earlier than animals that look like rhesus monkeys, that would suggest that the phylogeny was incorrect. Finally, biologists can compare the sequences of specific proteins and genes in the five animals.

According to cladistics, every taxon above the species level—whether it is a genus, a family, or a higher category—should be **monophyletic,** meaning that each taxon includes the ancestral species and all its descendants. Birds, for example, form a monophyletic taxon: all birds are descended from a common ancestor that was itself a bird, and all the descendants of this ancestor are birds.

Cladists reject any taxon that is **polyphyletic**—one that includes descendants of more than one ancestor. As we will see in the following chapters, a number of higher taxa appear to be polyphyletic. For example, protists have evolved from many different separate ancestors (Figure 19-6). They do not constitute a single lineage. For this and other reasons, cladists reject the traditional grouping of protists into the kingdom Protista.

Another group that are not monophyletic are the bony fishes. Traditional taxonomists group the lobe-finned lungfishes with all other fishes. But the descendants of the original lungfishes include not only modern lungfishes but also amphibians, reptiles, birds, and mammals. Because some of the descendants of lungfish ancestors are not fishes, cladists say the group "fishes," which includes bony fishes and lobe-finned fishes, is not monophyletic but "paraphyletic." A **paraphyletic** group contains some but not all of the descendants of a common ancestor. A strict cladistic interpretation would therefore say that the class of animals we call fish—including lungfishes but not mammals—does not exist. In a cladistic sense, we humans are a kind of fish.

Like other methods of classification, cladistics has some limitations. Some organisms cannot be classified using cladistics at all. In bacteria, for example, deciding which characters reveal ancestry is nearly impossible. Bacteria exchange genes and traits with such freedom that the usual lines of descent may not exist. As a result, bacteria are classified phenetically using DNA sequences.

Cladists who study the lineages of plants face similar problems, for separate species of plants often hybridize to form a new species that may or may not be reproductively isolated from the parent species. Even animals hybridize, confusing the search for one true lineage. The red wolf, if it truly is a hybrid, would not have a single, branching lineage but two lines of descent that separate, then come back together.

Cladists group organisms according to their relatedness. A cladogram is a hypothesized phylogeny, or family tree.

Do DNA and Protein Sequences Provide "Objective" Criteria for Classification?

As we saw in Chapter 15, the sequence of a protein evolves in much the same way as a morphological feature. For this reason, comparisons of amino acid sequences (or nucleotide sequences) suggest relationships between species. For example, all other things being equal, two species with identical genes for hemoglobin are more likely to be closely related than two species whose genes for hemoglobin differ by several bases.

Because DNA sequences are linear and unambiguous, it is easy to imagine ranking all genetic differences equally, as pheneticists suggest we do with morphological differences. Indeed, evolutionary biologists have embraced the molecular approach enthusiastically, partly for this reason. But the DNA sequences that specify proteins evolve in the same way as larger-scale characters, and distinguishing between homologous and homoplasious sequences may be even harder than with morphological traits. And, as mentioned earlier, the reversal of a mutation is impossible to detect. We don't know if two species share the same base sequence because it never changed in neither species, or if the sequence changed in one species and then changed back. Base sequences are extraordinarily valuable, but they are not more objective measures of relatedness than morphological characters.

One advantage of morphological data over molecular data is that morphological information can often be checked against the fossil record. If turtles and mammals share a particular protein, we can infer that this protein evolved before the two lineages split. But we cannot find the protein or its gene in the fossil record to check this hypothesis. In contrast, paleontologists can look in the fossil record to see if a certain bone evolved before or after the first appearance of a lineage of animals.

We know that morphological traits evolve at different rates. A structure such as the backbone may persist for 200 million years, while fur color may change dramatically in just a few generations. It makes more sense to group animals according to stable characters such as whether they have backbones than according to their external coloring.

In the same way, genes, and the polypeptides they specify, also evolve at different rates. Histone 4, a protein that binds tightly to nuclear DNA, appears to have remained nearly unchanged for over 500 million years, while other protein sequences diverge strongly over relatively short periods. Proteins, like morphological characters, evolve at different rates according to how much selective pressure they experience

(and what kind). It makes sense when classifying organisms to give priority to the proteins that have evolved most slowly.

If all genes and morphological features changed at constant rates and if convergent evolution and reversals never occurred, then cladists, pheneticists, traditionalist taxonomists, and molecular biologists might always agree on how to classify organisms. But the phenotypes of organisms change in response to natural selection and the chance appearance of new mutations. All classification is made difficult by natural selection, which continually shapes organisms and their genes in ways that sometimes obscure patterns of relatedness.

Like other traits, nucleotide and amino acid sequences may be homologous or homoplasious, stable or unstable. Molecular approaches to classification have opened up a rich, new vein of evidence, but the sequences themselves are not more "objective" measures of relatedness than morphological traits.

Key Concepts

- Naming organisms allows biologists to distinguish one species from another and to group related species.
- Biologists classify organisms into species and also into larger groups such as genera, families, and orders.
- Natural selection shapes organisms in ways that sometimes obscure patterns of relatedness.

Summary with Key Terms

How do biologists define species?

A **taxonomist** is a biologist who classifies organisms. Each classification of organisms is a **taxon**. Many biologists now use the word **systematics** for the naming and classification of organisms and **taxonomy** for the study of *how* organisms should be classified. Until this century, biologists usually viewed the members of a species as variants from a fixed underlying plan, in what is now called the **typological species concept.** Today, biologists usually define a species according to the **biological species concept,** which says that a species is a group of actually or potentially interbreeding organisms that do not interbreed with members of other such groups. Besides being **reproductive units, biological species** are also **genetic units** and **ecological units.** Every formal definition of a species has some practical or theoretical limitation. All the same, most species that have been identified are what taxonomists call "good" species, species that are obviously different from other similar species.

How do biologists group species into larger taxa?

Most of the time, morphologically defined species correspond to the biologically defined species. Classification of species into larger groups, however, may be difficult. Biologists use a **hierarchical** scheme, in which species are grouped into **genera**, genera into **families**, families into **orders**, orders into **classes**, classes into **phyla** (or divisions), and phyla into **kingdoms**. Kingdoms are further arranged into three **domains**—the Archaea, the Eubacteria, and the Eukarya. Each species has a two-word designation, called a **binomial**, which names both the **genus** and the species.

The biological species concept is not always useful. Classifiers may know nothing about breeding patterns, and many species interbreed with other species. **Phenetics** gives equal weight to every character, grouping organisms with similar clusters of characteristics. But not all common characters indicate relatedness. Similarities that are inherited from a common ancestor are said to be **homologous** (and are given more weight). Both similar traits that evolved separately (convergent evolution) and **reversals** are forms of **homoplasy**. Homology indicates relatedness but homoplasy does not.

Cladistics insists that the only scientifically meaningful classification scheme is one that reflects **phylogeny**, or evolutionary history. The goal of cladistics is to construct evolutionary trees, or **cladograms**, that show the relationship of groups of species. **Cladograms** are constructed using **shared derived characters** rather than **primitive characters**. Cladists argue that every recognized group should be **monophyletic**, that is, each group should include an ancestral species and all its descendants. But many traditionally accepted groupings are **paraphyletic** or **polyphyletic**.

Review and Thought Questions

Review Questions

1. Define a species according to the biological species concept. What two kinds of problems often make this definition impractical?
2. What two purposes are accomplished by Linnaeus's system of classification? Name the seven levels of taxa commonly used by taxonomists.
3. Define phenetics.
4. What is cladistics? Why do cladists require that a lineage be monophyletic to be considered part of a single taxon?
5. Why do cladists say that "the reptiles" (meaning turtles, crocodiles, snakes and lizards, and dinosaurs) is not a legitimate taxon?

6. Cladists consider the birds, class Aves, a legitimate taxon because birds are monophyletic. On the other hand, the very first bird was also most likely a dinosaur. Would a cladist argue that the group we call dinosaurs should include the birds? Explain why or why not.
7. Define "paraphyletic."
8. What do DNA and protein sequences add to the science of taxonomy?

Thought Questions

9. Do you think that the red wolf should be saved from extinction? Do you think that the decision to save it should be influenced or decided by whether it is a species, a subspecies, or a hybrid? Explain why you think what you do.
10. If you found a new species of bird, what would you name the bird? If you were in charge of classifying this bird, what system of classification would you use? Why? Now describe the process, step by step, by which you would classify your bird. You may make up characteristics.

BiologyNow Resources

Biology ⑧Now™

Active Figures

19-2: The connection between wolves and dogs
19-7: The process of classification of primates

Preparing for an exam? Take a diagnostic test on your BiologyNow CD-ROM.

Online materials relating to this chapter are at:
http://biology.brookscole.com/AAL3

About the Chapter-Opening Image
The red wolf illustrates the confusion that often arises in classifying organisms. How a species is classified can have important consequences. For example, the red wolf's status could profoundly affect the strength of the Endangered Species Act.

Prokaryotes: How the Other Half Lives

Key Questions

- How do bacteria acquire resistance to antibiotics?
- How do we know that an organism is a prokaryote?
- How do biologists classify prokaryotes?

Tracking Down a Killer

In 1983, a young scientist named Scott A. Holmberg showed for the first time that feeding antibiotics to farm animals can sicken and even kill people. Most of us have experienced food poisoning, the severe nausea and diarrhea that result from consuming meat, milk, or eggs contaminated with salmonella bacteria. Farm animals—whether cattle, pigs, or chickens—can carry salmonella, but animals fed antibiotics, Holmberg showed, carry especially deadly strains of the bacteria.

In February 1983, Holmberg was working in Atlanta, at the Centers for Disease Control (CDC), the nation's premier center for the study of all public health problems, from AIDS to smoking among teenagers. His boss, Mitchell Cohen, had received a report that 11 Minnesotans had been hospitalized with salmonella poisoning. Eleven people was hardly an epidemic: after all, 40,000 Americans are hospitalized with salmonella every year, and another 5 million suffer the severe diarrhea and vomiting associated with salmonella poisoning but don't need hospitalization. What was both disturbing and exciting about the Minnesota cases was, first, that most of the patients had been taking antibiotics when they came down with food poisoning and, second, that each patient was extremely sick. One man had died. Cohen strongly suspected that antibiotics were somehow causing or helping to cause the food poisoning. But how?

Paraskevas Photography

Cohen called Holmberg on a Saturday, and by the next morning, Holmberg was on a flight to Minnesota to find out. In Minnesota, Holmberg met with other **epidemiologists** [Greek, *epidemia* = how widespread a disease is]—researchers who study the incidence and transmission of diseases in populations. By that afternoon, Holmberg and state epidemiologist Michael Osterholm had two facts in hand.

First, all of the victims were infected with the same strain of *Salmonella newport*. This strain of bacteria, resistant to several common antibiotics, including penicillin and tetracycline, was unusual. The fact that all the victims had been made ill by the same rare form of *S. newport* suggested that they had all acquired the infection from a common source. In short, they must have all eaten the same thing. But what?

Second, 8 of the 11 victims had been taking antibiotics manufactured by the same drug company. At first glance, it looked like the antibiotics themselves were contaminated with salmonella bacteria. However, because 3 of the 11 patients had not been taking antibiotics at all, Holmberg and Osterholm ruled out that possibility.

Nonetheless, another clue suggested that antibiotics played an important role. One woman had been hospitalized with salmonella poisoning the day after she had begun taking amoxicillin for a sore throat. Her husband then developed a sore throat, too, and took some of his wife's amoxicillin. Within 48 hours, he was in the hospital with salmonella poisoning.

Holmberg and Osterholm hypothesized that the antibiotic had actually triggered the couple's salmonella poisoning (Figure 20-1). The couple must have been infected with antibiotic-resistant salmonella that suddenly flourished when the amoxicillin killed off all other competing bacteria. Without competition, the salmonella bacteria multiplied

DAY 1	DAY 3	DAY 4
Woman eats beef hamburger containing antibiotic-resistant *Salmonella newport* bacteria, which pass into the gut.	The woman takes an antibiotic for a separate infection, which kills off normal gut bacteria.	Antibiotic-resistant *Salmonella newport* multiply and she becomes ill.

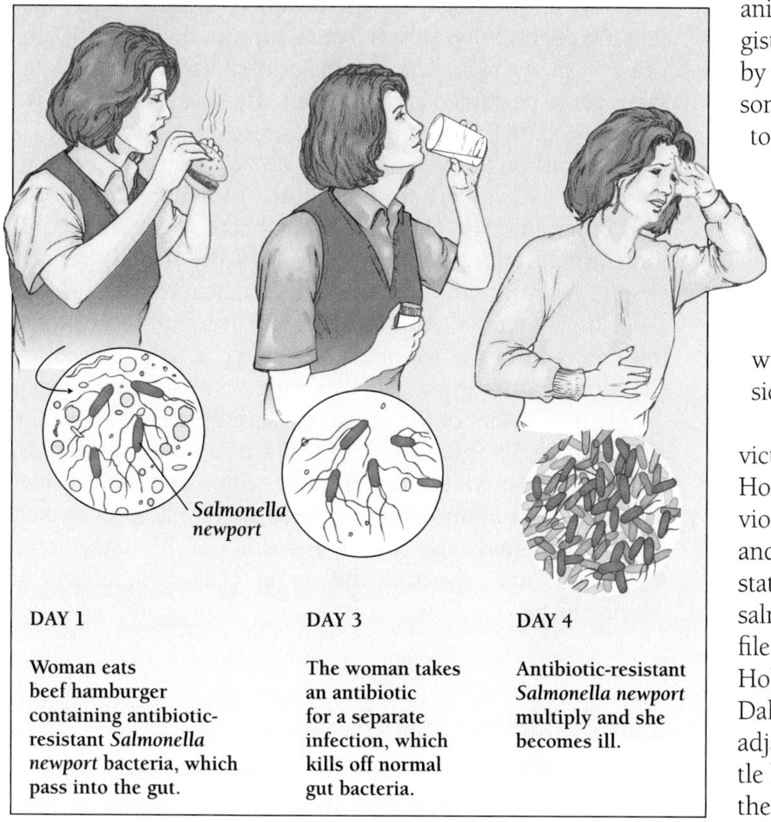

Figure 20-1

How antibiotics can trigger an infection. By eliminating competing bacteria in the gut, antibiotics create a perfect environment for the multiplication of antibiotic-resistant bacteria. If those bacteria are the kind that make us sick, we can become sicker much faster than usual. When our normal community of bacteria and fungi is unharmed, it keeps pathogens from taking over.

from a few thousand cells into a full-scale infection of millions upon millions of cells in just a few hours. But why were the couple harboring antibiotic-resistant salmonella bacteria in the first place? Where had they come from?

Holmberg spent the following two weeks searching for the common source of the food poisoning in the homes of salmonella victims. He looked through refrigerators and cupboards, searching for some sort of food they had all eaten—some common source for the antibiotic-resistant bacteria.

Finally, with still no answer, he returned to Atlanta frustrated and discouraged. He and his boss, Mitchell Cohen, knew that the most likely scenario was that all of the patients had consumed meat, milk, or eggs infected with resistant *S. newport*. In the United States, large numbers of farm animals are fed low doses of antibiotics to increase their growth rate. This practice kills bacteria that are sensitive to antibiotics, encouraging the proliferation of antibiotic-resistant bacteria. By 1983, Cohen strongly suspected that farm animals raised on antibiotics were making people sick, but no one had yet been able to prove it.

Cohen knew of a new technique, however, that might help Holmberg make a direct link between individual farm animals and individual salmonella victims. Molecular biologists had recently developed a way to "fingerprint" bacteria by characterizing their plasmids, simple loops of DNA that sometimes live inside bacteria (Chapter 12). With this new tool, every strain of bacteria could be distinguished from all others according to which plasmids it carried.

Cohen suggested that Holmberg find out where else in the United States *S. newport* existed. Holmberg wrote to epidemiologists all over the United States, and, within a month, he had a clue that would solve the case. The state epidemiologist in South Dakota wrote to say that antibiotic-resistant *S. newport* had recently sickened four people.

From Atlanta, Holmberg interviewed the South Dakota victims by phone. One of the victims, a dairy farmer, told Holmberg that his dairy cows had all had diarrhea the previous November, and that one dead calf had been autopsied and found to have died of salmonella. Researchers at the state agricultural laboratory that had analyzed the dead calf's salmonella told Holmberg that it had the same plasmid profile as the salmonella bacteria in the Minnesota patients. Holmberg soon discovered that three of the four South Dakota victims had eaten beef from cattle raised on a farm adjacent to the dairy farm. The beef farmer said that his cattle had been perfectly healthy, adding that he regularly laced their feed with the antibiotic tetracycline, a handful to every ton of grain. Even if the beef cattle had shown no symptoms of salmonella, however, they could have been carrying salmonella just the same.

"I knew what the story was," Holmberg later told *Science* magazine. "I knew that between this point here in South Dakota and that point there in Minnesota, there was a connection. Nature does not hand you coincidences like this. I was pretty sure that the beef herd was the source [of the *S. newport*]. . . . Now all I wanted to do was connect the two points. . . ."

Connecting the points, it turned out, was not so easy. In mid-January, just days before the salmonella outbreak in Minnesota, the beef farmer had sent all 105 head of cattle to slaughter. Since the beef cattle were gone, Holmberg had no way to test the cattle for salmonella.

Nonetheless, Holmberg was able to follow the path the tainted meat had traveled to supermarkets in Minnesota and Iowa, where 10 of the salmonella victims had shopped. In addition, Holmberg was able to show that the strain of *S. newport* from the dead dairy calf was truly distinctive. He examined 91 samples of different strains of *S. newport* from all over the United States, yet only the bacteria from the dead calf had the same plasmid profile as the salmonella in the sick Minnesotans. Holmberg no longer doubted that the beef cattle in South Dakota were the source of the antibiotic-resistant salmonella in Minnesota.

The plasmid fingerprinting technique had given epidemiologists their first clear evidence that feeding animals antibiotics can make the humans who eat them sick. Antibiotic-resistant bacteria are, it turns out, unusually dangerous. Holmberg showed that people are 21 times more likely to die from an infection by antibiotic-resistant salmonella than from an infection by nonresistant salmonella.

Amazingly, the practice of feeding antibiotics to farm animals continues today. Despite the clear hazards to public health, no laws regulate the practice. As a result of public pressure, only 20 percent of poultry growers now use antibiotics, the poultry industry reports. The beef industry recommends that its members use only antibiotics not used in humans. The pork industry continues to use human antibiotics without restraint.

How bacteria develop resistance to antibiotics, and how they pass that resistance among themselves, is a topic that reveals much about the lives of bacteria. In the rest of this chapter, we will learn a little about how bacteria are different from the rest of us and what kinds of bacteria populate the Earth. Along the way, we will see why bacteria are so good at becoming resistant to antibiotics.

20.1 Prokaryotes and People

The number of bacteria in your mouth at this moment exceeds the total number of people that have ever lived. You carry many more bacteria on your skin and in your gut than you have cells in your body. From a bacterium's point of view, your body is just another island of prime real estate to be colonized. Yet no human was even aware of the existence of bacteria until a Dutch cloth merchant named Antonie van Leeuwenhoek saw the first bacteria some 300 years ago (Chapter 4). And even after that no one realized that bacteria were important until about 150 years ago, when the

French biologist Louis Pasteur and others found that bacteria can cause disease (Chapter 6).

Prokaryotes, which are single-celled organisms with no nuclei, comprise two of the three domains of life—the Eubacteria [Greek, *eu* = true] and the Archaea [Greek, *archaios* = ancient]. About a third of the total biomass of life on Earth consists of prokaryotes, and some biologists think that there are huge numbers of prokaryotic organisms yet to be discovered in the depths of the oceans and even within the Earth's crust. Samples from ocean sediments hundreds of meters thick reveal as many as a billion cells per cubic centimeter of sediment. Of course, some biologists argue that since a lot of these cells are dead, they shouldn't count, and so the issue of exactly how much of the living biomass of the Earth is living prokaryotes remains surprisingly controversial.

When most people think about the diversity of life, they think of lions and wildebeests in Africa, trees, monkeys, butterflies, and even beetles in South America, or of the thousands of species of fish and marine mammals that populate the oceans. But life's diversity offers much more. Squash any passing insect and you'll discover hundreds of species of bacteria living inside. A healthy human gut contains about 600 different kinds of bacteria that live together in a functioning community. Others live in our mouths, and still others live on our skin. The vast majority of Earth's prokaryotes have yet to be named and classified. The few prokaryotes we know well are those that interact directly with humans.

But although we know most about bacteria that live in us or on us, bacteria are not limited to living with people and other animals. They live in the bodies of every kind of organism, even inside other cells. They live in soil, ice, and water. A pinch of ordinary soil contains something like 10 billion bacteria. And bacteria can live almost anywhere on Earth, from the ice fields of Antarctica to the steaming cauldron of a Yellowstone hot spring, from sediments deep below the bottom of the ocean to the driest desert soils. The smallest are 400 nm or less. (Recall that a molecule of DNA is only 20 nm wide.) The largest bacteria are about three-quarters of a millimeter across, as easily visible to the naked eye as the period at the end of this sentence. Bacteria are the most numerous, the most diverse, and the most ancient of all organisms.

How Do Bacteria Make Us Ill?

Most people think of bacteria as lowly pests—carriers of disease and spoilers of food. In fact, many bacteria are useful to humans in one way or another. The trillions of bacteria in a handful of soil, for example, help recycle carbon and other nutrients. All the organisms on Earth depend on bacteria to pull nitrogen from the air so that we can use it to build proteins.

Bacteria are useful to us as makers of cheese and yogurt and essential as the symbionts inside our bodies that help us digest our food. Some produce vitamins that we would otherwise have to get from special foods. And after that, they even clean up after us, as central players in sewage treatment plants that turn effluent into clean water.

But bacteria can also be **pathogens,** agents that cause disease. To cause a disease, a bacterium usually must accomplish two things:

1. It must invade the host, attaching and multiplying on or in the host's tissues and cells.
2. It must produce a toxin—a chemical that destroys or interferes with the host's normal processes.

Not all pathogenic bacteria are strangers. Many are normal residents of our bodies that are out of place or have multiplied when the body's defenses weaken. *Escherichia coli* in

our intestines, for example, are normal and useful members of the community of organisms in the body (Figure 20-2). But the very same organism in the bloodstream can kill in a matter of hours. In some cases, disease results just from the excessive growth of pathogenic bacteria. But in most cases, disease results from toxins made by the pathogen. These toxins may be **endotoxins,** toxic substances that are components of the bacterial cells themselves, or **exotoxins,** molecules that are secreted by the bacteria.

In the 1860s, physicians began to suspect that diseases were caused by bacteria and other microscopic pathogens. In the blood of sheep with the disease anthrax, biologists could see what we now know were anthrax bacteria. (Anthrax bacteria have been much in the news lately because of their potential for use as a biological weapon.) But the evi-

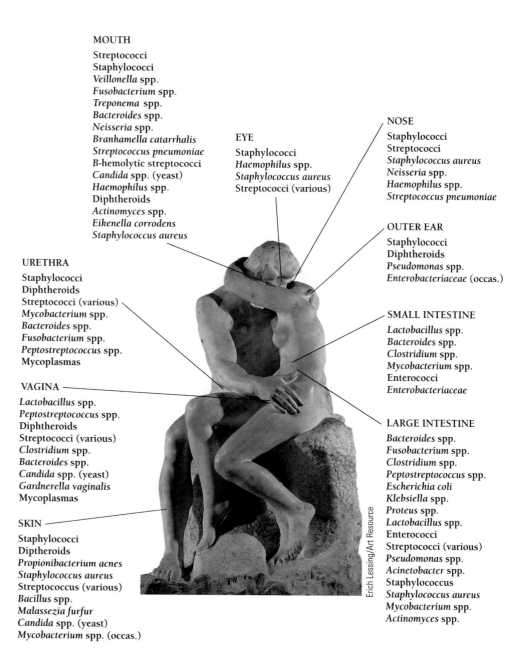

MOUTH
Streptococci
Staphylococci
Veillonella spp.
Fusobacterium spp.
Treponema spp.
Bacteroides spp.
Neisseria spp.
Branhamella catarrhalis
Streptococcus pneumoniae
B-hemolytic streptococci
Candida spp. (yeast)
Haemophilus spp.
Diphtheroids
Actinomyces spp.
Eikenella corrodens
Staphylococcus aureus

EYE
Staphylococci
Haemophilus spp.
Staphylococcus aureus
Streptococci (various)

NOSE
Staphylococci
Streptococci
Staphylococcus aureus
Neisseria spp.
Haemophilus spp.
Streptococcus pneumoniae

OUTER EAR
Staphylococci
Diphtheroids
Pseudomonas spp.
Enterobacteriaceae (occas.)

URETHRA
Staphylococci
Diphtheroids
Streptococci (various)
Mycobacterium spp.
Bacteroides spp.
Fusobacterium spp.
Peptostreptococcus spp.
Mycoplasmas

SMALL INTESTINE
Lactobacillus spp.
Bacteroides spp.
Clostridium spp.
Mycobacterium spp.
Enterococci
Enterobacteriaceae

VAGINA
Lactobacillus spp.
Peptostreptococcus spp.
Diphtheroids
Streptococci (various)
Clostridium spp.
Bacteroides spp.
Candida spp. (yeast)
Gardnerella vaginalis
Mycoplasmas

LARGE INTESTINE
Bacteroides spp.
Fusobacterium spp.
Clostridium spp.
Peptostreptococcus spp.
Escherichia coli
Klebsiella spp.
Proteus spp.
Lactobacillus spp.
Enterococci
Streptococci (various)
Pseudomonas spp.
Acinetobacter spp.
Staphylococcus
Staphylococcus aureus
Mycobacterium spp.
Actinomyces spp.

SKIN
Staphylococci
Diptheroids
Propionibacterium acnes
Staphylococcus aureus
Streptococcus (various)
Bacillus spp.
Malassezia furfur
Candida spp. (yeast)
Mycobacterium spp. (occas.)

Erich Lessing/Art Resource

Figure 20-2
A healthy community of microorganisms.
All of the microorganisms listed here live on the surfaces and in the interiors of normal, healthy human bodies. Most of these bacteria (and some fungi) live with us harmoniously. Indeed, we need them. Mice raised with no gut bacteria need to eat 30 percent more food than normal mice whose guts are full of bacteria. In addition, all these different species prevent any one microorganism from overgrowing. For example, *Candida albicans,* a kind of yeast, is a fungus that commonly lives in the mouth, gut, and vagina, as well as on the skin, usually harmlessly. (*Candida* is, of course, a eukaryote, not a prokaryote.) Yet when we take antibiotics that kill bacteria but not fungi, *Candida albicans* can begin growing and change its form to one that actually attacks the tissues of the body in a classic "yeast infection." A combination of plentiful bacteria and a healthy layer of body cells lining the mouth, gut, and vagina normally prevent this.

dence was weak; no one had yet provided any definite evidence that these tiny rod-shaped objects seen through the microscope actually *caused* anthrax, a disease that mostly affects livestock but also people.

To find out, the German physician Robert Koch began experimenting with anthrax bacteria, *Bacillus anthracis,* to see if by itself it could cause anthrax. Koch succeeded in showing that it did, and also discovered the bacteria that cause tuberculosis (TB), a deadly disease of the lungs, and cholera (a disease that kills by causing deadly bouts of diarrhea). Koch also demonstrated that wound infections are caused by specific bacteria. But Koch's most enduring contribution to science was a set of four rules for accurately identifying the pathogen for any disease.

The four rules, now known as **Koch's postulates** (Figure 20-3), are:

1. The same pathogen must be present in every person with the disease.
2. The pathogen must be capable of being grown in a pure culture—one uncontaminated with other organisms.
3. The pure cultured pathogen, introduced into an experimental animal, must cause the disease.
4. The same pathogen must be present in the infected animal after the disease develops.

More than 100 years later, Koch's postulates are still the accepted guidelines for identifying pathogenic organisms. In some cases, however, the postulates cannot be satisfied. For example, the bacterium *Treponema pallidum,* which causes the sexually transmitted disease syphilis, has yet to be cultured in the lab. In addition, bacteria that live inside other cells (such as the sexually transmitted bacterium *Chlamydia*) cannot grow in pure cultures because these bacteria always require the presence of host cells. Likewise, all viruses must grow inside other cells (Chapter 12), so pathogenic viruses such as HIV (human immunodeficiency virus) and influenza cannot be grown in a "pure culture."

Most pathogenic bacteria make us ill by releasing toxic chemicals that kill our own cells. Koch's postulates are guidelines for determining whether a microorganism causes a specific disease.

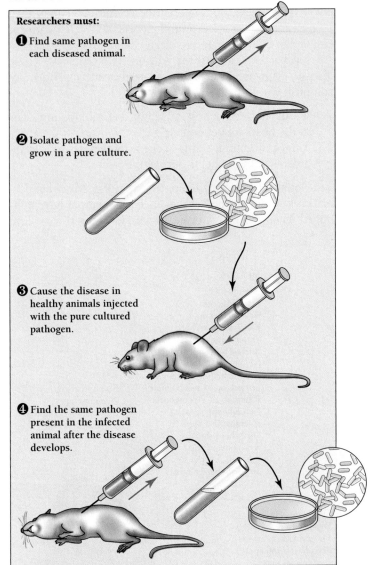

KOCH'S POSTULATES

Researchers must:

❶ Find same pathogen in each diseased animal.

❷ Isolate pathogen and grow in a pure culture.

❸ Cause the disease in healthy animals injected with the pure cultured pathogen.

❹ Find the same pathogen present in the infected animal after the disease develops.

Figure 20-3

Koch's postulates. ❶ The same pathogen must be present in every individual with the disease. ❷ It must be possible to grow the pathogen in a pure culture. ❸ The pure cultured pathogen must cause the disease when introduced into an experimental animal. ❹ The same pathogen must be detectable in the infected animal after disease develops.

How Do Bacteria Acquire Resistance to Antibiotics?

Whenever we are sick from a bacterial infection, we and our doctors want to eliminate the pathogenic bacteria as quickly as possible. The easiest way to do that is with an antibiotic. **Antibiotics** are drugs that kill bacterial cells without (much) harming our own eukaryotic cells. Since the first antibiotics came into use in the 1940s, medical researchers have discovered or synthesized more than 150 different antibiotics.

Amazingly, pathogenic bacteria now exist that are resistant to all antibiotics. In New York City, patients with certain strains of highly resistant tuberculosis must now be treated the same way tuberculosis patients were treated 50 years ago—by isolating them (Figure 20-4). No drugs can cure this form of drug-resistant tuberculosis. Health workers can only hope patients will get better on their own and try to prevent them from passing their resistant strains of tuberculosis to other people.

Where do bacteria acquire resistance to antibiotics? First we must understand how antibiotics work and where that resistance comes from. The first antibiotics discovered were molecules naturally synthesized by soil bacteria or

Figure 20-4

A tuberculosis isolation ward. In the 1920s, sick children were isolated on this ferryboat in New York's harbor. With the advent of antibiotics, hospitals no longer needed isolation wards for patients with tuberculosis. Today, because of strains of tuberculosis that are resistant to all known antibiotics, such wards have returned.

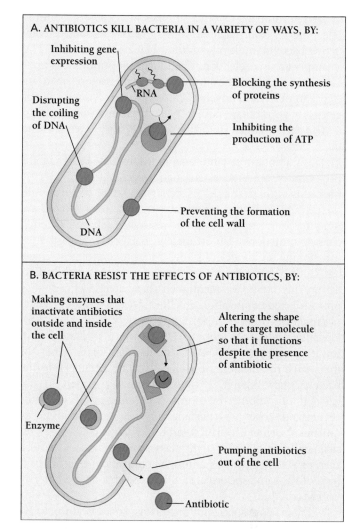

Figure 20-5

How does it work? A. Every antibiotic has its own way of preventing bacteria from reproducing. B. Bacteria have different ways of resisting the effects of antibiotics.

fungi. In fact, soil bacteria produce so many chemicals that are toxic to one another that pathogenic bacteria usually die when exposed to ordinary dirt.

The genes for resistance to these antibiotics come from the very organisms that make the antibiotics. A soil bacterium needs to be immune to its own toxins, after all, and they have evolved a variety of tools to fend off specific antibiotics.

What are these tools? That depends on the antibiotic. Some antibiotics, such as penicillin, kill bacteria by preventing the bacteria from synthesizing a cell wall. Since bacteria constantly break down and rebuild their cell walls, antibiotics that interfere with the rebuilding of the cell wall are potent killers of bacteria. Other antibiotics prevent bacteria from expressing their genes, interfere with DNA folding (which can disrupt gene expression and DNA replication), inhibit the synthesis of folic acid (important in nucleotide synthesis), or block protein synthesis.

Resistance to antibiotics can be classified into two main varieties (Figure 20-5A). Some resistance genes merely enable the bacteria to survive exposure to an antibiotic. Some resistance genes specify proteins that destroy the antibiotic as it enters the cell or that pump the antibiotic out of the cell. Still other resistance genes code for new proteins whose altered shapes make it impossible for the antibiotic to bind and do damage.

Other resistance genes enable the bacteria to release enzymes that actually destroy antibiotics in the environment (Figure 20-5B). This kind of resistance is the most danger-

ous. If, for example, a college student regularly took the antibiotic tetracycline to treat acne, he could gradually accumulate intestinal and other bacteria able to destroy tetracycline. If he then became infected by nonresistant *Chlamydia trachomatis,* for example, and his doctor prescribed tetracycline, his intestinal bacteria would destroy the tetracycline before it could kill the invading *Chlamydia* bacteria.

Resistance accumulates in populations of bacteria in several ways. In all populations of bacteria, a few individuals carry genes for resistance to antibiotics. Even baboons living in the wild, far from any humans, carry small numbers of intestinal bacteria that are resistant to some antibiotics. Humans have been exposed to antibiotics for decades, so we carry far greater numbers of resistant bacteria. One study of resistance in a group of students showed that one in 10 students carried large numbers of *E. coli* that were resistant to a combination of two antibiotics first introduced in the 1970s.

When humans or other animals are exposed to antibiotics, the resistant bacteria flourish in the absence of competing bacteria, multiplying until they are the majority. This is simply natural selection. The longer the antibiotic is present, the more likely the antibiotic-resistant bacteria are to increase in number. Thus, beef cattle regularly fed tetracycline gradually accumulate populations of bacteria that are resistant to tetracycline. Unfortunately, the story does not end there.

If antibiotic-resistant bacteria accumulated only in the presence of antibiotics, through selection, then society's problems with antibiotic resistance would be relatively minor. Strains of bacteria without resistance to a specific antibiotic could never acquire resistance except through chance mutations. Unfortunately, bacteria can pass fully functioning antibiotic resistance genes to completely different kinds of bacteria, so that resistance spreads rapidly. Earlier in this book, we mentioned that bacteria engage in a process called conjugation, in which one bacterium can pass plasmids and other DNA to another bacterium (Figure 12-8). Plasmids are simple loops of DNA carried separately from the bacterium's own DNA. Plasmids often carry genes for traits such as antibiotic resistance that are beneficial to the bacteria they inhabit. Because plasmids themselves pass genes to one another by means of jumping genes, plasmids accumulate genes for resistance the way a snowball accumulates snow as it rolls downhill. Some plasmids carry resistance to as many as eight classes of antibiotics. Bacteria then pass such plasmids, through conjugation, to other bacteria of the same species and to bacteria that are completely unrelated.

Unlike most species of organisms, bacteria are not discrete genetic entities, but a vast, integrated microbial world. As a result, multiple antibiotic resistance is the rule rather than the exception. In 1995, for example, hospital workers in Atlanta found that 25 percent of children fighting ordinary ear and respiratory infections by *Streptococcus pneumoniae* were infected with strains resistant to two or more antibiotics.

The tendency of plasmids to carry genes for resistance to several antibiotics has important consequences. Cattle treated with tetracycline, for example, can acquire bacteria that are resistant not only to tetracycline but to several other antibiotics as well. This is why the South Dakota beef cattle, which had been treated with tetracycline, carried *Salmonella newport* resistant to both amoxicillin and tetracycline. The plasmid that protected the bacteria from tetracycline also happened to carry genes for resistance to other antibiotics.

Genes for antibiotic resistance may allow a bacterium simply to survive exposure to an antibiotic or to remove or destroy any molecules of the antibiotic in its environment. Resistance genes accumulate in populations of bacteria, first, through natural selection, and second, by moving from one bacterial strain to another by means of plasmids and other mobile genes.

20.2 How Do We Know That an Organism Is a Prokaryote?

The defining characteristic of prokaryotes that sets them apart from other organisms is the absence of a true nucleus (Chapter 4). Prokaryotes lack not only a nucleus but also chloroplasts and mitochondria and all other membrane-enclosed organelles. Some prokaryotes do have extensively folded plasma membranes, however, and use these large surfaces for respiration, photosynthesis, and DNA replication.

Instead of neatly packing its DNA in a membrane-enclosed nucleus (as a eukaryote does), a prokaryote tucks its DNA into a concentrated mass called a **nucleoid.** The DNA in the nucleoid consists of a single molecule of double-stranded DNA that forms a closed loop. A prokaryote may also contain one (or more) small loops of DNA called plasmids (Chapter 12). Because a prokaryote's DNA forms a loop, it can replicate continuously, unlike eukaryotic DNA, which must replicate in pieces.

When prokaryotes divide, they do so through a process called fission. They do not require sex to reproduce, although they exchange genetic information so freely that they are never reproductively isolated. Unlike in most eukaryotes, a prokaryote "species" is neither a reproductive unit nor a genetic unit, and biologists must classify them according to similarities and differences in their DNA.

Prokaryotic cells come in a variety of shapes. The most common are the rod-shaped **bacilli** [singular, bacillus; Latin, = little rod], the spiral-shaped **spirilla** [singular, spirillum; Latin, *spira* = coil], and the spherical **cocci** [singular, coccus; Greek, *kokkos* = berry] (Figure 20-6). The bacilli include, for example, the gut bacteria *E. coli* and *Salmonella newport* (Figure 20-6A). The round coccus bacteria include the family of staphylococcus bacteria and the family of streptococcus bacteria that live on our bodies and sometimes cause boils, strep throat, or other infections (Figure 20-6C). Most bacteria are single-celled organisms, though some aggregate to form colonies of unspecialized cells (Figure 20-7).

The defining characteristic of prokaryotes is the absence of a nucleus or other membrane-enclosed organelles. The DNA of prokaryotes forms a single circle of double-stranded DNA.

Most Bacteria Have Complex Cell Walls

Like plant cells, most bacterial cells have rigid cell walls outside their plasma membranes. Like the wall of a plant cell, the wall of a bacterial cell gives the cell a shape and protects it from injury (Figure 20-8). However, while plant cell walls are made of cellulose, bacterial cell walls are made of a mixture of polysaccharides and polypeptides called **peptidoglycan.** The cell wall is a single, giant, bag-shaped macromole-

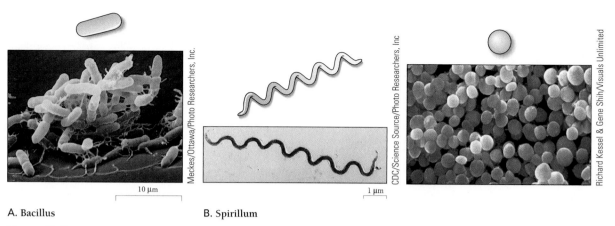

A. Bacillus   B. Spirillum

Figure 20-6

The shape of a germ. Bacteria come in many shapes, but three of the most common are the bacillus, the spirillum, and the coccus. A. A typical bacillus is *Salmonella newport*. B. The spirillum bacterium *Treponema pallidum* causes the venereal disease syphilis. C. The coccus bacterium *Staphylococcus aureus* lives harmlessly on the human body, but can cause serious infections when it invades lungs, wounds, or burns.

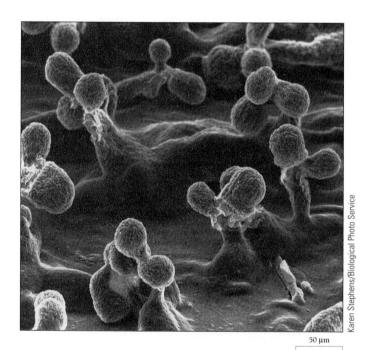

Figure 20-7

Some bacteria, such as this myxobacterium, are colonial. When cells of this species run short of nutrients, the cells cooperate to form a "fruiting body," which helps them disperse spores. With luck, at least some of the spores will land somewhere that is a nicer place for myxobacteria.

cule composed of a network of cross-linked peptidoglycan. The peptidoglycan cell wall is important to biologists for two reasons. It helps distinguish two major groups of bacteria, and it makes some bacteria susceptible to antibiotics. (Archaea, like eukaryotes, do not produce peptidoglycan.)

Biologists divide all of the bacteria into two major groups according to the way their cell walls soak up various dyes.

In the 1880s, the Danish bacteriologist Hans Christian Gram found that when he treated bacteria with a particular purple dye, some kinds of bacteria became permanently stained, while others could be washed clean. Those that became permanently stained he called "stain positive." The others were "stain negative." Today bacteriologists call these two categories of bacteria **gram-positive** and **gram-negative**, after Gram's stain (Figure 20–9).

What sets gram-negative and gram-positive bacteria apart from one another is the structure of the cell wall. Gram-negative bacteria—such as *E. coli* or the gonococcus bacteria that cause the sexually transmitted disease gonorrhea—have a thin, three-layered cell wall, whose outermost layer does not bind the Gram stain (Figure 20-9C). In contrast, gram-positive bacteria—such as the streptococci (which can cause strep throat) and the staphylococci (which can cause boils and other infections)—have a thick, single-layered cell wall containing at least 20 times as much peptidoglycan as the gram-negative bacteria. The peptidoglycan, itself arranged in up to 40 layers, binds the Gram stain so that it cannot be washed away.

Outside of the cell wall, some prokaryotes have other materials. The outer surfaces of some prokaryotes possess *pili* [singular, pilus; Latin, = hair], hairlike structures that attach the bacteria to surfaces and to other cells. The F-pilus, for example, provides a bridge for the transfer of DNA during bacterial conjugation.

The bacterium that causes pneumonia is enclosed in a **capsule,** a gelatinous layer of polysaccharides and proteins outside its cell wall (Figure 20-8). Capsules protect disease-causing bacteria from attack by the body's white blood cells, enabling the bacteria to overwhelm the body's immune system. Such bacteria are more deadly, or virulent.

Other bacteria possess a **slime layer,** a tangled web of polysaccharides that, like a capsule, lies outside the cell

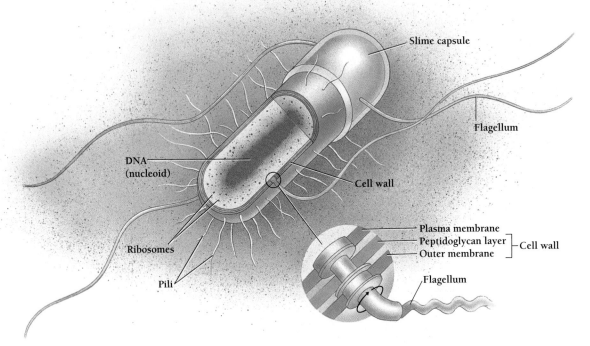

Figure 20-8

The basic structure of a bacterium. *E. coli* is a typical bacillus, with a slime capsule, an outer membrane, and a cell wall composed of a layer of peptidoglycan and a regular cell membrane (the inner membrane). Not all bacteria have an outer membrane, and one group, the mycoplasmas, have no cell wall. Prokaryotes have genes made of DNA but they don't keep the DNA in a nucleus (as eukaryotes do).

Biology⑤Now™ Learn more about the structure of prokaryotes by clicking on this figure on your BiologyNow CD-ROM.

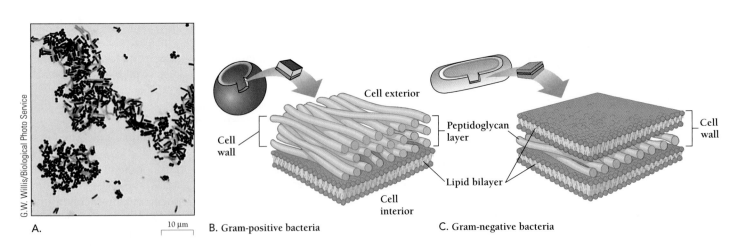

Figure 20-9

It's all in the wall. Bacteria absorb stain or not depending on what kind of cell wall they have. A. A mixture of gram-positive *(dark)* and gram-negative *(light)* bacteria. B. The gram-positive coccus bacterium *(left)* has a two-layered, peptidoglycan-rich cell wall that easily absorbs a colored stain. C. The gram-negative bacillus *(right)* has a three-layered, peptidoglycan-poor cell wall that resists staining.

Figure 20-10
Dental plaque bacteria. The slime we feel on our teeth in the morning is made by bacteria, which form dense colonies protected by a layer of slime. The five main bacteria that colonize the teeth are *Streptococcus sanguis, S. oralis, S. mitis, S. gordonii,* and *Actinomyces naeslundii. A. naeslundii* (rods) and *S. oralis* (cocci) typically begin colonization of the teeth first; the other bacteria multiply after the protective biofilm is established.

wall. A sticky slime layer protects the bacterial cell against dehydration, helps keep food from diffusing away, and also allows cells to attach to and move about on the surfaces on which they grow. Dental plaque is a slime layer, or "biofilm," made by the bacteria that inhabit the mouth. Plaque helps bacteria grow smoothly over the surfaces of our teeth (Figure 20-10). When bacterial plaque is not removed frequently it combines with the minerals in saliva to form hard deposits called tartar, which can lead to gum disease.

The presence of large amounts of peptidoglycan distinguishes the bacterial cell wall from the cell walls of archaea and eukaryotes. Gram-positive bacteria have a thick, peptidoglycan-rich cell wall.

How Do Bacteria Move?

Some bacteria move by secreting slime and gliding on it. Others use a propeller—the bacterial **flagellum** (Chapter 4). Bacterial and eukaryotic flagella are entirely different. The eukaryotic flagellum is an extension of the cytoplasm surrounded by membrane that beats back and forth. In contrast, the prokaryotic flagellum is a thin, rotating protein filament attached to the outside of the cell (Figure 20-8).

Many bacteria use their flagella for directed movement, or **taxis** [Greek, = to put in order]. (A taxicab performs di-

rected movements.) The direction of movement depends on the operation of the flagella. For example, when the flagella of an *E. coli* rotate in a counterclockwise direction, they twist around each other and propel the bacterium forward. When the flagella rotate in a clockwise direction, however, they separate and become uncoordinated. As a result, the bacterium tumbles aimlessly. When the flagella resume a counterclockwise spin, the bacterium moves in a new direction. By controlling the timing of clockwise and counterclockwise rotations, bacteria can move toward or away from food, oxygen, heat, light, and other stimuli. Some bacteria even use the Earth's magnetic field to guide them to the lower depths of the ocean. **Chemotaxis** is taxis in response to varying concentrations of specific chemicals.

Bacteria move in directed fashion, by gliding on slime or by rotating propellerlike flagella.

How Do Prokaryotes Grow So Rapidly?

Because bacteria keep their DNA in a nucleoid instead of a nucleus, bacteria can replicate their DNA continuously and divide very rapidly. In fact, the cells in well-nourished colonies of *E. coli* can divide every 20 minutes. Every time the bacteria divide, the colony doubles in size. That is, after one generation, one cell becomes two. After another generation the 2 cells become 4, then 8, 16, and so forth. After 20 such doublings (in less than 7 hours), a single cell would yield 2^{20}, or about a million, new cells; and after 30 doublings (10 hours), 2^{30}, or about a billion, cells. Such growth is "exponential" (Chapter 28).

If a single *E. coli* bacterium could be supplied with enough food to divide every 20 minutes continuously, the combined mass of the bacterium and all its descendants would exceed that of the entire planet after only two days. We can see that, in reality, exponential growth can last for only short times. Nonetheless, the rapid proliferation of prokaryotes means that a favorable new mutation such as antibiotic resistance can spread quickly to a huge number of progeny.

A streamlined metabolism and a simple means of replicating DNA together allow bacteria to grow and divide more rapidly than eukaryotes.

Prokaryotes Are Metabolically Diverse

We eukaryotes are monotonous from a biochemical point of view. Nearly all of us use glycolysis and respiration to produce energy. Nearly all of us need oxygen to live. Likewise, all photosynthetic eukaryotes photosynthesize in the same way. In fact, the relentless sameness of metabolic pathways in all eukaryotes strongly supports the idea that we are all descended from a common ancestor.

In contrast, prokaryotes use a great variety of biochemical pathways. Some, for example, perform glycolysis, but with end products other than ethanol or lactate (Chapter 6). Many of these end products contribute to the pleasant (or not-so-pleasant) odors of wines and cheeses. Some prokaryotes do not use glycolysis at all. Instead, they produce ATP by capturing the chemical energy of molecules in their environment. Methane-producing prokaryotes, for example, capture energy released when carbon dioxide (CO_2) reacts with hydrogen gas (H_2) to produce methane (CH_4). Some archaea capture the energy from the reaction of sulfur-containing compounds with hydrogen gas. Hydrogen gas, unstable in the atmosphere, nonetheless accumulates to high levels under the oceans and within the Earth's mantle.

Because prokaryotes are so diverse, they flourish in places where eukaryotes could never survive. Over billions of years, prokaryotes have evolved an amazing variety of adaptations that contribute to their continuing success. Some bacteria can form spores, dormant structures that are resistant to chemicals, heat, drying, radiation, freezing, and even the vacuum of space. Bacterial spores may survive intact for millions of years, an adaptation that allows the bacteria to survive extraordinarily hostile environments. For example, biologists have recovered and reanimated bacterial spores from the digestive tracts of bees entombed in amber for 25 to 40 million years and from salt crystals possibly hundreds of millions of years old.

Bacteria also make antibiotics that can kill competing microorganisms, and, as we've seen, they can exchange genes for resistance to antibiotics by means of mobile genes such as plasmids. Bacteria can obtain energy from the waste products of other species (for example, carbon dioxide and methane). The adaptations of prokaryotes in turn provide new opportunities for eukaryotic life. A two-meter-long tubeworm, called *Riftia pachyptila*, lives near hydrothermal vents on the ocean floor. *Riftia pachyptila*, sometimes called "just a bag of bacteria," is sustained by symbiotic bacteria that use the energy of inorganic sulfur compounds from the vents to make energy-rich compounds that the worms can live on (Figure 20-11).

The enormous success of prokaryotes depends on their metabolic diversity and their ability to transfer genes to one another.

20.3 How Do Biologists Classify Prokaryotes?

Because prokaryotes do not (usually) reproduce sexually, taxonomists cannot use the biological species concept, which defines a species as a reproductively isolated population (Chapter 19). Reconstructing the phylogeny of bacteria is equally difficult, as the fossil record for prokaryotes is nearly nonexistent. So biologists classify prokaryotes according to how they get their energy and building blocks (metabolism) and according to DNA and protein sequences.

So far, some 1,700 species of prokaryotes have been named and classified, although estimates of the number of

A.

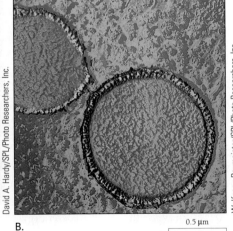

B.

0.5 μm

Figure 20-11

How eukaryotes profit from the special adaptations of prokaryotes. All animals have bacteria inside of them, but these tubeworms are hardly more than bags of symbiotic bacteria. A. The worms, *Riftia pachyptila*, live near hydrothermal vents deep in the ocean. B. The bacteria supply their protective worm container with energy from inorganic sulfide compounds.

species vary widely—from 5,000 up into the millions. Biologists agree on some broad groupings. In one classification, suggested in 1987, taxonomists divided the bacteria into 12 kingdoms; a more recent list included more than 25 kingdoms. Each kingdom's DNA sequences differ from those of the other kingdoms more than plant DNA differs from animal DNA.

Generally, microbiologists classify bacteria into broad categories according to the way they get food. **Autotrophs** [Greek, *autos* = self + *trophos* = feeder] obtain carbon atoms directly from carbon dioxide, just as plants do. Autotrophs include the **photoautotrophs,** which use light energy to make organic compounds from CO_2, and **chemoautotrophs,** which use energy from oxidizing inorganic substances such as hydrogen sulfide (H_2S) or ammonia (NH_3).

Heterotrophs [Greek, *heteros* = other + *trophos* = feeder], which include the great majority of prokaryotic species, get carbon and energy from organic molecules such as glucose made by other organisms (often autotrophs). Heterotrophs include the **photoheterotrophs,** which use light energy, and the **chemoheterotrophs,** which get energy from organic molecules instead of light. Heterotrophic prokaryotes may absorb nutrients from living organisms, from dead organisms, from the waste of organisms, or from particles from dead organisms.

Prokaryotes also vary in the way they use oxygen. Heterotrophs that must have oxygen for respiration are called **obligate aerobes** (because they are "obligated" to have air to live). Those that can grow either with or without oxygen are called **facultative aerobes.** Still others, for which oxygen is a deadly poison, are called **obligate anaerobes.**

How Do Biologists Classify the Archaea?

The prokaryotes include two large domains—the **Archaea** [Greek, *archaio* = ancient] and the Eubacteria ("true" bacteria). The Archaea are the oldest group of organisms and both the Eubacteria and the Eukarya (protists, fungi, plants, and animals) evolved from the ancestors of modern Archaea. But DNA sequences suggest that the Archaea are more closely related to us eukaryotes than to bacteria. Archaea probably separated from the Eubacteria about 3.5 billion years ago. (To put that in perspective, the Earth is only 4.6 billion years old.) Archaea differ from Eubacteria in the structure of ribosomal RNA and in the kinds of lipids in their cell membranes (see box). Archaea entirely lack peptidoglycan in their cell walls.

So far biologists have found only three kingdoms of Archaea—the Euryarchaeota, the Crenarchaeota, and the Korarchaeota (Figure 20-12). This last kingdom is recently discovered and its members are known only from their DNA. The Crenarchaeota include the **thermoacidophiles,** which inhabit hot sulfur springs such as those in Yellowstone National Park (Figure 20-13) or hot undersea volcanic vents called "hydrothermal vents." These archaea grow best in water that is boiling (100°C) and even as high as 130°C. (The temperature that hospitals use to reliably sterilize medical equipment is 120°C.) Some species flourish only in hot sulfuric acid that would destroy most organisms. The unusual enzymes these organisms use to survive and flourish in such conditions have attracted the attention of biotech companies. Finally, now that microbiologists know what to look for, they have begun to find Crenarchaeota in less dramatic settings—in ocean water and ordinary soil.

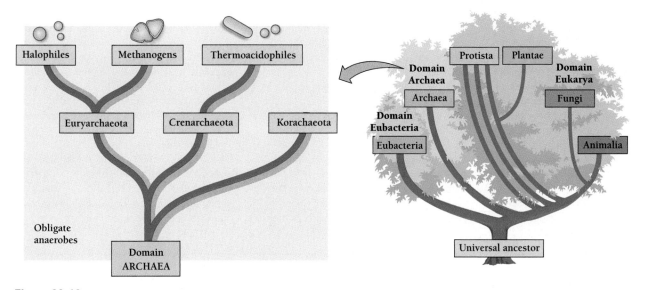

Figure 20-12

Phylogeny of the Archaea. The domain Archaea includes three distinct phyla. The domain Eubacteria includes at least 25 groups, now given kingdom status, of which a dozen are shown here. The Eubacteria are divided into heterotrophs and autotrophs. The autotrophs are further divided into photoautotrophs and chemoautotrophs.

Extreme Biology   Carl Woese and the Archaea

In the summer of 1996, a researcher working on an obscure microbe at The Institute for Genomic Research in Rockville, Maryland, announced a finding that immediately hit the television nightly news and the front pages of newspapers all over the United States. In sequencing the microbe's genome, the researcher, Carol Bult, had provided irrefutable evidence that all organisms, historically classified as either eukaryotes or bacteria, actually fall into three basic categories.

The microbe, named *Methanococcus jannaschii,* was only the fourth organism for which researchers had completed such a sequence, but the implications were profound. Bult's work not only supported the theory that the archaea are distinct from both bacteria and eukaryotes, it also provided independent support for two other ideas: (1) that we eukaryotes are more closely related to archaea than to bacteria and (2) that both eukaryotes and bacteria are descended from ancient organisms that resembled archaea.

Yet, despite the drama with which these important results were revealed, it was an oddly anticlimactic end to a very long story. Thirty years before, Carl Woese, a young molecular biologist at the University of Illinois, Urbana, began the painstaking task of working out the evolutionary relationships of the bacteria. Earlier researchers had made the attempt and failed. "The ultimate scientific goal of biological classification cannot be achieved in the case of bacteria," wrote the eminent microbiologist Roger Stanier.

Woese (pronounced woes) admired Stanier's own considerable efforts to classify the bacteria but did not accept that classifying the bacteria was impossible. Indeed, he had an ingenious plan for discovering their relations to one another. He knew that ribosomal RNAs are among the most conserved parts of all organisms. Because the work of translating RNA into proteins is fundamental, organisms with mutations in

Courtesy of University of Illinois at Urbana-Champaign News Bureau

Figure A
Carl Woese discovered the archaea in 1977.
The scientific community did not fully accept his results until 1997.

their ribosomal RNA nearly always die and ribosomes change hardly at all over millions of years. Ribosomes are, in short, some of the best sources of information about long-term evolutionary relationships among organisms.

For 10 years Woese catalogued short strings of nucleotides that he cut from the ribosomal RNAs of different microorganisms. It was mind-numbing, labor-intensive work that left him, he later said, "just completely dulled down." Yet it was fruitful work, for he gradually unraveled the relationships among some 60 different kinds of bacteria.

Then, in 1976, Woese's colleague Ralph Wolfe sent him some methanogens on which to work his magic. Methanogens were curious bacteria that came in a variety of shapes, but all produced methane as a by-product of their metabolism. No one knew where they fit in with the other bacteria. Woese set to work to find out. To his surprise, however, the methanogens lacked the characteristic sequences of ribosomal RNA that all other bacteria share. As he told

Wolfe, the methanogens didn't seem to be bacteria at all.

It was dramatic news, for it meant that in addition to the eukaryotes, and the bacterial prokaryotes, we humans were sharing the planet with a completely different domain of organisms—the Archaea—never before recognized. Yet, although the story hit the front page of *The New York Times,* the news faded away immediately. For most people, the news was abstract and meaningless. And most of Woese's own colleagues ignored his discovery. They simply didn't believe him. Although Woese provided the answers to many of the questions that most intrigued the eminent microbiologist Roger Stanier, Stanier, who has since died, never acknowledged Woese. For years, Woese says, he waited in vain for some friendly acknowledgment from Stanier.

R.G.E. Murray, the author of the leading microbiology textbook, didn't even include the archaea until 1986, and then only as a subgroup within the prokaryotes—even though Woese's work had demonstrated that the archaea were distinct from other bacteria. A handful of other researchers stood behind Woese, including Wolfe, arguing with their colleagues on his behalf. Yet, Woese's work was otherwise almost universally snubbed.

Woese merely retreated back to his laboratory, continuing his work, gathering new and better data, and publishing a series of landmark papers on the evolution and classification of bacteria. Today Woese is highly acclaimed. He has been elected to the prestigious National Academy of Sciences and awarded both a MacArthur Foundation "genius" grant and microbiology's top honor—the Leeuwenhoek medal.

Great discoveries in science frequently meet initial rejection. As Wolfe told *Science* magazine in 1997, "We can say, 'Oh, that's just the way science works.' But it was a personal experience for [Woese]. He's the one who had to live through it."

Figure 20-13
Thermoacidophiles. Thermoacidophiles may be chemoautotrophs (reaping energy from sulfur compounds or methane) or heterotrophs (reaping energy and carbon from small organic compounds made by other organisms). The best-known genus, *Sulfolobus,* lives in the hot sulfur springs of Yellowstone National Park. *Sulfolobus* can survive a temperature as high as 88°C (190°F) and an acidity as low as pH 1. Although *Sulfolobus* thrives in heat, it dies of cold at temperatures below 55°C (130°F).
Biology ⓔ Now™ Learn more about thermoacidophiles and related archaea by clicking on this figure on your BiologyNow CD-ROM.

The second major kingdom of Archaea, the Euryarcheota, includes the **methanogens,** shown in Figure 20-14, and the **halophiles,** which live in extremely salty environments (Figure 20-15). Methanogens derive energy from hydrogen gas by adding hydrogen to the carbon in carbon dioxide (CO_2) to make methane (CH_4). Methanogens are all obligate anaerobes and mainly live in two environments: (1) swamps and sewage treatment plants, and (2) the guts of animals, especially of cows, termites, and other animals that digest cellulose. Even the human gut harbors methanogens that produce methane, which, when released, is called flatulence. Methanogens play an essential role in the ecology of the Earth. Where methanogens flourish in swamps made anaerobic by the growth of other bacteria, they release methane, or "marsh gas," into the atmosphere. There, the methane is quickly oxidized to

carbon dioxide, which is then available for photosynthesis. If methanogens did not recycle carbon from organic molecules back into the atmosphere, other organisms would gradually run out of carbon. Natural gas, which we use to heat our homes and cook our meals, is about 98 percent methane.

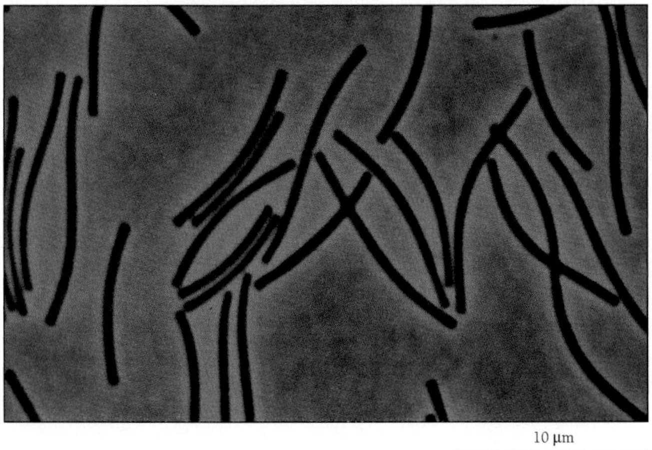

The archaea include three kingdoms. The most widespread and ecologically important archaea are the methanogens.

How Do Biologists Classify the Eubacteria?

Modern microbiologists classify the eubacteria according to DNA sequences that code for the RNA in ribosomes (see box). Because unrelated bacteria exchange genes freely, the phylogeny of bacteria is more like a wall of intertwined vines than any proper tree of life. By the late 1990s, taxonomists had begun to worry that it might never be possible to unravel the complex phylogeny of bacteria. But more recent work suggests that at least some parts of bacterial genomes remain relatively unaltered by gene transfers.

A preliminary phylogeny suggests three main lineages— (1) the spirochetes and chlorobi, (2) the proteobacteria and mitochondria, and (3) the cyanobacteria and firmicutes (Figure 20-16). The **spirochetes** are a phylum of bacteria with long, helical cells and flagella, which they use to propel themselves. Most spirochetes are free-living and anaerobic, but a few are pathogens, including *Treponema pallidum*, which causes the sexually transmitted disease syphilis (Figure 20-17). The **chlorobi,** or green sulfur bacteria, are a phylum of free-living bacteria that lack flagella (and the ability to propel themselves). They contain a form of the photosynthetic pigment chlorophyll (as plants do), but they pho-

10 µm

Figure 20-14
Methanogens. Methanogens, which *gen*erate the gas *methan*e, include bacteria of all three standard shapes—bacilli, spirilla, and cocci.

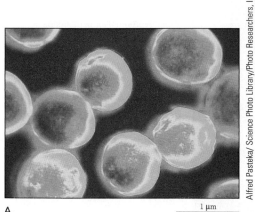

Alfred Pasteka/ Science Photo Library/Photo Researchers, Inc.

Kevin Schafer/CORBIS

A.   1 μm   B.

Figure 20-15

Halophiles. Most bacteria cannot grow in high concentrations of salt. Because salt prevents the growth of most bacteria, it acts as a preservative in ham, beef jerky, and other salty foods. Unlike any other bacteria, halophiles such as these *Halococcus* bacteria flourish in briny places the world over—from the Dead Sea to Utah's Great Salt Lake and California's Mono Lake. Halophiles, usually flagellated rods or cocci (like these), form a lavender-pink scum over their salty homes. The pink color comes from a purple pigment in the plasma membrane that captures light energy. A. Halophile bacterium. B. Evaporation salt ponds along the shore of San Francisco Bay.

Figure 20-16

Phylogeny of the Eubacteria. The domain Eubacteria includes something like 25 kingdoms. But here we combine them into just three main lineages. Because of both gene transfer among unrelated organisms and endosymbiosis (when prokaryotes such as chloroplasts and mitochondria take up residence inside of eukaryotes), the true history of life is more like a wall of interwoven vines than a simple tree. The common ancestor of all life was some form of archaean. The eukaryotes and eubacteria later split off from this archaean-like ancestor. Still later, bacteria joined the eukaryotic branch as mitochondria and chloroplasts.

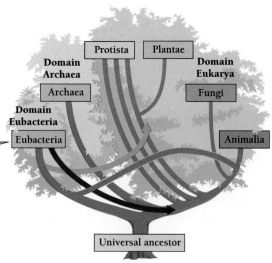

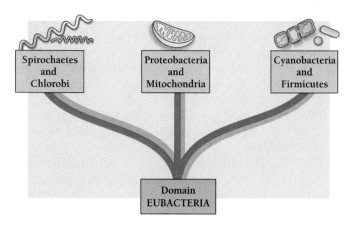

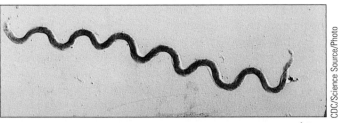

CDC/Science Source/Photo Researchers, Inc.

1 μm

Figure 20-17

Swimming syphilis spirochete. The most famous spirochete is *Treponema pallidum,* which causes the sexually transmitted disease syphilis. *T. pallidum* is transmitted from one person to another by means of direct contact between moist mucous membranes. Until the introduction of penicillin in the 1940s, syphilis was incurable, progressive, and fatal. All spirochetes contain internal flagella enclosed between layers of the cell wall, by means of which they move differently from any other bacteria. Although some live freely and some are parasites, all move by undulating quickly, even through the most viscous liquids. Most are finicky about their diets and *T. pallidum* is no exception. Its requirements are so specific that although it grows well inside the human body, biologists have been unable to grow virulent strains in the laboratory.

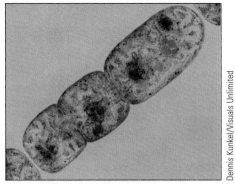

Dennis Kunkel/Visuals Unlimited

A.

Dennis Kunkel/Phototake

B.

tosynthesize only in the presence of sulfur, which donates electrons. (In plants, water donates the electrons for photosynthesis.)

Cyanobacteria have ancient lineages that human royalty can only envy. The oldest known fossils are 3.5-billion-year-old cyanobacteria. Free-living, aquatic bacteria that photosynthesize, cyanobacteria are one of the largest and most important groups of bacteria on Earth (Figure 20-18A). Because they photosynthesize, they generate oxygen as a waste product. Billions of years ago, cyanobacteria covered the Earth and pumped massive amounts of waste oxygen into the Earth's atmosphere, paving the way for the evolution of heterotrophic organisms such as ourselves. Without cyanobacteria we wouldn't be here.

As if that weren't enough of an accomplishment, cyanobacteria can even claim credit for inventing plants. The tiny organelles called chloroplasts that photosynthesize inside of plant cells can trace their ancestry back to ancient cyanobacteria (Figure 20-18 B and C). Sometime about half a billion years ago, ancient cyanobacteria moved inside of eukaryotic cells and began photosynthesizing, a cooperative process called "endosymbiosis." You can read more about the evolutionary history of cyanobacteria and endosymbiosis in Chapter 18.

Another phylum of bacteria on the same branch of the tree as the cyanobacteria are the firmicutes, which include the **mycoplasmas.** Mycoplasmas are unusually small bacteria that live in soil and in other organisms. They differ from all other bacteria in lacking a cell wall, which makes them baglike and shapeless. More important, many cause disease such as pneumonia in humans and other animals. Because they do not make a cell wall, they are unaffected by peni-

George Chapman/Visuals Unlimited

C.

Figure 20-18

Cyanobacteria and chloroplasts. The cyanobacteria were among the first photosynthetic organisms on Earth. A. A modern filamentous cyanobacteria *(Spirulina sp.)* B. Biologists think that ancient cyanobacteria came to live symbiotically inside eukaryote cells as chloroplasts. This closeup of a cyanobacterium *(Nostoc sp.)* shows some of the internal structure necessary for photosynthesis (see Chapter 7). C. Shown here is a chloroplast from a corn leaf, with its elaborate system of membranes for photosynthesis.

cillin and other antibiotics that interfere with cell wall synthesis (Figure 20-5). Some species colonize the urethra, where they metabolize urea (the major component of urine). Mycoplasmas cause a variety of diseases besides pneumonia, including some forms of pelvic inflammatory disease, postpartum fever (infections after delivering a baby), kidney disease, and some forms of arthritis.

One of the largest phyla of bacteria are the **proteobacteria,** which include "enterobacteria," such as *Es-*

cherichia coli (*E. coli*) and *Salmonella enterica,* which live in the guts of animals; the nitrogen-fixing bacteria; the purple sulfur bacteria and the myxobacteria. The proteobacteria are all gram negative, but they come in every shape, including rods, cocci, and spirilla. One group, the rickettsia, are tiny, rod-shaped proteobacteria that are commonly parasitic. For example, Rocky Mountain spotted fever is a rickettsia infection transmitted to humans by tick bites. The most interesting fact about rickettsia, however, is that it is the closest living relative of the mitochondria, the energy-producing organelle inside all eukaryotic cells. Just as cyanobacteria moved inside the ancestor of the first plant cells, some ancestor of rickettsia moved inside an early eukaryote.

The myxobacteria were at first classified as fungi because, like fungi, they form multicellular reproductive structures called "fruiting bodies." Myxobacteria are common in soils, where, like fungi, they break down organic matter, making small molecules available to plants and other organisms (Figure 20-7).

From the perspective of most other organisms on Earth, some of the most important bacteria may be the **nitrogen-fixing bacteria,** which provide nitrogen in forms such as ammonia and amino acids from which other organisms can make proteins (Figure 20-19). Although nitrogen gas (N_2) constitutes some 80 percent of our atmosphere, the triple bond that holds the two atoms of nitrogen together is impossible for most organisms to break. As a result, other organisms depend on nitrogen-fixing bacteria to provide nitrogen in a usable form. You can read more about nitrogen-fixing bacteria in Chapter 25.

Other medically notable proteobacteria include *Helicobacter pylori,* which causes stomach ulcers and stomach cancer (Chapter 1); *Neisseria gonorrhoeae,* which causes the sexually transmitted disease gonorrhea; *Bordetella pertussis,* which causes whooping cough; *Vibrio cholerae,* which causes cholera; and *Yersinia pestis,* which causes bubonic plague.

In this chapter we have briefly surveyed the two domains of prokaryotes and learned how the biology of bacteria helps them acquire resistance to antibiotics. In the next chapter, we survey the first two kingdoms of eukaryotes, the Protista and the Fungi.

Key Concepts

- Prokaryotes are successful because of their rapid growth rates and their metabolic versatility.
- Bacteria can transfer genes to one another via plasmids and viruses, which makes it easy for them to acquire resistance to antibiotics.
- Because prokaryotes do not breed and because they freely transfer genes among species, prokaryotes cannot be classified according to the biological species concept; but their evolution and phylogeny can be deduced from their DNA sequences.

Summary with Key Terms

How do bacteria acquire resistance to antibiotics?

Most **pathogenic** prokaryotes make us ill by releasing toxic chemicals that kill our own cells. These may be **endotoxins** or **exotoxins. Koch's postulates** are guidelines for determining whether a microorganism causes a specific disease, a topic of interest to **epidemiologists.** Resistance may allow a bacterium to survive exposure to an **antibiotic** or, alternatively, to actively destroy any antibiotic in its environment. Resistance genes accumulate in populations of bacteria, first, through natural selection, and second, by moving from one strain of bacterium to another by means of plasmids and other mobile genes.

How do we know that an organism is a prokaryote?

Prokaryotes tuck their DNA into a **nucleoid** rather than a true nucleus. They also lack other membrane-enclosed organelles. Most are single-celled organisms, though some aggregate to form colonies. They come in many shapes, but three forms dominate: the rod-shaped **bacilli,** the spiral-shaped **spirilla,** and the spherical **cocci.** Most have cell walls made of a **peptidoglycan.** Bacteriologists classify bacteria according to whether they are **gram-positive** or **gram-negative.** Some bacteria have specialized outer surfaces that form a **capsule,** a **slime layer,** or hairlike structures called pili. Many prokaryotes use their **flagella** for directed movement, or **taxis** (including **chemotaxis**).

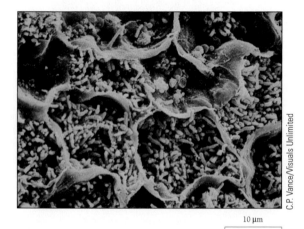

C.P. Vance/Visuals Unlimited

10 μm

Figure 20-19
Aerobic heterotrophs. Nitrogen-fixing bacteria growing in nodules on the roots of a plant. Some nitrogen-fixing bacteria live freely in soil and water, while others associate intimately with specific plants, living, for example, in tiny nodules on the roots of peas, beans, lupines, locusts, and other members of the pea family. The nitrogen-fixing bacteria provide the plant with a usable form of nitrogen, and the plant provides carbohydrates for energy. This relationship allows extraordinarily efficient fixation of nitrogen, as much as 100 times faster than that performed by free-living bacteria.

Prokaryotes grow continuously and rapidly, and usually reproduce asexually.

Prokaryotes are biochemically more diverse than eukaryotes. They may be either **autotrophs** or **heterotrophs.** **Photoautotrophs** obtain energy from sunlight, while chemoautotrophs obtain energy by oxidizing inorganic compounds. Some heterotrophic prokaryotes derive energy from light (**photoheterotrophs**) but also require organic compounds. **Chemoheterotrophs** depend exclusively on organic compounds for both energy and carbon atoms. Some heterotrophs require oxygen to live (**obligate aerobes**), others can live either in the presence or absence of oxygen (**facultative aerobes**), and some cannot live when oxygen is present (**obligate anaerobes**).

How do biologists classify prokaryotes?

Prokaryotes are mainly classified according the DNA sequences that specify ribosomal RNA. The **Archaea** include the **methanogens,** which make methane from carbon dioxide and hydrogen gas, the **halophiles,** which grow only in high concentrations of salts, and the **thermoacidophiles,** which derive energy by oxidizing sulfur compounds or methane.

The **Eubacteria** include about 25 phyla, but they can be classified into three main groups, the **spirochetes** and **chlorobi,** the **proteobacteria** and mitochondria, and the **cyanobacteria** and firmicutes (including the **mycoplasmas**). The proteobacteria bacteria are an immense group of bacteria that include the enterobacteria and the **nitrogen-fixing bacteria.**

Review and Thought Questions

Review Questions

1. What features of eukaryotes do prokaryotes lack?
2. What molecule makes up much of the cell wall of members of the Eubacteria but not of Archaea?
3. What is the difference between gram-positive and gram-negative bacteria? Which kind is susceptible to penicillin? Explain why.
4. Describe two ways in which antibiotics work against bacteria. Why do these mechanisms not affect eukaryotic cells?

5. Describe the two major ways that antibiotic resistance is spread through populations of bacteria.
6. What features allow prokaryotes to reproduce so much more rapidly than eukaryotes? What advantages does this give bacteria?
7. Why don't biologists classify prokaryotes in the same way they classify plants or animals?
8. What aspect of the biology of prokaryotes do biologists use to classify them?
9. Are humans and other eukaryotes more closely related to the Archaea or to the Eubacteria?
10. What is the origin of the eukaryotic organelle called the mitochondrion? To what kind of prokaryote is it most closely related?

Thought Questions

11. Are antibiotics helpful against a viral infection such as a cold or flu? Why or why not?
12. Animals breeders began giving animals antibiotics because studies showed that animals raised on antibiotics grew faster. Today, this effect is not as strong as it was 30 or 40 years ago. Why might antibiotics have this effect? Why might antibiotics no longer be so good at increasing growth rates?

BiologyNow Resources

Biology ⑤ Now™

Active Figures

20-8:   The structure of prokaryotes
20-13: Thermoacidophiles and related archaea

Preparing for an exam? Take a diagnostic test on your BiologyNow CD-ROM.

Online materials relating to this chapter are at:

http://biology.brookscole.com/AAL3

About the Chapter-Opening Image

Salmonella bacteria that infect hamburger typify our problems with antibiotic-resistant strains of bacteria, the theme for this chapter about bacteria.

21

The Grab Bag Protists and the Purebred Fungi

Key Questions

- What feature distinguishes the eukaryotes from the prokaryotes?

- What feature of the protists makes classifying them so much harder than classifying the fungi?

- What features distinguish fungi from other eukaryotes?

- What three important ecological roles do fungi play?

Dr. Paul A. Zahl/Photo Researchers, Inc.

The Fungus Fighters

In 1948, two government researchers set out to find a drug that would ultimately save thousands of human lives. In just two years, Elizabeth Lee Hazen and Rachel Brown succeeded in producing a drug that could safely cure deadly fungal infections in humans—a drug that is still a medical mainstay today. Both women had overcome great obstacles to obtain their scientific training. That they were highly educated, practicing scientists at all was something of a miracle. That they were able to make a major contribution to medicine was a testimony to their lifetime dedication to science and the respect they inspired in those around them.

Elizabeth Lee Hazen, born in August 1885 to a Mississippi cotton farmer and his wife, was an orphan by the time she was two years old. Raised by relatives, Hazen grew up in a large family in Lula, Mississippi. After high school, she enrolled in the Mississippi Industrial Institute and College, the first state-supported college for women in the United States. Without such an institution, she might never have gone to college, might never have studied physiology, anatomy, botany, zoology, and physics (Figure 21-1). Her grades were unremarkable. Like many people with great contributions to make, her strengths would not become obvious until she was much older.

After graduating from college, Hazen taught high school for six years. When she was 31, she enrolled as a graduate student at Columbia University, in New York, where one interviewer told her that she was so ignorant he doubted she had ever been to college. Nonetheless, she soon proved herself, and the following year Columbia awarded her a master of arts degree in biology. During World War I, she worked for the Defense Department diagnosing bacterial infections. After the war, she returned to Columbia, and in 1927, at the age of 42, she received a Ph.D., becoming one of a small number of women with doctorates in the United States. She stayed at Columbia until 1931, when the New York State Department of Public Health lured her away to supervise its large medical pathology lab. But she maintained a connection with Columbia. In 1944, when the health department needed an expert on fungal infections, she returned to study mycology, at the same time continuing her full-time work in the pathology lab.

For most of us, fungi are no more than a mild curiosity. Mushrooms are a nice addition to a pizza. Mold is an annoyance in the shower. Some of us may have had unpleasant experiences with a fungus, a bout of athlete's foot, a yeast infection, nothing that a simple antifungal medication could not treat. In most people, the immune system's white blood cells quickly engulf and digest any fungal spores that make it into the body, and fungi are not a problem.

Under the right circumstances, however, a fungus can destroy a living human being as thoroughly as it can rot a dead stump in the forest. Diabetics, for example, have tissues that are unusually high in sugar. The warm, sugary environment of a diabetic's lung apparently creates a perfect environment for mucormycosis, the lush growth of a fungus in the order Mucorales. A common, usually harmless food mold, some species of Mucorales can grow through the lungs into the heart. Alternatively, this fungus may infect the sinuses, then run hyphae, long strands of the fungal body, throughout a person's head, destroying nerves and brain function. People with damaged immune systems—burn victims, transplant patients being treated with immunosuppressive drugs, and those with immune diseases such as AIDS—are all susceptible to fungal infections. For such people, fungi are a potentially life-threatening problem.

In the 1940s, when the first antibiotics came into use, medical experts discovered that patients treated with antibiotics became highly susceptible to fungal infections. It seemed that many of the bacteria killed by the antibiotics were important in keeping fungi at bay. Yet no antifungal drugs existed, and doctors were helpless to stop the spread of the often-fatal infections.

Elizabeth Hazen was determined to find a drug that would help. By 1948, she had established an immense collection of pathogenic (disease-causing) fungi with which she and her team of technicians could diagnose fungal infections from people all over the eastern United States. Even more important, Hazen had for years been testing soil samples for bacteria that could kill infectious fungi. By 1948 she had found two different bacteria that produced substances toxic

Figure 21-1
Rachel Brown and Elizabeth Lee Hazen. The discovery and use of bacteria-killing antibiotics in the 1940s increased the number of dangerous fungal infections. In 1950, Elizabeth Hazen and Rachel Brown developed the first fungicide, which could safely be used to treat fungal infections in humans.

to fungi. But what were the substances? Hazen needed the help of a chemist to isolate the active ingredient. She took a train from New York City to the health department's main office in Albany, New York, 150 miles away. There she had an appointment with senior researcher Gilbert Dalldorf, himself a talented bacteriologist and physician. Dalldorf listened to Hazen's proposal, then took her to see chemist Rachel Brown.

Like Hazen, Brown had succeeded in science against long odds. She was born in 1898 and raised in Springfield, Massachusetts, in a small, middle-class family. In 1912, when Rachel Brown was 14 years old, her father abandoned his family. Rachel's mother, Annie, then supported herself, her two children, and both her parents on a secretary's salary. Despite her poverty, Annie Brown was determined that her two children should go to the best schools she could afford. By Rachel's senior year in high school, her heart was set on Mount Holyoke College, a small, private college for women. But Rachel was turned down for a scholarship, and her mother could not possibly afford the tuition.

Miraculously, a wealthy friend of Rachel's grandmother offered to pay all of Rachel's expenses for a full four-year college education. Brown earned a B.A. in chemistry from Mount Holyoke and then put herself through graduate school at the University of Chicago. In 1926, she completed a Ph.D. dissertation in chemistry and shortly after took a job in Albany with the New York State Department of Public Health, an institution that hired an unusually high number of women scientists. There, she worked steadily, isolating proteins and other molecules from infectious bacteria and protists that could be used for diagnostic tests or for vaccines. When Elizabeth Hazen and Gilbert Dalldorf stuck their heads in the door of her lab, it was the beginning of a fruitful collaboration and friendship that would last for more than 25 years.

Hazen explained that she had two strains of bacteria that could kill fungi, but that she needed to isolate the antifungal substances for further testing. In the months that followed, Brown and Hazen sent bulky package after bulky package between Albany and New York City. Hazen would send samples of bacteria, then Brown would extract different compounds for further testing by Hazen. The two researchers quickly isolated three fungicidal molecules from two kinds of bacteria. Two of these molecules were highly toxic to mice and therefore not useful as drugs. The third substance, however, was only mildly toxic and killed virtually any fungus it came into contact with; they called it fungicidin. It looked like a miracle drug, another penicillin.

At a meeting of the National Academy of Sciences in October 1950, Hazen and Brown presented their results. A reporter from *The New York Times* jumped all over the story, and within a day or two, the new lifesaving drug was big news. Pharmaceutical companies began calling the state lab, demanding to be let in on the project.

Dalldorf and Brown nearly panicked. Fungicidin had been tested in animals but never in humans. For clinical trials, huge quantities of the drug had to be manufactured cheaply and quickly. Only a pharmaceutical company could do that. But Dalldorf and Brown knew, from previous experience with another drug, that they needed to patent fungicidin quickly before someone else did. Without a patent, to make clear who owned the rights to the drug, no pharmaceutical company would develop it. And Hazen, Brown, and Dalldorf wanted the drug to be used to save lives, the sooner the better.

By February, the researchers had filed a patent application with the United States Patent Office; a month later, they licensed the patent to Squibb Pharmaceuticals. Fungicidin, it turned out, was a name already in use for another chemical, so Hazen and Brown renamed it nystatin (pronounced nye-stat-in), after New York State. Within five years, Squibb had found a way to produce nystatin cheaply and had successfully tested the drug in humans. As the first treatment for fungal infections in humans, nystatin was tremendously useful and still in use today.

Squibb sold so much nystatin that between 1955 and 1979 alone, royalties from sales of the drug came to $13.4 million. Half of this went to a nonprofit foundation that funded scientific research. The other half went to the nonprofit Brown–Hazen Fund, which supported research and training in the medical and biological sciences and awarded grants to women scientists. Neither Hazen nor Brown ever accepted any of the enormous wealth that resulted from their discovery. Brown apparently helped put several young women through college, as she herself had been helped. Both Hazen and Brown continued to work into their 70s, content in their achievements, honored with medals, honorary degrees, and elections to prestigious societies.

How Do We Know That Protists and Fungi Are Eukaryotes?

In this chapter we will survey the first two kingdoms of eukaryotes: the fungi and the protists. In contrast to all the prokaryotes, eukaryote cells keep their DNA inside a membrane-enclosed **nucleus.** Eukaryotes have other membrane-enclosed organelles, too, including mitochondria and chloroplasts. In fact, all eukaryote cells are defined by their internal membranes. In contrast, prokaryotes (Archaea and Eubacteria) have no internal membranes (Figure 21-2).

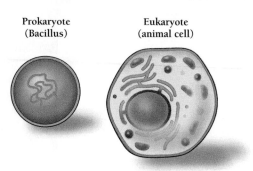

Prokaryote (Bacillus)   Eukaryote (animal cell)

Figure 21-2

How do eukaryotes differ from prokaryotes? Prokaryotes have no internal membranes and no nucleus. Eukaryotes keep their DNA in a nucleus, whose boundary is a membrane similar to the one surrounding a whole cell. Eukaryotes also have other organelles that are surrounded by such membranes, including mitochondria, chloroplasts, and vacuoles. In the last chapter, we looked at the prokaryote (the Archaea and Eubacteria). In this chapter, we turn to the first two kingdoms of Eukaryotes, the Protista and the Fungi.

Protists differ from fungi in several ways. Most fungi are multicellular and have complex internal structures and specialized cells that serve different purposes. In contrast, most protists are single celled. A few kinds of protists form colonies, but these are primitive collections of nearly identical cells. A big difference between fungi and protists is how easily fungi can be classified. Most of the fungi are remarkably alike and make an easily recognized kingdom of organisms. The fungi comprise a single, monophyletic lineage. But the protists are so diverse that many biologists believe that they should be divided into a dozen or more lineages or even kingdoms. Protists are grouped not by what they have in common, but by not belonging to any of the other three eukaryotic kingdoms (Figure 19-5).

21.1 What Are Protists?

If we look at a drop of pond water under a microscope, we discover an amazing world of tiny, wriggling creatures. Many of these microscopic creatures are members of the kingdom Protista [Greek, *protos* = first]. The **protists** range from the simple amebas to the most complicated cells on our planet (Figure 21-3). Protists flourish wherever there is moisture. In many marine ecosystems, photosynthetic protists such as red and brown algae trap nearly all of the solar energy that eventually reaches other organisms.

Photosynthetic protists are the basis of aquatic food chains. On land all animals eat either plants or other organisms that eat plants. But in oceans and lakes, animals eat either protists or other organisms that eat protists. Protists' abundance in the fossil record suggests that they have been successful for a long time.

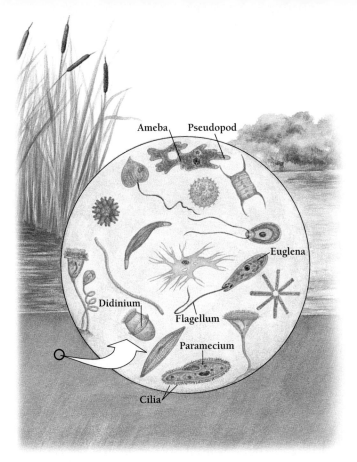

Figure 21-3

A drop of pond water may contain a community of protists. Thousands of species of protists live in oceans, lakes, ponds, and puddles. Protists include the photosynthetic euglenids, the flexible amebas, and ciliates such as *Paramecium* and *Didinium,* as well the beautiful dinoflagellates and diatoms. Some protists anchor themselves to some object or merely drift. Others propel themselves by waving flagella or cilia or by ameboid movements.

At a biochemical level, protists are more or less identical to plants, animals, and fungi: all derive energy from glycolysis and respiration, and, like plants, the photosynthetic protists all use water as a source of electrons. Protists lack the biochemical diversity of the prokaryotes, but they are champions of cellular diversity. No other kingdom includes so many different kinds of cells.

Protists include organisms that photosynthesize (like plants), organisms that eat other organisms (like animals), and organisms whose life cycles closely resemble those of molds and other fungi. Some are photosynthetic and have chloroplasts. Others are heterotrophic and possess vesicles and lysosomes for digesting other organisms. Some are shapeless bags, while others have elaborately sculpted shells. Some are free living, some parasitic. Some produce one-celled reproductive structures called **spores**; others do not. Some are single celled and others are multicellular like plants, animals, and fungi.

Some protists are **plankton** [Greek, *planktos* = wandering], organisms that float in the water, carried by currents. Others are **motile**, meaning that they move actively in search of food. Of the motile forms, some propel themselves with one or two long **flagella** (Figure 21-3). Some use hundreds of **cilia**, and others use protrusions called **pseudopodia** [Greek, *pseudos* = false + *pous* = foot], temporary extensions of the cytoplasm (Figure 21-3).

Protists that photosynthesize like plants are collectively called **algae**, although they aren't necessarily related to one another. Algae are simple, plantlike organisms that live in moist environments everywhere: in open water, ice, soil, and even small puddles or pockets of water in trees or rocks. Like plants, algae use chlorophyll and other pigments to capture light for photosynthesis. However, whereas true plants photosynthesize with both chlorophyll *a* and chlorophyll *b,* most protists do not have any chlorophyll *b.* Algal protists also differ from true plants in the structure of their cell walls and the arrangement of their flagella.

With so much diversity, we may ask what binds the protists together? The answer is, surprisingly little. What protists have in common is that they do not fit in with plants, animals, or fungi. At one time biologists classified all organisms as either plants or animals. But common pond-water protists of the genus *Euglena* illustrate why this classification never worked (Figure 21-4). In sunlight, most species of *Euglena* develop chloroplasts and photosynthesize. In the dark, however, *Euglena* live a heterotrophic life, absorbing nutrients from their surroundings, even sometimes losing their chloroplasts. *Euglena* is not a plant and not an animal, but it is a fine protist.

Another example of how confusing protists can be come from a green alga. *Chlamydomonas* is a single-celled organism capable, like *Euglena,* of both photosynthesis and directed movement. But *Chlamydomonas*'s close relative *Volvox* consists of colonies of hundreds to thousands of cells, each of which looks just like a free-living *Chlamydomonas.* The two kinds

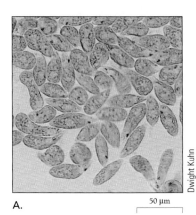

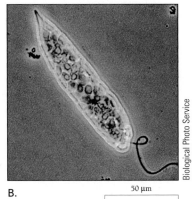

A.

B.

50 μm

50 μm

Dwight Kuhn

Biological Photo Service

Figure 21-4

Neither plant nor animal. A. *Euglena* in sunlight are green with chloroplasts. B. *Euglena* in a dark environment use their flagella to go looking for light.

of *Chlamydomonas* are identical except that one is single-celled and one is colonial. Biologists have had to accept that protists that seem very different from one another may be closely related and those that seem very similar may be unrelated. In this chapter, we survey a few of the phyla of protists, grouped mainly according to DNA sequences and other molecular clues.

> Protists are a grab-bag kingdom, including every eukaryote that is not animal, plant, or fungus. Protists show enormous diversity at the cellular level.

How Do Protists Reproduce?

All protists are eukaryotic. Protist cells have nuclei and other membrane-enclosed organelles, and almost all have mitochondria. Most (but not all) also have chromosomes and undergo mitosis, as described in Chapter 8. All can reproduce asexually by simply dividing in two, and some can also make haploid gametes, by means of meiosis, and reproduce sexually.

Every protist species goes through two phases in its life cycle—reproduction and dispersal. Dispersal may take many forms. Some free-living, motile protists swim or crawl. Others scatter seedlike **cysts,** enclosed structures that contain reproductive cells. Still others infect a host such as an insect, bird, or mammal that will disperse the protist.

Reproduction may be either asexual or sexual, depending on the species and environment. In humans and other animals, our cells are nearly all diploid ($2n$)—having two sets of chromosomes. When animals reproduce we make haploid ($1n$) egg and sperm cells that contain only one set of chromosomes. Some protists are, like us, diploid for most of their life cycle and produce haploid gametes (Figure 21-5). Others are always haploid and always asexual.

Still others are haploid for most of their life cycle but produce a short-lived diploid zygote. Some protists have a multicellular haploid phase and also a multicellular diploid phase, a life cycle pattern that is called **alternation of generations** (Figure 21-5). Plants also have alternation of generations.

Figure 21-5
Alternation of generations in a green alga. A. Some protists are haploid during most of their life cycle. B. Others alternate between a diploid ($2n$) form and a haploid (n) form. In this life cycle of a sea lettuce (a photosynthetic protist), we see that the diploid sporophyte makes haploid spores, which form the gametophyte. The haploid gametophyte in turn makes haploid gametes, which come together in a process of fertilization, similar to the way human eggs and sperm fuse at fertilization. The sea lettuce, genus *Ulva,* grows in shallow seas in the temperate zones. Each individual sea lettuce has a flat, leaflike "thallus," up to a meter long, and a rootlike "holdfast," which attaches the thallus to a rock or other object. C. Still other protists are primarily diploid, as humans are.

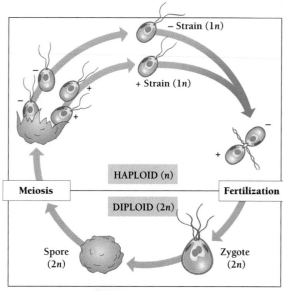

A. **Haploid**

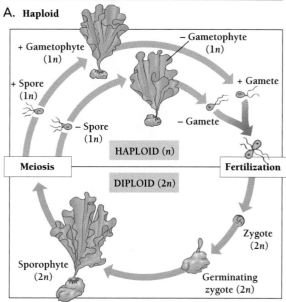

B. **Alternation of generations**

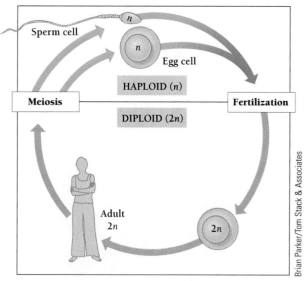

C. **Diploid**

Brian Parker/Tom Stack & Associates

21.2 What Are the Six Main Groups of Protists?

The classification of the protists is an area of active and controversial research, and it would hardly be honest to say that what follows is a definitive classification (Figure 21-6). However, DNA sequences suggest that the most common protists can be grouped into six groups: (1) the Discricristales (including the euglenids, flagellates, and acrasid slime molds); (2) the Chromalveolates (including the dinoflagellates, ciliates, water molds, diatoms, and various algae); (3) the radiolarians and foraminifera; (4) the amebas; (5) the red algae; and (6) the green algae.

Table 21-1
Which Protists Use the Same Chlorophylls as Plants?

Chlorophyll	a	b	c	Storage material
Plants	a	b		Starch
Chlorophytes	a	b		Starch
Rhodophytes	a			Starch
Euglenophytes	a	b		No Starch
Dinoflagellates	a		c	Starch or oil
Phaeophytes	a		c	No starch
Chrysophytes	a		c	No starch

Discricristales

The discricristales include the **euglenophytes** [Greek, *eu* = true + *glene* = eyeball], which take their name from their light-detecting eyespots. Their eyespots enable them to live a double life. In the dark, they live heterotrophically, digesting organic matter in the water around them. But when sunlight becomes available, euglenophytes use their eyespots, and a single flagellum, to swim into the sunshine (Figure 21-4). Euglenophytes photosynthesize with both chlorophyll *b* (a pigment also used by plants), which gives them their grass-green color, and chlorophyll *a*. Except for the green algae, no other algae or any other protists make chlorophyll *b* (Table 21-1). Unlike plants and green algae, however, euglenophytes have no cell wall and

store energy in a polysaccharide called paramylon, instead of starch (which plants and green algae use).

Euglenophytes are closely related to **trypanosomes**, flagellated protists that cause sleeping sickness in Africa and Chagas disease in South America. Trypanosomes, which have complex life histories, are transmitted to people and animals via insect bites (the tsetse fly in Africa and the "kissing bug" in South America).

Another related group within the discricristales are the **flagellates**, a diverse group of single-celled heterotrophs. All have at least one long flagellum, and some have thousands. Some flagellates are free-living in fresh or salt water, while others are parasites in animals. Some convert from a flagellated to an

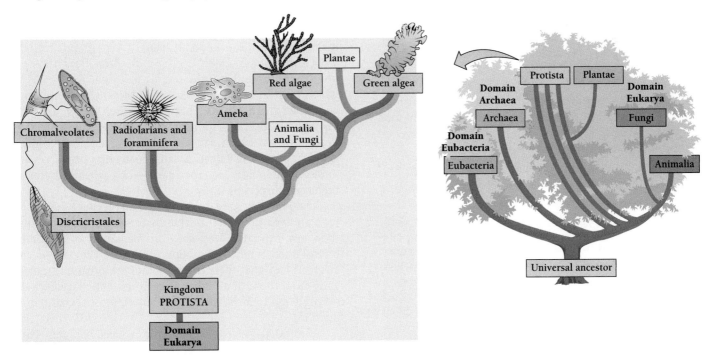

Figure 21-6
The diverse kingdom Protista. Most protists belong to just six major lineages—the discricristales, the chromalveolates, the radiolarians and foraminifera, the amebas, the red algae, and the green algae (closely related to plants).

Figure 21-7
The slug form of an acrasid slime mold.

ameboid form and back again, depending on the availability of food. A well-known flagellate is *Giardia lamblia*, a human parasite transmitted in water contaminated with human feces. *Giardia* can cause abdominal cramps and severe diarrhea. Species of *Giardia* differ from most protists and other eukaryotes in lacking mitochondria. However, biologists have discovered that the *Giardia* genome contains some of the same genes that mitochondria have, and biologists now think that *Giardia* once had mitochondria but exported the genes to the nuclear DNA and then lost its mitochondria.

Acrasiomycota [Greek, *acrasia* = confusing two things + *mukes* = fungus], or **acrasid slime molds,** are peculiar protists, animal-like at some stages of their lives and funguslike at others. They live in damp soil, in fresh water, on decaying vegetation, and especially on rotting logs. When nutrients are plentiful, acrasid slime molds flourish as separate amebalike cells, consuming bacteria and other food particles by phagocytosis. When food is scarce, the individual amebas aggregate to form a multicellular stage that looks a bit like a miniature garden slug (Figure 21-7). The slug form secretes a slime track and wanders about in search of food and light. When it finds food, it settles down and develops a mushroomlike **fruiting body** that produces spores. Each spore turns into an individual ameba. Reproduction is asexual; both the amebas and the slug form are haploid.

Euglenophytes can live as autotrophs or heterotrophs. The flagellates are single celled and use flagella to move. Acrasiomycota (acrasid slime molds) live as individual amebas and as multicellular slugs, depending on their food supply.

Chromalveolates

The chromalveolates are a huge and diverse group of protists that include two major lineages. The dinoflagellates, apicomplexans, and ciliates lie on one branch. The water molds, golden algae, brown algae, and diatoms lie on the other.

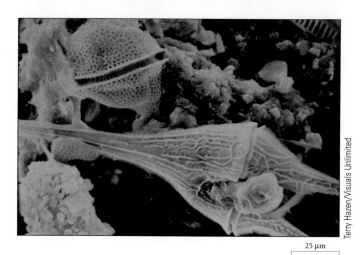

Figure 21-8
Two dinoflagellates. Dinoflagellates are renowned for their diversity and beauty. Shown here are just two kinds of luminescent dinoflagellates.

Dinoflagellates, Apicomplexans, and Ciliates

Some of the most flamboyant protists are the **dinoflagellates,** single-celled organisms that float freely as plankton in warm oceans (Figure 21-8). Most dinoflagellates have a rigid wall, or **test.** The test is crossed by two grooves, each containing a flagellum. The two flagella spin around like propellers, driving the tiny organism through the water and giving the dinoflagellates [Greek, *dinos* = whirling + *flagellum* = whip] their curious name. A few dinoflagellates form colonies, and some live symbiotically with corals, sea anemones, and clams. Dinoflagellates make up a major part of the phytoplankton, the basis of most aquatic food chains.

Of the thousands of species of dinoflagellates, the best known are those that grow rapidly to cause "red tides," in which population explosions, called blooms, of dinoflagellates color the ocean red, brown, or other colors, depending on the species (Figure 21-9). Some red tides release deadly toxins, which can kill fish and make filter-feeding shellfish poisonous to eat. Because many dinoflagellates are luminescent—glow in the dark—they create dramatic nighttime displays.

Dinoflagellates are a very old group. Because their hard tests are well preserved in the fossil record, we know that they have existed at least since the beginning of the Cambrian Period, some 570 million years ago. They are so ancient, in fact, that their chromosomes are organized differently from those of all other eukaryotes. Dinoflagellates do not undergo mitosis, meiosis, or any form of sexual reproduction.

The 4,000 species of **Apicomplexa** are all parasites. They are distinguished by a distinctive cell structure called an *apical complex,* which is made up of pipelike molecules called microtubules (Chapter 4). Apicomplexans reproduce by a combination of asexual and sexual processes and have unusually complex life cycles, involving two or more different hosts. For

example, four species of the genus *Plasmodium* cause malaria, a disease that annually infects about 500 million people and kills 2 to 3 million. Humans are one host for *Plasmodium*. But the four species of *Plasmodium* infect humans with the help of a second host—mosquitoes of the genus *Anopheles*.

A female mosquito needs a blood meal in order to reproduce. When she sucks blood from a human, she first injects a bit of saliva. The saliva contains an anticoagulant, a chemical that keeps the blood flowing freely and allows her to drink her fill (Figure 21-10). A mosquito infected by the *Plasomodium* parasite injects a diploid (2*n*) form of the parasite, called a *sporozoite*, along with her saliva. The sporozoite passes to the human liver and there forms *merozoites*. The merozoites enter red blood cells, divide rapidly, and release toxins that cause the fever and chills of malaria. Some of the merozoites form *gametocytes* that are sucked up by other mosquitoes. Inside the mosquitoes, the gametocytes form gametes that fertilize one another and form new sporozoites, so the cycle can begin again.

The complex life cycles of all apicomplexans ensure their survival. At any time, different phases of an apicomplexan are present in several forms in different hosts, making malaria difficult to eradicate. The malaria *Plasmodium*, for example, is present in the gut and salivary glands of mos-

Figure 21-9
More dinoflagellates! This is a satellite photo of a red tide off the east coast of Italy. Populations of dinoflagellates sometimes explode and form "red tides." Some red tide species make poisons that clams, mussels, and other shellfish accumulate in their bodies, making these ordinarily edible shellfish deadly. Some dinoflagellates secrete toxins that kill whole schools of fish, whose flesh the dinoflagellates then feed on. During a bloom of such dinoflagellates, toxic vapors over the water can make boaters temporarily ill.

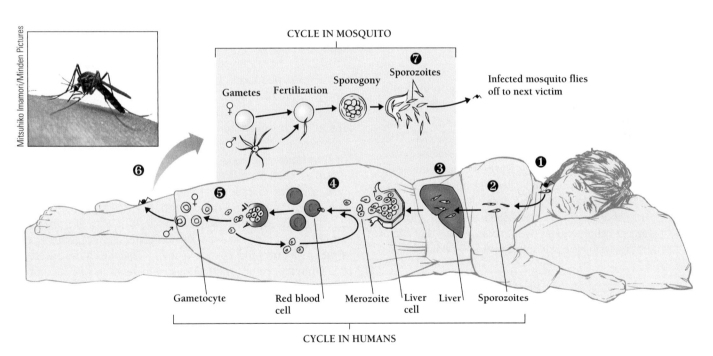

Figure 21-10
How *Plasmodium* causes malaria. ❶ A mosquito infected with the malaria parasite *Plasmodium* has a diploid (2*n*) form of the parasite, called a *sporozoite*, in the wall of its gut and in salivary glands. ❷ When the mosquito sucks blood, it injects a few of these sporozoites into the human host's blood. ❸ The sporozoites travel through the bloodstream to the cells of the liver, where the sporozoites divide and develop into *merozoites*, which are also diploid. ❹ The merozoites enter red blood cells and divide rapidly until the red cells break open and release merozoites and toxins into the blood, causing the fever and chills of malaria. ❺ Some of the merozoites develop into male or female *gametocytes*. When a new mosquito bites the same person, it sucks up the gametocytes, which make haploid (1*n*) gametes. ❻ Still inside the mosquito, the haploid gametes fuse to form diploid zygotes that embed themselves in the mosquito's intestinal wall. ❼ These develop into sturdy *oocysts* that divide to form thousands of new sporozoites, some of which migrate to the salivary gland, where they begin the cycle again. (The mosquito shown here is not in the genus *Anopheles*.)

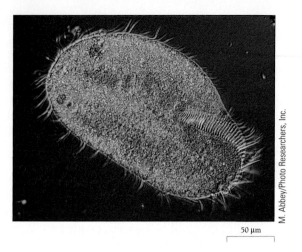

Figure 21-11
The cilia of a *Paramecium*. The cilia surround the oral groove and form oarlike arrangements for movement.

quitoes and in the liver and blood of humans. Malaria has been controlled in the United States through the elimination of mosquito-breeding areas in most parts of the country and the widespread use of potent insecticides such as DDT. Yet, worldwide, the number of new cases of malaria has more than doubled since the mid-1970s. The mosquitoes have evolved resistance to insecticides, and *Plasmodium* has evolved resistance to antimalarial drugs used in humans, and malaria is more of a health problem than ever.

The **ciliates** (Ciliaphora) include about 8,000 species of free-living, single-celled heterotrophs that live in both fresh water and salt water. Their name [Latin, *cilium* = eyelash] reflects the presence, on their surface, of thousands of cilia, shorter and more numerous than eukaryotic flagella. The cilia, which may be arranged in bundles or in sheets, help ciliates move and feed (Figure 21-11). The best-known ciliate is probably *Paramecium*, a common resident of pond water and long used in teaching and research. Most ciliates lack shells, although a few marine ciliates form shells of sand held together by organic cements. The fossil record shows that such shell-forming ciliates have existed for about 100 million years.

Dinoflagellates are among the oldest and most primitive of the eukaryotes. Apicomplexans are nonmotile, spore-forming protozoa, such as the malaria-causing *Plasmodium*, that live as parasites in animals. They are known for their apical complex and their complex life cycles. *Paramecia* and other ciliates are single-celled, free-living heterotrophs coated with cilia.

Water Molds, Golden Algae, Brown Algae, and Diatoms

The **oomycota**, including the "water molds," behave like real molds (fungi), but they are neither molds nor fungi. Like the true fungi that we'll discuss later in this chapter, the oomy-cota are either parasitic, deriving nourishment from living organisms, or *saprophytic,* deriving nourishment from dead organisms. And like true fungi, the oomycota extend hyphae-like threads into the tissues of their hosts, release digestive enzymes, and absorb the resulting organic molecules.

Oomycota reproduce sexually, and their life cycles include both haploid and diploid phases. As in animals, the haploid gametes differ greatly in size; the large ovum is the source of this phylum's name, oomycetes [Greek, *oion* = egg + *mukes* = fungus], meaning "egg fungi." Among the most common oomycetes are the water molds, which grow in fuzzy masses on the remains of algae or animals. In fresh-water ecosystems, water molds decompose and recycle nutrients from dead organisms.

A few oomycetes cause diseases in animals and plants. Historically, the most important of these is *Phytophthora infestans*, which causes blight in potatoes. In 1845 and 1849, this oomycete infected and destroyed the Irish potato crop. A million Irish died of hunger, and another 1.5 million fled to the United States. A similar blight on the German potato crop late in World War I helped force an end to the war.

Golden-yellow pigments give the **golden algae,** or **chrysophytes** [Greek, *chrysos* = golden + *phyta* = plant], their appealing name. Like dinoflagellates, many chrysophytes form hard tests containing silica and other minerals that have helped preserve them in the fossil record as far back as 500 million years ago. Each chrysophyte family contains both single-celled and colonial forms. The colonial forms may be spherical or loosely branched. Most chrysophytes are plankton, widespread in temperate lakes and ponds. Only one group lives in the oceans.

Chrysophytes reproduce asexually in one of two ways. A single cell, called a swarmer cell, may swim away from a colony to found another colony. Alternatively, the hundreds or thousands of cells of a mature colony may split into two groups, which then float away from each other to form two new colonies. Chrysophytes can also form cysts that resist cold temperatures and desiccation.

The **phaeophytes** [Greek, *phaios* = dusky brown], or brown algae, together with the unrelated rhodophytes (red algae) make up most of the organisms that we call seaweeds. Phaeophytes are all multicellular, and most live in the cool waters off temperate coasts. Phaeophytes include the biggest protists in the world—the sargassums and giant kelps, which in deep water may be up to 100 meters tall (Figure 21-12). In the Sargasso Sea, off the east coast of North America, are immense mats of brown alga called sargassum. Most members of this genus attach to rocks or sand, but those in the Sargasso Sea form free-floating mats. Another species, *Sargassum muticum*, from Japan, is an invasive exotic throughout the world. *S. muticum* now grows along both coasts of North America and the coast of Europe.

Most phaeophytes reproduce sexually. However, meiosis in phaeophytes produces one-celled reproductive structures called **spores,** instead of gametes. The spores, which are haploid, germinate into new multicellular organisms in

Figure 21-12
A Pacific kelp forest. Kelp forests such as this one are home to whole communities of organisms. In shallow coastal waters, the giant kelp *Macrocystus* may grow at enormous rates, producing masses of brown algae that provide both food and shelter for fish, sea urchins, sea otters, and other organisms. Although the giant kelps are simpler than true plants, they have a few specialized structures. Floating photosynthetic blades capture light for photosynthesis. The blades are kept near the surface by gas-filled bladders and anchored to the sea floor by holdfasts.

which all the cells are haploid. This multicellular haploid form is called the **gametophyte** because it makes gametes. In contrast, the diploid form, which makes the spores, is called the **sporophyte.** Such alternation of generations also occurs in green algae, red algae, fungi, and the true plants. Brown algae can also reproduce vegetatively, so that a single branch of *Sargassum muticum* can float hundreds of miles in a current of water and start a new colony.

By protist standards, phaeophytes are huge and complex, but their individual cells are much like those of other protists. Phaeophytes make chlorophylls *a* and *c* and produce other pigments like those of chrysophytes (Table 21-1). But, unlike true plants, phaeophytes never make chlorophyll *b* or starch.

The **diatoms** (Bacilliarophyta) are some of the most numerous and beautiful of all protists. Most of the 10,000 species of diatoms are single cells, although some form simple filaments or colonies. Each species has a unique and elaborate silica test (Figure 21-13). Most diatoms live as free-floating plankton in fresh or salt water. Some, however, can move on slime that they extrude from a slit between the two parts of the test. The tests of diatoms are well preserved in the fossil record, dating from the Cretaceous Period, between 65 and 135 million years ago.

Diatoms can reproduce sexually, forming gametes by meiosis. Usually, however, they also reproduce asexually, dividing their chromosomes through ordinary mitosis. At cell division, each half of the test stays with one of the daughter cells, which then assembles a second half.

Oomycota resemble fungi, but their spores have flagella. Chrysophytes (golden algae) have both single-celled and colonial species. Phaeophytes (brown algae), which are related to the chrysophytes, are the largest and most complex of the protists. Diatoms have elaborate tests and can reproduce sexually as well as asexually.

Radiolaria and Foraminifera

Like diatoms, **radiolarians** are among the most beautiful protists. Their glasslike skeletons can be found in the fossil record as far back as the early Cambrian, more than 500 million years ago. In fact, the distinctive shapes radiolarians have evolved over time help geologists identify the ages of rocks. Most radioloarians live as free-living plankton in oceans. They can reproduce sexually or asexually. Some are predators, while others feed on even smaller plankton, which they filter from the water (Figure 21-14A).

Figure 21-13
Diatoms. Diatoms' intricate tests give them an edge against the dinoflagellates in competing for the title of most beautiful protist.

50 μm

A.

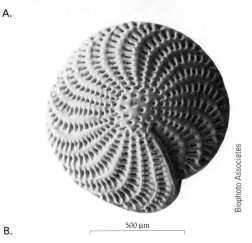

B.

500 μm

Biophoto Associates

Dennis Kunkel, Visuals Unlimited

C.

Ric Ergenbright/CORBIS

Figure 21-14
Radiolarians and Foraminera. A. The tests, or shells, of these salt water radiolarians illustrate the beauty and diversity of this group of protists. B. The intricate tests of foraminerans are rich in calcium. C. As marine foraminifera die and drop to the sea floor, their tests form thick sedimentary deposits hundreds of feet thick, as shown in this picture of the famous white cliffs of Dover, England.

The **foraminifera** [Latin, *foramen* = little hole + *ferre* = to bear] are marine organisms with tests made of organic materials reinforced with grains of sand or minerals and filled with tiny holes (Figure 21-14B). Foraminifera range in size from about 20 μm to several centimeters. Most are tiny and live in the sand or attached to other organisms or stones. Two large groups are free-floating plankton and serve as food for clams, mussels, and other invertebrates that filter food from water. The tests of foraminifera are abundant in the fossil record, especially in the last 230 million years. They make up much of the chalky sediment on the bottoms of the seas. The famous white cliffs of Dover, in England, as well as other cliffs like them, were created undersea as billions of foraminifera rained down on the ocean floor over millions of years (Figure 21-14C).

Foraminifera have complex sexual life cycles. Like plants, they have true alternation of generations, with both haploid and diploid phases. Some harbor within them photosynthetic protists such as dinoflagellates, chrysophytes, or diatoms.

The ancient radiolarians have beautiful glasslike tests. The foraminifera have chalky tests full of holes.

Rhizopoda

Rhizopoda, or amebas, are common throughout the world, in oceans, in fresh water, in soil, and as parasites of animals. Hundreds of millions of people in the tropics (and about 10 million in the United States) harbor parasitic amebas that infect the lower intestinal tract. Fortunately, only about 20 percent of such infections produce the symptoms of amebic dysentery, which range from mild discomfort to severe diarrhea and death. In the most severe cases, the ameba may cause abscesses that rupture the abdominal wall or penetrate the diaphragm and lungs.

An ameba continuously changes its shape, in an apparently chaotic way (hence the old name of one species—*Chaos chaos*). Amebas do this by means of **pseudopodia,** temporary extensions of membrane-enclosed cytoplasm. An ameba extends a pseudopod and then moves its cytoplasm into the pseudopod, so that the pseudopod becomes the body. The movement of amebas only appears chaotic, for they actually move in a directed manner toward sources of food. They then envelop the food in their pseudopodia (Figure 21-15).

All amebas reproduce through simple cell division. Some, such as the parasitic amebas that cause amebic dysentery, can form cysts that won't dry out and can't be digested by their hosts. A few construct distinctive tests by gluing together tiny grains of sand and other inorganic bits. Similar tests exist in the fossil record, with some dating back to the Precambrian, more than half a billion years ago.

A relative of the amebas are the myxomycota [Greek, *muxa* = mucus + *mukes* = fungus]—the plasmodial slime molds, whose appearance resembles the acrasid slime molds

Figure 21-15
Beautiful Ameba. Scanning electron micrograph of *Amoeba proteus*.

mentioned earlier. The slug form of a plasmodial slime mold differs from that of an acrasid slime mold in having no cell boundaries. It's just a mass of cytoplasm with many nuclei but no boundaries between cells. The plasmodial slime molds also differ from the acrasid slime molds in having a sexual cycle. The amebas of plasmodial slime molds are haploid, and two of them can join to form a diploid zygote. The nucleus of the zygote divides repeatedly, but the zygote itself does not divide. The result is a slug form with numerous nuclei that can feed and move about. It propels itself by growing more in one direction than another. When food or water becomes scarce, the slug forms a fruiting body, undergoes meiosis, and releases haploid spores.

Rhizopoda (amebas) use pseudopodia to move and ingest food. Myxomycota (plasmodial slime molds) live as amebas or as a multinucleated plasmodium.

Rhodophytes (Red Algae)

Although the brown algae include the biggest, most spectacular seaweeds, most of the world's seaweeds are **rhodophytes** [Greek, *rodon* = rose], or **red algae.** Nearly all of the 4,000 species of rhodophytes are marine, and most live in tropical seas, attached to rocks or other algae. Among the most beautiful of the rhodophytes are the coralline algae. These algae deposit calcium carbonate in their cell walls and contribute to the building of coral reefs. Calcium-depositing rhodophytes have been preserved in the fossil record as far back as 500 million years.

The cell walls of red algae occasionally contain cellulose but more often contain another polysaccharide, whose attached sulfate groups give the rhodophytes their characteristic slippery feel. Two such polysaccharides—agar and carrageenan—are useful to humans. Agar helps gel instant desserts, jellies, and baked goods, and is also the most widely used culture medium for growing bacteria in the laboratory. Carrageenan is used to stabilize oil and water emulsions in foods such as salad dressing, as well as in paints and cosmetics.

All rhodophytes reproduce sexually and have complex life cycles. Meiosis can take place right after fertilization, with no sporophyte stage. Alternatively, the zygote can develop into a diploid organism, which later produces haploid spores that grow into male and female gametophytes. The female gametophyte produces a large ovum, while the male produces a much smaller sperm.

Rhodophytes (red algae) have distinctive pigments and complex sexual cycles.

Chlorophytes (Green Algae)

The photosynthetic **chlorophytes** [Greek, *chloros* = green], or green algae, are more like plants than any other group of protists. In fact, plants almost certainly arose from a chlorophyte ancestor and we could reasonably place the chlorophytes in Chapter 22, with the plants. Chlorophytes are unlike the other photosynthetic protists in having both of the chlorophylls that plants have—chlorophylls *a* and *b*—and in making starch, just as plants do. Among the rest of the protists, only the euglenophytes have both chlorophylls *a* and *b*. Modern chlorophytes include several multicellular groups, such as these colonies of *Volvox* (Figure 21–16).

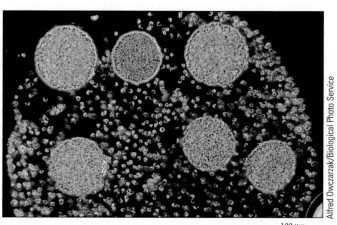

100 μm

Figure 21-16
Colonial green alga. Are colonial algae really multicellular individuals, or just blobs of identical cells? Members of the colonial genus *Volvox* may contain 500 to 60,000 cells arranged in a single-layered, hollow sphere. The cells in most colonies of green algae are identical. In *Volvox* and some other genera, however, there is some division of labor, suggesting a rudimentary form of multicellularity. For example, only some cells are able to reproduce. In addition, the cells may move smoothly through the water by coordinating the beating of their flagella, much as our heart cells coordinate their contractions.

Chlorophytes come in a great array of colonial and multicellular forms. They range in size and complexity from the single-celled *Chlorella* to the 25-foot long seaweedlike *Codium magnum*. Chlorophytes are mostly free living and aquatic, but some species live on the surface of snow, on tree branches, in the soil, or in close association with other organisms, including fungi, hydras, and other protists.

Among the most common of the green algae is the single-celled *Chlorella*, which lives in fresh water, salt water, and soil. Chlorella-like species also live inside of other organisms, including other protists and invertebrate animals such as sponges, hydras, and flatworms. A close association between two organisms is called **symbiosis** [Greek, *sumbios* = living together], and organisms living in a symbiotic relationship with one another are called *symbionts*. Chlorella individuals provide their symbiotic partners with energy-rich compounds produced by photosynthesis.

The photosynthetic chlorophytes (green algae) are more like plants than other protists and also the most diverse of the plantlike protists.

21.3 How Do True Fungi Differ from Other Organisms?

We love fungi. We put fungi in our salads and fungi in our sauces. We use fungi to ferment wines and cheeses and to leaven bread. What is more, we need fungi. A walk through a damp forest demonstrates the enormous ecological importance of fungi: they are prodigious decomposers, recycling the nutrients of dead plants and animals. Without fungi, dead trees would fall and accumulate, and the soil, a little bit of which washes away in every storm, would never be replaced.

But fungi also plague us in countless ways. *Cryptococcus neoformans,* for example, is one of many unpleasant fungi that torment humans and other animals. Distributed worldwide, it infects first the lungs, then the membranes surrounding the brain, and sometimes the kidneys, bones, and skin. The first sign of infection is a cough. Usually, the symptoms that bring *C. neoformans's* victims to a doctor, however, are headache, blurred vision, confusion, depression, or inappropriate behavior (such as odd speech or dress). Another infectious fungus attacks the sinuses, then the brain, producing intolerable headaches and ultimately consuming its victim alive.

Still other fungi destroy entire harvests of wheat, potatoes, and other crops. Fungi destroy in small ways as well as large. In the kitchen, fungi ruin bread, potatoes, jams, and sauces. In the bathroom, they attack our feet, producing athlete's foot, and grow on the walls of the shower and the porcelain surfaces of the toilet.

Fungi grow wherever there is sufficient warmth, moisture, and energy-rich organic molecules. Sometimes fungi find nourishment in surprising and bothersome places—the lining of a camera lens, clothing, shoes, or the wooden planks of a ship. During the American Revolution, the British lost more ships to dry rot (a fungus) than to the fledgling American Navy. In World War II, fungal infections of the skin took more men out of action in the South Pacific than did battle wounds.

Fungi are a little like plants and a little like animals. As long as biologists classified all organisms as either plants or animals, fungi posed a problem. Like animals, fungi are heterotrophic—they derive their energy from other organisms. But fungi enclose their cells in rigid walls as plants do. Fungi are distinct from both plants and animals, however: they take in food differently from animals, their cell walls are chemically different from those of plants, and they lack the embryonic stages of plants and animals. Biologists all agree that the fungi deserve their own kingdom.

How Can We Characterize the Fungi?

With the exception of the single-celled yeasts and a few others, most fungi are **molds** composed of masses of threadlike filaments called **hyphae** [Greek, *hyphe* = web; singular, hypha]. The hyphae contain many nuclei (Figure 21-17). In some fungi, the nuclei all lie in a common cytoplasm rather than in separate membrane-enclosed cells. Such fungi are said to be **coenocytic** [Greek, *koinos* = shared + *kytos* = hollow vessel]. In **dikaryotic** fungi (which have two nuclei per cell), walls called **septa** separate the nuclei within a filament. The septa, however, are perforated by large pores that allow the cytoplasm to flow along the hyphae. Hyphae grow, branch, and intertwine to form a mass, called a **mycelium** [Greek, *myketos* = fungus]. Growth may be so rapid that in the time it takes you to read this page a single mycelium can add meters of hyphae.

Fungi's ability to grow rapidly helps compensate for another odd characteristic of their cells—lack of flagella and

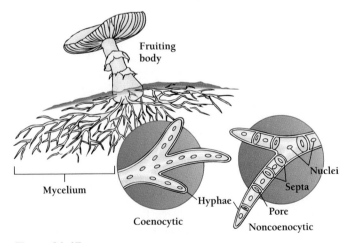

Figure 21-17

Two types of fungal hyphae. The nuclei within hyphae may all share the same cytoplasm (coenocytic), or they may be separated by septa.

Figure 21-18

A fairy ring. An underground mycelium grows outward over hundreds of years. When the mycelium "fruits," it forms mushrooms at its outer edge, which release spores. The grass within the circle of mushrooms is often stunted and pale. For this reason, people once believed that dancing fairies caused these rings.

cilia. Fungi can move in two ways only. The mycelium can move from place to place by growing, or the mycelium can produce spores that the wind scatters.

Another distinguishing characteristic of the fungi is their cell walls, which are made of a tough polysaccharide called **chitin,** the same substance in the hard shells of insects, spiders, and crustaceans. In plants, the major component of the cell wall is cellulose.

Recall that all fungi are heterotrophs, organisms whose food is organic molecules produced by other organisms. A fungus's food may be either dead or alive. Saprophytic fungi feed on the remains of dead organisms, while parasitic fungi feed on living organisms. Fungi obtain food, whether dead or alive, by infiltrating the bodies of organisms and other food sources with long, thin hyphae. The hyphae secrete digestive enzymes that break down the food, and absorb, rather than ingest, the resulting nutrients.

Fungi may reproduce asexually or sexually. Each cell of the hyphae is usually capable of generating a whole new mycelium, allowing most fungi to reproduce vegetatively (asexually). A single mycelium can produce separate individuals by means of simple cell division, by the breaking up of existing hyphae, or by budding. Fungi also reproduce by making and dispersing hardy spores capable of resisting heat, cold, drought, and lack of food. Spores are so small that the wind can scatter them all over the planet.

Spores, which are always haploid, may be made in a variety of intricate structures, called **sporophores.** Ordinary mushrooms are the sporophores of a large underground mycelium, which may be many meters in diameter (Figure 21-18). In general, asexual spores form when food is plentiful and the environment is otherwise welcoming. Fungi reproduce sexually when food is scarce or the environment has become unfavorable. Recall that sexual reproduction creates a greater diversity of genotypes in the next generation. This diversity increases the chances that a few individuals will survive a hostile environment. Each haploid spore can germinate and produce a new haploid mycelium. Fungi differ from both plants and animals in that they do not go through any embryonic stages. Spores form mycelia directly.

Fungi are filamentous, lack flagella, have distinctive cell walls, absorb food, and reproduce by means of spores.

How Do Mycologists Classify Fungi?

The three main phyla of fungi are the zygomycetes, ascomycetes, and basidiomycetes (Figure 21-19). These three phyla differ from one another in how they make spores and whether the hyphae have septa. Most fungi are capable of

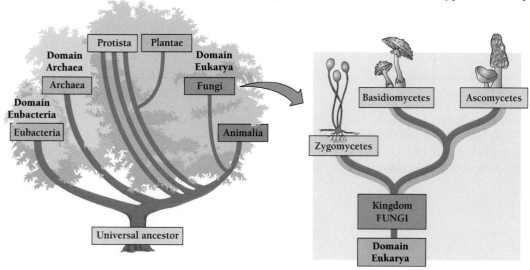

Figure 21-19
The kingdom of Fungi. A selection of phyla of Fungi.

Biology ⒺNow™ Learn more about fungi by clicking on this figure on your BiologyNow CD-ROM.

Health and Biology   Fungal Infections Today

Fungal infections have become increasingly common. Between 1991 and 1994, for example, fungal infections in one county in southern California increased from 400 per year to 4,500 per year. In fact, today nearly 40 percent of all deaths from hospital-acquired infections are due not to bacteria or viruses but to a handful of fungi.

Most hospital fungal infections are due to *Candida albicans,* a virulent relative of baker's yeast. *Candida* infects 70 percent of AIDS patients. Even if science could eliminate *Candida,* however, other fungi would quickly take its place. *Cryptococcus, Histoplasma,* and *Aspergillus* are three of the next most common fungi that infect patients with impaired immune systems. Even ordinary baker's yeast, a normal inhabitant of the human mouth and the stuff we use to make bread rise, can grow out of control in individuals who are already very sick.

Healthy people also fall victim to some seemingly ordinary fungi. In California's San Joaquin Valley, for example, many people become infected by the soil fungus *Coccidioides,* which causes symptoms ranging from mild flulike illnesses to attacks in which the fungus spreads throughout the body (Figure A).

Against this onslaught of rot, health care workers have only a few broadly effective drugs. Nystatin and two other drugs derived from it are still the most effective way to treat fungal infections. One of the most powerful antifungals is amphotericin B, a drug with side effects so nasty that patients sometimes call it "amphoterrible." It can cause fever, chills, low blood pressure, headaches, vomiting, inflammation of blood vessels, and kidney damage.

Safe antifungal drugs with few side effects are much harder to find than safe antibiotics. This is because antibiotics, which kill prokaryotes but not our own eukaryotic cells, work by exploiting a variety of differences in cell chemistry between prokaryotes and eukaryotes. But fungi are eukaryotes, like us. Their cells work like our own cells, and chemicals that are bad for fungi are usually bad for us, too.

Nonetheless, the chemistry of fungal cells does differ from that of our own, and any differences can be exploited by researchers who design drugs. For example, the three drugs most commonly used all work by attacking ergosterol, a lipidlike solid that helps form the cell membranes of fungi. Our own cell membranes contain cholesterol instead of ergosterol. Fungal cells also differ from ours in having a cell wall. Researchers' latest hope is to find a drug that attacks the cell walls of fungal cells without harming our own cells.

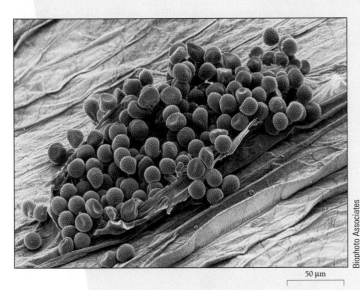

50 μm

Biophoto Associates

Figure A

Fungal spores are everywhere. The tiny haploid particles by which mushrooms and other fungi reproduce float in and out of our nostrils with each breath, land on bread and jams in our kitchens, and grow into dark streaks in our bathrooms. Outside our homes, a rain of fungal spores dusts trees and shrubs, lawns, dead leaves, and soil. Creeks, rivers, and lakes disperse fungal spores from one watershed to the next. And winds carry fungal spores from one end of the Earth to the other.

---

reproducing both sexually and asexually. Asexual reproduction may occur vegetatively (as when parts of the mycelium divide to form new individuals) or through the production of spores by mitosis. Sexual reproduction always occurs by means of spores.

Zygomycetes

The mycelium of a **zygomycete** [Greek, zygote + *mukes* = fungus], such as bread mold, is coenocytic: it consists of hyphae without dividing septa, and all the nuclei share a common cytoplasm. The only exceptions are two kinds of spore-forming structures, which form asexual or sexual spores. Most of the time, zygomycetes reproduce asexually by forming spores in spore-containing capsules called **sporangia.**

The common black bread mold, *Rhizopus stolonifer,* illustrates the sexual life cycle of a typical zygomycete (Figure 21-20). When zygomycetes reproduce sexually, they do so by forming a *zygosporangium,* a structure that produces haploid spores from a diploid zygote.

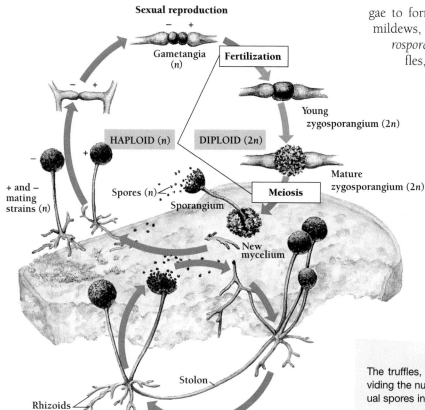

Sexual reproduction

Gametangia (n)

Fertilization

HAPLOID (n)   DIPLOID (2n)

Young zygosporangium (2n)

Meiosis

Mature zygosporangium (2n)

+ and − mating strains (n)

Spores (n)

Sporangium

New mycelium

Stolon

Rhizoids

Asexual reproduction

Figure 21-20
Zygomycetes make spores. Life cycle of the black bread mold *Rhizopus stolonifer*.

Biology ⓔNow™ Learn more about the life cycles of fungi by clicking on this figure on your BiologyNow CD-ROM.

Whether zygomycetes reproduce sexually or asexually, they always use spores. Both kinds of spores travel on air currents and can, in the right environment, divide to produce a new mycelium. Because spores drift randomly, their arrival on a hospitable host (a piece of bread, for example) is entirely a matter of chance. A few zygomycetes, however, have evolved a means of actively launching themselves into likely new environments.

Bread molds and other zygomycetes are coenocytic and usually reproduce asexually by means of spores.

Ascomycetes (Sac Fungi)

Ascomycetes are a large and diverse group, with some 30,000 species (Figure 21-21). They are monophyletic (one lineage) and include most of the fungi that combine with al-

gae to form lichens, as well as most yeasts, the powdery mildews, the molds, the cup fungi, the bread mold *Neurospora* (once a mainstay for geneticists), and the truffles, prized by chefs for their earthy flavor.

Ascomycetes [Greek, *askos* = wineskin + *myketos* = fungus], or sac fungi, are so named because their sexual spores are always produced in little sacs, called **asci** (singular, ascus). They have no sporangia for making asexual spores. Their hyphae usually have septa separating the nuclei. These septa, however, are perforated. As a result, molecules and even organelles can circulate throughout the cytoplasm. The first antibiotic used by humans came from a strain of the common ascomycete mold *Penicillium chrsyogenum*. Another famous ascomycete is *Saccharomyces cerevisiae*, the yeast used to make bread dough rise and beer ferment.

The truffles, yeasts, mildews, and other ascomycetes have septa dividing the nuclei of their hyphae and no sporangia. They produce sexual spores in little sacs called asci.

Basidiomycetes (Club Fungi)

The 25,000 species of **basidiomycetes**, or club fungi, include many of the most familiar fungi: the mushrooms, the bracket fungi, and the puffballs (Figure 21-22). These fungi take their name from the tiny, club-shaped **basidium** [Latin, = little pedestal; plural, basidia] that they use to produce spores. The basidia develop on the underside of a mushroom's cap, usually on the sides of the thin gills that radiate from the center like the pages of an open book (Figure 21-23). The mushroom itself, from which the basidia develop, is

Michael Fogden/DRK Photo

Figure 21-21
The familiar cup fungi are ascomycetes. These are bird's nest fungi.

Health and Biology Ergotism

After *Penicillium* mold, the most famous fungus may be the ascomycete *Claviceps purpurea*, or ergot, which infects the flowers of rye and other grains. This peculiar sac fungus produces tiny spikes, called sclerotia, or ergots, in the grain. The small, black or brown ergots—compact masses of hardened mycelium—replace a few of the seeds that would normally form in the seed head (Figure A).

Each ergot contains a witch's brew of deadly substances that can poison humans and other animals that eat the infected grain. In extreme cases, the symptoms of this poisoning, called ergotism, consist of hallucinations and fatal convulsions. In milder or chronic poisoning, sufferers may have headaches, muscle pains, numbness, and cold fingers and toes—all signs of constricted blood vessels that can result in gangrene, a usually fatal infection when not treated with antibiotics.

During the Middle Ages, ergotism was common in northern Europe and other areas where rye bread was popular. Also called St. Anthony's fire, ergotism could fell thousands of people in a season. In the year 994, for example, 40,000 people died from eating ergot-infected grain. Modern grains are cleaned of the dark masses of mycelium, and today ergotism is rare.

Ergot in small doses is medically useful. The main ingredient is an alkaloid—ergotamine tartrate. Ergotamine stimulates the smooth muscle of the peripheral and cranial blood vessels and also inhibits the neurotransmitter serotonin. As a result, ergotamine causes small arteries and smooth muscle fibers to contract. It is used to induce labor, to control bleeding from the uterus following childbirth, to treat high blood pressure, and—in conjunction with caffeine—to relieve severe migraine headaches. Two other ergot alkaloids,

bromocriptine and pergolide, are used in the treatment of Parkinson's disease. These two alkaloids help relieve the symptoms of Parkinson's disease by activating receptors for the neurotransmitter dopamine.

Ergot was also the original source of the psychoactive drug lysergic acid diethylamide (LSD). LSD induces mood changes (euphoria, anxiety, and depression), impaired judgment, and distorted perceptions of many sensations, including time, space, or self-image. Despite LSD's reputation as a powerful hallucinogen, true hallucinations are apparently rare.

Dennis Drenner

Figure A
Ergot. At left, the small black ergot (or sclerotium) in the seed head of a rye plant. At right, a normal, uninfected seed head.

J. Robert Stottlemyer/Biological Photo Service

Michael Fogden/DRK Photo

Figure 21-22
Basidiomycetes. Bracket fungi *(left)* and puffballs *(right)* have the same sexual life cycle as supermarket mushrooms.

Figure 21-23
Like the pages of a book. In basidiomycetes, the spore-producing basidia usually develop on the paper-thin "gills," which radiate from a central stalk like the pages of a book.

the equivalent of a fruit and biologists call it a **basidiocarp** [Latin, *carpus* = fruit].

In the common field mushroom, the mycelium spreads underground and may be up to 30 m in diameter (Figure 21-19). The basidiocarps (mushrooms) develop at the periphery of the mycelium in "fairy rings"—often overnight. Folk legends once held that mushroom fairy rings appeared at the bidding of fairies who were tired of dancing. In reality, mushrooms are able to grow so quickly because a huge underground mycelium can quickly mobilize vast amounts of material. As the secondary mycelium exhausts the nutrients in the center of the ring, however, the ring grows outward in search of more nutrients. Because the mycelium typically grows at about 30 cm per year, we can estimate that a fairy ring 30 m in diameter is a century old.

Another group of basidiomycetes, less charming than mushrooms but at least as important, are the rusts and smuts. Rusts and smuts ruin billions of dollars worth of wheat and other grains in the United States and elsewhere every year. As distinct from other basidiomycetes, however, rusts and smuts do not form basidiocarps. Instead, they are parasites with complex life cycles, usually involving hosts of two species.

Mushrooms, rusts, and other basidiomycetes are named for the club-shaped basidia that produce spores. They have septa and distinctive life cycles.

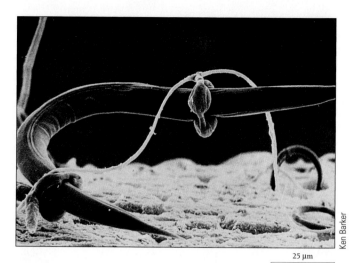

25 µm

Ken Barker

Figure 21-24
Not all fungi are passive. Here a fungus lassoes a passing nematode worm.

21.4 What Ecological Roles Do Fungi Play?

Some fungi are predatory (Figure 21-24). But the two most important ecological roles of fungi are as ecosystem recyclers. Both roles depend on powerful enzymes that can reduce substances as different as Wonder Bread and solid rock into small molecules that can be absorbed by fungi or other organisms. Using these enzymes, fungi decompose organic matter and make nutrients available to other organisms. Fungi, together with bacteria, recycle much of the world's organic matter from dead plants and animals, not to mention feces, leaves, and logs. Fungi also participate in close, mutually beneficial symbiotic associations with other organisms. We will discuss two kinds of important fungal associations, lichens and mycorrhizae.

What Role Do Fungi Play in Lichens?

Each of the 25,000 recognized species of lichens looks like a moss or simple plant growing on a rock or a tree trunk. **Lichens** may be leafy, shrubby, or crusty. They come in a variety of colors, including black, white, red, orange, yellow, green, and brown (Figure 21-25). Lichens are not plants at all, however, but associations of fungi with photosynthetic partners. In all but about 20 lichen species, the fungal partner is an ascomycete. The photosynthetic partner in a lichen is either a green alga or a cyanobacterium (Figure 21-26).

The association between a fungus and its photosynthetic partner is not entirely mutual. The various fungi cannot live without the help of photosynthetic partners.

Figure 21-25
Lichens are diverse. They are also able to absorb exquisitely tiny amounts of minerals and nutrients dissolved in rain and dew. While this ability is crucial to their survival in harsh settings, it makes lichens especially vulnerable to pollution. The reduced growth of lichens is a valuable indicator of deteriorating air quality.

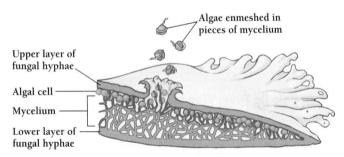

Algae enmeshed in pieces of mycelium

Upper layer of fungal hyphae

Algal cell

Mycelium

Lower layer of fungal hyphae

Figure 21-26
Lichen. Like a pie, a lichen is constructed with a protective crust of fungal hyphae above and below and a filling made of mycelium and algal cells.

The photosynthetic partners, however, are all capable of living without fungi. Most of the diversity of lichens comes from the fungi, for 90 percent of all photosynthetic partners in lichen are the same few organisms (from one of just three genera of green alga or of one genus of cyanobacteria). In contrast, the fungal partner is unique to each lichen. Nonetheless, the symbiotic association between fungi and their photosynthetic partners helps both parties.

The symbiotic association between lichens and their partners makes them especially tough. Fungi extract nutrients from rock and other harsh microenvironments. By doing so, fungi allow the algae or cyanobacteria to grow where they otherwise could not, while the photosynthetic algae provide a source of energy to their fungal partners. Lichens therefore grow where no plant could survive—bare rock, mountaintops, the Arctic tundra, even the seemingly lifeless valleys of Antarctica.

Lichens usually reproduce asexually, either by fragmentation or by sporelike starters called soredia, which are clumps of algae enmeshed in a bit of mycelium. The fungal and photosynthetic components can also reproduce independently, either sexually or asexually.

Lichens are mutually beneficial symbiotic associations of fungi and photosynthetic algae or cyanobacteria. In contrast to their photosynthetic partners, lichen fungi cannot live alone.

What Role Do Mycorrhizae Play in the Lives of Plants?

Fungi form close associations not only with photosynthetic bacteria and photosynthetic protists but also with true plants. **Mycorrhizae** [Greek, *mukes* = fungus + *rhiza* = root] are symbiotic associations between fungi (usually a zygomycete or a basidiomycete) and the roots of plants (Figure 21-27). At least 80 percent of all plant species have mycorrhizae. Some biologists now think that even more plants depend on mycorrhizae. In such associations, the fungus's hyphae penetrate the plant roots and also extend into the surrounding soil.

What are these fungi doing? Although some fungi obtain sugar from the plant's roots, the mycorrhizae are not parasites, for they help the plant flourish by supplying phosphate and metal ions. Indeed, what the mycorrhizae do for plants is more obvious than what the plants do for the mycorrhizae. In one experiment, plants whose mycorrhizae had been removed grew more slowly than plants that had mycorrhizae. Plants with no mycorrhizae were then provided with fertilizer, however, and they grew as well as the plants with mycorrhizae. This experiment suggested that the mycorrhizae help plants pull nutrients from the soil.

Mycorrhizae are fungi on the roots of plants that supply minerals to the plant.

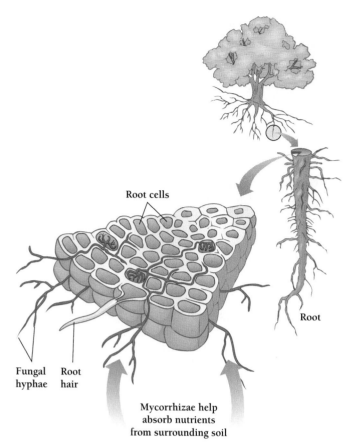

Root cells

Fungal hyphae

Root hair

Root

Mycorrhizae help absorb nutrients from surrounding soil

Figure 21-27
Symbiotic fungi and plants. Mycorrhizae grow inside of the roots of trees and other plants. The symbiotic association of fungi and plants probably played an important role in the evolution of terrestrial plants.

Where Did the Fungi Come From?

Because fungi have no hard parts, the fossil record of fungi is not rich. Nonetheless, fossil fungi can be found at least back to the Ordovician Period, 430 to 500 million years ago. The earliest fossils of land plants, which are about 400 million years old, contain petrified mycorrhizae, and some biologists now believe that plants could not have colonized the land without the aid of the fungi. Most mycologists think that the first fungi were zygomycetes, which have the simplest life cycle. Basidiomycetes probably derive from ascomycetes, which are in turn descendants of the more primitive zygomycetes.

Fungi evolved from protists at least 430 million years ago. The fungi may have been responsible for the colonization of land by plants.

In this chapter we have surveyed the simplest eukaryotes, the Protista, as well as the Fungi, the first kingdom that consists entirely of multicellular organisms. In the next

chapter, we will survey the plants, which, along with the fungi, are essential for humans and all other animals. On land, plants provide all the energy and simple organic materials that animals and fungi need to live. Plants and fungi are thus inextricably dependent on one another. Animals likewise absolutely depend on plants and fungi.

Key Concepts

- Protists and fungi are eukaryotes.
- Protists are polyphyletic and enormously diverse, particularly in cell structure.
- Protist plankton are important contributors to aquatic food chains.
- True fungi lack flagella and have cell walls made of chitin (a protein also used by insects).
- Fungi play important ecological roles as decomposers, as partners in lichens, and in association with the roots of higher plants.

Summary with Key Terms

What feature distinguishes the eukaryotes from the prokaryotes?
Unlike the prokaryotes, eukaryote cells keep DNA inside a membrane-enclosed **nucleus**. Eukaryote cells also have other membrane-enclosed organelles. Prokaryotes have no organelles.

What feature of the protists makes classifying them so much harder than classifying the fungi?
Protists are a polyphyletic (many lineages) group of eukaryotes that are not fungi, plants, or animals. Because the different lineages of protists are unrelated, they can be considered a "grab bag" kingdom. Most protists are single-celled organisms, but many are multicellular. Some protists are **plankton**, which drift. **Motile** protists propel themselves with **flagella, cilia,** or **pseudopodia**. Protists include **algae** (any protists that perform photosynthesis). Protist life cycles vary greatly from group to group. Each protist goes through two phases: reproduction and dispersal. Some are haploid throughout their life cycles and never undergo meiosis. Others are mostly diploid. Others undergo **alternation of generations**, like plants, passing through both haploid and diploid phases. The diploid phase is the **sporophyte**, which produces **spores** (sometimes encased in **cysts**), and the haploid phase is the **gametophyte**, which produces the gametes (eggs and sperm).

The most common protists can be grouped into six groups: (1) the **discricristales**, including the **euglenophytes, flagellates,** and **acrasid slime molds**; (2) the **chromalveolates**, including the **dinoflagellates, apicomplexa, ciliates, water molds (Oomycota), golden algae (Chrysophyta), brown algae (Phaeophyta),** and **diatoms**; (3) the radiolarians and foraminifera; (4) the amebas (Rhizopoda); (5) the red algae (Rhodophyta); and (6) the green algae.

Among the discricristales, the euglenas can live as either autotrophs or heterotrophs and are closely related to trypanosomes, flagellated protists that cause some human diseases. Acrasid slime molds live as amebas and as multicellular "slugs," which can produce fruiting bodies.

Among the chromalveolates, the old and primitive dinoflagellates are marked by distinctive tests (shells) and long flagella. Apicomplexans are nonmotile, spore-forming protists, which live as parasites in animals. Ciliates are single cells, coated with cilia. Chrysophytes have both single-celled and multicellular species, and diatoms can reproduce sexually as well as asexually. Phaeophytes (brown algae) include kelp and are the largest and most complex of the protists.

Radiolarians have glasslike tests and foraminifera have calcium tests pierced by little holes. Rhizopods (amebas) move and ingest food with pseudopodia. Rhodophytes (red algae) have distinctive pigments, but no flagella. Chlorophytes (green algae) are the most diverse of the algae and the most closely related to true plants.

What features distinguish fungi from other eukaryotes?

Fungi are heterotrophic (as are animals) and enclose their cells in rigid walls (as do plants). Unlike plants or animals, they do not undergo embryonic development. Fungi live mostly on land and, with a few exceptions, consist of a mycelium, a mass of intertwined filaments. The threads of mycelium predigest their food with secreted enzymes and then absorb the nutrients. The cell walls of fungi are made from chitin, the same material in the external shells of insects.

Most fungi are molds, that may reproduce asexually or sexually. Fungi may also reproduce and disperse as spores, which can resist harsh environmental conditions. Mycologists classify fungi by the way they produce spores (in sporophores) and by the appearance of their mycelia. Zygomycetes (conjugation fungi) are coenocytic, lacking septa between the nuclei in their hyphae. Ascomycetes (sac fungi) have septa (dikaryotic) but no sporangia. They produce spores in sacs called asci. Basidiomycetes (club fungi) have septa and distinctive life cycles. Each fruiting body of a basidiomycetes mushroom is called a basidiocarp. On the thin gills beneath the cap lie the basidia, which produce spores.

What three important ecological roles do fungi play?

Fungi play important ecological roles, both as decomposers and as participants in symbiotic relationships. Lichens are mutually beneficial associations of fungi and photosynthetic algae or cyanobacteria. Mycorrhizae are mutually beneficial associations between fungi and the roots of plants. Fungi evolved from protists at least 430 million years ago. Some biologists believe that plants could not have colonized the land without the aid of fungal mycorrhizae.

Review and Thought Questions

Review Questions

1. How do eukaryotes differ from prokaryotes?
2. Why is the kingdom Protista so much more diverse than the fungi, the plants, or the animals?
3. Name three kinds of organs that protists use to propel themselves through water. Biologists use what word to describe organisms that are self-propelled?
4. Define alternation of generations. How is it different from the life cycle of humans?
5. Name the six major lineages of protists.
6. Which lineage of protists is most closely related to plants? What three features together support this connection?
7. What are the two hosts of the malaria protist *Plasmodium*?
8. What gives the apicomplexa their name?
9. Why are apicomplexa difficult to eradicate?
10. What features characterize the true fungi?
11. To which phylum of fungi do common grocery store mushrooms belong? How is it that mushrooms can appear so quickly after a rain?
12. What are the two partners that compose a lichen? Which partner is more independent?
13. Describe the structure of a lichen.
14. What do mycorrhizae do for plants?

Thought Questions

15. If you wanted to invent a new fungicide that would kill fungal cells but not human cells, what differences between the biology of fungal cells and biology of human cells might you exploit for this purpose?
16. Paleontologists have speculated that vascular plants could not have evolved to the extent they have without the symbiotic relations with fungi they now depend on. How would you either test such a hypothesis or rephrase it in a way that would make it testable?

BiologyNow Resources

Biology ⓔ Now™

Active Figures

21-19: Fungi
21-20: The life cycles of fungi

Preparing for an exam? Take a diagnostic test on your BiologyNow CD-ROM.

Online materials relating to this chapter are at:

http://biology.brookscole.com/AAL3

About the Chapter-Opening Image

This mushroom represents one of the main focuses of this chapter on protists and fungi.

22

How Did Plants Adapt to Dry Land?

Key Questions

- How did plants adapt to life on land?
- How do bryophytes survive on land without vascular tissues?
- What are the advantages of a vascular system?
- How are seeds an adaptation to a dry, terrestrial environment?
- What roles do flowers and fruits play in angiosperm reproduction?

The Loves of Plants

How to classify organisms is often a matter of dispute. Prokaryotes are commonly classified according to their metabolic pathways; protists, according to their morphology; and fungi, according to their style of reproduction. When 18th-century biologists first began to classify plants, they could not agree on what traits to use. By 1799, naturalists had suggested at least 52 different systems for classifying plants. One system grouped plants according to the form of the flower, seed, and seed coat. Other systems gave priority to the shape of the fruit.

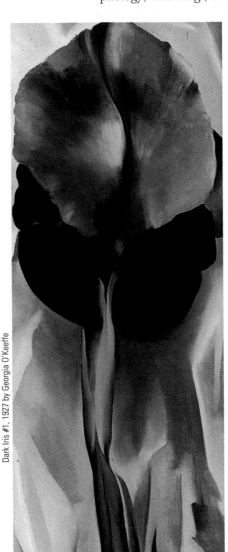

Dark Iris #1, 1927 by Georgia O'Keeffe

It was Carolus Linnaeus's system that ultimately came into use, but not without heated arguments from both scientists and clergy. Linnaeus based his classification solely on the reproductive structures of flowers. Indeed, he was fascinated by the "marriages of plants," as he called them (Figure 22-1). The son of a country parson, Linnaeus equated sex with marriage, and his descriptions of "vegetable coitus," as some called it, are simultaneously explicit and romantic.

As the historian of science Londa Scheibinger wrote, "Linnaeus emphasized the 'nuptials' of living plants as much as their sexual relations. Before their 'lawful marriage,' trees and shrubs donned 'wedding gowns.' Flower petals opened as 'bridal beds' for a verdant groom and his cherished bride, while the curtain of the corolla lent privacy to the amorous newlyweds."

Linnaeus divided plants first according to whether their marriages were public or private. Public marriages were those in which the flowers were visible to the naked eye; private marriages were those in which the flowers were too small to be seen. (The division corresponded to the public and private marriages then recognized in most of Europe. In England, the 1753 Marriage

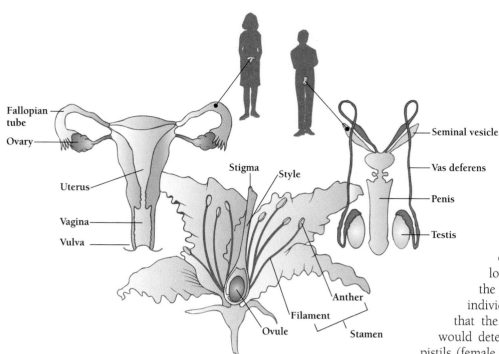

Figure 22-1

Linnaeus equated flower parts with the sexual parts of animals. He argued that the anthers are the testes; the filaments, the vas deferens; and the pollen, the seminal fluid. In the female parts, the stigma is the vulva; the style is the fallopian tube; the ovule is the ovary; and the seeds are the zygotes.

Act eliminated clandestine marriages by requiring a public proclamation before a marriage could be considered legal.)

Linnaeus further divided plants according to whether the male and female parts were together on the same flower, "in one bed," as he put it. Otherwise, if male and female parts were in separate flowers, they were in "two beds." Those in one bed, also called "hermaphrodites," were divided according to how many male parts (stamens) were present. These Linnaeus termed "husbands" to the single female pistil, or "bride." Flowers with one stamen were monandrian—having one man. (Linnaeus called his beloved wife Sara "my monandrian lily.")

As most flowers, however, have more than one stamen, indeed several, many naturalists and clergymen were horrified by Linnaeus's terminology. One St. Petersburg professor asked what kind of God Almighty would introduce such "loathsome harlotry" into the plant kingdom. The bishop of Carlisle wrote to the Linnaean Society in 1808 that "a literal translation of the first principles of Linnaean botany is enough to shock female modesty."

The Englishman William Smellie, who edited the first edition of the *Encyclopedia Britannica* (1771), rejected the "alluring seductions" of plant sexuality. He was especially repulsed by Linnaeus's description of plants that had reproductive structures of both sexes as "hermaphrodites." Smellie compared Linnaeus's work to that of the most "obscene romance-writer."

Most naturalists avoided Linnaeus's more vivid imagery. One exception was Erasmus Darwin, Charles Darwin's maverick grandfather. His *Loves of the Plants* (1789) is a collection of poems depicting flowers as practitioners of free love and even incest.

Linnaeus's taxonomy had other drawbacks. For example, in his system the number of stamens determined the class to which a plant species belonged. But in some groups of plants, the number of stamens can vary from individual to individual. He also decided that the number of stamens (male parts) would determine class, while the number of pistils (female parts) would determine order, the next taxonomic group down. That is, he deemed the stamens more important than the pistils. This system neither corresponded to the phylogeny (relatedness) of plants, nor was the system arbitrary. It was simply Linnaeus's way of importing into science traditional views about the relative positions of males and females.

Although many of Linnaeus's species designations—genus and species—remain in use today, his classification of groups above the genus level has been abandoned in favor of systems that more accurately reflect evolutionary relatedness. As botanists have learned more about the lives of plants, they have acquired a better understanding of how plants live and how they have evolved.

In the rest of this chapter, we will see that plants are amazingly diverse. Although the animal kingdom includes four times as many species as the plant kingdom, it is plants that first define a habitat—by their shape, size, and behavior. To a large extent, the animal inhabitants of grasslands differ from those of tropical forests because grasses differ structurally and chemically from trees. We can argue, then, that the plants lay the foundation for more habitat diversity than any other kingdom of organisms.

22.1 How Did Plants Adapt to Life on Land?

The tree is an aerial garden, a botanical migration from the sea, from those earlier plants, the seaweeds; it has a purchase on crumbled rock, on ground. The human, standing, is only a different upsweep and articulation of cells. How treelike we are, how human the tree.

—Gretel Ehrlich

The beauty of trees and flowers has inspired poets and gardeners alike. Plants delight the eye, tell us the season and even the time of day. They provide us with delicious foods, beautiful clothes, and strong building materials. Botanists like to say that plants provide "food, fiber, feed, fuel, and pharmaceuticals." But even this catchy list doesn't convey the importance of plants.

Through photosynthesis, plants (together with bacteria and protists) provide all of the energy we derive from food, whether that food is itself animal or plant. Plants living millions of years ago captured the energy that we now collect as coal, oil, or gas to heat our homes, to fuel our automobiles, and to drive our power plants. Plants provide the oxygen we breathe. Plants support the entire web of terrestrial life. Without plants, none of the other kinds of organisms could live on land.

A **plant** is a multicellular organism that performs photosynthesis and develops from an embryo. Almost all plants live on the land, and even those that live in water are descended from terrestrial plants. All plants, however, are descended from protists, and plants capture energy from the sun in much the same way as many of the marine protists. Nonetheless, the invasion of the land by plants nearly half a billion years ago depended on the evolution of structures different from anything found among plants' marine ancestors.

What Are the Advantages and Disadvantages of Life on Land?

The land offers three major advantages for photosynthetic organisms: more light (unfiltered by water), more carbon dioxide (which is more concentrated in air than in water), and a more reliable supply of minerals and other nutrients (which are more concentrated in soil than at the surface of a lake or sea).

Plants could not leave the water without solving a host of serious problems. To begin with, the seas provide a continuous supply of water. Water drives photosynthesis and provides a medium in which gametes and spores can disperse or meet for fertilization. Water also provides buoyancy for organisms. Even a 30-meter-tall brown alga can float. So aquatic plants, unlike terrestrial ones, require no skeleton or other structural support system.

To colonize the land, then, plants had to be able to obtain, conserve, and distribute water; to accomplish fertilization with little or no water; and to support their own weight. The evolution of plants depended principally on four adaptations:

1. A waxy **cuticle,** which coats the plant's exposed surfaces and reduces water loss through evaporation
2. The ability to absorb water from dew, rainwater, or groundwater (using roots, for example)
3. Enclosed reproductive organs called **gametangia** in which gametes form
4. Enclosed **sporangia** in which spores form

All plants use some version of a waxy cuticle, an absorption system, and gametangia. Plants that make seeds are especially well adapted to a dry existence; their sperm may be carried thousands of miles in dry pollen grains and their seeds are encased in hard shells that may live in a state of dormancy for millions of years.

Land offers plants more light, carbon dioxide, and nutrients than the sea. But plants living on land need adaptations for absorbing and retaining water, for fertilization with little or no freestanding water, and, in the case of larger plants, some form of structural support.

How Are Vascular Plants Different from Nonvascular Plants?

Despite important similarities among all plants, plants may be divided into two groups—the tracheophytes (vascular plants) and the bryophytes (nonvascular plants). Each group has different solutions for the problems of terrestrial life.

The **tracheophytes** [Latin, *trachea* = windpipe] include all the most familiar living plants, from ferns to fir trees and flowering fuchsias. All tracheophytes have pipelike tissues that conduct water and nutrients from one part of the plant to another. Because this system of conducting tissues resembles the vascular system of humans and other vertebrates, the tracheophytes are also referred to as the **vascular** plants. The tracheophytes' vascular system allows these plants to grow far larger than nonvascular plants, sometimes reaching enormous sizes.

The only true plants that lack such a vascular system are the **bryophytes** [Greek, *bryon* = moss], the mosses and their relatives. Because bryophytes lack a vascular system, they are much smaller than the tracheophytes. The bryophytes are also much less diverse, totaling only about 24,000 species, far fewer than the (approximately) 290,000 named species of tracheophytes.

Most botanists refer to the major groups of plants as divisions, equivalent to the phyla of animals that we will survey in the next chapter. However, the term phylum is also becoming acceptable. Each division, like each animal phylum, contains classes, orders, families, genera, and species.

Some botanists divide the plants into only two divisions—Bryophyta and Tracheophyta. That is, they recognize plants with specialized vascular systems and those without. Most botanists, however, count 10 or 12 divisions of living plants, of which all but one (the bryophytes) are vascular plants. The divisions of vascular plants vary in their modes of reproduction and dispersal.

Table 22-1 lists 10 divisions. Although it gives both a formal Latin name and a common name for many of the divisions, we will use the common name if there is one (for example, "conifers" rather than the Coniferophyta).

Bryophyta is the single division of nonvascular plants. The nine divisions of Tracheophyta are all vascular plants.

Table 22-1
The Divisions of Plants

Division	Number of species
Bryophyta (mosses, liverworts, and hornworts)	24,000
Psilotophyta	Small
Lycopodophyta	1,100
Equisetophyta (horsetails)	15
Pteridophyta (ferns)	13,000
Coniferophyta (conifers)	529
Cycadophyta (cycads)	100
Ginkgophyta (ginkgo)	1
Gnetophyta	70
Anthophyta (flowering plants)	275,000

How Do We Know About the Evolution of Plants?

The fossil record for plants, unlike that for protists, is excellent. Plant fossils are easier to find not only because plants are multicellular and therefore bigger but also because the vascular plants, at least, have many hard parts such as bark, cones, pollen, and leaf veins—all of which nicely survive the process of fossilization.

The oldest plant fossils date from the beginning of the Silurian Period, about 430 million years ago (Figure 22-2). These fossils are not only the oldest plant fossils but also the oldest terrestrial organisms. Until then, organisms apparently lived only in the seas. The first plants appeared on the land, probably in associations with fungi similar to present-day mycorrhizae (fungi that live symbiotically with plants). The successful invasion of the land by plants and fungi allowed the later evolution of terrestrial animals. Among these animals, the most populous and diverse were and continue to be the insects. As we will see, the evolution of the flowering plants (angiosperms) went hand in hand with the evolution of insects and fungi.

We can deduce some of the characteristics of the first plants by noticing what all living plants have in common. All show alternation of haploid and diploid generations, all contain both chlorophylls *a* and *b*, all have cell walls made of cellulose, and all are multicellular. Almost all plants store energy as starch. All develop from embryos that are surrounded by nonreproductive ("sterile") tissue. And, during cell division, all plant cells use a microtubular structure, the

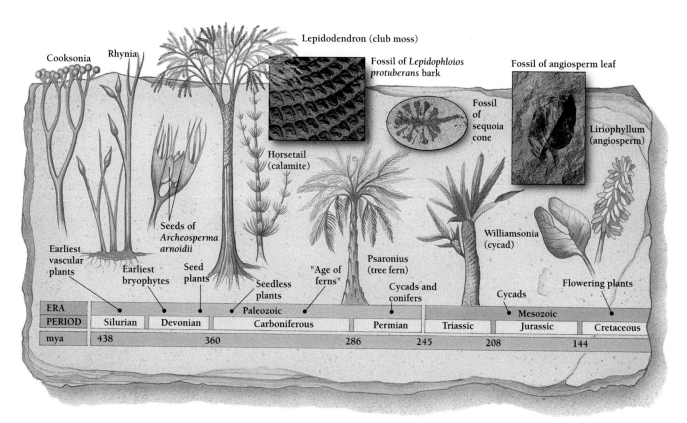

Figure 22-2
The evolution of plants. Notice the first appearance of vascular plants, bryophytes, seed plants, seedless plants, cycads and conifers, and flowering plants.

BiologyNow™ Learn more about characteristics of plants by clicking on this figure on your BiologyNow CD-ROM. (Louise K. Broman/Photo Researchers,Inc.; William E. Ferguson; McCutcheon/Visuals Unlimited)

phragmoplast, to build a new cell wall between the daughter cells. Plants share these striking characteristics with the green algae, suggesting that all plants are first cousins with the green algae.

True plants, however, have characteristics that set them apart from the green algae. In addition to a waxy cuticle, specialized structures for absorbing water, and specialized enclosed reproductive organs, plants also have **stomata** [Greek, *stoma* = mouth], tiny mouthlike pores that open and close to regulate the flow of carbon dioxide and other gases in and out of the plant.

Evidence for plant evolution comes from fossils and from comparisons of living species. The first plants probably evolved from a common ancestor that resembled a green alga.

What Does the Fossil Record Tell Us About the Evolution of Plants?

Although it is tempting to think of nonvascular plants as more primitive than vascular plants, the fossil record of vascular plants is actually older. The earliest bryophyte fossils date from the Devonian Period, about 400 million years ago, while the earliest vascular plants appear at the beginning of the Silurian Period, 430 million years ago (Figure 22-2).

By the time bryophytes first appeared in the early Devonian Period, several types of simple vascular plants were widespread, some of which are strikingly similar to modern species such as ferns and club mosses. Botanists divide the vascular plants into those that produce seeds, such as the flowering plants and the conifers, and those that do not. To the seedless plants, the ferns and their allies, we owe our warm houses and speedy cars. These plants, buried deep beneath layers of sediment, have turned into the famous fossil fuels, mainly oil and coal, on which we are now so dependent. Seedless plants began flourishing in the Carboniferous Period, from 345 to 280 million years ago, when the world's climate was warm and wet. Shallow seas and dense swamps covered most of the land.

The growth of the earliest seedless plants, including ferns, horsetails, and lycophytes, was so luxuriant that a large proportion of these plants did not decay completely before being buried under more dead plants and other sediments. Over millennia, these buried, compressed plants became the major component of all fossil fuels—coal, oil, and natural gas. The large amounts of carbon in these fossils give the Carboniferous Period [Latin, *carbon* = coal + *ferre* = to bear] its name.

During the Carboniferous Period, plants grew taller and taller in an adaptive "light war." The tallest plants had access to the most sunlight and shaded the plants below. The tallest were 40-meter lycophyte trees, as tall as large oaks. Far below this canopy of lycophytes spread vast stands of 18-meter horsetails and 8-meter ferns. Ferns, in fact, became so successful that the late Carboniferous Period is often called the "Age of Ferns."

Although a few Carboniferous plants had seeds, none had flowers. Seed plants, which first appeared in the late Devonian Period, were the *progymnosperms* [Greek, *pro* = first + *gymnos* = naked + *sperma* = seed], the ancestors of the modern gymnosperms. **Gymnosperms** are modern seed plants that do not have fruits or flowers. Gymnosperms include pine trees and other conifers, as well as the cycads, which look a little like palm trees (Figure 22-2).

Besides having seeds, the progymnosperms had more elaborate branching and more complex vascular systems than the earlier vascular plants. The Carboniferous swamps contained conifers and palmlike cycads, as well as lycophytes, horsetails, and ferns. As the climate became drier, the giant lycophytes, horsetails, and ferns began to disappear, and the conifers and the cycads began to dominate the landscape.

At the end of the Paleozoic Era, about 225 million years ago, the Earth's climate changed dramatically, and for 100 million years the dominant plants were the conifers and great cycads. Then, about 130 million years ago, a new kind of plant appeared—one with flowers.

Flowering plants, which have both seeds and fruits, became the most abundant plants on Earth. Today we live in the Age of Flowers. Of the present 315,000 species of plants, including all bryophytes and all tracheophytes, an astounding 275,000 are flowering plants. All of the flowering plants fall into two groups. These are the **eudicots,** such as roses, poppies, oaks, and others; and the **monocots,** such as grasses, lilies, onions, daffodils, and cattails. Of the once dominant gymnosperms, only 529 species remain today.

Vascular plants predate nonvascular plants. To the first seedless plants of the humid Carboniferous Period we owe our current supplies of oil and gas. As the climate dried, first ferns, then conifers and cycads dominated the landscape.

How Did Angiosperms Evolve?

The flowering plants, or **angiosperms** [Greek, *angion* = vessel + *sperma* = seed], make up by far the largest division of plants (Figure 22-3). The angiosperms enclose their seeds in a vessel that we know as a fruit. A peach is a fruit, and its pit is an enormous seed encased in a hard shell.

The first fossils of flowering plants date from the Cretaceous Period, about 125 million years ago. These early angiosperm fossils are leaves and pollen grains, rather than flowers.

In the absence of fossil flowers, evolutionary biologists who would like to know how angiosperms evolved from the gymnosperms have had to depend on comparisons among living plants. The magnolia, for example, is considered a

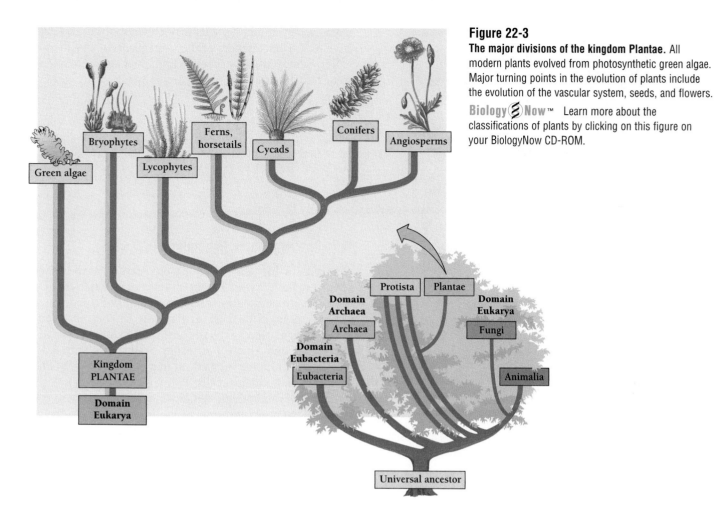

Figure 22-3
The major divisions of the kingdom Plantae. All modern plants evolved from photosynthetic green algae. Major turning points in the evolution of plants include the evolution of the vascular system, seeds, and flowers.

Biology⊜Now™ Learn more about the classifications of plants by clicking on this figure on your BiologyNow CD-ROM.

primitive plant because it has so few specializations. The magnolia has only simple leaves, woody stems, and insect-pollinated flowers, features that probably characterized the first angiosperms.

In the 125 million years since the first angiosperms appeared, they have diversified both morphologically and biochemically. They have improved their transport system and developed *deciduous* leaves, which fall off seasonally, so that modern angiosperms are better able to withstand periods of drought.

In addition, many modern angiosperm species have adaptations that prevent self-fertilization. Such species are more genetically diverse than gymnosperms and therefore better able to survive environmental changes. Angiosperm diversity and their dominance of the land attest to the success of their many evolutionary experiments.

Thanks to improved transport and deciduous leaves, many angiosperms have been able to colonize ever drier (and colder) habitats. Adaptations preventing self-fertilization have contributed to angiosperms' enormous diversity and success.

22.2 The Nonvascular Plants Include Mosses, Liverworts, and Hornworts

The nonvascular plants, or **bryophytes**, include the mosses, which give the division its name, and two less common classes—the liverworts and the hornworts. Mosses and other bryophytes generally live in wet environments, some tropical and some temperate (Figure 22-4).

Because all bryophytes lack specialized vascular systems, they depend on freestanding water for both photosynthesis and fertilization by free-swimming sperm. Although bryophytes can live only where freestanding water occurs, they nonetheless disperse well and grow all over the world. One species of moss even grows high on an Antarctic mountain just a thousand miles from the South Pole.

The bryophytes play important but little understood roles in the world's ecosystems, recycling carbon and absorbing nutrients and toxic metals. Peat moss alone covers up to 2 percent of all land on Earth, an area the size of the

Figure 22-4
Moss. Sphagnum moss surrounding a skunk cabbage in the Quinalt Rain Forest, Olympic National Park, Washington. Most mosses grow in habitats that are always wet or damp.

ment, however, many plants may crowd together to form a large mat. The closeness of the short plants allows them to retain water in the tiny spaces within the tangle. Because direct sunlight dries them out, bryophytes usually grow in shade.

> Although bryophytes lack a specialized vascular system, all parts of their bodies are adapted to absorb water, which gives them a spongy feel.

How Do Bryophytes Reproduce?

The cells of humans and most other animals are diploid (2*n*), containing two of every chromosome, while their gametes (sperm and eggs) are haploid (*n*), having just one set of chromosomes. In most animals, the haploid phase consists of single-celled eggs and sperm. This haploid stage is short-lived and barely noticeable. Sperm last only a few weeks, and although human ova may survive 45 years, they exist only as individual cells.

In plants and other kingdoms, the haploid stage may be multicellular and constitute a significant part of the life cycle. As mentioned in the last chapter, a sexual life cycle in which the haploid (1*n*) and diploid (2*n*) phases are both multicellular is called **alternation of generations.**

In bryophytes (and all other plants), the haploid stage is a multicellular **gametophyte** that makes (haploid) ova and sperm. The ova and sperm come together in the female gametophyte to form the diploid phase, called the **sporophyte** (Figure 22-5). The diploid sporophyte produces haploid **spores** by means of meiosis. But these haploid spores are not gametes; they are unlike any phase in our own life cycle or that of other animals. If the haploid spore finds itself in a welcoming environment, its cells begin dividing, through mitosis, and the bryophyte grows directly into a leafy, green gametophyte. The gametophyte (which makes gametes) is a mature individual, but it is haploid.

The haploid gametophytes may be male or female. Each sex develops *gametangia*, structures in which the male or female gametes develop (Figure 22-5). The male gametangium, called an *antheridium*, produces sperm with two flagella. The female gametangium, called an *archegonium*, produces and houses the ova and is generally shaped like a flask with a long neck. Some bryophytes bear archegonia and antheridia on the same plant.

To reach the ovum in the archegonium, the sperm must either swim through water adhering to the spongy mat or be splashed into the vicinity of the neck of the archegonium (Figure 22-5). Sometimes insects carry moss sperm, suspended in water, from male to female plant. Once sperm are near an archegonium, chemical attractants draw them into the neck, and the sperm swim through a fluid in the

United States. In some areas, bryophytes are the principal harvesters of solar energy, and all other organisms depend on them.

Bryophytes are also remarkable because the most obvious part of the bryophyte is not the diploid phase but rather the haploid phase. When we look at a moss-covered stump, we are looking at the haploid phase, in which each cell has only one set of chromosomes.

How Do Bryophytes Absorb Water?

Although some bryophytes have elongated cells that conduct water, they lack the specialized transporting tissues of the vascular plants. They anchor themselves to surfaces by specialized structures called *rhizoids* [Greek, *rhiza* = root]. Unlike the roots of vascular plants, rhizoids are not specialized for absorbing water and nutrients. Instead, all parts of a bryophyte are absorbent, and the bryophytes consequently have a spongy feel. Bryophytes can grow on rocks, tree trunks, and other hard surfaces that resist the roots of vascular plants.

Bryophytes are small, usually less than two centimeters high and less than 20 centimeters long. In a moist environ-

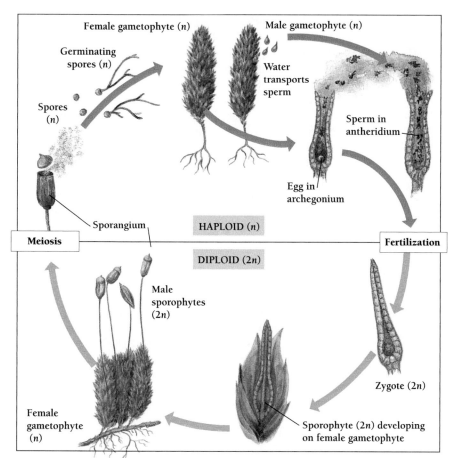

Figure 22-5
Life cycle of a moss. The mosses illustrate "alternation of generations," the life cycle in which an organism has both diploid and haploid phases that are multicellular. Diploid (2*n*) sporophytes growing from a haploid gametophyte make spores *(n)*, through meiosis. The haploid spores divide, through mitosis, to form haploid gametophytes. The haploid gametophytes make sperm and eggs (gametes). The fusion of the sperm and egg (fertilization) forms a diploid (2*n*) zygote. The diploid zygote grows from the haploid gametophyte, forming a new sporophyte (which makes haploid spores).

archegonium's neck to reach the ovum at the center of the swollen base.

Fertilization occurs within the archegonium. Immediately after fertilization, the diploid zygote begins to divide by mitosis to produce the sporophyte. The sporophyte, which is usually not photosynthetic, consists of a stalk, as tall as 15 to 20 cm, at the top of which develops a fruitlike **sporangium** [Greek, *spora* = seed + *angeion* = vessel]. Cells inside the sporangium undergo meiosis to produce the haploid spores, which can develop into a new mossy gametophyte.

Because bryophyte spores need to germinate in an environment that will sustain them, dispersal is essential to bryophyte reproduction. Some bryophytes propel their spores from exploding capsules. Some depend on splashing rainwater. A few bryophyte species live on dung, and flies carry the spores to fresh dung. Once in a hospitable environment, the haploid spores germinate, divide by mitosis, and develop into mature, leafy, male or female gametophytes.

Plants exhibit alternation of generations, a sexual life cycle in which haploid and diploid phases are both multicellular. Bryophytes' life cycles differ from those of vascular plants significantly and from those of animals profoundly: in a bryophyte's life cycle, the haploid gametophyte phase dominates.

22.3 The Vascular Plants Include Seed Plants and Seedless Plants

The tracheophytes (vascular plants) are both more diverse and more numerous than the bryophytes. Vascular plants include more than 290,000 named species, but some taxonomists say that the number of vascular plant species runs into the millions. We need only look around us to appreciate the success of the vascular plants—wherever we see green, we nearly always encounter a vascular plant.

Of What Advantage Is a Transport System to a Plant?

The ability of vascular plants to transport water and nutrients allows them to grow to huge sizes and to develop specialized tissues. Vascular plants also have more division of labor than the bryophytes, with distinctive tissues and organs that include roots, leaves, and stems. Their transport systems contain two distinct kinds of conducting tubes: **xylem** [Greek, *xylon* = wood], which carries water and minerals from roots to the photosynthesizing leaves, and **phloem,** which distributes the sugars and other organic molecules made in the leaves to the rest of the plant (Figure 22-6). Phloem consists of elongated, living cells. Xylem consists of the stiff walls of dead cells. Vascular plants also produce a hard, supporting material called *lignin*, which usually lines the walls of the xylem. Lignin, a major component of wood, provides the extra support that allows trees to attain such enormous sizes.

Plants' ability to transport water and nutrients has allowed them to grow larger and develop specialized tissues.

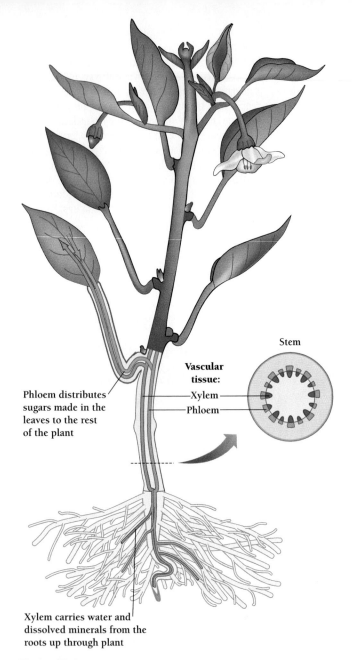

Figure 22-6
Vascular plant. Vascular plants have tissues that conduct fluids throughout their bodies much as our own veins and arteries conduct blood. Xylem *(blue)* primarily conducts water and dissolved minerals from the roots to the leaves. Phloem *(maple syrup color)* conducts sugar from the leaves to the rest of the plant.

Phloem distributes sugars made in the leaves to the rest of the plant

Stem

Vascular tissue:
Xylem
Phloem

Xylem carries water and dissolved minerals from the roots up through plant

How Do Vascular Plants Reproduce and Disperse?

Recall that in the bryophytes, the haploid gametophyte—the leafy, green part we call moss—is the more prominent part of the plant life cycle. In vascular plants, the diploid sporophytes are more prominent. The seed of a peach tree, for example, develops, in the right environment, into a mature sporophyte—a peach tree. In the spring, the peach tree blooms, producing flowers, which are reproductive structures. Parts of the flowers undergo meiosis to produce short-lived male gametophytes (pollen) and female gametophytes, which produce sperm and ova.

Vascular plants that produce seeds have at least one great advantage over the bryophytes. In colonizing new habitats, seed plants do not rely on the chance that a dispersed seed will fall into a benign and nourishing environment. Instead, the seed comes encapsulated in a tough coat that protects the embryo inside.

Within the safety of the plant's female reproductive organ, the plant embryo stops developing, and the mother plant packages the embryo, with a reserve of food, inside a protective seed coat. A seed is able to withstand long periods of cold and drought. When the seed finally encounters water and warmth, it germinates, and the embryo uses the stored food to finish developing into a photosynthetic seedling.

In vascular plants, the diploid phase dominates the life of the plant. The protective coat of a seed, and the food stored inside, allow a seed to wait indefinitely for suitable conditions for germination and growth.

Tracheophytes Lack Seeds

Four divisions of living vascular plants lack seeds: the pterophytes (ferns), psilotophytes, lycophytes, and equisetophytes. (Another four divisions of seedless plants exist only in the fossil record.) We will discuss three divisions of living tracheophytes: the ferns, lycophytes, and equ9isetophytes.

Pterophytes (Ferns)

Some two-thirds of the 13,000 species of ferns live in the tropics, and the rest inhabit the temperate zones. Ferns range in size from a few centimeters to giant 30-meter tree ferns, as tall as oak trees. Ferns are common in moist parts of shady forests. Their leaves (called *fronds*) often have a lacy appearance and consist of many *pinnae* [singular, pinna; Latin, = feather], small leaflets that extend from a central stalk (Figure 22-7).

In the large ferns of the tropics, fronds grow from a central stem. In most ferns, however, fronds ˙grow from a *rhizome,* a horizontal stem that spreads on or below the ground. Each frond develops from a characteristic coil called a fiddlehead.

Reproduction in ferns and other seedless plants resembles that in the bryophytes. Fertilization can occur only if the released sperm can swim to an egg (in a mature archegonium). Ferns, like other seedless plants, therefore require a wet environment to provide at least a film of water in which the sperm can swim.

As is true of other vascular plants, the sporophyte (the mature fern) is much more prominent than the gametophyte. The leaves of the fern sporophyte are also called **sporophylls,** because they are specialized for making spores, from which the gametophytes grow. Clustered on the

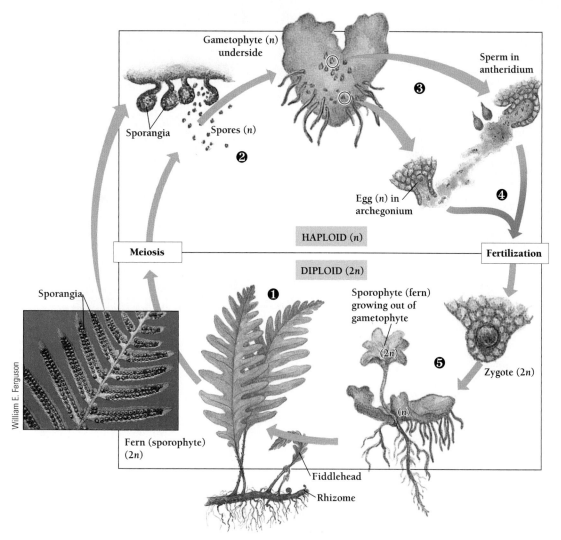

Figure 22-7
The life cycle of a fern. ❶ A fern (or sporophyte) releases haploid *(n)* spores. ❷ Spores mature into haploid *(n)* gametophytes. ❸ The heart-shaped gametophytes make sperm and eggs (gametes), which come together on the underside of the gametophyte. ❹ The fusion of the sperm and egg (fertilization) forms a diploid (2*n*) zygote. ❺ The zygote grows right out of the gametophyte, forming a new sporophyte, or fern.

underside of the sporophylls are groups of sporangia called *sori* [singular sorus; Greek, *soros* = heap] (Figure 22-7). The diploid sporangia produce haploid spores by means of meiosis. Some ferns catapult their ripe spores several meters into the air, where breezes carry them aloft. Such airborne spores may disperse over enormous distances. On landing, each spore germinates to form a tiny nonvascular gametophyte, which is typically flat, thin, heart-shaped, and less than a centimeter wide. This gametophyte is green, photosynthetic, independent, and fragile. It usually bears both male and female reproductive structures.

Lycophytes

Lycophytes (also called lycopods) have true roots, stems, and leaves, as well as a simple arrangement of xylem and phloem. As in ferns, some of the leaves are sporophylls, with sporangia on their upper surfaces. The sporophylls may be green and look like ordinary leaves, or they may be distinct and grouped into spore-producing cones, called **strobili** [singular, strobilus; Greek, *strobilos* = cone]. The most common lycophytes are the club mosses, which form evergreen, mosslike mats beneath temperate forests. Their strobili resemble little clubs (Figure 22-8).

Ferns have independent gametophytes.

Lycophytes have true roots, stems, and simple leaves.

Figure 22-8
Lycophyte. The club mosses are vascular plants, not true mosses (bryophytes).

Figure 22-9
Horsetails. Equisetophytes contain bits of silica that give them a rough feel.

Equisetophytes

The living equisetophytes are all members of the single genus *Equisetum,* or horsetails (Figure 22-9). These plants grow in moist places along the banks of streams or at the edges of woods. Like the lycophytes, the equisetophytes have true roots, stems, and leaves. Their jointed stems contain many discrete strands of xylem and phloem. The outer cell layer of the plant contains silica (the same material from which glass is made), which gives the plants a rough texture useful to campers and settlers for scouring pots. Indeed, another name for these plants is "scouring rush."

> Equisetophytes have true roots, stems, and complex leaves. Their stems are jointed, and their outer cell walls are reinforced with silica.

How Do Seed Plants Reproduce?

Seeds are of great value in a dry, terrestrial environment because they liberate a plant from the need for freestanding water. **Seed plants** manage fertilization without water in the same way we do, through a form of internal fertilization. The gametophytes have no independent lives but develop within the tissues of the sporophytes. The gametophytes of seed plants are tiny—even less prominent than in the seedless vascular plants.

Many seedless plants make just one kind of spore and one kind of gametophyte. In contrast, all the seed plants make two kinds of gametophytes (male and female), from two kinds of spores on two kinds of sporangia. The sporophytes make haploid microspores (male) and megaspores (female) in separate sporangia. In some species, the male and female sporangia occur together on the same plant, while in others, the two kinds of sporangia occur on male and female plants. In either case, the microspores and megaspores develop into distinct male and female gametophytes.

The female gametophyte stays inside the megasporangium, which lies safe within the main body of the adult plant (the sporophyte), awaiting the arrival of the sperm. The megasporangium is in turn covered by one or two additional layers of tissue, called integuments, which will later develop into the protective seed coat. The megasporangium and integuments together form the **ovule.**

The small, male gametophytes are grains of pollen, released from the microsporangium and carried, usually by wind or insect, to the female gametophyte. The male gametophyte then develops a slender pollen tube, which carries a sperm nucleus to the ovum produced by the female gametophyte. At fertilization, the sperm and ovum fuse to form a diploid zygote. The resulting zygote, still enclosed within the female gametophyte, begins to develop into an embryo.

After fertilization, the ovule and its contents become a seed. The **seed** includes (1) the new sporophyte embryo; (2) a female tissue that nourishes the developing embryo; and (3) the ovule from the previous sporophyte generation. Inside the seed coat are the embryo and a source of food for its later development. The seed coat protects the young embryo from drying out or from being eaten. Once the seed germinates, the embryo resumes its development and growth.

In the following sections, we will discuss the five divisions of living seed plants. The 771 named species of the first four divisions are gymnosperms, which produce seeds that are not enclosed in fruits. In contrast, the 275,000 species of angiosperms produce seeds that are enclosed in a fruit.

Seeds, which enable plants to colonize relatively dry environments, consist of a diploid zygote and a source of food encased in a seed coat.

Pat O'Hara/CORBIS

Figure 22-10
Sugar pine. As this sugar pine in northern California shows, gymnosperms may be immense organisms.

Gymnosperms: Seed Plants Without Flowers

The four divisions of gymnosperms—conifers, cycads, ginkgos, and gnetae—are all vascular plants with seeds, but without flowers and fruits. They are not closely related to one another. We briefly discuss the conifers, cycads, and ginkgos.

Conifers

The conifers [Greek, *konos* = cone + *phero* = carry] take their name from their characteristic cones, which are male and female strobili. The conifers include many of the most common trees in the Northern Hemisphere—the pines, firs, spruces, and cedars—as well as the magnificent redwoods of the California and Oregon coasts. Conifers grow well in the short growing seasons of northern latitudes and high altitudes. They are mostly evergreen, keeping their leaves throughout the year (Figure 22-10). Individual leaves may last through two or three winters. (Bristlecone pines, some of which live to be more than 5,000 years old, retain their needles for up to 45 years.) Conifer forests extend almost all the way across North America, Europe, and Asia.

Many conifers, including the familiar pines, have needlelike leaves that are well adapted for dry conditions, including, for example, northern winters, when the air is dry and the groundwater is frozen. A needle's small surface area and heavy cuticle minimize evaporation. Despite its simple shape, however, the needle usually has two vascular bundles. The stems of conifers are heavily reinforced with lignin, which gives trees the structural support to reach enormous sizes. The lignin in wood is also what makes it strong enough to support enormous trees and buildings.

Most conifers bear male and female cones on the same tree. The male, pollen-producing cones are generally small (one to two centimeters long) and consist of hundreds of sporophylls. Wind carries the pollen grains, often in huge yellow clouds, to the female cones. The female, or ovulate, cone consists of tough scales arranged in a spiral. Figure 22-11 shows the life cycle of a typical conifer, a pine tree.

Conifers produce male and female gametophytes in cone-shaped strobili.

Cycads

The cycads are large-leafed plants that look like palms (Figure 22-12). Some cycads, called "sago palms," are used as ornamental plants in homes and gardens. But true palm trees are flowering plants, while cycads have neither flowers nor fruits: they bear naked seeds on the scales of cones near the top of each plant. Individual plants bear either male or female cones, but not both. Some people eat the seeds and underground stems of cycads, but cycads contain toxins that can cause gastrointestinal problems, degenerative brain disease, and even paralysis.

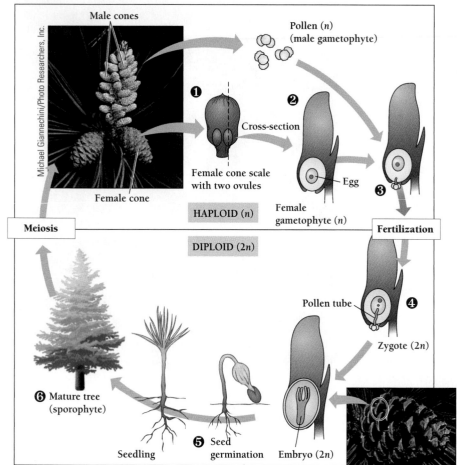

Figure 22-11

Life cycle of a pine. Like other conifers, pines have two kinds of cones: pollen cones and seed cones. ❶ In the seed cones, which are the familiar "pine cones," each scale holds two ovules. Each ovule in turn contains a female sporangium (2*n*). ❷ A cell within the sporangium undergoes meiosis to produce four haploid cells, one of which undergoes mitosis to produce the female gametophyte, which makes an egg. The female gametophyte in the ovule develops two or three "archegonia" each of which in turn makes a single giant ovum, loaded with carbohydrates and proteins. ❸ When pollen blows onto the female cones, a sticky fluid in the opening to the gametophyte captures the grains of pollen. ❹ After a few months, the pollen gametophyte develops into a pollen tube, which makes its way to one of the archegonia in the ovule. One of the cells of the pollen tube divides to form two sperm nuclei, which then enter the archegonium. As much as a year after pollination, one of these nuclei finally fuses with the ovum to form a zygote, which develops within the ovule to form a (2*n*) seed. ❺ If the seed falls in moist soil, it may germinate and develop into a tiny tree. ❻ With luck, it will survive gophers and deer and grow into a massive pine tree. Botanists call the tree a sporophyte.

The small male cones of cycads consist of microsporophyll scales, each of which contains many pollen-producing microsporangia. Female cones are much larger. The cone of one Australian cycad may be a meter long and weigh 90 pounds (40 kilograms). As in conifers, pollen travels by wind to the ovule, germinates to produce a pollen tube and sperm, and fertilizes the ovum.

Cycads produce male and female cones on separate plants.

Ginkgos

The ginkgo is the last living species in the division Ginkgophyta, the rest of whose members went extinct about 250 million years ago. The ginkgo, or maidenhair tree, is a native of China and Japan and a living fossil that has hardly changed in 80 million years. Ginkgo trees dominated Cenozoic forests. They may grow as tall as 40 meters and live for more than

Figure 22-12

Rare cycad. Linnaeus first described *Zamia pumila* L. in 1763. (The "L." stands for Linnaeus.) He thought it was a dwarf palm, not realizing that its reddish cones separated it from the true palms, which are angiosperms. This particular species of cycad is grown in gardens, but it has been eradicated in parts of its native range and now grows naturally only in Florida, Mexico, Central Cuba, and the Dominican Republic.

Figure 22-13
Asahikawa ginkgo tree. As beautiful as a cherry tree in bloom, this ginkgo tree in Hokkaido, Japan, has begun to turn a golden yellow as it prepares to drop its leaves in the fall.

1,000 years (Figure 22-13). Yet these beautiful trees, which still decorate many gardens and parks, are now extinct in the wild.

Like the cycads, ginkgo trees are either male or female. The male trees have cones (strobili), while the female trees carry fleshy (and smelly) seeds at the ends of short stalks. The seeds of the female tree have an odor so rancid that most people prefer to plant male trees only.

Ginkgo is widely advertised as an herbal remedy. In one study, chemical extracts from ginkgo enhanced attention, short-term memory, and reaction time in people with Alzheimer's disease. In other studies, ginkgo extracts seemed to impair memory. Herbal supplements makers are unregulated, unlike the pharmaceutical industry, which is required to demonstrate what a drug can do and what its side effects are. In addition, supplements vary widely in what herbs they contain and what active ingredients, no matter what the label says. For more about herbal supplements, see the box in Chapter 30.

Ginkgos resemble cycads in their life cycle and conifers in their growth patterns.

22.4 The Angiosperms Are Seed Plants with Flowers and Fruits

The diversity of the angiosperms dwarfs that of all other divisions of plants. Angiosperms, also called **Anthophyta** [Greek, *antho* = flower], range in size from the tiny duckweed *Wolffia,* less than a millimeter long, to *Eucalyptus* trees, taller than 100 meters. With some 275,000 known species, angiosperms are the most widespread and familiar of plants. We see them daily in our lawns, gardens, parks, fields, and forests. They provide us with furniture, clothes, medicines, and dyes. We stock our kitchens with their seeds, fruits, leaves, stems, and roots. Angiosperms provide us with wine, beer, coffee, and tea. We use the more brilliant angiosperms to brighten our lives and enliven our loves (Figure 22-14).

Despite their diversity, all angiosperms share two important adaptations: flowers, which promote fertilization, usually by insects or other animals; and fruits, which promote the survival and dispersal of seeds.

Of What Use Are Flowers?

Flowers are reproductive structures that ensure the distribution of pollen, either by animals or by wind. Some flowering plants rely on wind alone to distribute pollen. Grasses, rushes, sedges, plantain, birch, poplar, and oaks, for example, have small inconspicuous flowers and no scent. Because wind pollination is so inefficient, wind-pollinated flowers release mind-boggling amounts of pollen. Individual plants

Figure 22-14
Apricot tree in bloom, Pakistan. Fruit trees are the quintessential angiosperms. Their spectacular displays of blossoms attract pollinating bees in spring. By summer, the pollinated blossoms have developed into delicious, ripe fruit.

of the weed *Mercurialis annua* may release over a billion grains of pollen.

Other flowers attract insects (especially bees, beetles, butterflies, and moths), birds (especially hummingbirds), and even bats and other small mammals. All these animals' visits are rewarded with a sugary fluid called nectar, a share of the high-protein pollen, or nutritious petals or other flower parts. Animals learn to associate these delicious foods with the seductive perfumes and bright colors of flowers. Not all of a flower's attractions are necessarily attractive to humans: the eastern skunk cabbage (*Symplocarpus foetidus*), for example, attracts pollinating flies by emitting an odor like that of dead fish.

Most flowers are specialized to ensure pollination by just one group of animals. Flowers pollinated by moths, for example, are usually large, white, and overwhelmingly fragrant at night, when moths fly. In contrast, flowers pollinated by butterflies are small and brightly colored, since butterflies fly during the day and rely on sight more than smell. One orchid species produces flowers that look so much like female bees that male bees come to mate with the flowers. As the male bee attempts to copulate with the flower, he picks up pollen, which he then carries to another flower in another futile attempt to mate.

Insects are as well adapted to take advantage of flowers as flowers are adapted to take advantage of insects. One South African fly has a tongue nearly three inches long, which allows it to take nectar that is out of the reach of other insects. Insects and flowers provide a striking example of **coevolution,** the mutual adaptation of two separate evolutionary lines. Coevolution occurs when two species are significant factors in the lives of one another. In the case of insects and flowers, coevolution ensures a food supply for the insects and pollination for the plants. You can read more about coevolution in Chapter 27.

Flowers attract animals that distribute pollen.

Of What Use Are Fruits?

A fruit is a mature ovary that encloses and protects seeds. Many fruits promote seed dispersal by animals or wind. The most familiar fruits are those that attract by appearance and taste, for example, peaches, apples, and berries. An animal usually digests the fleshy part of such fruits, but the seeds pass unharmed through the digestive system. (In fact, some seeds cannot germinate until they pass through the digestive tract of an animal.) Seeds excreted by the animal can end up far from the plant that produced them, together with a nourishing supply of fertilizer.

Other fruits use different stratagems to transport their seeds. Thistles and dandelion fruits, for example, soar like kites in the slightest breeze, and maple fruits whirl through the forest like free propellers. Burrs stick to the fur (or socks) of passing animals, while coconuts have spread to countless tropical islands by floating across the ocean. Touch-me-nots (*Impatiens*) propel their seeds from fruits that explode on touch.

A fruit is a ripened ovary that encloses and protects the seeds and usually enhances their dispersal.

How Do Angiosperms Reproduce?

In angiosperms, as in all the other seed plants, the diploid sporophyte generation is more prominent than the haploid gametophyte. In fact, the female gametophytes of angiosperms are even tinier than those of the gymnosperms, and they live their whole lives inside the sporophyte's flowers. (The male gametophyte—the pollen—disperses.)

The flowers produce haploid microspores (male) and megaspores (female) in separate sporangia, from which the gametophytes derive. After meiosis, the microspores divide once to give rise to two haploid nuclei—the tube nucleus and the generative nucleus (Figure 22-15). The male gametophyte resumes development only after it arrives on the **carpel,** the flower part that contains the ovule. The pollen tube now grows, and the generative nucleus divides into two sperm nuclei. By the time the pollen tube reaches the ovule, it contains two sperm nuclei.

Meanwhile, within the ovule, the female gametophyte has developed. Meiosis yields four haploid megaspores, only one of which survives. That megaspore divides three times by mitosis to produce the embryo sac, the mature female gametophyte. The embryo sac has eight nuclei and seven cells. One cell (the central cell) contains two haploid nuclei (called the polar nuclei). Another cell becomes the haploid ovum.

Fertilization in angiosperms is different from that in any other organism. As we will discuss in more detail in Chapter 32, there are actually two separate fertilizations, together called a **double fertilization:** (1) one sperm nucleus fuses with the nucleus of the ovum to form the zygote, and (2) the other sperm nucleus fuses with the two nuclei of the central cell to form a triploid nucleus called the primary endosperm nucleus. The zygote then divides by mitosis and develops into an embryo. The primary endosperm nucleus divides to form an **endosperm,** which will provide food for the developing embryo and, in some cases, for the germinating seedling.

Flowers reproduce by means of double fertilization. Two sperm nuclei from the pollen grain fertilize two ova from the ovary, resulting in a diploid zygote and a triploid cell that forms the nutritious endosperm.

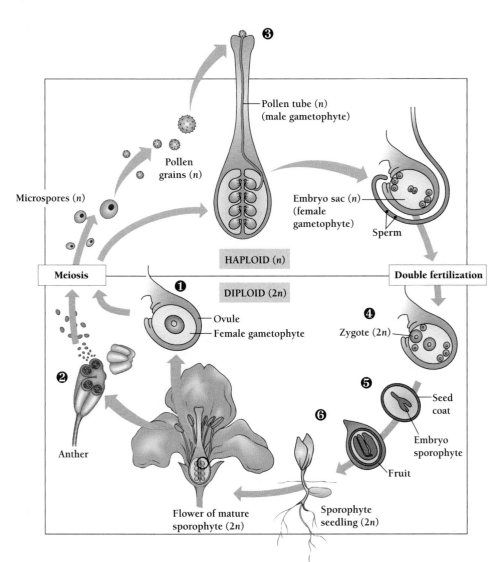

Pollen tube (n)
(male gametophyte)

❸

Pollen
grains (n)

Microspores (n)

Embryo sac (n)
(female
gametophyte)

Sperm

Meiosis

HAPLOID (n)

Double fertilization

DIPLOID (2n)

❶

Ovule

Female gametophyte

❹

Zygote (2n)

❷

Anther

❺

Seed
coat

❻

Embryo
sporophyte

Fruit

Flower of mature
sporophyte (2n)

Sporophyte
seedling (2n)

Figure 22-15
Life cycle of a flowering plant. ❶ Inside a flower (part of the sporophyte) haploid *(n)* spores in each ovule develop into a female gametophyte, and ❷ haploid spores in the anthers develop into male gametophytes (pollen). ❸ When a grain of pollen reaches a stigma (carried by an animal or the wind), the pollen grows a pollen tube down to the ovary, where a sperm from the pollen grain fertilizes an egg, creating ❹ a diploid (2*n*) zygote. ❺ From each ovule, a seed develops. Each seed consists of an embryo (sporophyte), a store of food, and protective seed coat. ❻ The ovary wall forms a fruit, which helps disperse the seeds. Eventually, the seed germinates and the new sporophyte begins to grow.

What Are the Parts of a Flower?

As in the gymnosperms, separate sporangia produce male spores and female spores in angiosperms. As in the case of the gymnosperms, the sporangia lie on modified leaves, or sporophylls. In angiosperms, however, both male and female sporophylls are flower parts instead of cones. The male sporophylls are **stamens,** and the female sporophylls are carpels**.**

A flower is a short piece of stem that usually ends with four whorls, or circles, of modified leaves. These are the carpels, the stamens, the petals, and the sepals (Figure 22-15). The innermost parts are the carpels. Each carpel (or pistil) encloses one or more female sporangia (ovules) in a swollen base called the **ovary.** Pollen does not reach the ovules directly but first falls on the stigma, the carpel's receptive area. The style, a slender tube through which the pollen tube enters the ovary, extends from the stigma to the top of the ovary. After fertilization and development, the ovule becomes the seed, and the ovary becomes the fruit.

Just outside the carpels lies the next whorl of modified leaves, namely the stamens, or male sporophylls. Each stamen

consists of an **anther,** a thick, pollen-bearing structure, and a **filament,** a thin stalk that connects the anther to the base of the flower. Within the anther are four sporangia, called pollen sacs, where the pollen grains develop (Figure 22-15).

Outside the stamens is a whorl, or **corolla,** of **petals.** In flowers pollinated by insects or other animals, the petals are usually large and brightly colored. In flowers pollinated by wind, however, the petals are reduced or missing altogether. Sometimes the petals fuse, so that the corolla is an uninterrupted tube.

The fourth and outermost whorl is the **calyx,** a whorl of leaflike parts called **sepals,** which enclose the rest of the flower and protect it before it opens. The sepals are usually small and green, but in some flowers they may be large and colorful.

Flower structures vary widely. Those that lack one or more of the four whorls are said to be incomplete. Those that lack either stamens or carpels are said to be imperfect (as well as incomplete). Flowers may have radial symmetry, so that any cut along the axis of the flower will produce two identical halves. Or they may have bilateral symmetry, so that only one cut through the flower's axis will produce identical (or mirrored) halves.

Fruits are as diverse as the flowers that produce them. Some develop from a single carpel (with one or many seeds) or from several fused carpels with a single ovary. Others, such as raspberries or strawberries, derive from many separate carpels. And still others, like the pineapple, come from many flowers.

> The male sporophyll is the stamen—including anther and filament—and the female sporophyll is the carpel (or pistil)—including style and ovary. A corolla of petals and a calyx of sepals generally surround the stamens and carpels.

Plants' colonization of the land 430 million years ago may have set the stage for the evolution of insects and other terrestrial animals. The terrestrial environment varies far more than the marine environment, because of extremes of temperature and rainfall, for example. As a result, terrestrial organisms—both plants and animals—are themselves more diverse than their marine cousins. Animals, which we survey in the next two chapters, are especially diverse. Most animal species are insects, and the evolution of insects is closely tied to that of plants.

Key Concepts

- A plant is a multicellular organism that photosynthesizes and develops from an embryo.
- The nonvascular plants lack vascular tissues for transporting water and nutrients.
- The evolution of a vascular system enabled plants to transport water and nutrients, to grow large, and to diversify.
- Seeds allow plants to withstand dry environments.
- Flowers and fruits have allowed plants' rapid diversification.

Summary with Key Terms

How did plants adapt to life on land?

Plants are mostly terrestrial, multicellular organisms that perform photosynthesis. They evolved from the ancestors of modern green algae and adapted to life on the land. Most plants are terrestrial. The land offers three major advantages for photosynthetic organisms: more light, more carbon dioxide, and a more reliable supply of minerals and other nutrients. The major drawback, however, is less water, so that, for plants to invade the land, they had to evolve mechanisms for obtaining and conserving water.

The first plants probably evolved from green algae at least 430 million years ago. The first plants with seeds (an adaptation to a dry environment) appeared by the late De-

vonian Period. Fossils of the first flowering plants date from the Cretaceous Period, about 125 million years ago. The evolution of plants depended principally on three adaptations: (1) a waxy **cuticle,** which coats the plant's exposed surfaces and reduces evaporation; (2) the ability to absorb water and minerals from dew, rainwater, or groundwater; and (3) jacketed reproductive organs. All plants also have **stomata** and life cycles characterized by **alternation of generations** between **gametophytes** and **sporophytes.** The **gametangia** of the gametophytes produce gametes, and the **sporangium** of the sporophytes produces **spores.**

How do bryophytes survive on land without vascular tissues?

The **bryophytes** have no vascular system for distributing water and nutrients and therefore cannot grow very large. Bryophytes include mosses, liverworts, and hornworts. The haploid (gametophyte) phase of bryophytes is more prominent than the diploid (sporophyte) phase.

What are the advantages of a vascular system?

Vascular plants are far more diverse and more numerous than the bryophytes. Vascular plants contain two kinds of conducting tubes: **xylem,** which carries water and minerals from roots to leaves, and **phloem,** which distributes sugars made in the leaves to the rest of the plant. In vascular plants, the diploid phase of the life cycle is more prominent than the haploid phase.

How are seeds an adaptation to a dry, terrestrial environment?

Most kinds of vascular plants disperse as **seeds,** which contain a diploid embryo, a food reserve, and a protective coat. A seed contains the embryo, the female gametophyte, the organ that produces the female gametophyte and layers of tissue called the integuments, together called the **ovule.** Many **seed plants** have flowers and fruits, which aid in fertilization and in the dispersal of seeds. **Tracheophytes** such as ferns, lycophytes, and equisetophytes lack seeds. Five divisions of tracheophytes make seeds, but four of these, collectively called the **gymnosperms,** do not have flowers.

Gymnosperms—conifers, cycads, ginkgos, and gnetae—produce male and female gametophytes in **strobili.** In conifers the strobili are separate cone-shaped structures on the same tree. Cycads produce male and female gametophytes in strobili that lie on separate plants. Ginkgos resemble cycads in their life cycle and conifers in their growth patterns.

What roles do flowers and fruits play in angiosperm reproduction?

Anthophyta, or **angiosperms,** are the most diverse group of plants. The gametophytes of angiosperms are even more reduced than those of the gymnosperms. Both male and female gametophytes are part of the flowers of the **sporophyte,** as are the male and female **sporophylls.**

A flower is a short piece of stem that usually ends with four circles of modified leaves: the **carpels, stamens, petals,** and **sepals.** Each carpel—including stigma, style, and ovary—encloses the female sporangia. The stamens are the male reproductive structures. Each stamen consists of an **anther** and a **filament.** A **corolla** of petals, often brightly colored and scented, attracts insects that carry pollen grains (male gametophytes) from flower to flower. A **calyx** of sepals encloses and protects the rest of the flower before it opens.

Fertilization in angiosperms is different from that in any other organism and consists of two separate fertilizations—a **double fertilization.** The zygote divides and develops into an embryo. The primary **endosperm** nucleus divides to form an endosperm, which will provide food for the developing embryo and, in some cases, for the germinating seedling. Flowering plants are divided into **monocots** and **eudicots,** according to the form the embryo takes.

A fruit is a ripened, or mature, **ovary** that encloses and protects the seeds. Flowers and fruits allow angiosperms to interact with animals. Plants have **coevolved** with animals, fungi, and other organisms.

Review and Thought Questions

Review Questions

1. Describe six traits that together set plants apart from members of the other kingdoms of organisms.
2. List several advantages and disadvantages to life on land.
3. What traits enable plants to live on land?
4. Compare alternation of generations with the life cycle of an animal, such as ourselves.
5. What group of plants that flourished 300 million years ago provided the fossil fuels oil, coal, and natural gas? What is the name of that geologic period?

6. How do bryophytes absorb water? What important roles do mosses play in the world's ecosystems?
7. What are xylem and phloem and what functions do they perform in vascular plants?
8. Describe the sporophyte phase of a pine tree.
9. Describe the gametophyte phase of a cherry tree.
10. What are the advantages of seeds to terrestrial plants?
11. What are the advantages of flowers and fruits to terrestrial plants? Name the parts of a flower.

Thought Questions

12. Why did plant taxonomists abandon Linnaeus's system of classification? What characteristics of plants reflect their evolutionary relatedness?
13. Why do these characteristics represent relatedness more than the number of stamens or the number of pistils?

BiologyNow Resources

Biology⬧Now™

Active Figures

22-2: Characteristics of plants
22-3: The classifications of plants

Preparing for an exam? Take a diagnostic test on your BiologyNow CD-ROM.

Online materials relating to this chapter are at:

http://biology.brookscole.com/AAL3

About the Chapter-Opening Image

Artist Georgia O'Keeffe is known for her depictions of flowers as sexual organs. Linnaeus had the same idea almost three hundred years ago.

23

Protostome Animals: Most Animals Form Mouth First

Key Questions

- What is an animal?
- What characteristics distinguish the sponges (Parazoa) from other animals (Eumetazoans)?
- What characteristics distinguish the radial animals from the bilateral animals?
- What characteristic distinguishes the protostomes from the deuterostomes?

Art Wolfe/Getty Images

Progress and the Burgess Shale

One day about 530 million years ago, a mud slide the size of a city block suddenly broke loose and buried tens of thousands of small marine animals living in the ooze at the bottom of a shallow sea. Acres of mud sealed off the animals so abruptly and completely that the creatures never had a chance to decay. Over millions of years, the mud slowly turned to rock, perfectly preserving the animals within. Geologic processes slowly lifted the rock thousands of feet high, and today the ancient mud slide, and the creatures entombed within it, sits high in the Canadian Rockies. It is called the Burgess Shale.

When the fossil remains of the tiny animals were finally discovered half a billion years later, they unleashed a bitter debate about the nature of evolution. Within the solid mass of the Burgess Shale, the perfectly preserved creatures revealed a remarkable diversity of animals, a diversity that suggested different things to different people.

The Burgess animals' first discoverer was paleontologist Charles Doolittle Walcott and his family. In 1909, Walcott, together with his wife, Helena Walcott, and several of their children were collecting fossils in the Canadian Rockies when they came across an isolated slab of rock crammed with the small, perfectly preserved organisms. Walcott quickly found the source of this slab, higher up the slope of the mountain, and the following summer, Charles, Helena, and their half-grown children returned and dug more than 80,000 fossils from the mountain.

Walcott was a distinguished leader of the early 20th-century American scientific establishment (Figure 23-1). However, his role as top administrator at the Smithsonian Institution, as well as his positions on the boards of other scientific organizations, left him little time for deep thought or serious research. As discoverer of the Burgess fossils, he was entitled to name and classify the new organisms. In his haste, however, he classified all of the Burgess animals as either trilobites (already well-known ancient organisms) or primitive members of still living groups (such as crustaceans, jellyfish, and polychaete worms). In short, Walcott mistakenly assumed that all his animals were members of already well-known phyla.

Walcott well knew how little time he had given the Burgess animals. For the rest of his life, he dreamed of studying them in detail. In

1926, he announced his plans to retire so that he could spend his last years with the Cambrian animals. But he died just three months short of his planned retirement, and even though he was one of the leading scientists of his day, he never fully appreciated his own remarkable discovery.

Finally, in the 1970s, three British paleontologists—Harry Whittington, Derek Briggs, and Simon Conway Morris—undertook to reexamine the Burgess fossils. The fossils were not merely flat impressions but fully formed stone animals whose ancient bodies could be dissected and studied. Using dental drills, Whittington, Briggs, and Conway Morris carefully chipped away layers of stone tissues, revealing the anatomies of some of the most bizarre animals ever found (Figure 23-2). Altogether, argued Whittington, Briggs, and Conway Morris, the Burgess animals represented no fewer than 25 basic body plans, of which only four survive in modern organisms.

In 1989, the Harvard paleontologist and popular writer Stephen Jay Gould wrote a book, *Wonderful Life,* describing the animals of the Burgess Shale and chronicling the work of Walcott, Whittington, Briggs, and Conway Morris. Gould argued that the diversity of life forms in the Cambrian dwarfed anything we know today. The range of anatomical designs among the animals that died in a block of mud 530 million years ago far exceeded that of all the creatures in today's oceans. The Burgess animals, wrote Gould, should have challenged the ablest classifiers of the time. Yet Walcott had classified these bizarre organisms into well-known groups of organisms such as jellyfish and shrimps.

Each of the body plans represented in the Burgess Shale, argued Gould, could have been assigned to a new phylum. Only because many of these types are represented by just a single species had taxonomists hesitated to give each one a whole phylum to itself. Charles Walcott had failed to recognize the true diversity of these creatures, Gould said, for two reasons. Walcott had assumed that the creatures he had found would fit into categories he already knew; and he assumed that evolution, by its nature, inevitably progresses from a few simple forms to a diversity of complex and modern forms (culminating in human beings). This idea—that evolution is progressive—is controversial, however, and depends on how the word "progressive" is defined.

Is evolution progressive? Does diversity increase? Do organisms become more complex? Gould argued that the diversity of the Burgess Shale provided evidence that evolution was *not* a progression from simple to complex, from few forms to

Figure 23-1
Charles Walcott. Paleontologist Charles Doolittle Walcott and his wife, Helena Walcott, discovered a rich deposit of ancient organisms in the Canadian Rockies that demonstrated that a great diversity of life existed 500 million years ago.

Brown Brothers

many. He argued that life is less diverse today than during the Cambrian explosion. The Burgess Shale, he said, illustrated that multicellular life began in great variety all at once. Since then, for reasons unknown, most of these forms became extinct. The organisms that survive today, he said, have been merely lucky survivors of various catastrophes in the history of the Earth.

Gould's arguments were not wholeheartedly accepted by other paleontologists and evolutionary biologists. Soon after the appearance of *Wonderful Life,* further studies of the Burgess fossils by Conway Morris and others suggested that the Burgess animals were not as novel as Conway Morris had initially believed. The animals could be comfortably grouped into smaller numbers of known phyla, although not necessarily the phyla that Walcott had selected.

Were the animals that lived 530 million years ago more diverse than those living today, as Gould had argued? Probably not. While paleontologists have recognized about 11 phyla of animals from the Cambrian, biologists have grouped living animals into 35 phyla. As far as is known, no whole phylum of animals has ever gone extinct. The dinosaurs, for example, were merely a group within the class Reptilia in the phylum Chordata, the phylum to which we humans belong. And the Chordata are represented in the Burgess Shale animals.

Certainly, the history of life suggests that diversity and complexity have increased since the first appearance of life 3.5 billion years ago. All living organisms, in all their current diversity, have descended from the same few simple ancestors. The complex creatures, such as ourselves, that now populate the world have evolved, one generation at a time, from simpler organisms.

In 1995, the philosopher Daniel Dennett attacked Gould, dissecting Gould's arguments about diversity and progress as aggressively as Gould had scrutinized Walcott's taxonomy of the Burgess animals. Dennett insisted that evolution is, in fact, progressive, in that evolution tends to increase complexity.

As organisms evolve adaptations in response to their environment and to other organisms, complexity increases. For example, the grinding teeth of a horse are an adaptation to tough grasses, which evolved toughness and lack of digestibility in response to grazing by plant-eating animals. In the pre-Cambrian, no land plants existed, so a horse could not have existed.

Increasing complexity may be described as progressive, even though no goal is involved. Such progressive change does not, however, imply that evolution is *designed* to pro-

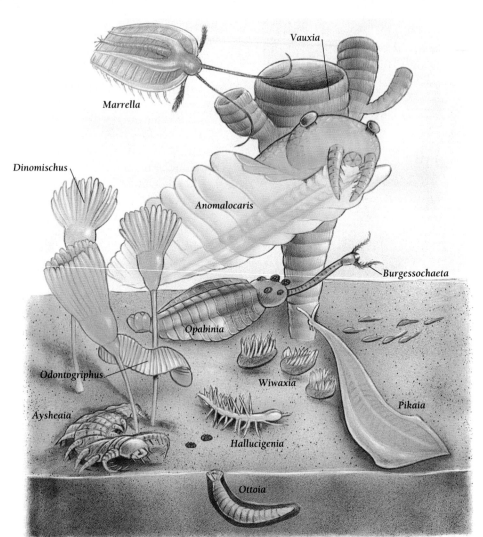

Figure 23-2

Some Burgess Shale animals. The large, green *Anomalocaris* grew up to a meter long. Just below is *Opabinia (brown),* smaller, but with five eyes and a front-facing "nozzle." *Pikaia* may be the oldest known Chordate—if not our direct ancestor, at least a relative of one. *Odontogriphus* (the pink, ribbonlike animal) was a flat, swimming creature. *Hallucigenia* (the green creature walking on the sea floor) had seven pairs of spines and seven tentacles. (Because of an initial misunderstanding by paleontologists, *Hallucigenia* is upside down in this figure.) The three reddish, flame-shaped creatures are *Wiwaxia,* which probably crawled along the sea floor. What look like green plants on stems are actually *Dinomischus,* a sessile animal. *Vauxia* was a sponge.

duce increasingly complex organisms. Nor does evolution *necessarily* produce increasing complexity. After all, we are surrounded by relatively simple organisms—including bacteria, amebas, ferns, and flatworms—survivors, like us, of the ravages of major catastrophes, like those that wiped out the dinosaurs, and the constant, local, minor ones that constitute natural selection. All these relatively simple organisms are descendants, as we all are, of a simple ancestor that lived more than 3.5 billion years ago. All organisms, both ancient and modern, are the product of the same evolutionary processes that have continued for billions of years.

23.1 What Is an Animal?

As animals ourselves, we tend to see our own kingdom as the most interesting. Many of us have spent hours at zoos, observing the beautiful and the bizarre. Even the best zoos, however, barely hint at the variety of animals that exist on Earth. A coral reef is a much better place to observe the tremendous di-

versity of animals. A coral reef swarms with fishes, shelled animals, worms, sponges, corals, sea anemones, and many other unfamiliar organisms. Most land dwellers are awed by the variety of shapes, domes, branches, fans, tubes, and stars, as well as by the range of colors, from dark red to luminous blue.

What Features Characterize the Animals?

An animal is a multicellular, heterotrophic organism that develops from an embryo. All animals are eukaryotes, and most reproduce sexually. A few, such as the corals, can reproduce by asexual processes, such as budding. Unlike plants, fungi, and protists, however, animals never show alternation of generations. With rare exceptions, the only haploid cells are the gametes (sperm and eggs).

Nearly all animals ingest their food—usually other organisms or their remains. A few animals—including many corals—appear to perform photosynthesis. Careful examination reveals, however, that these animals harbor symbiotic algae that are actually the ones performing photosynthesis.

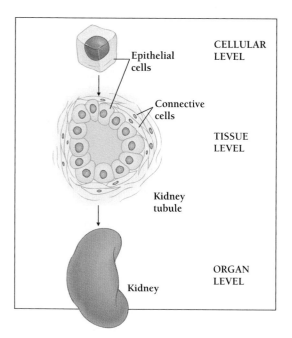

Figure 23-3
The structure of animals. A kidney is an organ, which is made up of two or more tissues. The two tissues in the kidney tubule, for example, are connective tissue and epithelial tissue, which are made of connective cells and epithelial cells, respectively.

Like plants, animals are highly structured. Almost all animals contain many different kinds of specialized cells. Groups of similar cells, along with the materials they secrete, called **matrix**, form **tissues**, units of structure and function. Two or more kinds of tissue in turn can form an **organ**, a structural unit with a distinctive function. For example, one class of cells can form an **epithelium**, a tissue that lines a surface (such as the outside of the body), while another kind of cell, along with its matrix, can form **connective tissue**, a network of loosely connected cells that may help support other body parts. A kidney, which contains both epithelial and connective tissues, is an example of an organ (Figure 23-3). The specialization of cells and the grouping of cells into tissues and organs allow animals to function efficiently.

Unlike plants, most animals can move about in search of food or mates. Some animals, such as the corals, are stationary but can move body parts to capture prey. Many have stages during development when they disperse. Almost all animals coordinate their movements in patterns that we can describe as behavior. For example, animals capture food, avoid predators, and breed. For all but the sponges, the coordination of such behavior depends on the functioning of a network of nerve cells (neurons).

An animal is a multicellular, heterotrophic organism that develops from an embryo and usually reproduces sexually.

How Do Zoologists Classify the Animals?

Zoologists (pronounced zo-OL-ogists), biologists who study animals, estimate that living animal species number about 4 million. Zoologists divide these millions of living animals into about 35 phyla (Figure 23-4). (Extinct animals form many more phyla.) The 40,000 species of vertebrates (animals with backbones)—including fish, frogs, snakes, birds, and mammals—are just a subphylum of one of these 35 phyla. Vertebrates include a mere one percent of all animal species. Most animals (so-called "invertebrates") have no backbones. Of these, the vast majority are insects (mostly beetles, in fact), snails, jellyfish, and worms.

How Do Biologists Use Symmetry to Classify Animals?

Zoologists divide all the creatures in the animal kingdom into two subkingdoms. The subkingdom **Eumetazoa** [Greek, *eu* = true + *meta* = middle + *zoa* = animals] includes almost all of the 4 million named species of living animals. Excluded are the 5,000 species of sponges, which make up the small subkingdom **Parazoa**. The sponges are the simplest animals, for they have no organs, as described earlier, and no regular symmetry. Their lack of symmetry and tissue organization makes them seem almost more like colonial protists. But they are definitely animals.

The Eumetazoa (all the other animals) have either of two kinds of symmetry. Animals with **radial symmetry**—the simple jellyfish and sea anemones, for example—can be rotated along their central axis without changing their appearance (Figure 23-5A). These animals have a top and bottom, but no front and back and no left and right. Many of these animals live most of their lives either floating passively in the water (jellyfish) or attached to a rock or some other support (sea anemones).

Most animals are **bilaterally symmetrical**, meaning that their left and right halves are (approximately) mirror images of one another (Figure 23-5B). Such animals have a defined **anterior** (front) and **posterior** (back end). The *dorsal* surface is the back, or top, which, in most animals, faces the sky. The *ventral* surface is the part facing the Earth. All the bilateral animals, from worms to snails to humans, belong in the "Bilateria."

Bilateral symmetry implies some sort of head and, ordinarily, a preferred direction of movement—usually "head" first. Such directed movement, in turn, requires an integrated system of muscles, nerves, and sense organs that the sponges and radially symmetrical animals lack. In general, the sense organs, which detect prey and predators, and the nerves, which coordinate behavior, lie near the front of the animal (Figure 23-5B). Reproduction, digestion, and excretion usually lie in the rear, the posterior end.

The Parazoa (sponges) have no symmetry and little tissue organization. The Eumetazoa are classified into bilateral animals (Bilateria) and radial animals.

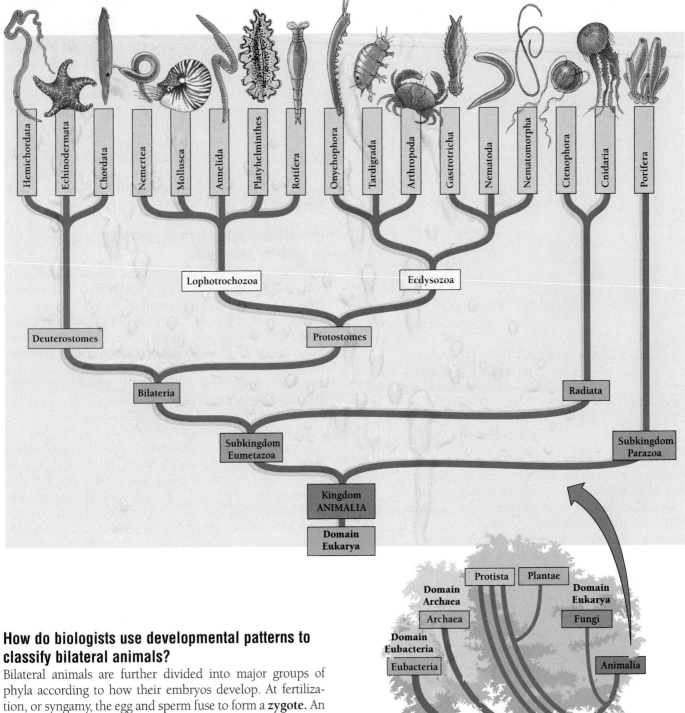

Lophotrochozoa

Ecdysozoa

Deuterostomes

Protostomes

Bilateria

Radiata

Subkingdom
Eumetazoa

Subkingdom
Parazoa

Kingdom
ANIMALIA

Domain
Eukarya

Hemichordata · Echinodermata · Chordata · Nemertea · Mollusca · Annelida · Platyhelminthes · Rotifera · Onychophora · Tardigrada · Arthropoda · Gastrotricha · Nematoda · Nematomorpha · Ctenophora · Cnidaria · Porifera

Protista · Plantae

**Domain
Archaea**

Archaea

**Domain
Eukarya**

Fungi

**Domain
Eubacteria**

Eubacteria

Animalia

Universal ancestor

**Figure 23-4
The major phyla of protostome animals.**

Biology ⒺNow™   Learn more about protostomes by clicking on this figure on your BiologyNow CD-ROM.

How do biologists use developmental patterns to classify bilateral animals?

Bilateral animals are further divided into major groups of phyla according to how their embryos develop. At fertilization, or syngamy, the egg and sperm fuse to form a **zygote.** An animal zygote divides by mitosis to produce a hollow ball of cells called a **blastula.** The blastula folds in on itself to form three cell layers surrounding a simple cavity (Figure 23-6). The three cell layers develop into distinctly different tissues in the adult. The innermost layer, which surrounds the cavity, is called the **endoderm** [Greek, *endon* = within + *derma* = skin] and gives rise to the intestines and other digestive organs. The outermost layer, called the **ectoderm** [Greek, *ecto* = outside], gives rise to skin, sense organs, and nervous system. The middle layer, the **mesoderm** [Greek, *mesos* = middle], gives rise to muscle, skeleton, and connective tissue. In illustrations, the ectoderm (skin and nerves) is always blue,

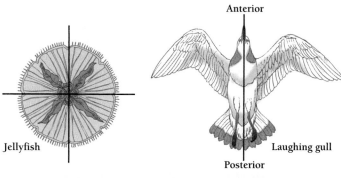

Figure 23-5

Symmetry. A. Radial symmetry in a jellyfish. B. Bilateral symmetry in a laughing gull. All animals with bilateral symmetry have anterior and posterior ends and dorsal and ventral surfaces. Shown here are the ventral (under) surface of a jellyfish and a gull.

the mesoderm (muscles) is always red, and the endoderm (gut) is always yellow.

All metazoans (except the radial ones) develop these three layers, which are like a tube within a tube. The inner tube becomes the digestive tract, with a mouth at one end and an anus at the other. The outer tube is the body wall. Between the two tubes (of most animals) is a fluid-filled space, in which the internal organs hang. In most bilateral animals (including ourselves), a layer of mesoderm cells lines this space, and the lined cavity is called a **coelom** [Greek, *koilos* = hollow] (Figure 23-7D). Animals with a coelom are called **coelomates**. These include, for example, snails, earthworms, insects, and all of the vertebrates.

The additional internal space of a coelom offers several advantages. For example: (1) it prevents muscle movement from interfering with the operation of the digestive and circulatory systems, (2) it can provide a hydrostatic skeleton against which muscles can work, and (3) it provides a pro-

tected space for the production of sperm and ova. For example, your lungs could not fill and empty so easily if each breath meant struggling against surrounding muscles instead of just filling an empty space within the chest cavity. In the same way, a coelom keeps the muscular movements of the intestinal tract from interfering with the movements of the surrounding muscles.

The simplest animals, including the radial jellyfish and the bilateral flatworms, have no coelom and are called **acoelomates** [Greek, *a* = without] (Figure 23-7A). In between are the **pseudocoelomates** [Greek, *pseudo* = false], which have an organ-containing cavity, but without the mesodermal lining of a true coelom. A **pseudocoel** lies between the endoderm and the mesoderm. Almost all the pseudocoelomates are tiny worms called nematodes.

Most animals are bilateral coelomates, which are further divided into two major groups according to how they develop as early embryos (Figure 23-6). At one end of the innermost cavity that forms the gut is an opening called the **blastopore.** In some animals, the blastopore eventually develops into the adult mouth and the anus forms later. In other groups of animals, the blastopore develops into the anus, and the mouth develops later. Those in which the mouth develops from the blastopore are called the **protostomes** [Greek, *protos* = first + *stoma* = mouth]. These include, for example, snails, earthworms, and all the arthropods—the spiders, insects, and their evolutionary relatives. In contrast, those in which the anus forms first, from the blastopore, and the mouth forms second are called **deuterostomes** [Greek, *deuteros* = second + *stoma* = mouth]. Most coelomates are protostomes. We vertebrates and our cousins the echinoderms (sea stars and sea urchins, for example) are deuterostomes.

Protein and DNA sequencing data supports the main division of the bilateral animals into protostomes and deuterostomes. Molecular data also confirms the traditional classification of the deuterostomes into subcategories

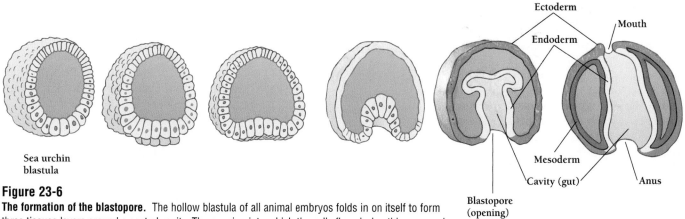

Figure 23-6

The formation of the blastopore. The hollow blastula of all animal embryos folds in on itself to form three tissues layers around a central cavity. The opening into which the cells flow during this process is called the blastopore. In some animals (protostomes), the blastopore becomes the mouth; in others (deuterostomes) it becomes the anus. In both kinds of animals, the hollow center becomes "the tube within a tube" that later forms the gut.

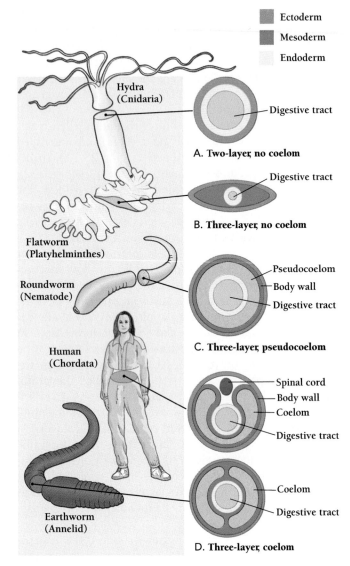

Ectoderm
Mesoderm
Endoderm

Hydra
(Cnidaria)

Digestive tract

A. Two-layer, no coelom

Digestive tract

B. Three-layer, no coelom

Flatworm
(Platyhelminthes)

Roundworm
(Nematode)

Pseudocoelom
Body wall
Digestive tract

C. Three-layer, pseudocoelom

Human
(Chordata)

Spinal cord
Body wall
Coelom
Digestive tract

Coelom
Digestive tract

Earthworm
(Annelid)

D. Three-layer, coelom

Figure 23-7
Tissue layers. A. A cnidarian has just two tissue layers and no coelom. B. The flatworms have three layers, but still no coelom. C. The roundworms have three layers and a pseudocoel. D. Most animals are true coelomates, including earthworms and humans. Coelomate animals have three tissue layers and a true coelom in which the gut is suspended.

In this chapter we will begin with the simplest animals, the sponges; move to the acoelomates, first those with radial symmetry, then those with bilateral symmetry; and then move to the pseudocoelomates. We will finish with an examination of the many groups of protostome coelomates. In the next chapter, we will briefly survey the deuterostome coelomates.

23.2 The Sponges Have No Organs

The Porifera (sponges) belong to the subkingdom Parazoa and differ from all other animals in having no symmetry and minimal organization and cell specialization. The sponges have only simple connective tissues and no organs. Fossil sponges date from the early Cambrian Period, 530 million years ago.

The 5,000 species of sponges (most of which are marine) come in many sizes, shapes, and colors. Almost all sponges are **sessile**, permanently anchored to rocks, logs, or coral. They range in size from a few millimeters to two meters. Some are shapeless blobs, while others resemble fans, cups, crusts, and tubes (Figure 23-8A). Like the synthetic "sponges" we use at the kitchen sink, the members of Porifera [Latin, *porifera* = hole-bearer] are full of holes.

The sponge's cells lie in three layers surrounding a central cavity. A sponge's outer, epithelial layer of cells is perforated with tiny holes, through which water enters the sponge. Each sponge pumps an enormous volume of water through its simple body. A sponge the size of a fingertip may pump 20 liters of water a day. This water, loaded with nutritious microorganisms and organic matter, enters through the pores, flows into the cavity, and exits through the sponge's large opening (or openings). The driving force for this flow comes from a layer of flagellated cells lining the sponge's central cavity (Figure 23-8B).

Between the outer epidermal layer and the flagellated cells lining the cavity is a layer of **mesenchyme** [Greek, *mesos* = middle + *enchyma* = infusion], loosely attached cells embedded within a jellylike substance. Within this simple tissue, amebalike cells capture food and shuttle it from the inside cells to the outer epithelial cells. These same amebalike cells also give rise to either sperm or eggs. Most sponges are *hermaphrodites:* a single organism is both male (like the Greek god Hermes) and female (like the goddess Aphrodite).

Although sponges are multicellular, they are extremely primitive. No nerve cells coordinate a sponge's responses, and each cell functions as an independent unit. Nonetheless, single sponge cells, separated from others by being forced through a piece of cheesecloth, will spontaneously reorganize themselves into a functioning sponge.

(mainly sea star-type animals on the one hand and vertebrates such as fish and mammals on the other). But molecular data does not always parallel the classification of the protostomes by form and developmental pattern. As a result, the classification of many protostome animals remains unsettled.

Zoologists divide the Eumetazoa according to the presence or absence of a coelom or pseudocoel and according to whether the blastopore develops into a mouth (protostomes) or an anus (deuterostomes).

The individual cells of sponges resemble protists. But unlike protists, sponges develop from embryos and produce sperm and eggs.

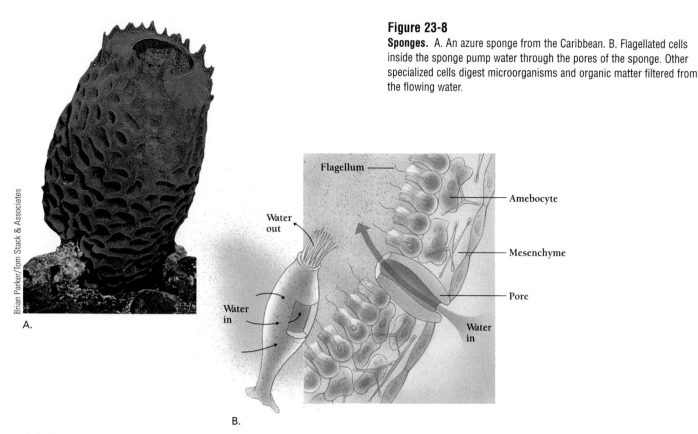

Brian Parker/Tom Stack & Associates

A.

B.

Figure 23-8

Sponges. A. An azure sponge from the Caribbean. B. Flagellated cells inside the sponge pump water through the pores of the sponge. Other specialized cells digest microorganisms and organic matter filtered from the flowing water.

23.3 Radial Animals

The 9,000 species of radially symmetrical animals fall into two phyla, Cnidaria (pronounced ni-DAR-ee-uh) and Ctenophora (pronounced ten-OFF-or-uh). The cnidarians, which are simpler and far more numerous, are among the oldest animals in the fossil record, present well before the Cambrian explosion. All the radial animals lack a coelom.

Cnidarians (Hydras, Jellyfish, and Anemones)

Zoologists recognize four classes of **cnidarians:** Hydrozoa (hydroids), Scyphozoa (jellyfish), Anthozoa (corals and sea anemones), and the jellyfishlike Cubozoans. Almost all are marine (with the exception of *Hydra* and a few other freshwater hydrozoans). By far the largest class is Anthozoa, with about 6,200 species. Hydrozoa include about 2,700 species, and Scyphozoa only about 200.

Anthozoans [Greek, *anthos* = flower] take their name from their flowerlike appearance. They include sea anemones, sea pansies, sea fans, and sea whips. Each anthozoan has a cylindrical body with a crown of tentacles. Symbiotic algae enhance their plantlike appearance and also contribute to their nutrition, often allowing them to grow in water that is poor in nutrients. Especially impressive are the corals, whose secreted skeletons of calcium carbonate are the foundation of gigantic coral reefs.

The anthozoans have the most complex behaviors of the cnidarians. Some species of sea anemones, for example, feed on crabs, mussels, or even fish. A sea anemone can catch and con-sume prey with its tentacles. Some sea anemones even attack the tentacles of other sea anemones that intrude too closely.

All cnidarians contain two layers of cells that function as true tissues. But cnidarians have no organs, making them the least complex of the Eumetazoa. Cnidarians come in two basic body plans—**polyps,** which resemble cylinders, and the bell-like **medusae,** or jellyfish (Figure 23-9). Both forms have a single mouthlike opening to a central cavity, and both use

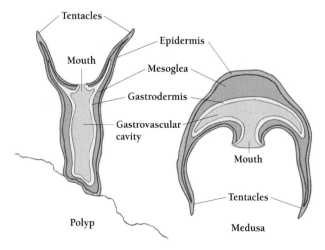

Figure 23-9

The two body plans of the cnidarians. Jellyfish and other cnidarians have just two body plans: the polyp and the medusa. The polyp is the sessile form exemplified by sea anemones and hydras. The medusa, exemplified by the jellyfish, is basically a polyp turned over.

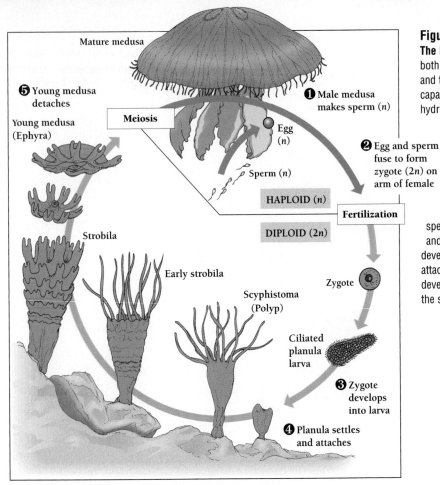

Figure 23-10

The life cycle of a hydrozoan. The class Hydrozoa includes both *Hydra*, a simple polyp often studied in biology classes, and the formidable Portuguese man-of-war *(Physalia)*, capable of stinging to death an adult human. Most hydrozoans are jellyfish (also called jellies) that have both polyp and medusa stages. In the polyp stage, hydrozoans are colonial, with many polyps budding around a branched central cavity. *Hydra* is an exceptional hydrozoan, for it lives as a solitary polyp, with no medusa stage. All hydrozoans shed sperm and eggs from testes and ovaries that project from the body wall. ❶ The male medusa makes haploid sperm and the female medusa makes haploid eggs. ❷ Egg and sperm fuse to form a diploid (2*n*) zygote. ❸ The zygote develops into a "planula" larva. ❹ The planula settles and attaches on a surface such as a rock and begins to grow and develop into a strobila. ❺ Small young medusas detach from the strobila.

tentacles to capture food. Polyps are usually partly sessile, attached to rocks or other surfaces, with mouths and tentacles pointed upward. Medusae generally float free with the mouth pointed down and tentacles dangling, like the fringe on the edge of an umbrella. Some cnidarians are polyps throughout their life cycles, some are only medusae, and some cycle from one form to the other (Figure 23-10). Most cnidarians start out life as free-swimming larvae and become sessile as adults.

Both polyps and medusae have an internal digestive cavity, which is surrounded by an outer layer of epidermis and an inner layer of gastrodermis. Like higher animals, cnidarians perform most of their digestion extracellularly—that is, within the digestive cavity. Digestion in the cavity, however, is not complete, and the cells lining the cavity engulf fragments of the small animals that compose the cnidarian's diet.

Between the two cellular layers is the **mesoglea** [Greek, *mesos* = middle + *glia* = glue], a jellylike material that contains few if any cells. In some medusae, such as the jellyfish, the mesoglea may be quite thick and gelatinous. In polyps, the mesoglea is usually thin, and in some cases (such as in the sea anemones) consists entirely of cells.

Cnidarians [Greek, *cnide* = nettle] take their name from specialized stinging cells, called **cnidocytes,** that lie on their tentacles (Figure 23-11). Each cnidocyte can use water pressure, at some 140 times the pressure of the atmosphere, to

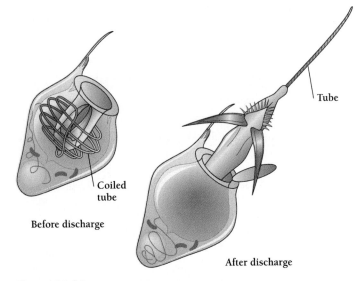

Before discharge

After discharge

Tube

Coiled tube

Figure 23-11

Cnidocytes. Cnidarians are named for their stinging cells, or cnidocytes. Coiled inside each cnidocyte is a tiny barbed harpoon, covered by a lid. When the cnidocyte fires, the lid opens and the harpoon ejects, turning inside out as it unfurls. If the barbs strike the flesh of another animal, they catch and inject a paralyzing toxin. When the victim stops moving, the cnidarian pulls its prey to its mouth and begins to digest it.

fire a tiny barbed spear called a **nematocyst**. As whalers use barbed harpoons threaded with rope to impale whales and drag the animals back to the whaling boat, cnidarians use nematocysts and their attached threads to capture their prey.

Cnidarians work much more quickly than whalers, however. Only a few milliseconds elapse between the detection of the prey by the tickling of the cnidocyte's flagellum-like cnidocil and the firing of the nematocyst. After it has pierced its prey, the nematocyst can discharge a poisonous protein. The nematocysts of one cnidarian—the Portuguese man-of-war—produces a neurotoxin potent enough to kill a human swimmer.

Once the nematocysts have attached their tiny tethers, the tentacles draw the prey mouthward. Pulling and pushing the prey requires coordination among the tentacles, which is controlled by simple muscle cells and nerve cells. A cnidarian's nerve cells are unlike those in more complex organisms, for they transmit impulses in both directions, rather than just one.

Cnidarians are radially symmetrical acoelomates with true tissues but no organs. Named for their specialized stinging cells, the cnidarians include the hydras, jellyfish, corals, and sea anemones.

Ctenophores (Comb Jellies)

The **ctenophores** [Greek, *ktenos* = comb + *phora* = motion], or comb jellies, are a small phylum of about 90 living species. Their name refers to the rows of comblike cilia that these beautiful animals carry in bands along their short bodies (Figure 23-12). Ctenophores seem to waltz through the water. By means of the coordinated action of their cilia, they move forward, mouth first, capturing prey with their sticky

Figure 23-12
A ctenophore, or comb jelly. These graceful animals are noted for their beauty.

(James R. McCullagh/Visuals Unlimited)

tentacles, while slowly rotating. They are hermaphrodites (simultaneously male and female) and usually shed both sperm and eggs into the open sea.

Because the combs are arranged symmetrically, and because one species of ctenophore has cnidocytes, biologists once classified the ctenophores as cnidarians. But ctenophores' paired tentacles lack the radial symmetry of true cnidarians, and modern zoologists have given the comb jellies their own phylum. The fossil record shows that the ctenophores have existed separate from the cnidarians for at least 400 million years.

Unlike the cnidarians, the ctenophores (comb jellies) are not radially symmetrical, for they possess paired tentacles. In addition, ctenophores move through the water by means of rows of cilia.

23.4 The Lophotrochozoan Protostomes

As mentioned, the bilateral animals are divided into two main taxa, the deuterostomes (in which the mouth forms after the anus) and the protostomes (in which the mouth forms before the anus). We'll discuss the deuterostomes in Chapter 24. In the rest of this chapter, we discuss the protostomes. Traditional morphological taxonomy divides the protostomes according to how the cells of their embryos divide.

More recent molecular evidence suggests possibly two main groups of protostomes—the **Lophotrochozoa** [Greek, *lopho* = crest or tuft + *trokos* = wheel, disk] (worms, snails, and other soft-bodied animals) and the **Ecdysozoa** [Greek, *ekdysis* = casting off or shedding] (crabs, barnacles, spiders, insects, and other animals with a hard external skeleton). The Lophotrochozoa include flatworms, ribbon worms, mollusks (snails and octopods), annelids (earthworms), and many other large groups of animals. The Lophotrochozoa also include the *rotifers*, tiny aquatic animals with saclike bodies (Figure 23-13).

Platyhelminthes (Flatworms)

One phylum of Lophotrochozoa, Platyhelminthes is an acoelomate (like the Cnidarians and Ctenophores). Flatworms include many species of parasites that infect hundreds of millions of people, as well as the pond planarian *Dugesia*, used in biology laboratories all over the world. Flatworms differ from more advanced animals in having only one opening to their digestive cavities. Mouth and anus are the same.

The 20,000 named species of **Platyhelminthes** (flatworms) live in a wide range of environments. Most free-living species are marine, although some (such as *Planaria*) thrive in fresh water, and some are terrestrial. One species is

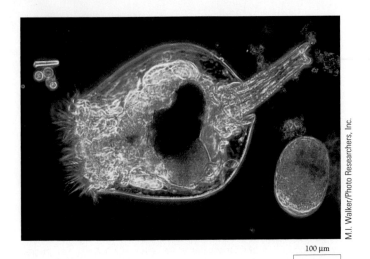

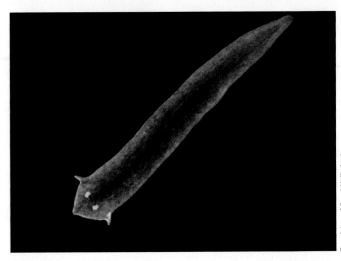

Figure 23-13

Rotifer. Brachionus female with eggs. The 2,000 species of rotifers are among the smallest of all animals, ranging in size from 40 micrometers (μm) to two millimeters. These common freshwater animals are often shaped like tiny trumpets, with a crown of cilia around the open end. The cilia direct water containing bacteria, protists, and small animals into the mouth. The whirling cilia make the crown look like a rotating wheel, giving the phylum its name [Latin, *rota* = wheel + *ferre* = to bear]. Rotifers also have tiny but effective jaws, which grind food after it enters the mouth. The rotifers' success depends on the rapidity of their reproduction: some species can double their population every two days. And females commonly reproduce without males, through parthenogenesis. Rotifers are a major source of food for other animals that live in fresh water.

A.

Digestive system

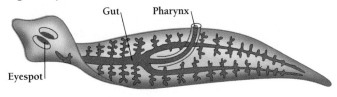

Nervous system

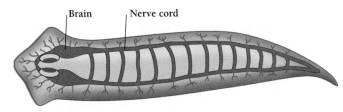

Reproductive system

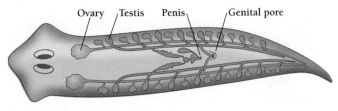

B.

Figure 23-14

Flatworm. A. Live planarian glides forward. B. Flatworms are simple animals with a one-way gut and no coelomic cavity separating the layers of their bodies. But, like all bilateral animals, they have a head with sense organs (note eyespots), as well as intricate digestive, nervous, and reproductive organs.

specialized for living in bat guano. Zoologists divide the flatworms into three classes: Turbellaria (free-living flatworms), Trematoda (flukes), and Cestoda (tapeworms). All three classes share the same acoelomate body plan.

All flatworms have three body layers—an endoderm that lines the digestive cavity, an ectoderm on the outside, and a mesoderm in between. They have feeding, digestive, and reproductive organs, a brain, and a simple system for regulating water balance.

One thing flatworms do not have is a circulatory system. Some flatworms nonetheless grow to 60 cm, and tapeworms may be as long as 10 meters (over 30 feet). How do such large animals distribute oxygen and nutrients to their cells? Flatworms survive without a circulatory system by keeping diffusion distances small. Their flattened shapes allow gases to diffuse rapidly to and from the outside environment, and the gut cavity of each flatworm is branched and extends throughout the body, so that no cell is far from the gut. The single opening to the flatworm's gut cavity serves as both mouth and anus. Likewise, the gut cavity itself serves both as a reservoir of nutrients and for wastes (Figure 23-14).

Turbellaria, the free-living flatworms, can reproduce either asexually or sexually. A free-living flatworm's simple nervous system allows it to move efficiently toward food and away from light. Some flatworms detect light with primitive

paired eyespots on their heads. They can also detect odor. A good way to collect flatworms is to put a piece of meat in the bottom of a muddy pond. Chemicals diffusing from the meat stimulate receptors on the flatworm's head. When a flatworm smells the meat, it turns its body and moves toward its future meal.

The differences among the three classes of flatworms relate to differences in the way they eat and reproduce. The flukes and tapeworms are parasites. Their outer layer of cells resists the digestive enzymes of their hosts, and they lack eyespots and chemical receptors for detecting food. Flukes take in food through a specialized mouth, while tapeworms lack a digestive system altogether and simply absorb digested food from their hosts.

Parasites are specialized for rapid and prolific reproduction in their hosts, and they often evolve stripped-down bodies, lacking many ancestral characteristics that are not useful for their specialized lives. Flukes and tapeworms illustrate many common adaptations of parasites: (1) rapid reproduction, (2) distinct stages that allow passage and dispersal through more than one host, (3) organs for attachment to their hosts, (4) specialized digestion (in the case of tapeworms, direct absorption), and (5) reduced sense organs. As a result, the more complex free-living flatworms are probably more similar to the original flatworm ancestors than are the streamlined flukes and tapeworms.

Trematoda (flukes) range in size from less than a millimeter to more than 8 centimeters. All flukes (more than 8,000 species) are parasites. Each attaches to its host (whether outside or inside) by a hook or sucker. Like the free-living flatworms, flukes have a digestive system with a single opening.

Many flukes are hermaphrodites and generally reproduce by mutual copulation. They are far more prolific than free-living flatworms, producing 10,000 to 100,000 times as many eggs. Some flukes live their whole lives on a single host. Most, however, have more complicated lives, involving two or more hosts. A liver fluke, for example, uses three distinct hosts to complete its life cycle—a mammal, a snail, and a fish (Figure 23-15).

Cestoda (tapeworms) are even more specialized for reproduction than are the flukes. An adult tapeworm has no digestive system at all. It lives in its host's gut, and its cells directly absorb passing nutrients.

A tapeworm maintains its easy life with a specialized attachment organ, called a **scolex,** at its anterior end. Behind the scolex lie a set of repeated segments, called **proglottids.** In the beef tapeworm, the proglottid chain may be as long as 10 meters (Figure 23-16). Each proglottid is a complete reproductive unit, with both male and female organs. If two or more tapeworms live together, they fertilize one another. Otherwise, a tapeworm self-fertilizes. In either case, the proglottids toward the rear end of the tapeworm are filled with embryos, each within a separate shell, which pass, with the feces, to the outside world. If cattle eat vegetation carrying these embryos, they acquire tapeworms. In the United States, about 1 percent of all cattle have tapeworms, an important reason for eating only inspected beef.

Platyhelminthes (free-living flatworms, flukes, and tapeworms) are the simplest animals with heads. They have three body layers, true organs and tissues, a brain, and an organ for regulating water balance. They do not have a one-way digestive system: they take in food and excrete wastes through the same structures. The parasitic members of this phylum possess fewer specializations than the free-living flatworms.

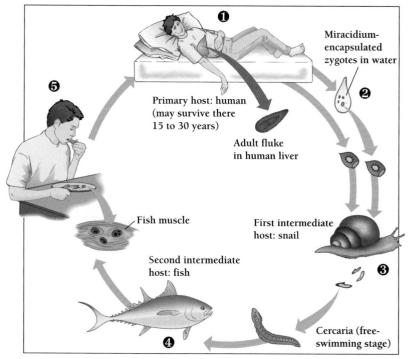

Figure 23-15
Parasitic flatworm. The elaborate life cycle of a human liver fluke. The liver fluke lives its adult life as a 1- to 2-cm worm in the livers of humans, pigs, cats, and dogs. ❶ Inside a host, each worm mates and reproduces, then deposits thousands of fertilized eggs (zygotes). ❷ If deposited on land, the zygotes may be eaten by a terrestrial snail. If deposited in water, the zygotes develop into swimming larvae, called miracidia, that penetrate the skins and body walls of aquatic snails. ❸ Inside the snail, the miracidia give rise (through a series of stages) to cercaria, which have a digestive tract, suckers, and a tail. ❹ The swimming cercarium then reenters and burrows into the muscles of a fish. Inside the fish's flesh, cercaria form the cysts, called metacercaria. ❺ When a human eats the fish, the metacercarium escapes from its cyst and develops into a sexual adult, which mates.

Biology ⓔNow™ Learn about the life cycle of another kind of fluke by clicking on this figure on your BiologyNow CD-ROM.

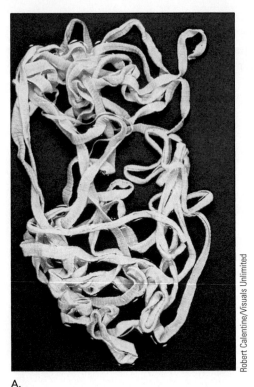

Robert Calentine/Visuals Unlimited

A.

Oliver Meckes/Photo Researchers, Inc.

Figure 23-16
Most tapeworms have two or more hosts. A. The broad fish tapeworm *(Diphyllobothrium latum)* infects tiny crustaceans called copepods, copepod-eating fish, and fish-eating mammals, including humans and bears. The broad fish tapeworm occurs worldwide, can reach 20 meters in length, and can shed a million eggs a day. A large tapeworm can live for its host's entire life. B. At the front end of a tapeworm is the scolex, which attaches to the inside of the host by means of hooks (top) and suckers (two holes).

250 μm

B.

Nemertea (Ribbon Worms)

Ribbon worms, or **Nemerteans,** are a group of about 1,150 species of free-living predatory worms, most of which live inconspicuously in the sea, ambushing smaller prey with harpoons and lassos. A few species live in fresh water or on land in the tropics. Their flat, velvety bodies may be less than a millimeter long or as long as 54 meters (Figure 23-17). Each has a characteristic snout, or **proboscis**—a long, sensitive, retractable, and sometimes venomous, tube. A ribbon worm uses its proboscis to explore its environment, defend itself, and capture prey. Ribbon worms are abundant in tidal mudflats, where they spend their days hidden in the mud, emerging at night to feed.

Ribbon worms somewhat resemble flatworms, but careful comparisons have persuaded biologists that these two groups are not closely related. Unlike flatworms, ribbon worms have a coelom. Ribbon worms also have a digestive tract with both a mouth and an anus and a circulatory system. The two-ended gut allows continuous eating, uninterrupted by the need to use the same opening for excretion. The second opening also permits more efficient extraction of nutrients from food, as it progresses in a single direction. The gut is like an assembly line, performing different operations at each stage. The anterior end breaks down the food, the posterior processes wastes.

The ribbon worm's circulatory system is relatively simple; in fact, some species have only two blood vessels. The blood is usually colorless, but may also be yellow, red, orange, or green. There is no heart to pump the blood in a particular direction: instead the vessels and the body wall contract irregularly, sloshing the blood through the worm's simple circulatory system.

Ribbon worms (Nemertea) have a two-ended digestive system, a circulatory system, and a proboscis.

In the rest of this section, we briefly survey two large phyla of Lophotrochozoa—the mollusks (which include snails, slugs, clams, oysters, squids, and octopods) and the annelids (which include earthworms, segmented marine worms, and leeches).

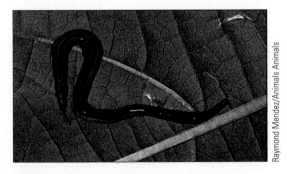

Raymond Mendez/Animals Animals

Figure 23-17
Ribbon worm (Nemertea). Some ribbon worms grow up to 54 m long. All have a two-ended digestive tract and circulatory system.

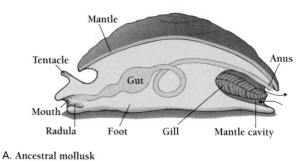

A. Ancestral mollusk

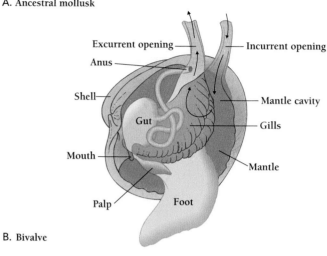

B. Bivalve

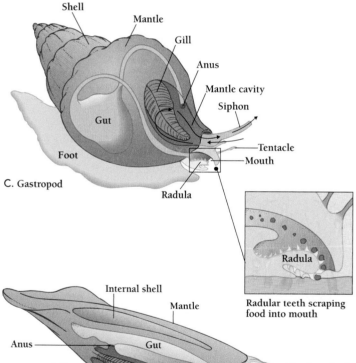

C. Gastropod

Radular teeth scraping
food into mouth

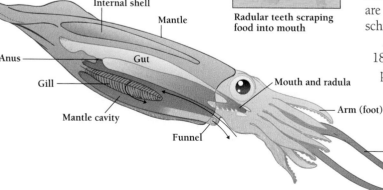

D. Cephalopod

Figure 23-18

The body of a mollusk. A. The basic mollusk has a radula, a foot for locomotion, and gills in a mantle cavity. Cilia on the gills drive water through the mantle cavity, from which the gills can absorb oxygen. The passing water also carries away carbon dioxide and other wastes. Land snails and slugs have no gills; instead, oxygen diffuses directly from the mantle cavity into the surrounding circulatory system. B. Bivalves have the foot, mantle, shell, cavity, and coelom of other mollusks, but they also have siphons for directing a current of water through the mantle cavity. C. Gastropods such as snails depart from other mollusks in having a twisted body and a distinct head with many sense organs, including touch receptors, chemical receptors, and eyes. D. Cephalopods have specializations not shared by other mollusks, including suckered tentacles and a mantle cavity into which the animal can pull water. This water is forcefully expelled through a funnel that can be pivoted to force a water stream in different directions, driving that animal forward. Some cephalopod species are quite large. The largest known fossil cephalopods had straight conical shells up to 5 m (16 ft) long.

Mollusks Share a Common Body Plan

If we take the diversity of species as a measure of a phylum's evolutionary success, then, among the animals, mollusks are second only to the arthropods (insects, spiders, and crabs). Biologists have named more than 100,000 species of living mollusks, in virtually every environment where animal life is possible. Mollusks are abundant in the seas, in fresh water, and on land. The number of land species of slugs and snails alone outnumber all species of reptiles, birds, and mammals combined. Mollusks may be as small as a grain of sand (about a millimeter long) or as long as a whale. (But the 20-meter giant squid is not nearly as heavy as a whale.) Most mollusks, whether snails or clams, live in protective shells. Indeed, one of mollusks' distinguishing characteristics is their shell. But the name **mollusk** [Latin, *molluscus* = soft] refers to this animal's soft body rather than its shell.

Mollusks are important to humans: oysters, scallops, squid, and other mollusks are among the great delicacies of the culinary world. Pearls and "mother of pearl" (from shells) have long been used to decorate people and their possessions, and the secretion of the marine snail *Murex* long served as the source of the dye royal purple, coveted by kings. Mollusks are also pests: some destroy crops and flowers, others damage docks and boats, and still others are intermediate hosts for such devastating parasites as schistosomes and other flukes.

All mollusks share a common body plan (Figure 23-18A). The intestinal tract, as well as the excretory and reproductive organs, lie within the main body of the mollusk, which is called the **visceral mass** [Latin, *viscera* = internal organs]. Attached to the visceral mass is the **foot**, a muscular extension of the body that the mollusk uses for sensing, grabbing, creeping, digging, and just holding on. The dorsal surface (top) of the visceral mass forms a **mantle**, specialized tissue that, in most mollusks, secretes a shell.

tle also enclose a space, called the **mantle cavity,** which contains the mollusk's breathing organs. Both the mouth and the anus usually open into the mantle cavity. Snails and other shelled mollusks can withdraw their bodies inside their mantle cavities, in some cases pulling behind them a protective flap called the **operculum,** which resembles a trapdoor.

In octopods, squid, sea slugs, and other aquatic mollusks, the mantle cavity contains gills—thin organs that transfer dissolved oxygen to the circulatory system. Air-breathing snails and slugs also use the mantle cavity to absorb oxygen.

Most mollusks have a **radula,** a rasping tongue covered with teeth made from chitin. Mollusks may use the radula to protect themselves, to capture prey, to scrape vegetation, or to tear food into tiny bits. A snail in an aquarium, for example, uses its radula to scrape algae from the aquarium's glass wall.

All mollusks have a specialized digestive tract, beginning with a mouth and ending with an anus. All have one or two kidneylike **nephridia,** tubular excretory organs that remove nitrogen-containing wastes from the coelomic fluid and regulate water and salt concentration.

All mollusks also have a circulatory system with a heart that receives oxygen-carrying blood from the gills and pumps it to other body tissues. Unlike humans and other vertebrates, most mollusks have open circulatory systems, in which the heart pumps blood through spaces between the tissues rather than exclusively through blood vessels. The only exceptions are the cephalopods, which have a closed, well-developed circulatory system.

Mollusks reproduce sexually. Usually, male and female gonads are in separate individuals, but some (such as land snails) are hermaphroditic (both male and female). In some species, including sea slugs and oysters, individuals change from male to female and back several times during a mating season.

Cross-fertilization is the rule, even among the hermaphrodites. Self-fertilization, however, does occur—an important adaptation for animals that sometimes move too slowly to find a mate. Among the land snails and slugs, fertilization is internal. In other mollusks, gametes are shed externally.

Mollusks are soft-bodied coelomates, with sophisticated organs for respiration, circulation, digestion, and excretion. All the mollusks share a similar body plan, including a foot, a mantle cavity, and a visceral mass. Most have shells and a rasping tongue called a radula. They reproduce sexually, and their embryos develop into a characteristic larval stage.

Zoologists divide the mollusks into seven or eight classes, of which we will briefly discuss three: the bivalves (which include clams, oysters, and mussels), the gastropods (which include snails, slugs, and abalones), and the cephalopods (which include squid, octopods, and the chambered nautilus).

Bivalves

Bivalves [Latin, *bi* = two + *valva* = part of a folding door] include clams, oysters, scallops, mussels, and other mollusks with paired shells. A ligament hinges the two shells (or valves) and a muscular "foot" protrudes from between them (Figure 23-18B). Two large muscles, called adductor muscles, pull the shells together. Although most bivalves lead sedentary lives attached to rocks, some move about a little more. A clam, for example, uses its foot to burrow into the sandy or muddy bottom of the sea or river. The most active bivalves are the scallops, which "swim" by using their adductor muscles to clap their shells together, simulating a human swimmer's "frog kick." A scallop's enormous adductor muscles, tender and delicately flavored, are familiar to seafood lovers (although some restaurants misleadingly give the name "scallops" to circular plugs of skate wing).

Bivalves such as clams, oysters, scallops, and mussels have two shells and usually lead sedentary lives.

Gastropods

The **gastropods** are the most diverse class of mollusks, with some 80,000 named species, including snails, whelks, conchs, limpets, abalone, and slugs. They have a distinctive twisted body and, except for the shell-less slugs, asymmetrical spiral shells.

Gastropods differ from all other mollusks in having a mysterious 180-degree counterclockwise twist of the body (viewed from above) that occurs during embryonic development. One result is that both the anus and the mantle cavity end up at the front end (anterior end) of the body, not far from the mouth (Figure 23-18C). Torsion may occur very rapidly, sometimes over the course of just a few minutes. Biologists have many hypotheses about what advantages torsion might confer on gastropods, but they have not yet agreed on an answer.

The shells of gastropods show great variation in form and color. Most gastropods are marine, but many inhabit fresh water or land. Unlike the quiet bivalves, gastropods are active creatures, equipped with a well-developed head and a sensitive nervous system. Gastropods are free living and consume plants, small animals, or decaying organic matter. A few species live as parasites.

Gastropods such as slugs, snails, and abalone are active animals with twisted bodies, single shells (or no shell), and well-developed heads.

Cephalopods

The **cephalopods,** which include nautiluses, squids, cuttlefish, and octopods, are the most active and intelligent mollusks (Figure 23-18D). Although thousands of extinct

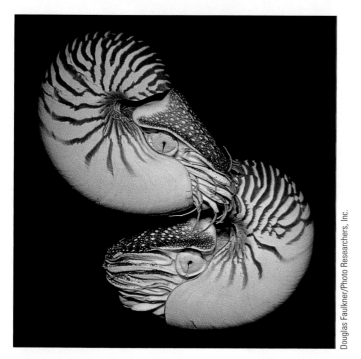

Figure 23-19
The chambered nautilus. The nautilus shell consists of a series of chambers arranged in a spiral. The nautilus occupies only the outermost chamber, and as the animal grows, it builds a new, larger chamber and moves in. The nautilus uses the remaining chambers to float or sink at will. In the same way that a balloonist makes a balloon rise by adding more hot air to the balloon, the nautilus injects gas into its shell to float higher in the water. To sink, the animal siphons off the gas. Today, only a few species of nautilus exist, all living in the western Pacific Ocean. But relatives of the nautilus, called ammonites, flourished for hundreds of millions of years, from the Devonian Period to the end of the Cretaceous Period.

cephalopods possessed well-developed shells, among the 600 living species only the nautilus possesses an external shell (Figure 23-19). Surrounding the mouth is a set of tentacles that have evolved from the foot of a cephalopod ancestor. Each species has a characteristic number of tentacles (or arms)—a squid has 10, an octopus 8 (hence its name), and a nautilus 80 or 90 tentacles. Cephalopods, which are all carnivorous, use their tentacles to capture prey, to pull themselves along the sea bottom, and to steer themselves in open water.

Considering the intellectual limitations of their relatives, the bivalves and the gastropods, cephalopods are amazingly intelligent. In a laboratory aquarium, an octopus can learn to distinguish objects of different shapes and even to find its way through a maze. Octopods are very good at escaping, and they think nothing of disassembling an aquarium to do so. Presented with stones or small bricks, an octopus can build a protected enclosure, hide behind it, and bob up and down to look around. An octopus also shows much planning and patience in stalking and luring its prey. Cephalopods engage in complex mating rituals (Figure 23-20).

Figure 23-20
A male squid caresses his mate with his many arms.

An octopus's interesting behavior is controlled by the most sophisticated brain of any invertebrate. Information about the outside world comes from touch receptors on the tentacles, as well as from large and complex eyes. When we consider that octopus eyes evolved independently of vertebrate eyes, it is remarkable how similar the eye of an octopus is to our own (Figure 23-21). This similarity is a result of convergent evolution.

Cephalopods respond to their environments not only with directed movements toward prey and away from predators, but also with a complex set of color changes. Most cephalopods have specialized skin cells, usually with three different pigments. The brain controls these cells in immediate response to the environment. An alarmed octopus, for example, may suddenly acquire dark stripes or spots, which presumably confuse potential predators. Yet another escape trick is to squirt a trailing cloud of dark ink. Some cephalopods, especially squids, can move amazingly fast by jet propulsion, squirting water from their mantle cavities.

Cephalopods are mollusks in which the foot has evolved into tentacles. They are predatory, intelligent, and active.

What Do All Annelids Have in Common?

Although earthworms and other annelids are common, they are not nearly as diverse as the mollusks. The 12,000 species of annelids live as both predators and scavengers in salt or fresh water and on land. They range in size from less than a millimeter to over three meters. They may be red, pink, green, brown, or purple; plain, striped, or spotted.

Annelids have long, segmented bodies consisting of identical (or nearly identical) sections called **metameres**. Segmentation (which also occurs in arthropods and verte-

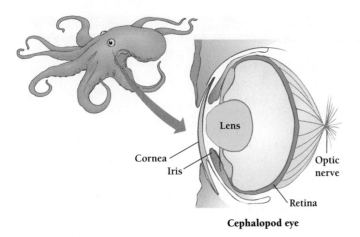

Cephalopod eye

Cornea
Iris
Lens
Optic nerve
Retina

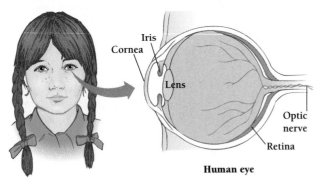

Human eye

Iris
Cornea
Lens
Optic nerve
Retina

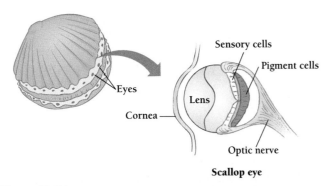

Scallop eye

Eyes
Sensory cells
Pigment cells
Lens
Cornea
Optic nerve

Figure 23-21

Convergent evolution of vertebrate and mollusk eyes. The cephalopod eye has long astounded biologists for its resemblance to the vertebrate eye. Cephalopod eyes can also be enormous. In 1933, a 70 feet-long squid whose eyes measured 15 inches across washed up on a New Zealand beach. Scallops have simple eyes that cannot focus. However, they have many of these eyes, which are probably useful for detecting the shadows of sea stars and other predators.

millimeter to over three meters. They may be red, pink, green, brown, or purple; plain, striped, or spotted.

Annelids have long, segmented bodies consisting of identical (or nearly identical) sections called **metameres.** Segmentation (which also occurs in arthropods and vertebrates) allows animals to achieve larger sizes by repeating an already successful organizational plan. Injuries to individual metameres are less likely to be lethal, since other segments perform the same functions. In addition, the segments can move independently, which gives the animal more flexibility.

Like mollusks, annelids have specialized excretory organs called *nephridia*. Like us, annelids have a closed circulatory system, with blood contained entirely within vessels. Several vessels carry blood from one end of the body to the other, with many fine capillaries within each segment. Some of the larger vessels serve as hearts and pump the blood along its complicated circuits. Annelids have no gills or lungs, however. Gases move directly through the skin to the blood.

Zoologists divide annelids into three classes: (1) polychaetes are free-living bristle worms, most of which are marine; (2) oligochaetes, which include the earthworms and related forms, live in fresh water or salt water or on land; and (3) hirudinea, or leeches, are mainly freshwater parasites that suck blood.

Annelids have segmented bodies, consisting of identical or nearly identical sections, and closed circulatory systems.

Polychaete Worms

The exotic and beautiful polychaete worms include many unusual and arresting species, including clamworms, plumed worms, scale worms, peacock worms, and sea mice (Figure 23-22). Most zoologists think that the earliest annelids were marine polychaetes. Each polychaete segment contains a pair of leglike paddles, called **parapodia,** which the worm uses to swim, crawl, or burrow, as well as to respire. On each parapodium is a bundle of bristles called **setae,** which give the polychaetes their name [Greek, *polukhaites* = many hairs, or bristles]. A polychaete worm's head is well developed and contains a variety of sense organs, including two to four pairs of eyes. Unlike oligochaetes (earthworms), polychaetes have separate sexes. A polychaete is always either a female or a male.

Polychaetes are primitive marine annelids. Each of a polychaete's segments has a pair of leglike parapodia with setae.

Oligochaetes

Oligochaetes include the common earthworms, the most familiar of the annelids. Each oligochaete segment, in contrast to that of the polychaetes, lacks parapodia altogether and contains just four pairs of bundled setae, leading to the class's name [Greek, *oligo* = few].

Oligochaetes are masterpieces of segmentation. An earthworm, for example, consists of 100 to 175 nearly identical segments, with a few altered segments at the front and

A.

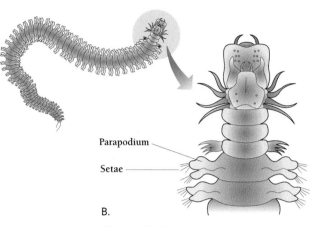

Parapodium

Setae

B.

Figure 23-22

Polychaetes. A. Many species of polychaetes—the tube worms—live in hidden places on the shallow sea bottom—under rocks, in burrows, or inside sponges, mollusks, or other animals. Some use their own secretions to build tubes where they live their entire lives. B. Polychaetes are annelids with legs, or parapodia, as shown here.

Figure 23-23

Earthworms. A single square meter of meadow soil may contain thousands of earthworms. As an earthworm moves through soil, it swallows particles of soil. The soil passes through the digestive tract, where any organic matter is ground finely and absorbed. Undigested clay, sand, and other matter pass through the digestive tract and out the anus as small cylindrical "castings." A single worm eats its own weight in soil every day. Because earthworms are so common, they serve to keep soil mixed, which greatly benefits other soil animals and plants.

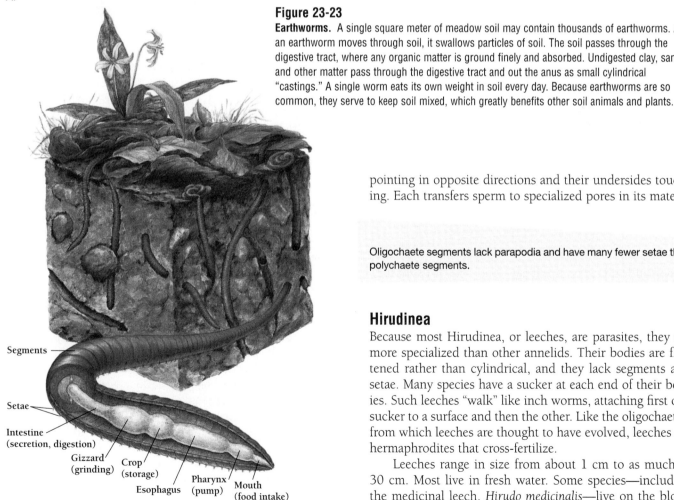

Segments

Setae

Intestine
(secretion, digestion)

Gizzard
(grinding)

Crop
(storage)

Esophagus

Pharynx
(pump)

Mouth
(food intake)

pointing in opposite directions and their undersides touching. Each transfers sperm to specialized pores in its mate.

Oligochaete segments lack parapodia and have many fewer setae than polychaete segments.

Hirudinea

Because most Hirudinea, or leeches, are parasites, they are more specialized than other annelids. Their bodies are flattened rather than cylindrical, and they lack segments and setae. Many species have a sucker at each end of their bodies. Such leeches "walk" like inch worms, attaching first one sucker to a surface and then the other. Like the oligochaetes, from which leeches are thought to have evolved, leeches are hermaphrodites that cross-fertilize.

Leeches range in size from about 1 cm to as much as 30 cm. Most live in fresh water. Some species—including the medicinal leech, *Hirudo medicinalis*—live on the blood

Extreme Biology   The Origin of Feces

Did the evolution of the one-way digestive tract fuel the Cambrian explosion? In a controversial new theory, geochemists have suggested that the development of one-way digestive systems by metazoans provoked drastic changes in the deep ocean, changes that allowed metazoans to colonize the bottom of the sea in an unparalleled adaptive radiation.

Until the Cambrian explosion, some 540 million years ago (mya), life consisted almost exclusively of single-celled bacteria and algae. As photosynthetic algae proliferated, they pumped increasing amounts of oxygen into the atmosphere. To see how this change in the composition of the atmosphere affected ancient life, a group of geochemists led by Graham Logan measured the chemical signature of fossilized organic matter in ancient ocean deposits.

To their surprise, the researchers found that at the beginning of the Cambrian, a dramatic change in the proportion of different carbon isotopes occurred. Before 590 mya, the sea floor deposits were relatively high in ^{13}C, a heavy isotope of carbon. Sometime between 590 and 540 mya, however, the deposits became lower in ^{13}C and higher in ^{12}C. This switch in "isotope signature" coincided closely with the flowering of multicellular life known as the Cambrian ex-

plosion. The researchers could not help asking, Were the two events related?

It seemed possible. Organisms tend to have higher ratios of $^{12}C{:}^{13}C$ than the atmosphere. During photosynthesis, plants incorporate into sugars and other new compounds a higher proportion of ^{12}C than ^{13}C. (This is because ^{12}C is lighter than ^{13}C and diffuses more quickly.) The result is that ^{12}C accumulates in photosynthetic organisms and all the organisms that eat them. Organic matter therefore has more ^{12}C than the atmosphere.

But nearly all organisms respire—a process that breaks down organic compounds and releases carbon dioxide. Respiration puts the ^{12}C back into the atmosphere. If photosynthesis and respiration are balanced, the ratio of ^{12}C to ^{13}C is also balanced. In that case, researchers would not see an excess accumulation of ^{12}C in ocean floor sediments.

But Logan and his colleagues had an idea that would account for the change in ratios. Before 590 mya, the researchers hypothesized, bacteria in the upper layers of the ocean decomposed nearly all organic waste, releasing organic ^{12}C back into the atmosphere. Early deep-sea sediments, then, were relatively high in ^{13}C.

Huge numbers of bacteria would have filled the upper layers of the ocean, and

their respiration would have depleted ocean waters of oxygen. The surface would have remained oxygenated, but the water near the bottom would have contained very little oxygen. Few organisms could have survived there.

But what if the bacteria were suddenly prevented from decomposing organic matter? Logan and his colleagues suggested that the evolution of a one-way digestive tract made all the difference. Animals with a one-way gut can package wastes in concentrated packets (feces), which fall to the ocean floor before surface-dwelling decomposers can absorb and digest them. With less decomposition occurring in the upper layers of the ocean, the overall oxygen content of the water would have risen. Ultimately, argue the researchers, oxygen would have diffused into the deeper waters of the ocean—allowing colonization by oxygen-requiring metazoans. That, the researchers say, paved the way for the Cambrian explosion.

Did the first fecal pellets fuel the Cambrian explosion? Many researchers remain skeptical of this tentative train of hypotheses. But the idea that the origin of all the major animal phyla depended on the evolution of feces has a certain appeal.

---

of humans and other mammals. Until the 19th century, leeches were used to treat such conditions as asthma, rheumatism, and drunkenness. Physicians (or barbers, who traditionally performed leech therapy as well as cutting hair) applied leeches, which sucked up to three pints of blood. When the patient fainted from blood loss, the physician removed the leeches. Today, physicians use leeches, in a more restrained manner, to drain excess blood from surgical wounds.

A leech's impressive capacity to consume blood depends on powerful muscles of the pharynx (the anterior end of the digestive tract), on sharp teeth, and on a chemical that prevents blood clotting, called an anticoagulant. Physicians use leech anticoagulant to treat heart attack, stroke, and cancer.

23.5 The Ecdysozoan Protostomes

The Ecdysozoa share a peculiar trait: they molt, shedding their outer skins or exoskeletons sometime during their lives. Some of these animals shed repeatedly. The name comes from the insect hormone ecdysone, which promotes, for example, the metamorphosis of a caterpillar into a butterfly. The Ecdysozoa include the nematode worms; the velvet worms (onychophorans); the Nematomorpha, or horsehair worms; and the Gastrotricha, which are microscopic, bottle-shaped animals that live in freshwater or marine environments; the tardigrades (water bears); and all of the arthropods—the insects, spiders, crabs, and scorpions. All of these groups of animals are interesting and even wonderful, but we will discuss just two large groups, the nematode worms and the arthropods.

Nematodes (Roundworms)

The 12,000 named **nematodes**—also called roundworms—are slender, cylindrical, and unsegmented. They range in length from a few millimeters to nearly a meter. Most are free living, however, and quite small. Free-living nematodes in the soil are actually aquatic animals, living in the film of water that surrounds each soil particle. They help to aerate the soil and circulate minerals and organic matter. Nematodes are among the most numerous of animals. A rotting apple in an orchard may contain 100,000 nematodes. An acre of soil contains billions of the little worms.

Many nematodes are parasites. They infect virtually all plants and animals. In fact, it has been said that if all organisms except the nematodes suddenly disappeared, a ghostly image of the world's former occupants would remain in the form of billions upon billions of nematode worms.

Nematodes lack a proper coelom. Instead, they have a cavity called a **pseudocoel** [Greek, *pseudo* = false, + *koilos* = hollow]—"false" because it lacks the mesodermal lining of a true coelom. The pseudocoel plays two important roles: (1) it is a distribution route for nutrients, gases, and wastes; and (2) it serves as a hydrostatic skeleton, a support for the body that depends on the rigidity of enclosed fluid under pressure.

Because nematodes lack a circulatory system, fluids move about the body within the pseudocoel. These fluids push against the walls of the pseudocoel and thereby stiffen the whole body in the same way that water stiffens a hose. The nematode's muscles work against this support in the same way that the muscles of a vertebrate pull against a bony skeleton. The hydrostatic skeleton allows the worm's body to return to its original shape after muscle contraction has stopped.

The nervous system of a nematode is more complex than that of an acoelomate. A free-living nematode has sense organs that detect movement and chemicals, nerve cords that run the length of its body, and a primitive brain made up of a ring of nerves just behind its mouth.

Nematodes reproduce sexually, with male and female usually separate. In most species, fertilization is internal and therefore requires copulation. Some species are hermaphroditic, and some can even fertilize themselves. Still others are parthenogenic, meaning that females reproduce without males. The unfertilized eggs develop into normal embryos and adults, just as a diploid zygote would. Nematodes are prolific breeders: a single female can produce more than 25 million eggs in her lifetime.

Nematodes (roundworms) are tiny, cylindrical worms that live everywhere in great numbers. They have a fluid-filled pseudocoel that acts as a hydrostatic skeleton and circulatory system, and they have sense organs and a primitive brain.

What Traits Do the Arthropods Share?

The **Arthropoda,** most of which are insects (some three-quarters of a million species), include more named species than any other phylum of organisms. Including insects, spiders, scorpions, and shrimps, arthropods are the most successful phylum on the planet. Named arthropod species number 1,113,000, and zoologists estimate that many more species remain to be described and named. The greatest number of species live in the tropics, but a student could probably collect hundreds (or even thousands) of species of arthropods on any college campus.

Besides being diverse, arthropods are also numerous. Zoologists estimate that the world today contains some 10^{18} (a billion billion) individual arthropods, hundreds of millions for every human being (a figure that might not startle a mosquito-bitten hiker). In a temperate climate, one square kilometer may contain 20 million arthropods. In the tropics, there are even more.

The enormous success of arthropods depends on both external and internal adaptations. All arthropods share two major external adaptations—an exoskeleton secreted by the cells of the epidermis and jointed limbs.

Jointed limbs, which give the arthropods their name [Greek, *arthron* = joint, + *podus* = foot], allow them to move agilely, despite the rigidity of their exoskeletons. Like the annelids, arthropods have segmented bodies. Indeed, both phyla are probably descended from the same segmented ancestor. Some, like the centipedes and millipedes, have legs on nearly every segment. But many of the legs on an arthropod have become adapted for other functions. Over millions of years of evolution, legs have changed into mouthparts, antennae, gills, pincers, claws, and egg depositors.

These specialized appendages often develop from fused segments, called **tagmata** [singular, tagma]. For example, the anterior segments of all arthropods are fused into a highly organized head. Other tagmata include the middle portion of an arthropod, called the **thorax,** and the rear portion, called the **abdomen.**

The **exoskeleton,** a hard external supporting framework, consists of layers of chitin and protein. It helps to protect the animals from predators and parasites and provides mechanical support. The exoskeleton also slows water loss, a feature that has allowed the arthropods to flourish away from water and dominate animal life on land since the end of the Devonian Period, more than 350 million years ago.

An exoskeleton has drawbacks, however. Because a hard exoskeleton prevents an animal from growing, arthropods periodically must shed their exoskeletons and secrete new, larger ones in a process called **molting.** The process of molting is dangerous. After an arthropod sheds its old armor and the new one is still hardening, the animal is highly vulnerable to predators. (Soft-shelled crab is a delicacy in many an East Coast restaurant.)

One solution is for an arthropod to do all its growing as a soft larva (as a caterpillar, for example). Once the animal

A.   Eggs   Larva   Pupa (inside cocoon)   Adult

B.   Eggs   Young nymph   Nymph   Adult

Figure 23-24
Complete and incomplete metamorphosis. A. During complete metamorphosis an insect, such as this isabella moth, hatches from an egg and develops as a legless larva that looks more like a worm than a mature insect. During development, a larva may go through many successive stages, called *instars,* before settling into a usually inactive stage called a *pupa,* or *chrysalis.* The pupa is usually encased in a protective cover, called a *cocoon.* The metamorphosis from larva to pupa and pupa to adult often involves dramatic changes in both external and internal structures. B. In incomplete metamorphosis, these changes occur more gradually. In the harlequin bug (shown here) early stages of development, called nymphs, look much like the adult, and molting functions mostly to increase size rather than to change form. Even in these metamorphoses, however, the juvenile forms lack wings.

has reached its adult size, it develops into the adult form in a process called **metamorphosis.** Some insects undergo gradual metamorphosis in stages that are not that different from one another (Figure 23-24). Others, such as flies, butterflies, beetles, and bees, undergo a complete transformation from a grub, caterpillar, or maggot to a flying insect. Metamorphosis and molting divide the life cycle of an insect into many distinct stages, each of which may have different food and lifestyle. In butterflies, for example, the caterpillar generally eats plants, while the adult butterflies drink nectar from flowers. In some arthropods, adults do not feed at all; their only role is to mate and die.

The internal anatomy of an arthropod is similar to that of an annelid. Both arthropods and annelids have a tubular gut that stretches from mouth to anus. The arthropod coelom is less prominent than that of an annelid and consists mostly of a cavity that encloses the reproductive organs.

The circulatory systems of arthropods are open. The heart pumps blood through the spaces of the body, and the blood returns through a series of one-way valves. The respiratory systems of arthropods vary according to taxonomic class and source of oxygen. Aquatic arthropods may have gills, with which they extract oxygen from the water in which they live. Spiders and some other air-breathing arthropods use stacks of modified gills, collectively called a **book lung,** to pull oxygen directly from the air. (All animals increase surface area for gas exchange in one of two ways. They either fold a surface outward to form gills or fold a surface inward to form lungs, as we discuss in Chapter 39.)

Most terrestrial arthropods, including the insects, however, have neither lungs nor gills. Instead, they take in air through regulated openings in the body wall called **spiracles.** Air passes from the spiracles to special ducts called **tracheae.** Each trachea branches into tiny tracheoles, which deliver oxygen to individual cells throughout the body.

Reproduction in arthropods is almost always sexual. Male and female are usually separate animals, though some species are hermaphroditic. For terrestrial arthropods, fertilization must be internal, since there is no external water through which sperm can swim. Millipedes, for example, accomplish fertilization in a multilegged embrace. Elaborate courtship rituals help males and females of the same species recognize one another. Male fruit flies, for example, do a little dance, during which the female decides whether she wants to mate with him. In an experiment where fruit flies were kept in the dark for several generations, however, the males abandoned the dance—which the females could not see anyway.

Because arthropods that are predaceous do not always distinguish between mates and meals, copulation can be a dangerous matter. Such animals may have especially protracted courtship rituals, which not only help them to recognize one another but also may serve to prevent one partner from eating the other.

Excretion of nitrogen-containing waste in most terrestrial arthropods depends on unique organs called **Malpighian tubules** (Chapter 40). These tubules absorb fluid from the blood and convert waste molecules into insoluble crystals of uric acid or guanine. The Malpighian tubules also reabsorb and recycle salts and water, so that little water is wasted.

The nervous systems of arthropods are often complex. Many arthropods can move, run, and fly quickly, and many have complicated sexual and social behaviors. The nervous system consists of a double chain of ganglia that runs along the lower surface of the body. At the head, the chain curls upward to form a brain—three pairs of fused ganglia. The sense organs—especially the eyes—of arthropods are more sophisticated than those of annelids. While some arthropods (such as the spiders) have simple eyes, most arthropod species have **compound eyes**, made up of numerous simple light-detecting units, called **ommatidia** [singular, ommatidium].

Arthropods include the greatest number of species of all animal phyla. External adaptations of the arthropods include the exoskeleton and jointed limbs. Internal adaptations of arthropods include circulatory, respiratory, reproductive, excretory, and nervous systems.

How Do Biologists Classify Arthropods?

The specialization of appendages, especially in the first few segments, is the basis of classification of living arthropods into three subphyla. (Extinct arthropods such as the trilobites may have their own subphyla.) In one subphylum—the **Chelicerates**—the first pair of appendages are mouthparts called chelicerae. The chelicerae serve as pincers or fangs, often associated with poison glands (Figure 23-25A).

In the other two subphyla, **Uniramia** and **Crustacea**, the first pair (or the first few pairs) of appendages are antennae, and the next pair are jaws, or mandibles (Figure 23-25B and C). These jaws differ from ours, which move up and down. Instead, an arthropod's mandibles crush and grind as they move from side to side. The Crustacea, which include water fleas, shrimps, lobsters, and crabs, are almost entirely aquatic, while the Uniramia, which include the insects, the centipedes, and the millipedes, are mostly terrestrial. The crustaceans all have branched (or biramous) appendages, while the Uniramia all have unbranched (or uniramous) appendages.

Biologists classify living arthropods into three subphyla according to the specialization of the appendages, especially the first two.

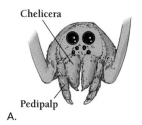

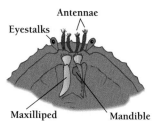

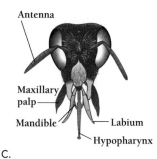

Figure 23-25

The mouthparts of arthropods. A. Chelicerae and pedipalps characterize chelicerates such as spiders and scorpions. B. Crustaceans such as this ghost crab have antennae, mandibles, and biramous appendages. C. The uniramia, including insects, millipedes, and centipedes, have antennae, mandibles, and uniramous appendages.

Chelicerates

The chelicerates are the only arthropods without antennae or jaws. The most anterior pair of appendages are the pincerlike chelicerae, and the second are the pedipalps, which perform different functions in the different classes (Figure 23-25A). The most familiar chelicerates are the horseshoe crabs and the spiders, of which there are some 35,000 named species. The spiders are just 1 of 11 orders of arachnids; arachnids are, in turn, just one of three classes of the chelicerate subphylum.

Spiders, mites, and other arachnids have six pairs of appendages (four pairs of legs, plus chelicerae and pedipalps, but no antennae). Like most other arachnids, spiders are predatory, capturing and eating insects and other small animals. Some spiders hunt and pursue their prey. Others trap their prey in elaborate silk webs. One spider genus, *Mastophora,* "fishes" for its prey with a silk line, tipped with a bit of glue. Spiders also use silk to make balloons (for making long trips), droplines (for making a quick exit), egg cases (for protecting their offspring), shrouds (for storing dead prey), and gifts from males (for luring attractive female spiders).

Spiders and other arachnids have no jaws and cannot chew. They digest only liquefied food. Once a spider has trapped its prey, it injects a paralyzing poison through its chelicerae, tears and grinds a bit of its prey with chelicerae and pedipalps, then regurgitates a potion of digestive enzymes into the body of its prey, which liquefies the prey. The spider then sucks its meal into its gut, where it completes digestion.

The scorpions are the oldest order of arachnids, extending back at least to the Silurian Period, about 425 million years ago. Scorpions differ from most chelicerates in having clearly segmented abdomens. This allows the scorpion to curl its abdomen, holding its stinger aloft. Other arachnids include the distinctive daddy longlegs (or harvestmen) and the mites, of which there are some 30,000 named species and perhaps a million more as yet unnamed. Mites, including the ticks, are usually less than 1 mm long, although the largest are 2 cm long. They live almost everywhere, eating plants, fungi, and animals.

Figure 23-26
Krill. Small marine crustaceans like these make up krill, the marine soup that feeds the great baleen whales.

William E. Ferguson

Chelicerates—such as spiders, mites, and horseshoe crabs—have no antennae and no jaws.

Crustaceans

The crustaceans include some 35,000 species of lobsters, crabs, shrimps, barnacles, and crayfish, as well as less conspicuous species such as water fleas, fairy shrimp, pillbugs, and krill (Figure 23-26). Crustaceans usually have three pairs of chewing appendages (including the mandibles), many pairs of legs, and, unlike other arthropods, two pairs of antennae. Their legs are biramous, or split into two parts at the tip.

The most numerous crustaceans are the copepods. These tiny crustaceans, each a few millimeters long, are a major component of plankton, the microscopic plants and animals that float near the surface of the oceans. Copepods eat marine algae and are a major food in the diet of whales, nurse sharks, and other filter-feeding marine animals. Next to nematodes, copepods are the most abundant animals on Earth.

The decapods, which include lobsters and crabs, are the largest crustaceans. Some lobsters may grow to be 60 centimeters long, and the Japanese spider crab is more than three meters across. The decapods [Greek, *deka* = ten + *pous* = foot] have five pairs of legs.

The oddest of the crustaceans are the barnacles. One wit described these creatures as "nothing more than a little shrimplike animal, standing on its head in a limestone house and kicking food in its mouth." Unlike other crustaceans, which are active, a barnacle lives its adult life in a single place, submerged in salt water and attached (by its head) to a submerged rock, piling, boat bottom, or even a whale. Barnacles use their appendages to direct passing food particles into their mouths. They secrete calcium-containing plates, which are firmly cemented to the underlying surface.

Crustaceans—such as lobsters, crabs, and shrimps—have jaws, biramous appendages, and, usually, two pair of antennae.

Uniramia

The Uniramia are the most diverse subphylum of arthropods, including 750,000 insects, 10,000 species of plant-eating millipedes, and 2,500 species of carnivorous centipedes. Almost all Uniramia take in air through tracheae, use Malpighian tubules to rid themselves of nitrogenous waste, and have unbranched (uniramous) appendages.

The myriapods [Greek, *myrioi* = countless + *pous* = foot], which include the millipedes [Latin, *mille* = thousand], the centipedes [Latin, *centum* = hundred], and two less familiar classes, are the most clearly segmented arthropods.

In insects, segmentation is clearest in the larvae, but adults are segmented as well. The body consists of three tagmata (fused segments): the head, the thorax, and the abdomen. The segments of the head are completely fused. The thorax consists of three segments, each of which carries a pair of legs, and, in many insects, one or two pairs of wings. The abdomen consists of 12 or fewer segments, none of which carries legs or wings.

Entomologists (scientists who study insects) divide the class Insecta into about 30 orders. These orders differ both in adult morphology and in their pattern of development. Some insects (for example, lice, fleas, and silverfish) have no wings, some (flies and mosquitoes) have one pair of wings, some (dragonflies) have two similar pairs of wings, and some (beetles) have two pairs, each with a distinctive structure. The front wings of beetles are hard shells that serve to protect the delicate hind wings, which alone are responsible for flight.

Uniramia have one pair of antennae and one pair of mandibles.

Zoologists divide the animal kingdom into the Parazoa (sponges) and the Eumetazoa (all the rest). The eumetazoans fall into two major groups, the Radiata and the Bilateria. The bilateral animals, in turn, can be divided into the protostome animals and the deuterostome animals. We have briefly surveyed many of the phyla within these groupings. These animals range from the simple sponges and jellyfish to animals as complex as insects or as alert as octopods. In the next chapter we will see that the deuterostome coelomates are also diverse, ranging in complexity and intelligence from sand dollars to humans.

Key Concepts

- An animal is a multicellular organism that eats other organisms and develops from an embryo.
- Zoologists divide animals into two subkingdoms—the Parazoa (sponges) and Eumetazoa (all other animals). The members of the Eumetazoa are grouped according to their symmetry (radial or bilateral).
- The bilateral animals can be grouped into the deuterostomes and protostomes. The protostomes are further divided into two main groups, the Lophotrochozoa (many of which are worms and mollusks) and the Ecdysozoa (all of which shed their skins or exoskelotons).
- Arthropods include the greatest number of species and the greatest numbers of individuals of all animal phyla.

Summary with Key Terms

What is an animal?

An animal is a multicellular, heterotrophic organism that develops from an embryo. Most animals reproduce sexually, producing a single-celled **zygote** from the fusion of an egg and a sperm. The developing embryo goes through a stage called a **blastula,** which is a hollow sphere made up of identical cells. The blastula folds into three tissue layers surrounding a central cavity that becomes the intestinal tract, or gut. The three cell layers are the **endoderm, ectoderm,** and **mesoderm.** These three layers develop into different types of **tissues** and **organs.** Tissues such as the **epithelial** and **connective** tissues are composed of specialized cells and their **matrix.** Organs are composed of two or more types of tissues.

What characteristics distinguish the sponges (Parazoa) from other animals (Eumetazoans)?

Zoologists divide animals into two subkingdoms, the **Parazoa** and **Eumetazoa.** The Parazoa consist mainly of the Porifera, or sponges, which have no symmetry and no tissues organized into organs. Sponge cells lie in three layers (epidermal cells, **mesenchyme** cells, and flagellated cells) surrounding a central cavity. Sponges are mostly **sessile.**

What characteristics distinguish the radial animals from the bilateral animals?

The eumetazoans include the **radially symmetrical** cnidarians and ctenophores and all the rest of the animals, which have **bilateral symmetry.** Like the sponges, the radial animals lack a **coelom.** Animals with bilateral symmetry generally have distinctive **anterior** and **posterior** ends. Animals with a coelom are called **coelomates.** Those with no coelom are called **acoelomates.** Cnidarians and ctenophores are radially symmetrical acoelomates. **Pseudocoelomates** (mostly nematodes) have an organ-containing cavity, but without the mesodermal lining of a true coelom.

Cnidarians, which include jellyfish, hydras, corals, and sea anemones, have tissues but no organs. They have two tissue layers sandwiched around a layer of **mesoglea.** They may have two body plans: **polyps** (which resemble cylinders) and **medusae** (which resemble bells). Both types have specialized stinging cells, called **cnidocytes,** which fire tiny barbs called **nematocysts. Ctenophores** (comb jellies) resemble the cnidarians but also have distinctive features, including bands of cilia.

What characteristic distinguishes the protostomes from the deuterostomes?

The bilateral animals (bilateria) are divided into the **protostomes** and the **deuterostomes,** according to whether the embryonic **blastopore** develops into the mouth (protostomes) or the anus (deuterostomes). The majority of bilateria are protostomes, including the annelid worms, mollusks, and arthropods. However, two major phyla of bilateria are deuterostomes—the echinoderms and the chordates, covered in Chapter 24.

The protostome animals can be divided into two main groups, the Lophotrochozoa and the Ecdysozoa. The **Lophotrochozoa** include many soft-bodied animals, including the flatworms, ribbon worms, mollusks, and annelids. **Platyhelminthes** (flatworms) are the simplest animals that have heads and organs. They consist of three layers of cells. They have a one-way gut and no circulatory systems. But their flat shape allows gases and molecules to diffuse rapidly. The flatworms include three classes—the **Turbellaria, Trematoda,** and **Cestoda.** The first class are free-living flatworms, while the other two include only parasites. All parasites are specialized for rapid and prolific reproduction in their hosts. The tapeworm consists of a **scolex** and many **proglottids. Nemertea** (ribbon worms) have a two-ended digestive system, a circulatory system, and a **proboscis.**

Mollusks (which include snails, slugs, clams, oysters, squids, and octopods) are unsegmented animals that typically have shells. Mollusks all have a **visceral mass**, a muscular **foot**, a **mantle**, a **mantle cavity**, a **radula**, and one or two kidneylike **nephridia**. Many shelled mollusks have an **operculum**. There are seven classes of mollusks, of which we discussed three: **gastropods, bivalves, and cephalopods.** Gastropods have single shells, well-developed heads, and twisted bodies. Bivalves have two shells and most lead sedentary lives. Cephalopods have forward-pointing tentacles instead of a foot.

Annelids (which include earthworms, segmented marine worms, and leeches) have segmented bodies that consist of identical or nearly identical **metameres.** Annelids have closed circulatory systems, with blood contained entirely within vessels. Annelids are divided into three classes: polychaetes, oligochaetes, and hirudinea. Each segment of a polychaete (or bristle worm) contains a pair of paddlelike projections, called **parapodia,** each with a bundle of bristles, or **setae.** Oligochaete segments lack parapodia. Hirudinea (leeches) are more specialized than other annelids and lack obvious segmentation.

The second major group of protostome animals are the **Ecdysozoa,** which all molt. They include the nematodes and arthropods, as well as many other interesting phyla. **Nematodes** (roundworms) are cylindrical and unsegmented, and have a one-way digestive system with both a mouth and an anus. The prominent **pseudocoel** of nematodes serves as a distribution route for nutrients, gases, and wastes and as a hydrostatic skeleton.

Arthropods include the greatest number of species and the greatest numbers of individuals of all animal phyla. External adaptations of the arthropods include an **exoskeleton** and jointed limbs. Arthropods have many specialized appendages that develop from **tagmata,** fused segments that form the head, the **thorax,** and the **abdomen.** The specialization of appendages, especially in the first few segments, is the basis of classification of arthropods into three subphyla: **chelicerates, crustaceans,** and **uniramia.** Because an exoskeleton prevents an animal from growing, many arthropods periodically **molt.** Other arthropods do all their growing in larval form, then change into adults during **metamorphosis.** Internal adaptations include circulatory, respiratory, reproductive, excretory, and nervous systems. Arthropods breathe by means of a **book lung,** or **spiracles** and **tracheae,** and they regulate excretion of nitrogenous wastes by means of **Malpighian tubules.** Some arthropods have simple eyes, but most have **compound eyes,** made up of numerous **ommatidia.**

Review and Thought Questions

Review Questions

1. How is an animal different from a plant or a protist? Why are sponges grouped with animals? What animal features do sponges and cnidarians lack?

2. What criteria do zoologists use to classify the Eumetazoa?

3. What is a coelom and how is it different from a pseudocoel?
4. What specializations set parasites apart from other members of their phyla?
5. What is the difference between a protostome and a deuterostome?
6. Members of which class of mollusks would make the most interesting pets? Why?
7. Which phylum of protostomes includes the greatest number of species of any phylum of organisms?
8. What specific characteristics of a grasshopper tell you that it belongs in the class Insecta?
9. How do millipedes differ from centipedes? Which class includes more species?

Thought Questions

10. What is progress? In what sense is evolution progressive?
11. Zoologists use the following traits (among others) to classify the animals: number of embryonic cell layers, presence or absence of a true coelom, and symmetry. Do you think that such characteristics truly reflect relatedness between different groups of organisms? If you were going to classify the animals, what criteria would you consider using?

BiologyNow Resources

Biology ⓔ Now™

Active Figures
23-4:   Protostomes
23-15: The life cycle of another kind of fluke

Preparing for an exam? Take a diagnostic test on your BiologyNow CD-ROM.

Online materials relating to this chapter are at:
http://biology.brookscole.com/AAL3

About the Chapter-Opening Image
The succulent katydid of Peru is a classic and beautiful invertebrate, the subject of this chapter.

24

Deuterostome Animals: Echinoderms and Chordates

Key Questions

- How are protostomes and deuterostomes alike and different?

- What traits distinguish the chordates and what traits distinguish the vertebrates?

- What environmental conditions stimulated the evolution of the swim bladder, the lung, and the muscular limbs in early fishes?

- How did the amniotic egg allow vertebrates to colonize the land?

Paraskevas Photography

Are We Upside Down?

In 1830, two French anatomists staged a public debate before the prestigious French Academy of Sciences on a topic so fundamental that historians of science have returned to it again and again. How many basic body types do animals have?

On one side, the politically powerful French anatomist and paleontologist Baron Georges Cuvier (1769–1832) argued that all animals should be divided into four distinct types. On the other side was Étienne Geoffroy Saint-Hilaire (1772–1844), a renowned comparative anatomist, who argued that animals were of just one type, not four.

Geoffroy was not a man to be intimidated even by a figure as powerful as Cuvier. During the French Revolution, Geoffroy had risked his life to save teachers and colleagues from execution. And when in 1807 Napoleon ordered Geoffroy to obtain collections from Portugal's museums by any means possible, Geoffroy slyly traded specimens from French museums instead of seizing the collections by force.

Geoffroy had long maintained that all vertebrates—mammals, birds, reptiles, and amphibians—had the same basic "body plan," a description of their symmetry and the arrangement of their tissues. In all vertebrates, for example, the backbone is at the back (**dorsal**) and the heart lies along the chest (**ventral**). The unity of the vertebrates was an idea that Cuvier and his followers had accepted.

But in 1830, Geoffroy Saint-Hilaire took his argument a huge step further, proposing that chordates (vertebrates and their relatives) have the same body plan as arthropods. In essence, he said, an insect and a mammal are similar. The single main difference, he said, was that we vertebrates are turned upside down: where a grasshopper's nerve cord runs along its belly, a human's nerve cord runs along its back (Figure 24-1). And where a human heart beats against the chest wall, a grasshopper's heart lies up along its back. As one modern biologist has put it, "If you lay down on your back and waved your arms, you would be doing what insects do when they walk."

Geoffroy's hypothesis, which he called *unité de plan* (singleness of plan), was provocative. Not only was it in direct opposition to Cuvier's theory of separate, unrelated groups, it also implied evolution. Cuvier firmly believed in the fixity of species and had publicly ridiculed Lamarck and other proponents of evolution. According to Geoffroy's hypothesis, however, humans and all other vertebrates were each a simple variation on a basic invertebrate body plan. Geoffroy's idea could be interpreted to mean that vertebrates had evolved from invertebrates.

For Geoffroy, *unité de plan* was a serious scheme, based on his dissections of thousands of animals. But during the debate, the pugnacious Cuvier made *unité de plan* look like a joke, listing, one by one, all of the differences between a duck and a squid.

Historians of science still argue over who won the debate on technical points. But the reality was that Cuvier's scientific and political influence was so profound that for 164 years no biologist resurrected *unité de plan* for serious dis-

cussion. In fact, a researcher risked his or her reputation even to talk about the discredited idea.

So the matter might well have rested. In the early 1990s, however, biologists working separately on frogs and fruit flies began to untangle the matrix of genes that underlies the development of the overall body plan. To their astonishment, they discovered that the frog genes and the fly genes were the same genes. Unwittingly, they had resurrected Geoffroy Saint-Hilaire's idea of a single body plan.

In fruit flies, for example, a gene called *dpp* codes for a signaling protein that somehow activates other genes necessary for the formation of dorsal structures—all the body parts that lie along the back of an animal. In addition, *dpp* seems to suppress the development of ventral structures—all the body parts that lie along the abdominal side of an animal. *Dpp* also suppresses the development of nerve cells. A second fruit fly gene, *sog*, counteracts the effects of *dpp*. *Sog*, which is expressed only in the ventral regions of a developing fruit fly, overrides *dpp* in the ventral parts, allowing nerve cells and other ventral structures to develop.

When researchers looked for similar genes in frogs, they found *bmp-4*. Like *dpp*, *bmp-4* codes for a signaling protein. But, oddly, it operated in reverse. *Bmp-4* activates genes that help form the ventral structures in the frog embryo and suppresses the formation of nerve cells in the dorsal regions (Figure 24-1).

No one made an explicit connection between the three genes, however, until September 1994, when two German

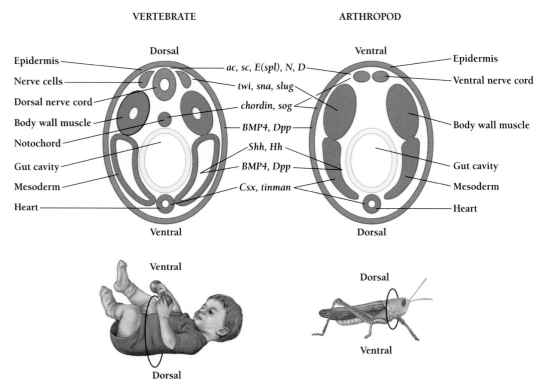

Figure 24-1
Opposites? The vertebrate body plan features a dorsal nerve and a ventral heart—an upside down version of the arthropod body plan, which has a ventral nerve cord and a dorsal heart. At left are cross-sections through a vertebrate embryo and an upside-down arthropod embryo, showing the similar expressions of proteins and genes in the two groups.

VERTEBRATE

Dorsal

Epidermis
Nerve cells
Dorsal nerve cord
Body wall muscle
Notochord
Gut cavity
Mesoderm
Heart

ac, sc, E(spl), N, D
twi, sna, slug
chordin, sog
BMP4, Dpp
Shh, Hh
BMP4, Dpp
Csx, tinman

Ventral

ARTHROPOD

Ventral

Epidermis
Ventral nerve cord
Body wall muscle
Gut cavity
Mesoderm
Heart

Dorsal

Ventral

Dorsal

Dorsal

Ventral

495

researchers pointed out that *dpp* and *bmp-4* do the same thing, only acting in opposite areas of frog and fly. Detlev Arendt and Katharina Nübler-Jung suggested that "the longitudinal nerve cords of insects and vertebrates derive from one and the same centralized nervous system in their common ancestor." After 164 years, they had revived Geoffroy Saint-Hilaire's *unité de plan.*

They were not ridiculed, however. Instead, researchers in California immediately uncovered a fourth gene. Edward De Robertis (at UCLA) and his colleagues found that the gene *chordin* stimulates the organized development of nerve cells in the dorsal region of a frog embryo, overriding *bmp-4*, just as *sog* overrides *dpp* in a fly.

If Geoffroy Saint-Hilaire and his modern supporters were right, then *dpp* and *bmp-4* were essentially the same gene, just expressed in different parts of the embryo. Likewise, *sog* and *chordin* had to be the same gene, separated by half a billion years of evolution.

In a series of dramatic experiments, De Robertis and several other researchers in California and Wisconsin demonstrated that the fly genes (*dpp* and *sog*) work in frogs, and the frog genes (*bmp-4* and *chordin*) work in flies. *Sog* mRNA injected into frog embryos, for example, promoted the development of dorsal structures. Likewise, *chordin* mRNA injected into flies induced ventral development, as *sog* normally does. Further, when researchers sequenced *chordin* and *sog*, they discovered that the base sequence of *chordin* (in frogs) is 47 percent identical to that of *sog* (in flies). And biologists have since uncovered several other developmental genes that are the same in both frogs and flies.

Has molecular biology redeemed *unité de plan*? Was Geoffroy Saint-Hilaire right, Cuvier wrong? Many biologists still hesitate to pronounce the matter settled. But nearly all are intrigued. If organisms as different as lobsters and linebackers share genes that help determine basic body plan, these organisms surely share a common ancestor.

De Robertis, for one, is eagerly seeking more genes shared by vertebrates and invertebrates that might shed light on the structure of that common ancestor. Until he and his colleagues succeed, however, even modern molecular biologists will hesitate a bit, reluctant to embrace an idea so long ridiculed.

24.1 Deuterostome Bilateria

As we discussed in the last chapter, the early development of the deuterostome coelomates differs from that of the protostomes. In all deuterostomes, including ourselves, the tiny hole in the embryo called the **blastopore** develops into an anus instead of a mouth. The mouth develops later (or second), hence the term deuterostome [Greek, *deuteros* = second + *stoma* = mouth].

The deuterostomes consist of three phyla (Figure 24-2): **Echinodermata** ("spiny-skinned" animals such as sea stars), **Hemichordata** (acorn worms), and **Chordata.** The Chordata include the vertebrates—the fish, the amphibians, the reptiles, the birds, and the mammals—as well as the less well-known sea squirts and lancelets. While sea urchins and humans differ radically as adults, their embryos share many basic features.

Phylum Echinodermata (Sea Stars and Others)

The 6,000 species of echinoderms fall into six classes of marine animals. Among these are two of the most familiar inhabitants of tidepools—sea stars (or starfish) and sea urchins. The other four living classes of echinoderms are the sea lilies and feather stars, the sea daisies, the sea cucumbers, and the brittle stars. Another 20 classes of echinoderms are now extinct. The echinoderms are an ancient lineage. Fossil sea stars, sea cucumbers, and sea lilies from the Burgess Shale suggest that the echinoderms had diverged from the protostomes some 530 million years ago.

Echinoderms are plentiful on the bottoms of intertidal zones (between high tide and low tide) and deep oceans. The largest is a sea star one meter across. The smallest are just a few millimeters. Most echinoderms have calcium-rich spines that project from their skin, giving the phylum its name [Greek, *echinos* = hedgehog, a small European mammal covered with spines, like a porcupine].

As adults, echinoderms are not bilaterally symmetrical. They often have five nearly identical parts arranged around a central axis. But their larvae are bilateral (Figure 24-3). Males and females shed sperm and eggs into sea water, where fertilization takes place. In the open ocean, the chances that an individual egg will meet a sperm of the right species is small. So echinoderms make a lot of gametes, to increase the odds of a sperm and an egg meeting. A single female may produce more than 2 million eggs in a breeding season.

Sea Stars (Starfish)

A sea star's body consists of a central region, with arms radiating out from the center. Under the center is the mouth, which opens into a digestive tract. Sea stars feed on all sorts of other animals, including sponges, corals, mollusks, crustaceans, worms, and even fish and oysters. However, a sea star prepares and eats an oyster rather differently from the way we would. The sea star first spreads its arms around the oyster and, like a human diner, it pries open the shell. But, in a surprising move, the sea star places its mouth near the gap in the oyster shell and extends its own gut through the opened shell, turning its stomach inside out. A sea star can perform this unusual feat even if the opening in the oyster shell is as small as a tenth of a millimeter. During the next few hours, the sea star completely

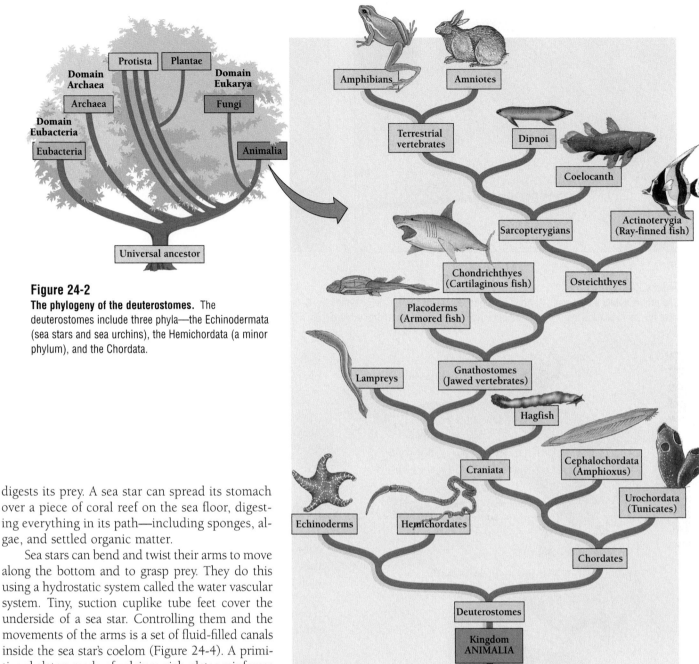

Figure 24-2
The phylogeny of the deuterostomes. The deuterostomes include three phyla—the Echinodermata (sea stars and sea urchins), the Hemichordata (a minor phylum), and the Chordata.

digests its prey. A sea star can spread its stomach over a piece of coral reef on the sea floor, digesting everything in its path—including sponges, algae, and settled organic matter.

Sea stars can bend and twist their arms to move along the bottom and to grasp prey. They do this using a hydrostatic system called the water vascular system. Tiny, suction cuplike tube feet cover the underside of a sea star. Controlling them and the movements of the arms is a set of fluid-filled canals inside the sea star's coelom (Figure 24-4). A primitive skeleton made of calcium-rich plates reinforces the whole body. Over this supporting structure is a delicate skin, with thousands of cells that are exquisitely sensitive to touch.

Sea stars and some other echinoderms can regenerate arms that are damaged or broken away. Some species of sea stars can regenerate a whole new individual from a single arm. In such asexual reproduction, the sea star breaks into two parts, and each regenerates the missing arms. Usually, however, echinoderms reproduce sexually, with males and females releasing gametes into the sea.

Echinoderms are ancient, spiny-skinned animals that are radially symmetrical as adults and bilaterally symmetrical as larvae.

Phylum Hemichordata (Acorn Worms)

The acorn worms are more similar to chordates like us than the echinoderms. Most of the 65 species of acorn worms live in U-shaped burrows on the bottom of the sea. They range in length from 2.5 cm to 2.5 m (nearly 8 ft). Zoologists classify these soft-bodied marine animals as hemichordates ("half chordates") because they resemble chordates in two respects: (1) they have **gill slits**, holes that directly connect the throat to the outside; and (2) in addition to a ventral nerve cord, they have the beginnings of a dorsal nerve cord (like our own spinal cord). As we

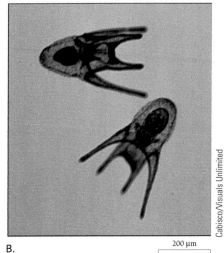

200 μm

A.

B.

Figure 24-3
Sea urchin adults and larvae. A. Adult echinoderms appear to have radial symmetry. B. But the larvae have the same bilateral symmetry as other deuterostomes. The adult form evolved from a bilateral ancestor.

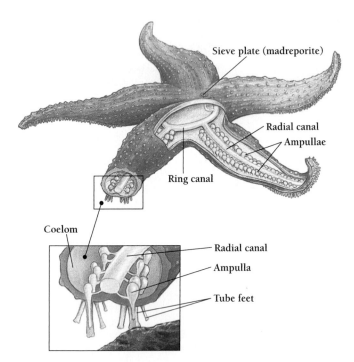

Figure 24-4
The sea star's water vascular system. Perhaps the most remarkable feature of sea stars and all other echinoderms is the water vascular system, which serves as both a circulatory system and an elaborate hydrostatic skeleton that allows complex movements of the tube feet. Water enters the system through a perforated sieve plate on the dorsal surface, which filters the water as it enters a tube. The filtered water flows down a tube into the ring canal, which surrounds the mouth (ventral) and from there into the radial canals that extend into each of the arms. In the arms, the radial canals branch to connect to long rows of tube feet. Each tube foot is also connected to a muscular fluid-filled sac, called an ampulla. In sea stars and other echinoderms, each tube foot ends in a sucker. The ampulla regulates fluid pressure, so that each tube foot may extend or retract, hang on or let go. Using the tube feet, a sea star can slowly pull itself along the bottom of the sea.

will discuss later, both of these features are fundamental characteristics of chordates.

Acorn worms have two traits in common with chordates, the phylum we belong to—gill slits and a dorsal nerve cord.

Phylum Chordata

The Phylum Chordata includes three subphyla—the Urochordata (tunicates), Cephalochordata (amphioxus), and the Vertebrata. Members of all three subphyla exhibit four hallmark characteristics at some time in their lives: (1) a flexible rod running along the back (dorsal surface), called the **notochord**, which gives the phylum its name; (2) a **dorsal hollow nerve cord**, running between the notochord and the surface of the back; (3) **pharyngeal slits** (sometimes called gill slits), holes in the sides of the body that run from the inside of the gut to the outside surface of the animal; and (4) a segmented body and a "postanal" tail, a tail that extends beyond the anus (Figure 24-5).

The notochord is a long column of fluid-filled cells, with the character of a stiff, but flexible, sausage. In the tunicates and cephalochordates, as well as in some fish and larval amphibians, the notochord provides back support for the adult animal. All other chordates, including humans and other mammals, also have a notochord. But it usually appears only briefly early in embryonic development, then disappears. In most chordates, the notochord is replaced by a bony or cartilaginous backbone. Cartilaginous fish (such as sharks) have both a notochord and a surrounding vertebral column, while adult reptiles, birds, and mammals have only a vertebral column.

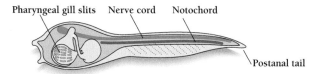

Tunicate larva (urochordate)

Pharyngeal gill slits   Nerve cord   Notochord

Postanal tail

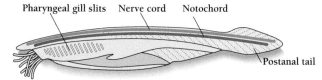

Amphioxus (cephalochordate)

Pharyngeal gill slits   Nerve cord   Notochord

Postanal tail

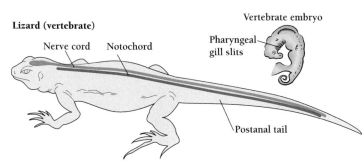

Lizard (vertebrate)

Nerve cord   Notochord

Vertebrate embryo

Pharyngeal gill slits

Postanal tail

Figure 24-5
The four features of chordates. The three chordate subphyla—the Urochordata (tunicates), the Cephalochordata (amphioxus), and the vertebrates and hagfish—share four distinguishing features: pharyngeal gill slits, a dorsal nerve cord, a notochord, and a postanal tail.

Dave Fleetham/Tom Stack & Associates

Figure 24-6
Our chordate cousin the tunicate. To the average person, a tunicate might look more like a sponge than a close relative of the vertebrates. Like sponges, adult tunicates feed by filter feeding. But unlike sponges, tunicates have a true digestive tract. Cilia in the pharynx pull water through an incurrent siphon into the mouth, past the pharyngeal slits, and out through an excurrent siphon. Sticky mucus traps food particles and carries them down into the digestive tract. The anus empties into the excurrent siphon.

The dorsal nerve cord of chordate embryos is hollow. As the nerve cord develops into the adult spinal cord and brain it remains partly hollow, its inner spaces filled with fluid. For comparison, the ventral nerve cords of invertebrates such as earthworms and spiders are solid. Gill slits are present in the embryos of all chordates (including humans), but not in adults (except in the fishes and a few amphibians).

All chordates have a notochord, a dorsal hollow nerve cord, pharyngeal slits, and a tail at some time in their lives.

What Do We Have in Common with Tunicates and Cephalochordates?

Adult tunicates do not look at all like other chordates (Figure 24-6). They are sessile marine animals that attach to rocks, boats, or the ocean floor. Their name derives from a tough outer coating, or tunic, made, amazingly, of cellulose. (Cellulose is common in the cell walls of plants, but rare in

animals.) While adult tunicates have pharyngeal gill slits, they have little else to recommend them as chordates. A tunicate larva, however, has a notochord, a dorsal hollow nerve cord, pharyngeal gill slits, and a postanal tail, most of which are lost when it metamorphoses into an adult. The common ancestor of the chordates probably resembled a tunicate larva.

Cephalochordates [Greek, *cephalo* = head], or lancelets, include about 45 species of small, segmented, fishlike animals. Most belong to the single genus *Branchiostoma,* whose common name is amphioxus (Figure 24-5). In these animals, the notochord, which is present in adults as well as in embryos, extends all the way through the head, hence their name. Sets of chevron-shaped (<<<) muscles pull against this internal skeleton. The segmentation of a lancelet is superficially similar to that of annelid worms and achieves the same advantages. Cephalochordates have all four chordate characteristics.

As larvae, tunicates have a notochord, a dorsal hollow nerve cord, pharyngeal slits, and a tail, all characteristics of chordates. Cephalochordates have notochords both as larvae and as adults.

Subphylum Vertebrata

The **Vertebrata**, or vertebrates, have many features that distinguish them from other chordates. The most important are a distinct head, with a cranium (or skull) and a brain, and a segmented vertebral column made of units called vertebrae. Arches of the vertebrae encircle and protect the dorsal nerve cord, which becomes the adult vertebrate's spinal cord (Figure 24-7).

Vertebrates also have characteristic internal organs—livers, kidneys, and hormone-secreting (endocrine) organs. The heart pumps blood through a closed circulatory system, including a system of fine capillaries that carry gases and nutrients to every cell in the body. The same system carries metabolic wastes away from cells. Liver, kidneys, and other organs closely regulate the chemical composition of the blood, ensuring a stable internal environment for all cells. You can read more about how these organs work in Part VII of this book.

Most modern vertebrates have bony skeletons. **Bone** consists of fibers of the protein collagen and crystals of calcium phosphate. Embedded in this extracellular material are living bone-making cells, blood vessels, and nerves. Bone is strong enough to support the huge bodies of the largest dinosaurs.

The embryos of all vertebrates have skeletons made of **cartilage,** also built from collagen, but softer and more elastic than bone, without calcium phosphate. Some parts of the adult skeleton retain cartilage, such as our ears and noses. But, except in the sharks and other cartilaginous fishes, bone mostly replaces cartilage in adults.

Vertebrate bone is so distinctive and so well preserved in the fossil record that biologists know more about the evolution of the vertebrates than about any other group. In this chapter, we survey the major groups of living vertebrates and some of those that are extinct. Some are fishes and some are four-footed animals, or **tetrapods** [Greek, *tetra* = four + *pous* = foot].

> Vertebrates have vertebral columns, skulls, and a well-developed brain. Most vertebrate skeletons are made of bone.

Not All Vertebrates Have Jaws

The most primitive vertebrates had spinal columns and skulls, but no jaws. Jawless fishes were common in the past. Paleontologists have uncovered the fossils of some 200 species of jawless fishes and named them **osteostracans** [Greek, *osteo* = bone] because their skin contained a distinctive shell-like skeleton of bony plates. The osteostracans lived half a billion years ago, during the Cambrian (Figure 24-8A).

Modern jawless fishes include only about 60 species of soft-bodied fish with round, jawless mouths, including the lampreys and hagfish. Lampreys and hagfish are hardly glamorous animals. Hagfish survive by eating polychaete worms and pieces of dead fish, scavenged from the muddy sea bottom. (Hagfish are not strictly vertebrates, as their backbones are a little different from ours. They belong to a group called the Craniata that includes all the vertebrates and all the hagfish.)

Lampreys are nearly all parasitic (Figure 24-8B). A lamprey lives by attaching itself to another fish with its circular jawless mouth, then tearing a hole in its living victim with its rasping tongue. The lamprey then sucks whatever fluids flow from the wound. A lamprey larva, like a lancelet, has a notochord and segmented muscle groups. It differs from a lancelet, however, in possessing a brain, a liver, and a kidney, just as we do. The lamprey is the closest living relative of the jawed vertebrates.

> Lampreys are the only living vertebrates that lack jaws.

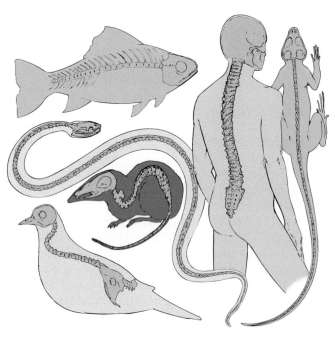

Figure 24-7

Which of these vertebrates has the strongest back? All vertebrates share a vertebral column composed of many vertebrae, as well as a thick skull that protects the brain. In most vertebrates, the vertebral column, skull, and skeleton consist largely of bone. In a few, such as the sharks and rays, these elements are flexible cartilage instead. The hero shrew *(center),* of Africa, is reputed to have the strongest backbone of any vertebrate. According to legend, a man can stand on this tiny insectivore's back without injuring the animal.

Placoderms, Now Extinct, Were Armored Fish with Jaws

At the beginning of the Devonian Period, some 405 million years ago, freshwater fish developed the first vertebrate jaws. These fish rapidly spread into the oceans, diversified, and

Figure 24-8
The jawless fishes. These fishes have neither working jaws nor paired fins. A. Extinct osteostracans.
B. A modern lamprey.

Courtesy Dr. Kiyoko Uehara

Berthoule-Scott/Jacana/Photo Researchers, Inc.

multiplied. Paleontologists sometimes call the Devonian the "Age of Fishes." Among the prominent jawed fishes in Devonian seas were a class of armored fishes called **placoderms** [Greek, *plak* = plate + *derma* = skin]. The placoderms had jaws, paired fins, and much less armor than the osteostracans.

Today, all vertebrates except the hagfish and lampreys have hinged jaws, with which they can seize, bite, and sometimes chew. Jaws permit a diverse diet and may have allowed placoderms and other jawed fishes to replace the ancient osteostracans. By the end of the Devonian Period, armored placoderms had disappeared, replaced by the Chondrichthyes and Osteichthyes. These two classes of fishes still dominate the aquatic regions of the Earth's surface. Vertebrates with jaws are called **Gnathostoma** [Greek, *gnatho* = jaw + *stoma* = mouth].

Placoderms were the first vertebrates with jaws.

Chondrichthyes: Fish with Jaws and Cartilaginous Skeletons

In most vertebrates, the soft cartilaginous skeleton of embryos is gradually replaced with bone. In sharks, skates, and rays, however, bone does not replace cartilage during development

(Figure 24-9). Adult Chondrichthyes have lightweight cartilaginous skeletons, giving the class its name [Greek, *chondros* = cartilage + *ichthys* = fish]. Chondrichthyes, with only about 750 species, are much less diverse than the 30,000 species of ray-finned fishes (Actinopterygia).

Chondrichthyes also lack a swim bladder, a balloonlike organ that helps other fish float. Instead, sharks and their allies increase their buoyancy by storing large amounts of oil in their livers. Nonetheless, the Chondrichthyes are heavier than water and sink to the bottom when they are not moving.

Although sharks and rays lack the bony scales of the placoderms and osteostracans, their skin is covered with small toothlike scales (denticles) that give it the feel of sandpaper. In addition, they have teeth (derived from their scales), which they use, along with their jaws and their ability to swim rapidly, to live as successful predators.

Sharks have particularly well-developed senses that help them find their prey. Among these senses are keen vision and smell. The lateral line system, a row of tiny sense organs along each side of the body, detects changes in water pressure. Other receptors, originally derived from the lateral line system, detect tiny electrical currents, including those generated by muscular contractions of potential prey.

The Chondrichthyes have cartilaginous skeletons and no swim bladder.

A.

B.

Figure 24-9

The Chondrichthyes. A. Rays and skates have evolved adaptations for a life on the ocean floor, where they feed mostly on mollusks and crustaceans. Their pectoral fins are enormous, giving these fish their characteristic appearance. B. Sharks are mostly fast-moving predators with streamlined bodies, tapered at both ends. Lacking a buoyant swim bladder, a shark keeps from sinking by swimming constantly. (Some sharks take breaks on the sea bottom or in a sea cave.) Paired fins (pectoral and pelvic) on the underside help keep the moving shark afloat.

Osteichthyes Have Bony Skeletons

The Gnathostomes (the vertebrates with jaws) split into three groups—the Placoderms, the Chondrichthyes, and the Osteichthyes [Greek, *osteon* = bone + *ichthys* = fish]. But all of the descendants of the osteichthyes, or "bony fishes," are not in fact fish. They include the amphibians, the mammals, and what we commonly call the reptiles and birds (Figure 24-2). The osteichthyes evolved two major lineages—the sarcopterygians, from which all terrestrial vertebrates evolved, and the ray-finned fishes, or Actinopterygii [Greek, *aktin* = ray + *pterygon* = wing].

The ray-finned fishes are extraordinarily diverse and live virtually wherever there is water (Figure 24-10A and B).

More than 30,000 species fill oceans and seas, lakes and streams all over the world. Adults range in size from about a centimeter to more than 20 feet long. They make up more than 95 percent of all fishes and half of all species of vertebrates. Ray-finned fishes are the best swimmers in the world.

Like the cartilaginous fishes, the ray-finned fishes first appeared in the Devonian Period, but in fresh water rather than the oceans. Because it's harder to float in fresh water than in salty sea water, the freshwater ray-finned fishes evolved a **swim bladder,** an air-filled sac that helps them control their buoyancy. The ray-finned fishes eventually returned to the sea, taking their swim bladder with them. The

A.

B.

C.

Figure 24-10

The two subclasses of bony fishes: the ray-finned fishes and the lobe-finned fishes. A. The fast-swimming barracuda typifies the highly evolved ray-finned fishes. B. A pike can achieve accelerations of 12 to 24 g (times gravity), reaching top speeds of 6 meters per second (equal to a 4-minute mile). To find out how fish are able to swim so fast, researchers at MIT have built a robotic pike (*Esox lucius*), named "Robopike," which swims by itself. C. The 350-million-year-old marine coelacanth *Latimeria* is a lobe-finned fish that was thought to be extinct—until museum curator Marjorie Latimer found one in a fisherman's catch in 1938.

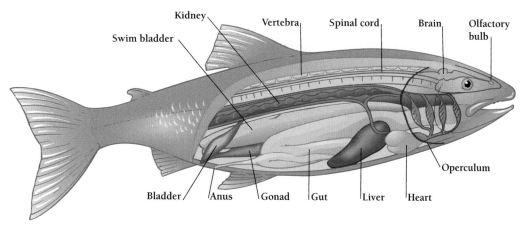

Kidney Vertebra Spinal cord Brain Olfactory bulb

Swim bladder

Operculum

Bladder Anus Gonad Gut Liver Heart

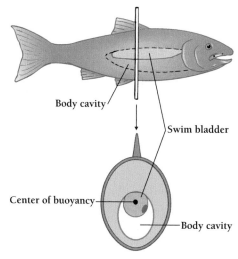

Body cavity

Swim bladder

Center of buoyancy

Body cavity

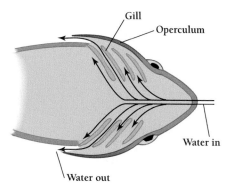

Gill

Operculum

Water in

Water out

Figure 24-11

The body plan of a bony fish.
Fish share many features with terrestrial vertebrates, including a brain, a liver, and true kidneys. In addition, a bony fish has a gas-filled swim bladder from which evolved the terrestrial lung. By regulating gases in the swim bladder, a fish can adjust its buoyancy, allowing it to float at the level it wants without exertion. Bony fishes also possess a bony operculum over the gills that allows them to pump water across the gills, facilitating the exchange of oxygen and carbon dioxide.

Biology ⑤Now™ Learn more about how fish absorb oxygen by clicking on this figure on your BiologyNow CD-ROM.

ray-finned fishes also differ from the Chondrichthyes in having bony skeletons and thin, bony scales, as well as paired fins supported by thin, bony rays, originally derived from bony scales (Figure 24-11).

The lobe-finned fishes, Sarcopterygii [Greek, *sarkodes* = fleshy] are the vertebrate lineage to which we belong. Unlike the fins of ray-finned fishes, the fins of lobe-finned fishes contain muscle as well as bone. The living descendants of the sarcopterygians include the rare coelacanth, as well as the lungfish (dipnoi) and the four-legged terrestrial vertebrates, or tetrapods. Modern lungfishes can use their sturdy fins to walk on land, and biologists believe that the first amphibians must have resembled lungfish whose muscular fins gradually evolved into limbs. Modern lungfish live in stagnant freshwater ponds, using their lungs to extract oxygen they get by gulping air from the surface. Lungfishes can radically alter their metabolism to survive long periods of drought by burying themselves in the mud at the bottom of a pond.

Ray-finned fishes (Actinopterygia) have bony skeletons, swim bladders, and paired fins. Lobe-finned fishes (Sarcopterygii) include the lungfishes, which can breathe air with simple lungs and walk on land.

24.2 What Adaptations Have Vertebrates Evolved for Life on Land?

By the end of the Devonian Period, fish had appeared in every conceivable aquatic niche—in seas, streams, and ponds. Except for a few lungfish adventurers, however, vertebrates had not found a way to live on the land. Plants and insects had already adapted to dry land. Just out of reach stood the wide world, all inviting, with great, lush forests and countless tasty insects and other arthropods. The terrestrial environment offered good meals and few predators.

For vertebrates, the movement from water to land was a dramatic event, accompanied by immense changes in their anatomy. To colonize the land, vertebrates had to solve several problems:

1. They needed to obtain and conserve water in a comparatively dry environment.
2. They needed to extract oxygen from air rather than water.
3. They needed strong skeletons to support their bodies—they could no longer depend on the buoyancy of water for support.
4. They needed to control fluctuations in body temperature, which can be much greater in air than in water.
5. They needed a moist environment for fertilization and early development.

Through natural selection, vertebrates evolved adaptations that solved all of these problems. Individual animals cannot solve problems, but lineages of animals can solve problems by evolving and adapting over great expanses of time. You can read more about how evolution works in Chapters 15–17.

The terrestrial vertebrates evolved adaptations that solved the problems of living on land.

Amphibians Are Terrestrial Animals That Begin Their Lives in Water

During the Devonian Period, the freshwater lakes and streams dried up frequently, leaving many aquatic organisms stranded in foul ponds and muddy streambeds. Only fish able to extract oxygen from the air, with a primitive lung,

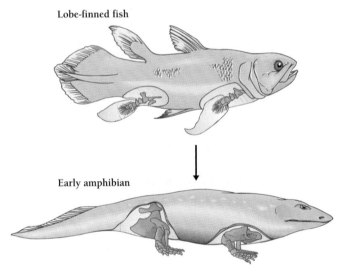

Figure 24-12
The first amphibians. The first amphibians evolved from lobe-finned fishes at the end of the Devonian Period.

were able to survive. Indeed, the Devonian Period saw the development of the two most important terrestrial adaptations: lungs and limbs.

Nearly all the freshwater fishes that survived the Devonian Period had a kind of lung, a simple sac enhanced with a rich supply of blood. (The swim bladder probably evolved from this simple lung.) The Devonian fish also developed **double circulation**, a kind of figure "8," in which blood circulates first between the heart and lungs and then between the heart and the rest of the body. This arrangement allows the heart to deliver oxygenated blood from the lungs to the body (Chapter 38).

In the Devonian Period, freshwater fish probably used their well-developed fins to walk across the bottoms of shallow pools and overland from one deep pool to another. Fish able to struggle across mudbanks to a deeper pool survived and passed on their primitive legs to their offspring. Some evolved highly effective legs and other adaptations to land.

The early amphibians were the first vertebrates to spend most of their time on land. Reptiles, birds, and mammals are all descendants of these earliest amphibians. The first amphibians somewhat resembled modern salamanders and lobe-finned fishes, and dominated the land for about 100 million years, until the rise of the dinosaurs in the Triassic (Figure 24-12). Some amphibians grew to lengths of three or four meters and were more terrestrial than modern amphibians. Modern salamanders, for example, have evolved several adaptations for moving about in shallow water, including a flat tail, that earlier amphibians did not possess.

Today, three orders of Amphibia exist (Figure 24-13): Urodela (salamanders), Anura (frogs and toads), and Gymnophiona (caecilians). Anurans lack tails; urodeles have them. The names Anura [Greek, *an* = without + *oura* = tail] and Urodela [Greek, *oura* = tail + *delos* = visible] describe this simple distinction. The frogs (anurans) are by far the most diverse amphibian order, including about 90 percent of the 3,000 amphibian species. The tropical Gymnophiona [Greek, *gymnos* = naked + *ophioneos* = snakelike], commonly called caecilians, lack feet and resemble snakes or earthworms. The

A.   B.   C.

Figure 24-13
Three orders of amphibians. A. Jordan's salamander, from North Carolina. B. A river frog from Florida. Frogs differ from toads in the quality of their skins: frog skin is smooth and moist, while toad skin is bumpy and dry. C. Gray caecilian. The caecilians are legless amphibians that live in the tropics.

least familiar amphibians, they live underground like moles, pushing through the soil and searching for small invertebrates to eat. Biologists have named about 160 species of these snakelike amphibians.

Adult toads and frogs have powerful leg muscles for jumping and well-developed ears and vocal cords for communicating. They have extendible, sticky tongues with which they can expertly capture flying morsels. Most amphibians can gulp air into their lungs, but they also absorb oxygen through their skins, which are moist and thin. In fact, some (lungless) salamanders breathe entirely through their skins and the linings of their mouths.

Amphibians are only partly adapted to life on land. They do not drink water, but instead absorb it through thin, damp skins. But such skin is not good at retaining water, so most amphibians must live in very moist environments or risk drying out.

An amphibian's life cycle generally consists of two stages, an aquatic stage and a terrestrial stage—hence their name [Greek, *amphi* = double + *bios* = life]. Most amphibians start life in water, when the male and female shed sperm and eggs together (Figure 24-14). Because amphibian eggs lack shells and special membranes to prevent water loss, their embryos must develop in a very wet environment.

The fertilized egg (zygote) develops into an aquatic larva, or tadpole, which, like a fish, obtains oxygen through gills and uses its long tail to swim. Later, in a process called **metamorphosis**, the tadpole loses its gills and develops lungs and limbs, and so becomes a terrestrial an-

imal (Figure 24-14). If it is a frog, it also loses its tail. The adult amphibian can now survive on land.

In a few species of frogs, salamanders, and caecilians, the zygote develops directly into the terrestrial form, bypassing the larval tadpole. Other species avoid the need for free-standing water for their larvae in other ways. They lay their eggs in moist places on land, in pockets of water in leaves, or in wet mossy places. Several species carry the developing eggs in special pouches on their backs or in the mouth (Figure 24-15). One South American frog (*Rhinoderma*) lays its eggs on moist ground, and groups of adults guard the developing embryos.

Michael Fogden/DRK Photo

Figure 24-15
Some amphibians care for their young. Here the reticulated poison dart frog carries two tadpoles to water.

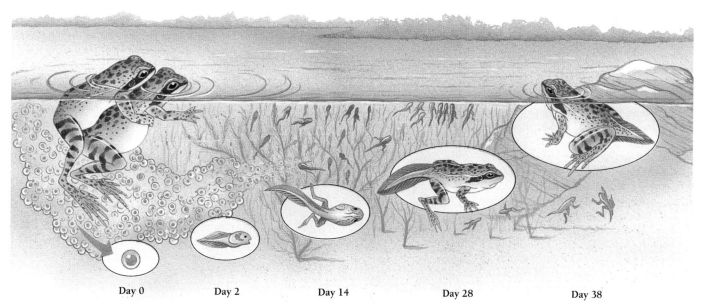

| Day 0 | Day 2 | Day 14 | Day 28 | Day 38 |

Figure 24-14
How tadpoles turn into frogs. Most amphibians lay their eggs in water. The eggs develop into tadpoles, which eat and grow under water. When the time is right, the tadpoles metamorphose into adult amphibians.

Biology (ⓔ)Now™ Learn more about the life cycle of frogs by clicking on this figure on your BiologyNow CD-ROM.

Extreme Biology Why Are Amphibians Disappearing?

Costa Rica's spectacular golden toad (*Bufo periglenes*) has not been seen since 1989, Australia's gastric brooding frog (*Rheobatrachus silus*) was last seen in 1979, populations of the common toad (*Bufo bufo*) have declined on Norwegian coastal islands, and many species of frogs in California have vanished from most of their historic ranges (Figure A). Throughout the world, amphibian species are declining in number and, in some cases, going extinct locally.

Is some unknown phenomenon exterminating toads and frogs?

The question first arose at the First International Herpetological Congress, held in 1989 in Canterbury, England. At that meeting, herpetologists recounted what each thought was a personal sorrow: the disappearance of an amphibian species from a study site. So many scientists reported the same kinds of losses that herpetologist David Wake organized a 1990 meeting in Irvine, California, to discuss whether there was a global decline in amphibian populations. At that meeting and at a follow-up symposium, the herpetological community shared information about declines in amphibian populations in Australia, Canada, the western and southeastern United States, Central America, the Amazon Basin, and the Andes.

Many frog species have boom-and-bust population cycles—populations explode one year and drop to almost nothing in another year. Such natural variation makes it difficult to interpret declines in some species. For most amphibian populations, scientists do not have long-term data on population cycles. Without information about natural population cycles, it is impossible to determine whether, for any given species or population, a dip in numbers is natural variation or a permanent decrease, possibly caused by human activities.

Nonetheless, herpetologists agree that an ongoing, worldwide decrease in amphibian populations is occurring. They further agree that the primary cause is the fragmentation and destruction of habitat, as is unfortunately true for many other organisms.

(You can read more about habitat fragmentation and extinction in Chapter 28.) Frogs may go the way of the dinosaurs. Species in habitats as different as Costa Rican cloud forest, Sonoran desert, and montane pools in the Sierra Nevada are all succumbing.

Herpetologists do not yet agree what other reasons might account for the worldwide disappearance of amphibians. Some researchers say amphibian declines are just part of the worldwide loss of biodiversity caused by destruction of forests, wetlands, and other habitats. The disappearance of amphibian species, they argue, is no different from the extinctions of thousands of other species.

Other researchers suspect that amphibians—with their permeable skin, permeable eggs, and complex life cycles—are particularly vulnerable to certain environmental changes. In particular, amphibians seem to be especially sensitive to increases in UV radiation and chemical pollutants. De-

Figure A
Golden toads.

creases in numbers of the North American tiger salamander (*Ambystoma tigrinum*) have been linked to pollution, especially acid rain. The Tarahumara frog (*Rana tarahumarae*) is extinct in the United States, and researchers have implicated high levels of the toxic metal cadmium. The massive die-offs of fertilized eggs of the Cascades frog (*Rana cascadae*) and the Western toad (*Bufo boreas*) may be due to increases in UV radiation stemming from the destruction of ozone in the stratosphere; the eggs of both of these latter species are low in *photolyase*, an enzyme that helps repair UV-damaged DNA.

A host of nonnative predators have been accused of decimating amphibian populations. In the western United States, bullfrogs (*Rana catesbeiana*), nonnative trout, bass, and sunfish, all predators of amphibians, have been introduced into countless waterways. Losses of the Yavapai leopard frog (*Rana yavapaiensis*), the mountain yellow-legged frog (*Rana muscosa*), and the Chiricahua leopard frog (*Rana chiricahuaensis*) have all been attributed to such introduced predators.

Some researchers argue that an accumulation of environmental stresses, rather than a single cause, may reduce or destroy a population. For example, a population of frogs may become fragmented and reduced by the loss of wetland habitat. If a subsequent drought kills many more individuals, the population may never recover. Or stress from increased UV exposure and pollution may make individuals more susceptible to disease—one explanation for the extinction of 11 distinct populations of Western toads (*Bufo boreas*) in western Colorado.

In 2002, biologist Tyrone Hayes, of UC Berkeley, published a landmark study showing that frogs are sensitive to the herbicide atrazine, heavily used in agriculture. Extremely low levels of atrazine cause males to form egglike gametes instead of sperm, which makes the males sterile. Atrazine is so common in some parts of the United States that it rains out of the sky.

When an embryo begins to move, a frog quickly takes it into its mouth. When the embryo finishes developing, the frog yawns, and a tiny froglet jumps out of his mouth.

A few salamander species do not complete metamorphosis. The mudpuppy and the axolotl, for example, retain their gills and continue to live exclusively in water even after they mature into reproducing adults. Under certain conditions, the axolotl matures into an adult with lungs instead of gills. But the mudpuppy never loses its gills and never develops lungs.

Amphibians live a dual life. The larvae are usually aquatic and the adults are usually terrestrial.

Amniotes Can Reproduce on Land

The first vertebrates to evolve an entirely terrestrial lifestyle did so by developing an egg that could be laid on dry land. This egg, called the **amniotic egg**, was the key to vertebrates' successful invasion of dry land and remains the link that ties mammals, reptiles, and birds together. We are all Amniota.

The **amniotes** include three main lineages—mammals, turtles, and "diapsids" (snakes and lizards, crocodiles and alligators, and all of the various kinds of dinosaurs, including the modern birds). For many years, biologists divided the amniotes into a different set of three lineages—the reptiles (Reptilia), birds (Aves), and mammals (Mammalia). All the amniotes that were not birds or mammals were lumped into one group, called the reptiles. But we now know that even though the word reptile is wonderfully evocative, the taxon Reptilia doesn't work as a formal classification because it doesn't reflect these creatures' evolution and relatedness (Figure 24-16). (Either birds would have to be considered reptiles, or the reptiles would have to exclude turtles, crocodiles, and all the dinosaurs.) Some of the most fascinating amniotes are now extinct, including the swimming ichthyosaurs, flying pterosaurs, the triceratops dinosaurs, and others.

The amniotic egg is so well adapted to a dry environment that even crocodiles and turtles, which live primarily in water, return to the land to lay their eggs. The *amnion* is a membrane that encloses the developing embryo in its own

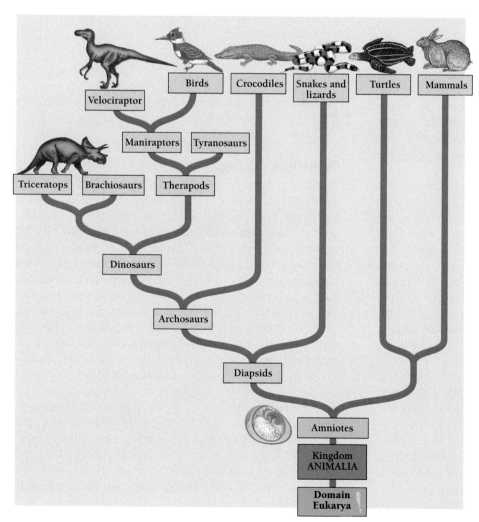

Figure 24-16

The amniotes. The vertebrates that make an amniotic egg completed the transition to a terrestrial life begun by the amphibians. The mammals and turtles split off the main line early in amniote evolution, followed by a diverse group called the diapsids. The diapsids, in turn, split into the crododiles and dinosaurs, which include modern birds.

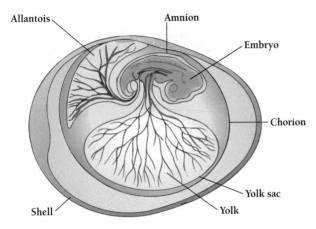

Figure 24-17

The amniotic egg. The most obvious feature of an amniotic egg is the yolk sac, which serves as a pantry—a storage area for fats, proteins, and other nutrients for the developing animal. But fish eggs and amphibian eggs also have a yolk sac. What distinguishes the amniotic egg are three other membranes: the amnion encloses the embryo in water; the allantois serves as a sort of outhouse, a separate area for wastes; and the chorion, the outermost membrane, encloses embryo, yolk, and allantois. The chorion helps regulate the exchange of oxygen and carbon dioxide between the embryo inside and the world outside.

little pond, eliminating the need for a separate aquatic stage (Figure 24-17). The amnion, in turn, lies within a porous shell that allows the exchange of gases with the surrounding air. Just beneath the shell is another membrane, called the *chorion* [Greek, = skin]. Also within the shell is the *yolk,* a rich food supply surrounded by a membrane called the *yolk sac.* A fourth membrane, the *allantois,* functions in both respiration and excretion. These four membranes are the defining characteristics of the amniotic egg.

In all amniotes, fertilization is internal, meaning that a male must inject its sperm into the female reproductive tract. In snakes, lizards, turtles, crocodiles, and birds, a hard shell forms around the amniotic egg before it emerges from the female. Because the shell would prevent a sperm cell from reaching the egg's nucleus, fertilization must occur before the shell forms. Even in mammals, in which development occurs inside the female, fertilization is internal.

To transfer their sperm, all male mammals, turtles, crocodiles, and lizards and snakes have a penislike copulatory organ. Mammals, birds, crocodiles, and turtles have a single penis, while snakes and lizards have two (Figure 9-8). In birds, some groups have one and others don't. Because bird groups that have a penis (ostriches and ducks) are older, less-evolved lineages than other birds, biologists believe that modern birds lost their penises during evolution.

In most amniotes, both sexes possess a **cloaca,** a common entrance and exit chamber for the digestive, urinary, and reproductive systems. In the female, the cloaca serves as an entrance for sperm from the male and an exit for the fertilized eggs. In most mammals, the females have three separate entrances for the digestive, urinary, and reproductive tracts, and the males have two. One exception are the monotremes—a primitive groups of mammals that includes the duck-billed platypus. Monotremes have fur, like other mammals, but they lay eggs and have a cloaca, like all the other amniotes.

As a group, the amniotes are entirely adapted to life on land. Although mammals and some snakes give birth to live young, most amniotes lay a shell-encased egg on bare rock or soil or in some sort of nest. Amniotes have evolved not only a remarkable egg and internal fertilization, but also more efficient kidneys for conserving water and legs better suited to walking on land than those of amphibians. Amniotes all have dry, watertight skin that conserves water, in the same way as the cuticles of insects and plants. Because amniotes cannot breathe through such tough skins, they depend entirely on their lungs for oxygen. Although a few amniotes—such as sea turtles, whales, crocodiles, and penguins—are secondarily adapted to aquatic life, they are all descended from terrestrial amniotes. They do not have gills and must breathe air.

The mammals, turtles, and three kinds of diapsids total five lineages of *living* amniotes:

1. Mammals
2. Turtles
3. Lizards and snakes
4. Crocodilians
5. Birds

The evolution of the amniotic egg has enabled amniotes to colonize the land.

Testudines (Turtles)

Turtles separated from one of the earliest lineages of amniotes and first appear in the fossil record about 200 million years ago. The order Testudines includes about 250 living species. With their rocklike carapace (top shell) and plastron (bottom shell), turtles are invulnerable to most predators (Figure 24-18). Their shell develops from and is part of the backbone and ribs. Turtles have no teeth and chew their food with beaklike jaws. (Some are vegetarians and some are carnivorous.) Turtles are deaf and dumb, but they can see well and in color and have a good sense of smell. They lay their eggs on land, and the temperature of the nest determines the sex of the hatchlings. Low temperatures produce males and high temperatures produce females.

Turtles are slow. They are heavy, breathe relatively inefficiently (because of their rigid carapace and plastron), and have a low metabolism. But even if they live a slow life, they last a long time. Some are believed to live more than 150 years. Today, many species of turtles are extinct or endangered, but until now their particular way of life was a successful one. Turtles have hardly changed in 200 million years.

Figure 24-18

Turtles. Galápagos tortoises in a pond on a volcanic island. Like other turtles, these Galápagos tortoises belong to one of the oldest lineages of amniotes. Turtles have not changed (evolved) much in their 200 million years of life on Earth. But on the 5-million-year-old Galápagos Islands, even turtles have evolved in small ways. In Charles Darwin's 1835 visit to the Galápagos Islands, he noticed that on each island, the tortoises were different. They varied from island to island in the lengths of their necks, the shapes of their shells, and other traits. Darwin later concluded that the tortoises on each island had evolved separately.

Squamata (Lizards and Snakes)

The order Squamata include the 3,800 species of lizards and 2,700 species of snakes, as well as 140 species of legless reptiles called the amphisbaenians, which are entirely adapted to an underground life, much like that of moles. Amphisbaenians are blind and burrow through soil hunting invertebrates. Another order of living reptiles related to the lizards and snakes includes just one species, the tuatara, a lizardlike animal that survives on a few islands off the New Zealand coast (Figure 24-19).

Crocodilia (Crocodiles and Alligators)

Along with the birds, the 22 species of Crocodilians are the last survivors of a huge taxon—the Archosaurs—that also included all the dinosaurs and dominated the Earth for hundreds of million of years. Crocodiles are the most advanced of the surviving reptiles (other than birds), with sophisticated social behavior and several specialized traits shared only by birds and mammals (Figure 24-20).

Crocodilians can stand up on their legs and gallop, more like mammals than lizards, and can reach speeds of 10 miles per hour. They have a bony "palate" on the roof of their mouth that separates the mouth from the breathing passages of the nose. A palate allows an animal to eat and

Figure 24-20

The crocodiles and alligators. Nile crocodile washing and releasing baby. The Crocodilia include the crocodiles, alligators, and caimans. Their physiology and behavior are different from that of the lizards and snakes.

A.

B.

C.

Figure 24-19

The lizards and snakes (Squamata). A. Crested dragon lizard, Malaysia. The Squamata comprise the lizards and snakes. B. Two male Panamint rattlesnakes in ritual combat, White Mountains, California. C. The lone species in the order Sphenodonta is the tuatara of New Zealand.

breathe at the same time. Both mammals and crocodiles have a bony palate, but lizards and snakes do not.

Crocodiles also have a four-chambered heart, similar to that of mammals. They are able swimmers and can even "tail walk" like dolphins, with their heads out of the water. Crocodiles' sight and hearing are excellent and they can vocalize, like birds and mammals. Finally, like mother birds and mother mammals, mother crocodiles care for their offspring. In some species, families stay together long after the young can care for themselves. Crocodiles did not evolve these adaptations recently; like the turtles, they've remained almost unchanged for 200 million years.

Crocodiles range in size from 5 to 25 feet long. Like cats, they hunt by lying in wait, then chasing down their prey with a short burst of speed. They have no stamina and cannot sustain a chase. Mostly they eat fish, but they are not fussy eaters; larger crocodiles and alligators will not ignore humans and other large prey.

How Do Terrestrial Vertebrates Regulate Body Temperature?

Life depends on the thousands of biochemical reactions that take place in every cell. The rate of a reaction depends on temperature, so the temperature of an organism is crucial to the way it lives. In aquatic animals, the temperature of the water around them usually establishes a fairly constant body temperature.

On land, however, air and ground temperatures fluctuate enormously. Animals that have left the water must therefore find a way to regulate temperature. Turtles, crocodiles, and snakes and lizards do so by moving into and out of the sun. Because they depend on external sources of heat, they are called **ectotherms** [Greek, *ektos* = outside + *thermos* = heat]. In contrast, mammals and birds warm their bodies by capturing the heat released by metabolism and are called **endotherms** [Greek, *endo* = within]. Endotherms keep nearly constant body temperatures by regulating both the heat produced by biochemical reactions and the heat lost by evaporation. Some evidence suggests that birds' cousins the velociraptors may have been endothermic as well.

Ectotherms allow their body temperature to fluctuate with air temperature; they regulate temperature to some extent by basking in the sun or moving into the shade. Endotherms maintain their body temperature within a few degrees of a constant temperature most of the time.

What Traits Characterize the Birds?

Birds are wonderfully diverse, with some 8,700 species (Figure 24-21). Most species have distinctive colored plumage and strange, haunting songs and calls. Yet birds are remarkably uniform. All are endothermic and all have feathers, and most can fly. The ability to maintain a constant temperature in cold environments allows birds to live where no crocodilian or squamate could. The Arctic tern, for example, spends half the year north of the Arctic Circle, then flies halfway around the world to spend the rest of the year in Antarctica. Both keeping a high body temperature and flying require enormous amounts of energy. In return for this energy expenditure, birds have access to vegetation, insects, and seafood that ectotherms can never reach. Flight also provides good protection from most predators.

Most birds find food and shelter on land, though many birds depend on streams, ponds, and seas for their food. Like the reptiles, birds produce amniotic eggs and do not need water for fertilization or early development.

Fertilization is always internal, but only a few species possess a penis. Ducks and flightless birds such as ostriches,

A. B. C.

Figure 24-21
Birds are both distinctive and diverse. Shown here are a great blue heron (A), a blue-throated hummingbird (B), and an emperor penguin and its young (C).

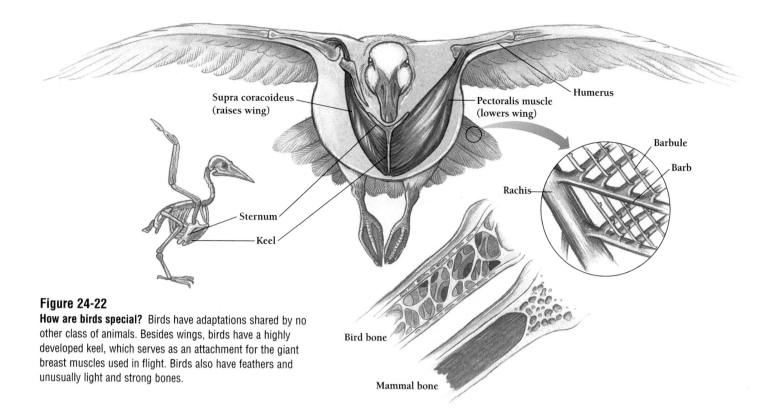

Figure 24-22

How are birds special? Birds have adaptations shared by no other class of animals. Besides wings, birds have a highly developed keel, which serves as an attachment for the giant breast muscles used in flight. Birds also have feathers and unusually light and strong bones.

Labels in figure: Supra coracoideus (raises wing); Humerus; Pectoralis muscle (lowers wing); Barbule; Barb; Rachis; Sternum; Keel; Bird bone; Mammal bone

as well as herons, flamingos, chickens, and turkeys, all have rudimentary penises. In most other birds, both males and females have a cloaca only. Birds accomplish fertilization by bringing together their cloacas. As in the tuatara, the sperm from the male's cloaca passes into the female's cloaca. Such intimacy often requires a long courtship, and many bird species have elaborate mating rituals.

Feathers are made of keratin, the same protein that forms a lizard's scales and our own fingernails and hair. Flight feathers are an amazing adaptation to flight, but down, which insulates the bodies of birds, is also important. Many biologists now believe that feathers were originally an adaptation for warmth. Flight came later, as did a host of adaptations that lightened the body and made birds still better flyers. These included the loss of teeth (which lessens the weight of the head), the hollowing of the bones (which lightens the whole skeleton), and the reshaping of the breastbone (which forms a keel to which flight muscles attach) (Figure 24-22).

Feathers, flight, and endothermy characterize the birds and together distinguish them from other amniotes.

What Traits Characterize the Mammals?

Mammals [Latin, *mammae* = breasts] take their name from their **mammary glands**, milk-producing organs in the female that characterize the mammals. No other vertebrates have mammary glands. Mother mammals nurse their young with warm milk, a rich mixture of fats, sugars, proteins, minerals, and vitamins, as well as antibodies that help defend newborns against infections.

Two other traits distinguish the mammals from other animals: hair and two sets of teeth ("baby" teeth and adult teeth). A mammal's hair, like a bird's feathers, helps conserve heat, a trait that is useful for an endotherm. Mammalian teeth, far more specialized than those of other vertebrates, help mammals get the tremendous amounts of energy they need to maintain a constant body temperature. Mammalian teeth may be specialized, for example, for grinding plants (horses and other herbivores), for shearing flesh (lions and wolves), for breaking bones (hyenas), or for seizing fish (dolphins and whales).

Mammalian teeth are extremely well preserved in the fossil record. Since a single tooth tells volumes about its owner, paleontologists know more about the evolution of mammals than is possible for other classes. In fact, some paleontologists now believe that the efficiency of the mammalian molar is at least partly responsible for the evolutionary success of the mammals since the end of the Cretaceous Period. When the dinosaurs died off, mammals radiated into about 30 orders, of which about two-thirds still have living representatives (Figure 24-23).

Mammalogists divide the living mammals into three subclasses—prototherians, metatherians, and eutherians. These subclasses have distinctive skulls and teeth that show that they first diverged from each other during the Mesozoic Era, even before the dinosaurs disappeared.

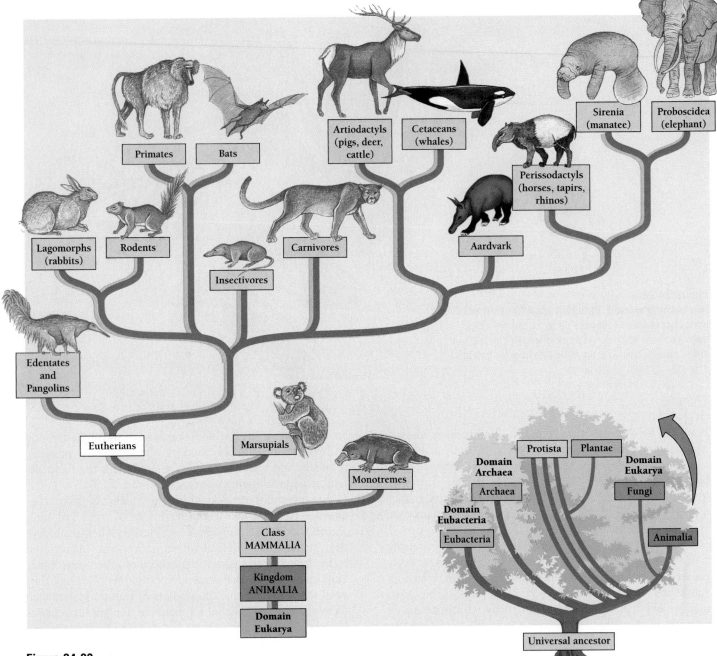

Figure 24-23
Some orders of the class Mammalia.

These three subclasses differ markedly in the ways they reproduce.

Most mammals are **eutherians** [Greek, *therion* = wild beast] or placental mammals. In humans, elephants, rats, rabbits, and every other kind of eutherian mammal, embryonic development takes place within the uterus of the mother. In fact, the word "placental" refers to the **placenta** [Latin, = flat cake], the flat structure that joins the lining of the uterus with the membranes of the fetus.

Metatherians [Greek, *meta* = beside, after + *therion* = wild beast], or **marsupials** [Latin, *marsupium* = pouch], such as the kangaroo and koala, reproduce differently from

eutherians (Figure 24-24). Newborn marsupials are not babies at all, but undeveloped fetuses. A newborn red kangaroo, for example, is no bigger than a bumblebee. Emerging from the uterus only 33 days after fertilization, the tiny creature painstakingly pulls itself along its mother's fur until it finds a nipple. The mother's nipples are in a furry pouch, where the young kangaroo sleeps and nurses for about six months. Even after it begins to wander a bit on its own, the young kangaroo regularly returns to the pouch until it is almost eight months old. Meanwhile, the mother may have another, younger baby in the pouch, as well as an embryo in her uterus.

Figure 24-24
A red kangaroo with its offspring. All marsupials carry their young in a special pouch that contains milk-secreting nipples. A female kangaroo may have a youngster at her side, another in the pouch, and a third in her womb.

The **prototherians** [Greek, *proto* = earliest + *therion* = wild beast], or **monotremes** [Greek, *monos* = single + *trema* = hole], are the strangest mammals (Figure 24-25). They are endotherms with hair and mammary glands, like other mammals. Like the reptiles from which they and all other mammals are descended, however, monotremes lay eggs that are similar to those of reptiles and birds. In addition, their digestive and reproductive systems empty into a cloaca, hence their name. The only living monotremes are

Figure 24-25
Duck-billed platypus. Unlike other mammals, monotremes lay eggs. They are considered the most primitive of all the living mammals.

the platypus and the two spiny anteaters (or echidnas), which live in Australia and Papua New Guinea. They are believed to be primitive mammals, but exactly how they are related to other mammals is still unknown. Mammals, both modern and ancient, are generally classified according to their tooth structure, and adult monotremes have no teeth.

Three characteristics set mammals apart from other amniotes—mammary glands, hair, and two sets of teeth.

In this chapter we have seen how creatures as seemingly different as sea urchins and rabbits can share traits that show they are related. This chapter concludes our survey of the six kingdoms of organisms. In the next section, we see how ecological forces shape these many living organisms.

Key Concepts

- Deuterostomes include both invertebrates and vertebrates.
- At some time during their development, all chordates have a notochord, a dorsal hollow nerve cord, pharyngeal gill slits, and a tail.
- Vertebrates have a segmented spinal column and a distinct head with a skull and a brain.
- Freshwater fishes evolved swim bladders, lungs, and limbs, adaptations that set the stage for the colonization of the land by amphibians and amniotes.

Summary with Key Terms

How are protostomes and deuterostomes alike and different?

Research in molecular genetics supports the hypothesis, first advanced in the 19th century, that chordates have an inverted version of the same body plan as most protostome coelomates. Structures that are **dorsal** in frogs and other vertebrates are **ventral** in arthropods. Deuterostomes, however, differ from protostomes in several ways, including, for example, the fate of the embryo's **blastopore**. Deuterostomes include three phyla: **Echinodermata**, **Hemichordata**, and **Chordata**. Echinoderms, such as sea stars, are bilaterally symmetrical as larvae and are descended from bilateral deuterostomes. Hemichordata (acorn worms) have pharyngeal slits and a dorsal nerve cord, both chordate characteristics, but not a notochord or a postanal tail.

What traits distinguish the chordates and what traits distinguish the vertebrates?

Chordates share four main features sometime during their lives: a stiff flexible rod, called a **notochord**; a **dorsal hollow nerve cord**; **pharyngeal slits** (or **gill slits**); and a postanal tail. Chordates include the **Vertebrata**, the **Urochordata** (tunicates), and the **Cephalochordata** (amphioxus). Tunicates resemble other chordates as larvae, but not as adults. Cephalochordates have notochords both as larvae and as adults. Vertebrates have skulls, segmented spinal columns, closed circulatory systems, and characteristic internal organs. The embryos of all vertebrates have skeletons made of **cartilage.** Most vertebrates replace the cartilage with **bone** during development.

What environmental conditions stimulated the evolution of the swim bladder, the lung, and the muscular limbs in early fishes?

Primitive vertebrates such as the **osteostracans** (extinct) and the hagfish and lampreys have no jaws. **Placoderms** (now extinct) were the first vertebrates with jaws (**Gnathostoma**). And most modern vertebrates now come equipped with jaws, including the **Chondrichthyes** (sharks and rays), in which bone never replaces cartilage, and the **Osteichthyes** (bony "fishes"), which have bony skeletons and a **swim bladder** or lung. The osteichthyes include the ray-finned fishes and the lobe-finned fishes. All are the descendants of fishes that colonized fresh water. Probably forced to survive in shallow pools, this lineage evolved swim bladders and lungs. One group, the lobe-finned fishes, also evolved muscular fins and legs. All **tetrapods,** or four-footed animals, are descended from this lineage.

Tetrapods have many special adaptations for land life, including **double circulation** and lungs, efficient kidneys, water-resistant skin, strong bony skeletons, and jointed limbs. Amphibians are semiterrestrial **ectotherms** that always begin their lives in water but later develop lungs and other adaptations to live on land during **metamorphosis.**

How did the amniotic egg allow vertebrates to colonize the land?

The **amniotes** can all reproduce away from water as a result of the development of the **amniotic egg.** The different parts of the amniotic egg—the amnion, the chorion, the allantois, and the yolk sac—all contribute to the egg's ability to regulate water balance on dry land. Amniotes also developed watertight skins and strong skeletons. Most amniotes have a common exit for digestive, urinary, and reproductive systems, called the **cloaca.** Birds added feathers, flight, and **endothermy,** while mammals added hair, two sets of teeth, endothermy, and **mammary glands** for nursing their young. The mammals are divided into the **eutherians,** or **placental** mammals; the **metatherians,** or **marsupials;** and the **prototherians,** or **monotremes.**

Review and Thought Questions

Review Questions

1. Describe three characteristics that distinguish the deuterostomes from the protostomes.
2. What characteristics do humans share with sea stars?

3. What four characteristics define the chordates? Can you locate all of these features in your own body? For each feature, explain where it is or where it once was.
4. Rank the following organisms according to how closely related they are to humans: tunicates, lampreys, and sea urchins.
5. Rank the following organisms according to how closely related they are to birds: lizards, velociraptors, mammals, and sea urchins.
6. Is the vertebral column, or backbone, actually made of bone in all vertebrates? What is bone?
7. Of what value are jaws?
8. Why must amphibians return to water to reproduce?
9. Which groups of animals have amniotic eggs?
10. What is endothermy? Why is it important?
11. What three mammalian characteristics distinguish a human from a turkey?
12. What is the function of a placenta?

Thought Questions

13. Compare the vertebrate body plan with the arthropod body plan. What advantage(s) might there be to flipping over the arthropod body plan? How would you test your hypothesis?
14. How would the day-to-day lives of humans be different if we were marsupials instead of placental mammals?
15. What did Cuvier do to win the debate with Geoffroy Saint-Hilaire? How might the history of biology have turned out differently if Cuvier had used a different debating style?

BiologyNow Resources

Biology ⓔ Now™

Active Figures
24-11: How the fish body plan is adapted to breathing
24-14: The life cycle of frogs

Preparing for an exam? Take a diagnostic test on your BiologyNow CD-ROM.

Online materials relating to this chapter are at:
http://biology.brookscole.com/AAL3

About the Chapter-Opening Image
This child illustrates the idea that the vertebrate body plan is the reverse of that of invertebrates such as lobsters and insects. As one biologist put it, "If you lay down on your back and waved your arms, you would be doing what insects do when they walk."

PART V ECOLOGY

Jim Brandenberg/Minden Pictures

Ecosystem Dynamics

Key Questions

- How do energy and nutrients flow among the individual members of ecosystems?

- How does photosynthesis provide the energy that drives all ecosystems? How is that energy lost at higher trophic levels?

- How are food webs both simple and complex?

- What are the results of human alteration of geochemical cycles such as the water cycle, the carbon cycle, and the nitrogen cycle?

Is the Biosphere a Superorganism?

In the early 1960s, the National Aeronautics and Space Administration (NASA) began planning its first explorations of the solar system. The space agency was especially keen to examine Martian soil for signs of life or conditions that could support life. But analyzing the Martian soil would require new instruments specially designed for the project.

For these, NASA turned to James E. Lovelock, an eccentric English chemist who had invented an ingenious device that detected tiny amounts of chemicals. Lovelock's device was widely used to measure trace amounts of pollutants such as the insecticide DDT.

In the early 1960s, NASA invited Lovelock to the Jet Propulsion Laboratory in California to help design instruments for detecting life on Mars. Lovelock's fascination with the search for extraterrestrial life and his insatiable curiosity soon led him beyond this technical chore to reconsider the much greater question of life on Mars. What if Martian or other life didn't look or act anything like terrestrial life? Could it still be detected? If so, how? How could life be described so that scientists could recognize it even in an alien form?

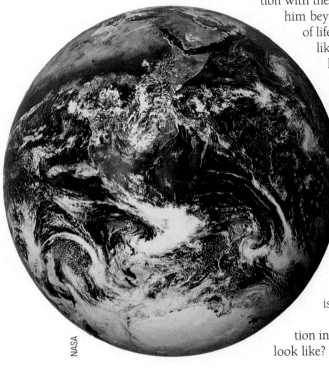

NASA

Lovelock approached these questions by considering the concept of "entropy." *Entropy* is a measure of disorder in "systems." A **system** is any collection of interacting parts or objects. If any system—whether a solar system or a dorm room—is left alone, its disorder, or entropy, will increase (Chapter 5).

Organisms, by contrast, create order and thus *decrease* local entropy. Plants, for example, convert carbon dioxide and water into highly structured leaves, stems, and flowers. And animals convert sugars, amino acids, and simple fats into muscle and brain tissue. But this decrease in local entropy runs on energy from the sun. And the decrease in entropy is only local. The universe as a whole is running down; its entropy is increasing.

A sure sign of life on Mars, Lovelock realized, would be a reduction in local entropy. But what in the world does a "reduction in entropy" look like? Lovelock couldn't stop thinking about the problem. He wondered how an alien being would recognize life on Earth. And it dawned on him that an alien would notice the peculiar chemical composition of Earth's atmosphere.

Earth's atmosphere is loaded with unlikely combinations of chemicals. For example, the atmosphere contains both methane and oxygen. Methane is a simple one-carbon hydrocarbon made by bacteria. Methane and oxygen readily react with one another to form carbon dioxide and water.

$$CH_4 + 2\ O_2 \rightarrow CO_2 + 2\ H_2O$$

This reaction constantly destroys methane. If no living organisms were present, methane would disappear from Earth's atmosphere. But bacteria in marshes, rice paddies, and cows' stomachs constantly make more methane, so the gas persists in our atmosphere.

Thinking about methane made Lovelock realize that the Earth's chemically unlikely atmosphere could be seen as an extension of our planet's **biosphere**—Earth's system of living organisms and their physical environment. Lovelock envisioned all of Earth's ecosystems, including the atmosphere, oceans, and soils, as constituting a single, immense living individual. Just as a snail's (nonliving) shell is considered part of its body, the atmosphere and oceans can be considered to be elements of a living biosphere. Lovelock further hypothesized that this immense organism acts to preserve itself. Lovelock named this entity Gaia, after the Greek goddess of the Earth.

The Gaia hypothesis, first published in the 1960s, vastly entertained the public. TV anchors glibly announced that the Earth was one big organism. Environmentalists seized on Gaia as a cultural metaphor, stressing the need to consider the biosphere and its environment—the atmosphere, the oceans, and the soils—as a whole. NASA's first breathtaking photographs of Earth as a cloud-swathed blue ball floating in space intensified Gaia's already enormous popular appeal.

Yet while biologists found the idea of Gaia amusing, most also found it deeply embarrassing. Despite Lovelock's impressive credentials as an inventor, many biologists viewed him as a crank. Academic ecologists of the 1960s didn't want their careful scientific work confused with the freewheeling political activity of environmental groups. But Lovelock's credentials as a scientist seemed to blur the line between ecology and the environmental movement. Evolutionary biologists, for their part, dismissed Gaia, arguing that the biosphere cannot be considered an organism because it neither reproduces nor evolves.

Lovelock now denies that he intended Gaia to be understood as an actual organism. He says Gaia is more like a **superorganism,** a group of individuals that interact almost like the cells of the body (Figure 25-1). Yet at times Lovelock has described Gaia in almost human terms, as though "she" had intentions and goals. He has referred, for example, to "Gaia's intervention" and "Gaian impatience."

The line between the Gaia metaphor and formal science blurred further when a biologist of national standing stood up for Lovelock. In 1974, Lynn Margulis, a distinguished

Figure 25-1
Do superorganisms really exist?
This Portuguese man-of-war consists of a tightly integrated colony of individuals. Each individual in the colony is a kind of polyp (related to the jellyfish) that cooperates with its comrades to form a complex superorganism. A bright blue, gas-filled float doubles as a sail. Wind drives the colony along the surface of the ocean. Beneath hang tangles of polyps bearing stinging tentacles that can reach 20 meters in length. The polyps quickly immobilize and digest fish or other creatures that come in contact with the deadly tentacles.

Kelvin Aitken/Peter Arnold, Inc.

professor of biology at the Amherst campus of the University of Massachusetts, began enthusiastically promoting a version of the Gaia concept—to the horror of her scientific colleagues.

In Margulis's view, Gaia is not a superorganism but a giant ecosystem, a system of all interacting organisms and their physical environment. Margulis, now a member of the prestigious National Academy of Sciences, proposed a related idea at the cell level that is now widely accepted (Figure 25-2A). She argued and eventually persuaded her fellow biologists that some of the tiny organelles inside our cells were once free-living bacteria. Specifically, Margulis proposed that mitochondria (which break down sugar) and chloroplasts (which photosynthesize) were once bacteria that came to live permanently inside other bacteria, to form the first nonbacterial, or eukaryotic, cells. In other words, our cells are a combination of bacteria that began cooperating billions of years ago. Such a cooperative arrangement is an example of **symbiosis** [Greek, *syn* = together + *bios* = life], any close association between two organisms, whether mutual or parasitic. Margulis asserts that just as different kinds of cells can live together to create successful forms of life, the different organisms of our biosphere interact in complex ways to create a functioning ecosystem. The view that organisms cooperate as well as compete is fundamental to the view that our biosphere is a tightly integrated whole (Figure 25-2B).

Most biologists still reject Lovelock's idea that our biosphere is a superorganism, but nearly all scientists now accept Lynn Margulis's view that the biosphere is a giant, tightly integrated ecosystem. Just don't call it Gaia.

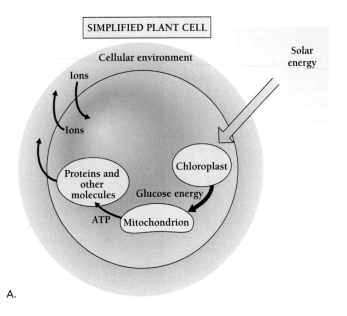

SIMPLIFIED PLANT CELL

Cellular environment

Solar energy

Ions

Ions

Proteins and other molecules

Chloroplast

Glucose energy

ATP

Mitochondrion

A.

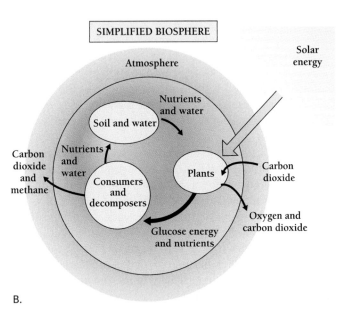

SIMPLIFIED BIOSPHERE

Atmosphere

Solar energy

Nutrients and water

Soil and water

Nutrients and water

Carbon dioxide and methane

Plants

Carbon dioxide

Consumers and decomposers

Oxygen and carbon dioxide

Glucose energy and nutrients

B.

Figure 25-2

Living organisms cooperate. A. Most biologists now agree that the chloroplasts and mitochondria that generate ATP and other energy-rich molecules were once prokaryotic cells that came to live inside other prokaryotic cells. The idea that eukaryotic cells are symbiotic associations of prokaryotic cells is the brainchild of biologist Lynn Margulis.
B. Margulis further argues that the different organisms of our biosphere interact in similar ways to create a functioning ecosystem. Shown here is a simplified biosphere.

25.1 What Is Ecology?

Ecology [Greek, *oikos* = home] is the study of the interactions of organisms with one another and with their physical environment, or "home." It is also the study of how those interactions determine the abundance and distribution of organisms.

How Are Individuals and Species Organized?

An individual organism is a member of a number of systems. For example, every individual is a member of: a **population,** an interbreeding group of individuals of the same species; a **community,** an interacting group of many species that inhabit the same area; and an **ecosystem,** an interacting group of species together with the nonliving parts of their environment (Figure 25-3).

All ecosystems taken together make up the biosphere. Because organisms live everywhere at the surface of the Earth and influence the chemical composition of the atmosphere, the biosphere is immense. Extending over the whole of the Earth's surface, the biosphere begins two miles (3 km) underground and extends into the highest reaches of the atmosphere. It ends where the atmosphere ends and space begins. Even in the deepest parts of the ocean, seven miles below sea level, living organisms persist.

The biosphere has many of the properties of living organisms that are listed in Chapter 1. For example, the biosphere obtains energy from its environment, transforms chemicals, and changes over time. But the biosphere differs from individual organisms in important ways. For example, unlike organisms, the biosphere does not reproduce itself. Further, the biosphere does something that no organism does. It recycles. No organism supports its own growth by eating its own waste, but the biosphere does.

What Is an Ecosystem?

An ecosystem consists of parts that are living, or **biotic,** and those that are nonliving, or **abiotic.** The living parts include all organisms from trees and whales to the tiniest microbes. The abiotic parts of an ecosystem include forces such as wind and gravity; conditions such as elevation, light intensity, humidity, temperature, salinity, or acidity; and physical substances such as soil, air, and water. The abiotic parts of the biosphere include both organic and inorganic substances. For example, soil usually contains both inorganic nitrogen, minerals, and water and organic material such as wood particles from trees and shrubs, chitin from the exoskeletons of dead insects, and disintegrating proteins from all kinds of organisms.

The living and nonliving components of an ecosystem interact with one another as a unit. The atmosphere, the water, and the soil both permit and limit life within an ecosystem. A freshwater lake, for example, provides all of the conditions necessary for certain fish and aquatic plants to flourish. Yet, the same lake would be inhospitable to plants and animals adapted to an estuary or a lake with different physical attributes, such as temperature, pH, or nutrient levels.

Just as the physical environment affects organisms, organisms affect their physical environment. Trees, for example, block sunlight, change the characteristics of soil, and re-

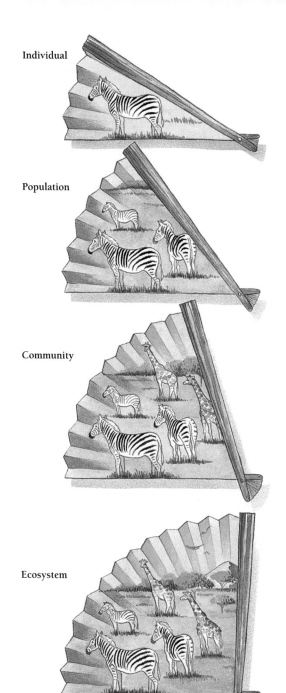

Individual

Population

Community

Ecosystem

Biosphere

lease oxygen into the atmosphere. Animals alter the physical structure of the environment. For example, beavers dam streams and elephants can uproot whole stands of trees in order to eat their leaves.

An ecosystem's boundaries are fuzzy. A pond may blend little by little into marsh, and then into a mixture of brush and open meadow. Streams bring nutrients and organisms from other areas into the pond and carry materials to other ecosystems downstream. All ecosystems interact with other ecosystems. Seeds disperse, animals migrate, and flowing water and air carry organisms, their products, and remains from one place to another.

Each ecosystem, community, or population may be part of a larger one. A pond ecosystem may be part of a meadow ecosystem, the meadow part of a forest, and the forest part of a two-thousand-mile drainage system such as the Mississippi River's. We may therefore speak of the frog population of the pond, of the meadow, of the forest, of the local drainage of streams, or of a continent-wide drainage system. How a population is defined depends on the mobility of the organism. Populations of interbreeding birds may cover whole hemispheres, while populations of frogs may be limited to a single valley. To some degree, then, ecologists choose the boundaries of the populations and ecosystems they study, always keeping in mind the larger picture.

In this chapter, we discuss ecosystems as entities whose parts interact in complex ways. In succeeding chapters of this section on ecology, we will see how:

- ecosystems and communities change over time
- human intervention affects ecosystems
- individual populations behave
- individual organisms behave.

Ecology is the study of the interactions of organisms with each other and with their physical environment. An ecosystem is a functional unit consisting of both biotic parts and abiotic parts. An ecosystem can be a subset of a larger ecosystem, and one ecosystem may blend smoothly into another. Nutrients and energy can also flow among different ecosystems.

Figure 25-3

Levels of organization. The biosphere is made up of different ecosystems, each ecosystem is made up of different communities, each community is made up of populations of different species, and each population is made up of individuals of the same species. An individual is a member of a population, a community, an ecosystem, and the biosphere.

  Learn more about levels of organization by clicking on this figure on your BiologyNow CD-ROM.

25.2 How Does Energy Flow Through Ecosystems?

Recall from the beginning of this chapter that life represents a local increase in order ("decrease in entropy"). This increase in order is powered by energy. The energy that drives ecosystems nearly all comes from the sun and passes from plants to plant eaters to carnivores in a series of steps. As energy passes from one level to the next, a certain amount of energy is lost, so that more energy is always needed. Without solar energy, all life on Earth would die out.

How Does Energy Enter Ecosystems?

Nearly all the energy that enters an ecosystem arrives as sunlight. Every square meter of the Earth's surface receives an average of 1.5 million Calories of light energy per year, the equivalent of about 3,000 McDonald's Big Macs. Of this, the atmosphere reflects 30 percent directly back into space as light, then absorbs another 20 percent, which it reradiates as heat. Of the remainder that reaches the actual surface of the Earth, most is absorbed by rocks, soil, and water. Only one or two percent of solar energy hitting the Earth actually enters ecosystems (about 60 Big Macs per square meter per year).

Different parts of the Earth receive different amounts of sunlight. Tropical regions at the equator receive nearly 2.5 times as much sunlight as areas near the poles. When the sun is low in the sky, as it is at the poles (and at dawn and dusk in all areas), more light reflects back into space. As a result, the poles receive far less sunlight than the equator. And while the equator receives the same amount of light all year around, the poles and the temperate regions, such as North America, receive more light in the summer and less in the winter. Such differences in energy supply profoundly affect the ecologies of these regions.

However much solar energy enters an ecosystem, it is nearly all first captured through **photosynthesis**, the process by which green plants and some bacteria use solar energy to transform carbon dioxide and water into sugar.

$$6\ CO_2 \ \overset{\text{light}}{+} \ 6\ H_2O \rightarrow C_6H_{12}O_6 + \ 6\ O_2$$
$$\text{carbon dioxide} \ + \ \text{water} \ \rightarrow \ \text{sugar} \ + \ \text{oxygen}$$

How Does Energy Flow Through Ecosystems?

Ecologists classify organisms into two groups according to where they get their energy. **Autotrophs** [Greek, *auto* = self + *trophe* = to nourish], or **primary producers**, are mainly plants and photosynthetic bacteria that produce their own food using solar energy. **Heterotrophs** [Greek, *hetero* = other + *trophe* = to nourish] are all the organisms that consume other organisms. Heterotrophs ultimately get all their energy from the solar energy captured by the autotrophs.

Ecologists further divide the heterotrophs into **consumers** (mostly animals and protists), which consume other organisms, and **decomposers** (mostly bacteria and fungi), which break down the dead. Consumers include both **herbivores,** such as deer, quail, and caterpillars, all of which eat plants, and **carnivores,** such as mountain lions, hawks, and wasps, all of which eat other animals.

Some animals feed from the whole carcasses of dead animals or from large pieces of dead plants and animals. Such **scavengers** include those that eat dead plant material, such as millipedes, sow bugs, earthworms, and termites, and flies, lobsters, and catfish, which often eat dead animals. Other organisms that consume and decompose the dead are **saprophytes** [Greek, *sapro* = putrid + *phyte* = plant], including many fungi, bacteria, and even some plants.

Many levels of consumers can exist. For example, hawks eat snakes, which eat ground squirrels, which eat lizards, which eat predaceous insects, which eat caterpillars, which eat plants. Such a linear sequence is called a **food chain.** Traditionally, ecologists have argued that food chains cannot have more than about three links (Figure 25-4). But in reality animals are enormously flexible about what they will eat, and chains of 6 to 11 links are common.

Ecologists assign the organisms in a food chain to different **trophic levels** [Greek, *trophe* = to nourish], depending on where they get their energy. For example, plants, which get their energy directly from the sun, are in the "first trophic level"; caterpillars, which get their energy from plants, are in the second; birds that eat caterpillars are in the third. Trophic level is simply an organism's position in a food chain.

But the different levels of a food chain are approximations only. Most predators eat from many levels of the food chain, depending on season; what food is available; and the size, age, and even sex of the predator. For example, female harrier hawks, which are larger than males, take larger prey than males do, and male carib hummingbirds, which are larger than their mates, take nectar from different flower species than the females. Many animals eat one thing as larvae and another as adults. Frogs, for example, are often herbivores as tadpoles but carnivores as adults. Some species mutually eat each other. The adult fish of species A may eat the young of species B, while the adults of species B eat the young of species A. Which species is at a higher trophic level? Clearly, neither.

Animals such as humans, bears, pigs, and crows that consistently eat almost anything they can digest are called **omnivores** [Latin, *omnis* = all + *vorus* = devouring]. One study of a desert ecosystem showed that nearly 80 percent of species were omnivorous. Most animals don't have the luxury of turning up their noses at food that isn't in the "right" trophic level. In many ecosystems, then, trophic level beyond the primary producers and strict herbivores is almost meaningless.

Every ecosystem has many interconnected food chains. The collection of all the food chains in an ecosystem is called

Figure 25-4
A simple food chain. In a simple food chain, the productivity of plants and other primary consumers goes to herbivores such as caterpillars, which are eaten in turn by predators such as robins.

a **food web** (Figure 25-5). Food webs are dynamic entities that change over time. In one year, a population explosion of oak moths means that bird and insect predators focus on oak moth caterpillars. In another year, oak moths are rare, and predators eat something else, perhaps mostly one thing, perhaps a diversity of insect prey.

One computer modeling study of food webs showed that most of the time only two trophic levels separate any two species, including the highest predator and the lowest herbivore. This result held up in predicting the behavior of seven real-life food webs, including desert, pond, and estuary ecosystems. Previous studies had assumed that food webs had many more distinct trophic levels and that the extinction of a single species would affect other species only

up to three steps away from its position in the food web. The new research suggests that there are no organisms three steps away. Losing a single species will affect all its predators and prey and all the organisms that interact with those predators and prey—in other words, most of the species in an ecosystem. Real-life food webs include many animals that eat members of their own species, their own predators, and surprising numbers of other species (Figure 25-5).

Every ecosystem includes one or more food chains that together form a food web. Primary producers transform energy from the sun into chemical energy, often sugar. Consumers obtain energy by eating producers, or organisms that have eaten producers. An organism's place in a food web is called its trophic level.

How Much Energy Flows Through an Ecosystem?

Every ecosystem is unique, whether it is a tiny pond or the vast Serengeti Plain of Africa. Yet similar ecosystems share fundamental characteristics, including climate, productivity, total mass of living organisms, and numbers of species. We'll discuss many of these characteristics in coming chapters. Here we can look at two important characteristics of ecosystems, both measures of how much energy flows through an ecosystem. These are *productivity,* the total energy captured in the chemical bonds of new molecules each year, and *biomass,* the total mass of all living organisms in an ecosystem.

Biologists can easily measure productivity by measuring plants' rate of photosynthesis and their rate of respiration, then subtracting one from the other. The result is the **primary productivity.** Primary productivity varies enormously from ecosystem to ecosystem. The most productive ecosystems are marshes. A marsh is twice as productive as a temperate forest, four times as productive as a wheat field, and 35 times as productive as a desert. In Chapter 26, we discuss some of the reasons that ecosystems differ in their productivity.

Another important characteristic of ecosystems is total **biomass,** the dry weight of all the organisms living in it—including producers, consumers, and decomposers. ("Dry weight" excludes water, so the dry weight of a 120-pound woman, for example, would be about 60 pounds.) Rain forests have more total biomass than other ecosystems, more even than the more-productive marshes.

In terrestrial ecosystems, plant biomass usually exceeds herbivore biomass, which exceeds carnivore biomass. A graph of the total biomass of organisms at each trophic level of a forest ecosystem looks like a pyramid, a **pyramid of biomass** (Figure 25-6). In aquatic ecosystems, this pyramid can be inverted (Figure 25-7). In a pond, for example, the biomass of algae may be less than the biomass of fish. How can the biomass of the consumers be greater than that of the producers? The answer lies in the flow of energy from one trophic level to the next.

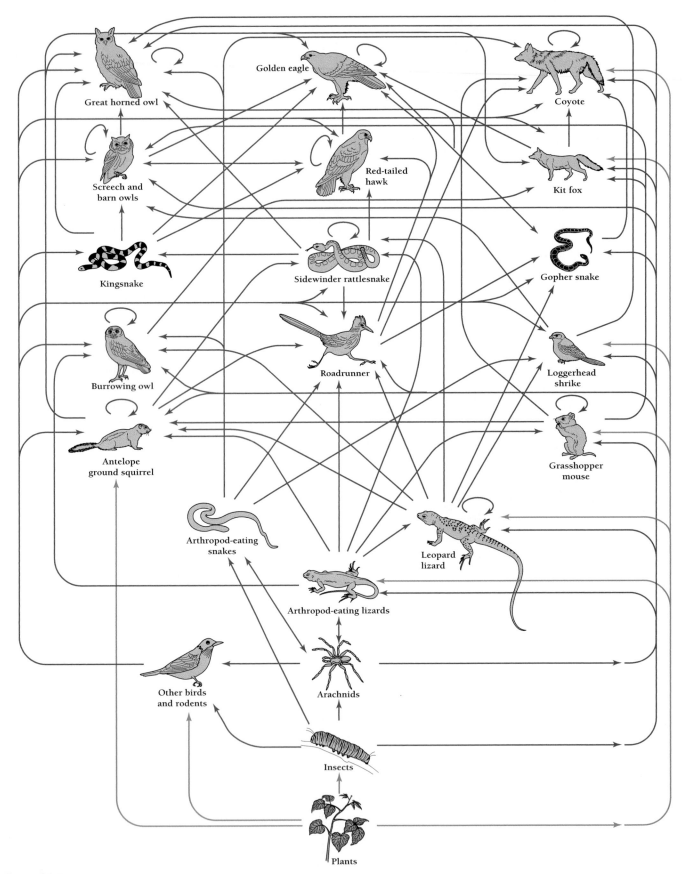

Figure 25-5

A food web. A food web is the sum of many food chains. The primary producers that capture sunlight are photosynthetic plants or bacteria. Organisms that only eat plants or photosynthetic bacteria occupy the next trophic level. Above that are organisms that eat other consumers, whose interrelationships may be complex. These could include, for example, bobcats that eat mice and rabbits or frogs that eat insects, which themselves may have eaten either plants or other insects. A frog may feed on insects that fed on a dead bobcat. Organisms at higher trophic level may feed from every other trophic level. Decomposers return nutrients back to primary producers. (After Gary Polis, 1991.)

Within the figure, the following labels appear:

Great horned owl · Golden eagle · Coyote · Screech and barn owls · Red-tailed hawk · Kit fox · Kingsnake · Sidewinder rattlesnake · Gopher snake · Burrowing owl · Roadrunner · Loggerhead shrike · Antelope ground squirrel · Grasshopper mouse · Arthropod-eating snakes · Leopard lizard · Arthropod-eating lizards · Other birds and rodents · Arachnids · Insects · Plants

Figure 25-6

A pyramid of biomass. In a grassland such as this National Bison Range in Montana, the biomass of the primary producers is greater than that of herbivores and carnivores—eastern meadow voles (*Microtus pennsylvanicus*) and bobcat (*Lynx rufus*). (Grassland, Richard J. Green/Photo Researchers, Inc.; voles, Gary Meszaros/Visuals Unlimited; bobcat, R. Lindholm/Visuals Unlimited)

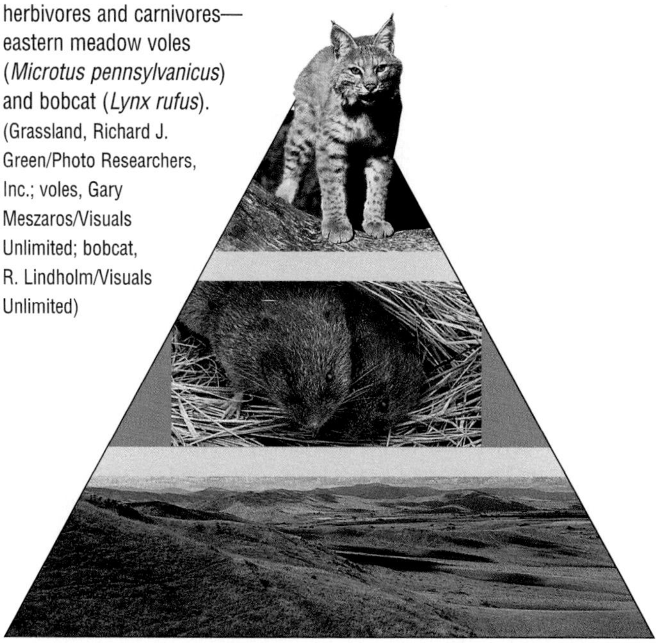

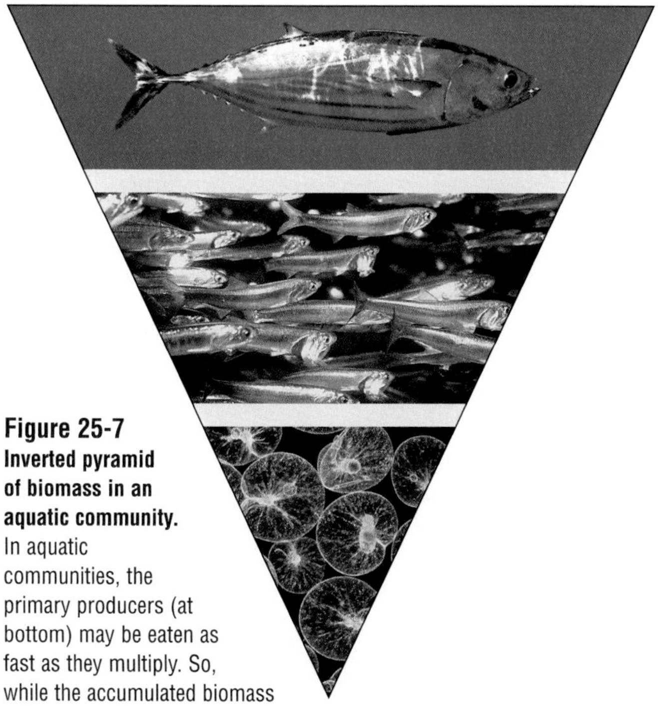

Figure 25-7
Inverted pyramid of biomass in an aquatic community. In aquatic communities, the primary producers (at bottom) may be eaten as fast as they multiply. So, while the accumulated biomass of the consumers is large, that of the producers may be small. Here schools of tuna feed on northern anchovies, which feed on plankton. (Tuna, David B. Fleetham/Tom Stack & Associates; anchovies, Norbert Wu/Peter Arnold, Inc.; plankton, D.P. Wilson/Science Source/Photo Researchers, Inc.)

During the course of a year, the algae in a pond grow faster than the fish. But the fish eat the algae as fast as it grows and convert only a fraction of it into fish. The rest is used to power movement and metabolism or is lost as heat.

Once we compare the energy flowing through the first trophic level with the energy flowing through higher levels, it becomes clear that the producers always process more energy than the consumers. This relationship holds in all ecosystems.

Every process releases energy in the form of heat (Chapter 5, Figure 5-4). As energy moves through the various trophic levels, from producers to herbivores to carnivores, more and more energy is lost, so that the amount of available energy in each trophic level looks like a pyramid. Such a **pyramid of energy** is always upright (Figure 25-8).

Of the approximately 1.5 million Calories that strike a square meter of land each year, only about one percent is actually available to herbivores such as deer and cattle. Of this 15,000 Calories, herbivores convert only ten percent (1,500 Calories) to tissues. Each trophic level passes on only about ten percent of the energy of the one below it; the rest is lost as heat. This generalization is called the **ten percent rule.** The ten percent rule explains why a pyramid of energy is always upright.

The ten percent rule also explains why ecosystems have so few trophic levels and so few individuals at the highest trophic levels. If on a square meter of land, herbivores supply 1,500 Calories per year, carnivores will have only 150 Calories

Figure 25-8
Pyramid of energy. The pyramid of energy is always upright, no matter whether for a forest or a pond. Here the grassland of the Masai Mara National Reserve in Kenya fixes carbon and energy on which zebras feed. These spotted hyenas, from Kruger National Park, South Africa, are top predators in a similar ecosystem. (Despite hyenas' reputation as scavengers, lions are actually more likely to steal prey from hyenas than the reverse.) (Masai Mara National Reserve, Renee Lynn/Photo Researchers, Inc.; zebras, Jeremy Woodhouse/DRK Photo; hyena, Gil Lopez-Espina/Visuals Unlimited)

per year to live on, about as many as are in a cup of spaghetti. So carnivores need to feed from a huge area.

The higher on a food chain an animal eats, the more area it needs to supply enough energy to live. Mountain lions, for example, roam over hundreds of square miles. All top carnivores are highly mobile animals, including, for example, sharks, killer whales, and eagles. Their travels force them to expend even more energy, which means that they need even more food. Top carnivores also tend to be "generalists," animals that eat a variety of foods. They can't afford to specialize too much.

Worldwide, humans use about 40 percent of the Earth's primary productivity. This includes not only wheat fields and farms for the food we eat and forests for the wood we use to build our houses, but also the areas on which we grow lawns and golf courses and areas that were formerly forests and marshes that are now covered by houses, roads, parking lots, factories, and shopping malls.

Two important measures of energy flow in ecosystems are primary productivity and biomass. A pyramid of biomass may not always be upright. But a pyramid of energy is always upright, because energy is lost as it moves from one trophic level to the next.

25.3 How Do Ecosystems Recycle Materials?

We can see that energy constantly escapes from the biosphere. In contrast, the biosphere recycles water, carbon, and other materials (Figure 25-9). As nutrients move from one trophic level to another, they may change form. But, unlike energy, nutrients rarely escape from the biosphere entirely. Sometimes these elements are part of organisms; sometimes they are part of the atmosphere (Table 25-1). A single carbon atom in your fingernail, for example, may have been, at different times, part of an apple, part of a bicarbonate ion in the ocean, and part of a lump of coal. Carbon and other materials pass through many forms—both biotic and abiotic—in a system called a **biogeochemical cycle.**

Oxygen, nitrogen, phosphorus, and other materials pass through biogeochemical cycles as well. Carbon dioxide moves from plants to consumers to water and air and back to plants and other producers. During photosynthesis, plants take up carbon dioxide and water, form chains of carbon and hydrogen (sugars), and release oxygen. Plants, animals, and other organisms then take that same oxygen from the atmosphere, use it to burn carbon compounds, and release carbon dioxide.

The biogeochemical cycles of materials such as carbon and oxygen involve the whole biosphere. A carbon atom

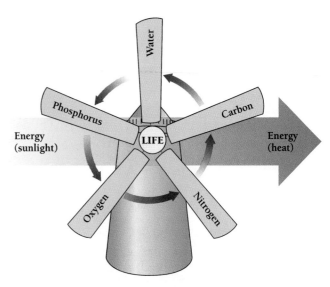

Figure 25-9
Ecosystems recycle material but lose energy forever. Sunlight drives ecosystems the way wind drives windmills. What's left is heat that passes into the atmosphere and, ultimately, out into space. In contrast, carbon, nitrogen, water, phosphorus, and oxygen cycle endlessly through the biosphere, like the vanes of a windmill.

may travel all over the world on currents of wind or ocean. In fact, atoms of oxygen and nitrogen become so widely dispersed that each time one of us takes a breath, we breathe in a few atoms from Julius Caesar's dying breath, in 44 B.C. Other materials, such as calcium and phosphorus, tend to stay put and cycle within smaller, more local ecosystems. We will discuss three biogeochemical cycles: those of water, carbon, and nitrogen.

Organisms recycle materials such as water, carbon, oxygen, and nitrogen, but they always lose energy through metabolism and respiration.

Table 25-1
Major Components of Earth's Atmosphere

Molecule	Percent in atmosphere
Nitrogen (N_2)	78.0
Oxygen (O_2)	21.0
Argon (Ar)	0.934
Water (H_2O)	0.10–1.0
Carbon dioxide (CO_2)	0.035
Neon (Ne)	Trace
Helium (He)	Trace
Methane (CH_4)	Trace

What Drives the Water Cycle?

All life is intimately connected to water. Organisms need a moist environment for the cells of their bodies and plenty of water for the insides of their cells. Vertebrates are 50 to 60 percent water, and plants even more. Plants use some 500 grams of water for every gram of biomass they produce. Some of this water is broken down to help form sugars during photosynthesis. But most of the water passes right through a plant unchanged. Plants pull water from the soil and release it as vapor through tiny pores (stomata) in their leaves, a process called **transpiration** (Chapter 31). On a hot summer day, an average maple tree transpires more than 50 gallons of water—enough to fill a large bathtub—every hour.

Flowing water also refreshes and flushes ecosystems. Running water brings in minerals, organic substances, and even organisms and carries away wastes. Water cycles through both biotic and abiotic parts of the biosphere (Figure 25-10).

More than 97 percent of the water on our planet is in the oceans, which cover 70 percent of the Earth's surface.

Sunlight evaporates large quantities of water from the oceans and the land. Of the water that evaporates from the land, plants transpire nearly 90 percent (Chapter 31). Despite all this "evapotranspiration" (evaporation plus transpiration), our huge atmosphere is only one-tenth to one percent water vapor. What water vapor there is, however, easily condenses into clouds that drop rain or snow onto the Earth's surface. Day after day, water evaporates, forms clouds, and rains in a rapid cycle that drops nearly a meter of rain over the entire surface of the Earth each year.

Overall, water moves from the oceans to the continents. Of the water that vaporizes into the atmosphere, only 17 percent comes from the land, and the rest comes from the oceans. Yet 25 percent of all rain falls *on* the land. Why does water move from the oceans to the land? Clouds over the ocean tend to stay warm, which allows them to hang on to their water. But when water vapor gets cold, it precipitates. Clouds blowing in off the ocean often encounter tall mountains, which push the clouds up higher into the atmosphere. There, cold temperatures cause the water vapor to

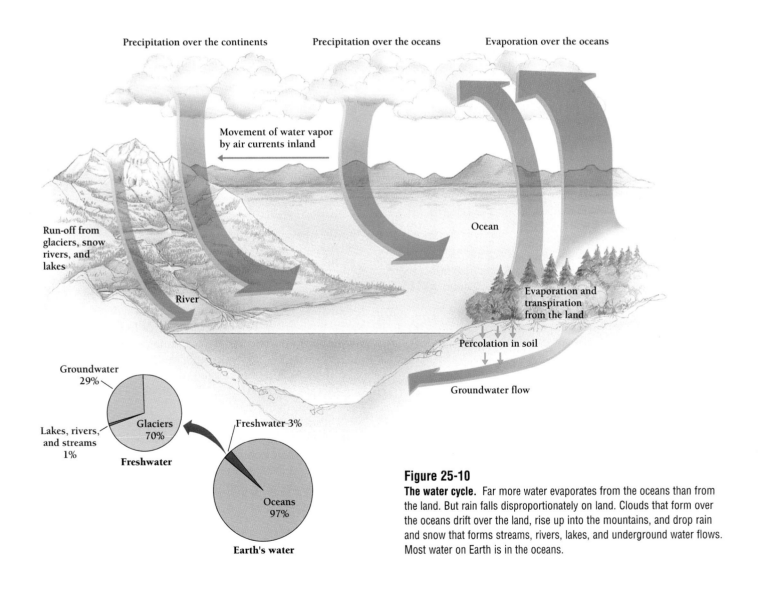

Figure 25-10

The water cycle. Far more water evaporates from the oceans than from the land. But rain falls disproportionately on land. Clouds that form over the oceans drift over the land, rise up into the mountains, and drop rain and snow that forms streams, rivers, lakes, and underground water flows. Most water on Earth is in the oceans.

condense into rain or snow, which drops onto the mountains and fills lakes, rivers, and underground aquifers (see Chapter 26, Figure 26-6). The water then returns to the oceans as runoff in streams and rivers.

Sunlight drives the evaporation of water from the oceans, the largest reservoir within the water cycle. This water then returns to Earth in the form of rain or snow and flows back to the oceans by way of rivers and streams.

Where Does Carbon Go?

Plants remove carbon dioxide from the atmosphere and transform it into sugars and other molecules. These forms of carbon eventually pass to herbivores, carnivores, scavengers, and decomposers, all of which release carbon dioxide back into the air.

Our atmosphere contains only trace amounts of carbon. Still, this adds up to about 635 billion tons, most of it in the form of carbon dioxide. But most carbon is not in the atmosphere or even in living organisms. The oceans contain nearly 50 times as much carbon as the atmosphere, mostly in the form of bicarbonate ions (HCO_3^-). Bicarbonate ions tend to settle to the bottoms of oceans, where they form thick sediments. Over millions of years, these immense layers of sediment turn into sedimentary rocks, which currently hold 16 million billion tons of carbon (Table 25-2).

By comparison, we organisms are a mere smear of carbon on the surface of the Earth. Living organisms contain only about 2,700 billion tons of carbon, one-tenth of what is dissolved in all of the oceans of the world. Each year, terrestrial plants convert about 12 percent of all atmospheric carbon dioxide into sugars and other organic molecules. This carbon then flows through a worldwide food web (Figure 25-11). Plants, herbivores, carnivores, and decomposers

break up these carbon compounds by oxidizing them to carbon dioxide. Eventually, the carbon returns to the oceans and atmosphere.

The largest reserves of carbon by far are in "fossil fuels," including coal, oil, and methane (natural gas). The 25 million billion tons of fossil fuels come from plants and other organisms that lived millions of years ago. Using the energy of the sun, these ancient organisms converted carbon dioxide to sugars and other chains of carbons. When these organisms died, however, their carbon compounds were not eaten and decomposed by other organisms, as usually happens, but transformed into thick oils that remained preserved deep underground for millions of years. Fossil fuels are fossilized sunlight, and there's only so much of this amazing resource. Current estimates are that we will have burned through most of our fossil fuel reserves in under 100 years. Burning fossil fuels releases huge amounts of energy, which can be used to run factories, warm houses, and power vehicles of every description (Figure 25-12). But it also increases the amount of carbon dioxide in the atmosphere, with disastrous consequences.

Carbon atoms cycle through oceans and atmosphere, with large reservoirs in living organisms and in fossil fuels.

How Have Humans Altered the Carbon Cycle?

Less than 200 years ago, humans began burning large amounts of coal, oil, and gas. In the process, we began returning fossil carbon to the carbon cycle. By also burning vast stretches of the world's forests, humans have increased the Earth's annual carbon dioxide production by 7.8 billion tons per year. As a result, total carbon dioxide in our atmosphere has increased by 30 percent since 1750.

To measure this change, scientists take ice samples from deep in the polar ice caps. The air trapped inside these frozen samples dates back thousands of years, and atmospheric scientists can measure the percentage of carbon dioxide in this ancient air. In about 1750, before the Industrial Revolution, the concentration of atmospheric carbon dioxide was about 280 parts per million (ppm). Now, 250 years later, the average concentration of carbon dioxide in the atmosphere is about 365 ppm. Most of the increase, however, has occurred in the last 50 years.

Because carbon dioxide traps solar energy, more carbon dioxide means higher temperatures. Indeed, global temperatures have increased by about 0.5°C in the last 100 years, about half of that in the last 40 years (Figure 25-13A). Warming in the 20th century was greater than at any time in last 1,000 years (Figure 25-13B). Of the 10 hottest years on record, 9 were between 1987 and 1997. No one questions that the world is heating up, and few now question that the cause is increasing carbon dioxide in the atmosphere.

Table 25-2
Where's the Carbon?

Coal, oil, and natural gas that are still in the ground contain more carbon than all other sources combined.

Where	Billions of tons	Percent
Fossil fuels	25,000,000	60.0
Sedimentary deposits	16,000,000	39.0
Oceans	30,000	0.07
Living organisms	3,000	0.007
Atmosphere	635	0.0015
Human carbon released/decade	80	

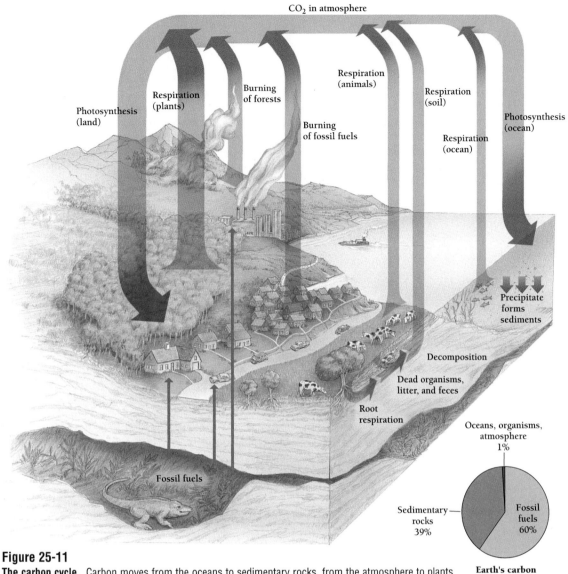

CO₂ in atmosphere

Photosynthesis (land)

Respiration (plants)

Burning of forests

Respiration (animals)

Burning of fossil fuels

Respiration (soil)

Respiration (ocean)

Photosynthesis (ocean)

Precipitate forms sediments

Decomposition

Dead organisms, litter, and feces

Root respiration

Fossil fuels

Oceans, organisms, atmosphere 1%

Sedimentary rocks 39%

Fossil fuels 60%

Earth's carbon

Figure 25-11

The carbon cycle. Carbon moves from the oceans to sedimentary rocks, from the atmosphere to plants and other organisms and back into the atmosphere, and from fossil fuels (ancient plants) back to the atmosphere. Most carbon is in the Earth's crust as sedimentary rocks or as fossil fuels (coal, oil, and gas).

The tendency of carbon dioxide, methane, and other gases to trap solar radiation is called the "greenhouse effect." The glass walls of a gardener's greenhouse allow light in. Once the light is inside, it degrades to heat. But the heat does not pass easily back out through the glass, so some of the

Figure 25-12

The 19-foot-long, 7-foot-tall Ford Excursion weighs more than 7,000 pounds. Sport utility vehicles (SUVs) now account for 40 percent of new-vehicle sales. Because these vehicles are technically trucks, brakes and other safety features are less strictly regulated than in cars, and they can legally burn more fuel and emit more pollutants per mile than passenger cars. The Ford Excursion, with the length and heft of a killer whale, is so big that the law requires no mileage tests.

Ron Kimball Photography, Inc.

heat is trapped inside the glass, where it warms the air inside the greenhouse. For a gardener in a cold climate, this is wonderful, but warming the whole Earth is not wonderful.

Scientists predict that unless we reduce carbon dioxide emissions, the average temperature of the Earth's surface may increase by 1.5° to 6°C over the next 100 years (Figure 25-14). The Earth's average temperature hasn't varied by more than about 1.0°C in the last 10,000 years, so an increase of even 2°C would cause havoc. Such an increase would melt the polar ice caps and raise sea levels worldwide by 18 to 20 inches, inundating coastal cities and destroying valuable wetlands. Already, the Arctic ice pack has lost about 40 percent of its thickness, the Antarctic ice cap is breaking up, and glaciers all over the world are melting away. Global sea levels have risen three times faster over the past 100 years than in the previous 3,000 years.

Extreme Biology   Measuring Carbon Dioxide Levels for 40 Years

The first hint that carbon dioxide might be a problem came in 1896, when a scientist noticed that high concentrations of "carbonic acid" in the air increased the temperature of the ground. By the 1930s, scientists had begun to believe that atmospheric carbon dioxide concentrations were increasing. Accurate measuring techniques designed in the 1950s removed all doubt.

In 1958, C. D. Keeling set up a system for measuring carbon dioxide concentrations in the air at Mauna Loa, a 4,169-meter active volcano on the island of Hawaii. The measurements were taken from towers erected on a barren lava field 3,400 meters above sea level. On the rocky slope, the local influences of vegetation and people were small, and Mauna Loa was and is one of the best places in the world for measuring undisturbed air. (When the volcano itself emits carbon dioxide, the researchers leave out those measurements.)

Keeling was not content to make his measurements for just a few years. He and his colleagues have made continuous measurements for more than 40 years, compiling the longest continuous record of atmospheric CO_2 concentrations anywhere (Figure A). Their precise records at Mauna Loa show a 15 percent increase in the annual concentration of CO_2, from about 315 parts per million in 1960 to about 365 ppm in 2000.

Amazingly, their methods and equipment have remained almost unchanged since 1959. The researchers collect air samples from air intakes at the tops of four 7-meter towers and one 27-meter tower. Four air samples are collected every hour and an infrared gas analyzer registers the concentration of CO_2 in the stream of air.

To ensure that the measurements are accurate, the flow is replaced by a stream of gas of known CO_2 concentration every 20 minutes. The gases used to calibrate the instruments are themselves calibrated against standardized gases whose CO_2 concentrations are measured even more accurately.

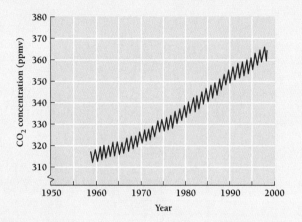

Figure A

Increasing carbon dioxide concentrations on Mauna Loa, Hawaii. Carbon dioxide levels rise smoothly from 1960 to 1999. The up-and-down squiggles represent seasonal variation in photosynthesis. In summer, plants photosynthesize more than they respire, taking up carbon dioxide and releasing oxygen. In winter, plants respire more, breaking down carbon-rich molecules and releasing carbon dioxide. Overall, carbon dioxide levels decrease during the growing season and increase during the rest of the year. (Source: Dave Keeling and Tim Whorf, Scripps Institute of Oceanography)

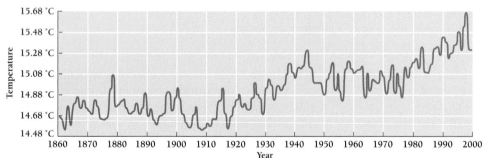

A.

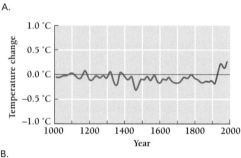

B.

Figure 25-13

Is it hot yet? The 20th century was the hottest century in the last 1,000 years. A. Since 1860, the average global temperature has increased about one degree Centigrade (almost two degrees Fahrenheit). B. This increase contrasts dramatically with average temperatures over the last 1,000 years. (School of Environmental Science, University of East Anglia)

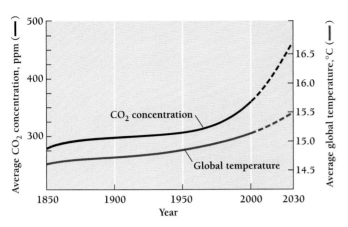

Figure 25-14

Atmospheric carbon dioxide and temperature. The close correlation between the increase in atmospheric carbon dioxide and global temperature suggests that carbon dioxide is causing global warming.

Biology ⓔ Now™ Learn more about global warming and the greenhouse effect by clicking on this figure on your BiologyNow CD-ROM.

Both plants and animals are responding to these changes. In the past 30 years, Europe's growing season has increased by 10 days. More than two dozen species of British birds have begun nesting earlier, and both bird and butterfly species are gradually moving north. In North America, several hot, rainless summers severely damaged wheat and corn crops. Yet, meanwhile, Alaska has become so warm that farmers there have been able to grow rye and even wheat. Meanwhile, researchers are worried that global warming could stop the Gulf Stream from bringing warm water north along the Atlantic Coast of the United States, leading to a far colder climate for the Northeast.

Nations around the world have leaped to halt the trend. In December 1997, 174 nations signed a treaty in Kyoto, Japan, agreeing to reduce overall carbon dioxide output. The "Kyoto Protocol" requires industrialized nations to cut their greenhouse gas emissions by five percent by the year 2010. Nearly all industrialized nations have promised to do so. Only Australia and the United States, the largest emitter of these gases, have so far refused. In 2003, the United Kingdom announced its commitment to reduce its carbon dioxide output by 60 percent over the next 50 years. The U.K. plans to do so partly by boosting energy output from renewable energy sources (e.g., solar power and wind generators) from three percent to 20 percent by 2020.

How Can We Get Rid of Carbon Dioxide?

Most proposed solutions to the problem of global warming do not include the obvious solution, which is to burn less fuel. This could be accomplished through a combination of energy conservation and the use of alternative sources of energy such as solar power, wind power, or nuclear power. But because most nations want to maintain or even increase fuel

consumption, researchers are looking for ways to remove carbon dioxide from the atmosphere.

Hide It in the Trees

One way is to plant trees that can mop up excess carbon dioxide through photosynthesis. In Canada, for example, plans are under way to plant billions of poplar trees over millions of acres of land. As these trees grow, they will take up carbon from the atmosphere and store it as wood. In Bolivia, a coalition of three American power companies, a U.S. environmental organization, and the Bolivian government are spending $10 million to protect tropical forests there.

Bury It

Meanwhile, a Norwegian oil company has begun pumping carbon dioxide into the ground under the North Sea at the rate of a million tons a year. And on Hawaii's Kona Coast, the Japanese, Norwegian, and U.S. governments together have built a 1.5-mile pipe that can pump liquid carbon dioxide 1,000 meters below the surface of the ocean.

Fertilize the Ocean

Perhaps the most dramatic plan is to stimulate the growth of ocean plankton by fertilizing the ocean with iron. In one study, iron fertilizer caused a 60 percent drop in carbon dioxide released into the atmosphere. Researchers hypothesize that the extra carbon goes into plankton. These either become part of the ocean food chain or die and sink to the bottom of the ocean, carrying their carbon with them.

Humans have increased carbon dioxide concentrations in Earth's atmosphere by about 30 percent by burning fossil fuels and cutting down forests. The apparent result is an increase in global temperatures that may dangerously alter the global climate. The Kyoto Accord is an agreement by 174 nations to reduce carbon dioxide in the atmosphere.

Why Is Nitrogen Both Common and in Short Supply?

Molecular nitrogen (N_2) is the most abundant element in the atmosphere, comprising 78 percent of the air we breathe (Figure 25-15). Nitrogen is also an indispensable part of the proteins and nucleic acids that all organisms need. Yet, ironically, usable nitrogen has been in short supply for millions of years. The huge quantities of atmospheric nitrogen are not directly available to animals, plants, and most other organisms because the two atoms of a nitrogen molecule are held together by a powerful, almost unbreakable triple bond. Plants, animals, and most other organisms cannot break this bond. They can only use **fixed nitrogen** such as **ammonium ions** (NH_4^+) or nitrate ions (NO_3^-).

Where does fixed nitrogen come from? One important source is lightning. During storms, the high temperatures generated by lightning—greater than the surface tempera-

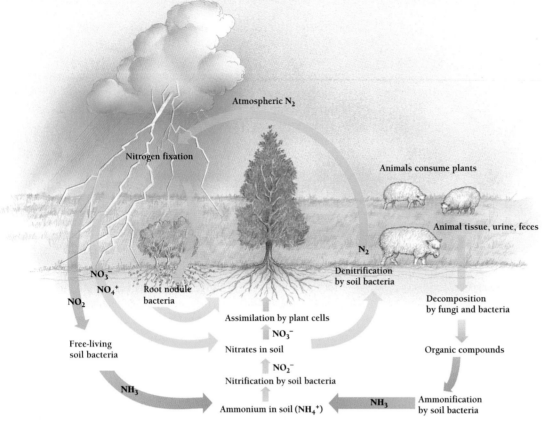

Figure 25-15
The nitrogen cycle. Nitrogen atoms in the atmosphere are fixed by bacteria and lightning and assimilated by plants. Nitrogen continuously cycles through the atmosphere, soils, oceans, and organisms.

ture of the sun—convert atmospheric nitrogen to NH_4^+ and NO_3^-. Rain washes the fixed nitrogen to Earth, supplying 5 to 10 percent of all usable nitrogen.

Amazingly, however, most nitrogen is fixed by bacteria. These tiny organisms have evolved powerful enzymes that can break the triple bond in N_2 and convert it to ammonia. Some of the **nitrifying bacteria** are free-living bacteria (such as cyanobacteria) that live in soil or water and convert nitrogen to ammonia. Others live in mutualistic relationships with certain plants. For example, plants in the pea family (peas and beans) play host to *Rhizobium* bacteria, which live inside tiny "root nodules" growing on the plants' roots. The *Rhizobium* bacteria convert atmospheric nitrogen into ammonia that the plants use. The plants, in turn, provide the bacteria with energy-rich sugars (an example of symbiosis).

Plants with symbiotic nitrifying bacteria can use ammonium ions directly in the synthesis of amino acids and proteins. Plants with no nitrifying bacteria survive on what is in the soil. Unfortunately, bacteria in the soil break down most of the free ammonium to nitrite (NO_2^-) and then nitrate (NO_3^-), a process called **nitrification.** Still other bacteria convert nitrate to N_2, a process called **denitrification.** Because these bacteria transform fixed nitrogen back into N_2, fixed nitrogen is usually in short supply.

Ecosystems also lose fixed nitrogen when it washes away. Nitrate is highly soluble, and heavy rains can wash it out of soils and into lakes and streams. Eventually, it ends up in the oceans.

Because nitrogen is often in short supply, plants soak up nitrate very easily and conserve it by removing it from dying leaves, stems, and roots. However, such conservation measures are often thwarted. For example, when caterpillars cut living leaves, the nitrogen-filled leaves fall to the forest floor, where their nitrates may wash from the ecosystem. Animals and other consumers also lose nitrogen. When animals break down proteins during respiration, they convert them to ammonia, urea, or uric acid. Bacteria obtain energy from urea and uric acid by converting them back into ammonia, a process called **ammonification.** The ammonia (or nitrite or nitrate) can then be used once again by plants, completing the cycle of fixed nitrogen among plants, animals, and decomposers.

In natural ecosystems, most fixed nitrogen moves locally from producers to consumers and decomposers and back to producers. Only small amounts of nitrogen are added to the system through nitrification or lost through denitrification.

Fixed nitrogen cycles locally among all the organisms of an ecosystem. Lightning and nitrogen-fixing bacteria fix nitrogen. Small amounts of fixed nitrogen washes out of ecosystems and into oceans. Soil bacteria also return nitrogen to the atmosphere.

How Have Humans Altered the Nitrogen Cycle?

The world's organisms have evolved over millions of years in an environment with limited amounts of available nitrogen. Yet, in the last few decades, human activity has doubled the amount of fixed nitrogen entering the biosphere and accelerated the movement of fixed nitrogen through ecosystems. In some areas, such as Antarctica, virtually no human-generated nitrogen enters ecosystems. But in highly developed areas such as northern Europe, the extra nitrogen is damaging ecosystems.

The extra fixed nitrogen comes from many sources. The largest single source is industrial fertilizers. The United States alone manufactures about 40 billion pounds a year for use on crops, but virtually every country in the world manufactures cheap fertilizer (Figure 25-16). Another source of extra nitrogen is the crops themselves. On nearly one-third of the Earth's surface, crops have displaced natural vegetation. Many of these crops are soybeans, peas, alfalfa, and other members of the pea family that harbor nitrifying bacteria, which fix atmospheric nitrogen. We humans also release large amounts of fixed nitrogen when we burn forests and grasslands, drain marshes, and burn fossil fuels in automobiles, factories, and power plants.

All this extra fixed nitrogen affects the atmosphere, the carbon cycle, and the functioning of ecosystems. All kinds of fixed nitrogen—including nitrous oxide, nitric oxide, and ammonia—are greenhouse gases that are helping to raise global temperatures. In addition, when nitric oxide dissolves in water, it becomes nitric acid, the major component of acid rain, which kills ponds, forests, and fish.

But excess fixed nitrogen in soils can cause profound changes to ecosystems. For example, in terrestrial ecosystems, excess nitrogen leaches nutrients from the soil. As ammonium accumulates in the soil, fungal decomposers lose out in competition with bacterial decomposers. The bacteria convert ammonium to nitrate and release hydrogen ions, which acidify the soil. Because the nitrates have negative charges, they attract lots of positively charged minerals, including calcium, magnesium, and potassium. Then, because the nitrates are water soluble, they leach away into streams and lakes, carrying away the calcium, magnesium and potassium. In this way, extra nitrogen that is sometimes used to fertilize the soil actually decreases overall soil fertility. In addition, as calcium leaches away and soil acidity rises, increasing amounts of aluminum dissolve in soil water. The aluminum poisons the roots of trees, and, washed into nearby streams, also kills fish.

Over time, soils become less and less fertile, and trees grow very slowly or not at all. In parts of northern Europe, the Front Range in Colorado, and the forests surrounding the Los Angeles Basin, excess nitrogen is so serious that forests are slowly dying off. In many ecosystems, the amount of nitrogen available controls the nature and diversity of plant and animal life and the cycling of carbon and minerals. Excess nitrogen decreases the diversity of species in an ecosystem, as a handful of species grow wildly out of control while others are crowded out.

Worldwide, 20 percent of all forms of human waste, including sewage and fertilizer, ends up in rivers. In aquatic ecosystems, excess nitrogen, two-thirds of which comes from farms, washes down rivers and accumulates in estuaries, creating "dead zones" in coastal areas such as the Gulf of Mexico and the Adriatic Sea. In a dead zone, massive blooms of algae draw millions of tiny grazing plankton. As the algae die and drop to the bottom, along with the grazers' fecal pellets, staggering numbers of bacteria bloom at the bottom and quickly consume all the oxygen in the water. Fish, shrimp, and crabs swim away from the deoxygenated water, but bottom creatures such as clams, snails, and worms slowly suffocate. In extreme cases, every living organism either swims away or dies.

Dead zones can be enormous. In the summer of 1999, researchers at Louisiana State University mapped a dead zone at the mouth of the Mississippi River that covered 20,000 square kilometers (7,700 square miles), an area the size of Massachusetts (Figure 25-17). By 2002, it had ballooned to 22,000 square kilometers. Although the Mississippi River dead zone is the largest in the Western Hemisphere, it is only one of many and there are even larger ones elsewhere in the world. For example, in the Baltic Sea a dead zone covers up to 38,000 square miles.

> Human activity has doubled the amount of fixed nitrogen entering the biosphere and accelerated the movement of fixed nitrogen through ecosystems. The extra nitrogen comes from industrial fertilizers, crops, burned tropical forests, and the burning of fossil fuels. In the long run, excess nitrogen decreases the fertility of soils, kills forests and aquatic ecosystems.

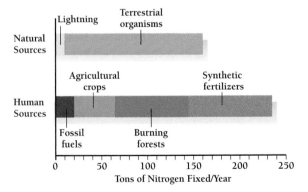

Figure 25-16

Sources of fixed nitrogen. Humans produce half again as much fixed nitrogen as all natural sources combined. As a result, available nitrogen in the biosphere has more than doubled. Excess nitrogen damages soils and kills forests and fish.

In this chapter, we have discussed the flow of energy and the cycle of materials within ecosystems. In the next chapter, we will survey a selection of ecosystems. Along the way, we will see how humans are damaging some of these ecosystems.

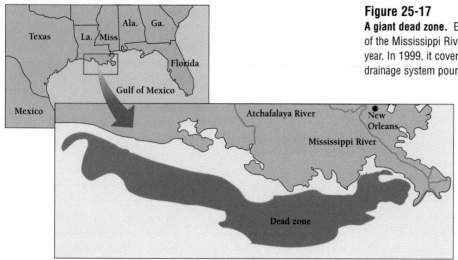

Figure 25-17
A giant dead zone. Each summer, a huge dead zone forms at the mouth of the Mississippi River. This dead zone seems to be getting larger each year. In 1999, it covered 20,000 square miles. The rivers in the Mississippi drainage system pour a total of about 1.6 million metric tons of nitrogen into the Gulf of Mexico each year. Blooms of algae and plankton stimulate blooms of bottom bacteria, which deplete the water of oxygen. Fish leave and other organisms die.

Key Concepts

- An ecosystem is an interacting group of species, together with nonliving components such as sunlight, soil, water, and air. All of the world's ecosystems, together with the atmosphere and the oceans, make up the biosphere.
- Food webs are the combination of all possible food chains within an ecosystem. Most of the time, only two links separate any two species.
- Plants and other photosynthetic organisms capture energy from sunlight, but only a small fraction of that energy is available to other organisms.
- The biosphere recycles water, carbon, nitrogen, and other nutrients but forever uses and loses energy.
- Without the effect of greenhouse gases, Earth would be a lifeless planet. But too much carbon dioxide, methane, and other greenhouse gases can overheat the planet.

Summary with Key Terms

In the 1960s James E. Lovelock proposed that the Earth's atmosphere, oceans, and soils were a single, immense, symbiotic **superorganism** called Gaia. Remnants of his idea persist in our current understanding of what the **biosphere** is and how it works.

Ecology is the science concerned with understanding the relations among organisms and their environment. Ecologists study **populations**, **communities**, and **ecosystems**. A **system** is an assemblage of interacting parts or objects. An ecosystem is the living **(biotic)** community together with the nonliving **(abiotic)** components of the community's environment, including soil, water, air, and weather. An ecosystem shares some properties of an organism, but not others. Ecosystems can be subsets of larger ecosystems, and one ecosystem can blend smoothly into another. Nutrients and energy can also flow among different ecosystems. All the world's ecosystems are bound together into a global biosphere by flowing energy and nutrients.

How does photosynthesis provide the energy that drives all ecosystems?

Almost all the energy that drives ecosystems comes from the sun. Green plants and some bacteria capture this solar energy through **photosynthesis**.

How are food webs both simple and complex?

The biotic part of an ecosystem consists of **autotrophs** (or **producers**) and **heterotrophs**. Heterotrophs include **consumers** (such as **herbivores, carnivores, scavengers**) and **saprophytes**. Autotrophs belong to the first **trophic level** in every **food chain**. Heterotrophs occupy different trophic levels, from the second level up, depending on what they eat. Animals that eat from several levels of a food chain, including humans, bears, pigs, and crows, are called **omnivores**. Food chains are usually interwoven into a complex **food web**. Food webs are full of loops. For example, many animals eat others of their own species or eat their own predators.

How do energy and nutrients flow among the individual members of ecosystems?

Energy flows through ecosystems. One way of thinking about energy flow is in terms of a **pyramid of biomass** (dry weight) or a **pyramid of energy**. Only the pyramid of energy is always an upright pyramid. This is because, according to the **ten percent rule**, each trophic level has access to only about ten percent of the **primary productivity** available to the one beneath it. Humans use about 40 percent of the Earth's primary productivity.

Ecosystems recycle water, carbon, nitrogen, and many other materials in **biogeochemical cycles**. Water enters the atmosphere from the oceans through evaporation and from the continents and islands through evaporation and **transpiration** from plants. Atmospheric water condenses into clouds, then precipitates as rain, sleet, or snow. Rain and snowmelt form streams and rivers that carry fresh water to the oceans. Along the way, fresh water carries minerals, organic materials, organisms, and wastes into and out of ecosystems.

Plants and other producers absorb carbon dioxide and turn it into sugar through photosynthesis. Producers, consumers, and decomposers all produce carbon dioxide by respiring—oxidizing carbohydrates back to carbon dioxide. In aquatic ecosystems, carbonates tend to precipitate to the bottom to form thick sediments. These sediments form rocks that are eventually lifted up and eroded. Large reserves of carbon also exist in the form of fossil fuels.

The Earth's atmosphere is 78 percent nitrogen. But this nitrogen is not available to most organisms until lightning or **nitrifying bacteria** transform it into **ammonium ions** or nitrate, which plants can use to make proteins. Soil bacteria break down ammonium to nitrite (NO_2) and then nitrate (NO_3) in a process called **nitrification.** Other bacteria convert nitrate back to N_2, a process called **denitrification,** and the nitrogen returns to the atmosphere.

Consumers and decomposers get such **fixed nitrogen** by eating plant proteins, or each other. Animals excrete waste nitrogen as ammonia, urea, or uric acid. Bacteria obtain energy from urea and uric acid by converting them back into ammonia, a process called **ammonification.** Ecosystems conserve nitrogen, but some is lost when rain washes it from fallen organic matter and carries it to the sea.

What are the results of human alteration of geochemical cycles such as the water cycle, the carbon cycle, and the nitrogen cycle?

Over the last 200 years, humans have returned fossil fuel carbon to the carbon cycle. By also burning continent-sized stretches of the world's forests, humans have increased the Earth's annual carbon dioxide production by 7.8 billion tons per year, so that total carbon dioxide in our atmosphere has increased by 30 percent since 1750. The parallel increase in global temperatures supports the idea that atmospheric carbon dioxide causes a "greenhouse effect." Human activity has also doubled the amount of fixed nitrogen entering the biosphere and accelerated the movement of fixed nitrogen through ecosystems. The extra nitrogen comes from industrial fertilizers, crops, burned tropical forests, and the burning of fossil fuels. In the long run, excess nitrogen decreases the fertility of soils, kills forests and aquatic ecosystems.

Review and Thought Questions

Review Questions
1. What is ecology?
2. What distinguishes an ecosystem from a community?
3. What properties do ecosystems share with organisms? How are ecosystems different from organisms?
4. Distinguish between autotrophs and heterotrophs. Give two examples of each.
5. Explain why the pyramid of energy is always narrowest at the top.
6. Explain how the biomass of aquatic producers can be smaller than the biomass of the consumers.

7. From how many trophic levels have you eaten in the last 24 hours? If there are consumers or decomposers such as animals, fungi, or bacteria on your list, draw a food chain, showing how they are interconnected.
8. Sketch a food web of the organisms that live in and around your house.
9. Where does the carbon dioxide in the atmosphere come from?
10. Humans are omnivores and eat from many levels of the food chain. Make a list of primary producers, herbivores, scavengers, and decomposers that humans commonly eat. Include a plant, an animal, a fungus, a protist, and a bacterium.
11. Draw and label a diagram of the nitrogen cycle.

Thought Questions
12. Describe the steps of the nitrogen cycle. How is nitrogen lost from ecosystems? How is it replaced? (Mention two ways.) Why does it need to be "fixed"?
13. Suppose a strawberry farmer treated his fields with a fumigant (poison gas) that killed all of the bacteria in the soil. Describe two different things the farmer could do to ensure that the strawberries received an adequate supply of nitrogen.
14. Suppose that all of the primary producers were removed cleanly and instantaneously from the biosphere. List five immediate consequences (e.g., declines in oxygen levels) and five indirect consequences.
15. A gallon of gasoline weighs only six pounds. Yet each gallon of gasoline burned releases almost 20 pounds of carbon dioxide into the atmosphere. Assuming gasoline is all octane, examine the reaction below and explain why the carbon dioxide released weighs more than the gasoline.

$$2 \, C_8H_{18} + 25 \, O_2 \rightarrow 16 \, CO_2 + 18 \, H_2O$$

16. Explain how a carbon atom in one of your front teeth could easily have been in the tooth of a *Tyrannosaurus rex* 65 million years ago.

BiologyNow Resources

Biology ❸ Now ™

Active Figures
25-3: Levels of organization
25-14: Global warming and the greenhouse effect

Preparing for an exam? Take a diagnostic test on your BiologyNow CD-ROM.

Online materials relating to this chapter are at:
http://biology.brookscole.com/AAL3

About the Chapter-Opening Image
NASA's first breathtaking photographs of Earth as a cloud-swathed blue ball floating in space reminded Earth's human population that Earth has definite boundaries and that there is only one Earth.

26 Terrestrial and Aquatic Ecosystems

Key Questions

- How do light, temperature, and humidity determine the characteristics of an ecosystem?

- How do mountains affect rainfall and temperature?

- What characteristics of aquatic ecosystems determine if they will be species rich and productive or species poor and unproductive?

- How do human activities affect ecosystems?

How Many Ecosystems Can You See in a Day?

One hundred fifty years ago, pioneers in covered wagons struggled to cross a flat, sunbaked desert, the last barrier before reaching fertile California. In the salt-encrusted flats of one 140-mile-long valley, not a single plant grew. Even on the slopes above the valley, plants grew only sparsely and water was rare. The oxen pulling the wagons died one by one, and the pioneers who made it to the other side called the place Death Valley.

In the summer, daytime temperatures in Death Valley, California, can reach an unbearable 134°F. Average yearly rainfall is a minuscule 1.66 inches. The long valley lies sunken along a geologic fault line, surrounded by rocky mountains nearly as dry as the desert itself. It is the lowest and hottest place in North or South America.

Lowest of all is Badwater, 282 feet below sea level. Badwater is the last place most people would want to go in the summer. Yet this salty pool is the starting line for a footrace that makes the Boston Marathon look like a jog around the block. Some 135 miles to the west and more than 15,000 feet higher lies the top of Mt. Whitney, the highest point in the contiguous United States, and the unofficial finish line for the grueling race. Each summer, a band of 20 to 80 runners attempts the 135-mile race from Badwater to Mt. Whitney (Figure 26-1). Race times range from 70 hours to as little as 26 hours.

Along the way, the runners pass through one of the most abruptly changing series of biological communities in the world, including most of the communities discussed in this chapter. The runners begin the race at 6 A.M., when the desert is still relatively "cool." Nighttime air temperatures sometimes drop no lower than 95°F. The flat floor of Death Valley itself supports little more vegetation than the moon. And ground temperatures regularly reach 200°F.

From Badwater, there is nowhere to go but up, and the runners climb out of Death Valley into the foothills and mountains of the Panamint Range, passing miles of widely dispersed shrubs. Desert salt bush has small, leathery leaves as salty as the saltiest corn chip. In the driest places, creosote bushes grow, each one 10

Matt Frederick

Mt. Whitney (14,494 ft)
Mile 150 – **unofficial finish line**

F

Mt. Whitney trail head (8360 ft)
Mile 135 – **official finish line**

E
D
C
B

Lone Pine
(3700 ft)
Mile 122

Townes Pass
(5000 ft)
Mile 59

A

Badwater
(-282 ft)
Mile 0
**Start
of race**

Sea level

Figure 26-1
Six ecosystems in a day. Runners in the punishing
Badwater 146 race pass through six ecosystems in just
over 24 hours on their way from Death Valley to the top
of 14,000-foot Mt. Whitney.

California

Matt Frederick

or 20 feet from the last. As in most **deserts,** only
10 percent of the ground has any vegetation whatever. Some of the runners may see animals such as a
rosy boa, a kit fox, or a roadrunner.

The human runners climb several thousand feet out of
Death Valley, and while temperatures continue to rise as the
sun rises overhead, average temperatures here are lower
than in the valley below. Creosote bush gives way to desert
sage, a species that tolerates the cold winters of high-altitude
deserts. By the time some of the runners reach the top of
Townes Pass (5,000 feet), it is evening and they are running
into a nauseatingly hot sun as it sets over still another range
of mountains.

By noon the next day the lead runners have reached the
broad Owens Valley. A hundred years ago, rich **grasslands**
and orchards surrounded the 15-mile-long Owens Lake.
Towns and tourist resorts dotted the edge of the lake, and a
steamboat ferried silver bullion to the south end of the lake.
Today, the lake bottom is dust; all but a film of water at the
center has been diverted to the millions of people living in
the cities of southern California. Owens "Lake" is so dry and
windy that its frequent dust storms are the largest single
source of airborne particulate pollution in the United States.

The runners push past the dead lake and north to the
town of Lone Pine, where those few who can resist stopping
to eat and rest continue on and turn west. For those who stop,
it's a good time to change shoes, as the runners' feet may swell
several shoe sizes. The runners have come 122 miles so far,
but towering before them is a 6,000-foot nearly vertical cliff
of solid granite, the grand eastern scarp of the Sierra Nevada
mountain range. Pacing themselves, they climb up through
the last of the desert sagebrush along a road cutting up the

edge of a deep canyon. Below them, a mountain stream slides
through narrow meadows and cottonwoods, its banks overgrown with thickets of willow, giving the runners their first in-
toxicating whiff of fresh mountain water. They have arrived in
a **temperate deciduous forest.**

As the runners climb higher, they pass through dense
thickets of prickly shrubs, or mountain **chaparral.**

Gradually, as the runners gain altitude, the chaparral of
small pinyon pines and mountain mahogany gives way to
taller and more open **coniferous forest** of Jeffrey pine, western white pine, and red fir (Figure 26-2). The air is cooler
and thinner, and breathing becomes difficult. But the runners have now reached Whitney Portal, the end of the paved
road up Mt. Whitney. Here, at 8,361 feet above sea level and
135 miles from Death Valley, the race is officially over. In
2002, Pam Reed, of Tucson, Arizona, won the race more
than four hours before the next runner. The next morning,
she went running, then headed up the trail to the top of
Mt. Whitney with a friend, while fellow racers were still attempting to finish the grueling marathon. Some runners
consider the real end of the race to be the top of Mt. Whitney. Those who continue on from Whitney Portal (or hike it
the next day) find that the coniferous forest soon thins out

Figure 26-2
Coniferous forest. At higher elevations and latitudes, cooler temperatures and moister soils support coniferous forests.

to a hot open **chaparral** of mountain mahogany, prickly Sierra chinquapin, and sagebrush.

As the trail briefly levels out in a small valley, the chaparral turns to another coniferous forest of lodgepole and fox-tail pines, with a cool understory of willow and a wildflower studded marsh along Lone Pine Creek. The runners (or hikers) climb up another rocky trail, through a willow-covered meadow, past a waterfall, and up a series of switchbacks through a garden of wildflowers. At timberline, the few sparse, stunted trees give way to alpine **tundra,** composed of grasses, nonwoody plants (herbs), and knee-high willows and huckleberries. Still higher, even these thin out to an occasional bunch of wildflowers trembling in the wind on the bare slopes of the mountain. The runners pant up a granite trail that was long ago blasted in the side of the mountain with dynamite. Looking back from among giant blocks of granite, the runners can see the Owens Valley 10,000 feet below. Even the very summit of Mt. Whitney, 14,494 feet high, is not quite bare rock. The summit is too high, too cold, and too exposed to wind and weather to support trees, shrubs, or even herbs. But here and there, lichens—symbiotic associations of fungi and photosynthetic algae—grip the cold rock.

Those who make it to the top collapse on the rocks, euphoric with accomplishment and oxygen deprivation. The rest of the runners are scattered across six distinct **ecosystems**— desert, temperate grassland, temperate deciduous forest, chaparral, coniferous forest, and tundra.

26.1 How Does Climate Determine the Nature of an Ecosystem?

In every region of the world, the physical environment and ecological succession lead to a characteristic community of plants and animals. Regions with similar climates tend to

have similar ecosystems. A desert ecosystem is the same in the Sahara and in the Mojave. The weather is dry and the plants and animals are adapted to extreme "aridity," or dryness. The deserts of the southwestern United States and northwestern Mexico, central Asia, Africa, and Australia all have plants with spiny or succulent leaves and burrowing animals that emerge only at night. Likewise, the northern parts of Canada, Alaska, Scandinavia, and Russia all share the same kind of sparse forest, called taiga (Figure 26-3).

But even though similar ecosystems look alike, they differ in the particular species they include. The grasslands of North America are dominated by deer, elk, and bison, while the grasslands of Africa are dominated by zebra, gazelles, elephants, and giraffes. Differences in the individual species

A.

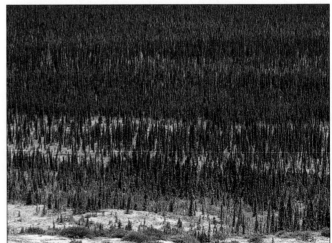

B.

Figure 26-3
Taiga here, taiga there. Although the particular tree species may vary, all taiga is characterized by sparse, stunted coniferous trees, whether in Canada, Scandinavia, or Russia. A. Siberian taiga. B. Spruce taiga in central Alaska.

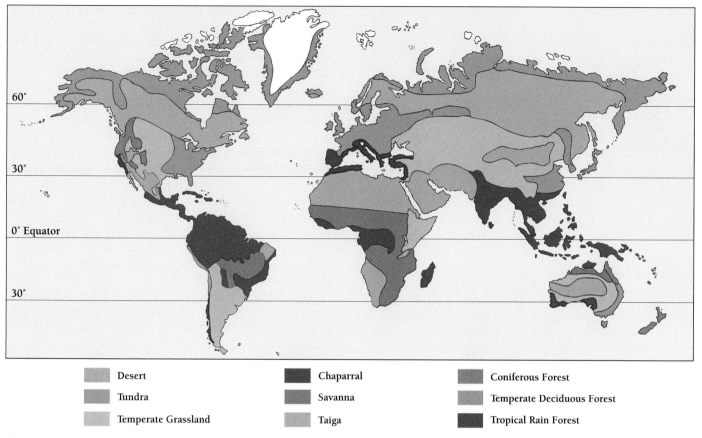

Desert		Chaparral		Coniferous Forest
Tundra		Savanna		Temperate Deciduous Forest
Temperate Grassland		Taiga		Tropical Rain Forest

Figure 26-4

World distribution of ecosystems. The sharp boundaries in this diagram are misleading, as ecosystems frequently grade together smoothly. In addition, each solid area actually represents an average of many vegetation types. Botanists divide California alone into 24 "floristic provinces." Notice that the distribution of ecosystems around the world matches rainfall (Figure 26-6).

Biology⒠Now™ Learn more about ecosystems by clicking on this figure on your BiologyNow CD-ROM.

in an ecosystem are largely the result of the unique evolutionary histories of each region.

In this chapter, we discuss eight terrestrial ecosystems and several aquatic ecosystems. Along the way, we will see what effects humans have on each of these ecosystems. We begin with a discussion of what factors determine climate in different parts of the world (Figure 26-4).

How Do Sunshine and the Earth's Movements Determine Climate?

The two factors that most determine what kinds of plants will grow in an ecosystem are temperature and moisture. These depend, in turn, on the intensity of sunlight and patterns of wind, rain, and ocean currents. All of these depend on the Earth's movements and the distribution of mountains and valleys.

The Earth's daily rotation constantly mixes the air, so that all the air at a given latitude has about the same temperature. But different latitudes have different temperatures and climates. This is because the latitudes closest to the poles receive less solar energy and experience more extreme seasons than those near the equator (Figure 26-5). As a result, the climate near the equator is both warmer and less variable than the temperate or polar climates.

Differences in air temperature cause air to move in currents that transfer heat away from the equator and toward the poles. Three rules determine the direction of these currents: (1) hot air rises and cold air falls, (2) hot air holds more moisture than cold air, and (3) the rotation of the Earth twists the moving air.

Figure 26-6 shows the pattern of air currents. Right at the equator, not much wind blows, and old-time sailors called this calm area the *equatorial doldrums* [Old English, *dol* = dull]. There, warm, moist air rises, which creates a region of low air pressure. The low air pressure pulls **trade winds** toward the equator from about 30° latitude (north and south). Because of the Earth's spin, the trade winds blow from the east. The trade winds were named when sailing ships used them to sail westward.

As the warm air from the equator rises, it cools. Because cool air can hold less moisture than warm air, the moisture

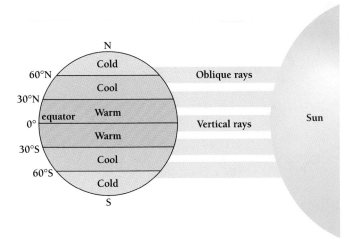

Figure 26-5
Temperature depends on latitude. Temperature and moisture are the two most important determinants of ecosystem type. The latitudes closest to the equator receive the most sun. The latitudes closest to the poles receive less solar energy than those near the equator. Because the sun's rays arrive at an oblique angle, more of the light is reflected by the atmosphere back into space and less arrives at the surface of the Earth.

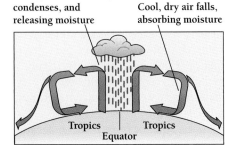

Figure 26-6
Air currents. A pattern of rising and falling air, combined with the Earth's rotation, causes air currents to flow from the east near the North and South Poles, to flow from the west in the middle latitudes, and to flow from the east along the equator. Rain falls at latitudes where moist air rises and cools.

condenses into clouds and falls as rain over the equatorial regions, creating the warm, rainy climate of the **tropics** (Figures 26-6 and 26-7). The equatorial air spreads north and south, raining and drying out as it goes. By the time this air reaches about 30° latitude, the air is quite dry and the lands beneath are nearly all deserts.

As the air loses water, it cools. The cool, dry air drops down toward the Earth's surface, and the Earth's spin gives the air a twist that makes it blow from the west. These **westerlies** blow at latitudes between 30° and 60° in both the north and south (Figure 26-6). As this air moves over the Earth's surface, it warms and picks up moisture. At about 60° latitude, the air again rises, cools, and rains (or snows). At this cool but wet latitude are most of the great coniferous forests. Once the air is dry again, it drops down to create a dry, cold region of Arctic tundra. Here, the winds, called the **polar easterlies,** come from the east.

The winds at each latitude depend on the overall movement of warm air from the equator toward the poles (Figures 26-6). The rotation of the Earth also changes the direction of the air movements, causing clockwise air currents in the Northern Hemisphere and counterclockwise currents in the Southern Hemisphere. The effect of rotation on movement, called the **Coriolis effect,** causes Northern Hemisphere winds (including hurricanes, typhoons, and tornadoes) to twist clockwise and Southern Hemisphere winds to twist counterclockwise. The Coriolis effect also induces ocean currents to turn in great circles—clockwise in the north and counterclockwise in the south.

Ocean currents often move along the edges of continents. The Gulf Stream, for example, carries warm water from the tropics up the eastern coast of the United States and then across the North Atlantic, where it warms Great Britain and much of northern Europe. The warm Gulf Stream gives northern Europe a far milder climate than would be expected from its latitude.

How Does Topography Influence Climate?

Mountain ranges also influence climate, creating "rain shadows." A **rain shadow** is a land of little rain on the downwind side of a tall mountain range. Mountains force warm surface air to rise to higher, colder altitudes, where moisture condenses into clouds and falls as rain or

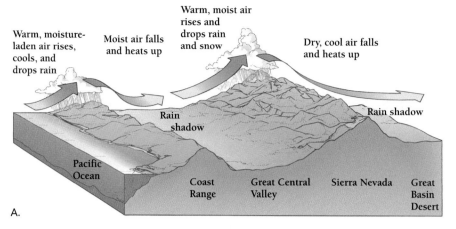

A.

B.

Figure 26-7
A rain shadow. Rain shadows occur in all parts of the world. One of the most dramatic examples is in the western United States. A. Rain clouds driven inland from the Pacific Ocean hit the Coast Range and ride up the slope, like skateboarders on a ramp. As the clouds rise, they cool and rain precipitates out, soaking the mountains. The clouds drop down the other side, warm up, and hold their remaining moisture. Over the Sierra Nevada and Cascade ranges, the clouds drop quantities of rain and snow. B. By the time the moving air reaches the Great Basin, it has lost nearly all its moisture and little rain falls in the desert beneath.

snow. Along the Pacific Coast, for example, prevailing westerlies drive warm, moist ocean air up and over two mountain ranges. When warm, wet air coming in off the Pacific Ocean hits the low Coast Range, the clouds rise, cool, and then dump up to 100 inches of rain a year on the coastal mountains (Figure 26-7A). As the air moves east, down the farther side of the Coast Range, it warms and absorbs moisture, drying the land. The same westerlies then hit the Sierra Nevada and Cascade mountains, which have some of the tallest peaks in the continental United States. Between 3,000 and 14,000 feet, the ocean air cools and drops rain and snow. By the time the ocean air reaches the far side of the mountains, however, it is so dry that all of the lands in the Owens Valley, Death Valley, most of Nevada, and parts of Oregon, Idaho, and Utah comprise a huge rain shadow called the Great Basin desert (Figure 26-7B).

Tall mountains have other effects on climate as well. As we climb a mountain such as Mt. Whitney, the climate changes in the same way it does as when we travel from low latitudes to higher ones (Figure 26-8). The Appalachian Mountains, the Sierra Nevada, and the Rocky Mountains all contain extensive coniferous forests, which turn to alpine tundra above the "tree line," the upper limit at which subalpine trees can grow. Even in the tropics, snow covers the highest mountains.

All of these geographic patterns of temperature and rainfall create just a few major types of climate. Within each climate, similar adaptations are useful. For example, the species that live in a desert are recognizably similar, whether the desert is in Asia, North America, Australia, or Africa. Likewise, a marsupial "wolf " from the forests of Australia has the same long, running legs and powerful jaws as the gray wolf of North American forests (Figure 17-16).

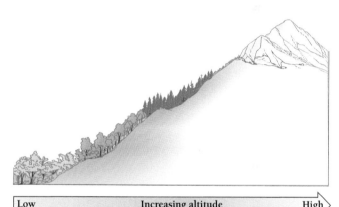

Low	Increasing altitude	High
Tropical rain forest	Temperate forest	Coniferous forest	Taiga	Chaparral	Tundra	Scarce vegetation (rock and glaciers)

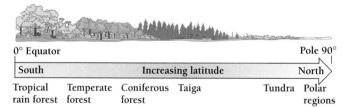

0° Equator		Pole 90°
South	Increasing latitude	North
Tropical rain forest	Temperate forest	Coniferous forest	Taiga		Tundra	Polar regions

Figure 26-8
The effects of latitude and altitude are similar. High-altitude ecosystems are similar to those at high latitudes.

26.2 How Do Ecosystems Differ?

The biosphere includes many kinds of communities and ecosystems that can be subdivided almost infinitely. In this chapter, we'll look at 12 ecosystems that illustrate both life's diversity and a few unifying themes (Figure 26-4). We will

Ecosystems with the least seasonality and the most rain are the most productive. Productivity is measured as the dry weight of new plant material that accumulates on a square meter of land in one year.

Ecosystem	Temperature (winter low/summer high)	Precipitation (cm/y)	Average Primary Productivity (g/m²/y)
Desert	warm/hot	<25	90
Open ocean	varied	aquatic	125
Tundra	cold/cold	<25	140
Lakes and streams	varied	aquatic	400
Chaparral	mild/mild	20–40	700
Grassland	cold/warm	25–75	600
Cropland	varied	irrigated	650
Taiga	cold/cool	50	800
Savanna	warm/hot	90–150	900
Temperate deciduous forest	cold/hot	75–125	1,200
Tropical rain forest	warm/hot	200–450	2,200
Coral reefs	warm/hot	aquatic	2,500
Swamp and marsh	warm/cold	aquatic	3,000

begin with the driest ecosystems, the warm, dry deserts and the cold, dry tundra, and then move through progressively wetter ecosystems, including chaparral, temperate grassland, and savanna; taiga, coniferous forest, and deciduous forest; and finally, wettest and warmest of all, the tropical rain forest (Table 26-1). At the end of the chapter we'll look at aquatic ecosystems, including estuaries and coral reefs.

Desert Ecosystems Are Surprisingly Delicate

Because they receive little rainfall and are low in nutrients, deserts have the lowest net primary productivity of all the ecosystems. These barren regions produce only about 90 grams dry weight of biomass per square meter per year, about the weight of 12 quarters. Deserts may be hot or cold, but all are dry, receiving less than 25 centimeters of rain per year. In the southwestern United States and northeastern Mexico, desert extends east from the Sierra Nevada to the Rocky Mountains and south to the Sierra Madre. Africa, Arabia, central Asia, central Australia, and parts of South America all have deserts.

What plants dominate a desert depends on the temperature. Sagebrush dominates the cold deserts of Nevada and Utah, while cacti and oily creosote bushes dominate the lower, warmer deserts of southeast Arizona and Mexico (Figure 26-9). Notice that the world's deserts cluster at about 30 degrees latitude, either north or south.

Desert plants have special adaptations to conserve water. Sagebrush, for example, has small leaves that minimize evaporation, while succulents (including the cacti) store wa-

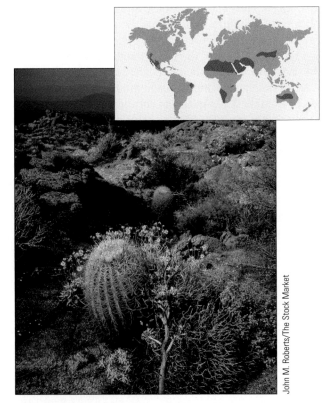

Figure 26-9
Deserts. Barrel cactuses punctuate the many deserts of the southwestern United States. All desert plants adapt to lack of water with leaves and stems that conserve water (by minimizing evaporation, for example). Cactuses also store water, making them attractive to thirsty animals. To protect themselves, cactuses arm themselves with dangerous spines.

John Shaw/Tom Stack & Associates

ter in fleshy tissue for future use. Cacti, with their valuable stores of water, grow prickly spines that protect them from thirsty animals.

Like desert plants, desert animals must also be able to survive extreme heat and dryness. In the summer, many are active only during the cool hours of dawn and dusk. Lizards and snakes thrive, as do birds such as cactus wrens and road-runners, and rodents such as kangaroo rats and mice. Many rodents **estivate,** retiring to underground burrows and entering a sleeplike state called *torpor* during the hottest and driest months of the summer. Round-tailed ground squirrels (*Ammospermophilus tereticaudus*) and several species of pocket mice estivate, as does the cactus mouse (*Peromyscus eremicus*). Laboratory experiments suggest that either lack of food or lack of water can trigger estivation in the cactus mouse.

Although deserts might seem tough, they are delicate. The wheels of off-road recreational vehicles destroy fragile desert soils, as well as plants, lizards, and tortoises. An even greater threat to deserts is the conversion of desert habitat into suburban developments and farms. Excessive irrigation can destroy any soil, but desert soils are especially vulnerable. Irrigation water from rivers and wells is loaded with dissolved minerals and salts. As the water evaporates from the hot desert soil, the minerals and salts are left behind as a deposit, a process called salinization. In time, the soil becomes so caked with minerals and salts that farming is impossible. By then, even hardy desert plants cannot grow in the poisoned soil.

Because deserts have little water and few nutrients, they are the least productive of all ecosystems. Desert plants and animals are adapted to conserve water.

How Can a Dry Ecosystem Be So Wet?

At the North and South Poles lie deserts of ice devoid of terrestrial plants (though not of algae and rich marine ecosystems). Just a few hundred miles south lies the cold **tundra** [Finnish, *tunturi* = arctic hill], a vast, open land of bogs, ponds, and lakes. Among the lakes are short grasses, shrubs, mosses, and lichens, nearly all less than 20 centimeters tall. The few small shrubs and trees grow on the banks of lakes or streams. Tundra sweeps round the North Pole, covering more than one-fifth of the Earth's land surface, including northern Canada, Alaska, Scandinavia, and Siberia (Figure 26-10). In addition, alpine tundra cloaks the upper slopes of high mountains and plateaus in temperate regions.

The tundra is cold. Freezing temperatures occur in every season, and **permafrost**—permanently frozen ground less than a meter below the surface—persists throughout even the warmest summers. Still, a one- or two-month summer growing season allows topsoil to thaw and plants to grow. As soon as the brief summer warmth and rains arrive, the plants' extensive underground roots allow them to quickly produce leaves, flowers, and fruits.

Tundra receives no more rain or snow than a desert. Each year, less than 25 centimeters falls, mostly as summer and autumn rains. In winter, the snowpack rarely exceeds 10 to 20 centimeters. Despite this low precipitation, the soil is perpetually wet, for the cold air slows evaporation and the icy permafrost keeps water from draining lower.

Mammals in the tundra include caribou (or reindeer), polar bears, weasels, foxes, and small rodent herbivores called lemmings. In the summer, large numbers of birds arrive from the south to nest and raise young. Swarms of flies and mosquitoes fill the air. The tundra supports an enormous amount of life in the summer, but the diversity of species is low.

In the past 30 years, oil wells, oil pipelines, and roads have begun to impinge on life in even this remote ecosystem. The noise of exploratory drilling has forced bowhead whales farther from shore, and oil field workers' garbage has increased populations of brown bears, foxes, ravens, and gulls that feed on the eggs and young of arctic birds, reducing their populations. Roads, off-road travel, and construction cause dust, thawing of the permafrost, flooding, erosion, and damage to vegetation. Because spring is so short in

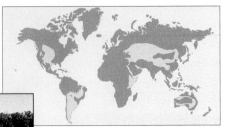

Tom McHugh/ Photo Researchers, Inc.

the polar regions, such damage can take hundreds of years to mend.

Tundra has a cold, dry climate and a short growing season. It is unusually sensitive to human activities.

Why Are There No Trees in Grasslands?

Temperate grasslands have well-defined seasons, with warm summers and cold winters. Annual rainfall is low, usually 25 to 75 centimeters per year, enough to keep grasslands from turning to deserts but not enough to sustain forests. In addition, grasslands usually have a dry period, during which fires are common. Wildfires and the lack of rain, together with heavy grazing by animals, keep trees from taking hold in this ecosystem.

Grasslands extend over much of the Earth's surface, mostly in the interiors of the continents. They include, for example, the steppes of Russia, the veld of South Africa, the pampas of Argentina, the puszta of Hungary, and the prairies of the central United States and Canada. Before European settlement, more than a million square miles of grassland covered central North America (Figure 26-11).

The types of grasses vary with the amount of rain and with season. In the eastern American prairie, for example, tall grasses, especially big bluestem, once dominated; in the drier western plains—in the rain shadow of the Sierra Nevada and Rocky Mountains—short grasses dominated.

Grasslands can support large numbers of mammals. For millions of years, African grasslands supported huge herds of wild gazelles and zebras, while the Asian steppes sup-

ported sheep and horses. All of these herbivores in turn supported populations of wolves, lions, humans, and other predators. When Lewis and Clark crossed the American prairie in 1804, some 60 million bison and 40 million pronghorn antelope grazed in vast herds. The explorers had to literally push their way through these herds of animals. Yet 19th-century hunters drove these populations to the brink of extinction in less than 75 years.

Today, in place of the bison and antelope live some 20 million humans, who have virtually eliminated the temperate grassland ecosystem. In place of hundreds of thousands of square miles of American and Canadian prairie, we have created the world's "breadbasket." Corn has replaced tall bluestem, and wheat has replaced the short grass, sustained by the use of industrial fertilizers and herbicides. Mass-produced pigs, cows, and beef cattle are the chief herbivores; humans, the top carnivores. A few thousand bison live in parks and reserves, and a few hundred thousand pronghorn antelope live in the plains adjacent to the Rocky Mountains. Their natural predators, however, the Great Plains wolf and the grizzly bear, are gone—shot, poisoned, and trapped to extinction.

Temperate grasslands have warm summers and cold winters. In North America, corn and wheat fields have replaced most temperate grasslands.

Chaparral Always Burns Eventually

Chaparral [Basque, *chabarro* = dwarf evergreen] dominates five widely separated temperate regions—California, central Chile, the shores of the Mediterranean Sea, southwestern Africa, and southwestern Australia. These regions share a "Mediterranean" climate, with mild, rainy winters and warm, dry summers. In the summer, prevailing westerly winds blow cool, moist sea air over the warm land. But the warmed air holds the moisture, so no rain falls.

Chaparral is characterized by dense shrubs and scattered, broad-leafed trees (Figure 26-12). The shrubs are typically spiny, thick, and dense. Each of the five chaparral regions of the world has its own species, but all chaparral plant species share common adaptations to summer drought and to periodic wildfires. Wildfires clear the way

Figure 26-12
Chaparral. Dense shrubs and scattered trees characterize Mediterranean climates.

for fresh growth and kill trees that would shade out the chaparral. The seeds of some plants actually require exposure to the intense heat of a brushfire before they can germinate. Perennial shrubs such as chemise and coyote brush are well adapted to fire. When fire burns away their dry branches, these species resprout from still-intact roots.

Chaparral supports some of the greatest diversity of species of any temperate community, including, to name a few, lizards, snakes, rabbits, chipmunks, squirrels, pack rats, kangaroo mice, bobcats, coyotes, thrushes, quail, sparrows, wrens, and hawks.

One of the greatest dangers to chaparral communities is overzealous fire prevention. By putting out small fires, humans ensure the accumulation of dry wood. Inevitably, a fire so big and hot that firefighters cannot put it out sweeps thousands of acres, incinerating the roots of the chaparral vegetation and destroying hundreds of suburban houses (Figure 26-13). Such hot fires also kill seeds and sterilize the soil, so that until the land is recolonized, nothing grows back. In addition, because of the way such superhot fires affect the soil, disastrous erosion may follow.

Chaparral ecosystems are characterized by warm, dry summers and mild, wet winters. Many chaparral species have special adaptations to fire.

Savannas Host Some of the Most Spectacular Grazing Animals

Savannas are grasslands with scattered trees or small clumps of trees (Figure 26-14). The trees are usually deciduous, dropping their leaves in the dry season. Like temperate grasslands, savannas have a regular dry season and periodic wildfires that limit the spread of trees. But savannas generally have more rain (usually 90 to 150 centimeters per year) and hotter climates than temperate grasslands. Savannas are often transitional ecosystems between temperate deciduous forest and prairies or between evergreen tropical rain forests and deserts. Savannas cover much of south and east-central Africa, much of Australia, parts of the Americas, and Southeast Asia.

Savanna grasses are often tall—2 meters or more in Africa, 1.5 meters in South America—and the trees short—usually less than 10 to 15 meters tall. Deep, dense underground networks of roots allow grasses to obtain water where trees cannot and so survive the dry seasons.

Animal life in savannas varies greatly from continent to continent. The best-known animals are those of the African plains. There, herds of zebras, wildebeest, and gazelles graze while elephants and giraffes browse on the trees. Among these herbivores, lions, cheetahs, hyenas, wild dogs, and other carnivores hunt and scavenge. Savannas also support large, flightless, grazing birds—the ostrich in Africa, the emu in Australia, and the rhea in South Amer-

Figure 26-13
Fire. Brush fires in chaparral communities destroy houses in minutes. Here a 20,000-acre fire sweeps through a suburban neighborhood in southern California.

Figure 26-14

Savanna. Grasslands and scattered trees in the Serengeti, in Tanzania.

ica. Termites consume up to one-third of all plant litter and serve as a major food source for birds and mammals. In some areas, termite mounds cover the savanna. These mounds may reach 9 meters in height and number 10 to 150 per hectare.

In most savannas, hunters and habitat destruction are driving many species to the brink of extinction. In East Africa, the seemingly infinite herds of grazing animals are becoming food for rapidly growing human populations. Elephants have become particular targets for poachers, not because of their food value but because of their valuable ivory tusks. Their slaughter continues despite worldwide efforts to outlaw trading in ivory. Nonetheless, the real threat to African savanna is the continuing conversion of savanna to farms and ranches that support the growing human population. Wildlife sanctuaries protect some of the African and South American savannas, but the end of the savanna ecosystem seems inevitable.

Tropical savannas are grasslands with scattered deciduous trees that support vast herds of grazing animals.

Taiga, the Spruce–Moose Ecosystem

South of the tundra, most northern lands are both slightly warmer and wetter. Here sparse conifer forests called **taiga** grow monotonously across thousands of miles of Siberia, Scandinavia, Canada, and Alaska (Figure 26-3). Winters are bitterly cold, and summers are short. The one or two species of trees (often spruce and pine) grow only 5 to 10 meters high. Ferns, mosses, and lichens grow sparsely on the forest floor. Shrubs are rare, except for blueberries and gooseberries that flourish in open areas where trees have temporarily died off.

Moose, mice, squirrels, porcupines, and snowshoe hares are the main vertebrate herbivores. Predators include wolves, grizzly bears, lynxes, and wolverines. Birds migrate to the taiga in summer but leave before the first snows fall. Except for garter snakes, reptiles cannot survive the long, harsh winters. Mosquitoes, black flies, and other insects, however, abound in the summer.

The taiga is so inhospitable that human populations are sparse. In some areas, logging has eliminated the original forest, and deer have moved in, replacing the moose. A common parasite of the deer—the nematode brain worm—kills moose and is hastening their disappearance.

One of the greatest threats to the taiga comes from the air. Factories, power plants, and automobiles produce particulate pollutants and oxides of sulfur and nitrogen that rise and travel thousands of miles to the taiga. The oxides dissolve in the clouds, dropping acid rain and snow on lakes and rivers. The result has been the wholesale death of fish and forests in many regions of the Northern Hemisphere, including Scandinavia and eastern Canada.

Lichens, in particular, accumulate sulfur dioxide, heavy metals, and other toxins. Ecologists use both the health and chemical composition of taiga lichens to monitor global air pollution.

Taiga has cold winters and short, mild summers.

Figure 26-15
Deciduous forest. In autumn, the leaves of deciduous trees turn color, as chlorophyll is pulled from the leaves. North County National Scenic Trail, Itasca State Park, Minnesota.

G. Alan Nelson/Dembinsky Photo Associates

Temperate Deciduous Forests

Temperate deciduous forests typically have warm, rainy summers and cold winters. They receive from 80 to 140 centimeters of rain and snow each year. However, temperature and precipitation vary more than in the tropics or the polar regions. As a result, deciduous forests may look rather different from one another. In the northern United States, for example, the dominant trees are maple, oak, hemlock, and birch, while in the southeastern states the dominant trees are pine, oak, and hickory.

The deciduous trees that make up these forests shed their leaves each autumn as a way to minimize water loss at times in the winter, when freezing temperatures make water unavailable (Figure 26-15). The plants of a temperate forest occupy four vertical layers: trees, shrubs, herbs (nonwoody plants), and ground cover. The trees occupy the highest layer, their tops touching to form a nearly continuous canopy. In the shade beneath the trees are shrubs and bushes whose branches lie closer to the ground. The herb layer may be especially rich and beautiful in the spring. Finally, the ground layer consists of mosses and liverworts, often living among the accumulated leaf litter.

Animal life is abundant in all the plant layers and in the soil beneath the litter (Figure 26-16). Trees and shrubs provide food and nest sites for many species of birds. Woodpeckers and chickadees, for example, nest in cavities in the trees, while warblers and thrushes build nests on their branches. Mammals such as chipmunks and squirrels; reptiles such as snakes, lizards, and wood turtles; and amphibians such as salamanders and tree frogs also find food and homes in the forest.

The eastern third of the United States, western Europe, Japan, Chile, and eastern China were once dominated by temperate deciduous forests. On every continent, humans have cut down much of the forest for lumber or firewood and to make way for farms, towns, and cities. Before European colonization 400 years ago, forests covered 46 percent of the United States, and most of this forest was ancient and deciduous. Although in the eastern United States, forests have returned, they still cover only 32 percent of the land and consist largely of young trees, all the same age and species. More important, these young forests have not brought back all the local species of plants, animals, and fungi that went extinct 200 years ago.

Much of the new forest is in small, isolated patches. These tiny patches of forest tend to hold far fewer species of both plants and animals than the extensive virgin forests they replaced (Chapter 28). In addition, such small areas do not provide the large ranges needed to support carnivores at the top of the food chain. A single mountain lion, for example, needs over 300 square miles. A single family of gray foxes needs a square mile.

Temperate deciduous forests have warm, rainy summers and cold winters.

Tropical Rain Forests Are the Most Diverse Ecosystems

Most **tropical rain forests** are within 10° of the equator—in the Amazon basin of South America; in central and west Africa; and in Malaysia, Indonesia, and Papua New Guinea (Figure 26-17). The weather in the tropical rain forests is more or less constant throughout the year, with average temperatures of about 27°C (80°F) every month of the year. Rainfall is heavy, ranging from 2 to 4.5 meters per year.

Warmth and moisture are so plentiful in a tropical rain forest that plants compete for light rather than water. And tropical plants capture light so completely that the forest floor is usually rather dim, with little vegetation. Visitors often compare a tropical forest to a cathedral, with its open floor, high vaulted ceilings, and soft filtered light.

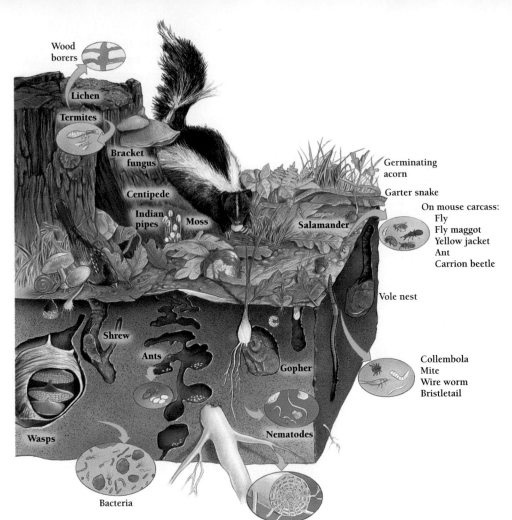

Figure 26-16
Beneath our feet. Biologist E. O. Wilson has written, "It is a failing of our species that we ignore and even despise the creatures whose lives sustain our own." The rich soils and leaf litter of a temperate deciduous forest supports hundreds of organisms from bacteria, fungi, and ferns, to worms and skunks. Shown here are just a few, including the mycorrhizal fungi on plant roots that fix nitrogen and the nests of ants, wasps, a gopher, a vole, and a shrew. The gopher primarily eats roots, the vole eats plants above ground, and the shrew eats worms, insects, and other invertebrates. A garter snake hunts invertebrates such as slugs and small mice and other vertebrates. While fungi such as the mushroom and bracket fungus feed off any organic matter in the soil or on the tree at left, a towhee (bird) digs through the leaf litter looking for many of the same prey the shrews eat. The skunk has a bird's egg in its mouth, but is considering the apple, tossed there by a human hiker, while a mouse carcass provides food for fly and carrion beetle maggots, which in turn feed a passing salamander.

Nutrients cycle rapidly through rain forest ecosystems. The soils of tropical rain forests are surprisingly poor in minerals and other nutrients. One reason is the heavy rainfall, which leaches nutrients from the soil. Another is the year-round work of ants, termites, fungi, and bacteria, which rapidly break down leaf litter to simple molecules. In most ecosystems, a dry season or a long, cold winter slows the breakdown of leaf litter and other organic matter, and soils tend to build up to depths of up to several feet. In the rain forest, only a thin layer of nutrients exists, just beneath the leaf litter at the very surface.

Rainforest trees use an extensive network of extremely shallow roots to sweep up whatever nutrients are available. The bulk of nutrients and energy in a tropical ecosystem are not in the soil but in the tops of trees and in the other living organisms.

Despite their thin soils, tropical rain forests are the second most productive ecosystem in the world, surpassed only by estuaries and marshes (Table 26-1). And tropical rain forests are first in biodiversity. Of all the Earth's species, 50 to 90 percent live in tropical rain forests. More

Figure 26-17
Tropical rain forest.
Terra-firma Amazonian rain forest, in Peru.

Extreme Biology How Does Acid Rain Alter Ecosystems?

Many people may have heard that severe acid rain can destroy lakes and whole forests. In the worst cases, fish, frogs, plants, and other organisms have disappeared from formerly productive lakes and streams. Yet, increasingly, ecologists are finding that acid rain also has the capacity to transform ecosystems in subtler ways.

Researchers have long known that acid rain increases the nitrogen available to plants. Indeed, some economists have argued that acid rain is a cure for global warming. They say that because nitrates from acid rain can stimulate plants to grow faster, plants in acid environments can take up more carbon dioxide, compensating for the carbon dioxide released when we burn fossil fuels.

Although this is a reassuring thought, ecological research does not support this hypothesis. In one study, researchers at the University of Minnesota applied nitrogen to plots of native Minnesota tall-grass prairie for 12 years to imitate the excess nitrates that result from both acid rain and fertilizer applications by farmers. Because native grasses use nitrogen efficiently, they flourish in places where the nitrogen content of the soil is low. In the nitrogen-treated plots, however, exotic weeds flourished, outcompeting and displacing native Minnesota grasses.

But that was only half of the bad news. As prairie grasses die, their tissues break down very slowly, and over time much of their carbon, nitrogen, and other constituents remain stored in the soil. Such long-term storage is, in fact, the key to the famous, rich soils of the American Midwest.

But in the Minnesota study, the weeds behaved differently from grasses. As the weeds died, their huge stores of nitrogen stimulated the growth of nitrogen-hungry soil microbes, which rapidly dismantled the exotic weeds into basic building blocks. The microbes released more nitrates into the groundwater and more carbon dioxide back into the air. The net effect was more carbon dioxide in the air—not less—and more nitrates in the soil, which stimulated the growth of more weeds.

In the spruce forests of Germany's Fichtelgebirge mountains, the excess nitrogen in acid rain supplies the trees with more nitrogen than they can use. Instead of stimulating growth, the acid rain destroys soil. Negatively charged nitrates raining down on forest soils bind with calcium, magnesium, and other positively charged cations in the soil. As the nitrates flow into lakes and streams, they take the calcium and magnesium with them. In some areas, such acid leaching has removed all of the calcium deposited in the soil for the last 500 years.

And, as we saw in the last chapter, as calcium leaches away and soil acidity rises, increasing amounts of aluminum dissolve in soil water. The aluminum poisons the roots of trees and kills fish.

In some areas, all the trees in a forest are dying (Figure A). In others, ecologists predict, trees deprived of calcium and magnesium will simply habituate and grow quite slowly, like the tiny bonsai trees that gardeners grow in pots. The result may be pygmy forests.

Simon Fraser/Photo Researchers, Inc.

Figure A
Acid rain can kill forests. Emissions from nearby steel plants and oil refineries, combined with salty sea air, have killed these trees near Port Talbot, Wales.

---

species of butterflies live in one rain forest in Costa Rica, for example, than live in all of the rest of North America. The diversity of species in a tropical rain forest can be staggering. One ecologist found 163 species of beetles living in just one tree in the Panamanian rain forest. Another ecologist found 445 species of trees on an area of Brazilian rain forest the size of a city block. Fifteen of those trees were new species.

More than two-thirds of the plants in a tropical forest are trees. Most are flowering, broad-leaved evergreens (Figure 26-18). Other plants common in tropical forests but rare in temperate forests are lianas, vines that are rooted in soil but climb into the tree tops, and epiphytes, plants such as orchids that grow entirely on other plants, with their roots exposed to air and rain but not embedded in soil.

As in a temperate forest, the vegetation forms discrete layers. Most of the trees form a continuous layer, called the **canopy**, 30 to 40 meters above the ground (Figure 26-18). Rising above the canopy are occasional, tall, isolated "emergents"—trees with umbrella-shaped crowns extending to a height of 50 meters or more. Below the canopy, a third layer of shorter trees 10 to 25 meters tall catch whatever light they can from gaps in the canopy. Finally, a short shrub layer consists of dwarf palms and giant herbs with especially large leaves, adapted to capture the faint light that filters down to the forest floor.

All these flowering plants provide a year-round supply of fruits and nectars for an array of birds, insects, snakes, bats, monkeys, and squirrels. Some live permanently in the canopy. Others live in the rivers or on the forest floor.

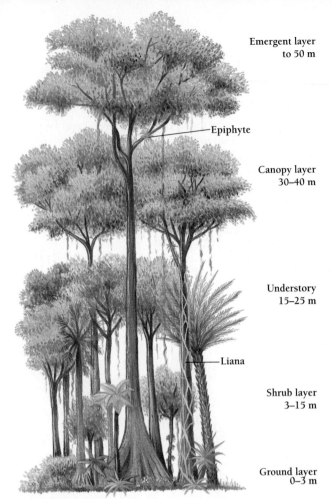

Figure 26-18
The layers of a tropical forest. Tropical rain forests are noted for their extensive layering. Shown here are the protruding emergent trees, the canopy, the understory, the shrubs, and the ground-level plants. Because the roots of tropical trees are so shallow, the trees often grow huge, supportive buttresses.

Emergent layer to 50 m

Epiphyte

Canopy layer 30–40 m

Understory 15–25 m

Liana

Shrub layer 3–15 m

Ground layer 0–3 m

Some eat underground plant parts, others, fruits and nectar. Ants and termites dominate this ecosystem. But anteaters and armadillos, pigs and their relatives, as well as poisonous and constricting snakes, all hold their own in the complex food webs of the tropical rain forest. The top carnivores of the forest floors used to be such large cats as the jaguar in South America and the tiger in Asia. But these big cats have all but succumbed to hunters and habitat destruction by humans.

How Are Tropical Rain Forests Being Destroyed?

Until recently, Earth contained 25 million square kilometers of tropical rain forests, an area greater than all of North and Central America. Today no more than eight million square kilometers remain. Estimates of the rate of rain forest destruction show that an area of tropical rain forest the size of the state of Washington is logged or burned every year.

Even this estimate may be too low. In 1999, field surveys of wood mills and forest burning in the Amazon showed that the true rate of destruction may be twice as high as previous estimates. As rain forest disappears, watersheds are destroyed, the local climate becomes drier, and remaining forests dry and become susceptible to fires. In short, rain forest destruction is an accelerating process. How is it happening?

Like other human populations, those in the tropics are doubling every few decades. To raise cash for these expanding populations, governments support the logging and burning of large areas of tropical rain forest. Tropical nations export timber and plywood from rain forest trees, metals mined from rain forest lands, and beef from cattle raised on clear-cut rain forest.

Once cleared, thin rain forest soils do not last long. Recall that the nutrients in tropical rain forest are in the vegetation, which has been carried away or burned. The average cattle ranch in Central America lasts only 6 to 10 years, after which the soil is too depleted of minerals to support grass for grazing cattle. Each hamburger made from the cattle raised on these ranches costs twenty-five square meters of tropical rain forest (about the area of a two-car garage), including about a half ton of trees, flowers, birds, mammals, snakes, lizards, frogs, butterflies, beetles, ants, and termites.

One of the greatest causes of rain forest destruction is small-scale farming. Individual farmers cut and burn a small area of land and raise a few crops. Once the soil is depleted, the farmers move on to another patch of land. This kind of farming is called "slash and burn" farming or "shifting agriculture." Because the individual fires set by farmers are so small, the damage is often invisible in satellite imagery until vast areas have been destroyed by thousands of farmers. In Chapter 28, we discuss how isolating small plots of forest creates "edge effects" that cause many species to go extinct.

Future textbooks may list the tropical rain forest as an ecosystem that has disappeared, for human activity threatens to eliminate all but a few parks within the next 20 to 30 years. Accompanying the destruction of the rain forest will be the disappearance of millions of species of plants, insects, and other animals found nowhere else. One-quarter of the world's 250,000 plant species and one-fifth of the 20,000 named butterflies will likely go. Yet most of the species that disappear will never even have been named.

As if that weren't bad enough, recall from Chapter 25 that the burning of forests releases tons of extra carbon dioxide into the atmosphere, increasing global warming, melting of the polar ice caps, and hastening other changes in climate.

Tropical rain forests have the greatest biodiversity of any ecosystem in the world. Unless major changes occur, however, 90 percent of it will be gone within 20 to 30 years.

26.3 What Are Aquatic Ecosystems Like?

Water covers 72 percent of the Earth's surface—70 percent saltwater oceans and two percent freshwater lakes, rivers, and streams. Like terrestrial ecosystems, aquatic ecosystems fall into just a few distinct types. Ocean ecosystems differ from one another according to temperature, depth, and distance from shore.

How Do Oceanographers Divide the Oceans?

Although ocean covers more than two-thirds of the Earth's surface and extends over nearly all of the Earth's latitudes, the oceans contain only about 10 percent of the planet's species. Because the oceans are all interconnected, they lack the many habitat islands that promote geographic isolation and speciation in terrestrial organisms (Chapter 17). Nonetheless, the oldest, largest, and most stable ecosystems on Earth are those of the oceans.

In the last few decades, **oceanography,** the study of the seas and their ecosystems, has revealed the shape of the seas. The ocean bottom, for example, contains vast mountain ranges and rifts, as well as a shallow continental shelf bordering each continent. The average depth of the oceans is about two miles. But the lowest spot on Earth, the Marianas Trench of the western Pacific Ocean, dips to seven miles below sea level—deeper than Mt. Everest is high.

Biologists distinguish **ocean ecosystems** according to ocean depth and distance from shore. The **benthic division** consists of all organisms that live on the ocean floor. These include those that live in the **intertidal zone,** which lies between high tide and low tide, and those that live at the bottoms of deep canyons in the **abyssal zone** (Figure 26-19). The **pelagic division** consists of all organisms that live in open water, above the bottom.

Organisms living in the intertidal zone are submerged at high tide and exposed to air and pounding waves at low tide. They must be able to withstand drying, hot sun, and crashing waves. Intertidal organisms anchor themselves to rocks or sand and close shells and other structures tightly to keep from drying out when the tide goes out. Tidepools reveal the rich diversity of the intertidal species (Figure 26-20).

In the pelagic division, microorganisms called **plankton** drift in the upper level of the water, moving only where currents take them. Fish, whales, and other free-swimming organisms feed on plankton or on each other. The pelagic division is further divided into the **neritic zone,** which is closer to shore, over continental shelf, and the **oceanic zone,** which is out beyond the continental shelf over the deepest water.

The neritic zone consists of shallow coastal areas near islands and continents. In the neritic zone, algae are the

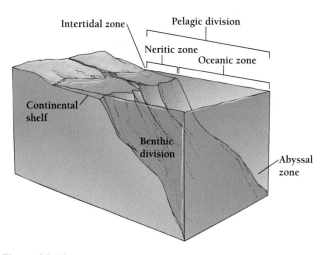

Figure 26-19
Dividing the ocean into zones. Oceanographers divide the ocean into divisions and zones according to the depth of the water and the distance from shore. The benthic division includes intertidal organisms and others that live on the bottom. The pelagic division includes the neritic zone (over the continental shelf) and the oceanic zone (deep water).

dominant primary producers. While the total area of the neritic zone is much smaller than that of the oceanic zone, it contains many more ecological niches and species.

The oceanic zone comprises most of the world's oceans. Here, photosynthetic cyanobacteria, diatoms, and dinoflagellates are the primary producers. The primary productivity of the open ocean is extremely low, however—comparable to that of a terrestrial desert. The reason for this low productivity is that the oceanic zone lacks

Figure 26-20
The intertidal zone. The intertidal zone is characterized by organisms that cling to rocks as the waves crash around them. Shown here is a Pacific Coast tidepool.

mineral nutrients, especially phosphorus and iron. Oceanic ecosystems recycle nutrients poorly, depending on a steady supply from terrestrial ecosystems. As organisms die, their remains fall from the surface layers to the ocean bottom. Once these nutrients drop to the bottom of the ocean, they remain there.

A few exceptions exist. Off the coast of Peru, the Humboldt Current draws nutrients up from the bottom in a process called **upwelling**. Upwelling creates a rich organic soup of nutrients that supports a rate of photosynthesis six times that found in other parts of the open ocean.

Where such currents bring nutrients, grazing protozoa and small animals feed on blooms of photosynthetic microorganisms. Small fish eat the tiny invertebrates, and larger fish eat the smaller fish. A marine food chain typically contains four or five trophic levels. This enormous productivity supports, in turn, huge populations of fish-eating birds, as well as a large human fishing industry.

Every year, around Christmas, the Humboldt Current reverses itself. Warm water flows down the coast, preventing the nutrient-rich cold waters from rising to the surface. Usually this reversal, called "El Niño" [Spanish, = the Christ child], is temporary and harmless. Every two to seven years, however, the current reversal lasts longer, with catastrophic results. In 1982 and 1983, for example, a persistent El Niño led to the wholesale starvation of fish-eating seabirds in the South Pacific and the collapse of the sardine industry.

Except in areas where upwelling bring nutrients from the ocean floor to the surface, ocean ecosystems are less productive and more homogeneous than terrestrial ecosystems.

Coral Reefs: The Tropical Rain Forests of the Sea

Coral reefs are found in tropical waters, usually within 30° of the equator. The combination of warm temperatures and lots of sunlight make coral reefs one of the most productive ecosystems in the world (Table 26-1). Biologists have estimated that one square kilometer of a coral reef can support 30 metric tons of fish. Coral reefs are also amazingly diverse, hosting all but one of the world's 33 animal phyla (major kinds of organisms).

Like tropical rain forests, the nutrients in coral reefs are tied up in living organisms. Photosynthetic algae living symbiotically with corals (which are living animals that build the reefs) provide the energy that drives the reef ecosystem. Together, the coral polyps (related to jellyfish) and some of their symbiotic algae create a substrate of calcium carbonate that forms the structure of the coral reef and provides food and shelter for hundreds of reef organisms. The biodiversity of coral reefs is comparable to that of tropical rain forests.

Human activities threaten the existence of coral reefs in many parts of the world. Pollutants poison the coral polyps, and dynamite and cyanide (used to kill and capture fish) damage the reefs and kill the living corals and other reef organisms. In some parts of the world, such as the Philippines,

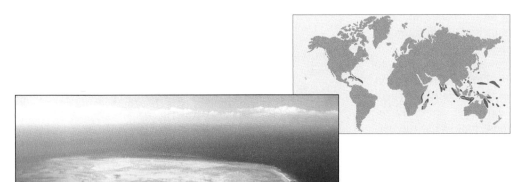

David Ball/Corbis

A.

B.

B. Jones/M. Shimlock/Photo Researchers, Inc.

Figure 26-21
Coral reef. Coral Reefs are important sites of biodiversity and some of the most productive ecosystems in the world. A. A tiny section of the Great Barrier Reef in Australia. B. Coral reef and sea fan in the Banda Sea, Indonesia.

most of the coral reefs have been completely destroyed (Figure 26-21).

Coral reefs are one of the most diverse ecosystems on Earth. Human activities threaten many of the world's reefs with extinction.

Why Are Estuaries and Marshes Valuable?

Estuaries [Latin, *aestus* = tide] are partly enclosed waters where freshwater streams or rivers meet the ocean. The water is less salty than the oceans, and the salinity is constantly changing. The back-and-forth flow of salt water and fresh water depends on the tides and rainfall.

Estuaries are the most productive of all ecosystems (Table 26-1). Tides and rivers provide a constant flow of water and nutrients (Figure 26-22). The primary producers in estuaries include plankton, algae, and large plants such as eelgrasses, sea grasses, marsh grasses, and mangrove trees.

Because of this high productivity, estuaries provide breeding and nursery grounds for many species of fish and shellfish. Countries that destroy their coastal wetlands ultimately destroy their fisheries. Estuaries also provide food and breeding grounds for mammals, birds, reptiles, and amphibians.

Estuaries are especially vulnerable to damage by humans. They receive pollution from streams and rivers, and they are prime targets for oceanfront development. All over the world, farms, condominiums, football stadiums, and industrial parks sit atop filled and drained wetlands.

In the United States, the destruction of wetlands has slowed in the past 20 years. Cities and states have ceased to regard marshes as mere sewers and breeding grounds for mosquitoes. Instead, people are beginning to realize that estuarine ecosystems are not only places of beauty but also irreplaceable contributors to the world food supply.

In 1906, Florida began draining large sections of the Everglades in an effort to control floods and provide fresh water for a growing population. In the 1960s, the U.S. Army Corp of Engineers diverted the Kissimmee River into a concrete canal, drastically reducing the flow of water into the Everglades. In places, the Everglades are now so dry that wildfires burn out of control. In other places, fish populations are so reduced that 90 percent of wading birds have vanished for lack of food. Excess phosphorus from farms has allowed cattails to take over large sections of the Everglades, crowding out species that formerly flourished there.

Fortunately, the state and the federal government are now struggling to reverse this damage and restore the Everglades. Florida farmers have changed the way they irrigate and spray their crops to minimize the runoff of both pesticides and excess phosphorus. Feed for cows and other animals is made with less phosphorus. (As one agricultural researcher put it, cows "can't excrete what they don't eat.") And giant artificial marshes now remove up to 80 percent of the phosphorus before it reaches the Everglades. Finally, Congress has given the National Park Service and the U.S. Army Corps of Engineers $8 billion to redirect water back into the Everglades. The Estuary Restoration Act of 2000 provides federal assistance in promoting estuary habitat restoration projects, as well as data collection throughout the United States and its territories.

Estuaries are the most productive ecosystems in the world.

Why Are Lakes, Ponds, Rivers, and Streams Susceptible to Pollution?

Limnology [Greek, *limne* = pool, marshy lake] is the study of freshwater lakes, ponds, rivers, and streams. As rivers and streams flow into lakes, they bring both water and dissolved nutrients. This material, both organic and inorganic, derives from the surrounding land, so the character of lakes and ponds varies from region to region. At the fringes of a lake, marshes and swamps often form transition zones to the adjacent terrestrial ecosystems.

As in ocean ecosystems, primary production in a lake depends on the amount of sunlight penetrating the lake's surface to photosynthesizing microorganisms and water plants. Heterotrophic protists and small animals such as rotifers and crustaceans graze on the photosynthesizing microorganisms, while birds, fish, and invertebrates feed on the larger plants. Fish also eat one another. Despite the efficient aquatic food web in lakes, organic matter falls to the bottom as detritus, where it is consumed by bacteria and invertebrates. Far more energy flows through the detritus food chains than through the chains involving grazing herbivores and carnivorous fish.

Michael P. Gadomski/Dembinsky Photo Associates

Figure 26-22
Estuary. Salt marsh at Assateague Island National Seashore, Maryland.

Biology⑧Now™ Learn more about the restoration of a famous estuary by clicking on this figure on your BiologyNow CD-ROM.

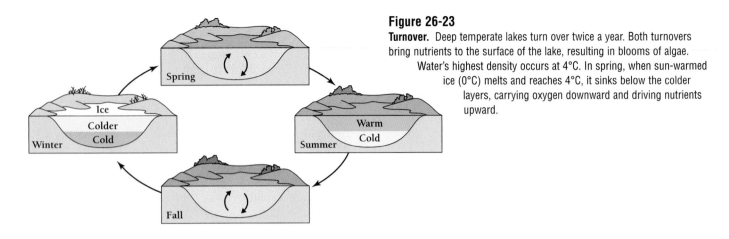

Figure 26-23
Turnover. Deep temperate lakes turn over twice a year. Both turnovers bring nutrients to the surface of the lake, resulting in blooms of algae. Water's highest density occurs at 4°C. In spring, when sun-warmed ice (0°C) melts and reaches 4°C, it sinks below the colder layers, carrying oxygen downward and driving nutrients upward.

In temperate regions, a lake's surface temperature varies greatly from season to season. In the summer, the warm air heats a layer of warm water as much as 20 meters thick (Figure 26-23). As winter approaches, the temperature of the surface layer falls until it is the same as the deeper layer. At that point, the waters of the two layers mix—an event called the *fall turnover.*

A similar turnover occurs in the spring as the winter ice melts. The two turnovers bring dissolved nutrients from the bottom sediments into the upper waters, where photosynthesis occurs. The result is a seasonal growth of algae. Turnover resembles the upwelling of nutrients that occurs in many coastal marine habitats.

A **eutrophic** lake is one with large amounts of minerals and organic matter. The surfaces of eutrophic lakes tend to be clogged with algae and cyanobacteria. As these organisms die and drop to the bottom, they stimulate the lake-bottom decomposers to multiply. During the summer, the decomposers so deplete the deeper layer of oxygen that the lake bottom can no longer support fish and other animals. As in the "dead zone" mentioned in the last chapter, few organisms can live in a eutrophic lake. In contrast, **oligotrophic** lakes have a limited supply of nutrients. These clear lakes have no permanent algal blooms, and they therefore contain more oxygen and support a more diverse community of organisms, including fish.

Eutrophication happens naturally over thousands of years, but humans often accelerate the process by dumping nutrient-rich sewage and other wastes into streams and lakes. Productivity increases at first, but diversity declines, and eventually most of the life in the lake dies.

Because rain washes both nutrients and poisons into rivers and streams, freshwater ecosystems easily accumulate pollutants.

We have seen how energy and materials move through ecosystems, as well as what kinds of ecosystems exist. In the next chapter, we will see how individual species interact and how ecosystems change over time.

Key Concepts

- Temperature and rainfall largely determine the character of an ecosystem.
- Latitude and mountain ranges largely determine climate.
- Areas with heavier rainfall support forests. Areas with less rainfall support grasslands or deserts.
- Human activity is degrading ecosystems worldwide.

Summary with Key Terms

How do light, temperature, and humidity determine the characteristics of an ecosystem?

Our planet contains a limited number of terrestrial and aquatic ecosystems, whose character is largely determined by temperature and moisture. The average temperature, the total annual precipitation, and the variability in each of these values help determine the character of an ecosystem.

At the equator, hot, moist air rises and flows north and south over the **tropics,** dropping rain. **Trade winds** blow cooler air from the tropics toward the equator. The **westerlies** are cool, dry winds that cross the Earth's surface between 30° and 60° latitude. These winds warm and pick up moisture as they go. In the Arctic, cold, dry **polar easterlies** blow from the east.

The climate of a region is determined by the total solar energy absorbed and by wind and ocean currents. These depend on latitude and the rotation of the Earth. The Earth's rotation creates the **Coriolis effect,** which causes winds moving north or south to rotate, clockwise in the Northern Hemisphere and counterclockwise in the Southern Hemisphere.

How do mountains affect rainfall and temperature?

Mountain ranges create desert **rain shadows** by forcing warm, moist air to drop most of its moisture on one side of the range. Ecosystems at high elevations resemble those at high latitudes. At high altitudes and latitudes, **coniferous forest** gives way to **tundra** at the tree line.

How do human activities affect ecosystems?

Dry **deserts** are characterized by widely spaced drought-tolerant plants, such as sage brush (in cold regions) and cacti (in hot regions). Some desert animals **estivate** during the hot, dry summers. The greatest human threats to deserts are off-road vehicles and agriculture. The cold, dry tundra is characterized by short, open grassland, bogs, ponds, lakes, and year-round **permafrost.**

Temperate grasslands are characterized by tall grasses in wet areas and short grasses in drier areas. Fire, insufficient water, and grazing animals all keep trees from becoming established in this ecosystem. Much of the world's grassland has been converted to agriculture, eliminating entire food chains.

Chaparral regions have mild, wet winters and warm, dry summers. They are characterized by dense shrubs and scattered, broad-leafed trees, mostly drought and fire tolerant. Suppression of natural brush fires leads to superhot fires that destroy houses, chaparral, and valuable soil.

Savannas are characterized by tall grasses and short, widely spaced trees. Savannas are hotter and wetter than temperate grasslands. The cold **taiga** is characterized by short, homogeneous pine or spruce forests.

Temperate **deciduous forests** are characterized by a heavy cover of deciduous trees and a limited understory. They are commonly associated with dense populations of humans that have cleared the forests repeatedly, and hunted to extinction most of the top predators.

Tropical rain forests are characterized by a dense **canopy** with a tall understory and little vegetation on the dim forest floor. Tropical countries are cutting down or burning their rain forests at a rate that will eliminate this ecosystem within a few decades.

Oceanography has revealed the depth and character of the oceans and the organisms that live in them. **Ocean ecosystems** can be divided into **benthic** and **pelagic divisions.** The benthic division includes the **intertidal** and the **abyssal zones.** The pelagic division includes the **neritic** and **oceanic zones.**

What characteristics of aquatic ecosystems determine if they will be species rich and productive or species poor and unproductive?

The primary producers are **plankton**, which grow best in areas of **upwelling** nutrients and sustain fish and other members of a marine food chain. The total number of species living in the oceans is far less than the number that live on land.

Coral reefs are one of the most diverse (and threatened) ecosystems. **Estuaries** are marshes with partly fresh, partly salt waters. Estuaries are the most productive ecosystems and also provide breeding grounds for many marine species. Humans have a long history of draining marshes and estuaries for farms or developments.

The study of freshwater lakes, ponds, rivers, and streams is called **limnology,** one of the oldest branches of

ecology. Lakes that are rich in organic matter are **eutrophic.** Those lower in nutrients support a more diverse community and are said to be **oligotrophic.** Humans frequently pollute fresh water with sewage and industrial wastes.

Review and Thought Questions

Review Questions

1. Explain why the polar regions experience more extreme seasons than the equatorial regions.
2. What kind of ecosystem exists in a rain shadow? Explain why.
3. How might the division of a forest or other habitat into small sections interfere with top predators?
4. What keeps grasslands from changing to forests?
5. Why do tropical plants make good houseplants?
6. What percentage of the Earth's surface is covered by water?
7. Why are estuaries ecologically and economically important?
8. What organisms are the primary producers in the oceanic zone? What organisms are the primary producers in the neritic zone?
9. What causes fall turnover, and what benefits does turnover have for freshwater organisms?
10. What causes a lake or pond to become eutrophic?

Thought Questions

11. Explain why the soils of tropical rain forests are poor, even though tropical rain forests are so productive.
12. Why do tropical rain forests have so many species?
13. Why is the open ocean species poor? Why do coastal areas support more life and more kinds of life?

BiologyNow Resources

Biology ⊛ Now™

Active Figures

26-8: Ecosystems
26-22: Restoration of a famous estuary

Preparing for an exam? Take a diagnostic test on your BiologyNow CD-ROM.

Online materials relating to this chapter are at:

http://biology.brookscole.com/AAL3

About the Chapter-Opening Image

Nothing says "desert" more clearly than an organ pipe cactus. Every ecosystem includes characteristic species. Desert ecosystems, for example, often host species of plants that store water and use spines to defend themselves against thirsty herbivores.

27 Communities: How Do Species Interact?

Key Questions

- How can the species of a community be both interdependent and independent?

- What kinds of specific interactions among species lead to adaptation, character displacement, and other forms of coevolution?

- What factors determine the species diversity of a community?

- What does it mean for a community to be stable?

- Why and how do communities change over time?

Volcano: An Ecosystem Is Destroyed

At 8:32 A.M. on May 18, 1980, a medium-sized earthquake shook Mount St. Helens in Washington State. Twenty seconds later, the volcano erupted, and the entire north side of the mountain slid away in a thundering roar. It was the largest landslide in recorded history. Explosions from the volcanic eruption tore through more than half a cubic mile of sliding debris and spewed rocks, ash, gas, and steam across adjacent valleys at velocities approaching the speed of sound. The series of blasts flattened an area the size of the city of Chicago—a forest with enough timber to build 300,000 houses. Every tree within 15 miles was either blown to bits, burnt to a crisp by 600°C gases, buried under tons of volcanic ash, or blown down. From the north face of the mountain poured a river of melted snow and ice, mud, boulders, and ash. In less than 10 minutes, this colossal river of rocks filled the valley below to a depth of 150 feet. A black plume of pumice and ash rose 15 miles into the air, and 520 million tons of ash began drifting eastward across the United States.

The death toll included an estimated 11 million fish, 27,000 grouse, 11,000 hares, 6,000 black-tailed deer, 5,200 elk, 1,400 coyotes, 300 bobcats, 200 black bears, 57 human beings, and 15 mountain lions. Three more explosions in May, June, and July dropped thick blankets of pumice and ash. In the end, every exposed slope within five miles of the crater was stripped of trees and other vegetation and buried under six feet of ash and rock. In places, the mud, ash, and rock piled up an eighth of a mile thick. The scene was as gray as a moonscape. For another 10 miles from the blast zone, thousands of full-grown fir and hemlock trunks lay flattened in the ash. It was hard to imagine how anything could have survived.

Yet, an enormous amount of life did survive (Figure 27-1). Along snow-covered ridges that faced away from the blast, saplings still bent by the weight of one or two meters of winter snow, as well as seedlings and shrubs

Figure 27-1
Life returns. Mount St. Helens with Spirit Lake in the foreground before the eruption in 1980 (top photo). Mount St. Helens in 1994 with Spirit Lake (bottom photo). By 1994, plants had begun to clothe even some of the most devastated areas. Ecologists predict that Mount St. Helens's ecosystem will return to mature coniferous forest in about 200 years.

buried under the snow, came through unscathed. Shrubs that were obliterated at the surface, but whose roots survived underground, sprouted and grew anew. From underground, burrowing animals such as pocket gophers and a variety of insects and other invertebrates emerged. Within weeks, plants began growing on the naked slopes of the mountain wherever rain had washed away the sterile ash to expose the old soil beneath.

Within 10 years, dogwood, elderberry, huckleberry, and other shrubs, as well as seedlings of fir, hemlock, and other conifer trees covered the slopes of Mount St. Helens. Ecologists say that by 2030, the area will begin to look like a young coniferous forest. In 200 years, the area should resemble an old-growth forest. The Mount St. Helens ecosystem—almost completely destroyed—will rebuild itself.

27.1 Are Communities Integrated or Individualistic?

The process by which Mount St. Helens renews itself is called **succession**, which is the predictable change in numbers and kinds of organisms in a particular area over time. How and why succession occurs has been one of the burning questions in ecology for more than a hundred years. All the organisms that live in a particular habitat are regarded as part of a **community**.

In Chapter 25 we discussed biologist Lynn Margulis's view that our biosphere is a tightly integrated whole that is composed of organisms that cooperate as well as compete. Surprisingly, the philosophical differences that fueled the debate over Gaia are old issues in the science of ecology. Do species interact in ways that create a cohesive, functioning environment? Or are species free agents that operate independently of other species? For nearly a hundred years, ecologists have been arguing about whether communities of organisms are "integrated" or "individualistic."

It all began in 1898, when one of America's first ecologists, Henry C. Cowles (1869–1933), published a groundbreaking paper on plant succession. Cowles was a warm, funny man of infectious high spirits who inspired generations of young ecologists. As a graduate student, he studied the succession of plants on sand dunes beside Lake Michigan.

Cowles saw that over thousands of years the water level of Lake Michigan has gradually dropped, leaving behind a broad, sandy shore. Fierce winds blow the sand from the former lake bottom into giant dunes. Grasses eventually cover the sand, growing long roots that stabilize the dunes. After that, bearberry, juniper, and other shrubs grow on the dune, displacing the grass. Further back from the shore, Cowles found jack pine and other trees growing among the shrubs. The tall pines create a moist, shady habitat that kills the shrubs, which are adapted to the dry open sand dunes, and encourages the growth of a black oak forest. Finally, in some places, the black oak forest gives way to beech and maple forest.

This sequence of communities starts right by the lake, where the newest sand dunes lie, and ends far up into the woods, where the oldest forests had grown up and matured (Figure 27-2). In every place around the lake, the same series of changes was taking place. Cowles realized that natural communities can change predictably over time and that the individual species have a natural relationship to one another. The tall pines couldn't come in until the grasses and shrubs added leaf litter and other organic matter to the sand. And the oak trees couldn't come in until the pines had shaded out the shrubs. But just as the shrubs on the Michigan dunes were overshadowed by larger plants, Cowles was himself soon overshadowed by a stronger personality.

Succession's most important booster was Frederick Edward Clements (1874–1945). Clements was Cowles's opposite—arrogant and distant, but a prolific researcher and

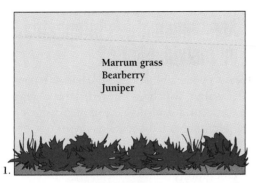

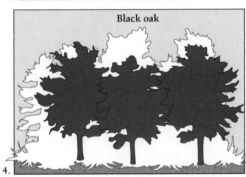

Figure 27-2

Sand dune succession. Henry C. Cowles showed that as the water level in Lake Michigan drops, the sand dunes left behind are gradually colonized by a succession of different plant communities. 1. Grasses and shrubs keep the sand dunes from blowing away and add organic matter that helps form a simple soil. 2. Cottonwoods grow in the sandy soil, shading out the grasses and shrubs. 3. Jack pines grow up through the cottonwoods, further adding to the soil and creating shade. 4. In the cool moist soil, oak trees germinate and grow, eventually shading out the jack pines. 5. Oak forest may give way to beech, maple, and hemlock.

writer all of his life and much revered by other scientists. In one of the first automobiles, and long before the construction of paved highways, Clements and his wife, Edith, drove thousands of miles across the roadless western United States, camping and studying plant communities. Clements wrote about every major ecosystem in North America, and, in 1916, published the definitive book on plant succession. Clements argued that in every geographic area, as defined by climate and soil, succession always created the same predictable community of plants, a **climax community.**

Clements argued that the climax community was an organism. Succession, he said, was a developmental process, like the embryonic development of an organism. Sounding rather like James Lovelock of Chapter 25, Clements wrote, "As an organism, the [climax] formation arises, grows, matures, and dies . . . each climax formation is able to reproduce itself, repeating . . . the stages of its development."

Clements's description is of an **integrated** community, one that consists of characteristic species that always interact with each other in predictable ways. In this view, all maple tree forests in a given area would have approximately the same species of plants and animals, and all these species would interact in similar ways.

Other ecologists found the integrated view compelling. Anyone can see that deserts usually contain cactuses, small mammals, lizards, and snakes, while ponds usually contain water plants, frogs, fish, and mosquitoes. Clements's view—that a community is an integrated collection of interdependent species—is one of the most important philosophical themes in ecology. But its simplicity is also its weakness. It invites criticism.

In 1926, a less-famous American ecologist named Henry Allan Gleason published a paper frankly challenging Clements. Gleason argued that communities are not integrated but composed of separate populations that merely happen to occupy the same habitat. Provocatively, Gleason wrote, "In conclusion, it may be said that every species of plant is a law unto itself. . . . [A plant] association is not an organism . . . but merely a coincidence." In Gleason's view, communities were **individualistic.**

Gleason's impolite attack on Clements so embarrassed other ecologists that most refused to even discuss Gleason's heretical ideas. As Gleason later recalled, "Not one believed my ideas; not one would even argue the matter. . . . For ten years, . . . I was an ecological outlaw."

In time, though, ecologists began to take Gleason's ideas seriously. A healthy and productive scientific debate ensued between Clements's integrated ecologists and Gleason's individualistic ones. Gleason and other individualists argued that every species has an independent distribution and that, in effect, every community is unique.

The individualistic view is supported by many careful studies of species distribution patterns that show no obvious clusterings of species. A survey of Wisconsin forests, for example, showed that no two tree species consistently share the same range (Figure 27-3A). Maple trees may be associated with beech trees in one place, elms in another. In the individualistic view, a beech-maple forest is not an organic entity or community but two species that just happen to share the same moisture requirements. Today, few ecologists accept Clements's rigid definition of a climax community. Most research shows that species composition varies from community to community.

On the other hand, the integrated view of ecological communities rests firmly on thousands of studies that demonstrate interactions among species. To take an obvious example, most woodpeckers and termites could not easily survive in a treeless grassland. Nor could many plants survive without the fungal mycorrhizae that help them draw nutrients from the soil (Chapter 21). But more subtle interactions also determine what lives where. For example, the presence of sugar maples appears to limit the distribution of basswood trees, perhaps through competition for water or shading (Figure 27-3B). So, even though the unity of a community may not be as well defined as Clements envisioned, ecologists still think of communities as functioning units whose member species all interact.

The integrated and the individualistic views are not mutually exclusive ideas but the extreme ends of a continuum. Just as the members of a family sometimes behave as individuals who only happen to live together at the same address and sometimes behave as an integrated whole, the species that make up a community can be both individualistic and integrated. A **community** is both an integrated set of interacting species and a group of organisms that occur together because their ranges happen to overlap. In the next section, we will examine a few of the ways in which different organisms in a community interact.

The individualistic and integrationist views of ecosystems represent extreme ends of a continuum of thought.

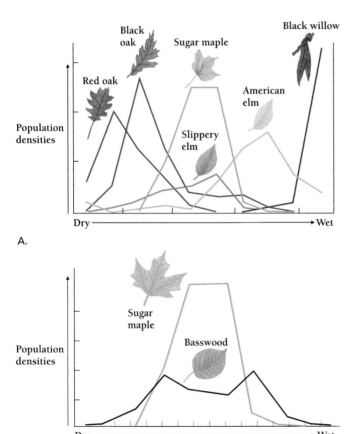

Figure 27-3

Are communities individualistic or integrated? Studies of trees in Wisconsin support both ideas. A. Oaks, maples, and other trees are distributed according to the dampness of the soil, independently of other trees. B. But basswood grows best where sugar maple is absent, even though it is capable of growing in the same soils.

27.2 How Do Species in a Community Interact?

We saw in Chapter 26 that the nature of a community depends partly on temperature, moisture, and sunlight. But within each community, the interactions among different species also help to establish the character of a community. These interactions include those between the consumers and the consumed and among competitors.

Who Eats Whom?

We usually think of consumers as being larger and stronger than the consumed. Hawks, for example, generally prey on mice, squirrels, and other animals smaller than themselves. Most predators are indeed larger than their prey, although carnivores that hunt in groups, such as wolves, regularly bring down prey far larger than themselves. But herbivores are often smaller than the individuals they consume. Hordes of tiny oak moth caterpillars can strip all the leaves from a 60-foot oak tree. And parasitic consumers, such as ticks, fleas, and tapeworms, are smaller than their hosts.

Consumers, then, come in many forms. The main consumers besides the decomposers and scavengers are herbivores, predators, and parasites. A **parasite**, by definition, consumes only part of the blood or tissues of its host and does not necessarily kill the host. In contrast, a **predator** usually (but not always) kills its prey and consumes most of its prey's body.

In some cases, the differences between predators and parasites may blur. For example, parasites do occasionally kill their host outright, and parasites frequently weaken the host so severely that the host dies of other causes. And some predators consume only part of their prey. For example, some predaceous fish bite chunks of flesh from other fish, without necessarily killing the victim.

Herbivores at first seem to fall outside these definitions, but they actually fit rather nicely. Leaf-eating caterpillars and deer, for example, browse on living plants, just like parasites. And seed-eating birds, which kill and consume an entire plant embryo, may be considered predators. In reality, ecologists rarely stress the parasite-predator distinction in herbivores.

How Do Plants and Animals Defend Themselves?

With notable exceptions, consumers' adaptations for finding and eating their favorite food exhibit a certain sameness: in the case of predators, swiftness, intelligence, acute senses, and sharp teeth; in the case of herbivores, patience and a good digestive system. In contrast, organisms have a diverse arsenal of defenses against potential consumers.

Camouflage

The subtlest defense against being eaten is to remain invisible. A rabbit simply dives down a hole or freezes so that it blends into a background of shrubs or grass. But many organisms go to far greater lengths to **camouflage** themselves by mimicking twigs, leaves, stones, bark, and other materials in their environment (Figure 27-4). A walking stick insect, for example, looks just like a twig. Some species of walking sticks even sway slightly, as though an errant breeze were blowing.

Chemical Defense

Plants defend themselves against consumers in a variety of ways. Trees protect themselves from termites and other wood-eating creatures with thick layers of bark. Roses and cactuses fend off herbivores with thorns. The single most important defense that plants use, however, is toxic chemicals. The unforgettable wallop of hot chili pepper, the sharp tang of raw broccoli, and the deadly toxin in poison hemlock all result from certain kinds of **secondary plant compounds**—chemicals that are not essential to a plant's normal metabolism but that often serve a defensive purpose (Chapter 30).

Chemical defense is not limited to plants. Bees, ants, and wasps inject a powerful acid into attackers. The bombardier beetle and the skunk both spray would-be predators with noxious chemicals (Figure 27-5). The famous poison-dart frog (*Phyllobates*) of South America and the skin and feathers of the pitohui bird of Papua New Guinea contain the same deadly toxin (Figure 27-6).

Warning Coloration

If a skunk looked like a rabbit, it would be attacked over and over again, in spite of its potent defense. But because a skunk has a striking and memorable black-and-white pattern, predators quickly learn to avoid the animal. Among all animals, chemical deterrents work best if accompanied by a bright, memorable design called **warning coloration.** Warning coloration, such as the yellow-and-black stripes of the aggressive, stinging yellow jacket, helps predators remember which prey to avoid. Predators will avoid any animal with the colors and patterns they associate with pain, illness, or other unpleasant experiences. Most warning col-

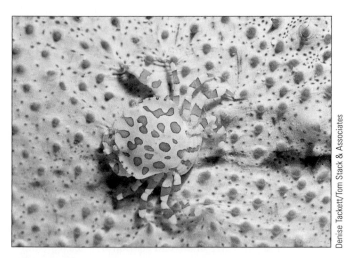

Figure 27-4
Camouflage. Many animals avoid being eaten by blending in with their background. This harlequin crab lives on the underside of a sea cucumber, camouflaged against the background of its host.

Denise Tackett/Tom Stack & Associates

Figure 27-5
Chemical warfare. Everyone knows that skunks fight back with noxious chemicals. So do many plants and insects, such as this bombardier beetle.

Thomas Eisner and Daniel Aneshansley/Cornell University

Figure 27-6

Warning coloration. In 1992, researchers discovered three species of birds in the genus *Pitohui*, whose skin and plumage contained the same toxin as that found in poison-dart frogs of South America. Like wasps, skunks, and poison-dart frogs, pitohui birds bear striking colors that may function to warn predators. Research has shown that inconspicuously colored birds taste better than flashy ones.

ors are vivid shades of red, yellow, orange, and white, combined with a contrasting shade of blue or black.

Because predators avoid animals with warning coloration, some species copy the warning coloration of toxic or dangerous species. A situation in which one animal *mimics* the color pattern of another is called **mimicry.** Some perfectly harmless and quite edible flies mimic the black-and-yellow stripes of yellow jackets and bees. A situation in which a harmless animal imitates a dangerous one is called **Batesian mimicry.**

In a second form of mimicry, called **Müllerian mimicry,** two or more dangerous species evolve similar colors. The similar colors signal a similar hazard to any predators the two mimics share in common. For example, monarch butterflies are toxic to birds because the caterpillars feed on milkweed, a plant that contains a toxic secondary plant compound. This toxin accumulates in both the monarch caterpillar and butterfly and makes it toxic to birds. A bird that eats a monarch quickly becomes ill (Figure 27-7A). After a few such experiences, the bird avoids monarchs, and the monarch's striking warning colors help the bird remember.

A second butterfly, the viceroy, is unrelated to the monarch but also contains a toxin that makes birds sick. The viceroy looks remarkably like a monarch, and biologists believe these two butterflies have evolved to resemble one another (Figure 27-7B). The advantage to each butterfly is that birds that have learned to avoid the monarch also avoid the viceroy and birds that avoid the viceroy also avoid the monarch.

In this way, Müllerian mimicry helps predators learn the same lesson from a variety of prey species. Bees and wasps both use a similar pattern of yellow and black stripes. Anyone who has been stung by a bee also learns to avoid wasps and hornets. Such encounters lead to greater safety for bees, wasps, and hornets. The more species that use these kinds of patterns and colors, the more this universal signal is reinforced.

Symbiosis: How Do Organisms Live Together?

A species may live in an intimate association with another species, an arrangement called **symbiosis,** meaning literally "living together." Symbiosis may be cooperative or antagonistic. Each species may affect the other positively, negatively, or not at all. Table 27-1 summarizes some of the possible interactions between any two species. In **parasitism,** one species benefits at another's expense. Tiny parasitic wasps, for example, lay their eggs inside of caterpillars. The wasp larvae literally eat the caterpillars alive as both animals mature. In **predation,** one species also benefits at another's expense.

Mutualism describes a symbiotic relationship between two species that mutually benefits each (Figure 27-8). For example, the bull's horn acacia (*Acacia cornigera*) of Central America has special enlarged thorns that house colonies of ants (*Pseudomyrmex ferruginea*). The acacia's nectar attracts the ants, which use it as their major food source. The acacia also provides nutritious fats and proteins in nodules at the

A.

B.

Figure 27-7

Müllerian mimicry. A. The monarch butterfly is so noxious that birds such as this jay quickly cough them up. B. Viceroy butterflies are unrelated but also toxic. Monarchs *(left)* and viceroys *(right)* have evolved amazingly similar color patterns. Each butterfly's message reinforces the other's: "Don't touch!"

Table 27-1
Interactions Among Organisms

Interaction	Effect on A	Effect on B	Example
Competition between A and B	Harmful	Harmful	Otter and mink
Predation by A on B	Beneficial	Harmful	Lion and gazelle
Symbiosis			
Parasitism by A on B	Beneficial	Harmful	Flea and dog
Commensalism of A with B	Beneficial	Harmless	Fish and anemone
Mutualism between A and B	Beneficial	Beneficial	Ants and acacia

Biology ⒺNow™ Learn more about predation interactions by clicking on this figure on your BiologyNow CD-ROM.

tips of its leaflets. In return, the ants protect the acacia from being eaten by caterpillars and other herbivorous insects. This arrangement benefits both species. One study showed that acacias grown without ants for 10 months weighed less than one-tenth of those with intact ant colonies.

Another kind of symbiosis is **commensalism,** an intimate relationship between two species that helps one but neither helps nor harms the other. For example, the anemone fish lives safely among the poisonous tentacles of sea anemones. The fish, which benefits from the protection of its host, is a commensal. The sea anemone, which catches and eats other species of fish, leaves the anemone fish alone, possibly because it secretes a protective slime that keeps the anemone from recognizing the fish as prey.

Species interact as competitors, as predator and prey, and as symbionts.

Figure 27-8
Mutualism. The cleaner shrimp lives safely near the mouth of this predaceous moray eel. The shrimp feeds on parasites it obtains while cleaning the eel, and the eel benefits from the removal of parasites.

Do Organisms Coevolve?

The mutualistic relationship between ants and bull's horn acacias demands special adaptations in each species. The acacia has evolved enlarged thorns with a tough, woody exterior and a soft, pithy center that is easy for the ants to excavate for nests. The acacia also has "nectaries" at the bases of its leaves that supply the ants with nectar and nodules of nutrients at the tips of its leaflets. The ants likewise have special adaptations. For example, most species of ants are active only at night or only during the day. But members of the ant species *Pseudomyrmex ferruginea* remain active and protect the acacia day and night. The ants not only chase off plant-eating animals but even protect the acacia from other plants that compete with the acacia for light, nutrients, or water.

These distinctive adaptations in the ant and the acacia appear to result from a long and mutual evolutionary history in which the needs of each species exert selective pressure on the other. The interdependent evolution of two or more species whose adaptations appear to be selected by mutual ecological interactions is called **coevolution.**

Biologists love to suggest likely examples of coevolution. For example, many pollinating insects specialize on flowers that seem, in turn, to be specially constructed to be pollinated by just one kind of insect. Mutualisms such as these present the most likely examples of coevolution. But demonstrating that coevolution has occurred in individual cases is difficult. To do so, a biologist must show that the actions of one species have brought about an evolutionary change in the other, and vice-versa.

All species that have ecological interactions may exert some selective pressure on one another. And virtually every species in a community interacts (however indirectly) with every other species. For example, soil bacteria that break down organic matter into nutrients that plants can absorb through their roots are as essential to herbivores and carnivores as to the plants. Coevolved relationships can be either very specific, as in the case of the acacias and the ants, or very diffuse. The concept of "diffuse coevolution" is another way of acknowledging that all organisms in a community affect one another.

Interactions between two species influence the evolution of both, a process called coevolution.

How Does Competition Affect Organisms?

Some species interact as consumers and consumed, and others as commensals or mutualists. Still others compete. In **competition**, one organism uses a resource that is in limited supply, negatively affecting another species' ability to survive and reproduce. For example, if two species of deer live in the same area and eat the same herbs, and the herbs are in limited supply, the two kinds of deer are in competition with each other.

Can Two Similar Species Share the Same Habitat?

Measuring competition in natural communities takes careful planning and months or years of research. But laboratory studies can quickly show how competition occurs in simple situations. The Russian ecologist G. F. Gause studied populations of two similar species of *Paramecium—P. caudatum* and *P. aurelia*. Each *Paramecium* species, cultured by itself, flourished on the bacterial food that Gause provided. But when Gause mixed the two species, *P. caudatum* disappeared from the culture within two weeks. Gause provided only one environment and one kind of food for the two species, and *P. aurelia* was evidently better adapted to Gause's conditions than was *P. caudatum*. One species thus drove the other to extinction.

These and similar experiments have led to a generalization called the **competitive exclusion principle,** which says that when two species compete directly for a resource that is in short supply, one species can eliminate the other. A resource in short supply is a **limiting resource.** For example, territory might be a limiting resource for a pride of lions, but air would not be. Similarly, water is often a limiting resource for terrestrial plants but not for aquatic plants.

The competitive exclusion principle predicts that one species always wins. But the same species does not win under all conditions. For example, in a 40-year series of experiments at the University of Chicago, Thomas Park and his colleagues studied the effects of different environments on competition between two species of grain beetles (*Tribolium castaneum* and *Tribolium confusum*). These beetles were allowed to grow in containers of flour, the favorite food of the larvae. When grown separately, both species did best in warm, moist conditions. When grown together in warm, moist containers, *T. castaneum* always drove *T. confusum* to extinction. In a cold, dry container, however, *T. confusum* outcompeted *T. castaneum*.

According to the competitive exclusion principle, no two species can simultaneously exploit the same resources in the same place. The place in which an organism lives is called its **habitat.** In contrast, the way an organism uses its environment is called its **niche.** The habitat of the desert tortoise is the desert. The niche that the tortoise occupies is that of a burrowing herbivore. If a habitat is an organism's address, the niche is its job.

The competitive exclusion principle states that no two species can indefinitely occupy the exact same niche in the same habitat.

How Do Competing Species Coexist?

In the laboratory, species can be made to compete for the same limiting resources. But in nature, two species occupying the same habitat can, in practice, split up the niche in an infinite number of ways. For example, the ecologist Robert H. MacArthur (1930–1972) tested Gause's competitive exclusion principle in field studies of five species of North American warblers. All five bird species hunt for insects and nest in the same kind of spruce tree. The niches of these five birds seem nearly identical, but MacArthur's studies showed that the five warblers had split up the spruce tree niche extremely finely. He showed that each species of bird hunts insects in a different part of the tree and nests at a separate time from the other warblers. As a result, the warblers' niches overlap without being identical. Ecologists call splitting a niche **resource partitioning.**

Another study, by ecologist Joseph H. Connell, shows how two species go about splitting a niche. In tidepools on the coast of Scotland live two species of barnacles. One species, *Balanus balanoides,* lives only on the lowest intertidal rocks, which are wet even during low tide. The other species, *Chthamalus stellatus,* lives higher up, in between the low-tide and high-tide lines.

To find out if the two species were competing, Connell removed *Balanus* from some tidepools and *Chthamalus* from others (Figure 27-9). Connell found that where *Balanus* was removed, *Chthamalus* moved down to occupy the newly opened area and survived there perfectly well. Likewise, where *Chthamalus* was removed, *Balanus* was able to survive on the higher, drier rocks.

Balanus dominated the lower levels because its heavy shell grows under the shell of *Chthamalus* and pries it up off of the rocks, preventing *Chthamalus* from growing on the lowest rocks. Higher up, however, *Chthamalus*'s higher tolerance for dry conditions allows it to prevail over the physically stronger *Balanus*.

Connell's two barnacle species illustrate the difference between a species' **fundamental niche,** its potential ability to utilize resources, and its **realized niche,** the resources that it actually uses in a particular community. The fundamental niche of *Chthamalus* included the lower rocks, but its

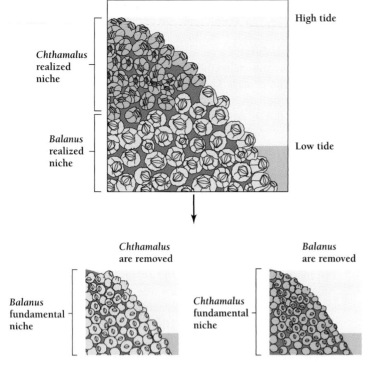

Figure 27-9
Niche splitting in barnacles. In a classic experiment, the ecologist Joseph Connell showed that the two barnacles *Balanus* and *Chthamalus* are capable of growing on the same rocks if the other species is removed. Otherwise they split the tidal niche, *Balanus* below and *Chthamalus* above.

realized niche did not. The fundamental niche of *Balanus* included the higher rocks, but its realized niche did not.

Similar experiments on rat parasites also illustrate competitive exclusion. Although the tapeworm and the spiny-headed worm both prefer the upper intestine of a rat, when the two worms are together, the spiny-headed worm takes the food-rich upper tract and displaces the tapeworm to the lower end of the intestine. All these experiments, and many more like them, suggest that competition in natural populations helps determine where species live and which ones can live together.

Competing organisms in natural environments may partition resources so that both species persist in the same habitat. Competition helps determine the species that can live in a community.

Exotic Species Are a Threat to Biodiversity and an Example of Competition

Species in the wild can easily drive one another to extinction through either predation or competition. In communities of species that have evolved together over thousands of years, we are unlikely to see such dramatic interactions. But when a brand new species arrives in an old established commu-

nity, anything can happen. Over the last few hundred years, humans have introduced thousands of new species to communities around the world. The result has been a dramatic increase in the rate of species extinctions. People introduce animals both on purpose and by accident, but whether the new species is a plant, a pig, a snake, or a bacterium, it's called an "exotic" because it's from somewhere else. An **exotic species** is one that has been introduced to a geographical area where it is not native. Exotic species have caused 39 percent of known species extinctions on Earth—more extinctions than are caused by any other single cause.

Throughout history, people have always brought cultivated plants and domestic animals wherever they traveled. The Polynesians introduced pigs to the Hawaiian Islands long before the first Europeans arrived. And Europeans introduced hundreds of animals to North and South America, Australia, New Zealand, and Hawaii—including, for example, domestic animals to eat, such as pigs and chickens; European wild boar to hunt; dogs and cats as pets or working animals; and English sparrows just for the pleasure of hearing their song. Even more plants arrived as crops and ornamentals. Still more species arrived (and continue to arrive) by accident. Weed seeds can be carried around the world in clothing, food, or mixed into crop seeds. Rats, mice, snakes, and mosquitoes stow away on board ships and airplanes. Mussels, snails, and fish travel in the ballast of ships. Disease and parasites come with their human or nonhuman hosts. In Hawaii, nearly all the native birds have been driven to extinction, killed by malaria brought by nonnative birds that came with humans.

Not all exotic species are a threat to natives. Nearly all vegetables grown in North America are exotic species, yet few of them are any hazard to native ecosystems. And gardeners know that some ornamental plants need constant nursing and attention to grow at all. These species are in no danger of taking over nearby ecosystems.

How is it that some exotics can be so destructive? Some exotics, by chance, are extremely well adapted to their new habitat. For example, when someone introduced the aggressive Minnesota northern pike to a lake in northern California sometime in the 1990s, that person unleashed what may turn out to be an ecological disaster. The pike have not only taken over this single lake but threaten to move downstream and upstream into all the waterways in California. In the absence of natural predators, the pike have been reproducing at exponential rates. In the years 2000, 2001, and 2002, state game workers captured 600, 6,358, and then 17,635 pike. Despite game workers' efforts to control the pike, populations are exploding, and there are more pike than ever. A single female may carry more than 80,000 eggs. The pike is not only an exotic, it is a highly **invasive** exotic, one that outcompetes local species and dominates its new habitat (Figure 27-10).

Invasive species often succeed so well because they do not share an evolutionary history with other organisms in the community. An invasive plant might contain a toxin that no native herbivore can tolerate. While native plants are be-

Figure 27-10

An invasive exotic snake. In the 1940s, the brown tree snake *Boiga irregularis* arrived on the island of Guam from Papua New Guinea as a stowaway on military ships and planes. In the 1970s, biologists began to notice an alarming rate of extinction of native birds and proposed various reasons. Julie Savage, who was then a graduate student at the University of Illinois, deduced that the brown tree snake was to blame. By 1990, all of the island's rain forest birds, its fruit bats, and most of its reptiles had been driven to extinction by a single exotic species.

Michael and Patricia Fogden/Corbis

ing eaten by native herbivores that have evolved the ability to eat them, the invasive exotics grow without restraint. The northern pike has no natural predators or competitors in California and is well adapted to the conditions in the lake.

Most interactions among species result in the coevolution of adaptations: prey evolve defenses against predators, which in turn evolve better mechanisms for capturing prey; competitors evolve morphological or behavioral adaptations that allow them to coexist. Such coevolution allows predators, prey, parasites, hosts, and competitors to coexist in the same community by limiting their ability to drive each other to extinction. Invasive exotics enter a community free from such limits. Prey are defenseless and edible; predators and parasites don't exist; and competing natives are at a disadvantage because they have to cope with predators and parasites.

Once established, most exotic species are impossible to eradicate. They are here to stay and likely to spread. In an effort to control the northern pike, sometimes called the "Water Wolf" for its voracious appetite, California has poisoned the entire lake, installed fish-grinding gates in the dam that allow the release of water but not fish, erected electrical barriers, and detonated explosives. In addition, officials patrol all adjacent lakes and reservoirs daily, scanning the lakes for pike. The state has already spent $15 million combating the fish, and yet their numbers continue to grow. It seems only a matter of time before the pike slip downstream to California's extensive waterways. At most immediate risk are salmon, trout, and other aquatic vertebrates. Yet fisheries experts say their greatest worry is that the fisherman who brought the pike in the first place will introduce it

to other lakes. DNA studies suggest that someone deliberately introduced a starting population of about 50 fish.

Preventing the introduction of potential pest species in the first place is the best solution, but humans continue to introduce new species wherever they go. Whether homesick for England, China, or Minnesota, people love to surround themselves with familiar and useful plants and animals.

Introduced exotic species are the leading cause of known biological extinctions.

Character Displacement: Does Competition Influence Evolution?

We have seen how competition between two species can lead to niche splitting. Species may evolve differences that make niche-splitting strategies permanent. For example, two species of caterpillars that both eat members of the rose family (which includes roses, apples, plums, and other species) might reduce competition by specializing, one on apple trees and one on roses. Any change in morphology or behavior that results from competition is called **character displacement.**

The finches of the Galápagos Islands provide some of the best-documented examples of character displacement. In the case of the Galápagos finches, character displacement has led to speciation. For example, the small ground finch *Geospiza fuliginosa* and the medium ground finch *Geospiza fortis* occur together on some islands, while on other islands only one of the two species lives. On islands where these birds occur separately, both species' beaks are medium sized, and they eat similar-sized seeds. But on islands where the two birds occur together, *G. fortis* individuals have larger beaks and specialize on larger seeds than *G. fuliginosa* (Figure 27-11). In other words, when the two species occur together, they split the niche and specialize on either big seeds or small ones. Where they occur separately, they each take the middle ground.

Recent studies have shown that the medium-sized beak is the best option under some circumstances. On islands where the two species occur together, they often hybridize, producing offspring with intermediate-sized beaks. When food is scarce (in times of drought), the hybrids reproduce poorly. When seeds are plentiful, however, the hybrids reproduce *better* than their parent species.

Competition and adversity appear to drive these two species apart, while lack of competition and easy pickings drive them together. Currently, they are separate species. Their distinctive adaptations almost certainly have resulted from previous competition. As Joseph Connell put it, character displacement is "the ghost of competition past."

Over evolutionary time, competition causes character displacement through natural selection.

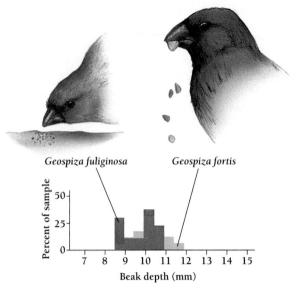

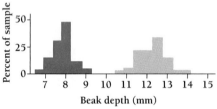

Figure 27-11

Character displacement in Galápagos finches. A. On islands where only one of two finches, *Geospiza fuliginosa* and *Geospiza fortis,* occur, the two species have beaks of similar, intermediate size, because each is a generalist, feeding on seeds of many sizes. B. But on islands where the two birds occur together, they evolve different-sized beaks that minimize competition between the two. *G. fortis* birds evolve large beaks and specialize on larger seeds, while *G. fuliginosa* birds evolve smaller beaks and specialize on smaller seeds. This is a classic example of both character displacement and niche splitting.

Biology ⊜ Now™ Learn more about Galápagos finches by clicking on this figure on your BiologyNow CD-ROM.

27.3 What Determines Biodiversity?

We have seen that for two competing species to coexist, they must exploit different niches. Both resource partitioning and character displacement increase the number of species in a community.

Each species occupies a unique niche, and each niche has a unique set of properties, such as food supply, sunlight, and other resources. All the unique but overlapping niches fit together to form a kind of multidimensional jigsaw puzzle—the community. The number of niches is a measure of the richness and complexity of a community.

How Can We Measure Diversity?

The number of species in a community is called **species richness.** For example, a single river in Amazonia may contain more species of fish than all the rivers of Europe. But species richness is only a crude measure of diversity, since some species may occur in great numbers while others are rare. To take an extreme example, the city of San Diego has an unusually large and well-stocked zoo, and so the city's species richness is high. The vast majority of organisms living in San Diego, however, consist of only a few species of animals—mostly humans, rats, mice, cockroaches, ants, and dust mites.

A better measure of diversity is **relative abundance,** which takes into account how common each species is. Ecologists use various mathematical summaries of relative abundance. Often, each species is assigned an "importance value" based on the number of individuals in an area, on biomass, or on primary productivity. The relative abundance of species in a tropical rain forest is extremely high. If we find 200 species of trees, most of them will be equally common. In contrast, a temperate deciduous forest is dominated by just a few species. In one West Virginia forest, for example, more than 83 percent of all trees were either yellow poplars or sassafras.

Careful study of a particular community can reveal the number and kinds of species that compose it. But what determines the particular species makeup of a community? Do latitude, rainfall, and other abiotic components determine the diversity of a biotic community? Or does the biological makeup of each community depend on its unique history—that is, on which plants and animals happen to arrive in what order? Are there any general rules that determine the diversity within a community?

How Do Biotic Factors Influence Diversity?

We have seen how competition and the resulting character displacement can increase the number of species in a community. Predators can also maintain diversity by preventing competitive exclusion. Some predators prevent common species from overwhelming less common ones. For example, when ecologist Robert Paine removed sea stars from a richly diverse intertidal community of algae and invertebrates, the entire area became overgrown by mussels, drastically reducing the number of species and the structural complexity of the community. The sea stars had kept the mussels in check by eating them, and the removal of the sea stars let the mussel population skyrocket. Paine argued that a predator that promotes diversity in this way should be called a **keystone species.**

Predators adjust their predation according to what is available. When one prey species becomes common, a predator will begin to prey preferentially on that species, ignoring one that has become rare. By switching from prey to prey, according to which is most common, a predator can keep the population of the dominant species at low enough densities that other species can survive.

How Do Abiotic Factors Influence Diversity?

The relative importance of biotic and abiotic factors in ecosystem diversity is still a subject of intense research. That abiotic factors play an important role, however, is certain. Species diversity generally increases from the poles to the equator, and from high elevations to low elevations (Figure 27-12). Physically varied areas (with hills, valleys, and streams) allow more unique niches than flat, featureless landscapes. Finally, the larger the size of an ecosystem, the more species it is likely to contain.

One way that ecologists have studied diversity is by studying the species diversity on islands. They have found that large islands, which have more habitat diversity and more niches, support more species than small islands.

Furthermore, the number of species on an island may remain stable—even as individual species appear and disappear. For example, in 1968, the Channel Islands, off the coast of southern California, had about the same number of bird species as they had had in 1917. But nearly one-third of the species were new.

In the 1960s, Robert MacArthur and another American ecologist, Edward O. Wilson, argued that the diversity of island species depends on a balance between the rates of immigration of new species and extinction of old species. According to the MacArthur-Wilson **theory of island biogeography,** a new island, devoid of already established species, should experience a high rate of immigration and a low rate of extinction, since there are fewer species to disappear. As more species arrive, the theory went, newer immigrants would have a harder time establishing themselves because of competition, and the rate of extinction would increase. MacArthur and Wilson also predicted that islands nearer the mainland would be easier for new immigrants to get to and should therefore have a greater rate of immigration and, therefore, more species than more distant islands.

Finally, MacArthur and Wilson formalized what ecologists as far back as Gleason had known—that the number of species in any habitat is proportional to its area. MacArthur and Wilson said the number of species (S) could be predicted by the following equation:

$$S = CA^z$$

where C and z are constants and A is the area of the ecosystem or island (Figure 27-13).

In 1969, Wilson and ecologist Daniel Simberloff devised an experimental test of the MacArthur-Wilson theory. They counted all the arthropod species they could find on several small islands (10 to 20 meters in diameter) in Florida, exterminated all the arthropods (mainly insects) with a chemical pesticide, and then monitored the gradual

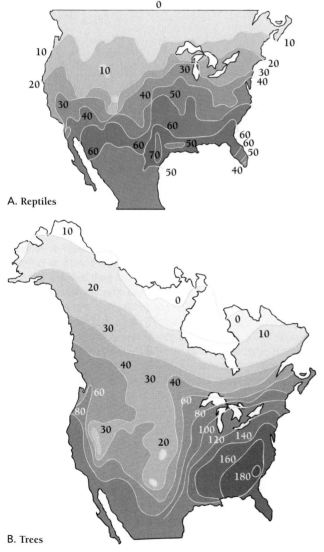

A. Reptiles

B. Trees

Figure 27-12
How does latitude affect species numbers? In North America, the number of species (numbers on maps) of both reptiles and trees increases from north to south. In general, species diversity increases as one travels from the poles toward the equator. (Source: R. L. Smith)

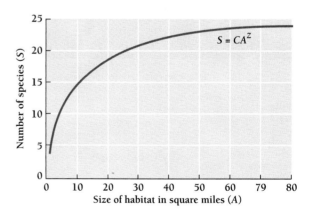

Figure 27-13
Larger habitats support more species. A graph of bird species on islands around Papua New Guinea shows that larger ecosystems support more species than smaller ones. We say that species number *(S)* is proportional to area *(A)*.

Table 27-2
Diversity on Islands Before and After Defaunation

Kind of Feeding	Number of species	
	Before	After
Herbivores	55	55
Scavengers	7	5
Detrivores	13	8
Wood borers	8	6
Ants	32	23
Predator	36	31
Parasite	12	9
Unknown	1	3
Total	164	140

nize an island may be a historical accident, but the *number* of species depends on the abiotic characteristics of the island. Of course, in this experiment, all of the trees, shrubs, and other plants on the island remained intact. These, too, probably played a critical role in creating a fixed number of niches for arthropods.

In 1984, Jorge R. Rey repeated this experiment on islands in Oyster Bay, Florida. Rey fumigated islands and then carefully counted the number of insects, spiders, and other arthropod species as they recolonized the islands (Figure 27-14). Rey's work confirmed what Wilson and Simberloff had shown—that an equilibrium number of species exists for each island and that the number of species depends on the area of the island. In addition, Rey showed that, as more species became established on an island, fewer species successfully immigrated to the island and more species on the island went extinct, thus establishing an equilibrium.

recolonization of the islands. Within six months, the islands had recovered their former arthropod diversity (Table 27-2).

Each island had approximately the same number of species as before it was fumigated. However, the kinds of species that arrived were often different from what had been there before. This study suggested that which species colo-

How Do Groups of Populations Interact?

Understanding the factors that govern species diversity on islands helps us understand species diversity on the continents. A forest surrounded by croplands or housing subdivisions stands as a kind of island, as does a meadow surrounded by forests or a mountain peak surrounded by valleys. Ecologists call these **habitat islands** (Fig-

Jorge Rey, University of Florida

A.

Figure 27-14
Does every island support a particular number of species? A. To find out, Jorge Rey first eliminated all of the invertebrate animals on islands in Oyster Bay, Florida, by covering the islands with giant tents and filling the tents with pesticides. After all the insects and other invertebrates were dead, he removed the tents and monitored the return of species to each island from the mainland. B. By about 20 weeks, each island was recolonized by a characteristic number of species. Bigger islands (top) accumulated more species than did smaller islands (bottom). Rey found that the species that colonized an island might not be the same ones that lived on the island before fumigation, but the *total number* of species on each island was about the same.

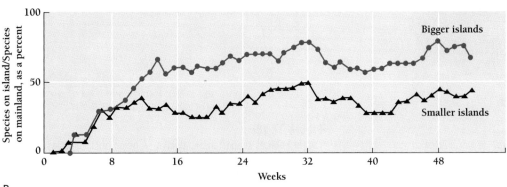

B.

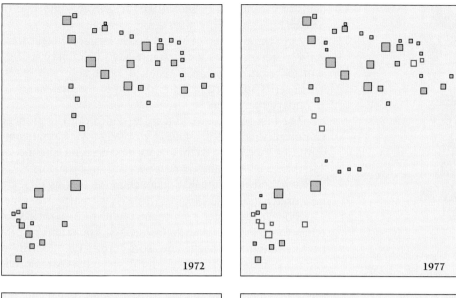

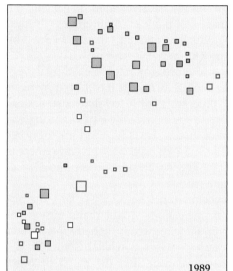

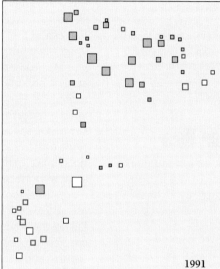

□ Populated

□ Newly extinct

□ Extinct

□ Recolonized

William J. Weber/Visuals Unlimited

Figure 27-15

Extinction and recolonization in a ghost town. In the desert between Nevada and California is an abandoned gold-mining town called Bodie. Living among the rocky diggings left over from the old 19th-century mines are small subpopulations of tiny rabbits called pikas. These subpopulations are in a constant state of change, with recolonizations and extinction. Bigger boxes indicate bigger populations. Between 1972 and 1977, about 10 subpopulations went extinct *(empty red squares)* but three new subpopulations appeared *(green-filled squares)* as pikas from other subpopulations recolonized empty habitats. Between 1977 and 1989, another 12 subpopulations went extinct and 2 new subpopulations appeared. Two years later, another nine subpopulations had disappeared and one area had been repopulated. Together, all these interacting subpopulations are called a metapopulation. (After Smith and Gilpin.)

ure 17-10). For many species, however, a small island of habitat cannot support a population of sufficient size for plants and animals to find mates or for animals at higher trophic levels to find sufficient prey. Sometimes chance events eliminate all the individuals in such small populations. Eventually, one or more species in these habitat islands goes "locally" extinct.

In the face of such local extinctions, the persistence of a population depends on migration from other habitat is-

lands. A group of local populations that alternately go extinct and recolonize vacant habitat islands is called a **metapopulation** (Figure 27-15). Just as with real islands, the sizes of islands, the number of islands, and their distance from one another all influence the rate of recolonization. In addition, the space between the habitat islands can also influence how easily and quickly species can migrate (Figure 27-16). We discuss extinction and population dynamics in more detail in Chapter 28.

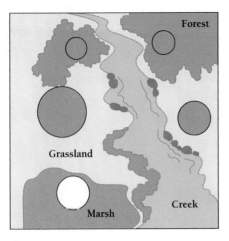

Figure 27-16

It's what's in between that counts. In MacArthur and Wilson's model of island biogeography, habitat islands are separated by uniform expanses of "matrix," and the likelihood of colonization is simply proportional to distance. In most communities, however, habitat islands *(circles)* are surrounded by a matrix of different microhabitats, where it may be easier for individuals to travel among islands that are far apart than among those that are closer together. For example, a frog traveling from the little pond in the forest in the upper left to the empty habitat *(white circle)* in the marsh at the lower left would find it easier to swim two miles downstream to the marsh than to hop a half mile through dry grassland.

27.4 Communities Change Over Time

We have seen now that the characteristics of communities depend in part on the geography and form of the habitat. Now we will look at how communities change over time.

Why Do Communities Change?

After Mount St. Helens blew up in 1980, the near-total destruction of the community of plants and animals that lived on its slopes initiated the gradual development of a new community. Windblown seeds from fireweed, grasses, and other plants first colonized the slopes of the volcano, rooting into the remaining soil wherever the thick volcanic ash had been washed away by erosion. Gradually, these weedy plants covered the barren slopes of Mount St. Helens. In time, shrubs such as dogwood and huckleberry, and small trees such as fir and hemlock, grew in among the first colonizers and replaced them. The trees would replace, or *succeed,* the shrubs, as the shrubs had begun to replace the weeds.

Each set of plants represents a stage in succession, the change in numbers and kinds of organisms in a community over time. Ecologists distinguish between **primary succession,** the invasion of a completely new environment such as a sandbar or new volcanic island, and **secondary succession,** the sequence of stages in a badly disturbed community. Examples of secondary succession include the reestablishment of a community in a burnt or logged-over forest and the return of natural communities to abandoned crop-

lands or vacant lots. In secondary succession, some kind of preexisting community plays a role in succession.

After the eruption of Mount St. Helens, some areas were entirely sterile. These areas were ready for primary succession to begin. Other areas retained fertile soil with seeds. These seeds sprouted and contributed to the first stages of secondary succession.

The first community in a succession is called a **pioneer community,** while later stages are called **successional communities.** According to the theory of succession, the mix of species in a community changes over time in a way that is cumulative and directional. Climax communities, as they are now understood by ecologists, themselves change over time.

Pioneer and later communities differ in a variety of other ways, summarized in Table 27-3. For example, pioneer communities usually consist of species that flourish in disturbed areas and breed rapidly. At the other extreme, climax communities often consist of species that breed slowly but gradually take over in undisturbed areas. In general, the short-lived weedy species of pioneer communities (thistles, for example) are gradually replaced, or succeeded, by longer-lived organisms (fir trees, for example).

Most descriptions of succession refer to changes in vegetation because succession is a concept introduced by plant ecologists. However, other kinds of succession exist. For example, a deer's carcass, a rotting tree, or a pile of dung can support a succession of insects, bacteria, and fungi over periods of two months or less (see box). In such cases, there is no climax community, and succession occurs on a small scale in both time and space.

Succession also occurs in patches. If a tree falls in a forest, it leaves a hole in the canopy. Bright shafts of sunlight fall to the forest floor, allowing weedy, light-adapted plant species to crowd out shade-adapted plants. Gradually, the shrubs and trees may take over, and the patch becomes shady forest again. But meanwhile other trees have fallen elsewhere in the forest. The result is a forest with patches of

Table 27-3
How Plant Community Traits Change During Succession

Trait	Early stages	Later stages
Biomass	Small	Large
Physical structure	Simple	Complex
Role of detritus	Minor	Major
Net primary production	High	Low
Stability (absence of change)	Low	High
Plant species diversity	Low	High
Seed dispersal by	Wind	Animal
Seed longevity	Long	Short

After Barbour, Burke, and Pitts 1980.

Extreme Biology   Succession on a Corpse

"The worms crawl in, the worms crawl out" goes the childhood ditty. But which "worms" and exactly when they crawl in and out can tell special criminal investigators how long ago, and often where, a person died. By studying the crawling creatures collected from a corpse, forensic entomologists, scientists who apply their knowledge of insects to legal matters, can solve otherwise baffling cases.

The father of American forensic entomology, Bernard Greenberg of the University of Illinois at Chicago, likes to recount the case of a missing nine-year-old girl. After her partly skeletonized body was discovered in an abandoned apartment building, Greenberg collected fly larvae and pupae from the body and took them back to the laboratory. There he raised them at temperatures like those in the apartment building over the days preceding the girl's discovery. He soon learned that the flies took 21 days to mature under those conditions, which meant that the fly eggs had to have been laid 21 days before the girl was found. The last time the little girl had been seen alive was 21 days earlier, near the apartment building in the company of a man in his 20s. As a result of Greenberg's testimony, this man was subsequently convicted of her murder.

Beetles, flies, and other arthropods colonize corpses in a predictable succession. "Flies are the best investigators of corpses," says Carl Olson, an entomologist at the University of Arizona. Metallic blue or green blowflies, drawn by odors that include a compound called cadaverine, can find an uncovered corpse within ten minutes of death. Blowflies and flesh flies lay eggs in wounds and moist natural body openings, such as eyes, ears, nose, and mouth. Inside the body, bacteria go to work decomposing the body, generating various gases in the process.

As the masses of maggots feed, their activity and that of the bacteria raise the temperature inside the corpse. "It's like when you walk into a room full of people—notice how hot it is?" Olson asks. "The maggots on a corpse create their own environment."

The maggots eventually puncture the skin, initiating what is called the decay stage and releasing what University of Hawaii researcher Lee Goff calls "strong, distinctly unpleasant odors." As the fly larvae complete their feast, they hop or crawl off to pupate in the soil. By that time, the fly larvae have created a hospitable environment for the next wave of insects, primarily beetles. It is a classic example of ecological succession. Any remaining larvae become food for the visiting beetles, which also finish off the flesh.

By the postdecay stage, only the skin and bones remain. A different set of beetles now arrive. These feed on dried skin and hair. They are the same insects that destroy fur coats and cashmere sweaters. When the body is reduced to bone alone, all of the carrion-feeding arthropods depart.

Figure A.
Burying Beetles *(Necrophorus marginatus)* lay their eggs on a mouse.

John Cunningham/Visuals Unlimited

---

habitat at many different stages of succession and many more species than if the forest were uniform.

Any kind of physical disturbance, whether a fallen tree or a collapsed bank along a stream, can create a patch in which secondary succession can begin. Even something as minor as a gopher mound in a meadow creates a disturbed area where a patch of different species can live. A disturbed area can be colonized by seeds already in the soil (a "seed bank") or by organisms from nearby areas.

Succession usually implies a progression through a series of transitional stages to a climax community. Succession can occur in disturbed areas, creating a patchwork of different stages of succession.

Why Does Succession Occur?

Clements argued that succession occurs because organisms degrade their own environment. For example, following a forest fire, the first plants to grow are those that are adapted to live in open sun. But the shade created by these sun-loving plants prevents their own seedlings from growing and favors shade-tolerant plants.

While many of Clements's scientific heirs still contend that each stage in a succession prepares the way for the next, other ecologists argue that this is not always true. Some of the changes in a succession, they say, merely reflect differing rates of growth. Shrubs, for example, appear to come before trees even when both types of seedlings start at the same time because the shrubs take less time to reach maturity.

In reality, **pioneer** species can facilitate succession, tolerate succession, or inhibit succession. In primary successions, however, pioneer organisms always facilitate succession. This is because most organisms cannot colonize bare rock or sand. The formation of soil from bare rock depends on the action of pioneer lichens and plants adapted to thin or nonexistent soils. Once these pioneer species have created a thin layer of soil, other species can grow.

> Pioneer or successional species change their environment in such a way as to allow colonization by succeeding species.

What Determines Which Community Is the Climax Community?

The climax community is, by definition, the one that persists the longest without changing. But what factors determine which community will persist? As we saw in Chapter 26, the character of a community depends heavily on such abiotic factors as climate, soil, and terrain. The dry, frozen tundra could never support a tropical rain forest. Early ecologists theorized that an area has only one possible climax community.

Communities assemble themselves flexibly, however, and their particular structure also depends on the specific history of the area. For example, giant sequoia forests in the Sierra Nevada include trees as old as 2,000 years. These forests appear to be a classic example of a long-lived climax community. In fact, however, these ancient forests persist only if periodic fires sweep through and kill young white fir and incense cedars growing beneath the mammoth sequoias (Figure 27-17). When humans regularly put out fires, the white fir and incense cedars grow up among the sequoias. Eventually, when a fire too big to put out arises, the smaller trees pass the flames up into the crowns of the giant sequoias, destroying the whole forest. Once that happens, a white fir and incense cedar forest springs up in its place.

> The character of a climax community depends on climate, terrain, and history.

Why Do Communities Stay the Same for Long Periods?

An ecologist who studies a pond today may well find it relatively unchanged a year from today. Individual fish may be replaced, but the number of fish will tend to be the same from one year to the next. We can say that the properties of an ecosystem are more **stable** than the individual organisms that compose the ecosystem.

Figure 27-17
A grove of giant sequoias. Such rare groves depend on frequent forest fires to prevent succession to white fir and incense cedar.

At one time, ecologists hypothesized that species diversity made ecosystems stable. They thought that the greater the diversity, the more stable the ecosystem. Support for this idea came from the observation that long-lasting climax communities usually have more complex food webs and more species diversity than pioneer communities. Ecologists concluded that the apparent stability of climax ecosystems depended on their diversity and complexity. To take an extreme example, farmlands dominated by a single crop are so unstable that one year of bad weather or the invasion of a single pest species can destroy the entire crop. In contrast, the many species that make up a native grassland will tolerate considerable damage from weather or pests.

But mathematical models of ecosystems suggested that a more complex system is more likely to break down than a simple one, just as a computer loaded with conflicting software is more likely to break down than a simple hand calculator. The question of what factors make communities stable remains controversial.

For now, ecologists cannot even all agree on what "stability" means. Stability can be defined as simply lack of change. In that case, the climax community would be considered the most stable, since, by definition, it changes the least over time. But, alternatively, stability can also be defined as the speed with which an ecosystem returns to a particular form following a major disturbance, such as a fire.

This kind of stability is also called **resilience.** In that case, climax communities would be the most fragile and the least stable, since they can require hundreds of years to return to their former state. As we saw in Chapter 26, tropical rain forests have the greatest biodiversity of any ecosystem and some of the highest primary productivity. Yet once destroyed, they do not quickly return.

Even simple lack-of-change stability is not always associated with maximum diversity. Maximum diversity in some areas is found not in the climax community but in midsuccessional stages. A mature redwood forest, for example, has fewer species than an immature forest. Even though ecologists know that the presence of certain species is essential to ecosystems, diversity by itself probably does not ensure stability.

Can Ecosystems Bounce Back?

Because ecosystems all over the world are being severely disturbed or eliminated by human activities, ecologists would like to know what factors contribute to the resilience of ecosystems. The severe destruction caused by a volcanic eruption such as Mount St. Helens is nothing in comparison to the huge areas of temperate and tropical rain forests that we humans have destroyed. We need to know what aspects of an ecosystem are most important to its resistance to damage, as well as its recovery.

Many ecologists now think that resilience comes not from diversity of species in one place but from the patchiness of the environment. An environment that varies from place to place supports more kinds of organisms than an environment that is uniform. In a community with many species, a local population that goes extinct is quickly replaced by immigrants from an adjacent community. Even if the new population is of a different species, it can approximately fill the niche vacated by the extinct population and keep the food web intact. Some studies suggest that the communities along rivers never exhibit a stable climax at any one location but instead consist of a mosaic of established, obliterated, and successional communities that result from banks collapsing, sandbars forming, and floods molding the landscape.

In addition to patchiness, other factors that contribute to an ecosystem's resilience include resistance to erosion and high primary productivity. As long as the soil is intact, seeds will quickly sprout, and shrubs, ferns, herbs, and grass will hold the soil in place, preventing erosion. Partly damaged ecosystems often support greater plant growth than mature ones, since younger trees grow more vigorously than trees in a darker climax forest.

Some of the factors that contribute to ecosystem resilience are patchiness of the physical environment, primary productivity, and resistance to erosion.

In this chapter we have seen how communities change over time and what factors, biotic and abiotic, influence the character of a community. In the next chapter, we will examine what factors affect populations and what happens when populations grow too big.

Key Concepts

- Over time, communities undergo predictable changes in species composition, a process called succession.
- Communities of species are both integrated and individual. The species of a community are both replaceable and bound to one another by complex relationships.
- Relationships among species, such as symbiosis, competition, and predation, coevolve over time.
- No two species can occupy the exact same niche and habitat indefinitely.
- Species diversity is determined both by abiotic factors such as latitude, geography, and habitat size and by biotic factors such as character displacement and the presence of keystone predators.

Summary with Key Terms

How can the species of a community be both interdependent and independent?

Henry Cowles did the first major American work on **succession,** and Frederic Clements published the first textbook describing succession. Clements believed that **communities** were **integrated**—real organic entities, with species playing the same roles in different communities. Clements also introduced the idea of a **climax community,** a single community of plants that, he said, characterizes a particular area and climate. In contrast, the ecologist Henry Gleason argued that communities were **individualistic**—mere associations of species that happen to occupy the same area.

What kinds of specific interactions among species lead to adaptation, character displacement, and other forms of coevolution?

The species in communities interact in many ways. They may compete or they may relate as **predator** and prey or as host and **parasite.** Organisms defend themselves from predators and herbivores by using **warning coloration, mimicry (Batesian** or **Müllerian), camouflage,** and **secondary plant compounds.** Species may share a **symbiotic** relationship, such as **parasitism, mutualism,** or **commensalism.** In evolutionary time, two or more species may **coevolve** mutualisms and other special relations with one another.

Species may also **compete** for a **limiting resource.** The kind of community an organism occupies is its **habitat.** The way the organism uses the resources in its habitat is called its **niche.** The **competitive exclusion principle** says that if two species occupy the exact same niche, one species will eliminate the other through competition. Competition from other species may reduce the **fundamental niche** of an organism to a smaller **realized niche**. In **resource partitioning**, species break up a large niche into several smaller ones. Competition between species can cause them to evolve different adaptations, a process called **character displacement**.

What factors determine the species diversity of a community?

Communities vary in the number of species they contain, or **species richness**, and in the **relative abundance** of each kind of species. Ecologists are still trying to discover what determines the diversity of a community. **Keystone species**, productivity, harshness of the environment, latitude, size of ecosystem, and historical accident all seem to play roles.

In the 1960s, Robert MacArthur and Edward O. Wilson proposed a **theory of island biogeography** to account for species diversity on islands of habitat. Tests of this theory suggest that the number of species on an island is a predictable consequence of the characteristics of the island, and not just a historical accident. **Habitat islands** do not have to be conventional oceanic islands. They can be mountaintops or ponds in a forest, for example. A group of local populations that alternately go extinct and recolonize vacant habitat islands is called a **metapopulation**.

Why and how do communities change over time?

Primary succession occurs when organisms colonize a previously unoccupied area, such as a volcanic island or a barren sand dune. **Secondary succession** occurs when organisms recolonize an area that has suffered some catastrophe, but which still has its soil intact. In secondary succession, the first **pioneer** plant species in a **pioneer community** are annuals, followed by perennials, shrubs, and trees. The final stage in development is called the climax community, but biologists no longer believe that the idea of a final stage makes sense. All stages can change, depending on circumstances. Intermediate stages are called **successional communities**.

What does it mean for a community to be stable?

Climax communities are either more **stable** than pioneer or successional ecosystems or less stable, depending on the definition of stability that is used. Patchiness of the physical environment, primary productivity, and resistance to erosion are some factors that seem to contribute to community **resilience**, the ability of a community to recover from fires or other cataclysms.

Review and Thought Questions

Review Questions

1. Contrast the integrationist view of communities with the individualistic one. Describe a piece of research that supports one view or the other.
2. Name a commensal and a parasite of humans.
3. Name a mammal, bird, reptile, amphibian, and insect that have warning coloration.

4. Define the competitive exclusion principle.
5. Give an example of a limiting resource for humans.
6. Define the terms "niche" and "habitat." Give an example of each for the same organism.
7. What factors determine the number of species living on an island?
8. From Figure 27-14, calculate the loss of subpopulations of pikas between 1972 and 1991 as a percent of the number present in 1972.
9. Give two definitions of "ecosystem stability." What factors determine the long-term stability of an ecosystem?

Thought Questions

10. Describe a climax community somewhere near where you live. What criteria did you use to determine whether it was a climax community? How would you go about testing your theory that it is a climax community?
11. Describe a series of small habitat islands that exist within 5 or 10 miles of your house and might be subject to MacArthur and Wilson's theory of island biogeography.
12. You have discovered two butterflies that look nearly identical but clearly belong to different families. Assuming that one is mimicking the other, think of an experiment that would show which one was the mimic and which one was the "original."
13. If succession occurs because pioneer or successional species improve conditions for subsequent species, what can we say about climax species?

BiologyNow Resources

Biology (S)Now™

Active Figures
27-11: Galápagos finches
Table 27.1: Predation interactions

Preparing for an exam? Take a diagnostic test on your BiologyNow CD-ROM.

Online materials relating to this chapter are at:
http://biology.brookscole.com/AAL3

About the Chapter-Opening Image
Destruction of an ecosystem. When Mount St. Helens erupted in 1980, it destroyed a 230-square-mile ecosystem, about the size of the city of Chicago.

Populations: Extinctions and Explosions

Key Questions

- How does habitat fragmentation lead to species extinctions?
- How do populations grow?
- What factors limit the growth of populations?
- How do energy subsidies increase food production?
- How do human overpopulation and development affect the biosphere?

Fighting for Life

In 1998, an eminent biologist gave up his job at the University of California. It was the second time Michael Soulé had left the security of a tenured faculty position—one of the strongest guarantees of lifetime employment in the United States. The academic life was, for Soulé, too much thought and too little action. He felt impelled to act, to literally save the world. The problem, as he saw it, was that everything was disappearing. All of the animals and special habitats he had studied and grown to love during his 40 years as a biologist were vanishing. His gardens of discovery were winking out one by one. "Every field biologist has had this experience," he told writer Robert Coontz. "You return to a study site you knew as a graduate student, or as an assistant professor, or where as a child you first became interested in biology, and it's been devastated."

Soulé began his career in the normal way of biologists. In his boyhood, he roamed the hills and canyons outside San Diego, catching lizards and snakes, watching birds, rabbits, and other wildlife.

Painting by Charles Knight. Field Museum of Chicago photo # Geo-CK-301c by Ron Testa.

In college he studied zoology, then went to graduate school at Stanford. His mentor there was population biologist Paul Ehrlich, whose books on human overpopulation written with his wife Anne Ehrlich have made Ehrlich far more famous than his decades of research on butterflies.

In 1967, he took a job as a professor at the University of California at San Diego, returning to his boyhood home. He taught biology and studied symmetry in lizards. All animals, including humans, become a little asymmetrical as they develop from embryos to adults. A high degree of symmetry, however, is an indicator of good health. In fact, many animals choose their mates partly on the basis of how symmetrical they are. We could say that symmetry is one biological definition of beauty. Asymmetry, on the other hand, often indicates poor health. Asymmetry, Soulé found, was also an indicator of pollution and other environmental stresses, and he was finding more asymmetry in his animals than he wanted to find.

Before his eyes, the canyons of San Diego County where he'd grown up were filling with houses. The fragments of land that survived between one housing development and the next were too small to support many individuals of each species. Not only in San Diego, but around the world, species were going extinct at an unprecedented rate.

These shrinking habitats drew Soulé's interest to the problem of inbreeding. He could see that once a population got small enough, the remaining individuals would have only a few others with which to breed. He knew that inbreeding would bring out genetic defects in the offspring. Fewer would reach maturity, and those that did would be less fit—less able to survive and reproduce themselves. He theorized that small populations spiraled downward toward an extinction from which they would never be able to escape. Soulé called this downward spiral the **extinction vortex** (Figure 28-1).

It was all interesting material for a researcher. Soulé published his experimental work in journals and presented his ideas at scientific meetings. The university promoted him from assistant professor to tenured associate professor and, finally, to full professor.

Yet something was missing. It was one thing to coolly document the slow destruction of study sites and beloved childhood haunts. It was quite another to do something about it. In those days, real biologists were not supposed to act on the grief they felt when yet another study site disappeared beneath a sewage treatment plant or shopping mall. In the mid-1970s, Soulé's fellow biologists accepted the wall that then divided biology from the environmental movement. For biologists to affiliate themselves with "ecofreaks" who chained themselves to redwood trees was to abandon any pretense of objectivity. At that time, a biologist who expressed sympathy with down-in-the-trenches environmentalists, let alone became one, would lose credibility as a serious biologist.

But Soulé took action. In 1978, he helped organize the First International Conference on Conservation Biology. It was the first time that field biologists acknowledged that they had both a responsibility and a special ability to do

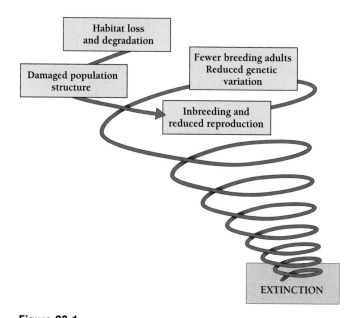

Figure 28-1

The extinction vortex. Biologist Michael Soulé hypothesized that when a population shrinks because of hunting or loss of habitat, the reduced population size causes inbreeding. Inbreeding, in turn, can cause ill health and poor reproduction in each new generation, accelerating the extinction process in a downward spiral. In addition, extinction of one species degrades the environment for other species in the habitat. The feedback loop thus involves more and more species as time passes.

more than just document the demise of the natural habitats they knew. Perhaps, they began to tell one another, their research could provide information that would help save some of these habitats.

By then, Soulé was in his early forties and had been on the UC San Diego faculty for 12 years. It's common for people this age to look for change, but most professors would content themselves with a new line of research, a sabbatical, or even a sports car. For Soulé, that wasn't enough. He quit his tenured job at UC San Diego and joined a Zen center in Los Angeles. But although he stayed at the Zen center for five years, he never abandoned biology. Instead, he was reinventing what it meant to be a biologist. In 1980, he helped put together a collection of essays on conservation biology, titled *Conservation Biology: An Evolutionary Ecological Perspective,* that galvanized field biologists everywhere and marked the informal beginning of **conservation biology** as a field separate from traditional ecology.

For the first time, a large number of biologists began to openly acknowledge that human overpopulation was creating a tidal wave of extinctions unprecedented in the history of humankind. In 1987, a group of research biologists formed the Society for Conservation Biology and began publishing a scientific journal about the science of conservation. Soulé returned to the University of California, this time to the Santa Cruz campus, to lead a department of environmental studies. When he eventually left again, it was to further advance the conservation of natural habitats.

28.1 How Do Populations Go Extinct?

Many factors cause species to go extinct. Biologists today lump all these factors into two categories. One category includes predictable events such as habitat destruction or deliberate extermination through hunting or trapping, called **deterministic** events. The other category includes unpredictable and infrequent events such as unusual storms, what biologists call **stochastic** (meaning chance or random) events.

How Do Deterministic Factors Drive Species to Extinction?

One deterministic factor that can drive species to extinction is competition from other species. In nature, one kind of tree may shade a smaller species, driving the local population to extinction. The larger tree has successfully competed with the smaller one for light. Usually, however, the result is extinction not of a whole species but of only a particular population. As we saw in the last chapter, individuals from another population can reestablish a population when opportunity arises.

Humans also drive local populations to extinction. But our large numbers and organized approach to taking over resources allow us to drive *every* local population of a species to extinction. The result is the destruction of whole species and ecosystems rather than a single subpopulation here or there. How do humans drive other species extinct?

One way is simply hunting and killing as many organisms as possible. In this way, humans have driven to extinction hundreds of animals, including California grizzly bears, Tasmanian wolves, passenger pigeons, dodos, and great auks.

Another way is **habitat destruction.** Sometimes we do this without even realizing we are doing it. If we remove all the trees and shrubs from a forest ecosystem and then plant wheat, we eliminate not only the trees and shrubs but also habitat for all the other organisms that live in the forest. If we repeat this pattern from Maine to Nebraska, we eliminate not only trees and habitat but also thousands of subpopulations of insects, spiders, worms, fungi, and bacteria.

In the 18th century, pioneer families cleared millions of acres of forest in the eastern United States. Most of the farms they built and tended are now long abandoned, and much of that forest has returned. But while many of the same species of trees and shrubs are back, hundreds of other species are gone forever. For many of these species, not a single population is left to recolonize the new forest. For others, the remaining subpopulations, in Canada or the Far West, are simply too far away to recolonize the new forest.

When humans drive other species to extinction, it is usually by destroying their habitats.

How Does Fragmentation Contribute to Habitat Destruction?

A subtler process that drives species to extinction is **habitat fragmentation.** Natural landscapes are a mosaic of different habitats. Forest blends gradually into savanna, chaparral, or grassland. Riparian (river) forests may run through drier habitats, connecting wetter forest ecosystems together. A flying animal such as a bird or insect can often travel easily from one patch of forest to another. If a population disappears from one habitat patch, individuals from another subpopulation soon arrive to recolonize.

For organisms that cannot fly, however, recolonizing is usually more difficult. Humans create sharply bounded patches of habitat in an uncrossable sea of pavement, houses, and crops. Many species cannot easily move from one patch to another.

Even a simple road can permanently divide a habitat for turtles, amphibians, and other slow-moving animals.

Fragmentation also reduces the total area of a habitat. We saw in Chapter 27 that numbers of species depend on the area of the habitat. According to MacArthur and Wilson's species-area curve, a 90 percent loss of habitat (area) results in a 50 percent loss in species. Most species simply cannot survive in a habitat that is too small. For example, numerous studies have shown that scarlet tanagers, pileated woodpeckers, ovenbirds, Acadian flycatchers, and many other birds rarely live in forests that are smaller than 40 acres. Large animals at the top of the food chain, such as hawks, need even larger areas.

Some studies suggest that in Brazilian rain forests, 500,000 acres is the minimum size needed to preserve local biodiversity. That is a little more than one-tenth the area of New Jersey. But rain forests in Peru and Venezuela, for example, have different sets of species and so would each need a separate 500,000-acre preserve.

Another important consequence of habitat fragmentation is **edge effects.** The smaller the pieces into which a habitat is cut, the more edge the habitat has—relative to the core, or interior. At the edges of a habitat, many interesting things happen. The edge of a forest or marsh has a microclimate different from (e.g., drier or windier) that of the center. At the edge of a forest, more sunlight enters and different kinds of plants grow. Safe nesting areas may be missing, and predators from the adjacent habitat may visit to take birds' eggs, kill young animals, or browse on young trees. In areas heavily populated with humans, these visitors may include dogs, cats, and, of course, humans themselves.

As a result of edge effects, many species can survive only in the very center of a habitat patch, far from the edge. A habitat that appears large enough to us may be very small for certain species, because little or no safe habitat exists that is far enough away from the edge. Because of differences in edge effects, fragmenting a habitat can be harmless for one species and disastrous for another (Figure 28-2).

One way humans destroy habitat is by breaking it up into small fragments with lots of edge and little central core.

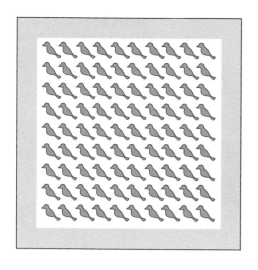
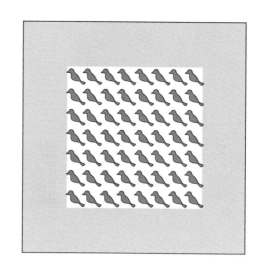

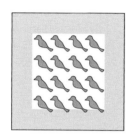

A. Species A

B. Species B

Figure 28-2

The edge effect. Species have different tolerances for a fragmented habitat. While some species can thrive at the edges of habitats, others do not. As natural habitats are fragmented by roads and other human development, the proportion of edge increases. A habitat patch that seems large to us and seems to support some species just fine is in fact too small for other species. In A, the population of blue birds (top) is reduced from 100 individuals to 64 when their habitat is fragmented (bottom). In contrast, the red birds in B need to stay farther from the edges of their habitat to survive. The same habitat fragmented in the same way reduces their numbers from 64 to one red bird per patch. All the red birds will go extinct because there aren't enough birds in each patch to mate and lay eggs.

Draining the Gene Pool

Recall that Michael Soulé's extinction vortex hypothesis suggested that inbreeding accelerates extinction. One source of support for this hypothesis is a 1998 report in the journal *Science* on a long-term study of greater prairie chickens. Beginning in the early 19th century, extensive habitat destruction pushed this native midwestern bird toward the brink of extinction. In Illinois, the greater prairie chicken is almost gone (Figure 28-3). During a three-decade study of these birds, the population size dropped from 2,000 birds in 1962 to 50 birds in 1994. Researchers also recorded a decline in hatching success (hatched eggs per nest) from a high of nearly 100 percent in the 1960s to less than 40 percent in 1990. Measures of genetic diversity showed that the Illinois population had less genetic diversity than larger populations in nearby states and that alleles present in the population before the 1970s later disappeared. The data suggest that as the population shrank, so did its genetic diversity. The data supported Soulé's prediction that as a population shrinks, so will its genetic diversity.

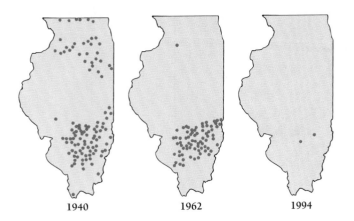

Figure 28-3
The end of the Illinois greater prairie chicken. Rust-colored dots indicate populations. One hundred years ago, greater prairie chickens *(Tympanuchus cupido)* lived abundantly throughout Illinois. Today, only two tiny populations survive in the southern part of the state.

Steve & Dave Maslowski/Photo Researchers, Inc.

1940   1962   1994

Biologists hypothesized that genetic defects due to inbreeding were keeping many chicks from hatching. Because most organisms have two sets of genes, a bad one often has no effect on the individual because a normal version of the gene on the other chromosome compensates for the bad one. When animals mate only with close relatives, however, it easy for an individual to end up with two copies of the same bad gene. Such individuals often have a multitude of subtle or dramatic genetic diseases.

When researchers brought in prairie chickens from larger, genetically healthier populations elsewhere in the United States, both genetic diversity and hatching success of the Illinois prairie chicken increased (Figure 28-4). Soulé's extinction vortex hypothesis predicted that declining population size and declining genetic variability would mutually reinforce each other and increase the likelihood of local extinction. Sadly for the prairie chickens, Soulé's prediction seems to be borne out by this study.

According to the extinction vortex hypothesis, loss and degradation of habitat reduce the size of a population, change its structure, and lead to a loss of genetic diversity. The loss of genetic diversity leads to inbreeding and reduced reproduction, which leads to smaller population size.

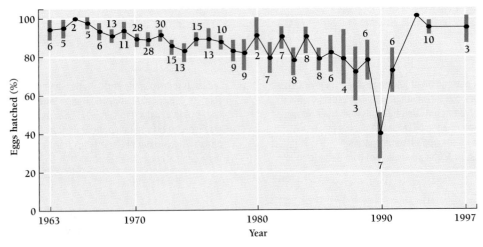

Figure 28-4
Inbreeding among greater prairie chickens. As the population of prairie chickens shrank, the percentage of eggs in a nest that successfully hatched dropped from 100 percent in 1963 to less than 40 percent in 1990. Biologists strongly suspected that inbreeding and the expression of deleterious genes accounted for the decline. Each data point shown here is the average of hatching success at all the nests in the population each year. Numbers refer to the number of nests per year. The vertical orange bars indicate variation in hatching success. Long bars mean more variation. Short bars mean little variation. Increased variation in any trait can be a sign of ill health in a population. Notice that the bars become longer over time, suggesting increasing health problems. In the late 1990s, biologists introduced prairie chickens from large, genetically diverse populations in other states. The result was a dramatic increase in hatch success in the Illinois prairie chickens and a decrease in variation, too. Biologists believe this was likely due to an increase in genetic diversity.

How Do Stochastic Factors Drive Species to Extinction?

Once a population is as ravaged as that of the Illinois greater prairie chicken, almost any chance event can drive it to extinction. In a population of 25 birds, a freak storm could kill several of the healthiest hens in one year. The next year, all the chicks could be males, just by chance. Chance, or **stochastic**, events such as these would not affect a larger population. A storm might kill a few hens, but many others would take their place. And in a population of 2,000, all the chicks would not be one sex. Finally, small populations can lose valuable alleles through genetic drift, as appears to have occurred among the greater prairie chickens of Illinois.

Chance events are more likely to lead to extinction in very small populations.

How Can We Preserve Ecosystems and the Species That Live in Them?

The work of conservation biologists has shown that, unless something dramatic is done soon, approximately half of the world's species will go extinct in the next decade or two. Humankind has now converted one-third of the Earth's land surface to agriculture and covered more with cities, suburbs, roads, and factories. The remaining land includes some of the harshest, least productive areas of the planet, including deserts, jagged mountain ranges, and arctic tundra and taiga. The rate of destruction is increasing. In tropical areas, about half of all nations have cut or burned 80 percent of their forests.

Twenty tropical nations have committed to saving 10 percent of their land in parks and other preserves. But because far more species live in the tropics than elsewhere, saving only 10 percent of all tropical rain forests means a likely loss of more than half of all the species on Earth.

In the spring of 1998, Michael Soulé expressed his disappointment at this trend in an article in the journal *Science*. He wrote, "Campaigns with targets in this range [10 percent] can create the unintended and false impression that such a paltry tithe to nature is enough to prevent a mass extinction of species."

Not long after the publication of this statement, Soulé left the university once and for all to join the Wildlands Project in Colorado. The ambitious goal of the Wildlands Project is to persuade people and their governments to set aside 50 percent of the North American continent for the preservation of biological diversity. "Our generation is the key," Soulé said in 1990. "Whatever we save in the next few decades is all future humanity will have to use and to appreciate. If we don't do it, it will be too late."

Conservation biologists have discovered several rules that allow preservation projects to save the greatest numbers of species. The best preserves are large areas, whose size minimizes edge effects in sensitive species. In addition, many conservation biologists recommend **corridors**, narrow strips of protected habitat that allow individuals to travel from one subpopulation to another (Figure 28-5). The most likely beneficiaries of corridors are small, slow-moving animals, such as frogs and salamanders that need to migrate overland to breeding ponds and streams. If a local population goes extinct, individuals from another preserve can then travel to the empty habitat and recolonize it. Other winners would be large predators, such as mountain lions and wolves that need large ranges. Corridors compensate for local extinctions and maintain genetic diversity.

We can preserve many more species if we set aside large preserves, connected by corridors and surrounded by buffer zones.

28.2 How Do Populations Grow?

We have discussed how populations go extinct. But all organisms have phenomenal powers of reproduction. Populations of bacteria, cockroaches, and rats sometimes seem to explode out of nowhere. Indeed, California sea stars have so many offspring that if all of the descendants of just 100 California sea stars survived and reproduced for 15 generations, the number of sea stars would exceed the number of electrons in the visible universe. Instead, most of the larvae are eaten by other animals and the sea star population remains at reasonable levels.

All organisms are capable of multiplying amazingly rapidly when conditions are ideal. Even elephants, which rarely have more than about 10 offspring in their 60-year life spans, can leave a lot of descendants. For example, if all of the 10 offspring of an elephant survived, and each of them had 10 offspring, and so on, the original elephant could have as many as 100,000 great-great-great-grandchildren in just 300 years (Figure 28-6).

In nature, we sometimes see such unrestrained growth when a species invades a new area, or, for example, when a protozoan discovers an uninhabited human gut. But neither elephants nor sea stars nor protozoans can multiply without end. Even though natural populations are capable of huge growth rates, they rarely grow so fast that they eat themselves out of house and home. Whenever populations begin to grow too fast, they eventually stop. When food or space runs low, or wastes degrade the environment, individuals tend to have fewer offspring. Sometimes, predators or disease bring a population under control.

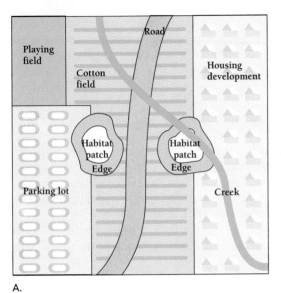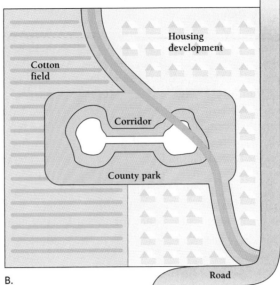

A.

B.

Figure 28-5

How can wildlife preserves help keep species from going extinct?
A. Two small wildlife preserves are bordered by houses, a playing field, and a parking lot. A road and a cotton field separate the preserves. Pesticides from the cotton field and people's backyards flow directly into the stream and then into the preserve on the right. Dogs, cats, humans, and other animals can enter and destroy plants, animals, nests, and dens. Edge effects are huge. Recolonization would be difficult or impossible for species that cannot easily cross the road and the cotton field. B. In the redesigned preserve, a corridor connects the two habitat patches. The road has been rerouted around the two patches, and the cotton field has been moved to one side. In this arrangement, if a species goes extinct in one habitat patch, individuals can recolonize from the other habitat patch. An inner buffer zone (purple) acts as a wide edge surrounding the corridor and both patches. In this buffer area, human use is restricted to a minimum. An outer buffer zone (green) allows more general human use but still no houses, crops, or industry. The community loses a parking lot and a ball field but gains a park with a small but intact ecosystem.

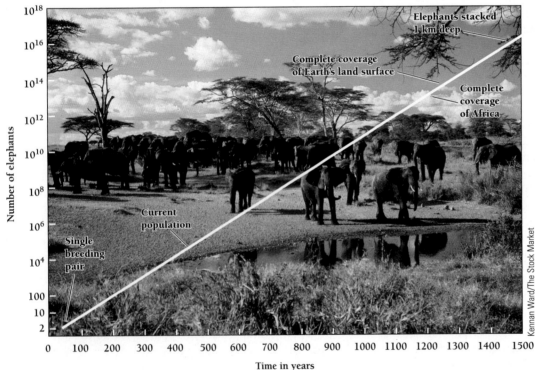

Figure 28-6

Unrestrained growth in elephants. Elephants, like the ones at this water hole, have the lowest r_{max}, or intrinsic rate of natural increase, of almost any species. Even so, one pair of elephants could (hypothetically) produce enough progeny to cover the Earth's surface in only 1,200 years. Fortunately, elephants rarely reproduce at the maximum rate.

Kennan Ward/The Stock Market

Natural populations may fluctuate from season to season and from year to year, but most are surprisingly stable over the long run. Two important questions in population ecology are, How do populations grow? and What factors limit population growth?

What Is Exponential Growth?

Under the most favorable conditions, the number of *Escherichia coli* in a nutrient-rich broth can double every 20 minutes. After the first 20 minutes, one bacterium forms two; after another 20 minutes, the two have formed four; after another 20 minutes, there are eight; and so on. We say that the bacteria have a **doubling time** of 20 minutes.

We can calculate the number of bacteria (*N*) from the number of times the population has doubled since the time at the beginning of the experiment (*t*) using the following equation:

$$N = 2^t$$

For example, if 60 minutes had elapsed, the population of *E. coli* would have doubled three times. So *t* would equal 3, and

$$N = 2^3 = 2 \times 2 \times 2 = 8$$

In this equation, *t* is an "exponent," so we say the bacteria's growth is "exponential." Exponential growth is a growth pattern in which a population regularly doubles every so often. But exponential growth also means that the larger the population, the faster it increases. For example, we know that 16 *E. coli* cells could double in 20 minutes to 32 cells. The population would increase by 16 cells. But a population of 16 million cells would also double in 20 minutes. That population would increase by 16 *million* cells over exactly the same time period. The large population and the small one would have the same doubling time, but the increase in actual numbers would be much larger in the second case.

When bacteria increase exponentially, a graph of their numbers looks like the letter J (an exponential curve). At the bottom of this **J-shaped curve,** the population size (*N*) increases rather slowly. Then it accelerates. When there are no limits to population increase, *N* increases faster and faster. The bigger the population, the faster it grows. Human populations are an example of a natural population with a J-shaped growth curve (Figure 28-7).

Figure 28-7
A classic J curve. The human population has been increasing exponentially for thousands of years. We are rapidly approaching the limit of what our planet can sustain. Already, we monopolize about 40 percent of all the solar energy absorbed through photosynthesis on Earth. The failure of our growth curve to level out is mainly due to closed sewers, refrigeration, cleaner water, and improved nutrition and hygiene. All of these factors, along with vaccinations, have increased the birth rate and decreased the death rate.

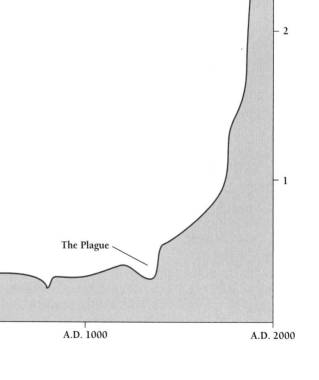

The Plague

1000 B.C.   0   A.D. 1000   A.D. 2000

The growth rate of a population is the difference between the average birth rate and the average death rate. **Birth rate** is the number of individuals born each year per 1,000 individuals. **Death rate** is the number of individuals who die each year per 1,000 individuals. The population will grow only if the birth rate exceeds the death rate. (We can express this as, $r = b - d$.) When the birth rate of a population, b, is at its maximum, and the death rate, d, is at its minimum, the growth rate, called r, is as large as it can be. This is called r_{max}, or the **intrinsic rate of natural increase**. For example, if b is 10 per 1,000 and d is 2 per 1,000, then r (which equals $b - d$) would be 8 per 1,000. In shrinking populations, d is greater than b, and r is a negative number.

All populations are capable of exponential growth. The larger they are, the faster they grow.

What Limits the Size of a Population?

A J-shaped growth curve is *always* temporary. We may see a J-shaped population curve when a bacterium has first colonized a container of nutrients. But even *E. coli* growing under ideal laboratory conditions soon find limits to their growth. The bacteria run out of food and space, and their own wastes begin to poison them. The number of bacteria gradually reaches a plateau. Their growth curve now looks more like a tilted and stretched out "S," and is called an **S-shaped curve** (Figure 28-8).

The population size at which a species tends to plateau is the upper limit of its population size, or the **carrying capacity** of the environment for that species. The carrying capacity is also called **K.** When a population reaches its carry-

ing capacity, the environment cannot support any more growth and the population has reached **zero population growth.**

The carrying capacity of the environment limits the size of a population.

What Determines Whether a Population Grows?

The carrying capacity of the environment for a particular species depends on the needs of the species and, therefore, on a multitude of factors—food supply, territories, predators, and competitors, to name just a few. Bees, for example, need nectar, pollen, and a safe place for a hive. Even if a given habitat has plenty of flowers, the carrying capacity for bees may be low if there are only a few good places for a hive within flying distance of the flowers.

Changes in a population size ultimately result from just four processes: birth, death, **emigration** (movement out of the population), and **immigration** (movement into the population). The net change in population size is the sum of births and immigration, minus deaths and emigration. (We can say that $r = b + i - d - e$, where i is immigration and e is emigration.)

Anything that influences any of these four factors can increase or decrease population size. Competition, disease, number of prey species, number of nesting sites, rainfall, soil fertility, and hundreds of other factors can all affect population size and the value of K.

Density-Dependent Factors

The rates of birth, death, immigration, and emigration all depend on the population density. As a population's density approaches the carrying capacity, its members must increasingly compete with one another for limited resources. For example, one study of a population of 25 pairs of tawny owls living at carrying capacity suggested intense competition. The 50 owls, capable of producing as many as 100 offspring per year, successfully fledged only 18 owlets. Because of limitations in the food supply—in this case, rodents—some owls were unable to feed their offspring, others did not bother to incubate their eggs, and still others did not even breed. Thus, competition for food depressed the birth rate. But even 18 new owls were too many. Only 11 territories opened up, as a result of deaths among the adults. So only 11 of the 18 fledglings could find territories on which to breed. The other seven young owls had to emigrate.

We can see how a population at carrying capacity can have a low birth rate. Animals run out of food, nesting sites, or territories. Plants competing with one another for water or light cannot get enough. High population density can also encourage emigration and discourage immigration (Figure 28-9). High population density can also increase the death rate from predation, infectious diseases, and parasites.

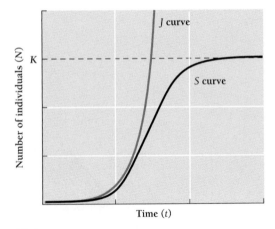

Figure 28-8

How populations grow. All animals are capable of exponential growth for short periods, as represented by the J curve. In natural populations, growth declines as population size increases and approaches the carrying capacity *(K)* of the environment, as represented by the *S*-shaped curve.

Figure 28-9

A plague of locusts. When the population density of certain species of locusts begins to rise, the individuals develop into longer-winged, more gregarious types that fly together. When the proportion of these gregarious types reaches a certain threshold, they rise in great masses and emigrate to a new area, consuming everything in their path—a classic plague of locusts.

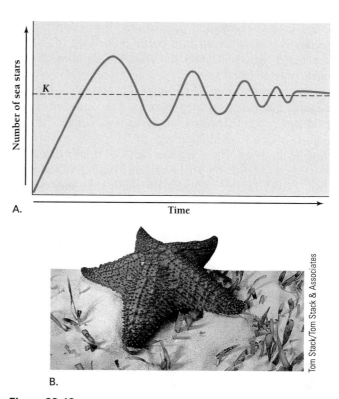

Figure 28-10

Sea stars approach *K*. Starfish produce so many offspring that their populations could easily skyrocket if other animals didn't eat so many of the larvae. A normal population in a new habitat would be expected to grow rapidly, then fluctuate around *K*, the carrying capacity.

The denser a population, the more slowly it grows. Factors that limit population growth according to density are called **density-dependent** factors.

Predation, parasitism, disease, competition, and emigration all limit growth in proportion to the density of the population.

Density-Independent Factors

Some populations can also be limited by **density-*independent*** factors, including fire, drought, storms, tornadoes, volcanic eruptions, and other natural disasters. For example, severe storms can all but eliminate a population of butterflies long before the caterpillar larvae run out of plants to eat. Changes in population density that occur without regard to population density do not follow the S-shaped curve. Density-independent factors are essentially random, or stochastic.

Do Natural Populations Actually Grow Exponentially?

The growth of bacteria in the laboratory follows the S-shaped curve closely. Wild populations also follow the S-shaped curve but usually fluctuate more. Often, population density will temporarily **overshoot** the carrying capacity (*K*) of the environment. For example, a population of mountain

sheep may wildly exceed the carrying capacity, then drop far below the carrying capacity, and then cycle back up to *K* again.

When a population exceeds the carrying capacity of its environment, it will, by definition, decrease. The population may then increase again, fluctuating up and down indefinitely. Many populations fluctuate up and down for a while, then gradually level off at the carrying capacity (*K*) (Figure 28-10). Whether the fluctuations continue depends partly on the extent of the original overshoot. The overshoot depends, in turn, on r_{max}. A species capable of growing rapidly, such as a sea star, is much more likely to overshoot its carrying capacity than a species such as an elephant, whose intrinsic rate of growth is low.

Natural populations do not usually follow the S-shaped curve exactly.

Why Do Some Populations Cycle Up and Down?

If populations rarely show uncontrolled exponential growth, what limits population growth? The answer seems to vary with the species and the circumstances. Predators sometimes control herbivore populations. For example, the predatory Australian ladybird beetle introduced to California orange groves in 1888 successfully controlled an outbreak of scale insects.

Many populations show regular cycles of population density, and biologists long assumed that variation in the food supply was the reason. In Alaska, for example, populations of rodents called lemmings cycle up and down every three to six years. These animals can start from a density of less than one lemming per hectare and reach levels of more than 25 per hectare. At peak densities, lemmings destroy their food supply and attract throngs of predators. With little food, the lemmings stop reproducing and their population plummets. The cycle then repeats itself. Many other animals, including snowshoe hares, arctic foxes, and ptarmigans, show similarly regular cycles.

Exactly why populations cycle, however, remains a mystery. Fluctuations in the food supply can sometimes explain cycling, but not always. In two groups of meadow voles—relatives of lemmings—researchers found that even though one population had 10 times as much food as the other population, it cycled as much as the population with less food.

> Population cycles probably result from interactions between species. So far, however, biologists have not discovered exactly how or why cycles occur.

28.3 What Factors Influence the Values of r and K?

Species differ greatly in their intrinsic rates of increase, r_{max}. Birds that lay six eggs per clutch, for example, have higher values of r_{max} than those that lay only two. Humans that begin reproduction at age 15 have a greater r_{max} than those who delay until age 30. And a tree that makes seeds for 200 years has greater r_{max} than one that reproduces for only 10 years. Number of offspring, age of first reproduction, and life span are all traits collectively called **life history traits.** Every species has a unique life history.

How Do Life Histories Affect Survivorship and Age Distributions?

Elephants invest energy in the production and care of a few, large offspring. As a result, the offspring have a good chance of living long enough to reproduce, usually several times. In contrast, when sea stars reproduce, most of the millions of larvae are filtered from the water and eaten by larger ani-

mals. Only a tiny fraction of sea star larvae grow up to be sea stars with offspring of their own.

One way of representing these two contrasting life histories is a **survivorship curve,** a graph that shows the fraction of a population that is alive at successive ages. The opposite of survivorship is **mortality,** the fraction of a population that dies at a given age. Figure 28-11A shows three kinds of survivorship curves. In a **convex** (or **type I**) survivorship curve, survivorship starts out high and decreases slowly with age. Then, at some point, survivorship begins to decline more quickly. Organisms with convex survivorship curves have a good chance of surviving until they have finished reproducing and caring for their young. Then the chance of dying in a given year increases dramatically. Elephants, dall sheep, adult fruit flies, and humans tend to show convex survivorship curves (Figure 28-11B).

Some organisms are as likely to die in midlife as when young or old. When survivorship is plotted logarithmically, birds and many other such organisms show a **diagonal** (or **type II**) survivorship curve—a straight, declining line (Figure 28-11C). Finally, many species are most likely to die early in life. Their survivorship curves are called **concave** (or **type III**).

Few organisms have survivorship patterns that fit any of these curves exactly. The commonest natural pattern is a high mortality rate among the youngest and oldest individuals and a low mortality among those in midlife (Figure 28-11D).

The **age structure** of a population is the fraction of individuals of various ages. We can represent the age structure of a population by a graph like the three in Figure 28-12, which show the number of people in each five-year age group in Sweden, the United States, and India in 2005. Each such group forms a **cohort,** the set of individuals that enter the population, or are born, at the same time. In this case, we show males and females separately, since their age distributions are slightly different. In all countries, more males are born than females. But in developed countries especially, more females survive in every age category (see box).

The age distribution in a population is not just a function of survivorship; it also depends critically on the birth rate. A rapidly reproducing population, such as that of India, has a greater fraction of young people than does a stable population such as Sweden (Figure 28-12). The age structure also depends on the ages of individuals who move into and out of the population (immigrants and emigrants). As a result, the age structures of human populations differ from country to country, often dramatically.

We can predict the future age structure of a given population if we have the following information: (1) the present age structure, (2) the mortality of each cohort (the chances that individuals of a given age will die each year), (3) the age structure of immigrants and emigrants, and (4) the **fertility** of each cohort (the number of offspring each individual in a cohort is likely to produce). Most such predictions are based

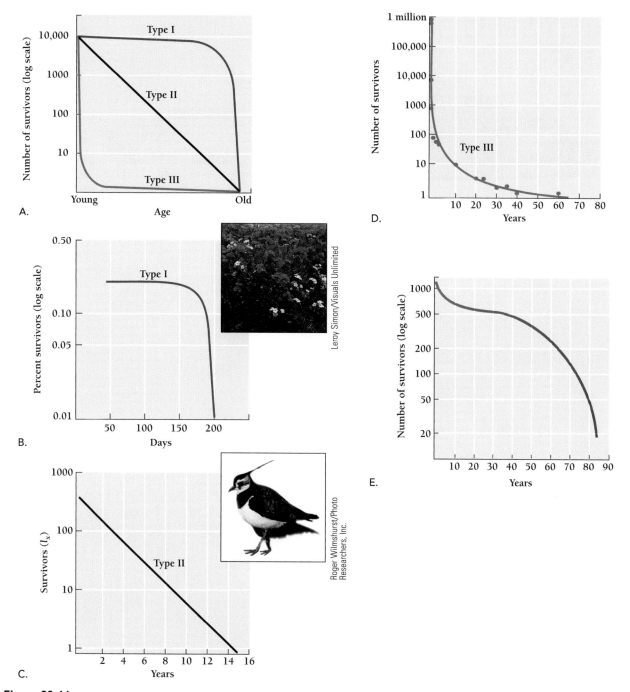

Figure 28-11

Life and death. We all die eventually. But some species are more likely to die young, and some are more likely to die old. Survivorship curves are a way of showing different patterns of mortality and survivorship. A. The members of a type I species survive until they are old and then rapidly die off. The members of a type II species are equally likely to die at every age. Most members of a type III species die soon after they are born or hatch. Those few that survive live a long time. B. The flower Drummond's phlox shows a classic type I survivorship curve, starting with germination. C. The European lapwing shows a classic straight type II survivorship curve. D. In the desert shrub *Cleome droserifolia*, only about 39 out of one million seeds survive to one year of age, resulting in a type III survivorship curve. E. Many natural populations show a combination of shapes, with, for example, high early mortality, a period of steady mortality, and then increased mortality late in life. The survivorship curve shown here is for humans living in the city of Breslau, Poland, in the 17th century. Even today, substantial mortality among infants and children is typical in many parts of the world.

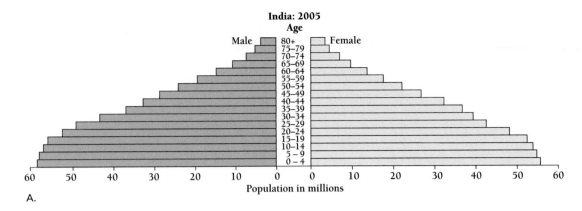

India: 2005
Age

A.

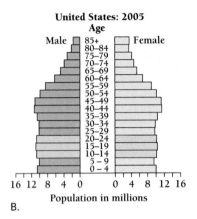

United States: 2005
Age

B.

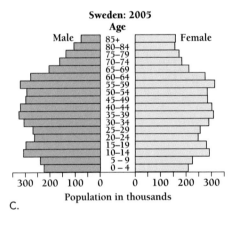

Sweden: 2005
Age

C.

Figure 28-12

Age distributions for three countries. A. India, which has a high birth rate, exhibits the classic pyramidal age structure of an expanding population. A huge proportion of the population is under 20 years old. But the increase has slowed dramatically compared to 40 years ago. B. In the United States, a slowly declining birth rate and an increasing life span are gradually transforming the age structure from a pyramid into a rectangle. The large bulge in the middle is the "baby boom" generation, those born after World War II. The baby boom, aged 41 to 59 in 2005, consists of about 72 million people. "Generation Y," born between 1979 and 1994, is the smaller bulge shown in tan. These are the offspring of the baby boom generation, numbering about 60 million. C. In Sweden (shown in thousands instead of millions), zero population growth rate results in stable population size, with a more rectangular age structure. (Source: U. S. Census Bureau, International Data Base)

Biology(🌐)Now™ Learn more about age distribution in populations by clicking on this figure on your BiologyNow CD-ROM.

Extreme Biology   Why Are Age Distributions of Males and Females Different?

Age distributions such as those in Figure 28-12 traditionally show males and females separately. We usually think of males and females as being represented in about equal proportions, but in many sexually reproducing organisms, large differences exist in the ratio of males to females.

In certain species of wasps, for example, most offspring are female. All fertilized eggs become females and all unfertilized eggs become males. Consequently, a shortage of males means that many eggs will go unfertilized and, in the next generation, more males will hatch. On the other hand, an excess of males means that most of the eggs will be fertilized and will therefore be female. Other organisms use still more complicated negative feedback mechanisms to maintain their sex ratios.

Ants, known for their cooperative behavior, are less cooperative when the subject is sex ratios. A queen ant is equally related to her female and male offspring and bears equal numbers of each. But the males, which develop from an unfertilized egg, are haploid, and the female workers, which develop from fertilized eggs, are diploid. As a result, female workers who share the same father are more related to one another than they are to their brothers.

When Finnish researchers studied the sex ratios of ants from 59 colonies of wood ants, they discovered that in colonies where all the workers share the same father, the males die off disproportionately. Even though the queen lays equal numbers of male and female eggs (the researchers determined the sex of 3,000 ant eggs), adult females far outnumber the adult males. What happens to the males? The researchers suspected that female workers either neglect the male larvae or kill them outright. Indeed, in the laboratory, eggs of males added to a nest of closely related females disappeared without a trace.

The situation was very different in colonies in which the queen had mated with two or more males. In such colonies, the females are less related to one another, and the ratio of male and female adult ants remains about equal.

What about humans? Are our sex ratios 50:50? Most people know that women live longer than men, on average. As a result, among older people, women outnumber men. Not everyone knows, however, that in all countries, more baby boys than baby girls are born each day. The difference is not great. In North America, for example, 105 boys are born for every 100 girls.

But by the time each cohort reaches prime reproductive age, in the early 20s, the sex ratio is nearly exactly 50:50. This adjustment occurs gradually. Baby boys are slightly more likely than baby girls to die of infections and other health problems. Older boys are slightly more likely than girls to die in accidents, and adolescent boys are dramatically more likely to die in all kinds of accidents than adolescent girls. For example, the leading cause of death for all Americans between 5 and 27 years of age is motor vehicle accidents. But a teenage boy driving a car is more than three times as likely to die in a car accident as a teenage girl driving a car. Among 21- to 24-year-olds, the ratio approaches 4 to 1. It might seem amazing to think that normal 20th-century behavior plays a role in natural selection. But while the kinds of risky behavior that teenage boys engage in may change from century to century, the result—a high death rate—remains the same.

---

on current conditions. But when mortality and fertility change, predictions may be too high or too low.

Differences in survivorship curves lead to differences in age distributions.

Demography: How Do Scientists Measure Populations?

Studying the growth and changes in the structure of human populations is the concern of **demography** [Greek, *demos* = people + *graphos* = measurement], the population biology of humans. It is important for economists, health planners, and insurance companies, as well as for biologists. Demographers use a variety of statistical measures to describe populations. Mortality is expressed in terms of the death rate, the number of individuals per 1,000 who die each year. Fertility is expressed in terms of the birth rate, the number of individuals per 1,000 who are born each year.

For example, the U.S. Census Bureau estimated that in 2003, in a U.S. population of about 290 million, 4.1 million babies would be born and 2.5 million people would die. In addition, immigration would add another 1 million, for a total increase in the population of 2.6 million.

4.1 million + 1.0 million − 2.5 million = 2.6 million increase

Demographers use several measures to express population growth. One is the **annual rate of increase**, the actual percentage by which a population increases each year. For example, the 2.6-million-person increase in the 2003 population of the United States came from a population of 290 million, for an annual rate of increase of 0.9 percent (9 per 1,000). The annual rate of increase for the entire planet in 2003 was expected to be about 1.17 percent (11.7 per 1,000), or 74 million more people (Figure 28-13).

Another measure of population growth is **completed family size**, the average number of children that reach reproductive age born to each family. For example, the esti-

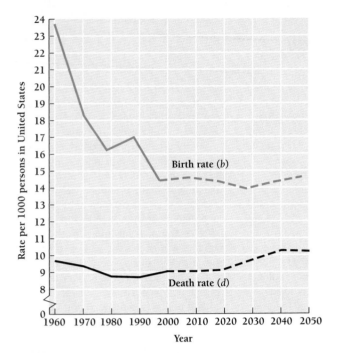

Figure 28-13
Birth and death in the United States. The dotted lines indicate projected birth and death rates in the United States. The death rate is increasing as the population ages, while the birth rate has leveled off. The natural rate of increase, *r*, equals birth rate minus the death rate. Based on this graph, what was *r* for the year 1980? What was *r* in 2000?

mated average completed family size in the United States in 2003 was 2.07 children. **Replacement reproduction** is the family size at which each couple is replaced by just two descendants. At the replacement level, a population neither grows nor shrinks. Replacement level in the United States is 2.1 children per family. The additional 0.1 makes up for the small number of children who die young or otherwise have no children themselves.

We can see that the population of the United States is not quite replacing itself each generation. But we can also see that the total population of the United States is still growing, at the rate of 0.9 percent per year. This paradoxical growth is not due only to immigration, as people sometimes argue. Without immigration, the annual rate of increase would still be about 0.6 percent. In the next section, we will see how population growth can continue despite an average family size of 2.0 or less.

Why Is Population Momentum Important?

The population of the United States continues to grow even though completed family size is below replacement level. In fact, even if tomorrow morning human couples the world over began having no more than two children each, the world population would continue to grow, increasing from about 6 billion today to 9.6 billion people by

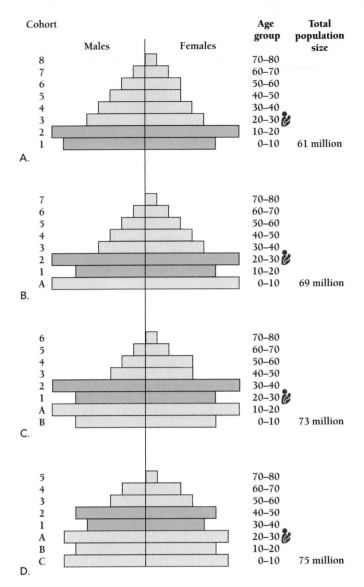

Figure 28-14
Population momentum. Because of population momentum, a population can keep growing even after individuals have only one child per person, replacement level. A. If a population has been growing rapidly, it will include a large proportion of children, all of whom will grow up and have enough children to replace themselves. For simplicity, we have assumed here that all reproduction occurs when a cohort is 20 to 30 years old. B. Starting here, two parents have two children only. But because Cohort 2 is so large, their reproduction causes the total population to increase from 61 million to 69 million. C. Replacement is still 2.0 children, but as Cohort 1 moves into its reproductive years, the population continues to grow, to 73 million. D. The many grandchildren of Cohort 2 are now coming into the world. In addition, more individuals are living into their 60s and 70s. Three generations after the population reaches replacement level (one child per person), it has reached 75 million and is still growing.

about 2050. This paradox is called **population momentum** (Figure 28-14).

The reason for population momentum is simple: in a rapidly growing population, a large proportion of individuals will be young. In fact, nearly 30 percent of the people in the world today are younger than 15 years old. That comes to 1.75 billion people. Even if they only replace themselves, they will increase the world population by 1.75 billion people. And the 1.75 billion people who are now youngsters will not begin to die until they are in their 60s, some 50 years from now. In the meantime, they will continue to live alongside their 1.75 billion children and their 1.75 billion grandchildren.

Even if human couples today began having no more than two children each, the world's population would continue to grow for another 50 years.

28.4 What Is the Earth's Carrying Capacity for Humans?

Demographers predict that if current population growth trends continue, the world population will increase by 3.3 billion people in the next 50 years. Where will population increase end and how? Just as sea stars cannot go on multiplying exponentially, human beings cannot continue to increase without limit (Figure 28-15). Our growing population raises many questions. Some questions concern our impact on other organisms. Some concern our impact on ourselves. How large, for example, can the human population become before it destroys the biosphere that supports us? If we decrease our birth rate, can we make our population level off or decline to a population size that the Earth can sustain? Or are we headed for a catastrophic population crash caused by famine, disease, or war?

Many people believe there is nothing to worry about and we shouldn't think about global overpopulation. Estimates of population growth made in the 1960s were based on population growth rates of the late 1950s, and because birth rates subsequently declined in so many countries, these estimates were, fortunately, too large. Some people have argued that because these estimates were too high, all estimates are too high and that we have nothing to worry about. But one reason that birth rates declined was that governments and other organizations worked to reduce population growth rates. Global population growth rates have slowed because people cared to do something about it (Figure 28-16 A and B).

Some people argue that AIDS, other epidemics, or wars will reduce world population growth. Other people argue that our planet can easily support 20 or 40 billion people. But the Earth's carrying capacity for humans depends on what else we want on Earth with us. If all the

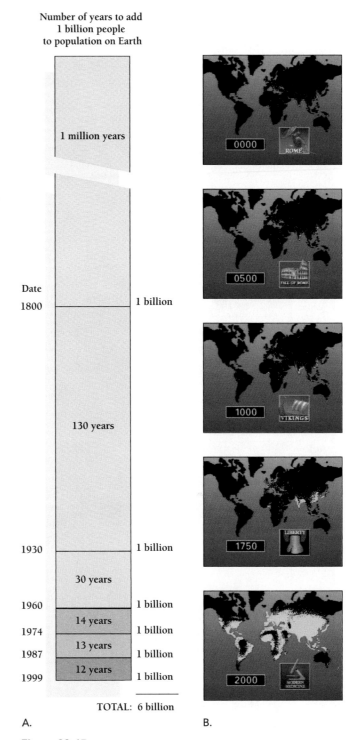

Figure 28-15

Human population growth. A. The number of years it takes to add a billion new people to our population has decreased from 130 years in the 19th century to just 12 years. B. Population estimates based on historical records show that large population increases began in India and China centuries earlier than in other countries. In just 250 years, the United States has gone from one of the least populous regions on Earth to the third most populous country in the world. (B, Reprinted with permission from Zero Population Growth, Inc.)

sunlight that strikes the Earth could be used to feed human beings, we might turn Earth into a giant feedlot for 40 billion or more humans, but it's likely that the biosphere would be heavily damaged, and it's not clear that even humans could survive under whatever conditions existed in such a scenario.

Whether we have yet exceeded the Earth's carrying capacity for humans is a matter of debate.

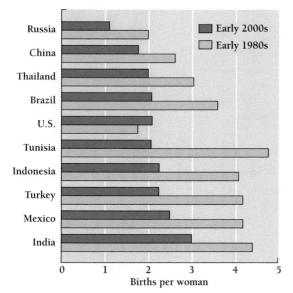

How Well Is Our Energy Supply Keeping Pace with Human Energy Consumption?

The vast majority of humans live in cities and suburbs, where little if any food is grown. Yet each city dweller consumes about a million Calories per year. The principal source of this food is farms and ranches that cover one-third of the land. Agriculture provides additional energy to, or **subsidizes,** cities and suburbs.

Agriculture is itself subsidized. The productivity of a cornfield depends not only on sunlight but also on the fossil fuels needed to run tractors, to pump irrigation water, and to make chemical fertilizers. A farm subsidized with fossil fuel produces about 10 times what an unsubsidized, natural ecosystem does. All of the extra energy is called an **energy subsidy.**

Urban humans not only consume a lot of food energy, they also use another 100 to 1,000 times more energy—for industry, transportation, heating, lighting, and other activities. This energy goes to buildings, roads, dams, bridges, airports, and all the goods and services we've grown used to. The overall energy flow through an urban area may be 1,000 times that of a pond or meadow. Much of this energy ultimately comes from fossil fuels pumped from wells in Alaska, the Middle East, or the ocean floor.

Because humans derive much of their energy and materials from areas far from where they live, the idea of carrying capacity, as applied to natural populations, doesn't quite

Figure 28-16

Declining birth rates. A. In nearly every country in the world, women are having fewer children. B. In some parts of the world, population growth is now very low. But very few countries have zero population growth, so the global human population continues to grow.

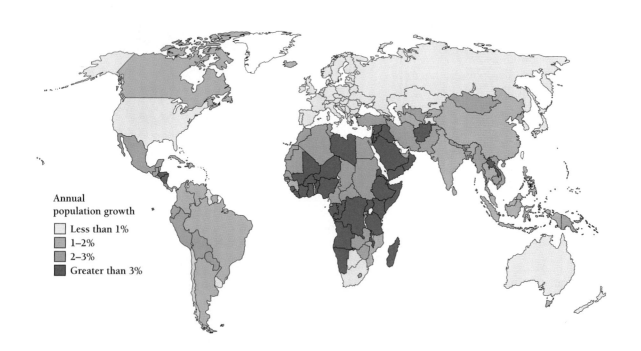

Extreme Biology   How Much Do We Have and Need?

"In the 21st century, we are going to pass through a bottleneck. The land required to maintain one American's standard of living, on average, is 12 acres; the amount required to sustain the current standard of living of someone in one of the developing countries, where 80 percent of the world's population lives, is one acre. Now here's the problem: The whole world wants to live like Americans. Thus we have the aftershock following the population explosion: not only too many people, but people attempting legitimately, understandably, to increase the quality of their lives. It's been estimated that to bring the whole world up to America's standard of living would require two more planet Earths. The results, politically, economically, and environmentally, of the rush to achieve that impossibility in the next century will be our greatest problem."—E. O. Wilson, Professor of Biology, Harvard University

Figure A
Two families, in India *(left)* and Germany *(right)*, with their possessions.

Peter Ginter/Material World

work. Ecologists talk about the carrying capacity of a particular habitat patch or ecosystem. But technology allows humans to exceed such limits, for humans now use resources from all over the world.

Some ecologists suggest that we think of the human use of resources in terms of **ecological footprint**, the area of land needed to provide all of the resources a single person uses. For example, the average person in the United States uses more than 10 times the energy that an average human in the world as a whole does. We use almost 10 times the paper, generate five times as much carbon dioxide, and use three times as much fresh water. To fulfill these demands, an aver-

age American needs about 5.1 hectares—our ecological footprint. In India, the ecological footprint of the average person is just 0.4 hectares, less than one-twelfth ours (see box).

When people talk about overpopulation, they often mention the nations with the largest populations—China, India, and the United States—or those with the fastest-growing populations, in Africa and the Middle East. For example, Saudi Arabia's growth rate is 3.7 times that of the United States. But Saudi Arabia has only 24 million people, to our 290 million. So while every year its population is increasing by nearly 800,000 people, ours is increasing by 2.6 million people.

When we consider total impact on the environment, however, we should look at both population size and ecological footprint together. India's 1.05 billion people each need about 0.4 hectares, or 0.42 billion hectares for the whole country (0.4 × 1.05 billion). But America's 290 million people each need 5.1 hectares, or 1.48 billion hectares for the whole country (5.1 × 0.29 billion).

The area of the whole United States is only 0.91 billion hectares, however, so we use 60 percent more land than we have to supply our ever-increasing demand for cars, clothes, and other goods. Even though we are far fewer in number, our impact on the world is more than three times greater than that of India.

Cities depend on energy subsidies and agriculture, which also depend on energy subsidies. Countries with the highest standards of living make the greatest demands on the world's resources.

How Well Has Agriculture Met the Human Demand for Food?

In the late 1960s, agricultural researchers recognized that many countries were outgrowing their food supplies. Every day, hundreds of millions of people were literally starving. To solve the hunger problem, researchers bred rice, wheat, and other crops that could produce up to 10 times as much food as older varieties. The effort, called the *Green Revolution,* was a boon to countries such as India, China, and Mexico, where crop increases temporarily outstripped human population growth.

Unfortunately, many of the Green Revolution's "miracle crops" depend on the use of chemical fertilizers, irrigation, and fossil-fuel-powered farm equipment. As American ecol-

ogist H. T. Odum wrote, the increased food supply is "partly made of oil." The world's poorest farmers cannot afford to buy the fertilizer, machinery, and fuel they need to grow the newer varieties of crops. But because few farmers now use the old crop varieties—bred over hundreds of years to grow well in particular regions—these old standby varieties are becoming unavailable.

Hunger continues to be a serious problem. Although food supplies increased in the 20th century, the world population grew even faster, and per capita grain production has now leveled off (Figure 28-17). About 800 million people are chronically hungry and living on inadequate diets that leave them susceptible to disease and premature death. According to the World Health Organization, about 10 million people die prematurely of hunger each year. If everyone in the world ate the way Americans do, the world food supply would support only about 2.5 billion people instead of 6 billion.

The Green Revolution increased the world food supply and so Earth's carrying capacity for humans. Today, however, population growth is outstripping the increased food supply.

28.5 How Is Human Overpopulation Polluting the Biosphere?

Humans are not only using up limited resources. We are also damaging air, water, and other renewable resources by polluting them with industrial waste, garbage, and sewage. The more of us there are, the more pollution we generate. In addition, by multiplying, we force more and more people to live in dangerous places. Studies show that

A.

B.

Figure 28-17
Grain production. A. Classic, terraced Asian rice fields. B. While world grain production is leveling off, the human population continues to grow. As a result, the amount of grain produced per person is declining.

as we become more numerous, natural disasters such as earthquakes and floods kill more of us and destroy more buildings and other property. Limited water supplies and farmable land may ultimately limit our population's ability to grow, just as these factors affect other natural populations.

Pollution Has Degraded the Water

Pollution is most obvious in freshwater communities, although marine ecosystems suffer from pollution as well. Historically, most people have disposed of sewage and other wastes by dumping them into the nearest body of water, to be carried away downstream to the ocean. But such polluted water is dangerous to human health, so government agencies have focused on cleaning up rivers and lakes. The worst damage has not been to people, however, but to other organisms. Several kinds of pollution damage ecosystems, including chemical and thermal pollution and sediments.

Thermal pollution occurs when factories dump hot or warm water into rivers and lakes. For example, when a power plant pumps cool river water through a hot nuclear reactor, the resulting hot water flows back into the river, killing some species and causing others to multiply too fast. Warm water can destroy aquatic ecosystems.

Sediments are particles of soil and sand suspended in water that settle to the bottoms of lakes, streams, rivers, and other bodies of water. They damage aquatic ecosystems by blocking light, burying organisms that live on the bottom, and filling in bodies of water. Lack of light prevents photosynthesis, limiting the productivity of ecosystems. In coastal waters, sediments can inundate coral reefs and shellfish beds.

Sediments result from the natural process of erosion. Indeed, natural erosion working over millions of years cut the Grand Canyon a mile into the Earth. And erosion created the fertile flood plains of the Mississippi and Nile rivers. But virtually all human activities—including the construction of houses and roads, agriculture, logging, and strip mining—enormously accelerate erosion. The result can be too much sediment building up too fast.

Chemical pollutants include a variety of molecules, not all of which are toxic. For example, *biodegradable pollutants* include carbon dioxide, sewage, fertilizers, and other wastes that organisms can consume. Like thermal pollution, these pollutants promote the growth of some organisms and inhibit the growth of others. Nitrogen and phosphorus in fertilizers used in agriculture and home gardening wash into rivers and lakes and promote "eutrophication" (Chapter 26).

Nondegradable pollutants include solid wastes, such as aluminum cans, plastics, glass, and hundreds of other human products that now accumulate in dumps. Even newspaper, technically biodegradable, does not degrade in most landfills. Up to 40 percent of all landfill is decades-old newspaper sports pages, stock market reports, and comics.

At sea, most waste is dumped overboard as a matter of course. Floating objects may stay on the surface for years. Plastic six-pack loops strangle birds and other animals. Lost fishing nets go on catching and killing fish and other animals long after the nets' owners have retired.

The hardest pollutants to deal with are outright poisons. These include heavy metals, such as mercury and lead; reactive gases from smog; radioactive waste; pesticides; and an unknown but growing number of toxic chemicals used in industry and agriculture. Recent research suggests that the peculiar atmospheric chemistry at the North and South Poles causes hundreds of tons of industrial mercury to precipitate out of the atmosphere each year and settle on the ice, from there entering Arctic and Antarctic food chains. Fish are now frequently contaminated with mercury. In coastal California, mercury blown across the Pacific Ocean from China sometimes rains down during winter storms.

Mercury poisoning causes fatigue in adults and damages the nervous systems of children and infants. About eight percent of U.S. women of childbearing age have enough mercury in their bodies to constitute a risk to any children they might bear. Little is known about the effects of thousands of other toxins. Still less is known about the interactions among these chemicals. For example, some substances can be relatively harmless by themselves but toxic in combination with other compounds.

One of the most interesting classes of toxins are *endocrine disrupters,* compounds that mimic or inhibit hormones such as estrogen and testosterone. Many plastics and other useful modern chemicals contain endocrine disrupters that interfere with normal development. Endocrine disrupters can cause both subtle changes in cell growth in reproductive tissues and gross malformations in the reproductive organs. You can read more about endocrine disrupters at the beginning of Chapter 36.

Many toxic chemicals become concentrated as they move up the food chain, a phenomenon called **biomagnification** (Figure 28-18). Plants concentrate a compound to many times that in the groundwater. Herbivores that eat the plants further concentrate the toxin in their flesh, and carnivores concentrate it even more. The highest concentrations are found in top carnivores, such as large freshwater fish, fish-eating birds such as pelicans and eagles, and, of course, humans. Trout in the Great Lakes, for example, have concentrations of pesticides and mercury that range from 1,000 to 14 million times that in the water in which they swim. The highest concentrations of the pesticide DDT ever found in humans came from men who regularly caught and ate catfish.

Water and soil pollution come in the form of thermal pollution, biodegradable pollutants, nondegradable pollutants, and poisons.

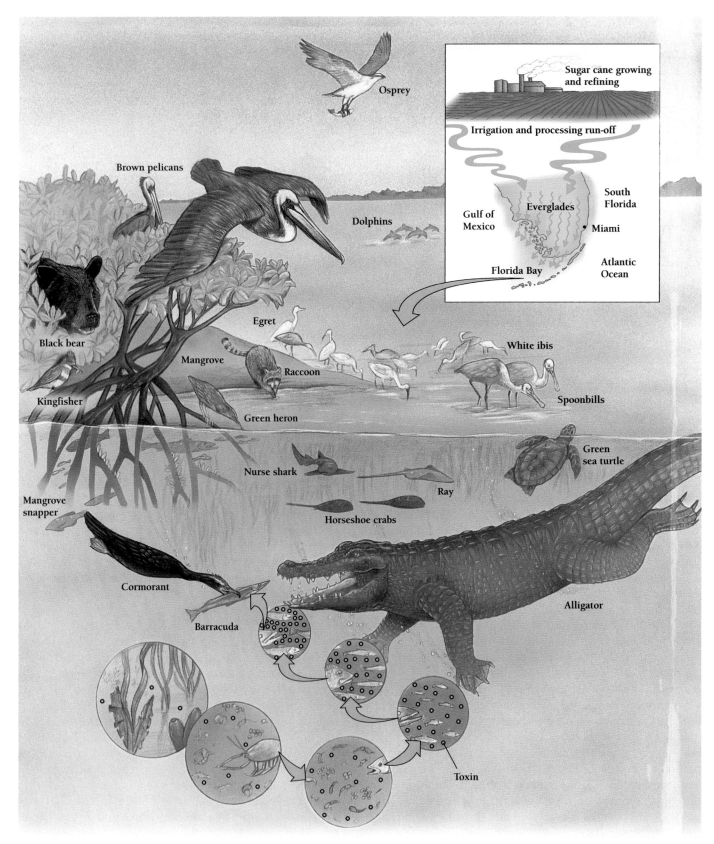

Figure 28-18

Biomagnification. As organisms consume food, they consume whatever toxins are in the food. Some toxins accumulate in the body, so that at each level of a food chain, the toxin becomes more concentrated—as illustrated here by the increasing concentration of blue dots.

Pollution Has Also Degraded the Air

Industrialization often means a darkening of the air and surrounding landscape as particles from incompletely burned coal and oil spew from factories and power plants. Tiny particles of smoke and soot cause lung disease in humans. But while soot may be the most visible and easily removed product of burning fossil fuels, the invisible chemical products of combustion have the most wide-ranging effects.

Nitrogen and sulfur oxides can travel thousands of miles with air currents before dissolving in rain and damaging or destroying forests and aquatic ecosystems. Sunlight converts nitrogen oxides in the atmosphere to the still more reactive compounds of smog. And, as we discussed in Chapter 25, burning fossil fuels has doubled the production of carbon dioxide in the atmosphere, contributing to global warming.

Another group of dangerous chemicals are the chlorofluorocarbons, chemicals that have eaten a hole in the ozone layer in the upper atmosphere. **Chlorofluorocarbons** are small molecules used as coolants in refrigerators and air conditioners, in the electronics industry as solvents to clean circuit boards, and as propellants in spray cans.

Ozone (O_3) is a highly reactive molecule made of three oxygen atoms instead of the two in the oxygen (O_2) we breathe. At ground level, ozone destroys living tissue. But 20 to 50 kilometers above the Earth, a layer of ozone called the **ozone layer** absorbs most of the ultraviolet light from the sun, shielding living things from potentially dangerous amounts of ultraviolet light (Figure 28-19). Ultraviolet light that reaches the Earth's surface hurts organisms by increasing mutations in DNA, destroying protein molecules, and increasing the number of chemically reactive molecules in the lower atmosphere. Some biologists suspect that excess ultraviolet light may have driven some species of frogs to extinction. For humans, the most obvious effect of increased ultraviolet light is higher rates of skin cancer. But ultraviolet light also breaks down the B vitamin folate, raising the risk of developmental defects in humans and other animals.

Almost 20 years ago, atmospheric scientists began to document a thinning of the ozone layer, particularly over Antarctica. They linked the disappearing ozone to the presence of chlorofluorocarbons in the upper atmosphere. By 1988, the evidence for the connection between increased chlorofluorocarbons and decreased ozone had become so clear that chlorofluorocarbon manufacturers agreed to replace these compounds with others.

Scientists initially expected that the hole in the ozone would soon return to normal. Yet despite dramatic reductions in the use of ozone-destroying chemicals, the ozone hole over Antarctica continued to enlarge and reached a record 27 million square kilometers in 1998—larger than all of North America. In the thinnest places, the ozone is only 70 percent of normal. Even over the United States, the ozone layer has thinned by two or three percent. Com-

puter models by atmospheric scientists at NASA suggest that the ozone layer will continue to thin for another 10 or 15 years.

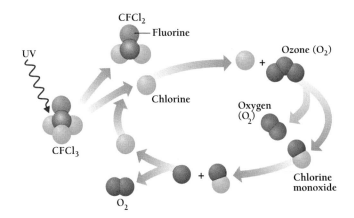

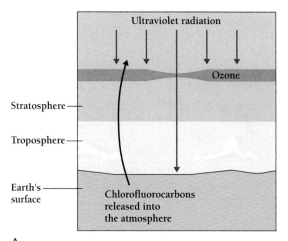

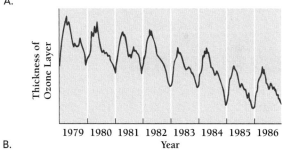

Figure 28-19

How do chlorofluorocarbons break down the ozone layer? A. Ultraviolet light from the sun breaks down $CFCl_3$ into $CFCl_2$ and chlorine. The chlorine reacts with ozone to form oxygen and chlorine monoxide, thus depleting ozone in the upper atmosphere. As the ozone layer thins, more ultraviolet light penetrates deeper into the atmosphere, harming both people and other animals. B. The ozone layer has been thinning steadily since the mid-1970s.

Biology 🄴 Now ™   Learn more about the breakdown of the ozone layer by clicking on this figure on your BiologyNow CD-ROM.

Air pollution includes visible particles of soot, invisible products of combustion, and chlorofluorocarbons, which damage the ozone layer.

Human population growth has had enormous effects on the biosphere. Our numbers have crowded out whole ecosystems, fragmented ecosystems, driven thousands of species to extinction, polluted the air and water, and altered the global climate.

Many researchers still optimistically believe that we can clean up the mess. We can remove pollutants in water and air, we can reduce the greenhouse effect, and we can slowly close the hole in the ozone layer. We can even stop multiplying so fast. Because of population momentum, however, it may take us several decades to halt our growth. One thing we cannot do is bring back the species we have driven to extinction. Once they are gone, they are gone for good. Biologists estimate that the evolution of replacement species will take 5 to 10 million years.

Key Concepts

- Populations may go extinct when habitat is damaged or destroyed. Chance events, including the disappearance of alleles from a gene pool, can hasten extinction.
- Natural populations grow exponentially only briefly before reaching a plateau, fluctuating, or crashing.
- Density-dependent factors decrease population size in response to population size.
- The growth of human populations can interfere with the growth and survival of other species and even whole ecosystems.

Summary with Key Terms

Although a conservation movement has existed in the United States for more than 100 years, 1980 marked the informal beginning of **conservation biology** as a field separate from traditional ecology.

How does habitat fragmentation lead to species extinctions?

Both **deterministic** and **stochastic** factors can drive species to extinction. Humans drive other species to extinction through **habitat destruction**, hunting, and the introduction of exotic species. Humans destroy habitat by breaking it up into small fragments with lots of edge and little central core. Roads, suburban developments, and agricultural fields all increase **habitat fragmentation** and **edge effects.** According to the **extinction vortex** hypothesis, loss and degradation of habitat reduce the size of a population, change its age and sex structure, and lead to a loss of genetic diversity. The loss of genetic diversity leads to inbreeding, reduced reproduction, and smaller population size. In a small population, chance, or stochastic, events are more likely to lead to extinction. We can preserve many more species if we set aside large preserves, connected by **corridors** and surrounded by buffer zones.

How do populations grow?

All living organisms are capable of **exponential growth.** Bacteria have a **doubling time** of as little as 20 minutes. Unrestrained exponential growth results in a rapid and limitless increase in total population size over time. A graph of this increase is sometimes called a **J-shaped curve.** The maximum possible rate of increase for a species is r_{max}, also called the **intrinsic rate of natural increase.**

What factors limit the growth of populations?

The growth of natural populations is limited both by **density-dependent factors,** such as space, resources, and predators, and by **density-independent factors,** such as weather and catastrophes. Density-dependent factors impose a limit to the growth of natural populations called the **carrying capacity** of the habitat, or K. When populations approach the carrying capacity of their habitats, the size of the population levels off, forming an **S-shaped curve.** When a population stops growing, it is said to have **zero population growth.**

Populations increase or decrease as a function of birth rate, death rate, immigration, and emigration. When populations overshoot their carrying capacity, they crash, then rise until the carrying capacity is reached or surpassed once more.

Ecologists classify species by **life history traits,** such as life span and number of offspring. One such measure is the **survivorship curve: convex,** or **type I; diagonal,** or **type II;** and **concave,** or **type III.** If we know the **age structure** of a population, which represents the size of each age cohort in a population, the age structure of immigrants and emigrants, and the fertility and mortality of each **cohort,** we can predict the future age structure of the population. **Demography,** the study of factors that influence human population growth, expresses **mortality** as death rate and **fertility** as birth rate, **replacement reproduction,** or **completed family size.** Demographers express the rate at which a human population increases as the **annual rate of increase.** The global population of humans has been increasing exponentially for thousands of years. **Population momentum** means that even if humans immediately and permanently reduced their annual rate of increase to zero, the world population would continue to grow for several decades.

How do energy subsidies increase food production?

Human agriculture is heavily **subsidized** with energy from fossil fuels. Despite the energy subsidy, global food production is beginning to fall behind increases in global population size.

How do human overpopulation and development affect the biosphere?

Human overpopulation increases water and air pollution, with attendant problems such as **biomagnification,** global warming, and the destruction of the **ozone layer** by chlorofluorocarbons. The impact of humans can be expressed in terms of **ecological footprint.**

Review and Thought Questions

Review Questions

1. The world population of humans is about 6 billion, and the doubling time is currently around 60 years. Using the equation $N \times 2^t$, calculate the world population after 30 years, 60 years, and 90 years. Graph your results, plotting N against time.

2. Using the information in the table (below), calculate r for Sweden. Show your work.

	Sweden	United States	India
b (births per 1,000)	10	14	23
d (deaths per 1,000)	11	9	8
r (rate of natural increase per 1,000)	—	6	15
Annual rate of growth (%)	0.1	0.9	1.5
Life expectancy (years)	80	77	63
Infant deaths per 1,000 live births	3	7	60
Total fertility rate (children per woman)	1.5	2.1	2.9
N (population in millions)	9	290	1,050

3. What factors make India's r value so much higher than that of the United States?

4. Why does Sweden have a higher death rate than the United States?

5. In your garage live a mated pair of healthy, mature Norway rats. Assume these animals can produce 10 young every 3 weeks (actually, Norway rats can produce up to 11). Assume also that each litter consists of five females and five males. Females are able to bear young at 12 weeks of age. If none of the adults or offspring die, and if conditions in your garage are so pleasant that none emigrate, how many rats will occupy your garage at the end of one year?

6. What is the intrinsic rate of natural increase of this pair of rats, expressed as the rate of increase per year?

7. What factors limit the carrying capacity of a habitat?

8. Give three examples of density-dependent limits to human population growth and three examples of density-independent limits.

9. What is replacement reproduction, and why is replacement reproduction for people in the United States 2.1 instead of just 2.0?

10. Why do urban populations require an energy subsidy?

11. Approximately how many people starve to death each year?

12. What type of survivorship curve does Sweden have?

Thought Questions

13. If a way could be found for the Earth to support 30 billion people, five times the current density, what other organisms and ecosystems would probably have to go to make room for these people? What could be retained? How would the world be better or worse than it is with a population of 6 billion?

14. Make a list of factors, both density-dependent and density-independent, that might limit the population of rats in your garage.

BiologyNow Resources

Biology *Now* ™

Active Figures

28.12: Age distribution in populations
28.19: Breakdown of the ozone layer

Preparing for an exam? Take a diagnostic test on your BiologyNow CD-ROM.

Online materials relating to this chapter are at:

http://biology.brookscole.com/AAL3

About the Chapter-Opening Image

The woolly mammoth—which went extinct 11,000 years ago—led a parade of modern extinctions that now includes millions of species.

29

The Ecology of Animal Behavior

Key Questions

- How does behavior develop?

- How does ecology influence the behavior of individuals and species?

- What are the advantages and disadvantages of sociality?

- How is altruism adaptive?

Why Is Sociobiology Controversial?

In 1979, a shy and bookish expert on the biology of ants was delivering a scientific talk in Washington, D.C., when a group of angry students charged the podium and doused him with water. E. O. Wilson, an eminent Harvard professor, had unwittingly placed himself at the center of a furious controversy by writing an ambitious textbook on animal behavior. The 1975 publication of *Sociobiology: The New Synthesis* almost overnight transformed Wilson into a notorious and, to some people, repugnant political figure.

Wilson became the target of angry public attacks by both scientists and nonscientists that left many people with the impression that sociobiology was promoting racism, sexism, classism, and even Nazism. "Sociobiology," the study of social behavior from an evolutionary perspective, got such a bad name, in fact, that biologists finally abandoned it in favor of **behavioral ecology**, the study of the adaptiveness of all forms of behavior, both social and individual.

The firestorm of debate that Wilson ignited was fueled by several separate but intermingling currents of thought, each with its own long history. These included the ongoing political debate over the role of genetics in shaping human behavior, the related "nature-nurture" debate, and a simmering feud between two groups of scientists.

Until about 1972, two separate disciplines for the study of animal behavior existed—the European science of **ethology**, the systematic study of animal behavior from a biological point of view, and the primarily American science of **experimental psychology**. Both ethologists and psychologists were interested in animal behavior, but their approaches, emphases, and philosophies differed dramatically.

Experimental psychologists focused on how animals learn. For example, the Russian physiologist Ivan Pavlov discov-

ered that he could teach a dog to salivate at the sound of a bell. First, Pavlov would ring a bell each time he fed the dog. When the dog saw the food, it salivated. In time, the bell alone would cause the dog to salivate, a behavior Pavlov termed a **conditioned reflex.**

Experimental psychologists were primarily interested in human behavior, and animals were only substitutes for humans. These researchers thought of behavior as being mostly learned. The adaptive value of behavior was of little interest to them, so all animals were considered more or less equivalent. As the famous experimental psychologist B. F. Skinner wrote in 1959, "Pigeon, rat, monkey, which is which? It doesn't matter."

Ethologists, by contrast, thought of behavior—however much of it was learned—as an adaptation that affected the survival of each individual and the evolution of species. Early ethologists preferred to study the normal behavior of animals living in the wild, with special emphasis on behavior that was inherited, or "innate." For example, the tendency of newborn babies to suck milk from a nipple is considered an innate behavior—even though practice improves a baby's ability to suck.

Experimental psychologists were suspicious of ethology for two reasons. First, they believed that because of humans' tremendous ability to learn, genetics and evolution were irrelevant. Second, ethologists emphasized the idea that behavior can be inherited, which reminded the psychologists of the 19th-century idea that whites were intrinsically (genetically) superior to all other races. In the 20th century, this idea was used to justify a variety of evils, including Hitler's human breeding program; the Nazis' mass murder of Jews, Gypsies, and others during World War II; and forced sterilizations in this country.

When E. O. Wilson's book came out in 1975, it brought these conflicts to a head. In his book, Wilson argued that by the year 2000 both ethology and experimental psychology would wither away, to be replaced by sociobiology and neuroscience, the study of the nervous system. Psychologists couldn't help but be offended. For them, acknowledging the value of the evolutionary viewpoint was one thing. Retiring meekly to the sidelines while neuroscientists and ethologists (turned sociobiologists) sorted out the problems of behavior was quite another.

Worse, Wilson speculated freely about the possible genetic basis of various aspects of human social behavior. Many people felt that Wilson's book provided ammunition that could be used to discriminate against the poor, the uneducated, minorities, and women.

A handful of biologists unfairly accused Wilson of being a "genetic determinist," someone who believes that all human behavior is determined by genes. But no biologist takes such an extreme view, and naturally Wilson denied the accusation. Today, the debate has lost much of its heat. Biologists do not argue about whether behavior is "controlled" by genes or by environment; all agree that behavior is always influenced by both.

29.1 How Is Behavior Acquired?

An animal's behavior contributes to its ability to survive and reproduce in the same way that its anatomical, cellular, and molecular characteristics do. Since genes can shape behavior, natural selection can operate on genes that influence behavior, just as it does on genes that influence structure. The adaptiveness of a given behavior depends to a large extent on ecology—the relationship of an individual to its environment. In this chapter, we will examine the ecological significance of behavior, as well as the role of natural selection in shaping behavior.

What Is Innate Behavior?

The human ability to walk is innate. Provided that we are physically able, we all begin walking at about one year. Even toddlers whose legs have been in casts during infancy begin to walk on schedule (Figure 29-1). **Innate** behavior is behavior that an animal engages in regardless of previous experience. The tendency of a dog to bark at strangers is innate. Most breeds bark at least occasionally. A dog's upbringing and psychological well-being may influence how much it barks and at whom, so that environment plays a role as well, but the dog's tendency to bark is inborn.

By contrast, a dog *learns* who is a stranger and who is familiar. **Learning** is a change in an animal's behavior in response to a specific previous experience. Mammals may be the most sophisticated learners, but essentially all animals—from flatworms to fish—can learn.

Different species of animals are able to learn different things. A grizzly bear is good at learning how to catch salmon in a stream, while a killer whale is good at learning how to catch salmon in the ocean. Every animal has innate

Figure 29-1
Programmed behavior. Humans all begin walking at about the same age. However, the style of walking, whether short steps or long, for example, is learned.

Figure 29-2
A modal action pattern. A greylag goose rolls a wayward egg back to her nest. If the egg is removed in the middle of this procedure, the goose continues the tucking movements until she is back on the nest.

tendencies to learn different things. We can see, then, that learned and innate behavior are not completely separable.

One interesting example comes from two races of the garter snake *Thamnophis elegans*. Garter snakes that live along the coast of California are known to relish slugs, while inland snakes will not touch them. To see if this difference in behavior was innate, researchers hatched baby snakes in the laboratory so the snakes would have no opportunity to learn about slugs, then offered the young snakes a menu that included slugs. Despite their inexperience, snakes from the coastal population readily ate the slugs, while those from inland populations refused them. This result suggested that genes might determine the snakes' taste for slugs. In a separate experiment, the researchers crossbred the two kinds of snakes. Among the hybrids, more snakes refused slugs than accepted them. The researchers suggested that a gene influencing slug refusal might be dominant.

Any behavior is likely to include both innate and learned aspects.

What Stimulates Innate Behavior?

A highly stereotyped innate behavior is called a **modal action pattern.** Examples of modal action patterns abound. All dogs scratch their ears in the same way, lifting a rear leg above a foreleg. This behavior is so stereotyped that if a human scratches a dog's ear, the dog's hind leg often will rise above the foreleg and move up and down in midair as if scratching. Parrots and many other birds also scratch their heads by bringing a foot over the wing. This mode of head scratching is not necessarily required by their anatomy, however. A parrot is perfectly capable of bringing a foot under the wing, and does so to clean its bill.

A modal action pattern can be partly fixed and partly flexible. For example, when an egg rolls out of the nest of a greylag goose, the goose can roll the wayward egg back into her nest by pushing it with the underside of her bill (Figure 29-2). This bill-tucking movement allows her to pull a loose egg back toward her, but she also must adjust to the sideways wobble of the egg by moving her head from side to side. If the goose is tucking the egg back into its nest and an experimenter suddenly removes the egg, the goose will continue tucking. But the bird's adjustments to the side-to-side movements of the egg will stop. The tucking movement is fixed, but the sideways adjustments are flexible.

Modal action patterns serve a purpose only when performed at the proper time and place. Some signal in the environment must stimulate a modal action pattern. In human infants, a pair of eyes stimulates a smile (Figure 29-3). A loose egg stimulates a greylag goose's chin tucking. A fly provokes the characteristic flick of a frog's tongue. In each case, the animal's senses detect a stimulus and the nervous system issues the order to act. The stimulus for a modal action pattern is called a **sign stimulus.**

Innate behavior often has a learned component. For example, the chick of a herring gull will peck at its mother's beak to beg for food the first time it sees its parent (Figure 29-4). Pecking is therefore regarded as innate. The pioneering ethologist Niko Tinbergen reasoned that since a herring gull chick always pecks at a red spot on its parent's bill, the red

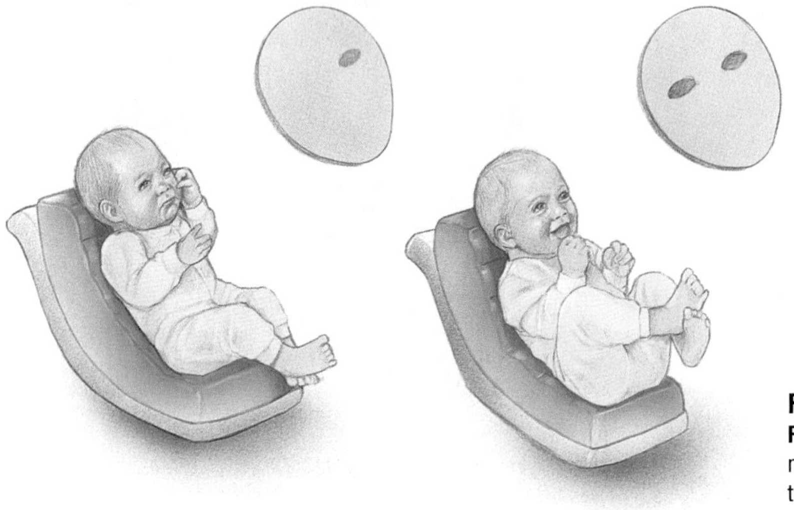

Figure 29-3
Releasing a smile. A flat, face-sized mask with one eye spot will not make a young baby smile. The same mask with two eye spots triggers a happy smile. The two eyes constitute a kind of sign stimulus. Older babies learn to prefer human faces.

A.   B.

Figure 29-4
Learning to beg. A herring gull chick begins begging for food soon after it hatches, a behavior that is partly innate and partly learned.

Figure 29-5
Tick sign stimuli. A female tick needs a blood meal to produce her eggs, but she must wait for a meal to come to her. For days or months, she sits unmoving at the end of a twig. At the moment a warm animal passes beneath her twig, she awakes, lets go of the twig, and drops. If she lands on a suitable animal, she inserts her feeding organ into the skin and sucks herself full of blood. If she lands on the ground, she climbs up onto a nearby plant and waits even longer. A tick responds to three separate sign stimuli. In response to light, a tick moves to the end of the twig. In response to either carbon dioxide (which animals exhale) or butyric acid (found in sweat), a tick lets go of the twig. Finally, in response to any warm surface—even a balloon filled with warm water—a tick will begin feeding. Light, carbon dioxide, butyric acid, and warmth all act as sign stimuli for a tick's behavior.

Butyric acid fumes and CO_2

spot may be what provokes the pecking behavior. To test this theory, Tinbergen presented chicks with models of gull bills with different colored spots. He found that the red spot elicited the pecking behavior better than any other colored spot, regardless of the shape of the model.

But even this behavior is partly learned. Later research by biologist Jack Hailman showed that a newly hatched gull chick pecks at anything that looks remotely like a beak, including the beaks of its siblings. Over the course of several days, however, the chick *learns* to recognize both its parents and food, so that it pecks in response to a much more limited range of stimuli.

Not all sign stimuli are visual. Touch, sound, and smell can also trigger stereotyped behavior patterns (Figure 29-5). A female silk moth emits a potent chemical, called bombykol, whose odor can attract males from over a mile away. When the male detects as few as 200 molecules of bombykol, he flies upwind until he loses the scent, flies in a zigzag pattern until he detects the bombykol again, and flies upwind again. He continues this pattern until he finds the female. Chemicals like bombykol that influence the behavior of another individual of the same species are called **pheromones.** Pheromones can activate modal action patterns.

Most sign stimuli are quite simple. Even with our own extraordinary brains and sensory systems, we simplify the complexity of the world. As a result, we can easily recognize public figures from simple line drawing caricatures, or close friends from just a glimpse in a crowd. Animals with less developed nervous systems than our own also respond to caricatures. When Tinbergen presented gull chicks with a stick with three red spots instead of just one, the chicks pecked at it even more than if it had been a real gull's bill. For a herring gull chick, three red spots constitute a **supernormal stimulus,** a stimulus even more stimulating than the normal stimulus.

A modal action pattern is a stereotyped behavior triggered by a sign stimulus. The sign stimulus can be seen, heard, felt, or smelled.

How Much Can Animals Learn?

In an early episode of the television show *Sesame Street*, Ernie is astonished that Bert has been able to teach Bernice the pigeon to play checkers. "Why, a pigeon that can play checkers!" Ernie exclaims. "That must be the smartest pigeon in the whole world!" "She's really not that smart," demurs Bert. "In the ten games we've played? She's only beaten me twice."

Pigeons may or may not be able to beat Muppets at checkers, but they are surprisingly capable learners. Pigeons and chickens can learn to distinguish between an underwater scene with fish and one without. They can even understand abstract ideas. Pigeons can learn to choose one of two disks according to whether it matches the color of a third disk.

Almost any animal can learn. Pigeons can beat humans at tic-tac-toe. Rats can learn not only to run a maze accurately and consistently but to run a mirror image of the maze, switching back and forth on cue. Octopuses can learn to open jars. Even planarians, among the simplest of all animals, can learn (Figure 29-6).

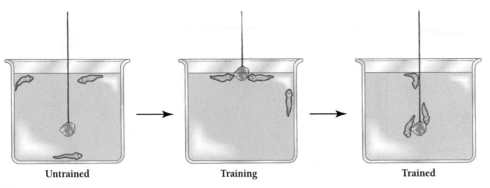

Untrained   Training   Trained

Figure 29-6
Learning in flatworms. The simple planarian, with barely a brain to call its own, learns to find food suspended from a string. Untrained worms search the surface or the bottom of a container for food and miss the meat hanging in the middle. During training, a piece of meat is suspended at the surface and then gradually lowered into the water. The flatworms learn to locate the string at the surface and then to follow it down to the meat. Trained worms quickly find food suspended in the water.

Biology⬧Now™ Learn more about types of learning in animals by clicking on this figure on your BiologyNow CD-ROM.

Animals pay close attention to one another's behavior. Experiments have shown that blue jays prefer to eat what they have seen other jays eating. Crows, upon seeing another crow sicken or die after eating poisoned bait, will shun the bait. One of the most striking examples of learning in natural populations of animals has occurred in at least 11 species of small English birds, including blue tits, titmice, and others. Starting in around 1920, a few birds learned to open the caps of milk bottles delivered to the front steps of English houses (see blue tit in chapter-opening photo). Beneath the caps was a rich meal of pure cream. The cream was so attractive that the birds attacked the bottles as soon as the delivery person was gone, and, in some areas, flocks of birds followed the milk trucks down the streets. If the bottles were capped with cardboard, the birds pulled the whole cap off or peeled it off in layers of paper. If the bottle was capped with foil, the birds poked a hole in the foil, then peeled the foil off in strips. This behavior first appeared in a few limited areas, then, over a period of years, gradually spread, suggesting that the birds learned from one another.

Animals can learn a great deal from their environment and from each other. Even behavior strongly influenced by genes is still open to modification through learning.

Can Old Dogs Learn New Tricks?

Just as humans learn languages most easily when young, many birds learn to sing their species song only when they are young. For many kinds of learning, timing is critical. A chaffinch (a kind of bird) raised in isolation so that it cannot hear adult birds singing never learns to sing an adult song.

Even if the chaffinch hears the right song as an adult, he cannot learn it. If, however, a young chaffinch hears the correct song during the first few months of life, he can learn to sing a proper chaffinch song—even if he cannot hear the song when he is actually practicing. The period of time during which an animal can learn a particular behavior pattern is called the **sensitive period.**

Sensitive periods exist for a variety of different kinds of behavior. For example, children born blinded by cataracts can learn to see if the cataracts are surgically removed before about age 10. If the cataracts are removed later in life, however, these people see only a random patchwork of colored shapes, which they cannot interpret. As infants we learn to interpret complex patterns of shading and color as a three-dimensional world. But for newly sighted people, the patterns may be meaningless.

The process by which an animal learns behavior during a sensitive period is called **imprinting.** The ethologist Konrad Lorenz provided the most famous example. Geese learn to recognize their mother by imprinting on the first moving object they see after hatching, usually the mother goose. But newborn geese will also follow a matchbox on a string, a rubber ball, a flashing light, or even a person (Figure 29-7).

Just as the development of normal vision depends on visual experience during a critical period, the development of normal social behavior depends on experience at sensitive periods. A young monkey isolated from other monkeys for six months anytime during the first year and a half of life does not develop social interactions with other monkeys when later given the chance. Instead, the deprived monkey crouches in the corner of its cage, rocking back and forth, much like a severely disturbed human child. An adult monkey subjected to the same isolation returns to its friends without such serious behavioral problems.

Figure 29-7
Are you my mother? Nobel laureate Konrad Lorenz showed that goslings imprint on the first moving object they see after hatching, including Lorenz himself.

Harry and Margaret Harlow, psychologists at the University of Wisconsin, attempted to find out what aspects of a young monkey's social life normally prevented such behavioral disorders. The researchers isolated young monkeys with and without a cloth-covered wooden dummy. The baby monkeys that had a wooden dummy, or "surrogate mother," showed fewer symptoms of isolation than the ones with no dummy.

The Harlows could completely prevent the symptoms of isolation by allowing the young monkeys to spend just 15 minutes a day with a normally socialized monkey that spent the rest of its time with the colony. Such limited social experience usually produced full socialization only if it occurred during the sensitive period in the first year and a half of life. In a few cases, however, monkeys reared in isolation could be partly "cured" by later exposure to persistently friendly monkeys. In monkeys, at least, positive experiences later in life can sometimes partly make up for a horrific early life.

For some complex behavior, then, the sensitive period is not the only time an animal can learn. We can look at hu-

mans for another example. Although children learn new languages faster and better than adults do, adults can learn new languages rather well if they are willing to work at it.

Many forms of learning occur only at particular times, called sensitive periods. Other forms of learning take place most easily during sensitive periods but can occur later as well.

How Do Genes Influence Behavior?

Behavior is the result of both genetic influences and experience (Figure 29-8). The nesting behavior of two small African parrots—the peach-faced lovebird and Fischer's lovebird—offers a perfect example. In both species of lovebird the female cuts long strips of bark or leaves for nest building. A peach-faced lovebird can carry up to six strips at a time by tucking them into her rump feathers, while a Fischer's lovebird carries these strips to the nest one at a time in her beak.

A hybrid of the two species tries both methods. When she builds her first nest, she starts out by tucking strips into her feathers, like her peach-faced parent. But she is inept, and the strips fall out before she can get to the nest. With time, she learns to carry just one strip at a time in her beak. After about three years, she gives up tucking strips into her feathers. However, just before she flies off with a strip in her beak, she still makes little sideways movements of her head, as if she were about to tuck a strip into her rump feathers.

The exact role of genes in influencing behavior is hard to pin down. In rare cases, a single gene can affect behavior. For example, in humans, a rare mutation in one gene causes Lesch-Nyhan syndrome, in which people compulsively hurt themselves, biting their fingers and lips and banging their heads. The vast majority of

Figure 29-8
Behavior is partly genetic. Specific types of behavior, such as smiling, are often shared by groups of related species.

physical and behavioral traits, however, are "polygenic," influenced by many genes. The complex genetics of all phenotypic traits—including behavioral traits—means that biologists cannot easily determine the exact degree to which a given behavior pattern is genetically or environmentally influenced. For example, despite years of research on the heritability of IQ scores—just one aspect of human intelligence— scientists can say only that IQ is somewhere between 30 and 70 percent inherited.

The balance is environmental. Nutrition, health, and environmental stimulation—in the womb, in infancy, and in early childhood—all affect intelligence to some degree. For example, one study found that premature baby boys fed breast milk had several more IQ points as older children than those who had been given formula. Yet in premature baby girls, the researchers found no difference.

29.2 How Does Ecology Influence Behavior?

Economists have debated whether people make rational economic decisions. For example, if food costs twice as much at one store as another, most people go out of their way to shop at the cheaper store. On the other hand, if travel costs are high, they will not spend more money getting to the cheaper store than they would save on the groceries. To a remarkable degree, then, people behave rationally. Animals, with far less intellectual capacity than humans, also behave surprisingly rationally about such decisions.

How Do Animals Make the Best Choices?

Crows open and eat thick-shelled sea snails, called whelks, by picking up the largest one they can find, carrying it 5 feet up in the air, and dropping it on the rocky shore. If the whelk breaks, the crow feasts. If the whelk stays intact, the crow carries it aloft and drops it again, repeating the procedure until the whelk's shell breaks.

Ecologists studying this tedious feeding strategy wondered if this was the best strategy to get the most food for the least energy expended. To see if the crows' strategy was "optimal," researchers themselves dropped variously sized whelks from different heights. The biologists soon found that the largest whelks were the most likely to break and that none of the whelks broke consistently unless dropped from a height of 5 meters or more. The crows' strategy of choosing only the largest whelks and carrying them to about 5 meters was indeed optimal. Further experiments revealed that if a whelk didn't crack open on the first try, persistence with a single whelk was more likely to succeed than starting over with a new whelk. So, a crow's decision to drop the same whelk over and over proved equally sensible. Overall,

the crows' feeding behavior is an efficient way of spending energy to get more energy. Such energy-efficient feeding is called **optimal foraging.**

When animals forage, they may efficiently maximize the amount of food they get for the energy expended. Ecologists call this process optimal foraging.

How Does Predation Influence Foraging Behavior?

Maximizing energy intake is not the only consideration that guides feeding behavior. When predators are present, an animal must adjust its behavior to increase its own chances of surviving. One study of chickadee feeding, for example, revealed that the chickadees would eat seeds at a researcher's feeding tray only if the tray seemed safe. When researchers mounted a model of a hawk nearby, the chickadees took the seeds away to a safer place. Taking the seeds away cost the birds time and energy, but given the apparent presence of a hawk, moving the seeds was a good way to both eat and survive.

How Does Competition Influence Foraging Behavior?

The presence of competing species can also change feeding behavior. For example, North American mink introduced into Sweden in the 1920s competed directly with European otters for food and caused otter populations to decline. However, the two animals continued to coexist. In America, the mink normally eat almost anything that moves, including crayfish, fish, muskrats, rabbits, birds, and small burrowing mammals. The European otters have a more specialized niche than the mink, feeding on fish, crayfish, crabs, and other aquatic invertebrates only. In Europe, the larger otters defend this narrow niche from the mink aggressively, sometimes killing mink that poach fish and crayfish. As a result, in areas where the otters survive, mink eat less fish and crayfish than those in America.

Foraging behavior depends, in part, on how an individual interacts with members of other species, especially predators and competitors.

How Do Animals Compete for Resources or Mates?

Animals of the same species may cooperate or compete with one another, according to the demands of their current environment and the constraints of their evolutionary histories. Among the most dramatic competitive kinds of behavior are the ritualized fights between pairs of bighorn sheep. The rams run together at full speed, bashing their heads together. The outcome of such combat determines which male

Figure 29-9
Agonistic behavior in rabbits. Fighting between males is not confined to large animals such as deer or lions. Here, two rabbits fight to establish dominance. Agonistic behavior also includes aggressive behavior and appeasement displays.

will get to mate with a female. The clash of horns is mostly symbolic: usually the fighters do not seriously injure each other. Other species use other symbolic contests—who can roar the loudest, stand the tallest, or stare the longest.

These contests are examples of **agonistic behavior** [Greek, *agonistes* = champion]—all the aspects of competitive behavior within a species, including aggression (outright attacks and fighting), aggressive displays, appeasement, and retreat (Figure 29-9). Ethologists use the term "agonistic behavior" instead of the word "aggression" to acknowledge that animals never exhibit pure aggression or pure fear. What most animals (humans included) show is a mix of fearful behavior and aggression, sometimes more of one, sometimes more of the other.

Some animals use agonistic interactions to establish a **dominance hierarchy,** a ranking of individuals that fixes which animals have first access to resources or mates. Dominance hierarchies establish rules that allow animals to live in the same area with limited resources. The classic example of a dominance hierarchy occurs in a flock of hens. Such a flock quickly establishes a "pecking order," which specifies who pecks whom without being pecked in return.

One way that animals compete with others of their own species for mates, resources, and rank is through agonistic encounters.

Territoriality

In many species, an animal engages in competitive behavior to establish and defend a **territory,** an area defended by an individual (or group of individuals), from which other indi-

viduals of that species are excluded—through displays, calls, chases, or fights. Some species, a few birds and fish, for example, use territories exclusively for mating. More commonly, species use territories for foraging or for raising young as well.

Establishing and defending a territory necessarily involves agonistic behavior. Animals use a wide variety of signals to mark their territories, including chemical scents, brightly colored plumage, or patrolling behavior. All of this signaling behavior is energetically expensive, most of all patrolling, calling, and fighting.

How do animals balance the benefits of defending a territory (in terms of food, mates, and a place to nest and rear young) against the costs? Behavioral ecologists who have studied the "economics" of territoriality have shown that some birds carefully adjust territory size according to costs and benefits. For example, a hummingbird on a patch of nectar-rich flowers will defend only a small territory when the flowers are putting out lots of nectar (and lots of other birds are trying to get the flowers). The same bird will defend a larger part of the same patch of flowers when the flowers are making less nectar (and not so many other birds are trying to get to the flowers). Some species are so sensitive to changes in food supply and "defense costs" that they adjust the size of their territories every day.

Animals defend territories for raising young, mating, or foraging—as long as defending the territory will increase their reproductive potential.

How does competition for mates influence the evolution of males and females?

In Chapter 15 we defined **natural selection** as the differential reproductive success of individuals with a particular set of genetically influenced traits. One special kind of natural selection is **sexual selection,** the differential reproductive success of individuals due to differences in their ability to acquire mates (Chapter 16). The enormous tail of a male peacock is a classic example of trait thought to result from sexual selection. The bird's gaudy tail attracts pea hens (females) but also probably increases a male's vulnerability to predators (Figure 29-10A).

Some biologists distinguish sexual selection from other kinds of natural selection, arguing that natural selection always results in the accumulation of adaptive structures and behaviors, while sexual selection often results in "maladaptive" structures and behavior, such as the peacock's tail. But all adaptations to the various demands of survival and reproduction are balanced against one another. Thoroughbred horses can run much faster than the wild horses of Mongolia. But the domesticated racehorses are much more likely to damage their longer, more delicate legs. The wild horses have balanced speed against sturdiness. They aren't as fast, but they are more likely to survive and reproduce in the wild.

A.

B.

Figure 29-10
The price of being male. A. The gaudy tail of a male peacock increases the bird's reproductive success despite increasing the bird's vulnerability to predators. B. Sparring male northern elephant seals. Larger males are more likely to win fights with other males. After driving another male away, the winner has access to more females and therefore produces more offspring than smaller males, who may not mate at all.

In the case of sexual selection, a selective force is coming from other members of the same species in a particular context, but the result is still the usual balance among conflicting selective forces.

Most well-documented examples of sexual selection are in males. The selective force in sexual selection in males comes either from other males, who are competing for mates (male competition), or from females, who are choosing a male with which to mate (female choice). A nice example of how competition among males results in selection comes from studies of elephant seals. Just a handful of males win all the fights, and a single big male may mate with up to 100 females (Figure 29-10B). The other source of selection pressure in sexual selection comes from the choices that females make. For example, in tungara frogs, which live in Panama, female frogs consistently prefer larger males to smaller ones. The females choose the larger males, however, not by comparing male physiques but by listening to the male frogs' mating calls. The song consists of a whine, followed by one or more low-pitched "chucks." Bigger frogs have lower "chucks," so female frogs prefer frogs with deep voices (Figure 29-11A). The lower the "chuck," the more the female frog is attracted, even when the "chuck" comes from a tape recorder.

Unfortunately, the male frog's deep "chucks" also attract the attentions of frog-eating bats, which, like the female frogs, prefer a big frog to a little one (Figure 29-11B). Bats force male frogs to balance survival against reproduction. To whatever extent variation in "chuck" pitch is under genetic influence, sexual selection will tend to favor the evolution of low-pitched chucks—but maybe not too low.

Competition for mates over evolutionary time results in sexual selection, the differential ability of individuals with different genotypes to acquire mates.

A.

B.

Figure 29-11
Cost-benefit analysis by male frogs. The female tungara frog of Panama prefers a male that makes a deep chuck when he calls. Larger males make deeper chucks. Males with the deepest voices have the best chance of attracting a female and, unfortunately, the best chance of being eaten by a frog-eating bat. A. A male tungara frog calling while floating in his pond. B. A hapless frog that attracted a bat instead of a female frog.

29.3 Some Animals Live in Social Groups

A tiger lives a solitary life. It meets other tigers rarely and for not much longer than it takes to mate. Yet the tiger's close relative the lion lives in a tightly knit group called a pride (Figure 29-12A and B). The members of a pride of lions are never far apart. They sleep together, hunt together, and raise their young together. Why do some animals live socially, while others live apart?

How Is Social Behavior Advantageous?

Living in groups affords distinct advantages. The greatest advantage to group living is that large groups of animals can more quickly spot predators and other dangers than can solitary animals. Researchers have found, for example, that the more pigeons that are in a flock, the sooner the pigeons will notice a hawk and the less likely that the hawk will catch one of the pigeons.

Groups of animals can also better defend themselves. When threatened, a herd of musk oxen forms a circle with their young in the middle and their horns pointed outward. And the thousands of honeybees in a hive can drive off a predator or an animal bent on stealing their honey where a lone bee would be helpless. Finally, group living can be an advantage in finding food. Predators that hunt cooperatively, such as lions and wolves, can kill prey far larger than themselves (Figure 29-12B).

Social living has disadvantages as well, however. Animals living in groups must compete with each other for food, mates, and breeding areas. Individuals waste time and energy fighting or threatening one another and searching for food in the same places. Groups of animals also transmit diseases to one another and attract predators.

Whether animals live in groups depends on the ecological costs and benefits. Many songbirds stake out territories in the spring and actively repel all but their own mates and offspring. Yet when autumn arrives, and food becomes harder to find, the same birds abandon their territories and roam the countryside in large communal flocks.

> Sociality allows animals to spot both prey and predators more easily. But groups of animals are themselves more visible to predators, and individuals in groups must constantly compete for food, mates, and breeding areas.

Honeybees Are an Extraordinary Example of Social Behavior

The coordination of individuals within a colony of social insects rivals that of the cells within an individual. We can view an insect colony as a superorganism, in which thousands or even millions of individuals function as a single unit. Social insects form enormous colonies. An African termite mound, for example, may contain 2 million workers. But the number of individuals alone is not what makes the social insects special. It is social organization that allows social species to take advantage of environmental resources unavailable to more solitary animals.

The familiar honeybee provides an excellent example. Humans long ago domesticated the honeybee to harvest the bees' useful honey and wax. Scientists since Aristotle have

B.

Figure 29-12
Loners and social animals. A. Tigers, which hunt alone, kill smaller animals. B. Lions, which hunt in groups, can kill animals much larger than themselves.

studied the workings of beehives and published tens of thousands of papers on bees. The fascination with bees has extended beyond the practical to the almost mystical. Humans admire the fact that the "busy bee" works so hard and loyally and ferociously defends its queen (long thought to be a king).

A honeybee hive consists of double-layered wax combs that contain thousands of hexagonal cells. In these wax cells, the bees store honey and pollen or house developing larvae. Each bee grows from an egg deposited in a separate cell. The egg develops into a larva, which then develops into a pupa and finally into an adult.

A hive may contain 20,000 to 80,000 workers, each engaged in a succession of tasks. At the beginning of her life, a worker feeds nectar and pollen to the larvae. Later, she makes wax to enlarge the comb, removes the dead and the dying, guards the hive against intruders, and scouts the surrounding territory for food and nesting sites. Finally, toward the end of her short, six-week life, she undertakes the most hazardous job of all, foraging for nectar and pollen.

All the workers in the hive are female, and all are sisters. None reproduce, however. Only a single female—the queen—lays eggs. Most of these eggs develop into sterile workers, but a few may become queens. Workers can create queens by feeding a few, select larvae a special rich diet. Normally, the hive's queen produces a "queen substance," a pheromone that prevents the workers from building special wax cells to raise new queens. In the spring, however, the queen's production of this chemical falls off, and the workers build the royal cells. Workers soon push the old queen out of the hive, and she flies off with a large swarm of workers to start a new hive (Figure 29-13).

About eight days after the old queen leaves the hive, the new queens begin hatching. Each one can form a swarm and fly off to start a new colony. If a queen elects to stay in the old hive, however, she first seeks and destroys rival queen larvae. She identifies these competitors by making quacking sounds outside the queen cells. If another queen responds to this challenge, a fight to the death ensues. There can be only one queen in each colony.

Once a queen has disposed of her rivals and established her queenship, she mates repeatedly in a series of "nuptial" flights. She leaves the hive, approaches a group of males, and releases small amounts of queen substance. The pheromone attracts males, one of whom will mate with her. The male releases his sperm into the queen's genital chamber and quickly dies. The queen may mate with as many as 17 males over a four-day period. In this way, she obtains enough sperm for a lifetime. She stores the sperm and uses it to fertilize her eggs, which she deposits one at a time into separate wax cells. A queen may live and lay eggs for more than five years.

All of the social insects, including termites, ants, wasps, and honeybees, have three traits in common: (1) they cooperate in raising their young; (2) only a few individuals re-

Figure 29-13
Swarming honeybees. A swarm of bees will stay in one place for days at a time while scouts look for a suitable site for a new hive. The scouts return to the swarm and dance the location of a hole in a tree or cave. Each scout, however, is dancing a different location. When all the scouts agree, the swarm sets off to build a new hive.

Biology ⊗ Now™ Learn more about bee communication by clicking on this figure on your BiologyNow CD-ROM.

produce, while sterile workers assist and defend the fertile queen; and (3) generations overlap, so that offspring help their mother.

Honeybees display an extraordinarily complex social structure that includes the division of labor and cooperation in rearing their young.

Is Altruism Adaptive?

In Chapter 16 we learned that natural selection favors traits that increase the likelihood that an organism's offspring will survive. Why, then, we must ask, do the sterile workers in a beehive persist in a behavior that can have no adaptive value for them individually? The workers increase the reproductive output of the queen but at the expense of their own reproduction. Biologists call such an arrangement **altruism,** behavior that benefits others at the expense of the animal that performs the behavior.

People sometimes jump into frozen lakes to rescue others. We know that this behavior is risky, because many of these would-be rescuers themselves die. Such behavior is clearly altruistic. Similarly, ground squirrels risk their own lives to warn one another of an approaching predator. And the young of many species of birds help their parents rear the next generation. The social insects, however, demonstrate the most extreme example of altruism.

If altruistic behavior has a genetic basis, then biologists must ask how it could possibly evolve. We would expect natural selection to favor sacrifice by parents to help their offspring but to act against the sacrifice of offspring to help their parents. Why do worker bees work so hard to help their mothers reproduce?

Does Altruism Arise by Natural Selection?

How could evolution produce workers, however industrious, if they leave no offspring? For a while, Darwin himself thought that this paradox was fatal to his whole theory of evolution by natural selection. To save his theory, Darwin argued that a family (the queen and her offspring, for example) must be the unit of selection, rather than the individual. By their self-sacrifice, the workers perpetuate genes that they share with their sisters, one of whom will become a queen. What is selected is the ability to be altruistic. The altruistic family is successful in passing on its genes.

In the 1960s, the British biologist W. D. Hamilton extended Darwin's idea, suggesting the theory of **kin selection.** Hamilton reasoned that an individual increases its reproductive output by helping relatives, which share its genes, to reproduce. Sisters (and brothers) are just as related to one another as they are to their parents. They share half of their genes. Thus, a worker bee helps ensure the continuation of her genes within her sisters and their descendants (recall that some of the sisters become queens).

Hamilton's work established the concept of "inclusive fitness." Fitness is a measure of selective advantage defined as the contribution to the next generation of one genotype in a population relative to the contribution of other genotypes. **Inclusive fitness** is the sum of an individual's genetic fitness (which includes that of its own direct descendants) plus all its influence on the fitness of its other relatives. Inclusive fitness, then, is the total success in passing alleles to the next generation, either by the efforts of a parent or by the altruistic acts of a relative.

One explanation for the adaptive value of altruism is inclusive fitness, which includes the concept of kin selection.

Why Might Intricate Societies Have Evolved among Certain Insects?

Intricate societies evolved at least 11 separate times among the bees, ants, and other members of the order Hymenoptera, yet only once (in the termites) among all other insects. Why the difference? One reason may be that the bees and wasps have an unusual kind of sex determination, called **haplodiploidy.** In haplodiploidy, males develop from unfertilized eggs and are therefore haploid, while females are diploid. Other than providing sperm, males contribute

nothing to the economy of the hive. Haplodiploidy dramatically alters the usual genetic relationships among parents and siblings. It means that a worker's inclusive fitness is greater if she devotes her energies to her siblings than if she reproduces herself. Let's see how this works.

First consider the usual case, say in humans. Half your alleles come from your mother, half from your father. Your brothers and sisters also share half of their alleles with your mother. What is the chance of sharing an allele with your sister? Half of all your alleles came from your mother, and the chance that you and your sister inherited the same allele is 0.5. So the chance that you each inherited a given allele from your mother is 0.5 × 0.5 = 0.25. Similarly, the chance that you each inherited the same allele from your father is 0.25. So the chance that you and your sister share any allele is 0.25 + 0.25 = 0.5, exactly the same as the chance that you share an allele with your mother or father. A similar calculation shows that you share one-eighth of your alleles with your first cousins. The British evolutionary biologist J. B. S. Haldane summarized these relationships by joking, "I would gladly lay down my life for two of my brothers or eight of my first cousins."

As in the case of ordinary sex determination, the queen passes half her alleles to her female offspring. However, since males are haploid, a male passes on all his alleles. That is, his daughters inherit all of his genes. As a result, the (female) workers share 75 percent (rather than 50 percent) of their alleles with one another. Because the queen mates with many different males, not all of her offspring share the same father. However, the average relatedness will be greater than 50 percent.

If the workers were to have their own offspring, they would pass on 50 percent of their alleles. They can therefore pass on more of their genes by putting their energies into making sisters than they can by reproducing and making daughters. To put it another way, the workers more efficiently perpetuate their genes by helping their mother, the queen, produce more offspring (including new queens) than by producing eggs themselves.

Haplodiploidy in some social insects probably facilitates kin selection.

Does Inclusive Fitness Always Explain Altruistic Behavior?

The haplodiploid social insects are not the only animals that help one another. Grazing animals, birds, and other animals that live in groups commonly help protect one another. In as many as 300 species of birds, "helpers" may assist a breeding pair feed and protect the nestlings. Whether such behavior can be explained by kin selection, however, depends on how closely the altruists are related to one another.

A few studies suggest kin selection as an explanation for altruism. For example, when a pair of Florida scrub jays breeds, as many as six other jays may help feed and protect

George D. Lepp

George D. Lepp

A.

B.

Figure 29-14

How kinship affects behavior. A. At Tioga Pass, California, students watch individual Belding ground squirrels marked with numbers written with hair dye. These ground squirrels live in colonies consisting of related females and their offspring. B. Older females sound a loud alarm when predators approach. Males take no such chances and scurry silently away.

the nestlings. These helpers are usually the breeding jays' offspring from previous clutches, offspring that have yet to mate themselves. Because territories suitable for breeding are in short supply, these helpers may not be able to breed effectively. Like the honeybee workers, the young jays may be better off helping their siblings than trying to breed themselves.

An even more persuasive example comes from Belding ground squirrel populations high in the Sierra Nevada, in California. In the summer, female squirrels defend territories where they raise their offspring. The two biggest dangers to Belding ground squirrels, besides winter snows, are predators and other squirrels. Unrelated or distantly related squirrels regularly kill other ground squirrels, killing up to eight percent of all pups. At the approach of a coyote or strange ground squirrel, some female squirrels stand up tall and sound a shrill alarm call that alerts other nearby squirrels (Figure 29-14). Other individuals who first sight a coyote or a stranger make no alarm and simply dive into the nearest burrow silently, leaving the other squirrels to fend for themselves. The squirrels that sound the alarm increase their own chance of being attacked, so their behavior is altruistic.

Paul Sherman, a biologist now at Cornell University, asked why these squirrels risk their own lives to help others, and why some squirrels behave altruistically, while others do not. Sherman discovered that most of the squirrels that sound the alarm are yearling and older females, who are warning their female relatives—sisters, mothers, and daughters, as well as juvenile sons. However, males and childless females rarely warn other squirrels. The older female incurs a risk that is counterbalanced by the chance that she will save the life of a close relative. Her altruistic behavior may contribute to her inclusive fitness.

In the past 20 years, many biologists, anthropologists, and other enthusiasts have tried to extend the concept of kin selection to virtually all altruistic behavior. But not all examples of altruistic behavior conform to the kin selection model. Dwarf mongooses, for example, help breeding pairs care for their young, but not all of the helpers are relatives. Kin selection does not account for such altruism.

There are other explanations for why animals behave altruistically, but one of the most interesting was first suggested by ethologist Robert Trivers. Trivers argued that animals capable of recognizing one another might engage in altruism on the understanding that the recipient would return the favor sometime in the future. Trivers called this idea **reciprocal altruism,** since the recipient reciprocates. Such a system at first seems too susceptible to cheaters to work. Biologists wondered, How can an animal ensure that the recipient of an altruistic act will return the favor? But research in the area of "game theory," a branch of mathematical logic applied to business, military, sociobiological, and other problems, suggests that altruism can pay off in spite of cheaters.

Kin selection may explain altruism in other animals besides the social insects. But other explanations, such as reciprocal altruism, are possible, as well.

Extreme Biology Tit for Tat and Prisoner's Dilemma

The classic game-theory analysis of the risks and benefits of cooperation and altruism is presented in the "game" called prisoner's dilemma. Imagine that you and a friend are arrested for committing a crime, and you are held in separate jail cells, unable to communicate. The police immediately commence interrogating the two of you.

Suppose you claim innocence, but your friend confesses and incriminates you.

You'll be in prison for a long time, but your friend will get off scot-free. On the other hand, you could get off yourself if you incriminate your friend and your friend pleads innocent. If you each confess (and incriminate each other), the state's job of convicting you will be easy, and your sentences will be light. But if you both plead innocent, the state will have little evidence against you, and your sentences will be even lighter.

What to do? Regardless of what your friend does, you are better off confessing, which is true for your friend, too. On the other hand, if both of you confess, you will be worse off than if both of you pleaded innocent. The safest solution is to confess. Even though pleading innocent might get you off scot-free, it's the riskiest solution to the problem.

The game of prisoner's dilemma seems to show that selfish behavior is safer than cooperative behavior. In 1981, however, a political scientist named Robert Axelrod and a biologist named W. D. Hamilton combined forces to explore what happens when the game is played over and over. They reasoned that in real life, people and animals interact repeatedly. Perhaps cooperation might pay off in the long run.

To make the endeavor more interesting, Axelrod invited game theorists from sociology, political science, economics, mathematics, and biology to design strategies to compete in an ongoing computer tournament. Predictably, players that always cooperated were taken advantage of. But strategies of constant betrayal also did badly.

The winning long-term scheme was a simple strategy called tit for tat. In tit for tat, the player always cooperates (pleads innocent) in the first round. After that, the player does whatever the opponent did in the last round. In short, you reward your friend for cooperating and punish your friend for ratting on you. Over time, tit for tat always won.

In the early 1990s, an even more successful strategy emerged from Axelrod's tournament. "Modified tit for tat" occasionally forgives other players for a betrayal, just often enough to nudge the game in the direction of cooperation and reciprocity. Although prisoner's dilemma is a simple game, it suggests that reciprocity may explain altruism in some situations.

Support for the tit-for-tat model of reciprocal altruism comes from studies of a surprising animal, the vampire bat (Figure A). Vampire bats *(Desmodus rotundus)* feed on the blood of cattle and other mammals by chewing through the skin and then injecting an anticoagulant that causes the blood to flow long enough for the bat to complete its meal.

But each night, about 7 percent of adults and 33 percent of young bats fail to find food. Without a meal, a bat will starve to death in about two and a half days. Amazingly, bats that have managed to get a meal help feed other bats that were not so lucky, so that few go hungry. Biologist Gerald Wilkinson, at the University of Maryland, noticed that bats often regurgitated part of their own meal and wondered if the bats were feeding only relatives—an example of kin selection—or nonrelatives, too—an example of altruism.

To find out, Wilkinson carefully observed who was feeding whom. Each day, 8 to 12 females and their offspring roost upside down in a hollow tree. But different bats come to the roost each day, so bats hang with both relatives and nonrelatives. Wilkinson found that bats regurgitated blood meals significantly more often to their kin than to non-kin. But he also observed that bats did regurgitate blood meals to non-kin. Not surprisingly, the non-kin bats that were fed were much more likely to associate with the bats that fed them than with the bats that didn't.

To test the tit-for-tat model, Wilkinson captured and caged some of the bats. Each cage housed unrelated females from two different populations, about 40 km apart. He then starved the bats, fed some of them, and looked to see which bats were fed regurgitated blood. Of the 13 blood feedings that Wilkinson observed, 12 occurred between bats from the same population, suggesting that the bats preferred to feed bats they already knew. And just as we humans are more likely to do a favor for someone who has done us one, a bat was much more likely to donate a meal to one that had previously given it blood.

Patricia and Michael Fogden/Corbis

Figure A
Give blood! In an example of kin selection and reciprocal altruism, vampire bats regurgitate blood meals, sharing with both kin and non-kin.

Key Concepts

- Animal behavior develops as a conversation between genes and the environment.
- Virtually all animals are capable of learning.
- The ecology of an animal determines the adaptive value of its behavior.
- Living in groups has both advantages and disadvantages.

Summary with Key Terms

How does behavior develop?

Experimental psychology is the laboratory study of basic behaviors common to many animals; **ethology** is the study of natural behavior particular to certain groups of animals; and **behavioral ecology** is the study of the evolution of all forms of behavior, both social and individual. All animal behavior is influenced simultaneously by both genes and environment. Humans and other animals learn from one another as well as from their environment.

Innate behavior includes a baby's first smile and the salivating of a dog over his dinner bowl, but all behavior can be modified through **learning**. A dog salivating at the sound of a can opener is a **conditioned reflex**. A **modal action pattern** is highly stereotyped innate behavior performed in response to a **sign stimulus**. An exaggerated sign stimulus is a **supernormal stimulus**. Stimuli may be sights, sounds, tactile experiences, or smells in the form of **pheromones**. Some forms of learning, called **imprinting**, occur only at special times in development, called **sensitive periods**.

How does ecology influence the behavior of individuals and species?

Animals seem to accurately weigh the costs and benefits of different foraging strategies in a process called **optimal foraging**. When animals compete for mates or resources, they engage in **agonistic behavior**, which includes aggression, appeasement, and retreat. Animals such as chickens, dogs, and primates use agonistic behavior to establish **dominance hierarchies**. Competition for mates and mate choice results in **sexual selection**. Many animals defend a **territory**, space and resources for reproduction.

What are the advantages and disadvantages of sociality?

Social groups allow animals to spot both prey and predators more easily. However, groups of animals are themselves more visible to predators than are solitary animals. Also, the individuals in a group must constantly compete with one another for food, mates, and breeding areas.

How is altruism adaptive?

The social insects display complex social structure and **altruism**. One explanation for the adaptive value of altruism is **inclusive fitness**, which includes the concept of **kin selection**. All of the social insects except the termites are **haplodiploid**, which means that sisters are more closely related to one another than to their own offspring. Such extreme relatedness facilitates kin selection and altruism. Another explanation for altruism is **reciprocal altruism**.

Review and Thought Questions

Review Questions

1. Define a modal action pattern and give an example.
2. What is a pheromone?
3. Give an example of a supernormal stimulus in advertising you have seen.
4. Give an example of a sensitive period.
5. Define and give an example of optimal foraging.
6. Distinguish between aggression and agonistic behavior.
7. Give an example of territorial behavior in dogs.
8. Why, according to the theory of inclusive fitness, are honeybee workers better off not reproducing?
9. Give an example of an innate response modified by learning in an adult human.

Thought Questions

10. Three pizza parlors offer specials on identical giant pizzas. One parlor will deliver the giant pizza for $25. The other two don't deliver. One is just around the corner, and its pizza is $20. The other is eight miles away and charges $12. It costs you 50¢ a mile to drive your 1980 Buick Skylark, and the car needs a new battery. If you break down in the neighborhood where the $12 pizza is, you might get mugged. What is the optimal solution to this foraging problem?

BiologyNow Resources

Biology Ⓔ Now™

Active Figures

29.6: Types of learning in animals
29.13: Bee communication

Preparing for an exam? Take a diagnostic test on your BiologyNow CD-ROM.

Online materials relating to this chapter are at:

http://biology.brookscole.com/AAL3

About the Chapter-Opening Image

English birds called tits learned from each other how to uncap milk bottles and sip the cream.

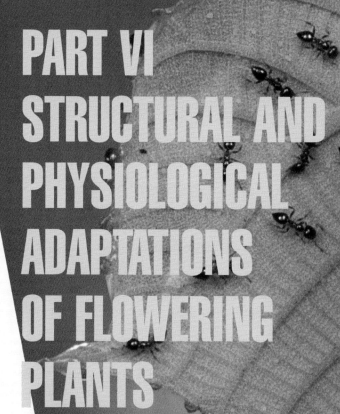

PART VI
STRUCTURAL AND PHYSIOLOGICAL ADAPTATIONS OF FLOWERING PLANTS

30

Structural and Chemical Adaptations of Plants

Key Questions.

- What physical constraints shaped the evolution of vascular plants?

- What kind of tissues make up vascular plants?

- How do plants defend themselves against herbivores?

Should We Eat Our Vegetables?

In 1983, a biochemist at the University of California, Berkeley, caused a stir in both the scientific community and in the health food industry when he asserted that enormous numbers of ordinary vegetables were loaded with mutagens, chemicals that cause DNA to mutate. Under the right circumstances, a mutagen can cause cancer. Natural chemicals, the biochemist argued, were as carcinogenic as anything manufactured by American industry.

Bruce Ames was already well known for his invention of the Ames test, an efficient method for testing the mutagenicity of chemicals. In the Ames test, plates of bacteria are treated with a chemical and the bacteria are then monitored for changes that indicate that their DNA has mutated. Because Ames's work had pointed out the hazards of various environmental toxins and mutagens, Ames was much admired by members of the environmental movement. In 1977, he had warned against the use of the pesticide ethylene dibromide, which he described as a "potent carcinogen." He also pointed out the hazards of a chemical used as a flame retardant in children's clothing and warned, more generally, against inadequately tested industrial chemicals.

But many people saw his 1983 paper as an about-face, in which he seemed to assert that pesticides, air pollution, and other environmental hazards were nothing compared to the toxins found in ordinary vegetables and meat. Indeed, Ames reviewed the scientific literature on mutagenic and toxic compounds in food and concluded, "The human dietary intake of 'nature's pesticides' is likely to be several grams per day— probably at least 10,000 times higher than the dietary intake of man-made pesticides." Ames listed scores of toxic chemicals found naturally in ordinary foods, including burned or browned meats and other foods, most fats— especially burned or rancid fats—and great numbers of compounds made naturally by plants.

Herbal teas made from plants in the family Euphorbiaceae, for example, contain potent promoters of carcinogenesis. Wilted carrots contain a mutagen. When we metabolize the alcohol in beer, wine, and spirits we make a compound that is both a mutagen and a teratogen [Greek, *terat* = monster + *gen* = to make], a compound that causes birth defects.

Royalty-Free/Corbis

As we will see later in this chapter, many plants have evolved the capacity to defend themselves chemically. Many plants respond to any kind of damage, usually inflicted by animals biting their leaves or bark, by secreting high levels of toxins, many of which are mutagenic, teratogenic, or carcinogenic. Ames's list of plant and other food toxins filled four pages in the journal *Science*.

Ames's fellow researchers were appalled, accusing Ames of ignoring the well-known hazards of smoking and of backing off from his stance against dangerous pesticides and the chemical pollution of air and water. If nothing else, they asserted, he was confusing people about what constituted a real hazard and what did not.

Ames argued that he still believed in the strict testing of industrial chemicals, but he also insisted that the natural components of a diet played a role in the incidence of major diseases such as heart disease, colon cancer, and breast cancer. A large part of his 1983 paper had been devoted to natural plant compounds, such as vitamin E, vitamin C, and β-carotene, that seem to prevent cancer rather than cause it. These compounds, called antioxidants, work by preventing the formation of free radicals, compounds formed during normal metabolic processes that can damage DNA. All plants manufacture antioxidants to protect their DNA and other important molecules from free radicals. For example, β-carotene, which functions in photosynthesis, also functions as an antioxidant in both the plants that make it and the animals that eat the plants.

Despite the many toxins in plants, their antioxidants make the argument for eating more fruits and vegetables a no-brainer. One study showed that smokers who did not eat fruit regularly had a 30 percent higher incidence of lung cancer than smokers who ate fruit. In later years, Ames argued that lack of antioxidant-containing fruits and vegetables in the diet was a major cause of the degenerative diseases of aging—including heart disease, cataracts, and cancer.

We can summarize Ames's argument by saying that if we don't smoke or drink too much, and if we eat our fruits and vegetables, we need not worry very much about trace amounts of pesticides or other contaminants in our food and drinking water—regardless of whether those contaminants come from factories or plants. Nutritionists now regularly advise Americans to eat more fruits and vegetables, not only for the vitamins, but also because of the high levels of antioxidants (Table 30-1).

Of course, people exposed to high levels of toxins—whether from toxic waste dumps near their homes or from working in industries such as agriculture or metals processing—continue to run serious risks. And, finally, compounds that mimic hormones may turn out to affect our biology even in small doses (Chapter 37). In general, however, most water and air pollution is more threatening to other organisms than to us.

In the rest of this chapter we discuss plant structure and the chemical compounds found in flowering plants.

Table 30-1
Foods That Are Good Sources of Three Antioxidants

Food	Vitamin A	% RDA Vitamin C	Vitamin E
Apples (1)	—	10	—
Apricots (3)	50	—	—
Blueberries (1 cup)	—	33	—
Strawberries (1/2 cup)	—	100	—
Cantaloupe (1/2)	>100	>100	—
Orange (1)	—	>100	—
Broccoli	90	200	—
Carrot (1)	400	—	—
Red pepper (1)	84	250	—
Winter squash (1 cup)	150	—	—
Sweet potato	500	—	—
Almonds (1/2 cup)	—	—	170
Mayonnaise (1 T)	—	—	80
Peanuts (1/2 cup)	—	—	60
Safflower oil (1 T)	—	—	45

RDA = Recommended daily allowance

30.1 What Physical Constraints Shape Plants?

Plants are photosynthetic. With few exceptions, they live by turning carbon dioxide and water into sugars using the energy from sunlight (Figure 30-1). To do this, they need water, light, and carbon dioxide (Chapter 7).

Until about 500 million years ago, most life on Earth was aquatic. The first photosynthetic organisms to invade the land left behind fossils—traces of themselves in the form of spores, bits of plant "skin," and bits of the plant "veins" that carry fluids. The first multicelled plants almost certainly descended from green algae. Some modern green algae associate with fungi to form lichens (Figure 21-25), and early plants may have evolved from a similar association. The algal partner could have provided sugars from photosynthesis while the fungal partner provided a way of getting water and minerals from the soil. Such "symbiotic" associations are so worthwhile that four-fifths of modern plants have fungal partners, called *mycorrhizae*, on their roots (Chapter 21).

From some such primitive form, recognizable plants evolved over 100 million years to become the dominant feature of land life. Plants make nearly all the food that supports terrestrial ecosystems and feeds all animals, including humans. The most diverse plants, with more than 275,000 species, are the angiosperms, or flowering plants.

Figure 30-1

Plants are photosynthetic. Using the energy in sunlight, plants transform water and carbon dioxide into sugars. A simplified equation representing photosynthesis would read: water + carbon dioxide = sugar. (More correctly: $6\ H_2O + 6\ CO_2 \rightarrow C_6H_{12}O_6 + 6\ O_2$)

Biology ⊜ Now™ Learn more about photosynthesis and water flow in flowering plants by clicking on this figure on your BiologyNow CD-ROM.

What Factors Drove Plant Evolution?

The land has distinct advantages over lakes and seas. Aquatic plants absorb water and dissolved carbon dioxide from the water that surrounds them. But they can only grow in shallow water because sunlight does not penetrate water very far and plants need sunlight for photosynthesis. In water, nutrients are also often highly diluted or even virtually nonexistent. But on land, plants can find plenty of light and nutrients, including a higher concentration of carbon dioxide—the basic building block of plants. The two big drawbacks of life on land are that (1) water is usually in short supply, and (2) without water to buoy plants up, gravity exerts a crushing force. To survive on land, plants had to

evolve new ways to get water and support their bodies against the pull of gravity.

Most terrestrial plants harvest energy from sunlight using their leaves. Because leaves are solar collectors, most plants have evolved broad, thin leaves that have a large surface for collecting sunlight. Inside leaf cells are tiny organelles called chloroplasts that use solar energy to turn carbon dioxide and water into sugar. Leaves are green because they contain chloroplasts, and chloroplasts are green because they are loaded with the green pigment chlorophyll, involved in the actual capture of solar energy (Chapter 7). Chloroplasts have their own DNA and most likely evolved from free-living photosynthetic bacteria that came to live inside early plant cells.

Terrestrial plants have also evolved roots that grow into the soil. Roots keep plants from falling over and also pull water from moist soil. Terrestrial plants also evolved stems (and branches and trunks), which support leaves, flowers, and fruits against the constant tug of gravity and also provide roadways for the transport of materials between the ground and the air (Figure 30-1). Inside the root and stem of every plant is a sophisticated system of pipes called the "vascular system" that carries nutrients to wherever they are needed.

The first plants with vascular systems, the **vascular plants,** did not evolve until about 435 million years ago. Until then, all plants were nonvascular plants, like the 24,000 species of mosses and other nonvascular plants (*bryophytes*) that still flourish today. But the vast majority of terrestrial plants are vascular—with true roots, stems, and leaves. Vascular plants include about 275,000 named species of flowering plants (e.g., daisies and oak trees), 13,000 species of ferns, and about 500 species of gymnosperms, such as the conifers (e.g., fir and pine trees). The basic form of vascular plants solves all of the major problems of terrestrial life and has remained unchanged for hundreds of millions of years. Vascular plants live virtually everywhere on the planet—anyplace that has sunlight, carbon dioxide, and any amount of water at all.

Terrestrial plants have evolved roots, stems, leaves, and a sophisticated vascular system to survive on land.

What Are the Structure and Function of Shoots and Roots?

Plants need to collect nutrients from both the ground and the air, so most plants consist of an aboveground **shoot system**—stems and photosynthetic leaves—and an underground **root system** that anchors the plant to the ground and absorbs water and minerals from the soil. Shoots and roots depend on one another utterly: roots could not live without the energy provided by photosynthesis in the leaves

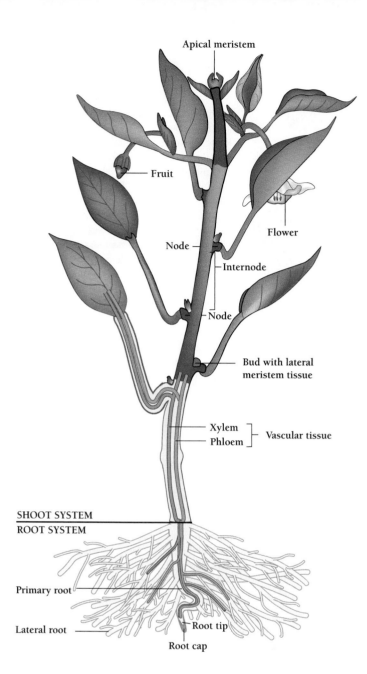

Apical meristem

Fruit

Flower

Node

Internode

Node

Bud with lateral
meristem tissue

Xylem
Phloem — Vascular tissue

SHOOT SYSTEM
ROOT SYSTEM

Primary root

Lateral root

Root tip

Root cap

Figure 30-2
Shoot and root. The main parts of a flowering plant include the root system and the shoot system. The shoot includes stems, leaves, flowers, fruits, and apical and lateral buds that contain meristem tissue. A cross-section through leaf, stem, or root reveals an outer layer of epidermis, some sort of central core of vascular tissue and intermediate ground tissue. The vascular tissue consists of xylem, which carries water and minerals, and phloem, which carries sugars made in the leaves.

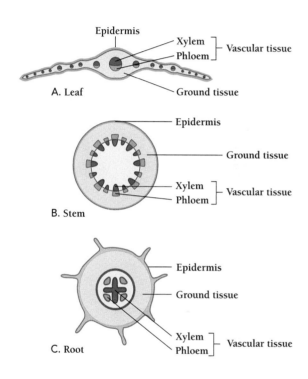

Epidermis

Xylem ⎤ Vascular tissue
Phloem ⎦

Ground tissue

A. Leaf

Epidermis

Ground tissue

Xylem ⎤ Vascular tissue
Phloem ⎦

B. Stem

Epidermis

Ground tissue

Xylem ⎤ Vascular tissue
Phloem ⎦

C. Root

of the shoot, and shoots could not live without the water and minerals supplied by the roots.

Materials travel between shoot and root in the plant's **vascular system** [Latin, *vas* = vessel], the network of conducting tubes through which fluids move (Figure 30-2). For example, sugars made in the leaves travel to the roots to power the molecular pumps that pull water from the soil, and water from the roots travels to the leaves to participate in the chemical reactions that form those sugars. The vascular system consists of two kinds of tissues—**xylem**, which transports water and minerals, and **phloem**, which transports the sugars made in the leaves during photosynthesis. The two kinds of vascular tissue run parallel to one another

in "vascular bundles," but they are quite different. Phloem, which consists of living cells, transports sugars from leaves to growing regions or any part of the plant that needs energy. Xylem mostly consists of dead cells. Besides transporting water and minerals from the roots, it provides mechanical support in large woody plants such as trees. Lumber, used for houses and furniture, consists almost entirely of xylem.

Vascular plants include a shoot, with stems and leaves, and a root. Connecting all the parts of a vascular plant is a vascular system composed of xylem, which carries water and minerals, and phloem, which carries sugars.

Health and Biology Herbal Supplements

Herbal supplements are big business these days, and Americans spend about $10 billion a year on herbs they hope will make them feel better. But do they necessarily get what they pay for? Scientific evidence suggests not.

One of the most popular supplements is the herb Echinacea, or purple coneflower, which is thought to boost immunity and shorten the duration of colds. But unlike the pharmaceutical industry, the food supplements industry is almost entirely unregulated. That means that when you buy a bottle of Echinacea (or any other herb), it may not contain what you think it does.

Biologist Christine Gilroy at the University of Colorado Health Sciences Center found this out the hard way while trying to study the health effects of Echinacea (as she reported in the March 24, 2003, *Archives of Internal Medicine*). Gilroy began by ordering the herbal supplement from three different suppliers, and then, being a careful scientist, she had some of each sample tested to make sure it contained what it was supposed to. None of the samples did. Growing in the United States are nine different species of Echinacea, each of which makes a slightly different array of secondary compounds. Only two species, *Echinacea pallida* and *Echinacea purpurea,* are known to offer health benefits. (Both species are threatened in some parts of the United States and protected.)

Figure A
Echinacea purpurea. The Eastern purple coneflower grows from Texas east to the Atlantic but is threatened through much of its range. Like many plant species used as herbal supplements, the collection of wild plants has decimated wild populations.

Gilroy's study ground to a halt when she found that one of the three supplements contained almost no Echinacea at all. Even the other two, bought directly from Echinacea growers, were full of other plants and

E. angustifolia, a species not known to offer any health benefits. Gilroy couldn't help wondering if previous researchers even knew what was in the supplements they tested.

Curious if her sample of three was typical, she ordered more Echinacea, from 59 different commercial suppliers. Six contained no Echinacea at all and 28 claimed to contain species that weren't present. Even those that contained Echinacea usually did not contain as much Echinacea as claimed. Gilroy concluded that any given study of one brand of Echinacea supplement would tell researchers nothing about another brand.

The U.S. Food and Drug Administration would like to ensure that supplements contain what they are supposed to. However, since no one is sure what chemicals (if any) in a given plant species are the ones that have pharmaceutical effects, there's no way for anyone to tell suppliers what *should* be in a bottle of Echinacea or ginseng. Plant tissues typically contain hundreds of compounds with perhaps dozens of active ingredients.

Besides species differences, plants also produce different secondary compounds depending, for example, on the maturity of the leaves; the environment of the plant, including temperature, availability of water, nutrients, and light; and whether the plant has been cut or eaten by herbivores. For now, herbal supplements are largely a matter of faith, faith in the herb and faith that the herb is in the bottle.

What Tissues Make Up Flowering Plants?

The cells of plants make up tissues, which in turn make up organs such as leaves, roots, and stems. Because each tissue type in plants is continuous throughout the plant, botanists use another level of organization, the tissue system. Three tissue systems occur in plants: the dermal, ground, and vascular tissue systems (Figure 30-2).

All three specialized tissue systems develop from regions of actively dividing undifferentiated cells called **meristem** [Greek, *meristos* = divided]. Like the "stem cells" of animals, meristem cells can differentiate to form many different types of tissue. Meristem tissue at the tips of roots and shoots are called **apical meristem** [Latin, *apex* = top]. Apical meristem tissue makes primary tissues through **primary growth**.

Many plants also possess **lateral meristem tissue**, cylinders of actively dividing cells within the roots and stems that

make "secondary tissue" through **secondary growth**. Primary growth helps plants grow taller or their branches and roots grow longer, while secondary growth helps trees and other plants grow larger in diameter. Secondary growth usually results in the development of wood, the familiar hard substance in our furniture and fireplaces (Chapter 32).

Covering and protecting the whole plant is the **dermal tissue system,** the equivalent of our skin (Figure 30-2). The outermost cell layer, the **epidermis,** secretes a waxy covering, called the **cuticle,** that keeps the aboveground parts of the plants from losing water. Most epidermal cells are simple building blocks like the bricks we use to build walls. But epidermis also includes glandular cells that secrete sticky or smelly substances and tiny prickles or hairs called trichomes. Most important, the lower surfaces of most leaves are covered with tiny holes that open and close to regulate the flow of carbon dioxide and water vapor. These tiny holes, called **stomata** [Greek, *stoma* = mouth], open and close like tiny mouths. When the stomata are open, water vapor evaporates out of the leaf and carbon dioxide moves inside the leaf to participate in photosynthesis. When the stomata are closed, usually as a way of conserving water, carbon dioxide cannot enter the leaf (Chapter 31).

The **ground tissue system** lies between the epidermal tissue and vascular tissue and makes up most of the plant (Figure 30-2). Ground tissue consists of three main types of tissue—parenchyma, collenchyma, and sclerenchyma. All three are simple tissues made of just one kind of cell. **Parenchyma** cells [Greek, *para* = beside] are delicate, living cells with large vacuoles, but only a thin primary cell wall. Parenchyma cells photosynthesize, store starch, and make up most of the softer tissues of plants (Figure 30-3).

Collenchyma cells [Greek, *kolla* = glue] are characterized by an unevenly thickened primary cell wall. Collenchyma

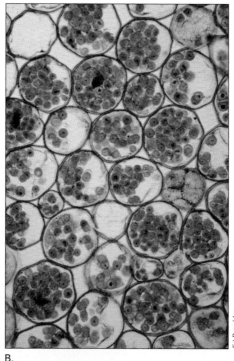

A.

Ed Reschke

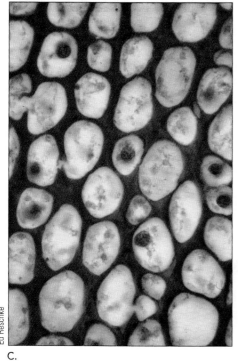

C.

Ed Reschke

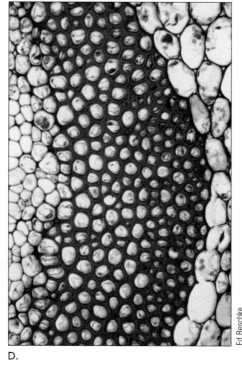

D.

Ed Reschke

Figure 30-3

Four plant cell types. Most of the parts of a plant are made of just a few cell types. A. The tough, waterproof epidermal cells protect the outside of the plant from injury and drying. Shown here are the brick-like cells of the epidermis of a spiderwort. Notice the stomata, which allow oxygen, carbon dioxide, and water to diffuse in and out of the plant. B. Inside, delicate parenchyma cells photosynthesize and store starches. (The parenchyma cells shown here are from the roots of a ranunculus and so do not photosynthesize.)Tough-walled collenchyma and sclerenchyma cells strengthen stems and leaves. C. Collenchyma cells from the stem of an elderberry. Notice the uneven thickening of the cell walls. D. Sclerenchyma cells from the stem of a *Helianthus sp.* flower.

Biology●Now™ Learn more about cell types and tissues in plants by clicking on this figure on your BiologyNow CD-ROM.

cells often form bands or sheets just inside the epidermis and provide mechanical support in the growing regions of stems and leaves. In celery, the bands of chollenchyma are easy to find; they are the long strings just under the surface (Figure 30-3).

Sclerenchyma cells [Greek, *skleros* = hard] have stiff cell walls that give mechanical support to the plant. Unlike parenchyma and collenchyma cells, sclerenchyma cells don't become fully functional until they are dead. Some sclerenchyma cells consist of long, thin cells that join together into strong plates or cords. Linen fabric, for example, is woven from strands of sclerenchyma from the flax plant.

Flowering plants are made of dermal, ground, and vascular tissues, all of which develop from meristem tissues. The epidermis and its cuticle cover a plant as our skin does, and stomata allow gas exchange. Ground tissue consists of parenchyma, collenchyma, and sclerenchyma.

How Do Flowering Plants Cope with Gravity?

Gravity is an ever-present challenge to plants. The more plants spread their leaves and branches to maximize their exposure to sunlight—and the more leaves they make—the stronger the stems and branches need to be to support the weight. If you hold a heavy book in one hand, you will see that it is harder to hold it out horizontally than to hold it down at your side or even straight up above your head. Plants that spread their leaves in the sun have the same problem. To hold leaves, stems, and branches out flat—horizontal to the Earth's surface—requires materials that are both strong and lightweight. In addition, the larger trees

grow, the stronger their tissues need to be to support so much weight.

A plant's answer to gravity is a rigid **cell wall** that surrounds each cell's plasma membrane. Like our own bony skeleton, a plant's cell wall supports the softer tissues against the tug of gravity. All plant cells have a **primary cell wall** outside the plasma membrane, and some cells also make a thicker **secondary cell wall** between the cell membrane and the primary cell wall (Figure 30-4). The primary cell wall consists of **cellulose,** other carbohydrates, and specialized proteins. For its weight, cellulose is the strongest material known. The mostly cellulose nutshell of the Australian macadamia nut, for example, is stronger than concrete, glass, brick, or commercial-grade aluminum; yet the density of the nutshell is less than half that of any of these other materials. A hard, stiff polymer called **lignin** adds strength to the secondary wall, and in some cases to the primary wall, as well.

The woody parts of trees and shrubs provide support against gravity. But what about plants that have no wood? Some plants, called *herbs* (or herbaceous plants), are entirely soft and green. Herbs have no woody shoot and can easily wilt and sag. (To a botanist, the term "herb" does not mean familiar kitchen seasonings, which may or may not be woody.)

Herbs handle gravity just fine as long as they have plenty of water. The cytoplasm of most plant cells fills only part of the space within the cell wall, often as little as 10 percent. The rest of the internal space consists of a large **central vacuole,** filled with a watery fluid. Water pressure in the central vacuole pushes the rest of the cell's contents and cell membrane against the cell wall. The cell wall prevents the cell from swelling and bursting. The pressure of the fluid contents of the vacuole against the hard cell wall, called "turgor," helps support herbs and other plants against the

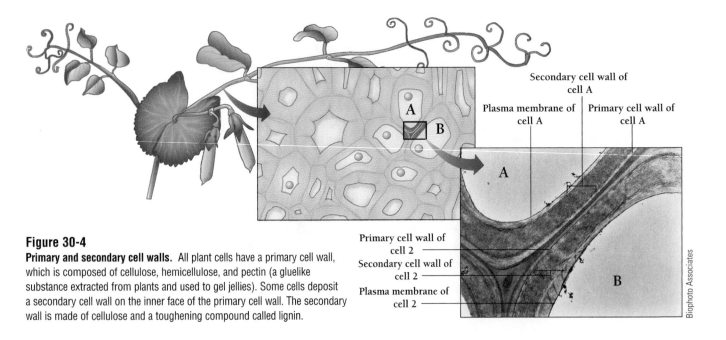

Figure 30-4
Primary and secondary cell walls. All plant cells have a primary cell wall, which is composed of cellulose, hemicellulose, and pectin (a gluelike substance extracted from plants and used to gel jellies). Some cells deposit a secondary cell wall on the inner face of the primary cell wall. The secondary wall is made of cellulose and a toughening compound called lignin.

pull of gravity. Plants stressed by a lack of water wilt because they have lost turgor.

Both turgor pressure inside a plant's cells and structural materials, such as cellulose and lignin, support plants against the pull of gravity.

What Kinds of Cells Make Up the Vascular Tissues?

The third tissue system, the **vascular tissue system,** consists of xylem, phloem, and some associated tissues (Figure 30-2). Xylem carries water and minerals in two kinds of cells, tracheids and vessel elements (Table 30-2). Both kinds of xylem cells have lignin-reinforced secondary cell walls tucked inside of primary cell walls, and both kinds of xylem cell carry

water only after they are dead and their cell membranes are gone. Although only the tracheids and vessel elements carry fluids, the xylem tissue also includes parenchyma cells, fibers, and other cells.

Tracheids are long, thin, spindle-shaped cells (Table 30-2). Water and dissolved substances move from one tracheid cell to another through tiny pores called **pits,** regions of the cell wall that have a primary cell wall but no watertight secondary cell wall. Without a secondary cell wall, the only barrier between two cells is their primary cell walls, which are permeable to water. Pits form at the tapered ends of each tracheid cell as pairs between adjacent cells, so that water can flow smoothly through the pits, from one hollow tracheid to the next (Figure 30-5).

Tracheids are the only conducting cells in the xylem of conifers and most other nonflowering plants. But flowering plants may also have **vessel elements,** which are shorter,

Table 30-2
Xylem and phloem cells.

Xylem, which carries water and minerals, is made of two kinds of cells—tracheids and vessel elements. Phloem, which carries sugary sap, is also made of two kinds of cells—sieve tube elements, which carry the sap, and companion cells, which regulate what goes in and out of the sieve tube elements.

Xylem		Phloem	
	TRACHEID CHARACTERISTICS: Dead at maturity Thin, elongated shape Tough secondary wall Pits with primary cell wall only FUNCTIONS; Slow seeping of water through pits on all sufaces of tracheid Mechanical support in woody plants		**SIEVE TUBE ELEMENT** CHARACTERISTICS: Living cells Lack most organelles Elongate shape Primary cell wall with intact cell membrane Perforate ends of cells (sieve plates) connect to form sieve tubes FUNCTION: Carry sap consisting of water, sugars, proteins, and RNA
	VESSEL ELEMENT CHARACTERISTICS: Dead at maturity Shorter and wider than tracheids Perforated end walls Primary and secondary cell walls FUNCTION: Rapid passage of water through perforated end walls		**COMPANION CELL** CHARACTERISTICS: Living cells with primary cell wall Variable shape Numerous plasmodesmata connect to sieve tube element FUNCTION: Nourish sieve tube element

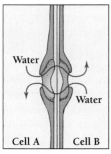

Tracheid

Pits

Figure 30-5

Pit pairs. Pit pairs are tiny openings in the secondary cell wall of tracheids that allow water to flow through the xylem from one tracheid to the next. A. In a simple pit pair, a hole penetrates the waterproof secondary cell wall, leaving only the water-permeable primary wall. In living cells, plasmodesmata connect the cytoplasm of the adjacent cells, providing a path for fluid flow. B. Simple pits weaken the xylem, so that it's likely to break, and some plants have evolved a stronger kind of pit pair, called the bordered pit pair, in which a tiny dome of secondary wall reinforces the wall over each pit. C. In conifers and some other plants, a bordered pit functions as a valve. When water pressure on one side of the bordered pit pair is higher than on the other, a buttonlike thickening of the primary cell wall pair blocks the pit, preventing water flow. During the growth of a shoot or root, these valves keep water in the growing regions.

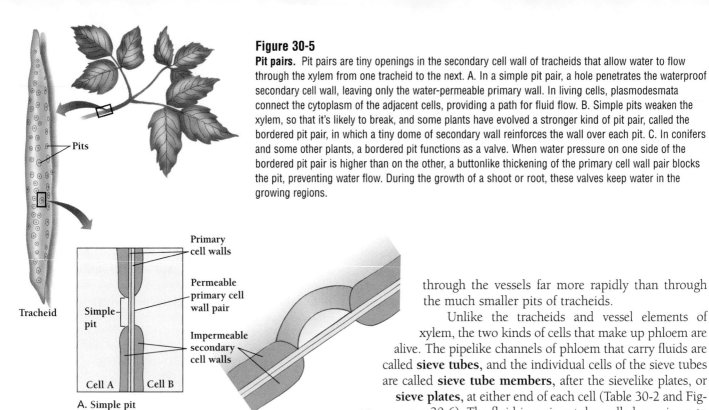

Primary cell walls

Permeable primary cell wall pair

Simple pit

Impermeable secondary cell walls

Cell A   Cell B

A. Simple pit

Primary cell walls

Bordered pit

Pit borders

Secondary cell walls

Cell A   Cell B

B. Bordered pit

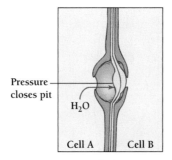

Water

Water

Cell A   Cell B

C. Conifer bordered pit: Water pressure equal in cells A and B.

Pressure closes pit

H_2O

Cell A   Cell B

Pressure higher in cell A than in cell B closes pit.

through the vessels far more rapidly than through the much smaller pits of tracheids.

Unlike the tracheids and vessel elements of xylem, the two kinds of cells that make up phloem are alive. The pipelike channels of phloem that carry fluids are called **sieve tubes,** and the individual cells of the sieve tubes are called **sieve tube members,** after the sievelike plates, or **sieve plates,** at either end of each cell (Table 30-2 and Figure 30-6). The fluid in a sieve tube, called **sap,** is up to 30 percent sucrose (table sugar). The rest of the sap consists of water, other nutrients produced in photosynthetic tissue, and messenger molecules such as plant hormones and RNA. The sieve plate at the end of each sieve tube member has more and bigger pores than the sides of the sieve tubes. Fluid therefore mainly flows lengthwise through a sieve tube, as if it were a pipe, rather than in and out through the sides.

Although sieve tube members are living cells, they are specialized for carrying sap, and they lack nuclei, ribosomes, and vacuoles. Without nuclei and ribosomes, they are incapable of synthesizing the proteins they need to live, so next to each sieve tube member is a **companion cell—** a fully functional cell that regulates what materials enter and leave the phloem and supplies the sieve tube member with proteins and energy-rich molecules (Table 30-2 and Figure 30-6). The companion cell is connected to the sieve tube member by cell junctions called **plasmodesmata,** channels of cytoplasm that connect adjacent cells through small pores. We'll learn more about how phloem works in Chapter 31.

Besides sieve tube members and companion cells, phloem also includes parenchyma cells, sclerenchyma fibers, sclereids, and other cell types, depending on the plant species. The cell walls of the sieve tube members do not contain lignin to strengthen the walls.

broader, and less tapered than tracheids. Vessel elements connect end to end to form long open channels called vessels (Table 30-2). The end wall of each vessel element is perforated with large holes, like those in the colander we use to drain spaghetti. These holes are wide open, and water flows

Xylem, which carries water and dissolved minerals, usually consists of two kinds of nonliving cells—narrow tracheids and wider vessel elements. Phloem, which carries sugary sap, consists of living cells called sieve tube members and companion cells.

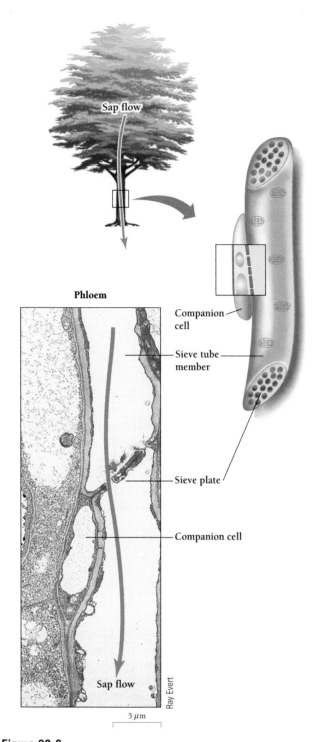

Figure 30-6

The living cells of phloem. Phloem differs dramatically from xylem in consisting of living cells. Like xylem cells, phloem cells come in two types. Sieve tube members are short and wide with flat ends, like vessel members. Modified plasmodesmata between adjacent sieve tube members allow sugar to pass from cell to cell. Nearby companion cells load sugars in and out of the sieve tube members and probably supply the sieve tube members, which lack nuclei, with energy and useful molecules.

How Do Eudicots and Monocots Differ in Structure?

Of the more than 300,000 named species of terrestrial plants now living on Earth, more than 90 percent are flowering plants, or angiosperms. Botanists traditionally divided the angiosperms into two groups based on their early development. When a seed first germinates, it has a root and a shoot. The very first simple leaves are called **cotyledons,** or seed leaves (Figure 30-7A). Some flowering plants, called **monocots,** have one cotyledon, while other flowering plants, called **dicots,** have two cotyledons.

All monocots evolved from a common ancestor, but DNA evidence suggests that the dicots evolved more than once. Most familiar dicots fall into a single lineage called the **eudicots.** But a few dicots seem to have evolved separately, including water lilies, star anise, and a handful of other flowering plants.

Monocots include all of the grasses, lilies, palms, and orchids—about 65,000 mostly herbaceous (nonwoody) species in all. Eudicots include almost all trees and shrubs (but not conifers and other gymnosperms, which are not flowering plants), as well as thousands of nonwoody plants, or herbs.

Monocots and dicots differ in the arrangement of vascular tissues within the shoots and roots, the number of flower parts, and the structure of the root system. The leaves of monocots and eudicots have characteristic patterns of large, visible veins (Figure 30-7B): in monocots these veins generally run parallel to one another (as in the leaves of corn and other grasses), while those of a eudicot usually form a branching pattern (as in a tomato leaf or a maple leaf). The flower parts of a eudicot usually come in multiples of four or five, while those of a monocot come in multiples of three (Figure 30-7D).

Inside a monocot's stem, the vascular bundles lie scattered throughout the stem, while those of a eudicot form a ring (Figure 30-7C). In many monocot roots, the vascular system includes a central nonconducting core, made of ground tissue. Inside the vascular cylinder of a eudicot's root, xylem and phloem often form a star shape (in cross-section) that runs the length of the root.

Monocots and eudicots differ also in the structure of their root systems (Figure 30-7E). Monocots generally have a delicate system of many small roots branching from the stem. These tiny, branching roots often form a thin, fibrous mat, familiar to anyone who has dug into a grassy lawn or field. In contrast, most eudicots produce one or several large roots, with much smaller roots branching off. In many eudicots, the entire root system connects to a single vertical root, called a taproot. A carrot is a taproot that has enlarged to serve as a storage organ.

Most flowering plants fall into two morphological groups that have characteristic differences in their embryos, flowers, leaves, roots, and stems: the monocots and the eudicots.

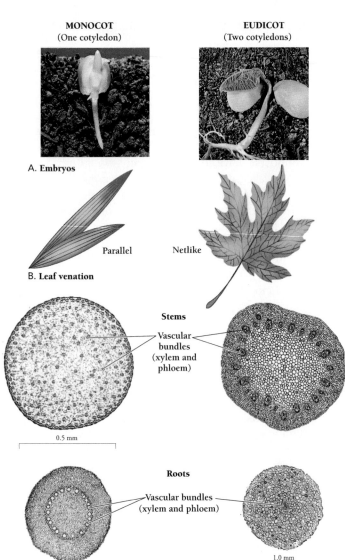

MONOCOT
(One cotyledon)

EUDICOT
(Two cotyledons)

A. Embryos

Parallel

Netlike

B. Leaf venation

Stems

Vascular bundles (xylem and phloem)

0.5 mm

Roots

Vascular bundles (xylem and phloem)

0.5 mm

1.0 mm

C. Stems and roots

D. Flowers

Fibrous

Taproot

E. Roots

Figure 30-7
Characteristics of monocots and eudicots. A. The monocot embryo (corn) has a single cotyledon, while the eudicot (bean) has two cotyledons. B. Monocots have parallel veins in their leaves, while eudicots have a more netlike arrangement of veins. C. Cross-sections through the stem and root of a corn plant (a monocot) and a buttercup (a eudicot). In monocots, vascular bundles in the stem are scattered throughout the ground tissue of the stem, but form a tight cylinder in the roots. In eudicots, the vascular bundles cluster near the outer surface of the stem, but crowd to the very center in the roots. D. The flower parts of monocots, such as this iris, as well as lilies, corn, and other grasses, come in groups of three. The flower parts of eudicots, such as this buttercup, more often come in groups of four or five. E. Monocots have masses of small roots. A eudicot may have a single, large taproot, with smaller branch roots. (A, left, Barry L. Runk/Grant Heilman Photography, right, Runk/Schoenberger from Grant Heilman; C monocot stem, © Dwight Kuhn, others, Runk/Schoenberger from Grant Heilman; D, left, John Gerlach/ Visuals Unlimited, right, John Gerlach/Tom Stack & Associates)

30.2 How Do Plants Defend Themselves Against Herbivores?

Plants are good to eat and they can't get away. Since the evolution of the first herbivores, plants have been fighting to stay alive. Bacteria eat plants, protists eat plants, fungi eat plants, and animals eat plants. Even other plants parasitize plants. How do plants defend themselves against this onslaught of hungry organisms?

Mechanical Defenses

Plants have numerous ways of preventing other organisms from taking a bite. One defense is the plant's waxy cuticle, which excludes both bacteria and fungi fairly well. Hairy leaves or very sticky leaves discourage caterpillars and other tiny herbivores (Figure 30-8). Spines (modified leaves), thorns (modified branches), and prickles (outgrowths of the stem) fend off larger mammalian herbivores.

Chemical Defenses

Plants' most effective defenses are probably not physical but chemical. Over hundreds of millions of years, plants have evolved a pharmacopoeia of chemical defenses against plant-eating animals. Most of these compounds are **secondary plant compounds,** those that do not participate

Figure 30-8
Mechanical defenses. Leaf hairs protect delicate plant epidermis from excessive sun and hungry insects, as well as from drying out and other dangers. Some leaf hairs secrete tarlike substances that make the leaf too sticky for caterpillars to walk on and too bitter for larger herbivores to eat.

directly in growth and development of the plant. "Primary compounds" include a mere handful of compounds that are essential for plants' day-to-day functioning, such as glucose, amino acids, proteins, ATP, and DNA. In contrast, secondary compounds are chemically and functionally diverse, and often unique to small groups of plants or even to a single species.

Plants manufacture thousands of defensive compounds, but most fall into three groups: terpenes, phenolic compounds, and alkaloids. The largest group are the lipid (fatty) **terpenes.** For example, pyrethroids from *Chrysanthemum* leaves are terpenes commonly used in insecticides. Peppermint, lemon, basil, and sage all derive their aromas from mixtures of terpenes called essential oils that repel or poison insects. Many plants store these repellent and toxic oils in glandular hairs on the surface of the epidermis, where the oils advertise the plant's toxicity to arriving insects. Cotton plants derive much of their insect, bacterial, and fungal resistance from the terpene gossypol, a toxic yellow pigment (under investigation as an anticancer agent).

Phenolic compounds are aromatic substances whose molecular structure includes a flat carbon ring (Chapter 3). Some phenolics repel herbivores, while others attract insect pollinators. Lignin is a phenolic that does double duty; it is a major constituent of plant cell walls (that provides both strength and waterproofing) that also makes plants indigestible to both animals and microscopic pathogens. Algae also make lignin, and biologists think that lignin is an ancient compound, whose original function, to repel pathogens, has expanded to include mechanical support for terrestrial plants. Some plants secrete phenolic compounds that poison any plants growing nearby. Desert plants, especially, resort to this unneighborly practice to eliminate competition for precious water.

Another class of phenolic compounds are the tannins, commonly found in woody plants. Like lignin, tannins reduce growth and survivorship in plant eaters. Cattle, deer, humans, and apes avoid eating unripe fruits and other plant parts with high levels of tannins because of the tannins' sharp, astringent taste. Yet, we humans prefer a bit of tannin in our tea, apples, blackberries, and red wine. Finally, the pigments responsible for most of the red, pink, purple, and blue colors of flowers, fruits, leaves, and other plant parts, called anthocyanins, are also phenolic compounds.

Alkaloids, found in 20 to 30 percent of all vascular plants, include many of the most powerful drugs known to humans. The list includes nicotine (the highly addictive active ingredient in tobacco), cocaine, morphine (from which heroin is made), the hallucinogen psilocybin, and the toxin strychnine, not to mention caffeine and theobromine (the stimulant in chocolate). At low doses, alkaloids such as caffeine, nicotine, cocaine, and morphine are used widely as stimulants and sedatives, for both medical and nonmedical purposes. But at high doses, many alkaloids are extremely toxic. Alkaloids in lupines, larkspur, poison hemlock, and other plants can kill livestock and humans ignorant enough to eat them. And cattle or goats that graze on such toxic plants sometimes give birth to offspring with dramatic birth defects, as may human mothers who drink the animals' milk.

Other secondary plant compounds besides the terpenes, phenolics, and alkaloids include the **mustard oil glycosides.** These compounds give cabbage, broccoli, radishes, and other plants in the mustard family their characteristic pungent odor and flavor. Many insects and mammals are repelled by mustard oil glycosides. But others have learned to cope with these compounds, and some insect herbivores specialize on plants in the mustard family (Figure 30-9). For these insects, the mustard oil glycosides act as an attractant. Insect herbivores that specialize on noxious plants avoid competition with other herbivores. Many compounds that originally evolved as repellents to herbivores now act as attractants for certain groups of animals.

Many secondary plant compounds repel or poison insects and other herbivores.

In this chapter we have briefly seen how the standard root-and-shoot form of a terrestrial plant allows the plant to collect water and minerals from the soil and to collect solar energy and carbon dioxide from the air. We have briefly discussed how the different tissues support plants against the pull of gravity and transport water and nutrients. Plants use secondary plant compounds to strengthen their bodies and protect themselves from herbivores and pathogens. In the next chapter, we discuss how plants raise water from the roots to the tops of trees that may be hundreds of feet tall.

Figure 30-9
Attractive insect repellent. Most insects avoid plants in the mustard family because the leaves contain unpleasant compounds. But cabbage butterflies prefer cabbage, broccoli, and other mustards.

Key Concepts

- Most flowering plants rely on underground roots to obtain water, minerals, and aboveground shoots to obtain carbon dioxide, and sunlight.
- A plant's vascular system consists of xylem, which carries water and minerals absorbed by the roots, and phloem, which carries sugars and other nutrients produced by photosynthesis.
- Plants produce secondary plant compounds that protect plants against herbivores, pathogens, and other plants; provide structural support; and/or attract animal pollinators and seed dispersers.

Summary with Key Terms

What physical constraints shaped the evolution of vascular plants?

Despite the enormous diversity of **vascular plants,** they share characteristic organizational features. Most consist of an underground **root system** and an aboveground **shoot system.** The shoot system usually includes stems, leaves, and flowers.

Materials travel between shoot and root through the plant's **vascular system,** a network of two kinds of transport tissues. The **xylem** transports water and minerals from root to shoot, and the **phloem** transports the energy-rich products of photosynthesis from leaves or storage areas to the rest of the plant.

Flowering plants fall into two morphological groups—the **monocots** and the **dicots**—that have characteristic differences in their embryos, flowers, leaves, roots, and vascu-

lar systems. The embryos of monocots have one **cotyledon** and the embryos of eudicots have two. Most dicots are **eudicots,** but a handful evolved separately.

What kind of tissues make up vascular plants?

In both classes of plants, roots, stems, and leaves consist of three types of tissue systems. The **dermal tissue** is the protective outer covering, including the **cuticle,** the **epidermis,** and its many **stomata.** The **vascular tissue system** lies inside, surrounded by the **ground tissue system.** Roots, stems, and leaves differ in the arrangement of vascular system and ground tissue. Tissues may consist of **parenchyma, collenchyma, sclerenchyma,** and other cell types.

Plants grow from undifferentiated cells that form **meristem** tissue. The **primary growth,** or lengthening, of a plant occurs only in specialized regions of dividing cells, called **apical meristem,** which lie at the tips of roots and shoots. **Secondary growth** is an increase in diameter of shoot or root that occurs in a cylinder of dividing cells called a **lateral meristem.**

Plant cells have **cell walls** made of **cellulose,** other carbohydrates, and proteins. All plant cells have a **primary cell wall** outside the plasma membrane. Some have a thicker **secondary cell wall** between the primary wall and the membrane. The polymer **lignin** strengthens many plant cell walls. Most of the internal space of a plant cell is occupied by a **central vacuole** that is filled with a watery fluid that presses against the cell wall. This water pressure in the vacuole of each cell gives plants an overall stiffness that allows them to resist the pull of gravity.

Xylem consists of two types of conducting cells—long, tapered cells, called **tracheids,** and shorter, flat-ended cells, called **vessel elements.** The xylem functions only after the cells die, leaving behind empty cell walls. Water and minerals move through the tracheids from dead cell to dead cell by way of pairs of **pits.** Water moves through large holes in the ends of the vessel elements. The **sieve tubes** of phloem are made of living **sieve tube members** connected by **plasmodesmata** to their **companion cells.** Each end of a sieve tube member is a **sieve plate** with holes in that allows **sap** to pass from cell to cell.

How do plants defend themselves against herbivores?

Plants defend themselves against herbivores with both mechanical and chemical defenses. Mechanical defenses include the waxy cuticle, sticky or prickly hairs, as well as larger prickles and thorns. **Secondary plant compounds** are chemicals not needed for the major metabolic pathways of a plant. Many secondary plant compounds repel or poison insects and other plant-eating animals. Plants manufacture thousands of secondary plant compounds, but most fall into three groups: **terpenes, phenolic compounds,** and **alkaloids.** Alkaloids, such as nicotine and cocaine, are nitrogen-containing secondary plant compounds found in 20 to 30 percent of all vascular plants. Other secondary plant compounds besides terpenes, phenolics, and alkaloids include the **mustard oil glycosides.**

Review and Thought Questions

Review Questions

1. What adaptations do terrestrial plants have that aquatic plants do not?
2. How is the root necessary for photosynthesis? How is the shoot necessary for photosynthesis?
3. What is meristem tissue and why is it important?
4. In what ways do monocots differ from eudicots?
5. Name the three tissue systems of plants.
6. Describe how companion cells interact with sieve tube members.
7. Is bread made from plants that are monocots or eudicots? Name six other plants you know that are monocots and six that are eudicots.
8. Name three general roles that secondary plant compounds can play in the life of a plant.
9. What does the word "meristem" mean literally and how does that meaning relate to what it is and does?

Thought Questions

10. Certain "prostrate" plants spread out over the ground without growing very tall. These plants can maximize their exposure to sunlight. Why don't all plants conserve energy and adopt this form? In other words, what are the advantages of growing tall?
11. When Bruce Ames first published his papers on the carcinogenicity of secondary plant compounds, many people felt that he was behaving irresponsibly. They worried that the public would misunderstand and (1) not bother to eat fruits and vegetables and (2) not bother to avoid toxins that are genuinely dangerous. Some of his critics thought that Ames was endangering the lives of innocent lay people. Do you think Ames's behavior was irresponsible? Are scientists more obligated to publish what they know is true or more obligated to limit what they publish, according to how the public will likely respond to such information?

BiologyNow Resources

Biology Now™

Active Figures

30-1: Photosynthesis and water flow in flowering plants
30-3: Cell types and tissues in plants

Preparing for an exam? Take a diagnostic test on your BiologyNow CD-ROM.

Online materials relating to this chapter are at:

http://biology.brookscole.com/AAL3

About the Chapter-Opening Image

This bunch of carrots reminds us that all terrestrial plants need roots to pull water and minerals from the soil and also that plants make powerful compound. Some, such as the carotene that carrots make, are good for us, but some are toxic.

31

How Do Plants Move Water and Sugars?

Key Questions

- How do terrestrial plants obtain water and minerals from the soil?
- What drives water up from the roots to the topmost leaves of a plant?
- How do a plant's stomata regulate water movement?
- How do plants transport sugar?

The Devastating Dry Leaf Creature

In the 1980s, disaster threatened the burgeoning California wine industry. For years, grape growers had been planting a kind of grapevine that had only weak resistance to a devastating insect pest. Ignoring the advice of European grape experts, California vineyard owners had planted 50,000 acres of fertile farmland with vines that were vulnerable to *Daktulosphaira vitifoliae,* a tiny sucking insect that feeds on roots and leaves. The insect's common name, phylloxera, means "dry leaf" in Greek [*phyllos* = leaf + *xeros* = dry], which is also the name of the disease they cause.

It was only a matter of time before these tiny relatives of aphids developed a way to overcome the vines' feeble resistance. And by the late 1980s, phylloxera was destroying whole vineyards. The pattern was always the same; the leaves of a few plants would dry out, turn red, and drop from the vines in midsummer. Within months, the rootstock died and the pest spread to nearby plants.

It wasn't as if California grape growers hadn't heard of phylloxera. In the 19th century, the infamous pest had devastated the European wine industry. In France alone, the insect destroyed some two and a half million acres of vineyards—a national disaster. Nor was the invasion confined to Europe. The tiny pest infested vineyards all over the world, including those in California. Indeed, 100 years ago, the California State Legislature directed the new University of California to create a department of viticulture ["grape growing"]—specifically to study phylloxera.

Don Ellis

The cure for the 19th-century Great Wine Blight, as it came to be called, came from a combination of American science and French technology. The American entomologist C.V. Riley worked out the complex life cycle of phylloxera, showing that it lived on wild grapes of eastern North America, and ironically, was accidentally introduced to France in 1860. The wild American grapes were adapted to the pest and flourished despite the phylloxera feeding on their roots. By the turn of the century, French viticulturists had put two and two together, solving the phylloxera problem in Europe by grafting high-quality French wine grapes onto wild American roots, or "rootstock." The French hailed Riley as a savior for helping to save their wine industry.

Figure 31-1
Phylloxera. These tiny insects attack the roots of grapevines, leaving the damaged root tissue susceptible to infection by bacteria and fungi.

How do phylloxera kill grapevines? The insects begin by clustering on the roots and piercing the outer layers of the root with their tiny mouthparts. Something in their saliva, some combination of enzymes and possibly even plant hormones, induces the roots to form a gall (a tumorlike ball of tissue) around the insects. Inside this protective gall, the insects safely feed, digesting the tissues of the roots (Figure 31-1).

Phylloxera do not take enough energy and materials from the vine to kill it directly. But the thousands of insects cutting into the roots make the plant susceptible to infections by fungi and bacteria. These "secondary infections" invade and destroy the vascular system in the roots. Unable to transport water and nutrients, the vine dies.

Nineteenth-century grape growers developed and tested thousands of different rootstocks. Some did well in the cool, wet climate of northern Europe. Some grew better in warm, dry soil. A few resisted phylloxera. Phylloxera-resistant rootstocks were found to secrete tannins and other substances that walled off the root galls and so prevented infections by bacteria and fungi. One rootstock that had only moderate resistance to phylloxera but produced heavy loads of fruit was named AXR1. It was tempting to use it.

Initially, the phylloxera left AXR1 alone. But first in Sicily, and then in France and Africa, grapes grown on AXR1 rootstocks that had flourished for a decade or two began to wither and die. In the 1920s scientists began advising growers against using the rootstock.

By then, however, Americans were in the midst of Prohibition, the 1919 law that prohibited the manufacture and sale of all alcoholic beverages—including wine. Phylloxera was the least of California grape growers' problems. They dug up their sprawling vineyards and planted apricots, peaches, and plums, or sold the fertile land for housing developments. As grapevines disappeared from California, phylloxera all but disappeared both from the land itself and from people's minds. When Prohibition ended in 1933, Cal-

ifornians began replanting vineyards. But they knew little about phylloxera.

In the 1950s, a researcher at the Davis campus of the University of California tested 18 rootstocks. Three of them yielded large crops of fruit and had other desirable traits for California vineyards. For simplicity, the University's Agricultural Extension Service chose to recommend just one rootstock, AXR1.

In the 1970s, American wine lovers recognized for the first time that California wines were surprisingly good and far cheaper than French wines. The California wine industry boomed. Soon everyone from retired dentists to multinational corporations was planting vineyards. Growers around Napa and Sonoma counties, the finest wine-producing area in the United States, planted nearly three-quarters of their vines on AXR1 rootstock. If California's phylloxera adapted to AXR1, or if the European strain of AXR1-munching phylloxera found their way into the state, the insects would eventually feast on nearly 50,000 acres of vines.

In the summer of 1983, the first sign of trouble appeared. Austin Goheen, a plant pathologist (an expert on plant diseases) from the University of California, Davis, was called out to a vineyard to inspect some mysteriously dying grapevines in Napa County. Goheen immediately recognized the telltale yellow colonies of tiny phylloxera insects on the roots of the stricken plants, but he couldn't be sure they were killing the vines. Small populations of the insects still existed in many regions of California, and almost every kind of rootstock harbored at least a few of the troublesome pests. Was this a new virulent strain of phylloxera that was sapping the life out of the roots, or was something else doing the damage?

Not until late in 1985 could researchers state with certainty that the vines harbored a new strain of phylloxera, one that could attack and kill AXR1 vines. But even then, many questions still remained unanswered. How fast would these new, virulent creatures spread? Should growers who lived miles from an infection site tear out their vines now, or wait until the insects moved closer? Would this new insect strain ravage vines in other grape-growing regions of California where soils and climate were different?

The researchers didn't know. Should they recommend that growers stop using AXR1? Tearing out the vines and planting new ones would cost the grape industry millions of dollars. The researchers felt they needed more information before issuing such an extreme warning. After all, they had found the insects in only a few vineyards. They studied and watched the phylloxera for another three years, while the insects spread relentlessly from site to site and the increasingly panicked growers clamored for guidance.

In 1989, the university finally issued a press release warning growers to stop using AXR1. Many growers immediately began tearing out their AXR1 vines and replacing them. But no one knew which rootstock to use instead. Some growers used four or five different rootstocks in a single vineyard, hoping some of it would withstand phylloxera.

The change was costing the wine industry a fortune, and many growers were furious with the university, blaming the researchers for recommending AXR1 in the first place. And why, they demanded, hadn't the university made its announcement earlier? Nurseries, stuck with millions of dollars worth of AXR1 they wouldn't be able to sell, threatened lawsuits if the researchers proved wrong.

Despite all the finger pointing, everyone shared some blame in the disaster. Growers and researchers alike knew that other nations didn't use AXR1. But as long as the rootstock had continued to do well in California, no one wanted to think about it or to pay to study it. The wine industry provides the funds for most research in viticulture, and it decides what kind of research to fund. As Jim Wolpert, the University of California's chief viticulturist, said, "If I'd put in a proposal in 1982 to the winegrape industry for funding to work on AXR1 and phylloxera, they would have said I was crazy." Not until phylloxera started killing vines did these minute creatures receive the attention they deserved. And by then scientists lagged far behind in their race to save the state's vineyards.

Phylloxera continues to infect northern California's vineyards. But newer rootstocks can wall off the tissue-piercing insects, so water coming into the roots flows smoothly up to the leaves and grapes in the hot sun above. And most vineyards are flourishing despite the insect's presence.

In this chapter, we'll see how this flow of water is as necessary to a plant's life as the flow of blood is to each of us. The flow of water both sustains photosynthesis and carries minerals and nutrients from the roots to leaves and from leaves to roots. In the first part of this chapter we will see how plants obtain water and minerals from the soil and what forces drive water and minerals from the roots upward. Later, we will see how sugars manufactured in the leaves are transported down to the roots and other parts of the plant.

31.1 How Do Terrestrial Plants Obtain Water and Minerals?

In a healthy plant, water and dissolved minerals flow from the soil into the roots, then move up through the stem to the plant's leaves. Some of the water contributes hydrogen atoms for making sugar during photosynthesis. But more than 90 percent of the water exits the leaves as water vapor and drifts away into the atmosphere.

Plants need other materials besides water from the soil and carbon dioxide from the air. To make amino acids and nucleotides, for example, plants need large amounts of nitrogen, sulfur, and phosphorus; smaller amounts of magnesium, calcium, and potassium; and tiny amounts of molybdenum, copper, zinc, manganese, boron, iron, and chlorine. Plants obtain these nutrients almost entirely as ions dissolved in the water absorbed by their roots. Roots are the major suppliers of both water and essential minerals.

Roots also anchor the plant to the ground, store carbohydrates, and produce several hormones that regulate growth. Finally, roots are important in *vegetative reproduction,* the process by which plants reproduce without sex (no flowers, fertilization, or seeds).

How Do Roots Transport Water?

Most of the water that enters a root comes through **root hairs,** each of which is the extension of a single epidermal cell (Figure 31-2). Huge numbers of root hairs form just behind the root tips, greatly increasing the surface area of the epidermis. A single rye grass plant, for example, may have 10 billion root hairs with a total surface area of more than 400 square meters, as much as the walls of a small house. The epidermis and the root hairs are coated with a slimy substance called **mucigel** that lubricates the root tips as they force their way between soil particles. Water enters the root hairs and flows into the cortex.

Roots vary greatly in their external form, but they have a common internal organization. A cross-section of a eudicot primary root shows the epidermis on the outside, the large cells of the ground tissue beneath the epidermis, and the vascular system in the center (Figure 31-2).

Most of a root's ground tissue is **cortex,** a tissue consisting of large, thin-walled parenchyma cells, which in some plants store starch—most dramatically in sweet potatoes and carrots. Water flows through the cortex to the **endodermis** [Greek, *endon* = within + *derma* = skin], the innermost layer of the cortex, which regulates the flow of water into the vascular system.

Water flows from the soil to the vascular system by two paths (Figure 31-3). Both paths flow through the cortex to the endodermis. The first path, called the **apoplast,** flows through the material of the cortex's cell walls. Water (and some dissolved minerals) soak into the cell walls much as they soak into the cellulose fibers of paper towels. The second path, the **symplast,** flows right through the cells of the cortex. Water and minerals pass directly from the cytoplasm of one cell to the cytoplasm of the next by way of connecting plasmodesmata.

Most minerals arrive at the endodermis by the way of the symplast rather than apoplast. In the root hairs, cell membranes actively transport ions from the soil into the cytoplasm of the epidermal cells. The dissolved ions then move (along with water) through plasmodesmata into the cells of the endodermis, surrounding the vascular tissue.

The endodermis (shown in purple in Figure 31-3) is key to a root's ability both to absorb water and to retain minerals. Each cell of the endodermis secretes into its walls a waxy, waterproof substance that makes the cell walls as waterproof as waxed paper (Figure 31-3). The boxlike cells have six sides, just like a cardboard box, but only four sides are waterproof. The other two sides have normal cell walls and membranes that control the flow of water and minerals through the endodermis. The barrier of waxy cells forms a waterproof cylinder within the root, called the **Casparian strip,** that prevents minerals and water from passing from

A.

Figure 31-2

Roots. A. Grasses grow dense masses of roots covered with tiny root hairs. Roots of Italian rye grass *(Lolium multiflorum)*. Root hairs increase the surface area of roots, so that plants can take up more water and minerals. B. A cross section through a root *(Smilax sp.)* shows the three main parts: epidermal tissue (outermost, orange ring), ground tissue (wide dark ring), and central vascular tissue (xylem and phloem).

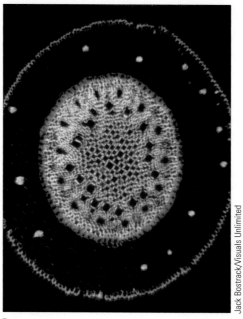

B.

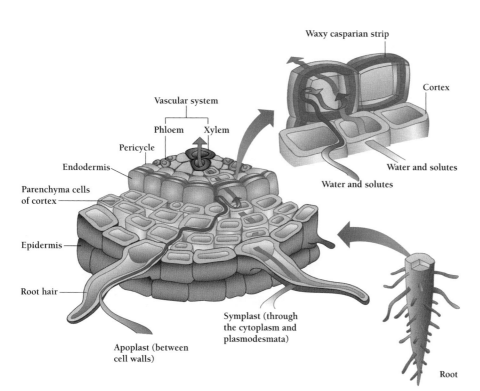

Figure 31-3

The two paths of water through the roots. Fluid in the *symplast* travels through the cytoplasm and plasmodesmata of the cells of the cortex. Water in the *apoplast* passes only through the extracellular spaces, including the cell walls, stopping at the endodermis. The endodermis is a waterproof layer of cells that surrounds the vascular cylinder. The Casparian strip prevents movement of water and dissolved minerals between the vascular cylinder and the cortical apoplast. Water and minerals can cross the endodermis only by entering endodermal cytoplasm, allowing the membranes of the endodermis to regulate the flow of solutes and water.

the apoplast into the center of the root. For water and minerals to move from the cortex to the vascular system, they must pass through the cell membrane and the cytoplasm of the cells of the endodermis. The Casparian strip and the rest of the endodermis also keep nutrients in the xylem from flowing back to the soil and undesirable molecules in the soil from entering the xylem.

Inside the cylinder formed by the endodermis is a sheath of lateral meristem, called the **pericycle** [Greek, *peri* =

around + *kykos* = circle], which gives rise to branch roots. Inside the pericycle are the phloem and xylem of the vascular system (Figure 31-3).

Water from the soil enters the root by way of root hairs, travels through the cells of the cortex, and then passes through the endodermis into the vascular tissue.

Stems Transport Water, Minerals, and Nutrients Between Roots and Leaves

The water that enters the root eventually moves up the xylem into a stem. Stems support leaves, flowers, fruits, and seeds. Stems also elevate all these organs toward pollinators, seed dispersers, and the sun; store nutrients and water; and preserve perennial plants through the stress of winter or summer. Some stems are photosynthetic.

As in the root, the stem's epidermis is on the outside, with the vascular system embedded in the ground tissue. The stem's epidermis differs from that of a root in three respects—(1) the absence of root hairs, (2) the presence of occasional stomata, and (3) the presence of a thick cuticle. The cuticle, which encases the stem and prevents water loss, is a waterproof layer made of wax and *cutin*, a polyester similar to that used in clothing.

In eudicots, the stem's epidermis encloses a ring of cortex (ground tissue) that itself surrounds a core of vascular tissue (Figure 31-4). We usually think of stems as long and thin, and the stem's cortex is often thin, sometimes no more than a few millimeters. But in some cacti, the cortex can be 20 to 30 cm thick. The outer parenchyma cells of the cortex, particularly in young stems, may be green and photosynthetic. The stems of most nonwoody plants do not have an endodermis or a pericycle.

What Do Leaves Do?

The main role of leaves is to make food through photosynthesis. Occasionally, leaves play other roles, including camouflaging a plant, protecting buds and flowers, and storing food for the embryo or young plant (Figure 31-5). But most leaves are organs of photosynthesis, and their most obvious characteristic is that their cells are packed with bright green **chloroplasts**—organelles that photosynthesize. The green color comes from the pigment chlorophyll, which captures light energy during photosynthesis.

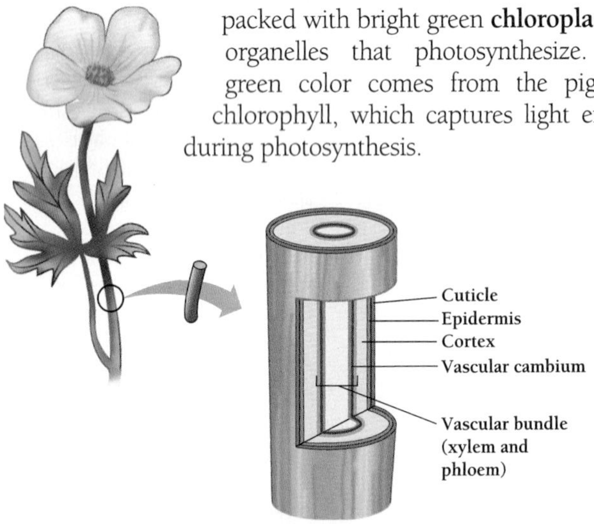

Figure 31-4

Anatomy of a stem. Shown in this eudicot stem are the outer cuticle and epidermis; the ground tissue, which is made of cortex; and vascular tissue, made of xylem and phloem.

Cuticle
Epidermis
Cortex
Vascular cambium
Vascular bundle (xylem and phloem)

Figure 31-5

A diversity of leaves. Leaves may be cotyledons, sepals, bud scales, or floral bracts. As organs of photosynthesis, they may be large or small, flat or round, waxy or hairy. Shown here are (A) the flowerlike leaves of the poinsettia, (B) the camouflaged leaves of the stoneflower, also known as stone plant (Genus *Lithops*, from Namibia), (C) pine needles, and (D) the bud scales and new leaves of a horse chestnut. (A, Lefever/Grushow from Grant Heilman; B, Kjell B. Sandved/Photo Researchers, Inc.; C, David Sieren/Visuals Unlimited; D, W. Ormerod/Visuals Unlimited)

Most leaves have a thin, flat part, called the **blade,** and a stalk, called a **petiole,** that connects the leaf to the stem (Figure 31-6). The arrangement of leaves on their stems, as well as the sizes and shapes of leaves, influences their ability to capture light. Many plants move their petioles to position their leaves, so that the blades of each leaf move out of the shade of other leaves.

In addition to their central role in photosynthesis, leaves also help move water from the roots through the xylem. One of the most obvious characteristics of leaves is a network of **vascular bundles,** commonly called veins, made of **xylem** and **phloem** (Figure 31-6). From the stem, the veins in the leaf branch into smaller and smaller veins. The xylem brings water to the leaf, and the phloem takes sugars back to the rest of the plant. Every part of a leaf is no more than a few cells away from the vascular system. Veins also act as mechanical struts, which, like the stiff metal spokes of an umbrella, help to keep the blade extended flat.

A leaf's epidermis is usually a single layer of cells covered with a layer of cutin and waxes. Plants that have evolved in dry climates usually have a thicker epidermis that slows water loss. The epidermis of many leaves also includes special-

Figure 31-6

Backlit maple leaves. These leaves from the maple tree *Acer rubrum* show many of the characteristics of leaves, including: a wide flat surface that increases the leaf's ability to gather light for photosynthesis; veins that carry water and minerals from the roots to cells in the leaf and sugars made in the leaves to other parts of the plants; and a petiole that joins leaf to stem.

The ground tissue of the leaf consists of **mesophyll** [Greek, *mesos* = middle + *phyllon* = leaf], green parenchyma cells that are responsible for most of a plant's photosynthesis. Most eudicot leaves have two kinds of mesophyll—**palisade parenchyma** and **spongy parenchyma** (Figure 31-7). The cells of the palisade parenchyma, which lie just under the leaf's upper epidermis, are elongated and packed with chloroplasts. Below

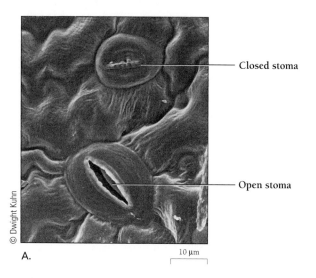

Closed stoma

Open stoma

A.   10 µm

ized hair cells, called trichomes, that have a variety of shapes and functions. Some are so thick and fuzzy that they shade the leaf from intense light; others prevent water loss or defend the plant against plant-eating animals.

To exchange gases with the atmosphere, leaves contain specialized pores, called **stomata** [singular, stoma; Greek, = mouth], through which carbon dioxide, oxygen, and water vapor enter and exit the airy interior of the leaf (Figure 31-7). The distribution of stomata on the leaf varies depending on the environment. Aquatic plants with floating leaves have stomata only on the upper surface. But the leaves of terrestrial plants are different; the lower surface of a leaf may contain tens of thousands of stomata per square centimeter, while the upper surface usually has few or even none.

Surrounding and defining each stoma are two *guard cells*—specialized epidermal cells that open and close the stoma. Stomata open when a plant needs carbon dioxide and can afford to lose water vapor through its leaves. Stomata close when the plant needs to conserve water. Stomata can also regulate a leaf's temperature. Just as an animal can cool itself off by sweating, an overheated leaf can cool itself off by opening its stomata more widely and allowing more water to evaporate.

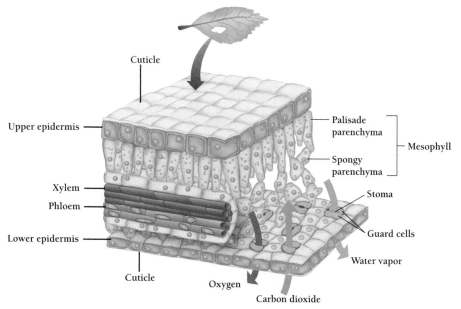

Cuticle

Upper epidermis

Xylem

Phloem

Lower epidermis

Cuticle

Palisade parenchyma

Spongy parenchyma

Mesophyll

Stoma

Guard cells

Water vapor

Oxygen

Carbon dioxide

B.

Figure 31-7

The paths of oxygen, carbon dioxide, and water vapor in the leaf.
Leaves consist of upper and lower epidermis enclosing mesophyll tissue that carries out photosynthesis and through which run the vascular bundles, or veins. A. Stomata in the surface of the leaf open and close to regulate the flow of water vapor and carbon dioxide in and out of the leaf. B. Inside the leaf, water passes from the xylem into the parenchyma cells, out to the layer of liquid water coating the cells, evaporates and then diffuses through the stomata. The net flow of water vapor is out of the leaf. Carbon dioxide in the atmosphere enters the plant through the stomata and is taken up by photosynthetic parenchyma cells. The net flow of carbon dioxide is into the leaf.

them are the rounder cells of the spongy parenchyma. The spongy parenchyma cells are only loosely associated with one another. Large air spaces separate the cells and allow the passage of carbon dioxide, oxygen, and water molecules.

Like the cells that line our lungs, these parenchyma cells are coated with a thin layer of water, which aids the diffusion of gases. Carbon dioxide entering a leaf through the stomata diffuses into the parenchyma cells. Water and oxygen leave the parenchyma cells and diffuse to the outside air, by way of the stomata. The shiny layer of water inside the leaf also creates many bright, reflecting surfaces. Sunlight bounces back and forth within this tiny hall of mirrors, increasing the likelihood that a chloroplast will absorb a photon for photosynthesis.

The top layer of cells in a leaf always shades those below. To allow at least a little light to reach the bottommost cells, leaves are usually as thin as they can be without collapsing. But, despite this adaptation, the cells in the top layers of leaves still get more light. So they are specialized for light absorption, while those nearest the underside of the leaf are specialized for the exchange of gases and water.

Leaves, the primary photosynthetic organs of plants, consist of an outer layer of epidermis and a central region of mesophyll through which run the vascular bundles, or veins. The epidermis is dotted with stomata.

31.2 What Drives Water Up to the Tops of Tall Trees?

We know that water moves up from roots to leaves through the xylem: if we place a plant cutting in water that contains a dye, we can see the path the dyed water takes through the plant. If the stems were transparent, we would be able to see the water move, for during the day the water ascends rapidly, at about 10 to 100 centimeters per minute.

What Is the Route of Water from Soil to Air?

Water enters the roots by way of the root hairs, moves through the cortex, crosses the endodermis, and enters the xylem. Once in the xylem, water ascends from the roots to the leaves through parallel and usually interconnected xylem vessels in the stem. The water then enters cells of the mesophyll and evaporates from the thin film of water on cell surfaces, diffusing through the air spaces of the mesophyll and exiting the leaf by way of the stomata. This evaporation and the passage of water vapor from leaf to air is called **transpiration.**

The rate of transpiration may be extraordinary: a tree may transpire 50 gallons of water in an hour, up to 30,000 gallons during a growing season. In a plant's lifetime, it typ-ically loses a hundred grams of water through transpiration for every gram of dry material it accumulates.

These tremendous quantities of water must sometimes move great distances. The tallest trees in the world are the California coast redwoods, *Sequoia sempervirens.* These trees are so large that entire churches can be built from a single tree. The tallest redwood of all is 367 feet, the same height as a 29-story building, and nearly a third the height of the Empire State Building. To pump water to such a height would require a pump exerting 180 pounds per square inch of pressure. But plants have no central pump. The cells of the xylem are themselves dead, so we know that these empty hulls exert no force.

What pulls water up? Plant scientists early in the 20th century proposed three possible mechanisms for water transport: (1) water is pushed up from the roots; (2) capillary action in the xylem pulls water in the same way that a paper towel pulls water out of a glass; or (3) the evaporation of water from the leaves creates suction that pulls water up the xylem. But which mechanism was the right one?

Root Pressure: Can the Roots Push Water to the Tops of Tall Trees?

Water tends to move from areas of purer water to areas with higher concentrations of solutes (Chapter 4). Thus, sugary raisins and cytoplasm-filled red blood cells both fill with water when placed in pure water. The movement of water across a membrane in response to a difference in solute concentration is called **osmosis** [Greek, *osmos* = push, thrust].

In the roots, the xylem usually has higher concentrations of solutes than the water in the surrounding soil. Water from the soil therefore flows through the cortex, across the endodermis, and into the xylem. As water flows into the xylem, the water pushes water already in the xylem up the roots and into the stem. The pressure that develops in the root xylem as a result of the inflow of water is called **root pressure.**

In small plants, root pressure can push water all the way up the stem and out of tiny holes at the margins of the leaves, in a process called **guttation** [Latin, *gutta* = a drop]. Early in the morning, gardeners often see the leaves of grass, tomatoes, strawberries, and other plants rimmed with tiny drops of water (Figure 31-8). These drops, which are not dew, result from root pressure. Guttation occurs on cool, humid mornings, when the air is too saturated for the water drops to evaporate from the leaves.

How far can root pressure push a column of water? Researchers can directly measure root pressure by attaching a pressure-measuring device to a cut stem (Figure 31-9). Root pressures measured in this way are usually less than one atmosphere, enough to support a column of water about 10 meters (30 feet) high—impressive, but not nearly enough to move water to the top of a redwood tree. In any case, many plants (including redwoods and other conifers) do not develop any root pressure at all.

Angelina Lax/Photo Researchers, Inc.

Figure 31-8
Guttation demonstrates the action of root pressure. The droplets at the margins of these strawberry leaves are not dew drops, but water forced by root pressure from tiny holes in the leaf.

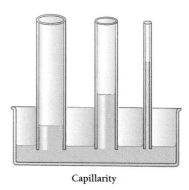

Capillarity

Figure 31-10
Is capillarity enough? Water rises higher in narrower tubes than in wide ones. In xylem, water rises by capillarity about half a meter, not nearly enough to get it to the top of a tall tree.

Another clue that showed plant physiologists that root pressure does not drive water to the tops of tall trees is that the water in the xylem is not usually under pressure. If we poke the xylem of a stem with a needle, water does not come spewing out, as if under pressure. Instead, air is sucked up into the xylem, indicating that the water in the xylem is being sucked up, as if by a vacuum.

Finally, root pressure moves water too slowly to account for the rapid transport of materials from roots to leaves (up to a meter per minute). The driving force for water transport must therefore come from some source other than root pressure.

The most pressure that roots can generate is enough to raise water some 30 feet, not the 300 feet or more of the tallest trees.

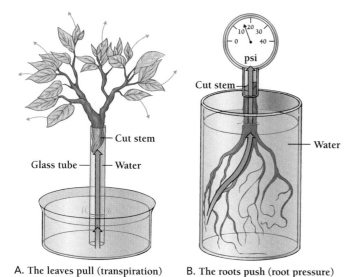

A. The leaves pull (transpiration)   B. The roots push (root pressure)

Figure 31-9
Root pressure. A. Cut flower stems pull water upward. B. Roots can also push water up through a cut stem. The pressure generated by the roots is a physical pressure that can be measured with a pressure gauge similar to that used to measure tire pressure.

Can Capillary Action Raise Water to the Tops of Tall Trees?

You can suck soda up a straw because water is **cohesive;** it sticks to itself very well, and columns of water do not break as they move up any narrow tube, or **capillary.** Water is also highly **adhesive,** having a tendency to cling to the surfaces of carbohydrates and other polar substances. Water's sticky adhesiveness is what causes it to soak into the cellulose of paper towels and the cell walls of the root cortex. Together, water's adhesiveness and cohesiveness cause it to move up the inside of most capillaries, including both glass tubes and xylem.

Could such capillary action drive water to the tops of tall trees? As it turns out, ordinary capillary action can pull water up only about half a meter, not nearly enough (Figure 31-10). Although root pressure and capillary action are not enough to drive water to the tops of tall trees, both effects are nonetheless important to water movement in plants, as we will now see.

Water's adhesiveness and cohesiveness enable it to move up capillaries, but not 300 feet up.

The Driving Force for the Ascent of Most Water Is Transpiration

If neither root pressure nor capillary action is enough to drive water up, then what does? The now-accepted explanation for the ascent of water is called the **cohesion-tension theory,** first suggested in 1914 by the Irish botanist Henry Dixon. The cohesion-tension theory states that transpiration in the leaves pulls water up the stem in continuous columns (Figure 31-11A). For decades after Dixon's proposal, many plant physiologists doubted that the theory was correct. Acceptance of Dixon's hypothesis rested on answering two questions. First, did transpiration generate enough tension to pull water up hundreds of feet from the soil? And, sec-

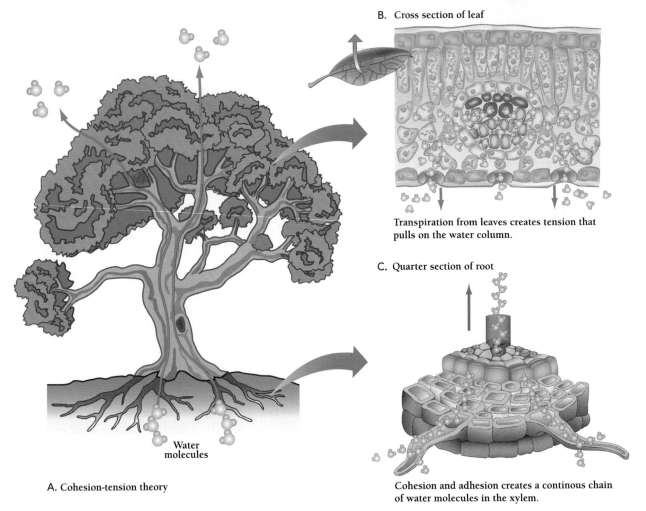

B. Cross section of leaf

Transpiration from leaves creates tension that
pulls on the water column.

C. Quarter section of root

Water
molecules

A. Cohesion-tension theory

Cohesion and adhesion creates a continous chain
of water molecules in the xylem.

Figure 31-11
Cohesion-tension theory. A and B. The evaporation of water molecules from leaf cells pulls water from
xylem in a continuous column from roots to leaves. C. Only the enormous cohesive strength of water
(a result of hydrogen bonds between adjacent water molecules) keeps the column from breaking. B. If a
bubble should fill a xylem cell and break a water column, the water inside will stop ascending to the leaves.

Biology 🅑 Now™ Learn more about transpiration and the cohesion-tension theory by clicking on this
figure on your BiologyNow CD-ROM.

ond, were long columns of water strong enough to be pulled
up a tree like pieces of rope? Wouldn't the narrow columns
of water in the xylem break under such tension?

The cohesion-tension theory raised two questions—one about the
amount of tension generated by transpiration and one about the cohesive strength of water.

How Can Transpiration Draw Water
to the Tops of Tall Trees?

We've seen that, through osmosis, water tends to move from
areas of lower solute concentration into areas of higher
solute concentration. A different way of saying the same

thing is that water tends to move into areas that have fewer
water molecules. So water molecules move into dry air as
readily as they move into sugary raisins or cells, This is because the concentration of water molecules in air is low
compared to the concentration of water molecules in a cell.

The tendency of water molecules to move along gradients of both concentration and pressure is called **water
potential.** Water potential is always relative to something
else. We have seen that pure water outside of a cell full of
solutes tends to flow into the cell. Because the pure water
has a strong tendency to flow into the cell, we say the pure
water has a "high water potential." Similarly, because water
has a strong tendency to evaporate into air, we say that water has a high water potential relative to dry air.

Inside the mesophyll of a leaf, water is evaporating from
virtually every cell surface, saturating the air spaces in the

mesophyll with water vapor. If the air spaces remained saturated, water potential would stop moving out of the mesophyll cells and transpiration would cease. Outside the stomata, however, is the whole atmosphere, which is almost never saturated.

Air can be dry, as in the Mojave Desert, or very humid, as in a rain forest. Even at 98 percent relative humidity, however, the tendency of water to evaporate into the atmosphere is very high. The water potential that results from the difference in water concentration between the humid air and the liquid water on the surface of mesophyll cells is enough to support a 300-meter column of water, nearly three times the height of the tallest trees. In principle, then, the transpiration of water from the air spaces of the leaves to the atmosphere could provide the driving force for the ascent of water (Figure 31-11B).

As water molecules move from the air spaces in the mesophyll to the drier atmosphere, more water evaporates from the film on mesophyll cells. Because of its cohesive properties, water leaving these cells pulls more water in behind, from the xylem in the leaves and stems and eventually from the xylem in the roots (Figure 31-11C). The result is the movement of water from soil to roots to xylem to leaves (all linked by an unbroken column of liquid water) and finally to air. The driving force for all this movement is water potential, the tendency for liquid water to evaporate into the air.

In humid conditions, water potential can raise water nearly three times as high as the tallest trees.

Does Transpiration Directly Affect Water Flowing Through the Stem?

It is one thing for a process to be theoretically possible and another for it to actually occur and contribute to the workings of an organism. Plant biologists therefore asked, Does water flow in the xylem increase as transpiration increases and decrease as transpiration decreases? In other words, are the two processes really linked? To find out, researchers heated small amounts of water in the stem and then measured how long it took the warm water to move up the stem. They found that fluid moves more quickly in the warmest part of the day, as transpiration increases.

Is Water Cohesive Enough to Sustain the Transpiration Pull in a Tall Plant?

The water columns in a plant are under tremendous tension; transpiration pulls each water column upward, while gravity pulls downward. This tug-of-war extends throughout the plant, with each water molecule clinging to its neighbors by way of hydrogen bonds. But transpiration depends on the water columns in the xylem remaining unbroken. If a water column should break, the top part of the column would be pulled upward and the bottom part would be left behind, with a gap left in between. Transpiration would cease.

Usually the water columns do not break. But large air bubbles can interrupt a water column, and air bubbles sometimes arise in the xylem. In winter, when the air and the plant are cold, gases stay in solution and bubbles rarely form. But in spring, warmer temperatures mean that gases in the xylem fluid can form bubbles and, if they're big enough, break individual water columns.

Woody plants have evolved a solution to this springtime hazard. Each spring, a burst of secondary growth creates new xylem routes. In many trees, most transport occurs in the newly formed xylem, in the outermost growth ring. The result is the familiar light and dark growth rings seen in cross sections of tree trunks.

The cohesive strength of water is usually sufficient to keep it from breaking under the tension generated by transpiration. If too many water columns break, transpiration slows.

Is Water in the Xylem Really Under Tension?

The greater the rate of transpiration, the greater the tension in the xylem as water is pulled upward. The tension in the xylem of a big tree is strong enough to narrow the tubular xylem cells—in the same way a straw collapses if you suck on it hard enough. Careful measurements show that during times of maximum transpiration tree trunks actually become slightly smaller in diameter. When transpiration slows, cell walls and the whole trunk spring back to a larger diameter.

31.3 How Do the Stomata Regulate Water Movement?

Stomata [Greek, *stoma* = mouth] are the gates through which carbon dioxide, water, and oxygen enter and leave the plant. In dry environments, plants risk losing more water through transpiration than is available from the soil. So plants must sometimes slow or stop transpiration (and water loss) by closing the stomata. Closing the stomata reduces water loss, but, of course, it also prevents the plant from collecting carbon dioxide. And without carbon dioxide a plant cannot photosynthesize. The tradeoff between saving water and photosynthesizing is called the **transpiration-photosynthesis compromise**.

To balance these needs, stomata usually open in response to light and to a decrease in levels of carbon dioxide in the air spaces of the leaf. When the air is hot and dry, and excessive transpiration threatens to wilt the plant, the stomata close. Without supplies of carbon dioxide from the outside air, however, most plants drastically reduce the rate of photosynthesis. Lack of carbon (because the stomata are closed) is one reason crop yields are low during a drought.

How Do Stomata Open and Close?

Although numerous, stomata are so tiny that they occupy only about 1 percent of the total surface of a leaf. Each stoma lies over an internal air space, which in turn connects to the rambling air spaces within the mesophyll. The size and distribution of stomata permit the efficient diffusion of carbon dioxide into the air spaces and of water vapor from the air spaces into the atmosphere (Figure 31-12A).

Two guard cells surround each stoma. These cells can change shape, opening and closing the stoma. As the guard cells gain water, they bulge with water and bend outward, opening the stoma (Figure 31-12B). When water exits the guard cells, they become flaccid and the gap between them closes.

Guard cells regulate the flow of water by actively pumping potassium ions (K^+) from adjacent cells of the epidermis (Figure 31-12C). As potassium ions flow into the guard cells, chloride ions follow them inwards. Water follows rapidly on the heels of these ions (by osmosis), and the cells swell, opening the stoma.

Several factors control the pumping of potassium ions and the consequent opening and closing of stomata, including light levels and water availability. For example, when water is scarce, guard cells lose turgor pressure and close. As water becomes less available,

the concentration of a plant hormone called **abscisic acid** increases. Abscisic acid inhibits the flow of potassium ions into the guard cells, the cells lose water, and the stomata close.

In response to light and low carbon dioxide, guard cells pump potassium ions from adjacent cells. Water follows the potassium into the guard cells, which opens the stoma, allowing carbon dioxide to enter. When plants are water stressed, they reverse this process and close their stomata.

How Can Plants Save Water Without Limiting Photosynthesis?

The transpiration-photosynthesis compromise prevents most plants from living in the driest deserts. In such places, most ordinary plants would wilt if they opened their stomata long enough to collect the carbon they need to fuel photosynthesis.

Plants such as the cacti, however, have evolved a biochemical adaptation that gets them around this dilemma. Unlike other plants, plants with **crassulacean acid metabolism (CAM)** can collect carbon dioxide by opening their stomata at night, when temperatures are cool and transpiration is low (Figure 31-13).

But photosynthesis cannot occur in the dark, so the plants need to store the carbon dioxide until daylight. CAM plants have evolved a biochemical trick for storing carbon dioxide collected at night for use in photosynthesis during the day. In CAM plants, carbon dioxide enters the stomata at

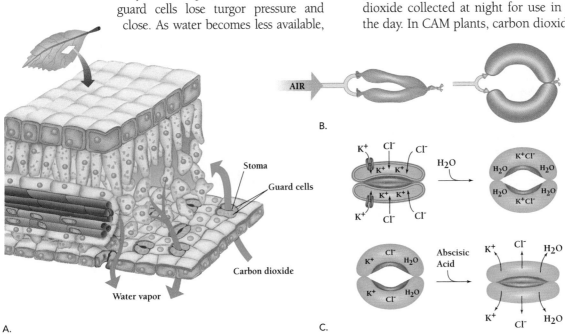

Figure 31-12

What causes stomata to open and close? A. When a stoma is open, water molecules from air spaces inside of the leaf diffuse into the surrounding atmosphere and carbon dioxide gas diffuses into the interior of the leaf. B. The birthday balloon model of stomata: Like a pair of balloons being inflated, guard cells stiffen and bend away from one another. C. The chemical model: the flow of ions and water cause stomata

to open and close. Each stoma has a pair of guard cells, which actively pump in potassium ions. Chloride ions and water molecules follow passively, filling the guard cells with water and causing the stoma to open. Under the influence of abscisic acid, potassium ions flow out of the guard cells, chloride ions and water molecules soon follow, closing the stoma.

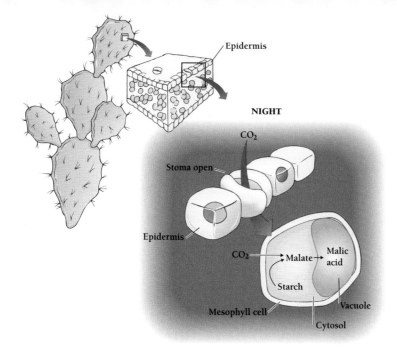

NIGHT

Epidermis

CO_2

Stoma open

Epidermis

CO_2 → Malate → Malic acid

Starch

Mesophyll cell

Vacuole

Cytosol

Stomata open and CO_2 enters plant tissue. In mesophyll cells, CO_2 is stored in vacuole as malic acid.

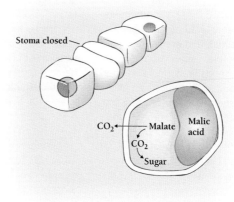

DAY

Stoma closed

CO_2 ← Malate ← Malic acid

CO_2

Sugar

Stomata close. Malic acid releases CO_2, which reenters the cytosol, where the CO_2 diffuses into chloroplasts for photosynthesis.

Figure 31-13

Dealing with dryness. CAM (crassulacean acid metabolism) plants such as this cactus have evolved a unique biochemical pathway that allows them to store carbon dioxide. These desert-adapted plants open their stomata to collect carbon dioxide at night, when temperatures are cool and water loss is minimal. CAM plants store the carbon dioxide as malic acid until morning. During the day, CAM plants close their stomata (thus saving water) and use the carbon dioxide in malic acid to photosynthesize in the chloroplasts.

night, diffuses into cells and combines with a breakdown product of starch or another carbohydrate to form an organic acid (Figure 31-13). The cell stores this acid, called malic acid, in a vacuole. During the day, the acid gradually diffuses back into the cytoplasm, where it releases carbon dioxide for photosynthesis. One can actually taste this process: the leaves of CAM plants are sour in the early morning, from the accumulated acid, and sweet at the end of the day, after the acid disappears. More than 23 families of flowering plants use the CAM pathway, including the succulent stonecrops, pineapple, and the houseplant "maternity plant."

Crassulacean acid metabolism gives some plants another way of solving the transpiration-photosynthesis dilemma.

31.4 How Do Roots Pull Water and Minerals from the Soil?

Water and minerals both usually come from the soil surrounding a plant's roots. Soils themselves result from both living and nonliving processes. They consist of particles of weathered rocks and partially decayed organic matter.

The best soils for plant growth hold enough water to provide a continuous supply after a rainfall, but not so much water that the roots cannot get the oxygen they need for respiration. The water capacity of soils depends in part on the sizes of the particles of weathered rock. Soils consisting only of small particles (less than 2 μm) are called **clays;** soils of medium-sized particles (from 2 to 20 μm) are called **silts;** and soils of large particles (from 20 to 200 μm) are called **sands. Loams** consist of mixtures of all three particle sizes. The ideal soil for most plants is a loam containing about 20 percent clay, 40 percent silt, and 40 percent sand.

In most soils, earthworms, insects, bacteria, and fungi quickly break down dead organisms, parts of organisms (such as fallen leaves) and feces. The residue of such decay is a black or brown material, called **humus.** (When we make it ourselves by piling up leaves and vegetable trimmings, we call it compost.) Good soil has a high capacity for positively charged ions, which bind to negatively charged particles of clay and humus. Humus does not provide the plant with organic matter (as was once thought), but it does provide a good reservoir of water and bound ions while also permitting oxygen to diffuse easily.

Plant scientists have determined what molecules and ions plants need by growing them with their roots in water instead of soil, a technique called hydroponic culture. By varying the minerals added to the water, researchers can tell which nutrients plants need for normal growth. The

Extreme Biology Adaptations to a Limited Water Supply

All plants need ample supplies of water, so dry climates require special adaptations. Biologists sometimes divide plant adaptations to dry environments into four strategies—escape, resist, avoid, and endure. Desert annuals, for example, escape from drought (Figure A). Their seeds lie dormant, indifferent to any drought, until enough rain falls to wet the soil. Then, in a matter of days, they germinate, grow to maturity, and produce new seeds. Such seeds can survive a single dry season or 20 years of drought.

Succulents and cacti resist drought by storing water (Figure B). CAM photosynthesis allows them to save water by keeping their stomata closed during the day. Shallow root systems quickly absorb water from the soil after each rainfall, and these plants store this water for future use.

Other plants, such as desert palms and mesquite, avoid water stress by growing only where rare water supplies exist. Palms grow only at oases, where their roots have access to relatively large amounts of groundwater. Mesquite bushes thrive by growing roots deep into the ground to water tables as much as 50 meters down.

Finally, some plants survive by enduring tremendous water losses. For example, while most plants die after losing 25 to 50 percent of their weight in water, the creosote bush can lose 70 percent of its weight in water and still survive (Figure C). How these plants survive such punishment is still unknown.

Even plants that grow in regions that ordinarily supply plenty of water have adaptations that allow them to survive water stress. The most common responses to drought are the slowing of cell growth and the reduction of transpiration by closing the stomata. Most plants can recover from temporary dry spells, although they may not make up for the lost period of growth.

In temperate climates, freezing temperatures make water unavailable for part of the year, and many species of plants cannot survive prolonged frosts. Surprisingly, the danger from frost turns out not to be from the ice itself. Ice crystals form only outside of cells and do not necessarily damage cells. Instead, the presence of ice reduces the concentration of liquid water, causing water to move out of cells by osmosis. The cells therefore lose turgor.

Many plants can tolerate freezing temperatures, in some cases –25°C or lower. Winter rye and winter wheat, crocuses, tulips, and daffodils all grow slowly but steadily during the winter months. These frost-resistant plants prevent water loss by producing their own "antifreeze," which in some cases just consists of a concentrated solution of an amino acid. When spring comes, these plants are able to grow more rapidly, taking advantage of unfiltered sunlight before trees and taller plants leaf out.

Plants that are resistant to water stress—winter wheat, for example—are said to be acclimated or hardy. As plant biologists breed hardier plants, they greatly increase world food production. For example, if winter wheat and winter rye could withstand temperatures just 2°C colder than they now can, they could replace large tracts of spring wheat and rye in the United States, Canada, and Russia. Because winter crops make better use of the abundant spring rains, such a replacement could increase grain harvests by an astounding 25 to 40 percent.

Figure A
Desert annuals mostly exist as dormant seeds. When rain comes, the seeds germinate, grow into mature plants, bloom, and go to seed—all in a period of days or weeks.

Figure B
Succulents, such as this agave, and cacti resist drought by storing water.

Figure C
The creosote bush endures punishing dehydration. It can lose up to 70 percent of its water and still live.

Table 31-1
Nutrients from the Soil

Nutrient	Concentration*	Symptoms of deficiency
Nitrogen	1–4%	Uniform loss of color in leaves, first on oldest leaves
Potassium	0.5–6.0%	Yellowing at margins of leaves
Calcium	0.2–3.5%	Terminal bud dies, new leaves hooked at tip. Tips and margins withered
Phosphorus	0.1–0.8%	Plants stunted, leaves abnormally dark green
Magnesium	0.1–0.8%	Yellow between veins of leaf
Sulfur	0.05–1.0%	Leaves pale green with dead spots
Chlorine	100–10,000 ppm	
Iron	25–300 ppm	In young leaves, large veins, green; rest of leaf yellow
Manganese	15–800 ppm	Dead spots scattered over leaf surface; only veins and veinlets remain green
Zinc	14–100 ppm	
Copper	4–30 ppm	
Boron	5–75 ppm	Petioles and stems brittle; bases of young leaves break down
Molybdenum	0.1–5.0 ppm	

*Typical concentrations in healthy plants in % weight or parts per million (ppm)

same experiments have demonstrated the symptoms that result from deficiencies of particular nutrients (Table 31-1).

Most soils are composed of some combination of clay, silt, sand, and humus. Hydroponic experiments have shown which minerals plants need.

Why Do Roots Need Oxygen from the Soil?

Groundwater is full of dissolved minerals, usually in the form of ions. But plants must actively transport these ions into their root cells and concentrate them. Because such active transport requires energy, root cells use large amounts of ATP. The ATP comes from oxidizing sugars made in the leaves. Roots, therefore, depend on the presence of oxygen in the soil. Because waterlogged soils hold less dissolved oxygen than air, most plants will eventually drown in waterlogged soils. (Plants grown hydroponically grow in aerated water, like that in fish tanks.)

Because root cells actively transport ions from the soil, root cells need oxygen to generate ATP by means of respiration.

Why Do Some Plants Obtain Minerals from Animals?

Many plants live in areas where minerals, especially nitrogen, are in short supply. In many swamps and wetlands, for example, highly acidic water interferes with plants' ability to

Figure 31-14
Hungry plants. Some plants, such as this Venus flytrap, *Dionaea muscipula*, live in environments that have insufficient supplies of nitrogen. Such plants may evolve techniques for trapping and digesting insects and other small animals.

William E. Ferguson

take up dissolved nitrates. A few carnivorous plants, such as the Venus flytrap, *Dionaea muscipula,* have solved this problem by trapping and digesting insects to extract nitrogen compounds, phosphates, and other ions (Figure 31-14).

31.5 How Do Plants Transport Sugar?

We have seen that roots depend on the products of photosynthesis, but cannot themselves photosynthesize. How do roots and other plant parts obtain the energy and materials for building and maintaining cells and their myriad processes? To find out, modern biologists have tracked the movement of the products of photosynthesis.

Which Way Does Sugar Move?

If researchers expose leaves to carbon dioxide containing radioactive carbon, carbon-14, the plant will take up the radioactively labeled carbon dioxide and use it to build sugars and other carbon-containing molecules. By studying the distribution of the radioactively labeled sugars, researchers have found that the products of photosynthesis move entirely within the sieve tubes of the phloem (not in the xylem).

In woody stems, functional sieve tubes are confined to the inner bark (Figure 31-15). Removing a strip of bark around a tree trunk, a technique called **girdling**, interrupts the phloem and blocks the transport of photosynthetic products. Girdling always kills the tree. Early colonists in North America girdled trees in the spring and cut them down in the fall to harvest partially dried firewood.

In the early 18th century, the Italian anatomist Marcello Malpighi and the English physiologist Stephen Hales studied how girdling kills trees. They noted that while the bark below the girdle shriveled and died almost immediately, the bark above the girdle remained healthy at first. But a girdled tree usually fails to leaf out in the following spring, and before long, the whole tree dies. What causes this pattern of death? The answer is that girdling first prevents sugars from flowing from the leaves to the roots. The leafy top of the tree remains alive for a while, but once the roots die, they no longer deliver water and minerals to the leaves, which then die.

Sugars are said to move from a **source** to a **sink**. A source can be a site of photosynthesis, such as the leaves where the sugars are made. Or a source can be a storage site, such as the roots, which release sugars by breaking down stored starches. Sinks are areas that are actively using sugars, including the growing tips of roots, shoots, young leaves, and fruits, as well as storage organs that are storing rather than releasing energy-rich molecules. A potato can be a sink in the summer, as it accumulates carbohydrates, and a source during the following spring, when it provides the plant with energy for the new growing season.

Sugars in plants move from sources, usually leaves or storage areas, to sinks, usually roots or storage areas.

What Is Inside a Sieve Tube?

Analyzing the contents of phloem was long a difficult problem for plant physiologists because the sieve tubes are small and the sap inside quickly seals any holes made in the sieve tubes. Amazingly, the most useful method for analyzing phloem contents came from a 1953 suggestion of two insect physiologists.

These researchers were studying the nutrition of aphids, relatives of the phylloxera discussed at the beginning of this chapter. Aphids drink sap, the sweet fluid contained in the phloem (Figure 31-16). An aphid has a specialized mouthpart, called a *stylet,* that pierces a sieve tube. The sap in the phloem is under pressure, so as soon as the aphid's stylet enters the phloem, the sap flows through the stylet and into the aphid.

By cutting off the aphid but leaving the stylet embedded in the sieve tube, plant physiologists have been able to collect the sap and study its composition. Sap is a concentrated solution of sugars, including up to 30 percent sucrose (table sugar). Sap also contains a few minerals, especially potassium ions, and carbohydrate derivatives, which vary considerably from plant to plant. Within the rose family (which includes apple, cherry, apricot, and pear trees), for example, sap contains little sucrose but lots of sorbitol, a sugar alcohol derived from glucose. Sap also carries molecular messages, including snippets of RNA and about 600 different proteins that regulate the expression of genes in cells all over the plant.

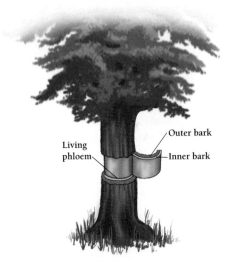

Figure 31-15
Sugar on the outside. A. In woody plants, living phloem is confined to the inner bark. B. Stripping the bark from a tree will kill it if the strip goes all the way around. Such girdling destroys the phloem and prevents the products of photosynthesis from reaching the roots.

Jerome Wexler/National Audubon Society/Photo Researchers, Inc.

Figure 31-16
An aphid in action. An aphid pokes its stylet into a sieve tube cell for a drink of sugar water. To collect the contents of the phloem, a researcher cuts the aphid off its stylet.

The Osmotic Flow of Water Moves Sap in the Phloem

By using aphid stylets to follow the movement of radioactively labeled sugars through the phloem, researchers have found that solutes move through phloem much faster than could be explained by diffusion alone. They concluded that the contents of the phloem undergo some kind of transport. The explanation for how this transport occurs is called the **pressure flow hypothesis,** first advanced in 1926 by the German plant physiologist Ernst Münch.

Münch's insight was that the concentration of sugar is higher near a source than near a sink. Because the concentration of water is lower in a sieve tube near a source, water follows the sucrose (by osmosis). Near a sink, sucrose leaves the sieve tube and again water follows (Figure 31-17). The difference between the pressure at the source and the pressure at the sink pushes the contents of the sieve tube, in bulk, in the direction of the sink.

The pressure flow hypothesis explains why sugars move from source to sink once they are in the phloem.

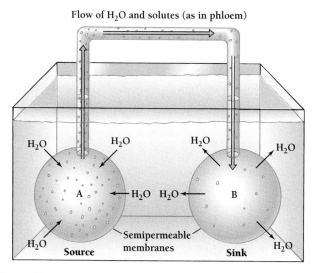

Flow of H₂O and solutes (as in phloem)

Figure 31-17
Münch's model for the pressure flow hypothesis. To test his hypothesis, Münch built a laboratory model that consisted of two selectively permeable membranes connected to each other by a tube. In this simplified diagram, the first membrane (labeled A) represents the source, and the second membrane (labeled B) represents the sink. Inside the first membrane, A, is a solution more concentrated than the solution outside of it. Water consequently flows by osmosis into A. The solution inside B is less concentrated than the solution outside. Osmosis moves water from B out into the surrounding solution. As a result of these osmotic processes, the water pressure in A increases and the water pressure in B decreases. The resulting differential in pressure causes the contents of the connecting tube (water and solutes) to flow from A to B.

Biology ⑧ Now™ Learn more about the pressure flow hypothesis by clicking on this figure on your BiologyNow CD-ROM.

What Draws Water and Sugar into the Phloem?

Once the sugar is in the phloem it flows away from the areas of highest concentration. But what prevents the sugar in leaf phloem from flowing back into the leaf mesophyll where it came from? And what concentrates the sugar in the phloem? The answer to both questions is active transport.

The cells of the mesophyll connect to one another by plasmodesmata, and sugars diffuse freely from cell to cell. However, few plasmodesmata connect mesophyll cells directly to sieve tube cells. Instead, sugar flows through plasmodesmata to bundle sheath cells, which lie adjacent to **companion cells,** specialized cells adjacent to the sieve tube elements (Figure 31-18). To get from the mesophyll into the phloem, sugar must flow into the bundle sheath cells and then be pumped across two cell membranes, that of the bundle sheath cell and that of the companion cell. This active pumping concentrates the sucrose in the sieve tubes of the phloem to levels several times that in the mesophyll. From the companion cells, the sucrose flows into the phloem, but cannot flow back.

Once the sieve tube contains a high concentration of sucrose, water flows into the phloem from the surrounding cells (by osmosis), greatly increasing the pressure. The pressure in the phloem near a sucrose source is more than 10 times that in a car tire, enough to push the sap through the sieve tubes at about one meter per hour, about the speed of the tip of the minute hand on a large classroom clock. (In contrast, water flows through the xylem about as fast as the clock's second hand.)

At a sink, the phloem unloads its sucrose (again aided by companion cells). The water again follows the sucrose, entering nearby xylem and flowing back to the mesophyll.

The pressure flow model for phloem transport works in a single direction—from a source, with a higher concentration of sucrose, to a sink, with a lower concentration. One of the reasons that plant physiologists were slow to accept this hypothesis is that phloem transport can often move in opposite directions—from leaf to root or from leaf to stem tip. But the flow of sap in a single sieve tube probably only goes in one direction at a time.

Cells in the mesophyll pump sugars by active transport from cell to cell and then into the phloem.

In this chapter we have seen how plants exploit the water potential in air to draw water and minerals from the ground up to the leaves; what processes open and close stomata; how sugars synthesized in the leaves are actively pumped into the phloem; and, finally, how those sugars move from areas of high sugar concentration to areas of low sugar concentration. In the next chapter, we discuss how plant embryos first develop a root and shoot.

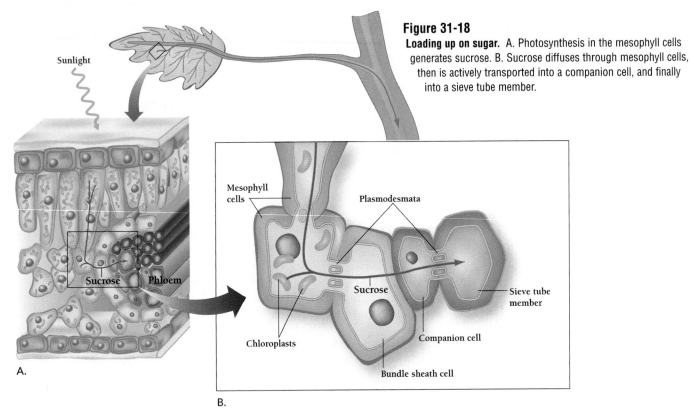

Figure 31-18
Loading up on sugar. A. Photosynthesis in the mesophyll cells generates sucrose. B. Sucrose diffuses through mesophyll cells, then is actively transported into a companion cell, and finally into a sieve tube member.

Sunlight

Mesophyll cells

Plasmodesmata

Sucrose

Phloem

Sucrose

Sieve tube member

Chloroplasts

Companion cell

Bundle sheath cell

A.

B.

Key Concepts

• In plants, water moves from soil to air through the vascular system.
• The main driving force for the ascent of water is transpiration, with upward transport depending on water's cohesive and adhesive properties.
• Stomata regulate the rate of transpiration by opening and closing.
• The bulk flow of water drives the transport of sugars in the phloem.

Summary with Key Terms

How do terrestrial plants obtain water and minerals from the soil?

Leaves are the major organs of photosynthesis. Most leaves are green because they are packed with bright green **chloroplasts.** Each leaf consists of a (usually) flat **blade** and a stalk, called a **petiole,** that attaches the leaf to the stem. The surfaces of a leaf usually contain specialized pores, called **stomata,** through which carbon dioxide, oxygen, and water vapor enter and leave. The interior of a leaf is composed of **mesophyll,** which is made of **palisade parenchyma** and **spongy parenchyma.**

Stems carry water and minerals from roots to leaves and nutrients from leaves to roots in **vascular bundles,** composed of **xylem** and **phloem.** The xylem lies toward the center and the phloem toward the outside of each bundle. The arrangement of vascular bundles in monocots differs from that in eudicots.

Water moves from soil to air by way of the xylem. Absorption and transport begin in the roots. Roots include vascular tissue (xylem and phloem), which is surrounded by ground tissue (**cortex**), all enclosed in a layer of epidermis. Water flows from the soil to the vascular system by way of the **mucigel**-covered **root hairs** and cortex. The innermost layer of the cortex is the **endodermis,** inside of which is a cylinder of meristem called the **pericycle.** The waxy, waterproof cell walls of the endodermis cells form the **Casparian strip,** which prevents water from flowing *between* the endodermal cells into the vascular tissue. Water and minerals can move through the root by two pathways, the **symplast** and the **apoplast.** The Casparian strip prevents movement of water and dissolved minerals between the vascular cylinder and the apoplast. Water must flow instead via the symplast, *through* the living cells of the endodermis. The membranes regulate the movement of both ions and water into the vascular tissue.

What drives water up from the roots to the topmost leaves of a plant?

Water moves into and out of living cells by **osmosis,** traveling from areas of low solute concentration to areas of high solute concentration, and from areas of high water concentration to areas of low water concentration. Water also moves from areas of higher pressure to areas of lower pressure. The tendency of water molecules to move along gradients of concentration and pressure is called **water potential.**

Increasing turgor pressure tends to cause cells to swell. In plant cells, the cell wall resists swelling and maintains cell shape, so that turgor pressure increases the stiffness of the plant and provides mechanical support. Solutes in the xylem cause water to flow into the roots, resulting in **root pressure**. Root pressure is beautifully demonstrated in **guttation**, the expression of droplets of water from the edges of leaves on cool, damp mornings.

Because of water's **adhesiveness** and **cohesiveness**, it tends to move up narrow tubes, or **capillaries**, a process called capillarity. But root pressure and capillarity cannot account for the movement of water to the tops of tall trees.

The major driving force for the ascent of water in the xylem is **transpiration** (passage of water vapor from plant to atmosphere). And the driving force for transpiration is low water potential in air. According to the **cohesion-tension theory**, water vapor passes from the leaves into the atmosphere through open stomata. The evaporation of water from the leaves pulls water up the xylem in continuous columns. Experimental measurements have established that water in the xylem is under tension. The integrity of the water columns in the xylem is critical and depends on the cohesive properties of water.

How do a plant's stomata regulate water movement?

Plants regulate transpiration and photosynthetic activity (requiring the passage of carbon dioxide from atmosphere to plant) by opening and closing their stomata, a process mediated by the hormone **abscisic acid**. During hot, dry periods, plants must balance the rate of transpiration against the rate of photosynthesis, a trade-off called the **transpiration-photosynthesis compromise**.

Many plants have special biochemical and anatomical adaptations that reduce water loss while permitting the entry of enough carbon dioxide to sustain photosynthesis. **Crassulacean acid metabolism (CAM)** plants collect carbon dioxide by opening their stomata at night, when temperatures are cool and transpiration is low.

Most plants obtain minerals as well as water from the soil. Roots actively transport ions from the soil through cells of the cortex into the xylem. **Loams** are different combinations of variously sized particles of **sand, silt,** and **clay**. Decayed organic matter in soils is called **humus** (or compost). Insectivorous plants obtain nutrients from the bodies of animals.

How do plants transport sugar?

Many parts of a plant do not perform photosynthesis, and so they require energy-rich molecules transported from the leaves. The movement of such molecules, mostly in the form of sucrose, occurs in the phloem. **Girdling** a tree or shrub interrupts the phloem and blocks the transport of photosynthetic products.

The osmotic flow of water, described in the **pressure flow hypothesis,** drives the flow of sugar water through the phloem. **Companion cells,** and the adjacent sieve cells, actively transport sugars into the phloem, and water follows by osmotic flow. At the other end of a sieve tube, sucrose leaves the phloem, and water follows. The resulting pressure differential between the sucrose **source** and sucrose **sink** powers bulk flow that carries the contents of the phloem—including sucrose—to all parts of the plants. Differences in water pressure move the fluids in both the xylem and the phloem.

Review and Thought Questions

Review Questions

1. Draw a diagram of the endodermis, including the Casparian strip, and explain how these structures regulate the flow of water and minerals into the root xylem.
2. In what ways does water coming into a leaf contribute to photosynthesis?
3. Which is higher—the water potential of a plant cell in air or the water potential of the same cell in a lake? In which direction would water move in each case?
4. Describe root pressure and guttation.
5. Explain why water's natural cohesiveness is an essential element of the cohesion-tension theory.
6. How would you test the cohesion-tension theory?
7. What problem do CAM plants solve?
8. Name the three main particle sizes in soil. Which is biggest and which is smallest? The most fertile soils, or loams, are generally made up of what combination of these three kinds of particles?
9. Describe the pressure flow hypothesis. Is a potato a source or a sink?

Thought Questions

10. Why should xylem be composed of dead cells while phloem must be composed of living cells? What differences in the way these two parts of the vascular system function could explain this difference?
11. Do you feel that any of the characters, whether scientists or grape growers, described in the story about phylloxera did anything wrong or unethical? Explain why or why not.
12. The chapter-opening story stated that funding for research on phylloxera came mainly from the wine grape industry itself. Do you think that researchers would or would not have realized the problem with AXR1 sooner if funding for such research had been available from the federal government? Explain your reasoning.

BiologyNow Resources

Biology *Now*™

Active Figures

31-11: Transpiration and the cohesion-tension theory
31-17: The pressure flow hypothesis

Preparing for an exam? Take a diagnostic test on your BiologyNow CD-ROM.

Online materials relating to this chapter are at:
http://biology.brookscole.com/AAL3

About the Chapter-Opening Image

Memorial at the School of Agriculture, Montpellier, France (1911). A sculptor has represented the rescuing of the French wine industry with American rootstock as a young American rescuing the ailing, old French vine.

Growth and Development of Flowering Plants

Key Questions

- How do plants reproduce sexually?

- How do plant embryos establish a body plan?

- How do plants build a shoot and a root through primary growth?

- How do plants build flowers?

- How do plants build outward through secondary growth?

- How do plants reproduce asexually?

The Origin of Corn

When Columbus left Spain in 1492, he was searching for a shortcut to China, Japan, India, and the rest of Asia. There, in "the Indies," Columbus hoped to trade European goods for spices and silk. Of course, he did not reach the Indies, but instead a whole new world, unknown to Europeans.

Columbus's most valuable discovery may have been corn. Corn, also called maize, was good to eat, easy to grow, and bountiful. And when Portuguese sailors finally reached the eastern coast of the real Asia, in 1516, they brought American corn to trade. Within 50 years of Columbus's arrival in the West Indies, corn grew in every corner of the world. Wherever corn was introduced, it replaced native crops as an indispensable part of the diet. Corn became so thoroughly adopted in every tiny village where it was introduced that peoples in Africa, India, China, and Turkey all came to believe that maize was native to their region.

Where did corn come from? Certainly it is American. Archaeologists have found corn (*Zea mays*) that is thousands of years old throughout North and South America. The oldest corn, from a site in Panama, is 7,000 years old. Most botanists and paleontologists believe that corn originated in Mexico. But what we know as corn is the product of intense artificial selection. Its unparalleled yields allowed human populations to grow far beyond those that subsist by hunting and gathering. The Western Hemisphere alone has thousands of varieties of corn grouped into some 300 races. Yet corn is almost unable to reproduce by itself; it is a plant that has become entirely domesticated (Figure 32-1). From what wild plant, then, is corn descended?

In the 1930s, the Harvard botanist Paul Manglesdorf crossed various varieties of corn—a process called backcrossing—to produce what he suggested was a model for wild-type corn. This approach would be similar to crossing several different breeds of dog to produce something that looked like a wolf. Manglesdorf found that as he backcrossed the corn, the cobs became smaller and the number

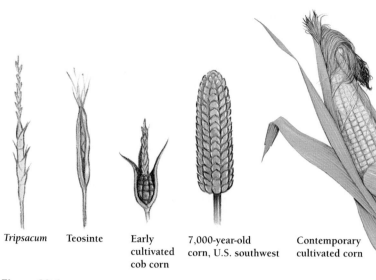

Figure 32-1

Breeding modern corn. The ancestor of corn looked more like wheat, with a few small loosely wrapped kernels. As early as 7,000 years ago, Native Americans had bred an almost-modern corn cob, whose many large kernels were suitable for popping.

Tripsacum Teosinte Early cultivated cob corn 7,000-year-old corn, U.S. southwest Contemporary cultivated corn

crossbreed the two plants, classical geneticists could tell little about them.

Based on his backcrosses and his students' Bat Cave discovery, Manglesdorf dismissed Beadle's teosinte hypothesis. The ancestor of corn is pod corn, he argued, and teosinte is a hybrid of ancient pod corn and another grass called *Tripsacum*. Modern corn, he suggested, is a mix of these three related grasses. *Tripsacum* did not have 20 chromosomes either, however, and the question of corn's ancestry remained unresolved. For 40 more years, Manglesdorf and Beadle, as well as their intellectual descendants, picked at each other's theories. By the 1970s, however, many botanists thought that an extinct teosinte would prove to be the ancestor of corn.

In 1978, Hugh Iltis, a professor of botany at the University of Wisconsin, sent a Christmas card to a colleague at the University of Guadalajara, in Mexico. The card bore a drawing of a fanciful extinct teosinte, something that might resemble the ancestor of corn. Amused, the Mexican botanist challenged her students to find such a plant in the wild. Young Rafael Guzman took the challenge and spent his Christmas vacation searching for the hypothetical plant. In the mountains near Jalisco, Guzman found a new species of teosinte, seemingly the model for Iltis's drawing.

Guzman sent Iltis some seeds, and Iltis found, to his delight, that the plants—called *Zea diploperennis*—had half the number of chromosomes that teosinte has. Like corn, *Zea diploperennis* has 20 chromosomes. The search for corn's ancestor, Iltis declared, was over. Indeed, in the 1980s, genetic research at the University of Minnesota showed that approximately five genes control the traits that distinguish corn from *Zea diploperennis*. Researchers hope to discover what these genes do.

Is the corn debate over? Most people now agree that corn is about 10,000 years old and probably descended from some form of teosinte from central Mexico. A likely ancestor for corn seems close at hand, yet paleobotanists and geneticists continue their comfortable bickering.

For most of us, it is a mess we can safely overlook. In comparison, the changes in corn's morphology under thousands of years of selective pressure are dramatic and revealing. Corn kernels are so tightly wrapped in the husk that the seeds can never disperse. Corn's human-selected morphology has made it entirely dependent on humans for reproduction. It is nearly alone in this odd defect. Nearly every other species of plant on Earth can reproduce and grow with no help at all.

In this chapter we will see how plants reproduce, how seeds form, and how the embryos in seeds develop into a mature plant.

of kernels inside the husk decreased until each corn kernel (a single seed surrounded by its fruit) was wrapped in its own husk. The ancestor of corn, he predicted, would be found to be an extinct grain that looked something like wheat, with one seed per pod. He called this hypothetical ancestral corn *pod corn*.

In the summer of 1948, two of Manglesdorf's students took a bus to Bat Cave, New Mexico. The cave had been occupied 3,000 years earlier by people living on the shore of an ancient lake. As the two young men dug, they found hundreds of pieces of corn. The deeper they dug, the smaller and more primitive the corncobs became. When they reached the bottom, they found tiny cobs of popcorn in which each kernel was enclosed in its own husk—in other words, Manglesdorf's pod corn.

Manglesdorf, who had given the two students $150 for bus fare and sleeping bags, was delighted. He later wrote, "Seldom in Harvard's history has so small an investment paid so large a return." Manglesdorf took a few kernels of the ancient Bat Cave corn, dropped them into a pan of hot oil, and they popped!

But while Manglesdorf's own students and other archaeologists found the pod corn hypothesis persuasive, geneticists had their own ideas. The geneticist George Beadle had proposed, in 1939, that corn was descended from a Mexican grass called teosinte (tay-o-SIN-tay). Nineteenth-century naturalists had first suggested that teosinte might be the ancestor of corn, and Beadle found that kernels of teosinte popped in hot oil were "indistinguishable from popped corn." Teosinte, however, has 40 chromosomes whereas corn has only 20, a difference that would prevent the two plants from interbreeding. Without being able to

32.1 How Do Plant Embryos Establish a Body Plan?

The life cycle of a flowering plant consists of two prominent phases: the diploid **sporophyte,** which is the familiar plant we all recognize, and the haploid **gametophyte,** which produces haploid eggs and sperm by mitosis. Recall that "diploid" means that a cell has two copies of each chromosome ("2n"), and "haploid" means that a cell has only one of each chromosome ("n" or "1n"). Humans are diploid, and, like plants, we make haploid eggs and sperm. When animals' and plants' haploid sperm and eggs join at fertilization, they form a diploid (2n) zygote that develops into an embryo and finally an adult.

In plants, the cycling of the sporophyte and the gametophyte is called **alternation of generations** (Figure 32-2). Most of this chapter concerns the diploid sporophyte generation, the dominant phase of the life of every flowering plant. But we begin by describing the development of the gametophytes, the production of haploid (n) gametes (eggs and sperm) and their union to produce a diploid (2n) zygote.

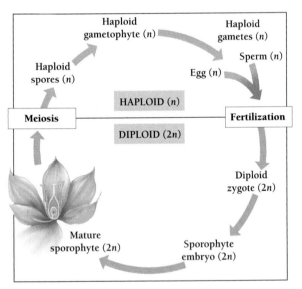

Figure 32-2
Alternation of generations in flowering plants. In flowering plants, the diploid sporophyte, which produces haploid spores, alternates with the haploid gametophytes, which produce haploid gametes. Both the sporophyte and the gametophyte are multicellular. The sporophyte is the conspicuous shoot and root, including most parts of the flowers. Inside the flowers are the gametophytes, which make the sperm and the eggs. Some plants have male and female gametophytes on separate plants; some have them on the same plant, but in different flowers; and some plants have both male and female gametophytes within the same flower.

Biology ⊗ Now™ Learn more about alternation of generations in another flowering plant by clicking on this figure on your BiologyNow CD-ROM.

How Do Plants Make Sperm?

In flowering plants, sperm are made inside tiny gametophytes called **pollen grains,** which are released from the flower to travel to other plants. Pollen is made in flower parts called **anthers,** which are attached to the flower's base with a **filament.** The anther and filament form the **stamen** (Figure 32-3A).

Within the anther are four pollen sacs in which diploid cells undergo meiosis to form haploid cells called **microspores** [Greek, *mikros* = small], each of which develops into a male gametophyte. Each microspore undergoes mitosis to produce, in most flowering plants, two haploid cells, one *generative cell* and one *tube cell.* The two cells are together enclosed within a common cell wall (that of the original microspore) to form a pollen grain. The outside of the pollen grain is often elaborately and distinctively sculpted (Figure 32-3B). Indeed, botanists often can recognize plant species by their pollen alone, even when the pollen is fossilized.

The anthers release the pollen, often immense quantities of it, and insects, wind, or other agents then carry some of it to the female part of a flower of the same species. But most pollen never reaches the stigma; instead it ends up in streams and lakes, on the ground, or in the hives of bees (Figure 32-3C).

Stamens produce haploid microspores, which develop into male gametophytes known as pollen grains.

How Do Plants Make Egg Cells?

The female reproductive part of a flower is called a **pistil.** A flower may have one pistil or several, and some flowers lack them altogether. The pistil is the most central of the flower's parts and consists of one or more **carpels.** The swollen base of the pistil is the **ovary** [Latin, *ovum* = egg], which produces the female gametophytes (Figure 32-4). When a pistil consists of several carpels, the carpels join to form a compound ovary. A projection, called the **style,** connects the ovary to the **stigma,** the structure that receives the pollen grains.

The wall of the ovary produces one or more **ovules,** structures that, after fertilization, form a seed. An ovule contains a diploid cell. Two cell layers called **integuments** surround the ovary (Figure 32-4). The integuments help to protect the developing embryo, eventually forming the seed coat.

The development of egg cells begins when the cell undergoes meiosis. One diploid cell in each ovule undergoes meiosis to form four haploid cells called **megaspores** [Greek, *megas* = large]. One of the four megaspore cells develops into the **embryo sac,** which is the mature female ga-

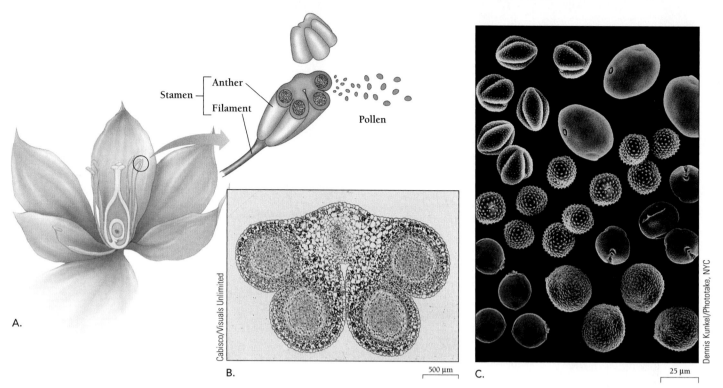

Figure 32-3

How do flowers make pollen? A. The anther is the pollen-producing part of the stamen. B. Each anther includes four pollen sacs, inside of which haploid microspores form. Each microspore divides and develops into a male gametophyte, a two-celled pollen grain. C. Pollen grains have distinctive textures and patterns. Shown here in this false-color SEM are pollen from ragweed, timothy brush, alder, and poplar.

metophyte, while the other three haploid cells degenerate. The megaspore nucleus undergoes three rounds of mitosis to form eight haploid nuclei. These eight nuclei form only seven cells, however, because one cell, the central cell, contains two nuclei (called polar nuclei). One of the other six cells of the embryo sac is the egg cell.

Carpels produce haploid megaspores, which develop into female gametophytes called embryo sacs. Each embryo sac has one egg.

How Does Fertilization Occur in Plants?

After a pollen grain lands on the stigma, the pollen grain's tube cell forms a **pollen tube.** The pollen tube grows down the style and through an opening through the two integument layers, finally reaching the embryo sac (Figure 32-4). As the pollen tube grows, the generative cell of the pollen grain divides mitotically to form two sperm cells. These cells ride the tip of the pollen tube as it grows. When the pollen tube reaches the embryo sac within the ovule, the tube tip opens, releasing the sperm cells.

Flowering plants differ from other organisms in undergoing **double fertilization.** Of the two sperm that arrive on the pollen tube, one fuses with the egg to form the diploid zygote ($n + n = 2n$). The diploid zygote develops into an embryo and eventually into a plant. The other sperm fuses with both haploid polar nuclei to form a triploid nucleus ($3n$) in the central cell. The resulting triploid central cell divides to form the seed's **endosperm,** a triploid tissue that nourishes the growing embryo (Figure 32-4).

Shortly after fertilization, both the zygote and the endosperm nucleus undergo repeated rounds of mitosis. The developing seed consists of an embryo, the endosperm and the integuments. In many species, the endosperm is used up during development, and the mature, or ripe, seed has only the embryo and the seed coat (derived from the integuments).

Flowering plants undergo double fertilization. One sperm fuses with the egg to form a diploid ($2n$) zygote and one sperm fuses with two haploid nuclei to form a triploid ($3n$) cell that develops into a nourishing endosperm. Together, the zygote and endosperm form the core of a plant's seed.

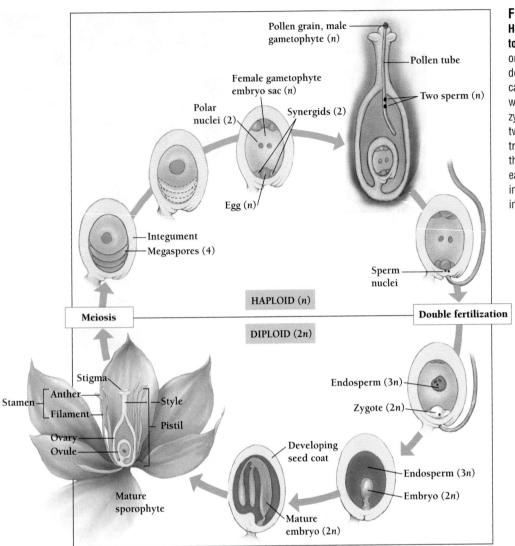

Figure 32-4

How do flowers bring egg and sperm together? When a pollen grain lands on the stigma, its pollen tube grows down the style and into the ovule, carrying two sperm. One sperm fuses with the egg cell to form the diploid zygote. The other sperm fuses with the two nuclei of the central cell to form a triploid ($3n$) cell that divides to become the $3n$ endosperm. After fertilization, each ovule becomes a seed, which includes the embryo, endosperm, and integuments.

How Does a Plant Form a Seed?

The first cell division in a zygote divides the cell into two cells, one of which becomes the embryo. The other cell develops into supporting structures, including a narrow stalk that, like an umbilical cord, attaches the embryo to the surrounding tissue, stimulating growth of and supplying nutrients to the embryo (Figure 32-5A).

With further development, the embryo attains a heart-like shape, the lobes of which become the "seed leaves," or **cotyledons** (Figure 32-5B). **Eudicots** such as soybeans have two cotyledons, while **monocots** such as corn have only one. The fundamental shoot-root polarity of the embryo, with an apical shoot meristem and an apical root meristem, is established early in plant development. The cotyledons absorb nutrients from the endosperm and later distribute them to the developing plant.

Opposite the cotyledons in the developing embryo lies the **radicle,** the beginning of a root complete with an apical meristem. Sometimes the cotyledons grow so long that they must curve backward to fit into the seed.

At this stage, most of the embryo's cells resemble the differentiated cells of a mature plant—with dermal tissue, ground tissue, and vascular tissue—except that their plastids do not yet contain chlorophyll. The meristem cells remain small and undifferentiated.

In many plants, by the time the embryo is mature the endosperm is fully consumed, while some plants retain the endosperm in the mature seed. In cereal grains (monocots), such as corn and wheat, most of what we eat is endosperm. In eudicot plants such as peas, the developing plant derives nourishment from the cotyledons (as do we).

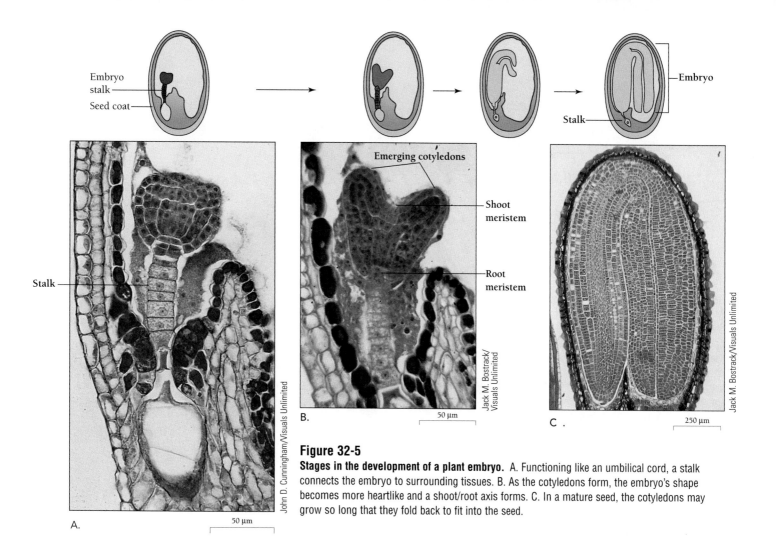

Embryo stalk
Seed coat
Embryo
Stalk
Stalk
Emerging cotyledons
Shoot meristem
Root meristem

B.   50 μm
John D. Cunningham/Visuals Unlimited
Jack M. Bostrack/Visuals Unlimited
Jack M. Bostrack/Visuals Unlimited

C .   250 μm

A.   50 μm

Figure 32-5
Stages in the development of a plant embryo. A. Functioning like an umbilical cord, a stalk connects the embryo to surrounding tissues. B. As the cotyledons form, the embryo's shape becomes more heartlike and a shoot/root axis forms. C. In a mature seed, the cotyledons may grow so long that they fold back to fit into the seed.

After cotyledons form, the seed dries and the seed coat hardens around the embryo, providing a tough and water-resistant covering. In most seeds, the embryo then becomes **dormant** [French, *dormir* = to sleep], and growth and development stop.

A dry seed contains a dormant embryo, with one or two cotyledons—and in some seeds some endosperm—surrounded by a tough seed coat.

How Does the Embryo Resume Development?

The resumption of seed growth is called **germination.** In the first stage of germination, the seed absorbs water and swells to double or triple its size during dormancy. Once the seed is hydrated, it awakes, with a burst of metabolic activity, cell division, and growth.

The first part of the embryo to break out of the seed coat is the radicle (Figure 32-6). Growing rapidly, the radicle immediately turns downward into the soil (toward gravity) and begins to form a root system. In many eudicots, a thin loop between the radicle and the cotyledons, called the hypocotyl, pushes up through the soil toward the sun. The hypocotyl pulls the cotyledons and small true leaves out of the soil (Figure 32-6). Once out of the ground, the hypocotyl straightens and the green cotyledons and true leaves spread out under the sun. In other kind of plants, such as corn, the cotyledons stay underground, but the growing shoot pushes the leaves and apical meristem straight through the soil toward the sun.

At germination, the seed swells with water, and embryonic development resumes with a burst of metabolic activity. The radicle breaks out of the seed coat and grows down into the soil. The stem grows toward the sun.

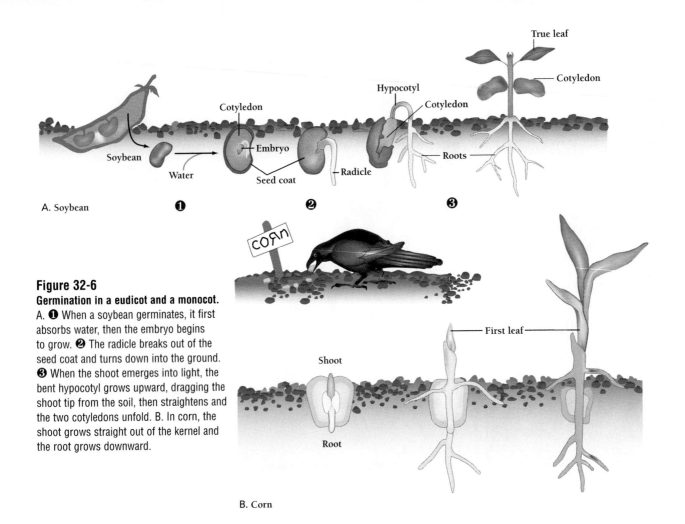

Figure 32-6
Germination in a eudicot and a monocot.
A. ❶ When a soybean germinates, it first absorbs water, then the embryo begins to grow. ❷ The radicle breaks out of the seed coat and turns down into the ground. ❸ When the shoot emerges into light, the bent hypocotyl grows upward, dragging the shoot tip from the soil, then straightens and the two cotyledons unfold. B. In corn, the shoot grows straight out of the kernel and the root grows downward.

How Does the Embryo Obtain Nourishment?

Cells in the dormant embryo maintain a low level of metabolism, but germination triggers a huge increase in biochemical activity. The energy and molecular building blocks for this activity comes from starches, oils, and proteins stored in the seed. The presence of these stores means that seeds are rich foods for animals.

Seeds release their stores using enzymes that are not present (or are present in much lower amounts) in the ungerminated seed. Biologists understand this process especially well in grains in part because grain sprouting is an important part of beer production. In cereals such as corn, barley, and wheat, the outer layer of the endosperm secretes enzymes, such as α-amylase, that break down the starches and other molecules stored in the endosperm. (An α-amylase in human saliva also breaks down starch into sugars as we chew.) It is the embryo itself that stimulates the production of these enzymes; if researchers remove the embryo, the cells of the outer endosperm secrete no enzymes and the endosperm remains undigested.

At germination in cereals, the outer endosperm secretes enzymes such as α-amylase that release nutrients to the growing embryo.

What Environmental Cues Trigger Germination?

We can see the obvious effect of environment on plant growth by keeping a germinating seedling in the dark. Both eudicots and monocots growing in the dark display a condition called *etiolation* [French, *etioler* = to blanch or whiten], with little or no chlorophyll production, a thin, spindly appearance, and poor leaf development (Figure 32-7).

A dormant plant embryo in a dry seed can suspend growth for days, months, years, decades, even centuries, depending on the species. For example, 1,000-year-old lotus seeds dug from a former swamp in Manchuria, China, amazed researchers by germinating when placed in water. Researchers have even succeeded in growing plants from lupine seeds found in 10,000-year-old frozen sediments in Alaska. Seeds do not seem to have a built-in clock that determines the time of germination, but instead depend on cues from their environment.

Plants use a wide variety of mechanisms to prevent premature germination. In many legumes, for example, a hard seed coat prevents the entrance of water or oxygen needed for germination. In these and other species, germination often requires *scarification*, harsh treatment to break the seed coat barrier. In nature, scarification conditions include fire;

Figure 32-7
Etiolation. Corn plants grown in the dark lack chlorophyll and appear spindly compared to those grown in normal light.

acid and enzymes in an animal's digestive system; scraping against sand or rocks; successive freezing and thawing; or the action of a fungus. In the laboratory, researchers can scarify seeds and so initiate germination with sandpaper, alcohol, or even sulfuric acid.

Other species prevent premature germination with chemical inhibitors, such as sodium chloride, cyanide, mustard oils, and other organic compounds. Rainfall then stimulates germination by washing away the inhibitors.

Light, water, temperature changes, stomach acid, and other environmental cues can all trigger germination.

32.2 How Do Plants Generate a Shoot and a Root?

Just behind the tip of both shoot and root are **apical meristems,** groups of *undifferentiated* cells that divide to produce the specialized cells of the shoot and root (Figure 32-8). Meristem cells resemble those in the early embryo and can continue to divide throughout the life of the plant. When meristem cells divide, some of the daughter cells remain meristem cells, that is, undifferentiated. Others divide several times, then differentiate into specialized cells.

Apical meristem can repeatedly generate a whole set of plant parts called a module, a sort of unit of construction that, in the shoot, can consist of stem, leaves, buds, and

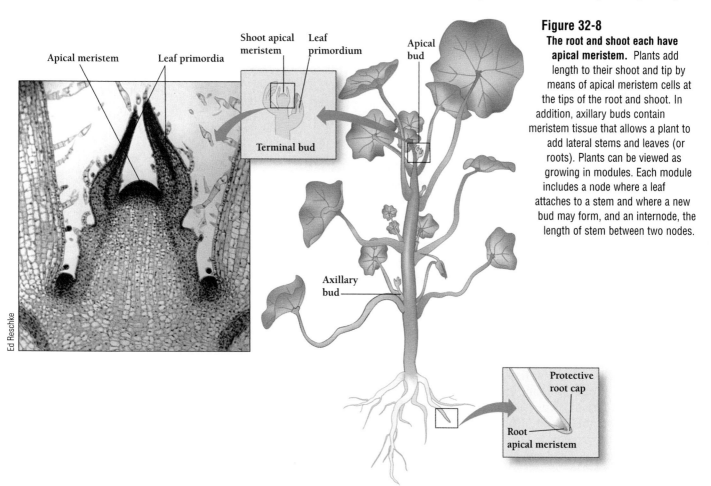

Figure 32-8
The root and shoot each have apical meristem. Plants add length to their shoot and tip by means of apical meristem cells at the tips of the root and shoot. In addition, axillary buds contain meristem tissue that allows a plant to add lateral stems and leaves (or roots). Plants can be viewed as growing in modules. Each module includes a node where a leaf attaches to a stem and where a new bud may form, and an internode, the length of stem between two nodes.

other organs (Figure 32-8). Each developing leaf attaches to the stem at a *node*. The length of stem between any two nodes is an *internode*. If we look from node to node, we see repeating modules, each consisting of a segment of stem, a leaf, a node, and a bud. A **bud** contains another shoot tip, with its own apical meristem, and a few tiny leaves below the meristem. In both shoot and root, cell division in the apical meristems causes the plant's primary growth. Most of the lengthening of shoots and roots, however, occurs because differentiated cells grow longer.

Each species of plant has a characteristic pattern of modular repetition. Modular organization, fixed repetition rules, and controlled outgrowth of buds underlie the branching patterns of stems. The modular organization of plants allows plants great flexibility in responding to their environment.

Plants form their bodies by building modular structures at the apical meristems at the ends of the shoot and root.

How Do the Daughters of Meristem Cells Elongate and Differentiate?

Some daughter cells leave the dividing meristem and stop dividing. These cells grow—mainly by taking water into a large central vacuole—and differentiate into any of a variety of cell types. Elongated differentiating cells, already arranged in columns, may be up to 50 times longer than those of the meristem.

Unlike animal cells, plant cells cannot move around and form new associations during development. After division ceases, they can only grow or gain and lose water pressure (turgor). Plant development therefore depends heavily on orderly cell division and cell expansion. Every cell has its place. As cells divide within the meristem, the planes of cell division and rates of expansion establish future spatial relationships.

Cell expansion depends mainly on the osmotic flow of water into the central vacuole. A cell builds its cell wall in such a way that the wall is weaker in one direction than in others. As a result, as the vacuole swells, the cell elongates in the direction of the weaker wall, instead of growing larger in all directions. The orientation of the cellulose fibers determines the direction of plant growth and depends on the action of plant hormones.

A hormone is a compound produced in one tissue and transported to cells in another tissue, where the hormone alters the behavior of cells. Plant hormones have dramatic effects on the pattern of cell division and expansion and influence the growth of an entire plant. For example, the plant hormone ethylene causes seedlings to grow short and fat, while another one, gibberellin, causes the growth of

tall, thin shoots. We'll discuss hormones in more detail in Chapter 33.

The development of form in plants depends on oriented cell division and cell expansion, which are influenced by hormones.

Root Tips Grow and Mature in Four Zones

Primary growth of roots occurs almost exclusively near their tips. In general, growth occurs in four overlapping zones—the zones of cell division, cell elongation, and cell differentiation, and the root cap (Figure 32-9).

The apical meristem near the tip of the root contains most of the roots' dividing cells. The actively dividing cells of the meristem have a small, uniform size. Farther back from the tip of the root, in the **elongation zone,** are cells that are growing longer. Even farther back from the meristem, in the **differentiation zone,** are cells that are becoming specialized to form cortex, endodermis or, vascular tissue, for example. Root hairs marks the beginning of the differentiation zone and fully mature root cells.

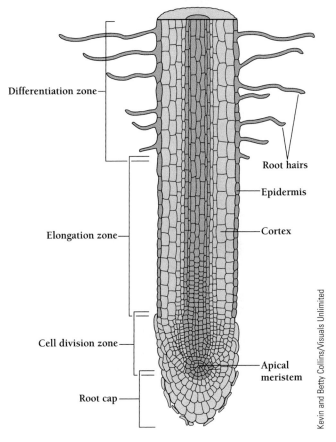

Figure 32-9
Growth in a root. In the root, plant biologists distinguish four growth zones—the root cap and three zones—of cell division, cell elongation, and cell differentiation.

At the very tip of the root is the **root cap**, whose cells make **mucigel**, a polysaccharide slime that lubricates the growing root as it pushes through the soil. The root cap protects the cells dividing in the apical meristem, which lies just behind the root cap. As the root grows, soil abrasion wears away cells in the root cap, but the root's apical meristem produces replacement cells.

Growing root tips include four zones: the slimy root cap, the cell-division zone, the cell-elongation zone, and the cell-differentiation zone.

Primary Growth in Shoots

Because shoot apical meristem must produce stem, leaves, branches, and flowers, primary growth is more complicated in shoots than in roots. The region of primary growth in a shoot may extend back as much as 10 to 15 cm from the shoot's tip and may include several nodes. As in the root, the shoot's apical meristem contains most of the dividing cells. Cells produced by division in the meristem grow longer, pushing the meristem upward. Farther from the meristem, cells become more mature, longer and, eventually, differentiated.

Shoot apical meristem occurs in buds, small growing— or potentially growing—extensions of the shoot. A plant has an **apical bud**, essentially the tip of the plant (though not necessarily the highest structure) and usually several **axillary buds**, just above the points where leaves join the stem (Figure 32-8). Each bud contains apical meristem and several minute shoots, each with a tiny node and internode, and one or more **leaf primordia**, small groups of undifferentiated cells to the side of the apical meristem that will eventually form leaves (Figure 32-8).

Many plant species show **apical dominance**, meaning that the apical bud suppresses the growth of nearby axillary buds by releasing a plant hormone that inhibits the growth of axillary buds. Axillary buds near the apical bud, at the top of the plant, are more inhibited than those lower down and farther away. That is why branches near the bottom of a conifer tree are much longer than those at the top. Removing the apical bud allows the axillary buds to grow, and many gardeners routinely produce bushier plants by pinching off the apical bud (Figure 32-10).

Each leaf starts as a tiny extension of the apical meristem. Apical buds suppress the growth of axillary buds, an effect called apical dominance.

How Does a Flower Derive from a Specialized Bud?

Some of the most interesting questions in plant biology concern the molecular and cellular events of flower development. How do plants control the development of these

Carolina Biological Supply/Phototake

Figure 32-10
Apical dominance in alpine firs (Abies lasiocarpa). Apical meristem tissue at the top of these trees exerts "dominance" over meristem tissue in the side branches, preventing them from growing. The branches closest to the top of the tree are most inhibited, while the branches closest to the bottom are least inhibited, creating the classic "Christmas tree shape." If the tip of the tree is removed by cutting or lightning, the side branches grow more, giving the tree a rounder, bushier shape.

beautiful patterned organs? To begin to understand how a flower forms, we need a clear picture of its parts and how it develops from a bud. Some buds are specialized to form flowers or a flowering shoot. A flower consists of up to four sets of parts arranged in **whorls**, concentric circles of modified leaves that develop into specialized flower parts (Figure 32-11A). The outer whorl (#1 in Figure 32-11A) consists of the **sepals**, the usually green parts at the base of a flower. Whorl #2 consists of the **petals**, the usually showy, colored parts of the flower. Whorl #3 consists of the male stamens. And whorl #4, the innermost whorl, includes the female pistils, each containing one or more carpels. The two inner whorls ultimately produce gametes—egg cells in whorl #4 and sperm cells in whorl #3.

Not all flowers have all four whorls. In willow trees and date palms, for example, some individuals have male flowers (with stamens but no pistils) and some have female flowers (with pistils but no stamens). Such plants are said to be *dioecious* [Greek, *di* = two + *oikos* = house], because the male and female plants live separately. In other plants, such as corn, such male and female flowers both occur on the same individual plant, so we say corn is *monoecious* [Greek,

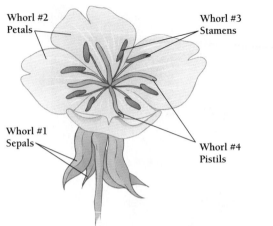

Whorl #2
Petals

Whorl #3
Stamens

Whorl #1
Sepals

Whorl #4
Pistils

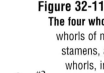

#4 #3

#2

#1

Figure 32-11

The four whorls of flowers. A. A flower consists of up to four whorls of modified leaves. These are sepals, petals, stamens, and pistils. This evening primrose has all four whorls, including both pistils and stamens. B and C. Every corn flower possesses two kinds of flowers. One set of flowers, the tassels, have stamens but lack pistils, and the other set, the silk, have pistils but lack stamens.

A. The four whorls of the evening primrose

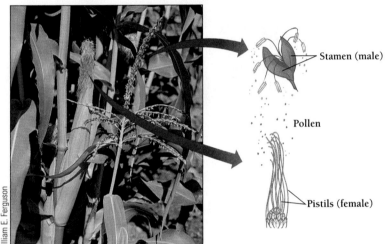

William E. Ferguson

Stamen (male)

Pollen

Pistils (female)

B. Corn tassels are flowers with stamens (♂),
corn silks are flowers with pistils (♀).

ilar ways. One of the most striking examples are the **homeotic selector genes**, which influence how and where organs form.

Very similar genes seem to regulate the development of body-part development in organisms as diverse as yeast and mammals. In fruit flies, for example, mutations in homeotic selector genes cause portions of a wing to grow where a balancing organ normally grows. A fly with three of these mutations develops a second set of wings.

In plants something similar happens. In Arabidopsis and snapdragons, each of three homeotic mutations changes the fate of one or more flower whorls, and a plant with all three mutations grows a "flower" made entirely of leaves. One interpretation is that meristems make leaves unless regulatory genes "tell" them to make something else.

Homeotic selector genes help regulate the development of flower parts in Arabidopsis.

monos = single + *oikos* = house]. (If we were to use this terminology for animals, we would say that humans, with our separate sexes, are dioecious.) But many plant species have just one kind of flower that has both stamens and pistils. Such a flower is neither monoecious nor dioecious.

The generalized flower develops from four whorls, which form sepals, petals, stamens, and pistils. Many flowers are missing one or more of these parts.

How Do Genes Regulate the Development of Flowers?

Studies of the tiny weed *Arabidopsis thaliana,* a member of the mustard family (like cabbage and broccoli), has shown that plant and animal embryos develop in surprisingly sim-

Secondary Growth Depends on Two Kinds of Lateral Meristems

The tallest and arguably the most magnificent of Earth's organisms are trees such as oaks and redwoods. These magnificent trees require both primary and secondary growth. **Secondary growth** refers to the thickening of stems, branches, and roots. Unlike primary growth, which occurs in apical meristem tissue, secondary growth arises from cell division in two kinds of lateral meristem tissue. The first, **vascular cambium,** produces secondary xylem and phloem, and the second, **cork cambium,** makes **cork,** the waterproof, insect-resistant covering on woody stems and roots (Figure 32-12). The only parts of a woody plant that lack vascular cambium and cork cambium are the tips of stems, roots, and branches that are still undergoing primary growth (elongation).

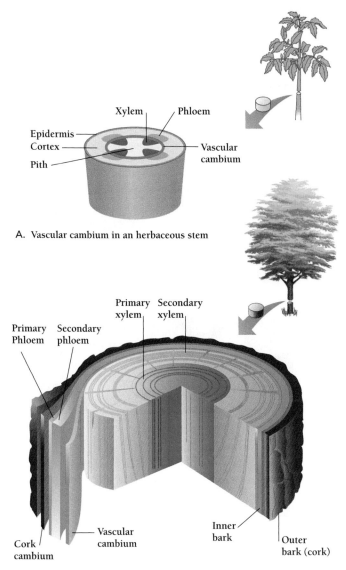

Xylem | Phloem
Epidermis
Cortex
Pith
Vascular cambium

A. Vascular cambium in an herbaceous stem

Primary Secondary
xylem xylem

Primary Secondary
Phloem phloem

Cork cambium | Vascular cambium

Inner bark | Outer bark (cork)

B. Vascular and cork cambium in a woody stem

Figure 32-12
Secondary growth. A. In a herbaceous stem, the cylinder of vascular cambium produces secondary xylem (to the inside) and secondary phloem (to the outside). B. In a woody stem, the vascular cambium produces secondary xylem (to the inside) and secondary phloem (to the outside). Everything outside of the vascular cambium is called bark—including the secondary phloem, the cork cambium and cork, and the epidermis. C. Cross-section through a woody stem showing two growth rings, each consisting of early wood (big cells) and late wood (small cells). The dark line at the outer edge of each ring indicates lack of growth during the winter.

Biology ⓢ Now™ Learn more about growth rings by clicking on this figure on your BiologyNow CD-ROM.

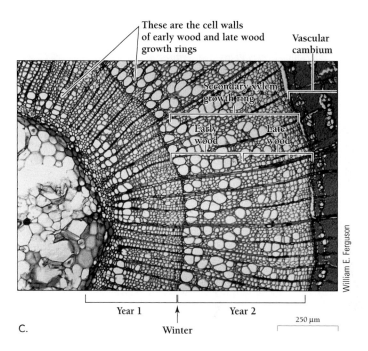

These are the cell walls of early wood and late wood growth rings

Vascular cambium

Secondary xylem growth ring

Early wood | Late wood

Year 1 | Year 2

Winter

250 μm

William E. Ferguson

C.

Vascular cambium, which lies between the primary xylem and the phloem, adds secondary xylem to the outside of the primary xylem and secondary phloem to the inside of the primary phloem. Each year the vascular cambium lays down new xylem and phloem, forming wood that pushes the primary xylem and phloem farther apart each year (Figure 32-12B). Secondary growth makes lumber for building houses and fences, pulp for paper, and root vegetables such as carrots, beets, turnips, and radishes.

As the vascular cambium increases the girth of a trunk or branch, the surrounding epidermis stretches and breaks. Just inside the epidermis, however, is another layer of meristem cells, the cork cambium. The cork cambium makes cells that grow outward toward the surface of the tree and other cells that grow inward toward the phloem. The outward-facing cork cells build waxy, fatty suberin and other secondary compounds into their cell walls. The cork cells then die, creating a thick layer of dead cork tissue that protects the tree from dehydration and attacks by herbivores and pathogens. **Bark** includes everything outside the actively dividing vascular cambium, including phloem, cork, and storage tissues.

As the tree grows, cork, cortex, and older phloem are stretched and crushed, so the living part of the bark makes up only a thin layer, a few millimeters thick. We can identify different tree species by the characteristic pattern of tears in the bark—including the smooth strips of the eucalyptus bark, the rough bark of the oak, and the regular cracks in the bark of ponderosa pine.

Secondary growth changes from season to season. In temperate zones, the vascular cambium is most active in the spring, producing large-diameter tracheids and vessel elements that, together, are called "early wood." In the summer, activity slows, and tracheids and vessel elements are much smaller, forming "late wood." Early wood is less dense and

Extreme Biology   Why Is *Arabidopsis thaliana* Well Suited to Genetic Analysis?

Arabidopsis thaliana, a small, common weed, also called "wall cress," is so well suited to genetic studies that it has been called the "botanical *Drosophila*." One major advantage Arabidopsis has over most plants is that a mature plant can be grown from a seed in as little as five weeks. In addition, Arabidopsis can self-fertilize, making it easy to isolate homozygous strains. It is so short and spindly that many plants can be grown easily and inexpensively in a small area, and its equally short genome has already been completely sequenced.

Geneticists have already identified hundreds of mutations in Arabidopsis. Some of these mutations lead to specific enzyme deficiencies, others to alterations in response to plant hormones, and still others to alterations in pattern. For example, one mutation leads to a missing shoot, another to a missing stem, another to a missing root (Figure A). Mutations affect mature plants as well as seedlings. One mutant completely lacks a shoot meristem, both in the intact plant and in tissue culture. Other Arabidopsis mutations affect the formation of specific types of tissue, while still others prevent the proper establishment of the embryo shoot-root axis.

Once a mutation has been found, it is relatively straightforward—if time-consuming—to identify and sequence the gene in which the mutation lies. The function of the gene can then be studied by putting it into normal or mutant Arabidopsis.

Researchers have performed such studies for a relatively small number of genes but have already learned much about Arabidopsis development.

(a)

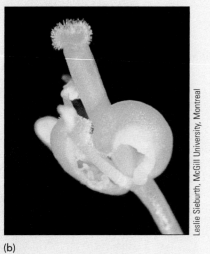

(b)

Leslie Sieburth, McGill University, Montreal

Figure A

Mutations in the flower *Arabidopsis thaliana*. (a) A normal flower with all four whorls: sepals, petals, stamens, and carpels. (b) The *apetala* 2 mutation converts sepals into carpels and petals into stamens.

---

lighter in color than late wood. The alternation of early wood and late wood creates distinctive rings of vascular tissue called **growth rings** that are obvious in a cross-section of a tree trunk or large branch (Figure 32-12C). Between growth rings, a tree produces little or no new tissue during the winter. Each ring represents one season, or year, and the number of rings in a tree trunk reveals the age of the tree.

> Secondary growth, which depends on vascular cambium and cork cambium, changes from season to season, resulting in characteristic growth rings in trees and other woody perennials.

32.3 Meristems May Produce Whole Plants

The same kind of meristem tissue that drives primary and secondary growth can produce whole new plants. Plants can reproduce asexually, without fertilization or seeds, in the process of **vegetative reproduction.** New strawberry plants form from root runners, and many ordinary people clone their favorite houseplants by means of vegetative reproduction. In some species, a stem cut from the main plant (a *cutting*), placed in water, generates new roots. These roots are said to be *adventitious* [Latin, = not properly belonging to], because they do not develop from a root apical meristem.

Some species that usually do not form such adventitious roots do so in the presence of the plant hormone auxin, which is sold in gardening stores. Amateur and professional gardeners often dip cuttings into a solution of auxin to enhance root formation. When the rooted cutting is planted, it grows into a clone—a mature plant that is genetically identical to the parent plant. In fact, farmers can use such cuttings to produce whole crops that are genetically identical.

One of the oldest and most commonly used means of artificial vegetative propagation is grafting. Horticulturists propagate several crops almost exclusively by grafting, including oranges, grapefruit, lemons and limes; avocados, apples, and plums; roses; and grapes. In **grafting,** a horticulturist grafts a cutting of one woody plant onto the root of another plant (Figure 32-13). The stem cutting is called the

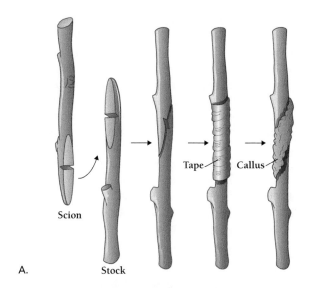

A.

B.

A.B. Joyce/Photo Researchers, Inc.

Figure 32-13
Grafting allows a scion of one variety to grow on the stock of another.
A. The process of creating a graft. B. A tree with many different apples growing on the same rootstock.

scion and the root is called the *rootstock* (or simply the *stock*). Recall from Chapter 31 that French grape varieties are universally grafted onto American rootstocks. Rather than try to breed a single hybrid with the best grapes and the best roots, grape growers simply graft a variety that makes good grapes onto a rootstock that resists pests or supplies nutrients effectively.

Most commercial fruit trees and roses are the products of such grafts. The trick in grafting is to bind scion and stock so that the vascular cambium of each plant is in contact with that of the other. Secondary growth establishes continuity both within the xylem and the phloem of scion and stock. The buds, leaves, flowers, and fruits all derive from the scion, while the root system derives entirely from the stock.

Another method of vegetative propagation is through tissue culture. Tissue culture has become increasingly important with the advent of genetically engineered plants. Carrots, cotton, tomatoes, and Douglas fir trees have all been propagated this way.

The starting material for plant tissue culture is a block of tissue taken from a plant and placed on a sterile nutrient medium that contains sugar, minerals, and hormones. The cells grow and divide to make an unorganized, undifferentiated mass. Under the right hormonal conditions, the little lump of cells will form shoots and roots, becoming a tiny plant, or "plantlet." This plantlet can grow and reproduce itself either sexually or vegetatively.

Meristem tissue allows plants to reproduce themselves asexually in a process called vegetative reproduction that does not involve fertilization or seed production. Cuttings, grafting, and tissue culture allow the artificial propagation of desirable plant varieties.

Key Concepts

- In developing seeds, plant embryos establish a body plan soon after fertilization and then pause after the seed matures.
- The primary growth and development of plants comes from apical meristems. Secondary growth comes from lateral meristems.
- Specific genes control the pattern of development in flowers.

Summary with Key Terms

How do plants reproduce sexually?

In **alternation of generations** in flowering plants, the diploid **sporophyte** produces haploid spores, and the haploid **gametophyte** produces haploid gametes (eggs and sperm). A flower may contain both male and female reproductive structures. **Stamens** (male structures), each made of an **anther** and a **filament**, produce haploid **microspores**, which develop into pollen grains (male gametophytes). **Pistils** (female structures) consist of one or more carpels. **Carpels** produce haploid **megaspores**, which develop into female gametophytes, the **embryo sacs**. The female gametophytes make female gametes (eggs) and the male gametophytes make male gametes (sperm). Each carpel contains a **style**, which connects the ovary to the **stigma**. Each **ovary** (made of one or more carpels) produces one or more **ovules**, which form the embryo sac and the **integuments** of the developing seed.

A **pollen grain** develops a **pollen tube** and two sperm cells. Flowering plants undergo **double fertilization**, in which one sperm fuses with an egg cell and a second sperm fuses with two polar nuclei within the carpel. The first fusion produces the zygote (2*n*), which grows into an embryo. The other gives rise to the **endosperm**, a triploid (3*n*) tissue that nourishes the growing embryo.

How do plant embryos establish a body plan?
The zygote divides to form an embryo that first resembles a sphere, then a heart. Development stops after the seed matures, when the embryo loses most of its water and becomes **dormant.** A seed consists of a dormant embryo, its endosperm, and a seed coat.

The resumption of growth is called **germination**, a process that begins when the seed swells with water. The first part of the embryo to emerge from the seed is the **radicle,** which contains the root's apical meristem and quickly forms a functioning root system. The shoot apical meristem lies above the point of attachment of the **cotyledons,** or seed leaves. Flowering plants are divided into two classes—the **eudicots** and the **monocots**—based on the number of cotyledons.

How do plants build a shoot and a root through primary growth?
A seed may maintain its dormant embryo for many years until the proper environmental changes bring about germination. Germination triggers a huge increase in biochemical activity for which the growing plant derives energy from stored starches, oils, and proteins. In grains, α-amylase and other enzymes break down starches in the endosperm.

The primary growth of both roots and shoots depends on **apical meristems.** The growing root tip has four distinct zones: the **root cap,** which secretes **mucigel;** the cell division zone (apical meristem); the **elongation zone;** and the cell **differentiation zone.** The growing tip of a stem makes stem, leaves, branches, and flowers. The place where the leaves join the stem is called a node. In between the nodes are the internodes. Each leaf originates as an extension of the apical meristem, called the **leaf primordium.** An embryonic shoot, with apical meristem and leaf primordia make up a **bud.** A bud at the tip of a shoot is an **apical bud.** Buds located just above the points where leaves join the stem are **axillary buds.** Many species show **apical dominance,** meaning that apical buds suppress the development of axillary buds.

How do plants build flowers?
The four **whorls** of a flower may form **sepals, petals,** stamens, and pistils. Mutations in several **homeotic selector genes** in *Arabidopsis* can result in interesting flower deformities, including a flower consisting entirely of leaves. Similar genes are found in animals (including vertebrates) and fungi.

How do plants build outward through secondary growth?
Secondary growth depends on two kinds of lateral meristems: the **vascular cambium,** which produces secondary xylem and phloem, and the **cork cambium,** which produces cork, the outer protective covering of woody stems and roots. **Bark** includes phloem, **cork,** and storage tissues. The

rate of xylem growth changes from season to season, resulting in characteristic **growth rings.**

How do plants reproduce asexually?

In **vegetative reproduction,** the same kind of meristem tissue that drives primary and secondary growth produces whole new plants. Many plants propagate in nature by producing adventitious roots or stems. Farmers, horticulturists, and researchers also propagate desirable varieties by stimulating adventitious root formation with auxin, by **grafting** a cutting of one plant together with the root or stem of another, or by allowing tiny plants to develop from undifferentiated tissue in artificial culture.

Review and Thought Questions

Review Questions

1. How does a plant's gametophyte different from its sporophyte?
2. What are the two products of double fertilization?
3. What determines when a seed germinates?
4. What happens during the germination of a seed?
5. Describe the four zones of growth and development in a root.
6. What is apical meristem tissue?
7. How is meristem tissue important to plants?
8. What happens if you remove the topmost tip of a growing plant?

9. Explain how a gene mutation can dramatically alter flower morphology.
10. Give some examples of vegetative reproduction. How is it different from sexual reproduction?

Thought Questions

11. Why would yeasts, plants, and vertebrates share similar genes for regulating development?
12. What do you think are the advantages of growing fields of identical plants, whether grapes, corn, or trees for timber?

BiologyNow Resources

Biology🧬Now™

Active Figures

32-2: Alternation of generations
32-12: Growth rings

Preparing for an exam? Take a diagnostic test on your BiologyNow CD-ROM.

Online materials relating to this chapter are at:

http://biology.brookscole.com/AAL3

About the Chapter-Opening Image

Not only do modern corn kernels "pop," so do the kernels of corn's presumed ancestors.

How Do Plants Respond to Their Environment?

Key Questions

- How do plants respond to their environment?
- How do hormones coordinate the activities of cells?

A Life Lived Full

In 1928, a young English chemist found himself on the job market in the middle of the worst economic depression in modern times. Unable to find an academic position in England, 25-year-old Kenneth Thimann landed a job as an instructor in biochemistry at the new California Institute of Technology (Caltech), in southern California. In 1929, Thimann and his artist wife, Ann, sailed for America on their first wedding anniversary. It was the beginning of a new life, a life that would profoundly influence American agriculture and even the course of the Vietnam War.

Like many a biologist's career, Thimann's rests on an intriguing observation by Charles Darwin. In the 1880s, Charles Darwin and his son Francis performed a series of experiments designed to explain **phototropism,** the tendency of plants to bend toward light (Figure 33-1A). The Darwins showed that something in the tip of an oat seedling causes the seedling to bend toward light. When Charles and Francis Darwin masked the tips of seedlings with a lightproof foil, the seedlings no longer bent toward the light (Figure 33-1B).

Forty-five years later, the Dutch plant physiologist Frits Went and his father demonstrated that the "something" was a chemical substance. In 1926, the younger Went was serving his compulsory military service during the day and spending his evenings working as a graduate student in his father's plant physiology laboratory at the University of Utrecht. Frits Went set out to isolate the growth-promoting substance from the tip of the oat seedling, using the strategy illustrated in Figure 33-2. On a tiny cube of agar, he placed as many plant tips as he could fit. After an hour, he removed the tips, reasoning that the hypothetical substance would have entered the cube of agar. He then placed the block on one side of a seedling stump that was growing in the dark and waited to see what would happen.

If the agar contained a substance that could alter growth, then the stumps should bend toward or away from the side with the cube of agar. At first nothing happened. By 3 A.M., however, the seedling had begun to curve away from the agar cube. Excitedly, Went ran home, burst into his parents' bedroom, and declared, "Father, come and see, I've got the growth substance!" Sleepily, his father suggested he repeat the experiment during the day. "If it is any good," the elder Went said, "it will work again, and then I can see it." Indeed, it worked again, and Frits Went named the mysterious growth substance **auxin** [Latin, *augmentum* = increase].

Helmut Gritscher/Peter Arnold, Inc.

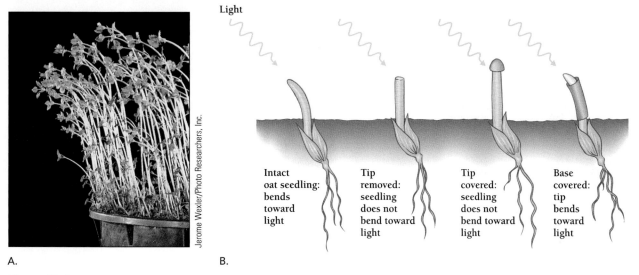

Light

Intact oat seedling: bends toward light

Tip removed: seedling does not bend toward light

Tip covered: seedling does not bend toward light

Base covered: tip bends toward light

A.

B.

Figure 33-1

What makes plants bend toward light? A. Seedlings bending toward light. B. Charles and Francis Darwin showed that the part of the seedling that responds to light is the tip. When the tip is covered, the plant will not bend. Later research showed that covering the growth region, below the tip, has no effect on bending.

Biology ⒺNow™ Learn more about phototropism by clicking on this figure on your BiologyNow CD-ROM.

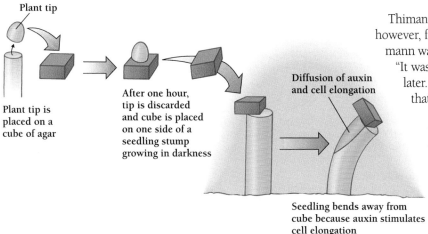

Plant tip

Plant tip is placed on a cube of agar

After one hour, tip is discarded and cube is placed on one side of a seedling stump growing in darkness

Diffusion of auxin and cell elongation

Seedling bends away from cube because auxin stimulates cell elongation

Figure 33-2

Does a substance cause plants to bend toward light? To answer this question, Frits Went placed the tips of seedlings on blocks of agar, guessing that a chemical substance would diffuse into the agar. He then balanced the blocks on the stumps of seedlings so that any substance in the agar would diffuse into only one side of the stem. The results confirmed his suspicions. Some substance causes the seedlings to bend. In addition, he learned that the seedling bends *away* from the substance.

When Kenneth Thimann arrived in California three years later, one of the first people he encountered was fellow instructor Herman Dolk. Dolk had been at the University of Utrecht with Frits Went, and he recruited Thimann to help him identify the chemical nature of auxin. Thimann and Dolk worked doggedly for two years, struggling to isolate the growth substance and solidifying a friendship.

Thimann's life seemed to couple tragedy with triumph, however, for in 1931, Dolk was killed in a car accident. Thimann was left to finish their work by himself (Figure 33-3). "It was kind of a lost feeling," he recalled almost 60 years later. "When Herman was killed, I was so deep into it that there was no choice but to go on."

Thimann went on to isolate the first known plant hormone, indole-3-acetic acid (IAA), which he showed was the Wents' "auxin." Then, in an inspired move, Thimann determined that by substituting other chemical structures for the indole in natural auxin, he could make synthetic auxins with the same properties as the natural hormone (Table 33-1). But these synthetic auxins differed from the natural ones in one important respect. Plants did not possess enzymes for breaking down the synthetic versions. The synthetic auxins, therefore, persist in the plant much longer than natural ones.

Thimann and others began a series of experiments to see how auxin affects plant growth. Auxin, they found, stimulates both rapid cell elongation and cell division, but also inhibits cell division in lateral buds.

In the 1930s, Frits Went left Holland to join Thimann at Caltech, and together they discovered that auxin also stimulates plants to grow roots. It was a tremendously valuable discovery. Today, virtually every nursery in the United States and Europe uses artificial auxin to induce the formation of roots in cuttings.

In the same year, Thimann and the Swedish-born biologist Folke Skoog, at the University of Wisconsin, showed that auxin could inhibit growth as well as promote it. Auxin

Figure 33-3
Kenneth Thimann. When asked for some advice to give to students, Thimann said, "Never give up."

is responsible for apical dominance, the tendency of the apical meristem to suppress growth in lateral buds. In addition, auxin inhibits leaves, fruits, and flowers from falling. Today, many fruit growers spray their orchards with synthetic auxin to prevent the fruit from falling before all of the fruit in the orchard is ready for harvest.

But it was Thimann's next discovery that influenced agriculture most dramatically. Thimann had attracted the interest of Harvard University, which invited him to join its faculty. At Harvard, he noticed that although a little auxin may either promote growth or inhibit growth, large amounts of auxin are fatal to plants.

During World War II, researchers at the University of Chicago developed Thimann's synthetic auxins for use as herbicides (products used to kill plants). The plan was to use the chemicals to defoliate areas hiding Japanese soldiers. In fact, the war ended before the herbicides were ever used.

The herbicides nonetheless came into heavy domestic use immediately after the war. Farmers no longer needed to plow under acres of weeds before planting a crop. Spraying herbicides was easier, and not plowing reduced erosion of valuable topsoil as well. By suppressing weeds with herbicides that killed eudicot weeds but not monocot crops, farmers growing corn and wheat were able to increase crop yields by 30 percent. Also, for the first time, road crews did not need to mow the edges of highways. They sprayed herbicide instead.

The most famous herbicides are the synthetic auxins 2,4-D and 2,4,5-T (Table 33-1). In the 1960s, American military strategists revived the idea of using herbicides as a weapon in the war against North Vietnam. American planes sprayed Vietnamese forests with heavy doses of herbicides to expose enemy positions. The sprayed trees dropped their leaves, revealing soldiers and civilians alike to American and South Vietnamese aircraft flying overhead. The herbicides

Table 33-1
Natural and Synthetic Plant Hormones and Their Effects

Hormone	Effects
Auxins (IAA)	Stimulate cell elongation and differentiation, apical dominance, development of fruit; act in tropisms

H_2C—COOH

Indoleacetic acid (IAA)

2,4-D	One of many synthetic auxins used as herbicides

O—CH_2—COOH

Cytokinins	Stimulate cell division, promote bud initiation, relieve apical dominance, inhibit root initiation, delay abscission and senescence

HN—CH_2—CH=$C$$\begin{smallmatrix}CH_3 \\ CH_2OH\end{smallmatrix}$

Zeatin

Gibberellins	Promote seed germination, stem elongation, and fruit development

Gibberellic acid (GA$_3$)

Abscisic acid	Closes stomata during water stress; causes dormancy in seeds

Ethylene	Triggers leaf and fruit abscission, fruit ripening, and defensive responses after injury; reorients cell expansion; alters gravitropic responses

were also used to destroy crops, greatly reducing the North Vietnamese food supply.

The U.S. government experimented with a variety of herbicide mixes, but the most commonly used was a mixture of 2,4-D and 2,4,5-T, called "Agent Orange." Chemical companies, under pressure to supply huge amounts of these synthetic auxins to the government, produced bulk quantities without the normal precautions. As a result, the herbicides were heavily contaminated with "dioxins," a large class of toxic chemicals, many of which are potently carcinogenic. In the 1980s, 27,000 Vietnam veterans sued five chemical companies for illnesses they suffered after returning from Vietnam. The chemical companies settled out of court for $180 million. (Vietnam and its people suffered even more extensive exposure to dioxin.)

Thimann took credit for the many uses to which his discoveries were put. He also insisted that his research was always basic research, not applied research. "Improvements that are based on pure research are more fundamental than research based on finding applications," he said. But he also distanced himself from the consequences of his research. Late in life, he argued that dioxin-laced 2,4,5-T was not dangerous, pointing to a fire in 1976 at a herbicide factory in Seveso, Italy. "A lot of people got spattered with 2,4,5-T that contained dioxin." Exposed workers developed chloracne, a severe but temporary pimpling that is characteristic of dioxin exposure. "But that's not cancer," said Thimann. "It heals by itself after a time, just like any other acne. They've done studies galore, studied their babies and everything. In the years since, nobody's shown any serious effects, and researchers know the workers were really dosed."

Thimann spent nearly 30 years at Harvard University, devoting himself to teaching and mentoring hundreds of students. When he was 58, he left Harvard to return to California. At the brand new University of California, Santa Cruz, he assumed the task of building a first-class science faculty and creating an environment in which students could come in close contact with their professors. His secretary once recalled that the two of them sometimes worked until midnight to get all their work done, but Thimann never refused to see a student. Neither his research nor the work of building a new university campus ever prevented him from putting students first. Even after he retired from teaching at the age of 68, he continued to supervise undergraduate research for more than 15 years. In his 90s, he still went to work every day.

In 1983, the Italian government awarded Thimann the Balzan Foundation Prize, one of the highest awards in biology. Great Britain's Royal Society and France's Academy of Science each elected him as a member. When Thimann's wife, Ann, died in 1991, he moved to Pennsylvania to be near his grown daughters. The University of Pennsylvania immediately provided him with an office, and he continued his involvement in research and academic life until he died in January 1997. He spent his last years studying aging (called "senescence") in plants. He never lost his love of science or of plants, once remarking, "We look on nature as a book—every now and then we can turn a page. It's a great thrill."

33.1 How Do Plants Respond to Their Environment?

Most animals move around to find new supplies of energy and nutrients, and they have evolved dozens of styles of locomotion, including swimming, running, and flying. In contrast, plants derive their energy from sunlight and their carbon from the air around them—an arrangement that makes locomotion unnecessary. In any case, plants obtain water from roots permanently embedded in ground, so they cannot just pick up and leave.

Plants are nonetheless active and responsive to their environments. Unlike animals, which have a permanent mature shape, plants change shape throughout their lives. A plant shaded by other plants grows toward any available light. A low-growing plant may spread along the ground, rerooting itself as it goes. A vine may grow up toward light or out across the ground. The roots of a plant growing in dry soil grow downward toward water. A plant exposed to more light than previously may gradually replace its old leaves with new ones that are less sensitive to light.

Not all plant activity is slow or accomplished by means of gradual growth. Plants continuously adjust the position of their leaves or flowers to track the sun from hour to hour as it moves across the sky (Figure 33-4). Flowers open and close, according to the time of day and other variables. Some plants forcefully eject seeds from their fruits. Species of *Mimosa* (called "sensitive plants") rapidly fold their leaves in response to a light touch.

Despite such activity, plants are limited in their ability to seek desirable environments or avoid undesirable ones. They must make the most of what is available, and environmental cues dictate the pattern of growth. In response to light and gravity, for example, the shoot grows up and the root grows down into the earth. In response to changes in day length, temperature, or soil humidity, plants may drop their leaves, go dormant, or burst into flower. Depending on environmental cues, the same cells that give rise to leaves and stems can give rise to flowers or roots.

Plants' ability to respond to their environment seems remarkable, since their cells, which are enclosed in a cellulose wall, are more rigid and more tightly glued to one another than those of animals. Yet when a leaf turns toward the morning sun, the cells on one side of its petiole must lengthen while those on the other side remain the same size. For the leaf to grow toward the sun, cell expansion must be perfectly coordinated.

In the rest of this chapter, we see some of the molecular and cellular mechanisms for coordinating the responses of plants to environmental cues.

Like other living organisms, plants are highly responsive to changes in their environment. Over time, a plant may alter its shape, its structure, and even the positions of its leaves and flowers in ways that increase its ability to survive.

Figure 33-4
Solar tracking. Flowers and leaves of some plants turn to face the sun as it moves across the sky.

How Do Plants Respond to the Pull of Gravity?

Growing toward or away from a stimulus is called a *tropism* [Greek, *trope* = turning], so plants' response to gravity is called **gravitropism** (Figure 33-5A). In response to gravity, roots generally grow down and shoots grow upward. Of course, side branches and roots may turn to grow horizontally. Gravitropism allows plants to respond vigorously to their environment; if we turn a young seedling on its side, the root turns and grows downward and the shoot turns and grows upward. How exactly does a plant "know" which way is up?

A simple experiment demonstrated the location of a root's gravity detector: removal of the root cap abolishes the root's gravitropism. The response of the root must depend on the cells of the root cap. But which component of the root cap cells detects gravity?

Large cells in the root cap (and in other tissues throughout the plant) contain organelles called **amyloplasts,** each of which contains several starch grains (Figure 33-5B). The amyloplasts are denser than other organelles, and, like sand in a jar of water, they consistently settle to the lowest part of the cell.

Plant physiologists long suspected that the amyloplasts detect gravity. To test this hypothesis, researchers treated roots with the plant hormones gibberellin and cytokinin, which caused the root cells to digest their amyloplasts. With no amyloplasts, the roots no longer responded to gravity.

Once a root detects gravity, large differences in the rate of elongation of cells on the top and bottom cause the root to bend. Since this response is so similar to the phototropic response studied by the Darwins and the Wents, we should not be surprised to learn that auxin plays a role in gravitropism as well. Auxin is made in the tips of shoots and is transported downward, creating a chemical difference that helps the plant define "up" and "down." Other plant hormones also participate in gravitropism, including ethylene and abscisic acid, which we discuss later. The initial trigger, however, seems to be calcium ions.

When a root is oriented horizontally, calcium ions accumulate along the lower surface of the root and trigger the transport of auxin to the lower side, where its high concentration inhibits elongation. (Oddly, although auxin promotes growth in stems, it inhibits the growth of root cells.) Meanwhile, the upper side, with less auxin, elongates more rapidly. As a result, the cells on the upper surface of the root grow more rapidly and force the root tip downward. When the root tip is oriented vertically once again, the asymmetrical distribution of calcium and auxin disappears, and the root grows straight.

> Gravitropism in plants depends on the settling of amyloplasts to the bottoms of cells. Auxin helps mediate the bending of roots and shoots in response to gravity.

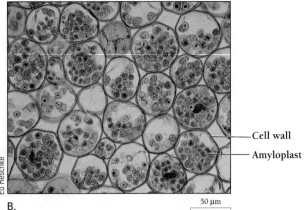

Cell wall
Amyloplast

50 μm

A.
B.

Figure 33-5
How do plants know which way is up? A. A fallen tree often sends new shoots upward, away from the pull of gravity. B. Amyloplasts in cells of the root cap drop to the bottoms of cells and help plants detect gravity.

Biology ⊜ Now™ Learn more about gravitropism by clicking on this figure on your BiologyNow CD-ROM.

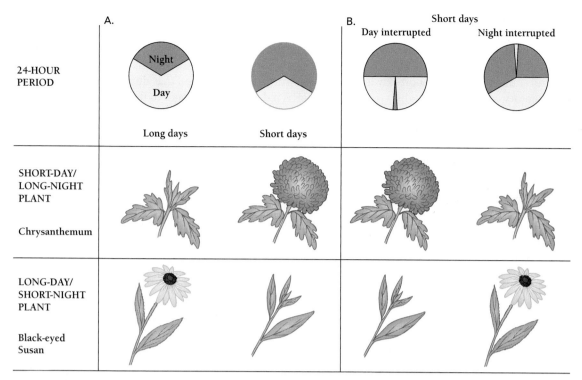

24-HOUR PERIOD	A. Long days	Short days	B. Short days Day interrupted	Night interrupted
SHORT-DAY/ LONG-NIGHT PLANT Chrysanthemum				
LONG-DAY/ SHORT-NIGHT PLANT Black-eyed Susan				

Figure 33-6

Short-day and long-day plants. A. Short-day plants such as chrysanthemums flower when days are shorter and nights are longer than a certain length. Long-day plants such as black-eyed Susans flower when nights are shorter than a certain ("critical") length. B. Plant physiologists Hamner and Bonner asked if plants were sensitive to day length, night length, or a ratio of day length to night length. When they interrupted the day with a period of darkness, it made no difference in whether either kind of plant flowered. But when they interrupted long nights with flashes of light, the short-day (long-night) chrysanthemum failed to flower, and the long-day (short-night) black-eyed Susan flowered. The two researchers concluded that in both kinds of plants, it is *only* the *length of the night* that determines flowering.

How Do Plants Detect Changes in Season or Daylight?

Every gardener, farmer, or hiker knows that different plants flower in different seasons and at different times of day. The flowers of crocuses and daffodils appear in the spring, and those of carnations and black-eyed Susans in the summer. Appropriate timing of flower and seed production determines how much sun and water will be available to new plants and which other organisms will be around to compete, to consume, or to pollinate, for example. In this section, we ask, How do plants detect and respond to seasonal changes?

In some species (such as cucumbers, peas, and tomatoes), the time of flowering depends only on maturity or size. Flowering may occur after a plant has grown for only a few weeks, or flowering may take months, or even years. In other species, flowering always happens at the same time of year, even if the plant reaches maturity much earlier.

Understanding the seasonal control of flowering is a practical as well as a scientific issue. Soybean farmers, for example, once tried to stagger their harvests by planting fields at two-week intervals. But they found that all the flowers appeared at once, late in the summer, and the whole harvest was ready at the same time regardless of the planting time.

For any species, the date of flowering varies with latitude and altitude. Spring-flowering plants in Massachusetts, for example, generally flower later than those in Georgia. From this observation, biologists conclude that flowering does not result from the operation of an internal annual clock. The flowering response turns out to be an example of **photoperiodism,** the response to the relative lengths of day and night as they change during the year. *Short-day plants* flower when days are short and nights are long, in the late summer, fall, or even winter. Examples include cocklebur, soybeans, and chrysanthemums (Figure 33-6).

In contrast, *long-day plants* require long days and short nights. These generally flower in the late spring or early summer and include spinach, sugar beets, and black-eyed Susans. Finally, in *day-neutral plants,* such as tomatoes, corn, and dandelions, flowering does not depend on day length. Plants from tropical areas are often day neutral: lacking both night-length variations and temperature fluctuations during the year, tropical plants have been under little pressure to regulate the seasons of their flowering. By contrast, plants closer to the poles have evolved responses that take advantage of the day-length signals. This adaptation ensures that plants will flower early enough to set seeds before cold weather strikes.

One of the first questions that researchers asked was whether flowering depended on the length of the day or the length of the night. To find out, they grew short-day plants—cocklebur in one laboratory, soybeans in another—in artificial cycles of light and dark. The plants flowered whenever the "nights" were longer than a critical length.

In 1938, Karl Hamner and James Bonner, plant physiologists at the University of Chicago, tried a different experiment. They interrupted the light period with a period of darkness or the dark period with a period of light (Figure 33-6B). For both short-day and long-day plants, interrupting the day made no difference in flowering. For short-day plants, however, interrupting the night that was longer than the critical length suppressed flowering. For long-day plants, interrupting a night that was longer than the critical length *promoted* flowering. The interruption did not need to be large—an ordinary light bulb turned on for 30 seconds was enough to prevent (or promote) flowering. The researchers concluded that both short-day and long-day plants measure the length of the night, not the day. Short-day plants should really be called "long-night plants," and long-day plants should really be called "short-night plants," but biologists continue to use the original terminology.

This early work on photoperiodism helped farmers control when their crops flowered and set seed and florists provide year-round supplies of flowers. But the chemical basis of photoperiodism remained a mystery.

In 1940, however, researchers at the U.S. Department of Agriculture Laboratory in Beltsville, Maryland, decided to study which wavelengths of light most effectively interrupted the long nights. They chose soybeans, which are short-day (long-night) plants. The researchers began by showing that illuminating a single leaf was enough to suppress flowering in the whole plant, which implied that the light effect occurred in the leaves.

They then built a large instrument that would light individual leaves in separate groups of plants with light of different colors. Their experiment showed that red light was most effective at interrupting the soybean's night (and suppressing flowering). The researchers concluded that the light detector must be a pigment that absorbs red light. When researchers finally identified the pigment, they named it **phytochrome** [Greek, *phyton* = plant].

The Beltsville group then made a remarkable discovery. They could cancel the effect of red light by immediately following the red flash with a light flash of longer wavelength. The most effective wavelength for the red-light effect was about 660 nm, and the most effective wavelength for the reversing flash was about 730 nm, in the "far-red" region of the spectrum. Sterling Hendricks, the plant physiologist who had built the instrument to determine the action spectrum, hypothesized that phytochrome must exist in two alternate forms, an inactive form that absorbs red light and an active form that absorbs far-red light. He called the inactive,

red-light–absorbing form P_r and the active, far-red–absorbing form P_{fr}.

Hendricks and his colleagues knew that sunlight contains much more energy in the red than in the far-red part of the spectrum. They proposed that during the day, abundant P_r absorbs red light and changes to its active form, P_{fr}. Then, during the night P_{fr} breaks down and levels fall. They hypothesized that P_{fr} promoted flowering in long-day plants and suppressed flowering in short-day plants (Figure 33-7).

Other plant physiologists initially doubted the existence of phytochrome. But Hendricks's hypothesis made a strong prediction: purified phytochrome should be a single compound that could switch from one form to another in response to light. The prediction stimulated a biochemical search that lasted nearly 40 years. When several research groups finally isolated pure phytochrome, it performed exactly as Hendricks had predicted. Under natural sunlight, both forms of phytochrome are present, but in darkness, in the absence of red light, P_{fr} breaks down, due to the activity of enzymes and the pigment's natural instability.

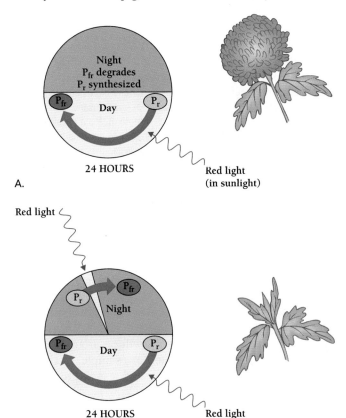

Figure 33-7

How do plants sense daylight? The flowering of long-day and short-day plants depends on the conversion of phytochrome from the inactive form, P_r, to the active but unstable form, P_{fr}. P_{fr} promotes the flowering of long-day plants and inhibits the flowering of short-day plants. A. During the day, the red light of sunlight changes P_r into P_{fr}. During the night, P_{fr} breaks down and its levels fall. At sunrise, sunlight begins converting P_r to P_{fr}, signaling the start of a new day. B. A flash of light at night converts P_r to P_{fr}, allowing the plant to sense the interruption of the night.

As P_{fr} breaks down at night, plants begin detecting night. Meanwhile, cells begin rebuilding P_r levels (when no red light is converting it to P_{fr}). At daybreak, sunlight begins converting P_r to the P_{fr} and the plant detects the end of night.

Besides acting as a daily clock, phytochrome is important in other processes, including flowering, germination, leaf formation, and the elongation of new seedlings. In almost every case, the process controlled by phytochrome depends on the conversion of P_r to P_{fr}. The red light in sunlight promotes these processes and its absence during the hours of darkness inhibits them. Phytochrome is a molecular switch that sunlight controls.

Plants detect the passing of the seasons by measuring the length of the night. During the day, red light from sunlight converts a pigment called phytochrome from its inactive form, P_r, to an active one, P_{fr}. At night, levels of the unstable P_{fr} fall.

33.2 How Do Hormones Coordinate the Activities of Cells?

A **hormone** is a compound produced in one tissue that alters the behavior of cells in another tissue. Plant hormones can move either directly from cell to cell or through the vascular system. Plant physiologists have identified five kinds of plant hormones: auxins, gibberellins, cytokinins, ethylene, and abscisic acid. There are three naturally occurring auxins, at least 60 gibberellins, and several cytokinins.

Hormones can control complex sequences of development and respond to an ever-changing environment because the same hormones affect different tissues differently; hormones have different effects at different concentrations; and hormones interact with one another in complex ways.

Auxin Coordinates Many Aspects of Plant Growth and Development

Frits Went explained phototropism by proposing that light on one side of a stem would cause auxin to flow to the other. There, auxin stimulates cell elongation, causing the stem to bend toward the light. But although Went had proposed this hypothesis, it had not yet been tested and verified. In the early 1960s, plant physiologist Winslow Briggs, who was then at Stanford University, decided to test Went's hypothesis (Figure 33-8).

Briggs thought up a way to measure the differences in auxin concentration between the side of a plant tip facing light and the side away from light. He sliced each seedling tip and its agar block in two and inserted a sliver of mica, a type of rock composed of layers of thin sheets. When the mica completely blocked auxin movement between the lighted side and the unlighted side, Briggs found no difference in the auxin levels of the two sides. When Briggs left some intact tissue near the very end of the tip, however, the auxin concentration on the shaded side rose to nearly twice that on the illuminated side. The light seemed to trigger the transport of auxin from the lighted to the unlighted side of the tip. The higher concentration of auxin on the unlighted side caused the cells on that side to elongate more rapidly, so that the seedling curved toward the light.

Briggs's experiments showed that auxin acts as the mediator for phototropism by moving through tissues in response to light and stimulating cell elongation. Further research

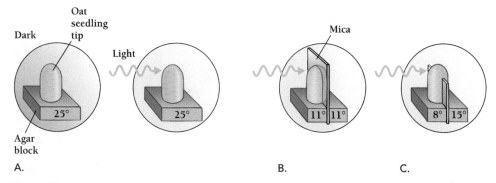

Figure 33-8

Does auxin move from one side of the seedling to the other? Or does the seedling synthesize extra auxin on one side? Winslow Briggs's experiment showed that light causes a redistribution of auxin in seedling tips. Briggs put oat seedling tips on blocks of agar, then measured the auxin content of the agar blocks by transferring them to oat seedlings, as Fritz Went had (Figure 33-2). The numbers on the blocks of agar are the approximate degrees of bending the blocks caused in oat seedlings. More degrees indicates more auxin. A. The auxin concentration in a seedling grown in the dark is the same as the concentration of that in a seedling grown in light. We can conclude that the seedling tip does not make more auxin in response to light. B. An oat seedling tip and its agar block is divided in half by an auxin-proof barrier and exposed to light on one side. But the auxin concentrations are the same on each side. Therefore, we know that extra auxin is not made on the shaded side. C. If the barrier is incomplete, so that auxin can move from one side to the other, the auxin concentration becomes much higher on the shaded side. We can conclude that auxin moves from the lighted side to the shaded side of the seedling tip.

showed how auxin causes cell elongation. Most of the time, turgor pressure pushes on the cell wall of a plant, but the rigid wall resists stretching. When auxin enters a cell, it stimulates the cell to secrete a protein, called *expansin,* that loosens the bonds among the molecules of the sides of the cell wall. As these bonds loosen, turgor pressure stretches the cell wall and the cell expands, like a long balloon being blown up.

Besides inducing cell elongation in stems, auxin affects the growth of many other plant tissues. For example, in the shoot, auxin inhibits the growth of lateral buds (apical dominance). In plants that show apical dominance, axillary buds do not develop into branches unless the terminal bud is removed. In the roots, the opposite happens. Auxin inhibits elongation of the main root, but promotes the growth of lateral roots. Auxin also promotes the "abscission" of leaves and fruits, in which their connection to a stem weakens and they fall. Some of auxin's effects occur rapidly, within 10 to 20 minutes of a plant cell's exposure to the hormone, and many occur because auxin affects gene expression in individual cells.

Finally, as we have seen already, high concentrations of auxin are effective as herbicides. Synthetic auxins are especially effective because plants cannot break down these substances, and so they persist for a long time. Three synthetic auxins—2,4-D, 2,4,5-T, and MCPA—have been particularly popular because they kill eudicots but not monocots. Farmers can spray an entire field of corn or wheat, killing all the eudicot weeds without harming the monocot crop. Similarly, ranchers can selectively kill sagebrush and mesquite, both eudicots, in grassy pastures or rangelands, and homeowners can spray a lawn and kill every eudicot growing there without killing any grass.

> Auxin causes plants to bend toward light by moving from the lighted side of a growing plant tip to the shaded side, where it stimulates cell elongation. Auxin also plays many roles, including inhibiting the growth and development of axillary buds. At high concentrations, auxins (including synthetic ones) are potent herbicides.

How Do Gibberellins Affect Cells?

Plants have more kinds of **gibberellins** than of any other hormone. All the gibberellins are complex molecules with 19 or 20 carbon atoms grouped into rings (Table 33-1). Most plant species have several different kinds of gibberellins, though probably not all are active forms.

Gibberellins can affect the overall growth of plants amazingly. In fact, biologists first discovered gibberellins because of their affects on stem elongation in rice plants infected by a fungus called *Gibberella fujikoroi.* The disease, first described by Japanese biologists in the 1890s, causes rice plants to grow so tall that they cannot support their own weight. The plants topple over into the water and rot. Japanese farmers called the disease *bakanae,* "foolish seedling." In the 1930s, two Japanese researchers identified the chemical

compound made by the fungus that is responsible for the extravagant growth of the rice seedlings, a gibberellin. In infected plants, the fungus raises gibberellin levels abnormally high. Subsequent research showed that a whole family of gibberellins existed. But because of World War II and poor communication, news of the discovery of gibberellins did not reach Western researchers until the 1950s.

Later research showed that plants normally make gibberellins. Some dwarf plant mutants—of peas, corn, beans, rice, for example—lack gibberellins altogether and are much shorter than their normal counterparts. When dwarf mutants are treated with gibberellins, in some cases with as little as a billionth of a gram per plant, they grow to normal size. Normally short plants treated with gibberellin also grow dramatically.

Like auxin, gibberellins seem to promote cell elongation by loosening cell walls, but with a more dramatic effect. If a segment of oat stem is treated with gibberellin and sucrose sugar (to provide energy), the oat stem will grow 15 times as long as an untreated stem.

Gibberellins also stimulate the breakdown of starches and sucrose to monosaccharides (simple sugars). The increased sugar concentration fuels the germination and early growth of seeds. In some cases, gibberellins can stimulate cell division—for example, in the apical meristem of the shoot. This effect is especially obvious in plants that keep their stems very short for the first year, forming a "rosette" (Figure 33-9). In the second year, when the plant flowers, the stem rapidly elongates (or "bolts"). Studies of these plants show that rosette plants are short because the stem

Courtesy of B.O. Phinney, University of California, Los Angeles

Figure 33-9
Gibberellin makes plant bolt. The hormone gibberellin causes stems to lengthen dramatically. This squat plant "bolts" when treated with gibberellin.

cells do not divide. Gibberellins—produced in the plant or applied experimentally—cause bolting by stimulating cell elongation and cell division. Finally, gibberellins can act to change the pattern of expression of specific genes, as in the case of α-amylase induction in the germinating barley seed (Chapter 32).

Gibberellins appear to be important in the normal growth of shoots, the germination of seeds, the flowering of plants, and the mobilization of food reserves from the endosperm of cereals and from the cotyledons of other plants. Because gibberellins are present at low concentrations in most tissues of a plant, their precise role has been difficult to understand.

Gibberellins regulate gene expression and promote cell elongation, cell division, and the breakdown of starches and sugars.

Figure 33-10
Witch's broom disease in a birch tree, Scotland. The fungus *Taphrina betulina* secretes cytokinins, plant hormones that can stimulate extreme lateral branching.

Cytokinins Stimulate Growth and Differentiation

Since the early years of this century, researchers have known that some substances, now called **cytokinins,** stimulate cytokinesis (cell division) in plant tissue culture. In all cases, however, these materials stimulated cell division only in the presence of the right amount of auxin. Thus, cytokinins and auxins together affect cell division.

In addition to stimulating cell division, cytokinins and auxins together help determine what kinds of tissues will form. The relative levels of cytokinins and auxins establish, for example, whether cells grown in the laboratory will continue to divide without differentiation or will differentiate into shoots or roots.

Although cytokinins and auxins work together, they also work in opposition to one another. Auxins promote root formation, while cytokinins suppress root formation. And while auxins promote bud dormancy and inhibit the formation of new buds, cytokinins release buds from dormancy and promote the formation of new buds. A plant infection called "witch's broom disease" provides an extreme example of the effects of cytokinins (Figure 33-10). The fungus that cause the disease secrete a cytokinin, which stimulates the increased lateral branching.

Auxins and cytokinins together coordinate the development of shoots and roots. The complementary actions of cytokinins and auxins depend not only on their different effects on cells but also on their differing distributions within the plant. As we noted earlier, auxins (which stimulate the formation of lateral roots) originate in shoot tips and move downward. In contrast, cytokinins originate in roots and move upward through the xylem. In a cutting, auxins therefore accumulate at the base and stimulate root formation. Conversely, cytokinins accumulate at the surface of a cut shoot and stimulate bud formation. Many trees sprout healthy shoots from cut stumps.

During the normal life cycle of a plant, cytokinins transported from roots to leaves by the xylem appear to prevent the aging of leaves. When a leaf ages, it loses chlorophyll and turns yellow, part of the process of **senescence,** the breakdown of cellular components leading to cell death in the leaf. Biologists can delay senescence in a cut leaf by applying cytokinins. Several fungi and two species of caterpillars use cytokinins to slow senescence, creating nutritious "green islands" on yellowing leaves on which these organisms can feed.

Naturally occurring cytokinins are present at exceedingly low levels. The first milligram of pure cytokinin isolated from plant tissue, for example, had to be extracted from 125 pounds (about 60 kg) of corn kernels. Modern techniques have made it possible to determine the amounts of specific cytokinins at levels as low as a billionth of a gram. And newer research suggests that cytokinins act by regulating patterns of gene expression in plant cells.

Together with auxins, cytokinins stimulate cell division, delay senescence, and determine how cells differentiate. Unlike auxins, cytokinins stimulate cell division and growth in axillary buds. Cytokinins function at extremely low concentrations.

How Did Researchers Identify Ethylene as a Plant Hormone?

Ethylene is a much simpler molecule than the other plant hormones, consisting of just two carbon atoms and four hydrogen atoms (Table 33-1). Unlike the other plant hormones, ethylene is a gas and is not transported in the vascular system.

The most dramatic, delicious, and economically important effect of ethylene is on ripening in fruits that have a distinct ripening phase. The ripening of a fleshy fruit (such as an apple or a tomato) is a complex process. Among the chemical changes that occur during ripening are the breakdown of chlorophyll, changes in the fruit's color, release of volatile compounds (which have a strong smell), the breakdown of starches into sweet sugars, and the breakdown of tough cell walls, making the fruit softer. Ethylene starts or speeds up all these changes (Figure 33-11).

In fruits, ethylene also triggers the increased production of more ethylene, so that once ripening starts, it spreads rapidly, both within a single fruit and from fruit to fruit. This positive feedback acceleration of ripening has practical consequences for fruit lovers. One can hasten the ripening of green fruits by putting them in an closed bag, which traps the ethylene, and by adding another piece of fruit that is already ripe and so a rich source of ethylene. For example, to ripen green peaches, one could put them in a paper bag with a ripe apple or a spotted banana.

Ethylene's effects also explain why "one rotten apple spoils the barrel." An overripe apple releases large amounts of ethylene and speeds up the ripening of those around it. These apples, in turn, release ethylene, and soon the whole barrel is filled with overripe apples, which soon rot.

Fruit growers and shippers knew how to hasten ripening long before they knew the chemical cause. The ancient Chinese ripened fruit in a room with burning incense, and Puerto Rican growers once built bonfires near their crops to stimulate flowering of their pineapples. Both these proce-

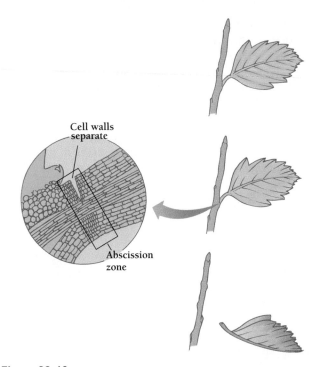

Figure 33-12

Abscission. A change in the balance between ethylene and auxin stimulates cells in the abscission zone at the base of a petiole to shrink and separate from one another. Finally, the fruit or leaf drops to the ground.

Figure 33-11

Peaches and apricots. Fruit ripening depends on the effects of the plant hormone ethylene. Ethylene stimulates the breakdown of the green pigment chlorophyll, the synthesis of red, yellow, or blue pigments, the release of aromatic compounds (that give the fruit a distinctive odor), and the breakdown of complex carbohydrates into sweet sugars. The resulting fruit attracts fruit-eating animals that help the plant by dispersing the seeds.

dures work because the incense and wood fires happen to produce ethylene.

Ethylene also causes plants to drop their leaves, a fact that led directly to the first identification of ethylene. Before the invention of electric power, many city streets were lit with gaslights. Gas lines carrying "illuminating gas" to each streetlight sometimes developed leaks and nearby trees lost their leaves. Workers soon discovered that the gas was the cause, and in 1901, a Russian graduate student named Dimitry Neljubov showed that the active component was ethylene.

Ethylene is present almost everywhere in a plant, dissolved in the plant's fluids. Under the right conditions, it causes plants to drop both leaves and fruit, a process called **abscission** (Figure 33-12). In addition, ethylene helps mediate other responses to the environment. For example, high levels of ethylene help seedlings find their way out from under rocks by triggering a "triple response" in the stems of seedlings grown in the dark—stems thicken, their elongation slows, and they begin to grow horizontally (Figure 33-13). When a tiny new seedling growing sideways out from under a rock or other object finally emerges from the soil, the amount of ethylene in and around the plant decreases (since there is no soil to hold the gas in), and the seedling resumes its upward growth.

Ethylene also plays an important role in signaling cell injury and in stimulating the plant's defensive responses.

A.

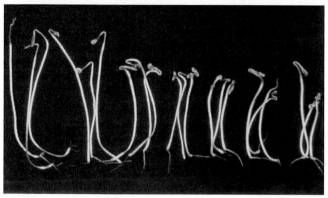

B.

Figure 33-13
The triple response to ethylene. In seedlings grown in the dark and exposed to ethylene grown in the dark, the stem thickens, elongation slows, and the stem begins to grow horizontally. A. Triple response of normal tomato seedlings with increasing ethylene from left to right. B. Response of mutant tomato seedlings that are relatively insensitive to ethylene, with increasing ethylene from left to right. (Yen, Lee, Tanksley, Klee, and Giovanoni, *Plant Physiology* [1995] 107:1343–1353)

Every injury stimulates ethylene production, and ethylene stimulates the formation of cork and the formation of lignin near a wound site, as well as defensive responses in other parts of the plant.

Ethylene is a gas that stimulates fruit ripening and the release of more ethylene. Ethylene also triggers the abscission of fruit and leaves, and the triple response in seedlings.

Abscisic Acid Promotes Dormancy in Buds and Seeds

Abscisic acid [Latin, *abscissus* = to cut off] has a confusing name because researchers initially thought it played a major role in abscission, the process by which leaves drop in the fall. But, despite its name, abscisic acid plays only a minor

role in abscission. We now know that a change in the balance between ethylene and auxin causes abscission.

What abscisic acid does do is play a major role in the suspension of bud and seed development, in signaling and preventing dehydration, and in regulating the opening and closing of stomata. Unlike the other plant hormones, abscisic acid is primarily an inhibitor. It antagonizes the effect of gibberellins, slows the growth of oat seedlings, inhibits the germination of wheat seeds, and prevents the synthesis of α-amylase by barley seeds. At a cellular level, it inhibits the transcription of some genes and stimulates the transcription of others. Abscisic acid also coordinates plant responses to a variety of environmental stresses, including drought, excess salt, waterlogging, cold, and mineral deficiency.

Abscisic acid antagonizes the effects of gibberellins and auxins, slowing growth of seedlings and inhibiting germination.

Are There Other Plant Hormones?

Research in the past decade has shown that beside the five basic plant hormones, a number of other signaling molecules help coordinate cellular activities in different parts of a plant. For example, when caterpillars chew on a leaf, the plant may produce *systemin,* which stimulates the synthesis of type of lipid called *jasmonic acid.* Jasmonic acid in turn stimulates the transcription of genes that promote defensive responses. One of these genes, for example, encodes a protein that inhibits the action of protein-digesting enzymes. The result is a bad case of indigestion for the caterpillar.

Another molecule recently identified as a plant hormone is *salicylic acid.* Salicylic acid derives its name from the willow tree [Latin = *salix*], whose bark was used for centuries by Native Americans and ancient Greeks to treat aches and fevers. Remarkably, salicylic acid, which we know as aspirin (acetylsalicylic acid), not only helps cure human ills but helps prevent plant infections. In plants, salicylic acid triggers defensive processes and promotes flowering.

One of salicylic acid's most striking actions is in the flowers of a number of heat-generating plants, such as those of skunk cabbage and the voodoo lily. In these flowers, salicylic acid stimulates the production of an electron carrier in the mitochondria that produces heat instead of ATP. The heat production is so dramatic that, on the day of flowering, the flower's temperature increases by 14°C (25°F). The high temperature vaporizes volatile chemicals that attract pollinating insects: a skunky odor in skunk cabbage and the odor of cow dung in voodoo lilies.

Many signaling molecules help plants respond to their environment by regulating patterns of gene expression and coordinating the behavior of cells.

In this last chapter of our section on plant anatomy and physiology, we have examined the ways that plants respond to their environments and the roles of plant hormones in plants' growth and development. We have reviewed classic experiments and some of the latest research in these areas. In the last section of *Asking About Life,* we study the anatomy and physiology of animals, with an emphasis on vertebrates.

Key Concepts

- Plants alter their activities and their development in response to environmental cues that include gravity and light.
- The internal signals in plants that help them regulate these activities include several kinds of hormones, the most prominent of which are the auxins, gibberellins, cytokinins, abscisic acid, and ethylene.

Summary with Key Terms

How do plants respond to their environment?

A plant's response to its environment depends on its ability to detect changes and conditions. For example, plants can detect gravity (**gravitropism**) by means of large cells in the root cap that contain starchy **amyloplasts**, which fall to the bottoms of cells. Plants detect day and night (**photoperiodism**) by means of the pigment **phytochrome**. Phytochrome is a molecular switch controlled by sunlight that regulates seasonal processes, such as flowering and seed production. Plants detect the passing of the seasons by measuring the length of the night. During the day, red light from sunlight converts phytochrome from an inactive form, P_r, to an active one, P_{fr}. At night, levels of the unstable P_{fr} fall.

How do hormones control the activities of cells?

Plant physiologists have identified five kinds of plant **hormones.** Each of these is a compound produced in one tissue or organ and transported to another tissue or organ, where it produces specific effects. The first plant hormone to be recognized was **auxin**, which coordinates **phototropism,** the growth of a plant toward light. Illuminating one side of the growing tip of an oat seedling causes auxin to move away from the lighted side. The higher concentration in the shaded side causes its cells to elongate, causing the plant to bend toward the light.

Gibberellins promote germination, trigger the mobilization of food reserves in germinating seeds, and stimulate growth in mature plants. **Cytokinins** stimulate cell division and growth and prevent leaf **senescence. Ethylene** promotes **abscission** of leaves and fruit, leaf senescence and fruit ripening, and also inhibits elongation of stems and roots. **Abscisic acid** promotes the breakdown of pigments and dormancy in buds and seeds.

Review and Thought Questions

Review Questions

1. Why does a plant need to detect changes in light?
2. Why does a plant need to detect gravity?
3. What kind of plants tend to be "day-neutral?" Why?
4. Define the word "hormone."
5. What are the primary effects of auxin?
6. How does phytochrome relay information about day length to plant cells?
7. How is senescence affected by cytokinins?
8. What is the triple response induced by ethylene?
9. Why might you expect the level of abscisic acid to increase in a plant subjected to environmental stress?

Thought Questions

10. Why do you think that some bacteria and fungi produce plant hormones? What advantage accrues to the fungus *Gibberella fujikoroi* in forcing rice seedlings to grow so tall?
11. If you were leading a committee charged with determining whether dioxin in Agent Orange likely caused health effects in military veterans, how would you design your study? (Experiments on humans would not be allowed.) What questions would you want answered?
12. Why do you think that auxin can both stimulate growth and inhibit it? Suggest a molecular mechanism by which this could occur.
13. Some plants become warmer on the day they flower and are ready to be pollinated. When women ovulate, their temperature also rises briefly. What might these two processes have in common?

BiologyNow Resources

Biology ⓔNow™

Active Figures

33-1:   Phototropism
33-5a: Gravitropism

Preparing for an exam? Take a diagnostic test on your BiologyNow CD-ROM.

Online materials relating to this chapter are at:

http://biology.brookscole.com/AAL3

About the Chapter-Opening Image

This slender seedling represents the tendency of a plant to bend toward light and the tendency of an engaged mind, such as Kenneth Thimann's, to bend toward a new question.

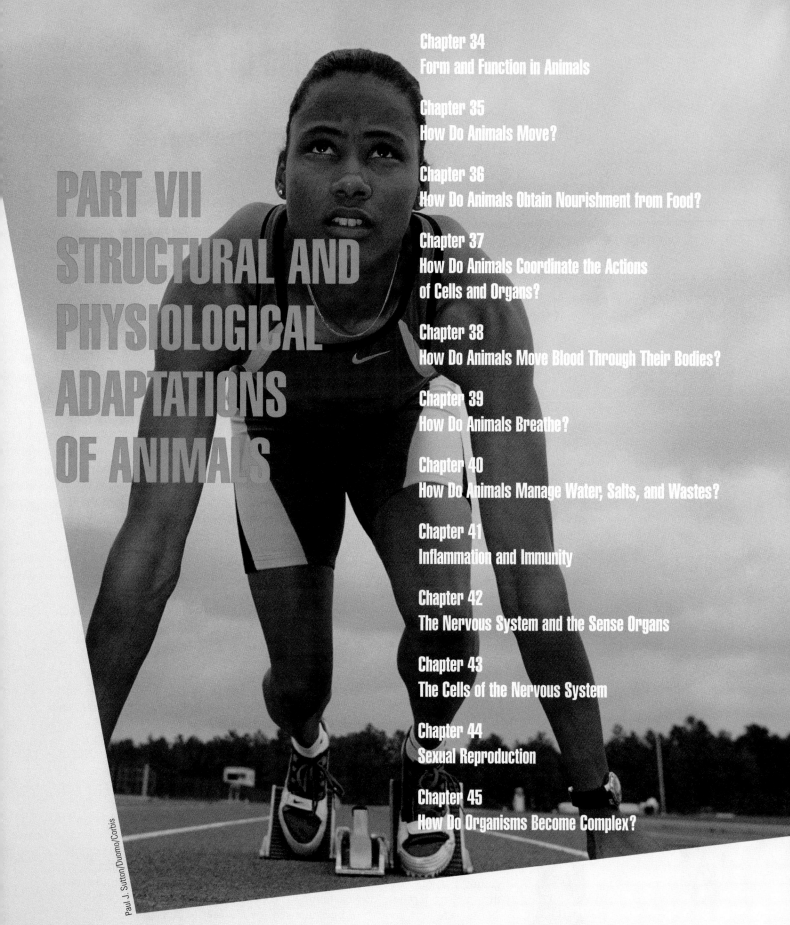

PART VII
STRUCTURAL AND
PHYSIOLOGICAL
ADAPTATIONS
OF ANIMALS

Paul J. Sutton/Duomo/Corbis

Form and Function in Animals

Key Questions

- What is the relationship between form and function?

- How are size, surface area, and shape related?

- How does locomotion influence shape?

- Why do animals sometimes conform to and sometimes resist changes in their environment?

Tyrannosaurus Wrecks: *T. rex* Treks, but Slowly

© John Sibbick

One of the high points of the movie classic *Jurassic Park* comes when an agile but stupid *Tyrannosaurus rex* pursues a Jeep at breakneck speed through a driving rain. Just when the *T. rex* seems ready to snap up actor Jeff Goldblum and down him like a dog biscuit, the gargantuan reptile runs out of steam and the three actors in the Jeep escape to (temporary) safety.

Could *T. rex* have run that fast? Zoologists and paleontologists say "no way." When an adult human runs, the force on each foot is hundreds of pounds—equal to the mass of the body times the acceleration of the foot against the ground. A 6.5-ton *T. rex* running at 45 miles per hour, say biologists, would exert so much force on its legs that they would snap like toothpicks.

Part of the drama of *Jurassic Park* stems from moviegoers' expectations that dinosaurs will be sluggish and clumsy. Indeed, the original reconstructions and paintings of dinosaurs from the early part of this century usually show dinosaurs moving very slowly or not at all. Early 20th-century paleontologists believed that dinosaurs were incapable of the kinds of sprints at which warm-blooded animals excel. Accordingly, dinosaurs were nearly always depicted with all four feet on the ground (Figure 34-1). Yet, the velociraptors and other predatory dinosaurs in *Jurassic Park* and its sequel, *The Lost World,* leap, run, whirl, cock their heads just like birds, and attack, swiftly and often intelligently. How can we tell which image is correct?

Because dinosaur skeletons share many features with those of modern reptiles—similar jaws, for example—early paleontologists assumed that dinosaurs resembled most modern reptiles in other ways as well. They assumed that dinosaurs laid eggs instead of bearing live young, that they were solitary rather than social, and that they were heterotherms—cold-blooded animals whose activity depended on air temperature. In the last several decades, all of these assumptions have been called into question.

In the 1970s, the maverick American paleontologist Robert Bakker electrified the imaginations of paleontologists, science illustrators, and small children by suggesting that dinosaurs might have been warm-blooded animals, more like birds than modern reptiles. After examining *T. rex*'s leg bones, Bakker concluded that the bones were long enough to have allowed the monster to run up to 45 miles per

Figure 34-1
Could a dinosaur outrun a human being? Early researchers assumed that dinosaurs were slow, heavy-footed walkers incapable of running. (K. Perkins/J. Beckett, courtesy of the Department of Library Sciences, American Museum of Natural History)

hour. But his quick calculation left plenty of room for further analysis.

British zoologist R. McNeill Alexander, an expert on the physics of animal movement, jumped into the argument with a more thorough study of dinosaur leg bones. Alexander concluded that although many dinosaurs had the long legs that would have enabled them to run fast, their legs were not strong enough to carry them at high speeds. Alexander estimated the strength of a bone by comparing the area of a cross-section to the length of the bone and the weight of the animal. He called this ratio a "strength indicator."

For example, the femur (or thigh bone) of an African elephant is surprisingly weak, with a strength indicator of only about 7. A running elephant moves at little more than 11 miles per hour. For comparison, the femur of an African buffalo, which is capable of running 30 to 35 miles per hour, has a strength indicator of about 22. Alexander calculated that the femur of *T. rex* had a strength of about 9, a little stronger, perhaps, than that of an elephant, but nowhere near that of a buffalo. At best, Alexander estimated, a *T. rex* could run no more than about 18 miles per hour.

Alexander's analysis seemed conclusive until 1995, when a paleontologist and a physicist at Indiana University took the whole argument one heavy step further. What would happen, asked James Farlow and John Robinson, if a 6.5-ton *T. rex* pursuing a Jeep at 45 miles per hour happened to trip and fall? The answer, derived from simple physics, was decisive.

In a running position, *T. rex*'s head would be about 11 feet above the ground. At 45 miles per hour, the five-foot-long head of a falling *T. rex*, with another 45 feet of its six-ton body coming from behind, would hit the ground with a deceleration equal to 16 times the pull of Earth's gravity—about 96 tons—enough to pulverize the creature's skull and excavate a crater eight inches deep. Then, the colossal corpse would have skidded some 50 feet until the mangled body finally overtook the head, snapping the monster's neck. Not a pretty sight.

Farlow and Robinson thus independently confirmed Alexander's calculations, estimating that *T. rex*'s maximum safe speed was probably 18 to 22 miles per hour. A Jeep would easily leave the dinosaur in the dust. As for Jeff Goldblum and the rest of us, it is a good thing the problem of outrunning *T. rex* and its six-inch teeth is purely theoretical. It's a rare human who can run that fast.

How can biologists know the anatomy, physiology, and habits of long-dead animals? Based on zoologists' extensive knowledge of the anatomy and habits of living animals, biologists can draw detailed conclusions from the skeletons and teeth of extinct animals. Before R. McNeill Alexander estimated the movements of dinosaurs, he spent years studying how living animals run and jump.

As we ask how animals move in the rest of this chapter, we will focus on general ideas that account for why their skeletons and their muscles look and act the way they do. We will see that the shape of a structure determines how it works. Zoologists sum up this idea by saying, "Form follows function."

34.1 Why Are Animals Shaped the Way They Are?

In this chapter, we introduced some basic principles that help determine shape and function with an overview of how animals move. As you read this chapter, keep in mind that form and function are closely related. If you know what the heart is supposed to do (pump blood), that will help you remember how it is shaped (hollow). And if you can think about how something is shaped, that will help you remember what it is for.

The rest of the chapters in animal physiology cover eating and digestion, circulation and waste management (the functions of the body that serve to keep the body operating smoothly), defenses that prevent the body from being overrun by other organisms, sexual reproduction and embryological development, as well as how cells communicate and keep the parts of the body working together. All of these topics deal with how the body is organized, or **anatomy,** and how its parts work together, or **physiology.**

We can study form at many levels. We can look at the overall shape of an animal or we can look at the shape of the parts of an animal. For example, the outside of a fish is streamlined, which eases its passage through water. We can look at the heart to see how it pumps, or we can look at the microscopic structure of heart muscle to see how it differs from other kinds of muscle tissue.

Every organ of the body is shaped in a way that reveals its function. Just as we know that a cookie sheet means someone is baking cookies and a muffin tin means muffins, the exact shape of a leg bone can tell us whether its owner is a runner, a jumper, a walker, or a swimmer.

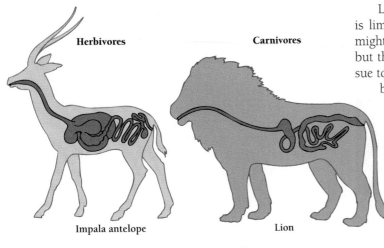

Herbivores   **Carnivores**

Impala antelope   Lion

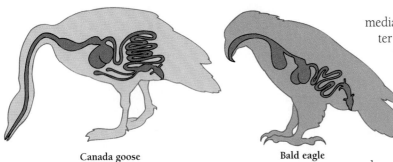

Canada goose   Bald eagle

Figure 34-2

Plant eaters have larger digestive tracts than meat eaters. Because plants, especially grasses, contain cellulose and other indigestible compounds, an animal that eats plants often has a larger digestive system than a meat eater. First, herbivores have to eat more material than carnivores to obtain the same amounts of nutrients. Second, for full digestion to take place, the food must remain in the digestive tract for a longer time, which again means it takes up more space. In contrast, meat is a richer source of protein and other nutrients, so animals that eat meat generally need to process their meals very little. As a result, their digestive systems tend to be simple and small.

The shape of a single tooth tells volumes about its owner's way of life. The large shearing teeth of lions reveal that they survive by tearing flesh from other animals. In contrast, the broad, flat grinding teeth of antelope tell us that they eat plants. The stomachs of lions and antelope are just as different (Figure 34-2). In cells as well, shape reveals function. We can see how the branched shape of a nerve cell is related to how it interacts with other cells (Figure 34-3). Even the shapes of molecules can tell us something about how they work.

As we discuss some of the requirements that shape animals, we need to remember that the form of an animal results from successive adaptations over long periods of time. Our hands are adapted from paws, which evolved from the lobed fins of long-extinct fishes. Every organ of the body bears within it clues both to its function and to its evolutionary history.

Likewise, the evolution of the different parts of the body is limited, or "constrained." For example, a five-toed foot might evolve into a wing or a hoof over long periods of time, but the same foot is unlikely to evolve enough nervous tissue to form a brain. Evolution is constrained by history and by what it has to work with.

As an animal evolves, a bone may take on a new function. A leg bone that formerly supported a small amount of weight may have to support a greater weight. In that case, natural selection will favor a thicker, heavier skeleton. So we say "Form follows function." That is, form evolves in response to the demands placed on the body part by natural selection—whether the "body part" is a leg bone or a cell membrane protein.

It is just as true that function follows form. The immediate function of an organ is limited by its shape. No matter how hard we flap our arms, we won't fly. We are limited by our form. So, while in evolutionary time, form follows function, in day-to-day time, function follows form (Figure 34-4).

What determines the form of an individual animal? Form can change over days, weeks, or years. Most form takes shape during the development of an animal from embryo to adult. The shape of an animal (or plant) results from the interaction of genes with the environment. In certain species of frogs, for example, the presence of predators in the water can cause the

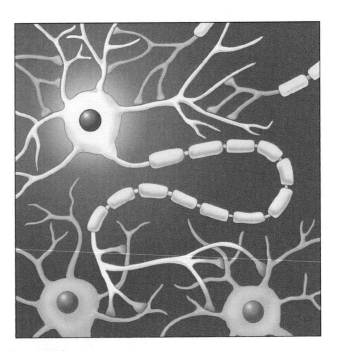

Figure 34-3

Form indicates function, even in cells. A nerve cell's primary function is to communicate with other cells. Its myriad fingerlike extensions allow it to connect to and talk to many other cells.

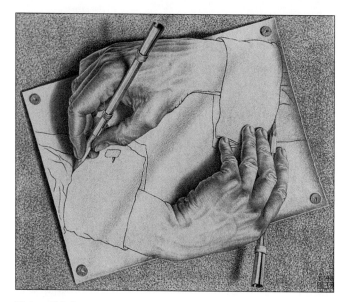

Figure 34-4
Form follows function. But function also follows form. (Drawing hands by M. C. Escher. By Cordon Art-Baam-Holland. All rights reserved.)

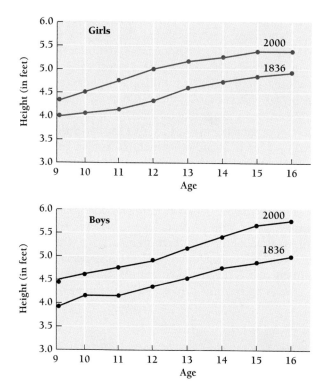

Figure 34-5
Children's heights. Average heights of children and teens working in British factories in 1836 compared to the average heights of modern American children and teens. Height is a good average measure of health, nutritional status, and standard of living. Height at different ages reflects an individual's nutritional history so far. Nutritionists say height is a measure of diet minus the demands of work, physical activity, and disease. In every country in the world, the average heights of children and adults are highly correlated with their expected life span and with the gross domestic product (GDP) of their country.

tadpoles to develop deeper-muscled tails that allow for sudden darts, while tadpoles in predator-free water swim and grow more slowly. In humans, good nutrition tends to result in large size, while poor nutrition and stress can limit growth (Figure 34-5).

34.2 How Are Size, Surface Area, and Shape Related?

When we compare a mouse and an elephant, the most obvious single difference is that one is big and one is small. But an elephant's size greatly limits its shape. To support its tremendous weight, its legs must be thick and straight, its feet wide and well padded. In contrast, the legs of a mouse are slender and flexed, a shape typical of smaller animals.

But animals differ in other respects as well. Depending on where they live (their habitat) and how they live (their niche), they may have organs with large surface areas or small ones; they may have complex digestive systems or simple ones.

How Does Size Affect Shape?

A mouse could not be as large as an elephant, and a polar bear could not be as small as a cat. For every type of animal, there is an appropriate range of sizes, and the evolution of a bigger or smaller size demands a change in form.

One reason for this change in form is gravity. As the linear dimensions (e.g., height and width) of an animal double, both the surface area of the animal and the cross-sectional

area of its bones increase four times. But its volume and weight increase eight times. For example, a hypothetical gazelle made twice as tall as normal would have bones four times as strong as a regular gazelle's, but would weigh eight times as much, so the animal's bones and muscles would bear a load that was twice as heavy. Its legs could not bear that much weight.

In general, we can say that the heavier the animal, the thicker its bones must be to support its weight. For the graceful little gazelle to become as big as a rhinoceros, it would have to evolve thick, heavy legs, so that every pound of weight had the same cross-sectional area of bone to support it. A little gazelle cannot simply evolve into a big gazelle. A big gazelle is a very different animal, and its whole lifestyle has to be different. It can no longer prance about and turn quickly to escape a predator. But being big has different advantages. A large animal can actively defend itself against many predators if it is willling to be aggressive (which a rhinoceros is).

Gravity affects the lives of large and small animals in other ways as well. Gravity will cause an ant (or any other object) dropped from an immense height to accelerate as it drops. But an ant's mass is so small that the force with which it hits the ground is tiny—not enough to hurt it. As a result, an ant dropped from the top of a ten-story building will likely walk away unscathed. But an adult human could not possibly survive such a fall without broken bones or worse.

We can draw interesting conclusions from this kind of discussion. For example, we can see that small animals are more suited to life in trees and other high places than large animals are. Even though monkeys and most other primates live in trees, humans, gorillas, baboons, and other large primates spend most of their time safely on the ground. Likewise, large, leaf-eating herbivores, such as elephants and giraffes, do not generally climb trees. Instead, they have evolved necks or trunks that allow them to reach into tall trees while keeping four feet firmly on the ground. And, of course, as we saw at the beginning of this chapter, large animals such as *Tyrannosaurus rex* must avoid not only heights but also running too fast.

Large animals must have proportionately stronger and thicker bones than small animals. Because large animals fall much harder than small animals, size determines lifestyle (and lifestyle determines size).

How Does Size Affect Metabolic Rate?

Just as gravity influences size, so size influences metabolic rate. Mice, shrews, hummingbirds, and other small, warm-blooded animals have proportionately greater metabolic rates than larger animals. Small animals use both more calories per gram of body weight and, likewise, more oxygen to fuel this higher metabolism. To survive, a mouse must eat about one-quarter its own weight in food every day. Of course, a mouse doesn't eat nearly as many calories as an elephant, just more per gram of body weight.

Large animals have slower metabolisms than small animals.

How Do Animals Use Expanded Surface Areas?

Animals of the same size can vary enormously in shape. For example, the webbed flippers of a sea lion look nothing like our own hands and feet. One of the many changes that can occur in an organ during evolution is an increase or decrease in surface area. As animals evolve, the surface area of limbs, teeth, lungs, digestive tracts, and the tiny vessels of the circulatory system can all increase or decrease.

Webbed feet are a way to increase surface area without increasing overall size. Aquatic animals from frogs to ducks must push large volumes of water to propel themselves through the water. Likewise, animals that fly must push

large volumes of air to keep themselves aloft. The wings of birds and bats are limbs whose surface area has increased enormously, by means of longer bones, longer feathers, and expanded membranes.

Animals also use large surface areas to speed the movement of molecules across membranes and to increase or decrease the transfer of heat (Figure 34-6). We animals diffuse oxygen (in the lungs), nutrients (in the digestive tract), and water and salts (in the kidneys). As we will see in the coming chapters, the lungs, the digestive tract, the kidneys, and the circulatory system all have enormous surface areas whose function is to move nutrients and wastes in and out of the body.

The rate at which oxygen and other nutrients diffuse across a membrane depends on many factors. One is the difference between the amounts of material on the two sides, called a **concentration gradient.** For example, how fast oxygen moves from our lungs into our blood depends on how much oxygen is on either side. If there is a lot of oxygen on one side and little on the other, diffusion will occur rapidly. If, on the other hand, both sides have nearly the same amount of oxygen, then diffusion will occur slowly. Likewise, if the membranes are thick, diffusion will occur slowly, and if the membranes are thin, diffusion will occur more rapidly.

If a membrane is thin enough and a gradient is steep enough, diffusion rates can be quite high. But even so,

Stan Osolinski/Dembinsky Photo Associates

Figure 34-6
Big ears, cool head. This blacktail jackrabbit rids itself of excess heat through its ears. Many animals cool themselves off by radiating heat through enormous ears or other expanded surface areas.

Biology ⊘ Now™ Learn more about how other animals regulate their temperature by clicking on this figure on your BiologyNow CD-ROM.

some animals need to move more material even faster. Many evolve a larger diffusion surface. In the digestive system, for example, thousands of folds in the lining of the intestines expand the total surface area and so speed the absorption of nutrients. In the circulatory system, miles of tiny capillaries likewise expand the total surface area and speed the distribution of those same nutrients.

Many parts of the body function by means of enormous surface areas, including the webbed feet of aquatic animals, the intestines, and the circulatory system.

How Do Animals Use Bulk Flow?

In single cells and in tiny animals such as the tardigrade, or water bear, diffusion is enough to move oxygen, nutrients, and wastes where they need to go (Figure 34-7). But diffusion only works over very short distances. Larger animals need to move masses of material over longer distances. For example, the muscles of the digestive system squeeze lumps of food into the stomach and down through the intestines; the muscular heart pumps blood through the circulatory system; and other muscles push large volumes of air in and out of the lungs. The movement of masses of material is called **bulk flow.**

Bulk flow and diffusion can work together to move materials. For example, the heart pumps blood in bulk to the lungs. From the lungs, oxygen diffuses into the blood. The blood in the circulatory system then carries the oxygen *in bulk* to the tissues of the body, where the oxygen diffuses into individual cells.

How Do Animals Use Countercurrent Systems?

Bulk flow and diffusion can work together in interesting arrangements called **countercurrent systems,** in which fluids or gases run past each other and exchange heat or materials. Animals use countercurrent systems in dozens of ways. Whales use countercurrent systems to keep their fins from radiating too much heat into cold Arctic water. Fish use countercurrent systems to absorb oxygen from the water around them.

Figure 34-7
Look Ma, no circulation!
Tardigrades, such as this one, as well as rotifers and other microscopic animals, rely on diffusion alone to move nutrients and wastes. They need no circulatory system.

Diane R. Nelson

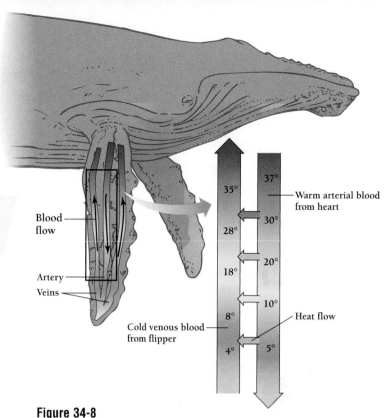

Figure 34-8
Countercurrent heat exchange. Whales and other marine mammals use countercurrent heat exchange to minimize heat loss from the flippers, whose large surface area is wonderful for swimming but a hazard in Arctic waters. As warm, arterial blood from the heart flows out into the flippers, it passes in close proximity to the returning venous blood, which is cold. The arterial blood becomes cooler and cooler as it enters the flippers, while the venous blood, warmed by the arterial blood, becomes warmer and warmer. In this way, body heat stays in the interior and the flippers remain cool.

Countercurrent systems allow animals to maintain steep gradients of temperature or molecules. Steep gradients in temperature make things happen faster. An ice cube placed in a hot frying pan melts faster than an ice cube in a glass of water because the difference in temperature between the skillet and the ice cube is so high. The same is true of concentrations of molecules (Chapter 4).

Dolphins and whales use countercurrent systems to keep the cold blood in their flippers from chilling the rest of the body (Figure 34-8). The body of a whale that swims in the freezing waters of the Arctic Ocean is well insulated against the cold. But its flukes and flippers, which are thin and flat, become extremely cold. To prevent cold blood from the flippers from chilling the body, the veins that carry the cold blood back to the body run close to the arteries that carry warm blood from the heart to the flippers. The warm blood leaving the whale's body heats the cold blood coming in from the flippers, while the cold blood leaving the fins cools the blood before it enters the flippers. This system is called a "countercurrent heat exchanger."

Countercurrent systems can move molecules as well as heat. Fish use a simple countercurrent system to extract oxygen

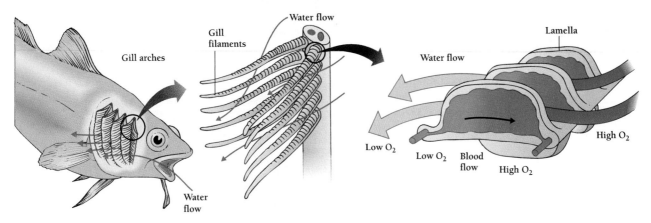

Figure 34-9
Countercurrent oxygen exchange. In fish gills, a countercurrent system ensures that the maximum amount of oxygen is transferred from water flowing past the gills to blood flowing in the opposite direction.

from water flowing past their gills. Water enters a fish's mouth and flows out of the body over the gills. The gills of fish consist of a huge surface folded into flat panels, called lamellae [Latin, *lamina* = thin plate]. Blood in the lamellae flows in the opposite direction from the water flowing past the gills. If we follow the water as it moves over the gills, the water passes blood that is ever poorer in oxygen. At every point along its path, the water carries more oxygen than the blood, so oxygen diffuses into the blood (Figure 34-9). The countercurrent arrangement of blood and water maximizes the steepness of the oxygen gradient between the blood and the water, and the steeper a gradient, the more quickly a substance will diffuse across it.

Countercurrent systems enable animals to establish steep gradients of temperature or concentration, and steep gradients hasten the diffusion of molecules and heat.

34.3 How Does Locomotion Influence Shape?

An animal's style of locomotion determines what sort of shape it evolves, just as size does. In order to propel themselves forward, animals must push against whatever is available. Aquatic animals push against water, terrestrial animals push against the ground or other objects, and flying animals push against the air. Both fast swimmers and flyers tend to be streamlined. But whereas most fish and dolphins use their tails for propulsion and their fins for guidance, birds use their wings for propulsion and their tails for guidance.

Terrestrial animals can slither, crawl, walk, run, or jump. Each style of movement uses the body in different ways. A deer and a kangaroo are similar in size and habits: both are medium-sized herbivores that escape from predators by running. But the kangaroo is adapted to hop on two legs while the deer is adapted to bound on four. These dif-

ferent styles of locomotion give the two kinds of animals very different anatomies.

Style of locomotion helps determine the overall shape of an animal.

What Kinds of Adaptations Do Swimmers Have?

Animals exclusively adapted to swim rapidly in water all have the same oblong shape also seen in fast cars and torpedoes. Seals, fish, and even squids all have a similar shape (Figure 34-10). Strong swimmers such as dolphins, tuna, and mackerel propel themselves by sweeping their tails back and forth or up and down. A swimmer's tail is therefore broad and flat, a shape (with an expanded surface area) that forces water to flow against the tail. The tail exerts forces against the water toward both the sides and the rear. The

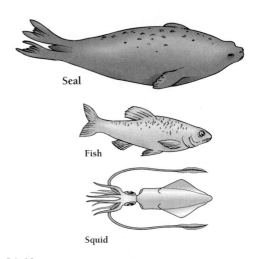

Figure 34-10
The shape of a swimmer. All strong swimmers share the same streamlined shape, including squid, fish, and seals.

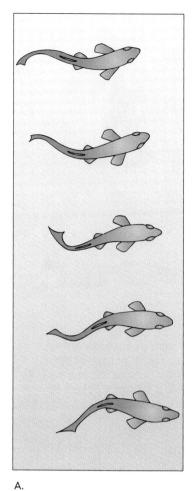

A.

B.

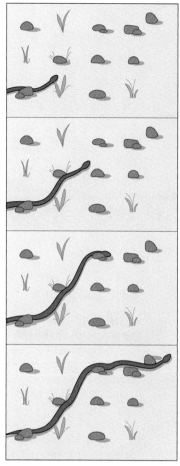

Figure 34-11

Swimming in water or on land. A. When a fish swims, it pushes alternately from side to side. The sideways forces cancel out, but the slight push backward against the water pushes the fish forward with each wriggle. B. When snakes, such as this blotched king snake, undulate from side to side through grass or across a bare rocky ground, they are doing the same thing as fish. The sideways forces cancel out, but the slight push backward against a rock or plant pushes the snake forward.

Biology ⑤ Now™ Learn more about vertebrate motion by clicking on this figure on your BiologyNow CD-ROM.

What Kinds of Adaptations Do Flyers Have?

Flying offers enormous advantages to all kinds of animals, from butterflies, locusts, and beetles to birds and bats. Flyers can quickly escape ground-dwelling predators; wander great distances in search of mates, nest sites, and food; and migrate seasonally, enjoying summer and spring all year round. Flyers can feed from the flowers at the tops of trees, as well as from any kind of food that is dispersed over a wide area. Some flyers even feed on other flyers. Bats and many birds eat insects. Some hawks and bats catch birds in flight. Finally, good flyers can disperse widely, so that a single species can colonize whole continents or hemispheres.

Birds can fly enormous distances and achieve amazing speeds. Thousands of golden plovers fly nonstop from the Aleutian Islands to the Hawaiian Islands every fall, neither eating nor sleeping for days. A peregrine falcon in a vertical dive was clocked at a speed of 125 miles per hour. And wind tunnel experiments with Lagger falcons suggest that this species may reach a velocity of 225 miles per hour.

Flying demands extreme specializations that can limit flyers when they are on the ground. The most obvious adaptation is wings, which are virtually useless for anything but flying. Because birds have such highly modified forelimbs, they walk on their hind feet only and also use those same feet for grasping food and twigs in the same way that four-legged animals use their forepaws.

Flyers have other specializations as well. First, they are extremely light for their size. Birds have hollow bones, and their skeletons are much lighter than those of ground dwellers. Birds also lack teeth, which tend to be heavy. Most flyers have reduced legs compared with their ancestors. The reproductive organs of flyers often shrink when the animal is not actually breeding.

Most birds eat foods that are high in calories and pass quickly through the digestive system, including insects, flesh, or seeds. All of these foods are high in protein and fat. Birds rarely eat leaves or grass, which contain a lot of difficult-to-digest cellulose. Because cellulose is hard to digest, herbivores (plant-eating animals) such as cattle and rabbits have relatively large, heavy digestive tracts in which large amounts of

forces to the side cancel each other out, however, and the net movement is forward (Figure 34-11A). Vertebrates less well adapted for fast swimming, such as sea lions, penguins, and platypuses, rely on various sorts of webbed flippers and feet to move themselves through the water.

Snakes undulate across the ground in much the same way that eels move through water. Each curve of a snake's body pushes against small objects and rough surfaces on the ground. As with the side-to-side sweep of a fish's tail, the lateral forces mostly cancel each other out, while the forces toward the rear push the snake forward (Figure 34-11B).

All fast swimmers have the same streamlined shape. Most swimmers undulate through the water. Snakes usually undulate like fishes.

plant material sit for long periods of time. Such heavy digestive systems are not for flying birds. The few flying birds that eat leaves—geese and quail, for example—are large-bodied birds that take flight reluctantly and with difficulty.

Flying animals have many specializations, including powerful wings, a light skeleton, hollow bones, reduced jaws and legs, reproductive structures that shrink when not in use, and streamlined digestive tracts.

How Do Animals Walk and Run?

Most animals walk by placing each foot on the ground and then successively rocking forward over each foot. The legs bend only a little. When an animal runs, however, the legs work vigorously, like pogo sticks. When the first leg hits the ground, the animal's body sinks down on the leg. Tendons in the legs work like rubber bands to store and release energy, propelling the animal upward and forward. Amazingly, nearly all animals run this way, not only horses and cheetahs, but ants, crabs, and cockroaches, too.

It doesn't matter whether an animal runs on four legs, six legs, or two, the pogo stick model applies to all. For example, intensive studies of cockroaches have shown that, using pogo stick locomotion, an American cockroach can run five feet per second (more than 3 mph). At top speed, a roach rises up off its front legs so that it is sprinting on just its hindmost two legs. Wind resistance keeps its surfboardlike body from falling back down, so that it can run bipedally. Such bipedal (two-legged) running has also evolved in many lineages of vertebrates, including lizards, dinosaurs and birds, not to mention apes.

What Kinds of Adaptations Do Runners Have?

Animals that run fast do so by maximizing two things—the length of each stride and the number of strides per unit of time. A "stride" is a full cycle of motion, two steps for a two-legged animal and four steps for a four-legged animal. A galloping horse covers nearly 23 feet per stride. A racehorse can manage well over 150 strides per minute, so that it can run up to 45 miles per hour. The much smaller cheetah has the same stride length as a horse, but it can sprint one and a half times as fast as a racehorse because it moves its legs one and a half times as fast. How can the cheetah have a stride as long as that of a racehorse? And how does the cheetah manage to move its legs so much faster than the horse?

Stride Length

The simplest way to make the stride longer is to make the legs longer (Figure 34-12A). The parts of the leg farthest from the body usually lengthen the most over the course of evolution. One reason that cockroaches run on two legs when they are sprinting is that their hind legs are the longest, which gives them a longer stride. Another way to

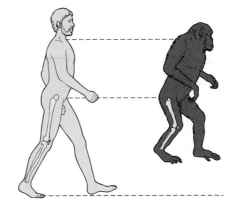

A. Proportionately longer legs

B. Flexible backbone

Figure 34-12
How to increase stride length. A. Humans are adapted for running compared to chimps. Compared with the rest of the body, our legs are much longer than those of chimps. Notice that in humans the bones of the calf are even more elongated than those of the thigh. B. Runners also frequently have flexible spines, which also increases the length of the stride.

increase stride length is to make the shoulder highly flexible. In most mammals, the collarbone (clavicle) anchors the shoulders firmly to the chest. But in cheetahs and other fast runners, the collarbone is either tiny and not connected to anything or absent altogether. Without a collarbone, runners can throw their forelimbs out much farther, lengthening the stride even more. Cheetahs, dogs, weasels, horses, and other fast runners also extend their stride length by alternately flexing and extending their spines (Figure 34-12B).

Stride Rate

One might think that animal runners would move their legs faster simply by contracting their muscles faster. But, in fact, large animals, which include most of the fastest runners, actually take more time to contract their muscles than small animals. It turns out that fast runners use the same trick to move their legs quickly that racing bicycles use to spin their wheels quickly.

The trick is gearing. Muscles that attach far from a joint have more leverage than those that attach close to the joint. In contrast, muscles that attach close to a joint move much faster than those that attach far away, but exert less force. You can demonstrate this idea with a pencil. If you hold the pencil at the eraser end, you can wiggle the graphite tip much faster than if you hold the pencil near the tip. On the other hand, you can exert more force with the pencil if you hold it close to the tip.

Animals use leverage and gearing in the same way. For example, moles that dig tunnels in forests and meadows use their forelimbs to dig forcefully through soil and rocks.

Moles don't need to move fast, but they need a lot of power. The muscles in their forearms therefore attach to the bones far from the joint, which gives the mole's arms more leverage, but little speed. In contrast, in the legs of runners such as cheetahs and gazelles, the muscles attach very close to the joint, which gives them speed, but little leverage.

Animal runners also increase stride rate by making their legs light. It is easier to move light legs quickly than heavy ones. Consequently, horses, gazelles, and other fast runners have more slender legs than slower animals. In all runners, the bones of the outer parts of its limbs tend to be reduced for lightness. Over evolutionary time, horses have lost more and more toes, so that their feet became relatively smaller (Figure 15-12). In runners, the bones of the legs, including the toes, are frequently fused together, which increases strength of the leg for the same amount of weight. The muscles that move these long light legs, although often large, are housed closer to the body, which also keeps the legs themselves light.

Animal runners increase overall speed by increasing stride length and stride rate. Long legs, specialized shoulders, and a flexible spine all increase stride length. Light legs and muscle attachments close to the joint increase stride rate.

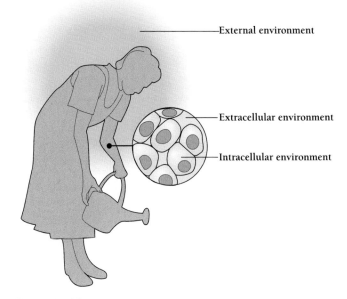

Figure 34-13
Inside out? Physiologists recognize three areas of the body—the external environment, the extracellular environment, and the intracellular environment. Animals regulate the insides of cells and the extracellular environment. But they do not regulate the external environment, which includes not only everything outside the skin, but also the contents of the gut.

34.4 Homeostasis and Tolerance

Two themes tie together many different aspects of physiology—**homeostasis** [Greek, *homeo* = same + *stasis* = standing still], the tendency of living organisms to maintain a constant internal environment, and **tolerance,** the (opposite) tendency of organisms to tolerate certain kinds of environmental change.

Homeostasis: How Do Organisms Self-Regulate?

In Chapter 1, we mentioned that organisms need to keep their insides separate from the outside world. The first step in doing this is to create a barrier. In the case of our bodies, it is our skin. In the case of our cells, it is the cell membrane.

In multicellular organisms such as animals, these two boundaries create three areas (Figure 34-13). Outside the body itself is the "external environment." Inside the body but outside the cells is the "extracellular environment." Finally, inside each cell is the "intracellular environment."

The intracellular environment is easy to visualize; it's everything inside the cell membrane. The contents of a cell are closely regulated. If pH and salt concentration are not just right, proteins change shape, which changes the way they behave, and normal cell functions grind to a halt.

The extracellular area is almost as closely regulated as the inside of the cell, because a cell's environment greatly affects the cell. In the space between our cells and in the blood in our veins, animals may regulate temperature, acidity, and salt balance.

The process by which an organism resists change, whether inside the cells or between the cells, is homeostasis. Homeostasis always involves two stages: (1) detecting change and (2) counteracting such change. Counteracting change is called **negative feedback.** If you are driving a car, you use negative feedback to regulate the speed of the car. If a speed limit sign says 55 mph and the speedometer says you are going 65, then you ease up on the gas and maybe even touch the brake to slow down.

In biology, the speed limit sign is called a **set point.** You may have heard that some physiologists believe that we have a set point for weight. Not everyone agrees with this idea, but we definitely have a set point for temperature. Normally, human body temperature ranges from 97° to 99° Fahrenheit (37°C). When we first get up in the morning, we are a bit cooler. Later in the day we get warmer. If we exercise strenuously we can become quite warm. But a healthy human rarely gets much warmer than 99°F or much cooler than 97°F.

All organisms resist change through homeostasis and negative feedback mechanisms.

How Much Change Can Animals Tolerate?

As we examine the different systems of the body in the coming chapters, we will encounter many examples of homeostasis. It is important to remember, however, that animals of-

Extreme Biology How Does the Body Regulate Temperature?

In vertebrates, a structure at the base of the brain called the hypothalamus detects changes in temperature. When the temperature is too low, the hypothalamus sends signals to other parts of the brain, which stimulate heat-seeking behavior, shivering, diverting blood from the skin, and increased heat production by the cells of the body. When we are cold, we also wrap our arms around our bodies and crouch, trying to reduce our surface area. Animals with fur contract tiny muscles at the base of each hair, which forces the hairs to stand up, forming a thick, insulating coat. Even nearly hairless humans get goose bumps, the visible contraction of those same muscles and a relic of our furry past. Birds ruffle their feathers similarly.

When an animal is too hot, the hypothalamus detects the change and signals the brain to initiate a different set of responses. Some animals flatten their feathers or fur. Some pant or sweat. Hot animals often become quiet and move as little as possible to minimize the production of excess heat.

Most of us have learned that mammals and birds regulate temperature and most other animals do not. But this is an oversimplification. Many other animals regulate body temperature, although usually not to the same degree.

The body temperature of most aquatic animals is close to that of the water around them—cool. But tuna, sharks, and some other fast-swimming fish can generate more power in their swimming muscles if they keep them warm. These animals use a countercurrent heat exchanger like the one in the fins of whales to keep their muscles as much as 14°C warmer than the water around them.

Some insects also heat parts of their bodies. Most insects are unable to fly in cold weather. We've all seen torpid flies on a dark windowsill on a cool, autumn morning. When the morning sun warms them, they seem to awaken and then fly away. But many wasps, bumblebees, and large moths and butterflies heat their flight muscles and are able to fly during cold weather

(Figure A). The large, hummingbird-like sphinx moth can fly at night in temperatures of only 10°C and yet maintain a body temperature of 41°C, greater than that of a human (37°C).

Joe McDonald/CORBIS

Figure A
White-line Sphinx moth. Some insects such as this sphinx moth have beeen described as "hot-blooded." Their bodies may be warmer than our own.

---

ten give in to environmental change instead of trying to resist it. For example, even as we humans regulate our core body temperature, we allow our arms and legs to warm up and cool down (Figure 34-14).

Even the most heat-tolerant animals have limits. Heat kills by changing the three-dimensional structure of proteins and by altering the behavior of cell membranes. Yet regulating temperature is energetically expensive. For many animals, a better solution is to adapt to or tolerate changes in temperature.

How much heat can an animal stand? The eggs of some shrimplike crustaceans can be boiled and yet still live and develop into healthy adults. Other animals can tolerate extreme cold. Some organisms can be frozen to −269°C, close to the temperature of space, then thawed out to live a normal life. Some polar intertidal invertebrates may freeze and thaw twice a day. When the tide goes out, they freeze in the subzero air. When the tide comes back in, the warmer water thaws them out and they resume their lives for a few hours until the tide goes out again.

Many animals cannot tolerate ice in their tissues, yet cannot generate enough heat to stay warm when the temperature drops. One solution is to have antifreeze for blood. Certain fish and amphibians, as well as many insects, contain glycerol or other substances that prevent their tissues from turning to ice when the temperature drops below

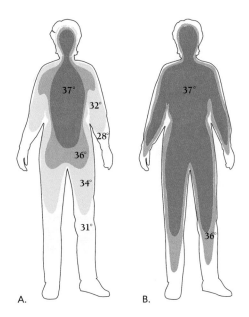

A.   B.

Figure 34-14
Tolerance in humans. A. In cold weather, the brain and the core of the body normally remain at approximately 37°C, while the temperature of the center of the legs and arms drops several degrees. Our skin temperature may fall even lower, approaching that of the environment. B. In warm weather, we let our whole body warm up by increasing circulation to the arms and legs. If it's even hotter, we open up the tiny blood vessels in our hands and feet to increase the radiation of excess heat.

freezing. Their body temperature drops below freezing, but no ice forms.

Many animals conform to their environment, a phenomenon called tolerance.

We have seen that the function of an organ determines its form and that the form of an organ determines its function. The next chapter in Part VII covers the structure and function of bone and muscle. In the remaining chapters of this last part of *Asking About Life,* we will see many more specific examples of how form follows function.

Key Concepts

- Form and function mutually determine one another.
- Increased surface area speeds the radiation of heat and the diffusion of molecules.
- Countercurrent systems are a way for organisms to establish useful gradients.
- Animals have adaptations for both homeostasis and tolerance.

Summary with Key Terms

What is the relationship between form and function?
Anatomy is how the body is organized, while **physiology** is how its parts work together. The shape of an animal depends on its size. Large animals must have proportionately stronger and thicker bones than small animals. Large animals also have lower metabolic rates than small animals.

How are size, surface area, and shape related?
Animals use expanded surface areas to increase the rate of exchange of heat and molecules, as well as to facilitate movement. Lungs, intestines, the circulatory system, and webbed feet and wings all function because they have expanded surface areas. One important use of expanded surface areas is **countercurrent systems,** which enable animals to establish steep gradients of temperature or concentration that help direct the movement of small molecules and temperature. (The difference between the amounts of material on the two sides of a membrane or from one region to another is called a **concentration gradient.**) To move large amounts of material from place to place, animals use **bulk flow.**

How does locomotion influence shape?
Style of locomotion also determines shape. Swimmers, which undulate through the water, have a streamlined shape. Flying animals have many specializations, including powerful wings, a light, rigid skeleton, hollow bones, reduced jaws and legs, and reproductive structures that shrink when not in use. Runners increase overall speed by increasing stride length and stride rate. Long legs, specialized shoulders, and a flexible spine all increase stride length. Light legs and muscle attachments close to the joint increase stride rate.

Why do animals sometimes conform to and sometimes resist changes in their environment?
Two themes in physiology are **homeostasis,** the tendency of living organisms to maintain their internal environment, and **tolerance** of environmental change. Most animals achieve homeostasis by means of **set points** and **negative feedback.** Animals regulate many aspects of the intracellular and extracellular spaces, but not all.

Review and Thought Questions

Review Questions
1. Give an example not in this chapter that illustrates the idea that how an organ works depends on its structure (function follows form).
2. Give an example not in this chapter that illustrates the idea that how the structure of an organ evolves depends on how the organ is used (form follows function).
3. Give two examples of how large size affects the shape of an animal.
4. Whose metabolic rate is higher, a mouse's or a monkey's?
5. Our lungs have as much surface area as a tennis court. How does this help us breathe?
6. What do biologists call the movement of masses of material, such as air in and out of the lungs or a meal through the intestines?
7. Explain what a countercurrent exchange system is. Give an example.
8. Why are fish, birds, and missiles streamlined?
9. Give three adaptations that allow cheetahs to run faster than other animals.
10. Define "homeostasis" and give an example.
11. Define "tolerance" and give an example.

Thought Questions
12. Imagine a giant, 12 feet tall. For his legs to be as strong as a normal man's, how thick would his femurs need to be?

BiologyNow Resources

Biology ⓔNow™

Active Figures
34-6: How animals regulate their temperature
34-11: Vertebrate motion

Preparing for an exam? Take a diagnostic test on your BiologyNow CD-ROM.

Online materials relating to this chapter are at:
http://biology.brookscole.com/AAL3

About the Chapter-Opening Image
Tyrannosaurus rex illustrates how much animals' bones can reveal about how animals move.

35

How Do Animals Move?

Key Questions

- What is the overall organization of an animal?

- What are the main building materials for an animal?

- How do animals use muscles and a skeleton to move?

- What environmental demands are placed on bone, muscle, and other tissues?

T. rex Redux

Leaving aside whether *Tyrannosaurus rex* could catch a fast-moving Jeep, biologists have continued to ask if *T. rex* could run at all. It is a question that biologists just can't leave alone, and in 2002, graduate student John R. Hutchinson and postdoctoral student Mariano Garcia, both at the University of California, Berkeley, published still another paper addressing this pressing question. To answer it, they studied the leg muscles of chickens. In the paper, published in the prestigious journal *Nature,* the two young researchers announced that *T. rex* probably could not run, at most managing a heavy-footed jogger's shuffle.

The question of how fast *T. rex* could run is a fun one for a biologist to tackle, and like a lot of scientific questions, there is more than one way to try to answer it. For example, ecologists have asked if there were any *T. rex* prey that were themselves fast runners. Unless large prey animals existed that could run more than 18 to 20 miles per hour, there would be no reason at all for *T. rex* to evolve legs massive enough to propel its 6.5 tons into even the slowest of sprints. Why bother?

The Field Museum/John Weinstein

Another approach is to look at the dinosaur's footprints. Sixty-five-million-year-old footprints from the behemoths suggest top speeds of about five miles per hour. If the fossil record is any indication, *T. rex* was a slacker who walked everywhere. On the other hand, the fact that the animals kept to a sedate walk on the mudflats in which their ancient footprints were fossilized doesn't mean they couldn't run on solid ground. But did they run? Could they run? McNeill Alexander had already calculated that *T. rex*'s leg bones weren't strong enough to support the forces generated if the beast's feet hit the ground running. The young researchers at Berkeley, Hutchinson and Garcia, wanted to calculate how big a *T. rex*'s muscles would have to be in order for it to run. The muscles of all animals are essentially the same, so mouse muscles contract in the same way as an elephant's or a lizard's. But the force a muscle generates depends on the area of its cross section. A muscle twice as thick in diameter will be four times as wide in area but also, unfortunately, eight times as heavy. Just as with bones, muscles that are twice as wide and twice as long still end up being weaker compared to the overall weight of the animal.

Hutchinson and Garcia calculated that domestic chickens have twice as much muscle as the minimum they need to run. For a chicken the size of *T. rex* to run, the chicken would need to be 99 percent leg muscle. Similarly, for *T. rex* to run, it would have to have had as much as 86 percent of its entire body mass just in the two extensor muscles that straighten the leg during running. Considering the beast's other leg muscles, immense body, long tail, massive skull, and six-inch teeth, this was impossible.

Furthermore, by examining the fossil leg bones of *T. rex*, Hutchinson and Garcia could estimate the area over which a muscle attached to the bone and such factors as the angle at which the muscles worked. Even with generous estimates of muscle mass, and even if *T. rex* used all of the muscles in its hind limbs, it still could not have generated enough force to loft its 6.5-ton body off the ground. For *Tyrannosaurus rex*, running was pretty much out of the question. Thanks to Hutchinson and Garcia, we know for sure now that any Jeep could easily out pace the monster.

More important to *T. rex*'s biology, if it couldn't run, it wouldn't have been able to catch smaller dinosaurs, many of which were clearly runners. *T. rex* most likely fed on dead animals that had been killed by other animals (carrion) or on other dinosaurs as large and slow as it was.

35. 1 The Geography of the Body

When we go on a journey, we often look at a map first, to see where we'll be going. As we explore the anatomy and physiology of animals, in Part VII of *Asking About Life,* it helps to keep in mind a map of the body.

How Are Animal Bodies Organized?

All vertebrates and their embryos share a common body plan—a tube inside a tube (Figure 35-1). The inner tube is the gastrointestinal (GI) tract, from mouth to anus, and hangs inside a cavity of the body called the **coelom.** Also inside the coelom are the heart, kidneys, and most of the body's other organs. The coelom separates the intestinal tract from the outer tube, which is the muscular wall of the body. In all vertebrates, from fish to birds, the coelom has two distinct parts: the **thoracic cavity,** which encloses the heart and lungs (when present), and the **abdominal cavity,** which encloses the stomach, liver, pancreas, kidneys, and most of the intestines.

The defining characteristic of vertebrates is the **backbone,** a column of partly hollow bones called **vertebrae** [singular, vertebra; Latin, *vertebratus* = jointed] that support the body and enclose the spinal cord. The backbone is also called the "spine," "spinal column," or "vertebral column," but the meaning is the same.

At the **anterior,** or front, end of the spinal column is the **skull,** the bony box that encloses and protects the brain; at the

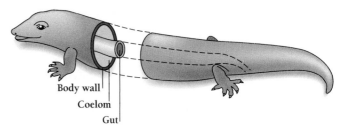

Figure 35-1
Vertebrate body plan. A vertebrate is a tube within a tube. The outer tube is the body wall. The inner tube is the gut.

posterior end of the backbone is the tail (Figure 35-2). Just behind the skull, the *pectoral girdle* supports the forelimbs—fins in fishes, front legs in horses, and arms in humans. Behind the pectoral girdle are the ribs, also attached to the backbone, and the *pelvic girdle,* which supports the pectoral fins in fish and the back legs in other vertebrates.

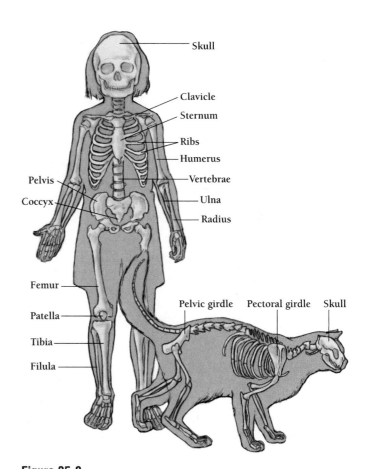

Figure 35-2
The vertebrate skeleton. In all vertebrates, the "axial skeleton" includes the skull, rib cage, and spinal column. The bones of the "appendicular system" include the shoulder, or pectoral girdle (clavicle and scapula), and arms, as well as the pelvic girdle and legs. In a few vertebrates such as snakes, many or all the bones of the appendicular system may be missing or vestigial.

The skull, rib cage, and backbone form the "axial skeleton." The arms and pectoral girdle (shoulder bones) and the legs and pelvic girdle (hip bones) form the "appendicular skeleton."

Most animals' bodies are organized into a four-tier hierarchy of (1) cells, (2) tissues, (3) organs, and (4) organ systems. Animals are made up of hundreds of different kinds of cells, such as nerve cells, skin cells, and muscle cells. Humans alone have some 200 different kinds of cells. A **tissue** is simply an assembly of similar cells that performs a specific function. For example, muscle tissue contracts and nervous tissue relays and stores information. Tissues include both cells and the materials the cells secrete, called **matrix.** For example, bone cells secrete a matrix of proteins and minerals that give bone its strength and flexibility.

Tissues may organize into **organs**—two or more kinds of tissues that together form a distinctive structure with a specific function, such as a kidney. Groups of organs together form **organ systems**. For example, the "urinary system" includes two *kidneys* that make urine, two *ureters* that carry the urine to the *bladder,* and a *urethra,* which empties the urine outside the body (Figure 35-3).

What Tissues Make Up Animal Bodies?

Biologists group most tissues into just four classes: epithelial, nervous, muscle, and connective. Each tissue type plays a particular role in how an organ works.

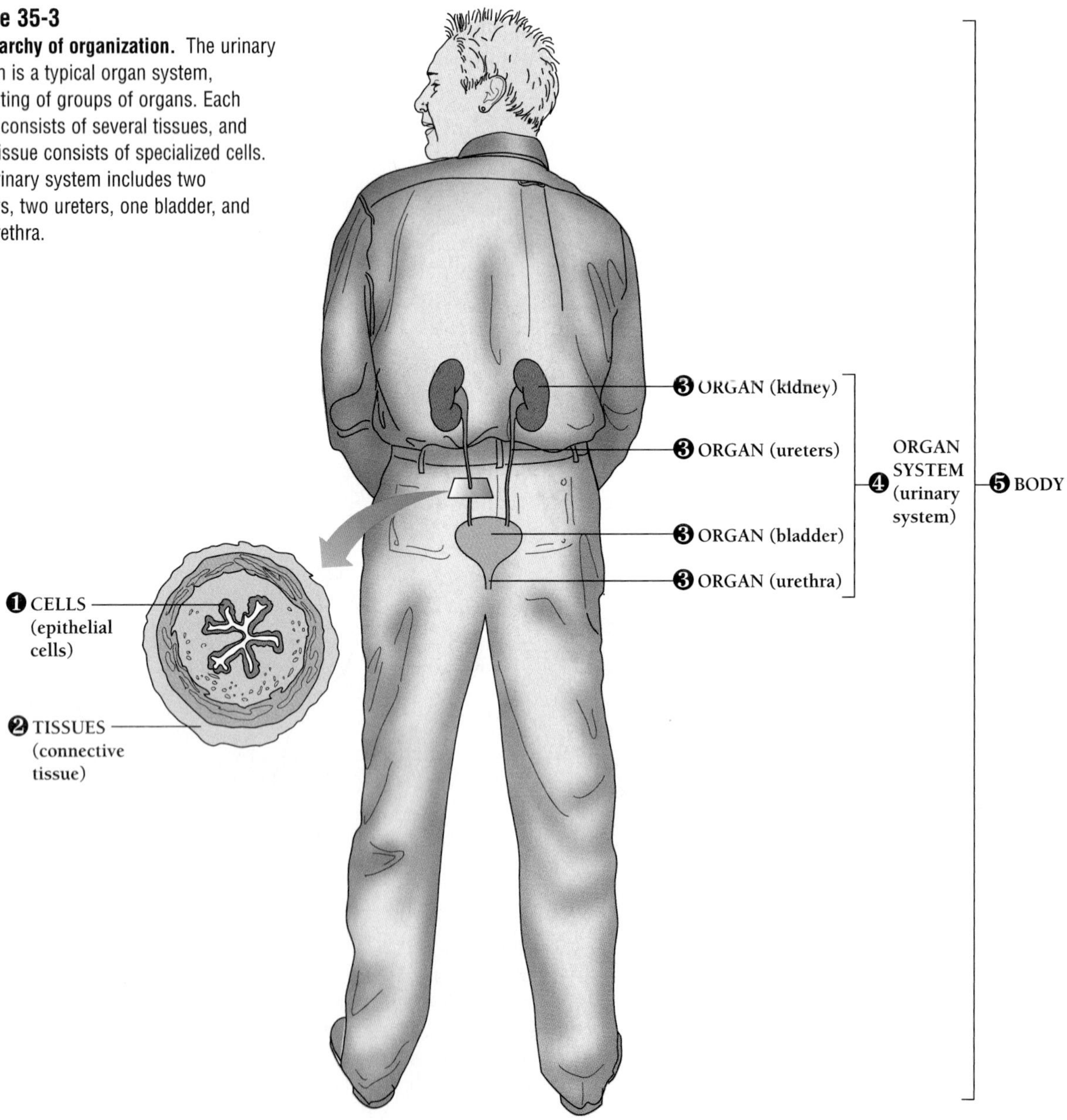

Figure 35-3
A hierarchy of organization. The urinary system is a typical organ system, consisting of groups of organs. Each organ consists of several tissues, and each tissue consists of specialized cells. The urinary system includes two kidneys, two ureters, one bladder, and one urethra.

❶ CELLS (epithelial cells)

❷ TISSUES (connective tissue)

❸ ORGAN (kidney)

❸ ORGAN (ureters)

❸ ORGAN (bladder)

❸ ORGAN (urethra)

❹ ORGAN SYSTEM (urinary system)

❺ BODY

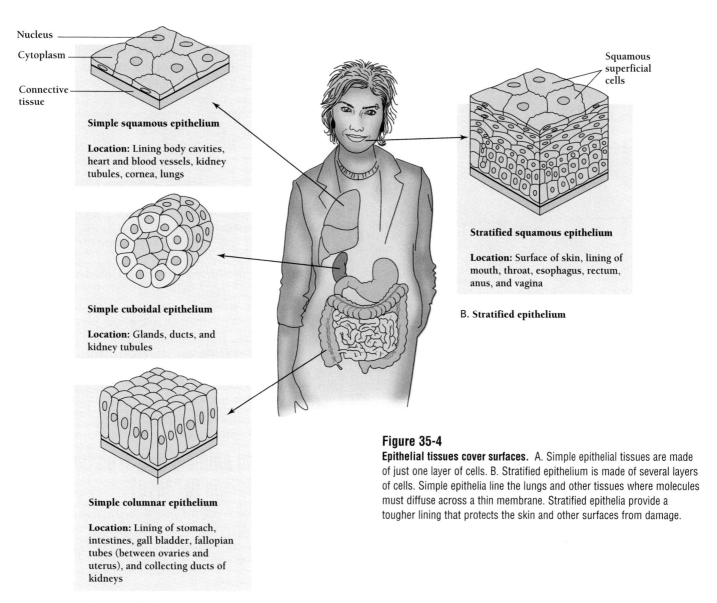

Nucleus
Cytoplasm
Connective tissue

Simple squamous epithelium

Location: Lining body cavities, heart and blood vessels, kidney tubules, cornea, lungs

Simple cuboidal epithelium

Location: Glands, ducts, and kidney tubules

Simple columnar epithelium

Location: Lining of stomach, intestines, gall bladder, fallopian tubes (between ovaries and uterus), and collecting ducts of kidneys

A. Simple epithelium

Squamous superficial cells

Stratified squamous epithelium

Location: Surface of skin, lining of mouth, throat, esophagus, rectum, anus, and vagina

B. Stratified epithelium

Figure 35-4
Epithelial tissues cover surfaces. A. Simple epithelial tissues are made of just one layer of cells. B. Stratified epithelium is made of several layers of cells. Simple epithelia line the lungs and other tissues where molecules must diffuse across a thin membrane. Stratified epithelia provide a tougher lining that protects the skin and other surfaces from damage.

Epithelial Tissue

Epithelial tissue, also called epithelium, forms sheets of tightly interconnected cells that line all inner and outer surfaces of the body, including those of the skin, mouth, nose, esophagus, stomach, intestines, lungs, kidneys, and reproductive organs (Figure 35-4A).

Epithelial tissues are named according to the shapes of the cells and the number of layers. A "simple" epithelium consists of just one layer of cells, while a "stratified" epithelium has several (Figure 35-4). The single layer of epithelial cells that lines the lungs, for example, is so thin that it is perfect for passing oxygen molecules through to the blood underneath. In contrast, the many layers of epithelium in the esophagus are tough enough to resist abrasion from half-chewed toast and crunchy peanut butter.

Nervous Tissue

Nervous tissue consists of two families of cells, the familiar branched nerve cells, or **neurons,** that carry electrochemical signals all over the body, and the various kinds of **glial cells,** which nourish and support the neurons. Together, these nerve cells form the tissues of the brain, the spinal cord, and the nerves. The defining characteristic of nervous tissue is that it carries information.

The shapes of neurons are a clue to what they do. Each neuron consists of a cell body and many fingerlike extensions called "dendrites" and "axons" (Figure 35-5). A **dendrite** usually receives a signal from another cell and transmits it to the neuron's cell body, while an **axon** usually carries a signal from the cell body to another cell. The cells relaying messages to and from neurons may be other neurons or different kinds of cells, such as muscle cells. Neu-

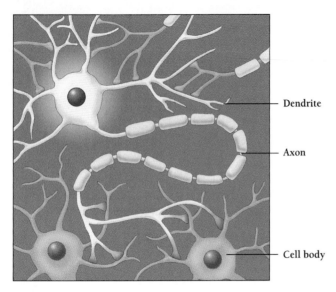

Figure 35-5

Talk to me. The primary function of nervous tissue is to send and receive messages. The dendrites of a neuron connect to hundreds or thousands of other cells, receiving incoming information by way of signaling molecules called neurotransmitters. The axon transmits its own message to other cells.

Labels in figure: Dendrite, Axon, Cell body

rons communicate with other cells, often by means of special signaling molecules, called *neurotransmitters*. A neuron releases a neurotransmitter, which conveys a signal to a second cell, which can be another neuron, a muscle cell, or even a hormone-producing cell. You can read more about the nervous system in Chapters 42 and 43.

Muscle Tissue

Muscle tissue consists of highly organized bundles of cells called "muscle fibers." The defining characteristic of muscle tissue is that it contracts. Like neurons, muscle cells burn a lot of energy (when they contract) and therefore contain large numbers of mitochondria. Vertebrates have three kinds of muscle tissue—**skeletal muscle** (in our arms and legs, for example), **cardiac muscle** (of the heart), and **smooth muscle** (in our intestines, blood vessels, and urinary and reproductive tracts). We'll discuss the different kinds of muscle tissue later in this chapter.

Connective Tissue

Most of us are familiar with the fibrous connective tissues—tendons and ligaments—that provide mechanical support to the skeleton. But several other kinds of tissues are also considered "connective," including fat and even blood. **Connective tissue** includes five major types of tissue: **loose connective tissue, fibrous connective tissue, cartilage, bone,** and **blood**. We'll discuss bone, cartilage, and fibrous connective tissue later in this chapter (Figure 35-6 B, C, and D).

Loose connective tissues, composed primarily of the protein collagen, bind together all the other tissues and organs of the body, "loosely connecting" our skin to the underlying muscle, for example, and also forming thin, tissuelike bags around each of the organs. The bags of connective tissue in which our organs are wrapped are like the plastic bags we use in a lunch box to enclose and separate a sandwich, some cookies, and some slices of an orange (Figure 35-6A).

Fat, or adipose tissue, is a special type of loose connective tissue that consists of tightly packed round cells that pad and insulate the body and store energy (Figure 35-6A). Each cell contains a drop of fat, a large drop when the body is storing energy and a small drop when the body has burned its stores.

Blood is also considered a kind of connective tissue. Its cells are red blood cells, which carry hemoglobin, and white blood cells, which are a major part of the immune system. Its matrix is a fluid called blood plasma that is made of water, salts, hormones, antibodies, and other proteins (Figure 35-6E). We discuss red cells and blood in more detail in Chapter 38 and the white cells of the immune system in Chapter 41.

35.2 How Do Animals Use Skeleton and Muscle to Move?

So far, we have surveyed the overall body plan of a vertebrate and the main types of cells and tissues in the body. In the rest of this chapter, we see how muscles and bone work together to move the body.

How Do Muscles and Bones Work Together?

All animals move by contracting their muscles against a rigid framework called a skeleton. The skeleton can be made of bone, like ours, or of other materials. It can be internal or external.

An earthworm has a **hydrostatic skeleton,** a fluid-filled coelom something like a water balloon that acts as a semirigid skeleton. To work its way through the soil, an earthworm uses a combination of circular and longitudinal muscles, which push and pull against its hydrostatic skeleton (Figure 35-7).

Most animals are arthropods—insects, spiders, and crabs, for example, and all arthropods have a rigid, external **exoskeleton** [Greek, *exo* = on the outside]. An exoskeleton resembles a hollow cylindrical tube. In some ways, this design is superior to the internal **endoskeleton** [Greek, *endo* = on the inside] of humans and other vertebrates. A hollow tube can support more weight than a solid rod of the same weight, so the exoskeleton of an insect can

A. **Loose connective tissue – Containment**

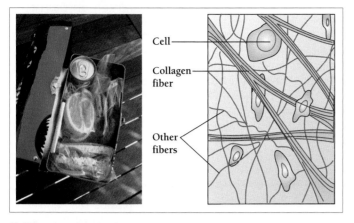

B. **Fibrous connective tissue –**
(Ligaments) – Hold joints together

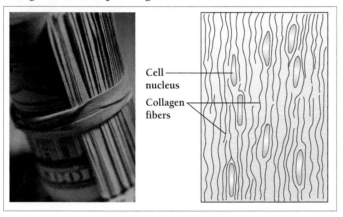

D. **Bone – Support**

Loose connective tissue (Fat) – Storage

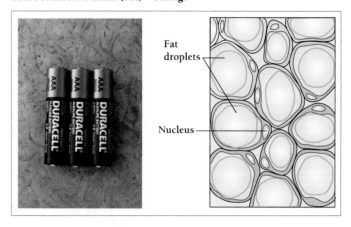

C. **Cartilage – Shock absorption**

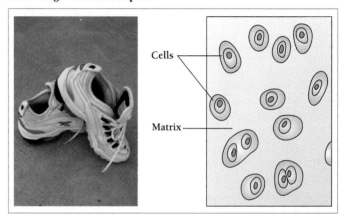

E. **Blood – Transport**

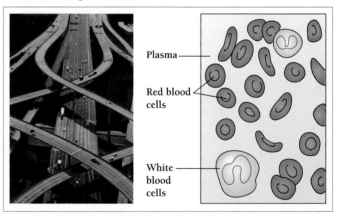

Figure 35-6

The five kinds of connective tissue. Connective tissues play surprisingly diverse roles in the body.
A. Loose connective tissues contain our various organs—as sandwich bags hold the parts of our lunch.
Another loose connective tissue, fat, stores energy—as batteries do. B. Fibrous connective tissues
such as ligaments and tendons transmit force or stabilize joints—as rubber bands hold things together.
C. Cartilage often acts as a shock absorber—as our athletic shoes do. D. Bones provide a solid
framework—as the steel girders of a building do. E. Blood transports material and information
throughout the body—as a freeway transports cars and trucks. (A and C, Kathleen Olson; B, LWA-Paul
Chmielowiec/CORBIS; D and E, Getty Images)

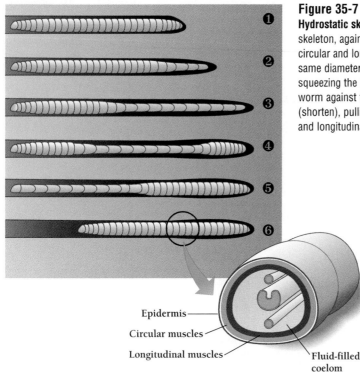

Figure 35-7

Hydrostatic skeleton. An earthworm's fluid-filled coelom provides a rigid hydrostatic skeleton, against which its muscles can pull. An earthworm uses a combination of circular and longitudinal muscles to inch through the soil. At rest, ❶ an earthworm is the same diameter from head to tail. To begin moving, ❷ & ❸ circular muscles contract, squeezing the body into a longer, thinner cylinder. ❹ Tiny hairlike feet then anchor the worm against the earth while its longitudinal muscles, running from head to tail, contract (shorten), pulling the body forward. ❺ & ❻ Waves of alternating contractions of circular and longitudinal muscles move the worm through the soil.

Epidermis

Circular muscles

Longitudinal muscles

Fluid-filled coelom

support more body weight, gram for gram, than a vertebrate skeleton.

Despite the differences between the two kinds of skeletons, the muscles of vertebrates and arthropods work in about the same way. Muscles can pull, but they cannot push. Imagine for a moment a whole, functional muscle sitting on a plate by itself contracting. If the muscle isn't attached to anything, it won't move anything except itself. Now attach one end of the muscle to a bone. The bone-muscle may wiggle a little, but it can't do any real work. But suppose the muscle is attached to two bones. Now if it contracts, it can pull the two bones together.

Now that we have a system for pulling the two bones together, we can add a second muscle that moves the bones apart. Skeletal muscles often work in pairs. For every muscle or set of muscles that bend a limb, another set of muscles unbends it. The bending and unbending of a limb involves a pair of **antagonistic muscles** that take turns pulling the limb one way or the other (Figure 35-8).

The human body contains hundreds of skeletal muscles that are attached to bones or to broad sheets of connective tissue (Figure 35-9). Together, these muscles make up more than 40 percent of the body's weight.

Many animals move by means of pairs of antagonistic muscles anchored on a skeleton. In the human body, skeletal muscles work our bodies by pulling against bone and each other.

The Vertebrate Endoskeleton

The basic vertebrate skeleton consists of a skull, a backbone, and ribs (the axial skeleton), as well as the bones of the pectoral and pelvic girdles and the limbs that usually go with

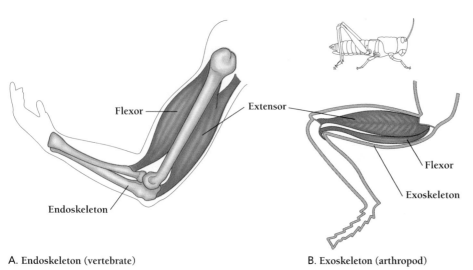

Figure 35-8

Antagonistic muscles. Skeletal muscles often work in antagonistic pairs. A. In vertebrates, antagonistic muscle pairs attach to the outside of an endoskeleton. B. In arthropods such as this grasshopper, antagonistic muscle pairs attach to the inside of an exoskeleton.

Flexor

Extensor

Endoskeleton

Flexor

Exoskeleton

A. Endoskeleton (vertebrate)

B. Exoskeleton (arthropod)

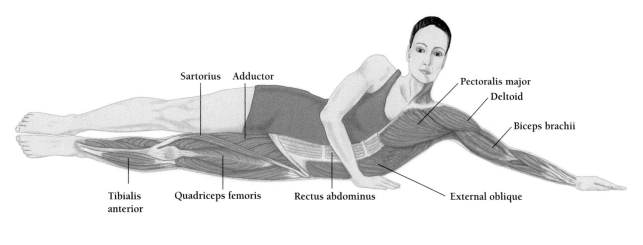

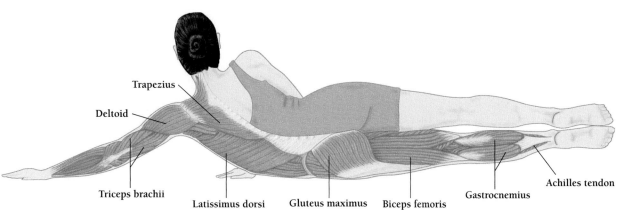

Figure 35-9
Muscles of the human body. Here are only some of the largest of the hundreds of muscles in the human body.

them (the appendicular skeleton). A human skeleton includes more than 200 major bones.

Most bones are attached to other bones at a **joint.** The immovable joints that fuse the separate bones of the skull into a rigid, protective helmet for the brain are called sutures (Figure 35-10A). But most joints allow bones to move relative to each other. Different kinds of joints allow different movements. For example, a knee joint allows the lower leg to swing back and forth (Figure 35-10B). This joint is almost like the hinge on a door, which swings in a single plane. In reality, the knee is capable of some lateral and twisting movement, one factor that leads to knee injuries. An array of flexible tendons and ligaments normally limit sideways or twisting motions. In contrast, the "ball-and-socket" joint between the thigh and the hip is more like a joystick. It moves in three dimensions and allows the thigh to swing in wide arcs in any direction (Figure 35-10C).

The bones of the vertebrate skeleton are joined together by a variety of joints, including sutures, hinge joints, and ball-and-socket joints.

What Forces Act on Bones and Other Connective Tissue Structures?

Most of the skeleton is made of bone. However, a few parts—such our noses and ears and the soft discs between our vertebrae—are cartilage. Cartilage also cushions the joints at the ends of the long bones, such as the tibia and femur of the leg. All the bone and cartilage tissues of the skeleton are bound together by tendons, ligaments, and muscles (Figure 35-11). **Ligaments** resemble rubber bands and wrap around the different parts of a joint, binding it together. In contrast, **tendons** attach muscles to bones.

Bones, cartilage, ligaments, and tendons must each withstand very different sorts of forces. Ligaments and tendons, which bind bones and muscles tightly together, must resist **tension,** the pulling action of two opposing forces. When we play tug-o-war, we exert tension on the rope and tension on our arms. Cartilage tolerates **compression,** the pushing action of two opposing forces. We feel compression in our legs when we jump up and down. Cartilage absorbs some of that shock, thus protecting the bones.

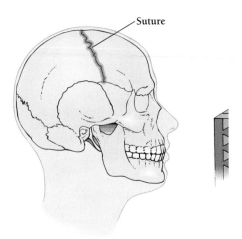

Suture

A. Sutures of the skull

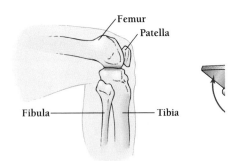

Femur
Patella

Fibula
Tibia

B. Knee: Hinge joint

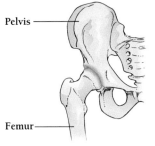

Pelvis

Femur

C. Hip: Ball-and-socket joint

Figure 35-10
How we are put together. All of our bones are connected by just a few types of joints. A. Most of the bones of the head lock together in sutures. These bones cannot move relative to each other. B. Most joints allow some degree of motion. Like a hinge on a door, a knee moves in one plane only. An array of flexible tendons and ligaments limit sideways or twisting motions. C. The thigh bone (femur) attaches to the pelvic girdle by means of a ball-and-socket joint. Like a joystick, the femur moves freely both from side to side and up and down.

Bones resist both tension and compression and also **shear,** the twisting action created by forces that are not opposite one another. We feel shear forces in our knees and hips when we turn suddenly while running. A *T. rex* might break its legs just taking a corner too fast. And even we ag-

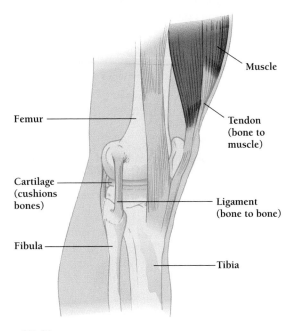

Muscle

Femur

Tendon (bone to muscle)

Cartilage (cushions bones)

Ligament (bone to bone)

Fibula

Tibia

Figure 35-11
Ligaments, tendons, and cartilage. Tough ligaments and tendons bind together the bones and muscles of the knee. Tendons connect muscle to muscle, ligaments connect bone to bone, and cartilage cushions the joint.

ile humans occasionally tear a ligament in a knee or ankle while playing sports.

Ligaments, tendons, cartilage, and bone must withstand a variety of forces, including tension, compression, and shear forces.

What Makes Our Skeletons So Strong?

Bones, tendons, and ligaments are enormously strong for their weight. For example, a tendon half an inch in diameter could support a two-ton car. All the parts of the skeleton are strengthened by a matrix of the protein **collagen.** This collagen matrix is responsible for the soft transparency of the cornea of the eye, the rocklike solidity of bone, and the toughness of a tendon. Collagen is the single most abundant protein in the extracellular matrix, and amounts to about 25 percent of all animal protein.

Collagen helps connective tissues resist both tension and twisting shear forces. Fibrils of collagen can twist together to form ropelike strands that resist tension or assemble in layers that have the shear strength of construction-grade plywood.

In bone and cartilage, collagen fibers are embedded in a flexible rubbery matrix that is somewhat flexible and resistant to tension (pulling). In bone, however, needlelike crystals of calcium phosphate make bone as hard as rock and help it resist both compression and shear forces.

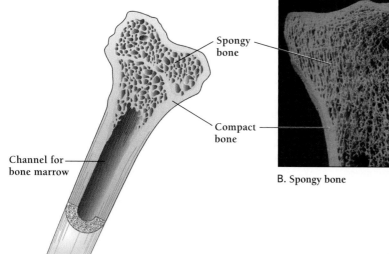

Spongy
bone

Compact
bone

Channel for
bone marrow

A. Structure of a long bone

B. Spongy bone

Don W. Fawcett/Visuals Unlimited

Figure 35-12
The structure of bone. A. Compact bone consists of dense layers of perpendicular fibers of collagen and crystals of calcium phosphate. At the core of some of the largest bones in the body is bone marrow, where red and white blood cells first appear. B. Spongy bone is made of the same hard material as compact bone, but in an open lattice structure.

The outer layer of most bone is hard, dense **compact bone tissue** that consists of dense layers of collagen and calcium phosphate crystals. Inside the compact bone and near the ends of long bones may be a layer of **spongy bone,** which is similar to compact bone, but with a light open-lattice structure (Figure 35-12).

Embedded in the spongy bone and sometimes also at the very center of the larger long bones is **bone marrow,** an aggregation of stem cells that give rise to red blood cells and white blood cells. Red blood cells carry oxygen throughout the body, and white blood cells are major components of our immune system. Without the stem cells in our bone marrow, we couldn't live.

Tough collagen gives bones, cartilage, tendons, and ligaments much of their strength. Bone is hardened by crystals of calcium phosphate. The stem cells in bone marrow make red blood cells and white cells.

How Does Bone Respond to Use?

We tend to think of our skeletons as solid and unchanging. Certainly bones are hard, but solid and unchanging they are not. We've already seen that bones are a regular network of holes and hollow passageways. Bones also grow and repair themselves against continual wear and tear and change shape according to how they are used.

Like other connective tissues, bone includes living cells that multiply and also mold the shape of a bone from day to day. The building and remodeling of bone is the task of two kinds of cells, the osteoblasts and the osteoclasts. **Os-**teoblasts [Greek, *osteon* = bone + *blastos* = bud] make bone matrix and **osteoclasts** [Greek, *osteon* = bone + *klastos* = broken] digest it. In all living bone, the osteoclasts continually break down and absorb bone, while the osteoblasts continually form new bone.

In young adults, the rate at which bone is created is approximately equal to the rate at which it is absorbed. In bones that are growing, or in bones that are repairing damage, the osteoblasts produce more bone than the osteoclasts destroy. But in older adults or younger people with some disease or other problem, the osteoclasts sometimes break down the bone faster than the osteoblasts can build it up. The result is the breakdown of bone, or **osteoporosis** (see box).

Bones adjust their size and density according to the amount of compression they experience. The bones of people who exercise become thick and strong, while those of couch potatoes (and astronauts living at zero gravity) tend to become thin, brittle, and decalcified. If a person limps to stay off an injured leg for a long period, the bone of the bad leg becomes thin and decalcified, while the healthy leg bone, which gets more use, becomes heavier and stronger. An unused bone can lose nearly a third of its mass. For this reason, physicians now encourage people with healing bone fractures to begin using the limb as soon as possible.

Osteoclasts and osteoblasts continually shape and reshape bones as we grow, heal, and change our habits and physiology.

The Three Kinds of Muscle

The bones of the skeleton support the body, but skeletal muscles hold the skeleton in position. Skeletal muscle is also called "striated muscle" because it has obvious stripes. As we'll soon see, the striations reflect the regular arrangement of protein filaments. Another kind of striated muscle is

Health and Biology Osteoporosis

When we are young, the osteoblasts outpace the osteoclasts, building bigger denser bones. By about age 30, however, the osteoblasts slow down and the osteoclasts nose ahead. Around this age, people begin a slow process of bone loss. In most people, for most of their lives, this loss is relatively harmless. But in others, this steady weakening of the bones is anything but harmless.

Some people's bones become so weak that minor stresses or falls result in fractures, often in the hip or the spine. Indeed, sometimes when old people "fall and break a hip," the hip has actually broken spontaneously and caused the fall, rather than the other way around. Such extreme bone weakness is called **osteoporosis**, technically defined as a decrease in bone mass with at least one fracture. Osteoporosis affects about 10 million Americans, 80 percent of them women. In addition, tiny fractures in the vertebrae cause the spinal column to bend, giving many older people a stooped posture and a noticeable hump in the upper back.

Our skeleton is like a retirement account. In our youth, we need to begin saving calcium for our old age and we should deposit as much bone into this account as we can. Once past age 30, however, all we can do is try to minimize withdrawals from the account. Osteoporosis is the nearly empty account that results when too little bone was deposited during our youth and too much was taken out later. By age 65, some women have lost half the mass of their skeleton.

Risk factors for osteoporosis include being old, female, Caucasian or Asian, sedentary, and thin, as well as smoking and drinking alcohol. But anyone can have osteoporosis. The people most prone to osteoporosis are those who begin adulthood with less bone than average. Children with chronic diseases such as cystic fibrosis, rheumatoid arthritis, and phenylketonuria all fail to build bone normally and are at risk for osteoporosis. Women are more likely to have osteoporosis than men for several reasons. First, adolescent boys build up more bone mass than adolescent girls, so they start out with more bone. Second, women lose more bone mass than men in the many years after menopause.

The hormone estrogen seems to inhibit the death of the bone-building osteoblasts, thus allowing for the build-up of bone. But during both menopause and lactation (nursing), estrogen levels decline and the osteoclasts break down bone faster than the osteoblasts can build it. Indeed, early studies showed that nursing mothers lost four to six percent of their bone mass in the first six months of nursing. More recent works suggests that women make up for this once they stop nursing. But the longer women nurse a baby, the longer it takes to return to their previous bone density.

Another hormone that contributes to bone loss during lactation and after menopause is the hormone prolactin, large amounts of which circulate in the blood of nursing mothers. Prolactin gets its name from its involvement in stimulating the production of milk. But prolactin plays many other roles, including stimulating parental behavior in males. In rats, prolactin was found to increase bone loss during lactation by directly inhibiting osteoblast activity and also by blocking estrogen receptors. Unlike most other hormones, prolactin secretion increases as we age, inhibiting the bone-building osteoblasts.

Another cause of osteoporosis is high-dose corticosterone (cortisone-type) therapy given to people with asthma or arthritis. High doses of cortisone cause the bone-building osteoblasts to die off and bone losses of up to six percent per year, with frequent bone fractures—classic osteoporosis.

The best way to prevent osteoporosis is to build up as much bone mass as possible before and during the 20s. And the best way to do that is to consume lots of calcium and vitamin D (which aids in the absorption of calcium) and to exercise a lot. Any exercise that puts stresses on the bones, such as walking, running, and jumping, will stimulate the osteoblasts to build bone in the legs, back, and hips. Weight-bearing exercises for the arms will likewise strengthen the upper back, arms, and wrists.

Good sources of calcium are, of course, dairy products such as milk and yogurt, but also broccoli and dark-green leafy vegetables such as kale, canned sardines and similar fish (eaten with bones), and calcium-fortified juice, bread, and cereal products.

Osteoporosis cannot be cured or even prevented completely, but various treatments can reduce its severity. Treatments include estrogen, which prevents the death of osteoblasts; calcitonin, a hormone that regulates calcium levels in the blood and inhibits the bone-destroying osteoclasts; and the drugs Fosamax, which also inhibits osteoclasts, and raloxifene, which mimics estrogen in some ways—preventing bone loss—but not in others. All of these treatments reduce bone loss but do not actually build bone.

Recently, the Food and Drug Administration approved a new drug, teriparatide, that does build bone. Teriparatide is a 34-amino-acid portion of human parathyroid hormone, which naturally regulates calcium and phosphate metabolism. The drug must be given in daily injections (along with vitamin D and calcium supplements). It can cause bone cancer in rats, though not, so far, in people. Teriparatide is not recommended for long-term use, for young people whose bones are growing, or for anyone with any risk for bone cancer.

Osteoporosis, bone loss with at least one fracture, affects four times as many women as men, in part because women start out adult life with less bone than men and then lose more, during lactation and after menopause.

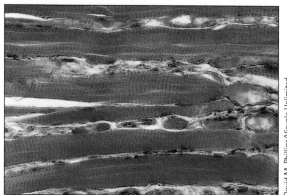

A. Skeletal muscle   50 μm

David M. Phillips/Visuals Unlimited

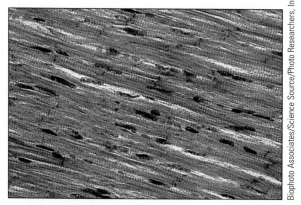

B. Cardiac muscle   25 μm

Biophoto Associates/Science Source/Photo Researchers, Inc.

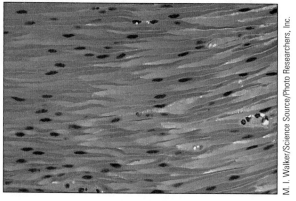

C. Smooth muscle   100 μm

M. I. Walker/Science Source/Photo Researchers, Inc.

Figure 35-13
The three kinds of muscle. A. The stripes, or striations, that enable skeletal muscle to contract are distinctive in micrographs such as this one. B. Cardiac muscle is also striated. C. Smooth muscle, which lacks striations, encloses the intestines, the urogenital organs, arteries, and other tubular structures.

cardiac muscle, responsible for making our hearts beat (Figure 35-13).

Smooth muscle, as its name suggests, looks smoother under a microscope than the striated cardiac muscle and skeletal muscle. Smooth muscle lines the walls of hollow in-

ternal organs such as the intestines, the uterus, and the blood vessels. Smooth muscles also contract and relax the iris of the eye, and contract into "goose bumps" to make our individual hairs stand on end when we are frightened or cold.

The difference between skeletal and smooth muscles lies both in their structure and in the types of nerves that control them. Skeletal muscles are generally under the control of a system of nerves called the **somatic motor system,** which controls our conscious movements. In contrast, the smooth muscles and the cardiac muscles are influenced by a separate system of nerves called the **autonomic nervous system,** which controls our reaction to anger or fear—the so-called "fight-or-flight" reaction. For example, when we are startled, the autonomic nervous system releases the hormone adrenalin, which speeds up the heart in preparation for running away from or fighting off a predator. At the same time, adrenalin shuts down the digestive system (which is smooth muscle), parts of the immune system, and other basic activities that we don't need in an emergency.

Smooth muscles and cardiac muscles are not attached to bones and are under the control of the autonomic nervous system rather than the somatic nervous system.

How Are Skeletal Muscles Organized?

Like other organs in the body, every skeletal muscle is enclosed in a tough sheet of connective tissue that separates it from the other parts of the body and protects the more delicate parts of the muscle (Figure 35-14). The muscle itself is composed of many bundles of muscle fibers (cells). Each bundle is enclosed in its own sheath of connective tissue, which carries blood vessels and nerves that supply the muscle. Surrounding each of the individual muscle cells are thinner sheaths of connective tissue. At each end of the muscle, the outer and inner layers of connective tissue come together and fuse into a tendon, which usually attaches the muscle to a bone.

Each muscle fiber in skeletal muscle consists of many threadlike **myofibrils** [Greek, *myos* = muscle + Latin, *fibrilla* = little fiber] that run the whole length of the fiber (Figure 35-14). Each myofibril itself consists of a string of short cylinders, called **sarcomeres** [Greek, *sarx* = flesh + *meros* = part].

Each sarcomere is marked by light and dark stripes. At either end of a sarcomere is a thin disc of protein, called a Z line. The Z lines of each sarcomere are perfectly aligned with the Z lines of all the other sarcomeres in a myofibril. This alignment explains why striated muscle fibers look striped.

The striped appearance of striated muscle derives from the ordered arrangement of striped sarcomeres. The light and dark lines of all the sarcomeres are aligned.

Every skeletal muscle consists of many parallel **muscle fibers,** or muscle cells. Muscle cells are called muscle "fibers" because they are so thin and long (up to a foot long). Each muscle fiber contains as many as 100 nuclei, as well as the long, thin proteins actin and myosin involved in muscle contraction.

Muscle tissue contains two kinds of fibers. **Glycolytic fibers** derive most of their energy from glycolysis, the oxygen-free breakdown of glucose sugar. **Oxidative fibers** derive most of their energy from cell respiration, the breakdown of small molecules such as glucose, which requires oxygen. (Chapter 6 reviews glycolysis and cell respiration.)

Oxidative and glycolytic fibers differ in several ways. Oxidative fibers are thin and packed with mitochondria, whereas glycolytic fibers tend to be thick and have only a few mitochondria. Oxidative fibers are endurance fibers that keep powering muscles over the course of a long run or walk. Glycolytic fibers power short bursts of exertion. Table 35-1 summarizes the characteristics of glycolytic and oxidative fibers.

Two characteristics of oxidative fibers give them a distinctive dark red color. Oxidative (endurance) fibers depend on a continuous supply of oxygen and are therefore well supplied with blood capillaries, which deliver oxygen-carrying hemoglobin. Oxidative fibers also contain lots of myoglobin, a relative of hemoglobin that moves oxygen off the hemoglobin in the blood and into the muscle fibers. The combination of myoglobin and a rich blood supply give oxidative fibers their dark red color. The "dark meat" in chicken and turkey legs is red because these muscles have a large proportion of oxidative fibers.

In contrast, the "white meat" of a chicken breast consists mainly of thick, white glycolytic fibers. Glycolytic fibers can deliver a lot of power but work only for short bursts. After a minute or so, their energy stores are exhausted. Glycolytic fibers are good for grappling with a heavy object or, in the case of a chicken, a short burst of flight to escape a predator.

Like bones, muscles grow or atrophy according to how much they are used. Brief, high-intensity exercise—such as weightlifting and sprinting—increases the diameter of glycolytic fibers, resulting in bulging muscles. In contrast, long-distance running increases the number of mitochondria in oxidative fibers and the circulatory system's capacity to deliver oxygen to these cells, but not the overall size of the muscles. As a result, the muscles of distance runners and other endurance athletes tend to remain thinner.

Table 35-1 The Two Kinds of Muscle Tissue

Oxidative	Glycolytic

Thin	Thick
Aerobic (uses oxygen)	Anaerobic (no oxygen)
Endurance	Short bursts
Many mitochondria	Few mitochondria
Myoglobin	No
Extra blood supply	No
"Dark meat"	"White meat"
"Wiry"	"Bulked up"

Ales Fevzer/CORBIS

Getty Images, Inc.

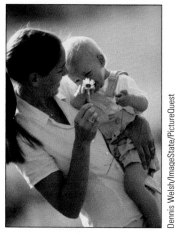

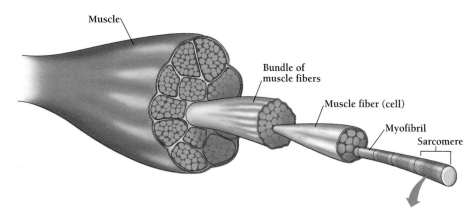

Figure 35-14

Building muscle. Muscle is one of the key characteristics of animals. Muscles allow animals to pursue of food, evade predators, and carry young. A single striated muscle consists of bundles of muscle fibers (muscle cells), which are made of bundles of myofibrils. Each myofibril consists of bundles of thick and thin filaments packed into barrel-shaped units called sarcomeres. The structure of sarcomeres is the key to how muscle contracts. At either end of a sarcomere is a thin disc of protein called a Z line, to which the thin filaments are attached. In the middle of a sarcomere is another thin disc of protein called the M line, to which the thick filaments are attached.

Biology⊜Now™ Learn more about muscle contraction by clicking on this figure on your BiologyNow CD-ROM.

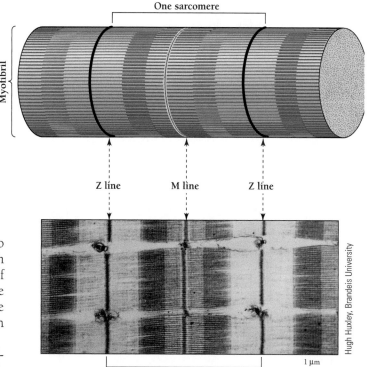

How Do Muscles Contract?

When a muscle fiber contracts, each sarcomere shortens, so that all the Z lines get closer together. What happens in each sarcomere determines what happens in the whole muscle. If each sarcomere shortens by 20 percent, then the whole muscle shortens by 20 percent. So, how does a sarcomere contract? The short answer is that two kinds of protein strands inside the sarcomere slide past each other.

Each sarcomere consists of hundreds of tiny, parallel filaments that come in two sizes—**thin filaments** and **thick filaments.** Thin filaments are mainly made of the protein **actin** and are anchored to the Z-disc ends of the sarcomere (Figure 35-15). In among the thin filaments are hundreds more thick filaments, all anchored in the middle to another protein disc, called the M line. Each thick filament is made of hundreds of thinner threads of the protein **myosin,** and each myosin molecule ends in a round head or **cross bridge** (Figure 35-15). The cross bridges are how muscles contract.

If you lift your head to stop reading for a minute, you contract muscles at the back of your neck. As each muscle fiber contracts, the cross bridges on the thick filaments reach out to and grab onto the thin filaments (forming temporary molecular bonds). All the cross bridges then pivot and pull the thick filaments through the surrounding thin filaments, let go of the thin filaments, and reach out further and pull again (Figure 35-15). This repeated grabbing and

letting go is something like the way a person climbs a ladder, repeatedly using both hands and feet to pull themselves up. In the same way, the thick filaments pull themselves along through the actin filaments. The whole process causes the thick and thin filaments to slide past one another in just a few millionths of a second (Figure 35-15). When a muscle relaxes, the thin filaments slide apart again. This model of muscle contraction is called the **sliding filament model.**

The signal for a set of muscle fibers to contract comes from a nerve cell. Groups of skeletal muscle fibers are under the control of a single neuron, which connects to the muscle in the middle of the fiber at a "neuromuscular junction." To make the muscle fiber contract, the neuron releases a

Relaxed sarcomere

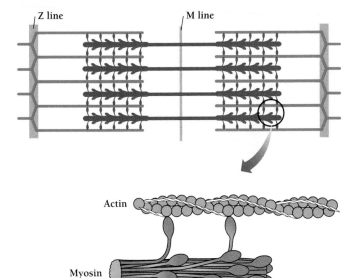

Contracted sarcomere

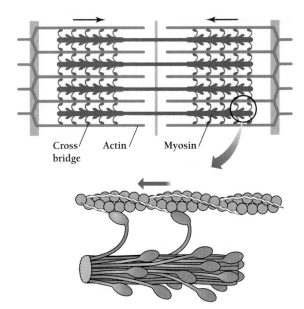

Figure 35-15

What makes the filaments slide? Actin thin filaments *(green)* and myosin thick filaments *(purple)* slide past one another during the contraction and relaxation of the muscle. Cross bridges on the thick filaments repeatedly grab onto and let go of the thin filaments, rapidly pulling the two sets of thin filaments together.

 **Biology⊗Now™** Learn more about sliding filaments by clicking on this figure on your BiologyNow CD-ROM.

neurotransmitter called acetylcholine that triggers a charge in the muscle cell membrane and a dramatic release of calcium ions inside the cell. The calcium binds to the thin filaments, exposing active sites to which the cross bridges bind, detach, and rebind.

> Muscles contract when a neuron stimulates the release of calcium, which allows cross bridges on the thick filaments to pull themselves along the thin filaments and so contract the sarcomere.

In this chapter, we have seen how animals use skeleton and muscles to move. In the next chapter we will find out how animals derive nourishment from food. We will examine both digestion, the breakdown of food particles to their constituent building blocks, and absorption of those building blocks by the gut.

Key Concepts

• A vertebrate body is organized as a tube within a tube.
• Organ systems, organs, tissues, and cells make up a hierarchy of organization.

• Most tissues are epithelial, nervous, muscle, or connective.
• Skeletal muscles move the body by pulling against bones, muscles, or other structures.
• The sliding filament model explains how individual muscle cells contract.

Summary with Key Terms

What is the overall organization of an animal?

In all vertebrate animals, the body forms a tube (the digestive tract) within a tube (the body wall). The space between the two tubes, called the **coelom**, includes the **thoracic cavity** and the **abdominal cavity**, which together contain many of the main organs of the body.

All animals have some kind of skeletal system. Earthworms and some other invertebrates work their muscles against a rigid, fluid-filled **hydrostatic skeleton**, arthropods have an external **exoskeleton**, and vertebrates have an **endoskeleton**. The defining characteristics of vertebrates are the **skull** and the **backbone**—a column of hollow bony segments called **vertebrae**. The skull lies at the **anterior** end of the body, and the tail is at the **posterior** end. The point of attachment between any two bones is called a **joint**. The knee is a hinge joint and the hip joint is a ball-and-socket joint.

What are the main building materials for an animal?

All animals consist of **tissues**, assemblies of cells and any associated **matrix**. Groups of tissues form **organs**, which together form **organ systems**. The four main classes of tissue are: **epithelial tissue, muscle tissue, connective tissue**, and **nervous tissue**. Epithelial tissue forms flat sheets that line the inner and outer surfaces of the body. Nervous tissue consists of **neurons** and **glial cells**. Neurons have a long **axon** and fingerlike **dendrites** that connect neurons to other cells and communicate with them by means of "signaling molecules."

The five main kinds of connective tissue are **fibrous connective tissue** (**tendons** and **ligaments**), **loose connective tissue, cartilage, bone**, and **blood**. Ligaments and tendons resist the pulling force called **tension**. Bones resist tension, **compression**, and **shear**. One of the main ingredients of bone, cartilage, and fibrous connective tissues is the protein **collagen**.

What environmental demands are placed on bone, muscle, and other tissues?

Adult bone consists of three major tissue types: **compact bone, spongy bone** (including **stem cells**), and **red bone marrow** (with stem cells). Specialized **osteoblasts** manufacture the bone matrix while their partners, the **osteoclasts**, continuously break it down. Osteoclasts and osteoblasts continuously shape and reshape bones as we grow, heal, and change our habits. The breakdown of bone by osteoclasts can lead to brittle bones, or **osteoporosis**.

How do animals use muscles and a skeleton to move?

Muscles arranged in **antagonistic pairs** move our bodies by pulling against bone and one another. Besides **skeletal muscle**, vertebrates also have **cardiac muscles**, which pump blood through the heart, and **smooth muscles**, which line the walls of hollow internal organs such as the intestines, the uterus, and the blood vessels. Smooth muscle and cardiac muscle are not attached to bones.

Skeletal muscles are generally under the control of a system of nerves called the **somatic motor system**, while the smooth and cardiac muscles are controlled by a parallel system of nerves called the **autonomic nervous system**.

Muscle fibers consist of bundles of myofibrils. Each **myofibril** consists of bundles of **thick filaments** and **thin filaments**. The functional unit of a myofibril is a **sarcomere**, a short cylinder of **myosin** thick filaments and **actin** thin filaments. According to the **sliding filament model** of muscle contraction, the thick filaments' **cross bridges** grab onto the thin filaments and pull the two parts together, so that the thick and thin filaments slide past one another and the muscle fiber contracts.

Review and Thought Questions

Review Questions

1. If a vertebrate animal is "a tube within a tube," what is the outer tube and what is the inner tube?
2. What limbs do the pectoral girdles of fish and humans support?
3. Name the four main classes of tissues.
4. What does each of these tissues do?
5. Would you expect an elephant's skin to be make of simple epithelium or stratified epithelium? Explain why.
6. What is the function of the dendrites on a neuron?
7. Which kind of connective tissue often encloses organs like a sandwich bag?
8. If you twist your knee during a soccer game, what kind of force has your knee been subjected to?
9. Name three differences between smooth muscle and striated muscle.
10. Give two examples of organs that have smooth muscle.
11. Describe the sliding filament model of muscle contraction in three sentences or less, mentioning thin filaments, thick filaments, and cross bridges.

Thought Questions

12. Give an example of something in everyday life that reminds you of the way muscle contraction works.
13. If you were an astronaut in space for six months, what would happen to your bones and muscles? What cells would be responsible for this effect?

BiologyNow Resources

Biology Now™

Active Figures

35-14: Muscle contraction
35-15: Sliding filaments

Preparing for an exam? Take a diagnostic test on your BiologyNow CD-ROM.

Online materials relating to this chapter are at:

http://biology.brookscole.com/AAL3

About the Chapter-Opening Image

Tyrannosaurus rex returns to show that the fossil bones of a long extinct animal can tell us something about both its muscles and its way of life.

36

How Do Animals Obtain Nourishment from Food?

Key Questions

- Why do animals need to eat?
- Why do animals need a digestive tract?
- How do mechanical and chemical digestion prepare food for absorption by cells?
- How does the intestine increase the rate of absorption?

Doctor Beaumont and His Reluctant Patient

In the summer of 1822, a shotgun accidentally discharged into a young French Canadian, blasting a huge hole in his stomach. Poor Alexis St. Martin's misfortune was physiology's gain, however, for the awful wound added immeasurably to our understanding of digestion.

Alexis St. Martin worked for the American Fur Company as a *voyageur,* transporting supplies among the company's remote outposts. That summer, St. Martin arrived at Mackinac Island, between Lake Huron and Lake Michigan, and jammed himself into a trading post cabin crowded with trappers, soldiers, and Indians. Most were there to buy and sell pelts, coats, and moccasins as well as to trade stories after a long and lonely winter. At the moment the shotgun went off, St. Martin was less than three feet from the muzzle, and the hole it made in his stomach was bigger than the palm of his hand. The horrified crowd of trappers fell back, and one of the store's clerks ran to Fort Mackinac to fetch a surgeon.

When Dr. Beaumont arrived, he saw a rib and some lung and St. Martin's stomach, all protruding from the enormous wound. The stomach had a hole in it, too, which, Beaumont later wrote, was "large enough to receive my forefinger, and through which a portion of his food that he had taken for breakfast had come out and lodged among his apparel. . . . I considered any attempt to save his life entirely useless." William Beaumont nonetheless cleaned and dressed the wound, "not believing it possible for him to survive twenty minutes."

In fact, St. Martin recovered amazingly. Within a year, he had regained enough strength to walk and care for himself and he ultimately lived into his 80s. But he would live the rest of his life with a hole in his stomach that leaked digestive juices when not plugged with a bandage. In his first year, he needed constant care, and Beaumont generously took him into his own home to care for him (Figure 36-1).

Despite St. Martin's amazing recovery, he was unable to return to work as a *voyageur,* and Beaumont hired him as a handyman to chop wood and help around

Merlin Tuttle/Photo Researchers, Inc.

Figure 36-1
Alexis St. Martin. William Beaumont and his family care for the wounded Alexis St. Martin in the 1820s.

The Granger Collection, New York

the house. By 1825, it had occurred to Beaumont that St. Martin's damaged stomach was a golden opportunity to find out how digestion works. How the stomach digests food was a standing question among physiologists of the era, and Beaumont began a series of experiments, tying food to a string and dipping it in through the hole in St. Martin's stomach to measure how long different foods took to break down. Beaumont even extracted juices from the young man's stomach to see if they would digest food outside the body.

St. Martin initially objected to these uncomfortable experiments. But they would bring Beaumont prestige in America's infant scientific community, and the ambitious Beaumont virtually kidnapped the reluctant young man. After the first round of experiments in 1825, St. Martin apparently fled, moving back to Canada to marry and start a family. But in 1829, Beaumont managed to lure his former patient back so that he could finish his experiments.

In 1833, Beaumont finally published a full description of his experiments and his conclusions in a book titled, *Experiments and Observations on the Gastric Juice and the Physiology of Digestion.* His work is a model of careful observation and experiment. For example, Beaumont showed that a piece of beef suspended in St. Martin's stomach on a string was digested over a period of several hours. Virtually the same changes occurred if a similar piece of beef was placed in a glass vial containing "gastric juice" from St. Martin's stomach. Beaumont also showed that the juice contained hydrochloric acid. Yet, he noted, hydrochloric acid alone would not digest the meat. What did it mean?

The answer only came later when physiologists discovered an enzyme called *pepsin,* which breaks down proteins in

the stomach. Pepsin digests protein far better than hydrochloric acid, but must work in an acid environment. Thus the acid's role in digestion is primarily to activate pepsin.

Beaumont listed 51 conclusions from his experiments and observations. His list was as important for the questions it raised as for the facts it summarized. Today, biologists are still studying the way we process food, looking, for example, at what makes us hungry and what makes us feel full.

36.1 Why and How Do Animals Eat and Digest Food?

All organisms need energy to power their bodies and raw materials to build and repair tissues and organs. Plants get energy from the sun and raw materials such as carbon dioxide and water from the air and soil in which they grow. With just a few simple materials, plants can build tens of thousands of macromolecules, including carbohydrates, proteins, nucleic acids, and fats.

Animals also need energy and raw materials, but they get them from plants, either by eating plants directly or by eating other animals that themselves ate plants. **Herbivores** [Latin, *herba* = vegetation + *vorare* = to swallow], such as caterpillars and cows, eat plants. **Carnivores** [Latin, *carn* = flesh], such as spiders and wolves, eat other animals. *Decomposers* eat the remains of plants and animals. **Omnivores** [Latin, *omnis* = all], such as humans, crows, and lobsters, eat both plants and animals.

Once an animal has acquired food, it immediately begins the process of **digestion,** reducing big chunks of food to small molecules such as sugars and amino acids. The two parts of digestion are **mechanical digestion,** in which the food is divided into tiny pieces, whether by teeth or other means, and **chemical digestion,** the enzyme-mediated breakdown of carbohydrates, proteins, and fats into simple building-block molecules, such as glucose, amino acids, and fatty acids.

The process by which cells in the intestinal tract take up these small molecules is called **absorption.** Finally, any material that was not digested and absorbed passes to the end of the digestive tract, which prepares the waste for **elimination.** The **digestive tract**—a long, narrow tube that runs from the mouth to the anus—is the organ system responsible for mechanical and chemical digestion, absorption, and elimination.

Once food has been broken down into simple molecules, the molecules are absorbed into the bloodstream, where the cells of the body can use them for either energy or raw materials. In the same way that wood can be used either to build a house or to heat it (by burning the wood), the molecules in our food can be used either to build new tissues or to fuel the body's activities. For example, our cells join sugars together to form carbohydrates, and they join amino acids together to form proteins (Chapter 3).

On the other hand, our cells can also extract energy from these same small molecules, or "burn" them, by breaking them down into even smaller molecules using oxygen. For example, in all animals, including humans and hummingbirds, the sugar glucose is a basic unit of energy. The blood carries glucose from the digestive tract to cells throughout the body. In the cells, the breakdown of glucose to carbon dioxide and water releases energy in the form of ATP molecules, which power most of an animal's activities (Chapter 6). The breakdown of molecules such as glucose to provide energy is called **metabolism**. Metabolism occurs inside cells and provides energy in the form of ATP or similar energy molecules. In contrast, digestion works on larger molecules, occurs in the digestive tract, and doesn't directly create ATP or other energy molecules.

Besides energy, our food also supplies essential building blocks that an animal cannot make itself, including essential amino acids, vitamins, and minerals. For example, most animals can make only about half of the 20 amino acids needed to build proteins, so they must get the others from their food.

Animals need food for energy and raw materials. Animals break down food by means of mechanical and chemical digestion, then absorb the resulting simple molecules, and discard the waste. Simple molecules such as glucose or amino acids can be used inside of cells either as fuel or as building blocks.

How Much Must an Animal Eat?

Food provides energy for all of an animal's activities. Physiologists usually estimate how much energy an animal is using by measuring oxygen consumption. An animal uses one liter of oxygen for every 5,000 calories that it burns. A calorie is the energy required to heat one gram of water 1°C. Both a "kilocalorie" and a "Calorie" (with a capital *C*) are 1,000 calories. Packaged food states the number of "calories" per serving, which are actually kilocalories. So we could say that an animal uses a liter of oxygen to burn every 5 Calories.

Calorie consumption is proportional to weight, so that, for example, a young woman of 120 pounds requires 1,800 to 2,400 kcal per day, while a young man of 160 pounds requires 2,400 to 3,200 kcal per day. Sedentary people need fewer calories than active ones, and older people need fewer calories than younger ones.

Our consumption of energy depends on what we are doing: running or making beds requires more energy than standing or sitting. But every animal requires a certain amount of energy just to stay alive and awake. This minimal energy requirement is called the **basal metabolic rate.** For humans, the basal metabolic rate is about 1,400 kcal per day for a 120-pound woman and about 1,700 kcal per day for a 160-pound man. The same amount of energy would keep a 75-watt bulb burning all day and night or would fuel a one-mile drive in a small car. For humans and other mammals, the basal metabolic rate is more than half the energy used in ordinary activities.

Animals must eat enough food for both basic metabolic needs and activities.

How Does an Animal Know How Much Food to Eat?

Healthy animals adjust their food intake to match their activities. For example, when researchers diluted ordinary rat food, the rats just ate more, so that they still acquired the same amount of energy. But damaging the hypothalamus, the part of the brain that controls appetite, can destroy a rat's ability to adjust its diet. One rat with a damaged hypothalamus may eat without stopping, while another, with damage to another part of the hypothalamus, may starve itself to death.

The hypothalamus exerts its effects by means of chemical signals called *hormones* (Chapter 37). For example, the hypothalamus secretes the hormone orexin, which stimulates intense hunger. Fat cells also "talk back" to the hypothalamus. When fat cells become full, they release the hormone leptin, which signals the hypothalamus to stop feeding behavior. Leptin binds to neurons that influence emotion and appetite and give us a sense of satiation. Pharmaceutical companies are now spending millions of dollars trying to find ways to control levels of these and other hormones, so that people can regulate their appetites using drugs.

An animal that takes in more food energy than it uses stores the excess as fat. Such **overnourishment** is common in developed countries such as the United States. Currently, more than half of Americans are overweight or obese. Being overweight contributes to a host of dangerous conditions, including high blood pressure, high cholesterol, diabetes, heart disease, and stroke.

But a much bigger problem worldwide is **undernourishment,** taking in too few calories. An animal that takes in less food energy than it needs must derive the extra energy from its own body. When energy stored as fat runs out, animals begin to break down their muscles and other tissues. Undernourishment affects about 13 percent of the world's population, or 840 million people. Many if not most of these chronically hungry people also suffer from specific deficiencies in their diet, including protein deficiency and different vitamin deficiencies.

Even in affluent countries such as the United States, where undernourishment is rare, some people still starve to death. In the year 2000, for example, 4,200 Americans died of some form of malnutrition. A fraction of these deaths occurred among the wealthiest Americans. In rare instances, teens with emotional problems and an obsession with weight loss eat so little that they actually die of starvation. This disease, called anorexia nervosa, is common among teenage girls, affecting about 1 in 100. But most girls do not die of this disease. Most actual deaths from anorexia nervosa—about 145 per year—occur among the elderly, both men and women 60 to 90 years old.

Animals usually regulate their food intake to match their activities, but this regulation can go awry either because of an inadequate food supply or because of a failure in normal regulation.

How Much Protein Do Animals Need?

Most people who are vegetarians know that they have to mix different kinds of food in order to get a "balance" of proteins. The reason for this is that of the 20 amino acids needed to make proteins, most animals can make only about 10. The ones that animals cannot make must come from their food and are called **essential amino acids.** Table 36-1 lists the essential amino acids for humans. Most of the essential amino acids are available in all the foods we normally eat. But several (lysine, methionine, and tryptophan) are found only in meat and a few plant foods.

Meat eaters don't think much about amino acids because meat has all the essential amino acids (as do eggs). But plant foods can have lots of one amino acid and hardly any of another. In order to get an adequate amount of lysine, methionine, and tryptophan, vegetarians need to eat a varied diet. And because animals do not store amino acids between meals, each meal must provide a balance of essential amino acids (Figure 36-2). An appropriate mix of food—one with **complementary proteins**—provides enough of each essential amino acid to sustain protein synthesis. For example, corn is low in the amino acid lysine but high in methionine, while beans are low in methionine but high in lysine. This makes corn and beans a good combination.

In parts of the world where food is limited and people depend on a single crop such as rice or corn, amino acid deficiencies are common. Where enough food is available, however, even the poorest people have developed healthful diets that contain complementary proteins—for example, corn and beans, wheat and chickpeas, or rice and lentils.

How Do Animals Keep Dangerous Digestion from Damaging Cells?

Digesting food requires harsh conditions. Humans help this process by cooking food or by marinating it in an acid solution such as vinegar, wine, or lemon juice. Like most other mammals, we also break food into bits by using our teeth to bite, puncture, tear, and grind the food with our teeth. In our stomachs and intestines, we squeeze and agitate food and expose it to strong acid and a powerful enzyme, which

Table 36-1
Nine Essential Amino Acids for Humans

Amino acid	Poor sources	Good sources
Histidine	—	Almost any good protein source: legumes, cereal grains, nuts, dairy, meat, eggs
Isoleucine	—	Any good protein source
Leucine	—	Any protein source
Lysine	Grains, nuts, corn	Legumes (peas, beans, lentils), milk, eggs, meat
Methionine	Legumes (peas, beans, lentils)	Corn, cereals, nuts, milk, eggs, meat
Phenylalanine	—	Any good protein source
Threonine	—	Any good protein source
Tryptophan	Corn, peanuts, almonds	Legumes (peas, beans, lentils), grains, milk, eggs, meat
Valine	—	Any good protein source

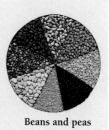

Corn and cereals

Essential amino acids:

Methionine
Valine
Histidine
Threonine
Phenylalanine
Leucine
Isoleucine
Tryptophan
Lysine

Beans and peas

Amy Ellis Dunleavy

Figure 36-2

A nutritious diet must provide complementary amino acids. Some foods such as corn are poor in lysine, while others such as beans are poor in methionine. Eating corn and beans together in the same meal delivers both of these essential amino acids, as well as all the others we need to build proteins.

together begin to break up the proteins. In the small intestine, detergents from the liver break the fats and oils into tiny droplets that can be attacked by enzymes.

But the harsh conditions of digestion are a two-edged sword. They prepare the tissues in food for further digestion, but they can also damage the tissues of the animal. To

Health and Biology Vitamins and Minerals

The main building blocks that animals require are sugars, amino acids, nucleotides, fatty acids, and glycerol. But animals also need small amounts of other materials, specifically vitamins and minerals. We need some minerals (calcium, phosphorus, potassium, sulfur, sodium, chlorine, and magnesium) in relatively large amounts and others (iron, manganese, and iodine) in smaller amounts, and still others (copper, zinc, molybdenum, cobalt, and selenium) in only in trace amounts. Table A lists a few of the minerals that are essential for humans, symptoms of insufficiency, and good food sources. Although humans need these 26 elements, any of them can be dangerous in large amounts.

Some animals cannot make organic molecules that are critical to their health. The most well-known example is vitamin C, also called ascorbic acid [Latin, *a* = not + *scorbutus* = scurvy], which plays a role in the biochemical reactions that build collagen in connective tissues. Humans become very sick without a good supply of vitamin C. Most people who get any raw fruits or vegetables at all do okay, but sailors far from shore for months at a time used to come down with *scurvy*, a disease characterized by bleeding gums, loosening teeth, and slow wound healing. The British navy supplied lemons to their crews to prevent scurvy, switching to limes in 1865, forever earning British sailors the title "limey." Al-

though the cure of vitamin C deficiency was understood in the early 19th century, the structure and biochemical function of ascorbic acid were not known until the early 20th century, almost 200 years after its first use to prevent scurvy. Besides helping to build collagen, ascorbic acid also serves as an *antioxidant,* limiting the damage caused to cells by oxygen and its products.

Some other vitamins have been harder to identify because they are present at high enough concentrations in the ordinary diet that deficiency diseases are rare. Despite these difficulties, we now know of 13 essential vitamins required in the human diet (Table B), which fall into two broad groups: *water-soluble vitamins* and *fat-soluble vita-*

Table A
Minerals Important in the Human Diet

Mineral	Good Sources	Symptoms of Deficiency	Functions
Sodium	Normal diet, table salt	Muscle cramps	Water balance; operation of muscles and nerves
Potassium	Fruits, vegetables, grains	Irregular heartbeat, fatigue, muscle cramps	Operation of muscles and nerves; acid-base balance
Chlorine	Table salt	Unlikely	Acid formation in stomach; nerve signaling
Calcium	Dairy, bony fish, leafy green vegetables; dried legumes	Osteoporosis	Formation of bone and teeth; clotting, nerve signaling
Phosphorus	Dairy, meat, cereals	Bone loss, weakness, lack of appetite	Formation of bone and teeth; energy metabolism (e.g., ATP); membrane phospholipids
Magnesium	Nuts, greens, whole grains	Nausea, vomiting, weakness	Enzyme action; nerve signaling
Sulfur	Protein foods	None, when protein needs are met	Proteins

ATP = adenosine triphosphate.

avoid such "self" digestion, animals perform most digestion in special areas outside the cell. Single-celled organisms use a special organelle whose membrane separates the digestive enzymes from most of the cell's own proteins. Spiders, starfish, and many other animals digest their food extracellularly by using the bodies of their prey as a digestion vessel. A spider catches a fly, then secretes digestive enzymes into the body of the fly. After the enzymes have digested the fly, the spider ingests the liquefied contents of its prey for further digestion. Venomous snakes likewise inject a venom

that partially digests the body of their prey before eating the prey. Starfish turn their stomachs inside out onto surfaces they want to digest, including coral and dead fish.

Even single cells keep digestive enzymes separate from their delicate cytoplasm. In *endocytosis,* a cell enfolds material outside the cell into a vesicle that fuses with a *lysosome,* whose digestive enzymes break down the material into small molecules that are released into the cell (Figure 36-3A and Chapter 4).

Multicellular animals digest food particles too in an extracellular digestion chamber. The simplest digestion cham-

mins. The water-soluble group includes vitamin C (ascorbic acid) and the B-complex vitamins, which have a common designation only because they usually appear in the same foods. Each of the water-soluble vitamins is a precursor of a particular *coenzyme,* a small organic molecule that binds to an enzyme and plays a role in catalysis. The fat-soluble vitamins, including vitamins A, D, E, and K, play diverse roles: vitamin A, for example, is the precursor of retinal, the light-sensitive molecule in the visual pigments, whereas vitamin K participates in blood clotting.

The listed substances are essential for humans, but not for all animals. Vitamin C appears to be essential only for humans, monkeys, and guinea pigs; other species can produce it themselves. Similarly, the fat-soluble vitamins are essential only for vertebrates.

Table B
Essential Vitamins in the Human Diet

Vitamin	Good sources	Functions
Fat-soluble		
A (retinol)	Leafy green and yellow vegetables, liver, eggs, milk	Visual pigment; development of bone and teeth; and maintenance of skin, gut, and other epithelia
D	Synthesized in skin exposed to sunlight; eggs, fortified milk	Bone growth; calcium absorption
E	Meat, milk, vegetable oils, whole grains	Prevents oxidative damage to cells; maintains levels of vitamin C
K	Green leafy vegetables; gut bacteria	Clotting; electron transport
Water-soluble		
B_1 (thiamin)	Meat, milk, eggs, grains, legumes	Coenzyme in production of carbohydrate and nucleic acids; development of connective tissues
B_2 (riboflavin)	Milk, meat, grains	Coenzyme in energy metabolism (e.g., FAD)
Niacin	Meat, bread, potatoes	Coenzyme in energy metabolism and biosynthesis (e.g., NAD and NADP)
B_5 (pantothenic acid)	Milk, meat, eggs, yeast	Coenzyme in synthesis of fatty acids and steroids
B_6 (pyridoxine)	Meat, potatoes, spinach	Coenzyme in metabolism of amino acids
Folic acid (folate)	Vegetables, cereals, eggs, meat; also gut bacteria	Coenzyme in metabolism of amino acids and nucleic acids
B_{12} (cobalamine)	Milk, meat	Coenzyme in metabolism of nucleic acids
Biotin	Legumes, nuts, eggs, liver	Coenzyme in metabolism of fats, glycogen, and amino acids
C (ascorbic acid)	Citrus fruits, potatoes, green leafy vegetables	Coenzyme in carbohydrate metabolism and in formation of connective tissue

---

bers, such as those of a hydra or flatworm, have a single opening that serves as both mouth and anus. A hydra's arms move food into the chamber, and enzymes then break the food into smaller pieces (Figure 36-3B). The surrounding cells take them up by means of endocytosis. The hydra discharges any indigestible food through the same opening.

Animals more complex than hydras and flatworms have digestive systems with two ends (Chapter 23). Food enters through the mouth and passes through an extended tube, called the **digestive tract,** the *gastrointestinal tract,* or the **gut** [Middle-English, *guttes* = entrails] (Figure 36-4). The undigested and unabsorbed material leaves through the **anus,** the opening at the posterior end of the digestive tract. The remnants of the food, together with bacteria that inhabit the tract, form the **feces** [Latin, *faeces* = dregs].

All organisms digest their food either outside of their own cells or in special compartments within their cells.

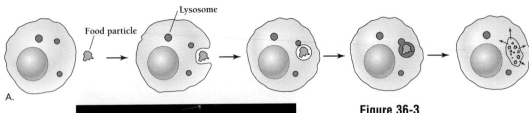

A.

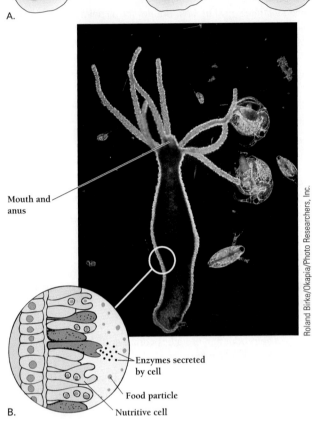

Mouth and anus

Enzymes secreted by cell

Food particle

B.   Nutritive cell

Roland Birke/Okapia/Photo Researchers, Inc.

Figure 36-3

Digestion in a single cell and in a simple organism. A. A single cell engulfs material in a vesicle that fuses with a lysosome full of digestive enzymes. After the enzymes digest the material, the resulting small molecules are released into the cytosol of the cell. B. A hydra is a simple animal that uses its tentacles to catch organisms even smaller than itself. As in other animals, digestion begins inside its large, primitive "stomach." After that, the cells lining the stomach engulf and digest the remaining particles by a process similar to that described in A.

What Are the Advantages of a Two-Ended Digestive Tract?

A digestive tract with two ends allows animals to process food more efficiently than a one-ended digestion cavity. The evolution of the two-ended digestive tract has the same advantages as Henry Ford's first industrial assembly line, in which workers, machines, and equipment were arranged so that a product such as a car passed consecutively from station to station until it was completely assembled. As food passes down the digestive tract, it undergoes a series of successive

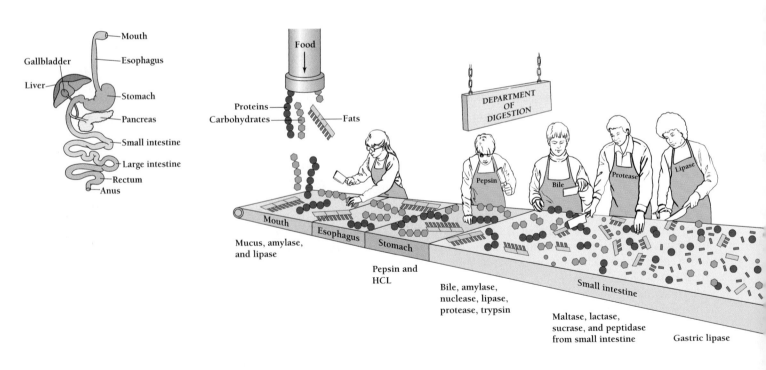

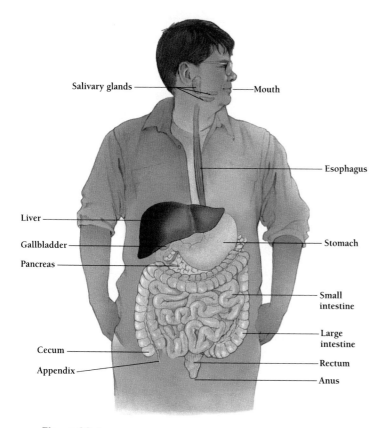

Figure 36-4
The human digestive tract, with accessory organs.

reactions that disassemble it into it component parts. We can think of the digestive tract as a *disassembly line* (Figure 36-5).

A two-ended digestive tract allows the digestive tract to work like a disassembly line.

36.2 How Do Vertebrates Digest Food?

The digestive system actually performs seven functions: (1) movement—agitating the food within specialized regions and pushing food through the system; (2) **mechanical digestion**—the physical breakdown of large pieces of food; (3) secretion—the production of acids, detergents, enzymes, and lubricants; (4) **chemical digestion**—the enzymatic breakdown of large molecules to their component building blocks; (5) **absorption**—the taking up of small molecules by cells; (6) storage—processing a meal a bit at a time, so animals do not have to feed continuously; and (7) elimination—the disposal of waste from the disassembly line (Figure 36-5).

The **lumen** [Latin, = opening, light] of the vertebrate digestive system, or "gut," is the inner space within the tract through which food passes. The lumen is continuous from mouth to throat to esophagus to stomach to small intestine

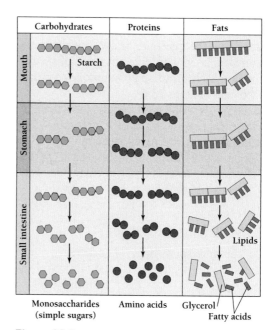

Figure 36-5
The digestive tract as a "disassembly line." Each region of the tract makes distinctive contributions to the process of digestion and absorption. Here simple sugars (monosaccharides) are shown as hexagons, amino acids as circles, and triglycerides as three-tooth combs. At the end of the gut, the colon retrieves water (wheelbarrow at far right).

Biology Now™ Learn more about the digestive tract by clicking on this figure on your BiologyNow CD-ROM.

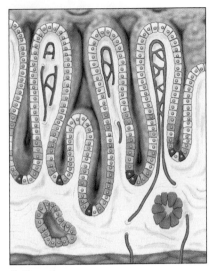
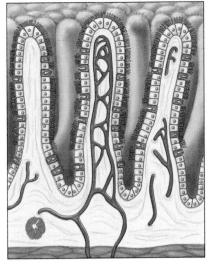
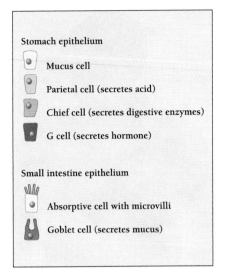

Stomach epithelium

    Mucus cell

    Parietal cell (secretes acid)

    Chief cell (secretes digestive enzymes)

    G cell (secretes hormone)

Small intestine epithelium

    Absorptive cell with microvilli

    Goblet cell (secretes mucus)

A. Stomach

B. Small intestine

Figure 36-6

Different gut surfaces. Each section of the gut has a different kind of lining. A. The stomach is lined with epithelial cells that secrete mucus, hydrochloric acid, digestive enzymes, and hormones. B. The small intestine is lined with epithelial cells that secrete mucus or absorb simple sugars or amino acids. Tiny fingerlike protrusions called villi (that are themselves covered with hairlike microvilli) greatly increase the surface area of the small intestine, increasing its ability to absorb.

to colon to rectum to anus. Just below the stomach, the pancreas, gall bladder, and liver empty digestive enzymes and bile into the small intestine.

At each end of the lumen—at the mouth and at the anus—are **sphincter muscles,** rings of contractile muscle that can open or close the lumen. Another sphincter muscle opens and closes the lumen below the stomach to control the passage of food into the intestines below. The entire gut, from mouth to anus, is lined with different kinds of specialized epithelial tissues, each composed of a unique combination of specialized cells (Figure 36-6).

To visualize the progress of a meal through the human digestive tract, imagine that Lexie St. Martin, the great-great-great-great-great-great-granddaughter of Alexis St. Martin, whom we discussed in the beginning of the chapter, is a student at the University of New Hampshire. Lexie is about to enjoy one of the hamburgers she has just barbecued at a school picnic.

How Does the Mouth Initiate Digestion?

Lexie's first bite contains some bun, some lettuce, and a chunk of meat. As she chews, her teeth tear and grind the bite into little bits. This mechanical digestion is the first step in the digestive process. Mechanical digestion increases the surface area of the food that is exposed to enzymes. This increased surface area speeds chemical digestion when the bite of hamburger arrives a bit lower in the gut.

Most animals feed on other organisms, whether plants or animals, whole organisms or just a few select parts. Our food usually consists of intact tissues that must, first of all, be broken down into smaller pieces. Reptiles and amphibians mostly eat and digest their food whole, but mammals and birds are adapted to begin breaking their food into chunks soon after it is eaten. Most mammals (though not all) break up food in the mouth, with teeth, tongue, and cheeks; a bird does the same with its **gizzard,** a muscular region of the gut, specialized for grinding food.

Mammals are the only vertebrates that can actually chew their food. To do this, mammals have specialized teeth, unlike those of other animals. For comparison, look at the unspecialized teeth of a crocodile (Figure 36-7A). Now look at human teeth, which include *incisors,* whose chisel-like edges cut bites of food; *canines,* whose pointed crowns grab and tear; and *premolars* and *molars,* whose flatter, ridged surfaces grind and crush (Figure 36-7D). Humans are omnivores, so we have an assortment of different teeth. Herbivores and carnivores have more specialized teeth. Herbivores have incisors (for cutting) and molars (for grinding), while carnivores have large canines (for grabbing) and knifelike "carnassials" for slicing meat (Figure 36-7B and C). In mammals that eat plants, the jaws and their muscles allow horizontal movement, for sideways grinding between the surfaces of the molars and premolars. Cheek and tongue muscles help out by moving the food into position for grinding.

Not all mammals chew, of course. For example, anteaters have no teeth and only weak jaws, but they use specially adapted tongues and salivary glands that secrete quantities of viscous saliva to lap up ants. Baleen whales have an elaborate filtering apparatus, called baleen, that filters small crustaceans (krill) from ocean water.

But for humans like Lexie, chewing is required for good digestion. The presence of food in the mouth, or even the smell or the mere thought of food, stimulates the first secretions of the digestive tract. The **salivary glands** secrete both enzymes and **mucus,** a viscous, slippery substance that coats the food particles and lubricates their movements within the

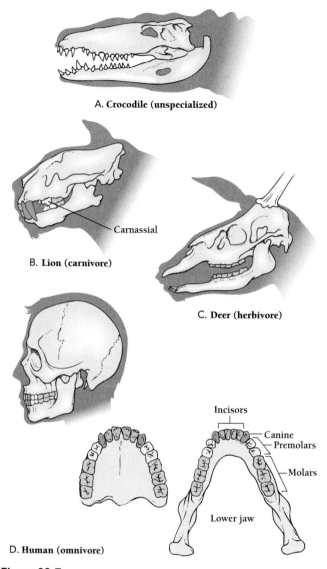

Figure 36-7

Specialization of teeth. A. The unspecialized teeth of a crocodile are good for grabbing and swallowing. Mammals have highly specialized teeth. B. Carnivores are known for their long canine teeth and large shearing carnassials, suitable for seizing and tearing flesh. C. Deer specialize in flat molars, suitable for grinding tough plant parts. D. Humans, pigs, bears, and other omnivores tend to have an assortment of teeth, so they can handle any kind of food that comes along.

mouth and the rest of the digestive tract. Mucus consists of water, salts, and *glycoproteins* (carbohydrates attached to proteins). The grinding action of the teeth exposes the food to the salivary enzyme **amylase,** which breaks the starch molecules in the bun into shorter fragments, some of them sugars. If Lexie chews for a minute, the hamburger bun will become noticeably sweeter as starch is converted to sugar. The tongue also secretes **lingual lipase,** which begins the breakdown of some of the lipids. These two enzymes in the mouth begin the digestion of carbohydrates and fats, but the proteins in Lexie's bite of hamburger remain intact until they reach the stomach.

Lexie's tongue shapes the chewed food and the lubricating mucus into a ball, called a **bolus,** which her tongue and cheek muscles push back into the throat, or **pharynx.** The pharynx is the common entryway both for food into the digestive tract and for air into the lungs. This being so, how does Lexie prevent the bolus from moving into her airways instead of her gut?

The answer lies in the intricate workings of the muscles of the pharynx. The narrow opening into the *larynx* (which leads to the respiratory tract) is guarded by a lid of elastic cartilage, called the **epiglottis.** When Lexie swallows, the muscles of the pharynx raise the larynx and the epiglottis folds over and covers the entrance to the larynx (Figure 36-8). For a second, Lexie can't breathe. At the same time, other pharyngeal muscles reflexively push the bolus past the larynx and down into the **esophagus,** a long tube that carries food to the stomach. Once the bolus is well into the esophagus, the larynx drops, the epiglottis reopens, and Lexie can breathe again. If you swallow in front of a mirror, you can see your larynx, or Adam's apple, rise briefly, then drop down again.

Although we initiate swallowing consciously, once food is in the back of the throat, reflexes take over. Swallowing involves the coordinated response of both smooth and striated muscles in the pharynx and esophagus, as well as the inhibition of muscles responsible for breathing. A swallowing center in the lower part of the brain (in the medulla) manages the operation through a network of nerves that connect to these muscles.

Digestion begins in the mouth with the mechanical digestion of the food, the chemical digestion of carbohydrates and lipids, the formation of a bolus, and the movement of a bolus into the esophagus.

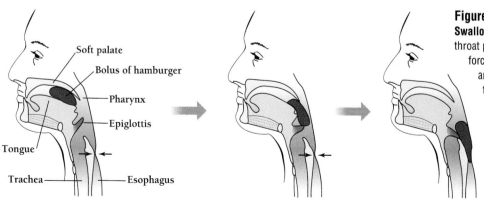

Figure 36-8

Swallowing, part 1. The muscles of the tongue and throat push a bolus of food back toward the throat, forcing the epiglottis down across the windpipe and closing it. Muscles in the throat then push the bolus down into the esophagus.

Biology Now™ Learn more about swallowing by clicking on this figure on your BiologyNow CD-ROM.

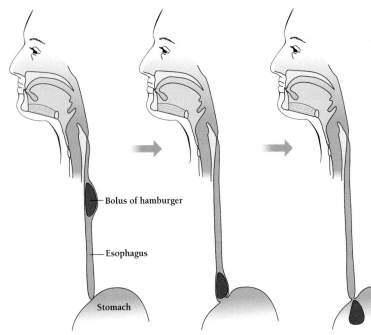

Figure 36-9
Swallowing, part 2. Peristaltic waves of contracting smooth muscle push the bolus down the esophagus and into the stomach. The intestines also push food through the gut by means of peristalsis.

- Bolus of hamburger
- Esophagus
- Stomach

What Do the Esophagus and Stomach Contribute to Digestion?

Smooth muscles propel the bolus through the esophagus. **Peristaltic waves,** coordinated contractions of smooth muscles, move the bolus down the esophagus toward the stomach (Figure 36-9). The alternating contractions of circular muscles behind the bolus and longitudinal muscles ahead of it operate something like those the earthworm uses to move through soil, as discussed in Chapter 35. During a wave, the longitudinal muscles in front of the bolus contract, shortening the esophagus ahead of the bolus. Meanwhile, the circular muscles behind the bolus contract, pushing the bolus forward toward the stomach. The esophagus goes straight down the chest cavity and into the stomach. Despite the downward course of the esophagus, gravity is unnecessary for moving food into the stomach: thanks to peristalsis, Lexie could swallow her burger while standing on her head.

The **stomach** [Greek, *stoma* = mouth] is the most muscular section of the digestive tract (Figure 36-10). If Lexie hasn't eaten for a while, her shrunken stomach holds only about two ounces, but it can stretch enough to hold far more—one or two liters of hamburger, green salad, milk, and chocolate chip cookies.

It has been six hours since Lexie had breakfast, and she's hungry, so bolus after bolus of hamburger arrives in her stomach. After each bite, the stomach's entrance squeezes shut to prevent the food from moving back up into the esophagus. Sometimes, particularly after eating a heavy meal or drinking alcohol, the muscles that close the top of the stomach relax, and the acidic contents from the stomach bubble back up into the esophagus, producing a burning feeling in the upper chest called "heartburn," or "reflux." Heartburn can be barely noticeable or so painful that it is mistaken for a major heart attack. Stomach acid is strong enough to dissolve steel. The stomach itself is usually protected from its acidic contents by mucus secreted by the cells lining the stomach.

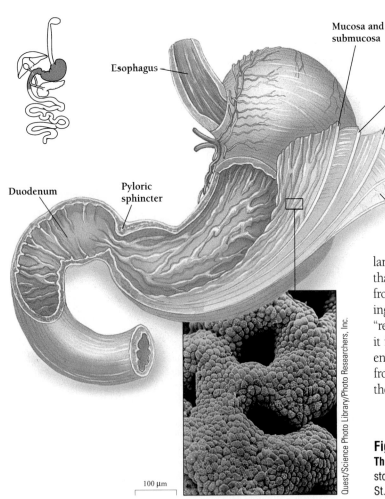

- Esophagus
- Mucosa and submucosa
- Muscle layers
- Serosa
- Duodenum
- Pyloric sphincter

100 µm

Quest/Science Photo Library/Photo Researchers, Inc.

Figure 36-10
The tissues and cells of a stomach. The deep folds (villi) give the stomach lining the velvetlike appearance that Beaumont noticed in Alexis St. Martin's stomach.

In response to food in the stomach, other cells in the stomach lining secrete **gastrin,** a hormone that calls the stomach to action. Gastrin stimulates the secretion of hydrochloric acid and pepsinogen, an inactive form of the enzyme **pepsin.** The acid activates pepsin, and the pepsin breaks the hamburger proteins into short lengths of polypeptides.

While the acid and pepsin are breaking down proteins, the acid denatures (or unfolds) the amylase and lipase that were secreted in the mouth and they stop digesting carbohydrates and lipids. The lettuce remains undigested, and pieces of it remain dispersed within the stomach contents. Meanwhile, the muscles of the stomach, stimulated by gastrin, churn everything together, continuing the mechanical digestion begun in the mouth. The result is a creamy, acidic liquid called **chyme** [Greek, *khumos* = juice]. Peristaltic waves move the chyme through the stomach's exit, another sphincter muscle, called the pyloric sphincter, which opens only briefly, allowing a small spurt of chyme to pass into the small intestine. (If Lexie happens to get food poisoning at this picnic, reverse peristalsis some hours later could cause the small intestine to begin ejecting chyme back into the stomach, which would in turn eject the chyme into the mouth, a process we all know as *vomiting.* But let's hope that the hamburger was thoroughly cooked.)

The muscles of the stomach mechanically digest food, while hydrochloric acid and pepsin chemically digest proteins. Food leaves the stomach as semi-liquid chyme.

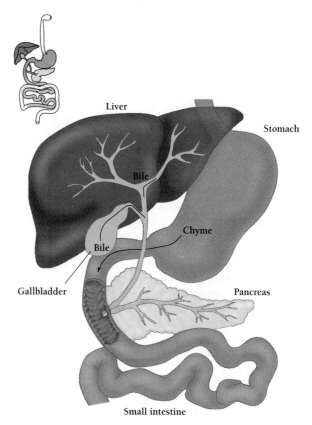

Figure 36-11

At the start. At the top of the small intestine, just where the first chyme from the stomach appears, the liver, gallbladder, and pancreas dump bile and digestive enzymes into the chyme as it starts down the intestinal tract.

What Does the Small Intestine Contribute to Digestion?

The chyme now enters the **small intestine,** the long, coiled section of the digestive tract between the stomach and large intestine where most absorption and chemical digestion take place. Two large glandular organs, the liver and the pancreas, help the small intestines finish the process of digestion begun in the mouth and the stomach (Figure 36-11). The **liver** secretes **bile,** a detergentlike fluid that helps enzymes break down fats. Embedded in the liver is a pear-shaped organ called the **gallbladder,** which stores and concentrates bile until it is needed. The **pancreas** produces digestive enzymes and an alkaline solution that neutralizes stomach acid, preventing the stomach acid from damaging the delicate surface of the small intestine and its digestive enzymes.

In the first few inches of the small intestine, the chyme from the stomach mixes with the secretions of the liver, the gallbladder, the pancreas, and the intestine itself. This first section of the small intestine is called the **duodenum** [Latin, *duodeni* = containing twelve] because the duodenum's length is about equal to the width of 12 large fingers.

The digestion that began in the mouth and stomach continues in the small intestine. But for the first time since Lexie took a bite of hamburger, small energy molecules and

building blocks now become available for absorption. The small intestine completes the digestion of macromolecules to small building-block molecules, and then absorbs 90 percent of them.

Both digestion and absorption depend on the thorough mixing of the chyme in the small intestine. The small intestine vigorously agitates its contents in much the same way a clothes washer agitates clothes. The inner layer of the small intestine, called the *mucosa,* consists of an absorptive epithelium, surrounded by a layer of connective tissue, blood vessels, and nerves, and a layer of smooth muscle (Figure 36-12). Surrounding the mucosa is another layer of connective tissue and two more layers of smooth muscle, the ones that agitate the chyme. Finally, another layer of epithelial and connective tissue, the *serosa,* surrounds the entire intestinal tract.

The smooth muscles of the small intestine move in two patterns. First, the intestines churn the chyme back and forth, more or less in place. This action thoroughly mixes the chyme with enzymes and other digestive secretions from the liver, pancreas, and intestines. The mixing also helps increase absorption through the intestinal lining by continuously exposing different parts of the chyme to the lining of the small intestine. After the intestine has absorbed most of the nutrients in a meal, its pattern of movement changes to peristaltic

Extreme Biology   What Kinds of Stomachs Do Other Animals Have?

In some animals, such as birds, the esophagus ends in a stomachlike sac, called the crop, which holds food for later digestion or to feed to nestlings. In pigeons, the crop produces a "milk" substance that the parents feed to their babies.

In birds, which have no teeth, whole chunks of food pass from the crop to a stomach that, like ours, secretes acids and enzymes and then to a muscular stomach called the gizzard, the inner surfaces of which are coated with an abrasive material. A bird also picks up and swallows small stones, which lodge in the gizzard and take the place of teeth in mechanically digesting the bird's food. Gizzards also contain pepsin and hydrochloric acid, so that the chyme that leaves a bird's gizzard is ready for the next steps of digestion in the intestine.

The stomachs of cattle—and those of moose, buffalo, and other related animals—have special adaptations that enable them to extract energy efficiently from the hard-to-digest grasses that make up most of their diets (Figure A). These animals are all ruminants—hoofed, horned, or antlered herbivores that can regurgitate partly di-

gested food for further chewing. A ruminant's stomach contains four distinctive chambers, which in a large cow may hold as much as 200 liters of grass and fluids, enough to fill a bathtub. Bacteria and protists in the first two chambers, the rumen and the reticulum, produce enzymes that break down cellulose, a feat that no animal can do by itself. The cow then regurgitates the partly digested

food, called cud, and chews it some more to further break it down. The cow then swallows it into the omasum, which absorbs water, and from there it goes to the abomasum and the intestine. In contrast to the chyme from Lexie's stomach, the chyme entering a cow's intestine contains no undigested cellulose, but only energy-rich molecules ready to be absorbed.

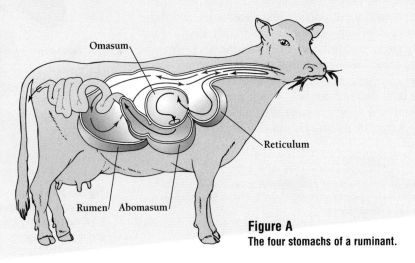

Figure A
The four stomachs of a ruminant.

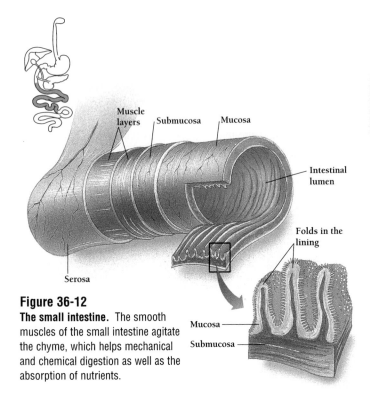

Figure 36-12
The small intestine. The smooth muscles of the small intestine agitate the chyme, which helps mechanical and chemical digestion as well as the absorption of nutrients.

waves (like those in the esophagus) that push the meal to the large intestine, the last region of the digestive tract.

Smooth muscles agitate the contents of the lumen and propel them onward.

How Does the Small Intestine Digest?

Once Lexie's meal enters the small intestine, chemical digestion begins in earnest. Proteases from the pancreas chop the peptide fragments from the stomach into lengths of a few amino acids. Other enzymes, called peptidases, break the polypeptides into amino acids. Meanwhile, pancreatic amylase attacks whatever starches survived the salivary amylase of the mouth. Maltose, sucrose, and other sweet disaccharides accumulate in the small intestine. Enzymes on the epithelium break the disaccharides into simple sugars, ready for immediate absorption. Nucleases secreted by the pancreas break down the DNA and RNA in Lexie's burger.

Lipids are not soluble in water and so are more of a challenge to digest. Because fats form large drops in the watery environment of the digestive tract, their oily interiors are protected from digestive enzymes. The liver solves this problem by secreting a detergent solution called bile that is bitter, alkaline, and a distinctive green-yellow or brown-yellow color. Bile (also called "gall") may move directly into the duodenum or enter the gallbladder where it is stored until a meal arrives.

Just as dishwashing detergent breaks up the oils left on a dinner plate, so bile breaks up the fat droplets within the chyme into much tinier droplets. The churning of the small intestine mixes the chyme with bile to form an *emulsion,* a smooth mixture of water and oil. The oil droplets in the emulsion are so tiny that lipases secreted by the pancreas can easily break the lipid molecules into fatty acids and glycerol that can be absorbed later by the small intestine.

By the time the remains of Lexie's meal have left the small intestine, essentially all the proteins, carbohydrates, lipids, and nucleic acids have been reduced to small molecules and absorbed. Table 36-2 lists the major digestive molecules of the small intestine. Only the cellulose from the lettuce remains intact, because Lexie has no enzyme that will break it down. It will pass through undigested.

Digestion in the small intestine depends upon enzymes produced by the pancreas and the intestine itself, as well as on bile produced by the liver and stored in the gallbladder.

How Does the Small Intestine's Large Epithelium Help Absorb Nutrients?

The shape of the small intestine helps it to both digest and absorb. First, its long, thin shape slows the passage of the chyme, giving enzymes more time to work. Second, for the intestine to most quickly absorb nutrients, the absorbing surface (the epithelial lining of the small intestine) must have a large surface area. In humans, the small intestine is only about an inch in diameter (which is why it is called the "small" intestine), but it is 15 to 20 feet long, which greatly increases its surface area.

To further increase its absorptive surface, the lining of the small intestine is covered with tiny fingerlike extensions, called **villi** [Latin, = shaggy hair; singular, villus]

(Figure 36-13A). Each villus extends about a millimeter into the lumen (Figure 36-13B). The epithelium covering it is made of many individual cells, each one of which is itself bristling with even tinier fingers called **microvilli** (Figure 36-13C). The villi and microvilli greatly increase the total area of the small intestine. If the surface of the human small intestine were completely smooth, it would have about the same area as the screen of a 28-inch television. With its villi and microvilli, it is instead about the area of a tennis court.

The individual cells of the epithelium absorb nutrients both passively and actively. The epithelial cells pass the sugars, amino acids, and other nutrients into the circulatory system, which carries them to the liver, which adjusts their concentration before they are dumped into general circulation.

The epithelial cells of the small intestine not only absorb small molecules but also prevent their entry into the body. Each epithelial cell is polarized, with the "top" facing the lumen and the "bottom" in contact with connective tissue. Joining the sides of the epithelial cells near the top are **tight junctions** that keep the chyme from leaking between the cells into tissues or blood (Figure 36-14). All materials must instead pass *through* the epithelial cells, giving them a chance to regulate what passes into the blood. The tight junctions functionally seal off the intestine's lumen (Figure 36-14). Below the tight junctions are "adhering junctions," which strengthen the connections between cells, and "gap junctions," which allow small molecules and ions to move among adjacent cells.

The overall process of absorption is so efficient that, normally, by the time the chyme appears in the large intestine, few nourishing molecules remain. What's left is mainly water, ions, and cellulose, together with mucus that helps the material continue moving through the intestine. On average, an adult human consumes about two liters of food and drink and secretes about seven liters of mucus, acid, and other fluids into the digestive tract each day. But the small intestine absorbs 95 percent of this, so that only about a pint of fluid enters Lexie's large intestine on a typical day.

The shape of the small intestine increases its surface area and allows time for both digestion and absorption. Villi and microvilli increase the surface area for absorption.

Table 36-2
Digestion in the Small Intestine

Carbohydrates (starch)	*pancreatic amylase* →	Two-sugar molecules (disaccharides)	*maltase, sucrase, lactase* →	Single-sugar molecules (monosaccharides)
Proteins (polypeptides)	*trypsin* →	Smaller polypeptides	*peptidases* →	Amino acids
Nucleic acids DNA & RNA	*nucleases* →	Nucleotides	*other enzymes* →	Nitrogenous bases, sugars, and phosphates
Fat	*bile* →	Fat droplets	*lipase* →	Fatty acids and glycerol

Figure 36-13

Cell specialization of the small intestine. Three electron micrographs. A. A cross section of small intestine shows how fingerlike villi increase the surface area of the intestine. B and C. Each of the millions of villi is covered with even tinier microvilli. (A, G. Shih-R. Kessel/Visuals Unlimited; B, From R. G. Kessel and R. H. Kardon, *Tissues and Organs: A Text-Atlas of Scanning Microscopy,* 1979, W. H. Freeman & Co.; C, David M. Phillips/Visuals Unlimited)

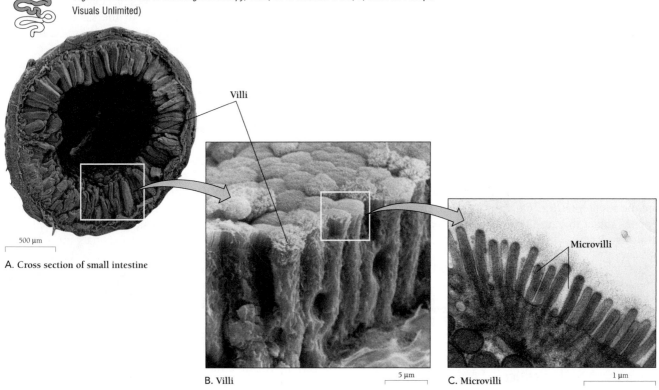

A. Cross section of small intestine

500 μm

Villi

B. Villi

5 μm

Microvilli

C. Microvilli

1 μm

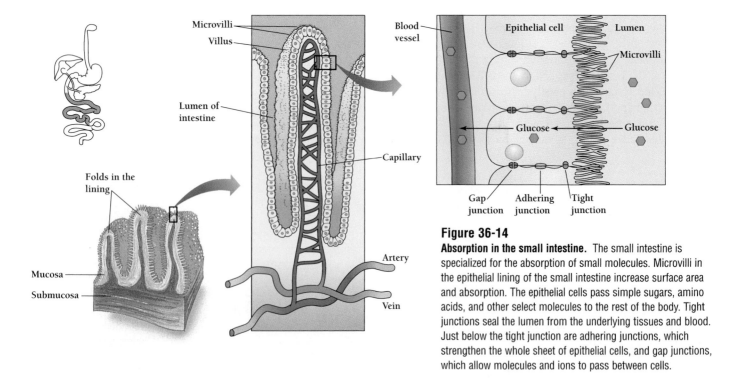

Figure 36-14

Absorption in the small intestine. The small intestine is specialized for the absorption of small molecules. Microvilli in the epithelial lining of the small intestine increase surface area and absorption. The epithelial cells pass simple sugars, amino acids, and other select molecules to the rest of the body. Tight junctions seal the lumen from the underlying tissues and blood. Just below the tight junction are adhering junctions, which strengthen the whole sheet of epithelial cells, and gap junctions, which allow molecules and ions to pass between cells.

What Does the Large Intestine Contribute to Digestion?

The **large intestine** reduces the volume of the chyme by more than 85 percent—by extracting water, nutrients, and dissolved ions—compacting what is left into feces. The large intestine is about five feet long and three inches wide, much shorter and wider than the small intestine. Most of the length of the large intestine is colon and is specialized for absorption. Although the colon has no villi, it has extensive microvilli. The colon ends in a straight portion, called the **rectum** [Latin, = straight], in which feces are stored until their elimination through the anus.

The large intestine removes water by pumping ions across its epithelium. Water follows, moving from a solution in the lumen with low solute concentration to a solution with high solute concentration. If the colon is irritated, its contents may move too fast for efficient absorption, resulting in **diarrhea,** or watery feces. On the other hand, if the intestine's contents move too slowly, the epithelium absorbs too much water, resulting in **constipation,** the inability to expel excessively dry feces. Undigested cellulose from fruits and vegetables provides *roughage,* also called *fiber* or *bulk,* which holds water and stimulates peristalsis. Insufficient roughage or water can cause constipation.

The colon contains large numbers of bacteria, which make up about half the dry weight of the feces. These bacteria live on mucus and unabsorbed molecules in the chyme. For example, indigestible carbohydrates such as those found in beans arrive in the colon virtually undigested. These complex molecules provide a rich food source for bacteria, which release breakdown products such as methane gas (which escapes through the anus as flatulence). Bacteria in the colon also synthesize three essential vitamins—vitamin K, biotin, and vitamin B$_5$ (pantothenic acid), which the colon absorbs.

The human colon may host hundreds of kinds of harmless or helpful bacteria. But disease-causing bacteria, viruses, or protists can also infect the small intestine and the colon, causing diarrhea. For example, cholera bacteria bind to the intestinal lining and release toxins that stimulate the intestine to release massive amounts of fluid. A cholera victim can die of dehydration in just a few hours.

Biologists are only now beginning to study the ecology of the bacteria and yeasts that live in our guts. But what we eat influences which species flourish from day to day, and the exact combination of species may determine, for example, whether we have gas, as well as subtler aspects of health. Germ-free mice, which have no bacteria at all, must eat significantly more food than mice with a normal "gut flora." One of the reasons that some antibiotics make us feel ill and cause diarrhea is that these drugs kill many of the normal bacteria in our intestines.

The first part of the large intestine, at the junction between the large and small intestines, is called the **cecum**

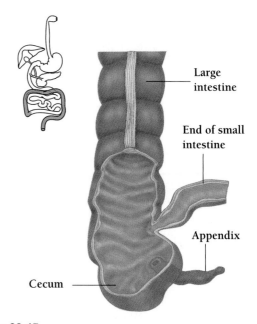

Figure 36-15
Nearing the end. The small intestine dumps chyme into the entrance to the large intestine close to the cecum and the appendix. From there the chyme passes through the different regions of the colon, gradually losing water and minerals as it approaches the rectum.

[Latin, *caecus* = blind], a pouch that stores material from the small intestine (Figure 36-15). Projecting from the cecum is a small, fingerlike projection called the *appendix,* which is part of the lymphatic system. The appendix is best known, however, for its tendency to become inflamed or infected (in *appendicitis*).

In contrast to the human cecum, the cecum of other mammals, especially herbivores, may be relatively large. In rabbits, for example, a large cecum accommodates bacteria and protozoans that digest cellulose in the rabbit's diet of plants. Undigested cellulose from the small intestine enters the cecum, where additional digestion can occur. The newly digested material, now containing energy-rich sugar (glucose molecules) and other nutrients, leaves the cecum and travels through the large intestine. But the large intestine is not able to absorb these nutrients, and they pass out in the feces. Rabbits, however, have evolved an effective (if unappetizing) recycling mechanism: they actually produce two types of feces, one of which is waste and the other of which includes sugars from the cecum. Rabbits eat the sweet ones, extracting the valuable molecules during their second passage through the gut.

The large intestine removes ions and water. Bacteria grow in both the colon and the cecum; in some species, bacteria in the cecum digest cellulose.

36.3 How Does an Animal Coordinate Digestion?

The digestive system must coordinate processes performed by cells that are far from one another, including movement, secretion, digestion, and absorption. When food arrives in the stomach, for example, there must be acid and pepsin; and when chyme arrives in the intestine, there must be pancreatic enzymes, bile, and bicarbonate. What coordinates all this activity?

The brain, spinal reflexes, and hormones all help coordinate the digestive system. Part of the control of the digestive system depends on a complex network of nerve cells within the digestive tract itself. This network continuously receives information about the status of the tract from specialized receptors on the epithelial membranes. These receptors report on the contents of the lumen—how full it is, its total solute concentration, its pH, and the concentration of specific digestion products. The nerve network processes this information and sends appropriate commands to cells within the tract. For example, when the small intestine is full, the nerve network increases the activity of its smooth muscles to increase mixing, digestion, and absorption. Meanwhile, other signals slow smooth muscle activity in the stomach, preventing more material from entering the intestine too soon.

The intestinal nerve network itself receives information from the brain and from hormones circulating in the blood. When we see, smell, taste, or even think of food, the brain tells the salivary glands of the mouth to secrete saliva and the stomach to secrete mucus, pepsinogen, hydrochloric acid, and the digestive hormone **gastrin.** Gastrin stimulates the secretion of more pepsinogen and hydrochloric acid. Our emotions can affect this process. For example, anxiety and stress inhibit the secretions of the stomach and also its movements, so that we find it hard to digest food when we are upset. (Recall from Chapter 35 that the autonomic nervous system responds to stress by speeding up the heart and shutting down digestion and parts of the immune system.)

As food arrives in the stomach and then the small intestine, gastrin continues to coordinate the action, and other hormones also come into play. For example, the arrival of lipids and carbohydrates in the duodenum stimulates the secretion of the hormone **CCK** (cholecystokinin). CCK inhibits both the movement of the stomach and the secretion of acids and pepsinogen, so that the meal comes into the small intestine more slowly, which gives the intestine more time to digest the lipids. As a result, a fatty meal stays in your stomach longer than a low-fat one—giving you a feeling of satiety. Hormones also control glucose levels in the blood and coordinate the absorption of blood glucose into the cells, as we'll discuss in more detail in Chapter 37.

Nerve networks, the brain, and hormones all help to coordinate digestion.

Key Concepts

- Animals depend on food for energy and raw materials.
- An animal breaks macromolecules into their component building blocks and absorbs the resulting small molecules.
- In vertebrates, digestion and absorption occur in a digestive tract that starts at the mouth and ends at the anus.
- The parts of the digestive tract are specialized for different functions, including mechanical and chemical digestion; movement; mucus, bile, and enzyme secretion; and elimination.

Summary with Key Terms

Why do animals need to eat?

Animals **digest** chunks of food into small molecules such as glucose and amino acids by means of both mechanical and chemical digestion, then absorb the resulting small molecules for use in the cells of the body. Small molecules can be used either as fuel (in **metabolism**) or as building blocks for macromolecules. The **digestive tract**, or **gut**, is responsible for movement, **mechanical digestion, chemical digestion,** secretion, storage, **absorption,** and **elimination.**

The amount of food that an animal needs to consume depends on its **basal metabolic rate**, the minimum energy needed to stay alive and awake, on its size, and on its activities. Failure to obtain enough of particular raw materials—including the **essential amino acids,** minerals, and vitamins—leads to **malnourishment. Overnourishment** is rare worldwide except in developed countries. Vegetarian diets require **complementary proteins.**

Why do animals need a digestive tract?

Individual animal species differ in their diets; **herbivores** eat only plants, and **carnivores** eat only animals. Humans are **omnivores,** eating plants, animals, and other organisms. Acids, detergents, and powerful enzymes are necessary for digestion, yet these harsh conditions can damage cells, so digestion always takes place outside of the cell's cytosol—in lysosomes, in a digestive cavity, or in a digestive tract that, in most animals, runs from mouth to **anus.**

How do mechanical and chemical digestion prepare food for absorption by cells?

In vertebrates, food enters the **lumen** of the digestive tract at the mouth, which is specialized for the initial preparation of food. In mammals, the teeth and jaws begin mechanical digestion. (Birds begin mechanical digestion in the **gizzard.**) The **salivary glands** secrete **mucus,** which lubricates the movement of the food in the mouth, and **amylase,** which begins to break down starches into sugars. **Lingual lipase** begins to break down fats. The mouth shapes the food and mucus into a **bolus,** to be swallowed by the **pharynx.** The **epiglottis** guards the entrance to the lungs and prevents choking.

Smooth muscle contractions within the **esophagus** move the bolus down to the stomach by means of **peri-**

staltic waves. Under the influence of the stomach hormone **gastrin,** the **stomach** agitates the food with more mucus and secretes hydrochloric acid and **pepsin,** a protease that works best in the acid environment of the stomach. The stomach forms **chyme,** a soupy liquid, and squirts the chyme into the **duodenum** of the **small intestine.** A **sphincter muscle** in the duodenum normally prevents chyme from moving back up into the stomach. **Bile,** enzymes, and other digestive secretions from the **liver, gallbladder, pancreas,** and intestine digest the chyme while the small intestine continues the physical mixing of the chyme. As the digested chyme passes through the small intestine, it absorbs the resulting small molecules.

How does the intestine increase the rate of absorption?

The lumen of the gut is lined with epithelium that is specialized for different functions in different regions. In the small intestine, the epithelium and immediately surrounding tissue layers are folded into **villi,** and the epithelial cells themselves have extensively folded plasma membranes, or **microvilli. Tight junctions** between the epithelial cells keep the chyme from leaking between cells, so all nutrients go through the gut epithelial cells to the circulatory system.

The **large intestine** reduces the volume of the chyme by more than 85 percent. The large intestine removes water and ions as the remaining chyme travels to the **rectum** and forms **feces.** Bacteria growing within the colon contribute about half the volume of the feces. Irritation of the colon can decrease absorption, leading to **diarrhea,** and insufficient bulk in the diet can lead to **constipation.** The first part of the large intestine, the **cecum** is, in some species, a sac in which bacteria break down cellulose.

The brain and a network of nerves and hormones, such as **gastrin** and **CCK,** help to coordinate muscle activity, secretion, digestion, and absorption within the digestive system.

Review and Thought Questions

Review Questions

1. For what two purposes do animals need food?
2. Which of the following are processes that the digestive tract performs?
   A. Absquatulation
   B. Mechanical digestion
   C. Chemical digestion
   D. Elimination
   E. All of the above
3. Why must vegetarians worry about balancing complementary proteins?
4. Give an example of a herbivore, a carnivore, and an omnivore. How do the teeth of herbivores and carnivores differ?
5. Why is digestion always outside the cytosol? Describe the three major ways that organisms keep digestion away from the cytosol of the cell.
6. What two enzymes begin chemical digestion in the mouth and what macromolecules do they digest?
7. How do hydrochloric acid and pepsin interact in the stomach?
8. Explain how peristaltic waves move food through the digestive tract. Draw a picture illustrating this process.
9. Explain what bile is, where it comes from, and how it functions in the digestive tract.
10. What part of the digestive tract is most responsible for absorbing nutrients?
11. In what part of the digestive tract is vitamin K made? What organism makes it?

Thought Questions

12. Why must birds have a gizzard?
13. If your pancreas no longer secreted its digestive enzymes, what food groups would you have to give up?
14. Orexin is a hormone that stimulates hunger. What does the word "orexin" have in common with the word "anorexia"?

BiologyNow Resources

Biology ⊘ Now™

Active Figures
36-5:   The digestive tract
36-8:   Swallowing

Preparing for an exam? Take a diagnostic test on your BiologyNow CD-ROM.

Online materials relating to this chapter are at:
http://biology.brookscole.com/AAL3

About the Chapter-Opening Image
All animals eat food and many of them seem to enjoy eating. This fruit bat shown in these four photos is eating a fig while hanging upside down. We've turned the last image upside down to better appreciate the bat's appreciative expression.

37

How Do Animals Coordinate the Actions of Cells and Organs?

Key Questions

- Why are signaling molecules important?
- How do hormones communicate with their target cells?
- How can the same hormone convey different meanings to different cells?
- How do hormones regulate blood glucose levels and coordinate our responses to stress?
- How do hormones regulate other hormones?

Environmental Effects on Sexual Development

In the early 1990s, researchers announced a trend that grabbed headlines and the attention of men everywhere. Western men, smirked TV newscasters, make half the number of sperm as men of the 1930s. As one biologist quipped at a Congressional hearing, "Every man in this room is half the man his grandfather was."

The news media had a field day. "Sperm counts down? Penises shriveled?. . . don't blame it on the feminists," declared *Newsweek* magazine. The culprits, researchers said, were environmental pollutants, called **endocrine disrupters,** that mimic or block the effects of natural hormones. Hundreds of industrial and agricultural chemicals used both in homes and in the natural environment are endocrine disrupters.

Endocrine disrupters include herbicides (for killing plants), insecticides (for killing insects and other animals), and "dispersants" (chemicals that prevent fluids from forming droplets, used in everything from dishwashing detergents to pesticides). Many endocrine disrupters are plastics or plastics softeners used in plastic jugs for juice, water, and milk; toys; teething rings; baby bottles; plastic tubing used in hospitals; and freezer containers. Many cosmetics and other beauty products contain a class of endocrine disrupters called phthalates that are associated with shortened pregnancies in women and low sperm counts in men. All of these compounds send confusing messages to cells throughout the body.

At the heart of research on endocrine disrupters is the question of how natural hormones work. A **hormone** is a substance—secreted by the cells of a particular organ or tissue—that moves to other tissues of the body by way of the circulatory system. Hormones work by binding to protein *receptors* either inside of or on the surfaces of cells. Once a hormone binds to a receptor, the cell receives a message that tells it to do something

© Stan Osolinski 1993/FPG

different. It may begin dividing or it may begin secreting some product, such as milk or another hormone. A single hormone can bind to more than one kind of cell and to more than one kind of receptor. Hormones and other signaling molecules send countless messages, in response to everything from what other animals are around to the time of day.

Endocrine disrupters interfere with these messages by mimicking natural hormones or blocking receptors for natural hormones. For example, a 1980 spill of the insecticide DDT and related compounds in a Florida lake reduced the birth rate of alligators there by 90 percent. The young male alligators that survived this toxic spill grew up with smaller-than-average penises, as if they had been exposed to large amounts of estrogen. Likewise, Atlantic salmon exposed to the dispersant nonylphenol (used in pesticides sprayed on eastern Canadian forests) experienced wide regional population declines because the nonylphenol kept them from developing normally from freshwater fish into saltwater fish—again, as if they had been exposed to large amounts of estrogen.

For many years, biologists have assumed that steroid hormones were active only in vertebrates (fish, amphibians, reptiles and birds, and mammals). But recent studies have shown that a sea slug (called the sea hare) carries a gene for an estrogen receptor. If this gene turns out to be widespread, it could mean that great numbers of invertebrate animals are also receiving confusing messages from the many synthetic endocrine disrupters that humans release into the environment.

One of the most dramatic and recent studies showed that the widely used herbicide atrazine lowers testosterone in male frogs and causes them to make eggs in their testes, hopelessly compromising their ability to reproduce. Researchers could trigger this effect in laboratory frogs with atrazine levels only one-third of what is considered safe for drinking water. Wild frogs in ponds and streams from California to Illinois showed the same abnormalities. And the more atrazine in the water, the worse the problem. Unlike some other endocrine disrupters, atrazine does not itself mimic or block a hormone. Instead it boosts levels of an enzyme that normally transforms testosterone into estrogen in the body. With high levels of this enzyme (called aromatase), testosterone levels drop and estrogen levels rise.

Can atrazine and other hormone disrupters affect humans? In the cases just discussed, probably not, although researchers wonder about the possible effects on the babies of farmworkers who have been chronically exposed to large amounts of atrazine. On the other hand, frogs, fish, alligators, and other aquatic animals are vulnerable to endocrine disrupters because they are immersed in pollutant-laced water throughout their development. Humans and other land dwellers probably expel the pollutant very quickly. And while sperm counts are down among human populations, most men still produce far more sperm than they need. Men's fertility—as measured by the number of children they conceive and by other measures—is fine.

But common endocrine disrupters in our environment do affect mammals. Bisphenol A (BPA), an estrogen mimic first developed in the 1930s as a possible synthetic estrogen (to be used as a drug), is routinely used to soften plastics used in plumbing pipes, food and beverage containers, and dental sealants to protect children's teeth. BPA leaches from all these products at high enough levels to cause mouse breast and uterus cells to behave as if they had been exposed to large amounts of estrogen.

Many endocrine disrupters at first seemed to have little or no effect when tested alone. But evidence is accumulating that some of them can have dramatic effects when combined. For example, environmental contaminants called PCBs (polychlorinated biphenyls) applied to turtle eggs individually had no obvious effect on turtle development. When two different PCBs were combined, however, their estrogenic effects turned males into females. Another study suggested that the combined effects of three chemicals used on U.S. soldiers in the 1991 Gulf War—the insect repellant DEET, the insecticide permethrin, and an anti–nerve-gas drug—may have caused the infertility and sexual dysfunction some veterans are experiencing. Rats exposed to low levels of these three compounds had tissue abnormalities in both the testes and areas of the brain that control muscle strength and movement, balance and coordination, and memory and emotions.

How can minuscule amounts of synthetic chemicals evoke powerful biological responses? The short answer is that industrial chemists have made molecules with sizes, shapes, and charge distributions similar to the hormones and other molecules that animals use for internal signaling. Understanding how a molecule can imitate a hormone requires a knowledge of natural signaling molecules, the subject of this chapter.

37.1 How Do Chemical Signals Coordinate the Behavior of Cells?

For animals to survive and reproduce, their cells must work together. Whether we are making eggs or sperm, hunting or shopping for dinner, responding to an old friend, or escaping a predator, our cells must be in constant communication with one another. Every kind of behavior or structural change in our bodies is influenced by signaling molecules.

What Kinds of Signaling Molecules Do Animals Use?

Cells can communicate with one another in a variety of ways. They can communicate directly by means of gap junctions that connect the cytoplasm of one cell with the cytoplasm of another. But cells not directly connected can use any of three types of signaling molecules: hormones,

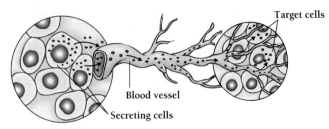

A. Hormones

Target cells

Blood vessel

Secreting cells

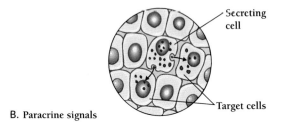

B. Paracrine signals

Secreting cell

Target cells

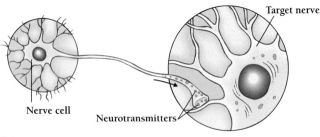

C. Neurotransmitters

Target nerve

Nerve cell

Neurotransmitters

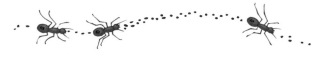

D. Pheromones

Figure 37-1

Types of chemical signaling. A. Hormones travel long distances within the body. B. Paracrine signals travel short distances. C. Neurotransmitters are specialized types of paracrine signals, though some experts classify them separately. D. Pheromones work between individuals. Ants lay down a trail of pheromones, which their nestmates follow and reinforce.

paracrine signals, and pheromones. **Hormones** are signaling molecules made and released by a specific organ or structure that travel to other tissues of the body by way of the circulating blood. Every hormone influences the behavior of a set of "target" cells, those that respond to the hormone (Figure 37-1A). For example, as we saw in Chapter 36, the hormone gastrin influences the behavior of cells in the stomach. Our focus in this chapter will be on hormones that are secreted by the tissues and organs of the **endocrine system** [Greek, *endo* = within + *krinein* = to separate].

A hormone signal is like a message broadcast over the radio. Anyone who can detect the signal can hear the message. Cells in the body detect a hormone's "message" by means of receptor proteins that bind to the hormone and re-

lay the message to the cell. Like a radio broadcast, the same hormone message can evoke different responses in different cells. For example, the hormone *secretin* stimulates the secretion of bile by liver cells and the release of digestive enzymes by pancreas cells, but inhibits the secretion of acid by the stomach (Chapter 36).

Paracrine signals [Greek, *para* = beside] are molecules that send signals only to cells in the immediate area (Figure 37-1B). Unlike hormones, paracrine signals do not travel through the blood, but act only at much shorter distances. Paracrine signals include growth factors, which stimulate cell division and keep cells from dying, and **prostaglandins,** which, just for example, maintain epithelia such as those in the stomach, help blood to clot, coordinate inflammation and fever, and cause contractions (including cramps) in smooth muscle. **Neurotransmitters,** molecules that carry information between adjacent nerve cells and between nerve cells and muscle cells, are also considered paracrine signals by many experts (Chapter 43).

Many neurotransmitters are identical to or closely related to hormones. But hormones can reach all the tissues of the body, while nerves carry signals to only certain cells (Figure 37-1C). If a hormone signal is like a message broadcast over the radio, a neural signal is more like a signal sent over phone lines. The neural message goes from a one particular phone (or cell) to another. (We will say more about neurotransmitters in Chapter 43.)

Pheromones are substances secreted by one organism that influence the behavior and physiology of another organism. For example, ants walking from a picnic back to their nest leave a chemical trail that allows other ants to find the picnic (Figure 37-1D). To attract mates, female moths release pheromones so potent that males can detect and respond to just a few molecules. Female hamsters release the pheromone aphrodisin, which prompts male hamsters to mate.

Researchers are so far aware of only two pheromones in humans, although others probably exist. Both pheromones are involved in synchronizing the monthly cycles of groups of women, so that they all ovulate (release eggs) together. One of the two pheromones is secreted in the armpits (of women) before ovulation and shortens the monthly cycle of other women exposed to it. The second pheromone is secreted around the time of ovulation and lengthens the monthly cycle of women exposed to it. Other studies have shown that the presence of women can increase beard growth in men and that vaginal secretions produced during or near the time of ovulation can raise men's testosterone levels. Pheromones are thought to be responsible, but surprisingly little is known.

Animals coordinate cell and organ activities using various signaling molecules, including paracrine signals, hormones, and pheromones. Hormones are chemical signals that travel through the entire body and stimulate specific responses in different tissues.

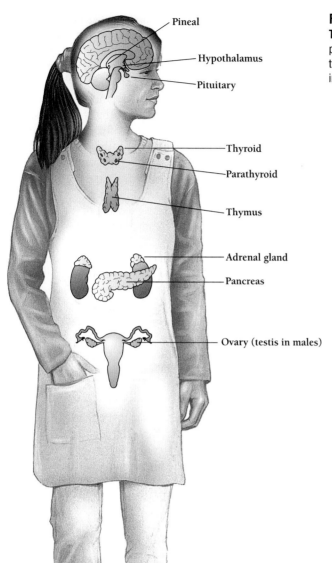

Pineal

Hypothalamus

Pituitary

Thyroid

Parathyroid

Thymus

Adrenal gland

Pancreas

Ovary (testis in males)

Figure 37-2
The major endocrine glands in a human. Although the pineal, hypothalamus, pituitary, thyroid, thymus, adrenals, pancreas, ovaries, and testes are considered the major endocrine glands, other tissues also secrete essential hormones. These include tissues in the heart, liver, intestine, stomach, and fat.

How Do Cells "Hear" a Hormone's Message?

All hormones bind to a specific protein in or on the target cell called a **receptor**. A cell receptor protein is like a portable radio that picks up a signal, such as a song or a news broadcast, and plays it in a way we can understand. When the hormone arrives, it binds to the cell receptor, which causes the cell to do something different; the cell has "understood" the hormone's message. What message a cell "hears" and how a cell responds to a hormone depends on which receptors the cell has.

Most hormones are produced in **endocrine glands**, which are composed mostly of epithelial cells specialized for secreting hormones into the blood. The major endocrine system organs are the pituitary gland, the pineal gland, the hypothalamus, the thyroid and parathyroid glands, the thymus, the adrenal glands, the pancreas, and the ovaries and testes (Figure 37-2). But many other organs in the body have cells that secrete hormones, including the stomach, intestines, heart, kidneys, liver, and adipose tissue (fat).

Two kinds of hormones send their signal in very different ways. **Lipid-soluble hormones** (such as estrogen, testosterone, and cortisol) pass through the cell membrane of the target cell and bind to receptors inside. **Water-soluble hormones** (such as insulin and epinephrine) bind to receptors in the cell membrane and do not enter the cell (Figure 37-3).

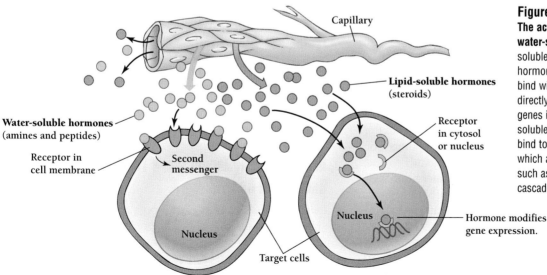

Capillary

Lipid-soluble hormones (steroids)

Water-soluble hormones (amines and peptides)

Receptor in cytosol or nucleus

Receptor in cell membrane

Second messenger

Nucleus

Nucleus

Target cells

Hormone modifies gene expression.

Figure 37-3
The actions of lipid-soluble and water-soluble hormones. Lipid-soluble hormones such as sex hormones cross the cell membrane, bind with receptors in the cytosol, and directly regulate the expression of genes in the cell nucleus. Water-soluble hormones such as epinephrine bind to receptors in the cell membrane, which activate a second messenger such as cyclic AMP, which creates a cascade of effects inside the cell.

The most well-known lipid-soluble hormones are the **steroids,** made from the building-block molecule cholesterol (Chapter 3). Steroid hormones work by slipping through the cell membrane, entering the nucleus of a cell, binding to the DNA, and turning on or off the expression of particular genes. The steroids include the stress hormones, or glucocorticoids (such as cortisol), the mineralocorticoids (such as aldosterone), and the *sex steroids* (such as estrogen).

The water-soluble hormones cannot pass through the cell membrane. Instead, they bind to protein receptors at the cell surface. In some cases, the receptor for a water-soluble hormone is an enzyme that changes the cell membrane. Usually, however, the receptor merely passes on the hormone's message to another signaling molecule inside the cell. This second signaling molecule, called a **second messenger,** alters the behavior of the cell.

One water-soluble hormone is **epinephrine,** also called *adrenaline* because it is made by the adrenal gland. When epinephrine (adrenaline) binds to the surface of a liver cell, it triggers the production of the second messenger **cyclic AMP,** which activates an enzyme that breaks down glycogen. Cyclic AMP amplifies the epinephrine's weak signal into a big one that causes the formation of thousands of glucose molecules. The effects of cAMP and other second messengers differ from cell to cell. In the intestine, for example, cAMP stimulates the pumping of sodium ions and water out of cells. In odor-sensitive tissues in the nose, cAMP triggers the firing of nerve cells that relay information about dinner to the brain.

Hormones affect cells by first binding to a receptor. Lipid-soluble hormones such as the steroids pass through the cell membrane and into the nucleus. Water-soluble hormones bind to cell-surface receptors that send a "second messenger" inside the cell.

How Do Synthetic Molecules Imitate Hormones?

Molecules that resemble a natural hormone can bind to the same receptor and initiate the same response. For example, the asthma drug isoproterenol binds to the same receptors as epinephrine (adrenaline), and either compound can be used to open the airways and ease the congestion caused by allergies and asthma.

Molecules that resemble a hormone may also bind to the receptor and block the natural hormone from binding. For example, the blood pressure drug propranolol binds to and blocks the same receptors in the heart that epinephrine normally binds to. By preventing epinephrine from binding to its own receptor, propranolol prevents the heart rate and blood pressure from increasing. For this reason, doctors often prescribe the drug to people who have high blood pressure.

Compounds that mimic other hormones are widely used in medicine. For example, doctors prescribe compounds that resemble estrogen and progesterone as birth control pills and for "hormone replacement therapy" in post-menopausal women. Synthetic thyroxin is prescribed to people whose thyroid glands don't make enough of this metabolic hormone.

But, as we saw at the beginning of this chapter, not all hormone mimics are useful drugs. Phthalates, a group of plastics softeners, block the receptors for testosterone and other *androgens* [Greek, *andro* = man], the generic term for testosterone-like steroids. In rats, phthalates can lower testosterone levels, shrink the sperm storage organ (the epididymis), cause testicular tumors, or even prevent the testicles from developing at all. These effects occur at levels similar to those experienced by humans exposed to normal amounts of plastic.

Experts estimate that 10 to 20 percent of male high school athletes and 30 percent of male college athletes use androgens, also called *anabolic steroids,* to increase muscle mass, endurance, and aggression. In humans, the use of natural or synthetic androgens increases the risk of reduced testicles and similar pathologies, as well as heart failure and stroke, depression of the immune system, jaundice, tumors of the liver, enlargement of the prostate gland, and obstruction of the urinary tract.

Many companies sell a variety of "performance enhancers," some of which are androgens and some of which are not. For example, the asthma drug clenbuterol mimics the hormone epinephrine, increasing the diameter of the respiratory passages and speeding blood flow to the skeletal muscles. Heavy use of clenbuterol, which is sometimes falsely sold as an androgen, can cause headaches, tremors, insomnia, and abnormal heartbeat.

Another "performance" drug sometimes sold as an androgen is gamma-hydroxybutyrate (GHB). GHB was developed as an anesthetic in the 1960s, but because it can cause seizures, it never came into medical use. In the 1990s, long after the usual 17-year patent had expired, the drug was reincarnated as a "health supplement" and "diet aid" and sold in health food stores. It is a dangerous drug, capable of causing confusion, hallucinations, seizures, and even coma.

Natural or synthetic molecules that resemble a hormone can bind to the same receptor and initiate or prevent the same response.

Hormones May Target More Than One Kind of Receptor

Different kinds of cells can respond differently to the same signal. If, by analogy, two different people heard on the radio that a storm was coming and that a certain road might close soon, one person might rush home early in case they couldn't get home later, while another person might rush to

Extreme Biology How Do Researchers Identify Hormones and Glands?

Endocrinology, the study of hormones, has traditionally asked three questions about a suspected hormone: (1) What organ is responsible for its synthesis? (2) Does the suspected hormone move through the circulation? and (3) What is its chemical identity?

To answer the first question, a researcher would determine whether a product of one organ affected the function of another. For example, a researcher would destroy a suspected endocrine organ and note the resulting changes. An often-cited example is the effect of removing the testes of a juvenile animal. The result is the failure to develop the secondary sexual characteristics that appear in males at sexual maturity. In roosters, a large red comb is a secondary sexual characteristic, and removing the testes results in the withering of the comb.

The second question—whether the suspected hormone travels through the circulation—distinguishes a hormone from a secretory product or a short-range paracrine signal. Researchers have addressed this question by restoring the suspected endocrine organ to another place in the body. In the case of a castrated rooster, transplanting testes into the abdominal cavity restores the comb. This result shows that the hormone responsible for the secondary sexual characteristics moves through the circulation rather than through a particular duct.

The third question—the chemical identity of the signaling molecule—is the hardest to answer. Researchers begin by demonstrating that an extract of the endocrine gland (for example, of a rooster testis) can produce the same effect as the intact gland. The next task is to isolate and identify the active compound and to determine its chemical structure. Finally, researchers synthesize the suspected hormone in the laboratory and show that the synthetic substance has the same effect as the suspected hormone. Again, in the case of the rooster's comb, pure testosterone can restore the comb just as well as an extract of rooster testes. The ability of a pure chemical, made in the laboratory, to mimic a hormone's effect establishes that hormone's chemical identity. The action must depend on that single hormone rather than on some unknown contaminant in the original extract or on two or more hormones.

This straightforward approach has not always worked. Endocrine organs sometimes make more than one hormone, so that researchers must replace a number of substances in order to reverse the effects of organ removal. The pancreas, for example, makes both insulin and glucagon. And the pituitary gland releases at least eight different hormones.

In vitro cultures of target cells or target organs provide another way that contemporary endocrinologists identify and study hormones, particularly in complex systems that involve many different signals and targets. Using both the traditional surgical methods with whole organisms and the more recent approaches with cultured cells and organs, endocrinologists have identified more than 50 hormones and more than 15 hormone-producing structures in vertebrates and similar numbers in invertebrates.

---

the airport to make sure they didn't miss their flight. The same message caused two different responses. Hormones can similarly have very different effects in different tissues. For example, in female mammals, the hormone prolactin stimulates milk production by the milk-producing glands. But in male mammals, the same hormone helps regulate the production of testosterone by the testes and also stimulates paternal behavior, possibly by binding to receptors in brain. How can one hormone do so many different things?

Part of the answer comes from studies of synthetic drugs that mimic hormones and other signaling molecules. Over many years, researchers discovered hundreds of synthetic compounds that bind to the same receptors as epinephrine. But different tissues responded differently to the synthetic compounds. For example, natural epinephrine causes the heart rate to speed up and the smooth muscles of the gut to relax, but the asthma drug isoproterenol only causes the heart rate to speed up. It does not cause the gut to relax.

At first, researchers wondered how a drug could mimic a natural hormone in one situation but not in another. But the answer turned out to be simple. Natural epinephrine binds to several different receptors. Some of these bind synthetic compounds and some don't. Further studies have shown that almost every natural signaling molecule binds to several different receptors.

Most hormones bind to different receptor types in different tissues.

37.2 How Do Hormones Coordinate Physically Distant Organs?

Animals coordinate all the organ systems of the body with a relative handful of hormones and signaling molecules. Here we briefly discuss a few examples.

How Do Hormones Help Keep Blood Glucose at a Nearly Constant Level?

In mammals, the liver, pancreas, small intestine, muscle, and other tissues that use glucose all influence the concentration of glucose in the blood. Shortly after a meal, when

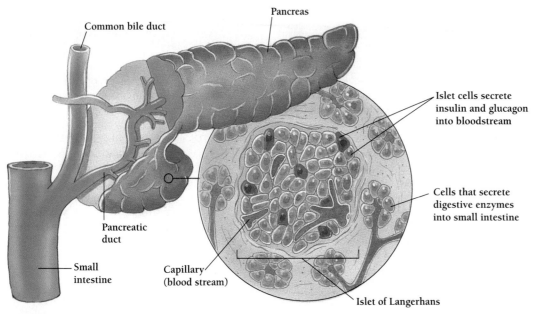

Figure 37-4

The pancreas. Endocrine cells in the islets of Langerhans secrete insulin and glucagon, which together maintain the right balance of glucose in the blood, as well as two other hormones.

blood sugar is high, the body is said to be in an **absorptive state,** in which cells take up glucose, make glycogen (which they use to store energy for instantaneous use), and increase the synthesis of fats and proteins. All this absorption reduces the concentration of glucose (and other molecules) in the blood. The body then goes into a **postabsorptive state,** during which liver cells break down glycogen to glucose and dump it into the bloodstream.

Two hormones—insulin and glucagon—coordinate the switch between the absorptive and the postabsorptive states. **Insulin** is a protein hormone produced by specialized cells in the pancreas (Figure 37-4). When the insulin-producing cells detect high levels of glucose, they make more insulin, which stimulates cells throughout the body to take up glucose, decreasing the glucose concentration in the blood. Insulin also stimulates the liver, the muscles, and adipose (fat) tissue to increase the synthesis of fats, proteins, and glycogen.

After the small intestine has finished absorbing glucose from a meal, the concentration of glucose in the blood decreases and insulin secretion stops. Other endocrine cells of the pancreas then begin to secrete **glucagon,** a second protein hormone that, together with the low levels of insulin, signals the beginning of the postabsorptive state. Cells then reduce their uptake of glucose and begin to break down glycogen, fats, and proteins (Figure 37-5).

Insulin and glucagon are produced by clusters of cells in the pancreas called the *islets of Langerhans* (Figure 37-4). High blood glucose levels stimulate some cells to make insulin, which stimulates glucose absorption from the blood (thus lowering blood glucose levels). Low blood glucose levels stimulate other pancreas cells to make glucagon, which promotes the production of more blood glucose from glycogen (thus increasing blood glucose levels).

The two protein hormones insulin and glucagon control blood sugar (glucose) levels. When blood sugar increases, pancreas cells make insulin, which promotes the synthesis of glycogen from glucose. When blood sugar decreases, pancreas cells make glucagon, which inhibits glycogen synthesis and stimulates its breakdown into glucose.

How Do Hormones Coordinate the Response to Stress?

Stress—the response to physical trauma, intense heat or cold, infection, pain, and fright—is part of the life of any animal. Faced with any of these stressors, mammals respond initially with an intense, short-lived alarm reaction, called the **fight-or-flight response.**

A key coordinator of the fight-or-flight response is epinephrine (adrenaline). By binding to different receptors in different tissues, epinephrine speeds the heart, redirects blood flow to the heart and other organs (by dilating some blood vessels and constricting others), and boosts the liver's production of glucose from glycogen (Figure 37-6). But epinephrine is a short-term hormone. After a few hours, the longer-acting hormone **cortisol** takes on a more important role.

A stressor, whether an approaching exam or an approaching grizzly bear, stimulates the brain to send signals to the **adrenal glands** [Latin, *ad* = to + *renes* = kidneys], which lie just above the kidneys. Each adrenal gland con-

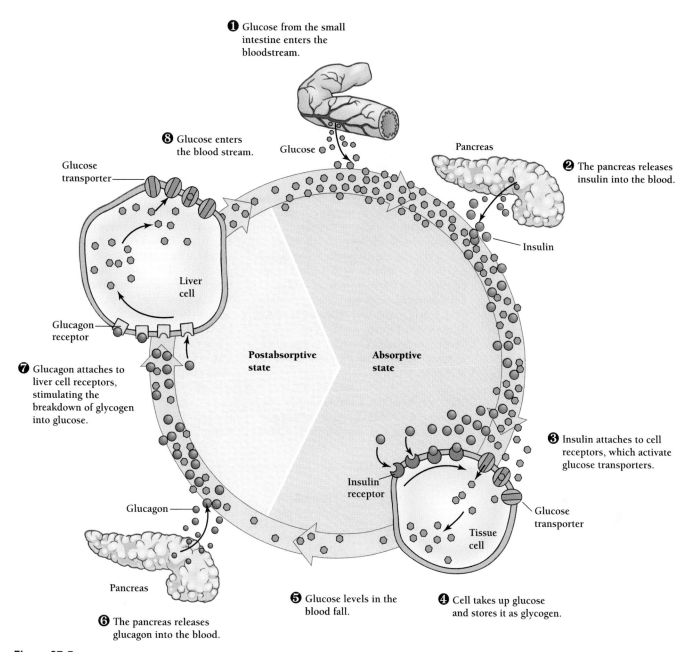

❶ Glucose from the small intestine enters the bloodstream.

❽ Glucose enters the blood stream.

Glucose

Glucose transporter

Liver cell

Glucagon receptor

❼ Glucagon attaches to liver cell receptors, stimulating the breakdown of glycogen into glucose.

Postabsorptive state

Absorptive state

Pancreas

❷ The pancreas releases insulin into the blood.

Insulin

❸ Insulin attaches to cell receptors, which activate glucose transporters.

Insulin receptor

Glucose transporter

Tissue cell

Glucagon

Pancreas

❻ The pancreas releases glucagon into the blood.

❺ Glucose levels in the blood fall.

❹ Cell takes up glucose and stores it as glycogen.

Figure 37-5

The pancreas helps regulates blood sugar level. This diagram presents a simplified version of how the pancreas regulates blood sugar. High blood sugar triggers the release of insulin during the absorptive state. Low blood sugar triggers the release of glucagon during the "postabsorptive state." In reality, regulating blood sugar is more complex. Other factors that stimulate insulin secretion besides high glucose levels in the blood include amino acids, fatty acids, and the digestive hormones gastrin and secretin. Factors that inhibit the secretion of insulin include the hormones epinephrine, norepinephrine, and somatostatin.

Biology 𝓔 Now™ Learn more about the pancreas and hormones by clicking on this figure on your BiologyNow CD-ROM.

sists of two parts—the outer *adrenal cortex* and the inner *adrenal medulla* (Figure 37-6). The adrenal cortex makes cortisol (and several other hormones), and the adrenal medulla makes epinephrine (as well as other signaling molecules). The signal to make cortisol comes from hormones made in the hypothalamus and anterior pituitary in the brain. These travel through the blood, a relatively slow process. In contrast, the signal to make epinephrine comes more quickly—also from the brain, but by way of the peripheral nervous system (Chapter 42).

The adrenal gland produces stress hormones in response to hormone signals from the brain. The adrenal cortex produces cortisol, and the adrenal medulla produces epinephrine (adrenaline), the major mediator of the fight-or-flight response.

In 1889, an observant animal caretaker noticed flies gathering around the urine produced by a dog whose pancreas had been removed. Without a pancreas, we now realize, the dog was unable to produce insulin, and glucose accumulated in its blood and urine. The observation was in fact responsible for the discovery that the pancreas is an endocrine organ.

Similarly, people suffering from **diabetes mellitus** [Greek, *diabetes* = siphon + Latin, *mellitus* = sweet] have an excessive concentration of glucose in their blood, which leads to its presence in the urine and an excess production of urine. Diabetics suffer from this condition because their bodies fail to produce normal levels of insulin or to respond to insulin, keeping the body in a permanent postabsorptive state.

About 10 to 20 percent of people with diabetes suffer from *insulin-dependent diabetes mellitus* (IDDM), also called *type 1 diabetes* or *juvenile diabetes* because it generally begins early in life. People with

juvenile diabetes fail to make any insulin because their immune systems have mistakenly attacked and destroyed the insulin-producing cells of the pancreas. Normal life depends on daily injections of insulin.

Most people with diabetes, however, do make insulin, but their target cells fail to respond to it. This condition is called *type 2 diabetes, non–insulin-dependent diabetes mellitus* (NIDDM), or adult-onset diabetes. Type 2 diabetes usually manifests itself in middle age, and it can usually be controlled by diet and by oral medicines. Type 2 diabetes has become more common, even among children, as more people throughout the world indulge in rich diets and sedentary lifestyles. According to a recent survey by the U.S. Centers for Disease Control and Prevention in Atlanta, for example, the prevalence of diabetes in the United States increased a whopping 61 percent between 1990 and 2001, and more than 8 percent just from 2000 to 2001. Almost 8 percent of all Americans now suffer from some form of diabetes.

The trend is particularly alarming because of the many negative effects diabetes can have on a person's health. According to the American Diabetes Association, diabetic adults are two to five times more likely to die of heart disease or suffer from stroke. Seventy-three percent of all diabetics have high blood pressure. In addition, diabetes can cause blindness. Indeed, diabetes is the leading cause of new cases of blindness in adults 20–74 years old. Diabetes is also responsible for certain forms of kidney disease, nervous system disease, dental disease, amputations, and pregnancy complications. In 1999, diabetes was the sixth leading cause of death in the United States and a contributing factor to many other deaths.

On the bright side, several studies have shown that lifestyle changes can prevent or delay the onset of type 2 diabetes among high-risk adults. These changes include improvements in diet and moderate exercise, such as walking for a couple of hours each week.

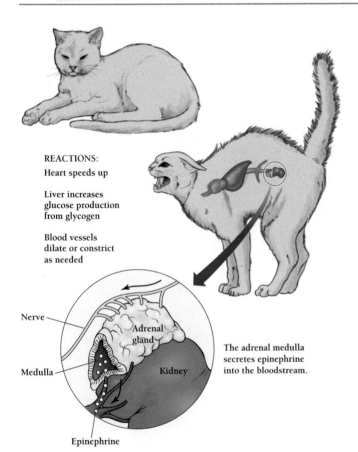

REACTIONS:
Heart speeds up

Liver increases glucose production from glycogen

Blood vessels dilate or constrict as needed

Nerve

Adrenal gland

Medulla

Kidney

Epinephrine

The adrenal medulla secretes epinephrine into the bloodstream.

Figure 37-6
Fight or flight. In cats, people, and other vertebrates, the water-soluble hormone epinephrine mediates the short-term "fight-or-flight" response.

How Do the Pituitary and Hypothalamus Regulate Other Endocrine Organs?

The pituitary gland is a small structure, about the size of a pea, at the base of the brain (Figure 37-7). Its name [Latin, *pituita* = slime] came from the mistaken belief that it produced the mucus of the nose. In humans, the pituitary has two main parts, called the posterior pituitary and the anterior pituitary.

The **posterior pituitary** is actually part of the brain itself; it consists of the ends of nerve cells that lie in a part of the brain called the **hypothalamus** [Greek, *hypo* = under + *thalmos* = inner chamber] (Figure 37-7). These nerve cells release two main hormones, vasopressin and oxytocin. *Vasopressin,* also called *antidiuretic hormone* (ADH), stimulates water reabsorption in the kidney (Chapter 40). *Oxytocin* acts at childbirth to stimulate the uterus to contract. Later, during lactation, the same hormone stimulates the mammary glands to eject milk. Oxytocin also increases during sexual arousal and orgasm in both sexes.

The **anterior pituitary** is not part of the brain but a separate gland. It makes several water-soluble hormones, includ-

Extreme Biology   How Do Hormones Control Insect Metamorphosis?

Among the most dramatic results of hormone action is the development of a butterfly or moth from a wormlike caterpillar. During this transformation *(metamorphosis),* the insect rebuilds every one of its organs. In a butterfly or moth, the fertilized egg develops first into a *larva* (a "caterpillar"), which is specialized for eating. As the larva grows, it undergoes several molts, each time casting off its outer skin and emerging as a slightly larger larva. The final larva undergoes a major transfiguration, or metamorphosis, into a *pupa,* an apparently dormant stage, which is enclosed in a silken cocoon. Finally, the pupa dissolves its cocoon, and the adult butterfly or moth emerges. The adult has a strikingly different body from that of the larva or the pupa: it is specialized for reproduction and, in some species, cannot even eat (Figure A).

Insects are particularly useful experimental subjects because their bodies are so hardy. Tying a loop of fine string around a larva, for example, results in pupa formation in just the front (anterior) half of the larva, while the rear (posterior) half stays a larva. Similar experiments allowed biologists to discover that molting and metamorphosis

depend on at least three hormones: *juvenile hormone,* a small molecule, derived from a fatty acid that prevents molting and metamorphosis; *ecdysone,* a steroid that promotes molting and metamorphosis; and *prothoracicotropin* (PTTH), a small protein that stimulates ecdysone production. Metamorphosis of larva into pupa occurs when juvenile hormone is low and ecdysone is high; metamorphosis of pupa into adult oc-

curs when PTTH production—triggered by cold temperatures, extreme drought, or short days—stimulates ecdysone production in the absence of juvenile hormone.

Metamorphosis in amphibians also depends on hormone action, this time thyroid hormone. Similarly, the steroid sex hormones—testosterone and estrogen—trigger the metamorphosis-like transitions of puberty in adolescent humans.

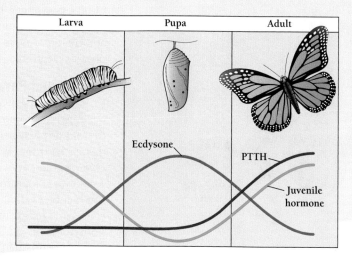

Figure A
Metamorphosis of a caterpillar into a butterfly. At least three hormones regulate insect metamorphosis—juvenile hormone, ecdysone, and prothoracicotropin (PTTH).

---

ing: *corticotropin,* which stimulates the production of cortisol in the adrenal cortex; *endorphin,* a natural pain suppressor; *thyroid-stimulating hormone* (TSH), which stimulates the production of thyroxin in the thyroid gland; *growth hormone* (GH), which stimulates tissue and skeletal growth; *prolactin,* which, among other things, stimulates milk production and helps regulate testosterone secretion; *melanocyte-stimulating hormone* (MSH), which is secreted by some animals and stimulates pigment production in specialized skin cells; and *follicle-stimulating hormone* (FSH) and *luteinizing hormone* (LH), together called *gonadotropins,* because they regulate sex hormone and gamete production in the gonads of both males and females (Chapter 44). The release of hormones by the anterior pituitary depends on other hormones produced by the hypothalamus.

The Hypothalamus

The hypothalamus produces at least seven regulatory hormones that control the release of hormones by the anterior pituitary. Five of these hormones stimulate the release of a specific hormone and two inhibit the release of hormones.

For example, *gonadotropin-releasing hormone* (GnRH) stimulates the secretion of FSH and LH by the anterior pituitary and *prolactin-inhibiting factor* (PIF) inhibits the pituitary's secretion of prolactin.

Hypothalamic hormones such as GnRH can enter the general circulation, but special veins also flow from the hypothalamus to the anterior pituitary, where they divide and form a second capillary bed. This diversion gives the hypothalamic hormones more direct access to their targets in the anterior pituitary.

The regulation of hormone release often involves **negative feedback;** when hormone levels rise, the hormone indirectly inhibits its own secretion. In male mammals, for example, GnRH stimulates the release of LH, which stimulates testosterone production in the testes. Testosterone from the testes then enters the circulating blood. When it reaches the hypothalamus, testosterone inhibits the production of GnRH. The lowered GnRH in turn reduces the release of LH, which reduces the secretion of testosterone by the testes. One important effect of androgen supplement use (by athletes, for example) is the suppression of GnRH in the hy-

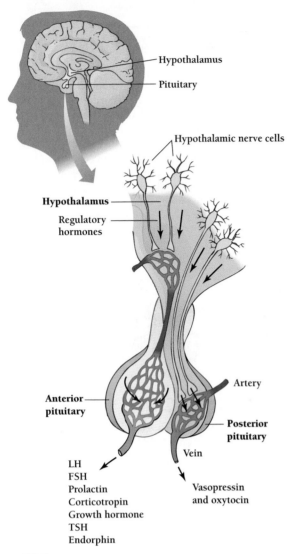

Figure 37-7
The hypothalamus and pituitary and their hormones. This diagram
emphasizes the relationships of the hypothalamus and pituitary: the
posterior pituitary consists of neuronal processes from the hypothalamus,
while the anterior pituitary receives signals from the hypothalamus via the
circulation.

Biology ⑤ Now™   Learn more about the hypothalamus and pituitary by
clicking on this figure on your BiologyNow CD-ROM.

pothalamus, which results in the suppression of the body's
own testosterone production. The suppression of GnRH release (and therefore LH and testosterone) can become permanent.

In female mammals, the regulation of estrogen works
similarly, but with a twist. High concentrations of estrogen
inhibit GnRH, leading to less estrogen. But very high estrogen concentrations can actually stimulate a burst of GnRH,
leading to a surge of LH, a process called **positive feedback.**
The surge of LH triggers ovulation (Chapter 44).

The hypothalamus integrates information both from the
endocrine system and from the nervous system and exerts

powerful control over hormone production in the adrenal
cortex, the gonads, thyroid, and liver.

The hypothalamus is the major mediator of information between the
brain and the endocrine system.

What Other Endocrine Organs Do Mammals Have?

Chemical signaling between organs is common, and many
organs are now known to have endocrine components (Figure 37-2). In addition to the organs we have already
mentioned—pancreas, adrenal glands, hypothalamus, and
pituitary—we can list the gonads, the thyroid and parathyroid glands, the pineal gland, the intestines, the placenta,
and the thymus, as well as the heart, kidneys, liver, and adipose tissue (Table 37-1).

Thyroid and Parathyroid Glands

In humans, the two lobes of the **thyroid gland** are in the
neck, just in front of the windpipe. In other vertebrates, the
thyroid lies just below the windpipe. The thyroid produces
thyroid hormone, which regulates metabolism and growth.
Overproduction of thyroid hormone, called *hyperthyroidism,*
results in uncontrolled, rapid metabolism, whereas under-
production, called *hypothyroidism,* may cause such symptoms as lethargy, weight gain, and intolerance of cold. Hypothyroidism during fetal life may result in abnormal
development, mental retardation, and low metabolic rate.
The thyroid gland also produces *calcitonin,* which stimulates
tissues to remove calcium ions from the blood.

Adjacent to the thyroid gland are the **parathyroid glands.**
These glands produce *parathyroid hormone,* which increases the
concentration of calcium ions in the blood—exactly the opposite of what calcitonin does. Between them, the two hormones
keep blood calcium relatively constant. Low blood calcium triggers the parathyroid to release parathyroid hormone, and high
blood calcium triggers the thyroid to release calcitonin.

Hormones from the thyroid gland regulate metabolism, growth, and
calcium levels in the blood. Parathyroid hormone also regulates calcium levels.

Kidneys

The kidneys release two important hormones (other than the
hormones released by the adrenal glands atop the kidneys).
These are erythropoietin, which stimulates the production of
red blood cells, and calcitriol, another hormone that regulates
calcium levels. **Calcitriol,** one of a family of related steroid
hormones collectively called vitamin D, increases available
calcium in the blood by stimulating the absorption of calcium
and phosphate from the gut and the release of calcium ions

Table 37-1
Major Endocrine Organs in Humans

	Hormones (examples)	Chemical Nature	Major known targets	Major known effects
Pancreas				
α cells	Glucagon	Protein	Muscle, liver	Stimulates glycogen and fat breakdown, glucose production, release of glucose and fatty acids into bloodstream
β cells	Insulin	Protein	Many targets	Stimulates glucose and amino acid uptake; lowers blood glucose levels; enhances glycogen, fat, and protein synthesis
Adrenal Gland				
Cortex	Cortisol, corticosteroids	Steroids	Many targets	Increase fat and protein breakdown, glucose production, tissue repair; inhibit immune function
	Aldosterone	Steroid	Kidney	Increases sodium retention and potassium excretion; maintains blood pressure and volume
	Androgen and estrogen	Steroids	Many targets	Regulate prenatal development of males, bone growth, sex drive
Medulla	Epinephrine, Norepinephrine	Amines**	Many targets	Increase heart rate, blood pressure, blood glucose; regulate blood vessel constriction and dilation; etc.
Hypothalamus				
	Releasing hormones	Peptides*	Anterior pituitary	Stimulate release of anterior pituitary hormones (including corticotropin (CRH), thyrotropin (TSH), growth hormone, and gonadotropins)
	Inhibiting hormones	Peptide,* amine**	Anterior pituitary	Inhibit release of anterior pituitary hormones (including prolactin, growth hormone, and thyroid stimulating hormone)
Pituitary Gland				
Posterior	Oxytocin	Peptide*	Uterus, breast	Stimulates uterine contractions, milk secretion from mammary glands
	Vasopressin (antidiuretic hormone, ADH)	Protein	Kidney	Prevents water loss
Anterior	Corticotropin (adrenocorticotropic hormone, ACTH)	Protein	Adrenal cortex	Stimulates glucocorticoid (including cortisol) secretion and growth of adrenal cortex
	Thyroid-stimulating hormone (thyrotropin, TSH)	Protein (glycoprotein)	Thyroid gland	Stimulates thyroxin secretion and growth of thyroid gland
	Growth hormone (GH)	Protein	Liver	Stimulates widespread tissue growth by stimulating the liver to produce insulin-like growth factors
	Melanocyte-stimulating hormone (MSH)	Peptide*	Pigment-producing cells	Stimulates pigment production in skin, hair, or feathers
	Follicle-stimulating hormone (FSH)	Protein (glycoprotein)	Ovary, testis	Promotes egg production and estrogen secretion in females; promotes sperm production in males
	Luteinizing hormone (LH)	Protein (glycoprotein)	Ovary, testis	Stimulates ovulation and progesterone production in females; stimulates testosterone production in males
	Prolactin	Protein	Breast, testis	Stimulates milk production in females; increases LH sensitivity and testosterone secretion in males
Gonads				
Ovary	Estrogen	Steroid	Many targets	Stimulates female sexual development, regulates menstruation and pregnancy, prepares breasts for milk production
	Progesterone	Steroid	Uterus, breast	Regulates menstruation and pregnancy, prepares breasts for milk production

Table 37-1
Major Endocrine Organs in Humans—cont'd

	Hormones (examples)	Chemical Nature	Major known targets	Major known effects
Gonads—cont'd				
Testis	Testosterone	Steroid	Many targets	Stimulates sex drive, male sexual development, sperm production, and bone and muscle growth
	Inhibin	Protein	Anterior pituitary	Inhibits secretion of FSH
Thyroid Gland	Thyroxine	Amine**	Many targets	Stimulates growth and metabolism
	Calcitonin	Protein	Bone	Regulates calcium uptake
Parathyroid Gland	Parathyroid hormone	Protein	Bone, small intestine, kidney	Stimulates movement of calcium into blood
Pineal Gland	Melatonin	Amine**	Brain	Regulates daily rhythm; may regulate the timing of puberty
Thymus	Thymosins	Protein	T lymphocyte	Regulates T-lymphocyte function

*A peptide hormone is a short polypeptide.

**An amine hormone is a small molecule usually based on an amino acid.

from bone. Calcitriol also stimulates the formation of osteoclasts (the cells that break down bone) and affects the white cells of the immune system and certain kinds of skin cells.

Erythropoietin, also known as **EPO,** is a hormone released by the kidneys when they detect low oxygen levels in the blood. EPO stimulates the stem cells in bone marrow to make more red blood cells. The increasing number of red blood cells increases the total amount of blood and improves the transport of oxygen from the lungs to the tissues.

Some endurance athletes, such as marathon runners and bicyclists, take EPO to increase their blood's ability to deliver oxygen, for which this hormone works very well. However, too much of an increase in red blood cells in the blood makes the blood heavier and thicker, which can put a strain on the heart. In the early 1990s, 18 cyclists died of heart failure, deaths attributed to abuse of EPO.

Ovaries and Testes

The gonads (ovaries and testes) produce sex steroids. The **ovaries** produce *estrogen* and *progesterone* and the **testes** produce *testosterone*. The sex steroids are responsible for the development and the maintenance of the secondary sexual characteristics as well as for the proper development of reproductive cells. We cover sex steroids in Chapter 44.

Pineal Gland

The pineal gland, which lies deep in the brain, makes the hormone melatonin from the neurotransmitter serotonin in response to darkness. In humans, the pineal gland detects light by way of nerve tracts that lead from the eyes. Melatonin production is low during the day and highest at night. Melatonin helps set daily sleep and activity cycles, called "circadian rhythms," and also inhibits the production of GnRH by the hypothalamus.

Thymus

Beneath the sternum, or breastbone, is the thymus gland, where an array of hormones called thymosins promote the maturation of immune cells called T lymphocytes (white cells). The thymus is quite large in babies and children. During puberty, adolescents grow faster than their thymus gland, and during adulthood the gland actually shrinks.

Key Concepts

- Animals use chemical signals to coordinate the activities of many cells and organs.
- Hormones are chemical signals made in specific tissues that move throughout the body and affect processes in target tissues.
- Hormones and other signaling molecules bind to specific receptor proteins in target cells. The interaction of a chemical signal and its receptor triggers characteristic responses in the target cells.

Summary with Key Terms

Why are signaling molecules important?

An animal must coordinate the activities of many tissues in the course of living and reproducing. Far-flung cells can communicate by means of several kinds of signaling molecules: **hormones,** chemical signals that are carried by the circulation and act at long distances; **paracrine signals,** which act only at short distances, such as **neurotransmitters** and **prostaglandins;** and **pheromones,** which influence the activities of other individuals of the same species. Hormones are made by **endocrine glands** of the **endocrine system** and cause characteristic responses in target cells.

How do hormones communicate with their target cells?

Hormones include those that are lipid soluble (such as **steroids**) and those that are water soluble (such as epinephrine). **Lipid-soluble hormones** bind to receptors inside a target cell and regulate the expression of genes. **Water-soluble hormones** bind to receptors in the cell membrane, which release a **second messenger** such as **cyclic AMP** that alters cell behavior.

How can the same hormone convey different meanings to different cells?

A single hormone often binds to many different receptors, with different kinds of receptors on different kinds of cells. As a result, different cells can receive different messages from the same hormone. **Endocrine disrupters** can mimic natural hormones, bind to and block the receptor, (preventing the natural hormone from functioning), or interfere with the normal synthesis of the hormone.

How do hormones regulate blood glucose levels and coordinate our responses to stress?

Insulin and **glucagon** help regulate blood glucose levels. Insulin establishes an **absorptive state**, in which cells take up glucose from the blood and make glycogen. Glucagon establishes a **postabsorptive state**, in which the liver produces more glucose and breaks down glycogen. **Cortisol** and **epinephrine** secreted by the **adrenal glands** contribute to the stress response. Epinephrine stimulates the **fight-or-flight response**.

How do hormones regulate other hormones?

Among the endocrine organs are the pancreas, stomach, small intestine, adrenal glands, **thyroid, parathyroid,** and gonads (**ovaries** and **testes**). The **anterior pituitary** is distinct from the **posterior pituitary,** which is actually an extension of the **hypothalamus,** a part of the brain. The hypothalamus and pituitary together regulate the production of hormones by other endocrine organs. The hypothalamus makes hormones that regulate the release of hormones by the anterior pituitary. The anterior pituitary produces hormones that regulate hormone production elsewhere in the body, in a manner often subject to **negative feedback.** In cases of **positive feedback,** an increase of one hormone can precipitate a feedback loop that further increases the production of that hormone.

The steroid hormone **calcitriol** (vitamin D) increases available calcium in the blood. When the kidneys detect low oxygen levels in the blood, they release the hormone **erythropoietin (EPO),** which stimulates the stem cells in bone marrow to make more red blood cells.

Review and Thought Questions

Review Questions
1. What is a hormone?
2. How do hormones, paracrine signals, and pheromones differ from one another?
3. How are the nervous system and the endocrine system both alike and different?
4. How does a hormone arrive at its target cell and what does the hormone bind to when it arrives?
5. How did compounds that mimic epinephrine help biologists understand how the same hormone could have different effects in different tissues?
6. Describe three ways in which an endocrine disruptor can interfere with normal hormone signaling.
7. How can the same hormone send different messages to different cells?
8. What are the two main classes of hormones and how do they differ in the way they work?
9. How does diabetes mellitus (type 1 diabetes) differ from type 2 diabetes?
10. What is the main function of the anterior pituitary? Name a few of the hormones released by the anterior pituitary and tell what each one does.

Thought Questions
11. In this chapter, we've seen that the same hormone can trigger different responses in different types of cells within the body. Do you think that the same process would allow similar hormones to trigger different responses in the cells of males and females or in the cells of different species?
12. In the 1950s and 1960s, physicians prescribed a synthetic estrogen called diethylstilbestrol, or DES, to pregnant women. In the 1970s, researchers noticed that the daughters of such women had a greatly increased rate of reproductive cancers. Suggest a mechanism by which DES might cause cancers of the reproductive organs in the grown daughters of women given DES. How would you test your hypothesis?

BiologyNow Resources

Biology ⑧ Now™

Active Figures
37-5: The pancreas and hormones
37-7: The hypothalamus and pituitary

Preparing for an exam? Take a diagnostic test on your BiologyNow CD-ROM.

Online materials relating to this chapter are at:
http://biology.brookscole.com/AAL3

About the Chapter-Opening Image
Some industrial and agricultural pollutants mimic the molecular structure of sex hormones. These compounds profoundly interfere with the normal development of alligators, fish, frogs, and probably other animals.

38 How Do Animals Move Blood Through Their Bodies?

Key Questions

- What do blood and the extracellular fluid do to keep animals alive?
- What are the essential components of blood and extracellular fluid?
- How do the heart and circulation move blood?
- How does the lymphatic system work with the circulatory system?
- What mechanisms control blood flow and heart rate?

Saving Blood, Saving Lives

One spring night in 1950, four doctors began an all-night drive from Washington, D.C., to Tuskegee, Alabama. They were on their way to help out at an annual free clinic where black physicians from all over the United States converged each year to treat people who had no other access to medical care. One of the four, Charles Drew, was, at 45, one of the most famous physicians in America: his work on blood banking had saved thousands of lives during World War II (Figure 38-1). But he never made it to Tuskegee.

Drew had spent a busy day in Washington, performing surgery and attending meetings at Howard University's College of Medicine, where he taught surgery. He must have been tired. After taking his turn at the wheel just south of Richmond, Virginia, Drew fell asleep and went off the road. The car rolled several times, forcing Drew against the steering wheel, crushing his chest, breaking his neck, and severing the major vein that delivers blood to the heart.

Drew was near death, but his friends found him an ambulance, which hurried him to the emergency room of Almanance General Hospital in Burlington, North Carolina. Like other hospitals in the South in 1950, Almanance treated blacks only in emergencies. But the physician on duty knew who Drew was and did all he could to stop the bleeding and replace the lost blood in his famous patient. Despite the doctor's best efforts, Drew was dead in less than two hours. In the weeks and months after Drew's death, many of his grief-stricken admirers were angry that the doctors at Almanance had not found a way to save Drew.

It was a painful irony that Charles Drew should bleed to death, having devoted his best years to improving the science of blood transfusion. His work had saved thousands of other people from bleeding to death. During World War II, German bombs devastated London, leaving tens of thousands of people with crushed limbs, broken bones, and horrendous burns. Although doctors could treat these injuries, their patients often died suddenly from *shock,* a life-threatening collapse of the circulatory system. Blood transfusions could save people from shock, but blood was in short supply. No one could keep blood cells from dying and spoiling after about a week, and

American Red Cross

British medical workers could not keep a steady supply of fresh blood. Thousands of bomb victims were dying from shock.

Fortunately, the liquid part of the blood, the *plasma* (without cells), can prevent shock almost as well as whole blood. Because plasma keeps for weeks, plasma collected in the United States could be shipped to Europe to save the lives of both civilian bomb victims and soldiers wounded in battle. In 1940, the United States had not yet joined in the effort to defeat Hitler's armies, but American plasma was helping Britain withstand nightly air raids.

Drew was a leading authority on blood storage and processing and a member of New York City's Presbyterian Hospital, associated with Columbia University. So it fell to him to launch and direct a program to develop better ways to collect, store, and ship plasma to Britain. Although plasma lasts longer than blood cells, it is loaded with glucose, proteins, and other nutrients that, unfortunately, make it an excellent growth medium for bacteria. In order to safely store and ship plasma, Drew had to find a way to prevent bacterial growth and to detect bacteria if they should slip into the blood supply. He worked out a foolproof system that prevented bacterial contamination as blood moved from a donor's arm into a collecting vessel, from the collecting vessel into a centrifugation system that removed the blood cells, and from the centrifuge into a sterile shipping bottle. To quickly separate the blood cells from the plasma, Drew modified a machine normally used to separate cream from milk. His invention allowed him to greatly increase the amount of blood processed each day.

Impressed by the success of Drew's program and procedures, the Red Cross set up a new blood bank at Presbyterian Hospital in 1941, this time for the U.S. Army and Navy, with Drew as assistant director. Within weeks, the armed forces, which were still segregated, announced their decision to refuse blood from African-American donors. Officials believed that some white soldiers would be afraid of transfusions if they thought the blood could have come from black donors. Both whole blood and plasma from African Americans could have saved lives, but while the Red Cross publicly stated that there was no biological basis for the decision, the agency gave in to pressure from the military and began turning away African-American donors who wanted to help with the war effort. In the face of this offensive policy, Drew silently resigned his position as assistant director and returned to teaching surgery to black medical students at Howard University.

Figure 38-1
Charles Drew. Charles Drew's work in blood banking helped save the lives of thousands of soldiers and civilians, both during World War II and since then.

Alfred Eisenstaedt/Time Life Pictures

Drew's greatest contribution in the last 10 years of his life was his work in educating the next generation of African-American physicians. His goal was to change the image of African-American doctors in the United States, who were then regarded as "just country practitioners . . . not particularly interested in advancing medicine." In Drew's day, for example, the American College of Surgeons would not admit Drew or any other African American of his generation to its membership, and many chapters of the American Medical Association also barred the membership of black physicians.

Drew remains a hero to this day. Amherst College, from which Drew graduated and where he ran track and played football, has established the Charles Drew House to promote the study of African and African-American culture. The Martin Luther King–Charles Drew Medical Center (associated with UCLA) in Los Angeles also honors Drew's contributions to medicine and his inspiration to generations of African-American physicians and medical students.

38.1 Why Do Animals Need Blood?

All the cells of the body are bathed in **extracellular fluid**, a watery fluid containing proteins and other substances. Cells depend on the extracellular fluid to provide food, oxygen, water, and salt, and to otherwise shield them from variations in the external environment. The concentrations of different molecules in the extracellular fluid must sustain but not harm the cells of the body. The extracellular fluid is a central player in homeostasis and is also both the external environment of the cells and the internal environment of the body.

Bathed in this fluid, cell membranes regulate the insides of cells, ushering in nutritious molecules, denying entry to others, and actively evicting toxic wastes. Nutrients pass from the extracellular fluid to the cells and wastes pass from the cells to the extracellular fluid (Figure 38-2).

Blood is an indispensable intermediary between the extracellular fluid and the environment outside the body. In mammals, for example, blood brings oxygen from the lungs and fuel and building blocks from the gut. Blood simultaneously takes carbon dioxide waste from the cells to the lungs and nitrogenous wastes to the kidneys. Blood also serves several other functions: it distributes hormones, it transports the molecules and the cells of the immune system, and

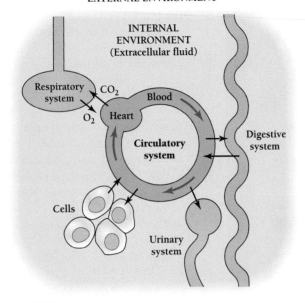

INTERNAL
ENVIRONMENT
(Extracellular fluid)

Respiratory
system

CO_2

O_2

Blood

Heart

Circulatory
system

Digestive
system

Cells

Urinary
system

Figure 38-2

The extracellular fluid is the intermediary between cells and the circulatory system. Extracellular fluid serves as a way station for incoming nutrients from the blood and outgoing wastes and signaling molecules from the cells. Blood carries materials from the extracellular fluid to other organs and tissues.

it conducts heat to all parts of the body. Like an urban subway system, which carries people from place to place throughout a city, the circulation picks up nutrient and waste molecules in one place and drops them off in another.

If the circulation is like a subway system, the extracellular fluid is like a platform at a subway station, where people getting off the train mingle with people getting on (Figure 38-3). In the same way, the extracellular fluid contains substances from both the blood and the cells, including, for example, glucose and other energy-rich molecules, building blocks such as amino acids, wastes such as carbon dioxide, and signaling molecules such as hormones. In addition, the extracellular fluid contains compounds such as bicarbonate and phosphate ions, which prevent the extracellular fluid from becoming too acidic or too basic. (Compounds that regulate acidity are called *buffers*.) Like a seedy subway station, the extracellular fluid may also contain material the body needs to get rid of—such as viruses, bacteria, and bits of cellular garbage. Along with extracellular fluid, this debris flows into the vessels of a second circulation, called the lymphatic system, which carries the debris in a fluid called **lymph** to lymph nodes scattered throughout the body. There, cells of the immune system destroy and digest any bacteria, viruses, and cellular garbage, as we'll discuss again briefly later in this chapter.

Circulating blood removes wastes and delivers nutrients to the extracellular fluid, or lymph, and also moves signaling molecules throughout the body. Cells remove nutrients from the extracellular fluid and expel wastes into it.

Figure 38-3

How is the extracellular fluid like a subway platform? Like a circulatory system, a big-city subway carries thousands of people to different destinations each day. On a subway platform, people getting off a train mingle with people getting on. Similarly, in the extracellular fluid, molecules of oxygen and glucose diffusing *out* the capillaries mingle with molecules of carbon dioxide that will diffuse *into* the capillaries.

What Is Blood?

Blood is an unusual connective tissue that is a combination of fluid and cells that carries all the substances that enter or leave the extracellular fluid. The fluid part of blood is called **plasma** [Greek, = form or mold], which is very similar in composition to the extracellular fluid.

In vertebrates, blood contains both plasma and three kinds of cells—red cells, or **erythrocytes** [Greek, *erythros* = red + *kytos* = receptacle]; white cells, or **leukocytes** [Greek, *leukos* = white]; and **platelets** [little plates], which are actually small fragments of cells (Figure 38-4). By far the most numerous cells in mammalian blood are red cells. A milliliter of human blood, just a few drops, contain about 5 billion erythrocytes, and a whole adult human contains about 25 trillion.

The first thing anyone notices about blood is how red it is. The bold color comes from the oxygen-binding protein **hemoglobin** [Greek, *haima* = blood + Latin, *globus* = ball] inside the red cells. Every red cell contains about 30 million hemoglobin molecules. Buried within each hemoglobin molecule are four iron-containing "heme groups" that help hemoglobin carry oxygen (Figure 39-10).

The white cells, or leukocytes, comprise several kinds of immune cells that defend the body against infections and tumors. Platelets are fragments of cells in the blood that help blood coagulate and so prevent blood loss. We discuss white cells and platelets in Chapter 41, on the immune system.

In all vertebrates, red cells, white cells, and platelets all originate from ordinary cells with nuclei. In mammals, however, red blood cells lose their nuclei, and platelets are membrane-enclosed bits of cytoplasm from larger cells, so that only the leukocytes are true cells. Researchers have suggested different reasons why nucleus-free red cells evolved in mammals. One hypothesis is that the loss of the nucleus allows more room for hemoglobin, thus maximizing the cell's capac-

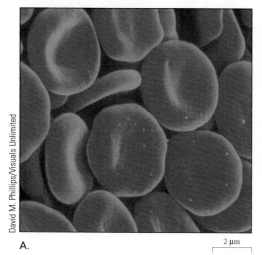

A.

2 μm

David M. Phillips/Visuals Unlimited

B.   Platelet   2 μm

Meckes/Ottawa/Photo Researchers, Inc.

Figure 38-4
Blood cells. A milliliter of blood, just a few drops, can contain about 5 billion red blood cells. A. Red blood cells have a characteristic flattened pillow shape. B. Platelets (such as the one attached to the red blood cell on the lower right) are actually fragments of membrane-enclosed cytoplasm from larger cells.

ity to carry oxygen. An alternative explanation is that the absence of a nucleus makes red blood cells more supple, so they are better able to squeeze through the narrow passages of the tiny blood vessels called capillaries. A red blood cell is so flexible that it can slip through a capillary half its diameter.

Vertebrates (mammals, birds, reptiles, amphibians, and fish) all use hemoglobin to carry oxygen, and so all have red blood. But invertebrate animals may transport oxygen using other proteins. For example, crustaceans (shrimp and crabs), spiders, and mollusks (snails and octopods) carry oxygen with the protein hemocyanin. Because hemocyanin contains copper instead of iron, it has a beautiful blue color. In invertebrate blood, oxygen-binding proteins float free in the blood rather than being carried inside of cells.

Vertebrate blood consists of plasma, red cells, white cells, and platelets.

38.2 How Do Animals Circulate Blood?

Many animals less than a millimeter in diameter (smaller than a period) get by without a circulatory system. These animals distribute nutrients, oxygen, wastes, and salts by diffusion alone. But diffusion across larger distances takes too long to be practical, and larger animals need a circulatory system with some kind of pump, or heart, to push materials around the body by means of bulk flow (Chapter 34). Circulating blood carries nutrients to cells and cell wastes to the outside world. Pipelike **blood vessels** usually carry the blood from one place to another, while some sort of pump, or **heart,** pushes the blood along (Figure 38-5).

Any muscular organ that pushes blood may act as a heart. In some animals, the squeezing of the blood vessels during body movements moves the blood. Valves in the

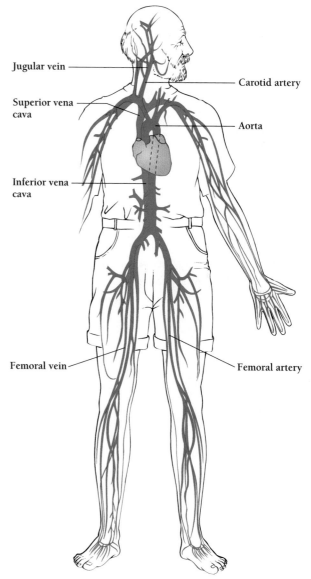

Jugular vein

Superior vena cava

Inferior vena cava

Femoral vein

Carotid artery

Aorta

Femoral artery

Figure 38-5
The major vessels of the human circulatory system. The circulation comprises the heart; the arteries, arterioles, and capillaries, which carry oxygenated blood from the heart; and venules and veins, which carry deoxygenated blood back to the heart. The aorta is the major artery through which blood leaves the heart and travels to the rest of the body. The venae cavae are the two major veins that carry blood back to the heart.

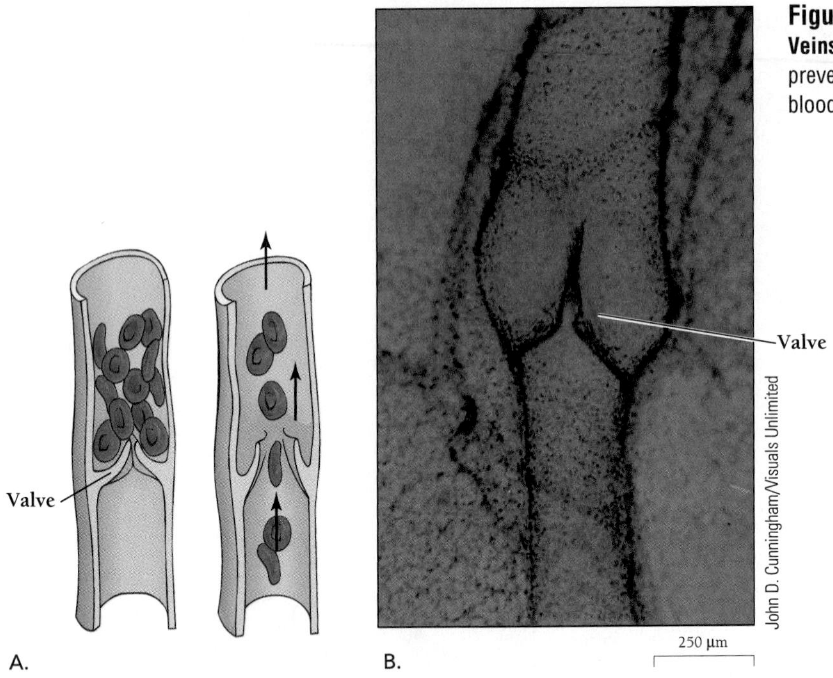

Figure 38-6

Veins contain one-way valves. A. Tiny flaps inside of veins prevent blood from flowing back toward the capillaries but allow blood to move toward the heart. B. A closed valve.

Valve

A.

Valve

B.

250 µm

John D. Cunningham/Visuals Unlimited

blood vessels ensure that blood flows in a single direction by preventing backflow of blood (Figure 38-6). Some species may have more than one heart, up to 10 in earthworms.

In vertebrates, the regular contractions of a single muscular heart do most of the pumping, with blood leaving the heart through **arteries** and returning to the heart through **veins.** The blood, the arteries and veins that carry the blood, and the heart that pumps it together make up the **cardiovascular system.** The route the blood takes through the body is called the **circulation** [Latin, *circus* = circle].

Invertebrates such as arthropods (spiders and insects) and mollusks (snails, slugs, and octopods) may have an **open circulation,** in which blood flows into a large open space that can occupy as much as 40 percent of the body's volume. In open circulation, blood pressure is low, but the blood directly bathes the cells of the body (Figure 38-7).

In a **closed circulation,** such as our own, blood moves under pressure through closed vessels in a continuous circuit—from heart to arteries to veins and back to the heart. The total volume of blood in a closed circulation is only about 5 to 10 percent of the volume of the body, far less than the 40 percent in an open circulation. A closed circulation has several advantages over an open one. For example, by changing how hard the heart pumps, an animal can easily increase or decrease the flow of blood through the whole body. And by varying the diameter of blood vessels, an animal can control the amount of oxygen, nutrients, wastes, and heat that reach different parts of the body.

Mammals and birds have a "double circulation," in which blood flows through two adjoining circuits, similar to a figure 8. In one circuit, deoxygenated blood from the veins passes to the heart and then to the lungs, where the blood picks up oxygen (Figure 38-8). The oxygenated blood then returns to the heart and passes into a second circuit, which delivers the oxygen-rich blood from the lungs to all the tissues of

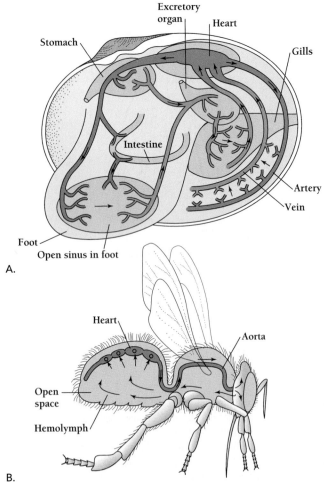

A.

B.

Figure 38-7

Open circulatory systems. A. In a mollusk such as the clam shown here, blood flows from the heart through closed arteries to contained open spaces and to the fluid, or hemolymph, that bathes all the internal organs; blood returns to the heart through closed veins. B. In an insect, such as this bee, the heart pumps blood through a blood vessel that runs along the top of the bee's body. Blood then flows into an open space and ultimately returns to the dorsal blood vessel through small openings.

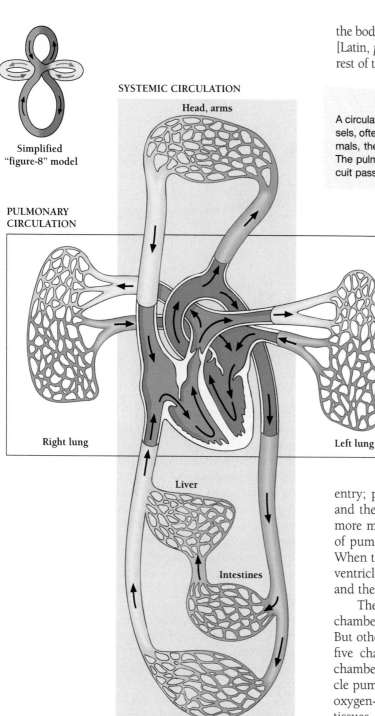

Simplified "figure-8" model

SYSTEMIC CIRCULATION

Head, arms

PULMONARY CIRCULATION

Right lung

Left lung

Liver

Intestines

Lower body

Figure 38-8
The figure-8 model of the circulatory system. The blood from the veins empties into the right side of the heart, which pumps it to the lungs. The blood returns from the lungs and then enters the left side of the heart, which pumps the oxygenated blood to the arteries.

the body. The route through the lungs is called the **pulmonary** [Latin, *pulmo* = lung] **circulation,** and the route through the rest of the body is called the **systemic circulation.**

A circulatory system consists of a closed or open system of blood vessels, often with a pump, or heart, to move the blood. In birds and mammals, the heart pumps blood through a figure-8 "double circulation." The pulmonary circuit passes through the lungs and the systemic circuit passes through all the other tissues of the body.

What Are the Advantages of a Four-Chambered Heart?

In order for the heart to pump blood through both halves of this figure 8, the heart needs two halves. Birds, crocodiles, and mammals have double hearts, with one-half of the heart pumping blood through the pulmonary circulation and the other half pumping blood through the system circulation. The two halves pump in synchrony, together squeezing blood through both halves of the heart once every second in humans.

Each half of a mammalian heart consists of two chambers. Blood first flows into a thin-walled **atrium** [Latin, = courtyard or entry; plural, atria], which holds the blood for a moment and then pumps it through one-way valves into the much more muscular **ventricle.** The ventricle does the real work of pumping the blood, either to the body or to the lungs. When the ventricles are full, pressure from the blood in the ventricles forces the valves between atria and ventricles shut, and the ventricles then pump the blood out.

The hearts of birds, crocodiles and mammals have four chambers—with two atria and two ventricles (Figure 38-9C). But other animals may have hearts with two, three, or even five chambers. Bony fishes, for example, have only two chambers, one atrium and one ventricle. The single ventricle pumps blood to the gills, where it picks up oxygen. The oxygen-rich blood then flows directly from the gills to the tissues, where it provides oxygen and other nutrients (Figure 38-9A).

The hearts of snakes, lizards, turtles, and amphibians contain three chambers—two atria and one ventricle (Figure 38-9B). In these animals, the oxygen-rich blood from the lungs often mixes with the oxygen-poor blood from the tissues and veins. These hearts are not inefficient versions of the four-chambered heart, however, but more flexible versions. For example, blood passing through a turtle's three-chambered heart can bypass the lungs when a turtle is diving and oxygen supplies are low.

For birds and mammals—which need to maintain a constant body temperature, highly developed brains, and

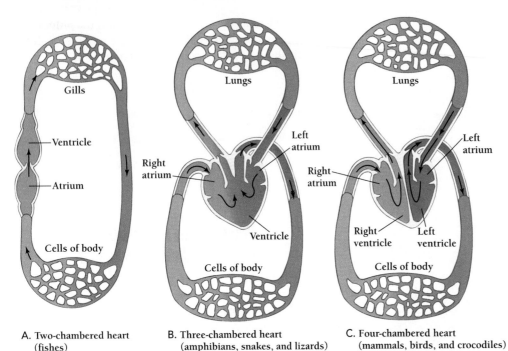

Figure 38-9
Evolution of the heart. A. Most fishes have a two-chambered heart, which pumps a single stream of unoxygenated blood forward into the arteries that feed the gills. B. Three-chambered hearts, intermediate between those of fishes and mammals, are found in lungfish, amphibians, and lizards and snakes. C. The four-chambered hearts of mammals, crocodiles, and birds pump two streams of blood, one oxygenated and one deoxygenated.

A. Two-chambered heart (fishes)

B. Three-chambered heart (amphibians, snakes, and lizards)

C. Four-chambered heart (mammals, birds, and crocodiles)

quick muscular responses, all of which require large supplies of oxygen—a three-chambered heart doesn't supply enough oxygen to the tissues. The four-chambered hearts of mammals and birds (and crocodiles) prevent the oxygen-poor blood from the tissues from returning to the tissues before it has acquired oxygen in the lungs, always keeping the two kinds of blood separate.

Another advantage of the four-chambered heart is that it allows animals to evolve to much larger sizes. Large animals, such as crocodiles and elephants, need high pressure in the left ventricle to pump blood to the rest of the body. If the pulmonary circulation were not separate, such high pressure would damage the delicate tissues of the lungs. In a four-chambered heart, the right ventricle is smaller and thinner-walled and pumps blood gently into the lungs.

The four-chambered hearts of crocodiles, birds, and mammals keep deoxygenated blood from mixing with oxygenated blood. Such an arrangement helps guarantee that oxygen-rich blood flows to all of the tissues of the body.

Which Way Does Blood Move Through the Human Heart?

In humans, the heart is about the size of a clenched fist and is located in the chest just under the breastbone, or **sternum** [Greek, *sternon* = chest]. Blood first enters the heart from the veins, specifically huge twin veins called the **superior vena cava** (which brings blood from the arms and head) and

the **inferior vena cava** (which brings blood from the legs and trunk). The blood from these veins, which is oxygen poor and therefore very dark, almost bluish, enters the **right atrium** of the heart (Figure 38-10). The right atrium pumps the blood through the right atrioventricular valve, or **right AV valve**, into the **right ventricle.** The right ventricle pumps the blood through the **pulmonary semilunar valve** to the **pulmonary artery,** which carries the blood to the lungs. In the lungs, the blood picks up oxygen (and becomes redder) and then returns to the heart by way of the pulmonary vein.

The **pulmonary vein** dumps oxygen-rich blood from the lungs into the **left atrium,** which pumps the blood into the **left ventricle,** by way of the **left AV valve.** The left ventricle, the most muscular of the four chambers of the heart, then pumps the blood through the **aortic semilunar valve** into the largest artery in the body—the **aorta,** which is more than an inch in diameter.

The aorta arches over the top of the heart, and its first branch sends blood to the head, while other branches supply the rest of the body (Figure 38-5). Almost one-quarter of the blood goes directly to the kidneys to be cleaned. The hard-working heart also needs its own blood supply, which comes by way of the *coronary* [Latin, heart] *arteries.* These branch off the aorta and feed capillary beds in the heart muscle with blood almost straight from the lungs.

In humans, blood enters the right atrium, which pumps the blood into the right ventricle, which pumps it out to the lungs. The blood returns from the lungs and enters the left atrium, which pumps the blood into the left ventricle, which pumps it into the aorta.

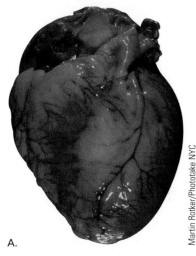

A.

Figure 38-10

The human heart. Blood enters and leaves the heart by way of several major blood vessels. Blood enters the right atrium from two giant veins (the superior and inferior venae cavae) and enters the right ventricle, which pumps blood to the lungs through the pulmonary arteries. Blood from the lungs returns to the left atrium (through four pulmonary veins), drops into the left ventricle and exits through the aorta. A. Photograph of a human heart. B. Diagram showing organization and blood flow.

Biology ⓔ Now™ Learn more about the human heart by clicking on this figure on your BiologyNow CD-ROM.

Martin Rotker/Phototake NYC

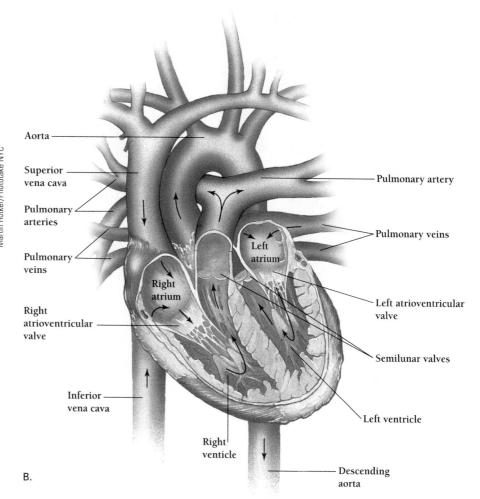

B.

38.3 How Does Blood Move Material To and From the Tissues?

The heart is the noisiest and most obvious part of the circulatory system, but what happens far from the heart in the tissues of the body is at least as important, even if silent and invisible to the naked eye. When an oxygen-laden red blood cell hurtles out of the heart through the aorta, it enters a 50,000-mile maze of blood vessels. What happens in that maze determines both how we feel at any given moment and our long-term health.

As the arteries leave the heart, they branch many times, and blood flows into smaller vessels called **arterioles.** From the arterioles, the blood spreads out into a network of millions of fine vessels called **capillaries,** only half the diameter of a single blood cell. The network of microscopic capillaries is so extensive that if all of the blood vessels in your body were placed end to end, they would extend two times around Earth's equator (Figure 38-11A). No cell in the body is more than about three cells away from a capillary.

Although arteries and veins do not normally leak at all, capillaries leak constantly. Except in the brain, the cells that

make up the capillaries are so loosely connected that small molecules easily diffuse between adjacent cells. Because of the pressure of the blood inside the capillaries, plasma oozes through these gaps (leaving all the blood cells behind) to form the extracellular fluid. Even some large molecules can move through the gaps between the vessels' cells. The sievelike walls of the capillaries filter blood cells and some of the larger molecules from the extracellular fluid, or lymph, so that it contains considerably less protein than the blood plasma.

The thin, leaky walls of the capillaries likewise allow them to absorb cell wastes from the extracellular fluid. As the capillaries feed into tiny veins called **venules,** which converge to form larger veins for the return to the heart, the blood inside carries whatever wastes seeped in from the tissues.

Arteries carrying blood from the heart branch to form smaller arterioles, which branch to form extensive capillary beds. In these capillary beds, plasma and nutrients ooze from the capillaries to form the extracellular fluid, and cell wastes enter the capillaries to be carried into venules and veins, which carry blood back to the heart.

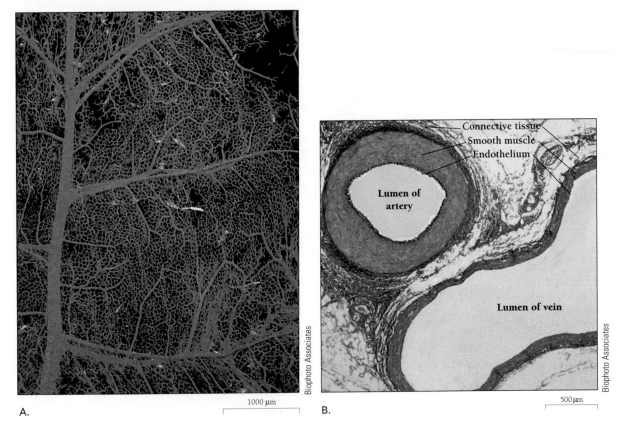

A. 1000 μm

B. 500 μm

Biophoto Associates

Connective tissue
Smooth muscle
Endothelium

Lumen of
artery

Lumen of vein

Figure 38-11
Arteries, veins, and capillaries. A. Arterioles branch into capillaries, so fine and so numerous that no cell is more than three cells away from a capillary. B. Veins, arteries, and capillaries are all lined with endothelial cells. Veins and arteries are further surrounded by connective tissue and smooth muscle, which contracts and relaxes to change the diameter of the vessel. Compared to veins, the walls of arteries are muscular and thick, which helps them withstand the high blood pressures generated by the heart.

How Does the Structure of Blood Vessels Help Them Carry Blood?

Arteries, veins, and capillaries are all hollow tubes, with a blood-filled space called the **lumen** [Latin, = light], surrounded by a thin layer of cells called the **endothelium** (Figure 38-11B). Unlike the epithelial cells of the skin or gut, which are tightly bound together, the endothelium of the capillaries is leaky and only one layer thick. As we've seen, the thin, loosely connected cells of capillary walls permit the easy passage of substances from the lumen to the surrounding extracellular fluid.

In contrast, the linings of arteries and veins are thick and impermeable. Surrounding the endothelial layer are supportive layers of elastic fibers, collagen, and smooth muscle. Blood pressure in the veins is much lower, and they generally have thinner and less muscular walls than arteries. In veins, movements of the body and one-way valves keep the blood moving slowly toward the heart (Figure 38-12).

The elastic walls of both arteries and large veins allow these vessels to narrow or expand in response to changes in blood pressure or in response to signals from the auto-

nomic nervous system to the smooth muscles. But the blood vessels—especially the arteries in which blood pressure is highest—must not be too elastic. If they stretch too much, they can balloon out and flatten the capillaries in the surrounding tissues. To prevent such ballooning, collagen fibers, like those found in connective tissue and bone, strengthen and stiffen the vessel walls.

If the collagen sheath of an artery fails and ballooning occurs, the result is an **aneurysm.** The aorta, for example, is normally between 1.6 and 2.8 cm wide. If it balloons to more than 5.5 cm, it can easily burst. In that case, 90 percent of people will bleed to death very quickly, rarely even reaching a hospital (see the box). Even if they do get to a doctor in time, it takes time for a doctor to deduce the cause, by which time it may be too late. And in about 35 percent of cases, the doctor does not recognize what the problem is. Aneurysms of both the aorta and the blood vessels in the brain kill 32,000 Americans each year. The thinner-walled veins are even more susceptible to such swelling, although the consequences are far less dangerous. Weakening of the walls of a vein near a valve and the weakening of the valve

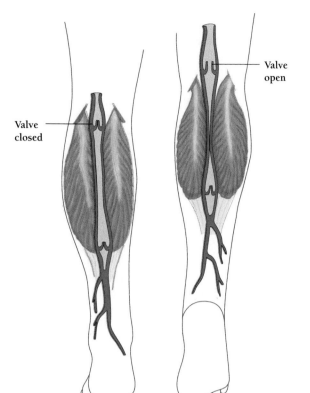

Figure 38-12

Valves in veins. When skeletal muscles, such as those in the lower leg, contract, the muscles squeeze veins, which pushes the blood upward toward the heart. One-way valves prevent the blood from flowing backward toward the capillary beds, so the blood can flow only toward the heart. Taking a walk speeds the blood back to the heart and lungs and contributes to our overall health.

itself can cause the blood to pool and distend the vein in "varicose veins." These can be both unsightly and painful.

The wall of a capillary is only one cell thick and leaky. The walls of the larger arteries and veins are thick, nonleaky, and surrounded by elastic fibers, collagen, and smooth muscle.

How Do the Lymph Nodes Clean the Extracellular Fluid?

Not all the extracellular fluid reenters the circulatory system through the veins. Some fluid reenters the circulation by way the vessels of the **lymphatic system** (Figure 38-13). The lymphatic system is a central part of the immune system. Without

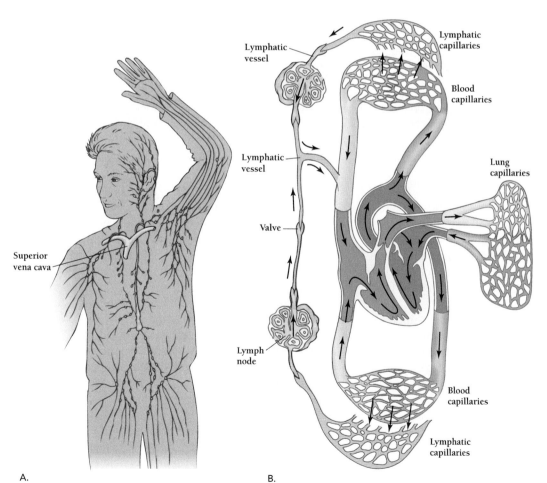

Figure 38-13

The lymphatic system. The lymphatic system carries the small amount of extracellular fluid that does not make its way back into the vascular capillaries. The lymphatic system also carries wastes from the extracellular fluid to the lymph nodes. A. Scattered throughout the body, the small lymph nodes filter dead cells, foreign cells, and other organic waste from the lymph. One-way valves keep the lymph draining in one direction. B. All the cleaned lymph drains into veins at the base of the neck that enter the superior vena cava and flow into the heart.

Business and Biology   Should People Be Tested for Aortic Aneurysms?

We think of the blood vessels of the body as quietly doing their jobs without any fuss, and most of them do for most of our lives. But as we age, the blood vessel walls can stretch, harden, and even break. In the United States, bursting aortic aneurysms in the chest or abdomen kill about 18,000 people each year, more people than are killed by AIDS (about 15,000). Eighty percent of burst aortic aneurysms occur in men, especially those who smoke, have high blood pressure, or atherosclerosis. Aneurysms are also more common in anyone with a parent or sibling who has one. The majority of aneurysms are harmless, but if an aortic aneurysm bursts, it is fatal 90 percent of the time.

One reason aortic aneurysms are so often fatal is that most people don't get to a hospital in time. But another reason is that many physicians are unfamiliar with the symptoms. Aortic aneurysms are misdiagnosed about 35 percent of the time (compared to only 5 percent for heart attacks). So even patients who get to a doctor in time may be sent home to die because the doctor doesn't realize something life threatening is occurring. When comedian John Ritter died of a burst aneurysm at the age of 54 in 2003, physicians interviewed by the media described a burst aneurysm as rare and undetectable.

In fact, aortic aneurysms are easy to detect and easy for a surgeon to permanently fix. Yet most people never have the $50 ultrasound test that could save them from a sudden, premature death. Because neither national Medicare nor private insurance companies normally pay for aneurysm tests, few doctors run this test, even on patients obviously at risk. Recently, some doctors have begun urging the federal government to cover the cost of aneurysm screening and set an example for insurance companies.

Would it be worth it to change this policy? In order to save thousands of people from fatal aortic aneurysms, we would have to screen millions of people. One aneurysm expert has recommended that all men over 60 and all women over 60 with a family history of aneurisms be screened. A British study of 61,000 men showed that screening those between 65 and 74 years old reduced the death rate from aortic aneurysms by an astounding 42 percent. In the United States, such screening could save 7,500 lives per year.

In many medical situations, doing the right thing saves money. For example, immunizing children against infectious diseases not only prevents the needless suffering of children, it also saves everyone the expense of caring for sick children and employers the cost of having parents take time off from work. Everyone gains.

But aortic aneurysms aren't like that. Because the vast majority of aneurysms are harmless and because those that do burst are usually immediately fatal, an undiagnosed aneurysm costs an insurer or Medicare almost nothing. There's no savings to counteract the expense of preventing them. And the expense could be enormous; screening millions of people could cost a billion dollars or more, depending on the number of people who are ultimately screened. Given the cost, it is difficult for insurers to justify screening even though it would save lives.

In the next few years, Medicare will have to decide whether to spend the money to prevent deaths from aortic aneurysms. For now, most people who discover they have an aortic aneurysm learn about it because they were screened for a tumor or some other problem that was producing symptoms. Each year, about 200,000 aortic aneurysms are diagnosed this way, most of them harmless. Surgeons operate on and repair about 50,000 of these, replacing the ballooned section of artery with a length of fabric or plastic tubing. If someone in your family has had an aortic aneurysm, it might make sense to get the test and pay for it yourself. It's cheaper than a pair of nice sneakers.

---

it, we would be dead in 24 hours. All mammals, birds, reptiles, amphibians, as well as many fishes have a lymphatic system.

Extracellular fluid in the tissues drains into the lymphatic vessels, where the fluid is called **lymph**. Lymph contains proteins and other large particles that cannot directly enter the capillaries, including bacteria, viruses, dead or foreign cells, and parts of cells.

Extracellular fluid enters the lymphatic system through lymphatic vessels, whose endothelial cells loosely overlap and act as one-way valves that allow fluid to seep into the vessels but not back out. The lymphatic system has no pump to move the fluid. Instead the movements of the skeletal muscles squeeze the lymph through the lymphatic vessels, and one-way valves prevent the fluid from moving backward. In people who don't move much, the lymphatic fluids move less than in people who walk and move about normally.

The lymph in the lymphatic vessels flows to **lymph nodes,** which filter bacterial cells and other large particles of waste from the lymph. Inside the lymph nodes are phagocytic (cell-eating) cells that engulf the material trapped in the lymph nodes. Dense concentrations of immune cells in the lymph nodes also monitor the lymph for signs of infection. During an infection, the lymph nodes frequently become swollen with increased numbers of immune cells. A doctor often checks the lymph nodes in the neck, under the arms, or in the groin for this sign of infection. We discuss these cells and their role in immunity in more detail in Chapter 41. The lymphatic system also includes the tonsils (in the wall of the throat); the thymus (in the neck), where immune cells mature; and the spleen, whose phagocytic cells clean the blood in the same way that the lymph nodes clean the lymph.

Once the lymph has been filtered by the lymph nodes, the purified lymph empties into two large lymphatic vessels that join the blood in two branches of the superior vena cava, below the shoulders.

Blood and lymph interact closely. One example of the interaction between the lymphatic system and the blood circulation occurs during inflammation. If the skin is damaged by a mosquito bite or a scrape, the surrounding tissues become hot, red, swollen, and painful or itchy. These symptoms are the direct result of dilation of the arterioles in the damaged tissue, which cause a temporary increase in blood flow to the tissue, which causes the heat and redness. While the small blood vessels are dilated, the gaps between the cells of the capillaries increase and even more plasma leaks out into the extracellular space than usual, causing the tissues to expand, which we call swelling. Meanwhile, white cells from other parts of the body home in on the damaged tissue, arriving by way of the circulation. The white cells slip out between the cells of the blood vessels and into the damaged tissues, where the white cells attack foreign cells or proteins and remove dead and dying cells.

Dead and foreign cells and proteins in the extracellular fluid flow into the lymphatic system. In the lymph nodes, particles of viruses, dead cells, and other waste are filtered from the lymph and ingested by phagocytic white blood cells.

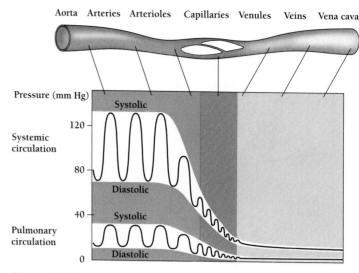

Figure 38-14

Pressure in the blood vessels. Blood pressure is highest in the aorta as the blood leaves the ventricles. In the arterioles and capillaries, the mean pressure drops from 100 to about 20 mm Hg. Finally, as the blood flows into the veins, it moves sluggishly, under virtually no pressure at all.

Biology Now™ Learn more about blood pressure by clicking on this figure on your BiologyNow CD-ROM.

38.4 How Does the Heart Beat?

A healthy heart coordinates the contractions of the atria and ventricles. In a resting adult, the heart beats about 70 times a minute. During each beat, the right and left atrium contract almost simultaneously, followed by the two ventricles.

The pumping of the heart generates pressure that drives blood through the circulatory system. This pressure in the arteries changes from moment to moment as the heart contracts and relaxes. We can feel the fluctuations of pressure in the larger arteries, such as in the one that crosses the wrist, where we can feel our pulse. Each beat of the pulse corresponds to an expulsion of blood from the heart. Medical workers usually measure blood pressure in the arteries of the upper arm, at the same level as the heart.

The pressure within the vessels decreases as the arteries divide into finer and finer vessels (Figure 38-14). In the capillaries, the pressure barely varies, and capillary blood flows smoothly. The capillary blood is still under pressure, however. This pressure drives some of the plasma from the capillaries into the surrounding extracellular space. Blood in the veins is under still less pressure and fluctuates hardly at all.

What Are the Phases of a Heartbeat?

Each heartbeat consists of alternating periods of contraction and relaxation. The period when both the atria and the ventricles are relaxed is called **diastole** [Greek, *dia* = between

+ *systellein* = to contract]. During the other part of the cycle, the atria contract, followed by the ventricles. The period of contraction lasts about a third of a second and is called **systole** [Greek, *systellein* = to contract]. During systole, the blood pressure momentarily increases. Blood pressure is therefore always expressed as two numbers, such as 120/80, referring to the high pressure generated during systole and the lower pressure during diastole. The units of pressure are mm Hg—the height (in millimeters) of a column of mercury (Hg) that could be supported by that pressure. You can read more about blood pressure at the beginning of Chapter 40.

Inside the heart, a system of valves ensures that the blood flows in one direction only. As we saw earlier, the valves between the atria and ventricles are called atrioventricular (AV) valves; those between the left ventricle and the aorta and between the right ventricle and the pulmonary arteries are called semilunar valves. Each valve consists of flaps of half-moon-shaped connective tissue, which allow blood to move in one direction but not the other. Blood from the ventricles cannot reenter the atria, and blood from the arteries cannot reenter the ventricles.

You can hear the heart working if you listen to your own heart with a stethoscope or to a friend's heart by putting your ear on his or her chest. You will hear that each heartbeat consists of two separate sounds—a low-pitched "lub" and a higher-pitched "dub." The first sound is that of the atrioventricular valves closing and the second is the closing of the semilunar valves (Figure 38-15).

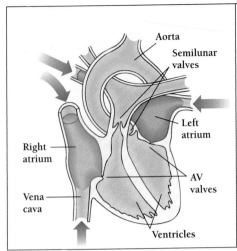
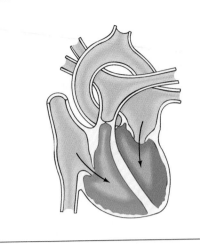
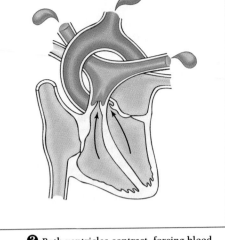

❶ Atria and ventricles relax. Blood flows into atria.

❷ Both atria contract, forcing blood into ventricles. AV valves then close.

❸ Both ventricles contract, forcing blood into aorta and pulmonary arteries. Semilunar valves then close.

A. Diastole

B. Atrial systole ("Lub")

C. Ventricular systole ("Dub")

Figure 38-15
What's that sound? The heart beats in three parts. ❶ During diastole, the atria and ventricles relax and blood fills the atria. ❷ During atrial systole, the two atria contract, forcing blood into the ventricles. As the atrioventricular valves close behind the blood, they make a low-pitched "lub." ❸ During ventricular systole, the two ventricles contract, forcing blood into the aorta and the pulmonary artery. As the semilunar valves close noisily behind the blood, they make a short "dub" noise.

A trained person can detect some heart abnormalities just by listening to the heart. A defective valve, for example, can produce a **heart murmur,** a low, hissing sound in addition to the "lubdub." This sound comes from the movement of blood in the wrong direction, for example, from a ventricle back into the atrium. Such a defect may result from abnormal development of the valve or from an infection such as rheumatic fever. Modern heart surgeons can replace a defective heart valve with an artificial one.

The period when the heart contracts is called systole, and the period in between contractions is called diastole. One-way valves allow blood to flow from the atria to the ventricles but not back, and from the ventricles to the arteries but not back.

What Coordinates the Beating of the Heart Muscle?

The action of the ventricles in pumping blood to the lungs and the rest of the body is essential to life. To pump blood, the muscle fibers (cells) of the ventricles must all work together—contracting together, relaxing together, and pausing together.

Like skeletal muscles, heart muscles are striated, and contraction depends on sliding protein filaments (Chapter 35). But unlike skeletal muscle fibers, heart muscle fibers are branched and tightly connected end to end, with special mechanical connections (desmosomes) and communication channels (gap junctions), so that the heart fibers are almost perfectly coordinated. If one muscle fiber in the ventricles contracts, all of the other fibers in the ventricles also contract, in a wave starting from the first fiber. The same is true of the atrial cells. But the atrial cells are physically separated from the ventricle cells by fibrous tissue, so stimulating the atrial cells to contract does not necessarily cause the ventricles to contract. The two groups of cells are, however, connected by a special tissue called the **atrioventricular (AV) node,** which conducts electrical signals from one part of the heart to the other.

About one percent of the heart muscle cells generate spontaneous, rhythmic electrical activity. These cells do not themselves contract but stimulate other cells to contract. One group of these special cells form a structure called the **SA (sinoatrial) node,** or **pacemaker,** which begins contraction before other cells in the heart, and so sets the pace of contraction for the rest of the heart. In mammals, the pacemaker lies near the top of the right atrium, where blood first enters the heart. The pacemaker emits electrical signals that travel to the left atrium. From there the signal goes to the AV node, which relays the signal to the ventricles through gap junctions (Figure 38-16). The pacemaker thus determines the rhythm of the heart and, with the AV node, coordinates the action of both atria and ventricles.

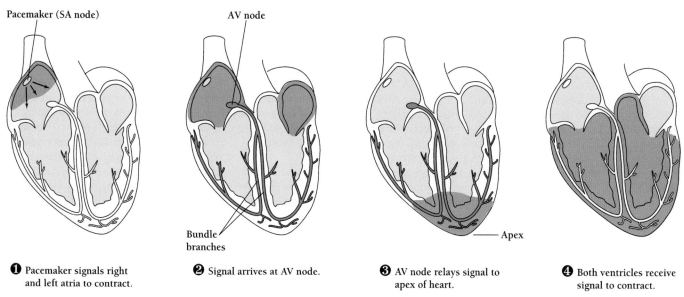

Pacemaker (SA node)   AV node

❶ Pacemaker signals right   ❷ Signal arrives at AV node.   ❸ AV node relays signal to   ❹ Both ventricles receive
   and left atria to contract.   apex of heart.   signal to contract.

Figure 38-16

Cardiac conduction. ❶ The heart's natural pacemaker, which lies in the right atrium, signals the right and left atria to contract and send blood to the two ventricles. ❷ The pacemaker's signal then arrives at the AV node, ❸ which transmits the signal through special muscle cells called "bundle branches" to the heart's apex. ❹ From there the signal to contract goes to both ventricles simultaneously and these two muscular cavities contract with enough force to drive blood to the brain, the hands and feet, and the lungs.

Ordinarily, the SA node sets the resting heart rate at about 70 beats per minute. If the pacemaker is damaged, however, the AV node can take over as a pacemaker, but at a slower rate—setting the heart rate at about 50 beats per minute. If both nodes are damaged, then other conducting cells set the frequency at only 30 beats per minute, too slow to deliver enough oxygen. Fortunately, surgeons can implant an artificial pacemaker, which emits electrical signals that stimulate a normal heartbeat.

Occasionally, the muscles of the heart begin working against one another, so that many of the fibers are contracting when they should be relaxed. The result is **fibrillation**, continuous, disorganized contractions. Fibrillation can affect the atria or the ventricles. If the atria fibrillate, blood continues to flow into the ventricles, so that people with otherwise healthy hearts can live with fibrillating atria. When the ventricles begin to fibrillate, however, death follows immediately and almost inevitably. During ventricular fibrillation, the different parts of the ventricles no longer contract simultaneously. Within a few moments, they cannot pump any blood at all. Blood continues to fill the ventricles from the atria, and the ventricles become distended with blood. Because the ventricles are not pumping blood, however, their own blood supply, which comes from the coronary arteries, is cut off. Within 90 seconds, the ventricles become so weak from lack of oxygen that they cannot contract. A strong electric shock can sometimes bring about *defibrillation*: fibrillation stops, and ventricles can reestablish normal, organized contractions.

The cells of the atria contract almost as one, as do the cells of the ventricles. Electrical signals for such contractions are conducted between the two parts of the heart by the atrioventricular (AV) node. A natural pacemaker, the sinoatrial (SA) node, sets the rhythm for the heart's contractions.

38.5 What Controls the Flow of Blood?

The total amount of blood pumped by the heart per minute—the "cardiac output"—depends on two factors: (1) the **heart rate,** the number of beats per minute, and (2) the **stroke volume,** the volume of blood delivered by each ventricle. Because the blood travels in a circuit, the volume that passes through the left ventricle must equal the volume that passes through the right ventricle. In a resting adult, the heart beats about 70 times a minute with a stroke volume of about 70 ml. So the volume of blood pumped each minute is about 70×70 ml $= 4,900$ ml, or about 5 liters. This is approximately the total volume of blood. In other words, each of us, while sitting still, pumps nearly all of our blood through the heart once a minute.

Health and Biology Atherosclerosis, Heart Attacks, and Strokes

Nearly 40 percent of all the people who died in the United States in 2000 died from thickening and hardening of the arteries, or **atherosclerosis.** During atherosclerosis, high levels of cholesterol-rich lipoproteins in the blood stimulate circulating white cells to remove the lipids. The white cells become bloated, attach to the arterial walls, and stimulate smooth muscle cells to divide, so that the artery wall becomes thickened. The mass of cells, along with fat, forms *plaque* that blocks the artery. Plaque deposits can damage the wall of the artery to such an extent that gaps appear between the cells of the endothelium and *blood clots* form. Blood clots are semi-solid clumps of blood that normally form to prevent excessive bleeding by blocking holes in the circulatory system. Unfortunately, blood clots further block the artery, which interferes with blood flow, or even stops it altogether. Either way, the cells that normally receive blood run short of oxygen, glucose, and other nutrients.

Arteries can be blocked anywhere in the body. Blocked arteries in the brain and heart are the most dangerous because even a

brief halt in the blood supply to these two organs can cause death or permanent health problems. Blocked or broken arteries in the brain cause *strokes* (sometimes called brain attacks), while those in the heart cause *heart attacks.*

More women than men die of cardiovascular disease. But, confusingly, the *death rate* is higher for men, which is the figure most sources report. The total number of deaths from all cardiovascular diseases (including both heart attacks and strokes) in the year 2000 was 501,000 women and 436,000 men, a total of 937,000 people. Women, who have more heart attacks later in life, are more likely to die of them. But men are more likely to have heart attacks when they are a decade or so younger, in their 40s and 50s.

One form of cardiovascular disease is a heart attack, which each year kills about 200,000 people. A **heart attack** is what happens when the coronary arteries or arterioles become blocked or broken, so that blood no longer reaches an area of heart muscle. Without oxygen, sugars, and other nutrients, the blood fibers cease to function

and even die, creating a nonfunctioning area of the heart, called an *infarct.* A heart attack is also called an *acute myocardial infarction* [acute = sudden + *myo*= muscle + *cardio* = heart]. If the block occurs in a major coronary artery, the whole heart may immediately stop beating. If the block occurs in a smaller vessel, the damage may be local and the person may survive, but with a damaged and malfunctioning heart.

Many deaths from atherosclerosis are preventable. Of the 900,000 or so annual deaths from atherosclerosis, about 300,000 per year are due to smoking cigarettes. People who smoke can reduce their risk by quitting smoking (see the box on page 764). People who don't smoke can reduce their risk of an early death from heart disease or stroke by exercising (even a little is better than none), by eating fruits and vegetables, the more the better (but even a little is better than none), and by reducing their intake of saturated fats (see the box on page 52). Many people at risk for heart attacks or strokes take small doses of aspirin to reduce the risk of blood clots, which can block important arteries.

When we dance, play basketball, or give a talk in front of a crowd, both heart rate and stroke volume increase, and the amount of blood pumped by the heart may go as high as 35 liters (about 9 gallons) per minute. In other words, all our blood goes through the circulatory system seven times a minute. Our physical and emotional states affect both the heart rate and the stroke volume. Running up two flights of stairs will dramatically increase both heart rate and stroke volume. But subtler influences, such as a disturbing thought, can also influence the heart. Both nerves and hormones regulate the beating of the heart.

The heart varies the amount of blood it pumps depending on our activities and moods. The amount of blood pumped is a function of how fast the heart beats and how much blood it moves with each beat.

What Hormones and Nerves Control the Heart?

Two systems of nerves influence heart rate—the **sympathetic** [Greek, *sym* = together + *pathos* = suffering] **nervous system** and the **parasympathetic** [Greek, *para* = be-

side, next to + sympathetic] **nervous system.** Sympathetic nerves mobilize the body in times of stress, speeding up the heart and opening or closing capillaries in different parts of the body. A sympathetic nerve acts on the heart's pacemaker by secreting **norepinephrine,** a signaling molecule closely related to epinephrine (Chapter 37). Parasympathetic nerves slow the heart by secreting a different molecule, **acetylcholine.**

Both norepinephrine and acetylcholine are **neurotransmitters,** molecules that transmit signals from nerve cells either to other nerve cells or to muscles or glands. Under the influence of acetylcholine, as in a person at rest, the heart beats about 70 times per minute. Under the influence of norepinephrine, the heart may speed up to more than 120 times a minute.

A third molecule, **epinephrine,** also speeds the heart during the "fight or flight" response in times of danger. Both epinephrine and norepinephrine increase stroke volume by causing the veins to constrict. As the veins narrow, the blood flowing through them comes under pressure, which increases the pressure on the blood flowing into the heart. As more blood flows into the heart, it reacts by pumping more blood out into the arteries. The net effect is to increase the volume of blood pumped, or cardiac output.

> Epinephrine and norepinephrine, associated with the "fight or flight" response, increase the amount of blood pumped through the heart, while acetylcholine reduces blood flow.

How Does the Body Deliver Blood to Some Areas But Not Others?

When we are running to catch a bus, we need to increase the blood supply to the muscles of the legs as well as to the heart itself. On the other hand, we don't need to increase the blood supply to the brain or the intestines. The body is remarkably good at delivering oxygen-rich blood where it is needed. Changes in the diameters of specific arterioles route the blood to the specific organs or muscles that are most active at a given time. In fact, at any moment, only 5 to 10 percent of all the capillaries in the body are fully open to the passage of blood. The local constrictions of arterioles restrict flow into the other 90 to 95 percent of the capillaries.

How does the body open and close arterioles? One form of control depends on the local concentration of oxygen, carbon dioxide, hydrogen ions, or other molecules that reflect metabolic activity. When the oxygen level in a tissue begins to drop, arterioles in that tissue dilate. Ordinarily, the smooth muscles that surround the arterioles remain contracted. Decreased oxygen in the surrounding tissues, however, causes the arterioles to relax. The result is **vasodilation,** the opening of the arterioles, which allows more blood to flow and more oxygen to arrive. Conversely, high oxygen levels stimulate **vasoconstriction,** the contraction of the arteriole's smooth muscle and the reduction of blood flow and oxygen delivery.

The endothelium that lines the inside of each blood vessel releases the signaling molecule NO (nitric oxide), which causes vasodilation. Drugs such as Viagra, used to sustain penile erections, work by stimulating the release of NO in the penis (Chapter 2).

Blood flow throughout the body is also regulated by the nervous system and by hormones and other substances. The sympathetic nervous system, for example, can rapidly override local control of the arterioles' dilation by releasing norepinephrine, which causes vasoconstriction. The effects of the sympathetic nerves vary from place to place in the body, according to physical and emotional states. For example, fear increases norepinephrine release in the area of the skin arterioles, causing the arterioles to contract and the skin to pale. A fever, or just an embarrassing moment, can decrease the activity of the same sympathetic nerves, causing vasodilation. In that case, blood flow increases and the skin reddens. If the temperature of the body rises, the sympathetic nervous system dilates the arterioles in the skin, allowing them to carry more blood to the surface of the body and radiates heat to cool the body.

Many substances influence the expansion and contraction of the vessels of the circulatory system, including, for example, *prostaglandins, bradykinins,* and *histamine.* Anyone who has ever suffered from an allergic reaction or an insect bite has encountered one of these. **Histamine** is a small molecule released by cells in damaged tissues that increases the blood flow and inflammation described earlier in this chapter. The increased blood provides nutrients that sustain the process of tissue repair as well as white blood cells that combat infection. "Antihistamine" drugs available in any store provide relief from itching and swelling by preventing histamine from acting. You can read more about inflammation in Chapter 41.

> The body regulates blood flow to different areas of the body by means of local control over vasodilation and vasoconstriction and the nervous and endocrine systems.

In this chapter we have discussed the importance of blood in helping to provide and maintain a constant internal environment. We have touched on some of the factors that control the rate of blood delivery. In the next chapter, we learn how the lungs take up oxygen.

Key Concepts

- Blood continuously refreshes the fluids of the internal environment, providing new materials and removing wastes.
- A mammal's heart pumps blood through two connected circuits.
- In a mammal's heart, a pacemaker coordinates muscle contractions in four separate chambers.
- Animals other than mammals have different patterns of blood circulation.

Summary with Key Terms

What do blood and the extracellular fluid do to keep animals alive?

Circulating blood removes wastes and delivers nutrients to the **extracellular fluid** and also moves signaling molecules throughout the body. Cells take nutrients from the extracellular fluid and expel wastes into it.

What are the essential components of blood and extracellular fluid?

The blood of invertebrates lacks cells, but the blood of vertebrates contains both **plasma** and large numbers of **erythrocytes, leukocytes,** and **platelets.** The **hemoglobin** in vertebrate red blood cells enables these cells to take up oxygen efficiently and also gives blood its bold red color. White cells and platelets play important roles in protecting the body from injury and infection.

How do the heart and circulation move blood?

The blood, the **heart**, and the **blood vessels** together make up the **cardiovascular system**. The route of the blood is called the **circulation**. **Arteries** carrying blood from the heart branch to form smaller **arterioles**, which branch to form extensive **capillary** beds, which converge to form **venules**, which converge into **veins**, which pour blood back into the heart. Arthropods and mollusks have an **open circulation**, and the blood travels part of its circuit through open spaces. Vertebrates, as well as other invertebrates, have a **closed circulation**, in which blood runs only within enclosed vessels. In birds and mammals, the heart pumps blood through a **figure**-8 "double circulation." The **pulmonary circulation** passes through the lungs, and the **systemic circulation** passes through all the other tissues of the body. The four-chambered hearts of birds, crocodiles, and mammals keep deoxygenated blood from mixing with oxygenated blood.

In humans, the heart is about the size of a fist and is located just under the **sternum**. The muscular walls of the heart form four chambers, two thin-walled **atria** and two more-muscular **ventricles**. From the **superior vena cava** and the **inferior vena cava**, blood enters the **right atrium**, which pumps the blood through the **right AV valve** and into the **right ventricle**, which pumps it out to the lungs by way of the **pulmonary semilunar valve** and the **pulmonary artery**. The blood from the lungs returns through the **pulmonary vein**, passing through the **left AV valve** to the **left atrium**, which pumps the blood into the **left ventricle**, which pumps it out the **aortic semilunar valve** into the **aorta**.

Each blood vessel consists of a thin layer of cells called the **endothelium**, which surrounds a hollow **lumen**. The walls of **capillaries** are leaky and only one cell thick, while the walls of the larger vessels are nonleaky and thicker. If the collagen sheath of an artery weakens, the artery wall balloons outward in an **aneurysm**. In veins, such a swelling near a valve can cause varicose veins.

How does the lymphatic system work with the circulatory system?

Extracellular fluid carrying debris and dead cells flows into the **lymphatic system** as **lymph**. Immune-cell-rich **lymph nodes** filter out dead cells, bacteria, and other wastes and then return the purified fluid to the circulatory system.

What mechanisms control blood flow and heart rate?

The period when the muscles of the heart are contracted is called **systole**, and the period in between, when they are relaxed, is called **diastole**. A defective heart valve that allows blood to flow in the wrong direction can cause an audible **heart murmur**.

The cells of the atria contract almost as one, as do the cells of the ventricles. Electrical signals for such contractions are conducted between the two parts of the heart by the **AV (atrioventricular) node**. A pacemaker, called the **SA (sinoatrial) node**, sets the rhythm for the heart's contractions. **Fibrillation** is the disorganized contraction of the muscles of the heart.

The amount of blood that the heart pumps varies depending on activities and moods. The amount of blood pumped is a function of how fast the heart beats and how much blood it moves with each beat. The amount of blood pumped by the heart (the cardiac output) equals the **heart rate** times the **stroke volume**.

The **sympathetic** and **parasympathetic nervous systems** both influence heart rate. Sympathetic nerves mobilize the body in times of stress, speeding up the heart—by secreting **norepinephrine** around the cells of the heart's pacemaker. Norepinephrine also causes **vasoconstriction**—the contraction of the smooth muscles surrounding the arterioles. Heat, illness, or embarrassment can all decrease the activity of the sympathetic nerves, allowing **vasodilation.** Parasympathetic nerves slow the heart by secreting **acetylcholine.**

Norepinephrine and acetylcholine are **neurotransmitters**, while **epinephrine**, which acts like norepinephrine, is a hormone. **Histamine** dilates capillaries and increases their tendency to leak fluid, increasing both blood flow and swelling.

Review and Thought Questions

Review Questions

1. What advantages do you think an open circulatory system has over a closed system?
2. What is the role of the extracellular fluid? How is it like a subway platform?
3. How would you expect your blood pressure to change (a) if it is measured in your leg rather than in your arm? (b) after strenuous exercise? (c) after severe blood loss? (d) after you have been frightened? (e) after a big meal? (f) after you stand up following a nap?
4. How does acetylcholine affect the heart?
5. Why does an isolated heart keep beating, but an isolated arm muscle does not contract by itself?
6. How does the structure of a capillary relate to its role in the circulation?
7. Why does the heart need its own arteries?
8. What ensures that blood flows in one direction only? What would happen otherwise?
9. What does a four-chambered heart do that a three-chambered heart does not?
10. The chapter opening story says that Charles Drew died because the "major vein that delivers blood to the heart" was severed in the accident. What is the name of that vein?

Thought Questions

11. Suppose you cut a blood vessel in your arm. How could you tell whether you have severed a vein or an artery?
12. How does the circulatory system illustrate the idea that form follows function?
13. The heart is a soft tissue not likely to be fossilized, so no one knows what kind of heart dinosaurs had. Give two reasons why dinosaurs might have had a four-chambered heart. Give different reasons favoring a three-chambered heart.
14. Why couldn't we mammals send blood from veins to heart to lungs, managing with a simple two-chambered heart the way fishes do?
15. The box on p. 750 says that more women than men die of cardiovascular disease, but that the death rate for men is higher. Both facts are true. Explain this statistical paradox.

BiologyNow Resources

Biology(⑤)Now™

Active Figures
38-10: The human heart
38-14: Blood pressure

Preparing for an exam? Take a diagnostic test on your BiologyNow CD-ROM.

Online materials relating to this chapter are at:
http://biology.brookscole.com/AAL3

About the Chapter-Opening Image
A blood poster from World War II. Charles Drew developed methods for shipping blood plasma overseas, saving thousands of lives during the war.

How Do Animals Breathe?

Key Questions

- How do the physics of gases and liquids determine the structure of gills and lungs?

- How do we breathe?

- How do lungs resist infection, damage, and collapse?

- How does hemoglobin pick up oxygen in the lungs and drop it off in the tissues?

- How do blood and the lungs rid the tissues of waste carbon dioxide?

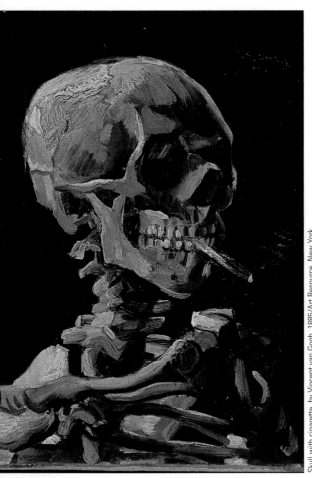

Skull with cigarette, by Vincent van Gogh, 1885/Art Resource, New York

Stanton Glantz and the Tobacco Industry

In 1995, a committee of the U.S. Congress singled out a scientist from among tens of thousands of others and canceled his research grant, astonishing medical researchers all over the United States. For Congress to assail a particular scientist in this way was virtually unheard of.

Normally, Congress gives the different government-funded research agencies a budget to distribute at their discretion. These agencies include the National Science Foundation, the National Aeronautics and Space Administration (NASA), the National Institutes of Health (NIH), and several others. The National Cancer Institute (NCI), by far the largest of the institutes at NIH, had awarded a three-year grant to Stanton Glantz, a professor of medicine at the University of California, San Francisco, to study the effects of public policy on tobacco use.

In particular, he had focused on how the tobacco industry works to keep Americans smoking. He then published a series of scientific papers in *JAMA*, the prestigious *Journal of the American Medical Association*, describing how the tobacco industry had for 30 years carefully concealed its knowledge that tobacco products are both deadly and addictive. Glantz's research also showed that elected officials who receive campaign contributions from tobacco companies were statistically more likely to vote in ways that helped the tobacco industry, a finding not likely to please some elected officials.

The Appropriations Committee of the U.S. House of Representatives argued that the NCI had no right to fund Glantz's work because it involved social science and political science. But Glantz argued that the tobacco industry was much like the mosquitoes that carry the disease malaria from person to person. Researchers say that a mosquito is a "vector" for malaria, and Glantz insisted, "You need to understand the vector of a disease in order to control it. The tobacco industry is the vector for lung cancer and heart disease."

For Glantz, the cancellation of his grant was only the latest in a series of conflicts with the tobacco industry. Much of the turmoil had begun on May 12, 1994, when a mysterious box arrived at Glantz's office. The box contained 4,000 pages of secret internal memos and research reports from a major tobacco company. The documents, Glantz later learned, had been stolen from the company by a paralegal who was astonished by what he was read-

Business and Biology   The Green Ball

In the 1930s, the makers of Lucky Strikes found to their dismay that women were passing up the company's cigarettes. The company suspected that the problem might be the cigarette's new green packaging, which clashed with most women's clothes. Green just wasn't popular. What to do?

The American Tobacco Company turned to Edward Bernays, the world's most famous public-relations man. Bernays's solution was to make green the most popular color in the United States. He held a "Green Ball" for well-connected women, and he enticed European fashion designers to create a collection of fashions, all in green. He held a Green Fashions Luncheon for fashion editors from major newspapers and magazines. He invented a Color Fashion Bureau that sent out press releases announcing that green was the color of the season. In only six months, Bernays succeeded in making green the top fashion color of 1934. Women no longer hesitated to buy Lucky Strikes because of the color, and sales increased.

Seventy years later, the situation is the same—and yet different. Cigarette manufac-turers today continue to pursue women aggressively, pushing cigarettes as glamorous and fashionable. In the summer of 2003, Legal Cigarettes paid the avant-garde fashion group As Four $20,000 to have its young models walk down the runway at an upscale New York fashion show—smoking cigarettes. Afterwards, everyone in the audience was to receive free cigarettes. But when the fashion models got wind of the plan, they refused to participate. As Four's four designers had to walk the runway smoking the cigarettes themselves.

The Granger Collection, New York

Figure A
Actress Jean Harlow in 1932—wearing green and smoking Lucky Strikes cigarettes. When this advertisement and others like it failed to attract female smokers, American Tobacco hired promoter Edward Bernays to convince American women that green was the fashion color (and Lucky Strikes, the fashion cigarette).

---

ing. The secret papers proved that the company, Brown and Williamson, had known since the 1960s that smoking is both deadly and addictive.

Brown and Williamson immediately went to court for the return of the documents, demanded the names of any people who had read them, and, allegedly, ordered a stakeout of the university library, where the documents were stored.

But it was too late. Copies of the documents had also arrived at several major newspapers and television networks, and the story became national news. A judge eventually ruled that the papers were in the public domain. In July 1995, the University of California at San Francisco posted the Brown and Williamson documents on the Internet, and Glantz's scientific papers analyzing the contents of the documents began appearing in *JAMA*. Within weeks of the first paper's publication, the House Appropriations Committee canceled Glantz's grant.

Why Was the Tobacco Industry Fighting So Hard?

If Brown and Williamson fought hard for the return of their papers, they had good reason. For the makers of Kools, Viceroys, and other popular cigarettes, the documents were acutely embarrassing. But more important, the documents could (and would) be used to attack the whole tobacco industry. A multibillion-dollar American industry, one that remains central to the economies of a half-dozen states, was in serious danger.

Until the late 19th century, tobacco smoking was viewed as a dirty habit, and only a few people smoked, nearly all of them men. In the 1880s, however, tobacco manufacturers discovered a way to process tobacco that made it possible to inhale without coughing. This allowed smokers to inhale deeply, increasing their exposure to nicotine, which is addictive. For many people, what had once been an occasional cigarette became a regular habit. During World War II, cigarettes were distributed to GIs, along with K rations and letters from home. Hollywood movies featured virile actors and alluring actresses lighting one another's cigarettes. Smoking was glamorous, seductive, romantic. More than just accepted, smoking became expected.

By the 1960s, cigarette sales in the United States had skyrocketed from 2 billion cigarettes per year in the 1880s to over 600 billion. And as cigarette sales skyrocketed so, eventually, did lung cancer death rates (Figure 39-1).

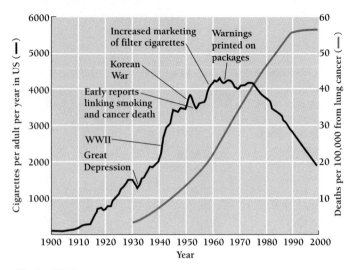

Figure 39-1

A rising death rate. As cigarette sales increased in the United States so did the death rate from lung cancer. In 1930 the annual lung cancer death rate was about 3 deaths per 100,000 people. Today, it is nearly 60 deaths per 100,000. Smoking is gradually going out of style in the United States, and the cancer death rate is finally leveling off. But tobacco companies have turned their sights on international markets, so smoking and smoking-related deaths are skyrocketing elsewhere.

The correlation between smoking and cancer was apparent as early as the 1880s, when a physician published a report suggesting that smoking caused cancer. Back then, when so few people smoked, doctors noticed that patients with cancer of the mouth, throat, or lungs were nearly always smokers. Researchers now know that smoking causes 90 percent of all cases of lung cancer.

But the first big surge in lung cancer deaths among American men didn't begin until the 1930s. Because lung cancer takes decades to develop, a 20-year lag appears between an increase in smoking and increase in lung cancer deaths.

In fact, in the United States, the lung cancer epidemic of the 1960s and 1970s was a public health disaster far greater in magnitude than the AIDS epidemic of the 1980s and 1990s. Doctors were horrified. Lung cancer is among the deadliest of all cancers, accounting for one-third of all cancer deaths (Figure 39-2). Untreated lung cancer patients live an average of eight months. Fewer than 15 percent of all lung cancer patients survive even five years. For comparison, 80 percent of women who get breast cancer survive five years, and many live much longer.

In 1964, the Surgeon General of the United States issued a report stating that cigarette smoking was hazardous to health. For the first time in 75 years, cigarette sales leveled off. Two years later Congress passed a bill requiring cigarette manufacturers to label their product as hazardous. For the first time ever, cigarette sales actually dropped.

It was a wakeup call for tobacco companies. Desperately, the tobacco industry began trying to develop a "safe"

cigarette. In the meantime, they did what they could to keep people smoking. Billions of dollars in sales and thousands of jobs were at stake.

Publicly, tobacco companies insisted that the scientific evidence for the cigarette-cancer link was weak. They blitzed television viewers with advertisements showing healthy young people smoking cigarettes against scenic backdrops—green meadows, waterfalls, and soaring mountains. The Brown and Williamson documents show, however, that the company's own research confirmed that cigarettes were both addictive and carcinogenic. Indeed, the quality of the tobacco industry's research was superior to that done by university researchers of the time.

In the late 1960s, the Federal Communications Commission (FCC) ruled that broadcasters must air one public service warning of tobacco's hazards for every four cigarette commercials. Viewers were suddenly besieged with shocking ads from the American Cancer Society, and cigarette sales dropped precipitously. Finally, in 1970, to get rid of the antismoking ads on television, cigarette companies pulled their own ads off the air. But by then millions of people had quit smoking. Between 1965 and 1979 alone, the percentage of smokers dropped from 42 percent of all adults to just 32 percent. Today, about 26 percent of men and 22 percent of women smoke.

The downward trend, however, has leveled off. And within certain groups, smoking is actually on the rise. For example, from 1980 to 1997, the percentage of high school seniors who smoked in the past month increased almost 20 percent (see box). Glanz, who is now director of the Center for Tobacco Control Research and Education at the University of California, San Francisco, thinks that one of the problems is that cigarette advertisements encourage young people to become addicted.

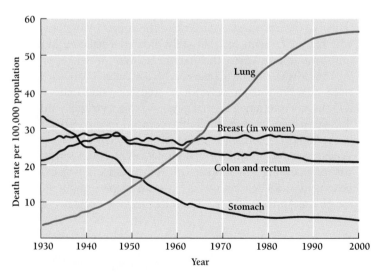

Figure 39-2

Lung cancer's separate path. Over the last 70 years, nearly every form of cancer has remained stable or declined. Only lung cancer has increased.

Business and Biology   Who Smokes and Why?

Smoking is the leading preventable cause of early death in the United States, by far. Each year, smoking kills nearly 500,000 people in the United States. More people die from smoking than die from all of the following combined: AIDS, car accidents, homicide, suicide, alcoholism, and all illegal drugs.

Tobacco smoke causes death in many ways—including cancer of the mouth, throat, larynx, esophagus, cervix, bladder, kidneys, and pancreas. Smoking also causes emphysema, a disease of the lungs in which people gradually suffocate to death. Tobacco-related lung cancer deaths add up to about 130,000 per year. But the great majority of smoking deaths are caused by cardiovascular disease, mainly heart attacks and strokes. Of the million deaths from cardiovascular disease each year, one-third (about 330,000) are caused by smoking.

In an effort to garner support from non-smokers, antismoking campaigns sometimes emphasize the effects of smoking on nonsmokers. It's true that about 3,000 non-smokers who die from lung cancer and an estimated 10,000 to 60,000 others who die of cardiovascular disease each year can probably blame secondhand smoke. And it's true that smokers are sick more often than nonsmokers—staying home sick 40 percent more often than nonsmokers. Their absence from work and their medical care together cost the United States some $65 billion a year.

But smokers are at far greater risk to themselves than to others. Their risk of a fatal illness is 30 or 40 times greater than that of nonsmokers. And although smokers cost society about 33 cents per pack of cigarettes, that figure is more than offset by the 52 cents per pack they pay in federal and state taxes. In short, it could be argued that nonsmokers reap nearly 20 cents a pack on smokers' self-destructive habit.

But the main reason smokers are not a drain on society's coffers is that smokers die young. Smokers work hard and pay for Social Security and Medicare benefits while they are working. But millions of nonsmokers live another seven years, on average, collecting benefits paid for by the work of smokers who are long gone.

The death rate among people who smoke more than a pack a day is more than twice that of nonsmokers. Yet, amazingly, smokers who are able to quit can cut their risk of death from heart disease by 50 percent—in the first year alone. After 10 or 15 years, even a former smoker's risk of lung cancer is only slightly higher than that of a lifelong nonsmoker.

Why do people take up smoking and keep smoking? It is illegal to sell tobacco products to anyone younger than 18. And yet adults rarely take up smoking, and those who do soon quit. Ninety percent of people who begin smoking after age 21 later quit. Most lifelong smokers—about 75 percent—begin smoking between the ages of 12 and 17.

We can conclude that if tobacco companies did not market cigarettes to those under 18, the market for cigarettes would soon disappear. Who would take up smoking? Certainly not adults. In a book about the tobacco industry, *The New York Times* reporter Philip Hilts wrote, "If it were true that the [tobacco] companies steer clear of children, as they say, the entire industry would collapse within a single generation."

Why do people keep smoking? Once people begin smoking, they find it difficult to quit. This is because an important component of tobacco smoke, nicotine, is a drug similar to heroin in its addictiveness. A cigarette is a system for delivering nicotine to the body; each puff on a cigarette delivers between 0.5 and 2.0 milligrams of nicotine to the bloodstream. Because the nicotine is inhaled, it reaches the brain within six or seven seconds, twice as fast as an injected drug.

Smokers say nicotine makes them feel good. The drug relaxes the muscles and increases alpha waves, brain waves that characterize a pleasant, relaxed state. If you want to know what it feels like to increase alpha waves in the brain, sit down in front of the television for a few minutes.

Within only days or weeks of starting to smoke even occasionally (less than daily), the symptoms of addiction can begin to surface in adolescents. Ninety-five percent of smokers are physically addicted, meaning that they experience physical withdrawal—intense physical discomfort—if they try to stop smoking. People who have been addicted to both heroin and cigarettes say that cigarettes are harder to give up. Most people who succeed in giving up smoking try to quit several times before they succeed.

By far the greatest surge in smoking is occurring in developing countries, where tobacco consumption is rising by 3.4 percent each year. In China, the fraction of men who smoke has reached an alarming 67 percent. If current trends continue, worldwide tobacco-related deaths could reach 10 million per year by the late 2020s, with more than 70 percent of deaths occurring in developing countries.

Efforts to curb this trend are under way. In January 2003, the World Health Organization completed a draft of a global treaty to reduce tobacco consumption. The treaty includes measures to control tobacco advertising, sponsorship, and smuggling.

Improving people's understanding of the healthy functioning of the lungs and the cardiovascular system, and how they can be altered by tobacco, may also help. In this chapter we will see how healthy lungs (and gills) work to absorb oxygen and rid the body of carbon dioxide and how the circulatory system moves oxygen and carbon dioxide between the lungs and the cells of the body. Along the way, we'll see how the lungs, heart, and blood vessels are all interconnected.

39.1 How Do Animals Exchange Carbon Dioxide and Oxygen?

Animals usually get most of their energy from cellular respiration, a biochemical process that requires oxygen (Chapter 6). Because cellular respiration takes place in almost every cell of the body, animal cells need to be able to acquire oxygen and distribute it to all the cells of the body. Because cell respiration generates carbon dioxide, a toxic waste product, animals must also be able to rid themselves of carbon dioxide.

Aquatic animals such as fish exchange oxygen and carbon dioxide across the surfaces of gills. Terrestrial animals such as birds, mammals, and reptiles all use lungs. In both gills and lungs, oxygen diffuses passively into the blood and carbon dioxide diffuses from the blood to the lungs or gills. The circulatory system carries oxygen from lungs or gills to the different tissues of the body and carbon dioxide from tissues to lungs or gills.

Lungs and gills both take up oxygen and release carbon dioxide.

Can Animals Take Oxygen Directly from the Air?

Oxygen is far more plentiful in the air than in water. In air, 21 percent (or approximately 1 of every 5 molecules) is oxygen, whereas in water, only 0.0006 percent (or 1 in 170,000 molecules) is oxygen.

Land animals have flourished so well, in part, because they are able to take oxygen directly from the atmosphere. For oxygen to be useful, however, it must first dissolve in water, since only dissolved oxygen can participate in the biochemical reactions of a cell. A land animal's first task, then, is to create a thin, wet surface on which oxygen can dissolve. Aquatic animals use oxygen that is already dissolved in the water around them. But the oxygen in water ultimately comes from the air.

But both land animals and aquatic animals need oxygen that has dissolved in water. Oxygen is far more plentiful in air than in water.

How Do Gases Exert Pressure?

In a gas such as air, the average distance between molecules is more than 10 times the distance between the molecules in a liquid. Because the molecules in a gas are so far apart, they interact very little with one another. Instead, each molecule moves about at random, with an average speed that depends on the temperature: the higher the temperature, the faster each molecule moves.

When a molecule in a gas strikes the surface of a liquid, it will usually bounce off into the space above the liquid. Each

time a gas molecule strikes a surface, it exerts a tiny force. All the gas molecules together exert a pressure on the surface of the liquid. The total pressure depends on the number of gas molecules that strike the surface and on the speed with which each molecule strikes the surface (that is, the temperature).

Each molecule contributes independently to the total pressure. But air is made up of different kinds of molecules. About 21 percent of the molecules are oxygen (O_2), about 0.03 percent are carbon dioxide (CO_2), and almost all the rest are nitrogen (N_2). Each kind of gas exerts a separate pressure, depending on the concentration and temperature of the molecules. Physiologists describe the pressure exerted by a type of molecule in a mixture of gases as the **partial pressure**, the pressure exerted by that one type of molecule (Figure 39-3). The total pressure is the sum of all the partial pressures. Air pressure is measured in "millimeters of mercury" (mm Hg) because columns of mercury can be used to measure pressure. In air, the partial pressure of oxygen is about 160 mm Hg, that of nitrogen about 600 mm Hg, for a total of about 760 mm Hg (normal air pressure). Carbon dioxide makes up such a small part of our atmosphere that its partial pressure is a barely noticeable 0.2 mm Hg.

Each kind of molecule in a mixture of gases has a different partial pressure.

How Much Dissolved Gas Will a Solution Hold?

We said earlier that when a gas molecule strikes the surface of a liquid, it usually bounces off. Sometimes, however, the gas molecule penetrates the surface and becomes part of the liquid. We say the gas "dissolves" in the liquid. Such a dissolved gas molecule can escape from the solution and be-

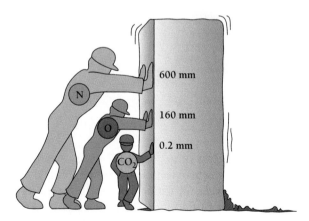

Figure 39-3
Partial pressure. The pressures of the different gases in the atmosphere add up to about 760 mm. About 78 percent of air is plain nitrogen, so it contributes a partial pressure of 600 mm Hg. Oxygen has a partial pressure of about 160 mm. The minute amounts of carbon dioxide gas in our atmosphere contributes a minuscule partial pressure of just 0.2 mm.

come a gas molecule again. When the number of gas molecules entering and leaving the solution is the same, we say the molecules have reached **equilibrium.**

The concentration of a molecule in a solution at equilibrium depends on the partial pressure of the gas and the temperature of the air and water. The higher the partial pressure of oxygen in air, the harder and faster the molecules hit the water and the more of them enter and dissolve, raising the concentration in the water.

Most molecules of a gas bounce off the surface of a solution, but some enter and dissolve. The amount of dissolved gas in a solution depends importantly on the partial pressure of the gas.

amount of water in a narrow-necked bottle. Similarly, oxygen exposed to water in an open pan will dissolve and equilibrate much more rapidly than oxygen exposed to the same amount of water in a narrow-necked bottle.

The same holds true for cells. Being small increases the ratio between surface area and volume, a cell's **surface-to-volume ratio** (Chapter 4). A one-celled organism such as an ameba can absorb all the oxygen it needs (Figure 39-4A). Even tiny multicelled animals such as rotifers or nematodes—less than the size of a period—can obtain enough oxygen by diffusion alone. But larger animals need special surfaces through which oxygen can diffuse. All animals increase surface area for gas exchange in one of two ways. They either

Why Is Surface Area Important in Gas Exchange?

If we know the temperature and partial pressure of oxygen, we can estimate its eventual concentration in a solution such as blood. What we don't know is how long it will take for the oxygen to reach that concentration.

The rate at which molecules of a gas enter a solution depends on three factors: temperature, partial pressure, and surface area. Both the molecules of a gas and the molecules dissolved in a solution move randomly in a process called diffusion (Chapter 4). In the lungs, oxygen spontaneously diffuses from a region of high partial pressure (the air in the lungs) to a region of low partial pressure (the blood).

The amount of oxygen that moves across a boundary in a given time (per second, for example) is proportional to the difference in partial pressures on the two sides of the boundary. That is why hospital patients are sometimes given high concentrations of oxygen to breathe; at high concentrations, the partial pressure of oxygen is higher and the oxygen molecules more quickly dissolve in the fluid covering the lungs.

Finally, large surface areas also speed the exchange of molecules. We understand intuitively that water in an open pan will evaporate more quickly than the same

Figure 39-4

Gills and lungs. Animals need a high surface-to-volume ratio to speed the uptake of oxygen. A. Organisms like this ameba that are smaller than a millimeter, about the size of a period, have high surface-to-volume ratios and need not expand their surface area. B. In larger aquatic animals, such as this sea star, gills evaginate and increase the surface area for oxygen exchange. C. The extensive gills of fish can extract a lot of oxygen in a short time. D. Spiders develop internal gill-like structures, called "book lungs." E. The lungs of amphibians, reptiles, birds, and mammals arise by invagination (infolding) rather than evagination (outfolding).

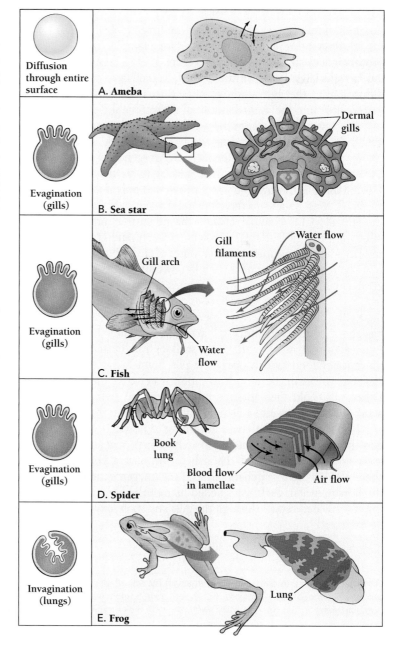

fold a surface outward (evagination) or fold a surface inward (invagination). Evaginated breathing structures are called **gills,** and invaginated breathing structures are called **lungs** (Figure 39-4).

> The rate of gas movement into a cell or an organism depends on the surface area through which the gas can move. Evaginated breathing structures are called gills, and invaginated breathing structures are called lungs.

39.2 How Do Aquatic Organisms Obtain Dissolved Oxygen?

Aquatic animals larger than about a millimeter obtain dissolved oxygen through gills. A sea star, for example, has tiny protuberances all over its surface (Figure 39-4B). A sea scallop has specialized gills within its shell. In each case, oxygen diffuses from the surrounding water across the expanded surface area into the animal's internal environment.

As oxygen diffuses from the water into the surface of the gills, however, the surrounding water loses oxygen. To exchange the oxygen-poor water for new, oxygen-rich water, aquatic animals either move their gills or move the water. Moving the gills through the water works well only if the gills are small. Most organisms therefore move water over the gills. In the scallop, for example, cilia on the gill surface move the water past the gill surfaces. Lobsters, crayfish, and crabs all use large internal paddles to move water over their gills.

The oxygen requirements of a fast-swimming fish are far greater than those of most invertebrates and slower-moving fish. So fast-swimming fishes have enormous surface areas on their gills (Figure 39-4C). The speedy mackerel, for example, has about 50 times as much gill area as a slow-moving bottom feeder such as a flounder. Like scallops and lobsters, fish move water continuously over their gills. Some fish use muscles and valves in the mouth to pump water over the gills. Others just hold their mouths partly open and swim constantly. Water flows into a fish's mouth and out past the gills.

Fish adjust their behavior according to how much oxygen they need. In one study, researchers kept mackerel in moving water that forced the fish to swim at a constant speed. When the researchers reduced the oxygen content of the flowing water, each fish opened its mouth wider, allowing more water to pass through the gills and more oxygen to be extracted.

> Dissolved oxygen can diffuse into an organism by way of gills. Many animals replace oxygen-depleted water with oxygen-rich water by pumping water across the gills or by swimming with their mouths wide open.

Why Can't Land Animals Breathe with Gills?

Why do fish die when they are taken from the water? Why can't gills extract oxygen from the air? Actually, some fish survive quite well out of water. An eel, for example, can breathe if it stays cool and moist. Eels can even crawl over land from one body of water to another. But, in air, most of an eel's oxygen comes through its skin, not through its gills.

The problem with gills out of water is that they collapse and flatten. Because gills lack mechanical rigidity, their surfaces fold and stick together when out of water, so that they cannot provide an expanded surface area for the diffusion of gases. Most gills also dry out. As they dry, they lose the thin layer of water necessary for oxygen to dissolve into.

One solution to this problem is to keep the gills inside the body. Spiders, for example, have stacks of internal plates called a "book lung" (Figure 39-4D). Book "lungs" are not actually lungs. They evaginate outward, like the pages of a book, not inward, like a lung. Anatomical comparisons suggest that they evolved from gills. Supporting bars keep the plates from collapsing.

Another solution is to breathe through the skin or some other broad surface. Some kinds of lungless salamanders, for example, breathe entirely through their skin, which is thin and rich in blood vessels. And some amphibians absorb air through the skin inside their mouths. But most land animals, including most amphibians, have inward-folded lungs (Figure 39-4E).

> Out of water, gills both dry out and collapse, which decreases their effective surface area.

39.3 How Do Air-Breathing Organisms Obtain Oxygen?

Aside from spiders, air-breathing organisms nearly all obtain oxygen from infolded (invaginated) surfaces that do not collapse easily. Land snails, many amphibians, and all reptiles, mammals, and birds depend on lungs, localized organs of gas exchange that are always associated with the circulation (Figure 39-4E). Lungs are extensions of tubes within an animal's body. They lie within the body cavity, where they are more protected than gills and less likely to collapse and dry out.

> Almost all air-breathing vertebrates depend on moist lungs to acquire oxygen.

By What Path Does Air Enter the Lungs?

Lungs do not function by themselves. They are one part of a **respiratory system**, which consists of all the structures responsible for the exchange of gases between the blood and the external environment. In vertebrates, the main components of the respiratory system are the lungs, the airways, and the muscles that move the air (Figure 39-5A). In addition, the circulatory system, which carries both oxygen and waste carbon dioxide, is also a critical part of respiration.

Air enters the body through the mouth and nose. These two openings lead to the throat, or **pharynx**, a common passage for food and air. Food and air diverge into two separate branches. The *esophagus* leads to the stomach, and the *larynx* leads to the lungs. The entrance to the larynx, called the glottis, is shielded by a flap called the epiglottis, which prevents food from entering the air tube during swallowing (Figure 39-5A).

The larynx contains the *vocal cords,* folds of membrane that vibrate as air passes over them, producing sound. The larynx leads directly into a long tube, called the **trachea**, or windpipe. The trachea enters the chest, where it forks into two smaller tubes, the **bronchi**, which lead to the two lungs. Inside the lungs, the bronchi branch into smaller and smaller extensions, with thinner walls. The smallest of the bronchi are the **bronchioles**, of which there are millions (Figure 39-5C and D). No gas exchange takes place in the larynx, the bronchi, or the bronchioles: they are only passages to the working surfaces of the lungs.

The surfaces on which oxygen dissolves are called **alveoli** [singular, alveolus; from Latin, *alveus* = a hollow], hollow sacs that are richly supplied with blood. In the alveoli, red blood cells in the passing blood pick up oxygen and carry it to the rest of the body. Each lung of a healthy young adult contains about 300 million alveoli. The fine branching of the air passages and alveoli greatly increases the surface over which oxygen molecules come into contact with cell surfaces. If the inside of each lung were like a smooth, hollow sphere instead of spongy mass, the surface area of the lungs would be not much larger than a regular sheet of paper. Instead, the lungs have a surface area similar to that of a tennis court.

The mammalian respiratory system consists of finely divided airways leading to hundreds of millions of blind sacs (alveoli) where oxygen is absorbed.

How Do We Breathe?

In mammals, the alveoli are dead ends. Air leaves each alveolus the way it arrived, through a bronchiole. Only by actively breathing can we move air in and out of the alveoli. At rest, before we inhale, our lungs contain about 2,400 milliliters (ml) of air. As we inhale (while sitting and reading) we regularly draw in and expel another 500 ml of air, so that the total fluctuates between 2,400 ml and 2,900 ml. The 500 ml of air we draw in and out during relaxed, quiet breathing is called the tidal volume. But during vigorous exercise, the maximum amount of air inhaled and exhaled, called the **vital capacity,** can reach 3 to 5 liters (Figure 39-6).

We can never expel all the air in our lungs and fill our lungs with completely fresh air. If we exhale forcefully as much air as we possibly can, we can expel all but about 1,200 ml of air, which remains in the alveoli. This air that we cannot expel (unless our lungs collapse) is called the residual volume. Freshly inhaled air always mixes with this residual air, which has already taken up some carbon dioxide and lost some oxygen.

Whereas amphibians and some reptiles push, or gulp, air into the lungs using muscles and valves in their mouths, mammals, birds, and most reptiles pull air into the lungs using muscles of the closed **thoracic cavity,** or chest cavity. Surrounding the thoracic cavity of humans are 12 pairs of ribs, supported by the spinal column and the sternum, or breastbone. Muscles and elastic connective tissue interconnect all the ribs and form the walls of the chest. At the bottom of the thoracic cavity is the **diaphragm,** a thick sheet of muscle that separates the thoracic cavity from the abdomen (Figure 39-5A). The lungs themselves are surrounded by a fluid-filled sac attached to the walls of the chest by connective tissue.

During **inspiration,** or inhalation, the diaphragm contracts and moves downward (Figure 39-6). This increases the space inside the chest cavity, decreasing the pressure in the lungs. The rib muscles contract, which expands the chest even more. The air pressure of the outside air now exceeds that inside the lungs, so air flows into the lungs. During **expiration,** or exhalation, the diaphragm relaxes upward into the chest cavity. When it is relaxed, it has a steep dome shape. The ribs resume their relaxed positions, which further contributes to a decrease in the size of the cavity. As the volume decreases, the pressure increases, and air flows outward. At rest, we have only to relax our muscles and air flows outward.

During exercise, we increase the flow of air by forcefully expelling (and then inhaling) more air. We do this by contracting muscles between the ribs and in the abdomen that boost the lungs' internal pressure.

Breathing causes tides of air to move in and out of the alveoli. Mammals pull air into the lungs by expanding the thoracic cavity. They push air out of the lungs by relaxing the muscles of the thoracic cavity.

What Keeps the Airways Moist and Clean?

Tightly connected epithelial cells line both the nonabsorbing surfaces of the airways and the absorbing surfaces of the alveoli. As we might expect from their differing roles, the cells of the airways differ from those of the alveoli (Figure 39-5 B and C).

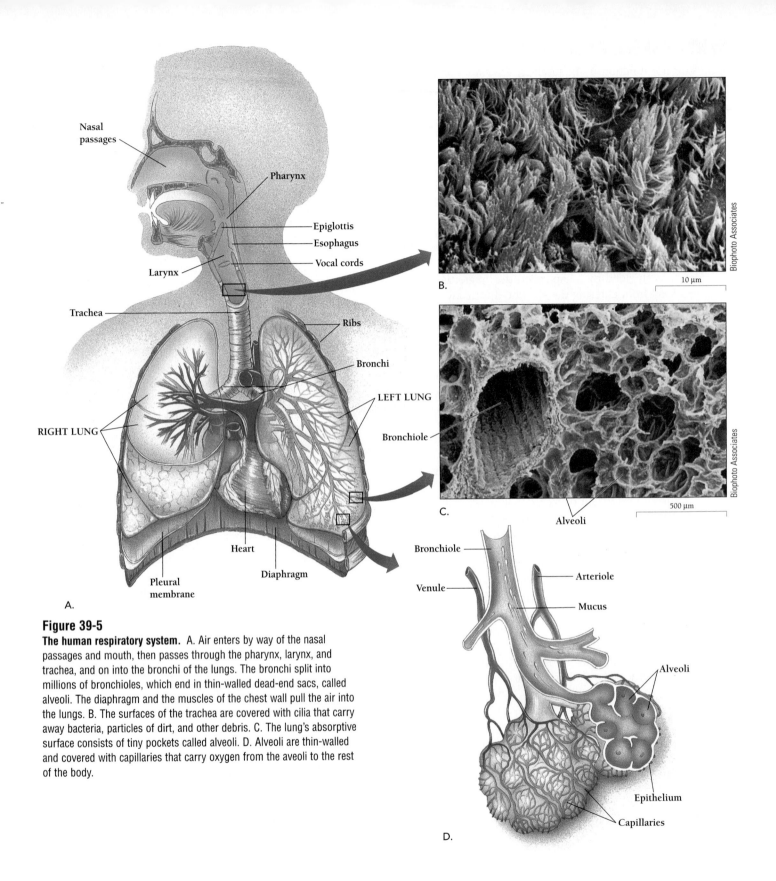

Figure 39-5

The human respiratory system. A. Air enters by way of the nasal passages and mouth, then passes through the pharynx, larynx, and trachea, and on into the bronchi of the lungs. The bronchi split into millions of bronchioles, which end in thin-walled dead-end sacs, called alveoli. The diaphragm and the muscles of the chest wall pull the air into the lungs. B. The surfaces of the trachea are covered with cilia that carry away bacteria, particles of dirt, and other debris. C. The lung's absorptive surface consists of tiny pockets called alveoli. D. Alveoli are thin-walled and covered with capillaries that carry oxygen from the aveoli to the rest of the body.

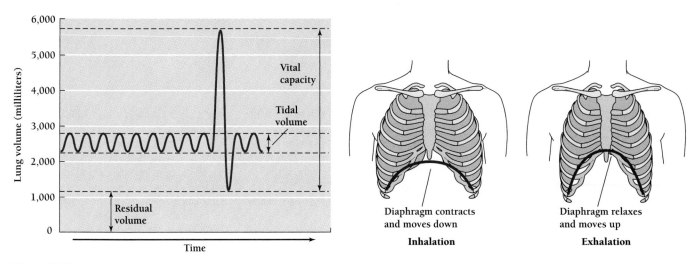

Figure 39-6
Take a deep breath. No matter how hard we exhale, about 1,200 ml of residual air is always left in the lungs. At rest, we take small breaths, and the tidal volume of air entering and exiting the lungs is about 500 ml. The harder we exercise, the larger the volume of air we take in and expel. A person in good condition can inhale and exhale about 5,000 ml per breath.

Biology🅔Now™ Learn more about respiration by clicking on this figure on your BiologyNow CD-ROM.

Most of the cells that line the trachea, bronchi, and bronchioles (but not those of the alveoli) have fringes of constantly beating **cilia** that carry away dirt, bacteria, viruses, and mucus. Many other cells in the airway linings secrete **mucus**, a thick solution of glycoproteins like that secreted by the membranes of the nose. Mucus keeps the surface of the airways wet, which in turn keeps the air within the airways moist and prevents the surfaces of the alveoli from drying out.

Mucus also helps protect the lungs from becoming clogged with dust particles or infected with bacteria. Dust particles, often carrying bacterial passengers, stick to the mucus, and the beating cilia drive the dust-laden mucus out of the airways into the throat. Swallowing then pushes the mucus into the digestive tract. The ciliated epithelium of the airways serves as an escalator for particles that would otherwise accumulate in the blind alveoli, where they could interfere with gas exchange and cause infection. The airways also contain scavenger white blood cells called macrophages that engulf the accumulated debris in the airways.

The mucus in the respiratory tract traps dust and bacteria, which cilia propel towards the mouth. Macrophages also engulf debris in the airways.

How Does Tobacco Smoke Damage the Lungs?

The three most dangerous substances in tobacco smoke are nicotine, carbon monoxide, and tar. Tar—a brown, oily substance like that in a chimney or a car's exhaust pipe—damages the lungs. Nicotine and carbon monoxide work together to damage the heart and arteries (see box on page 764).

Lungs need to be clean in order to collect oxygen effectively. So tar interferes with lung function immediately. Within minutes of the first puff on a cigarette, tobacco smoke paralyzes both the macrophages and the cilia so that they cannot clear the airways. One cigarette paralyzes the cilia for about an hour. Regular smoking destroys them outright. At the same time, the irritating smoke causes the lungs to secrete extra mucus, which accumulates in the lungs and clogs the alveoli. The tar in the smoke settles throughout the lungs, irritating the lungs and impairing the lungs' ability to absorb oxygen.

The only way to clear the lungs is for a smoker to cough, the classic "smoker's cough." The lungs, clogged with mucus and tar and unable to rid themselves of bacteria and viruses, become infected and inflamed. Smokers have frequent respiratory infections, especially bronchitis and even pneumonia. These chronic infections, combined with continual coughing, further damage the lungs.

The constant coughing of chronic bronchitis actually breaks the walls of the individual alveoli. Nitric acid and sulfuric acid in the burning tobacco further weaken the walls of the alveoli. As the walls of the alveoli break, the many small, bubblelike chambers become a few larger chambers, and the lung gradually loses its enormous surface area.

In the disease *emphysema,* so many alveoli are lost that the lungs can no longer absorb enough oxygen to support life. In the last weeks or months of life, a person with emphysema must breathe from an oxygen tank. Finally, even that is not enough.

Tobacco smoke tar also causes cancer, the unregulated growth of the body's own cells. Tobacco smoke contains at

Health and Biology How Does Tobacco Smoke Damage the Cardiovascular System?

Two of the three most dangerous substances in tobacco smoke—nicotine and carbon monoxide—damage the heart and arteries rather than the lungs. Nicotine is a stimulant with a broad range of effects. It increases the heart rate and constricts the blood vessels, causing the blood pressure to increase. Nicotine also stimulates the release of antidiuretic hormone (ADH), a hormone that causes the kidneys to retain water, which can further raise blood pressure. In addition, nicotine increases the tendency of the blood platelets to form clots. High blood pressure and abnormal clots are two factors that can trigger a heart attack or stroke.

Nicotine contributes to hardening of the arteries, a major cause of heart attacks and strokes. By increasing blood pressure, nicotine causes damage to the arteries, which then become inflamed and scarred. Finally, nicotine increases levels of cholesterol in the blood, which, in response to the inflammation, forms deposits on the arterial walls. These deposits further narrow the arteries and increase the likelihood that a blood clot will plug an artery and cause a heart attack or stroke.

While nicotine's addictiveness maintains the cigarette habit and raises blood pressure, the carbon monoxide in tobacco smoke interferes with the body's ability to supply oxygen to the cells of the body. Carbon monoxide is an odorless and toxic gas. Cigarettes are not the only source of carbon monoxide. Exhaust from automobiles and furnaces, for example, also release carbon monoxide. Government regulations outlaw levels higher than 10 parts per million (ppm) in homes or at work. Yet, cigarette smoke contains about 1,600 ppm carbon monoxide.

In the blood, carbon monoxide from cigarette smoke binds to hemoglobin and displaces oxygen. Inhaling carbon monoxide reduces oxygen supplies to every cell in the body, but the effects are most dangerous in the heart and brain. Oxygen deprivation in the brain impairs judgment, vision, and the ability to distinguish sounds, and oxygen deprivation in the heart reduces its ability to pump blood. Unfortunately, the high blood pressure induced by nicotine means that the heart needs to work harder than usual. The deadly combination of nicotine and carbon monoxide in cigarettes cause long-term damage to the heart. Cardiovascular diseases caused by cigarettes kill roughly 330,000 Americans each year.

---

least 50 different known *carcinogens,* chemicals that cause cancer. Because tobacco smoke ruins the lungs' ability to cleanse themselves, these carcinogens permanently coat the lungs. They can damage genes that control cell division, so that the cells divide without restraint. They can also damage enzymes that help regulate cell division. Some carcinogens enhance the effect of the other carcinogens. In time, small tumors develop, which eventually metastasize, or spread, to other areas of the body. Invasion of nearby nerves by cancer cells, for example, may cause partial paralysis.

How Do Alveoli Overcome the Surface Tension That Resists Expansion?

Each inhaled breath pulls air into the alveoli, expanding the volume of each of these tiny sacs. The small size of each alveolus enables the lungs to have lots of them, increasing the area available for gas exchange. But the small size of each sac also increases the difficulty of expanding it, just as small balloons are harder to blow up than giant ones.

A liquid's resistance to an increase in surface area is called surface tension. To overcome this resistance, the alveoli produce a **surfactant,** a detergent that reduces surface tension and allows expansion. Infants born prematurely often lack surfactant. These babies can breathe only with great difficulty and are said to be suffering from hyaline membrane disease. The prospects for survival are excellent, however, if the baby's doctor supplies a substitute surfactant.

Surfactants within alveoli reduce surface tension and allow the lungs to expand more freely.

How Do the Lungs Foster Gas Exchange?

Endotherms, animals that keep their bodies warm with the heat released by metabolism, burn more calories than other animals. Birds and mammals are endotherms, and therefore have high metabolic rates, requiring large amounts of oxygen and generating large amounts of carbon dioxide.

We have seen that the large surface area created by the alveoli helps speed the diffusion of oxygen and carbon dioxide to and from the blood. Other structural characteristics of the lungs also speed diffusion. The epithelial cells lining the alveoli are so thin that molecules of oxygen and carbon dioxide easily pass through. A dense network of tiny capillaries surrounds the alveoli, increasing the total blood surface area. Each capillary is so narrow that the red blood cells must squeeze through one at a time, their passage lubricated by a thin layer of blood plasma.

The same plasma leaks out of the capillaries and into the alveoli, helping to wet the inside surface of the lungs. Oxygen dissolves in this liquid and diffuses across the thin epithelium of the alveoli and across the thin walls of the capillaries to the blood (Figure 39-7). Once in the blood, the oxygen molecules diffuse into the red blood cells and bind to hemoglobin. The total distance diffused—from the inside of an alveolus to the inside of a capillary—is about 1/100th the thickness of a hair.

The blood carries the oxygenated hemoglobin through the pulmonary veins to the heart for distribution to the rest of the body. Carbon dioxide follows roughly the reverse path, from blood to alveoli to exhaled air.

The cellular structure of the alveoli promotes the diffusion of oxygen and carbon dioxide to and from the blood by providing a large, thin surface area associated with a rich blood supply.

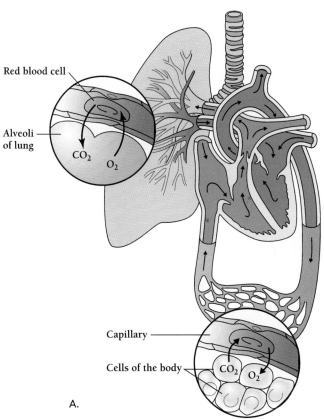

Figure 39-7

How we breathe. The human heart pumps blood into the lungs, where red blood cells pick up oxygen diffusing from the alveoli. Oxygen molecules pass through the epithelium of the alveolus and the endothelium of the pulmonary capillary. Inside the capillary, the oxygen enters the red blood cell and binds to hemoglobin. The red blood cells carry the oxygen back to the heart, which pumps it to all the tissues of the body. Any cells that need oxygen unload the precious cargo from the red blood cells. Meanwhile, the same circulating blood picks up carbon

Additional Ways to Exchange Gases

Among all the vertebrates, birds are the best at absorbing oxygen from the air. Many birds routinely fly long distances at altitudes above 6,000 meters (nearly 4 miles), far above where most mammals can live. Birds' great advantage is that they move air through their lungs in one direction. Mammals move air in and out, and much of the air in the lungs is residual air that is not fresh.

When a bird takes a breath, the air passes into a group of sacs called the posterior air sacs, which, like our own bronchi, do not allow gas exchange (Figure 39-8 ❶). When the bird exhales, the contents of the posterior air sacs move forward into the lungs ❷. In the next breath, air moves out of the lungs into another set of sacs, the anterior air sacs ❸. Finally, in the second exhalation, the contents pass back out into the atmosphere ❹. Air moves in a single direction through the lungs, and the air that enters the lungs is fully oxygenated.

In all vertebrates and most invertebrates, circulating blood distributes dissolved oxygen. Insects have a bloodlike fluid, which carries glucose and other materials, but, unlike our blood, it does not transport oxygen or carbon dioxide. Nor do insects have lungs or gills. Instead, insects move oxygen through a network of open passages called **tracheae** that carry oxygen directly to the tissues and carry carbon dioxide back out of the body (Figure 39-9). Air enters the body through openings called **spiracles,** which lead directly to the tracheae. This separate system for distributing oxygen more than compensates for the limitations of insects' open circulatory systems (Chapter 38). Until recently, biologists

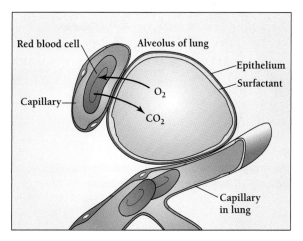

dioxide waste and carries it back to the lungs. Carbon dioxide exits the lungs by moving from the blood cells and plasma through the walls of the capillary and the alveolus. Once the carbon dioxide is in the alveolus, we eventually breathe it out.

Biology ⒮Now™ Learn more about gas exchange by clicking on this figure on your BiologyNow CD-ROM.

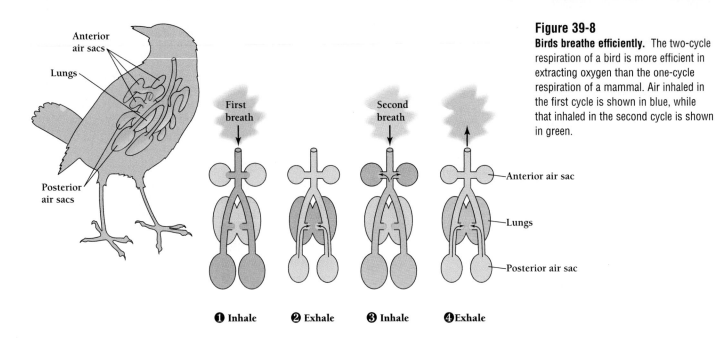

Figure 39-8
Birds breathe efficiently. The two-cycle respiration of a bird is more efficient in extracting oxygen than the one-cycle respiration of a mammal. Air inhaled in the first cycle is shown in blue, while that inhaled in the second cycle is shown in green.

Anterior air sacs

Lungs

Posterior air sacs

First breath

Second breath

Anterior air sac

Lungs

Posterior air sac

❶ Inhale   ❷ Exhale   ❸ Inhale   ❹ Exhale

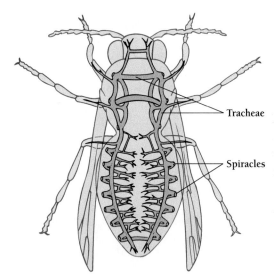

Figure 39-9
Insects breathe through tracheae. Instead of distributing dissolved oxygen, insects pump air itself through an extensive system of tracheae.

Tracheae

Spiracles

thought that the oxygen moved passively through the tracheae, but x-ray videos have revealed that many insects actively breathe by expanding and contracting the tracheae, much as we do our lungs. Crickets and ants can exchange up to 50 percent of the air in their tracheae every second.

In a bird's lungs, air flows in a single direction, allowing birds to take up oxygen more rapidly than mammals. Insects pump air through tracheae to all parts of their bodies.

39.4 How Does Oxygen Get to the Tissues?

In most animals, the blood carries oxygen to tissues throughout the body. If the blood consisted only of salts and water, it would not carry much oxygen, for little oxygen dissolves in plain water. So most animals have special oxygen-binding proteins. The presence of these proteins can increase the oxygen-carrying capacity of blood up to 50 times.

How Do Red Blood Cells Transport Oxygen?

Hemoglobin carries oxygen molecules in the blood much as a raft transports passengers down a flowing stream. In invertebrates, the oxygen-carrying pigments, such as hemocyanin, are usually dissolved directly in the blood. In vertebrates, however, hemoglobin molecules are packed into red blood cells.

The red cells are only containers for hemoglobin. Each red cell is like a transport raft, with no life of its own. It lives only a few weeks or months and lacks mitochondria and, in mammals, even a nucleus (Figure 39-10A). But a red cell may hold tens of millions of hemoglobin molecules.

Each molecule of hemoglobin consists of four polypeptide chains folded together. Each polypeptide enfolds a small molecule, called a **heme,** at the center of which is a single atom of iron (Figure 39-10B). One oxygen molecule can bind to each heme, so that each hemoglobin molecule can bind up to four oxygen molecules. Hemoglobin with one or more bound oxygen molecules is called **oxyhemo-**

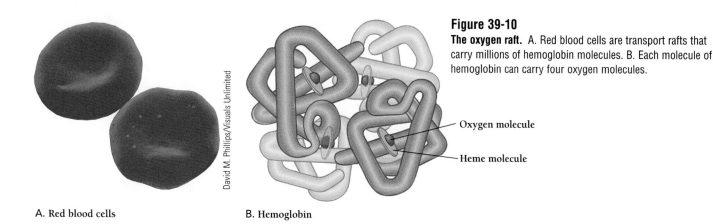

David M. Phillips/Visuals Unlimited

Figure 39-10

The oxygen raft. A. Red blood cells are transport rafts that carry millions of hemoglobin molecules. B. Each molecule of hemoglobin can carry four oxygen molecules.

Oxygen molecule

Heme molecule

A. Red blood cells

B. Hemoglobin

globin. Hemoglobin with no oxygen is called deoxyhemoglobin, or simply hemoglobin.

Oxygen-binding proteins greatly increase the oxygen-carrying capacity of blood.

How Does Hemoglobin Load and Unload Oxygen?

The partial pressure of oxygen in the alveoli is about 100 mm Hg, while the partial pressure of the deoxygenated blood coming into the lungs is only 40 mm Hg. As a result, the oxygen molecules in the alveoli passively diffuse into the blood (Figure 39-11). Because the surface area of the lungs is large, the oxygen molecules quickly reach equilibrium, and the partial pressure of oxygen in blood leaving the lungs is as high as that in the alveoli—100 mm Hg.

Once the oxygen is inside the capillaries of the lungs, the red blood cells quickly carry it to the heart to be pumped to the different tissues of the body. Tissue cells usually have a partial pressure of about 40 mm Hg, so the oxygen in the blood diffuses down the gradient into the tissues.

In the oxygen-rich tissues of the lungs, hemoglobin binds oxygen and takes it away. Indeed, by the time blood leaves the lungs, nearly all the hemoglobin carries oxygen on all four hemes. Physiologists say it is "saturated." Yet, in tissues with little oxygen hemoglobin easily unloads its cargo of oxygen. For hemoglobin to deliver oxygen, it must not only take up oxygen but also let it go.

How can a hemoglobin molecule respond so flexibly? The answer lies in its ability to change back and forth between two alternate three-dimensional structures. In one form, two of its four polypeptide chains move close together and the whole molecule easily binds oxygen (and does not easily let it go). In the other form, the four chains separate from each other and hemoglobin is only moderately likely to bind an oxygen.

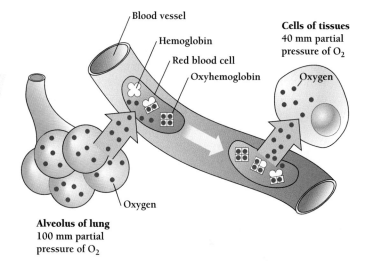

Blood vessel

Hemoglobin

Red blood cell

Oxyhemoglobin

Cells of tissues
40 mm partial pressure of O_2

Oxygen

Oxygen

Alveolus of lung
100 mm partial pressure of O_2

Figure 39-11

How does hemoglobin carry oxygen? Hemoglobin molecules associate with oxygen in the lungs, where the partial pressure of oxygen is high (100 mm Hg). Oxyhemoglobin, in red blood cells, moves through the circulation. In oxygen-consuming tissues, such as muscles, the partial pressure of oxygen is lower (40 mm Hg) and oxyhemoglobin dissociates into deoxyhemoglobin and oxygen. The released oxygen diffuses into the surrounding tissues.

Physiologists say that hemoglobin changes its "affinity" (or fondness) for oxygen. Each time a hemoglobin molecule binds an oxygen molecule, the hemoglobin becomes more "fond" of oxygen. In contrast, as hemoglobin unloads its oxygen molecules, it loses its affinity for oxygen.

Hemoglobin changes its structure according to the partial pressure of oxygen around it (Figure 39-11). In the blood in the lungs, the partial pressure of oxygen is high. So the hemoglobin molecules bind oxygen and also increase their affinity for oxygen. Because of their high affinity for oxygen, the hemoglobin molecules do not let any of their oxygen go.

In the tissues of the body, the partial pressure of oxygen is much lower, however. So, after leaving the lungs, saturated

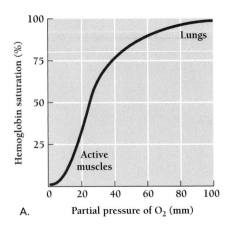

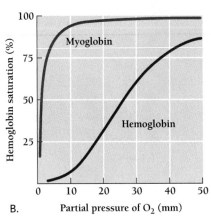

Figure 39-12

Oxygen dissociation curve. A. Hemoglobins changing affinity for oxygen. At low partial pressures of oxygen, like those found in active muscles, hemoglobin readily releases oxygen molecules to the tissues. At high partial pressures, like those found in the lungs, hemoglobin binds oxygen tightly and releases little. B. Myoglobin is an oxygen-binding protein similar to hemoglobin that, in animals, is primarily found in muscle. Myoglobin downloads oxygen from the blood into the muscle cells. Like hemoglobin, myoglobin is fully saturated at 100 mm. But myoglobin is also fully saturated at 40 mm, and nearly so at 20 mm. Myoglobin could never work well as an oxygen carrier in the blood because it does not let go of oxygen easily.

hemoglobin easily loses its fourth oxygen. But under most circumstances, the other three oxygen molecules stay bound to hemoglobin. They form a sort of emergency supply of oxygen. Only if the blood passes through tissues that are severely depleted of oxygen—such as the vigorously exercising muscles of a runner—does hemoglobin unload the other oxygen molecules—and then very quickly (Figure 39-12).

Researchers can easily study hemoglobin's oxygen binding because oxyhemoglobin is redder than deoxyhemoglobin. The oxygen-loaded blood of the arteries is bright red, while the oxygen-depleted blood of the veins is darker. Using a spectrophotometer, an instrument that measures light absorption, researchers can tell from the exact color of the blood how much oxygen is bound to hemoglobin at different partial pressures of oxygen. The amount of oxygen bound to the hemoglobin graphed against oxygen partial pressure is called an oxygen dissociation curve (Figure 39-12).

In adult humans, the percent of oxyhemoglobin increases from a partial pressure of about 10 mm Hg to about 60 mm Hg. Above 60 mm Hg, the percent of oxyhemoglobin does not increase much: the hemoglobin is effectively saturated with oxygen.

Hemoglobin has two alternate structures. One has a high affinity for oxygen and the other has a lower affinity for oxygen. In lungs, hemoglobin assumes the high-affinity state. In the tissues, hemoglobin assumes the low-affinity state.

What Other Factors Determine Hemoglobin's Affinity for Oxygen?

Besides being regulated by the partial pressure of oxygen, hemoglobin's affinity for oxygen is affected by pH, or acidity. Cells that are especially active often exceed their ordi-

nary supply of oxygen. In that case, they depend get energy (ATP) from glycolysis, which causes an accumulation of lactic acid, which increases acidity. The acidity decreases hemoglobin's affinity for oxygen, increasing the release of oxygen.

Cells that are especially active also release lots of carbon dioxide as they burn sugar (combine oxygen and glucose). When hemoglobin binds the waste carbon dioxide, its affinity for oxygen decreases. So hemoglobin in active cells releases its oxygen more easily.

Still another adaptation of hemoglobin helps increase oxygen delivery. In many mammals, hemoglobin binds to an abundant three-carbon molecule called BPG (2,3-bisphosphoglycerate). When BPG binds to hemoglobin, it decreases hemoglobin's affinity for oxygen, allowing the hemoglobin to release more oxygen in the tissues. The body has several ways of controlling BPG production to satisfy its changing oxygen needs. For example, hormones involved in coordinating a variety of functions, including epinephrine, testosterone, growth hormone, and thyroxin, stimulate red blood cells to produce BPG (see box).

Hemoglobin's affinity for oxygen changes according to the partial pressure of oxygen, pH, and the amount of carbon dioxide and BPG.

How Does Blood Carry Carbon Dioxide?

While oxygen is entering our blood, carbon dioxide is leaving. Carbon dioxide is the waste product of cell respiration, the means by which our cells make ATP (Chapter 6). Just as oxygen in the blood comes to equilibrium with oxygen in the alveoli, carbon dioxide in the blood comes into equilibrium with carbon dioxide in the alveoli. The difference between the partial pressure of carbon dioxide in the blood and the alveoli is very small compared to the difference for

A fetus has a special problem in obtaining oxygen. Since it cannot get oxygen from its own lungs, it must get oxygen from its mother's blood. Maternal blood comes into close contact with fetal blood only in the *placenta,* a structure in the womb that conveys nutrients from the mother to the fetus. In the placenta, the maternal and fetal blood pass in close proximity without actually flowing together. Still, oxygen and other nutrients can pass out of the mother's blood and wastes can pass from the fetus to the mother.

In order for the fetus to sop up oxygen from the maternal blood, the fetus's hemoglobin must have a higher affinity for oxygen than the mother's hemoglobin. And, indeed, a fetus makes a different kind of hemoglobin, called hemoglobin F.

Amazingly, however, when researchers studied the oxygen-binding characteristics of hemoglobin F, they found that it bound oxygen with exactly the same affinity as adult hemoglobin, hemoglobin A. For decades, this lack of difference in oxygen affinity baffled researchers: how did fetal blood do its job?

The answer came from the discovery of the action of BPG (2,3-bisphosphoglycerate). In humans, the major difference between hemoglobins A and F is that hemoglobin F does not bind BPG. When BPG binds to hemoglobin A, it decreases the molecule's affinity for oxygen. As a result, the hemoglobin can more easily give up oxygen to tissues that need it. Because hemoglobin F doesn't bind BPG, its affinity for oxygen is high, and the fetal hemoglobin absorbs oxygen from maternal blood. The hemoglobin F can still unload its oxygen in the fetal tissues because they have a lower oxygen partial pressure and lower pH than does the blood in the placenta (Figure A).

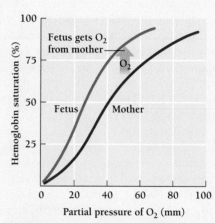

Figure A
Oxygen dissociation curves for fetal and maternal hemoglobin. Because fetal hemoglobin has a higher affinity for oxygen than does maternal hemoglobin, the maternal hemoglobin gives up oxygen to the fetal hemoglobin.

---

oxygen, but even so, the carbon dioxide readily diffuses into the alveoli.

Carbon dioxide enters the blood in three forms. A small amount, 5 to 10 percent, dissolves directly into the blood as a gas. Another 20 to 30 percent combines with hemoglobin inside the red blood cells. It does not compete with oxygen, however, because it attaches to a different place on the hemoglobin molecule than oxygen. Lastly, 60 to 70 percent of carbon dioxide combines with the water in blood to form a hydrogen (H^+) ion and a bicarbonate (HCO_3^-) ion—some of it inside red blood cells and some in the plasma.

Carbon dioxide readily diffuses from the blood into the alveoli. The blood carries most carbon dioxide in the form of bicarbonate ions.

39.5   What Factors Contribute to Oxygen Homeostasis?

Climb a tall mountain and you'll find yourself gasping for breath. Sprint to catch a bus or plane and you may wish you could transport oxygen a little faster. Happily, the body has adaptations that maintain oxygen delivery under most conditions.

How Can Blood Deliver Oxygen at High Altitudes?

At high altitudes, the difference in the oxygen partial pressures in the lungs and muscles is not as great as at sea level. When a person climbs a mountain, one of the first changes in the blood is an increase in the level of BPG, which binds to hemoglobin and decreases its affinity for oxygen. As a result, hemoglobin more easily delivers its cargo of oxygen to the brain and other tissues.

The body also adjusts to high altitude by building up a bigger supply of hemoglobin-rich red blood cells. Nearly a quarter of the oxygenated blood from the lungs and heart goes directly to the kidneys. In response to low oxygen levels, the kidneys release a protein hormone, called erythropoietin (EPO), that stimulates the bone marrow to make more red blood cells.

In humans, low oxygen in the tissues triggers an increase in BPG, which increases oxygen delivery by hemoglobin and the secretion of EPO, which increases the production of red blood cells.

How Do Animals Regulate Breathing?

The lungs and the blood, working together, accomplish both oxygen delivery and carbon dioxide removal. But how does an animal respond to lung failure, blood loss, or increased muscle activity, all of which may challenge the blood's ability to deliver sufficient oxygen?

As tissues use oxygen for respiration, they also produce carbon dioxide as a waste product. Increased cellular respiration always leads to an increase in carbon dioxide. In mammals, an increase in the partial pressure of carbon dioxide in the blood signals the need for deeper or more rapid breathing—a homeostatic response that helps blow off excess carbon dioxide.

Several factors control breathing rate, but by far the most powerful is the acidity of the blood. As mentioned earlier, when carbon dioxide reacts with water it releases hydrogen ions (H^+). When specialized receptors in the brainstem, aorta, and carotid arteries detect hydrogen ions, they transmit a signal to a special nerve complex in the brainstem that regulates breathing. High levels of carbon dioxide increase the number of hydrogen ions in the blood, which stimulates the brainstem to signal faster and deeper breathing. Nerves trigger the contraction of the diaphragm and other muscles needed for inspiration. As the lungs expand, stretch receptors report back to the brainstem, the contraction signals to these muscles cease, and expiration occurs. Researchers seeking to develop safer anesthetics, as well as treatments for breathing disorders such as sleep apnea, are trying to understand more about how this neural circuit works.

The brainstem also controls the circulatory system, changing the heart output according to the needs of the body. The maintenance of a constant internal environment means coordinating the actions of heart, blood vessels, and lungs to control the levels of oxygen and carbon dioxide as the status of the tissues changes.

Key Concepts

- Most animals use gills or lungs to obtain oxygen and dispose of carbon dioxide.
- Gills and lungs have large, wet surface areas that increase the exchange of oxygen and carbon dioxide.
- In mammals, the diaphragm and the muscles of the rib cage pull air into airways that lead to oxygen-absorbing surfaces.

Summary with Key Terms

How do the physics of gases and liquids determine the structure of gills and lungs?

For oxygen to participate in biochemical reactions, it must first dissolve in water. The final concentration of dissolved oxygen at **equilibrium** depends on its **partial pressure** and on the temperature. The speed with which oxygen dissolves depends on surface area.

Both lungs and gills are folded surfaces with large **surface-to-volume ratios. Lungs** are folded inward, **gills** are folded outward. An animal's **respiratory system** extracts oxygen from the environment. The circulatory system then distributes the oxygen throughout the body, often employing an oxygen-binding protein such as hemoglobin. Insects, however, distribute air directly to the tissues of the whole body from **spiracles** into a system of **tracheae.**

How do we breathe?

The respiratory system in mammals consists of the lungs, the airways, and the muscles that move air. Air moves into the lungs through the **pharynx,** the larynx (which contains the vocal cords), the **trachea** (or windpipe), the **bronchi,** and the **bronchioles.** The bronchioles lead to the **alveoli,** the major sites of gas exchange.

Contractions of the **diaphragm** and the muscles of the chest are responsible for the **inspiration and expiration** of air. Muscle contraction expands the **thoracic cavity,** pulling air into the lungs. Muscle relaxation expels air. At rest, each breath draws in and expels about 500 ml of air, called the tidal volume. Fresh air mixes with about 1,200 ml of residual air remaining within the alveoli after the last breath. At rest, most air is residual, but during heavy exertion, the lungs can move three to five liters of air with each breath, the **vital capacity.**

How do lungs resist infection, damage, and collapse?

The cells of the respiratory tract are specialized for several different functions. Some cells secrete **mucus,** which traps dust, bacteria, viruses, and other impurities. Other cells secrete **surfactants,** which allow the lungs to expand smoothly and freely. Most cells of the respiratory passages have **cilia** that push dust-laden mucus out toward the pharynx. The cells of the alveoli form a specialized epithelium, well adapted for gas exchange. Oxygen and carbon dioxide diffuse through the epithelium of the alveoli and the endothelium of the capillaries into the blood plasma.

How does hemoglobin pick up oxygen in the lungs and drop it off in the tissues?

Hemoglobin is made of four polypeptide chains and four **heme** groups. Each heme can bind one oxygen molecule, for a total of four oxygen molecules per hemoglobin. Oxygenated hemoglobin is called **oxyhemoglobin.** Hemoglobin binds and releases oxygen according to the partial pressure of oxygen in the surrounding tissues. More active muscles have lower oxygen partial pressures, causing a greater release of oxygen. The low pH and carbon dioxide released by active muscle also favor the release of oxygen.

As hemoglobin binds oxygen, its affinity for oxygen increases. Conversely, as hemoglobin loses oxygen, its affinity for oxygen decreases. This arrangement is at the heart of hemoglobin's ability to bind oxygen in the lungs and release it into the tissues of the body.

How do blood and the lungs rid the tissues of waste carbon dioxide?

The respiratory system also serves to dispose of carbon dioxide. In the lungs, carbon dioxide passes through the alveoli and exits from the body. Accumulated carbon dioxide in the tissues stimulates neurons in the brainstem to command an increase in breathing rate.

Review and Thought Questions

Review Questions

1. How does the partial pressure of a gas influence how much of it dissolves in a fluid?
2. In what ways does a lung differ from a gill?
3. What characteristics do gills and lungs have that maximize the rate of oxygen diffusion?
4. Since atmospheric air is richer in oxygen than water, why do fish suffocate when out of water?
5. Imagine air passing through the respiratory system from the nose to the alveoli. Name the parts along the way. Which part enables you to speak?
6. How do the lungs rid themselves of dust and microbes?
7. How does the diaphragm muscle help expand and contract the chest cavity? Draw a diagram.
8. Name the parts along the path that oxygen takes from the alveoli to the hemoglobin inside a red blood cell.
9. What changes hemoglobin's affinity for oxygen?

10. How does the body cope with the low partial pressure of oxygen at high altitudes?
11. In what three forms does carbon dioxide exist in the blood? Which is most common?
12. Considering what you know about both the circulatory system and the respiratory system, explain why a drug that is inhaled reaches the brain twice as fast as one that is injected.

Thought Questions

13. Explain what a vector is. What did Stanton Glantz mean when he said that the tobacco industry is the vector for heart disease and lung cancer?
14. Chemotherapy often inhibits the body's ability to produce new red blood cells. What symptoms might this cause? Describe how stimulating the production of red blood cells by the medicine Procrit, a synthetic form of the hormone erythropoeitin, could help reverse these symptoms.
15. Carbon monoxide (CO), present in cigarette smoke and engine exhaust, binds to the same part of a hemoglobin molecule as oxygen, but with a higher affinity. Explain how exposure to CO could affect respiration.
16. In the year 2000, about 2.4 million Americans died. Nearly 20 percent of those deaths were premature and resulted from smoking. Why do people smoke?

BiologyNow Resources

Biology ⑤ Now™

Active Figures

39-6: Respiration
39-7: Gas exchange

Preparing for an exam? Take a diagnostic test on your BiologyNow CD-ROM.

Online materials relating to this chapter are at:

http://biology.brookscole.com/AAL3

About the Chapter-Opening Image

In 1885, Vincent van Gogh painted this portrait of a skeleton smoking. Physicians of the 19th century already recognized that smoking was dangerous to health.

How Do Animals Manage Water, Salts, and Wastes?

Key Questions

- How is the regulation of fluid and salt balance an example of homeostasis?

- In what ways do freshwater, marine, and terrestrial animals face different problems in maintaining fluid balance and excreting nitrogenous wastes?

- How do vertebrate kidneys clean the blood and balance water and sodium?

- How do hormones regulate kidney function and blood pressure?

Will Corn Chips Raise Your Blood Pressure?

In the spring of 1996, a team of researchers from the University of Toronto published a paper in the *Journal of the American Medical Association*. The paper's conclusions ran counter to decades of authoritative nutritional advice. Julian Midgley, Andrew Matthew, and their colleagues wrote that for most people an extremely low-salt diet would not lower blood pressure and in some people might actually cause a heart attack. At a time when salt consumption in the United States had fallen 30 percent and supermarket aisles were jammed with low-salt crackers, low-salt chips, and low-salt soups, the researchers' stance dramatically contradicted accepted wisdom.

Americans love corn chips, potato chips, salty buttered popcorn, hot dogs, pizza, and greasy hamburgers. The more salt and fat, the better these foods seem to taste to us. Indeed, nutritionists uniformly agree that Americans consume far too much fat and far too much salt. Fat, especially animal and other highly saturated fat, contributes to both obesity and atherosclerosis, or hardening of the arteries. Both of these are major risk factors for heart attacks and strokes. Salt, we have been told, causes high blood pressure, a major risk factor for heart attacks, strokes, and kidney failure (Figure 40-1).

Even with declines in salt consumption, the average American consumes twice the recommended daily intake of salt. The minimum sodium requirement for normal day-to-day health is 115 milligrams (mg). (For comparison, a teaspoon of salt has about 2,000 mg of sodium.) Government nutritionists, in an effort to recommend something Americans can actually live with, suggest a diet containing no more than 2,400 mg per day, 21 times the minimum. Yet most of us take in about 4,000 mg of sodium each day, nearly 35 times what we need.

Amazingly, young people's kidneys easily deal with this daily excess. A typical 20-year-old can maintain normal blood pressure even after eating, over the course of a day, five cafeteria

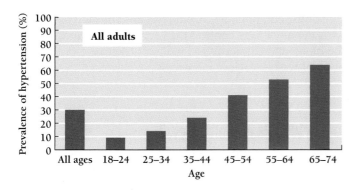

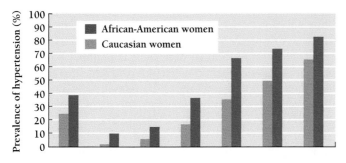

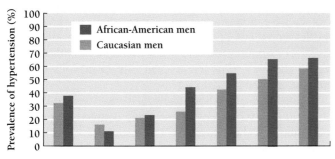

Figure 40-1

Prevalence of hypertension. Hypertension increases with age. African Americans experience more hypertension than Caucasian Americans, and men generally have more hypertension than women. However, African American women older than about 45 have the highest rate of any group.

One study of 32 countries found a consistent relationship between sodium consumption and high blood pressure, or **hypertension.**

But, according to Midgley and Matthew, a host of papers show that the connection between sodium and hypertension is indirect. To begin with, some researchers argue that a closer examination of people's behavior usually shows that those who consume a lot of sodium also drink more alcohol and tend to be overweight. Alcohol and obesity each cause hypertension, and these factors, rather than the salt itself, say the researchers, might be the cause of hypertension.

Researchers have long known that only 5 to 10 percent of Americans with high blood pressure clearly benefit from a low-salt diet. In these "sodium-sensitive" individuals, a high-salt diet causes blood pressure to rise, while a low-salt diet causes blood pressure to drop. Sodium sensitivity gradually increases in those older than 45 or 50 years of age. Apparently, as we age, our kidneys lose some of their ability to regulate salt and fluid levels in the blood. Considering the international data showing the strong association between sodium consumption and blood pressure, it was reasonable for health experts to recommend that Americans cut back on salt.

Midgley and Matthew's reanalysis of data from more than 50 previous papers suggested that a low-sodium diet lowers blood pressure only in people older than 45 years of age who have high blood pressure. Even then, blood pressure drops only modestly. And a low-sodium diet has virtually no effect in people with normal blood pressure or young people with high blood pressure.

The researchers also questioned the assumption that lowering blood pressure in older hypertensives actually prevents heart attacks and strokes or saves lives. Indeed, a super-low-sodium diet seems to be dangerous in some people. In one study, men with high blood pressure on extremely low-sodium diets had four times the risk of a heart attack as men consuming more sodium.

Should we eat as much salt as we can? Definitely not. Even for those of us whose kidneys are not sensitive to salt, too much salt burdens the kidneys and may possibly contribute to bone thinning, or osteoporosis, later in life. Will a high-salt diet raise blood pressure? It is beginning to appear that in most people, it does not, at least not immediately in healthy young adults. However, a high sodium intake is associated with an increased risk of stroke and heart attack, even where blood pressure is normal. Apparently, sodium makes blood vessels less elastic and may cause or worsen hardening of the arteries. Finally, a 2002 study showed that newborn babies with a preference for salt had higher blood pressure than other newborns.

In time, researchers will work out exactly how the high-salt diet so many people favor affects health. For now, it's enough to know that 115 mg of salt per day is all we really need, and anything over 2,400 mg is truly excessive.

pancakes, with a total of 4,000 milligrams of sodium; two slices of pizza, with 2,000 milligrams of sodium; and a can of spaghetti and meatballs, with 1,000 milligrams of sodium (for a grand total of 7,000 milligrams).

But nutritionists still suggest limiting salt intake because, generally, salt causes the body to retain water, which raises blood pressure. And physicians have long suspected that, long term, increased sodium increases blood pressure, leading to higher risk of heart attacks and stroke. The strongest evidence that increased sodium actually leads to increased blood pressure has come from comparisons of high blood pressure in different countries. In countries where sodium consumption is high, blood pressure is also high, while in countries where sodium consumption is low, average blood pressure is also lower.

40.1 Why Do Animals Need to Regulate Water and Salt?

Water is the major component of living organisms: in humans, it makes up half or more of our body weight. Water is the major component of both cells and the extracellular fluid in which they are constantly bathed. Even tiny changes in the concentrations of certain molecules can dramatically change protein structures, the rates of biochemical reactions, and other cellular processes. So cells and organisms must closely regulate the mix of water, salts, and other molecules both inside and outside of cells. To keep the right mix of water, salts, and other materials, organisms must detect changes in the normal concentration of these materials, then readjust the concentration, just one more example of homeostasis, the capacity of organisms to maintain a stable internal environment.

Organisms in different habitats face different types of challenges to proper water balance. Whether in deserts, oceans, or ponds, organisms evolve adaptations that help keep them from gaining or losing too much water (Figure 40-2). In dry, terrestrial environments, keeping enough water in the body and its cells is a constant challenge. In freshwater environments, keeping water *out* is the challenge, since the fresh water tends to move into the more-concentrated solutions of cells. The marine environment is more like a terrestrial one than a freshwater one. Because the ocean is saltier than the cells of marine fish, their cells tend to lose water through osmosis. So, like terrestrial animals, marine fish must expend energy to keep from losing water.

In vertebrates, the organ system that maintains both water and salt concentration is the urinary system. The **urinary system** includes the two **kidneys**, which produce urine, the twin **ureters**, which carry the urine to the **urinary bladder** for storage, and the **urethra**, which carries the urine outside the body. The vertebrate urinary system disposes of urea that results from the breakdown of proteins (and other nitrogen-containing compounds), maintains constant salt concentrations in body fluids, controls the volume and composition of the blood and extracellular fluid, and takes care of the disposal of wastes, toxins, and excess water and salt.

Which Way Does Water Flow?

To understand the flow of water between an organism and its environment, we can think of an organism as a bag of watery fluid. For a single-celled organism, this description is literally true: the cell membrane surrounds the cytoplasm, a dilute solution of salts and organic molecules. Outside the cell is more water, with more or less salt. Outside the individual cells of multicellular organisms is a similar dilute solution of salts and organic molecules, this time the *extracellular fluid*. In a human, the cytoplasm makes up about 40 percent of the body's volume, the extracellular fluid about 20 percent, and the blood plasma about 4 percent.

Women average about 50 percent water and men about 60 percent. The difference in water content results from differences in the amount of muscle and adipose tissue (fat). Women have more adipose tissue (which is 10 percent water) and less muscle (which is 75 percent water) than men. As a result a 170-pound man has just a few more pounds of solid weight than a 120-pound woman. But he has 42 pounds more water than she does.

Water can often flow freely between the extracellular fluid and the external environment and between the extracellular fluid and the cytoplasm. The direction in which water flows depends on osmosis. In Chapter 4, we discussed *osmosis* [Greek, *osmos* = impulse, thrust], the flow of water from dilute solutions into more concentrated ones (Figure 4-16). Net water flow between two solutions stops when the total concentrations of the two solutions are equal.

Water flows passively from a solution with a low concentration of ions and molecules to a solution with a high

A.   B.   C.

Figure 40-2

Desert, sea, or fresh water. Healthy kidneys maintain correct blood pressure and water and salt balance regardless of the external environment. Thanks to adaptable kidneys, reptiles such as these tortoises and turtles have evolved to live in environments in which the amount of salt and water vary enormously. A. Terrestrial: An endangered desert tortoise (Mojave Desert). B. Marine: A green sea turtle (Hawaii). C. Freshwater: An eastern chicken turtle (North Carolina).

concentration of ions and molecules. Although the movement of water happens passively, by osmosis, organisms often work to create concentration gradients to drive the transport of water or ions. You can read more about osmosis and concentration gradients in Chapters 4 and 34.

> The movement of water, ions, and other molecules between the extracellular fluid and the insides of cells depends on the concentration of ions and molecules in each fluid.

Water and Salt Balance in Aquatic Animals

When the total concentrations of ions and molecules in two solutions is the same, they are said to be *isotonic*. For example, the body fluids of most marine invertebrates are isotonic with seawater, so there is no net flow of water between these animals and their environment. In contrast, freshwater animals are saltier than their environments, and their body fluids are *hypertonic*. A cell in a solution of plain water takes in water, sometimes to the bursting point (Figure 4-17). Freshwater animals can reduce the relentless inflow of water with water-resistant skins. But to obtain oxygen, fish must constantly flood their gills with water, which, along with oxygen, moves into the gills. Freshwater fish must dispose of this excess water in their urine.

Producing a lot of urine, however, cannot by itself solve all the problems of water and salt balance. Urine,

which comes from the extracellular fluid, includes lots of salts and organic molecules. An animal that excreted large amounts of urine without reabsorbing these molecules would lose too many essential molecules. So freshwater animals pump the ions and other molecules out of the urine before excreting the urine, which is now very dilute, or *hypotonic*.

Saltwater fishes have body fluids with a salt concentration about one-third that of seawater. Saltwater fishes must therefore work to keep from losing water, actively pumping salt ions into its urine, disposing of the salt and keeping as much water as possible (Figure 40-3).

To compensate for water loss, a marine fish drinks lots of seawater. (In contrast, a freshwater fish does not drink at all.) But because the water contains lots of salt, these fish must remove the extra salt, mostly at the gills. Sea turtles and seabirds, which share this problem, have special organs that excrete excess salt (Figure 40-4).

Marine mammals generally do not drink seawater, but instead get most of their water from the organisms they eat. In addition, marine mammals can derive water from the breakdown of glucose or fat by cell respiration. (Recall from Chapter 6 that the breakdown of one molecule of glucose produces six molecules of water.)

> Freshwater vertebrates rid themselves of excess water by excreting dilute urine. Marine vertebrates tend to excrete salts and other wastes in minimal amounts of water.

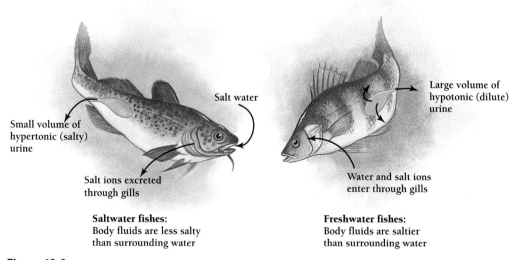

Saltwater fishes:
Body fluids are less salty
than surrounding water

Freshwater fishes:
Body fluids are saltier
than surrounding water

Figure 40-3
Salt water and fresh water present different problems. The kidneys of both saltwater and freshwater fishes must excrete water, salts, and nitrogenous wastes. But saltwater fishes are less salty than the surrounding seawater. Because the animals' cells tend to lose water and gain salt, they must expend energy to retain water while excreting salts. In contrast, the bodies of freshwater fishes are saltier than the surrounding water, so they tend to gain water and lose salt. Freshwater fishes must expend energy to excrete water without losing precious salts.

Figure 40-4
Salty tears. Like many sea animals, this sea turtle excretes excess salt in glands located near the eye.

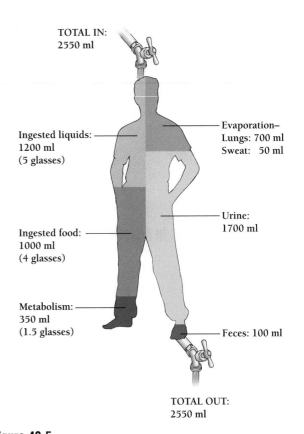

TOTAL IN:
2550 ml

Ingested liquids:
1200 ml
(5 glasses)

Ingested food:
1000 ml
(4 glasses)

Metabolism:
350 ml
(1.5 glasses)

Evaporation–
Lungs: 700 ml
Sweat: 50 ml

Urine:
1700 ml

Feces: 100 ml

TOTAL OUT:
2550 ml

Figure 40-5
Water in, water out. An average human ingests and excretes 2,550 ml of water each day. Water comes into the body as food and drink and leaves in urine, feces, sweat, tears, or vapor from the lungs, mouth, nose, and other moist surfaces.

How Do Land Animals Balance Water and Salt?

The major advantages of life on land are a rich supply of oxygen to breathe and a rich supply of plants to eat. The major disadvantage is getting and keeping water. Frogs, salamanders, and earthworms solve the problem by living partly in water and absorbing water through their skins. Some frogs, for example, have special abdominal skin patches that take up water, and some insects can also absorb moisture directly from the air.

But most terrestrial vertebrates obtain water through drinking, eating, and metabolism. An average adult, for example, drinks about 1,200 milliliters (ml) of fluids a day (about five cups) and obtains another 1,000 ml in food. Metabolizing the food yields another 350 ml of water, for a total intake of about 2,500 ml (more than half a gallon). Normally, the body loses about the same amount of water it takes in (Figure 40-5). Water leaves the bodies of land animals by three main routes: urine, feces, and evaporation. Evaporation takes place both from the outer surface of the body (the skin) and from the mucous membranes such as those in the nose and mouth, as well as from the lungs. Most water (1,700 ml) leaves as urine. An additional 100 ml per day is lost in the feces, and about 50 ml in sweat. Most of the rest leaves as evaporation from the lungs.

Species differ in the total amount of water that they process each day. Desert animals, such as kangaroo rats, have access to almost no water, so they conserve water by staying underground during the day, not sweating, concentrating their urine, and extracting extra water from the metabolic breakdown of macromolecules in their food (mostly dry seeds).

In all animals, water balance varies from day to day. A human may drink plenty of water one day, but not get enough to drink the next, sweat profusely during vigorous exercise, or lose too much water because of diarrhea or vomiting. Humans and other animals must therefore regulate water flow. We do so mostly by drinking more or less water (whether as water or in fruits and vegetables, juices, or other drinks) and by producing more or less concentrated urine. We animals are so good at this that, under most circumstances, our water content varies less than one percent. The failure of such regulation leads either to **dehydration** (in which too much water flows out of the body) or to "water intoxication" (in which too much water flows into the body) and **edema** [Greek, = swelling] (in which water accumulates in the tissues).

Land animals acquire water from drinking, food, and metabolism. They lose water in urine, feces, and evaporation from lungs and skin.

40.2 How Do Animals Dispose of Nitrogen-Containing Wastes?

In land animals, water and salt regulation are related to the disposal of wastes, especially **nitrogenous** (nitrogen-containing) **wastes**, the breakdown products of proteins, nucleic acids, and other nitrogen-containing compounds. Nitrogenous wastes come from excess protein in food and from the normal turnover of cell components.

Fish and most aquatic animals excrete excess nitrogen as **ammonia** (NH_3). Ammonia-loaded urine flows rapidly into the surrounding water. But ammonia reacts rapidly with organic molecules and is toxic to animals. Aquatic animals can dilute the ammonia to harmless concentrations, but land animals cannot afford to excrete that much water.

Land mammals solve this problem by using liver enzymes to convert ammonia into **urea**, a relatively nontoxic organic compound (Figure 40-6). Urea is less toxic, but, like ammonia, it must be dissolved in some water. Terrestrial animals therefore secrete urea in a minimum amount of water each day—in humans, at least 500 ml.

Insects, land snails, most reptiles, and birds require even less water than mammals. Instead of disposing of nitrogenous wastes as urea, they produce a nearly insoluble organic compound called **uric acid** (Figure 40-6). Even mammals like ourselves excrete the breakdown products of nucleic acids (DNA and RNA) partly as uric acid. Because uric acid is nontoxic and dissolves poorly in water, many animals excrete it in some sort of solid form. Some insects, as well as the embryos of reptiles and birds, go one step further. They squeeze as much water out of it as they can and then deposit dry uric acid within their own bodies.

> Animals rid themselves of nitrogenous wastes by excreting ammonia, urea, and uric acid.

40.3 How Do Kidneys Work?

Vertebrate kidneys do three things. They (1) filter the blood, (2) reabsorb useful molecules, and (3) excrete what's left (urea, salts, and toxins) as urine. The kidneys filter about one-fifth of all of the blood in the body every minute; regulate the balance of water, salts, wastes, toxins, and other organic molecules in the body; and, in vertebrates, excrete **urine**, a mixture of water, urea, salts, and wastes. Kidneys are paired structures that lie against the middle back (Figure 40-7). Blood brings wastes from all over the body, arriving at the kidneys by way of the **renal arteries** [Latin, *renes* = kidney] and leaving by way of the **renal veins**. The two kidneys clean the blood and excrete urine into a pair of tubes called the **ureters**. The two ureters carry the urine to a single **urinary bladder**, which holds the urine for storage and eventually empties it into the **urethra**, which goes to the outside of the body.

In most birds, reptiles, and amphibians, the ureter runs into a **cloaca**, a common exit chamber for the digestive, excretory, and reproductive systems. In male mammals, the urethra joins the reproductive tract so that urine and sperm both exit the body through the penis. In female mammals, the urethra empties urine only.

How Does a Kidney Filter the Blood?

The smooth, oblong shape of a kidney reveals little about how it works. But the interior is intriguingly intricate. If you cut through the middle of a kidney, you'll see an outer region, called the **cortex**, and an inner region, called the **medulla** (Figure 40-7).

Each of the twin renal arteries enters the center of one kidney and then branches into hundreds of thousands of fine arterioles in the cortex. Each arteriole forms a tangled network of capillaries, called a **glomerulus** [Latin, = little

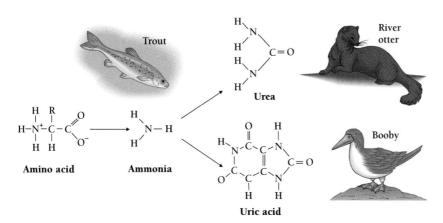

Figure 40-6
Ammonia, urea, uric acid. Animals excrete nitrogenous wastes in different forms. Trout and other aquatic animals can excrete nitrogen as ammonia (which is toxic) because they have access to plenty of water with which to dilute the ammonia. Otters and other mammals excrete urea, which is much less toxic. Boobies and other birds—as well as insects, land snails, and most reptiles—excrete nitrogen in the form of uric acid, which is a solid.

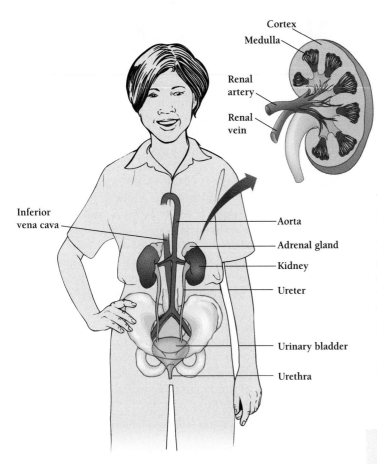

Cortex
Medulla
Renal artery
Renal vein
Inferior vena cava
Aorta
Adrenal gland
Kidney
Ureter
Urinary bladder
Urethra

The capillaries of the glomerulus are so leaky that together they filter all the body's plasma about every five minutes.

Tightly layered over the capillaries of the glomerulus, like a glove on a hand, is the inner layer of Bowman's capsule. The partly filtered plasma from the capillaries is filtered again as it passes through tiny slits in this first layer to enter the **capsular space** (Figure 40-10A). The pressure within this space is only about one-quarter that of the blood, but it is enough to drive the filtered plasma, or **filtrate**, out of the capsule and down into the renal tubule, where the next two stages of urine formation occur (reabsorption and excretion). The filtrate is similar in composition to the blood's plasma, but without cells or proteins. Besides water, it contains glucose, amino acids, salts, urea, and other small molecules.

> The glomerulus and Bowman's capsule keep cells and proteins in the blood plasma from entering the renal tubule. The filtrate that enters the tubule consists of salts, glucose, amino acids, and urea.

ball; plural, *glomeruli*] (Figure 40-8). Surrounding each glomerulus is a bulb called **Bowman's capsule,** which is actually the first part of a long, narrow tube, called the **renal tubule.** The glomerulus, Bowman's capsule, and the tubule together form a **nephron** [Greek, *nephros* = kidney], the functional unit of the kidney. Each human kidney contains about 1.25 million nephrons, whose combined length is about 85 miles.

The number of nephrons varies from person to person, and a recent study showed that people with high blood pressure have about half the number of nephrons as people with normal blood pressure. Although people seem to be born with all their nephrons, some evidence suggests that mothers who eat a nutritious diet while pregnant may increase the number of nephrons their babies have and so reduce the likelihood that their children will develop high blood pressure.

All the nephrons work in about the same way, so if we can understand how a single nephron works, we can understand how the kidney works. Each nephron *filters* the blood, *reabsorbs* water, sodium, and glucose, and *secretes* certain large molecules into the urine (Figure 40-9). The glomerulus and Bowman's capsule are responsible for filtration, the first stage of urine formation. Blood pressure in the capillaries of the glomerulus forces plasma out through tiny pores in the capillary walls, leaving behind all cells and most proteins.

How Does a Kidney Reabsorb Water and Salt From the Filtrate?

The renal tubule **reabsorbs** salt, water, glucose, and other organic molecules from the filtrate and returns them to the blood. The reabsorption of glucose is so efficient that, in a healthy person, no glucose at all escapes in the urine. Left behind in the gradually forming urine is all the urea, a few wastes, and a limited amount of water.

The kidneys clean the blood in the same way that most people clean out a cluttered drawer—by emptying the drawer out and putting back only what is worth keeping. Kidneys empty nearly everything from the blood, then put back the glucose and valuable organic molecules, as well as salt and water as necessary. As we'll see later in this chapter, hormones help determine how much water and salt the kidney tubules send to the bladder and how much goes back into the blood.

Each renal tubule consists of a long U-shaped loop with three main regions: (1) the wide **proximal convoluted tubule;** (2) the narrow **loop of Henle** (pronounced HEN-lee), whose *descending limb* drops down into the medulla and whose *ascending limb* ascends back into the cortex;

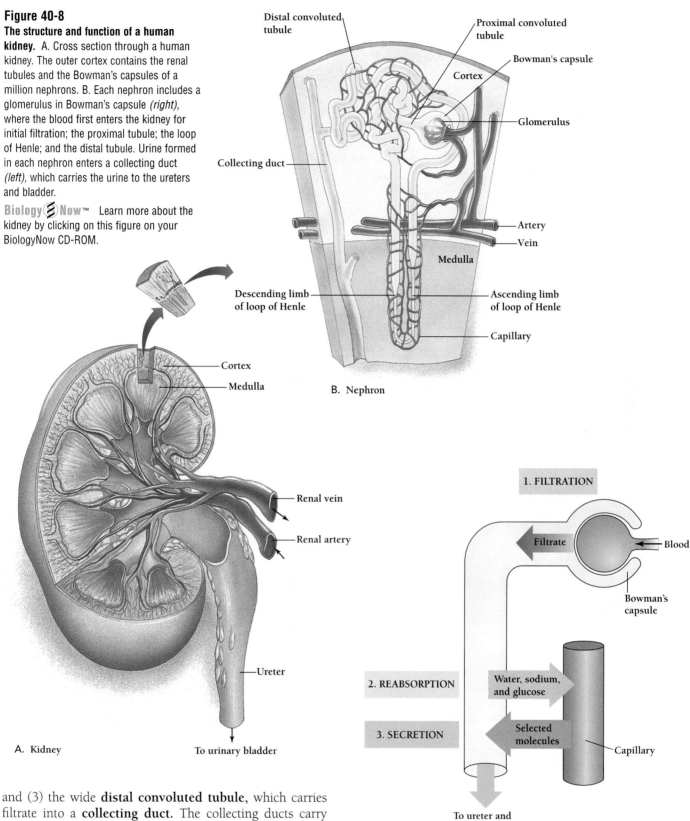

Figure 40-8

The structure and function of a human kidney. A. Cross section through a human kidney. The outer cortex contains the renal tubules and the Bowman's capsules of a million nephrons. B. Each nephron includes a glomerulus in Bowman's capsule *(right)*, where the blood first enters the kidney for initial filtration; the proximal tubule; the loop of Henle; and the distal tubule. Urine formed in each nephron enters a collecting duct *(left)*, which carries the urine to the ureters and bladder.

Biology ⊜ Now™ Learn more about the kidney by clicking on this figure on your BiologyNow CD-ROM.

Distal convoluted tubule

Proximal convoluted tubule

Bowman's capsule

Cortex

Glomerulus

Collecting duct

Artery

Vein

Medulla

Descending limb of loop of Henle

Ascending limb of loop of Henle

Capillary

B. Nephron

Cortex

Medulla

Renal vein

Renal artery

Ureter

A. Kidney

To urinary bladder

1. FILTRATION

Filtrate

Blood

Bowman's capsule

2. REABSORPTION

Water, sodium, and glucose

3. SECRETION

Selected molecules

Capillary

To ureter and urinary bladder

Figure 40-9

Overview of filtration, reabsorption, and secretion. Each nephron *filters* cells and large molecules from the blood, *reabsorbs* salts and water, and *secretes* urea and water and waste materials.

and (3) the wide **distal convoluted tubule,** which carries filtrate into a **collecting duct.** The collecting ducts carry urine into one of the two ureters.

In each of the three parts of the renal tubule, different epithelial cells line the inner surface. In the proximal convoluted tubule, great numbers of fuzzy microvilli increase the surface area of the tubule. Physiologists have named this fuzzy surface

Figure 40-10

The glomerulus in Bowman's capsule. A. A mammalian nephron includes a network of capillaries called the glomerulus. Blood pressure forces fluid out of the capillaries of the glomerulus and into the capsular space inside Bowman's capsule. The filtered fluid in Bowman's capsule, called filtrate, flows into the renal tubule for processing. B. Scanning electron micrograph of the surface of the glomerulus. C. Brush borders in the proximal convoluted tubule and in the upper ascending loop of Henle increase surface area, which helps the nephron absorb sodium ions and other materials from the filtrate.

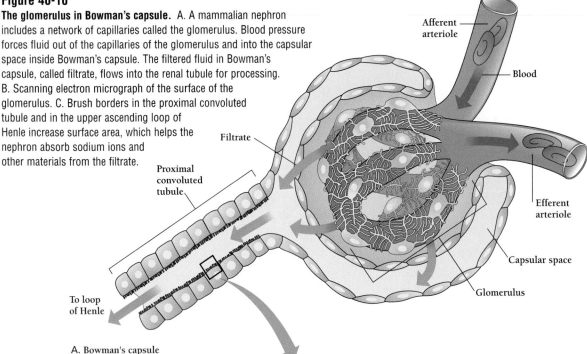

A. Bowman's capsule

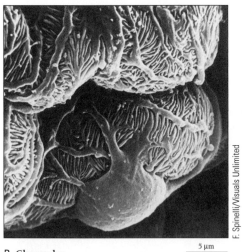

B. Glomerulus
5 μm
F. Spinelli/Visuals Unlimited

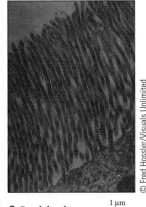

C. Brush border
1 μm
© Fred Hossler/Visuals Unlimited

a *brush border* (Figure 40-10C). The brush border increases the surface area over which sodium pumps work. As the pumps in the brush borders pump sodium out of the tubule, chloride ions and water follow. By the time the filtrate leaves the proximal tubule and enters the loop of Henle, about 70 percent of the salts and water have returned to the blood by reabsorption. The remaining filtrate has more urea, but the same salt concentration as when it started.

The loop of Henle absorbs half of the remaining water and two-thirds of the sodium and chloride ions in the filtrate. The kidney does this by setting up a concentration gradient between the outside of the loop of Henle and the inside. This gradient helps concentrate the filtrate and form urine.

The descending limb of the loop of Henle and the deepest part of the ascending limb are smooth inside and have no brush borders. This part of the renal tubule does not actively pump materials out of the filtrate. But the medulla at the bottom of the loop of Henle is so concentrated with urea and salts that water in the filtrate moves out through the walls of the tubule by osmosis, concentrating the filtrate (Figure 40-11).

The filtrate, which is now a more concentrated solution of water and urea, is becoming urine. It enters the upper part of the ascending limb, the distal tubule, and the beginning of the collecting duct, where more brush borders actively pump sodium out of the urine. The wall of the renal tubule here is waterproof, however, and so no more water escapes the lumen (Figure 40-11).

The filtrate leaving the distal tubule is not yet urine, for it still contains too much valuable water. When the urea-loaded filtrate reaches the far end of the collecting duct, urea diffuses out, raising the concentration of the fluid surrounding the deep part of the loop of Henle. The high urea content of this fluid, as well as its saltiness, is the driving force that pulls water passively out of the urine at the bottom of the loop. How much water and salt are removed depends on variables that we discuss later in this chapter. The urine that leaves the duct can attain the same concentrations of sodium, chloride, and urea as the surrounding extracellular fluid in the deepest region of the loop of Henle. For humans, this means a final maximum salt concentration almost four times that in blood plasma—a necessity for anyone regularly eating pizza and other salty foods. In the kangaroo rat, the loop of Henle is even more effective. The salt concentration in the urine of a kangaroo rat can be 14 times that in the an-

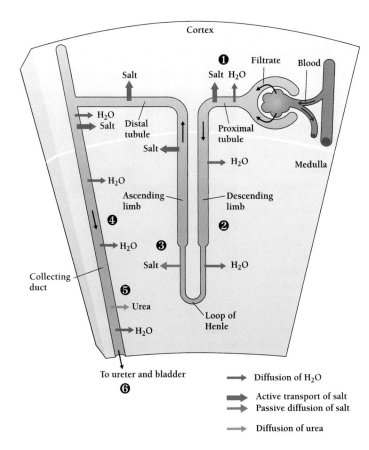

Cortex

Salt

Salt H₂O   Filtrate   Blood

❶

H₂O   Distal
Salt   tubule

Salt   Proximal
tubule

Salt   H₂O

H₂O   Medulla

Ascending   Descending
limb   limb

❹   ❷

H₂O

H₂O   ❸

Salt   H₂O

Collecting
duct

❺

Urea   Loop of
Henle

H₂O

To ureter and bladder

❻

→ Diffusion of H₂O

⟹ Active transport of salt
⟹ Passive diffusion of salt

→ Diffusion of urea

Figure 40-11

Countercurrent concentration of solutes in a nephron. A nephron is a countercurrent system that filters the blood and concentrates certain ions and molecules to form the urine. ❶ The proximal tubule actively pumps sodium ions out of the filtrate. The filtrate, which consists of water, urea, and other molecules, descends into the loop of Henle, which is surrounded by the increasingly salty medulla. ❷ Water passively diffuses out of the loop of Henle *(blue arrows)*. ❸ Beyond the turn of the loop, the tubule's epithelium is impermeable to water and the water remains trapped in the tubule. But the same region transports sodium ions out of the tubule, so that the fluid entering the collecting duct is dilute—lots of water, not much salt. ❹ As the water and urea go back down through the salty medulla, water flows passively out, concentrating the urea in the tubule. (❺ Some of the urea escapes into the lower part of the medulla.) ❻ The remaining fluid, the urine, flows from the collecting duct to the ureters and then to the bladder.

Biology ⑤ Now™ Learn more about the countercurrent system by clicking on this figure on your BiologyNow CD-ROM.

imal's blood plasma. For a desert animal that needs to keep water but get rid of salt, this is a life-saving adaption.

In addition to reabsorbing water and salt from the filtrate, the nephron also reabsorbs glucose, amino acids, proteins, and vitamins and returns these to the blood. Normally, any glucose or other nutrients in the blood either move passively out of the filtrate or are transported out, and returned to the blood. If the concentration of one of these molecules is unusually high, however, some of it may remain in the filtrate to be excreted in the urine. This is exactly what happens in cases of untreated diabetes mellitus. People with this disease excrete urine that is very sweet because it contains glucose. In diabetics, the concentration of glucose in the blood can be so high that it does not all pass out of the filtrate.

A kidney uses a concentration gradient to produce urine that is high in urea and other waste materials but low in glucose and other valuable nutrients. The concentration of urea increases until the filtrate reaches the far end of the collecting duct, where the urea diffuses into the surrounding tissues and draws water from the filtrate in the loop of Henle.

What Does a Kidney Secrete into the Urine?

Kidneys also secrete potassium and hydrogen ions, ammonia, and organic acids and bases into the urine. Secretion is a backup for filtration, a way to put into the urine wastes

that didn't leave the bloodstream in the glomerulus and Bowman's capsule. Materials that are frequently secreted into the filtrate, mainly at the proximal and distal tubules, include hydrogen, potassium, and ammonium ions; creatinine, a breakdown product of the energy molecule creatine; drugs, such as antibiotics; and other toxins. The secretion of hydrogen and ammonium ions is an important way for the kidneys to control the acidity of the blood. Minor changes in acidity can be fatal, so acidity is regulated very tightly.

40.4 How Do Animals Regulate Water and Salt Balance?

Humans and other animals consume different amounts of water and salt from day to day, and they eliminate different amounts in their feces, sweat, and urine. Maintaining a constant level of water and salt concentration in the blood (homeostasis) means that the kidneys must constantly adjust urine concentration and flow.

The body controls urine concentration and flow by a variety of complex mechanisms, including neural and hormonal control of fluid intake and the regulation of blood flow through the kidneys.

How Does the Brain Control Fluid Intake?

One way animals control water balance in the body is by varying how much water they drink. In mammals, the brain's hypothalamus contains a "thirst center" that has receptors that detect changes in salt concentration. An increase in salt concentration in the extracellular fluid of the hypothalamus stimulates these *osmoreceptors*, and we experience the sensation of thirst until we drink some water.

Once we drink, we do not feel thirsty for about 15 minutes (30 minutes on a full stomach). If by then the salt con-

Health and Biology The Artificial Kidney

Infectious diseases, drugs, poisons, genetic diseases, tumors, and physical trauma can all severely damage the kidneys. In fact, about 35,000 people die of kidney failure in the United States each year, about 1.5 percent of all deaths. When a person's kidneys begin to fail, urea and other substances build up in the blood. This condition is called uremia [Greek, *ouron* = urine + *aima* = blood] ("urine in the blood").

Many patients with kidney failure stay alive with the help of a device called a kidney dialysis machine. The patient's blood runs into a chamber containing a membrane that allows the passage of salt ions, as well as urea and other small molecules (Figure A). On the other side of the membrane is a large volume of a solution whose composition is the same as that of normal blood plasma. Sodium, chloride, and potassium ions, as well as glucose and other molecules, move passively across the artificial membrane and come to the same concentrations on each side. Urea and other small waste molecules diffuse across the membrane into the large volume, so that their concentration falls nearly to zero. The kidney dialysis machine thus restores the blood plasma to a nearly normal composition.

Patients who depend on dialysis machines cannot regulate their water balance with hormones, as the rest of us do. Instead, they must carefully balance their intake of food and drink to match losses from evaporation and elimination. Depending on the remaining abilities of their damaged kidneys, these patients produce little or no urine. Even the best dialysis machine, then, cannot approach the feats of the simplest kidney.

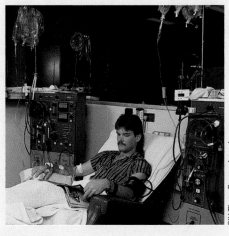

Figure A
Kidney dialysis. A young man with damaged kidneys waits as a dialysis machine cleans urea and other small waste molecules from his blood.

---

centration in the blood has not gone down enough, we get thirsty again and drink until the thirst center detects the right concentration of salt. If we lose a lot of water very quickly, say, by exercising strenuously on a hot day, it's possible to become dehydrated without becoming thirsty right away.

Most of us have read that we should drink eight glasses of water a day whether we are thirsty or not. That's eight cups, or a half gallon—a lot of water for smaller people. Many people who carry a water bottle in hand all day long, trying to get down that half gallon, are probably drinking more water than they need. There's little harm in that; it's nearly impossible to drink enough water to do yourself harm, and the (rapidly growing) bottled water industry appreciates your business.

But eight glasses is a vague benchmark, based on the approximate needs of an active young adult (Figure 40-5). The eight-glasses-of-water rule also assumes that you are getting almost no fluids from fruit, vegetables, soup, milk, coffee, tea, juice, or other drinks. But in reality, everything you eat and drink contains some water, so in most situations you don't need to drink a half a gallon of water a day. Unless you are sick, exercising strenuously, or suffering through unusually hot weather, you are unlikely to become seriously dehydrated if you drink when you feel like it.

When salt concentrations in the blood increase, the brain's hypothalamus detects the increase and triggers a sensation of thirst.

What Influences the Rate of Filtration?

The kidneys receive 20 to 25 percent of what the heart pumps each minute. Blood flow to the kidneys is much higher and more constant than blood flow to other areas of the body. But the arterioles sending blood to the glomeruli can open and close to regulate the amount of blood passing through the kidneys each minute. When the concentration of salt or nitrogenous wastes rises, blood flow to the kidney increases. If, for example, we eat a slice of pizza, the salt concentration in the blood will go up, causing the arterioles to relax and allow more blood to the glomeruli. The kidneys can then filter more and remove more salt.

An increase in salt concentration in the blood causes an increase in the blood flow to the kidneys.

How Do Hormones Regulate Kidney Function?

Suppose you play soccer on a hot afternoon and forget to drink water. Sweat is running down your shirt and you are becoming dehydrated. What can your kidneys do to maintain the correct balance of fluids and salts until you drink something? One thing they can do is hoard every bit of water in your body and not excrete any more than absolutely necessary. Another is to move some salts out into the urine heading for your bladder or, alternatively, to hoard salts if you are losing salts. Finally, because you are losing water, the total volume of blood in your body is dropping, so the blood vessels in the body must compensate by constricting a little. This helps maintain blood pressure in the arteries and arterioles.

How does the body make all this happen? The short answer is hormones. A few of the hormones that regulate water and salt balance are aldosterone, antidiuretic hormone (ADH), atrial natriuretic factor (ANF), and angiotensin II. We'll talk about each of these briefly.

If the concentration of sodium in the blood decreases, if the volume of blood decreases, or if the blood pressure decreases, then the **adrenal glands** [Latin, *ad* = toward + *renes* = kidneys], which lie on top of the kidneys, secrete the steroid hormone **aldosterone.** Aldosterone helps the body hang onto both sodium and water by stimulating the kidneys, sweat glands, salivary glands and intestines to absorb water and salt.

People who cannot secrete aldosterone because of malfunctioning adrenal glands lose too much salt and water in the urine. They can compensate by eating salty foods and drinking extra water. Aldosterone normally prevents dangerous dehydration and salt depletion. When we lose too much water and salt, the total volume of blood drops and blood pressure drops. When blood pressure drops too low, the heart cannot supply enough blood to the tissues, resulting in shock and sometimes death.

Aldosterone's effects are most dramatic in the presence of a second hormone, **antidiuretic hormone,** or **ADH,** made in the hypothalamus (in the brain). Like aldosterone, antidiuretic hormone prevents water loss and raises blood pressure in response to a fall in blood pressure or a rise in the saltiness of the blood. As its name suggests, antidiuretic hormone is the opposite of a *diuretic,* a substance, such as caffeine or alcohol, that stimulates water loss. ADH causes the body to retain water by increasing water permeability in the collecting ducts of the kidneys. Water then returns to the blood through the walls of the collecting ducts and the urine becomes more concentrated.

As we saw earlier, cells in the thirst center of the hypothalamus respond to changes in the salt concentration of the blood. After we eat a bag of corn chips, the concentration of salt in the blood rises and the thirst center orders the secretion of ADH into the blood. When the ADH reaches the kidneys, it increases water uptake in the collecting ducts. As a result, the urine becomes more concentrated, the blood becomes more dilute, and the salt concentration of the blood returns to normal. Other factors also influence ADH secretion. Pain, fear, cold, and stress can all decrease ADH levels and increase urine production.

Angiotensin II stimulates the release of both aldosterone and ADH, thus increasing the retention of water and sodium. Angiotensin II also stimulates thirst and elevates blood pressure by causing the blood vessels to constrict. The combination of salt and water retention, which increases the total volume of blood, and the constriction of the blood vessels, increases blood pressure.

When blood pressure drops or the concentration of salt in the blood falls, aldosterone and ADH cause the kidneys to retain water and sodium, raising blood pressure. ADH also prevents water loss and raises blood pressure. Low blood pressure and concentrated blood also stimulate the formation of angiotensin II, which stimulates the release of aldosterone and ADH and constricts arterioles, further increasing blood pressure.

Atrial Natriuretic Factor (ANF)

As we saw earlier, kidney function depends on blood pressure to provide the force for filtration and flow of the filtrate through the tubules. We have seen that an increase in ADH lowers urine volume and raises the volume of blood in the vessels. The result is an increase in blood pressure. But the body also needs a way to limit increasing blood pressure. One organ perfectly situated to monitor blood pressure is the heart.

In response to excessive blood pressure, cells in the smaller chambers of the heart (the atria) produce a hormone called **atrial natriuretic factor** [Latin, *atrium* = vestibule, referring to the heart chamber, + *natrium* = sodium + Greek, *ouron* = urine], or **ANF.** ANF counteracts all of the hormones we've just discussed that increase blood pressure. ANF inhibits angiotensin, aldosterone, and ADH, relaxes smooth muscles, reduces thirst, and increases the kidneys' elimination of sodium ions and water by closing channels in the collecting ducts. ANF thus increases water loss and lowers blood pressure.

In response to high blood pressure, the heart produces ANF, a hormone that increases water loss and lowers blood pressure.

The kidneys release another hormone, **erythropoietin (EPO)** in response to low oxygen levels in the blood. As mentioned in Chapters 38 and 39, EPO stimulates the production of red blood cells in the bone marrow to compensate for whatever factors may be causing blood oxygen levels to fall. These could be heavy exercise; a visit to the mountains, where oxygen levels are lower; an injury that causes blood loss; or a disease affecting the lungs. The increase in red blood cells not only improves the transport of oxygen, it also increases blood volume, which increases blood pressure.

40.5 How Do Invertebrates Regulate Water, Salts, and Wastes?

Animals in different environments must solve different problems of salt and water balance. Their solutions differ in ways that at least partly reflect their distinct evolutionary histories.

The simplest excretory structures are the "flame bulbs" of flat-worms, which help these simple freshwater animals excrete excess water. Water and materials from the extracellular fluid enter a flame bulb. Inside, beating cilia move water into tubules that drain into tubes that run the length of the flatworm's body and empty to the outside world through tiny pores. Under the microscope, the cilia in the flame bulbs appear to be flickering, an illusion that gives these cells their curious name.

Although we think of the earthworm as a terrestrial animal, it spends most of its life in damp tunnels in the soil. Its body fluids therefore have a higher concentration of molecules than the surrounding water, and the earthworm has the same problem as a freshwater animal—getting rid of excess water. Like mammalian kidneys, earthworm *nephridia* (singular, nephridium) both filter and reabsorb. Each body segment contains a pair of nephridia. Fluid in the earthworm's body cavity, or coelom, is under pressure. The pressurized fluid pushes through an opening in the tubular part of the nephridium, where salts and valuable molecules are reabsorbed. The dilute urine left behind passes from each nephridium through a pore in the body wall.

Crabs and most other marine invertebrates have body fluids with about the same concentration of salts as seawater. Many of these animals allow the concentrations of ions and molecules in their body fluids to fluctuate with that of their environment, an example of tolerance (Chapter 34).

Even though water and salt balance are less of a problem for marine invertebrates than for other animals, ridding themselves of nitrogenous wastes remains a problem. They accomplish this task with excretory organs that work by filtration, absorption, and secretion and eliminate excess nitrogen in the form of ammonia. Some invertebrates have arrangements of excretory organs very different from ours. Lobsters and other crustaceans, for example, perform excretion in their heads, through *green glands,* which function much like other invertebrate kidneys. And the abalone's two kidneys differ in function—the left kidney specializes in reabsorption and the right in secretion.

The excretory system of insects differs from all the other systems we have discussed. Insect excretory organs, called Malpighian tubules, are blind outpocketings of the gut. These tubules may number from two to several hundred. Their blind ends are bathed in the insect's body fluids (Figure 40-12).

Fluid flows into the tubules, but not because of a pressure difference (as in earthworms and marine invertebrates). Instead, the tubules actively pump potassium ions. Water, with dissolved nitrogen wastes, passively flows into the tubules by osmosis. The contents then flow into the gut, where the residues of digestion (the feces) mix with the products of excretion (the urine).

The insect's hindgut and rectum are especially efficient at removing water, so that the combination of feces and urine can be extremely dry, depending on the insect's diet. An insect that lives on fresh leaves gets plenty of water and so excretes lots of liquid urine. But a termite, which dines on wood, excretes dry little pellets with the consistency of the

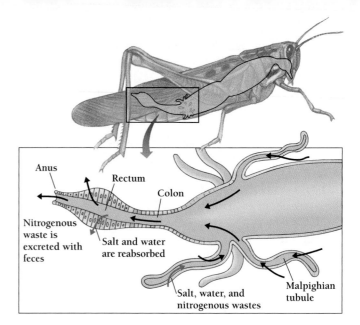

Figure 40-12
Malpighian tubules. In insects, the excretory organs, the Malpighian tubules, are blind outpocketings of the gut. The Malpighian tubules excrete salt, water, and nitrogenous wastes into the rectum, where it is excreted with the feces.

compacted sawdust "logs" that city people buy at the supermarket for their fireplaces.

In this chapter we have seen why animals need kidneys and how they excrete water, salts, and nitrogenous wastes. In the next chapter we will see how animals defend themselves against infection by microorganisms, promote healing, and minimize blood loss—all important functions of the immune system.

Key Concepts

- Animals must regulate water and solutes both inside each cell and in the extracellular fluid.
- Animals excrete nitrogenous wastes, usually diluted with water, by means of kidneys or other excretory organs.
- In mammals, water and small molecules from the blood flow into the kidney, which then "puts back" essential ions and molecules.
- Hormones coordinate salt and water excretion by altering blood flow and water permeability.

Summary with Key Terms

How is the regulation of fluid and salt balance an example of homeostasis?
Maintaining the right balance of water and salts as well as excreting nitrogenous and other wastes is a problem of homeostasis that animals solve in various ways. Freshwater animals

tend to gain water and lose salts, while marine animals tend to lose water and gain salts. The major problem of land animals is to acquire and conserve water. Every animal has characteristic adaptations for regulating salt and water, for preventing **dehydration** or **edema**. The failure to regulate fluid balance can also cause **hypertension**, or high blood pressure.

In what ways do freshwater, marine, and terrestrial animals face different problems in maintaining fluid balance and excreting nitrogenous wastes?

Animals must rid themselves of **nitrogenous wastes** from the proteins in their food. Marine invertebrates and fish excrete excess nitrogen in the form of **ammonia**. Mammals mostly convert the waste ammonia to **urea**, which is less toxic than ammonia. Birds, insects, and other animals excrete **uric acid**.

How do vertebrate kidneys clean the blood and balance water and sodium?

The vertebrate **urinary system** includes two **kidneys**, which secrete **urine** into paired **ureters**, which send the urine to a single **urinary bladder**, where the urine is stored until it can be excreted through the **urethra** or **cloaca**. The kidney regulates the composition of the urine by means of filtration, reabsorption, and secretion. Blood carries wastes to and from the kidneys by way of paired **renal arteries** and **veins**.

The kidney consists of separate functional units, called **nephrons**. Each of the million or more nephrons consists of a filtering unit—the **glomerulus** and **Bowman's capsule**—and a **renal tubule** that reabsorbs and secretes, ultimately draining into the ureters. In mammals, the wide **proximal convoluted tubule** of the nephron bends into the narrow **loop of Henle** and exits the kidney through the wide **distal convoluted tubule**, an arrangement that allows the mammalian kidney to produce a concentrated urine. Blood plasma squeezes out of the glomerulus, through the first layer of Bowman's capsule into the **capsular space**, forming a **filtrate** that enters the renal tubule. The renal tubule and **collecting duct** together regulate the excretion of water, sodium, and urea. Glucose is returned to the blood. Bowman's capsule and both convoluted tubules lie in the kidney's outer **cortex**, while the loop of Henle drops into the kidney's inner **medulla**.

How do hormones regulate kidney function and blood pressure?

Several hormones regulate the operation of the kidney. **Aldosterone**, from the **adrenal glands**, and **antidiuretic hormone (ADH)** stimulate the reabsorption of sodium and water in the nephrons. **Angiotensin II** stimulates the release of aldosterone and ADH and also the constriction of blood vessels, thus causing the kidneys to retain water and blood pressure to rise. **Atrial natriuretic factor** (**ANF**), secreted by the heart, counteracts the effects of these other hormones, lowering blood pressure. The kidneys release the hormone **erythropoietin (EPO)** in response to low-oxygen blood coming into the kidneys. EPO stimulates the development of extra red blood cells that can carry more oxygen.

Review and Thought Questions

Review Questions

1. Why do marine fish drink continuously, but freshwater fish not at all?
2. Sketch a nephron and collecting duct. How do the cells of the epithelium vary from segment to segment? How does the transport of ions vary from segment to segment? How does the epithelium's permeability to water vary?
3. Why do terrestrial animals need a special way to excrete nitrogenous wastes?
4. How does urea get to be more concentrated in the urine than in the blood?
5. How does glucose get to be less concentrated in the urine than in the blood?
6. Why are people with kidney problems also likely to have high blood pressure?
7. Suppose you drink two extra glasses of water at dinner. What effects do you expect the extra water to have on the secretion of hormones by the hypothalamus, adrenal glands, and heart?
8. Suppose you eat an exceptionally salty meal, without drinking any extra water. What effects do you expect the extra salt to have on the secretion of hormones by the hypothalamus, adrenal glands, and heart? How will blood pressure be affected?
9. What triggers the release of erythropoietin (EPO)? What does EPO do?

Thought Questions

10. Why do seabirds and sea turtles have salt glands? Why can a seagull survive by drinking seawater, but a shipwrecked sailor cannot?
11. Whales and seals do not have salt glands, but live well without fresh water. How do you think they might manage this?
12. America has some of the cleanest tap water in the world. Why do so many Americans today drink bottled water instead of tap water?

BiologyNow Resources

Biology⑤Now™

Active Figures
40-8: The kidney
40-11: Countercurrent system

Preparing for an exam? Take a diagnostic test on your BiologyNow CD-ROM.

Online materials relating to this chapter are at:
http://biology.brookscole.com/AAL3

About the Chapter-Opening Image
The salt box symbolizes the huge amount of salt that Americans consume, forcing their kidneys to work overtime removing it from the blood.

Inflammation and Immunity

Key Questions

• How does the body keep microorganisms from colonizing the body?

• What gives the immune system specificity, diversity, memory, and self-nonself recognition?

• How do immune cells recognize the body's own cells?

The Killer Cold

In the spring of 2003, a mysterious disease epidemic came out of China, spreading like influenza, but more deadly. People caught the pneumonia-like disease from their relatives at home, from strangers in hotel elevators, and on international flights. But most frighteningly, doctors, nurses, and other health care workers caught it from their patients. In the first few weeks, 90 percent of new cases were among health care workers.

On March 7, 2003, a very sick man entered a Toronto emergency room with a fever and shortness of breath. Doctors diagnosed 43-year-old Tse Chi Kai with pneumonia and kept him overnight in the ER. In the morning, Mr. Tse was worse and he was put in an intensive care ward. Hospital workers learned that his mother had recently traveled from Hong Kong, had been ill with the same symptoms, and had died on March 4. But it wasn't much to go on. The rest of the family was now sick, as well, and patients who had been in the emergency room with Tse were also coming down with the apparently deadly illness.

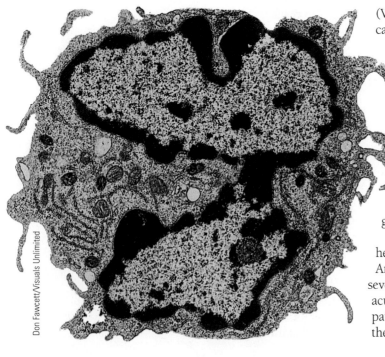

Don Fawcett/Visuals Unlimited

Then, on March 12, the World Health Organization (WHO) announced an epidemic of "atypical pneumonia," calling it a "world wide health threat." The following day, Tse died, and Toronto doctors understood with shock that they were witnessing the first few days of what could be a devastating epidemic if they didn't act fast. They knew they had to find everyone who had been exposed to Mr. Tse, and everyone who had been exposed to those people. Within a few days, hundreds of health care workers and thousands of others were asked to isolate themselves at home until they were sure they did not have the disease. Two entire hospitals were closed to all new patients, and all health care workers began to wear masks, gloves, and protective gowns.

Anyone with a fever of more than about 100.5°F, headache, malaise, and body aches needed to be watched. Anyone who also developed a dry cough after three to seven days probably had the disease, now called "severe acute respiratory syndrome," or SARS. About 10 percent of patients became so ill that they could no longer breathe on their own.

The Toronto public health agency searched Tse's home to find out what flight his parents had been on. They found the tags with the flight number still on the luggage and quickly tracked down all the passengers on that flight. Fortunately, none had become ill. By late March, three people had died in Toronto, one of them Tse, and increasing numbers of people were coming down with the illness. But the good news was that by acting decisively, Toronto health officials had slowed the outbreak, preventing far more infections and deaths. The warning from WHO had been critical in saving lives, but it had come at great cost.

Just a month earlier, WHO's Dr. Carlo Urbani had been working in Hanoi, Vietnam, when colleagues called him in to check out an unusual case. Urbani was a past president of the Italian chapter of the international organization Doctors Without Borders and an expert on communicable diseases, especially parasitic infections. But even though Urbani was "a worm guy," as his colleagues affectionately put it, no one was better at diagnosing illnesses. So they called him (Figure 41-1).

Urbani arrived to find that an American businessman, named Johnny Chen, had entered the hospital with what looked like the flu. But Urbani quickly realized that Chen did not have the flu. Furthermore, whatever he had was highly contagious; everyone around Chen was getting sick. Fifty-six percent of hospital workers exposed to Chen ultimately became ill with the disease. Urbani immediately alerted the WHO.

Afterward, Urbani and another WHO doctor persuaded the Vietnamese Health Ministry to begin isolating people with SARS symptoms and anyone else exposed and also to begin screening travelers. Meanwhile, with Urbani's warning, WHO was able to issue its March 12 worldwide alert and begin monitoring the spread of the disease in other parts of the world. In Toronto, the warning made all the difference.

But by then, Urbani had been taking care of SARS patients for two weeks. On March 11, he left Vietnam for a conference in Thailand on treating schoolchildren for parasitic worms. He wasn't feeling well, and by the time he arrived in Bangkok, a colleague who met him at the airport knew it was time to call an ambulance. His wife flew to Bangkok to be with him, but could only to talk to him through an intercom. His colleagues tried desperately to save him, but it was no use. He was on a respirator and

Figure 41-1
Dr. Carlo Urbani. Dr. Urbani of Doctors Without Borders was one of the first people to alert the World Health Organization (WHO) about the dangers of the SARS epidemic. After caring for SARS infected patients, Urbani himself came down with the disease and died within a few weeks.

Guido Picchio/AP/Wide World Photos

mostly unconscious until he died on March 29.

Carlo Urbani had played a key role in helping WHO stem the outbreak of a new and dangerous disease. But the research community was engaged in something almost as amazing. They were cooperating at an unprecedented level. Although the disease wasn't officially recognized by WHO until March 12, within six days of the formal announcement, researchers at a hospital in Hong Kong had already identified a paramyxovirus that might be causing the disease. A few days after that, researchers at a nearby hospital had suggested a second possibility— a round, spiky virus usually associated with common colds, called a coronavirus (Figure 41-2).

Researchers succeeded so rapidly by applying a combination of high technology and a new, more cooperative way of doing science to identify two likely culprits in just days. As soon as WHO formally announced the SARS epidemic, the agency also organized laboratories around the world to find: (1) the virus causing the disease, (2) a test, and (3) a treatment. In a matter of days, WHO organized 12 laboratories in eight countries that would share tissue samples, data, and results in lightning-fast emails and twice-daily conference calls. WHO set up a web site where researchers could post information about everything that was going on in their lab— whether discoveries, setbacks, or new questions.

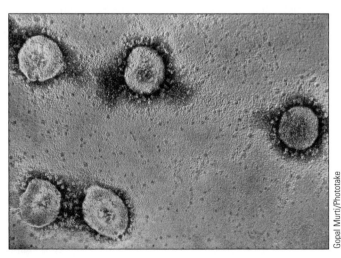

Figure 41-2
The SARS coronavirus. Related to viruses that cause ordinary colds, the SARS coronavirus is uncharacteristically deadly.

Gopal Murti/Phototake

One of those labs, at the U.S.'s Centers for Disease Control and Prevention, immediately placed its bets on the coronavirus, though until someone could prove that this wayward cold virus caused SARS, no one could be certain. A reliable test for either the virus itself or antibodies to the virus would allow health care workers to quickly separate infected people from those who have ordinary flu or colds and free quarantine people from a virtual house arrest. And a test for the virus would solve a logical problem.

Until researchers could identify the virus that was causing the disease, they could not be sure who had the disease. Since the early symptoms were so like that of an ordinary cold or flu, the people identified as having the disease might not all have had it. So if, for example, a coronavirus believed to be the culprit was not found in some patients, researchers couldn't be sure whether the patients didn't have SARS or if the virus they are looking for wasn't what was causing SARS. An accurate test for the virus would at least be a first step in breaking out of this problem.

By April 2003, WHO's cooperating labs had already come up with several possible tests. None was perfect. One never identified the coronavirus when it wasn't there (no "false positives"), but sometimes failed to identify it when it was (a few "false negatives"). Another test for antibodies to the virus was useful only after the patient had been sick for three weeks. These tests clearly weren't going to help the patients, but they could tell who had coronavirus and who didn't. For a little over a month's work, it was phenomenal progress—a triumph of international cooperation and a tribute to the dedication of Urbani and his many colleagues.

41.1 The Cast of Characters

Our biosphere teems with viruses, bacteria, fungi, protozoans, and parasitic worms, many of which will eagerly colonize humans and other living things. Some, like the SARS virus, seem to appear out of nowhere. Such pathogens already exist among birds, pigs, and other animals that humans come in contact with when a new version evolves that can infect and sicken people. The evolutionary success of pathogens depends on their ability to infect old hosts and new ones.

But our own evolutionary success depends on our ability to prevent and combat invasions of these tiny pathogens. In this chapter we review several lines of defense with which mammals protect themselves from pathogens—including the skin, the inflammation and blood-clotting systems, and the B and T cells of the immune system.

The skin keeps out pathogens as effectively as a castle wall keeps out human invaders. However, any opening in such a barrier—whether the mouth or a cut or puncture wound—allows pathogens access to the body. Should even the tiniest vessel of the circulatory system be cut, **blood clots** quickly plug any holes. At the same time the process of **inflammation** begins, with white cells swarming to engulf and consume bacteria, viruses, debris, and other elements that do not belong in the body. Both of these defenses are general; they don't depend on what kind of organism is attacking the body.

In contrast, the **immune system** can distinguish among individual pathogens, responding more vigorously to some than others and tolerating the body's own cells in some cases and not in others. The immune system consists of cells that can recognize and attack toxins, viruses, bacteria, and other microorganisms, and even cells of the body that have been infected. Some of the immune cells eat (phagocytose) invading viruses or cells, some punch holes in pathogens (lysis), and some secrete **antibodies**—molecules that both mark certain cells, viruses, or molecules for destruction and even disable them directly. Together, the antibodies and the various cells can incapacitate virtually any kind of pathogen (Table 41-1).

All immune cells are white cells, or **leukocytes** [Greek, *leukos* = clear, white], which come from stem cells in the bone marrow. *Stem cells* are immature cells with the potential to become any of a range of different cells. Stem cells can either differentiate into various kinds of mature, specialized cells or divide to produce more stem cells. Bone marrow stem cells form both white cells and the red blood cells that carry oxygen.

Biologists classify these many white cells into those that respond in the same way no matter what is attacking the body (nonspecific responses) and those that distinguish one pathogen from another (specific responses). Four kinds of cells mainly participate in the nonspecific responses—neutrophils, eosinophils, basophils, and mast cells, and two other kinds of cells, called **lymphocytes**, participate in the specific responses. These lymphocytes are known as B cells and T cells. In addition, two more—natural killer cells and **macrophages** [Greek, *makros* = large + *phagein* = to eat], the giant cell-eating white cells—can participate in immune responses, for a total of eight kinds of white cells (Table 41-2).

The nonspecific cells react to any injury or infection and respond to signals from the lymphocytes. But the B cells, which make antibodies that recognize individual proteins, can distinguish a coronavirus from a polio virus and a pollen protein from a peanut protein. The B cells remember almost any pathogen our immune cells have encountered in our lives. T cells, whose surfaces are studded with antibody-like receptors (called "T cell receptors"), also recognize specific pathogens.

The body defends itself against invading pathogens with three lines of defense: (1) the skin, (2) blood clots and inflammation, and (3) the immune system. Eight kinds of white cells coordinate both nonspecific inflammation and specific immune responses.

Table 41-1
How to Kill a Pathogen

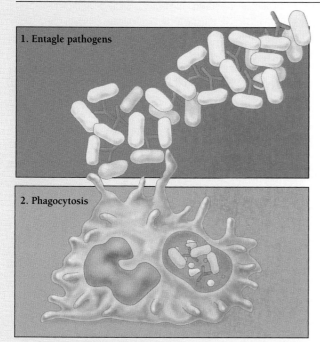

1. Entagle pathogens

2. Phagocytosis

3. Lysis

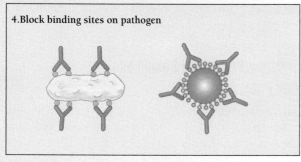

4. Block binding sites on pathogen

Table 41-2
Some Types of White Cells (Leukocytes)

Nonspecific (Inflammation Response)
Neutrophils—Fast acting. Phagocytic on bacteria and larger parasites

Eosinophils—Attack bacteria and larger parasites

Basophils—Reside in the bloodstream and release histamines, which trigger inflammation

Mast cells—Reside in the tissues and release histamines, which trigger inflammation and attract neutrophils in response to IgG antibodies.

Natural killer cells (NK cells)—Break open and kill infected or cancerous cells.

Macrophages—Slow acting. Phagocytic on microorganisms and debris that are marked with antibodies, present antigen, and release four kinds of intercellular signals called cytokines.

Specific
B cells—Develop into plasma cells, which make antibodies, and memory cells.

Helper T cells—In response to infected and altered cells, stimulate the production of B cells and other T cells by means of molecular signals.

Cytotoxic T cells—Recognize and kill infected cells and cancer cells.

41.2 How Does the Skin Keep Pathogens Out?

The skin is a large, stretchable barrier that repels most pathogens. In humans, the skin makes up about 15 percent of the body's weight. Like the linings of the gut, mouth, and other internal surfaces of the body, the skin consists of epithelial cells. But unlike the thin, internal epithelia, the skin epithelium consists of a thick pile of cell layers, which together make up the skin (Figure 41-3).

The outermost layer is the **epidermis** [Greek, *epi* = on + *derma* = skin]. At the very top is a thin layer of nondividing dead and dying cells that protect the dividing cells beneath. These outer cells are continually exposed to scraping, drying, infection, and other injuries. On average, these cells live only about a month. As the outer cells die, still-dividing cells from the lower layers replace them. Beneath the epidermis is the **dermis,** a thicker layer of living cells, and a layer of **hypodermis.** Embedded within these layers are hairs, nerve endings, muscles, sweat glands, oil glands, blood vessels, and receptors for heat, cold, and pressure.

At the top of the epidermal layers, the oil and sweat secreted by glands in the skin create a slightly acid environment that is hostile to the growth of many microorganisms. The skin nonetheless swarms with microorganisms that are adapted to these conditions. Most are benign bacteria and yeasts that, as a community, cause no damage. Indeed, many actively compete with pathogenic organisms, reducing the chance that pathogens can enter the body or that any one microorganism grows out of control. (Sometimes, antibiotics kill off too many of our natural

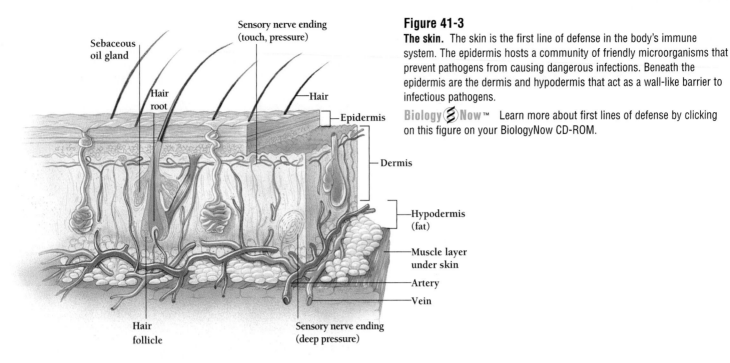

Figure 41-3

The skin. The skin is the first line of defense in the body's immune system. The epidermis hosts a community of friendly microorganisms that prevent pathogens from causing dangerous infections. Beneath the epidermis are the dermis and hypodermis that act as a wall-like barrier to infectious pathogens.

Biology⊘Now™ Learn more about first lines of defense by clicking on this figure on your BiologyNow CD-ROM.

bacteria and leave us open to infections by yeasts and other fungi.)

The skin and its community of bacteria, yeasts, and other organisms are so effective a barrier that most microorganisms cannot enter the body unless they do so by way of the mouth, nose, eyes, or breaks in the skin. Even those that do enter on particles of food or dust usually die because few survive exposure to **lysozyme**, an enzyme found in saliva and tears that digests the cell walls of many microorganisms.

A few microorganisms reach the digestive tract or respiratory tract. In the respiratory tract, cilia sweep mucus and trapped microorganisms up into the throat. From there they are swallowed and pass down the esophagus into the stomach, where they join a host of other microorganisms.

The stomach's acid environment is enough to kill most microorganisms. Some—*Salmonella* bacteria or poliovirus, for example—survive and pass into the digestive tract. Vast numbers of microorganisms flourish in the digestive tract. Indeed, more individual bacterial cells inhabit the gut than there are cells in the body. As in the case of the microorganisms of the skin, however, most are benign and usually compete with pathogenic bacteria for space and resources. As a result, under normal conditions, foreign microorganisms cannot survive in the gut.

The first barrier to invasion by microorganisms is the skin. Lysozyme in saliva and tears kills many foreign microorganisms that enter the eyes, mouth, or nose. Microorganisms that manage to enter the intestinal tract must compete with millions of others—almost all benign.

41.3 How Does the Body Defend Itself When the Skin Is Broken?

The skin provides the first line of defense against infection or invasion, and relatively few microorganisms breach its barrier. But wounds can open the way to infection and also damage blood vessels. Damaged blood vessels pose two dangers: blood loss and the rapid dispersal of microorganisms throughout the body by way of the circulating blood.

How Does the Body Limit Blood Loss and Keep Pathogens Out?

Mammals have evolved two adaptations—wound healing and "hemostasis"—each of which minimizes both bleeding and infection. When a wound heals, normal skin cells divide as they lose contact with other cells and stop dividing once they regain contact. But growing new cells to fill the gap is a process that takes time—days or even weeks. A shorter-term response that can take place in just seconds is **hemostasis** [Greek, *haima* = blood + *stasis* = standing], the collection of mechanisms that slow blood loss and the entry of microorganisms.

Hemostasis involves at least three interconnected mechanisms (Figure 41-4): (1) the contraction of smooth muscles in the damaged vessels, called **vasoconstriction**; (2) the clumping of small cellular elements in the blood, called **platelets**, at the site of injury; and (3) the formation of a blood clot, a lump of coagulated blood.

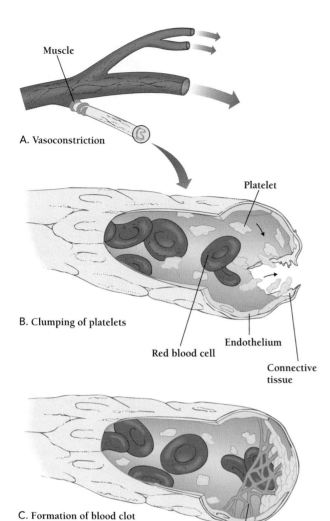

Figure 41-4
Hemostasis. If the skin is broken, the circulatory system also works to keep out pathogens. The circulation has three ways of preventing blood loss. A. Vasoconstriction. B. Clumping of platelets. C. Formation of blood clots.

Any injury immediately triggers the contraction of nearby blood vessels. Such vasoconstriction sometimes seals the leaky vessel completely, but at the least it reduces the loss of blood. Injury to the wall of a blood vessel also exposes collagen, the major structural protein of connective tissue. Platelets in the blood adhere first to the collagen and then to each other, forming clumps of platelets that can completely plug tiny openings in the wall of the blood vessel. Such minute ruptures occur hundreds of times every day, only to be sealed by platelets. In addition, the platelets stimulate further vasoconstriction and initiate clotting, the mechanism by which the blood forms a gell-like lump that blocks blood flow.

Although platelets can help initiate clotting, clotting is strictly a property of the plasma, the liquid, noncellular part of the blood. Clotting can occur in the absence of any whole blood cells, even in a tube of plasma outside the body.

The body closes breaks in the skin by means of wound healing, which prevents microorganisms from entering. Breaks in the circulatory system are sealed by means of vasoconstriction, the aggregation of platelets, and blood clots, all of which prevent blood from leaking out.

How Does the Body Prevent Damaging Blood Clots?

A serious wound can open up hundreds of blood vessels, including major veins and arteries. Halting the bleeding is critical for immediate survival, and the clotting system does that. But blood clots are also extremely dangerous. A single clot can block an arteriole in the brain (causing a stroke), in the heart (causing a heart attack), or in the lungs (causing a thrombosis). Any one of these events can cause permanent damage or death. The body therefore limits the formation of clots with a complex array of molecular checks (Figure 41-5).

A blood clot consists of a network of fibers made of the protein fibrin. Fibrin is not present in unclotted blood; otherwise, clots would form all the time and kill us. Fibrin forms from another protein called fibrinogen. When the enzyme thrombin digests away part of the fibrinogen polypeptide, it becomes fibrin. Thrombin is not normally in the blood either. Otherwise it too would cause the blood to clot all the time. Instead of thrombin, blood plasma contains prothrombin, a precursor to thrombin. At the site of tissue damage, another enzyme, called factor X ("factor ten"), converts inactive prothrombin to active thrombin, initiating the clot. Factor X is itself normally inactive, but several other enzymes can convert it to an active form at the site of damage.

What initiates the first step in this cascade is ordinary collagen—the main protein in connective tissue (Figure 41-5). When the connective tissue in blood vessel walls is damaged, the exposed collagen stimulates the clotting cascade. At each succeeding step, a single molecule of one enzyme activates many molecules for the next step. The result is a tremendous multiplication of the initial effect, so that a minor injury that exposes just a little bit of collagen causes the formation of huge amounts of fibrin.

The clotting cascade is so powerful that the body needs a way to quickly destroy fibrin so that all the blood does not clot. This anticlotting system consists of a multiplying cascade of enzymes similar to the clotting cascade. The last enzyme in the cascade is *plasmin*, which digests the bonds between fibrin molecules and dissolves the clot (Figure 41-5). Plasmin comes from *plasminogen*, which is activated by *tissue plasminogen activator (tPA)* or other enzymes. Some tissues have more tPAs than others. The uterus, for example, contains high levels of a tPA, which inhibits the clotting of menstrual blood.

Physicians treat heart attack victims with synthetic tPA, which, when given intravenously, stimulates the production of plasmin. The plasmin breaks up the clots that are block-

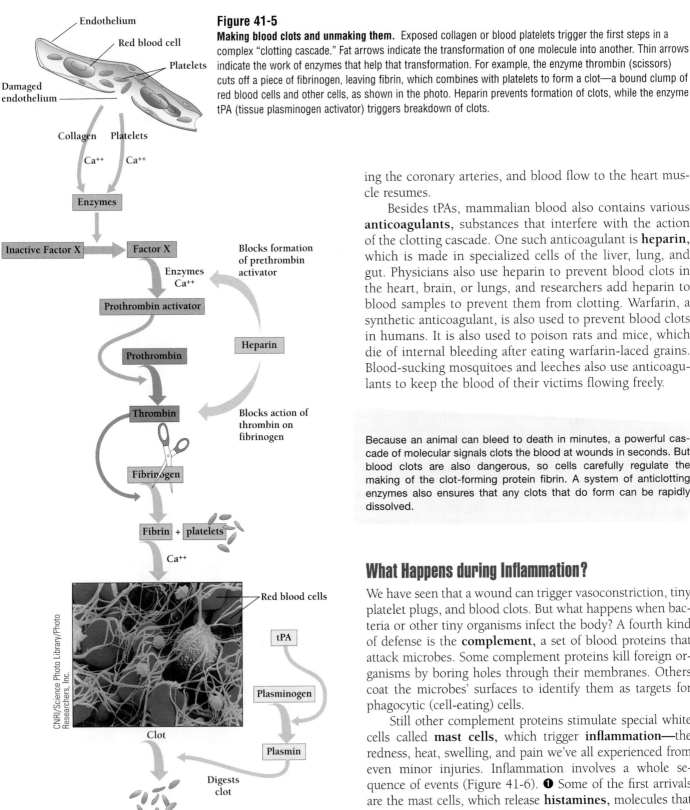

Figure 41-5

Making blood clots and unmaking them. Exposed collagen or blood platelets trigger the first steps in a complex "clotting cascade." Fat arrows indicate the transformation of one molecule into another. Thin arrows indicate the work of enzymes that help that transformation. For example, the enzyme thrombin (scissors) cuts off a piece of fibrinogen, leaving fibrin, which combines with platelets to form a clot—a bound clump of red blood cells and other cells, as shown in the photo. Heparin prevents formation of clots, while the enzyme tPA (tissue plasminogen activator) triggers breakdown of clots.

ing the coronary arteries, and blood flow to the heart muscle resumes.

Besides tPAs, mammalian blood also contains various **anticoagulants,** substances that interfere with the action of the clotting cascade. One such anticoagulant is **heparin,** which is made in specialized cells of the liver, lung, and gut. Physicians also use heparin to prevent blood clots in the heart, brain, or lungs, and researchers add heparin to blood samples to prevent them from clotting. Warfarin, a synthetic anticoagulant, is also used to prevent blood clots in humans. It is also used to poison rats and mice, which die of internal bleeding after eating warfarin-laced grains. Blood-sucking mosquitoes and leeches also use anticoagulants to keep the blood of their victims flowing freely.

Because an animal can bleed to death in minutes, a powerful cascade of molecular signals clots the blood at wounds in seconds. But blood clots are also dangerous, so cells carefully regulate the making of the clot-forming protein fibrin. A system of anticlotting enzymes also ensures that any clots that do form can be rapidly dissolved.

What Happens during Inflammation?

We have seen that a wound can trigger vasoconstriction, tiny platelet plugs, and blood clots. But what happens when bacteria or other tiny organisms infect the body? A fourth kind of defense is the **complement,** a set of blood proteins that attack microbes. Some complement proteins kill foreign organisms by boring holes through their membranes. Others coat the microbes' surfaces to identify them as targets for phagocytic (cell-eating) cells.

Still other complement proteins stimulate special white cells called **mast cells,** which trigger **inflammation—**the redness, heat, swelling, and pain we've all experienced from even minor injuries. Inflammation involves a whole sequence of events (Figure 41-6). ❶ Some of the first arrivals are the mast cells, which release **histamines,** molecules that stimulate inflammation. ❷ Capillaries open, or *dilate,* in the injured region. As the walls stretch, gaps open up between the cells that make up the walls, so that white cells and complement proteins in the blood can easily slip out of the capillaries and into the tissues. ❸ Attracted by complement proteins, large white cells called **neutrophils** migrate to the site

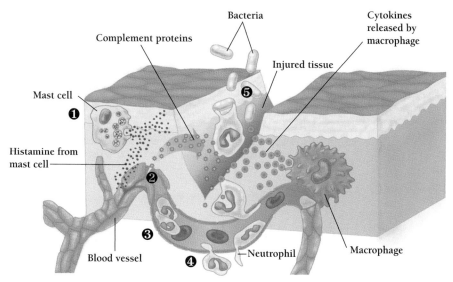

Bacteria

Complement proteins

Cytokines released by macrophage

Injured tissue

Mast cell

❶

❺

Histamine from mast cell

❷

Blood vessel

❸

❹

Neutrophil

Macrophage

Figure 41-6

A nasty cut. Tissue damage such as a cut ❶ stimulates the granules in a mast cell to release histamine, causing inflammation. ❷ Capillaries dilate and release complement proteins and white cells into the injured tissue. ❸ Attracted by complement proteins signaling molecules, ❹ neutrophils move into the damaged tissue and help initiate the (slower) specific immune response, ❺ eliminating bacteria or other pathogens.

of infection or damage. ❹ The neutrophils move out through the gaps in the capillary wall and in amongst the cells of the damaged tissue. ❺ Finally, the neutrophils bind to and phagocytose (engulf and digest) any microorganisms in the damaged tissue. Neutrophils can only bind to particles that have been marked for phagocytosis, either by complement proteins or antibodies.

The mast cells that initiate all this activity are large, round cells found in connective tissues throughout the body. Stuffed with histamine-filled vesicles, mast cells release their histamine in response to even the smallest paper cut. Allergy sufferers know a few of histamine's distinctive effects—the swelling and itchiness in the eyes, nose, throat, and other tissues. Histamines also trigger similar effects in response to infection by cold viruses. *Antihistamines* are drugs that block the effects of histamines. An antihistamine is useful for treating cold symptoms, allergies, and even insect bites, which also trigger the release of histamines.

Both neutrophils and macrophages ingest cells, viruses, and other materials. But neutrophils are more common (making up about 60 to 75 percent of all white cells), are the first to the scene of a wound or infection, and work only briefly. The enormous macrophages are less common, take longer to arrive, but continue ingesting microorganisms and debris over longer periods (Figure 41-7). Macrophages are found in connective tissue in the liver, brain, spleen, lungs, and lymph nodes.

Macrophages also perform an important role that neutrophils do not. As a macrophage digests invading microorganisms, it moves recognizable proteins from the pathogen out to its own cell membrane. Immunologists call this process "presenting antigen" because the macrophages cover themselves with the proteins that antibodies will respond to (Figure 41-8). Macrophages also send molecular signals that stimulate other immune cells. Cells "present" antigen in the same way that a bloodhound's handler gives a dog the shirt

Engulfed material

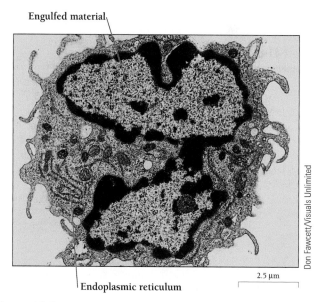

2.5 μm

Don Fawcett/Visuals Unlimited

Endoplasmic reticulum

Figure 41-7

A macrophage. This macrophage has engulfed something large, perhaps a bacterium. Also, notice at left the endoplasmic reticulum and Golgi apparatus, which produce and secrete signaling molecules such as cytokines.

of a lost child to sniff. From the distinctive odor the child has left on the shirt, the dog knows who to look for. Likewise, from the distinctive shape of the antigen presented by a cell, the immune system "knows" which antigen to look for.

Macrophages also release small messenger proteins called **cytokines** that regulate cell division, protein synthesis, and migration in other white cells. Cytokines include *interferons, tumor necrosis factor-α* (TNF-α), and various kinds of *interleukins*. TNF-α acts locally, stimulating macrophages and neutrophils to increase their phagocytic

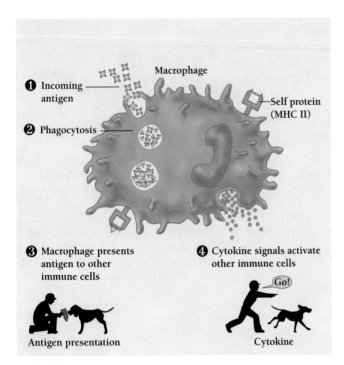

Figure 41-8

Presenting antigen. Macrophages are like the human tracker who waves a missing child's shirt in front of a bloodhound's nose. The dog knows to look for that particular smell out of all others. In the same way, ❶ a macrophage engulfs an antigen such as a cold virus, ❷ digests the virus's proteins into small pieces, then ships these pieces to the surface of the macrophage. ❸ There, the viral parts are displayed, or "presented," in the context of special "self" proteins. Like bloodhounds, specialized B and T cells respond by making antibodies and cells that will find this particular virus. ❹ Macrophages also release cell-signaling molecules called cytokines that further stimulate the activity of T and B cells. This is like the human tracker telling the bloodhound, "Go!"

---

rifice also prevents the cell from making and releasing any more viruses.

Complement proteins and mast cells react nonspecifically to damaged body tissues. Mast cells release histamines, which promote inflammation. Inflammation dilates blood vessels, increasing the delivery of white cells and complement proteins to damaged tissues. Macrophages phagocytose pathogens and also send cytokine signals that stimulate inflammation, fever, and the activity of other immune cells.

41.4 How Do We Become Immune to Disease?

Most of us have had contact with someone infected with chickenpox. If we have not had chickenpox, or been vaccinated against it, we may come down with chickenpox. If we've already had it, though, we say we are "immune to" chickenpox. What is immunity and how do we become immune? And how does a vaccine make us immune?

Four attributes characterize immunity: *specificity, memory, self-nonself recognition,* and *diversity.* We say the immune system is *specific* because it can tell a chicken pox virus from a cold virus. The specificity of the immune response distinguishes it from the **nonspecific** responses of clotting, complement proteins, inflammation, and interferons, for example. We say the immune system has *memory* because once the immune system has developed defenses against the chickenpox virus, it can attack the virus again and again— even decades later. If chickenpox virus shows up again, the immune system's response is usually even faster and stronger than the first one.

The immune system usually ignores the body's own cells as long as they are healthy. The immune system's apparent ability to distinguish the body's own cells from foreign cells and other particles is called *self-nonself recognition.*

The immune system is capable of recognizing and attacking an incredible *diversity* of pathogens, including parasitic worms, fungi, bacteria, viruses, and plant pollen, as well as purified proteins and other molecules. Our immune system can even respond to molecules that have been synthesized by organic chemists. How can an animal have specific responses to so many different substances, even to ones neither the animal nor any of its ancestors likely encountered in all of evolutionary history?

The answer is antibodies. The immune system makes millions of different protein **antibodies**, each of which binds to a specific **antigen** [an *anti*body *gen*erator]. Antigens are usually foreign proteins or other molecules, recognizable by the unique shape and electrical charge of subunits called **epitopes** [Greek, *epi* = upon + *topos* = place] (Figure 41-9). An antigen may have many epitopes, each of which maybe recognized by a different antibody.

---

activity. Interleukins [Latin, *inter* = between, among + Greek, *leukos* = white] carry messages among different white cells throughout the body, transmitting information by binding to cell-surface receptors. Interleukins can trigger both general inflammation responses and specific immune responses.

Interleukin-1 (IL-1), for example, attracts phagocytic cells to the site of injury, stimulates the T and B cells of the immune system, and causes fever by stimulating the cells in the brain's hypothalamus that regulate body temperature. Mild fevers help the immune system work better. A fever of about 102°F (39°C) slows the growth of bacteria, increases complement proteins, and causes T and B cells to multiply faster. A very high fever (105°F or higher) is dangerous, however, especially in adults.

Cells infected by viruses produce still another set of cytokines, called interferons. Interferons stimulate phagocytosis and interfere with the replication of viruses by slowing protein synthesis in cells infected by viruses. The shutdown of protein synthesis may kill the body's own cell, but the sac-

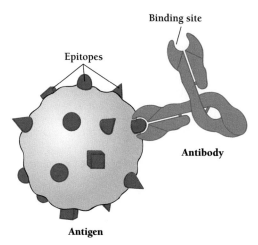

Figure 41-9
Antibody, antigen, and epitopes.
Each of the millions of antibody molecules that the body makes binds to a specific epitope—a subunit of the antigen that has a distinctive shape and charge.

Antibodies activate the complement system and block the activity of viruses and other pathogens in different ways. Antibodies can bind to and block the receptors the virus uses to infect a host cell, tangle large numbers of viruses or bacteria into clumps, or mark cells for later destruction by various kinds of killer cells. Macrophages, for example, recognize an antibody marker as an invitation to destroy the antigen as surely as a meter reader recognizes an expired parking meter is an invitation to give a parking ticket (Figure 41-10).

The immune system is characterized by specificity, memory, diversity, and its ability to distinguish self from nonself. Antibodies are responsible for specificity and diversity.

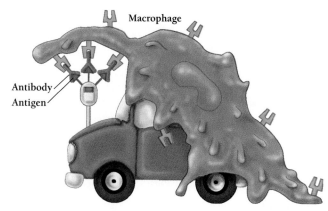

Figure 41-10
A macrophage engulfs anything marked with antibodies. Well, maybe not a car. But antibodies do mark a cell for destruction by macrophages in the same way that an expired parking meter marks a parked car for a parking ticket.

How Does An Antibody Recognize Its Antigen?

All antibodies are proteins called **immunoglobulins**. Immunoglobulins are made of two kinds of polypeptide chains—light chains and heavy chains—and come in five main shapes, called IgG, IgM, IgA, IgD, and IgE (Figure 41-11). These are pronounced "immunoglobulin G," "immunoglobulin M," and so on. The most common antibodies are the IgG antibodies, which consists of four polypeptide chains shaped like a Y (Figure 41-11). IgG antibodies act against viruses, bacteria, and bacterial toxins.

Antigens bind at "binding sites" at the top of the arms of the antibody. The different immunoglobulins have different numbers of binding sites, which gives them the power to glue bacteria and viruses into submission. An antibody with several binding sites can bind to the same site on two different bacterial cells. Because hundreds of antibodies can bind to different bacteria, the bacteria become linked together into a tangled mass. These masses of antibody-covered bacteria cannot easily spread through the body and they are easy targets for phagocytic cells.

Each of the five kinds of antibodies plays a different role. Some activate complement proteins, which then destroy any marked cells. Others protect the surface of the body such as the intestinal, respiratory, and urogenital tracts and are abundant in tears, saliva, and milk.

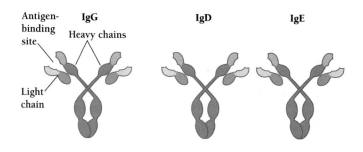

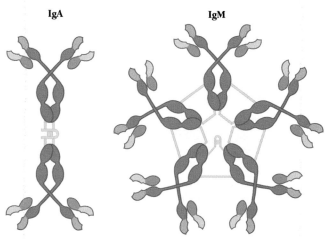

Figure 41-11
The five classes of antibodies. The five kinds of immunoglobulins differ in their heavy chains. Notice that IgG, IgE, and IgD have 2 binding sites each, while IgA has 4 and IgM has 10.

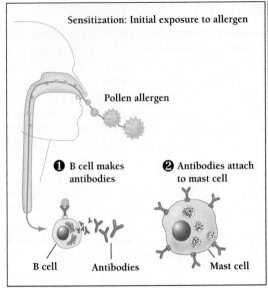

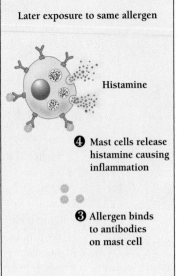

One group, the IgE antibodies, plays a role in allergies. Many people make lots of IgE antibodies in response to antigens that are not dangerous at all, including pollen, nuts, shellfish, or dust mites. This IgE response is the basis for most *allergic* reactions. When one of these antigens enters the body of a sensitive person, IgE molecules attach themselves to mast cells. When the antigen shows up again, it stimulates the mast cells to release histamine (Figure 41-12). The histamine triggers the inflammatory response, increased mucus secretion, and contraction of the airways, causing congestion, sneezing, a runny nose, and difficulty in breathing. An extreme response, called anaphylaxis, can result in death.

Antibodies are proteins called immunoglobulins with two or more binding sites for the epitope of a specific antigen. IgE antibodies are involved in allergies.

B Cells: Where Do Antibodies Come From?

Antibodies don't just appear out of nowhere. They are made by a special class of white cells called lymphocytes. **Lymphocytes** are white cells found in the lymphoid tissues (including lymph nodes, spleen, thymus, and tonsils) that develop from stem cells in the bone marrow (Figure 41-13).

The body makes two kinds of lymphocytes—B lymphocytes, or **B cells**, which make antibodies, and T lymphocytes, or T cells, which do other things. B cells mature in the Bone marrow, hence their name. Animals can make hundreds of millions of kinds of B cells. But each kind of B cell makes just one kind of antibody. We'll talk about the other major group of lymphocytes, T lymphocytes, or **T cells,** a

little later in this chapter (Figure 41-13). T cells play a different role and mature in the Thymus, hence their name.

During development, the stem cells that will form the various B cells give rise to a library of different B cells, each one making a different antibody (see box). Each B cell in the library and all the cell's descendants are called a "cell line" or a "clone," because they are identical. But since each line of B cells can make only one kind of antibody, how do the right B cells "know" when to spring into action, dividing and producing antibodies when their particular antigen shows up?

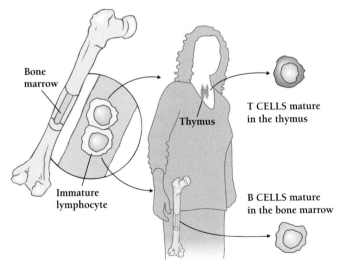

Figure 41-13
Where do B and T cells come from? B cells and T cells—the cells responsible for the immune system's *specificity, diversity,* and *memory*—first develop in bone marrow from stem cells. Some of these cells stay in the bone marrow and mature into B cells *(yellow),* which make antibodies. Others migrate to the thymus and mature into T cells *(purple).*

Extreme Biology   How Do B Cells Make So Many Different Antibodies?

The immune system is able to field 100 million or more kinds of antibodies, each with a unique three-dimensional shape able to bind to a single epitope. How does the immune system make antibodies with so many different shapes? Each set of B cells that produces a particular antibody is called a "clone," a group of identical cells. So if the body makes 100 million different antibodies, these are made by 100 million different kinds of B cells.

If every different antibody protein needed its own gene, the body would need about 100 million genes. But humans only have about 30,000 genes and many of those are unrelated to the immune system. How can a few thousand genes carry the information to make millions of unique antibody proteins?

Two facts make this problem solvable. First, as we have seen, every antibody is made from two polypeptides, so that different polypeptides can be combined in ways that multiply their differences. Just as we can combine five shirts and two pairs of jeans to make 10 unique outfits, so lymphocytes can combine 300 versions of one polypeptide with four versions of another to make 1,200 unique polypeptides. Second, it turns out that two different genes can be spliced together, with surprising results.

In 1965, two researchers suggested that the antibody genes in different B cells might mutate and become different from each other during B cell development. The different regions of these genes could then recombine to form even more kinds of antibody genes. In other words, these researchers were proposing that antibody diversity was not inherited at all, but arose during the development of the individual. It was a radical suggestion.

The idea was intriguing. But for a long time, biologists had no way of knowing if this could really happen. Then in 1976, Japanese-born biologist Susumu Tonegawa and colleagues working in Switzerland discovered that lymphocytes have unique arrangements of antibody genes found in no other cells in the body. Tonegawa showed that in B cells, separate DNA fragments join in unique ways to make new genes that coded for a single antibody polypeptide. Tonegawa's discovery was so revolutionary that in 1987 he received a Nobel prize.

We now know that each light chain is formed from three different gene sequences that can combine in 3,600 different ways and that each heavy chain is formed from four different gene sequences that can combine in 130,000 ways. Together, the 3,600 light chains and 130,000 heavy chains can potentially form almost half a billion different antibodies. As if these half a billion possibilities were not enough, immunoglobulin genes also mutate at high rates, further increasing antibody diversity. Mammals can probably make more than a billion different kinds of antibodies.

To answer that question, we need to know how B cells respond to antigen. The surface of each B cell is studded with a special form of its own antibody called a **B-cell receptor**. When an antigen such as a cold virus first appears in the blood and binds to the B-cell receptors, the B cells begin dividing and differentiating along two separate pathways. One line of B cells become **plasma cells,** which make and secrete antibodies. The other line of B cells become **memory cells.** Memory cells do not make antibodies but act as the immune system's memory after the first infection is over. In the absence of the antigen for that particular cold virus, the memory cells do nothing. But if the antigen shows up again, the memory cells divide rapidly to make huge numbers of plasma cells, which make antibodies (Figure 41-14).

During the first exposure to a virus or other antigen, the immune system reacts slowly because the number of plasma cells and antibodies is very low. This first slow reaction is called the **primary immune response.** But during the primary response, the immune system makes lots of memory cells so that it will be ready the next time. On a second exposure to the same antigen, the many memory cells ensure a much faster and stronger response, called the **secondary immune response** (Figure 41-14). The secondary response may be so strong that the body completely fights off the infection without the person ever knowing about it. The multiplication of memory cells explains the decades-long immunity that children acquire after exposure to a disease such as chickenpox. But the secondary immune response doesn't only happen the second time we encounter a virus or other pathogen. It can happen three times or fifty times if a pathogen keeps returning.

Antibodies are very specific, but not perfectly so. Because the epitopes that trigger an immune response are only small parts of antigens, antigens that share similar epitopes can trigger immune responses for each other. For example, strep bacteria, which cause strep throat, share antigens similar to those on our own heart muscle. In some people, a strep infection can trigger the production of B cells whose antibodies not only attack the strep bacteria, but also heart muscle, a dangerous condition called rheumatic fever (see box on page 799).

When an antigen binds to B-cell receptors, the B cells divide and differentiate into (1) plasma cells, which make and secrete antibodies, and (2) memory cells, which make more plasma cells if and when the antigen ever appears again. The secondary response is usually faster and stronger than the primary response.

How Do Vaccines Work?

A vaccine is an antigen that stimulates the B cells to make memory cells but doesn't actually make us ill. For example, heat-killed polio virus cannot cause polio. But it contains

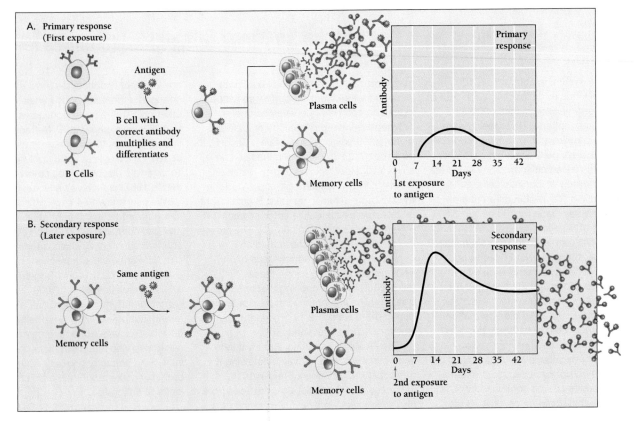

Figure 41-14

Making memories. When B cells encounter antigen, they make both plasma cells and memory cells. The plasma cells make antibodies to the antigen right away, but it takes a few days for the body to make enough plasma cells to fight off an infection. This is the typically slow primary immune response. But meanwhile, the memory B cells are also multiplying. If the same antigen shows up again at a later date, the many memory cells quickly produce lots of plasma cells, mounting a faster and more vigorous immune response—the secondary immune response.

the same protein antigen as the live virus and stimulates the same B cell line. So an injection with the inactivated material stimulates the B cells to make a population of memory cells against polio.

Some vaccines are made from live viruses that are "attenuated," which means they are weak versions of the real thing that don't make people very sick. The vaccine against measles, for example, is made with attenuated viruses. Still other vaccines are made from purified proteins from viruses or bacteria. Efforts to make such a vaccine against AIDS have so far failed.

Vaccines trigger the same kind of immunity that an infection does—called active immunity because the body makes memory cells. But there is another way of acquiring both antibodies and lymphocytes, called passive immunity, that will defend a person against an infection. For example, an injection of antibodies against measles can prevent infection in someone who has already been exposed to the disease. The antibodies are not "homemade," made by the body itself, but they mark the measles virus for destruction by the person's own macrophages and perform all the other services that their own antibodies would have. Sometimes these injections of antibodies are called gamma globulin. Another form of passive immunity occurs when mothers breast-feed their babies. Breast milk contains both antibod-

ies and immune cells, so babies who are breast-fed have more resistance to infections than formula-fed babies.

Vaccines stimulate B cells to make memory cells for a specific antigen, without making a person ill. Vaccines may consist of dead pathogens, live but attenuated pathogens, or of a purified protein from the pathogen. Breast milk and injections of antibodies can both provide passive immunity.

T Cells: When Good Cells Go Bad

So far we've focused almost exclusively on antibodies and the B cells that make them. But antibodies have an important limitation. They do not recognize the body's own cells even when those cells harbor viruses or other pathogens. Because colds, influenza, AIDS, and many other infections are intracellular (inside the cells), the body needs a way to recognize when its own cells have "gone bad." T lymphocytes, or **T cells**, recognize and destroy body cells that have become infected or cancerous.

Like B cells, T cells initially arise from stem cells in the bone marrow. But unlike B cells, T cells migrate to the thymus, a gland just below the neck, to mature. The thymus is like an elite school for immune cells. There the T cells learn

Health and Biology   Why Does the Immune System Sometimes Attack Itself?

Although the immune system is largely successful in distinguishing self from nonself, about five percent of the population suffers from some "autoimmune disease," in which the immune system attacks the body's own tissues, causing tissue damage or malfunction.

In the disease myasthenia gravis, for example, the body's own antibodies block muscle cell receptors that normally respond to acetylcholine, the chemical signal that triggers muscle contraction. When antibody levels are high, the muscles do not contract, resulting in weakness and even paralysis. One treatment is to remove the antibodies from the patient's blood by removing the blood plasma, removing the antibodies, then recycling the blood back into the body.

In insulin-dependent diabetes mellitus (juvenile diabetes), which affects about 0.5 percent of the U.S. population, T cells attack and destroy the body's own insulin-producing pancreas cells. Unable to make insulin, diabetics must inject themselves with insulin every day.

Until the early 1970s, biologists believed that in healthy people, the immune system always eliminated any B and T cells that reacted against the body's own cells and that autoimmune diseases were the result of some failure to eliminate such autoimmune cells. But it turns out that all healthy adults contain some autoimmune cells. In most people, such autoimmune cells are held in check, but in those with autoimmune disease, something goes wrong. But what?

Biologists have just a few clues. For example, the streptococcal bacteria that cause strep throat have surface antigens that resemble antigens on the cells in heart muscle, a happenstance called molecular mimicry. Antibodies and T lymphocytes recognize both antigens and attack not only the bacteria but also the heart muscle cells. Such an attack on the heart cells is called rheumatic fever, a serious and often fatal disease that was quite common in the days before antibiotics quickly cured strep throat infections.

In the case of insulin-dependent diabetes, T cells that respond to an antigen on a common virus (coxsackie virus) also recognize an enzyme made in insulin-producing pancreas cells. Researchers hypothesize that the viral infection causes the immune system to mount a permanent attack (aided by memory cells) against the pancreas.

One of the more surprising possible factors in autoimmune diseases may be pregnancy. Nearly 80 percent of people with autoimmune diseases are women, and researchers have only recently begun to ask why. One reason women are disproportionately affected may be "microchimerism," in which a woman carries stem cells belonging to other individuals, including her own mother and her own children.

During pregnancy, mother and baby can exchange stem cells—unspecialized cells that can form anything from skin cells to liver cells. Researchers have even discovered that some women can carry stem cells from

an older sibling. During the mother's pregnancy with a brother, say, the stem cells entered the mother's blood and continued to divide and live in her body. During a subsequent pregnancy, the descendants of the brother's cells enter the new baby. Blood from the umbilical cord may contain stem cells from two or more individuals. Men can pick up stem cells from their mothers and older siblings, too, but the likelihood of carrying several cell lines is greater in women.

Because women with documented cases of microchimerism have a higher risk of autoimmune diseases, some researchers have hypothesized that some sort of immunological battle among different cell lines is triggering autoimmune diseases in women. If the fetal stems cells are very similar to those of the mother, her immune system may tolerate them, and they differentiate to form skin, white cells, and various other tissues of the body. Later in life, the hypothesis goes, the immune system somehow recognizes the fetal cells and attacks them.

But so far no one has shown that microchimerism actually causes disease. In fact, some evidence suggests just the opposite—that fetal stem cells can rescue the diseased tissues of the mother. In one amazing case, a biopsy from a woman with liver disease revealed that her liver was partly composed of cells from a son she had conceived 20 years earlier. Her son's cells had apparently persisted for years, then differentiated to repair her liver.

---

to recognize the body's own cells. T cells that do not recognize body cells don't graduate. In fact, they are killed. T cells that learn their lesson enter the lymphatic system and blood and move around the body, destroying delinquent body cells.

Because the T cells are specialized for killing body cells, it's critically important that they be able to tell good cells from bad ones. It would be a disaster if the T cells began attacking normal healthy cells, for T cells are, like snipers, frighteningly good at what they do. For example, when people touch poison ivy or poison oak, oil from the plant penetrates the skin and marks healthy cells for destruction by T cells. The T cells may begin their attack within hours (Figure 41-15). T cells are also responsible for the body's rejection of organs surgically transplanted from other individuals.

T cells do not recognize foreign cells. T cells recognize only those cells that are like others in the body but some-

how different. Immunologists say that T cells recognize **altered self.** They do so with the help of molecules on the surfaces of T cells, called **T-cell receptors**, that recognize foreign antigens as altered self—as long as the antigens are in the context of a body cell.

The genes for T-cell receptors are much like the B cells' immunoglobulin genes. And, just as for antibodies, rearrangements of T-cell receptor genes form a vast T-cell library. Each line of T cells makes one kind of T-cell receptor, which recognizes just one antigen. In fact, T-cell receptors are even more diverse than the B cell's antibodies.

T-cell receptors recognize body cells that are cancerous or infected by a pathogen.

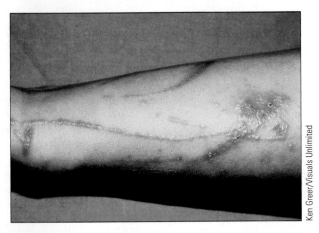

Figure 41-15

Poison ivy and cytotoxic T cells. An oil from poison ivy (and poison oak) happens to bind to proteins in skin cells, marking them as "foreign." Cytotoxic T cells "think" that the cells are infected with some pathogen and attack and destroy these otherwise healthy skin cells. The trick is to get the oil off the skin as soon as possible. Soap will work in the first few hours; special solvents can work in the first 24 hours or so. After that, the cells are marked for destruction and the T cells will arrive and attack. Cortisone and other similar steroid creams, as well as antihistamines, can help suppress the T_C cell response, but do not prevent it.

How Do T Cells Recognize Self?

T cells come in several types, but we'll focus on helper T cells, or T_H **cells**, which mainly manage other immune cells, and cytotoxic T cells, or T_C **cells,** which kill cells, as their name implies. T_C cells kill cells in more than one way. Some drill holes in their victim's cell membrane while others simply order body cells to die in the process of apoptosis, or programmed cell death (Chapter 45). T_H cells are central players in the immune response (see box on page 802). They release cytokines that tell B cells to make antibodies and also stimulate the multiplication of the killer T_C cells (Figure 41-16). T_H cells also bind antigen and present it to the B cells, much as macrophages do.

T cells recognize the body's own cells by means of special cell-surface proteins called **major histocompatibility complex,** or **MHC proteins.** The genes for MHC proteins vary enormously, so that (not counting identical twins) every individual's particular mix of MHC alleles is unique. It is on the basis of these unique MHC proteins that transplant surgeons "match" donated organs. The MHCs of a donated organ need not be identical to those of the organ recipient, but the MHCs must be similar, or else the recipient's immune system will attack and destroy the transplanted organ, a process called "rejection."

Two kinds of MHCs do similar but very different things. Both are plasma membrane proteins that present antigen fragments for recognition by T-cell receptors. **MHC Class I** proteins are present on the surface of almost every cell in the body and bind to T_C cells. In contrast, **MHC Class II** mole-

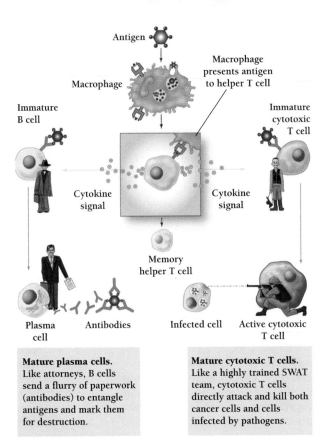

Mature plasma cells. Like attorneys, B cells send a flurry of paperwork (antibodies) to entangle antigens and mark them for destruction.

Mature cytotoxic T cells. Like a highly trained SWAT team, cytotoxic T cells directly attack and kill both cancer cells and cells infected by pathogens.

Figure 41-16

Tracking down the bad guys. An overview of the work of B cells and T cells. Once a macrophage or other cell presents antigen to helper T cells, the helper T cells send signals that activate the B and T cells to both hunt down the antigen and remember it. The B cells make plasma cells that make antibodies to mark antigen wherever it appears. Like the lawsuits, court orders, appeals, and other entangling paperwork produced by a good law firm, antibodies both mark the pathogen (defendant) and immobilize it, reducing its ability to cause disease. The B-cell "law firm" also creates a paper trail of memory cells that will immobilize the pathogen even faster if it ever shows up again. Meanwhile, helper T cells also alert the hyperaggressive cytotoxic T cells. Once "activated," cytotoxic T cells start killing any body cells infected by the pathogen. Cytotoxic T cells also have their own memory cells.

Biology ⊜ Now™ Learn more about the immune response by clicking on this figure on your BiologyNow CD-ROM.

cules occur only on the surfaces of antigen-presenting cells such as macrophages. When an antigen-presenting macrophage consumes a virus or microorganism, the macrophage covers its surface with fragments of the digested pathogen bound to MHC II proteins. The combination of MHC II and antigen alerts the T_H cells to the presence of cell-infecting viruses or microorganisms. The T_H cells then signal the B cells and T_C cells to take action (Figure 41-17).

T-cell receptors recognize "altered self" by means of MHC proteins. T cells only recognize antigens that are bound to MHC II proteins, found on the surfaces of special antigen-presenting cells.

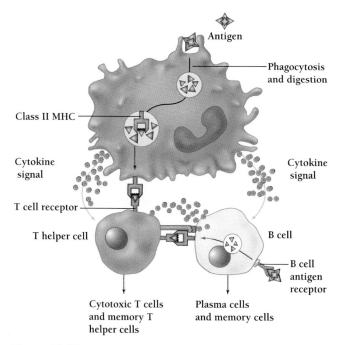

Figure 41-17

How self molecules help present antigen. Macrophages and other phagocytic cells present antigen chemically bound to Class II MHC "self" proteins. A helper T cell with the matching antigen receptor binds to the MHC II-antigen complex, which activates the helper T cell. The helper T cell activates a B cell that is displaying the same antigen and the B cell then multiplies to make plasma and memory cells. Activation of the B cell occurs partly through physical contact and partly through cytokine messages. Meanwhile, the helper T cell also activates the cell-killing cytotoxic T cells, which destroy any cells presenting a complex of antigen and MHC I self protein (not MHC II).

Individuals Have Distinctive Antigens

Every individual (except for identical twins) has a unique set of MHC proteins on their cells' surfaces. Individuals may also differ in other surface molecules. For example, human red blood cells have a surface glycoprotein that comes in three varieties, each genetically determined. These identifying molecules mean that anyone receiving a blood transfusion needs to get blood from the right "blood group."

The most important classification is that of the ABO blood groups. Individuals of blood group A have an antigen called A on the surfaces of their red blood cells; those of blood group B have another antigen, called B; those of group AB have both A and B; those of blood group O have neither A nor B. A type B or type O individual does not make A antigen and has antibodies to A antigen. They have these antibodies even if they have never been exposed to type A blood, probably because similar epitopes are present on the surfaces of common gut bacteria. Similarly, individuals who

do not have B (people with blood type O or A) have antibodies against B.

Before physicians understood these differences, blood transfusions were dangerous procedures. If a person with type A blood received a transfusion of type B blood, antibodies would attack the red cells, link them together in big clumps, and clog the blood vessels, often killing the patient.

Still another antigen on the red blood cells of some people is Rh factor—named after the rhesus monkeys in which the factor was first identified. Some people have the antigen and some don't. But unlike the unusual situation with the A and B antigens, people who lack Rh antigen do not generally make antibodies against it unless they are first exposed to Rh-positive blood.

Exposure to Rh blood can happen in two ways: after a transfusion of Rh-positive blood into an Rh-negative recipient—a situation that doctors try to avoid by testing the blood—or after a pregnancy in which an Rh-negative woman carries an Rh-positive baby. In that case, the mother may develop antibodies against Rh factor. Such antibodies are not usually present at high concentrations during the first pregnancy with an Rh-positive fetus. But during the delivery of the baby, the mother is likely to be exposed to the baby's Rh antigens, triggering the creation of memory cells in her own blood.

A later pregnancy with another Rh-positive baby can provoke a strong secondary response in the mother. The mother's memory B cells make plasma cells, which make anti-Rh antibodies that attack the fetus's red cells, sometimes killing the fetus. Physicians can prevent this dangerous immune response by treating Rh-negative mothers during their first pregnancy and all subsequent pregnancies with antibodies, giving her passive immunity.

In essence, the treatment prevents the immune system from learning about Rh-positive cells by killing the fetal cells in her blood. Anti-Rh antibodies injected into the mother's blood destroy Rh-positive fetal cells in her blood, so the mother makes no anti-Rh antibodies, and her B cells make no memory cells against Rh-positive cells that may appear in later pregnancies.

MHC proteins and other cell-surface antigens make the cells of every individual unique. Blood for transfusions must match the ABO and Rh blood types of the recipient's blood.

In this chapter we have seen how the body defends itself against pathogenic viruses, bacteria, and good cells gone bad. The power of the immune system—its specificity, memory, diversity, and self-nonself recognition—comes from the ability of its cells to communicate with one another. In the next chapter we will find out how the cells of the nervous system communicate with one another and with the muscles and organs of the body.

Health and Biology AIDS—Acquired Immunodeficiency Syndrome

In 1996, the number of new cases of AIDS, or acquired immunodeficiency syndrome, in the United States declined for the first time since the disease became known in 1980. Researchers heralded the six percent drop, from 60,620 new cases to 56,730, as proof of the success of both preventive programs and treatments. The good news in America is that fewer people are contracting AIDS and fewer people are dying of AIDS. Indeed, the annual death rate has dropped 70 percent, from 51,670 in 1995 to only 15,603 in 2001. The bad news is that there is still no cure and the total number of people in the United States who are living with AIDS is over 400,000, with another half million people carrying the HIV virus but not ill.

In the United States, 75 percent of people living with AIDS are men, among whom the majority contracted the disease through homosexual relations or intravenously injecting illegal drugs using contaminated needles. Education programs promoting safe sex and clean needles have dramatically reduced the transmission of AIDS among homosexual men and intravenous drug users.

Today about 20 percent of AIDS patients acquire the virus through heterosexual contact, up from 10 percent. This apparent doubling has caused alarm among some people, but the total number of new cases among heterosexuals has remained steady at around 12,000 per year for several years. It's a greater percentage of the total only because new cases of AIDS among homosexuals and drug users have declined so dramatically, which is good news.

AIDS is caused by HIV, the human immunodeficiency virus—an RNA virus that infects cells by reverse transcribing its RNA into DNA inside the cells it infects (see Chapter 13). While influenza viruses infect cells in the respiratory tract, for example, HIV's specialty is the immune system's helper T cells (T_H cells). These are the T cells whose cytokine messages mobilize the immune system by activating B cells and cytotoxic T cells. By attacking and killing T_H cells, the HIV virus renders the immune system almost helpless to defend itself. As the T_H cells die, AIDS patients are unable to fight off infections or certain kinds of cancers (Figure A). As a result, AIDS patients tend to die of pneumonia, fungal infections, and other common diseases that most people's immune systems fend off.

If the AIDS epidemic in the United States has slowed, the disease is spreading relentlessly in other parts of the world. Five million people became infected in 2002, and three million people died of the disease in that year. As of the end of 2002, some 42 million people were living with AIDS worldwide, nearly 30 million in sub-Sahara Africa, where AIDS is spread primarily through heterosexual contact. The World Health Organization predicts that another 45 million people will contract the disease by 2010.

Yet even amidst such grim statistics, many AIDS programs offer hope. Programs that primarily educate and treat young people have paid off in places like Ethiopia, South Africa, Uganda and Zambia. In Addis Ababa, for example, HIV infection rates among young women declined by more than one-third between 1995 and 2001. And in South Africa, HIV infections rates among pregnant teens dropped by a quarter between 1998 and 2001.

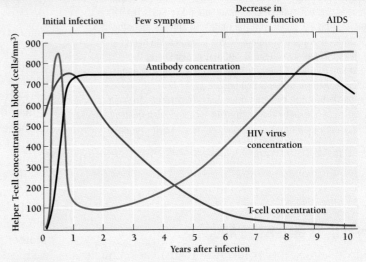

Figure A
Decline in T cells as AIDS infection progresses.

Key Concepts

- The skin is the first barrier against invading bacteria, viruses, fungi, and other pathogens.
- Inside the body, inflammation, complement, white cells, and signaling molecules provide nonspecific internal defenses against infection.
- The immune response demonstrates specificity, memory, diversity, and the ability to distinguish self from nonself.
- B cells and their antibodies recognize and help eliminate foreign molecules and cells. T_C cells recognize infected or cancerous body cells by means of "self" cell surface proteins.

Summary with Key Terms

How does the body keep microorganisms from colonizing the body?
The body defends itself against invading pathogens by means of both **nonspecific** defenses and the specific defenses of the **immune system**. Nonspecific defenses include the skin, the populations of microorganisms on the skin and in the gut, and **blood clots** and **inflammation**. The skin consists of the **epidermis**, the **dermis**, and the **hypodermis**. Saliva and tears contain **lysozyme**, an enzyme that dissolves the protein in bacteria and other pathogens.

The body deals with breaks in the circulatory system by means of **hemostasis**, which includes **vasoconstriction**, the clotting of the blood, and the aggregation of **platelets**. All of these processes prevent excessive blood loss and help prevent microorganisms from entering the circulatory system. Because blood clots are dangerous, the body regulates the synthesis of the fibrin in clots by means of a complex cascade of regulation. A second system of anticlotting enzymes destroys the fibrin in clots. Mammalian blood also contains **anticoagulants**, such as **heparin**.

All immune cells, whether specific or nonspecific, are white cells, or **leukocytes**. **Complement** proteins and **mast cells** react nonspecifically to damage to body tissues. Mast cells release **histamines**, which promote inflammation, by increasing blood circulation and attracting and mobilizing phagocytic cells. During inflammation, **neutrophils**, attracted by chemicals released by mast cells, attach to the capillary walls, move through the capillary wall into the damaged tissue, and phagocytose any microorganisms in the damaged tissue. **Macrophages** consume microorganisms and cellular debris and releases signaling molecules called **cytokines** that increase inflammation.

What gives the immune system specificity, diversity, memory, and self-nonself recognition?

Immune responses depend on two kinds of **lymphocytes: B cells**, which make antibodies (and more B cells, called memory cells), and **T cells,** which recognize antigens on the surfaces of the body's own cells. **Antibodies** recognize specific pathogens and not others by a pathogen's **epitope**-bearing **antigens. B cells** give the immune system its **memory.** Antibodies are proteins called **immunoglobulins**. A single B cell can make only one kind of antibody, but the immune system as a whole can produce more than a billion antibodies.

B cells respond to particular antigens by means of membrane-bound forms of antibodies called **B-cell receptors.** When a B cell encounters antigen it makes **plasma cells**, which make antibodies, and **memory cells**, which later make plasma cells. T cells are of two kinds: **T cytotoxic (T_C) cells**, which kill cells recognized as **altered self**, and **T helper (T_H) cells**, which send messages to nearby B cells and to T cytotoxic cells. Antigen recognition by T cells depends on **T-cell receptors** on the surfaces of T cells.

The immune system's initial response to a new pathogen is called the **primary immune response.** The reappearance of a pathogen stimulates a faster, stronger **secondary immune response** that is based on the multiplication of memory cells during the first exposure. The immune system's ability to distinguish the body's own cells from others is called self-nonself recognition.

How do immune cells recognize the body's own cells?

T-cell receptors do not recognize free antigens, but only antigens bound to the proteins of the **major histocompatibility complex (MHC). MHC Class I** proteins present antigens to T_C cells, and **MHC Class II** proteins present antigens to T_H cells (from the surfaces of macrophages and other antigen presenting cells).

Review and Thought Questions

Review Questions

1. Name the body's three lines of defense against pathogens.
2. Name the four special characteristics of the immune system. How do antibodies contribute to these?
3. What is hemostasis?
4. What prevents all of the blood from clotting at once?
5. Give an example of a cytokine and explain what it does.
6. Explain the difference between memory B cells and plasma B cells.
7. How do B cells make so many different kinds of antibodies?
8. How do antibodies recognize a specific antigen?
9. Explain the difference between passive immunity and active immunity.
10. Explain how B cells differ from T cells.
11. Explain the difference between helper T cells and cytotoxic T cells.
12. MHC Class I and Class II proteins are cell surface markers that perform two different roles. What are those roles?

Thought Questions

13. Sperm are one of the few kinds of cells that lack MHC Class I markers on their cell surfaces. What might happen if sperm had these identifying markers?
14. HIV, the AIDS virus, infects and kills helper T cells. Why do AIDS patients tend to die of pneumonia, cancer, and fungal infections?

BiologyNow Resources

Biology ⑤ Now™

Active Figures

41-3: First lines of defense
41-16: The immune response

Preparing for an exam? Take a diagnostic test on your BiologyNow CD-ROM.

Online materials relating to this chapter are at:

http://biology.brookscole.com/AAL3

About the Chapter-Opening Image

Most coronaviruses cause mild coldlike symptoms. But in the winter of 2002–2003, a deadly coronavirus appeared in China and quickly spread to other parts of the world, killing thousands of travelers and healthcare workers.

The Nervous System and the Sense Organs

Key Questions

- How does the nervous system respond to its environment?

- What does a brain do?

- What does the peripheral nervous system do?

- How do we sense?

A Single Mind or a Group of Minds?

A student research volunteer notices a momentary lapse in her ability to choose the right form of a verb. Another volunteer finds it hard to pay attention to objects in his left field of view. Yet another has difficulty recognizing that a face is angry, not happy. Researchers have discovered that they can trigger these and many other startling (but temporary) mental deficiencies by pressing a magnetic coil to a person's head and inducing an electric current in the brain tissue—a technique called transcranial magnetic stimulation (TMS). Where on a person's head the researchers place the coil determines which mental processes break down.

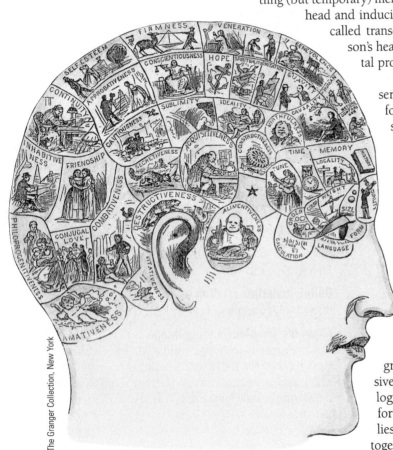

The Granger Collection, New York

TMS studies are only the latest in a long chain of observations that show that the brain processes different information in different areas of the brain. For example, researchers at Harvard University found that by applying a train of electric pulses to a brain region called the "left prefrontal cortex" they could stifle a person's ability to conjugate verbs, while sparing their ability to put nouns in the singular or plural forms.

Does this striking division of labor imply that the brain is a collection of individual modules acting independently from one another? Or does the brain work as an integrated whole? The question is amazingly similar to one posed in Chapter 27: do species in an ecological community act as independent agents in an individualistic manner, or do they form integrated communities? Like ecologists, neuroscientists have argued about this issue for many years.

Scientists now appreciate, however, that the integrated and individualistic views are not mutually exclusive and both are needed to explain how the brain and ecological communities work. The brain has areas specialized for processing certain types of information, but it also relies on *distributed processing,* in which different areas work together. So although verbs and nouns may map to different parts of the brain, speaking, understanding spoken language, reading, and writing all rely on the integrated activities of several areas of the brain at once.

The idea that the brain is subdivided into specialized parts first gained prominence in the early 19th century, when the German doctor and neuroanatomist Franz Joseph Gall tried to establish a connection between the size of various areas of the brain and a person's particular emotional or intellectual capacities. Gall's approach, which was to look at the shapes of people's skulls, was eventually discredited. But, in the early 1820s, a young scientist named Pierre Flourens decided to test Gall's ideas more rigorously. Flourens damaged specific parts of the brains of experimental animals in the hopes of learning what behavior each part of the brain controlled. Flourens found, for example, that destroying the *cerebellum,* a structure in the lower part of the brain, led to a loss of coordination. But removing slices of the *cerebral cortex,* the folded surface of the brain, led not to the loss of any specific traits, but instead to a vague, general loss of mental function.

Other neurologists continued to try to associate specific functions—such as touching or hearing—with specific regions of the brain. In the late 1800s, for example, the French neurologist Pierre Paul Broca described a patient who could not speak. After the patient's death, Broca found damage in a region in the left hemisphere of the brain. He concluded that this region was responsible for language.

Broca was mostly right, but he incorrectly assumed that language is an indivisible function localized in a single region of the brain. The German neurologist Carl Wernicke showed that language consists of separate functions when he described a patient who could speak but who could not understand what he himself was saying. After the man died, Wernicke dissected his brain and found that the cerebral cortex had been damaged in a single place, probably by a stroke. Wernicke knew about the area Broca had described and its importance for language. He also knew that the cerebral cortex contained another area where damage led to a general failure of hearing. Wernicke concluded that understanding words must involve several different elementary processes, each of which occurred in a different region of the brain (Figure 42-1). His work became the basis for the now-accepted idea of distributed processing.

Contemporary neuroscientists accept the view that individual regions of the brain perform specific functions and that complex functions of the brain require cooperation and information flow between several regions. One exciting application of this idea is in the development of devices that may, in the future, allow paralyzed patients to control prosthetic limbs via brain signals. For example, researchers at Duke University have used electrodes to listen in on neurons firing in the cerebral cortex in monkeys. By correlating these patterns of activity with the monkeys' movements, they can predict when and roughly how a monkey will extend its arm to move a joystick or reach for a piece of food. They can then send these deciphered signals to a computer control-

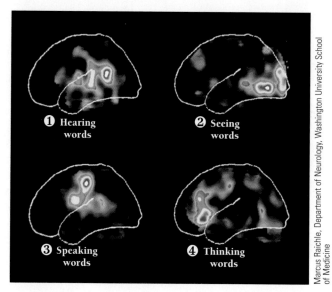

Marcus Raichle, Department of Neurology, Washington University School of Medicine

Figure 42-1
Different mental tasks involve different brain regions. These positron emission tomography (PET) scans show activity in different parts of the brain by tracking oxygen levels in blood flowing through the brain. In these images of the side of the brain, front of the brain at left, the hotter colors (red and yellow) indicate regions of high metabolic activity, while cooler colors (blue and green) show regions of low activity. These images show activity just after a subject has *(left to right)* ❶ heard someone else speak words, ❷ read words, ❸ spoken words, and ❹ thought words. Researchers can also probe brain function by tracking blood flow. PET and other brain imaging techniques have revealed patterns of brain activity associated with many mental activities, including language, attention, perception, memory, emotions, and movement.

ling a robotic arm and make the robotic arm mimic the monkey's movements. Someday, the researchers hope, paralyzed patients will be able to use similar devices to control the movements of prosthetic limbs.

In this chapter, we will focus on the basic principles underlying the organization and function of the nervous system. We will describe the roles of individual brain regions, particularly those involved in **sensation**—the detection of external stimuli by vision, hearing, touch, taste, and smell. We will also discuss how different components of the nervous system act together to generate integrated behaviors.

42.1 How Does the Nervous System Respond to Its Environment?

Nervous systems coordinate behaviors of all kinds by integrating and interpreting information. Vision, touch, hearing, smell, and taste report sensory information about the outside world to an animal's nervous system. At the same time, the nervous system also collects information about the body's interior world by means of receptors that monitor, for example, temperature, pH, oxygen supply, and levels of individual hormones.

The nervous system needs to be able to (1) detect changes in the body's internal and external environments, (2) interpret the meanings of those changes, and (3) respond by sending appropriate signals to the organs of the body (including the muscles). Our eyes, nose, ears, and other sense organs detect change, while our brain and spinal cord interpret signals from the sense organs and command the muscles and endocrine systems.

Nerves carry signals from the sense organs to the spinal cord or brain, and from the brain or spinal cord to the muscles, organs, and endocrine system. The intricate connections among nerve cells—in the brain, spinal cord, and nerves—determine how animals respond to a variety of stimuli.

An animal's nervous system detects changes in the external and internal environment, integrates and interprets the information, and coordinates an appropriate response.

Do Simple Animals Have Brains?

The simplest organisms that have nervous tissue are jellyfish, hydra, and other cnidarians (Figure 42-2A). A hydra, for example, has a web of interconnecting neurons, called a **nerve net,** that carries information to all parts of its body. A stimulus anywhere on the hydra's body triggers reflexive muscle contractions. In jellyfish, nerve nets coordinate complex swimming movements. But cnidarians have no brains, and, although capable of simple habituation (no longer responding to an often repeated stimulus), they do not otherwise learn.

The simplest animals capable of learning are the flatworms, whose nervous system resembles that of more complex animals (Figure 42-2B). The flatworm's nervous system consists of two parallel nerve cords, with nerves extending outward to the muscles. The two nerve cords converge near the front of the body to form a very simple "brain." Like other bilaterally symmetrical animals, flatworms have a well-defined direction of movement. They have light-sensitive organs, called eyespots, at the front end, and they process sensory information in clusters of nerve cells called *ganglia.*

In even a simple brain, the convergence of two nerves in a ganglion provides many opportunities to generate complex patterns of neural activity and behavior. A simple brain also allows animals to modify such patterns by learning through association. Flatworms can, for example, be taught to move in a variety of patterns in order to find food (Figure 29-8). More complex animals have greater concentrations of sense organs and ganglia in the head.

Flatworms are the simplest animals that can learn.

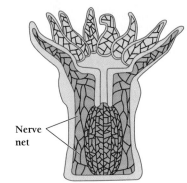

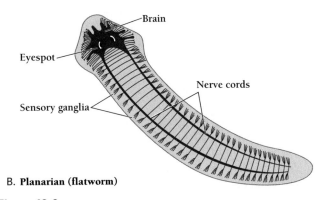

Figure 42-2
Invertebrate nervous systems. A. Nerve net in a sea anemone. Hydras, sea anemones, and jellyfish have no brain, but only an integrated nerve net. B. The nervous system of a flatworm, the simplest animal that has a brain and is able to learn.

How Is the Vertebrate Central Nervous System Organized?

Vertebrates show the most developed brains and the most sophisticated sense organs. (The only impressively smart invertebrate is the octopus.) Most neurons are concentrated in the **central nervous system (CNS),** which consists of the brain and the spinal cord (Figure 42-3). The system of nerves that connects the brain and spinal cord to the sense organs, muscles, and other organs is called the **peripheral nervous system.** In most vertebrates, the brain lies within a thick bony structure, called the *cranium,* and the spinal cord passes through a canal made by stacks of bony vertebrae. Both the cranium and the vertebrae help protect the CNS from traumatic damage.

Also protecting the brain and spinal cord is a set of three membranes, called the *meninges.* Both the brain and the spinal cord are hollow and filled with a fluid, called the *cerebrospinal fluid.* This fluid also surrounds the brain and spinal cord, filling the space between the two innermost meninges and cushioning the brain and spinal cord against injury.

The cerebrospinal fluid derives from plasma, the liquid part of the blood. But the capillaries that supply blood

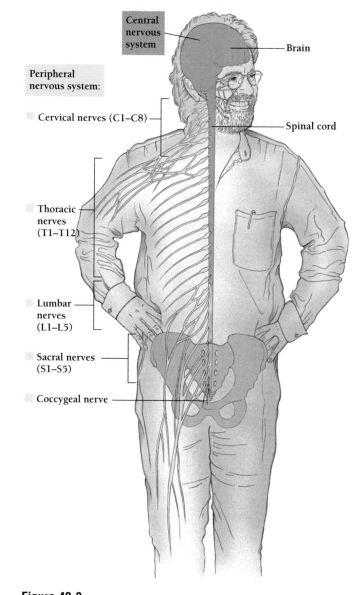

Among the molecules that easily cross over are alcohol, nicotine, heroin, and cocaine. Any drugs designed to treat neurologic or psychiatric problems must also be able to cross the blood-brain barrier, which makes designing such drugs that much more of a challenge.

The vertebrate central nervous system, the brain and spinal cord, is surrounded and protected by bones, membranes, and cerebrospinal fluid. The special capillaries supplying the CNS form a tight blood-brain barrier.

How Does the Spinal Cord Mediate Movements?

The **spinal cord** is a long bundle of specialized nerve fibers that runs from the brain to the tail (Figure 42-3). It receives sensory information from the peripheral nervous system about the position of the limbs, touch, temperature, and pain, and sends motor instructions to the muscles of the limbs and trunk. The spinal cord also influences heart rate, the movements of the intestines, and other nonvoluntary (autonomic) functions.

Neurons in the spinal cord called spinal **motoneurons** directly stimulate muscle contraction. The nerve fibers from the brain that send signals to the motoneurons cross from left to right (or from right to left) as they pass through the base of the brain, so that the left side of the brain controls muscles on the right side of the body.

The spinal cord can itself direct relatively complex behaviors even without a connection to the brain: a person or animal with a severed spinal cord can learn to move his or her legs in a coordinated rhythm and even to lift a leg over an obstacle.

Movement mediated by the spinal cord is only minimally influenced by conscious thought, and many of our most automatic behavior patterns, or **reflexes,** are mediated by the spinal cord. In a reflex, motor activity is initiated in direct response to a sensory stimulus. Physicians sometimes test reflexes by tapping the knee gently with a hammer to see if the lower leg will kick up reflexively (Figure 42-4). The hammer hits a tendon that connects the thigh's quadriceps muscle to a bone in the lower leg. As the physician's hammer strikes the knee, the force stretches the quadriceps muscle and stimulates stretch receptors in the muscle, triggering nerve impulses. These impulses travel by the sensory neurons to the spinal cord, where they deliver their message to motoneurons, which signal the quadriceps to contract, which jerks the lower leg. The information passes through just one connection between nerve cells, or *synapse,* so the stretch reflex is said to be *monosynaptic.* The normal function of this reflex is to restore the position of the lower leg after a rapid movement in the other direction.

Impulses from the sensory neurons also travel to a second group of spinal cord neurons that inhibit the hamstring muscles at the back of the thigh. Because these muscles

Figure 42-3

Nervous system of a human. The nervous system of a human resembles that of a flatworm in being linear and highly segmented. Our central nervous system consists of the brain and the spinal cord. The peripheral nervous system is everything else and consists of the peripheral nerves, which run to the arms, legs, and surface of the body. For example, the cranial nerves connect the brain to muscles, sense organs, and glands primarily in the head and neck. The spinal nerves carry motor and sensory signals to and from the skin, muscles, and joints of the limbs and trunk.

to the spinal cord and brain are different from those that supply the rest of the body. The cells of the CNS capillary walls are so tightly connected to one another that few molecules can pass out of the capillaries and into the cerebrospinal fluid. The capillary walls are said to form a *blood-brain barrier.*

A few special molecules are actively transported across the barrier, and small, nonpolar molecules can pass through.

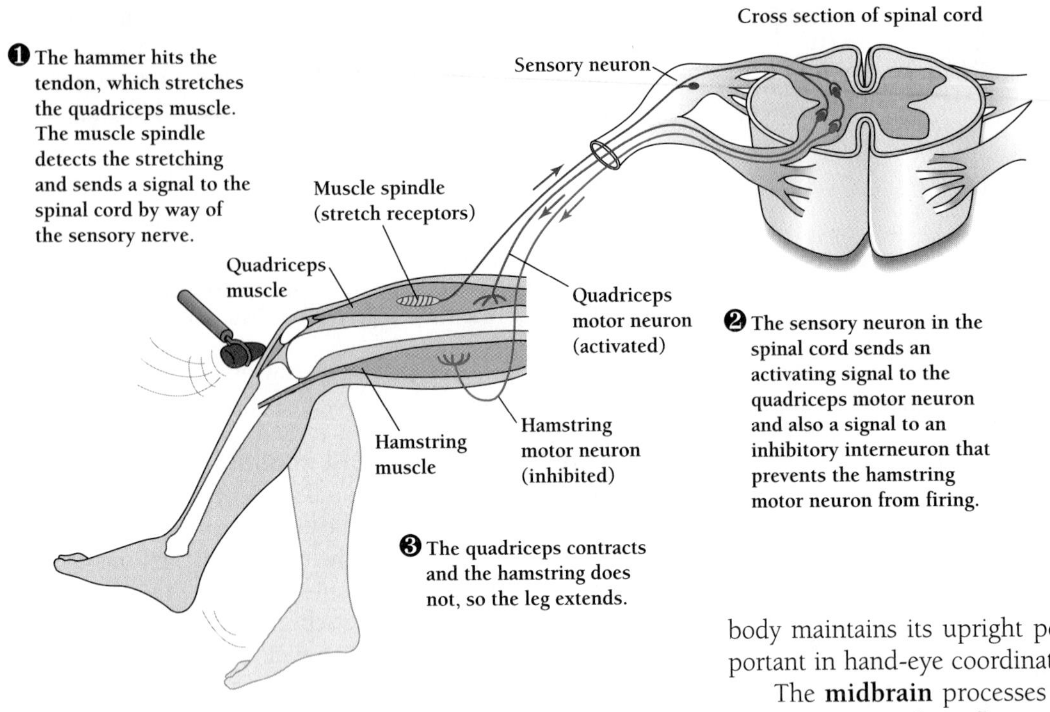

❶ The hammer hits the tendon, which stretches the quadriceps muscle. The muscle spindle detects the stretching and sends a signal to the spinal cord by way of the sensory nerve.

Cross section of spinal cord

Sensory neuron

Figure 42-4
The knee jerk reflex. The knee jerk reflex is mediated by both a monosynaptic reflex and a two-synapse reflex.

Muscle spindle (stretch receptors)

Quadriceps muscle

Quadriceps motor neuron (activated)

❷ The sensory neuron in the spinal cord sends an activating signal to the quadriceps motor neuron and also a signal to an inhibitory interneuron that prevents the hamstring motor neuron from firing.

Hamstring muscle

Hamstring motor neuron (inhibited)

❸ The quadriceps contracts and the hamstring does not, so the leg extends.

---

counter the quadriceps, inhibiting the hamstrings allows the quadriceps to contract, contributing to the knee jerk reflex.

The spinal cord receives somatic sensory inputs about the position of body parts, sends motor instructions to voluntary muscles, and coordinates autonomic functions. In a simple reflex, input from a sensory receptor directly stimulates a motoneuron and indirectly inhibits another.

The Vertebrate Brain

Paleontologists' studies reveal that the brains of jawless fishes that lived almost half a billion years ago consisted of three major divisions: (1) a hindbrain, (2) a midbrain, and (3) a forebrain. Although vertebrate brains have gone through many evolutionary changes since then, these same major divisions appear in humans and other modern vertebrates (Figure 42-5A).

Hindbrain and midbrain. The hindbrain and midbrain, shown in Figure 42-5A, take care of some of our most basic needs, including breathing, standing upright, and staying awake. The **hindbrain**, which consists of three parts—the **pons**, the **medulla oblongata**, and the **cerebellum**—integrates sensory information and motor instructions from the spinal cord as well as information from the rest of the brain. For example, the medulla regulates breathing and blood circulation, and the cerebellum coordinates information coming from the senses with outgoing motor instructions. This coordination ensures that the limbs move smoothly and the

body maintains its upright posture. The cerebellum is important in hand-eye coordination.

The **midbrain** processes visual and auditory information, such as in the reflexes that orient the head and neck toward a visual stimulus. One midbrain region, called the *substantia nigra,* helps initiate movement. Parkinson's disease—which results from the loss of many of the cells of the substantia nigra—severely limits a person's ability to control movements. Together, the midbrain, the pons, and the medulla are called the *brain stem.* Running through the brain stem is a network of ganglia that receive information from all other parts of the brain and regulate attention, sleeping, and dreaming.

The hindbrain integrates and processes both sensory information and motor instructions. The midbrain relays sensory information, helps organize movement, and regulates attention.

Forebrain. In humans, the **forebrain** is the locus of almost all conscious activity. It plays crucial roles in sensation, perception, movement, learning, memory, and mood. The forebrain consists of two major divisions: the diencephalon and the cerebrum (Figure 42-5B). The **diencephalon** consists of the **thalamus** and **hypothalamus,** which integrate information about our internal and external worlds—including sensory information about body temperature, eating, drinking, growth, stress, threats, mating, and pleasure.

An important part of the **cerebrum** is a thin surface layer called the **cerebral cortex** (Figure 42-5B). The cerebral cortex consists of two almost identical halves, connected by a thick bundle of nerve fibers called the *corpus callosum.* A large cerebral cortex is a hallmark of mammals. In humans, the thin cerebral cortex (only four to six cells deep) has become so large that it is packed around the rest of the

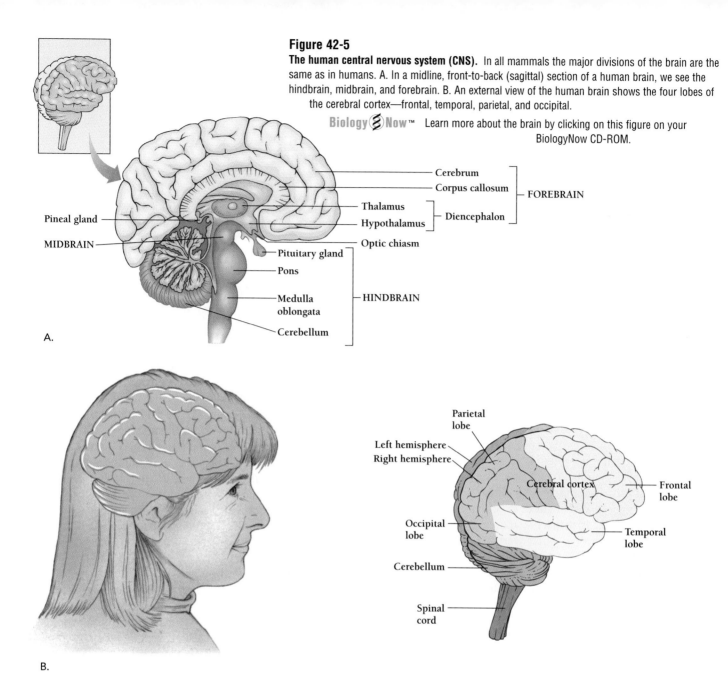

Figure 42-5

The human central nervous system (CNS). In all mammals the major divisions of the brain are the same as in humans. A. In a midline, front-to-back (sagittal) section of a human brain, we see the hindbrain, midbrain, and forebrain. B. An external view of the human brain shows the four lobes of the cerebral cortex—frontal, temporal, parietal, and occipital.

Biology Now™   Learn more about the brain by clicking on this figure on your BiologyNow CD-ROM.

brain like a crumpled sheet, almost completely covering the brain. The folds in the cerebral cortex provide landmarks for dividing it into four areas—the *frontal, parietal, temporal,* and *occipital lobes.* Because the brain has two halves, it has two of each lobe, a left and right frontal, a left and right parietal, and so on.

Each pair of lobes receives sensory signals (mostly via the thalamus), processes that information, and transmits instructions to some other part of the brain or to the spinal cord. The lobes also have regions specialized for integrating information called the **association cortices,** which help process both sensory input and motor output. In a mouse, only five percent is associative and the rest is sensory and motor. In humans, more than half of the cortex is associative, a fact that may account for our unusual intelligence.

Each pair of lobes receive a particular type of input: the frontal lobes mainly deal with movement and smell; the parietal lobes, with somatic sensation (e.g., pain, temperature and our sense of position and movement); the occipital lobes, with vision; and the temporal lobes, with hearing. In addition to receiving and processing this information, the lobes generate complex mental activities. The frontal lobes, for example, help us plan for the future and define our personalities. The parietal lobes, on the other hand, are important in paying attention. The temporal lobe, including the

hippocampus, is involved in recognition and the formation of memory.

Underneath the cerebral cortex lie other structures—the amygdala, which coordinates autonomic (involuntary) responses to emotional states, especially fear, and the basal ganglia, which help establish patterns of movement.

The forebrain plays important roles in sensation, perception, movement, learning, memory, and mood. In humans, the cerebral cortex is divided into four lobes, each of which plays distinctive roles in mental functions, including sensation, recognition, and planning.

How Do Humans Learn?

The association cortex, especially in the temporal lobe, is intimately involved in memory. **Memory** is the storage of knowledge about the world. Memories form in stages. **Short-term memories**—of experiences within the previous few minutes—can disappear rather easily. But the brain can stabilize memories into **long-term memories** that may persist for decades.

Memories may involve different degrees of consciousness. A *procedural* memory is one that has a reflexlike quality—automatic and not dependent on awareness. Skills (like riding a bicycle) and habits (like twirling a lock of hair) are examples of procedural memories. In contrast, a *declarative* memory is a recalled thought or experience, a memory that a subject can summarize in a declarative sentence. Some skills—such as driving a car—may start out as conscious, declarative memories, and then later become unconscious, procedural memories.

Different brain structures seem to be involved in procedural and declarative memories. Damage to one part of the cerebellum may destroy the ability to perform a particular task, while damage to the temporal lobe may interfere with the formation of long-term declarative memories.

The hippocampus has an important role in the formation of long-term memory from short-term memories and is also an important site in the generation of epileptic seizures. One of the most forceful illustrations of the role of the hippocampus came from a young man, called "H.M.," whose two hippocampi were removed in an effort to stop persistent seizures. After his operation, H.M. could still speak and understand language. He had memories from before his surgery, and he could remember for seconds or minutes. But he lost the capacity to form new long-term memories. Studies of both humans and other animals all point to the hippocampus and surrounding regions as a crossroads of short-term and long-term memory.

One of the most amazing discoveries about memories came from tests performed during neurosurgery. Wilder Penfield, a Montreal brain surgeon, stimulated the temporal lobes of several of his patients. These patients vividly experienced past events. For example, one older woman heard a childhood melody each time Penfield stimulated a particular spot in her brain. Despite the incredible amount of information gathered by physicians, psychologists, and neuroscientists, however, we are still far from understanding the basis of memory.

Memories form in stages, with short-term memories including experiences of the preceding few minutes and long-term memories including experience from many years before. The hippocampus plays an important role in the formation of long-term memories.

How Is the Vertebrate Peripheral Nervous System Organized?

The **peripheral nervous system,** which consists of all the nerve cells that are outside the central nervous system (although some cells can be part of both), connects the central nervous system (CNS) to the sense organs, muscles, and other organs. There are two types of peripheral nerves: the *cranial nerves* and the *spinal nerves.* The cranial nerves link the brain to muscles, sense organs, and glands primarily in the head and neck. The spinal nerves carry signals to and from the skin, muscles, and joints of the limbs and trunk.

Each of these nerves can be further classified as either afferent, efferent, or mixed, depending on the signals they carry. **Afferent** nerves carry sensory information *to* the CNS. **Efferent** nerves carry instructions *from* the CNS to other organs. Many nerves handle two-way traffic and so can be both afferent and efferent.

The efferent nerves that carry information to the skeletal muscles make up the somatic, or voluntary, nervous system. Other efferent nerves, those of the autonomic, or involuntary, nervous system, carry instructions to the smooth muscles of the endocrine, digestive, reproductive, excretory, respiratory, and vascular systems, as well as to the cardiac muscles of the heart.

The peripheral nervous system connects the CNS to sense organs and muscles. It includes nerves of both the somatic nervous system, which carries information to voluntary muscles, and the autonomic nervous system, which carries instructions to endocrine organs, smooth muscles, and cardiac muscles.

How Does the Autonomic System Work?

The autonomic system consists of two distinct sets of nerves—the **sympathetic** and the **parasympathetic** nerves (Figure 42-6). These nerves use different neurotransmitters, which act on different receptors in target

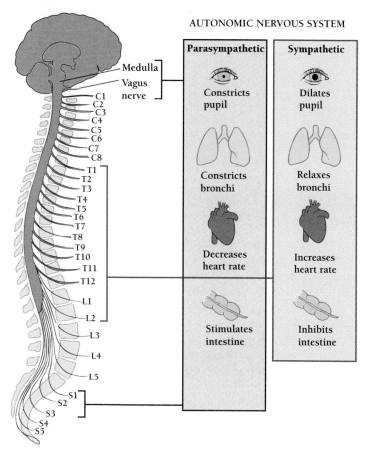

AUTONOMIC NERVOUS SYSTEM

Parasympathetic	Sympathetic
Constricts pupil	Dilates pupil
Constricts bronchi	Relaxes bronchi
Decreases heart rate	Increases heart rate
Stimulates intestine	Inhibits intestine

Figure 42-6

The human autonomic nervous system. The autonomic nervous system of mammals consists of the nerves that control involuntary activity such as the beating of the heart, the contraction of blood vessels, and the movements of the intestines. The autonomic nervous system includes the sympathetic nerves, which generally prepare the body for action, and the parasympathetic nerves, which generally prepare the body for digestion and calm.

cells. The sympathetic nervous system, which usually uses norepinephrine as a final neurotransmitter, helps coordinate the "fight-or-flight" reaction (Chapter 37). This response involves the mobilization of energy stores in preparation for vigorous action in the face of danger or other stressors. In contrast, the parasympathetic nervous system, which uses acetylcholine as a neurotransmitter, helps conserve energy by slowing the heart and increasing digestive activity. The sympathetic system is activated when you face a pop quiz without having studied, for example, while the parasympathetic system kicks in when you settle down to watch a favorite TV show.

Cells in the *thoracic* and *lumbar* sectors of the spinal cord give rise to nerves that drive sympathetic responses, while cells in the brainstem and *sacral* sector of the spinal cord give rise to those coordinating parasympathetic responses (Figure 42-6). Many organs, including the heart,

salivary glands, bladder, and sex organs, receive both sympathetic and parasympathetic inputs. As expected from their primary functions, these inputs often have effects that oppose each other. For example, sympathetic signals speed up the heart, while parasympathetic signals slow it down.

A single type of nerve, however, can also do two opposite things. Sympathetic nerves alone, for example, can control both the dilation and constriction of blood vessels. When the firing rate of these nerves increases, the blood vessels they control constrict. Conversely, when their firing rate decreases, the vessels dilate.

Although autonomic responses are usually involuntary, in some cases they can be consciously regulated. Using meditation, *biofeedback,* or other relaxation techniques, some people can learn to control their blood pressure, heart rate, or muscle tone. This ability can help treat ailments such as hypertension and stress.

The autonomic nervous system consists of the sympathetic nerves, which prepare the body for danger, and the parasympathetic nerves, which prepare the body for rest and digestion. Organs may receive input from both kinds of nerves. And the same nerves can transmit signals with opposite effects.

How Does Information Move Through the Nervous System?

The nerves of the peripheral nervous system are further classified according to what kind of information they carry. The auditory system, for example, contains cells responsible for hearing, and the visual system contains cells responsible for seeing.

The cells that detect the properties of the outside world are called **sensory receptor cells.** They initially encode this information in the form of electrical voltages across their plasma membranes. Information from receptor cells is passed to **relay neurons,** which carry information in the form of nerve impulses to particular sets of neurons in the brain (Chapter 43). Each part of the nervous system and almost every nerve cell receives information from many sources. Ultimately, however, all external information comes from the sensory systems.

The electrical properties of nerve cells allow neurobiologists to probe individual cells and the interconnections among cells of the nervous system. Researchers can see how the brain responds to external stimuli by recording electrical currents from the surface of the brain, from individual cells, and even across minuscule patches of nerve cell membranes.

Receptors detect characteristics of the internal and external world and generate signals. Relay neurons transmit the information to the brain as patterns of nerve impulses.

42.2 How Do Animals Sense Their Environment?

Many individual parts of the brain contribute to *sensation*, the detection of external stimuli. And far more contribute to **perception,** the conscious recognition and interpretation of those stimuli. But animals—human or otherwise—can respond to sensation in ways other than conscious perception. Sensory stimuli can help control movements, regulate internal organs, and maintain alertness, without our being conscious of making any changes. For example, changes in light intensity lead to changes in pupil dilation, of which we are mostly unaware.

What Senses Contribute to Sensation and Perception?

Neuroscientists recognize six senses—hearing, vision, taste, smell, balance, and touch. Touch actually includes four independent senses—pressure, pain, temperature, and sensing limb motion and position. Some animals detect additional qualities of the external world, such as electric fields (some fish), magnetic fields (some migrating birds), or ultraviolet light (insects and many birds).

Sensation begins in sensory receptor cells, specialized cells that detect a particular property of the internal and external world. Each receptor cell contributes to only one sense: receptors in the eye respond to light, those in the ear to vibration, and those in the taste buds and nose to particular chemicals. In vertebrates, sensory receptors for hearing, vision, taste, and smell are in specialized **sense organs**—the ears, eyes, taste buds, and the olfactory epithelium of the nose. But touch or other body (somatic) sensation, on the other hand, depends on sensory receptor cells, dispersed in skin and muscles, which report separately on pressure, pain, temperature, and limb position.

In most cases, sensation involves three stages:

1. **Transduction** translates the energy of a stimulus (such as light) into a change in the chemistry of the receptor cell.
2. **Transmission** carries a signal from the receptor cells to the central nervous system.
3. **Integration** combines the signals from many receptors to ultimately provide a unified perception of the outside world.

Each sensory system—including sensory receptors and the connections within the central nervous system—responds quantitatively to three characteristics of a stimulus—*intensity, duration,* and *location.* Sensory receptors in the ear, for example, produce larger electrical voltages in response to greater vibration, which we perceive as louder sounds.

Sensory receptors also respond to other qualities of a stimulus. The ear, for example, distinguishes among different frequencies, which are perceived as high notes and low notes. Similarly, the eye distinguishes among different wavelengths of light, which we perceive as different colors.

Sensation depends on signal transduction by the sensory receptor, the transmission of a signal to the central nervous system, and the integration of signals to form a unified perception of the outside world.

What Path Does Sensory Information Take to the Brain?

Sensory receptor cells feed information to relay neurons. Some relay neurons simply transmit the information from a single sensory receptor to the next step of the neural chain. Other relay neurons do more: they integrate information from several sensory receptors or from several other relay neurons.

Relay neurons usually send information first to the thalamus and then to particular regions of the cerebral cortex. Although all routes pass through the thalamus, they pass through different regions of the thalamus. And although all routes end in the cortex, they end in different regions of the cortex: visual information ends in the occipital lobe, auditory information in the temporal lobe, and somatic sensation in the parietal lobe (Figure 42-5).

Information from the olfactory system takes a different route, going first to other parts of the brain before arriving at the thalamus and cortex. The direct connection of the olfactory system to the parts of the brain dealing with emotion probably reflects its ancient evolutionary origins and helps explain both our difficulty in naming odors and our intense emotional responses to certain smells such as the scent of a parent or lover.

Each sensory receptor feeds information into a characteristic route. Each route is called a *labeled line* because the brain recognizes it as corresponding to a particular sense. Stimulation of a particular receptor, therefore, always evokes the same sensation, even when the stimulation is not entirely appropriate. For example, someone who receives a blow to the head may "see stars" because the blow mechanically stimulates receptors along the visual pathway. Similarly, a person who suffers deafness because of a defect in the ear will "hear" a tone when a researcher electrically stimulates the relay neurons in the auditory pathway. (This ability to "hear" an electrical stimulus has allowed the development of electrical devices, called *cochlear implants,* that allow thousands of deaf people to hear.)

After a sensation signal first arrives in the cortex, the primary sensory region further processes information and sends it to another region of the cortex. In the case of the visual system, there are at least 32 areas of the cortex that have representations of the visual world. Researchers do not yet know the significance of all these representations, but each is thought to extract different features of the visual image. One region, for example, appears to be highly specialized for the recognition of faces: people with damage to that area cannot identify faces but have otherwise normal vision and memory.

The functional organization of sensory systems shows that the brain is an *active* agent in sensation; it is emphatically not a passive recipient of information from the sense organs. The brain extracts only certain features of the external world. The brain also communicates back to some sensory receptors and relay neurons, actively modifying even the very beginnings of sensation.

Sensory information from different receptors passes through the thalamus and converges in different regions of the cerebral cortex. The cortex actively interprets this information.

42.3 How Do Animals Hear?

Hearing, like the detection of light or touch, depends on the ability to translate, or **transduce**, energy into electrochemical events in neurons. To understand how this transduction takes place, we must first understand the physical nature of a stimulus.

What we perceive as sound is actually pulsations of the molecules of the air. If we pluck the string of a guitar, it vibrates with a characteristic frequency. As it moves, it alternately pushes molecules together and apart, so that the surrounding air contains waves of pressure.

We can describe a sound wave in terms of the *frequency* of vibration, which we ultimately perceive as pitch, and of the *amplitudes* of the pressure differences, which we perceive as loudness. While this simple description works well for simple tones, such as those produced by a vibrating string, we will also want to understand the perception of more complicated sounds, such as those of a word. Fortunately, the same description applies, for any sound can be regarded as the sum of many tones of different frequencies and amplitudes. If we can understand how the ear transduces simple vibrations, then we will also understand how it transduces speech or music.

How Does the Mammalian Ear Detect Frequency and Amplitude?

The mammalian ear consists of three parts, called the **outer ear,** the **middle ear,** and the **inner ear** (Figure 42-7). The outer ear channels sound through the **auditory canal** to the **tympanic membrane,** or eardrum. The tympanic membrane vibrates in response to variations in air pressure.

Behind the tympanic membrane is the middle ear, which, like the outer ear, is filled with air. This air comes into the ears by way of two tubes (one on each side), called the *Eustachian tubes,* from the throat. The Eustachian tubes are normally closed, so that the air in the middle ear is isolated from the air outside. Because the tubes are usually closed, the air in the middle ear can become pressurized relative to the outside air. A person who rapidly gains altitude, in a plane or even on an elevator, sometimes experiences painful pressure on the eardrum. Fortunately, swallowing, chewing, or yawning open the Eustachian tubes and allow the middle ear to attain the same pressure as that outside.

Inside the middle ear, three small bones transmit and amplify the movement of the tympanic membrane to the fluid-filled inner ear (Figure 42-7A). The middle ear bones amplify the vibrations of the tympanic membrane, so that the fluid in the inner ear vibrates with less amplitude but far more forcefully than if it were directly linked to the eardrum.

Inside the inner ear is a curled-up organ called the **cochlea** [Latin, = snail], an enclosed, fluid-filled tube that detects vibrations coming from the eardrum and turns them into signals. The cochlea translates, or transduces, vibration into signals that we ultimately perceive as sounds. The part of the cochlea that actually transduces sound signals is the **organ of Corti** (Figure 42-7B). When the fluid in the cochlea vibrates, two membranes in the organ of Corti move with respect to each other. **Hair cells** sandwiched between the two membranes detect the movement (Figures 42-7). When the membranes move, the hair cells generate a neural signal. Each sound frequency stimulates a different set of hair cells, and the signals pass from the organ of Corti to relay neurons to the brain.

As in the visual and the somatic (body) sensory systems, the brain simultaneously processes several separate representations of sound. Each representation provides different information. One set of cells, for example, appears to register differences in the timing of signals from the two ears; another set of cells registers differences in sound intensity. These two kinds of information are especially useful in locating the source of a sound.

Sound vibrations transmitted by the eardrum move the fluid in the inner ear, and hair cells in the organ of Corti then detect the movement of the fluid. The brain interprets the patterns of signals from the organ of Corti as sounds.

How Does the Inner Ear Help Maintain Balance?

Besides responding to sound vibrations, hair cells also respond to acceleration or gravity. Consequently, many animals, including humans, use hair cells to determine the movement and orientation of their heads.

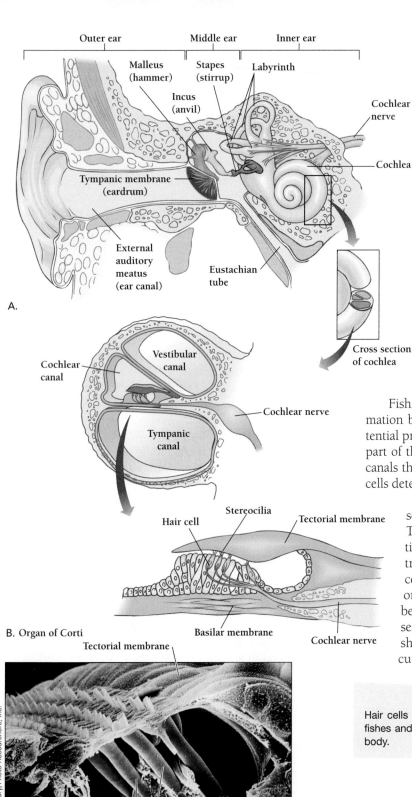

Outer ear | **Middle ear** | **Inner ear**

Malleus (hammer)
Stapes (stirrup)
Labyrinth
Incus (anvil)
Cochlear nerve
Tympanic membrane (eardrum)
Cochlea
External auditory meatus (ear canal)
Eustachian tube

A.

Cross section of cochlea

Vestibular canal
Cochlear canal
Cochlear nerve
Tympanic canal

Stereocilia
Hair cell
Tectorial membrane

B. Organ of Corti
Basilar membrane
Cochlear nerve

Tectorial membrane

G. Bredberg/Science Photo Library/Photo Researchers, Inc.

C.
Hair cells
25 μm

Figure 42-7

Inside the human ear. The ear consists of the external ear (the flap of cartilage and skin attached to the side of the head plus the ear canal), the inner ear (the ear drum plus three little bones that transmit sounds to the inner ear), and the inner ear (which senses both sound and our position relative to the ground). A. The organization of the three parts of the ear. B. Cross section of the organ of Corti in the inner ear, which detects vibrations in the fluid of the inner ear. C. Scanning electron micrograph, showing hair cells in the organ of Corti.

Biology (S) Now™ Learn more about the human ear by clicking on this figure on your BiologyNow CD-ROM.

Fish and frogs use hair cells to provide sensory information both about their own movements and those of potential predators and prey. In fish and frogs, the hair cells are part of the **lateral line system**, which consists of two long canals that run along the two sides of their bodies. The hair cells detect changes in water movements through the canals.

Inside the mammalian inner ear is a **labyrinth**, a set of fluid-filled canals that contain more hair cells. These cells detect fluid motion due to gravity, vibration, or movements of the head. The pattern of electrical activity from these hair cells tells the brain's cerebellum and hindbrain how the animal's head is oriented and in what direction it is moving. The cerebellum then integrates this information with other sensory information to determine how the muscles should move to maintain proper posture or to execute a particular movement.

Hair cells in the labyrinth of mammals and the lateral line system of fishes and frogs help establish the position and the movement of the body.

42.4 How Do Animals See?

Animals have evolved many different kinds of light detectors. In all cases, the primary event in vision is the action of light energy on a pigment molecule that changes from one form to another. This change takes place in a specialized light-detecting cell called a **photoreceptor**.

Different kinds of eyes and different visual systems use different strategies to deliver light to the receptor pigment and to present the resulting information to the rest of the nervous system. Planarians, a type of flatworm, have photoreceptors within a sense organ called an *eye-cup*. The receptors detect a shadow at the cup's edge, thereby identifying the direction of a light source. But the planarian's eye does not form any image of the outside world, and the flatworm lives in a world of light and shadow only.

In contrast, the eye of every vertebrate animal forms a distinct single image of the outer world. This image is projected against the back of the eye to a sheet of cells called the **retina** in much the same way that an image is projected against the film at the back of a camera. And just as photographic film responds to different frequencies and intensities of light from an image, photoreceptor cells in the retina contain pigment molecules that detect images before us.

In insects, the eye forms an image in a piecemeal fashion. The insect eye is really a collection of thousands of individual eyes, each of which forms a part of the full image. Together, the parts of an insect's *compound eye* compose an accurate image of the outside world. Although an insect eye sees less detail than the best vertebrate eye, insects are extraordinarily good at detecting movement—especially of large objects. Since larger animals often eat insects, their ability to detect such movement helps them stay alive.

The primary event in the detection of a visual image is the capturing of light energy by a light-sensitive pigment within a photoreceptor cell.

How Does a Vertebrate's Eye Form an Image of the Outside World?

The vertebrate eye is roughly spherical (Figure 42-8). Encasing and protecting the eye is a tough connective tissue, called the *sclera*. Light enters the front of the eye through a transparent region called the **cornea.** Behind the cornea is a muscular, donut-shaped disk called the **iris** that gives the eye its color. The iris surrounds the **pupil,** the dark, central opening through which light reaches the retina. The iris controls the size of the pupil in the same way that the iris of a camera controls the aperture of the lens. In dim light, the iris opens the pupil to let in more light. In bright light, the iris closes the pupil, limiting the amount of light that reaches the retina.

Just behind the pupil is a clear **lens,** which helps the cornea focus images on the rear surface of the retina. The problem in both a camera and an eye is that the point in space where an image comes into focus depends on how close or far away an object is. Faraway objects form images immediately behind the lens, while nearby objects form images some distance further. A photographer deals with

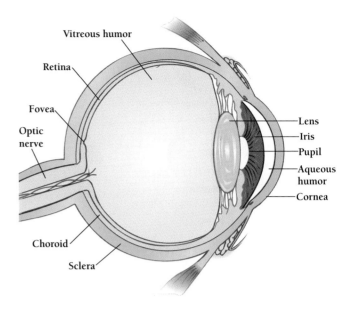

Figure 42-8
The human eye. The lens focuses light images on the retina, which contains photoreceptor cells that transmit the information to the brain.

this problem by focusing, moving the lens farther from the film for nearby objects and closer to the film for faraway objects.

The eyes of fish and amphibians work like a camera, moving the lens relative to the retina to focus objects of different distances. The mammalian eye, however, deals with the problem quite differently: it actually changes the shape of the lens according to whether objects are nearby or far away. Because a thicker lens bends light more, it forms the image closer to the lens. When focused on nearby objects, the mammalian lens thickens to form a closer image; when focused on faraway objects, the lens thins. The focusing of objects at different distances is called **accommodation.** The eyes alter the thickness of the lens by means of tiny muscles that pull on the lens. In people over about 40, the lens hardens and the muscles can no longer alter the shape of the lens. Seeing very close is usually no longer possible without reading glasses.

In a vertebrate eye, the lens focuses an image on the surface of the retina.

Why Must Many of Us Wear Glasses?

A person who is **nearsighted** (myopic) either has eyeballs that are too long from front to back or corneas that are too curved. The images of faraway objects come to a focus too soon in the space in front of the retina, so that the image on the retina is out of focus. A corrective concave lens (glasses

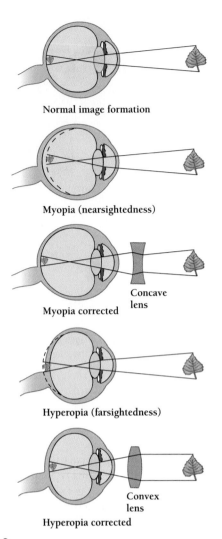

Figure 42-9

Why do we have to wear glasses? People with myopia (those who are nearsighted) usually have elongated eyeballs, so that images focus just short of the retina. The image that hits the retina is out of focus. People who are farsighted usually have shortened eyeballs, so that images focus behind the retina. Again, the image that hits the retina is out of focus. In both cases, lenses will correct vision.

or contacts) works by spreading the light before it enters the eye, so that images converge on the retina (Figure 42-9). A person who is **farsighted** usually has eyeballs that are too short, so that images of nearby objects hit the retina before they have come to a focus. A corrective convex lens (glasses or contacts) works by bringing the light to a focus on the retina.

As we age, cloudy spots, called **cataracts,** sometimes form on the lens, obscuring our vision. At one time, cataracts were a major cause of blindness in older people. But today, an eye surgeon can remove the damaged lens and replace it with an artificial one. Someone who has had to wear contacts all her life may find herself with 20:20 vision at age 80.

The eye's ability to focus also depends on the fluids through which light passes before and after it travels through the lens. The large space between the lens and the retina contains a transparent, jellylike substance, called the *vitreous humor.* A more watery substance called the *aqueous humor* fills the space between the lens and the cornea. Too much pressure within the aqueous humor—caused by blockage of the ducts that drain the cavity—is called *glaucoma,* a condition that damages the eye and can cause blindness.

Defects in the focusing apparatus cause eye disorders.

How Does the Retina Detect an Image?

The retina is a thin sheet of cells behind the vitreous humor. It consists of five types of cells arranged in distinct layers (Figure 42-10). The most numerous of these cells are the photoreceptors, which actually lie in the bottom layer of the retina, farthest from the light entering the eye. Light from the outside world must pass through the cornea, the lens, and the four other cell layers, including overlying nerve cells, before it reaches the photoreceptors. It is the photoreceptors, however, that convert the image into signals that the brain can interpret.

The retina contains two kinds of photoreceptor cells: cylindrical **rod cells** and cone-shaped **cone cells.** The rod cells are most dense slightly away from the edges of the retina and are phenomenally sensitive to light. When we are outside on a dark night and the pupil is opened up all the way, the 100 million rod cells in the human eye can detect even slight amounts of light and movement. The image that we see, however, is poorly defined. Because the rods are primarily around the periphery of the retina, we can actually detect faint objects more successfully by looking just to one side of the object, so that the image is projected not to the center of the retina, but to the edge.

The cone cells are made for full-color vision in very dim (one-quarter starlight) to bright light. Unlike the rods, the 6 million cones of the human retina are concentrated in the central portion of the retina, called the **fovea** (Figure 42-8). As a result, we see color most fully when we look directly at an object. We perceive the whole visual world as colored both because some cones also lie outside the fovea and because we constantly move our eyes so that different images impinge on the fovea.

Both rods and cones contain special pigments that absorb visible light and transmit a signal. Rods contain a single kind of pigment called *rhodopsin,* which is most sensitive to blue-green light. Cones come in three varieties, each of which has a distinct type of pigment; one of these is most sensitive to blue light, one to red, and one to green. The brain compares the images produced by the

Direction
of light

Direction of light

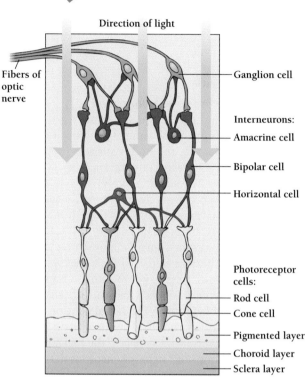

Fibers of
optic
nerve

Ganglion cell

Interneurons:

Amacrine cell

Bipolar cell

Horizontal cell

Photoreceptor
cells:

Rod cell

Cone cell

Pigmented layer

Choroid layer

Sclera layer

Figure 42-10
The organization of the retina. Light must pass through the outer layers of the retina before encountering the photoreceptor cells, the rods and cones. Rods, which help us see movement, faint objects, and shape but not color, have a single type of light-detecting pigment. Cones have one of three different pigments and allow us to see color. Information passes from the rods and cones to a series of interneurons—the amacrine cells, bipolar cells, and horizontal cells—which process the visual signal before relaying it to the optic nerve (which carries information to the brain).

different kinds of cones and interprets the differences as colors.

Photoreceptor cells contain pigments that absorb visible light. The absorption of light alters the cell's electrical properties and initiates a series of events that transmits visual information to the brain.

How Does Information Travel from the Retina to the Brain?

We know a lot about the path that carries visual information from the photoreceptors to the brain. It is clearly impossible to study each of the hundreds of millions of cells in the

body, but neuroscientists have managed to study thousands of them. The principles of information processing in the visual system seem to apply to other sensory systems.

At the beginning of this chapter we said that every sensory system consists of three stages: transduction, transmission, and integration. Sometimes, however, these three stages are not completely separate. In the visual system, the process of integration actually begins during transmission.

We have seen that the retina's rods and cones respond to images and transduce these images into a neural signal. The major pathway of transmission of visual information is from the rods and cones to intervening neurons, called **interneurons.** The interneurons in turn transmit information to ganglion cells to the optic nerve to the brain (Figure 42-10). But the interneurons not only transmit signals; they also *process* the signal before it reaches the ganglion cells, which feed directly into the optic nerve. Because of the interneurons, information from more than 100 million photoreceptor cells in the retina funnels down to a million ganglion cells. The ganglion cells receive a processed, or *integrated,* version of the visual signal from the interneurons.

The photoreceptor cells of the retina transmit image signals to interneurons, which process these signals as they transmit them to ganglion cells.

There Is Much More to Vision Than Meets the Eye

The axons of all the ganglion cells of the retina together form a thick stalk called the **optic nerve.** The optic nerves from each eye converge in a structure called the *optic chiasm* (Figure 42-11). There, some of the fibers cross to the opposite side of the brain and some continue on the same side. The fibers are arranged so that the left side of the brain receives information about the right half of the visual field, from both eyes. Similarly, the right side of the brain receives information from both retinas about the left side of the visual field.

Most of the optic nerve fibers in humans end in a region of the thalamus called the *lateral geniculate nucleus.* Each neuron in the lateral geniculate nucleus integrates information from several ganglion cells. Neighboring cells in the lateral geniculate nucleus connect to neighboring ganglion cells, so the overall organization of the image is maintained.

The signals that enter the lateral geniculate nucleus continue on to the *primary visual cortex,* a region at the back of the brain (Figure 42-11). The cells of the visual cortex respond to still more integrated versions of the visual information. As in the case of the retina, connections between adjacent cells—both directly and through interneurons—increase the contrast in the image.

Information from the lateral geniculate nucleus and from the primary visual cortex also flows to other parts of

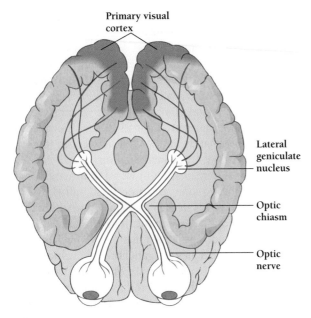

Figure 42-11

How do we perceive what our eyes see? Image information from the two eyes passes to the optic nerve into the brain to the lateral geniculate nucleus of the thalamus. From there the information passes to the primary visual cortex on each side of the brain. Many other areas of the brain also receive visual information.

the brain. In each case, the pattern of electrical activity reflects the spatial organization of the visual image, but different features are emphasized or downplayed depending on the brain areas involved. Ultimately, the interneural connections of the retina and brain establish not only what we see but also how we respond to it.

> The axons of the ganglion cells form the optic nerve, which carries visual information to the brain. Optic nerve fibers end in the thalamus, which serves both as an integrator of visual information and as a relay station.

42.5 Other Sensory Pathways

Other sense organs also report to the brain. In each case, the corresponding perception consists of a pattern of neural activity. Depending on which brain cells are active, we interpret these chemical and electrical changes differently—as sound or smell, touch or taste.

Perception refers to the interpretation of these neural events as the conscious experience of objects and events in the external world. This definition raises a basic philosophical question: just who is doing the interpreting? Some people envision the sensory pathways as television or telephone cables that carry electrical representations of pictures and

sounds to some inner observer. The hard thing to understand about the brain is that there is no such inner observer sitting at a switchboard or video monitor. The brain does not convert the electrical impulses back into "real" images. Perceptions are purely electrochemical patterns. The marvel is our ability to interpret these patterns and to respond to them. A perception may stimulate a person to run, to talk, or just to feel good.

Somatic Sensations

The visual system processes information only from the external world. In contrast, the somatic sensory system reports on information from three sources—from the external world, as it impinges on the body's surface; from the position of body parts with respect to one another; and from the interior of the body. The somatic (body) sensory system has several types of receptors—including those for touch, vibration, pressure, hair movements, position, heat, cold, and pain.

Each kind of receptor converts information into electrical signals, which, like the signals of the visual system, travel to the central nervous system. For example, receptors in the skin convert pressure into electrical activity. The receptors send signals to the neurons in the spinal cord or in the part of the brain called the medulla. The neurons in the spinal cord or medulla in turn send signals to the thalamus. Finally, the thalamus neurons send signals to neurons in the cerebral cortex.

As in the visual system, the pathways of information roughly maintain the spatial organization of the receptors, so that the sensory cortex has a complete map of the whole body. Parallel pathways carry corresponding information from different kinds of receptors—for example, for touch or for position—to adjoining columns of cells in the sensory cortex. The map in the cortex is somewhat distorted because some parts of the body, such as the mouth, hands, and genitals, deliver more sensory information than other parts, such as the back and legs.

> Sensory receptors throughout the body detect touch, position, temperature, and pain. Somatic sensory information passes to the spinal cord or to the medulla, through the thalamus, to the cerebral cortex.

How Do We Detect Molecules as Smells and Tastes?

Perception of smell and taste depends on sensory receptors that respond to molecules from the external environment. Similar receptors are responsible for determining the chemical status of the internal environment. These initial receptor molecules are proteins that may bind to the small molecules responsible for smell or taste. Like the receptors for hor-

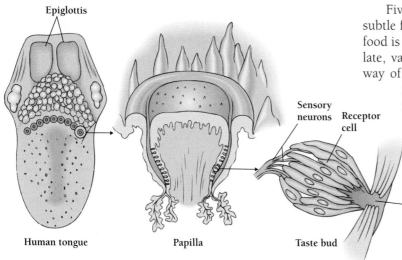

Epiglottis

Human tongue **Papilla** **Taste bud**

Sensory neurons **Receptor cell**

Gustatory pore

Figure 42-12
Taste buds on the tongue. The epithelium of the tongue and adjacent tissues contains taste buds, which contain taste receptors.

mones and neurotransmitters, binding initiates a cascade of cellular events that ultimately result in a signal being sent to the brain (Chapters 37 and 43). Taste information detected by taste buds flows to the medulla. Like sensory information from the eyes, ears, and skin, the mouth's contributions funnel through a distinct part of the thalamus before reaching the cerebral cortex.

Taste sensations depend on taste buds on the tongue and on the lining of the top of the digestive tract—the mouth, pharynx, larynx, and upper esophagus (Figure 42-12). A single taste bud harbors between 50 and 100 taste cells. All are modified epithelial cells. Some cells respond to dissolved molecules, while some provide support.

Biologists classify tastes into five major qualities—sweet, sour, bitter, salty, and umami. The amino acid glutamate, used as a flavor additive in the form of monosodium glutamate, or MSG, is responsible for the umami taste. A popular belief is that different parts of the tongue detect different tastes, but scientists have found that, in fact, all five qualities can be sensed by any region of the tongue that contains taste buds. Even individual taste cells respond to multiple qualities.

So how does the brain sort out the mixed messages it receives from the taste buds? Most scientists now think that the information that defines a particular taste is encoded by unique patterns of electrical activity generated by many scattered taste cells. No single neuron is capable of distinguishing between different qualities; it is the joint behavior of many cells as a group that represents a particular taste. A salty potato chip, for example, generates a pattern of responses across the mouth that is different from the pattern generated by a sweet candy bar. Conversely, foods that taste alike generate similar patterns of activity.

Five tastes are not, of course, enough to convey the subtle flavors of food. What we usually call the "taste" of a food is actually both taste and smell. The aromas of chocolate, vanilla, garlic, and lemon all come to our brains by way of our noses, not our tongues. That is why a stuffy nose can impair our ability to taste what we are eating. Smelling, or olfaction, is far more sensitive than taste. Devoted wine tasters, for example, can distinguish more than 100 different flavors—mixtures of taste, smell, and textures—in wine alone. Altogether a well-trained human can distinguish about 10,000 odors. Yet this is nothing compared with the olfactory powers of dogs. Dogs can distinguish far more odors and at much fainter concentrations, which makes them useful for searching for lost children and the victims of disasters.

Researchers now suspect that humans have from 100 to 1,000 distinct types of odor receptors. The smell receptors lie in the olfactory epithelium of the nose (Figure 42-13). Smell receptors, unlike taste receptors, are true neurons that connect directly to the brain's **olfactory bulb,** which extends into the space next to the nose. Each receptor neuron contains a single type of olfactory receptor protein, which can bind a range of related molecules, thereby triggering neural activity. The pattern of electrochemical activity of individual receptor neurons in the olfactory bulb reflects the specific type of odor. Unlike the relay stations for vision, hearing, and touch (all of which are in the thalamus), the olfactory bulb does not reflect the spatial organization of the outside world. As a result, our sense of smell provides no clues about the location of an odor. We can sniff in different directions to determine where a smell is strongest, but the information is not included in our perception of the smell itself.

When we taste and smell, receptors bind a specific protein, which triggers changes in nerve activity in the brain, which we perceive as taste or odor.

Sir John Eccles, who won a Nobel prize for his studies of the synapse, once quipped, "The brain is so complicated that it staggers its own imagination." It seems unlikely that we will ever completely understand exactly how the brain accomplishes its uncounted activities. We cannot predict how much biology may reveal about the processing of sensory information, of the control of muscle movements, and of the modulation of mood, memory, personality, mind, and soul. One thing, however, is certain: intelligent, wondering women and men will continue to use these marvelous organs to ask questions about ourselves and our fellow creatures.

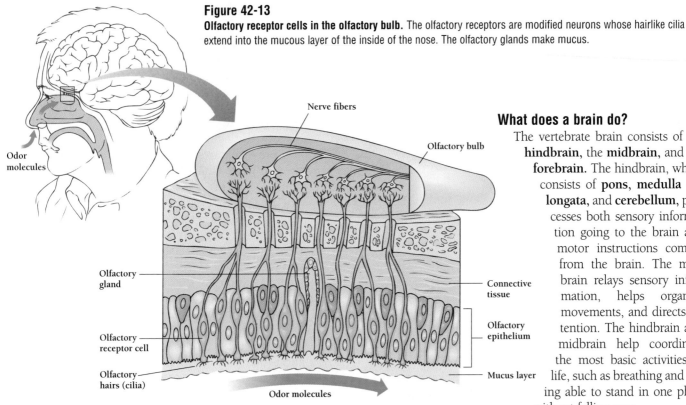

Figure 42-13
Olfactory receptor cells in the olfactory bulb. The olfactory receptors are modified neurons whose hairlike cilia extend into the mucous layer of the inside of the nose. The olfactory glands make mucus.

Nerve fibers

Olfactory bulb

Olfactory gland

Connective tissue

Olfactory epithelium

Olfactory receptor cell

Olfactory hairs (cilia)

Mucus layer

Odor molecules

Odor molecules

Key Concepts

- The nervous system regulates internal organs and movements.
- The nervous system is also responsible for sensation, cognition, learning, emotion, mood, and consciousness.
- Most senses depend on a sense organ with special receptor molecules, and on special neural pathways within the brain. The main senses that inform an organism about the exterior world are vision, hearing, touch, smell, and taste.

Summary with Key Terms

How does the nervous system respond to its environment?

The nervous system is a cellular network that senses and responds to changes in an animal's external and internal environment. Cnidarians have simple **nerve nets**, but most animals have nerve ganglia and clusters of ganglia for processing information. Vertebrates have a **central nervous system (CNS)** connected to the limbs, organs, and sensory organs by way of the **peripheral nervous system. Motoneurons** in the **spinal cord** directly stimulate muscle cells and participate in **reflex** movements that don't involve the brain. Sensory information about touch, position, temperature, and pain moves from receptors to the spinal cord and then up to the brain.

What does a brain do?

The vertebrate brain consists of the **hindbrain**, the **midbrain**, and the **forebrain**. The hindbrain, which consists of **pons, medulla oblongata**, and **cerebellum**, processes both sensory information going to the brain and motor instructions coming from the brain. The midbrain relays sensory information, helps organize movements, and directs attention. The hindbrain and midbrain help coordinate the most basic activities of life, such as breathing and being able to stand in one place without falling over.

The forebrain, which consists of the **diencephalon** and **cerebrum** (with its two hemispheres), is the site of activities that include sensation, perception, movement, learning, memory, and mood. Within the diencephalon, the **thalamus** integrates information about the external world, while the **hypothalamus** is the major regulator of the internal world. Each cerebral hemisphere consists of a **cerebral cortex** and underlying structures, including the **hippocampus**, amygdala, and the basal ganglia. In humans, the cerebral cortex is highly folded, with many ridges and grooves. The cortex receives sensory information and also processes that information, largely in the **association cortex**, which is intimately involved in higher mental functions, such as planning, attention, and **memory**. **Short-term memories** are less stable than **long-term memories**, whose formation depends on the workings of the hippocampus.

What does the peripheral nervous system do?

The **peripheral nervous system** consists of both **afferent** nerves, which carry sensory information to the CNS, and **efferent** nerves, which carry motor instructions from the CNS to both the voluntary muscles and to the internal organs. The autonomic nervous system contains both the **sympathetic** and the **parasympathetic** nerves.

How do we sense?

Sensation refers to the detection of a stimulus in the external world, and **perception** refers to its conscious recognition and interpretation. Both sensation and perception depend on **sensory receptor cells** and on the central nervous system.

In vertebrates, hearing, vision, taste, and smell are in specialized **sense organs.** Touch, or somatic sensation—including pressure, position, motion, pain, and temperature—depends on individual receptors that lie throughout the body.

Sensation depends on transduction, transmission, and integration. **Transduction** translates an external signal into a change in the chemistry of the sensory receptor cell. **Transmission** carries a signal from the receptor cell to the CNS, often by way of **relay neurons. Integration** processes and integrates the information in the signals from many different receptors, a task often accomplished by **interneurons.** Sensory information from each sense usually flows separately through the thalamus and converges in the cerebral cortex.

The auditory system detects and interprets differences in frequency, which we perceive as pitch, and differences in pressure, which we perceive as loudness. In mammals, sound passes from the **outer ear,** through the **auditory canal,** to the eardrum (or **tympanic membrane**), which transmits the vibrations through the bones of the **middle ear** to the fluid of the **inner ear.** In the **cochlea** of the inner ear is the **organ of Corti,** whose **hair cells transduce** vibrations into neural signals. Each frequency stimulates a different set of hair cells, and the brain interprets the patterns of stimulation as different patterns of sound. Hair cells in the **labyrinth** of mammals or in the **lateral line system** of fish and amphibians detect the body's position and motion.

Light enters the eye through the transparent **cornea** and passes through the **pupil,** whose opening depends on the action of muscles in the **iris.** The **lens** of the eye focuses images onto the **retina,** the thin sheet of cells at the back of the eye. In the retina, **photoreceptors** detect light. **Cones,** which detect bright light and color, are concentrated in the central **fovea. Rods** detect very faint greenish-blue light and are concentrated around the edges of the retina. Both rods and cones contain light-absorbing pigments. Misshapen eyeballs or corneas may result in **nearsightedness** or **farsightedness.** A cloudy spot on the lens is called a **cataract.** The ability to alter the shape of the lens to focus near or far is called **accommodation.**

The photoreceptor cells of the retina transmit information to the **interneurons,** which process and transmit the information to ganglion cells. The axons of the ganglion cells form the **optic nerve,** which carries visual information to the brain. Optic nerve fibers end in the lateral geniculate nucleus of the thalamus, which serves both as an integrator of visual information and as a relay station.

The primary event in taste and smell is the binding of a small molecule to a specific protein. This event triggers changes in nerve activity in the brain. Taste information is encoded by patterns of electrical activity across many taste cells that are relayed through the thalamus to the cerebral cortex. Olfactory receptor cells, however, themselves lie within a brain structure, the **olfactory bulb,** and the signals that result from olfaction do not directly reach the cerebral cortex.

Review and Thought Questions

Review Questions

1. What does a flatworm have that a jellyfish doesn't in terms of a nervous system?
2. What part of the brain is larger in mammals than in most other vertebrates? What does that part of the brain do?
3. What are the main functions of the midbrain and hindbrain?
4. What kind of cell detects and transduces the signal received by the sense organs?
5. How do sense organs relay signals to the brain or the spinal cord?
6. Explain how the knee jerk reflex works.
7. How does olfaction differ from other senses?
8. How do taste cells encode information?
9. What is the difference between an afferent neuron and an efferent neuron?
10. Briefly describe how the hair cells in the inner ear detect movement and vibration.

Thought Question

11. Explain in what ways the brain is integrated (a single mind) and in what ways the brain is a group of separately operating units (a confederacy of minds).
12. The chapter opening image shows a 19th-century map of the brain based on "phrenology," which is described below. Why can wrong ideas sometimes stimulate good questions?

BiologyNow Resources

Biology ☉ Now™

Active Figures

42-5: The brain
42-7: The human ear

Preparing for an exam? Take a diagnostic test on your BiologyNow CD-ROM.

Online materials relating to this chapter are at:

http://biology.brookscole.com/AAL3

About the Chapter-Opening Image

In the mid-19th century, scientists and the public began to realize that different parts of the human brain were responsible for different mental activities. Franz Joseph Gall estimated the strength of a trait in someone not from the anatomy of the person's brain, however, but from the ridges and bumps on the skulls of his living subjects. This practice, known as phrenology, captured the imaginations of many 19th-century intellectuals. Phrenology was more quackery than science, but it helped stimulate interest in the physical basis of the mind.

The Cells of the Nervous System

Key Questions

- How does a nerve cell carry information?
- How do nerve cells talk to one another?

Rita Levi-Montalcini

When Rita Levi-Montalcini (1909–) was a 21-year-old medical student, she asked her professor to suggest a suitable research project. A neuroanatomist at the Turin School of Medicine, in Italy, he suggested that she find out how the folds of the human brain were formed during development.

"It was a really stupid question, which I couldn't solve and no one could solve," Levi-Montalcini recalled at age 82. When the professor called her into his office a few months later to see how she was doing, he pronounced her attempts "real trash" and told her she was not cut out for scientific research. Neither of them could guess that she would eventually win a Nobel prize and become one of the most influential neuroscientists of the century. Nor could either of them know that her subsequent work would suggest new therapies for spinal cord injury and Alzheimer's disease.

Yet the 22-year-old didn't let her professor discourage her. After her initial failure, Levi-Montalcini quickly found a more manageable problem and never looked back. Levi-Montalcini gradually taught the professor to respect her, and he ultimately proved a faithful friend and mentor until his death many years later. All her life, Levi-Montalcini has skillfully navigated obstacles that would stop most people. Even in her 80s, she continued to inspire and charm new generations of biologists.

Born into an intensely patriarchal Italian-Jewish family, Levi-Montalcini rebelled against her parents' assumption that she and her twin sister Paola would marry straight out of high school. She begged, in vain, to be sent to a high school that would prepare her for college (Figure 43-1). Instead she was sent to a high school for girls that offered no math or science.

When Levi-Montalcini was just 20, however, her beloved governess, Giovanna, died of stomach cancer, and Levi-Montalcini resolved to become a doctor. She knew that she had not had the education that she needed to go to medical school, so she decided to find a tutor to prepare her for the entrance exams. To make the endeavor more fun, she persuaded her cousin Eugenia to join her. Eight months later they both passed the entrance exams with flying colors. In 1930, they entered the Medical School at Turin, Italy. By 1936, both young women had earned degrees in medicine with top honors. And after the false start with the human brain, Rita began work that fascinated her. "For the first time," she wrote in her autobiography, "I became passionate about research . . . ."

It was a passion that would be severely tested. Beginning in the 1930s, Mussolini, who had allied himself with Hitler, prohibited Italy's Jewish teachers and doctors from working.

In 1939, when Levi-Montalcini could no longer work at the university at Turin without endangering both herself and her non-Jewish colleagues, she left school.

For a year, she lived quietly with her mother, her sister Paola, and her brother Gino. Her father had died several years earlier. She and Eugenia spent much of their time treating Jews who had fallen ill but could not be treated by non-Jewish doctors.

Then one day she received a visit from an old friend from medical school who demanded to know what sort of research she was doing. Under the wartime conditions, it hadn't occurred to her that she could do any research, and she said nothing. Distressed by her apathy, he scolded her, "One doesn't lose heart in the face of the first difficulties. Set up a small laboratory and take up your interrupted research. Remember Ramón y Cajal, who in a poorly equipped institute, in the sleepy city that Valencia must have been in the middle of the last century, did the fundamental work that established the basis of all we know about the nervous system of vertebrates."

It was all the impetus Levi-Montalcini needed. She set up a laboratory in her bedroom and decided to study chick embryos, as chicken eggs were easy to come by. Partly inspired by the thought of the famous Spanish anatomist Santiago Ramón y Cajal, she chose to study the development of the nervous system.

Always practical, she would carefully remove the tiny chick embryos from the eggs after she had finished an experiment and turn the eggs into omelets in the kitchen downstairs, a practice that nauseated her brother Gino. Her research into how embryonic nerve tissue differentiates into specialized cells went splendidly. She described her fascination:

Now the nervous system appeared to me in a different light from its description in textbooks of neuroanatomy, where its structure is described as rigid and unchangeable. Only by following from hour to hour in different specimens, as in a cinematographic sequence, the development of nerve centers and circuits, did I come to realize how dynamic these processes are; how plastic and malleable is the entire nervous system . . . .

During development, the nerves of the peripheral nervous system appear to grow out from the spinal cord toward the limbs. The biologist Viktor Hamburger, at Washington University in St. Louis, had shown that embryos whose limb

Figure 43-1
Rita Levi-Montalcini as a young woman. As girls, Rita Levi-Montalcini and her sisters, Paola and Nina, hoped to emulate the Brontë sisters, the famous 19th-century trio of English writers. As they grew older, Rita became a Nobel prize–winning neuroscientist and Paola a famous painter.

buds have been amputated do not develop nerves to supply the limbs. He proposed that the limbs produce some special substance that tells the unspecialized nerve cells in the spinal cord what kind of cells they should turn into, as well as in what direction they should grow.

Levi-Montalcini's research suggested an alternative. She thought that the nerve cells in the embryonic spinal cord do specialize, or differentiate, forming nerves that can grow outward to the limbs, but that they soon die without some special substance from the limb buds to sustain their growth. Her discovery of this substance revolutionized the understanding of neural development.

But Levi-Montalcini's work was once again interrupted and this time her very life endangered. In 1943, German tanks approached Turin, and Levi-Montalcini and her family fled. It would be two years before she could resume her research.

After the war, Hamburger invited Levi-Montalcini to come to Washington University, in St. Louis, to work with him. She gladly accepted, and together, they found that her hypothesis that some substance sustains developing nerve cells was correct. In the 1950s, she and biochemist Stanley Cohen isolated this substance, which they called *nerve growth factor.* Levi-Montalcini joined the faculty at Washington University, and in 1986, she and Cohen shared the Nobel prize for Medicine and Physiology.

Their discovery led not only to a new understanding of neural development but also to the discovery of a whole set of related growth factors. These proteins stimulate the growth of neurons during early development and help sustain them during adult life. Growth factors, many researchers believe, may prevent cell death—and even promote regeneration—in spinal cord injuries, Alzheimer's disease, Parkinson's disease, and other neurodegenerative diseases.

Levi-Montalcini, now in her 90s, continues to make her mark on science. As of February 2002, at age 93, she was still chairing the Levi-Montalcini Foundation and keeping in close contact with the foundation's beneficiaries. She and her late sister Paola established the foundation in 1992 to help the careers of young people, focusing most recently on funding scientific education for African women.

"You never know what is good, what is bad in life," Levi-Montalcini told the magazine *Scientific American.* "I mean, in my case, [working in isolation during World War II] was my good chance."

823

43.1 How Are Nerve Cells Special?

Even a generation after biologists had accepted that organisms are made of cells (the cell theory), biologists were still not certain that the nervous system was made of cells. When they examined the tissues of the brain under the microscope, they did not see distinct cells, only dots, blurs, and dark spots that resembled nuclei. The problem, we now know, was that the brain contains so many intertwined cells that the only way to even see all of a single cell is to stain just one cell while leaving its neighbors unstained. In the 1880s, Camillo Golgi developed just such a method for staining only a few cells in a sample of brain tissue (Figure 43-2).

Shortly afterward, the Spanish neuroanatomist Santiago Ramón y Cajal used Golgi's method to study the nervous system in a variety of species. His most important contribution was clear evidence that the nervous system consists of cells. Ironically, Golgi himself did not accept Ramón y Cajal's conclusions. When in 1906 the two men jointly accepted a Nobel prize for their work, each man's acceptance speech contradicted the other's.

A hundred years have now passed, and neuroscientists are in agreement about much of the biology of nerve cells. Most nerve cells have the same kinds of organelles as other cells, though sometimes more of them. For example, because neurons use immense amounts of energy, they have great numbers of mitochondria, the organelles that provide power to cells. And because neurons also make lots of proteins, they likewise contain extensive endoplasmic reticulum.

The most familiar nerve cells are **neurons**, which carry information. But the nervous system also includes **glial cells**, which sustain the neurons and outnumber neurons in the brain by at least three to one. Nerve cells come in a variety of shapes and sizes that help them carry and transmit information to neurons, muscles, or glands. But most neurons are merely variations on a basic plan (Figure 43-3). Most neurons have a distinct center, called the *cell body,* which is about the same size as most other cells. Numerous short extensions, called **dendrites**, usually relay signals from other cells to the cell body. Longer, thicker extensions, called **axons**, usually carry signals away from the cell body and to other cells. Axons may be a meter or more in length. They may branch at their ends to make contact with target cells, but the number of axon branches is small compared with the number of branches in some dendrites. One kind of cell in the brain's cerebellum, for example, has enough dendrites to receive information from more than 100,000 other cells.

Although the glial cells do not carry information, they are as important as the neurons. The most numerous glial cells in the brain are the star-shaped *astrocytes.* These help regulate the levels of ions and chemical signals in the extracellular fluid so that neurons can transmit their signals reliably. They also provide structural support and serve as guides and scaffolding during the development of the nervous system. In addition, astrocytes play a role in the *blood-brain barrier,* which strictly controls the flow of substances from the blood into the brain. Other glial cells wrap around nerve cell extensions in a *myelin sheath.* As we will discuss later, the myelin sheath speeds the transmission of signals through the nerves.

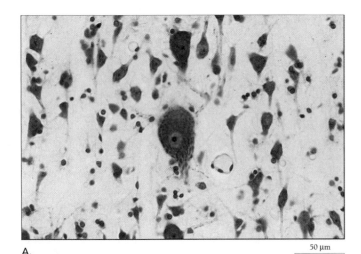

A.   50 µm

Figure 43-2

A network of single cells. In the late 19th and early 20th centuries, anatomists debated whether the brain was made of individual cells, as the rest of the body was known to be. A. With standard staining techniqes, it is impossible to tell where one cell ends and another begins. B. Camillo Golgi developed a silver staining technique (shown here) that allowed him to see that the nervous system is composed of individual cells.

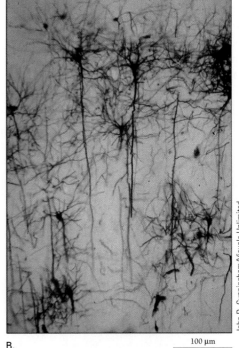

B.   100 µm

John D. Cunningham/Visuals Unlimited

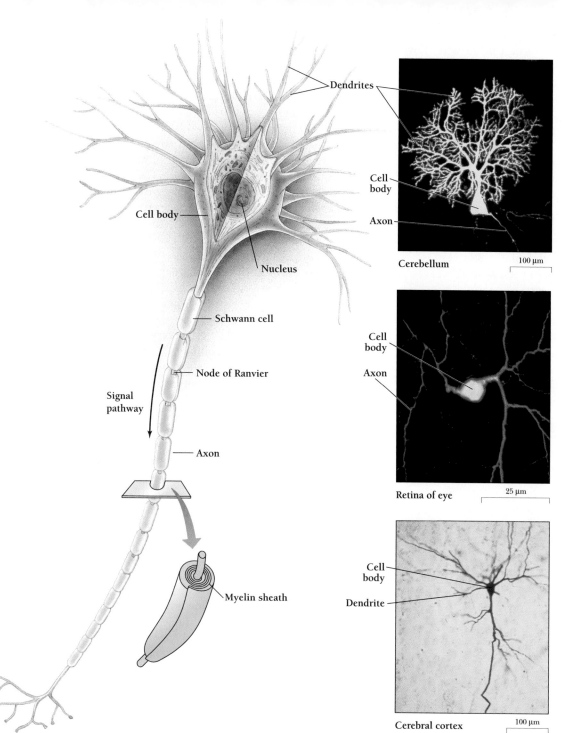

Dendrites

Cell body

Nucleus

Schwann cell

Node of Ranvier

Signal pathway

Axon

Myelin sheath

Figure 43-3
Nerve cells come in all shapes and sizes. The number of processes depends on how many other cells they connect to. The schematic drawing of a nerve cell shows the cell body with dendrites and an axon enclosed in an insulating myelin sheath. The photographs at right show different types of nerve cells, including cells from the eye's retina and the brain's cerebellum and cerebral cortex. (Photos, retina and cerebellum, David Becker/Tony Stone Images; cerebral cortex, Carolina Biological/Phototake, NYC)

Biology⊛Now™ Learn more about nerve cells by clicking on this figure on your BiologyNow CD-ROM.

Cell body

Axon

Cerebellum   100 μm

Cell body

Axon

Retina of eye   25 μm

Cell body

Dendrite

Cerebral cortex   100 μm

43.2 How Does a Neuron Carry Information?

Neurons carry different kinds of information. Whether we cut ourselves or catch sight of a loved one, the information about the event is carried to the brain by nerves, and the brain's response is transmitted to our organs and muscles by nerves. How do nerves do it?

Scientists have known since the late 18th century that nerves respond to electrical impulses. A small electric shock applied to the neurons that control a frog's leg, for example, will cause the leg to kick. Even though the electric current stimulates the frog's nerves, however, the electrochemical pulses carried by nerves are entirely different from the electric current that flows in the wires of your house. Household electric current, which consists of electrons jumping from atom to atom in a metal wire, flows much more rapidly than nerve impulses, which are carried by ions moving back and forth across cell membranes. And while electric current gradually loses energy as it flows passively through a wire, the strength of a nerve impulse is constantly renewed as it passes from nerve cell to nerve cell.

Another important difference between electric current and nerve impulses is the standard size of a nerve impulse. Electricity can flow in millions of volts or in tiny millivolts. But a nerve impulse always carries the same voltage. If a stimulus is sufficient to trigger a nerve impulse at all, the impulse will be the same size no matter how strong or weak the stimulus was. On the other hand, if the stimulus is too weak, the neuron will not react at all, which is called a **threshold** response.

If a neuron has threshold response, how can animals detect the full range of differences in temperature and light intensity? Nerves convey differences in intensity in two main ways. First, as the intensity of a stimulus increases, the number of impulses per second also increase. Second, neighboring neurons have different thresholds, so a slight touch may stimulate only a handful of neurons, but a sharp poke will stimulate many. As a result, the number of impulses reaching the brain each second increases because each receptor cell is sending more impulses and more receptor cells are firing off impulses.

Unlike electrical lines, nerves carry energy impulses that are always the same size and that do not dissipate during transmission. A neuron transmits an impulse only when stimulated by a signal that exceeds a minimum threshold. Nerves convey differences in signal strength by increasing the number of impulses.

How Does a Neuron Send an Impulse?

If nerve impulses are not simple electric currents, what are they? In the rest of this chapter we will try to understand how neurons generate and transmit impulses and how these impulses are then passed from one neuron to the next.

The chemical composition of the inside of a cell is always different from outside the cell because the cell's membrane allows some molecules and ions to pass in or out of the cell while preventing the passage of others. Membrane **channels** allow certain molecules and ions to pass through the membrane. Some of these channels have **gates** that open and close, regulating the passive movement of molecules and ions through the membrane. Other membrane proteins actively **pump** molecules and ions across the membrane using the energy from ATP.

All these different channels and pumps create a difference in the concentration of molecules and ions on the two sides of a cell membrane, or **concentration gradient.** Such a gradient has potential energy. If a concentration gradient also results in a net electrical charge on one side of the membrane, then an electrical gradient also exists, which also has potential energy.

An electrical gradient is called **voltage.** Voltage is what drives electric current. If we compare electricity to a waterfall, then the **current** is the amount of water, or the number of electrons, flowing per minute. The voltage is comparable to the height of the waterfall. The greater the voltage, the greater the work that can be performed.

Electricians measure voltage with a voltmeter, a device that determines the tendency of electrons to flow between two electrodes in units called *volts.* The voltages across cell membranes are much smaller than those in wall sockets. Even the small batteries used for toys and household appliances have voltages much greater than those of a cell. A standard radio battery, for example, generates about 1.5 volts, while most cells generate voltages of a few one-thousandths of a volt, or *millivolts.*

A neuron sends a signal to another cell by means of an electrical impulse, or **action potential,** a brief change in the voltage across its membrane. In a living cell, the voltage across a cell's membrane results from the difference in charge between one side of the membrane and the other. At rest, neurons maintain a voltage of about 70 millivolts across the plasma membrane, with the inside of the cell negative with respect to the outside. This voltage is called the **resting potential.** The action potential and the resting potential are called "potentials" because they refer to potential energy.

During an action potential, the cell membrane *depolarizes* and the cell's net charge actually reverses itself for just a fraction of a second. Such an impulse moves along the length of the neuron's axon membrane like a wave. Once the wave passes, the membrane *repolarizes,* returning to its resting potential. Like a camera's flash unit, which stores an electric charge for the flash's lightbulb, a cell's resting potential is potential energy that can be used to pass an impulse.

For many years neurobiologists could not easily measure the tiny voltages of nerve cells because standard metal electrodes were too big, and nerve cells were incomparably tiny and delicate. But the invention of microelectrodes and the discovery that squids have gigantic neurons solved both problems. *Microelectrodes* are tiny glass tubes that contain salt solutions and are small enough to maintain contact with a single cell. Neurobiologists record the tiny voltages picked up from microelectrodes with an *oscilloscope,* a device with a screen that displays variations in voltage over time (Figure 43-4). When a physiologist pokes a microelectrode into a neuron, the oscilloscope registers a negative voltage of about 70 millivolts, the resting potential. When stimulated, the neuron can suddenly and repeatedly change its voltage, generating an action potential that the electrodes also detect and register on the oscilloscope.

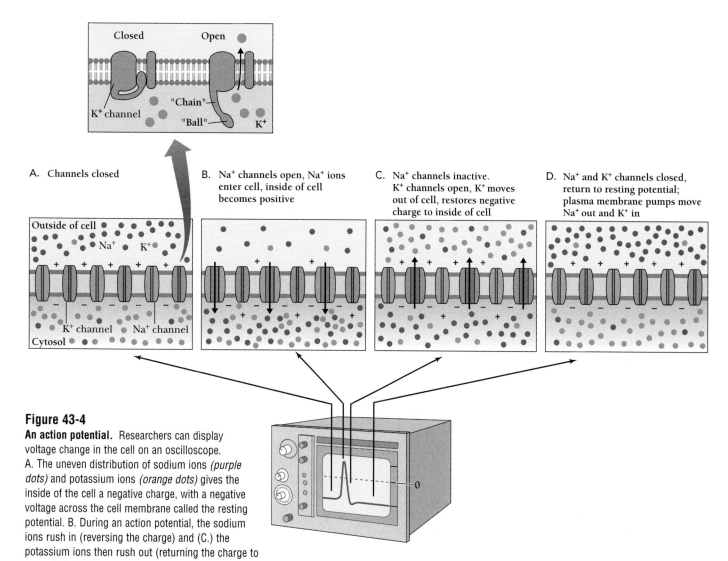

Figure 43-4

An action potential. Researchers can display voltage change in the cell on an oscilloscope. A. The uneven distribution of sodium ions *(purple dots)* and potassium ions *(orange dots)* gives the inside of the cell a negative charge, with a negative voltage across the cell membrane called the resting potential. B. During an action potential, the sodium ions rush in (reversing the charge) and (C.) the potassium ions then rush out (returning the charge to normal). D. At the end, the channels close and membrane pumps restore the ion distribution to what it was in A. Each voltage-gated potassium channel is a transmembrane protein with a "ball and chain" gate, or valve, that blocks the opening in the channel *(top left)*.

Biology ⑧ Now™   Learn more about potassium and sodium channels by clicking on this figure on your BiologyNow CD-ROM.

How Does a Neuron Maintain a Resting Potential and Generate an Action Potential?

A neuron's resting potential comes from a difference in the distribution of both negative and positive ions on the two sides of the plasma membrane. During the generation of a nerve impulse, the negative ions stay put, while the positive ions potassium (K⁺) and sodium (Na⁺) jump back and forth across the membrane.

Just before an impulse, during the "resting state," the inside of a neuron contains fewer sodium ions and more potassium ions than the outside of the neuron (Figure 43-4A). Sodium-potassium pumps in the cell membrane push sodium ions out and pull potassium ions inside, and the cell membrane maintains this uneven distribution of ions.

When something stimulates the neuron, sodium channels open, and sodium ions rush through to the sodium-poor interior of the neuron, like commuters rushing into an empty subway car whose doors have just opened. As the sodium ions rush in, the inside of the cell gains a slight positive charge, at least near the open sodium channels (Figure 43-4B). (Only a small fraction of the cell's sodium and potassium ions move during these events, and the voltage changes across the membrane are localized around the tiny patch of membrane where the channels are opening and closing.)

A few milliseconds later, the sodium channels close, the potassium channels open, and the potassium ions rush out of the cell (Figure 43-4C). The period of time when the sodium channels close and briefly resist opening again is called the **refractory period**. As the potassium ions flow out of the cell, they restore the cell's normal −70-millivolt resting potential and the neuron is ready to transmit another pulse (Figure 43-4D). Then the ATP-powered sodium-potassium pumps again go to work to push sodium out and pull potassium in.

The main reasons that the resting potential exists are the nerve cell membrane's sodium-potassium pumps, which establish the difference in sodium and potassium ion concentration, and the membrane's relative permeability to potassium (K^+) ions and impermeability to sodium (Na^+) ions.

Most of what we now know about how a neuron generates a nerve impulse came from studies of the giant neurons of squid. These neurons are responsible for the signals the squid uses to produce a jet of water to propel itself through water. In the 1930s and 1940s, English physiologists Alan Hodgkin, Andrew Huxley, and collaborators measured changes in voltage in the squid axon while changing the ion concentrations inside and outside the cell.

For example, the researchers found that the resting potential depended on the gradient of potassium ions, but the action potential depends on the gradient of sodium ions. In one experiment, Hodgkin and Huxley replaced all the sodium ions in the extracellular fluid with a larger ion that could not enter the sodium channels. The neuron's resting potential was normal, but it could no longer generate an action potential. This experiment showed that neurons generate an action potential through temporary changes in membrane permeability to sodium ions.

A neuron has a resting potential of −70 millivolts because membrane pumps create and maintain differences in sodium and potassium ion concentration across the membrane. The resting potential depends mostly on the gradient of potassium ions across the cell membrane. The action potential depends on the rapid flow of sodium ions into the neuron.

What Opens and Closes the Channels?

At the normal resting potential, sodium channels are closed; but when the cell depolarizes, the sodium channels open briefly and admit a small horde of sodium ions. Because the channel's permeability to sodium ions depends on the voltage across the membrane, sodium channels are said to be **voltage gated,** meaning that they open or close according to the voltage.

The potassium channels are also voltage gated. In one kind of voltage-gated potassium channel, the gate works like a "ball and chain" valve (Figure 43-4). When a cell is at rest, the ball plugs the channel entrance. When the membrane becomes depolarized, during an action poten-

tial, the ball (on its molecular "chain") pulls away from the channel entrance, allowing potassium ions to flow through.

An action potential depends on the opening and closing of voltage-gated sodium and potassium channels, whose "gates" open and close according to the membrane voltage.

How Does a Nerve Impulse Move Along an Axon?

The properties of voltage-gated channels explain both how action potentials occur and how they can move for long distances along neuron membranes. We can list six consecutive events in the production of an action potential in a single spot in the membrane:

1. An alteration in ion distribution in the neuron membrane causes a local depolarization that is large enough to exceed the threshold.
2. Sodium channels open and sodium ions flow into the cell.
3. As sodium ions flow inward, the inside of the membrane becomes locally positive.
4. The decreased polarization of the membrane causes more channels to open, increasing the positive charge on the inside of the membrane.
5. The sodium channels close spontaneously.
6. Potassium channels open and potassium ions then flow outward, restoring the membrane to the resting potential.

Notice that the pumping of sodium and potassium ions does not occur during the action potential. This pumping serves to maintain the distribution of ions responsible for the resting potential *in preparation for* an action potential.

Here's how the action potential propagates itself (Figure 43-5). As the membrane's inside surface becomes more positive at one spot, positive ions diffuse, and the positive charge spreads to an adjacent patch. As an adjacent patch becomes depolarized, its sodium channels open, leading to further depolarization and further spreading. In this way the action potential propagates itself down the axon, regenerating itself as it goes.

Once an action potential begins to move, it moves in just one direction. Recall that once the sodium channels have opened, they cannot immediately open again. As a result, an impulse cannot pass through again until after the refractory period. By that time the impulse has moved on. This prevents an impulse from moving backward. Each action potential passes through each membrane region only once.

An action potential propagates because a region of depolarization can move along the cell membrane, opening adjacent sodium channels and leading to further depolarization. Because each channel has a refractory period, the charge can move in only one direction.

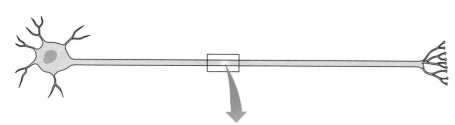

Figure 43-5
Wave action. The action potential moves along
the axon membrane as a wave of depolarization
(A and B) and repolarization (B).

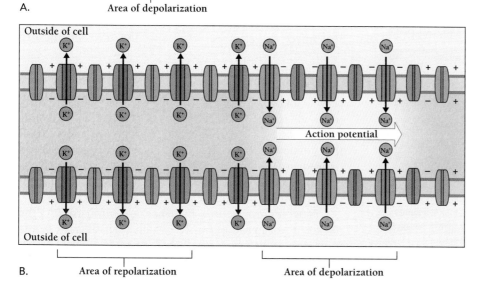

A. Area of depolarization

B. Area of repolarization   Area of depolarization

How Do Nerves Speed the Transmission of Action Potentials?

Although the thick axons of the squid conduct impulses at speeds up to 10 meters per second, most neurons have much smaller axons that would normally conduct impulses much more slowly. Given the expected rate of conduction in the sensory nerves leading from our feet, for example, it would take a painfully long time for us to realize that we had stepped on a nail or come too close to a campfire. Fortunately, most vertebrate nerves are wrapped in **myelin**. Myelin is a specialized, glistening sheath made of multiple layers of tightly packed glial membrane punctuated by gaps known as **nodes of Ranvier** (Figure 43-6).

The myelin sheath speeds the conduction of nerve impulses in several ways. Like the insulation on a hot water pipe, the myelin sheath insulates axons, preventing the passage of ions across the axon's membrane, except at the nodes of Ranvier. This insulation has two effects. First, action potentials are generated only at the nodes, which requires the movement of fewer ions than if the impulses were generated throughout the axon's length. Second, the insulation makes axons less leaky, and consequently, better conductors. Nerves wrapped in myelin are said to be **myelinated**.

The myelin sheath also helps isolate the ions that enter at the nodes of Ranvier from the influence of the charges outside the axon. Separated by a greater distance, the charges outside the cell have less of a tendency to attract the opposing charges inside the cell, which would keep them parked nearby (like the magnets on two sides of a piece of paper). In this way, the voltage resulting from the entry of ions can spread more quickly.

In this type of conduction, in which a nerve impulse jumps from one node to the next, is called **saltatory conduction** (Figure 43-6). A nerve impulse can jump along a myelinated axon up to 100 times faster than along an unmyelinated axon. An ordinary nerve conducts at about 1.2 meters per second, or about two and a half miles per hour, the speed of a normal walk. In contrast, a myelinated nerve conducts at up to 120 meters per second, or about 250 miles per hour. Not surprisingly, diseases in which myelin function is disrupted can be devastating. For example, in multiple sclerosis, a disease in which the myelin sheaths of the central nervous system deteriorate, people can have problems with vision, movement, and speech.

Myelin is present only in vertebrates. Invertebrates speed conduction by having axons of large diameters, such as those of the giant neurons of the squid.

Nerves wrapped in myelin conduct action potentials rapidly because nerve impulses jump from node to node.

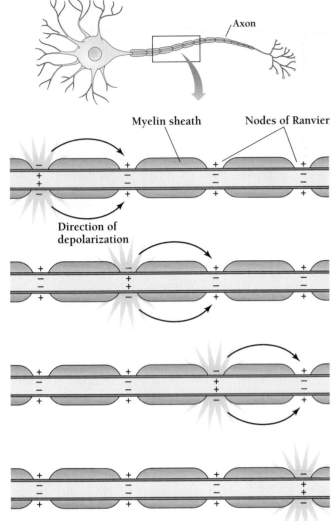

Myelin sheath

Nodes of Ranvier

Direction of depolarization

Figure 43-6
Fast Connection. A nerve impulse moves much more quickly down a myelinated axon than an unmyelinated axon because the impulse jumps from node to node. Ion currents can only flow where the myelin sheath is interrupted, at the nodes of Ranvier.

43.3 How Do Neurons Communicate with One Another?

Ramón y Cajal's studies showed that the nervous system consists of individual cells. Since then, studies with the electron microscope have revealed that there is a distinct connection point between one nerve cell and the next. The connection between two communicating neurons is called a **synapse.** The neuron sending information is the **presynaptic** neuron, and the neuron receiving information is the **postsynaptic** neuron. We have seen that a neuron can send information long distances along an axon. How does a presynaptic neuron trigger an action potential in a postsynaptic neuron?

The simplest synapse is the **electrical synapse,** which joins presynaptic and postsynaptic neurons through gap junctions, channels through the membranes of adjacent cells that allow ions and small molecules to pass freely from one cell to the next. An electrical synapse allows an action potential to continue traveling to the postsynaptic cell in the same manner and at about the same rate as it traveled down the presynaptic axon.

All electrical synapses are **excitatory,** meaning that action potentials in the presynaptic cell stimulate action potentials in the postsynaptic cell. Electrical synapses occur in a variety of invertebrate and vertebrate nerve circuits, where fast conduction between cells is advantageous and subtle modifications of a signal are unimportant. For example, electrical synapses occur in the vertebrate heart, where they coordinate the synchronous contraction of heart muscle cells.

The more common type of synapse, however, is the **chemical synapse,** in which presynaptic and postsynaptic membranes do not join (Figure 43-7). Instead, the presynaptic and postsynaptic cells are separated by a narrow **synaptic cleft.** Communication across such a gap requires the diffusion of one or more **neurotransmitters,** small signaling molecules made in the presynaptic cell that affect the electrical charge of the postsynaptic cell membrane.

Chemical synapses may either excite or inhibit, depending on the particular neurotransmitter and on the receptors on the postsynaptic cell. So far, biologists have identified at least 20 different neurotransmitters, including acetylcholine (Chapter 37). Many neurotransmitters are either amino acids (glutamate and glycine) or else molecules derived from amino acids (GABA, derived from glutamate; serotonin, derived from tryptophan; and dopamine, norepinephrine, and epinephrine, all derived from tyrosine).

A nerve impulse in a presynaptic neuron triggers the release of a neurotransmitter, which then diffuses across the synaptic cleft and binds to a receptor. A neurotransmitter can either excite or inhibit the activity of the postsynaptic neuron, but the effect is always delayed by the time needed for the presynaptic cell's machinery to release the neurotransmitter and for the neurotransmitter to diffuse across the gap. No such delay occurs in an electrical synapse. On the other hand, chemical synapses have two important advantages over electrical synapses. First, they may be either excitatory or **inhibitory** (meaning that the neurotransmitter reduces the probability of an action potential in the postsynaptic cell). Second, they can amplify signals—a weak electric current generated by a small presynaptic neuron can depolarize a large postsynaptic cell through the release of thousands of neurotransmitter molecules.

A synapse may be electrical or chemical. Electrical synapses are always excitatory, but chemical synapses may be excitatory or inhibitory.

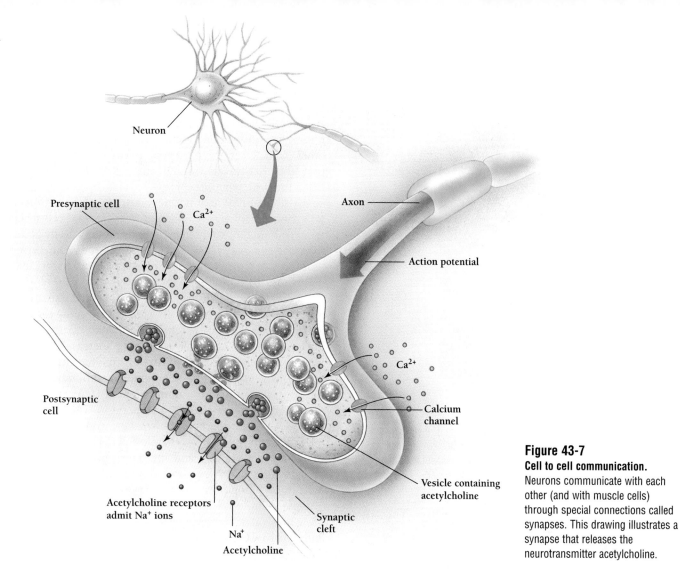

Neuron

Presynaptic cell

Ca²⁺

Axon

Action potential

Postsynaptic cell

Ca²⁺

Calcium channel

Acetylcholine receptors admit Na⁺ ions

Vesicle containing acetylcholine

Synaptic cleft

Na⁺

Acetylcholine

Figure 43-7
Cell to cell communication.
Neurons communicate with each other (and with muscle cells) through special connections called synapses. This drawing illustrates a synapse that releases the neurotransmitter acetylcholine.

How Does a Chemical Synapse Work?

Understanding how chemical synapses work is important not only to our understanding of the brain but also to medical and social problems. Essentially all the medications that affect the functioning of the brain act on synapses. Among these substances are antidepressants, sleeping pills, tranquilizers, antipsychotic medicines, and narcotics such as codeine, heroin, and cocaine. Each of these chemicals acts by mimicking or interfering with the production, release, or action of some neurotransmitter.

Let us look more closely at the structure of a chemical synapse. The best-understood chemical synapse is the vertebrate *neuromuscular junction,* the synapse between a motoneuron and a voluntary muscle cell (Figure 43-8). These synapses are similar to those between neurons (Figure 43-7), but they are larger and easier to study. On the presynaptic side, the electron microscope reveals tens of thousands of tiny vesicles, each about 50 nm in diameter and enclosed by a membrane. These vesicles are full of the neurotransmitter **acetylcholine.**

Knowing this much, we now want to know (1) what triggers the release of neurotransmitter and (2) how the released neurotransmitter causes the postsynaptic cell to fire an action potential. The trigger for release of the acetylcholine appears to be an increase in intracellular calcium (Ca^{2+}) ions. The concentration of calcium ions in the cytoplasm is much lower than in the extracellular fluid. Calcium ions will always flow *into* a cell—if there is a way in.

During an action potential, calcium ions flow into a presynaptic neuron through special calcium channels that open in response to the change in voltage during an action potential. Like sodium channels, these calcium channels are voltage gated and close soon after they open. And because they are present only near the synaptic cleft, they change the cell's calcium concentration only locally. Yet this very localized change immediately triggers the release of acetylcholine because the machinery that releases neurotransmitters is right next to the calcium channels.

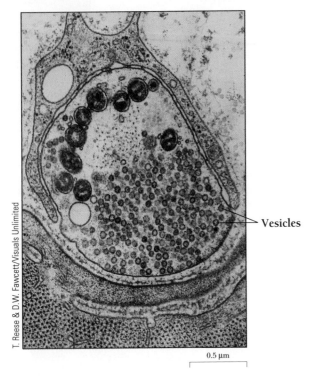

Vesicles

Figure 43-8
A neuromuscular junction. In this electron micrograph of the synapse between a motoneuron and a muscle cell, notice the clusters of small vesicles full of acetylcholine. When the motoneuron fires, some of these vesicles will release their acetylcholine into the synaptic cleft, as illustrated in Figure 43-7.

0.5 μm

Once the acetylcholine is released, it diffuses across the synaptic cleft to the postsynaptic membrane, where it binds to a receptor molecule. This acetylcholine receptor is itself an ion channel that allows the passage of sodium and potassium ions. Like the voltage-gated sodium channels and the voltage-gated calcium channels, this ion channel is gated. But it is acetylcholine rather than voltage that opens the gate. A molecule that specifically binds to another molecule is called a ligand, so the acetylcholine receptor is an example of a **ligand-gated channel.**

At neuromuscular junctions, an increase in the local concentration of calcium ions triggers the release of acetylcholine into the synaptic cleft. When acetylcholine binds to its receptor, it opens a channel that allows the passage of sodium and potassium ions.

Neurotransmitters May Either Excite or Inhibit Postsynaptic Neurons

When sodium ions pass through the channel of the acetylcholine receptor, they depolarize the membrane, making the inside of the membrane less negative than before. This volt-age change is called a **postsynaptic potential.** The postsynaptic potential triggered by acetylcholine excites the cell and is therefore called an **excitatory postsynaptic potential (EPSP).**

If an EPSP is sufficiently depolarizing, the postsynaptic cell fires an action potential. When the postsynaptic cell is a muscle cell, this action potential triggers contraction. On the other hand, when the postsynaptic cell is a neuron, the action potential travels through the cell body and axons to reach the next synapse.

A single EPSP may not depolarize a cell enough to fire an action potential. Whether the postsynaptic cell reaches the threshold for firing often depends on the activity of more than one synapse. Many presynaptic neurons can converge on a single postsynaptic cell. Some may be excitatory, while others may be inhibitory.

Inhibitory neurotransmitters open channels that allow the passage of potassium or chloride ions, leading the cell to increase their negative potential (to "hyperpolarize") and to be less likely to fire. This temporary increase in negative voltage is called an **inhibitory postsynaptic potential (IPSP).** The most common inhibitory neurotransmitter in the brain is **gamma-aminobutyric acid (GABA).** Virtually every neuron in the brain can respond to GABA, and some 30 percent of all brain neurons make GABA.

The activity of a postsynaptic cell depends on the summing of the excitatory and inhibitory potentials of all the presynaptic cells. A single postsynaptic neuron may integrate information from up to 100,000 other cells. Each postsynaptic neuron effectively serves as a microcomputer that itself fires or not according to the balance of depolarizing and hyperpolarizing responses to input from other neurons.

The firing pattern of the postsynaptic cell depends on the summing of the responses to both excitatory and inhibitory signals.

Neurons May Respond to the Same Neurotransmitters in Different Ways

The same neurotransmitter can have different effects in different postsynaptic neurons. For example, acetylcholine receptors come in two varieties, *nicotinic receptors* and *muscarinic receptors*. Skeletal muscle has nicotinic receptors, and acetylcholine triggers them to fire. Smooth muscle and heart muscle, on the other hand, have muscarinic receptors, and acetylcholine inhibits the production of an action potential. As a result, acetylcholine reduces the heart rate and slows the peristaltic contractions of the smooth muscles in the intestines (Chapter 37).

A single neurotransmitter can convey different messages to different cells, depending on the kind of receptor the cell has.

T. Reese & D.W. Fawcett/Visuals Unlimited

How Are Neurotransmitters Cleared from the Synapse?

We have seen how a neurotransmitter alters the postsynaptic neuron. But after the neurotransmitter has delivered its message, the postsynaptic neuron needs to be able to end the conversation, or hang up the phone. To end a message, neurotransmitters must be cleared quickly from the synaptic cleft to make way for the next set of signals. Neurons rely on three main strategies to accomplish this: (1) wait for the neurotransmitter to diffuse away from the synaptic cleft; (2) actively pump the neurotransmitter out of the synaptic cleft, either into the presynaptic neuron or into surrounding glial cells, a process called **reuptake**; and (3) destroy the neurotransmitter in the synaptic cleft using extracellular enzymes.

The time during which a neurotransmitter acts is limited by its diffusion and by its active removal by enzymes or transporters.

Many Psychoactive Drugs Act on Chemical Synapses

Not all the molecules that bind to neurotransmitter receptors are neurotransmitters. A variety of substances—including, for example, nicotine (from tobacco), curare (a frog toxin used in poison darts), and α-bungarotoxin (from snake venom)—all bind to the nicotinic acetylcholine receptor. Nicotine activates the receptor, while the others block the receptor and prevent the initiation of muscle contraction. All of these drugs are said to be *psychoactive,* meaning they alter mood and perception. And all work by either mimicking or disrupting the action of a neurotransmitter.

Many compounds that bind to neurotransmitter receptors have proven to be useful medications. For 2,000 years, for example, physicians have treated diarrhea with belladonna, an extract of a tall, bushy herb called deadly nightshade (*Atropa belladonna*). We now know that the extract worked because one of the compounds made by deadly nightshade, *atropine,* blocks muscarinic acetylcholine receptors. Once these receptors are blocked, acetylcholine cannot make the smooth muscles of the intestines contract. When the intestines are relaxed (instead of cramped), diarrhea is inhibited.

Other naturally occurring or chemically synthesized compounds affect other neurotransmitter receptors. Certain pesticides (malathion and parathion, for example) inhibit the breakdown of acetylcholine, leading to the overstimulation of insects' muscles and ultimately death. Benzodiazepines, commonly prescribed antianxiety drugs, bind to GABA receptors and increase the effectiveness of GABA in opening the chloride channel.

Still other substances influence the release, reuptake, or inactivation of specific neurotransmitters. Some of these are effective as painkillers, others are tranquilizers, still others are stimulants. Some drugs, such as cocaine, bring about euphoria and anesthesia, possibly by blocking the reuptake of the neurotransmitter dopamine. Among the most widely used medicines are the *selective serotonin reuptake inhibitors (SSRIs),* which include fluoxetine (Prozac) and sertraline (Zoloft). These drugs elevate mood and fight depression by blocking the reuptake of the neurotransmitter **serotonin** from certain synapses. Serotonin mediates both excitatory and inhibitory responses in various regions of the brain. Alterations in its levels may contribute to many psychiatric

Health and Biology What Is the Physical Basis of Drug Action?

Research into the way that neuroactive drugs work has led to important basic discoveries about the brain. One of the most startling of these discoveries occurred in the late 1970s, as researchers tried to understand why *opiates*—morphine, heroin, and related compounds—are such powerful drugs. Researchers realized that opiates work because the nervous system itself communicates information about pain using opiatelike neurotransmitters.

They reasoned as follows: (1) if opiates act powerfully on the brain, they must bind to specific receptor molecules to trigger changes in the responses of particular brain cells; (2) the binding of labeled opiates in the brain should reveal where the opiate receptors were; and (3) the specific distribution and properties of the opiate receptors should suggest that they must ordinarily bind to some endogenous brain compound. In the 1970s, several research groups identified three classes of such natural opiates, called *enkephalins, endorphins,* and *dynorphins*—polypeptides that act as natural pain suppressors and regulators of mood.

In the early 1990s, after more than a decade of trying, molecular biologists were able to identify recombinant DNA that specified the primary structures of opiate receptors. Understanding how these receptors work may allow us to understand the neural basis of pain and to develop ways of lessening it.

disorders in addition to depression, including anxiety, obsessive compulsive behavior, and schizophrenia.

Psychoactive compounds work by either mimicking or disrupting the action of neurotransmitters.

Key Concepts

- The nervous system consists of neurons, which carry electrical signals, and glial cells, which provide metabolic and structural support to the neurons.
- Neurons can carry nerve impulses over long distances.
- Connections between neurons may be either electrical, mediated by gap junctions, or chemical, mediated by the release of signaling molecules called neurotransmitters.
- Neurotransmitters act by binding to receptors in the plasma membrane of the target neurons.
- Networks of neurons mediate simple and complex behaviors, such as learning.

Summary with Key Terms

How does a nerve cell carry information?

The operation of the nervous system depends on the properties of individual nerve cells and the way in which they connect with one another. The cells of the nervous system include **neurons,** which carry information in the form of chemical and electrical signals, and **glial cells,** which provide mechanical and metabolic support. A neuron's **axon** transmits a signal, while the **dendrites** receive signals from other cells.

Like all other cells, neurons have different concentrations of molecules and ions inside than outside. The selective action of **channels, gates,** and **pumps** results in a greater concentration (a **concentration gradient**) of sodium ions outside the cell and of potassium ions inside, as well as a **voltage** of about -70 millivolts across the cell membrane. This **resting potential** can suddenly change when a neuron is stimulated, giving rise to an **action potential,** which can travel over large distances. Action potentials move more rapidly through axons that are surrounded by **myelin** or that have a large diameter.

The generation (and propagation) of an action potential depends largely on **voltage-gated** sodium channels in the membrane of a nerve cell. These channels open when the voltage across the plasma membrane exceeds a **threshold,** and the electrical **current** into the cells increases. The **refractory period** of the channels ensures that action potentials move one way down an axon. In **myelinated** axons, ion flow across the cell membrane occurs only at the **nodes of Ranvier,** giving rise to rapid **saltatory conduction.**

How do nerve cells talk to one another?

Nerve cells communicate with one another at **synapses,** which may be either **electrical** or **chemical.** Electrical synapses allow nerve impulses to propagate in much the same way as they travel down axons. At a chemical synapse, the **presynaptic** cell responds to the arrival of an action potential by releasing a store of **neurotransmitter** (such as **acetylcholine, GABA,** or **serotonin)** contained in synaptic vesicles.

The released neurotransmitter diffuses across the **synaptic cleft** to act on receptors on the **postsynaptic** neuron where its action may be either **excitatory,** evoking an **excitatory postsynaptic potential (EPSP),** or **inhibitory,** evoking an **inhibitory postsynaptic potential (IPSP).** These receptors may be **ligand-gated** ion channels, as in the case of the nicotinic acetylcholine receptors in muscle cells and GABA receptors in neurons. The action of neurotransmitters is terminated by diffusion, **reuptake,** or by degradative enzymes. Many psychoactive drugs act by mimicking or inhibiting the production, release, degradation, or action of neurotransmitters.

Review and Thought Questions

Review Questions

1. How are glial cells different from neurons? What role does each kind of cell play in the nervous system?
2. What is a resting potential? How does a neuron generate a resting potential?
3. What is an action potential? How does a neuron generate an action potential?
4. Why does an action potential travel in only one direction down an axon?

5. How does an electrical synapse differ from a chemical synapse?
6. What is a neurotransmitter?
7. Give one example of an excitatory neurotransmitter and one example of an inhibitory neurotransmitter. How do they each affect the postsynaptic neuron?
8. How does a chemical synapse work?

Thought Questions

9. What consequences—behavioral or otherwise—would you expect to see in an animal missing a gene that allows the synthesis of serotonin?

BiologyNow Resources

Biology⑤Now™

Active Figures

43-3:   Nerve cells
43-4:   Potassium and sodium channels

Preparing for an exam? Take a diagnostic test on your BiologyNow CD-ROM.

Online materials relating to this chapter are at:

http://biology.brookscole.com/AAL3

About the Chapter-Opening Image

The freshly hatched chick refers to Rita Levi-Montalcini's studies of the development of chick embryo nervous systems, carried out in her bedroom during World War II.

Sexual Reproduction

Key Questions

- How do mammals form sperm and eggs?
- How do the sperm and egg meet and fuse?
- How do hormones control reproductive events?
- What happens during pregnancy and lactation?
- How can humans limit reproduction?

Why Did Sex Evolve?

Textbooks tend to say that the purpose (or selective advantage) of sex is to recombine genes in novel ways, thereby increasing genetic diversity in each new generation. In Chapter 9, for example, we saw how the shuffling of chromosome regions during meiosis (when sperm and eggs are produced) enables a single human couple to produce countless unique embryos. In truth, however, no one knows why sex evolved, why it persists in the majority of multicelled organisms, or why some organisms seem to do fine without it. In addition, biologists have asked themselves why most organisms have just two sexes. Why not 13 different sexes, as certain slime molds have? What is the purpose of the sex chromosomes and how have they evolved?

Nonscientists and scientists alike also wonder about some other aspects of sex. Why are men and women different from each other? Why are men bigger than women, and why are female hawks bigger than their male mates? Which differences in behaviors and attitudes are "hardwired" during embryonic development? Which are the result of different levels of hormones such as estrogen and testosterone and might be changed with a few injections? Which differences are learned? What does it mean to learn to behave a certain way? Are women better drivers and men better navigators? If so, why? What attributes do they share with one another or with members of the opposite sex? And why? Discussions of all these questions nearly always lead to heated arguments.

Art Resource, New York

Amazingly, biologists cannot confidently answer a single one of these questions. Instead, they themselves continue to argue about the most fundamental aspects of sex, wrangling as fiercely as any group of nonscientists. Nonetheless, scientists have some fascinating insights into a few of these questions. Not all of them accept the idea that sex evolved to increase genetic diversity. For one thing, such increased diversity is not always a good thing. Most organisms live in environments similar to those of their parents. If one generation survived and reproduced successfully in that environment, why shake up the genome on the off chance that something better will come up? Sex can erase adaptive traits as easily as it can create them. If it ain't broke, why fix it?

One alternative theory comes from Richard E. Michod of the University of Arizona. Michod argues that sex evolved to repair damaged genes. Michod compares the genome to a vintage car that needs a steady supply of spare parts to keep it running. Any car collector knows that the best place to get parts is from another car of the same model. Chances are good that what is broken or worn out in one car will be usable in the other. From two broken cars, one whole one can be reconstructed.

In the same way, organisms can reconstruct a damaged genome using spare DNA. If a single strand is damaged, the adjoining strand can be used as a template for repair. But what if both strands are damaged? In that case, the correct sequence must be obtained from somewhere else.

Michod's research showed that bacteria can survive DNA damage (from mutagens such as ultraviolet light or chemicals) by swapping their own DNA with that of dead bacteria of the same species. In fact, Michod has found that only bacteria with damaged DNA actively scavenge for spare DNA, while those with healthy DNA do not bother. Bacteria that were best at recombining were most likely to survive.

As multicellular eukaryotes evolved from bacteria, says Michod, they may have used the same kind of recombination when making sperm and eggs. During meiosis, the two chromosomes, one from each parent, line up, duplicate, and swap DNA. As a result, each chromosome is a mix of genes from both parents. Recombinations, adds Michod, are most likely to occur at breaks in the DNA, often a sign of damage.

Such an exchange costs an individual, however. Thanks to sex, each individual passes on only half of its genes to each of its offspring. A female, for example, who could reproduce asexually could pass on all of her genes to every offspring, doubling her genetic representation in the next generation. Female whiptail lizards, aphids, and other organisms reproduce without sex, as do plants of both sexes (Chapter 9).

But offspring inheriting both sets of chromosomes from one parent run the risk of getting a double dose of damaged DNA. Sex keeps harmful recessive mutations masked, says Michod. Still, he does not explain why Chihuahua whiptail lizards, all of which are female, manage without spare parts from males (Figure 44-1). Michod's ideas are well known, but they are not yet accepted.

Figure 44-1
Chihuahua whiptail lizard. This unusual animal reproduces by parthenogenesis, the process by which eggs develop without any paternal genes. Males of this species do not seem to exist.

44.1 How Do Mammals Form Sperm and Eggs?

Few subjects occupy more attention, energy, wonder, and anxiety than sexual reproduction. For many species, such as the mayfly, adult life consists only of mating and dying, sometimes without even the diversion of feeding. Like annual plants, such animals die before their young even appear. Humans and other mammals have a longer and more complex adult life, with considerable energy devoted to the care of the next generation.

In mammals, successful reproduction involves many steps—production of eggs or sperm, mating and fertilization, nurturing the embryo within the uterus, delivery, nursing and cleaning, and continued care even after nursing has stopped. In both males and females, many organs contribute to these processes, all influenced by hormones.

Where Does Internal Fertilization Take Place?

In this chapter we examine the specific tissues and organs that allow sexual reproduction, focusing on human reproduction. We begin with the **gonads,** the paired organs where the eggs or sperm, together called "gametes," form and mature. Eggs, or **ova** [singular, *ovum;* Latin, = egg], form in the **ovaries** and move into the **oviducts** [Latin, *ovum* = egg + *ductus* = duct] toward the **uterus** [Latin, = womb] (Figure 44-2). Sperm form in the **testes** [Latin, singular, *testis;* diminutive, testicle]. Both eggs and sperm are haploid: they have just one set of chromosomes each. When an egg and sperm come together in the oviducts, the nuclei of the two gametes fuse, in the process known as **fertilization,** syngamy, or conception (in humans). The resulting **zygote** is diploid—it has two sets of chromosomes.

If all goes well, the one-celled zygote begins to divide and continues to make its way down one of the oviducts to the uterus. After about a week, the multicelled embryo embeds itself in the wall of its mother's uterus (Figure 44-2). When the embryo has developed the basic features of the organism it is to become, the bare outlines of all its organs, limbs, and so on, it is a **fetus.** In humans, the embryo is called a fetus nine weeks after fertilization.

At the bottom of the uterus is a tight ring, called the **cervix** [Latin, = neck], which encloses a narrow passage from the uterus to the vagina. The cervix is a sphincter, a muscular ring that normally stays contracted, leaving only a small opening connecting the uterus and vagina. At the end of a pregnancy, when the baby is ready to be born, the cervix opens to about 10 cm and the uterus contracts to push the baby out through the cervix and the vagina and into the wide world.

The vagina connects the genital tract to the outside. During sexual intercourse, the vagina encloses the penis and receives the ejaculated sperm. At birth, the baby passes out through the vagina, which stretches amazingly but may sometimes tear during this violent process.

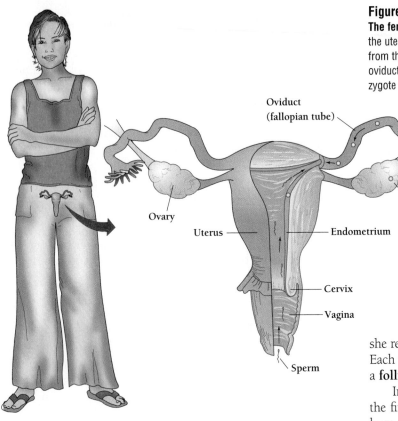

Figure 44-2
The female reproductive tract in humans. Two ovaries supply oocytes to the uterus. Oocytes (eggs) from the ovaries pass into the oviducts. Sperm from the male pass into the vagina, through the uterus, and up into the oviducts. In the oviducts, sperm may fuse with the oocyte. The resulting zygote travels to the uterus and develops.

Oviduct (fallopian tube)

Ovary

Uterus

Endometrium

Egg

Cervix

Vagina

Sperm

How Do Female Mammals Produce Ova?

The female sex organs, or ovaries, are solid, almond-shaped organs, about 4 cm long. The ovaries lie on the sidewalls of the lower abdominal cavity, one on each side, where they produce ova, or eggs, and also sex hormones. Much of the process of **oogenesis,** the production of eggs, begins before birth.

In both females and males, mature germ cells (eggs or sperm) descend from **primordial germ cells,** which are cells that have not differentiated to form skin or nerve cells. Such "undifferentiated" cells are also called *stem cells,* a broad term for any cell that has the potential to become more than one kind of cell. Embryonic stem cells can become any of the 200 kinds of cells in the body. More-differentiated stem cells, such as those in the bone marrow, may have somewhat more limited capacity to differentiate, although how much is not completely understood yet. (We discuss some of the controversy surrounding stem cells in the box on page 873.)

During the development of a female fetus, the primordial germ cells migrate to the outer surfaces of the ovaries, and multiply, forming diploid cells called **primary oocytes.** Each primary oocyte can divide by meiosis to form a haploid ovum, or egg (Figure 44-3).

A baby girl is born with an average of about 750,000 primary oocytes, each one partway through meiosis. As the baby grows up, many of these oocytes die, so that by the time she reaches puberty she has about 200,000 primary oocytes. Each primary oocyte, together with surrounding cells, forms a **follicle** [Latin, *folliculus* = small ball] (Figure 44-4).

In the ovary's follicles, the primary oocyte goes through the first division of meiosis, producing two haploid cells—a large **secondary oocyte** and a small **polar body.** Polar bodies are tiny cells that, like sperm, contain a nucleus but almost no cytoplasm (Figure 44-3B). Nearly all of the cytoplasm goes to the oocyte. Polar bodies are just a way for the oocyte to get rid of chromosomes; polar bodies do not themselves participate in fertilization. The polar body may divide again, but it cannot develop into an egg. Doctors sometimes use polar bodies to select genetically healthy embryos (Chapter 14).

Meanwhile, still in the follicle, the secondary oocyte enters meiosis II and divides again to form the ovum and a second polar body. In humans this final division does not occur until after the egg and sperm fuse.

The mature egg contains a single set of chromosomes tightly packed into its nucleus and a large amount of cytoplasm, including all of the organelles and structures that most cells contain. In addition, the egg also contains lots of high-energy yolk particles, which supply energy to the developing embryo.

Over a woman's lifetime, usually one primary oocyte will mature into an ovum every 28 days or so. Of the initially 200,000 present in the ovaries, only about 400 will complete meiosis before ovulation ceases (during menopause) at around age 50. All the rest of the oocytes degenerate. Of the 400 or so oocytes that mature, only a tiny fraction will ever meet a sperm.

The ovaries produce diploid oocytes. In humans, only an oocyte that is actually fertilized completes meiosis and forms a haploid primary oocyte (egg), plus three tiny polar bodies.

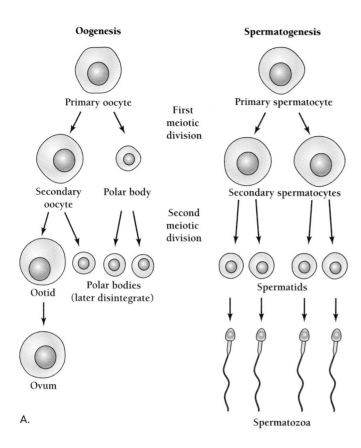

Oogenesis

Primary oocyte

First meiotic division

Secondary oocyte   Polar body

Second meiotic division

Ootid   Polar bodies (later disintegrate)

Ovum

A.

Spermatogenesis

Primary spermatocyte

Secondary spermatocytes

Spermatids

Spermatozoa

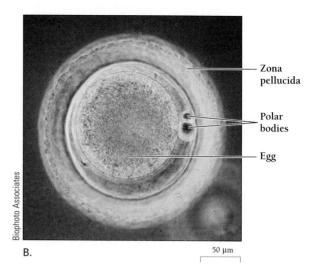

Biophoto Associates

B.

Zona pellucida

Polar bodies

Egg

50 μm

Figure 44-3

Oogenesis and spermatogenesis. A. The diploid primary oocyte and spermatocyte divide by meiosis (which is two divisions) to produce four haploid cells—called ootids and spermatids. One of the ootids contains virtually all of the cytoplasm from the original oocyte and becomes the ovum, or egg. The other three cells, called polar bodies, fall away. In contrast, all of the spermatids mature into sperm. B. In this photo of a rabbit zygote, two polar bodies still cling to the surface at about 3 o'clock.

Where Do Oocytes Go After They Leave the Ovaries?

Beginning at puberty, a few follicles begin developing and expanding with fluid each month. Usually only one of the several developing follicles in the two ovaries continues to expand each month. The single maturing follicle expands so much, however, that it bulges from the side of the ovary and bursts open, releasing the oocyte. This release of the immature egg is called **ovulation** (Figure 44-4).

When the oocyte has erupted from the follicle in the ovary, it enters the abdominal cavity and then finds its way into the opening of one of the two oviducts that carry the egg into the uterus (Figure 44-4). (In humans, each oviduct is called a fallopian tube.) The oviducts do not enclose the ovary, and the oocyte often floats free in the abdominal cavity, at least for a time. But the flared end of each oviduct contains hairlike cilia that draw fluid into the tube's lumen and sweep the oocyte inside. One might think that the oocytes would not enter the oviduct very reliably, but they do up to about 98 percent of the time. (Even women missing one ovary and the opposite oviduct can still easily conceive children. For this to happen, an oocyte from the functioning ovary must cross the abdominal cavity to the opposite oviduct.) Inside the oviduct, more cilia propel the oocyte toward the uterus, the chamber in which—if the oocyte meets a sperm along the way—the fertilized egg may develop into an embryo and then into a fetus.

The egg and sperm have at least two ways of finding one another. First, the egg or oviduct releases odorous mol-

ecules that attract sperm, which have chemical receptors similar to those in our noses. Second, both the egg and sperm move to an area of higher temperature within the oviduct. This warm spot is like a prearranged meeting place for a blind date. After the egg and sperm meet and fuse, the zygote begins dividing and developing as it travels toward the uterus (Figure 44-4).

The uterus, which lies in the middle of the pelvis, above and behind the bladder, is about the size and shape of an upside-down pear. It has thick muscular walls and a specialized lining called the **endometrium** [Greek, *endon* = within + *metro* = mother]. An early embryo can embed itself in the wall of the endometrium and tap into the rich supply of blood vessels for nutrients (Figure 44-4).

The secondary oocyte moves from the ovary, through the oviduct, into the uterus.

How Do Male Mammals Produce Sperm?

In mammals, the male gonads, or testes, are contained within the **scrotum** [Latin, = bag], a pouch that lies outside the body (Figure 44-5). The temperature within the scrotum is usually a few degrees below body temperature, which is ideal for sperm development. Sometimes, in fact, hot baths or tight clothes can temporarily interfere with sperm matu-

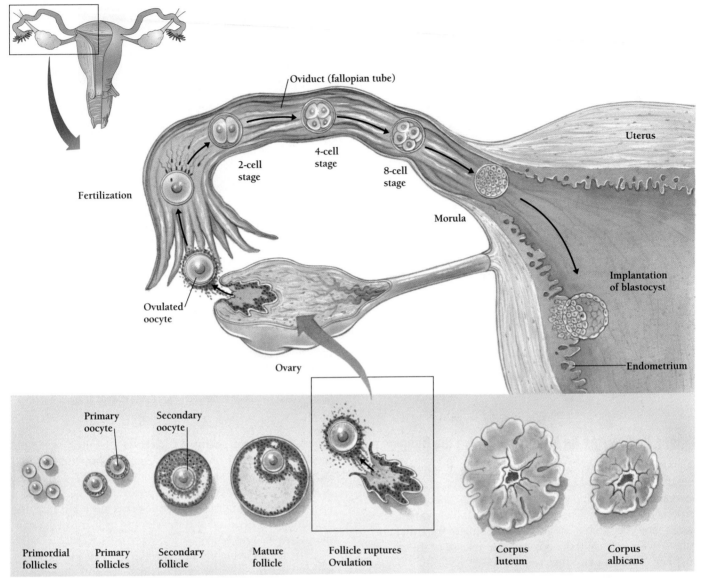

Figure 44-4
Development and maturation of primary oocytes and follicles. Inside the ovary, the primordial follicles *(bottom left)* enlarge to form secondary follicles. Inside the follicles, the primary oocytes mature into secondary oocytes. After the follicle ruptures (ovulation), the secondary oocyte migrates into the oviducts. If the oocyte fuses with a sperm, it finishes meiosis II and divides to form a mature oocyte and a third polar body. The zygote begins dividing. After about a week, the embryo embeds itself in the wall of the uterus. In the ovary, the ruptured follicle meanwhile matures into a corpus luteum (then a corpus albicans).

ration (although not reliably enough to be used as a form of birth control).

Each egg-shaped testis is 4 to 5 cm long, about the size of a golf ball, although not so heavy. A testis consists of hundreds of separate chambers filled with tightly coiled ducts, called **seminiferous tubules** [Latin, *semen* = seed + *ferre* = to bear], in which the sperm mature (Figure 44-6). Each mature sperm cell consists of a head, a body, or midpiece, and a tail. The head contains a dense nucleus, with a tightly packed haploid set of chromosomes. The midpiece and tail specialize the sperm for rapid movement. The midpiece contains dense concentrations of mitochondria—sources of ATP power for the journey to the egg—and the tail consists of a long flagellum, which, powered by the mitochondria, whips about and drives the sperm cell forward (Figure 44-7). The sperm also contains a special structure at its front—the **acrosome,** a huge lysosome whose enzymes help the egg and sperm fuse.

Sperm, like eggs, are descendants of primordial germ cells. During the process of **spermatogenesis,** primordial germ cells in the seminiferous tubules multiply (through mitosis), then undergo meiosis to produce four haploid

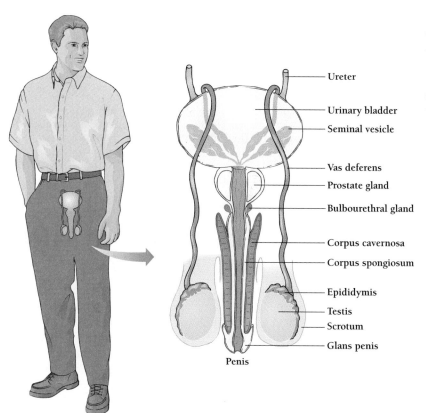

Ureter

Urinary bladder

Seminal vesicle

Vas deferens

Prostate gland

Bulbourethral gland

Corpus cavernosa

Corpus spongiosum

Epididymis

Testis

Scrotum

Glans penis

Penis

Figure 44-5
The male reproductive tract. Sperm form and mature in the testis, then pass into the epididymis until ejaculation, when they pass into the vasa deferentia and the penis. Fluids from the paired seminal vesicles and bulbourethral glands, as well as the single prostate gland, activate the sperm and speed them on their way.

spermatids (Figure 44-3). These divisions are just like the ones that produce the egg and its three polar bodies. Each spermatid matures into a tailed sperm as it moves into the seminiferous tubules (Figure 44-7).

In humans, the process of spermatogenesis takes more than two months. During this time, the developing sperm are surrounded by two kinds of specialized cells: **interstitial cells,** which synthesize the male sex hormone testosterone, and **Sertoli cells,** which nourish the developing sperm, regulate the passage of nutrients from the blood, and secrete hormones (Figure 44-6).

As the spermatids mature, they separate from the Sertoli cells and pass into the seminiferous tubules. From there, the sperm pass into the **epididymis,** an interconnected network of coiled ducts that leads to a single exit tube (Figure 44-6). The passage through the epididymis takes 12 to 20 days, during which the sperm complete their maturation. The entire process of sperm formation takes 60 to 70 days.

During ejaculation, smooth muscles in the epididymis contract and push the mass of mature sperm toward the **vas deferens** [Latin, *vas* = vessels + *deferre* = to carry down],

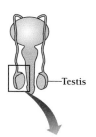

Testis

Figure 44-6
The development and maturation of sperm.
Inside the testis are hundreds of seminiferous tubules. On the inside wall of each tubule are diploid cells that divide by meiosis to form four spermatids. Over a period of two and a half months, the spermatids mature into sperm. Sertoli cells supply nutrients to the sperm.

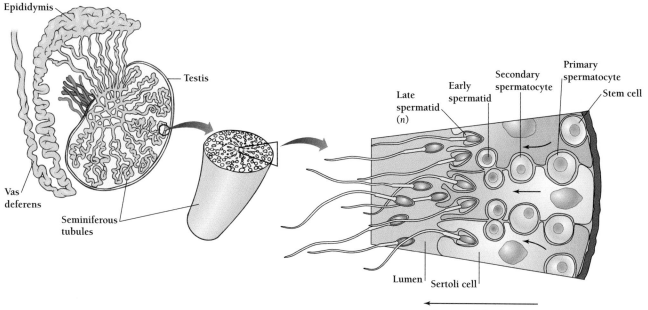

Epididymis

Testis

Vas deferens

Seminiferous tubules

Late spermatid (*n*)

Early spermatid

Secondary spermatocyte

Primary spermatocyte

Stem cell

Lumen

Sertoli cell

Direction of development

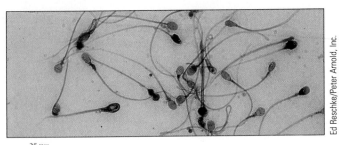

25 µm

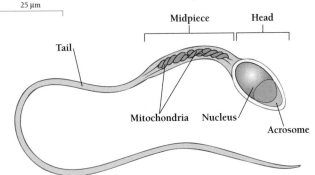

Figure 44-7
Human sperm. A mature sperm consists of a chromosome-packed head, a mitochondrion-rich midpiece, or body, and a long flagellum, or tail.

a large, thick-walled duct that carries the sperm into the **urethra,** the tube that carries urine from the bladder to the penis (Figure 44-5). Each ejaculation releases about 200 million sperm. But every hour, 200 million more sperm mature, far more than can be used. Unused sperm are absorbed by other cells in the testes.

In males, diploid germ cells divide through meiosis to form haploid spermatid cells, which mature into sperm cells in the seminiferous tubules.

How Do Sperm Cells Travel from the Testes to the Penis?

As the sperm leave the vasa deferentia for the urethra, two **seminal vesicles** secrete sugars and other nutrients around the sperm, after which the **prostate gland** secretes a thin, milky alkaline fluid into the urethra. Next, the **bulbourethral glands** inject a small amount of lubricating mucus into the mix (Figure 44-5). The sperm cells, together with the fluid from the seminal vesicles, the prostate gland, and the bulbourethral glands, make up the **semen** [Latin, = seed]. The many molecules in semen provide energy for sperm motility, help the sperm cling to the walls of the vagina and cervix, stimulate contractions of the female reproductive tract, reduce the viscosity of (female) cervical

mucus, and neutralize the acidity of the vagina—all of which help sperm travel through the female tract toward the egg.

Sperm cells move from the seminiferous tubules, through the vas deferens, to the urethra, and out through the penis. Secretions from several glands prepare the sperm for their travels in the female reproductive tract.

44.2 Sexual Intercourse: How Do the Egg and Sperm Rendezvous?

In Chapter 9, we saw that many animals lay eggs that are fertilized externally (for example, many invertebrates, fish, and amphibians). But in amniotes (reptiles, birds, and mammals), as well as in many other animals, fertilization occurs inside the body of the female. Reproduction therefore requires that the male deliver mature sperm into the female genital tract. Internal fertilization, or sex, is simply how the sperm and the egg meet.

> Said an ovum one night to a sperm,
> "You're a very attractive young germ.
> Come join me, my sweet,
> Let our nuclei meet
> And in nine months we'll both come to term."
> —Isaac Asimov

The External Genitalia

The human penis contains a single exit tube, the urethra, which ends in a slitlike opening. Surrounding the urethra is a spongy cylindrical tissue, the **corpus spongiosum,** enlarged at each end (Figure 44-8A). At the far end of the penis is a smooth cap, called the **glans.** Above this tissue lie two more spongy cylinders, each called a **corpus cavernosum,** which run along the length of the penis. All three spongy cylinders fill with blood during an erection, as we will discuss shortly. Covering the outside of the penis, including the glans, is a loose layer of skin called the foreskin. The glans and the foreskin are full of sensory nerves that make the penis sensitive to mechanical stimulation. In many cultures, the foreskin is surgically removed in the first days of life in a procedure called **circumcision.**

The external genitalia of the female are collectively called the **vulva** [Latin, *volvere* = to wrap] (Figure 44-8B). Two thin skin folds, called the **labia minora** [Latin, = small lips], surround the mouth of the vagina and the end of the urethra. These join at the front and form a hood over the **clitoris,** the small female homologue of the male penis. The clitoris includes spongy cylinders like those of the penis and

Figure 44-8
The male and female reproductive tracts in humans.

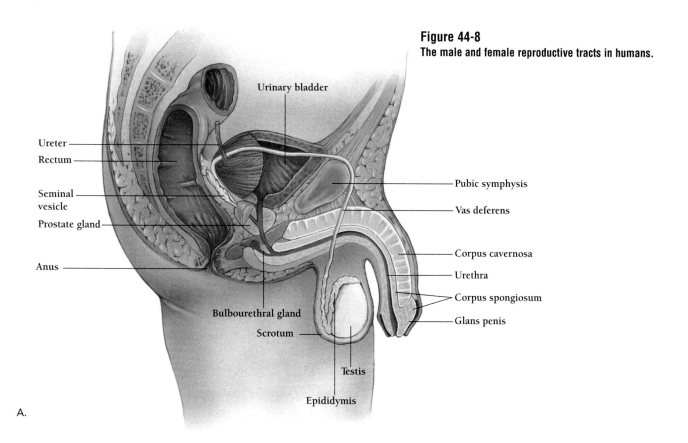

Urinary bladder

Ureter

Rectum

Seminal vesicle

Prostate gland

Anus

Bulbourethral gland

Scrotum

Testis

Epididymis

Pubic symphysis

Vas deferens

Corpus cavernosa

Urethra

Corpus spongiosum

Glans penis

A.

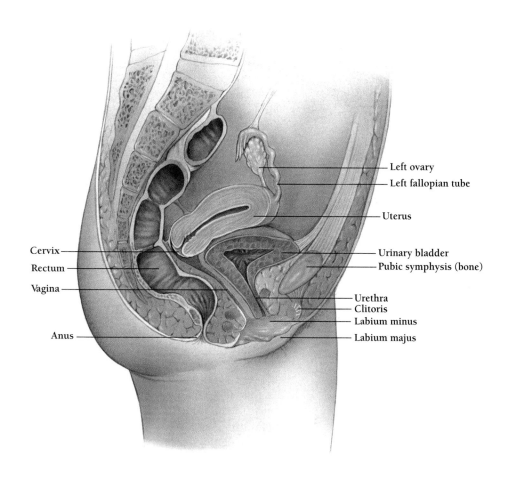

Left ovary

Left fallopian tube

Uterus

Urinary bladder

Pubic symphysis (bone)

Cervix

Rectum

Vagina

Urethra

Clitoris

Labium minus

Labium majus

Anus

B.

a glans that is especially sensitive to stimulation. On either side of the labia minora are two more skin folds, called the **labia majora** [Latin, = large lips]. Just inside the mouth of the vagina is a thin membrane of irregular ragged shape, called the **hymen.** The hymen often impedes penetration by the penis, but only at first. The first sexual intercourse nearly always tears a hymen that closes the vagina, so an intact hymen was long regarded as the only proof of virginity. Other events can also break the hymen, however, including infection, a fall, or vigorous exercise.

Some cultures cut off the clitoris in young girls or even fuse the labia by cutting them and sewing them together, leaving only a small exit hole for urine and menstrual blood. These various practices, collectively called **female circumcision,** may cause long-term health problems and pain, depending on the nature of the surgery.

Despite the striking differences between the male and female genitalia, these organs actually share many features in common. Testes and ovaries develop from the same embryonic structures. The labia majora and scrotum develop from the same structures and are supplied with the same sets of nerves. The surface of the labia minora likewise corresponds to the ventral surface of the penis and is similarly sensitive (Figure 44-9). The external male genitalia are most simply described as a female genitalia that have grown together and fused at the midline, with the clitoris enlarged to include the urethra.

> The external genitalia of males and females—the penis and scrotum in males and the clitoris and vulva in females—develop from the same embryonic precursors.

The Male and Female Sexual Responses Each Consist of Four Stages

Human sexual responses involve changes in blood flow and in the contractions of smooth and skeletal muscle. Physiologists divide the sexual response cycle into four phases: excitement, plateau, orgasm, and resolution.

The first phase, **excitement,** is marked by increased blood flow to the clitoris, the labia minora, and the breasts in the female and to the penis and testes in the male. In each sex, the affected tissues become engorged with blood, causing **erection** of both penis and clitoris and enlargement of the testes and breasts.

Erection results directly from an increase of blood flow and may happen quickly, sometimes within 5 to 10 seconds. Blood fills the tiny spaces within the spongy tissues of the clitoris and penis, causing them to enlarge and harden. In females, erection simultaneously engorges the clitoris and tightens the tissues around the base of the vagina. During sexual intercourse, this tightening squeezes the penis.

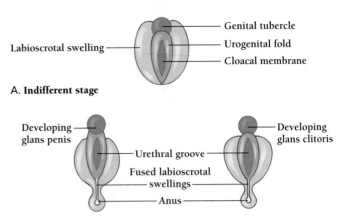

A. Indifferent stage

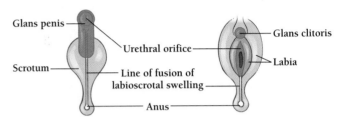

B. About 9 weeks development

C. About 12 weeks development

Figure 44-9

The common origin of male and female external genitalia. A. From four to seven weeks' gestation, the genitals of male and female human embryos are indistinguishable. The "indifferent" genitals consist of a genital tubercle, a urogenital fold, and paired labioscrotal swellings on either side of the urogenital fold. B. By nine weeks, the anus has formed and the labioscrotal swellings have begun to fuse along the midline at top and bottom. The structures of females and males begin to diverge, or differentiate. The labioscrotal swelling of males, for example, have fused more than those of females. C. By 12 weeks, the labioscrotal swellings have fused completely in males to form the scrotum. In females, the genital tubercle which forms the glans has shrunk, and the labioscrotal folds have completed fusion at top and bottom and developed into the labia majora and labia minora.

The bulbourethral glands of the penis release a small amount of mucus, which lubricates the head of the penis and eases entry into the now tightened entry of the vagina. Further lubrication comes from the female's **Bartholin's glands,** located beneath the labia minora and derived from the same structures as the bulbourethral glands in the male.

Erection in both females and males can occur as a result of physical stimulation of any part of the urogenital region, including both the sexual organs and the urinary tract. Stimulation of the glans, in particular, triggers a reflex response in which the blood vessels dilate.

Because nerve pathways from the brain control blood flow, sights, thoughts, or feelings can all cause erection and excitement even without physical stimulation. Just as the brain can stimulate erection and excitement, however, it can also inhibit excitement. Anxiety, fear, anger, or depression, for example, can all repress the sexual response.

AIDS (acquired immune deficiency syndrome) is probably the most talked about and feared STD today. Countless articles, TV programs, and educational materials describe how the immune systems of AIDS patients break down leaving them vulnerable to infections and cancers.

But although AIDS is the most deadly STD, in the United States, it is by no means the most common. And other STDs are responsible for many serious health problems, including infertility, pregnancy complications, and cervical cancer. Over 15 million Americans contract STDs every year, with more than two-thirds of cases occurring in teenagers and young adults. Some experts call it an epidemic. Yet most Americans underestimate their risk of infection. According to a 1998 survey, only 14 percent of men and 8 percent of women consider themselves at risk, even though at least 25 percent of Americans will be infected with an STD during our lifetimes.

STDs can be caused by viruses, bacteria, or single-celled organisms called protozoans. One of the most well known STDs,

herpes, is caused by a virus. Herpes sores increase the risk of catching AIDS. One of the most common viral STDs affecting young people is human papilloma virus (HPV). Although HPV sometimes causes genital warts, often it causes no visible symptoms and many people do not know they are infected. Unfortunately, some strains of HPV can cause cervical cancer. Although there is no cure for HPV, a simple test called a Pap smear detects cancerous or precancerous lesions, which can then be treated. Thanks to the Pap smear, cervical cancer is rarely fatal (Chapter 8).

The most common bacterial STD in the United States is chlamydia. Like HPV infections, chlamydial infections often go unnoticed because many infected people have few or no symptoms. In women, untreated chlamydial infections can result in infection of the oviducts, also known as pelvic inflammatory disease, which can cause severe pain and lead to infertility. Gonorrhea and syphilis are two other bacterial STDs that can also cause major health problems, such as increasing a person's risk of con-

tracting AIDS. Before the age of antibiotics, syphilis caused insanity and death. Today, both gonorrhea and syphilis are curable with antibiotics. But the sooner they are treated, the better for everyone.

One of the most common protozoan infections is trichomoniasis. In women, trichomoniasis often causes vaginal itching, a yellow-green discharge, and pain during sexual intercourse or urination. In addition, like gonorrhea and syphilis, it can increase the risk of contracting AIDS. Trichomoniasis can also be treated with antibiotics.

The best way to deal with STDs is prevention. After abstinence, the most effective methods are the male latex condom and having sexual intercourse only with a single, faithful, uninfected partner. Since most STDs can be treated, people who suspect they may have been exposed to an STD—whether because they have had multiple partners or believe their partner has—should ask their health providers to do a blood test for STDs. Often the test will be negative, but being tested regularly is the responsible thing to do.

During the second, or **plateau**, phase, breathing and heart rate increase in response to continued stimulation. The end of the vagina dilates and the uterus pulls upward. The third stage is **orgasm**, which refers to a whole complex of changes—smooth muscle contractions in the genital tract, skeletal muscle contractions throughout the body, and feelings of intense pleasure. Orgasm is typically brief, usually lasting three to five seconds.

In females, orgasm increases the probability of fertilization, but is not required. In males, however, full sperm delivery requires **ejaculation**, the propulsion of sperm out of the penis. During ejaculation, contractions of the male genital ducts and muscles in the penis force sperm out of the end of the penis.

During female orgasm, the smooth muscles in the uterus and the outer part of the vagina contract rhythmically and the cervix drops toward the pool of sperm. These contractions are, like those of the male, a reflex mediated by the spinal cord. Some research suggests that women who have orgasms less than a minute before the man, or up to 45 minutes after, retain most of the sperm in the reproductive tract, while those who have no orgasm, or have one more than a minute earlier, lose more sperm. Orgasm-mediated contrac-

tions of the uterus and oviducts, as well as ciliary currents in the oviducts, seem to propel the sperm upward toward the egg. However, fertilization can easily occur without female orgasm.

In the last phase, **resolution**, blood flow and muscle tension return to normal. In males, the penis ceases to be erect, and a second erection is not possible for a period of time that ranges from minutes to hours.

Physiologists divide the sexual response cycle in both men and women into four phases: excitement, plateau, orgasm, and resolution. Orgasm plays an important role in fertilization in both males and females.

How Does a Sperm Reach and Enter an Egg?

Once inside the vagina, many of the 200 million or so sperm will be destroyed by vaginal acid or will simply leak out. The rest will pile up against mucus in the cervix. At the right time of the month, the mucus is clear and thin and eases the passage of sperm into the uterus and oviducts. But even then, many sperm fail to wiggle through the cervical opening and end their days in the upper end of the vagina. The few hundred that make it through the cervix fan out into the uterus and up into the oviducts. In most instances of sexual intercourse, the sperm find no oocyte. At the right time of the month, however, hundreds or thousands of sperm may arrive in one of the oviducts and encounter a mature oocyte.

Even though the mammalian sperm may arrive near the egg within minutes of sexual intercourse, the sperm are incapable of fertilizing an egg until they have been in the female reproductive tract for a few hours. The fluids of the female reproductive tract dissolve substances coating the sperm's head, wash away factors in the semen that block fertilization, and make the sperm more motile, enabling them to reach and penetrate the outer layers of the egg.

When the sperm reaches the egg, the process of fertilization begins. Between the sperm's nucleus and the outer membrane is a vesicle called the acrosome that is packed with digestive enzymes. Under the influence of the hormone progesterone, the outer membrane of the acrosome fuses with the membrane of the sperm head and releases the enzymes, which break through the outer layers of the egg. Once the sperm passes through, the egg's membrane undergoes a series of changes that makes it impenetrable to any other sperm.

Substances in a woman's reproductive tract activate the sperm, so the sperm can reach and fuse with the egg. Once a sperm has passed through the outer layers of the egg, no other sperm can enter.

44.3 How Do Hormones Control Gamete Production?

Hormones regulate both the production of sperm in males and the production of ova in females. Both sexes use the same set of related hormones. What controls the production of the steroid hormones by the gonads? How does the body ensure, for example, that estrogen and progesterone are made in the right amounts and at the proper times?

The answer is that another endocrine organ, the **anterior pituitary gland**, makes and releases two hormones that together control the secretion of the sex steroids—**estrogen** and **progesterone** in females and testosterone in males. The pituitary gland, embedded in the base of the brain, consists of two lobes, anterior and posterior. The pituitary gland lies directly beneath the **hypothalamus** [Greek, *hypo* = under + *thalamos* = inner room], a part of the brain that regulates the expression of many hormone systems (Chapter 37).

The anterior pituitary produces and releases seven or eight distinct molecules, all of them small proteins or polypeptides. Two of these polypeptides are hormones that

act on the gonads in both males and females. Their names, **follicle-stimulating hormone (FSH)** and **luteinizing hormone (LH)**, come from their major effects in females. In males the concentration of FSH and LH are relatively constant after puberty. In females, however, the concentrations of FSH and LH change in a monthly cycle.

How Do Hormones Control Sperm Production?

The testes secrete **testosterone**, the best known of a class of androgens, or male hormones (Figure 44-10). Testosterone activates sperm production and boosts sex drive. Once testosterone is released into the blood, it circulates for only 15 to 30 minutes before it is bound by tissue receptors or degraded. As a result, testosterone levels can change hourly.

Like the ovaries, however, the testes also secrete estrogen, about one-fifth as much as is secreted by an adult female. Recent studies suggest that estrogen plays a central role in male reproduction. Like testosterone, estrogen stim-

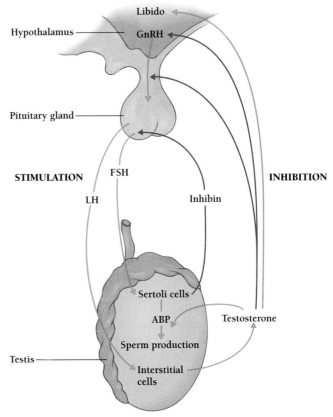

Figure 44-10
Hormonal regulation of sperm production in mammals. Here are just a few of the complex feedback systems that regulate male sperm production (and sexual behavior) in humans and other mammals. LH and FSH indirectly stimulate *(green arrows)* sperm production by stimulating the secretion of testosterone and androgen-binding protein. Through negative feedback loops, inhibin, testosterone, and other androgens hold *(red arrows)* sperm production in check. In the central nervous system, testosterone also stimulates *(green arrows)* libido (sex drive).

ulates sexual behavior in a variety of male mammals, including primates, even after individuals have been castrated. In addition, the male brain converts testosterone to estrogen, and estrogen levels in the brain rise and fall in synchrony with blood testosterone levels.

Hormones regulate the production of both sperm and testosterone by means of negative feedback. In the brain, the hypothalamus secretes pulses of **gonadotropin-releasing hormone, GnRH,** which stimulate the pituitary gland to produce LH and FSH. LH stimulates the interstitial cells of the testes to produce testosterone and other androgens, while FSH stimulates the Sertoli cells to produce androgen-binding protein (ABP). In the presence of ABP, testosterone stimulates the production of sperm (Figure 44-10).

When testosterone and other androgens released by the interstitial cells reach a certain level, they inhibit the production of GnRH in the hypothalamus and (probably indirectly) LH in the pituitary. When LH production increases, androgen levels increase, limiting the production of LH and, therefore, of androgens. Similarly, the Sertoli cells secrete the hormone *inhibin,* which enters the circulation and suppresses the secretion of FSH by the pituitary, decreasing ABP and sperm production. During the day, testosterone levels vary by about 25 percent. But the negative feedback system between the testes and the hypothalamus maintains testosterone levels at a relatively constant level from day to day.

Sperm production and testosterone production are thus intimately tied together. Not coincidentally, testosterone increases sex drive and aggressive behavior in general in both males and females. Among women, testosterone levels are higher among college students, attorneys, and managers than among teachers, clerical workers, and housewives. Among men, football players have higher testosterone levels than ministers. What is not known is whether men are successful at football, for example, because they have high testosterone levels, or whether they have higher testosterone because they play football.

Some evidence supports the latter hypothesis. For example, men's testosterone levels rise when they win games as different as football and chess and fall when they lose. Testosterone even falls when the team a man is rooting for loses. Research in this area generally suggests that violence is positively correlated with testosterone levels in both men and women and that family happiness is negatively correlated with testosterone levels in men. That is, the men with the best relationships with their wives and children also had lower levels of testosterone.

Human males do not begin substantial testosterone production until they are about 10 years old. By 13, they are usually sexually mature and testosterone production increases rapidly until about age 20, after which the hormone begins a steady decline (Figure 44-11). By the late 40s or 50s, most men begin to experience a mild decrease in sexual function similar to that which women experience during menopause. (Women's testosterone levels also decrease as

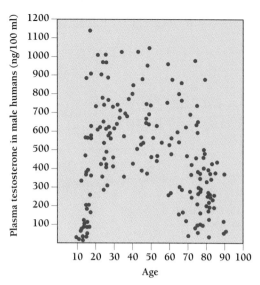

Figure 44-11
Increase and decrease in testosterone as a function of age. The average concentration of testosterone in the blood of human males increases dramatically at puberty, then gradually drops beginning at about age 20. Men vary enormously in testosterone levels. A 90-year-old man can have the same testosterone levels as a 19-year-old.

menopause approaches.) In some cases, the drop in male testosterone can cause the same kind of hot flashes that often characterize women's transition into menopause. Both testosterone and estrogen treatments have been shown to reduce hot flashes in men.

In men, the hypothalamus secretes GnRH, which stimulates the pituitary gland to produce LH and FSH. By boosting the production of testosterone and ABP, LH and FSH increase sperm production.

What Are the Main Events of the Female Reproductive Cycle?

Beginning at puberty (between ages 10 and 17), human females undergo reproductive cycles. Each cycle simultaneously produces a mature oocyte and prepares the wall of the uterus for **implantation,** the process in which an embryo burrows into the lining of the uterus. If an embryo does not implant, or implants but dies, the uterine lining is sloughed off in a process called **menstruation** [Latin, *mens* = month]. Each cycle is coordinated by several hormones, many of the same ones that mediate sperm production in males. Physicians have categorized the changes into two cycles—the ovarian cycle and the uterine cycle. The **ovarian cycle** traces changes in the ovarian follicles that give rise to the eggs, while the **uterine cycle,** also called the menstrual cycle, describes the changes in the lin-

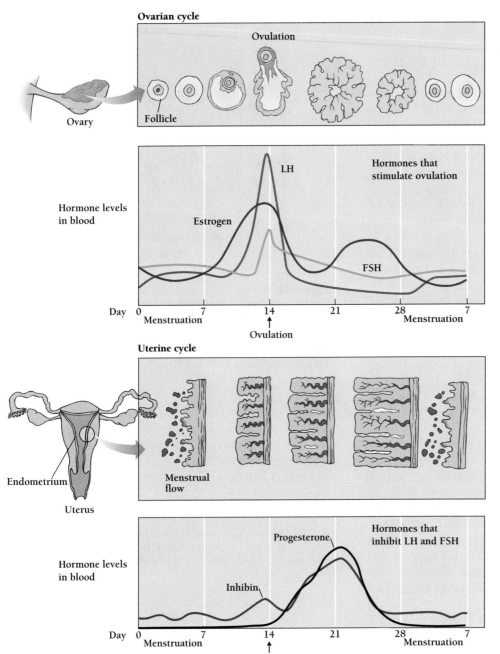

Ovarian cycle

Ovulation

Ovary

Follicle

Hormone levels in blood

LH

Estrogen

Hormones that stimulate ovulation

FSH

Day 0   7   14   21   28   7

Menstruation   Menstruation

Ovulation

Uterine cycle

Endometrium

Uterus

Menstrual flow

Hormone levels in blood

Progesterone

Hormones that inhibit LH and FSH

Inhibin

Day 0   7   14   21   28   7

Menstruation   Menstruation

Ovulation

Figure 44-12

Hormonal regulation of the ovarian and uterine cycle in humans. Estrogen, LH, and FSH dominate at ovulation. Progesterone and inhibin, on the other hand, dominate after ovulation, inhibiting LH and FSH.

Biology Now™ Learn more about how hormone levels affect ovulation and conception by clicking on this figure on your BiologyNow CD-ROM.

ing of the uterus (Figure 44-12). But the two cycles are not actually separate processes.

In 90 percent of healthy adult women, the reproductive cycle ranges from 21 to 38 days, but 28 days is typical. The very young tend to have irregular cycles, which vary in length from one cycle to the next. As young women mature, their cycles tend to become more regular. Women who live together ovulate and menstruate at approximately the same time of the month. Pheromones appear to regulate this coordination. Women who live with men have more regular cycles than those who live alone. Menstruation and ovulation cease at menopause, which usually occurs in the late 40s or early 50s.

Most variation in the length of the menstrual cycle occurs in the period between the onset of menstruation and ovulation, which makes predicting the day of ovulation difficult. The period between ovulation and menstruation is much less variable and is nearly always about 14 days. If a woman knows when she ovulated, she can estimate when she will menstruate. But knowing when she began menstruation will not necessarily tell her when she will next ovulate.

The female reproductive cycle includes the ovarian cycle and the uterine, or menstrual, cycle.

How Do Hormones Regulate the Ovarian Cycle?

As in males, a woman's hypothalamus secretes GnRH, which stimulates the pituitary to produce LH (luteinizing hormone) and FSH (follicle-stimulating hormone), which together stimulate the ovary to develop mature oocytes and to produce estrogen and other steroid hormones (Figure 44-13). In the days before ovulation, an increase in FSH (from the pituitary) stimulates 5 to 12 follicles in the ovaries to grow and mature (Figure 44-12). FSH causes growth and development of the oocyte and, together with LH, causes the follicle cells to release increasing amounts of estrogen, which promotes the growth of the follicle. This **positive feedback** cycle causes a buildup of estrogen in the blood in the days before ovulation. All but one of the follicles stop growing and ultimately degenerate, usually leaving just one follicle to develop a mature egg.

As the dominant follicle matures, it continues to secrete increasing amounts of estrogen. The rising levels of estrogen trigger a second positive feedback cycle. Right around the time of ovulation, the high levels of estrogen stimulate the hypothalamus to release a pulse of GnRH, which causes a large surge of LH and FSH (Figures 44-12 and 44-13). The LH spike triggers ovulation.

After the oocyte bursts from the follicle, the follicle collapses inside the ovary. There, stimulated by LH, it enlarges to form the **corpus luteum** [Latin, = yellow body], a structure that secretes progesterone and a little estrogen. The progesterone helps prepare the lining of the uterus for implantation of the embryo and also prevents further ovulation by signaling the hypothalamus to stop producing GnRH, which reduces LH and FSH. The corpus luteum likewise produces inhibin, which also inhibits the production of LH and FSH.

If the egg is not fertilized, the corpus luteum degenerates in 8 to 10 days. Inhibin and progesterone levels drop, and the pituitary begins producing FSH and LH once more. If the egg is fertilized and the embryo implants in the uterine wall, the implanted embryo produces *chorionic gonadotropin,* a hormone similar to LH that maintains the corpus luteum. The presence of **human chorionic gonadotropin, or HCG,** in blood or urine is the basis of most pregnancy tests. During early pregnancy, the corpus luteum enlarges and produces even more progesterone until about the fourth month of pregnancy.

Estrogen, then, participates both in negative feedback and—once each cycle—in positive feedback. In contrast, progesterone only promotes negative feedback. Together these two feedback systems bring about the regular cycles of ovaries and uterus.

In the days before ovulation, LH and FSH foster the maturation of an ovarian follicle. After ovulation, the corpus luteum secretes progesterone, which helps prepare the uterus for implantation, and, along with inhibin, inhibits the secretion of LH and FSH.

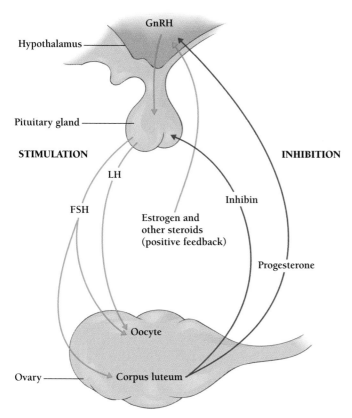

Figure 44-13

Hormonal regulation of egg production in mammals. Here are just a few of the complex feedback systems that regulate female egg production (and sexual behavior) in humans and other mammals. As in males, LH and FSH released by the pituitary stimulate *(green arrows)* oocyte production. Progesterone and inhibin inhibit *(red arrows)* the production of LH and FSH.

How Do Hormones Regulate the Uterine Cycle?

The lining of the uterus undergoes changes that parallel the changes that occur in the ovaries (Figure 44-12). While the follicles are growing, the lining of the uterus, called the endometrium, becomes two to three times thicker. After ovulation, the corpus luteum forms and its progesterone stimulates the endometrium to develop further. Blood vessels increase, and the endometrium secretes glycogen and other materials that will facilitate implantation and nourish the embryo.

If no fertilization occurs, however, the lining of the uterus disintegrates. During menstruation, the blood vessels coil and constrict, and 50 to 150 ml (about half a cup) of blood, mucus, vaginal secretions, and endometrial tissue washes out through the vagina. This bleeding may include bits of membranous tissue and blood clots. Although light bleeding may occur after fertilization, several days of medium to heavy bleeding usually means that fertilization either has not occurred or it has occurred but the embryo has died.

> The uterine cycle involves the thickening of the uterine lining, the endometrium, and its subsequent preparation for embryonic implantation. If no embryo implants, the endometrium breaks down and is washed away in a flow of blood, called menstruation.

44.4 What Happens During Pregnancy and Lactation?

Most animals whose eggs are fertilized internally—including reptiles, birds, and monotreme mammals (platypuses and echidnas)—lay shelled "eggs" that actually contain a developing embryo, not a single unfertilized egg cell. But many other animals, including some sharks, snakes, and most mammals, keep the developing embryo inside for up to 18 months (elephants) and then give birth to live young, a process known as **viviparity.**

How Can a Woman Tell if She Is Pregnant?

Pregnancy is the state in which a woman is carrying a zygote, embryo, or fetus inside her body. It begins with the fusion of the two nuclei of the egg and the sperm. At that moment, two cells become one cell.

Confusingly, doctors say pregnancy begins 2 weeks *before* fertilization, at the beginning of the last menstrual period. Thus, a doctor will tell a woman carrying a two-week-old embryo that she is "four weeks" pregnant. A normal pregnancy lasts about 38 weeks, but doctors say that it is 40 weeks, because, by tradition, they count from the last menstrual cycle. In this book, we refer to the actual number of weeks rather than adding the extra two weeks.

About a week after fertilization, the ball of cells that is the human embryo burrows into the lining of the mother's uterus ("implantation") and begins secreting human chorionic gonadotropin (HCG). HCG causes the corpus luteum to double in size and produce more progesterone. Progesterone prevents menstruation and triggers the secretion of thick cervical mucus, which forms a plug that keeps bacteria out, protecting the embryo from infection.

The embryo quickly develops the same set of four membranes found in all "amniotes," animals that have amniotic eggs (Chapter 24). The amniotic eggs of reptiles, birds, and mammals contain an amnion, yolk sac, allantois, and chorion, which we discuss in Chapter 45. From the chorion, mammals develop a **placenta,** a flat organ rich in blood vessels that conduct nutrients from the mother's blood to the embryo and wastes from the embryo to the mother. Connecting the embryo to the placenta is a thick "umbilical cord."

As the embryo implants and begins releasing hormones, many women notice nothing at all. Others, especially those who have been pregnant before, may recognize the subtle signs immediately. At one week, the woman may feel a slight cramping from implantation and sometimes light spotting. Not long after, other changes may begin, including sleepiness, dizziness, nausea, breast tenderness, bloating, and frequent urination.

The most obvious early sign of pregnancy, however, is the absence of a menstrual period. In the absence of pregnancy, menstruation occurs almost exactly two weeks after ovulation. When a period is about a week late, most women guess that they may be pregnant. To be sure, they get a pregnancy test, a test for the presence of HCG in either urine or blood.

Lab tests can detect HCG in the blood as early as 8 days after fertilization, and even inexpensive drugstore versions can detect it in the urine as early as 14 days after fertilization. These tests are highly reliable, but not perfect. Occasionally, a pregnancy test can give a "false negative," indicating that a woman is not pregnant when she is. One reason for a false negative is taking the test too soon. A woman who takes the test before the embryo has begun to release HCG will get a negative result. If the woman still thinks she may be pregnant, she can try again in a week. False positives are rare but can also occur.

> After one week, the embryo begins secreting HCG, which causes the corpus luteum to enlarge and secrete ever more progesterone and estrogen, which prevents menstruation, and is the basis of early pregnancy tests.

What Happens Between Conception and Delivery?

Physicians divided the 38 weeks of pregnancy into three "trimesters," of about three months each. In the first trimester, a woman probably won't "show" much but may gain a few pounds. Estrogen causes increases in the size of the breasts and the uterus, and sometimes intense nausea (morning sickness) and fatigue. By the end of nine months, the breasts will be about two pounds heavier and the uterus will be 22 times its usual size. Progesterone prevents the uterus from contracting and, along with estrogen, stimulates the development of milk-producing cells in the breasts.

The fetal placenta also produces a **human chorionic somatomammotropin (HCS)**, which makes glucose sugar available to the fetus, especially in the second and third trimesters. HCS works by making the mother resistant to insulin so that her blood glucose levels increase. The glucose crosses the placenta more readily than amino acids or fats, so that much of the glucose is redirected from the mother to the fetus. Simultaneously, HCS increases the metabolism of fat in the mother's body, so that she has access to energy-rich fatty acids to replace the glucose. HCS's effects on insulin sensitivity may contribute to *gestational diabetes*, a disorder that affects one to two percent of pregnant women causing excessive glucose in the blood.

In the second trimester, the corpus luteum degenerates and HCG levels decline. But the fetal placenta produces even

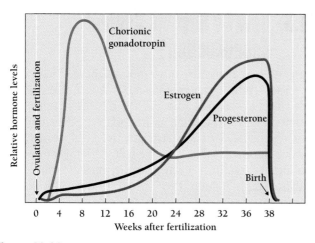

Figure 44-14

Hormone levels during pregnancy. Early in pregnancy, the embryo secretes human chorionic gonadotropin (HCG). HCG causes the corpus luteum to enlarge and secrete progesterone and estrogen, which prevent the uterus from expelling the embryo. In the latter half of pregnancy, HCG levels decline, but the fetal placenta secretes ever higher levels of estrogen and progesterone. The sudden drop in these hormones precipitates birth.

Biology ⊜ Now™ Learn more about how hormone levels change during pregnancy by clicking on this figure on your BiologyNow CD-ROM.

more estrogen and progesterone, which prevents the uterine and ovarian cycles from starting (Figure 44-14). The embryo becomes a fully formed fetus, and the woman may feel its first kicks at about four months (14 to 16 weeks). Traditionally, this stage is called "quickening." Her abdomen swells, her morning sickness usually goes away, and her appetite often increases. Most women feel at their best during the middle trimester.

By the third trimester, the mother's blood volume has increased by about 30 percent, or one or two liters. The growing fetus begins to impinge on the mother's internal organs. For example, the weight of the fetus compresses the bladder, contributing to frequent urination and even incontinence. In addition, the mother's diaphragm no longer moves freely, so some mothers find breathing difficult. On average, pregnant women gain between 25 and 35 pounds, of which approximately seven pounds is the fetus and four pounds is amniotic fluid, placenta, and fetal membranes. In addition, blood and extracellular fluid add another six pounds, and the breasts and uterus each increase in weight by about two pounds. Beyond that, many women gain several pounds of fat, which they will need when they begin nursing the baby.

Between conception and delivery, a complex web of hormones helps balance the needs of the fetus and mother.

Parturition: What Happens on D(elivery) Day?

The baby itself helps decide the moment of birth, called **parturition,** by sending chemical signals to the mother. Late in pregnancy, the fetus begins to secrete **cortisol,** a steroid hor-

mone associated with stress that increases the likelihood of labor. The mother's posterior pituitary secretes the hormone **oxytocin,** which (1) stimulates the muscles of the uterus to contract and (2) stimulates the fetal membranes to secrete prostaglandins, which also increase contraction. **Prostaglandins** are chemical signals that increase inflammation and smooth muscle contraction. The pressure of the baby on the inside of the uterus, especially the cervix, also increases the tendency of the uterus to contract. Over a period of days, contractions may begin, then stop, repeatedly. But at some point, contractions begin to come regularly.

As labor progresses, the contractions come with greater intensity and frequency. Contractions push the baby's head against the cervix, which stimulates the release of more oxytocin, and consequently more frequent and more intense contractions, an example of positive feedback. Because of positive feedback, contractions come closer and closer together. Depending on levels of oxytocin and prostaglandins, the progression from a few irregular contractions to intense frequent ones ranges from 2 to 48 hours.

The pain associated with labor results in part from lack of oxygen in the uterine muscles. Near the end, the mother must push the baby out using these same muscles, as well as the abdominal muscles. As the baby descends through the vagina, intense stretching and pressure can cause severe pain. Some physicians attest that the pain of labor without anesthesia can be equal to the pain that would occur with amputation of the fingers.

At parturition, cortisol, oxytocin, and prostaglandins combine to drive the uterus into a positive feedback cycle of stronger and faster contractions that ultimately expel the baby.

Nursing the Baby

A few minutes after a woman delivers her baby, she also delivers the placenta. Someone must physically cut the umbilical cord, which connects the baby to the placenta. Then, within a few hours, a woman's body begins reverting to its former state. Excess fluid is excreted and the uterus begins to shrink. The 11 pounds that make up the baby, the placenta, and the amniotic fluid are lost immediately. Within the first week, the six pounds of excess blood and extracellular fluid vanish as well.

What do not normally revert to their former state are the breasts. In the first few days, the breasts secrete **colostrum,** a fluid similar to milk but with less fat. Colostrum also contains high levels of antibodies. As the baby suckles, the stimulation of the nipple causes the mother's pituitary to secrete the hormone **prolactin,** which stimulates the milk glands to produce milk. After about three days, the mother's milk usually "comes in." Human milk contains fats that promote the development of the nervous system and immune cells and antibodies that help keep the baby from becoming ill.

The more often the baby sucks, the more prolactin is released and the more milk the breasts produce. A hungry baby may not be satisfied with the milk produced on one day. But if it is allowed to suck as long and as often as it likes, the next day will bring more milk. By the same token, an infant who leaves milk in the breasts all day will find less milk the next day.

Although the baby must suck hard to stimulate the nipples, a process that can be painful to the mother for the first few days or weeks, the milk ducts actually eject the milk into the baby's mouth with considerable force. Sucking also triggers the release of oxytocin (which also stimulated birth). Oxytocin causes tiny muscle cells surrounding the milk glands to contract forcefully and shoot the milk into the hungry baby's mouth, a process called **milk ejection,** or the "let-down" reflex.

Nursing a baby places great metabolic demands on a mother and drains calcium from her bones. Most mothers make up for this loss within a couple of years after they stop nursing, as bones take up calcium more than usual in the period after lactation.

During the first few days after birth, babies get a low-fat version of human milk called colostrum. The baby's suckling stimulates the release of oxytocin, which causes the milk glands to eject milk, and prolactin, which ensures the synthesis of milk for subsequent meals.

44.5 How Do Humans Control Reproduction?

Humans, more than most other animals, have sexual contact for reasons other than reproduction. And unlike most female mammals, female humans are sexually receptive even at times when fertilization is impossible. Some biologists have speculated that sex strengthens social bonding, increasing the chance that a couple will work together to nurture their young. In this way, sexuality may contribute indirectly to reproduction even when it does not lead to conception.

Humans do not necessarily want to bear as many children as possible, however, so **birth control,** the conscious regulation of reproduction, is an important issue for many sexually mature humans around the world (Figures 44-15 and 44-16). Worldwide, 62 percent of married or regular partners use some form of **contraception,** the prevention of conception—70 percent in developed countries, 60 percent in less developed regions. Ninety percent of couples that use contraception use modern methods of contraception such as birth control pills and condoms. Only 10 percent rely on traditional methods such as the rhythm method and withdrawal. The most commonly used modern methods are sterilization (20 percent), intrauterine devices, or IUDs (15 percent), and oral pills (8 percent). But there are other methods of preventing pregnancy.

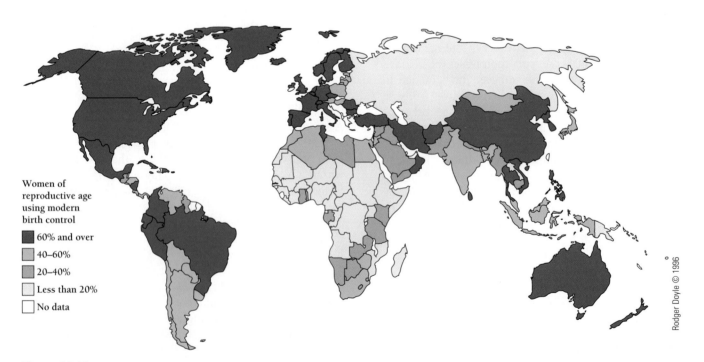

Women of reproductive age using modern birth control

- 60% and over
- 40–60%
- 20–40%
- Less than 20%
- No data

Rodger Doyle © 1996

Figure 44-15

Birth control around the world. Different countries rely more or less heavily on birth control, depending on culture, laws, economic factors, and education.

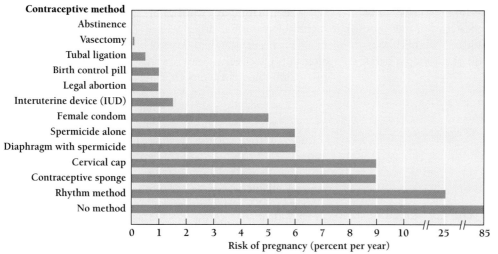

Figure 44-16
Some methods of birth control. Listed here are some of the most common methods of birth control, listed from most effective to least. A small fraction of surgical abortions fail. The failure rate is higher in pharmaceutical abortions. You can find more information in Table 44-1.

Abstinence

The most certain contraceptive method is not to have sexual intercourse, called "abstinence." **Abstinence** is also the best way to avoid contracting sexually transmitted diseases (STDs). Abstinence is common among teens and unmarried adults. In 2001, 65 percent of U.S. high school students were not currently sexually active and 54 percent had never had sexual intercourse. Many adults also abstain for periods of months or years. Anyone can abstain from sex at any time for any length of time. Virginity is not a prerequisite. On the other hand, abstaining generally means abstaining from any form of sexual intimacy. People who refrain from actual penetration but are sexually intimate in other ways are not abstaining and may conceive.

Abstinence is the most effective form of birth control.

Sterilization Provides Effective but Irreversible Birth Control

After abstinence, **sterilization** is the most effective form of birth control. It is almost 100 percent effective. (In extremely rare cases, a sterilization can fail.) Quick and relatively inexpensive surgical methods of male and female sterilization are the most widely used forms of birth control worldwide. Although sterilization can sometimes be reversed, it is usually a permanent change, so many people choose it after they have already had children and know they want no more or are certain that they never want to have any.

Vasectomy, the cutting and tying of the vasa deferentia, prevents sperm from entering the urethra (Figure 44-17). Vasectomies are extremely safe, do not require hospitalization, and are performed in less than an hour with local anesthesia. The risk of death from a vasectomy is about one per

million vasectomies, compared with about one in 6,000 for driving a car each year.

Tubal ligation, the cutting and blocking of the oviducts, or oviducts, prevents eggs from reaching the uterus after ovulation (Figure 44-17). Although safe, it is a more complicated operation than vasectomy. Because the oviducts lie within the abdomen, a tubal ligation requires one or more surgical incisions into the abdomen and a

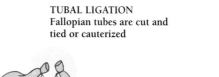

TUBAL LIGATION
Fallopian tubes are cut and tied or cauterized

VASECTOMY
Vasa deferentia are cut and tied or cauterized

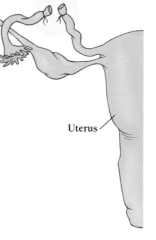

Uterus

Testis

Figure 44-17
Tubal ligation and vasectomy. In tubal ligation, the oviducts are either cut and tied, cauterized (sealed by burning), closed off with very tight rubber bands, which kill the tissue, or clamped with a tight clip. A tubal ligation requires abdominal surgery. Cutting and tying the vasa deferentia prevents sperm from reaching the penis. It is a much safer and simpler procedure than tubal ligation. Neither procedure should be considered reversible.

hospital-like setting. While modern techniques make this operation short and relatively uncomplicated, full recovery can take up to several weeks. The risk of fatal complications is about 1 in 38,500, lower than the risk of carrying a baby to term, about 1 in 10,000.

Sterilization—vasectomy and tubal ligation—is highly effective and should be considered irreversible.

Hormonal Contraceptives and Intrauterine Devices

Researchers are now trying to develop chemical means of preventing sperm maturation. Indeed, they have shown that monthly testosterone injections, which reduce sperm production through negative feedback, are as effective as condoms. But these injections are not yet commonly available. In contrast, a great variety of *systemic* contraceptives, those that go into the bloodstream, exist for women.

Chemical birth control can prevent ovulation, fertilization, or implantation. One version of the female **birth control pill** consists of a combination of estrogen and progesterone. Together these hormones act through negative feedback to prevent the anterior pituitary from secreting LH and FSH, thus preventing ovulation.

If used correctly, birth control pills are almost entirely effective in preventing pregnancy, with pregnancy rates of less than 1 percent per year. But the hormones slightly increase the risk of high blood pressure, strokes, and heart attacks. These risks are higher among smokers than among nonsmokers. The chances of fatal complications, however, are far lower than the mortality associated with pregnancy itself. For many couples, the pill is the preferred method of birth control. (It does not prevent the transmission of STDs, however.)

Another group of birth control pills contain only progesterone, with no estrogens. These pills (sometimes called "minipills") have few of the side effects of the combination pills, but they have a slightly higher pregnancy rate. They appear to work by inhibiting the movements of both ovum and sperm, as well as by interfering with implantation.

Another chemical method that interferes with ovulation involves the release of a substance that mimics progesterone from a capsule surgically placed under the skin. A single such implanted capsule, called Norplant, may last for up to 10 years.

Another hormonal contraceptive is a progesterone derivative called DMPA, or **Depo-Provera.** DMPA is taken as an injection, which appears both to suppress ovulation and to inhibit implantation for three months at a time. The risks associated with DMPA, however, are still the subject of debate. Some people argue that it is safer than the available birth control pills, while others point to serious problems, including depression and abnormal menstruation.

The **morning-after pill** is an oral contraceptive taken in higher-than-normal doses within 72 hours of unprotected intercourse. Like standard doses of birth control pills, morning-after pills effectively interfere with both fertilization and implantation. Side effects, such as nausea and cramps, are unpleasant but transient. In December 2003, a U.S. Food and Drug Administration (FDA) advisory body recommended that morning-after pills be available without a prescription throughout the United States, just like drugs such as aspirin and cough syrup. The FDA expects to make a decision sometime in 2004. In some states, morning-after pills are already available with a prescription, but women have to ask a pharmacist—and know to ask.

Another method for preventing implantation is the **intrauterine device,** or **IUD,** typically a small piece of plastic that is placed into the uterus, where it interferes with implantation.

Hormonal contraceptives suppress ovulation and inhibit implantation. The IUD also inhibits implantation.

Induced Abortions Terminate Pregnancies After Implantation

Abortion is the termination of a pregnancy. Approximately 50 to 75 percent of all embryos abort spontaneously during the first trimester in **spontaneous abortions.** In most cases, these embryos have fatal chromosomal or developmental defects. An **induced abortion** is the deliberate removal of an embryo or fetus from the uterus. The ease of the procedure decreases with the age of the embryo or fetus. In the first trimester (the first three months), the procedure is brief and can take place in a clinic or doctor's office. The embryo is usually removed from the uterus by gentle suction between 3 and 10 weeks after conception. Early abortions, between three and five weeks after fertilization are extremely safe, with a risk of death of only 1 in 263,000. Abortions between 5 and 10 weeks after fertilization are done by dilation and extraction (D&E) and are 10 times as safe as having a baby, with a risk of death of only 1 in 100,000. Some doctors now use the drug **RU 486,** also called **mifepristone,** along with prostaglandins, to induce abortions within the first nine weeks after conception. RU 486 blocks progesterone, thus allowing menstruation.

Induced abortions in the second trimester (the second three months) require different methods, which are performed only in a hospital. The most common method is to induce uterine contractions by injecting a solution of salt or of prostaglandin. RU 486 is also given in conjunction with prostaglandins. Induced abortions in the third trimester are extremely rare and usually only done to save the mother's life.

Until the late 19th century, abortion was legal in the United States before the "quickening" of the fetus, at four to five months. In 1873, however, Congress passed the Comstock Act, which banned all information about birth control, as well as the practice of abortion or any form of birth control. Even condoms were illegal. Similar laws were passed by individual states over the succeeding decades. In 1973, the U.S. Supreme Court declared all of these laws unconstitutional. The decision, called *Roe v. Wade*, declared that abortions in the early stages of pregnancy were legal in the United States, but that states could restrict abortions in the third trimester as long as the mother's health was not at risk. Today, abortion is legal in all parts of the United States, but remains deeply controversial.

An induced abortion is the deliberate removal and destruction of an embryo or fetus.

Barrier Methods Prevent the Union of Sperm and Ovum

Couples who may want to have children some day, but not yet, choose temporary forms of contraception. Most couples have access to contraceptives that are far more effective and reliable than withdrawal or the rhythm method. The easiest of these methods to understand are those that physically prevent the union of sperm and ovum by imposing a **barrier** between them (Table 44-1). A **condom** is a thin, elastic rubber sheath that covers the penis and prevents sperm from entering the vagina. By preventing actual physical contact between the male and female genital tracts, condoms prevent not only pregnancy but also the transmission of sexually transmitted diseases (STDs), such as herpes, human papilloma virus, acquired immunodeficiency syndrome (AIDS), chlamydia, syphilis, and gonorrhea (see box on page 845).

The major drawbacks of condoms are that they can break, leak, and sometimes fall off. In addition, if the con-

Table 44-1
Contraceptive Methods and Their Effectiveness

Method	% pregnant in a year* (ideal)	% pregnant in a year* (typical)	Serious health consequences	Protection from STDs?	Mortality rate (per 100,000)
No method	85	85	Pregnancy and childbirth	No	13
Abstinence	0	0	None known	Yes	0
Vasectomy	0.10	0.15	None known	No	0.3
Tubal ligation	0.5	0.5	Infection, increase in heavy bleeding	No	1.5
Condom (no spermicide)	2–3	12–21	Latex allergy, rare	Latex, yes	0
Birth control pill	<1	3-5	Heart disease, stroke (especially in smokers)	No	1.6†, 6.3‡
Depo-Provera	0.3	0.3	Possible increase in osteoporosis	No	—
Norplant	0.05	0.05	Infection at implantation site, abnormal vaginal bleeding	No	—
Morning-after pill	<1	25	Nausea, cramps	No	—
RU-486/Mifepristone	<1	5–8	Nausea, cramps	No	—
Intrauterine device (IUD)	<2	1–3	Pelvic inflammatory disease	No	1
Diaphragm with spermicide	6	18–20	Latex allergy or toxic shock syndrome, rare	Some	0
Cervical cap with spermicide	6	~18	Latex allergy or toxic shock syndrome, rare	Some	0
Contraceptive sponge	9	20#	None known	Some	0
Female condom	2–5	12–21	None known	Yes	0
Spermicide alone	3–6	15–26	Allergies, rare	Some	0
Rhythm method (partial abstinence)	6–10	20–25	None likely	No	0
Withdrawal	4	19	None known	No	0
Legal abortion within 7 weeks after fertilization	<1	<1	Infection, cramps	No	0.2

*Expressed as percent per year.
#Women who have borne children experience higher failure rates.
†In nonsmokers.
‡In smokers.

Business and Biology   What Services Do Fertility Clinics Offer Infertile Couples?

In the United States, about one couple in six, or 2.5 million couples, is infertile, meaning that they do not conceive within the first year of having sexual relations without contraception. About half the time, fertility problems lie with the man. Male infertility results from the failure to produce and deliver sperm that can fertilize an egg. An inability to produce enough healthy sperm may result from infections (such as mumps), poor nutrition, scrotum overheating (from tight clothing or hot tubs), sperm-damaging chemicals (including alcohol and other drugs) and radiation, or age. Problems of sperm delivery may be due to blockage or scarring in the genital tract or from impotence (inability to maintain an erection during intercourse) and premature ejaculation.

Female infertility likewise results from age, infections and scarring of the genital tract (including STDs; see the box on page 845), or hormonal imbalances that interfere with ovulation or maintaining pregnancy. In addition, chromosomal abnormalities that result in spontaneous abortions also increase as a woman and her partner age (Chapters 9 and 14).

Overall infertility rates have not changed over the years and many infertile couples still elect to adopt children. But the marketing of "assisted reproductive technologies," such as *in vitro* fertilization, has created a huge demand for fertility services. Current spending on fertility services in the United

States now exceeds a billion dollars a year. In just four years, from 1996 to 2000, the number of babies resulting from infertility treatments rose 73 percent.

Fertility clinics offer a wide selection of services. Women with unresponsive ovaries can be treated with FSH or combinations of FSH and LH to boost the number of mature follicles in the ovaries. Men with low sperm counts can ask a doctor to use *artificial insemination,* a procedure in which the physician uses a catheter to introduce the man's sperm into his partner's cervix before or on the day of ovulation.

One of the most costly, invasive, and effective techniques for overcoming infertility is *in vitro fertilization* (IVF), in which a woman's oocytes are removed surgically from her ovaries and fertilized with sperm in a dish. IVF can helps couples with a range of fertility problems, including blocked oviducts, certain sperm deficiencies, and female hormonal deficiencies. In IVF, a physician first induces the maturation of many ovarian follicles—either in the mother herself or in a woman who is donating eggs—using FSH or a mixture of FSH and LH. A surgeon later removes all the mature ova from the ovaries and they are placed in a dish with sperm.

After fertilization and some subsequent development, the physician transfers the embryos back into the uterus of the intended mother, where they can, hopefully, develop normally. Many fertility clinics im-

plant several embryos because most or all are likely to die after transfer to the uterus. Occasionally several will survive. And the previously "infertile" couple goes home with triplets or even quintuplets. In such cases, premature birth is extremely likely. Prematurity can seriously impair the health of the baby, and hospital costs are often astronomical. In the 1990s, the total cost of each *in vitro* baby averaged $60,000 to $100,000. The costs for a single premature baby can exceed a half million dollars.

Through careful selection of the number of implanted embryos, however, the chances of having more than one or two successful implantations can be minimized. In addition, if too many embryos implant and begin to grow, parents and their physician fertility expert may opt to abort a subset of the embryos. Extra embryos are usually frozen for later use in case the first implantation fails or if the couple want more children.

Because of the hormone treatments and surgery involved, it's harder to find a woman who will donate eggs than it is to find sperm donors. But because fertility clinics have created a valuable market for reproductive services, human eggs have now acquired a dollar market value. Infertile couples will typically pay $3,000 to $5,000 for an egg donation. And for eggs from an intelligent, blond athlete attending Stanford University, one fertility clinic advertised a $100,000 fee.

---

dom is put on too late, sperm may leak into the vagina before ejaculation. A condom must be put on as soon as the penis is erect, and it must be removed *immediately* after ejaculation, so that the sperm does not leak out from around the flaccid penis. Although usually less effective at preventing pregnancy and STDs, the female condom, a sheath that fits over the cervix and covers the vulva, has the advantage of not requiring the penis to be erect. (A man can ejaculate even when his penis is not erect.)

Another method for preventing sperm from reaching the ovum is a **cervical cap**, a thimble-shaped rubber or plastic cap, about an inch in diameter, that fits tightly over the cervix. A **diaphragm** is larger than a cervical cap but works in the same way, serving as a barrier to the entrance of sperm. A **spermicide**, a cream or jelly that kills sperm, is an essential part of the effectiveness of both cervical caps

and diaphragms, and can also be used with condoms. A woman may insert a diaphragm filled with spermicide up to several hours before intercourse. A cervical cap may stay in place for several days. However, additional doses of spermicide must be inserted before each instance of intercourse.

When properly used, condoms, cervical caps, and diaphragms are effective contraceptives, with failure rates of only three to five pregnancies per 100 women per year. With all three methods, however, improper usage and breakage lead to actual pregnancy rates of about 20 per hundred women per year (Table 44-1). One reason for these increased rates is the use of oil-based lubricants or spermicides that break down the latex material of the barrier.

The contraceptive sponge also provides a barrier to the entry of sperm into the uterus. Like the cervical cap and the diaphragm, the sponge fits over the cervix, but, unlike them, it requires no special fitting. Sponges have high failure rates—about 20 per hundred women who have never given birth per year, more in women who have borne children.

Some couples attempt to prevent fertilization with spermicides alone. These include jellies, foams, creams, and suppositories. Each of these methods, by themselves, ideally, gives pregnancy rates of about six per one hundred women per year, but in real day-to-day use, closer to 26 percent. Spermicides alone are not a reliable method of contraception. Finally, some women attempt to prevent pregnancy with a **douche** [French, = shower], the washing of the vagina immediately after intercourse. This method is highly ineffective, with a pregnancy rate of about 40 per one hundred women per year.

Barrier methods such as condoms and diaphragms are intended to prevent the sperm from reaching the oocyte. These methods are best used with spermicides.

The Rhythm Method

The **rhythm method** is a form of modified abstinence in which a couple refrains from sexual intercourse for about a week each cycle. For fertilization to occur, the egg in the oviduct must encounter a sperm at a time when both the sperm and the egg are in their prime. Eggs survive only 12 to 24 hours after they are released from the ovaries, so the sperm needs to encounter the egg within a few hours after ovulation. In essence, the sperm has to be ready and waiting.

Sperm can survive in the female reproductive tract for up to a week, but they are usually only able to fertilize an egg for three or four days. In reality, then, conception usually happens only when intercourse occurs in the few days before ovulation or on the day of ovulation itself. And although the maximum period of fertility for a woman is about eight days (the seven days before ovulation plus one day after ovulation), in practice the period when a woman can have intercourse and conceive is more like three to five days.

However, avoiding sexual intercourse during this narrow window of a woman's cycle is not a reliable way of preventing pregnancy. It's often quite difficult for a woman to predict when that three- to eight-day period will begin. Since her most fertile time is in the few days before ovulation, effective use of the rhythm method requires that she be able to predict the day of ovulation, which is difficult, especially in teens, whose ovulation cycles tend to be irregular. The less accurately a woman can predict ovulation, the longer the period of abstinence needs to be.

When practiced with complete commitment, the rhythm method only fails 6 to 10 percent of the time. That is, out of 100 women practicing the rhythm method, 6 to 10 will get pregnant in one year (Table 44-1). As most people practice the rhythm method, however, as many as 25 become pregnant in one year. For comparison, among sexually active women who practice no birth control, 85 out of 100 become pregnant in a year.

Several factors can increase the success of the rhythm method. Some women use a modified rhythm method, using contraception during the eight days before (expected) ovulation. If a woman has regular and predictable cycles, she has a better chance of correctly guessing the time of ovulation. In addition, if she knows when she ovulates each month, she has a better chance of estimating when it will occur the following month. One clue is a one-half to one-degree increase in body temperature on the day of ovulation. Another clue is a change in the cervical mucus, which becomes noticeably clear and elastic a couple of days before ovulation. Finally, on the day of ovulation, many women feel a slight pain on the side where ovulation is occurring, called *mittelschmerz* [German, *mittel* = middle + *schmerz* = pain, pain in the middle of the month].

All of these clues are useful to a woman trying to get pregnant. But *none* of them is immediately useful to a woman trying to avoid pregnancy because they all manifest themselves less than four days before ovulation. However, they are useful for pinpointing the day of ovulation and, therefore, for estimating the day of ovulation in the following cycle.

Monthly abstinence (the rhythm method) is one of the least effective forms of birth control.

Withdrawal

An even less reliable technique is **coitus interruptus** [Latin, *coitus* = sexual intercourse], or withdrawal, in which a man attempts to avoid conception by removing his penis from the vagina shortly before ejaculation. Although this method is widely practiced in some parts of the world, it is one of the least reliable forms of birth control (Table 44-1). One reason is that the secretions from the man's bulbourethral gland sweep small numbers of sperm into the vagina well before ejaculation. In addition, withdrawal before orgasm requires the man to exert considerable resistance to the stereotyped pattern of the sexual response. In other words, he does not always have enough willpower to withdraw.

Key Concepts

- Oogenesis and spermatogenesis are parallel processes that produce haploid oocytes in females and sperm in males.
- In vertebrates, the joining of a haploid sperm and a haploid egg to form a diploid zygote depends on sexual intercourse.

- Hormones regulate the production of ova and the maintenance of pregnancy and lactation in females, as well as the production of sperm in males.
- In both sexes, the anterior pituitary gland makes and releases hormones that control the secretion of these hormones.
- In mammals, male and female external genitalia develop from the same embryonic tissues, while internal genitalia develop from separate embryonic tissues.

Summary with Key Terms

How do mammals form sperm and eggs?

Mammals form haploid gametes (eggs, or **ova**, and sperm) for sexual reproduction from **primordial germ cells** in the **gonads**—the **ovaries** in females and the **testes** (testicles) in males. During **oogenesis**, diploid germ cells in the ovaries form haploid **primary oocytes**. Each month, several primary oocytes develop into follicles, which swell with fluid. Just before **ovulation**, a primary oocyte completes meiosis I and forms a **polar body** and a **secondary oocyte**, which enters meiosis II but does not complete it unless it is fertilized.

In the male's **scrotum** are the **testes**. Inside the testes, the **seminiferous tubules** make sperm, a process called **spermatogenesis**. The seminiferous tubules are surrounded by **Sertoli cells**, which nourish the developing sperm, and **interstitial cells**, which secrete **testosterone**. On the inside walls of the seminiferous tubules, diploid cells divide through meiosis to form four haploid **spermatids**, which mature into sperm, with a head, a midpiece, and a tail-like **flagellum**. The front of the head contains an enzyme-packed lysosome called the **acrosome**.

How do the sperm and egg meet and fuse?

The fusion of the egg and sperm is called **fertilization**, syngamy, or conception. The sperm moves into one of the **oviducts**, where it encounters an egg and fuses with it. The egg and the polar body divide once more, forming a mature ovum and two or three polar bodies. The nucleus of the sperm and egg fuse, and the resulting diploid **zygote** begins dividing as it moves down one of the two oviducts to the **uterus**.

After about a week, the developing embryo burrows into the **endometrium** of the uterus. At about nine weeks, all the major organs have formed, at least in general outline, and the embryo is called a **fetus**. When the fetus has matured at nine months, it is said to be full term, and is expelled through the **cervix** by muscular contractions of the uterus.

During **ejaculation**, sperm leave the testes by way of the **epididymis**, the **vas deferens**, and the **urethra** of the penis. Along the way, fluids from the **seminal vesicles**, the **prostate gland**, and the **bulbourethral glands** contribute to the formation of the **semen**. The penis consists of the **corpus spongiosum**, paired **corpora cavernosa**, and the **glans**, all well supplied with blood vessels and nerves. A loose layer of skin, called the foreskin, covers the penis and glans in males who have not been **circumcised**.

The female external genitalia consist of the **vulva**—including the **labia majora** and **labia minora**—and the **clitoris** (all of which may be removed or damaged in the procedure known as **female circumcision**). Like the penis, the clitoris includes a glans and paired corpora cavernosa. Inside the entrance to the vagina is a membrane called the **hymen** that is usually opened during first intercourse.

In humans, sexual response consists of four stages—**excitement**, **plateau**, **orgasm**, and **resolution**. During the excitement stage, increased blood flow to clitoris and penis results in **erection** of the penis and clitoris and tightening of the vaginal walls. In addition, the **Bartholin's glands**, in the female, and the bulbourethral glands, in the male, contribute lubrication. Enzymes in the female reproductive tract break down a coating on the sperm and increase its motility and its capacity to fuse with the egg.

How do hormones control reproductive events?

Gonadotropin-releasing hormone (GnRH) secreted by the **hypothalamus** regulates the secretion of **luteinizing hormone (LH)** and **follicle-stimulating hormone (FSH)** by the **anterior pituitary gland**. In males, hormones regulate the production of sperm and testosterone by means of negative feedback. LH stimulates the production of testosterone and other androgens. FSH stimulates the Sertoli cells in the testes to make androgen-binding hormone (ABH), which, in the presence of testosterone, stimulates the production of sperm. The Sertoli cells also secrete inhibin, which suppresses the secretion of FSH, shutting down sperm production.

In women, LH and FSH regulate the **uterine** and **ovarian cycles**. As in men, the hypothalamus secretes GnRH, which stimulates the secretion of LH and FSH. These stimulate the ovaries to develop mature oocytes (and **follicles**) and to produce **estrogen** and other steroid hormones. The growth of the follicles causes a buildup of estrogen in the blood in the days preceding ovulation, which triggers the release of more GnRH, and therefore LH and FSH, a **positive feedback** system. LH triggers ovulation and development of the **corpus luteum**. **Progesterone** and estrogen from the corpus luteum inhibit the production of GnRH and cause the thickening of the endometrium. If no embryo **implants**, the endometrium breaks down and washes away in a flow of blood, called **menstruation**.

What happens during pregnancy and lactation?

Like sharks and many snakes, humans and most other mammals are **viviparous**; the young develop inside the mother. After fertilization, an embryo's secretion of **human chorionic gonadotropin**, or **HCG**, doubles the size of the corpus luteum and increases the secretion of estrogen and progesterone, which prevents menstruation and ovulation and sustains the pregnancy. Testing for HCG is the basis for pregnancy tests. The embryo constructs a set of four membranes and a **placenta**, an organ for transporting nutrients from the mother. The placenta also secretes the hormone human chorionic somatomammotropin (HCS), which helps redirect glucose from the mother to the embryo.

Physicians divide pregnancy into three trimesters. During the first trimester, the mother may experience fatigue and nausea. During the second trimester she will feel the first movements of the fetus, her nausea will probably go away, and she will feel more energetic. In the third trimester, the increasing size of the fetus will impinge on her own organs and make her distinctly uncomfortable.

At **parturition, cortisol, oxytocin,** and **prostaglandins** combine to drive a positive feedback cycle of faster and faster uterine contractions. These contractions, combined with pushing by the mother, force the baby (and later the placenta) out through the cervix and vagina. After birth, the mother's breasts secrete **colostrum,** a low-fat version of human milk, for the nursing baby. After about three days, the colostrum is replaced by milk. The infant's suckling stimulates the mother's pituitary to release oxytocin, which stimulates **milk ejection,** and **prolactin,** which stimulates the synthesis of more milk.

How can humans limit reproduction?

Birth control measures include, in approximate order of effectiveness: **abstinence, sterilization (vasectomy** and **tubal ligation), birth control pills, intrauterine devices (IUDs),** a variety of **barrier** methods, the **rhythm method,** and **coitus interruptus.** Barrier methods include two kinds of **condoms,** the **cervical cap,** and the **diaphragm.** Cervical caps and diaphragms work much better with **spermicides.** (A **douche** is not a method of birth control.) Systemic contraceptives are synthetic hormones that prevent ovulation, fertilization, or implantation. These include the birth control pill, **Norplant, Depo-Provera,** and the **morning-after pill.** All of these chemical contraceptives are for women only. Forms of birth control that prevent conception are **contraceptives.**

When contraceptives fail, some women arrange to have a doctor perform an **induced abortion,** the intentional destruction and removal of the embryo or fetus. An induced abortion is distinct from a **spontaneous abortion,** the death and expulsion of an embryo due to a chromosomal abnormality or other damage. Abortions may be induced physically, or chemically, using **RU 486,** or **mifepristone,** for example.

Review and Thought Questions

Review Questions

1. Compare oogenesis with spermatogenesis. What is the main difference between these two processes?
2. What is the difference between a primary oocyte and a secondary oocyte? Which fuses with the sperm and what happens then?
3. What is inside of a mature oocyte, or egg? What is inside a sperm?
4. Draw a diagram with labels showing a cross section of a seminiferous tubule that illustrates how the sperm form and develop.
5. Draw a diagram of the female reproductive tract that shows what happens where, with egg, sperm, and arrows indicating directions of movement.
6. What glands contribute fluids that make up the semen?
7. Which glands in women and men secrete mucus during sexual intercourse?
8. What are the four stages of the human sexual response?
9. How do GnRH, LH, and FSH regulate the formation of mature oocytes and sperm? Draw a diagram showing where these hormones are made and how they interact with the gonads. What do the initials stand for?
10. If a woman has regular 28-day cycles and her period is one week late and a pregnancy test gives a positive result, how old is the embryo?
11. What hormones regulate milk production? How does a baby's suckling stimulate milk ejection?

Thought Question

12. In all mammals except monotremes, the testes are external to the body cavity and sperm cannot form effectively inside the warmth of the body cavity. Why do you think this might be so? Did mammals evolve external testes and then evolve sperm that were temperature sensitive or did the temperature sensitivity evolve first? Why would either happen? Why are we different from birds and other vertebrates, which make healthy sperm inside of their bodies?

BiologyNow Resources

Biology ⒺNow™

Active Figures

44-12: Hormone levels, ovulation, and conception
44-14: How hormone levels change during pregnancy

Preparing for an exam? Take a diagnostic test on your BiologyNow CD-ROM.

Online materials relating to this chapter are at:

http://biology.brookscole.com/AAL3

About the Chapter-Opening Image

This Fabergé egg symbolizes the female's investment of energy and materials in the oocyte.

How Do Organisms Become Complex?

Key Questions

- How does a zygote become an adult?

- How does a human embryo develop?

- Can differentiated cells be totipotent?

- How do molecular signals direct cell differentiation?

How Do We Become Complex?

Of all the wonders of life, none is more amazing than embryological development. Each of us has developed from a single cell, barely visible to the naked eye, into a complex adult organism. How does such complexity arise from the simplicity of a single egg? The Greek philosopher Aristotle argued that the female's egg provides formless substance and the male's semen gives the egg shape, molding it into an individual in the same way that a potter forms a bowl from clay. How sperm might accomplish such crafting, Aristotle did not say.

Eighteenth-century biologists explained the development of complexity by denying that simplicity had ever existed. In this view, all the parts of the adult organism already exist at the earliest stages of life. The adult is *preformed* in the sperm or the egg.

The French biologist Charles Bonnet formalized this theory, called preformation, in 1745. He argued that each egg contains a complete embryo, and each embryo contains more complex embryos inside itself, and so on, like a set of Russian dolls. Many of Bonnet's contemporaries argued that the complete embryo was in the sperm, not in the egg, but the idea was the same. Some microscopists even claimed to see a tiny creature, called a homunculus, curled up inside the sperm head, shown at left. Preformation implied that the whole organism resulted from the growth of a preformed miniature.

The doctrine of preformation avoided the difficult problem of explaining development from a single cell and also provided a simple view of inheritance. A child resembled its parents because it was already preformed in the egg or sperm of the parent. But the doctrine of preformation also presented a serious problem. If a child was already preformed, how would both parents contribute to the makeup of the child? Anyone could see that children took after their mothers *and* their fathers, not one or the other.

Preformation also left little room for evolution. Because all future generations must be contained within past generations, only a limited number of generations could be contained. Eventually, the tiniest embryos inside embryos would be too tiny to exist.

And preformation worked as a theory only as long as people believed that the Earth was only 4,000 years old. As geologists discovered that life was millions of years old and chemists discovered that matter was not infinitely divisible, the underpinnings for the theory of preformation crumbled.

It was an Estonian embryologist, however, who provided the evidence that finally disproved preformation. In the 1820s, Karl Ernst von Baer described the gradual development of a mammalian zygote. He saw that a single-celled zygote divided and formed three layers of tissue, from which the embryonic organs and tissues then developed. It was a gradual process, like the one that Aristotle had suggested 2,000 years before.

By the middle of the 19th century, it was clear that Bonnet's theory of preformation was incorrect. Yet von Baer's work raised a new question: if a single cell could divide and each descendant cell could specialize into different kinds of cells, how did the cells become specialized? Was each cell capable of developing into any kind of cell, or was each cell committed to becoming a certain kind of cell?

In order to answer this question, the German biologist Wilhelm Roux conceived an experiment that would tell whether or not embryonic cells were **totipotent**, capable of developing into any kind of cell, or **determined**, capable of developing into only one or a few kinds of cells.

Roux reasoned that as the cells divided, each one got only some of the genetic information present in the zygote. If that were true, and if only half of the embryo were allowed to develop, the result would be an embryo with half its structures missing. In 1888, Roux took a two-celled frog embryo and destroyed one of the two cells by piercing it with a hot needle (Figure 45-1A). The remaining cell formed only half an embryo. Roux concluded that embryonic cells were not totipotent. Instead, he argued that at each cell division the daughter cells receive half of the information in the dividing cell. As the zygote divides into more and more cells, these cells receive smaller and smaller shares of the genetic information in the zygote's original nucleus.

Within four years, however, Roux was challenged by the German biologist Hans Driesch. Driesch performed a similar experiment, but with dramatically different results (Figure 45-1B). Driesch used sea urchin embryos, and, instead of killing one cell and leaving it in place, he shook the two-celled sea urchin embryos until the pairs of cells separated and fell apart. As he watched, each of the two cells grew into a complete sea urchin. Driesch's result disproved the idea that each of the cells of an embryo had different information. An embryo, apparently, was not a mosaic of information. Instead, each cell might be totipotent, capable of forming all the structures of the adult.

Driesch's demonstration of totipotency in embryonic cells conclusively discredited the theory of preformation. If each cell of an embryo could develop into a separate adult, then neither the egg nor the sperm could be said to exclusively harbor the individual.

But why had Roux's experiment suggested preformation? The German embryologist Hans Spemann wondered if Roux's embryos developed abnormally because the dead cell

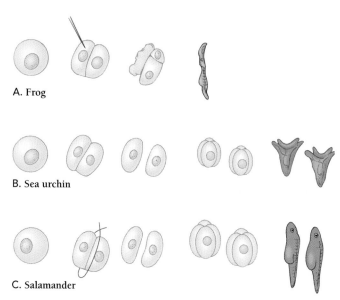

Figure 45-1

Totipotency and preformation. Is each cell of an embryo capable of forming a whole tadpole? A. To find out, Wilhelm Roux killed one cell of a two-cell frog embryo and got half a tadpole. B. But Hans Driesch shook apart the two cells of a two-cell sea urchin embryo and got two complete sea urchins—the first known artificial clones of an animal. C. Hans Spemann repeated Roux's experiment. Instead of killing one cell, however, he separated the two cells of the salamander embryo with a fine hair. Both cells developed into normal salamander larvae, thus showing that each cell was totipotent, capable of forming all the parts of the whole organism.

interfered with the development of the remaining embryo. Spemann repeated Roux's experiment, but instead of killing one cell, he used a fine hair (from a baby) to gently separate the cells of a two-cell salamander embryo (Figure 45-1C). And like Driesch, Spemann saw each cell develop into a whole, normal embryo. His result showed that Roux had misinterpreted his results, and, more importantly, solidly confirmed Driesch's view that each cell of a two-cell embryo can give rise to a whole adult.

The experiments of Driesch and of Spemann permanently laid to rest the theory of preformation and suggested that each cell inherits all of the genetic information contained in the zygote, not just part of it. But more questions remained. Where does complexity come from? How do cells become different from one another? How do organs and tissues form? How do the different parts of the embryo know how to arrange themselves?

As we will see later in this chapter, Spemann and his students went on to answer many of these questions, but each answer provoked still more questions. Near the end of his career, in 1936, Spemann gave a series of lectures at Yale University, in which he despaired of ever understanding the mechanisms of development. If he were alive today, he might be happy to learn that in recent years biologists have made great strides in understanding how organisms develop complexity and have solved many of Spemann's most perplexing puzzles.

45.1 How Does Fertilization Initiate Embryonic Development?

In all animals, fertilization, or syngamy, joins the haploid ovum and sperm to form a diploid zygote and initiates a remarkable series of events. For example, within a minute after a sea urchin sperm binds to an egg, the egg increases its rate of oxygen consumption. Within 10 minutes, the resulting zygote begins intense metabolic activity, using energy at a much higher rate.

Protein synthesis increases dramatically, and the embryo begins to divide. Earlier protein synthesis occurs as the egg's ribosomes (which come from the mother) read messenger RNA (mRNA) also synthesized by the mother's cells. Such mRNA, called **maternal mRNA,** is produced in the diploid egg cell (before meiosis) from the genes of the mother, rather than those of the zygote. The very early development of a zygote is controlled by maternal mRNA. In sea urchins and frog embryos, embryos lacking active nuclei still divide.

Fertilization is both a genetic and a developmental event. Protein synthesis and development in early embryos is directed by maternal mRNA.

45.2 How Do Cells of the Embryo Give Rise to Cells of the Adult?

Developmental biologists view the body plan of a vertebrate as "a tube within a tube." The outermost tube, or **ectoderm** [Greek, *ecto* = outside + *derma* = skin], forms the part of the animal that is in contact with the outside world—the epidermis, or outer skin layer, the nervous system, and the sense organs (Figure 45-2). The innermost tube, or **endoderm** [Greek, *endo* = inside + *derma* = skin], gives rise to the gastrointestinal tract, together with associated organs such as the pancreas and liver. Between the ectoderm and the endoderm is the **mesoderm** [Greek, *mesos* = middle + *derma* = skin], which forms bones, blood, muscles, and tendons, as well as the heart and kidneys.

These three layers eventually form the adult vertebrate. To see how this happens, we can study the early events of development and ask when the three layers first appear.

Biologists Separate Vertebrate Development into Eight Stages

Although the process of development is continuous, biologists often divide it into eight separate stages (Figure 45-3).

1. **Gamete formation**—the production of sperm and eggs (Chapter 44).

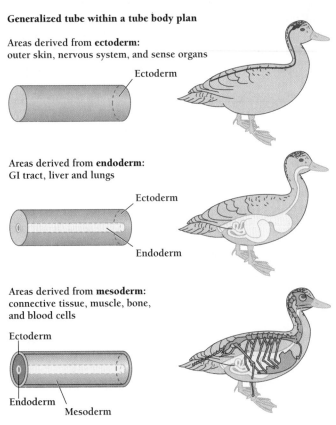

Generalized tube within a tube body plan

Areas derived from **ectoderm:**
outer skin, nervous system, and sense organs

Ectoderm

Areas derived from **endoderm:**
GI tract, liver and lungs

Ectoderm

Endoderm

Areas derived from **mesoderm:**
connective tissue, muscle, bone, and blood cells

Ectoderm

Endoderm

Mesoderm

Figure 45-2

A tube within a tube. An animal consists of a tube of endoderm *(yellow)* within a tube of ectoderm *(blue)*. In between the two tubes the mesoderm *(red)* develops. In vertebrates such as this duck, the ectoderm forms the outer skin, nervous tissue, and sense organs; the mesoderm forms muscle, bone, blood cells, and connective tissue; and the endoderm forms the intestines, liver, and most other internal organs.

2. **Fertilization**—the fusion of the haploid egg and sperm to form a diploid zygote.
3. **Cleavage**—the division of the one-celled zygote into many smaller cells; in many vertebrates these cells form a hollow ball of cells called a **blastula.**
4. **Gastrulation**—the rearrangement of embryonic cells to form the three germ layers (ectoderm, mesoderm, and endoderm) of a **gastrula.**
5. **Organ formation**—the migration and differentiation (specialization) of cells to form organs such as the heart, kidneys, and stomach. In humans, once an embryo has developed the rough outlines of all its organs, nine weeks after fertilization, it is called a **fetus.**
6. **Growth**—the increase in size of an organism after the organs and body plan are established; growth involves both cell division and the production of extracellular materials (such as bone, cartilage, and hair) and converts the collection of beginning organs into an adult.
7. **Metamorphosis**—a series of changes in form in which the larval form (for example, caterpillar or tadpole) changes into an adult form (for example, butterfly or

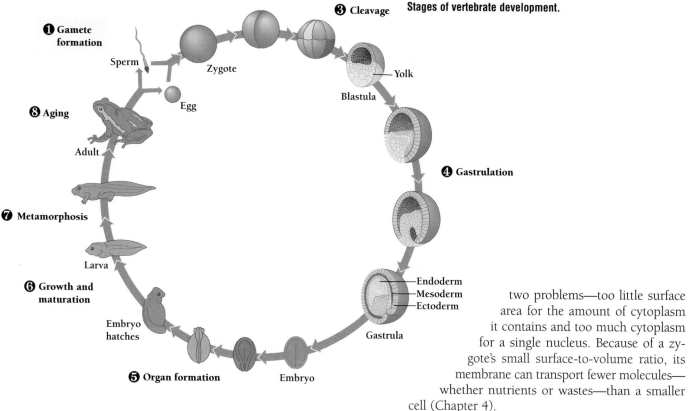

Figure 45-3
Stages of vertebrate development.

❷ Fertilization

❶ Gamete formation

❸ Cleavage

Sperm

Zygote

Yolk

Blastula

❽ Aging

Egg

Adult

❹ Gastrulation

❼ Metamorphosis

Larva

Endoderm
Mesoderm
Ectoderm

❻ Growth and maturation

Embryo hatches

Gastrula

❺ Organ formation

Embryo

frog), which has a different lifestyle; not all animals undergo metamorphosis.

8. **Aging**—once the adult form is established, cells die, extracellular tissues change, and organs and organ systems degenerate.

Why Does a Zygote Divide?

A zygote is often much larger than the average body cell. In many species, the zygote contains large stores of yolk—a mixture of proteins, lipids, and carbohydrates that nourishes the embryo until it can feed itself. Such a large cell has two problems—too little surface area for the amount of cytoplasm it contains and too much cytoplasm for a single nucleus. Because of a zygote's small surface-to-volume ratio, its membrane can transport fewer molecules—whether nutrients or wastes—than a smaller cell (Chapter 4).

The zygote's nucleus presents a similar problem. The nucleus of a large cell and a small cell are the same size, containing the same amount of DNA and the same amounts of regulatory proteins. Yet the nucleus of a zygote has far more cytoplasm and membrane to regulate than a smaller cell. A single nucleus cannot make enough mRNA for a large, rapidly developing embryo.

The division of the zygote into many smaller cells is one way to solve both of these problems (Figure 45-4). The series of rapid cell divisions, called cleavage, increases both the surface area and the number of nuclei for the embryo's cytoplasm. As each nucleus assumes influence over a separate cell, the cells can become different from one another, each cell potentially expressing different genes (making dif-

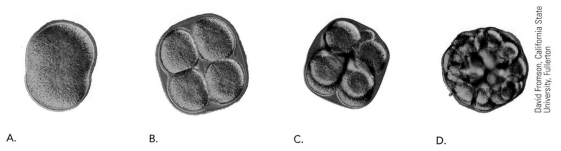

David Fromson, California State University, Fullerton

A. B. C. D.

Figure 45-4
Cleavage in a sea urchin embryo. A. A zygote beginning first cleavage. B. After two divisions, the embryo consists of four cells. C. The embryo continues to cleave, forming an 8-cell and then a 16-cell embryo. D. A 32-cell embryo.

ferent mRNA). Such differences in gene expression are the basis of cell specialization and differentiation.

Cleavage increases total cell surface area and the amount of DNA per unit of cytoplasm. Cleavage also converts the embryo from a single large cell to many smaller cells that can become different from one another.

How Do Zygotes Cleave?

Just as there are lots of ways to cut up a cake, there are lots of ways for a zygote to cleave, and different groups of animals have evolved different "cleavage patterns." The different cleavage patterns depend partly on the distribution of yolk inside the egg. Eggs with little yolk, such as those of the sea urchin, divide more or less equally, symmetrically, and quickly (Figure 45-4). In the sea urchin, each division takes less than an hour. In other species, eggs that contain a lot of yolk may either divide very slowly or divide incompletely. For example, yolky bird and fish eggs confine their divisions to a tiny disc of incompletely separated cells floating on the surface of the yolk (Figure 45-5). In mammals, which are descended from animals with yolky eggs, cleavage occurs slowly, each division taking 12 to 24 hours, and also forms a flat disc.

After dividing repeatedly into many smaller cells, animal embryos form a hollow ball of cells called a **blastula.** In mammals, the dividing cells form a **blastocyst** [Greek, *blastos* = germ + *cystos* = cavity]. The hollow cavity, in both blastula and blastocyst, is called a blastocoel (Figure 45-6A).

Embryos from different taxonomic groups divide at different rates and in different ways depending on the amount of yolk in the egg. Sea urchins and amphibians divide to form a hollow ball called blastula. Mammals divide to form a blastocyst.

How Does Gastrulation Set Up a Three-Layered Embryo?

During the process of **gastrulation,** the hollow ball of cells (blastula or blastocyst) folds into three layers and forms "a tube within a tube." The innermost tube will become the gut. The outermost tube is the wall of the body. For an easy way to imagine the conversion of a hollow blastula into this three-layered structure, think about pushing in one side of a soft, hollow ball with your finger (Figure 45-6B). The pushing-in process is called invagination. (The process is different in birds and mammals, as we discuss below.)

If you kept pushing, your finger would eventually reach the other side of the ball. In real life, your finger would not go through the two layers of the ball. But in embryos, at the spot where your finger reached the other side, the two layers fuse and open up, so that the ball of cells has two holes, one where your finger went in and one where it came out. In vertebrates, the second hole eventually becomes the mouth, while the first opening becomes the anus. Recall from Chapter 23 that most animals are classified according to whether the first hole becomes the mouth (protostomes such as insects and slugs) or the second hole becomes the mouth (deuterostomes such as vertebrates and sea urchins). One of the reasons that embryologists study sea urchins is that they are deuterostomes, like us.

To continue with the metaphor, the place where you first push your finger into the side of the ball is called the **blastopore** [essentially the little hole in the ball]. The innermost space, created by your finger, will eventually become the space inside the gut, or **archenteron** [Greek, *arche* = beginning + *enteron* = gut]. The archenteron is lined with endoderm cells (yellow in Figure 45-6). The cells on the outside of the ball will become the ectoderm (blue). In the gastrulating embryo, a group of cells sitting above the

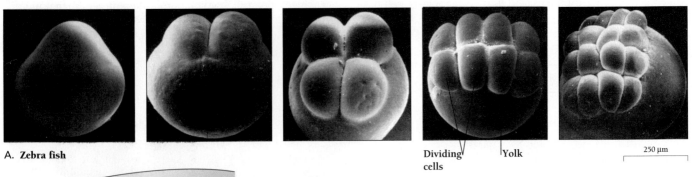

A. **Zebra fish**

Dividing cells | Yolk

250 μm

B. **Bird** — Blastula — Yolk

Figure 45-5

Cleavage in yolky eggs. A. In the yolky embryo of a zebra fish, cleavage is confined to a small disc on the outer surface of the yolk. B. Cleavage in birds is similar, resulting in a flattened blastula. Mammal eggs have lost their big yolky eggs during evolution, but they too have a flattened blastula. (A, from Beams and Kessel, 1976.)

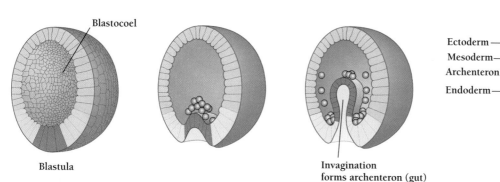
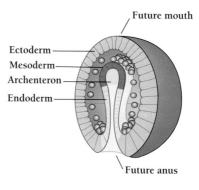

Future mouth

Ectoderm —
Mesoderm —
Archenteron —
Endoderm —

Future anus

Blastocoel

Blastula

Invagination
forms archenteron (gut)

A.

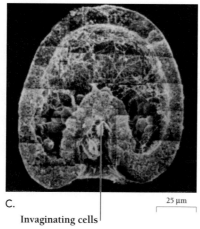

B.   25 μm
Invaginating cells

C.   25 μm
Invaginating cells

Figure 45-6
Gastrulation in a sea urchin. A. The sea urchin blastula is a
hollow ball of cells. Cells from the bottom of the blastula
move into the blastocoel and eventually become mesoderm.
Endodermal cells invaginate smoothly in through the
blastopore to form a structure called the archenteron. As
invagination proceeds, individual cells at the tip of the
archenteron detach. These cells form contacts with the wall of
the blastocoel and actually pull the archenteron up to the wall.
Then they disperse into the blastocoel and proliferate to form
most of the mesoderm. B. Scanning electron micrograph of
early sea urchin gastrula C. Cross section of sea urchin
gastrula. (B and C, from Morrill and Santos, 1985.)

invaginating archenteron migrates down (into the blastocoel) to form the mesoderm (red).

> During gastrulation, a blastula invaginates, giving rise to the endoderm,
> ectoderm and mesoderm. The inside of the invagination is the archen-
> teron, the inside of the future gut.

How Do Mammals Gastrulate?

As mentioned earlier, a mammalian embryo divides and
forms a **blastocyst**. The blastocyst includes two groups of
cells—the inner cell mass and the trophoblast. The **inner
cell mass** will eventually grow into the embryo itself, including a sac, called the *amnion,* filled with *amniotic fluid,*
which protects the embryo (and later the fetus) from mechanical shocks. The **trophoblast** [Greek, *trephein* = to
nourish + *blastos* = germ] is an outer cell layer that,
together with cells from the lining of the uterus, forms
the **placenta**, the organ that connects the embryo to
its mother. The blood vessels of the placenta deliver
nutrients from the mother to the embryo and carry

waste from the embryo to the mother, for her organs to
dispose of.

The inner cell mass separates into two layers, the
hypoblast, which gives rise to the yolk sac, and the **epiblast,**
which gives rise to the amniotic cavity (Figure 45-7). Where
these two sacs touch each other is the site where the embryo
itself will develop, from a flattened blastula called the
blastodisc.

The embryo forms the three germ layers by means of
complex movements of different groups of cells. The most
striking is the formation of a **primitive streak**, a mammal's
equivalent of the blastopore (Figure 45-8). Like a crowd
pushing through the front doors of an auditorium to get to
the best seats first, the cells piled up on the primitive streak
squeeze through a groove in the streak and into the space
between the epiblast and hypoblast, later separating into endoderm (yellow) and mesoderm (red). Cells left behind in
the epiblast layer become the ectoderm (blue).

> The mammalian embryo forms from a flattened disk of cells, and the
> cells invaginate through the primitive streak, to form the three-layered
> gastrula.

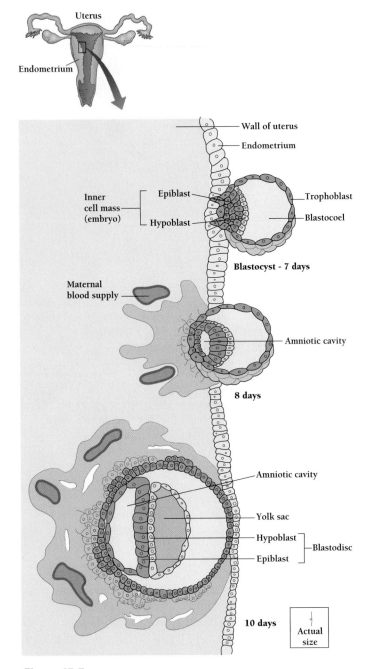

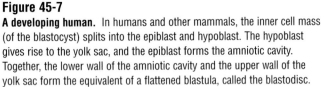

Figure 45-7
A developing human. In humans and other mammals, the inner cell mass (of the blastocyst) splits into the epiblast and hypoblast. The hypoblast gives rise to the yolk sac, and the epiblast forms the amniotic cavity. Together, the lower wall of the amniotic cavity and the upper wall of the yolk sac form the equivalent of a flattened blastula, called the blastodisc.

Organ Formation: How Does the Nervous System Form?

Once an embryo has established three germ layers, it next begins to form organs. An organ may derive from a single germ layer or may come from two different germ layers. In each case, organ formation consists of two major processes:

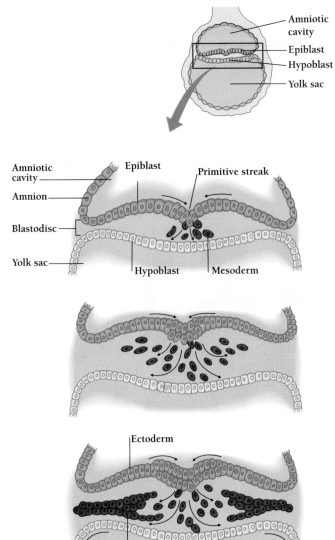

Figure 45-8
Mammals form three cell layers. In humans and other mammals, the epiblast cells invaginate along the "primitive streak," forming the mesoderm. The epiblast cells also enter the hypoblast to form the endoderm.

morphogenesis, the creation of form, and **differentiation,** the specialization of cells. The formation of the brain in higher vertebrates illustrates many of the main features of organ formation.

In vertebrates, one of the first organs to form is the **notochord,** a supportive rod that runs from head to tail, beneath the dorsal (back) surface. All animals in the phylum Chordata (including all vertebrates) have a notochord. The notochord, made from mesoderm, serves both as an internal skeleton for all vertebrate embryos and as an organizer for further embryonic development.

Just above the notochord, the ectoderm rearranges itself to form the nervous system. The cells along the central line of the embryo thicken to form a flat plate called the **neural**

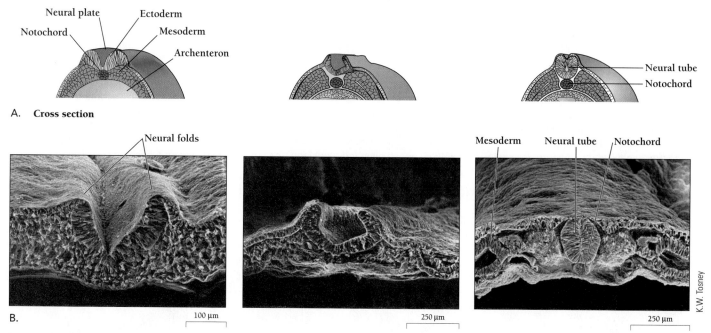

Neural plate   Ectoderm
Notochord   Mesoderm
               Archenteron

A.   Cross section

Neural folds

Mesoderm   Neural tube   Notochord

Neural tube
Notochord

B.

100 μm   250 μm   250 μm

K.W. Tosney

Figure 45-9

Making a neural tube. A. Cross sections through the middle of an embryo, showing the folding of the neural plate into the neural tube, which will develop into the brain and spinal cord. B. Micrographs showing the neural folds coming together and fusing at the top.

plate, which curls up into a **neural tube,** which forms for the brain and spinal cord (Figure 45-9). The cells left on the surface, above the neural tube, become skin (epidermis). The cells between the tube and the surface, called **neural crest cells,** migrate to other parts of the embryo and differentiate into nervous tissue, pigment cells of the skin, and other tissues. The entire process of forming the neural tube and the neural crest cells is called **neurulation.**

Once the neural tube is in place, it forms the different regions of the brain and spinal cord. The front of the tube enlarges and then bulges on each side to form the cerebral hemispheres of the brain. Just behind, another pair of bulges form the optic vesicles, which will become the eyes, as described in the box.

The notochord induces the formation of the neural plate, which folds into a hollow neural tube and forms the different regions of the brain and the spinal cord. Neural crest cells migrate to other parts of the embryo to form nerve and other tissues.

Programmed Cell Death Contributes to Normal Development

Growth and development usually suggest an increase in size of an organism and an increase in the number of cells. But the death and elimination of certain cells is also a major part of development. In the last decade, researchers have come to appreciate how much development depends on "programmed cell death," or **apoptosis** [Greek, *apo* = from + *ptosis* = fall]. Cells undergoing apoptosis are distinctive: their nuclei condense, their DNA is systematically digested, and other cells methodically engulf them.

Apoptosis is particularly common among cells of the vertebrate immune and nervous systems. For example, the "pruning" of neurons in the brain during adolescence is a major part of human development. But one of the most easily appreciated examples is the apoptosis that forms the embryonic digits (fingers or toes). In the early embryo, a thin tissue lies between the digits (Figure 45-10). This tissue per-

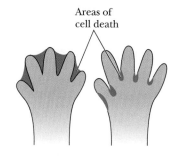

Areas of
cell death

Figure 45-10

Sculpture and development. Like a sculptor chiseling stone, the programmed death, or apoptosis, of individual cells generates form in a developing animal. The human hand first forms as a finlike paddle that lacks both fingers and thumb. As cells die between the fingers, their shape emerges. Later, the specific tissues—bones, muscles, nerves, and blood vessels—will develop within each finger. The same process occurs in the feet of frogs, lizards, and other vertebrates.

Extreme Biology   How Does the Eye Form?

The eye forms as a result of a simple set of cell movements (Figure A). When the optic vesicle comes into contact with the overlying skin ectoderm, the ectoderm thickens to form a plate that will develop into the lens. Once the lens thickening occurs, the optic vesicle folds back on itself (or invaginates) to form a double-walled optic cup, and the lens plate curls and pinches away from the skin ectoderm to form the lens vesicle. The

stalk that attaches the cup to the rest of the brain becomes the optic nerve, which will eventually carry visual information from the retina to the brain.

After this complex set of thickenings, foldings, pouchings, and pinchings, the cells of the lens, the retina, and the overlying skin ectoderm begin to differentiate. The cells of the lens make specialized proteins, called crystallins, that give the lens its optical

properties. The cells overlying the lens change character to become the cornea, the transparent covering of the eye. Finally, the cells of the retina develop into light-detecting rod and cone cells. Through these events a seemingly uniform layer of cells in the neural tube develops into an eye, our most prized contact with our surroundings.

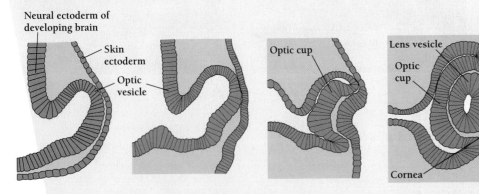
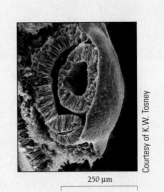

Courtesy of K.W. Tosney

250 μm

Figure A
Formation of the eye. A sheet of neural ectoderm folds to form the optic vesicle and optic cup. The skin ectoderm responds to the optic vesicle by forming a lens, as shown in the micrograph on the right. In humans, development of the eye first begins about 22 days after fertilization.

---

sists as the familiar webbing of a duck's feet. In nonaquatic vertebrates, the cells of the webbing die off, neatly separating the digits.

Programmed cell death, or apoptosis, is a central mechanism in embryonic development.

45.3 How Does Human Development Proceed after Implantation?

In humans and other mammals, fertilization occurs in the oviduct (Chapter 44). The zygote begins dividing while it is still moving down the oviduct toward the uterus. After about seven days, the embryo implants in the wall of the uterus.

Nine to 16 days after fertilization, the inner cell mass gastrulates. By 31 days, the placenta, which joins embryo to mother, becomes fully functional, moving nutrients and wastes between the embryo and its mother.

The placenta can also transport viruses, such as rubella ("German measles") or HIV (which causes AIDS); alcohol, nicotine, antibiotics, and other drugs; and even cells. Some of these substances may be **teratogens,** chemical substances or infectious agents that cause abnormal development (Figure 30-9). In the late 1950s, for example, doctors discovered that thalidomide, a sleeping pill and tranquilizer widely prescribed by doctors in Germany and England, was a teratogen. Some 7,000 pregnant women who were given this drug by their doctors gave birth to babies whose arms or legs were abnormal stumps (resembling a seal's flippers).

But the vast majority of embryos that make it past the first six to eight weeks develop into healthy, normal fetuses. An embryo becomes a **fetus** when every organ appears as a recognizable **rudiment** [Latin, *rudimentum* = beginning], or initial stage, from which the final form will develop. In humans, all these organs are in their proper places by the end of eight weeks. On the first day of the ninth week, the embryo is said to be a fetus.

But the fetus is itself rudimentary. The final form of each organ requires extensive cell specialization and growth. Each human hand and arm, for example, consists of 43 muscles, 29 bones, and hundreds of nerve pathways. None of these are present in the embryonic rudiment of an arm. In fact, except for the rudimentary heart, which begins pumping blood on day 22, the organs of an early fetus are largely nonfunctional. In addition, a nine-week human fetus is just a little over an inch long. Only with further growth and development does an early fetus become a baby.

Different parts of the body grow at different rates. A nine-week fetus, for example, has a head nearly the same size as all of the rest of its body (Figure 45-11). Although the head will never again be as proportionately large, it continues to grow and develop in advance of the other parts of the body. The brain, eyes, jaws, lungs, stomach, intestines, and kidneys grow and develop most rapidly, while the legs and feet lag behind. Between 8 and 12 weeks the fetus more than doubles in length, the external genitalia appear and begin to develop, and the kidneys excrete their first urine (into the amniotic fluid).

Between 12 and 16 weeks, a human fetus doubles its length again. By 16 weeks, the ovaries are differentiated and the primary follicles contain oogonia. Between 17 and 20 weeks, the mother feels the first fetal movements. By 20 weeks, the fetus is about 10 inches long, and, in males, the testes have begun to descend.

By 22 weeks, the lungs, intestines, and kidneys are sufficiently developed that it is sometimes possible—with intensive medical care—for the fetus to survive outside the womb. Between 20 weeks and 26 weeks, the fetus begins to fill out (Figure 45-11). At 24 weeks, the lungs begin to secrete a *surfactant*, a detergent that allows the lungs to inflate with air (Chapter 39).

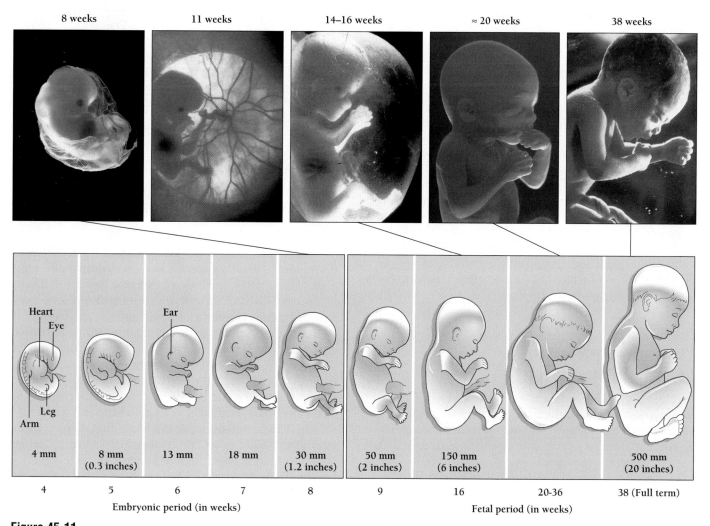

Figure 45-11

Stages of fetal development in humans. In humans, organ formation occurs in the first eight weeks, when the embryo is less than an inch long. At nine weeks, the fetal period begins; the organs gradually become functional over many weeks, and the fetus begins growing. (Photos from left to right, Science Pictures Ltd./SPL/Photo Researchers, Inc.; Petit Format/SPL/Photo Researchers, Inc.; Petit Format/Nestle/Photo Researchers, Inc.; James Stevenson/SPL/Photo Researchers, Inc.; Petit Format/Nestle/Photo Researchers, Inc.)

After 26 weeks, a fetus born prematurely has a good chance of surviving—because its respiratory system is usually capable of functioning. Between 26 weeks and 36 weeks, the fetus gains weight in the form of fat. By 36 weeks, the circumference of the baby's abdomen equals the circumference of its head. If it is born, it is not considered to be premature.

Physicians usually describe the stages of human development in terms of **trimesters,** periods of about three months each. Most of the important developmental changes occur during the first trimester. The first three months are the time when there is the greatest chance of a **miscarriage,** or **spontaneous abortion.** Approximately 50 to 75 percent of embryos abort spontaneously during the first trimester.

An embryo forms the rough outlines, or rudiments, of all of its organs long before those organs are functional. At the beginning of nine weeks, a human embryo possesses rudiments of all the major organs and is said to be a fetus, but only the heart is functional.

45.4 Embryonic Cells Become Increasingly Differentiated

Hans Driesch's experiment showed that each cell of a two-cell sea urchin embryo has the ability to develop into a whole (rather than a half) sea urchin. With this and similar results in mind, embryologists have come to distinguish the **fate** of a cell—what it becomes during development—from its **potency**—what it *could* become if allowed to develop in another environment. In the case of the two-cell sea urchin embryos, the *fate* of each is normally to become the right or left half of the sea urchin, but each has the *potential* to become a whole organism.

What Kinds of Experiments Distinguish Potency and Fate?

At early stages of development and for relatively simple organisms, we may follow the fates of individual cells just by watching and photographing them. To track more complicated events, embryologists can mark individual cells with fluorescent dyes.

Defining the potential of a cell is more difficult. One way of determining the development potential of embryonic cells is to transplant them from one embryonic environment to another. In 1918, Hans Spemann performed a now-famous transplant experiment in two species of newts. One species was darkly pigmented and the other was pale.

Spemann transplanted pieces of the dark, pigmented embryo into the pale, unpigmented "host" embryo. Because of the pigment, he could distinguish easily between structures formed from the host cells and those formed from the transplanted cells. Having studied the normal development of the newt, Spemann knew which embryonic cells should form neural plate and which should form skin.

Spemann transplanted cells he knew would become neural plate cells into an area that would become skin cells. If he transplanted the undifferentiated neural plate cells at the early gastrula stage, they formed skin instead of a neural plate. The transplanted cells developed according to their new environment. But, if Spemann waited until the late gastrula stage, the neural-cells-to-be developed into neural plate, even though they were in the wrong environment. Spemann concluded that later in development, cells become **determined**—that is, they are *committed* to their normal developmental fate, gradually losing the capacity to become something different. Although neural-plate-cells-to-be look just like skin-cells-to-be, they are already determined, meaning they can no longer differentiate in response to signals from their environment.

The fate of a group of cells is the tissue they will become during normal development. The potency of a group of cells is all the tissues that they could become in a different cellular environment.

Are Differentiated Animal Cells Totipotent?

Spemann's experiments raise two important questions: (1) How does a cell's environment influence its development? (2) How does a cell become determined, so that its environment no longer influences its development?

One hypothesis for how cells lose potency and become determined is that they literally lose genes during differentiation. In the 1950s, two developmental biologists, Robert Briggs and Thomas King, decide to test this hypothesis. They removed the nucleus from a leopard frog egg and replaced it with a nucleus from a cell from another frog, a technique called "nuclear transplantation." The egg developed under the influence of the transplanted nucleus. Briggs and King took nuclei from cells at different stages of development to see which nuclei could still direct development (Figure 45-12). A nucleus from a blastula cell was totipotent: it could often direct the development of the whole frog. But a nucleus from a neurula stage or later embryo implanted in an egg could no longer guide the development of the egg into a healthy, swimming tadpole. In leopard frogs at least, nuclei seemed to lose the potential to direct complete development.

Later researchers found that nuclei from the intestines of a different tadpole *could* generate a normal adult. Although most of the resulting zygotes did not develop normally, about one percent developed into normal adult frogs whose cells contained nuclei identical to those of the donor tadpole. A fundamental question remained, however: were

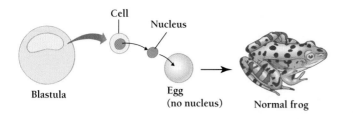

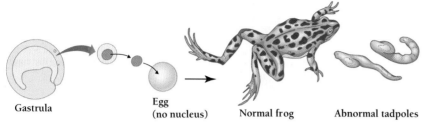

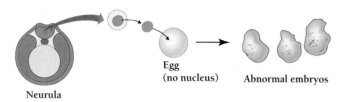

Figure 45-12
Do the nuclei of developing embryos remain totipotent as they differentiate? The experiments of Briggs and King suggested that as cells differentiate, their nuclei become less able to direct complete development. A nucleus from a blastula cell could direct the development of an egg into a normal leopard frog; a nucleus from a gastrula could direct development in some cases; but a nucleus from a neurula cell could never direct complete development.

fully differentiated animal cells totipotent? Until 1997, this question remained unanswered.

Early transplantation experiments did not make clear whether differentiated animal cells were totipotent.

Can a Differentiated Nucleus Direct Development from Egg to Adult?

One consequence of Briggs and King's nuclear transplantation experiments was that, for the first time, biologists could produce a group of genetically identical animals. Because the nuclei from the blastula cells were identical (all resulted from the cleavage of a single zygote), the resulting leopard frogs were clones.

For the first time, zoologists could clone an animal. Gardeners had been cloning plants for centuries. They could do so because virtually every plant cell is totipotent, so each cell can potentially develop into a complete adult. But no one had ever before cloned an animal. One important reason was that as animal cells differentiate, they seem to lose the ability to direct development. But no one knew why.

Animal breeders and biologists have long wanted to clone animals. A farmer with a prize dairy cow would love to be able to clone such an animal. A whole herd of prize dairy cows would not only ensure a steady supply of milk for the life of the individual cows but it would do so indefi-

nitely. More alluring still, a farmer with such a herd could sell the clones to other farmers.

Breeders would like to be able to clone adult animals with proven traits. No one can know an animal's worth until it is mature. Animal breeders can clone cells from embryos, but an embryo is basically an unknown: it is not a productive milk cow, a Derby-winning racehorse, or even a lab animal with a well-understood genotype and phenotype. Until recently, animal breeders had never succeeded in cloning cells from adult animals, and biologists' efforts with mice had failed over and over.

Then, in February of 1997, a team of biologists at the Roslin Institute, in Edinburgh, published a mind-bending result in the British journal *Nature*. The team, led by Ian Wilmut and Keith Campbell, had successfully reared a Scottish mountain sheep grown from an egg cell containing a nucleus transplanted from a mature cell from the udder of an adult sheep. In other words, the sheep, named Dolly, had been cloned from an adult animal. Dolly's impassive face appeared on the front pages of newspapers and magazines around the world. If biologists could clone sheep, commentators asked, why not people? Governments hastily passed laws banning the cloning of humans, although no one knew if the cloning of humans would ever be possible.

Geneticists were equally excited. If they could clone lab animals, they realized, they could study the subtle effects of environment on whole colonies of genetically identical lab animals. They could study, for example, the genetics of development as well as genetic diseases. More lucratively, biotechnologists could generate herds of genetically engineered animals, such as goats that secrete useful proteins in their milk (Chapter 13).

But developmental biologists viewed the research in a different light. From their perspective, one of the most exciting aspects of Wilmut and Campbell's work was that it seemed to answer the question about totipotency. If Dolly's nuclei had come from a fully differentiated cell, then, at least in sheep, it was possible to conclude, for the first time, that the nuclei of differentiated cells are totipotent.

The cloning of a Scottish mountain sheep in the 1990s suggested that differentiated nuclei are totipotent.

How Does Totipotency Change as a Result of Differentiation?

The first question that many people asked Wilmut and Campbell was, How did you do it? How had they prompted a differentiated nucleus to guide a full course of development starting in an egg cell? The answer was anything but clear. And even as researchers have followed up on the work—using many different techniques to clone sheep, cattle, goats, rabbits, pigs, cats, and mice—doubts remain about whether Dolly was truly cloned from a mature, differentiated nucleus.

The trick, Wilmut and Campbell argued, was to make the DNA of the donor cells behave more like the inactive DNA of an egg or sperm. As we saw earlier, the more differentiated a cell is the less able it is to direct development. Differentiated cells are unable to express all the genes that a totipotent cell can. Cells suppress gene expression by various means, including, for example, proteins that prevent ribosomes from transcribing certain genes, enzymes that alter the twisting of the DNA, and the addition or deletion of methyl groups (Chapter 3).

Wilmut and Campbell attempted to make differentiated udder cells more like egg cells by growing the udder cells in culture and starving them of essential nutrients. Gradually, the cultured cells passed into a dormant state in which many genes shut down and replication was impossible. Once the DNA in the donor nuclei was dormant, the team of biologists began transferring the nuclei into egg cells. Painstakingly, they transplanted 277 nuclei, of which only one successfully developed (Figure 45-13).

Subsequent attempts have done only a little better. On average, less than four percent of cloned embryos survive to become live young. Most die during fetal development or within the first 24 hours after birth. Even those four percent that live longer often suffer from some type of abnormality. To say that cloned animals are unhealthy is a huge understatement. They are lucky to live at all.

Some species are easier to clone than others. One reason seems to be that the later a transplanted nucleus starts expressing its genes, by making mRNA, the better chance it has of guiding normal development. For example, mice are harder to clone than sheep, probably because mouse embryos begin expressing their nuclear DNA after just one cell division, whereas sheep embryos don't begin until after three divisions. In monkeys, key proteins required for cell division are tightly attached to the egg's nucleus. When the egg's own nucleus is removed to make room for a donor nucleus, the proteins are lost.

Some studies suggest that not all cloned animals were actually cloned from differentiated cells. Most tissues contain mixtures of differentiated cells and a very few stem cells, totipotent cells that can differentiate into any kind of cell. Some cases of "successful cloning" may actually have resulted from researchers accidentally transferring a nucleus from a stem cell rather than an adult differentiated cell. Finding stem cells in a population of differentiated cells is nearly impossible, like looking for a needle in a haystack.

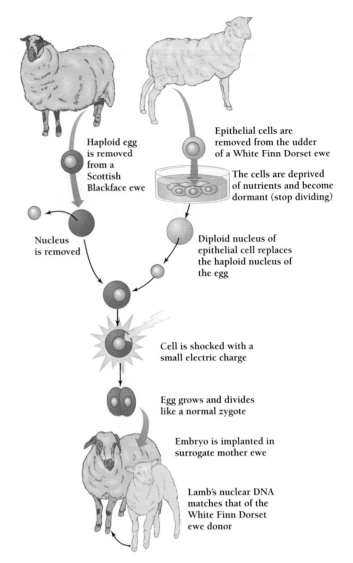

Figure 45-13

Cloning Dolly. The trick to cloning Dolly was to make differentiated cells less differentiated. By depriving the cultured udder cells of nutrients, the researchers induced the nuclei to enter a dormant state, then transplanted the nuclei into eggs surgically removed from another sheep. In February 2003, Dolly died of a lung disease apparently unrelated to her being a clone. She has been stuffed for display at a science museum in Edinburgh.

Biology ⊜ Now™   Learn more about the cloning of Dolly by clicking on this figure on your BiologyNow CD-ROM.

But researchers who transfer enough nuclei may sometimes get lucky without even knowing it.

For animal breeders, the question of whether a clone is derived from a fully differentiated cell or from a stem cell doesn't matter. For developmental biologists, however, the question of whether a fully differentiated cell is totipotent is as compelling today as it was in 1950. A paper published in 2002 may have finally put the question to rest. The researchers transplanted nuclei from differentiated cells of the immune system called lymphocytes. Lymphocytes rearrange parts of their DNA to produce antibodies and other proteins important

Business and Biology   Why Is Research on Human Stem Cells Controversial?

All the cells of a fetus and of an adult derive from the dozen or so cells of the inner cell mass. At the blastocyst stage, these cells are interchangeable and totipotent. Biologists have learned to grow such totipotent cells in artificial culture. Like other stem cells (discussed earlier), embryonic stem cells can divide to produce more stem cells and differentiate into any kind of specialized cell. Since the 1980s, biologists have been introducing genes into mouse stem cells and then putting them back into embryos to produce genetically altered mice—transgenic mice, knockout mice, and even knockin mice (Chapter 13).

In late 1998, biologists from the University of Wisconsin published a paper in the journal *Science*, reporting the isolation of *human stem cells.* These cells—like their mouse counterparts—not only proliferate in artificial culture but also differentiate into a variety of cell types, including cartilage, muscle, bone, skin, nervous tissue, and pancreas tissue.

The *Science* paper represented a minor milestone in biology, but it hit society as a whole like an earthquake. For many people—including the biomedical community and people with relatives suffering from diseases of defective cells such as diabetes and Parkinson's disease (so-called "patient advocates")—human embryonic stem cells are the holy grail, perceived not as a minor technical advance but the starting point for a revolutionary new kind of medicine.

Some biologists think that stem cells will make it possible to replace any damaged or diseased human cells, potentially curing any disease. Others hope to genetically engineer stem cells so that they do not provoke an immune response after transplantation. Such cells might allow physicians to replace worn-out heart muscle, insulin-producing pancreatic cells, or damaged brain cells. A few biologists even imagine using human embryonic stem cells to genetically engineer the human germ line, adding or deleting genes that could be passed on to an individual's children. Some of the most persis-

tently optimistic boosters of stem cell research think it will be possible to create customized stem cell lines that are genetically identical to a prospective patient's own cells.

But many outside the patient advocate and biomedical research communities are anything but pleased with these ideas. Some people object to research on cells from human embryos; they worry that the eventual success of any treatments would create a market for aborted embryos. A market for embryos could pressure women to sell embryos as they currently sell their eggs to the reproductive technologies industry. So even as scientists and physicians were initially celebrating the potential of stem cell research, some politicians called for a complete halt to any such research.

Patient advocates such as former actor Christopher Reeve (of "Superman" fame)—whose consuming interest is the cure of paralytic injuries—continue to be enthusiastic about the potential of stem cell research. But current laws severely restrict stem cell research in the United States (and parts of Europe). For example, the federal government only funds research on human embryonic stem cells if they have come from one of about a dozen cell lines established before August 9, 2001. And scientists and some politicians say many of these lines are not available and may not even exist. Experimenters who would like to use stem cells cultured from more recent embryos may not legally do so in the United States.

The biotech community's initial forays into this field have been almost abandoned. Declines in the stock market and the realization that stem cell investments could take years or decades to pay off have nearly dried up investments by venture capitalists. Biotech companies that funded early work on embryonic stem cells are laying off workers, struggling for survival, and often getting out of stem cell research altogether. One exception is a company that owns 5 of the 15 human stem cell lines that the U.S. government allows researchers to work with. And

even this company has narrowed its focus to creating pancreatic stem cells to treat diabetes, the fastest growing disease in the developing world (see the box on page 730).

Stem cell research is a scientific issue that deserves and even requires thoughtful input from nonscientists. Stem cell therapies may have the potential to treat serious diseases such as diabetes, Parkinson's disease, spinal cord injuries, and Alzheimer's disease. But such therapies can be developed only if biomedical researchers are allowed to experiment with cells that ultimately come from human embryos. Society as a whole has to decide where to draw the line, how to balance the many ethical issues involved.

It's an ethical debate that will not go away anytime soon. The patient advocacy, scientific, medical, and biotech communities are all eager to remove restrictions on stem cell use. And if politicians change their minds about stem cell research, the fertility industry is ready to help. A 2003 survey partly funded by the fertility industry showed that fertility clinics currently hold about 400,000 frozen embryos. Of these, nearly 90 percent are being held for "possible future use" by the couples who own them, at least for now. But such couples are aging and will eventually have to decide what to do with these embryos. Of the rest of the frozen embryos, 9,000 are slated for destruction, 9,000 are to be donated to other couples, and 11,000 could be available to science if the restrictions on stem cell research are eased.

Some researchers are actively seeking ways to sidestep the whole embryonic stem cell debate by finding ways to isolate stem cells from other sources. Early successes with the blood in the umbilical cords of newborn babies, so-called "cord blood," later turned to disappointment. The latest enthusiasm is for stem cells from adult bone marrow. With luck, marrow cells will offer real cures for the people who need them—but without the painful ethical side effects of embryonic stem cells.

for defending the body against disease (Chapter 41). Since these rearrangements occur late in differentiation, they are found only in mature lymphocytes, not in the stem cells that become lymphocytes. Mice cloned from mature lymphocyte nuclei carried the rearranged genes in all their tissues, demonstrating that the clones were derived from mature lymphocytes, not stem cells. Finally, embryologists could say with confidence that a fully differentiated cell could be totipotent, just as Dolly's cloning had suggested.

Cell differentiation results in changes in chromosome structure and chemistry that determine a cell so that it loses potency. In rare cases, nuclei from fully differentiated cells are totipotent and can be transplanted into eggs that develop, though not usually normally.

45.5 How Does Developmental Fate Depend on Chemical Signaling?

Cell differentiation depends on changes in patterns of gene expression. Differences in the patterns of gene expression are in turn influenced by the cellular environment. So how do cues from the embryo influence the developmental fates of individual cells?

Can Cells and Molecules Explain the Primary Organizer?

Arguably the most spectacular experiment in the history of developmental biology was one performed by Hans Spemann's student Hilde Mangold in 1924. Recall that Spemann had established that cells become determined during gastrulation. Mangold and Spemann asked *how* cells become determined.

The two researchers correctly guessed that determination of the ectoderm into neural tissue or into skin depended on contact between the ectoderm and the underlying mesoderm, which derives from the cells that invaginate during gastrulation (Figure 45-6). Mangold asked what would happen if the invaginating cells—the dorsal lip of the blastopore—were transplanted to another region of the embryo. Amazingly, the transplanted dorsal lip not only invaginated, it also induced the tissues around it to form a nearly complete second embryo.

Spemann named the dorsal lip the **primary organizer,** meaning that it established the entire organization of the embryo. He was so astonished by Mangold's result that he compared the action of the primary organizer to the workings of the mind—suggesting that it could not really be understood in chemical and physical terms.

Subsequent work, however, has demonstrated that the primary organizer's mechanism, as well as that of other tissues that help establish developmental patterns, is chemical. Researchers do not yet fully understand how pattern formation occurs, but they have identified many molecules that

guide this process within whole embryos and within individual organs. For example, in amphibians, cells in the primary organizer secrete proteins that promote, or at least help maintain, neural differentiation. They accomplish this by blocking the action of other proteins that induce the formation of epidermis. Some of these proteins also play roles in setting up patterns in other regions of the embryo.

Widely separated taxonomic groups often rely on similar chemical signals. Proteins that contribute to neural development in frogs, for example, also help regulate neural development in fruit flies. But while these signals promote the development of ventral structures in vertebrates, they favor the development of dorsal structures in invertebrates (Chapter 23).

Most of the proteins that affect development affect the transcription of specific genes. Often a signaling protein binds to a receptor on the surface of the target cell and stimulates a cascade of events that regulate the activity of a transcription factor. The altered transcription factor stimulates the production of new mRNAs and proteins, which in turn change the character of the target cell—for example, stimulating an undifferentiated precursor cell to become a nerve cell.

Hilde Mangold's transplantation experiments in amphibians showed that the primary organizer can establish the overall orientation (axis) of an embryo. Recent work has shown that the primary organizer contains pattern-forming proteins used by many species.

How Do Cells "Know" Where They Are and What to Do About It?

Development requires both that individual cells express appropriate genes and that cells be organized into functioning tissues and organs. The development of a whole limb, for example, requires the differentiation of the cells of the muscles, skin, and bone into a characteristic pattern, distinct from the organization of the same cell types of a specific muscle or part of the limb. How does the developing organism achieve this pattern?

Part of the answer to pattern formation lies in the interactions between layers of cells. For example, the skin ectoderm of a chicken can give rise to several different kinds of structures—fully tufted feathers, partly tufted feathers, scales, claws, or even teeth (an extraordinary finding, since birds do not have teeth) (Figure 45-14). The developmental fate of the ectoderm depends upon the source of the underlying mesoderm. For example, wing ectoderm placed over foot mesoderm will form claws instead of feathers.

But unlike the differentiation of specific tissues, the overall pattern of a limb, its large-scale form, does not depend on chemical signals between adjacent cell layers. Instead, concentration gradients of chemical signals called **morphogens** [Greek, *morphe* = form + Greek, *genes* = born] establish a map of the body to be. The concentration of a particular mor-

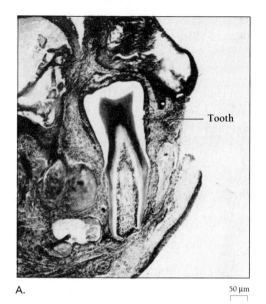

A.

50 µm

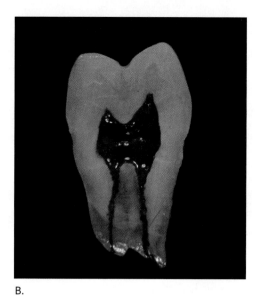

B.

phogen tells a cell whether it will become a bone cell in a finger or a bone cell in the elbow. The formation of patterns during development depends on each cell's "knowing" where it is in the body. It also depends on the interpretation of this information: each cell must also "know" *what* it is.

In vertebrates, one example of a morphogen is a small molecule called retinoic acid. In chicks, salamanders, newts, and frogs, retinoic acid induces the formation of limbs—complete with humerus, radius, and ulna. Indeed, one can induce chick limb buds to form extra digits by adding retinoic acid. In a regenerating newt limb, retinoic acid is present in a concentration gradient, with the highest concentration in the part of the limb farthest from the body.

Like a steroid, retinoic acid apparently directly influences gene expression by interacting with a transcription factor. Recall from Chapters 11 and 37 that steroid hormones such as progesterone and testosterone bind to receptor proteins inside a cell. The steroid-receptor complex regulates the transcription of specific genes. Similar receptors bind to retinoic acid. Which genes are expressed, however, differs from cell to cell. The cells' response depends on already established differences among the cells.

Researchers have identified a number of genes responsible for pattern formation in Drosophila. The first of those identified were two similar genes that—when mutated—led to a couple of bizarre flies called bithorax, with two sets of identical wings rather than two different sets, and antennapedia, in which legs grew where antennae should have (Figure 45-15). Later studies revealed that the proteins encoded by the genes *bithorax* and *Antennapedia* act as transcription

A.

B.

C.

Figure 45-15
Genes control pattern formation. A. A normal fruit fly, *Drosophila melanogaster.* B. A fly with the *Antennapedia* allele has legs where its antennae should be. C. A fly with the *bithorax* allele has two identical pairs of wings instead of two unlike pairs.

factors. *Bithorax* and *Antennapedia* are two of a family of **homeotic selector genes,** genes whose expression affects overall body plan.

Pattern formation seems to result from the action of morphogens such as retinoic acid. Retinoic acid influences the transcription of genes and therefore the differentiation of cells.

positional information that had already evolved before the Cambrian explosion. If flies and mice share these genes, then probably all arthropods and vertebrates and many other groups share them.

Animals as different as mice and fruit flies use the same gene products to establish the position of the parts of the body.

Mammals and Flies Use Many of the Same Transcription Factors to Establish Patterns During Development

The discovery of homeotic genes in Drosophila immediately led researchers to ask whether mammals had similar genes. Mammalian counterparts, called *Hox* genes, were easy to find, for their sequences are similar to those in Drosophila. In fact, the *Hox* genes of flies and mice are even arranged on their chromosomes in the same order. And the order of genes on the chromosome matches the pattern of action within the embryo. Genes that affect head development lie at one end of the cluster, while those that affect tail development lie at the other. How the order of genes on chromosomes affects their transcriptional regulation is now under intense investigation.

The most recent common ancestor of Drosophila and mice lived some 600 million years ago. We can conclude that during the last 600 million years flies and mice have used—with very few modifications—a system of specifying

45.6 Postembryonic Development
Metamorphosis Converts a Larva into an Adult

The first stage in which an animal has an independent life may not at all resemble the sexually mature adult. A frog embryo, for example, develops into a **larva** called an aquatic tadpole. Like a fish, a tadpole propels itself with a large-finned tail and absorbs oxygen from water through gills. After a period of growth, the tadpole **metamorphoses** into an adult frog, breathes air with lungs, and jumps on four legs. The frog also switches from a vegetarian diet to a carnivorous diet of flies and other invertebrates (Figure 24-14).

Metamorphosis is more common among invertebrates than vertebrates. The changing of a caterpillar into a butterfly is one of the most spectacular examples of metamorphosis. Many insects—including the beautiful butterflies, the more somber moths, and all the flies, beetles, and wasps—undergo **complete metamorphosis,** meaning that none of the tissues of the adult come from the larva (Figure 45-16).

Figure 45-16
Metamorphosis in insects. Complete metamorphosis in the monarch butterfly. A caterpillar feeds on milkweed, then forms a pupa. Inside, the pupa metamorphoses into a butterfly. (Photos, left to right, 1 and 3, Lior Rubin/Peter Arnold, Inc.; 2, Ed Reschke/Peter Arnold, Inc.; 4, Don Riepe/Peter Arnold, Inc.)

Biology ⊜ Now™ Learn more about the metamorphosis of butterflies by clicking on this figure on your BiologyNow CD-ROM.

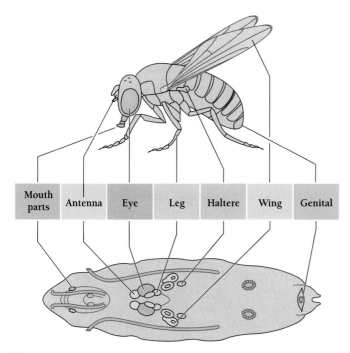

Figure 45-17
Imaginal discs. Imaginal discs in Drosophila and other insects are larval regions bound to develop into specific adult tissues.

Instead, the adult develops from groups of specialized larval cells called **imaginal discs** (Figure 45-17).

In insects that undergo complete metamorphosis, the larval stages are specialized for feeding and growth, while the adults are specialized for dispersal and mating. For example, butterflies may spend several weeks as caterpillars feeding on leaves and growing. Then the caterpillars (larvae) form an inactive phase, called a **pupa.** During the pupal stage, which can last a few days or a few months, the tissues of the caterpillar break down, and the cells of the imaginal discs form the tissues and organs of the adult. Finally, the adult butterfly inside breaks out and emerges as a sexually mature adult. The adult butterfly, unlike the caterpillar, has a strawlike proboscis for sucking nectar from flowers, wings for finding flowers and mates, and an avid interest in butterflies of the opposite sex.

Not all metamorphosis is complete. Grasshoppers, cockroaches, and bed bugs, for example, undergo **gradual metamorphosis,** in which the embryo hatches into an immature form, called a **nymph,** which is more or less a miniature adult. The nymph then undergoes a series of molts, each time getting bigger and gradually looking more like the adult (Figure 23-25).

Do humans and other mammals undergo metamorphosis? Some people consider puberty a variation on the theme of gradual metamorphosis. Before adolescence, mammals are dependent, sexually immature creatures, with size and body proportions different from those of adults. After a period of gradual change in proportions, adolescents undergo a burst of growth and develop characteristics that specifi-

cally facilitate reproduction. All these changes, like those that occur during metamorphosis in insects and amphibians, are controlled by hormones.

Many amphibians and insects undergo metamorphosis—which can be complete or gradual.

45.7 And So We Say Goodbye

In this final chapter of *Asking About Life,* we have surveyed the main events of developmental biology. We have seen, for example, that although sea urchins, amphibians, and mammals may gastrulate differently, the result is the same—three germ layers that form skin, internal organs, and gut. We have also seen that the cells of early embryos are totipotent: each cell contains all the genetic information needed to construct a fully formed adult organism. The cloning of Dolly the sheep, along with other research, suggests that the nuclei of differentiated adult cells are also totipotent.

Does this mean that each cell in our bodies contains everything needed to make a new person? Could we scrape a cell from a woman's cheek and recreate her? The answer is no, for each of our cells and each of our nuclei is the product not only of the genes it carries and expresses but also of the environment in which it lives and develops.

The cytoplasmic environment of a nucleus influences which genes are expressed and, therefore, the phenotype of the cell. For example, biologists can now induce a nucleus from a differentiated cell, such as an udder cell from Dolly the sheep, to "dedifferentiate" and direct development, but *only* in the context of an egg cell. The cytoplasm of the original udder cell probably cannot sponsor embryological development.

In a similar way, we know the environment of a cell influences its fate. Ectoderm in contact with the notochord becomes nervous tissue, while ectoderm in contact with the optic vesicle of the developing eye becomes lens tissue.

We can take another step back and see that the environment of a whole organism partly determines its phenotype as well. Wet soil produces a plant with a different phenotype than dry soil; a calorie-rich, well-balanced diet produces an animal with a different phenotype than a diet deficient in calories and specific nutrients; and a tadpole reared among predacious dragonfly larvae develops a different phenotype than a tadpole reared without predators.

Tongue in cheek, the British developmental and molecular biologist Sidney Brenner distinguished between two broad classes of developmental processes, those that occur by the "European Plan" and those that depend on the "American Plan." In the European Plan, Brenner argued, cells "do what their parents say," whereas in the American Plan, they "do what their neighbors say." Both plans operate—development indeed requires genetic programs but it also depends on environment.

This lesson is a fitting place to end this book, since this conclusion leads to a slew of further questions: How does a set of genes specify a process that unfolds in time? Which features of the environment are important in affecting a particular process? How does a particular environmental trigger stimulate or inhibit the expression of the contributing genes? The answer to each question opens the door to still more questions.

Asking questions and seeking answers is one of the foremost ways of engaging with life. An unanswered question, like an unclimbed mountain or an unattained trophy, is always beckoning, calling us to do our best. We hope that our book has provoked you to ask your own questions about life.

Key Concepts

- During development, an animal changes from a single cell to a complex organized multicelled organism by means of cell division, cell movement, cell specialization, and pattern formation.
- At the beginning of development, each cell contains all the genetic information necessary to produce an entire animal. As development proceeds, however, most cells become increasingly limited to a particular developmental pathway.
- The pattern of gene expression of individual cells and their participation in the formation of organs and body parts depend on chemical signals, many of which act by regulating transcription of specific genes.
- Virtually all animals appear to use the same mechanisms to achieve pattern formation and programmed cell death.

Summary with Key Terms

Can differentiated cells be totipotent?

Over the years, embryologists have wondered whether each cell of a developing embryo is **totipotent,** that is, capable of developing into any type of cell, or if each cell is fully **differentiated,** meaning it has a predetermined fate. The **fate** of a cell is the kind of cell it will become during normal development. The **potency** of a cell is all the kinds of cell that it could become in a different cellular environment. Cells whose fate cannot be changed are said to be **determined.** The cells of early embryos and adult stem cells tend to be totipotent.

How does a zygote become an adult?

Animal development begins with a single-celled zygote that appears simple and homogeneous. All vertebrates pass through the same stages: **gamete formation, fertilization, cleavage, gastrulation, organ formation, growth, metamorphosis** (in only some species), and **aging.**

Embryonic life begins with fertilization, after which cleavage divisions lead to the formation of a hollow ball of cells. In amphibians, cleavage results in an asymmetrical ball of cells called a **blastula.** Mammalian cleavage results in a solid ball of cells that develops into a hollow sphere called a **blastocyst.** In mammals, the blastocyst includes a **trophoblast,** which forms the tissues of the **placenta,** and an **inner cell mass,** which forms the embryo itself. The early development of a zygote is influenced by proteins translated from **maternal mRNA** transcribed during oogenesis from the genes of the mother.

During **gastrulation,** the hollow blastula invaginates to form a **gastrula,** with three germ layers—**ectoderm, mesoderm,** and **endoderm**—and a hollow space called the **archenteron,** which is the primitive gut. The ectoderm forms the epidermis, the nervous system, and the sense organs, while the endoderm gives rise to the gastrointestinal tract and associated organs. The mesoderm forms the bones and muscles, as well as organs such as the heart and kidneys.

In amphibians and sea urchins, gastrulation begins when an indentation, called the **blastopore,** forms on the side of the blastula. Mammals gastrulate when the inner cell mass separates into two layers, the **hypoblast** and the **epiblast.** Together, the hypoblast and epiblast form a blastodisc, which then invaginates along the **primitive streak** to form the three germ layers—ectoderm, mesoderm, and endoderm.

In vertebrates, part of the mesoderm forms the **notochord,** a supportive cord that runs from head to tail that also serves as an organizer for further embryonic development. In particular, the notochord induces **neurulation.** The notochord induces the formation of **neural plate** tissue, which folds into a hollow **neural tube.** The neural tube bends to form the different regions of the brain and the spinal cord. **Neural crest cells** from the surface of the neural tube migrate into distant parts of the embryo to form nerve, skin, and other tissues.

The optic vesicles of the neural tube induce the overlying ectoderm to form the lens and cornea of the eye. Organ formation of this sort depends both on differentiation, the acquisition of specific proteins and subcellular structures, and on **morphogenesis,** the creation of form.

How does a human embryo develop?

In nine-week-old human embryos, all of the major organs appear as recognizable, although nonfunctional, **rudiments.** The embryo is then called a **fetus.** Most **miscarriages,** or **spontaneous abortions,** occur in the first **trimester.** **Teratogens,** such as drugs and viruses, can cause malformations of the developing embryo. Programmed cell death, or **apoptosis,** contributes to normal development of the hands and feet and many other organs.

How do molecular signals direct cell differentiation?

In an embryo, cells differentiate according to their position. Transplantation experiments have shown that, in amphibians, tissue from the dorsal lip of the blastopore, called the **primary organizer**, can establish the overall orientation of the embryo. Chemical signaling between different parts of the embryo provides positional information to individual cells, leading to the formation of harmonious patterns of organs and body parts. Substances whose concentration provides positional information are called **morphogens**. For example the morphogen retinoic acid is a transcription factor that influences the transcription of genes.

Studies of mutants that affect the early development of Drosophila have revealed that the establishment of pattern depends on the production and distinctive distribution of a cascade of transcription factors. The genes for these transcription factors are called **homeotic selector genes.** Many of the proteins responsible for pattern formation in Drosophila are virtually identical to mammalian proteins, whose genes are called *Hox* genes.

Many animals **metamorphose** from a **larval** form to an adult form. In **complete metamorphosis**, tadpoles turn into frogs or salamanders, caterpillars develop into **pupae** and then into butterflies or moths, and maggots turn into flies. In flies and many other insects, the adult form derives from **imaginal discs** in the larvae. In **gradual metamorphosis**, a succession of **nymphs** gradually change from a larval form to an adult form.

Review and Thought Questions

Review Questions

1. What are the eight stages of development in a vertebrate?
2. How did embryologists disprove preformation?
3. What does it mean for a cell to be totipotent? What does it mean for a cell to be determined?
4. Compare cell fate with cell potency.
5. Why can early embryos develop for a while even without any nucleus at all?
6. How does the zygote's dividing into many smaller cells (cleavage) help the embryo?
7. What happens during gastrulation?
8. What happens during neurulation?

9. What is apoptosis?
10. How long after fertilization does a human embryo implant in the wall of the uterus of the mother?
11. What is the difference between a fetus and an embryo? How long after fertilization does a human embryo become a fetus?
12. At approximately how many weeks after fertilization can a human fetus survive outside the uterus?
13. What happens during nuclear transplantation? How is it different from cloning a plant cell?
14. What happens to the genes when a cell differentiates?

Thought Questions

15. Why is it hard to clone humans and other mammals? If you could clone someone, whom would you choose to clone? Would you expect them to be completely identical in every respect? What would you hope to accomplish by this?
16. Experiments in invertebrates suggest that the effects of maternal mRNA and other factors in the egg can influence phenotype for several generations. Design an experiment that would distinguish such maternal effects from those of the genotype. You may use any organism you like.

BiologyNow Resources

Biology ⑧ Now™

Active Figures
45-13: The cloning of Dolly
45-16: The metamorphosis of butterflies

Preparing for an exam? Take a diagnostic test on your BiologyNow CD-ROM.

Online materials relating to this chapter are at:
http://biology.brookscole.com/AAL3

About the Chapter-Opening Image
If every individual were already preformed within an egg or sperm then every egg or sperm would carry within it all of its descendants.

Glossary

3′ end The end of a polynucleotide at which the 3′ carbon of the nucleotide is not attached to another nucleotide but to a phosphate or a hydroxyl group.

5′ end The end of a polynucleotide at which the 5′ carbon of the nucleotide is not attached to another nucleotide but to a phosphate or a hydroxyl group.

α (alpha) carbon In an amino acid, the carbon atom to which the carboxyl group and the amino group are both attached.

α (alpha) helix A common secondary structure in proteins in which every carbonyl group of the polypeptide backbone is hydrogen bonded to the amino group four amino acids farther down the polypeptide chain.

abdomen In a mammal, the part of the body between the chest (thorax) and the pelvis; the rear portion of an arthropod.

abdominal cavity The part of the coelom that encloses most of the intestinal tract, as well as the liver, pancreas, and kidneys.

abiotic Nonliving; without life.

abscisic acid A plant hormone involved in dormancy in buds and seeds and the closing of the stomata.

absorb To take up into cells and into the circulation.

absorption spectrum The relative amounts of light of different wavelengths that a substance or solution absorbs.

absorptive state The state of an animal shortly after a meal, during which cells in many organs take up glucose and other small molecules to make glycogen, fats, and proteins.

abyssal zone Ocean waters deeper than 1,000 meters.

acetyl CoA A compound that consists of the two-carbon acetic acid (acetate) linked to a larger molecule called coenzyme A; acetyl CoA is the end point of glycolysis and a starting compound for the citric acid cycle.

acetylcholine A neurotransmitter used in the neuromuscular junction to stimulate muscle contraction; acetylcholine is also used by parasympathetic nerves to slow the heart.

acid A molecule (or part of a molecule) that can give up a hydrogen ion.

acoelomates The simplest animals, including the jellyfish and flatworms, that have no coelomic cavity.

acquired immunodeficiency syndrome (AIDS) A disease characterized by a deficiency of certain helper T cells; AIDS patients cannot mount effective immune responses to infections or to certain kinds of cancers.

acrosome A specialized lysosome in a spermatozoan; it contains enzymes that allow the sperm to enter the egg.

actin The protein that makes up actin filaments in nonmuscle cells and the thin filaments in muscle cells.

actin filaments The most flexible elements of the cytoskeleton; formed by the association of actin molecules into long filaments about 7 nm in diameter.

action potential (nerve impulse) A rapid, transient, and self-propagating change of voltage across the membrane of a neuron or a muscle cell; action potentials allow long-distance signaling in the nervous system.

action spectrum The relative effectiveness of different wavelengths in promoting a specific light-dependent process.

activation energy The minimum energy needed to initiate a chemical reaction or other process.

active site A groove or cleft on an enzyme's surface to which a substrate binds.

active transport Movement of a substance across a membrane in a manner that does not occur spontaneously but requires an expenditure of energy.

adaptation Any trait acquired through natural selection that helps an organism survive and reproduce in its environment.

adaptive radiation The multiplication of a single species into many species, each with a separate ecological niche.

adenosine diphosphate (ADP) A nucleotide that consists of adenosine, ribose, and two phosphate groups; ADP is the product of the hydrolysis of a single phosphate from ATP.

adenosine triphosphate (ATP) A nucleotide that consists of adenosine, ribose, and three phosphate groups; ATP is the universal energy currency, providing energy for many biochemical processes in all organisms.

adhering junction A molecular assembly on the surface of an animal cell that connects the cell surface to actin filaments on the inner surface of the plasma membrane.

adhesiveness The tendency to cling to a surface.

adrenal glands Endocrine glands that lie on top of the kidneys.

aerobic Requiring oxygen, usually for aerobic respiration.

afferent Leading toward, as blood vessels entering an organ or nerve fibers carrying sensory information from peripheral nerves or sense organs to the central nervous system.

age structure In a population, the fraction of individuals of various ages (or cohorts).

aging Progressive developmental changes in an adult organism.

agonistic behavior All the aspects of competitive behavior within a species, including aggression, aggressive displays, appeasement, and retreat.

alchemy An ancient study, one of whose aims was to turn "base" metals, such as lead or copper, into silver or gold.

aldosterone A hormone, made by the adrenal glands, that stimulates sodium and potassium reabsorption in the distal tubules and collecting ducts of the kidneys.

algae Plantlike protists that perform photosynthesis using chlorophyll *a*.

alkaloid A nitrogen-containing secondary plant compound.

allantois One of the membranes that surrounds an amniotic embryo; the allantois functions in both respiration and excretion.

allele One of several variant versions of the same gene.

allergy An inflammatory response to a harmless antigen; allergies involve the production of IgE antibodies and the release of histamine from mast cells.

allopatric speciation Species formation by geographical isolation.

altered self In the immune response, nonself antigens presented in the context of an MHC molecule. These occur on the surface of a body cell when the cell is infected by a virus, bacterium, or other microorganism.

alternation of generations A sexual life cycle in which haploid and diploid phases alternate.

altruism Behavior that benefits others at the expense of the animal that performs the behavior.

alveoli [singular, **alveolus**] The expanded surfaces at the ends of the smallest passageways of the lungs; they provide the huge surface area for the exchange of oxygen and carbon dioxide.

Alzheimer's disease A disease, associated with aging, in which patients suffer massive memory loss and disorientation.

amino acid A small molecule that contains both amino and carboxyl groups; the building block of polypeptides and proteins.

amino group A functional group that consists of one nitrogen and two hydrogen atoms (NH_2).

ammonia A small molecule that consists of one nitrogen and three hydrogen atoms; it is secreted as a nitrogen-containing waste by many cells and organisms.

amniocentesis Sampling of cells—usually with a syringe—in the amniotic fluid.

amnion One of the membranes that surrounds a reptilian, avian, or mammalian embryo.

amniote A vertebrate that makes an amniotic egg, one with four characteristic membranes including an amnion; includes reptiles, birds, and mammals.

amniotic egg An egg in which the embryo is surrounded by an amnion within a porous shell that allows the exchange of gases with the surrounding air.

amniotic fluid The fluid that surrounds the fetus.

amphibians Class of four-legged vertebrates, such as frogs and salamanders, that usually reproduce in water.

amphipathic Having both a hydrophilic and hydrophobic region.

amylase An enzyme found in saliva and pancreatic juice that hydrolyzes the polysaccharides of food into shorter fragments of maltose and glucose.

amyloplast In a plant, a plastid that contains large granules of starches.

anabolism The synthesis of complex molecules from smaller ones.

anaerobic Without oxygen.

anaerobic photosynthetic bacteria Photoautotrophic bacteria that live without oxygen.

anaphase The stage of mitosis during which sister chromatids separate.

analogy An older term for similar biological structures that have evolved separately through convergent evolution. See homoplasy.

aneuploid Having an abnormal number of chromosomes.

aneurism Abnormal ballooning of the wall of a blood vessel.

angiosperms The flowering plants.

angiotensin A hormone that stimulates the constriction of blood vessels.

Animalia The kingdom of animals—multicellular, heterotrophic organisms that undergo embryonic development.

annual rate of increase The proportion by which a population increases each year.

annuals Plants that complete their life cycles, from seed to mature plant, in a single growing season and then die.

antagonist A molecule that prevents a chemical signal from binding to its receptor and triggering its characteristic response in target cells.

antagonistic pair Muscles that pull in opposite directions.

antenna complex In a chloroplast, an association of chlorophyll and carotenoids that traps light and transfers the energy to the chlorophyll molecules that actually participate in photosynthesis.

anterior Toward the front of an animal.

anterior pituitary gland An endocrine organ that makes and releases a number of peptide hormones.

anther A thick pollen-bearing structure that is part of the stamen.

anthocyanins The phenolic pigments responsible for most of the red, pink, purple, and blue colors of flowers, fruits, leaves, and other plant parts.

Anthophyta Angiosperms; flowering plants.

Anthozoa Sea anemones, corals, and related species.

anthropoid Of the suborder Anthropoidea, which includes monkeys, gibbons, and great apes (including humans).

antibiotic A substance that kills (or interferes with the growth of) microorganisms.

antibody A blood protein that forms complexes with molecules (antigens), such as those on the surfaces of microorganisms.

anticodon A sequence of three nucleotides in transfer RNA that forms specific base pairs with the corresponding codon sequence in mRNA.

antidiuretic hormone (ADH or vasopressin) A hormone, released by the posterior pituitary gland, that prevents water loss in the kidneys and constricts blood vessels.

antigen (an *anti*body *gen*erator) A molecule that stimulates the production of an antibody or of another immune response; a molecule that binds to an antibody or other recognition molecule within an animal's immune system.

antioxidant Compound such as vitamin E, vitamin C, or β-carotene that prevents the damaging effects of oxidation.

anus The opening at the far end of the digestive tract through which the undigested and unabsorbed material leaves.

aorta The artery that carries blood from the left ventricle to the rest of the body.

apical dominance In plant development, the suppression of lateral bud development by terminal buds.

apical meristem In a plant, a self-renewing group of undifferentiated cells, just behind the tip of the shoot or root, that generates differentiated structures.

apoptosis The organized death of specific cells, often as a normal part of development (programmed cell death).

aqueous Watery; dissolved in water.

Archaea (Archaebacteria) A domain of single-celled organisms, separate from the Eubacteria, that often live under extreme environmental conditions.

archegonium In ferns and mosses, the female reproductive organ that produces and houses the ova.

archenteron The "primitive gut," the innermost tube of an animal embryo; it is lined with endoderm and will become the digestive tract.

aromatic Having a benzene-like planar ring of atoms.

artery One of the vessels that carry blood away from the heart.

arterioles Smaller vessels that branch off from arteries.

artificial selection The process by which humans select the most desirable individuals for breeding to make new types of plants and animals.

asexual reproduction A process that produces offspring with genes from a single parent.

assortative mating Nonrandom mating.

aster A starlike object visible in most dividing eukaryotic cells (other than those of vascular plants); the aster contains the microtubule organizing center.

atherosclerosis Hardening of the arteries.

atom The smallest unit of matter that still has the properties of an element.

ATPase An enzyme that breaks ATP into ADP and phosphate.

ATP synthase A molecule embedded in the inner membranes of chloroplasts and mitochondria that makes the energy molecule ATP.

atrial natriuretic factor (ANF) A hormone produced in the atria of the heart that increases water loss and lowers blood pressure.

atrioventricular (AV) node A special tissue that conducts electrical signals from one part of the heart to the other.

atrioventricular valves The valves between the atria and ventricles of the heart.

atrium One of the smaller thin-walled entrance chambers of the heart, through which blood enters the adjacent ventricle.

australopithecine An early hominid.

autonomic nervous system Involuntary nervous system; it coordinates the responses of smooth muscles, cardiac muscles, as well as organs of the endocrine, digestive, excretory, respiratory, and cardiovascular systems.

autonomous replication Propagation of a virus or plasmid in which the replicating DNA remains separate from the DNA of the host cell.

autosomes The chromosomes that do not differ between males and females.

autotroph An organism that can make its own organic molecules from simple inorganic compounds (such as carbon dioxide, water, and ammonia).

auxin A plant hormone that affects the growth and development of plants, including phototropism, stem elongation, apical dominance.

axillary bud (lateral bud) A bud that forms just above the point where a leaf joins the stem and which can develop into a branch.

axon A type of neuronal process that usually carries signals away from the cell body and connects to other cells.

β (beta) structure A common secondary structure in proteins in which the amino and carbonyl groups of the polypeptide chain are hydrogen bonded to other polypeptide chains or to distant regions of the same chain folded back on itself.

B lymphocyte Lymphocytes that when stimulated divide to form antibody-producing cells or memory cells.

bacilli [singular, **bacillus**] Rod-shaped prokaryotic cells.

backbone A column of hollow bony segments called vertebrae.

bacteriophage (or **phage**) A virus that infects bacteria.

balanced polymorphism A balance of different alleles in a population.

basal metabolic rate The minimum amount of energy required just to stay alive and awake.

base A molecule (or part of a molecule) that can accept a hydrogen ion.

base substitution A type of point mutation in which one base (nucleotide) is replaced by another.

basidiocarp The fruiting body of a mushroom.

basidiomycetes (or **club fungi**) Include many of the most familiar fungi: the mushrooms, the bracket fungi, and the puffballs.

basidium [plural, **basidia**] A tiny, club-shaped structure that produces spores.

Batesian mimicry A deception in which one species mimics another's warning coloration for protection.

behavioral ecology (formerly called **sociobiology**) The study of behavior from an evolutionary perspective.

behavioral isolation Reproductive isolation that results from different mating behaviors.

benthic division All the organisms that live on the ocean bottom.

biceps The major muscle on the front side of the upper arm.

bilateral symmetry Structural symmetry such that one cut through the axis will produce identical (or mirrored) halves.

bile A detergent solution, secreted by the liver; bitter, alkaline, and an ugly green-yellow or brown-yellow color.

binary fission The process of cell division (in prokaryotes) in which a cell pinches in two, distributing its materials and molecular machinery more or less evenly to the two daughter cells.

binding site The region of a protein molecule to which a substrate or a chemical signal binds.

binomial A two-part name that includes both the genus and species.

bioassay A method that estimates the concentration of a substance by measuring its biological activity.

biochemistry The study of the structures and reactions that actually occur in living organisms.

biogeochemical cycle The movement of a substance, such as carbon or nitrogen, through many forms—both biotic and abiotic.

biogeography The study of the past and present distribution of plant and animal species.

biological species The largest unit of a population of similar organisms in which gene flow is possible.

biological species concept A definition of species that says species are groups of actually or potentially interbreeding populations, which are reproductively isolated from other such groups.

biology The science of life.

biomagnification Toxic chemicals are concentrated at higher levels in the food chain.

biomass Aggregate dry weight of all organisms in a community or ecosystem.

biome A geographical region with a distinctive landscape, climate, and community of plants and animals. See ecosystem.

biosphere The system of living things that covers the Earth.

biotic Living.

bipedal Consistently walking on two rather than four feet.

birth control The conscious regulation of reproduction.

birth control pill A formulation of estrogen and progesterone that prevents the anterior pituitary from secreting LH, thereby stopping ovulation and preventing conception.

birth rate The annual number of births per 1,000 individuals in a population.

bivalves Includes clams, oysters, scallops, mussels, and other mollusks with two shells—a right shell and a left shell.

bladder A balloonlike organ that stores urine.

blade The thin, flat part of a leaf.

blastocoel In an animal embryo, the interior of the blastula.

blastocyst A modified blastula in which the cells do not lie within a single layer, but do enclose an internal cavity.

blastopore An opening at the end of the archenteron.

blastula A stage of an animal embryo that consists of a sphere with cells on the surface and fluid inside.

blending inheritance The pattern of inheritance in which offspring appear to have characteristics intermediate between those of their parents.

bolus In digestion, a ball of macerated food and lubricating mucus.

bone The hard connective tissue of a vertebrate; bone consists of fibers of collagen and crystals of calcium phosphate.

book lung Stacks of modified gills that provide a surface of gas exchange in some animals, including spiders.

bordered pit pair A region of the border between two plant cells in the xylem where the adjacent secondary wall overarches the pit membrane and reinforces the wall of a tracheid.

bottleneck The reduction of a large population to a few individuals.

bottleneck effect The restriction of genetic diversity of a population that results from the reduction of a large population to a few surviving individuals by a random disaster or harsh selection pressure.

Bowman's capsule In a kidney, the bulb that surrounds each glomerulus.

BPG (2,3-bisphosphoglycerate) A molecule in mammalian red blood cells that binds to hemoglobin, decreasing its oxygen affinity and allowing it to release more oxygen in the tissues.

branchial arches In fishes and amphibians, the tissues between the gill slits.

breathing center A nerve complex on which the coordination of breathing depends.

bronchi In the mammalian respiratory system, the two tubes that branch from the tracheae and carry air to and from the lungs.

bronchioles In the respiratory system, the smallest branches of the bronchi.

bryophytes The mosses and their relatives.

bud In a plant, the precursor of a leaf or a flower, consisting of several leaf primordia, separated by internodes whose cells have not elongated.

buffer A molecule that easily converts between acidic and basic forms by donating or accepting one or more hydrogen ions.

bulbourethral glands The glands that inject a small amount of mucus into the semen.

bulk flow The movement of masses of material through a living body. Examples include circulating blood and the movement of food through the digestive tract.

calcitriol A form of vitamin D; a steroid hormone that increases available calcium in the blood (from the gut and the bones) and stimulates formation of the bone-forming osteoclasts; also affects immune and skin cells.

callus Undifferentiated tissue that forms at the cut surface of a plant.

calorie A measure of energy; the amount of energy needed to raise the temperature of one gram of water 1°C. One thousand calories is a kilocalorie, or Calorie, a measure used in calculating the energy available in food.

calyx The fourth and outermost whorl of leaflike parts called sepals.

Cambrian explosion The burst of diversification—unmatched in the history of the world—in which all of the modern animal phyla that have fossilizable skeletons appeared.

camouflage A defense against being eaten in which organisms mimic materials in their environment in an attempt to be invisible.

canopy In a tropical rain forest, the continuous layer of trees 30 to 40 meters above the ground.

capillary One of the minute blood vessels that brings blood in closest contact with tissues; capillaries connect the finest arterioles with the finest venules.

capsid A coat of protein on the outside surface of a virus.

capsule A gelatinous layer of polysaccharides and proteins outside the cell wall; in disease-causing bacteria, the capsule protects from attack by the host's white blood cells, enabling the bacteria to overwhelm the body's immune system.

carbohydrate A compound that contains the equivalent of one water molecule (one oxygen atom and two hydrogen atoms) for every carbon atom; includes sugars and polysaccharides.

carbonyl A functional group that consists of a carbon atom attached to an oxygen atom by a double bond (C=O).

carboxyl A functional group that contains one carbon atom and two oxygen atoms; the carbon atom forms a double bond with one oxygen atom and a single bond with the other.

carcinogen A chemical that causes cancer.

cardiac muscle The muscles that pump blood through the heart.

cardiovascular system The blood, the heart, and the blood vessels together.

carnivore An animal that eats only other animals.

carotenoid A plant pigment, yellow or orange in color.

carpel In a flower, a female reproductive structure.

carrying capacity The plateau value of the S-shaped curve of population growth. The maximum number of individuals of a species that a habitat can support.

cartilage A type of connective tissue that cushions joints; cartilage consists of collagen, but without calcium phosphate.

Casparian strip In plants, a waxy wall that extends around the walls of endodermal root cells; it restricts the movement of water and dissolved solutes.

catabolism The reactions of metabolism that break down complex molecules, such as those in food.

catabolite activator protein (CAP) A positive regulator of transcription in *E. coli;* CAP binds to the promoter region of the *lac* operon and increases transcription whenever cyclic AMP is present at sufficiently high levels.

catalyst A substance that accelerates a chemical reaction and is not itself consumed in that reaction.

catarrhines Monkeys that have downward pointing nostrils and often no tail.

catastrophism The view that the discontinuities in the fossil record result from a series of catastrophes; some catastrophists argued that life arose anew after each catastrophe, rather than deriving from previously existing species.

cation A positively charged ion.

cecum A blind sac of the digestive tract, near the junction between the large and small intestine.

cell body The center of a neuron, including its nucleus, as distinct from its axon or dendrites.

cell cycle The orderly sequence of events that accomplish cell reproduction.

cell division The process by which a parent cell gives rise to two daughter cells that carry the same genetic information as the parent cell.

cell plate In a dividing plant cell, the precursor of a new cell wall between two daughter cells.

cell theory The summary of the cellular basis of organisms: (1) All organisms are composed of one or more cells; (2) cells, themselves alive, are the basic living unit of organization of all organisms; and (3) all cells come from other cells.

cell wall An external rigid structure surrounding all plant cells and most prokaryotes.

cellulose A polysaccharide made by plants; the major structural material of wood, cotton, and paper.

central cell A cell within the ovule of a flower; it contains two haploid nuclei, derived from the megaspore, and develops into endosperm after fertilization.

central dogma The frequently violated principle that "DNA specifies RNA, which specifies proteins."

central nervous system (CNS) The brain and spinal cord.

central vacuole A large membrane-enclosed space within a plant cell.

centrifugation A method for separating macromolecules and subcellular structures according to size (and shape) by subjecting them to high gravitational fields in a spinning tube.

centriole A pair of small cylindrical structures, each about 0.2 μm in diameter and 0.4 μm long, that lie at right angles to one another; centrioles are present at each pole of the mitotic spindle in animal cells and in some other eukaryotes.

centromere The point at which the two chromatids of a single chromosome are joined.

centrosome A microtubule organizing center that (in many cells) also contains a distinctive organelle called the centriole; the centrosome plays an important role in cell division.

cephalization The concentration of sense organs and ganglia in the anterior end of an animal.

cephalochordates (lancelets) About 45 species of small, segmented, fishlike animals. Most belong to the single genus *Branchiostoma*, whose common name is amphioxus.

cervical cap A birth control device; a thimble-shaped rubber or plastic cap, about an inch in diameter, that fits tightly over the cervix and prevents fertilization.

cervix The narrower, lower part of the uterus.

Cestoda Tapeworms.

channel A membrane protein that allows the passage of specific molecules or ions.

chaparral The dense growth of shrubs that is characteristic of the American Southwest and other regions, all on the west coasts of continents, that share a "Mediterranean" climate, with mild, wet winters and hot, dry summers.

chaperone One of a number of special proteins that prevent promiscuous interactions by binding to unfolded polypeptides and catalyzing correct folding.

character displacement In evolution, a change in morphology, life history, or behavior that results from competition.

chelicerae The first pair of appendages in spiders and other arachnids; they may serve as pincers or as fangs that are associated with poison glands.

chemical equilibrium A balance between forward and reverse chemical reactions.

chemical reaction A transformation in which different forms of matter combine or break down.

chemical synapse A distinct boundary between two neurons through which neurotransmitters diffuse.

chemiosmosis The linking of chemical and transport processes.

chemistry The study of the properties and the transformations of matter.

chemoautotroph An autotroph that derives energy by oxidizing such inorganic substances as hydrogen sulfide (H_2S) or ammonia (NH_3).

chemoheterotroph A heterotroph that uses no light, relying exclusively on organic molecules for both energy and carbon atoms; in contrast to a photoheterotroph.

chemotaxis The movement of a cell toward a higher (or, in some cases, a lower) concentration of a particular chemical.

chiasma [plural, **chiasmata**] The sites of exchange of DNA between homologous chromosomes during meiosis; chiasma are visible during prophase of meiosis I.

chitin A tough polysaccharide in the cell walls of fungi and the exoskeletons of arthropods.

chlorofluorocarbon (CFC) A gas used as a refrigerant and aerosol propellant known to interfere with the formation of an ozone layer in the stratosphere.

chlorophyll The green pigment of plants and protists; chlorophyll is responsible for absorbing light for photosynthesis.

chlorophytes Green algae.

chloroplast A large, green, membrane-enclosed organelle that performs photosynthesis.

cholesterol A small molecule that consists of four interconnected rings of carbon atoms; a

component of cell membranes and the starting compound in the synthesis of steroids.

Chondrichthyes Sharks and rays—fish with cartilaginous skeletons.

Chordata All vertebrate animals—fish, amphibians, reptiles, birds, and mammals—and animals that have notochords, including sea squirts and lancelets.

chorion One of the membranes that surrounds an embryo; it lies just beneath the shell within an amniotic egg.

chorionic villus sampling A procedure for sampling cells of the early embryo that uses fetal cells present in the placenta.

chromatid One of the two separate but connected bodies that make up a chromosome at the beginning of mitosis, when the chromosomes first become visible; a chromatid contains a single long molecule of DNA.

chromatography A method for separating molecules according to their relative affinities for a stationary support, called the stationary phase, and a moving solution, called the mobile phase.

chromoplast A plastid that contains the pigments of fruits and flowers.

chromosomal mutation A change in a relatively large region of a chromosome.

chromosome A discrete complex of DNA and proteins, visible with a light microscope within a dividing cell; originally used only for eukaryotic cells, but now also used to mean a single large molecule of DNA that contains the genes of a bacterium or a virus.

chromosome banding The distinctive pattern, visible in a light microscope, that results from the selective binding of certain dyes to individual chromosomes.

chrysophytes Golden algae and diatoms.

chyme A creamy, acidic liquid that passes into the small intestine.

cilium [plural, **cilia**] A protein assembly, consisting of microtubules, that can move a cell through a liquid medium (or a liquid medium over a cellular surface); a single cell usually contains many cilia, often arranged in rows; cilia have the same organizational plan as eukaryotic flagella but cilia are much shorter.

circulation The route of the blood throughout the cardiovascular system.

circumcision The surgical removal of foreskin.

citric acid cycle (Krebs cycle) A set of reactions in the mitochondria that help generate the high-energy molecule ATP from the breakdown of glucose sugar; converts the carbon atoms of acetyl CoA into carbon dioxide.

cladistics An approach to taxonomy whose first goal is to describe the groupings of organisms in a way that shows their phylogeny.

cladogram A family tree based on shared derived characters. The cladogram itself, which represents the relations among species, is the basis for classifying organisms, not the characters.

class A taxonomic group that consists of one or more orders; a subdivision of a phylum (or, for plants, of a division).

clay A soil that consists only of small particles (less than 2 mm).

cleavage A series of rapid cell divisions following fertilization in early animal embryos; cleavage divides the embryo without increasing its mass.

cleavage furrow A groove formed from the cell membrane in a dividing cell as the contractile ring tightens.

climax community The long-lived community at the end of a succession.

clitoris A highly sensitive erectile tissue that is part of the female genitalia; it is homologous to the penis.

cloaca A common entrance and exit chamber for the digestive, urinary, and reproductive systems in many vertebrates.

clone A population of genetically identical individuals or cells descended from a single ancestor.

closed circulatory system A circulatory system in which blood runs only within enclosed vessels.

cnidarians One of two phyla of radially symmetrical acoelomates; cnidarians include three classes: Hydrozoa (hydroids), Scyphozoa (jellyfish), and Anthozoa (corals and sea anemones).

cnidocytes Specialized stinging cells that lie on the tentacles of cnidarians.

coacervate Discrete tiny droplet into which proteins and polysaccharides can spontaneously concentrate.

cocci [singular, **coccus**] Spherical-shaped prokaryotic cells.

codominant Two alleles that each contribute to the phenotype of a heterozygote.

codon A group of three nucleotides that specifies a single amino acid.

coelom A cavity lined by a layer of mesoderm cells.

coelomates Animals with a coelom.

coenocytic Fungi in which the nuclei all lie in a common cytoplasm rather than in separate membrane-bound cells.

coenzyme An organic molecule (but not a protein) that is a necessary participant in an enzyme reaction.

coevolution The mutual adaptation of two separate evolutionary lines.

cohesion An attraction between molecules of the same substance.

cohesion-tension theory Theory that describes how evapotranspiration from the leaves of plants pulls continuous chains of water molecules up from the roots.

cohort The set of individuals that enter the population (or are born) at the same time.

collagen A fibrous protein, secreted by fibroblasts, that provides mechanical strength in cartilage, ligaments, tendons, and bones; collagen is by far the most abundant protein in the extracellular matrix.

collagen helix A regular structure found principally in the structural protein collagen; it consists of three polypeptide chains wound around each other.

collenchyma Plant cells that have a thick primary cell wall and often provide mechanical support; collenchyma are usually in the growing regions of stems and leaves.

colon The lower end of the large intestine; it is specialized for the absorption of water and ions.

commensal The species that benefits from the protection of its host in a commensalistic relationship.

commensalism An association between individuals of two species in which one organism benefits without harming the other one.

community An interacting group of species that inhabit a common area.

compact bone tissue The hard, dense bone that surrounds the spongy interior.

companion cell In plants, a long, nucleated, fully functional parenchyma cell that supplies a sieve tube member with proteins and energy-rich molecules.

comparative anatomy The study of morphological similarities and differences among organisms; it often provides clues to evolutionary relationships.

competition The interaction between organisms or species that depend on the same limited resource; competition occurs when one organism or species uses a resource in a way that limits the availability of that resource to others.

competitive exclusion principle When two species compete directly for exactly the same limiting resources, the more efficient species will eliminate the other.

competitive inhibitor A molecule whose inhibitory effects on an enzyme can be overcome by increased substrate concentration.

complement A set of blood proteins that attack microbial invaders.

complete metamorphosis The pattern of development in which there are distinct larval, pupal, and adult forms; this pattern is characteristic of many insects, including Drosophila, fleas, flies, beetles, wasps, moths, and butterflies.

completed family size The average number of children that reach reproductive age born to each family.

complex transposon A transposable element that contains other genes besides transposase, for example, a gene for antibiotic resistance.

compound A substance produced, by a chemical reaction, from two or more different elements.

compound eye An eye that consists of numerous simple light-detecting units; this type includes the eyes of most arthropods.

compound microscope A light microscope that contains several lenses.

compression The pushing action of two opposing forces.

concave (or **type III**) **survivorship curve** Description of a life history in which individuals have the greatest chances of dying early in life.

concentration The number of molecules in a given volume.

concentration gradient A graded difference in the concentration of a substance.

conception Fertilization or syngamy. The union of an egg and a sperm to form a zygote.

condensation (or **dehydration condensation**) **reaction** The linking of two building blocks, accompanied by the removal of a water molecule.

condom A thin rubber sheath that covers the penis and prevents sperm from entering the vagina; a condom is used both as a birth control device and to prevent the transmission of sexually transmitted diseases.

conjugation A process in which two temporarily attached bacteria exchange genetic material.

connective tissue A relatively sparse population of cells within a bed of extracellular matrix; includes bone and cartilage.

consumer An organism, usually an animal, that obtains energy by eating producers or other consumers.

contact inhibition of cell division The ability of cells to stop dividing when neighboring cells touch each other.

contractile ring A bundle of actin filaments that surrounds a dividing cell and pinches the cytoplasm in two during cytokinesis.

control The part of an experiment that acts as a standard against which the rest of the experiment is compared in order to test a hypothesis.

convergent evolution The independent evolution of similar features in separate groups of organisms.

convex (or type I) survivorship curve Description of a life history in which survivorship starts out high and decreases slowly with age until a certain point when survivorship begins to decrease more rapidly.

Coriolis effect The twisting effect of rotation on movement, particularly of wind or of water.

corolla In a flower, the whorl of petals.

corpora cavernosa In the male reproductive system of mammals, two spongy cylinders that run the length of the penis and fill with blood during erection.

corpus luteum A structure within the ovary that develops from the ruptured follicle; it secretes progesterone and some estrogen.

corpus spongiosum A spongy cylindrical tissue that surrounds the urethra.

corridors Narrow strips of protected habitat designated for the purpose of allowing a species to travel from one preserve to another. Depending on the size of the corridor, the usual beneficiaries are small, slow moving animals such as frogs and salamanders or large predators such as mountain lions and wolves.

cortex The outer region of a cell, a tissue, or an organ.

cortical microtubules Microtubules that encircle a dividing plant cell; they determine the placement of a new cell wall.

corticotropin (adrenocorticotrophic hormone or ACTH) A peptide hormone, made in the anterior pituitary, that stimulates the production of corticosteroids in the adrenal cortex.

cotransport The coupled transport of two substances across a membrane; cotransport depends on specific transmembrane protein molecules called cotransporters.

cotyledons (seed leaves) The very first leaves that sprout from a seed; present in the embryos of seed plants, they contain stores of nutrients for use after germination.

countercurrent system A system in which heat, fluids, or gases run past each other in a manner that allows the exchange of heat or matter.

covalent bond Shared arrangement of electrons that holds atoms together in molecules.

crassulacean acid metabolism (CAM) A variant of the C_4 pathway of the light-independent reactions of photosynthesis; CAM is used by succulent plants such as cacti.

creationism A religious belief that life was created very recently, as described in the Christian Bible.

Cro-Magnon An early race of *Homo sapiens*.

crop In the digestive tract of birds, a saclike extension of the esophagus; the crop can store food for later digestion.

cross bridges Extensions of myosin molecules that perform the work of muscle contraction by pulling the thin and thick filaments over one another.

crossbreeding (or crossing) The interbreeding of two genetically distinct organisms.

crossing over One type of genetic recombination; it involves the breakage and rejoining of single chromatids of homologous chromosomes.

Crustacea A subphylum of arthropods with two pairs of antennae and eyes at the ends of stalks.

crystal A solid that is enclosed by geometrically regular faces; the starting material point for x-ray diffraction analysis.

ctenophores Comb jellies, a small phylum of about 90 living species.

current The movement of electrical charges.

cuticle A waxy covering that keeps the above-ground parts of plants from losing water.

cyanobacteria Photosynthetic bacteria; previously called "blue-green algae"; the most ancient photosynthetic organisms known and the probable ancestors of chloroplasts.

cyclic AMP (cAMP) Adenosine monophosphate in which the same phosphate group is linked to carbon atoms 3 and 5 of ribose; it is used as a "hunger" signal in bacteria and protists and as a second messenger in mammalian cells.

cyclic photophosphorylation The production of ATP from light energy by a series of electron transfers that regenerate the absorbing chlorophyll; in plants, cyclic photophosphorylation depends on the flow of electrons from excited P_{700} in a cycle that regenerates P_{700}, which is then able to absorb another photon of light.

cyclins Proteins involved in regulating the cell cycle.

cyst An enclosed structure that contains reproductive cells.

cytochrome One of a set of heme-containing electron carrier proteins that change color as they accept or donate electrons.

cytokine One of a number of small proteins that regulate proliferation and protein synthesis in cells that participate in inflammation and in the immune response.

cytokinesis The division of the cytoplasm and formation of two separate plasma membranes.

cytokinin A plant hormone that influences the rate of division and differentiation.

cytoplasm In eukaryotes, that part of the cell outside the nucleus but inside the plasma membrane.

cytoskeleton A network of protein fibers that runs through the cytosol of eukaryotic cells; it consists of microtubules, actin filaments, and intermediate filaments, along with other associated proteins.

cytosol In eukaryotic cells, the part of the cytoplasm not contained inside organelles.

cytotoxic T cell An immune cell that kills altered body cells.

daughter cells The cells produced from a single parent cell after cell division.

dead space The volume of air in the air passages that does not come into contact with the surfaces of the alveoli in the lungs.

death rate The number of people who die each year per 1,000 in the population.

deciduous Shedding leaves annually, as in many trees and shrubs.

decomposer An organism, such as a bacterium or fungus, that lives on the energy in the complex molecules of dead organisms.

dehydration A physiological state in which the body lacks sufficient water.

deletion A type of mutation; the removal of one or more nucleotides.

demography The statistical study of populations.

denatured A protein that has lost its native, three-dimensional structure and its functional activity, while still having an unaltered primary structure.

dendrite A type of neuronal process that usually carries signals to the cell body.

denitrifying bacteria Bacteria that convert nitrate to atmospheric nitrogen.

density-dependent Referring to population growth, limited in proportion to population density—the denser the population, the more slowly it grows.

density-independent Referring to population growth, limited by factors other than population density—such as fire, drought, and other natural disasters.

deoxyhemoglobin Hemoglobin that has no bound oxygen.

deoxyribonucleic acid (DNA) The long thin molecules in which organisms store and transmit genetic information.

depolarizing A voltage change across an excitable membrane that makes the inside of the cell less negative and more apt to produce an action potential.

Depo-Provera (DMPA, depo-medroxy progesterone acetate) A birth control pill; a progesterone derivative taken as an injection; it suppresses ovulation and inhibits implantation for three months at a time.

descent with modification Evolution; change in organisms over time.

desert A dry, relatively barren region, with lower productivity than most other biomes.

desmosome An anchoring junction that consists of transmembrane proteins that attach to a cell's intermediate filaments.

detergent An amphipathic molecule that interacts both with water and with hydrophobic molecules.

determined In development, having a limited potency. A determined cell can no longer develop in accordance with new environmental signals.

detritus Particles of dead organic matter.

deuterostomes Those animals in which the anus forms first, from the blastopore, and the mouth forms secondarily.

diabetes mellitus A disease characterized by the overproduction of sweet (sugar-containing) urine.

diagonal (or **type II**) **survivorship curve** Description of a life history in which survivorship decreases in proportion to age, so that the curve is a straight, declining line.

diaphragm (1) A sheet of muscle beneath the lungs that separates the thoracic cavity from the abdomen. (2) A birth control device. A membrane of latex that prevents sperm from entering the cervix.

diastole The half of the cycle of the heart in which both the atria and the ventricles are relaxed.

dicots Flowering plants that form two cotyledons (seed leaves) during germination, including about 170,000 species; in contrast to monocots (grasses and lilies), which make only one cotyledon. Most dicots are eudicots.

differential gene expression The production of different proteins (and different amounts of proteins) in different types of cells at different times during the life of an organism.

differentiation The process by which tissues and cells become specialized and different from one another.

diffraction The scattering of electromagnetic waves by regular structures so that their interference produces a pattern of lines or spots; the diffraction of x rays by a crystal produces a pattern of spots that can be analyzed to reveal the structure of the molecules within the crystal.

diffusion Random movements of molecules that lead to a uniform distribution of molecules both within a solution and on the two sides of a membrane.

digestion The process of hydrolyzing large molecules into smaller units such as glucose, amino acids, fatty acids, and glycerol.

digestive tract (the **gastrointestinal tract**, the **alimentary canal**, or the **gut**) The tube, extending from mouth to anus, in which animals accomplish digestion, absorption, and elimination.

dikaryotic Having two nuclei per cell, as in certain fungi.

diploid Having two sets of chromosomes.

directional selection In evolution, selection that shifts the frequency of one or more traits in a particular direction.

disaccharide A sugar composed of two simple sugars, such as two glucose molecules or one glucose and one fructose.

disjunction In meiosis, the moving apart of two homologous chromosomes during anaphase I and of two sister chromatids in anaphase II.

dispersal The spread of a species or group of species from one area to another.

disruptive selection In evolution, the opposite of stabilizing selection: it increases the frequency of extreme types in a population, at the expense of intermediate forms.

distal tubule In the kidney, the wide tubule from which the urine flows into the collecting duct and then to the bladder.

disulfide A covalent bond between two sulfhydryl groups; disulfides are a common cross link between two cysteine side groups

within a single polypeptide or between two polypeptides.

diuretic A substance, such as caffeine or alcohol, that stimulates water loss.

divergent evolution The separation of one species of organisms into two (or more) species.

DNA ligase An enzyme that joins fragments of DNA together, essential in natural DNA replication and recombinant DNA technology.

DNA polymerase The enzyme that strings together nucleotides into DNA.

domain The highest taxon; three domains—the Archaea, Eubacteria, and Eukarya—include all living organisms.

dominance hierarchy A ranking of individuals that fixes (at least temporarily) who may dominate whom.

dominant Refers to an allele that alone determines the phenotype of a heterozygote.

dormant Referring to a stage in the development of a seed in which growth is suspended until restarted by environmental cues.

dorsal Refers to the top, or back, of an animal.

dorsal hollow nerve cord A flexible rod containing neurons and their processes that runs between the notochord and the surface of the back.

double circulation A pattern of circulation in which blood circulates separately between the heart and lungs and between the heart and the rest of the body.

double fertilization In flowering plants, the simultaneous fusion of two sperm nuclei from the pollen tube with three nuclei to form a zygote and an endosperm.

doubling time The time it takes a population to double.

Down syndrome A genetic disorder, almost always caused by trisomy of chromosome 21, that causes mental retardation and abnormal development of the face, heart, and other parts of the body.

downstream In DNA or RNA, toward the 3′ end.

duodenum The first section of the intestine, where most of digestion occurs.

duplication A mutation in which a sequence of DNA is duplicated so that it occurs twice. The two versions of the sequence may subsequently evolve separately.

Echinodermata "Spiny-skinned" animals such as sea stars.

ecological footprint The area of land needed to provide all of the resources a single person uses.

ecological isolation A barrier to reproduction that results from differences in the ways that species live.

ecology The study of the interactions of organisms with one another and with their physical environments.

ecosystem A community of organisms together with the nonliving parts of the community's environment.

ectoderm The outermost cell layer of an embryo, including, in vertebrates, the future skin, sense organs, and central nervous system.

ectotherm An animal that depends on external sources of heat.

edema The accumulation of water in the tissues.

edge effects A consequence of habitat fragmentation. Edge effects, such as incursions by light and predators, prevent a species from fully occupying a habitat island.

efferent Leading away, as blood vessels leaving an organ or nerve fibers carrying information from the central nervous system to the muscles and other organs.

egg (an **ovum**) A female gamete.

ejaculation The propulsion of sperm out of the penis.

electrical synapse A connection between two neurons through gap junctions.

electrochemical gradient A double gradient composed of a chemical gradient (the difference in hydrogen ion concentration, or pH) and an electrical gradient (the difference in charge).

electromagnetic radiation A form of energy that includes x rays, ultraviolet light, visible light, infrared radiation, microwaves, and radio waves.

electron A subatomic, negatively charged particle; its charge is exactly equal to that of a proton, but its mass is much smaller.

electron carrier One of the molecules that carry electrons from high-energy, reduced compounds (NADH, NADPH, and $FADH_2$) to oxygen.

electron microscope A microscope, with higher resolution than a light microscope, that uses beams of electrons to reveal subcellular structure.

electron transport chain The pathway of electrons in oxidative phosphorylation or photophosphorylation.

electronegativity A measure of the tendency of an atom to gain electrons.

electrophoresis A method for separating molecules (such as proteins and nucleic acids) according to their size and charge.

electrostatic interaction The attraction or repulsion of charges.

element A substance that cannot be reduced to simpler substances by chemical means.

elimination The disposal of the remains of digested food.

elongation In genetics, the adding of additional nucleotides or amino acids to a growing polynucleotide or polypeptide.

elongation zone A region of the growth zone of a plant's root.

embryo A set of early developmental stages in which a plant or animal differs from its mature form.

emergent property A characteristic that arises only at complex levels of organization.

emergents In tropical forests, trees with umbrella-shaped crowns extending to a height of 50 meters or more.

emigration Movement out of the population.

emphysema A disease of the lungs characterized by irreversible enlargement of the air spaces.

endemic Found nowhere else in the world.

endergonic Refers to a process in which free energy increases.

endocrine Cells or organs that are specialized for secreting specific signaling molecules into the general circulation.

endocrine gland An organ that is specialized for secretion of a hormone into the general circulation.

endocrinology The study of hormones.

endocytosis The process of taking in materials from outside a cell in vesicles that arise by the inward folding ("invagination") of the plasma membrane.

endoderm The innermost cell layer of an embryo, including the future gastrointestinal tract and associated organs such as the pancreas and liver.

endodermis In vascular plants, the innermost layer of the cortex in roots and stems.

endometrium The inner lining of the uterus.

endoplasmic reticulum (ER) An extensive and convoluted network of membranes within a eukaryotic cell.

endorphin One of three classes of polypeptides that act as natural pain suppressors.

endosperm In flowering plants, a storage tissue that develops during double fertilization.

endosymbiosis The close association of two organisms, one of which lives inside the other.

endosymbiotic theory The theory that mitochondria and chloroplasts are descended from prokaryotes that came to live inside ancient eukaryotic cells.

endothelium In the circulatory system, the epithelial cells that line the capillaries.

endotherm An animal that warms its body by capturing the heat released by metabolism.

endotoxin A toxic substance that is a component of a bacterial cell or other pathogen.

energy The capacity to perform work, to move an object against an opposing force.

energy subsidy The extra energy, usually provided by fossil fuels, that increases the productivity of a farm.

entropy A formal measure of disorder; entropy has a high value when objects are disordered or distributed at random and a low value when they are ordered.

enzyme A large molecule, almost always a protein, that accelerates the rate of a specific chemical reaction.

enzyme-substrate complex The association of enzyme and substrate that forms in the course of catalysis.

eon One of several very long periods of time into which geologists and paleontologists have divided the history of life; eons are much longer than eras; a billion years or more.

epicotyl The axis of a developing plant above the attachment point of the cotyledons.

epidemiologist A researcher who studies the incidence and transmission of diseases in populations.

epidermis In animals, the outer layer of the skin; in plants, the outermost layer of the embryo and the plant.

epididymis In the male reproductive tract of mammals, an interconnected network of coiled ducts in which sperm are stored.

epinephrine (adrenaline) A hormone, made in the adrenal medulla, that speeds the heart, dilates the blood vessels, and increases the liver's production of glucose from glycogen.

epiphyte A plant that grows entirely on other plants.

episome A virus or other genetic system that can propagate either autonomously or as an integrated part of the host's chromosome.

epithelial tissue Cells tightly linked together to form a sheet with little extracellular matrix.

epithelium A tissue that lines a surface (such as the outside of the body or the inside of the lungs).

epitope A specific shape and charge distribution that antibodies recognize.

epoch A relatively short period of geologic time; one of the subdivisions of the Tertiary and Quaternary Periods of the Cenozoic Era.

equilibrium In a chemical reaction, the point at which no further net conversion of reactants and products takes place.

era A period of time, ranging in length from 65 million to several billion years, into which paleontologists usually divide the history of life; eras are shorter than eons.

erection The rigid state of the penis or clitoris, caused by the blood filling the tiny spaces within spongy tissues.

erythrocyte A red blood cell.

erythropoietin (EPO) A hormone released by the kidneys in response to low levels of oxygen in the blood; EPO stimulates the bone marrow to make more red blood cells. Hazardous to health if taken as a performance supplement.

Escherichia coli (E. coli) A bacterial resident of the human gut and favorite experimental organism.

esophagus The part of the digestive tract that connects the mouth and pharynx with the stomach.

essential amino acids The amino acids that an animal cannot itself produce.

essentialism The view, originally argued by Plato, that individuals, whether chairs, daisies, or people, are only distorted shadows of an ideal, or essential, form.

estivate To pass time in a sleeplike state (torpor) during the hottest and driest months of the summer.

estrogens Any of a group of steroid hormones that regulate the female reproductive cycle, as well as the growth and development of female reproductive structures.

estuary A partly enclosed body of water where a freshwater stream or river meets the ocean.

ethology The study of animal behavior.

ethylene A plant hormone; a two-carbon molecule containing a double bond.

etiolated Having a thin, spindly appearance, poor leaf development, and no chlorophyll production.

Eubacteria The commonly occurring prokaryotes that live in water and soil or within larger organisms.

eudicot See dicot.

euglenophytes Protists that have distinctive light-detecting eye spots.

Eukarya The domain that encompasses the Animalia, Plantae, Fungi, and Protista kingdoms.

eukaryotic Cells that contain a central nucleus and other membrane-enclosed organelles.

Eumetazoa The larger of two subkingdoms of animals, including all of the animals except the sponges.

eutherians (placental mammals) Animals whose embryonic development takes place within the uterus of the mother.

eutrophic A lake that has excessive minerals and organic matter and insufficient oxygen.

evergreen Keeping leaves throughout the year.

evolution The process by which species arise and change over time. The idea that all organisms have descended from common ancestors.

excitatory In the nervous system, leading to the production of action potentials.

excitement A phase of sexual arousal that is characterized by increased blood flow to the clitoris, the labia minora, and the breasts (in the female) and to the penis and the testes (in the male).

exergonic Refers to a process in which free energy decreases.

exocytosis The export of molecules from a cell by a process that is approximately the reverse of pinocytosis; molecules to be exported are surrounded by membranes that move to the cell surface.

exon A segment of a gene (or of a pre-mRNA) that is also present in mature mRNA; most exon sequences encode polypeptide segments.

exoskeleton External skeleton.

exotoxin A toxic substance secreted by bacteria.

exponential growth A growth pattern in which a population repeatedly doubles in some constant period of time.

expiration Exhalation.

expressivity The variability in phenotype associated with a given genotype, usually in reference to particular alleles.

extension The unbending of a limb.

extinction vortex The theory that inbreeding brings out genetic defects that hasten the extinction of a population.

extracellular matrix A network of proteins and polysaccharides found in connective tissue.

F1 (or first filial) generation The initial progeny of a cross.

F2 (or second filial) generation The progeny of the F1 generation.

facilitated diffusion An increased rate of passive transport; it depends on the action of specific transporter molecules within the membrane.

facultative aerobe A heterotroph that can grow either with or without oxygen.

facultative anaerobe A microorganism that can live either anaerobically (by fermentation) or aerobically (using oxidative phosphorylation).

FAD (flavin adenine dinucleotide) An electron acceptor in oxidative phosphorylation.

FADH$_2$ The reduced form of FAD.

fall turnover In a lake, the annual mixing of the waters of epilimnion and hypolimnion.

fallopian tube Human oviduct.

family A taxonomic group that consists of one or more genera; a family is a subdivision of an order.

fat A triacylglycerol that is solid at room temperature; the component fatty acids are usually saturated.

fate What a cell or a tissue becomes during development.

fatty acid A small molecule consisting of a hydrocarbon chain ending in a carboxyl group; a component of phospholipids and triacylglycerides.

feces Waste matter of digestion, discharged through the anus; feces consist of the remnants of food together with bacteria that inhabit the intestinal tract.

female choice A courtship pattern in which the female assesses and chooses the best male.

fermentation The anaerobic extraction of energy from organic compounds.

fermenting bacteria Anaerobic heterotrophs that derive energy and carbon atoms from a variety of organic compounds in the absence of oxygen.

fertility The number of offspring each individual in a cohort is likely to produce.

fertilization The union of two haploid gametes to form a diploid cell or zygote.

fetal masculinization The development of male characteristics in a fetus carried by a woman treated with male hormones during pregnancy.

fetus In mammals, a stage of development in which all organs have formed.

fibril A threadlike structure, made of smaller filaments; the term is used to refer to cross-linked cables of collagen molecules and to assemblies of actin and myosin filaments.

fibrillation In the heart, continuous disorganized contractions.

fibroblast A flat, irregularly shaped cell found in connective tissue.

fibrous protein A protein with an elongated shape; fibrous proteins include most structural proteins.

filament (1) A small protein fiber; (2) in plants, a thin stalk that connects the anther to the base of the flower.

filtrate In the kidney, the water, urea, and other small molecules in the blood that freely pass into Bowman's capsule from the renal artery.

First Law of Thermodynamics The statement that the total amount of energy stays constant in any process; that is, energy is neither lost nor gained—it only changes form.

fitness A measure of selective advantage; it is defined as the contribution to the next generation of one genotype in a population relative to the contribution of other genotypes.

fixed nitrogen Nitrogen in the form of ammonia (NH_3) or nitrate (NO_3).

flagellum [plural, **flagella**] In eukaryotes, a protein assembly, consisting of microtubules, that can move a cell through a liquid medium (or a liquid medium over a cellular surface); in prokaryotes, a protein assembly that moves like a propeller.

flexion The bending of a limb.

fluid mosaic model The accepted model of biological membrane structure; the model stresses that proteins and phospholipid molecules can move within each leaf of the lipid bilayer unless they are restricted by special interactions.

fluidity A measure of the ability of substances to move within a membrane.

follicle In the ovary, the granulosa cells surrounding a primary oocyte.

follicle-stimulating hormone (FSH) A peptide hormone, made in the anterior pituitary that in females promotes the maturation of the follicle during the menstrual cycle and in males stimulates testosterone production.

food chain The sequence in which consumers eat either producers or other consumers.

food web The collection of all the interacting food chains of an ecosystem.

foot In a mollusk, a muscular extension of the body that the mollusk uses for sensing, grabbing, creeping, digging, and holding on.

foreskin A loose layer of skin around the outside of the penis, which ends in a flap and folds over the glans.

fossil An object, usually found in the ground, that represents the remains or imprint of past life.

founder effect The restriction of genetic diversity of a population that results from the founding of a new population by a small subset of a larger population.

frame-shift mutation A mutation (an insertion or deletion) that alters the groupings of nucleotides in subsequent codons.

free energy A measure of available energy under the conditions of a biochemical reaction; the term is abbreviated G, after the American thermodynamicist J. Willard Gibbs.

free radical A compound that is highly reactive because it contains an unpaired electron; it is biologically important because it can lead to DNA damage and to cell death; free radicals are formed during normal metabolic processes in both animals and plants.

fruiting body A mushroomlike growth on cellular slime molds that produces spores.

functional group A standard small grouping of atoms that contributes to the characteristics of an organic molecule.

fundamental niche A species' potential ability to utilize resources.

Fungi The kingdom that includes heterotrophic organisms, both multicellular and single-celled organisms.

G_0 The state of a cell that has withdrawn from the cell cycle.

G_1 The period of the cell cycle that represents the gap between the completion of mitosis and the beginning of DNA replication; it is also called the first growth phase.

G_2 The period of the cell cycle that represents the gap between the completion of DNA synthesis and the beginning of mitosis (of the next cell cycle).

gametangia In plants, the enclosed reproductive organs in which gametes form.

gamete A specialized reproductive cell through which sexually reproducing parents pass chromosomes to their offspring; a sperm or an egg.

gamete formation The production of sperm and eggs.

gametophyte The haploid form of a life cycle characterized by alternation of generations.

gap junctions Protein assemblies that form channels between adjacent animal cells.

gastrula A stage of an animal embryo in which the three germ layers have just formed.

gastrulation The process of forming a gastrula.

gate The part of a channel protein that opens and closes the channel in response to environmental signals such as voltage or the binding of a neurotransmitter.

gene The unit of inheritance. A DNA sequence that is transcribed as a single unit and encodes a single polypeptide, a set of closely related polypeptides, ribosomal RNA, or transfer RNA.

gene flow The movement or flow of genes from one population to another by means of interbreeding.

gene pool All the alleles of all the individuals in a population.

gene rearrangement The cutting and splicing of segments of specific genes; it is known to occur within the immune system.

genetic code The relationship between nucleotide sequence in mRNA and amino acid sequence in polypeptides.

genetic linkage The tendency of two or more genes to segregate together.

genetic map (linkage map) A summary of the genetic distances between genes.

genetic recombination Associations of genes that occur in offspring that did not exist in the parents.

genetics The study of inheritance.

genome The collection of all the DNA in an organism.

genotype The genes present in a particular organism or cell.

genus [plural, **genera**] A group of species that share many morphological characteristics; a taxonomic grouping of one or more species; a subdivision of a family.

germ cells (or **germ line**) Gametes and the cells from which they arise.

germ layer Any of the three layers that form in a vertebrate embryo—ectoderm, mesoderm, or endoderm.

germ layer formation The movement of embryonic cells to give the three germ layers (ectoderm, mesoderm, and endoderm); gastrulation.

germ line theory The view that antibody diversity results from genetic information that is already present in the zygote.

germination The resumption of growth by a seed.

gibberellin A plant hormone.

gill slits Openings that directly connect the throat to the outside.

gills Evaginated breathing structures.

girdling Removing a strip of bark around a tree trunk.

gizzard A region of the intestinal tract of birds specialized for the grinding of food.

glans The smooth cap formed by the tissue at the far end of the penis.

glial cell A cell within the nervous system that does not itself transmit electrical and chemical signals, but which provides metabolic and structural support for neurons.

globular protein A relatively compact protein that is roughly spherical in shape.

glomerulus A tangled network of capillaries in the cortex of a kidney.

glucagon A protein hormone produced in the pancreas; a signal for the postabsorptive state, glucagon inhibits glycogen synthesis and stimulates its breakdown into glucose.

glucose A simple, six-carbon sugar made by plants during photosynthesis (and metabolized by animals for energy).

glycerol A polar three-carbon molecule with three hydroxyl groups; a starting compound for triacylglycerides and phospholipids.

glycogen Polysaccharide used in animals for long-term energy storage.

glycolysis A set of ten chemical reactions that is the first stage in the metabolism of glucose.

glycolytic fiber A large muscle fiber that derives most of its energy from glycolysis and has few mitochondria.

glycoprotein A protein that contains covalently attached carbohydrates.

glycosidic bond The covalent bond between two sugar molecules in a polysaccharide or oligosaccharide.

goblet cells Specialized cells that make mucus.

Golgi apparatus In eukaryotic cells, a set of flattened discs of membrane, usually near the nucleus, involved in the processing and export of proteins.

gonad A gamete-producing organ; an ovary or testis.

gonadotropin One of two hormones (FSH and LH), made in the anterior pituitary, that act on gonadal tissue.

gonadotropin releasing hormone (GnRH) A polypeptide hormone, made by the hypothalamus, that stimulates the secretion of FSH and LH by the anterior pituitary.

gossypol A terpene, made by cotton plants, that is responsible for resistance to insects, bacteria, and fungi; it also works as a contraceptive in human males.

gradualism The idea that species evolve gradually and continuously through the steady accumulation of changes, rather than through sudden changes.

granum [plural, **grana**] A stack of thylakoids within a chloroplast, enclosed by the thylakoid membrane.

gravitropism The response of a plant to gravity.

greenhouse effect The warming of the Earth as the result of the absorption of heat by carbon dioxide and other gases.

ground tissue In plants, the cells that occupy most of the interior of the embryo and the plant.

growth control A regulatory mechanism that prevents cell division by allowing the cell cycle to proceed under some conditions and to stop under others.

growth factor One of a number of protein paracrine signals that stimulate cell division and cell survival.

growth hormone (GH) A peptide hormone, made in the anterior pituitary, that stimulates tissue and skeletal growth, milk production, and other processes in humans, cows, and other mammals.

gut The gastrointestinal tract, the alimentary canal, or the digestive tract.

guttation The process by which root pressure can push water all the way up the stem and out of tiny holes at the margins of the leaves.

gymnosperms Seed plants that do not have fruits or flowers, including, for example, pine trees and other conifers, as well as palmlike cycads.

habitat The place in which an organism lives, along with the set of environmental conditions that characterize that place.

habitat destruction The process by which humans make a habitat uninhabitable for a species. Habitat destruction can range from bulldozing and paving over the area to the elimination of a single, important (keystone) species.

habitat fragmentation The breaking up of natural habitat into fragments of land too small to support intact ecosystems or populations.

habitat island A fragment of habitat isolated from other similar fragments.

habituation The decrease in a behavioral response following repeated exposure to a harmless stimulus.

half-life The time required for half the atoms of a radioactive isotope to decay.

halophiles The salt-loving bacteria, one of the three phyla of Archaea.

haplodiploidy An unusual kind of sex determination in which males develop from unfertilized eggs and are therefore haploid, while females are diploid.

haploid Having a single set of chromosomes.

Hardy-Weinberg equilibrium A stable distribution of genotype frequencies maintained by a population from generation to generation.

heart The muscular organ responsible for pushing the blood through the circulatory system.

heart attack A sometimes fatal failure of the blood supply to the heart that causes the heart to beat irregularly or to stop beating.

heart rate The number of contractions (beats) per minute.

heat The form of kinetic energy contained in moving molecules.

heme An iron-containing organic molecule that gives hemoglobin and the cytochromes their red color; heme may donate or accept electrons, as its iron atom changes its charge between Fe^{2+} and Fe^{3+}.

Hemichordata Acorn worms; soft-bodied marine animals (up to 8-feet long) that resemble chordates in having gill slits and a dorsal nerve cord.

hemizygous In male mammals, having one (instead of two) copies of a gene because the Y chromosome lacks the gene present on the X chromosome.

hemoglobin The oxygen-binding protein that makes red blood cells red.

herbaceous plant (an **herb**) A plant that has no woody parts.

herbivore An animal that eats only plants.

heterotroph An organism such as a lion or a mushroom that lives off the organic molecules made by other living organisms. See autotroph.

heterozygous Having two different alleles for a single gene (in a diploid organism).

hierarchy An arrangement in which larger groups include smaller groups, which include still smaller groups.

high-energy bond A relatively unstable chemical bond that gives up energy as new, more stable, bonds form.

histamine The amino acid histidine minus the carboxyl group, a major stimulus for the inflammatory response; it is released by cells in damaged tissues; it dilates capillaries and increases their tendency to leak fluid.

histone One of a set of small, positively charged proteins that bind to DNA in eukaryotic cells.

homeostasis The tendency of organisms to maintain a stable internal environment in the presence of a changing external environment.

homeotic mutation A mutation that causes the cells of an embryo to give rise to an inappropriate structure in the adult, for example, to legs instead of antennae.

homeotic selector gene In a plant, a gene that establishes the fate of one or more whorls of a developing flower.

hominid A member of the human family, Hominidae; the only living hominid is *Homo sapiens*.

hominoid Hominids and apes; all have large skulls and long arms and tend to walk at least partially erect.

homologous chromosomes The two matching chromosomes that align during meiosis I.

homologous structures Similar structures in two or more species. Homologous structures may perform different or similar functions.

homoplasy Similar traits that have evolved separately. The possession by two species of a similar trait that is not derived from a common ancestor. "Homoplasy" has replaced the older word "analogy," which requires that two similar traits perform the same function. In contrast, similarities that are inherited from a common ancestor are said to be homologous.

homozygous Having two copies of the same allele (in a diploid organism).

hopeful monsters Individuals with macromutations such as chromosome rearrangements whose deformities have adaptive value. A term coined by biologist Richard Goldschmidt to describe a mechanism for a saltationist theory of speciation.

hormone A substance, made and released by cells in a well-defined organ or structure, that moves throughout the organism and exerts specific effects on specific cells in other organs or structures.

Hox genes Mammalian counterparts of the fruit fly (Drosophila) homeotic selector genes. Both kinds of genes help regulate the overall formation of the anterior-posterior axis of the body.

human chorionic gonadotropin (HCG) A hormone made after conception; the basis of the most common tests for pregnancy.

human immunodeficiency virus (HIV) The retrovirus that causes AIDS; HIV infects and kills T_H lymphocytes by first binding to the CD4 protein on the cell surface.

humus In soils, the residue of decayed dead organisms; a black or brown material that decays slowly.

hybrid The progeny of a cross of two genetically distinct organisms.

hybridoma A hybrid cell line derived from the fusion of a cancer cell (a lymphoma) to another cell, such as an antibody-producing cell.

hydrocarbon chain A chain of connected carbon atoms, with hydrogen atoms sharing other available outer shell electrons.

hydrogen bond A weak attraction between a hydrogen in one molecule that has a slight positive charge and a negatively charged atom in another molecule.

hydrogen ion (H^+) The result of a dissociation of a water molecule; it is also called a proton, though it is actually a hydronium ion (H_3O^+).

hydrolysis Breaking the bond between two building blocks by adding a water molecule, reversing the dehydration-condensation reaction.

hydronium ion (H_3O^+) A water molecule that has acquired an extra proton and a charge of $+1$; the proper name for a hydrogen ion.

hydrophilic Water-loving; refers to a molecule (or a part of a molecule) that is soluble in water by virtue of its interactions with water molecules.

hydrophobic Avoiding associations with water; nonpolar.

hydrophobic interaction The association of nonpolar molecules.

hydrostatic skeleton A rigid fluid-filled space (the coelom) that provides mechanical support in invertebrate coelomates.

hydroxide ion (OH^-) A water molecule from which a hydrogen ion has dissociated; it has a charge of -1.

hydroxyl The OH functional group, which allows molecules that contain it to form hydrogen bonds.

hymen A thin membrane of irregular ragged shape just inside the mouth of the vagina.

hypertension High blood pressure.

hypertonic Having a total concentration of solutes higher than that within a cell.

hyphae [singular, **hypha**] The threadlike filaments of a fungus.

hypocotyl The axis of a developing plant below the attachment point of the cotyledons.

hypothalamus A part of the brain that regulates the expression of many hormone systems.

hypothesis An informed guess—for example, about the way a process works or a structure is organized.

hypotonic Having a total concentration of solutes lower than that within a cell.

imaginal disc In insect development, a group of larval cells from which adult structures later develop.

imago Adult stage of an insect.

immigration Movement into a population.

immune response The defense system by which animals resist microorganisms and other foreign tissues, including cancer.

immunoglobulin An antibody; one of the members of a group of globular blood proteins called globulins.

implantation In mammals, the process in which a zygote burrows into the wall of the uterus.

imprinting The process by which an animal learns behavior during a sensitive period.

in vitro In a test tube.

in vivo In a living organism.

inbreeding Mating among close relatives, which greatly increases the number of homozygotes.

incipient species Subspecies that may be in the process of becoming separate species.

inclusive fitness The sum of an individual's genetic fitness (which includes that of its own direct descendants) plus all its influence on the fitness of its other relatives.

individualistic The view that every species has an independent distribution, and that, in effect, every community is unique.

induced abortion The deliberate removal of an embryo from the uterus.

induced fit A change in the conformation of an enzyme brought about by the binding of the substrate.

induction In embryonic development, the process by which one cell population influences the development of neighboring cells.

inflammation A set of responses to local injury; characteristics of injured tissue include redness, heat, swelling, and pain.

initiation The start of synthesis.

initiation site The first nucleotide actually transcribed from DNA into RNA.

innate Behavior that an animal engages in regardless of previous experience.

inner cell mass In a mammalian embryo, a small group of cells within a blastocyst that will eventually grow into the embryo itself and subsequently into the adult.

insertion A type of mutation; the addition of one or more nucleotides.

inspiration Inhalation.

insulin A protein hormone, produced by the pancreas, that regulates glucose uptake; a signal for the absorptive state, it promotes the synthesis of glycogen and inhibits its breakdown.

integrated In ecology, the view that a community consists of characteristic assemblages of species that interact with each other in predictable ways.

integrated replication A method for viral propagation in which the virus's DNA becomes integrated into the host cell's DNA and is replicated along with that of the host.

interferon A cytokine that interferes nonspecifically with the reproduction of viruses.

interleukin-1 (IL-1) A cytokine produced early during inflammation; it stimulates responses both in leukocytes and in organs distant from the site of infection.

intermediate filament In eukaryotic cells, a component of the cytoskeleton that consists of filaments 8 to 10 nm in diameter, thinner than microtubules but thicker than actin filaments.

intermediate form Fossils that grade from one form to another—for example, the fossils intermediate in both shape and in time from the tiny, horselike animal eohippus to the modern horse.

internode In a plant, the region of a stem between nodes.

interphase The part of the cell cycle in which the chromosomes are not condensed and the cytoplasm is not dividing.

interstitial cells In the testes, the matrix amongst the seminiferous tubules that synthesizes the male sex hormone testosterone.

intertidal zone The zone that lies between high tide and low tide.

intrauterine device (IUD) A birth control device; a small piece of plastic or other material that is placed into the uterus, where it interferes with implantation.

intron (intervening sequence) A segment of a gene (or of a pre-mRNA) that is transcribed into RNA but excised before the primary transcript matures into functional mRNA.

invagination The local folding of a cell layer to form an enclosed space with an opening to the outside.

inversion A mutation in which a segment of a chromosome is turned 180° from its normal orientation.

involuntary muscle A muscle that cannot be consciously controlled; smooth and cardiac muscles.

ion An atom (or a molecule) with a net electrical charge, the result of a different number of electrons and protons.

ionic bond A bond formed by ions with opposite charges.

isomers Molecules that contain the same atoms arranged differently.

isotonic Having a total concentration of solutes that is the same as a cell's interior.

isotopes Forms of an element that have different numbers of neutrons.

J-shaped curve A graph of the exponential growth of a population; the curve resembles the letter J.

joint The point of attachment between two bones or two parts of an exoskeleton.

karyotype The chromosomal makeup of a cell.

keystone species A predator or other organism whose presence promotes diversity or controls the structure of a community.

kidney In vertebrates, the major organ of excretion and of salt and water balance; regulates the composition of the urine.

kin selection The tendency of individuals in some species to increase their reproductive output by helping relatives, which share their genes, to reproduce.

kinetic energy The energy of moving objects.

kinetics The study of the rates of reactions.

kinetochore A specialized disc-shaped structure that attaches the mitotic spindle to the centromere.

kinetochore microtubules A subset of polar microtubules that run from a pole of the mitotic spindle to a kinetochore.

kingdom The next taxon down from domain; including Animalia (animals), Plantae (plants), and Protista (protists).

Koch's postulates Rigorous criteria for identifying the pathogen for a given disease.

krummholz A forest of stunted trees that grow near timberline on a mountain.

kwashiorkor Protein deficiency disease.

labia majora In the external genitalia of human females, the skin folds on either side of the labia minora that enclose small amounts of fatty tissue.

labia minora In the external genitalia of human females, the two thin skin folds that surround the mouth of the vagina and the end of the urethra.

lac repressor A protein that binds to the *lac* operator and prevents the expression of the *lac* operon.

lactose operon (or *lac* operon) A region of *E. coli* DNA that encodes three proteins used to derive energy from lactose: the enzyme β-galactosidase (which splits lactose into galactose and glucose) and two other proteins, called permease and acetylase.

lagging strand In replicating DNA, the newly made strand that is extended discontinuously.

larva A feeding form of an animal distinct from the later adult.

larynx The modified upper part of the trachea that contains the vocal cords, folds of membrane that vibrate as air passes over them, producing sound.

lateral bud (axillary bud) A bud that forms just above the point where a leaf joins the stem and which can develop into a branch.

lateral meristem A cylinder of actively dividing cells within a root or stem.

leading strand In replicating DNA, the newly made strand that extends continuously, with DNA polymerase adding nucleotides to its 3′ end.

leaf primordium A tiny extension of the apical meristem; it grows into a leaf.

learning Modification of neural activity and behavior as the result of experience.

leukocyte White blood cell.

ligament A band of connective tissue by which joints may be joined together.

ligand A molecule that binds to a specific binding site in a protein.

lignin A major constituent of the walls of cells specialized to provide mechanical support or transport water.

limnology The study of freshwater ecosystems—lakes, ponds, rivers, and streams.

linkage group A set of genes that do not assort independently because they are physically close to one another on the same chromosome.

linkage map A summary of the genetic distances between genes; a genetic map.

lipid Fats, oils, and related organic compounds found in the cell membranes of living organisms; function as energy-storage materials.

lipid-soluble signal A chemical signal that can enter a target cell by passing directly through the plasma membrane.

liposome An artificially produced vesicle that is surrounded by a phospholipid bilayer.

loam A soil that consists of a mixture of clay, silt, and sand.

locus [plural, **loci**] The position of a gene on a chromosome.

lophophore In three phyla of marine invertebrates, a filtering apparatus that catches food.

low-density lipoprotein (LDL) A carrier protein in the blood that binds to cholesterol.

lumen A space enclosed by a membrane (as in the endoplasmic reticulum) or epithelium (as in the gut).

lungs Invaginated breathing structures; localized organs of gas exchange that are always associated with the circulation.

luteinizing hormone (LH) A peptide hormone, made in the anterior pituitary, that in females induces ovulation and stimulates estrogen production and in males increases testosterone production.

lymph Fluid containing dead or foreign cells and waste proteins that passively moves through the lymphatic system.

lymph nodes Regions in the lymphatic veins where filterlike tissue separates cells and other detritus from the lymph.

lymphatic system The network of lymphatic vessels and nodes that provides a secondary route for fluids from the extracellular space to the bloodstream.

lymphocyte One of a class of white blood cells that develop within the lymphoid tissues (including lymph nodes, spleen, thymus, and tonsils); the cells responsible for the immune response.

lymphocyte library The collection of B lymphocytes, each containing a unique random rearrangement of immunoglobulin genes.

lysis The breaking down of a plasma membrane.

lysogenic cycle The reproduction of a virus along with its host without causing the host cell to lyse.

lysogenic virus A virus that reproduces either in a lytic cycle, in which it destroys its host, or along with the host; the virus can lie in a dormant (lysogenic) state and can be activated to enter a lytic cycle.

lysosome A small membrane-enclosed organelle that contains hydrolytic enzymes that break down proteins, nucleic acids, sugars, lipids, and other complex molecules.

lysozyme An enzyme found in tears, saliva, and other bodily fluids that weakens the cell walls of some bacteria.

lytic cycle Viral reproduction that destroys the host.

lytic virus A virus that destroys its host cell by lysing it.

macroevolution The evolutionary processes by which species and higher groupings (taxa) of organisms originate, change, and go extinct.

macromolecules Large molecules formed by the polymerization of smaller building blocks.

macrophage A large phagocytic cell, widely distributed throughout the body.

major histocompatibility complex (MHC) A set of 40 to 50 closely linked genes, some of which encode proteins on the surface of every somatic cell; MHC proteins present antigens to the immune system.

male competition Competition among males, usually for territory or access to females.

malnourishment A deficiency in one or more essential nutrients.

Malpighian tubules Excretory organs in arthropods.

mammary glands Milk-producing organs in the female that characterize the mammals.

mantle In mollusks, a specialized tissue that secretes a shell onto the dorsal surface of the visceral mass.

mantle cavity In mollusks, an enclosed space, formed by folds in the mantle, that contains the mollusk's breathing organs.

marsupials (metatherians) Mammals whose young are born in a fetal stage; marsupials carry the young in a pouch until they can fend for themselves.

mass A measure of the amount of matter.

mass number The total number of protons and neutrons in an atom's nucleus.

mast cell A large round cell; mast cells are distributed throughout the connective tissues and are filled with small histamine-containing vesicles, which are released at the sites of tissue damage.

maternal mRNA Any mRNA made in the egg from maternal DNA, by definition before fertilization.

matrix The intercellular substance within a tissue, or the interior substance of a mitochondrion.

matter Any substance.

mean The average value of a set of numbers, also called the arithmetic mean.

mechanical isolation A reproductive barrier that derives from incompatibilities between male and female reproductive organs.

medulla A part of the brain that regulates the rate of breathing; the inner region of a kidney or adrenal gland.

medusa The bell-like body plan that represents one of the two basic body plans of the cnidarians.

megaspore In a flower, the haploid cell within an ovule that develops into the embryo sac, the mature female gametophyte.

meiosis The process by which haploid gametes arise from diploid cells; meiosis distributes chromosomes so that each of four daughter cells receives one chromosome from each homologous pair.

meiosis I The first of the two divisions of meiosis, during which homologous chromosomes pair and are distributed into two daughter cells.

meiosis II The second of the two divisions of meiosis, during which sister chromatids are distributed to daughter cells.

membrane envelope A membrane that surrounds a virus particle.

memory cell In the immune system, a cell that can later be stimulated to produce cells that make particular antibodies.

menstruation The monthly shedding of blood, mucus, vaginal secretions, and endometrial tissue out through the vagina.

meristem In plants, a region of undifferentiated, actively dividing cells.

mesenchyme Loosely attached cells in a jellylike substance found in the embryos of vertebrates and the adults of some other animals. In vertebrate embryos, mesenchyme cells later form connective tissue, bone, cartilage, blood, and the lymphatic system.

mesoderm The middle layer of a vertebrate embryo, including the future connective tissues (bones, muscles, and tendons) and the cells of the blood.

mesoglea In cnidarians, the jellylike material that lies between the two cellular layers.

mesophyll Green parenchymal cells that are responsible for most of a plant's photosynthesis.

messenger RNA (mRNA) The RNA molecules that carry information from DNA to the ribosomes, where the mRNA is translated into a polypeptide.

metabolism All the chemical reactions occurring within an organism.

metamere One of the identical (or nearly identical) sections that make up an annelid's long, segmented body.

metamorphosis In many animal species, a series of dramatic changes in form leading from a larva to an adult.

metaphase The stage of mitosis or meiosis during which chromosomes move halfway between the two poles of the spindle, where they accumulate in the metaphase plate.

metaphase plate A disc formed during metaphase in which all of a cell's chromosomes lie in a single plane at right angles to the spindle fibers.

metapopulation A group of local populations, which may alternately go extinct and recolonize one another's vacant habitat islands.

metastasize Referring to cancer cells, to spread to other parts of the body.

metatherians (marsupials) Animals whose newborns are relatively undeveloped fetuses.

methanogen A methane-producing bacterium, one of the three phyla of Archaea.

micelle A cluster of amphipathic molecules.

microelectrode An electrode that is small enough to enter or maintain contact with a single cell.

microevolution Changes in the frequencies of alleles of genes in a population.

micronutrient A substance required in minute amounts for the life of an organism; examples include molybdenum, copper, zinc, manganese, boron, iron, and chlorine.

microtubule In eukaryotic cells, the largest elements of the cytoskeleton; they consist of tubulin molecules assembled into hollow rods, about 25 nm in diameter and of variable lengths, up to several μm.

microtubule organizing center (MTOC) The region of a eukaryotic cell from which microtubules emanate.

microvillus (plural, **microvilli**) A membrane-covered extension of an epithelial cell.

midbody In a dividing plant cell, the thin connection between daughter cells that persists until the end of cytokinesis; the midbody is packed with microtubules from the spindle apparatus.

midpiece In a spermatozoan, a section that contains a microtubule organizing center and dense concentrations of mitochondria—sources of ATP for the journey to the egg.

mimicry Pretending to be something else.

mineral An inorganic substance, especially one required for the life of an organism.

minipill A birth control pill that contains only progesterone, with no estrogens.

missense mutation A change in DNA that alters the codon for one amino acid into a codon for another amino acid.

mitochondrion [plural, **mitochondria**] In eukaryotes, the major energy-producing organelle; in cells that do not directly harvest sunlight, mitochondria produce nearly all of the ATP the cell needs to power the chemical reactions of the cell.

mitochondrial matrix The compartment surrounded by the inner mitochondrial membrane.

mitosis The process of the equal distribution of chromosomes during cell division.

mitotic spindle An elongated structure that develops outside the nucleus during early mitosis; contains the microtubular machinery that moves the chromatids apart.

mittelschmerz Pain during ovulation.

mobile gene A gene whose chromosomal address changes.

modal action pattern A highly stereotyped innate behavior.

modern synthesis The union of genetics and evolutionary biology in the 1930s that demonstrated the genetic basis of natural variation and natural selection.

mold A protist or fungus that grows as a downy coating on animal or vegetable matter.

molecular genetics The branch of genetics that studies how DNA carries genetic instructions and how cells carry out these instructions.

molecular weight The mass of a molecule relative to that of a hydrogen atom.

molecule A specific combination of individual atoms held together by covalent bonds.

mollusk A soft invertebrate; may be inside a shell (snail) or not (slug, octopus).

molting The shedding of feathers, fur, skin, or exoskeleton and the secretion of new ones.

monocotyledon (or monocot) One of a large group of flowering plants that have one cotyledon. Monocotyledons consist of about 65,000 species, including all of the grasses, lilies, palms, and orchids. Most are herbaceous.

monoculture The raising of a single crop.

monophyletic A taxon that includes an ancestral species and all its descendants.

monosaccharide A simple sugar with three to nine carbon atoms; the building block for polysaccharides.

monotreme (prototherian) An egg-laying mammal; like other mammals, monotremes have hair and mammary glands.

morning-after pill A birth control method that is taken after unprotected intercourse.

morphogen A substance that specifies the position of a cell within a pattern.

morphogenesis The development of form.

mortality The probability that an individual of a given age will die each year.

morula A solid ball of cells in an early mammalian embryo.

motor cortex A region of the brain responsible for movements, with each part of the body specified by a distinct area.

motor neuron (motoneuron) A neuron that directly connects with a muscle and commands muscle contraction; the final common pathway of instructions that direct movement.

mucigel In plants, the slimy substance that the epidermis and the root hairs secrete that enhances the absorption of water and minerals and helps root hairs penetrate the soil.

mucus A viscous, slippery substance that coats the food particles and lubricates their movements within the mouth and the digestive system.

Müllerian mimicry The resemblance of two or more equally dangerous species to each other; the similarities in color or form represent similar dangers to common predators, which therefore avoid all the mimicking species.

multicellular Made of many cells.

muscle fiber A giant muscle cell that contains the proteins responsible for contraction.

mustard oil glycosides Compounds that give cabbage, broccoli, radishes, and other plants in the mustard family their characteristic pungent odor and flavor.

mutagen An agent that increases the rate of mutation; a mutagen can be a chemical or a form of radiation.

mutation A change in the nucleotide sequences of DNA.

mutualism An association between organisms of two species from which both organisms benefit.

mycelium In a fungus, the mass into which hyphae grow, branch, and intertwine.

mycorrhizae Symbiotic association between the root of a plant with the mycelium of a fungus.

myelin An insulating structure around a nerve fiber; myelin consists of extensions of the plasma membrane of a Schwann cell or an oligodendrocyte.

myocardium The muscular walls of the heart.

myofibril A thread, consisting of actin and myosin and about 1 to 2 μm in diameter, that runs the length of a muscle fiber.

myoglobin An iron-containing muscle protein that pulls oxygen from the blood.

myosin A long, two-headed protein that interacts with actin and generates movement in an ATP-dependent manner.

NAD$^+$ (nicotinamide adenine dinucleotide) The major electron acceptor in oxidative phosphorylation.

NADH The reduced form of NAD$^+$.

natural selection The differential reproduction of individuals due to inherited traits; the major mechanism for evolution.

Neanderthal One of the lineages of *Homo sapiens*.

negative feedback The process of neutralizing external changes.

negative regulator A protein that reduces transcription of a particular gene or operon.

nematocyst In cnidarians, a tiny barbed spear fired with water pressure by cnidocytes.

nephridia (singular, **nephridium**) Tubular excretory organs that remove nitrogen-containing wastes from the coelomic fluid and regulate water and salt concentration.

nephron The functional unit of the kidney formed by the glomerulus, Bowman's capsule, and the renal tubule together.

neritic zone The zone of the ocean that is out from shore, over the continental shelf.

nerve growth factor (NGF) A protein paracrine signal necessary for the growth and differentiation of specific nerve cells.

neural crest A set of embryonic cells derived from the roof of the neural tube; neural crest cells migrate to different locations in the embryo and develop into a variety of different cell types in the adult.

neural plate A flat plate above the notochord in a vertebrate embryo; the future nervous system.

neural tube A hollow tube that forms from the neural plate, above the notochord in a vertebrate embryo; the precursor of the central nervous system.

neuromuscular junction The chemical synapse between a motor neuron and a voluntary muscle cell.

neuron A nerve cell specialized for the conduction of electrical and chemical signals.

neurotransmitter A signaling molecule that transmits signals from a nerve cell either to another nerve cell or to a muscle or a gland.

neurulation The process of forming the neural tube and neural crest.

neutrons Neutral particles without electric charge found in the nucleus of an atom.

niche The way an organism uses its environment.

nicotine An addictive and toxic alkaloid in tobacco used as a stimulant drug and insecticide.

nitrification The process by which bacteria convert ammonia to nitrite (NO_2) and nitrate (NO_3) to extract energy.

nitrogen fixation The conversion of atmospheric nitrogen gas into ammonia, which makes nitrogen available to organisms; nitrogen fixation is carried out only by specialized prokaryotes, some of which live within the roots of some plants.

nitrogen-fixing bacteria Bacteria that break the triple bond in atmospheric nitrogen to make ammonia or other nitrogen compounds that plants and animals can use.

nitrogenous wastes Nitrogen-containing products of protein breakdown.

node The region of the stem to which a petiole attaches.

node of Ranvier A gap between myelin wrappings along a nerve axon.

noncompetitive inhibitor A molecule whose inhibitory effects on an enzyme cannot be overcome by increased substrate concentration; a noncompetitive inhibitor usually binds to an enzyme at a location other than the active site.

noncyclic photophosphorylation The production of ATP from light energy by a series of electron transfers that do not directly regenerate the absorbing chlorophyll; in plants, noncyclic photophosphorylation occurs within photosystem II, with electron flow ultimately depending on both photosystems.

nondisjunction The failure of homologous chromosomes or sister chromatids to move apart during meiosis; nondisjunction results in a gamete having too many or too few chromosomes.

nonpolar Having an approximately uniform charge distribution.

nonsense mutation A change in DNA that changes the codon for an amino acid into a nonsense or termination codon and thereby leads to a truncated polypeptide.

norepinephrine A neurotransmitter derived from tyrosine; norepinephrine is used by the central nervous system and the sympathetic nervous system.

normal distribution A distribution of numbers (data) where equal numbers of data points fall on each side of the mean, forming a bell shaped curve in which about 68 percent of the data will fall within one standard deviation of the mean, and about 95 percent will fall within two standard deviations.

notochord A rod that runs from the front to the rear of a vertebrate embryo, beneath its back (dorsal) surface.

nucellus In a flowering plant, the central portion of the ovule in which the embryo develops; the megasporangium.

nuclear envelope The boundary of a nucleus; it consists of a double membrane separated by about 20 to 40 nm.

nuclear fission The breakup of large atomic nuclei, of uranium or plutonium, for example.

nuclear fusion The fusion of two atomic nuclei to form a larger one— formation of a helium nucleus from two hydrogen nuclei, for example.

nuclear pore An interruption of the nuclear envelope that forms channels between the contents of the nucleus and the cytosol.

nuclear reactions The high-energy transformation of individual atoms; occurs in stars and in nuclear reactors.

nuclear transplantation A technique for moving a nucleus from one cell to another.

nuclease An enzyme that catalyzes the hydrolysis of a nucleic acid.

nucleic acid A macromolecule (DNA or RNA) formed by the polymerization of nucleotides.

nucleoid In the prokaryotes, the restricted part of a cell that contains the cell's DNA; it is not surrounded by a membrane.

nucleolus [plural, **nucleoli**] A conspicuous structure within the nucleus, in which ribosomal RNAs are made.

nucleosome A DNA-histone complex, about 11 nm in diameter; each nucleosome contains a 146-nucleotide-long stretch of DNA and eight histone molecules.

nucleotide A small molecule that consists of a nitrogen-containing aromatic ring compound, a sugar, and one or more phosphate groups.

nucleus In eukaryotic cells, the membrane-enclosed structure that contains most of a cell's genetic information in the form of DNA.

obligate aerobes Organisms that require oxygen to live.

obligate anaerobe A microorganism that can grow only in the absence of oxygen.

oceanic zone The zone of the ocean that is beyond the continental shelf over the deepest water.

oceanography The study of the seas and their ecosystems.

octet rule The generalization that an atom is particularly stable and chemically unreactive when its outermost shell is full, meaning (usually) that it contains eight electrons.

oil A triacylglycerol that is liquid at room temperature; the component fatty acids are usually unsaturated.

Okazaki fragment In DNA synthesis, a stretch of DNA that is to be added to the lagging strand.

oligonucleotide A short length of DNA containing about ten nucleotides.

oligotrophic Referring to a pond or a lake that lacks nutrients; oligotrophic lakes are clear and have no permanent algal blooms; they contain more oxygen and support a more diverse community of organisms than eutrophic lakes and ponds.

ommatidia (singular, **ommatidium**) The simple light-detecting units that make up compound eyes.

omnivores Animals that eat plants, herbivores, and other carnivores.

oncogene A gene whose product can change normal cells into cancerlike cells.

oocyte A precursor of an egg cell.

oogenesis The production of mature eggs.

open circulatory systems A circulatory system in which blood mixes freely with extracellular fluids and bathes the organs of the body.

operator The DNA sequence in an operon to which a repressor protein binds.

operculum In fish, a protective bony flap that covers several gill arches on each side of the head; in mollusks, a protective flap that covers the mantle cavity.

operon In prokaryotes, a set of genes transcribed into a single mRNA.

opiates Morphine, heroin, and related compounds.

optimal foraging theory The theory that models how animals optimize their energy intake, safety, and other factors while foraging for food.

orbital A limited portion of an atom's space through which an electron moves.

order A taxonomic group that consists of one or more families; a subdivision of a class.

organ A structural unit with a distinctive function formed by two or more kinds of tissue.

organ formation During embryonic development, the movement and specialization of tissues and cells to produce functioning organs such as heart, kidneys, and the nervous system.

organelle A subcellular structure that performs a specialized task; in eukaryotic cells, many organelles are enclosed by membranes, which isolate the contents of the organelle from the rest of the cytoplasm.

organic Carbon-containing; the term refers to all carbon-containing compounds, even when they have nothing to do with organisms.

organic chemistry The study of the structures and reactions of carbon compounds.

organism A living thing such as a bacterium, plant, or animal.

orgasm A complex of changes that often accompany sexual intercourse—smooth muscle contractions in the genital tract, skeletal muscle contractions throughout the body, and feelings of intense pleasure.

origin of replication A DNA sequence at which replication begins.

oscilloscope A measuring instrument that displays voltages on a televisionlike screen, showing variations with time.

osmosis The flow of water across a selectively permeable membrane as a result of concentration differences.

osmotic pressure The pressure exerted by osmosis.

Osteichthyes A taxon of fish that have bony skeletons.

osteoblasts Specialized cells that resemble fibroblasts and manufacture the bone matrix.

osteoclasts Specialized cells that digest collagen and bone matrix.

ostracoderms Armor-plated jawless fishes, all of which are extinct.

outbreeders Species with complex physical or behavioral adaptations that promote cross-breeding with individuals who are not closely related.

ova [singular, **ovum**] Eggs.

ovarian cycle Refers to the events of the menstrual cycle within the ovaries—the production of the oocytes and the growth of the corpus luteum.

ovary The organ (in an animal or a flower) that produces female germ cells (or gametophytes); a female gonad.

overlapping genes A single stretch of DNA that contains the information for distinct polypeptides in different reading frames.

overshoot Population growth that exceeds the carrying capacity.

oviduct One of two long tubes that lead to the uterus; in humans, it is called a fallopian tube.

ovulation The release of the oocyte from the follicle.

ovule The structure in a carpel that, after fertilization, forms a seed.

ovum [plural, **ova**] A female gamete; an egg.

oxaloacetate A four-carbon compound that can combine with the acetyl group from acetyl CoA to produce citric acid, beginning the citric acid cycle.

oxidative fiber A thin muscle fiber that derives most of its energy from respiration.

oxidative phosphorylation The process that couples the oxidation of NADH and $FADH_2$ to the production of high-energy phosphate bonds in ATP.

oxidizing agent The electron acceptor in a redox reaction.

oxyhemoglobin Hemoglobin that is bound to oxygen.

oxytocin A peptide hormone, released by the posterior pituitary, that stimulates contractions of the uterus during childbirth.

ozone A highly reactive molecule made of three oxygen atoms, instead of the two in atmospheric oxygen.

ozone layer The layer of ozone gas 20 to 50 kilometers above Earth.

pacemaker A group of the cells of the heart muscle capable of rhythmic spontaneous contractions that set the pace of contraction for the rest of the heart

paleontology The study of ancient life.

palisade parenchyma In plants, the mesophyll cells that lie just under the leaf's upper epidermis and are elongated and packed with chloroplasts.

paracrine signal A chemical signal that acts only in the immediate region of its production.

paraphyletic Refers to a taxon that contains some but not all the descendants of the ancestral species.

parapodia In polychaetes, a pair of leglike paddles in a single segment; parapodia are used in respiration and to swim, crawl, or burrow.

parasite An organism that consumes parts of a larger organism; it does not necessarily kill the host.

parasitism A symbiotic relationship in which one species benefits at another's expense.

parasympathetic nervous system A division of the autonomic nervous system; it generally acts to husband resources, for example by slowing the heart and increasing intestinal absorption.

Parazoa The smaller of two subkingdoms of animals, which contains two phyla: the Placozoa and the Porifera (sponges); the parazoa differ greatly from all other animals in showing no symmetry and minimal organization and cell specialization—they possess only simple connective tissues and no organs.

parenchymal cells Plant cells with thin cell walls, chloroplasts, large vacuoles, and the machinery to perform photosynthesis.

parental generation (P generation) The original parents in a genetic cross.

passive transport Movement of a substance across a membrane that occurs spontaneously, without the expenditure of energy.

pathogens Agents that cause disease.

pattern formation The creation of a spatially organized structure during embryonic development.

pedigree A family tree; pedigrees are often used to show the inheritance of a disease within a family.

pelagic division The open waters of the ocean and the organisms within them.

penetrance The fraction of individuals with a particular genotype that show a corresponding phenotype.

pepsin The stomach enzyme that hydrolyzes proteins into smaller fragments each containing a few amino acids.

peptide A molecule that consists of amino acids linked by peptide bonds.

peptide bond The covalent bond between two amino acid molecules in a polypeptide; these bonds are formed by the carboxyl group of one amino acid attaching to the amino group of another amino acid.

peptidoglycan The cell-wall material of many prokaryotes; peptidoglycan may consist of a single covalently linked molecule formed from polypeptides and polysaccharides.

perennial A plant that lives and produces seeds for two or more years.

perforation plate In the vascular system of a plant, the end wall of each vessel element; the perforation plate contains one or more holes through which water flows.

pericardium A fibrous sac that encloses the heart itself within a watery lubricating fluid.

pericycle In plant roots, a sheath of parenchyma cells just within the endodermis, but outside the vascular tissue.

period In the geological time scale, a time interval of 30 to 75 million years that contains distinctive forms of life in its fossil record; a period is longer than an epoch, but shorter than an era.

peripheral nervous system (PNS) Nerve cells outside the central nervous system (CNS); the PNS carries information from the CNS to muscles and organs and to the CNS from sense organs.

peristalsis Waves of contractions in smooth muscle that, for example, push food through the digestive tract.

permafrost Permanently frozen ground less than a meter from the surface that underlies the soil during even the warmest summers.

petal One of the showy, usually colored, parts of a flower.

petiole The stalk that connects the leaf to the stem.

pH The concentration of H^+ ions in a solution; a pH below 7 is acidic and a pH above 7 is basic.

phage A virus that infects bacteria.

phagocytosis A type of cellular ingestion in which the cell's membrane surrounds a relatively large solid particle, such as a microorganism or cell debris.

pharyngeal slits (gill slits) Holes in the sides of the body that run from the inside of the pharynx to the outside surface of an animal.

pharynx The throat; a common entryway for food into the digestive tract, for air into the lungs, and for water into the gills.

phenolic compounds Aromatic substances that play a variety of roles in plants; some phenolics repel herbivores and pathogens, some attract pollinators or fruit dispersers.

phenotype The collection of all the properties of an individual organism.

phenotypic plasticity Variability in phenotype associated with a given genotype, usually due to environmental influences; used in reference to whole organisms and whole genomes rather than particular genes (see expressivity).

phenylketonuria (PKU) A human disease that results from the absence of the enzyme phenylalanine hydroxylase, which converts phenylalanine to tyrosine.

pheromone A substance secreted by one organism that influences the behavior or physiology of another organism of the same species; a pheromone can also activate modal action patterns.

phloem Conducting vessels that distribute the sugars and other organic molecules made in the leaves to the rest of the plant.

phosphate The ion (PO_4^{-3}) formed by the dissociation of hydrogen ions from phosphoric acid (H_3PO_4).

phosphodiester bond The links between nucleotides formed by the phosphate group of one nucleotide attaching to a carbon atom in the sugar component of another nucleotide.

phospholipid An amphipathic derivative of glycerol in which two hydroxyl groups attach to fatty acids and the third to a phosphate ester; phospholipids are principal components of biological membranes.

photoautotroph An organism that supplies itself with energy by means of photosynthesis.

photochemical reaction center A complex of chlorophyll molecules and proteins that converts captured light energy to chemical energy.

photoheterotroph A heterotroph that uses light energy but also requires organic compounds.

photon A package of energy; a light particle.

photoperiodism The response of a plant to the relative lengths of day and night.

photorespiration In plants, an oxygen-dependent process that does not produce useful energy molecules.

photosynthesis Process in which sugars are synthesized within the chloroplast and temporarily stored there.

photosystem I One of two distinct but interacting sets of electron transfer reactions responsible for storing light energy in high-energy chemical bonds; photosystem I best absorbs and uses light with wavelengths of about 700 nm.

photosystem II One of two distinct but interacting sets of electron transfer reactions responsible for storing light energy in high-energy chemical bonds; photosystem II best absorbs and uses light with wavelengths of about 680 nm.

phototropism The bending of a plant toward light.

phragmoplast A set of microtubules that extends between two dividing plant cells at right angles to the cell plate.

phylogeny Evolutionary history.

physiology The study of how living organisms are organized and how their parts work together.

phytochrome A plant pigment involved in many processes that depend on the timing of dark and light, including flowering, germination, and leaf formation.

pigment A molecule that absorbs visible light and has color to human eyes.

pilus (plural, **pili**) In bacteria, a long appendage that serves as the means of attachment of conjugating bacteria and a conduit for the transfer of DNA.

pinocytosis A type of endocytosis; the nonspecific uptake of bits of liquid and dissolved molecules.

pioneer community The first community in a succession.

pistil In a flower, a female reproductive structure.

pith In plants, the unspecialized tissue within central nonconducting core of the vascular system.

pituitary gland A pea-sized structure at the base of the brain that releases at least nine hormones.

placenta Tissue that passes nourishment, oxygen, and wastes between the mother and the developing embryo in placental mammals.

placental mammals Mammals whose embryonic development takes place entirely within the uterus of the mother.

placoderms Extinct armored fish with jaws.

plankton Organisms that float in the water, carried by currents.

Plantae The kingdom that includes plants, autotrophic multicellular organisms that undergo embryonic development.

plasma The fluid part of blood.

plasma cell A cell specialized for the production of antibodies.

plasma membrane The membrane that surrounds a cell.

plasmid A circular DNA that can replicate autonomously.

plasmodesmata [singular, **plasmodesma**] Fine intercellular channels between plant cells, derived from vesicles trapped in the growing cell plate.

plastid A plant organelle surrounded by a double membrane; plastids include chloroplasts, chromoplasts, and amyloplasts.

plate In reference to the Earth's crust, a large block that moves with respect to other blocks at the rate of a few centimeters a year.

platelet A small, membrane-enclosed element that is a component of mammalian blood; it is formed as a cytoplasmic fragment of a precursor cell in the bone marrow; platelets contribute to clotting.

Platyhelminthes The phylum of flatworms.

platyrrhines Monkeys whose flat noses have widely separated nostrils that point sideways.

pleiotropy The capacity of a single gene to affect many aspects of phenotype.

point mutation A change in a single nucleotide pair of DNA.

polar Having uneven distributions of electrical charge; having positive and negative ends (or poles).

polar body The smaller daughter cell that results from the unequal meiotic division of an oocyte.

polar fronts Regions of low pressure, where the westerlies end and polar easterlies begin.

pollen tube An extension of a pollen grain, which carries the sperm nuclei into the ovule.

polygenic Traits governed by many genes that vary smoothly and continuously within a population.

polymer A large molecule that consists of smaller identical (or nearly identical) subunits, called monomers.

polymerase chain reaction (PCR) A method that specifically and repetitively copies a segment of DNA between two defined nucleotide sequences.

polymerization The assembly of many small molecules into a larger one.

polymorphic Having two or more distinct phenotypes within a population; in genetics, having two or more alleles of a given gene.

polynucleotide A chain of nucleotides such as DNA or RNA held together by phosphodiester bonds.

polyp A body plan that resembles a cylinder; one of the two basic body plans of the cnidarians.

polypeptide A chain of amino acids held together by peptide bonds.

polyphyletic Refers to a taxon that includes descendants of more than one ancestor.

polyploid Having three or more complete sets of chromosomes, as in some plants.

polysaccharide A carbohydrate macromolecule formed by the polymerization of simple sugars (monosaccharides).

polysome Several ribosomes attached to a single piece of mRNA.

polyunsaturated A hydrocarbon chain (or fatty acid) with more than one double bond.

Pongidae Gorillas, chimpanzees, and orangutans; all the great apes except humans.

population A breeding group of individuals of the same species that inhabit a common area.

population genetics The quantitative study of the processes by which variation is generated and passed on within populations.

population momentum The continued growth of a population even though completed family size is below replacement level.

positive assortative mating Choosing mates on the basis of similarities to the individual's own genotype.

positive regulator A protein that increases the rate of transcription.

postabsorptive state The state of an animal in which cells derive energy and building blocks by breaking down stored glycogen, fats, and proteins.

posterior The back (tail-end) of an animal.

postsynaptic Refers to the neuron that is "downstream" from a synapse; a postsynaptic neuron responds to the presynaptic neuron.

postsynaptic potential A voltage change induced by the action of a neurotransmitter.

posttranscriptional processing A series of chemical modifications that convert the primary transcript of a gene to a mature RNA.

posttranslational Occurring after translation is completed.

postzygotic barrier Barrier to gene flow which makes a zygote either inviable (certain to die) or sterile.

potency What a cell or tissue could become during development if it were allowed to develop in another environment.

potential energy A general term for energy that can ultimately be converted into kinetic energy.

prebiotic evolution The evolution of organisms from nonliving matter.

Precambrian Referring to fossils that date from more than 590 million years ago, that is, before the beginning of the Cambrian Era.

predator An organism that usually (but not always) kills its prey and consumes most of its prey's body.

preformation The theory, now discredited, that all the parts of the adult organism already preexist at the earliest stages of life.

pre-mRNA The primary transcript of a protein-coding gene; the term is also used to refer to the partly modified, not yet mature mRNA precursor.

pressure flow hypothesis The accepted explanation for the transport of sap through the phloem.

presumptive Having a specified developmental fate.

presynaptic Refers to the neuron that is "upstream" from a synapse; an action potential in the presynaptic neuron can trigger the release of neurotransmitter into the synapse.

prezygotic barrier A barrier to gene flow that prevents the fusion of the sperm and egg to form a zygote (syngamy).

primary cell wall The wall that lies outside the plasma membrane.

primary growth In a plant, growth from apical meristem tissue; the extension of length.

primary lysosome A nearly spherical lysosome that has not yet fused with an endosome.

primary oocyte A precursor of an ovum; a diploid cell that is capable of dividing, by meiosis, to form a haploid ovum.

primary organizer The term used by Spemann to describe the dorsal lip of the blastopore, meaning that its action established the organization of the entire early embryo.

primary productivity The productivity of the first trophic level, the energy captured in the chemical bonds of new molecules each year for each square meter.

primary structure The linear sequence of amino acids in each polypeptide chain of a protein.

primary succession The invasion of a completely new environment such as a sandbar or new volcanic island.

Primates The order of mammals that includes humans, apes, monkeys, and lemurs.

primer An already existing polynucleotide to which additional nucleotides are added; all known DNA polymerases add nucleotides to a primer.

primitive characters Traits shared by all the members of a lineage and also shared with more distant relatives. In members of the cat family, for example, long canine teeth are a primitive trait because they are shared with other members of the order Carnivora, such as canids (dogs) and bears.

primordial germ cells The precursors of both sperm and eggs, which arise early in development.

principle of independent assortment The principle that the alleles for one gene segregate independently of the alleles of another gene.

principle of segregation The principle that a sexually reproducing organism has two genes for each characteristic, and these two copies segregate (or separate) during the production of gametes.

proboscis A long, sensitive, retractable, tube-like snout or mouthpart; in elephants and tapirs; in some insects, which use it to drink; in ribbon worms, in which the proboscis may be venomous.

producers Organisms, such as plants, that harvest energy directly from sunlight or, rarely, from inorganic molecules.

progesterone A steroid hormone secreted by the ovaries and placenta in females that stimulates growth of the uterine lining for pregnancy and the mammary glands for lactation.

proglottid One of the repeated segments of a tapeworm; a proglottid contains both male and female reproductive organs.

programmed cell death See apoptosis.

prokaryotic Cells that contain neither a nucleus nor other membrane-enclosed organelles.

prolactin A peptide hormone, made in the anterior pituitary, that stimulates milk production.

prometaphase Previously called early metaphase, the stage of mitosis during which the nuclear membrane disappears and the chromosomes attach to the spindle fibers.

promoter A DNA sequence that specifies the starting point and direction of transcription; it is where RNA polymerase first binds as it begins transcription.

prophage The dormant form of a bacteriophage in which the phage DNA has integrated into that of the host cell.

prophase The first phase of mitosis, when the diffusely stained chromatin resolves into discrete chromosomes, each consisting of two chromatids joined together at the centromere.

prostaglandin One of 16 paracrine signals derived from a 20-carbon fatty acid called arachidonic acid.

prostate gland In the reproductive tract of a male mammal, a gland that secretes a thin milky alkaline fluid into the lumen of the urethra.

protease A digestive enzyme that catalyzes the hydrolysis of peptide bonds; used to digest proteins.

protein A macromolecule consisting of one or more polypeptides.

proteinoids Polypeptides thought to have been the first to appear in prebiotic evolution; proteinoids are almost certainly without specific sequences of amino acids.

Protista The kingdom that consists mostly of single-celled organisms but that also contains some related multicellular species; protists include algae, water molds, slime molds, and protozoa.

protobiont A primitive cell—thought to represent a stage of prebiotic evolution—that could concentrate organic molecules and begin to evolve the first metabolic pathway.

proto-oncogene (cellular oncogene) The cellular counterpart of a viral oncogene; proto-oncogenes participate in the normal control of growth and differentiation.

proton A positively charged particle found in the nucleus of an atom. A single hydrogen atom consists of one proton and one electron. When a hydrogen atom is stripped of its electron, it becomes a hydrogen ion (H^+), or proton.

proton channel The route that protons (or hydrogen ions) follow as they flow down an electrochemical gradient.

proton gradient The difference in proton concentration across a membrane.

protostomes Those animals in which the mouth develops first, from the blastopore; protostomes include mollusks, annelids, and arthropods.

proximal tubule The tubule that lies just next to Bowman's capsule, in the kidney's outer cortex.

pseudocoel A cavity in certain invertebrates that lacks the mesodermal lining of a true coelom.

pseudopodia Temporary extensions of a cell's cytoplasm.

pulmonary Relating to the lungs.

pulmonary artery The artery that carries deoxygenated blood from the heart to the lungs.

pulmonary circulation Circulation from heart to lungs to heart.

pulmonary veins The veins that carry oxygenated blood from the lungs to the heart.

punctuated equilibrium The idea that species change very little most of the time (stasis), and that most anatomical or other evolutionary change in individual species occurs during a geologically brief period at the time of speciation.

pupa An inactive phase, following a larval stage, of insect development; a pupa may be enclosed in a cocoon.

purine A nitrogenous base that contains a particular nine-membered double ring, with five carbon and four nitrogen atoms; adenine and guanine are purines.

pyramid of biomass A graphic presentation of the total mass of organisms at each trophic level of an ecosystem.

pyramid of energy A graphic presentation of the amount of energy used at each trophic level in a specified time (usually a year).

pyrimidine A nitrogenous base that contains a particular six-membered ring, with four carbon and two nitrogen atoms; thymine, uracil, and cytosine are pyrimidines.

pyruvate A three-carbon compound that is the end point of the first stage of glycolysis.

quaternary structure The relationship among separate polypeptide chains in a protein.

R group (a side chain) Each amino acid's characteristic group of atoms.

races (or subspecies) Morphologically distinct subpopulations that can interbreed.

radial symmetry Structural symmetry in which rotation along the central axis doesn't change the appearance.

radula In mollusks, a rasping tongue covered with teeth made from chitin.

rain shadow The area adjacent to a mountain range, away from the prevailing winds, where little rain falls.

random drift Changes in gene frequency not caused by selection, mutation, or immigration, but by random events.

reabsorption In the kidney, the transport of specific substances such as salt, water, and glucose from the filtrate back into the blood.

reading frame The grouping of nucleotides into codons that specify an amino acid sequence.

realized niche The resources that a species uses in a particular community.

receptor A protein to which a chemical signal first binds.

receptor-mediated endocytosis A cell's uptake of specific substances that are recognized by receptor proteins on the plasma membrane.

recessive The term referring to an allele that does not contribute to the phenotype of a heterozygote.

reciprocal altruism An explanation for apparently selfless behavior; the hypothesis that animals engage in altruism with the expectation that the recipient would return the favor some time in the future.

recombinant Containing a combination of genes not found in nature.

rectum The straight portion at the end of the descending colon in which feces are stored until their elimination through the anus.

reducing agent The electron donor in a redox reaction.

reductionism The effort to understand the whole in terms of the parts.

reflex The most automatic behavior pattern, with motor activity directly responding to a sensory stimulus.

refractory period In an axon, the period during which the further excitation does not result in channel opening and impulse propagation.

regulatory cascade A set of reactions that amplify a signal from a relatively small number of molecules into a response that affects many more molecules.

relative abundance The proportional representation of a species in a sample or a community.

renal artery The artery that carries blood to each kidney.

renal tubule In the kidney, a long narrow tube leading away from Bowman's capsule.

renal vein The means by which blood leaves the kidneys.

renin An enzyme secreted by the kidneys.

replacement reproduction The family size at which each couple is replaced by just two descendants.

replication Copying of a single DNA molecule (or a single set of DNA molecules) into two copies.

replication fork A Y-shaped region of DNA where the two strands of the helix have come apart during DNA replication.

replication unit In eukaryotic DNA replication cells, a group of 20 to 50 origins of replication that form replication forks at the same time.

reproductive unit A species as the group of potential mates.

reproductively isolated The term referring to populations that are unable to interbreed.

reptile An air-breathing, egg-laying vertebrate with an external covering of scales or horny plates; reptiles include snakes, lizards, crocodiles, turtles, and dinosaurs. Birds are descended from dinosaurs and are therefore considered within the same taxonomic group as lizards, snakes, and dinosaurs.

resilience One definition of ecosystem stability; the speed with which an ecosystem returns to a particular form following a major disturbance such as a fire.

resistance factor (R factor) A plasmid that contains a gene for an enzyme that inactivates an antibiotic.

resolution (1) The minimum distance between two objects that allows them to form distinct images; (2) the part of the sexual response cycle in which blood flow and muscle tension return to normal after orgasm.

resource partitioning Splitting an ecological niche.

respiration The oxygen-dependent extraction of energy from food molecules.

respiratory system All the structures responsible for the exchange of gases between the blood and the external environment.

resting potential The voltage across the cell membrane when a cell is not producing action potentials.

restriction enzyme An enzyme that cuts DNA at a particular sequence.

retrovirus An RNA virus that uses reverse transcriptase for viral replication.

reuptake The pumping of transmitter from the synaptic cleft either into the presynaptic neuron or into surrounding glial cells.

reverse transcriptase An enzyme that copies RNA into DNA; present in HIV and RNA tumor viruses.

reversion A mutation that restores the previous version of a gene sequence.

rhizoid A thin, rootlike filament that enables mosses, liverworts, and fern gametophytes to anchor and to absorb nutrients.

rhizome A horizontal stem that spreads on or below the ground from which shoots grow.

rhodopsin A protein that detects light in the retina.

rhythm method A birth control method; a form of modified abstinence, in which a couple refrains from sexual intercourse at times in the monthly cycle when conception is likely to occur.

ribose A five-carbon sugar that is a constituent of RNA (ribonucleic acid).

ribosomes Complex assemblies of RNAs and proteins, about 15 to 30 nm in diameter, that are responsible for carrying out protein synthesis.

ribozyme An artificial RNA enzyme.

RNA polymerase The enzyme responsible for transcribing DNA into RNA.

root The part of a plant below the ground.

root cap A region of the growth zone of a plant's root tip.

root hairs Tiny projections, each of which is the extension of a single epidermal cell, through which most of the water that enters a root comes.

root pressure Osmotic pressure generated by the difference in solute concentrations between root tissues and soil solution.

root system An underground system that anchors the plant and absorbs water and minerals from the soil.

rough ER Endoplasmic reticulum that is associated with ribosomes.

RU 486 A drug used to induce abortions within nine weeks of conception.

rudiment Initial embryonic stage of an organ, from which the final form will develop.

ruminants Hoofed, horned herbivores that can regurgitate partly digested food, called cud, for further chewing.

S (for DNA synthesis) The period of the cell cycle during which DNA replicates.

salivary glands Exocrine organs in the mouth that secrete both enzymes and mucus.

saltationism The view, held by many 19th-century biologists, that evolution occurred by fits and starts.

saltatory conduction The jumping movement of an action potential down a myelinated axon, from one node of Ranvier to the next; saltatory conduction occurs at rates up to 100 times faster than conduction down an unmyelinated axon.

sample A representative selection of a population whose statistical qualities give information about the whole population.

sands Soils made up of large particles (from 20 mm to 200 mm).

sap A watery liquid containing sugars, minerals, and signaling molecules that moves through the phloem of plants.

saprophyte A plant that decomposes dead material.

saprophytic Deriving nourishment from dead organisms.

sarcomere In muscle, a small repeating cylinder within a myofibril; each sarcomere is about 2.5 μm long; the contractile unit of a striated muscle.

saturated A hydrocarbon chain (or fatty acid) in which all the bonds between carbon atoms are single bonds; the chain is called saturated because it contains the maximum number of hydrogen atoms.

savanna A tropical or subtropical grassland punctuated by solitary trees or small clumps of trees.

scanning electron microscope (SEM) A type of electron microscope in which electrons are reflected from the surface of the observed object.

scavenger An animal that feeds from whole carcasses.

scientific method A formal manner of formulating and testing hypotheses.

sclerenchyma A plant tissue that consists of cells with thick cell walls; may be fibers or sclereids.

scolex A specialized attachment organ at the anterior end of a tapeworm.

scrotum A pouch that lies outside the body that holds the testes.

Second Law of Thermodynamics The statement that, while the total energy in the universe does not change, less and less energy remains available to do work.

second messenger An intracellular molecule whose synthesis and degradation depends on the action of an extracellular signal and which stimulates changes inside a target cell.

secondary cell wall The wall that lies between the cell membrane and the primary cell wall.

secondary growth Growth at the lateral meristems that increases the thickness of a shoot or a root.

secondary immune response A robust immune response triggered by the reappearance of a previously encountered antigen or microbe.

secondary oocyte A precursor of the ovum; a haploid cell produced from the primary oocyte by meiosis.

secondary plant compounds Chemicals that are not essential to a plant's normal metabolism, but which often serve a defensive purpose.

secondary sexual characteristics Hallmarks of sexual differentiation that appear at sexual maturity; in humans these include an increase in body size, a deepening of the voice, hair growth, and the development of sexual drive.

secondary structure Regular local structures resulting from regular hydrogen bonding within adjacent stretches of a polypeptide backbone; secondary structures include α helices and β structures.

secondary succession The sequence of stages in a community that has suffered serious damage.

secretion In the kidney, transport into the filtrate; secreted substances include potassium and hydrogen ions, ammonia, and organic acids and bases; the process by which cells make and discharge substances.

secretory vesicles In eukaryotic cells, small membrane-enclosed vesicles that transport loads of newly made glycoproteins to the cell membrane for secretion outside the cell.

sedimentary rock Rock formed by pressure from succeeding layers turning layers of mud, sand, and other sediment into rock.

seed plant Any plant that bears seeds, including the gymnosperms and angiosperms. Seeds are an adaptation to terrestrial life.

segregation The separation of two homologous chromosomes during meiosis.

selectively permeable Allowing the passage of some ions and molecules (especially of water) much more rapidly than others.

semen The sperm cells, together with the fluid from the seminal vesicles, the prostate gland, and the bulbourethral glands.

semiconservative The pattern of DNA replication in which half of each parent molecule (one strand) is present in each daughter molecule.

seminal vesicles In the male reproductive tract, organs that secrete sugars and other nutrients into the semen.

seminiferous tubules The tightly coiled ducts in the testis in which the sperm matures.

senescence Aging; deterioration; in plant cells, the breakdown of cellular components leading to cell death.

sensitive period The period of time during which an animal can learn a particular behavior pattern.

sensitization An increased behavioral response to a noxious stimulus.

sepal One of the parts at the base of a flower; sepals are usually small and green, but in some flowers they may be large and colorful.

septa In a fungus, the walls that separate the nuclei within a filament.

serotonin A signaling molecule; a CNS neurotransmitter derived from the amino acid tryptophan involved in sleep, alertness, and mood; also constricts blood vessels and stimulates inflammation.

Sertoli cells In the testes, specialized cells that surround the developing sperm; they nourish the developing sperm, regulate the passage of nutrients from the blood, and secrete a fluid that fills the lumen.

sessile Used to describe an animal permanently anchored to a solid object.

seta (plural, **setae**) In annelids, a stiff, bristlelike projection.

sex chromosomes The chromosomes that differ between males and females; in humans, the 23rd pair of chromosomes.

sex steroid A steroid hormone that stimulates, maintains, and regulates reproductive organs and secondary sexual characteristics.

sex-determining region In male mammals, a gene on the Y chromosome that encodes a regulatory protein that promotes the sexual differentiation of the embryonic gonads into testes.

sex-linked gene A gene that has a different pattern of inheritance in males and females because the gene lies on a sex chromosome (usually the X chromosome in mammals).

sexual dimorphism A phenotypic difference between the sexes.

sexual reproduction A process that produces offspring that have inherited genetic information from two parents rather than one; because the genes from each parent are likely to differ, sexual reproduction provides new combinations of genes.

sexual selection The differential ability of individuals with different genotypes to acquire mates.

shared derived characters Traits shared by a lineage of organisms but not shared with more distant relatives. Among mammals, for example, fur is a shared derived character because other vertebrates do not have fur. In primates, fur is a primitive character, one that is shared with all other mammals. Cladists construct maps of relatedness, called cladograms, using shared derived characters.

shear The twisting action created by forces that are not opposite one another.

shell In an atom, a group of orbitals whose electrons have nearly equal energy.

shoot The part of a plant above the ground.

shoot system Stems, photosynthetic leaves, and flowers and other organs of reproduction.

sibling species Two species that are extremely similar but reproductively isolated from one another.

sickle cell anemia A blood disorder that gets its name from the curled appearance of red blood cells in sickle cell patients; a genetic disease caused by a mutation in the gene encoding the β-polypeptide of hemoglobin.

sieve plate In sieve tubes of a plant's phloem, the top and bottom end walls of each sieve tube member.

sieve tube members In the phloem of a plant, the cells that actually conduct fluid.

sieve tubes In the phloem of a plant, pipelike channels that carry organic matter from the leaves.

sign stimulus The key aspect of an object that triggers a modal action pattern.

silent mutation A change in DNA that has no effect on phenotype.

silt A soil of medium-sized particles (from 2 mm to 20 mm).

simple diffusion The random movement of like molecules or ions from an area of high concentration to an area of low concentration.

simple pit pair A region between plant cells where the secondary wall is absent, allowing small molecules to pass.

simple transposon The simplest transposable element; a short length of DNA that can move from place to place in a genome.

sink The site of storage or consumption of a substance.

sinoatrial (SA) node The pacemaker, which lies near the top of the right atrium in the mammalian heart.

sinus An open cavity; in insects and other arthropods and in most mollusks, it is a space through which blood travels.

sister chromatids The two chromatids that make up a single chromosome; the sister chromatids are duplicate copies of the same genetic information.

skeletal muscles The voluntary, striated muscles of vertebrates, which are often attached to bones.

skull The bony (or cartilaginous) case that encloses and protects the brain; it lies at the

forward (or, for humans, the top) end of a vertebrate's skeleton.

sliding filament model The accepted view of muscle movement, which holds that movement results from changing the relative positions of thick and thin filaments rather than from filament contraction.

slime A thick, slippery web of polysaccharides secreted by many organisms; usually protective.

smooth ER Endoplasmic reticulum that is not associated with ribosomes.

smooth muscle The involuntary muscles that line the walls of hollow internal organs such as intestines and blood vessels.

sociobiology (renamed **behavioral ecology**) The study of behavior from an evolutionary perspective.

sodium-potassium pump An active transporter that transports sodium ions (Na^+) out of cells and potassium ions (K^+) into cells, using the energy of ATP.

solute A substance that dissolves within a solvent.

solvent Any fluid in which other substances dissolve.

somatic cells In a multicelled organism, the cells that do not give rise to germ cells or gametes.

source The site of production (for example, of sugars in a plant).

speciation The process by which new species form.

species A biological species is a group of actually or potentially interbreeding populations that are reproductively isolated from other such groups. Species may also be defined differently, for example, according to phenetics or cladistics.

species diversity A measure of diversity that takes into account how common individuals of each species are.

species richness The number of species in an ecosystem.

sperm (spermatozoan) A male gamete.

spermatid The four haploid cells produced during the two meiotic divisions (meiosis I and II).

spermatogenesis The development of spermatogonia into mature sperm.

spermicide A cream or jelly that kills sperm; an essential part of the effectiveness of both cervical caps and diaphragms.

sphincters Rings of muscles that surround a tube.

spiracle In the insect respiratory system, an opening through which air enters the body; a spiracle leads directly to the tracheae.

spirilla [singular, **spirillum**] Spiral-shaped prokaryotic cells.

spirochetes Anaerobic heterotrophs that have a distinctive corkscrew shape.

spongy bone tissue Tissue that generally lies at the ends and inside the bones.

spongy parenchyma In a plant leaf, the mesophyll cells that lie just under the palisade parenchyma that are irregular but rounded in shape and separated by numerous air spaces.

spontaneous abortion (miscarriage) In mammalian development, embryonic or fetal death, usually resulting from congenital defects.

sporangium In plants and fungi, a structure within which cells undergo meiosis to produce the haploid spores.

spore A reproductive cell that can develop into a new individual, produced by various organisms including seedless plants, algae, fungi, and some protists; often with a protective covering that allows prolonged dormancy.

sporophyll A leaf or leaflike structure that is specialized for the production of spores; in flowering plants, the term refers to carpels and stamens; in ferns sporophyll refers to a leaf that bears sporangia.

sporophyte The diploid form of a life cycle characterized by alternation of generations; a sporophyte produces haploid spores that give rise to haploid gametophytes by meiosis.

sporozoans Parasitic protists, including those that cause malaria; sporozoans undergo alternation of generations.

stabilizing selection Selection that tends to act against extremes in the phenotype, so that the average is favored.

stamen In a flower, a male reproductive organ; the male sporophyll.

standard deviation An estimate of the variation in a set of data calculated from how much the data deviate from the mean.

starch Polysaccharides used in plants for long-term energy storage.

start The "point of no return" in the G_1 phase of the cell cycle; once a cell proceeds beyond Start, it proceeds through the rest of the cycle, including mitosis and cytokinesis.

stasis Lack of evolutionary change over long periods of time.

statistics A branch of mathematics; the analysis and interpretation of numerical data in terms of samples and populations.

statolith Gravity detector in a plant or animal.

stem cell An undifferentiated cell from which differentiated cells can develop.

steric inhibition Inhibition of an enzyme by a molecule whose shape resembles that of the substrate.

sternum The breastbone.

steroid A lipid-soluble molecule derived from cholesterol; many steroids are used as hormones.

stoma [plural, **stomata**] In a plant leaf, a tiny mouthlike pore that opens and closes to regulate the flow of carbon dioxide and other gases to the interior of the leaf.

stomach The most dilated and most muscular section of the digestive tract.

stop codon A codon in mRNA that signals the end of a polypeptide chain.

stroke Failure of the blood supply to the brain.

stroke volume In the action of the heart, the volume of blood delivered by a ventricle.

stroma The region within a chloroplast enclosed by the inner chloroplast membrane.

suberin The waxy substance, which is impermeable to water, that surrounds each cell of the endodermis.

subsidize Referring to the growth of crop plants, to provide additional energy.

substrate A reacting molecule in an enzyme reaction; a substrate is usually (but not always) much smaller than an enzyme.

succession Progressive, predictable change over time in kinds of species and numbers of species in a community.

successional communities Intermediate stages of succession.

sucrose Table sugar; a disaccharide consisting of two monosaccharides, glucose and fructose, linked together.

sugar A simple carbohydrate; a molecule that has the equivalent of one molecule of water (that is two hydrogen atoms and one oxygen atom) for every atom of carbon.

supernormal stimulus A stimulus even more stimulating than anything normally encountered in nature.

superorganism A group of individuals that cooperate more closely than the individuals of a colony and almost as closely as the cells of an individual.

surface tension The tendency of a liquid to form a smooth round surface.

surface-to-volume ratio The amount of surface area for each bit of volume.

surfactants Compounds that reduces the surface tension of liquids; occur naturally in the lungs; synthetic surfactants may be endocrine disrupters.

survivorship curve A graph that shows the fraction of a population that is alive at successive ages.

suture An immovable joint that fuses separate bones, as in the skull.

swim bladder An air-filled sac that helps fish control their buoyancy.

symbiosis A close association between two organisms.

sympathetic nervous system A division of the autonomic nervous system; the sympathetic nervous system initiates the "fight or flight" reaction.

sympatric speciation The splitting of one species into two without geographical isolation.

synapse The distinct boundary between two communicating neurons.

synapsis The pairing of homologous chromosomes in prophase I.

synaptic cleft A space of about 20 nm that separates the presynaptic and postsynaptic cells in a chemical synapse.

synchronous cell populations Cells that are all at the same stage of the cell cycle.

syngamy The coming together of an egg and a sperm at fertilization.

systematics The scientific study of the kinds and diversity of organisms and of any and all relationships among them.

systemic circulation The route of the blood through the body, minus pulmonary circulation.

systole The part of the heartbeat in which the atria and ventricles contract.

T cell (T lymphocyte) A kind of immune cell that matures in the *t*hymus gland; recognizes and destroys body cells that have become infected or cancerous.

tagmata (singular, **tagma**) Fused segments of an arthropod exoskeleton.

taiga The broad band of coniferous forest that extends across Canada, Alaska, Scandinavia, and Siberia.

tannin A secondary compound commonly found in woody plants.

taproot A single vertical root found in many dicots.

tar An oily, viscous material, consisting of hydrocarbons, found in cigarette smoke and the insides of chimneys.

target organ A structure upon which a hormone acts.

taxis Directed movement.

taxon [plural, **taxa**] A general term for any group of organisms at any level of the classification hierarchy.

taxonomy The naming and grouping of organisms.

T-cell receptor A protein on the surface of a T cell that can bind to a complex of an antigen and MHC molecule, so-called "altered self."

telophase The last phase of mitosis, during which the mitotic apparatus (including kinetochore, polar, and astral microtubules) disperses and the chromosomes lose their distinct identities.

temperate deciduous forests Temperate forests that receive 80 to 140 centimeters of precipitation each year, composed of trees that lose their leaves in the winter.

temperate grasslands Regions with well-defined seasons, with hot summers and cold winters; low annual rainfall keeps grasslands from turning to deserts but is not enough to sustain the growth of trees.

template A guide for the assembly of a complementary shape; in the context of DNA or RNA synthesis, one strand acts as a template for the assembly of a complementary sequence.

temporal isolation Reproductive isolation that results from differences in the times at which two populations reproduce.

ten-percent rule The generalization that the organisms of any trophic level provide the next higher trophic level with only 10 percent of the energy that they have assimilated from the lower trophic level.

tendon A type of connective tissue that attaches the muscles to the bones.

tension The pulling action of two opposing forces.

teratogen A compound that causes birth defects.

terminal bud A bud that lies at the tip of a shoot.

termination The ending of chain growth.

terpene One of a class of unsaturated hydrocarbons that form the largest class of secondary compounds in plants.

terrestrial Living on land.

territory An area occupied by an individual (or group of individuals), from which other individuals of that species are excluded.

tertiary structure (or **conformation**) The three-dimensional spatial arrangement of all the atoms of a polypeptide.

testable A hypothesis is testable if there exists an experiment that would disprove the hypothesis if it were incorrect.

testicles Human testes.

testis (plural, **testes**) A male gonad, which produces sperm.

testosterone A steroid hormone produced in the testicles in males involved in sperm production, sex drive, and development of secondary sexual characteristics.

tetrad A united chromosome pair visible during meiosis I consisting of four chromatids.

theory A system of statements and ideas that explains a group of facts or phenomena.

theory of island biogeography MacArthur and Wilson's theory that species diversity on an island is a function of the size of the island and its distance from a source of colonizing species, whether the mainland or other islands.

thermoacidophiles One of the three phyla of Archaea; the bacteria that inhabit hot sulfur springs such as those in Yellowstone National Park.

thermodynamics The study of the transformations and relationships among different forms of energy.

thick filament In striated muscle, a myosin filament, which is 14 nm in diameter.

thin filament In striated muscle, an actin filament, which is 7 nm in diameter.

thoracic cavity Chest cavity; that part of the coelom that encloses the heart and lungs.

thorax The middle portion of an arthropod.

thylakoid A flattened disc surrounded by the innermost membranes of a chloroplast.

thylakoid membrane A membrane, within the stroma of a chloroplast, that delineates stacked vesicles, called grana.

thymine dimers Two adjacent T nucleotides linked together in the same DNA strand; they are formed as a result of exposure to ultraviolet light.

thyroid stimulating hormone (TSH) A peptide hormone, made in the anterior pituitary, that stimulates the production of thyroxin in the thyroid gland.

tidal volume The amount of air drawn in and then expelled in a single breath.

tight junction The fusion of the membranes of adjacent cells into an impermeable sheet, so that materials cannot pass amongst the cells.

tissue A group of similar cells and associated intercellular material that performs one or more specific functions; tissues are often integrated with other tissues to form an organ.

tolerance In the immune system, a state of induced unresponsiveness to proteins and other molecules; tolerance prevents immune attacks on the body's own components.

totipotent Able to develop into a whole organism.

trachea [plural, **tracheae**] A tube, which arises by invagination, that carries air within an animal's body; in vertebrates, the windpipe.

tracheids Long, thin, spindle-shaped cells in the xylem of plants.

tracheophytes Vascular plants; tracheophytes include all the most familiar living plants.

tradewinds Winds that blow from about 30° latitude, steadily toward the equator.

transcription The production of RNA from DNA.

transcription factor A protein that binds to DNA and regulates transcription.

transfer RNA (tRNA) A small RNA molecule that serves as an adaptor in protein synthesis; each tRNA contains an anticodon that allows it to bind to a codon in mRNA and each becomes linked to a specific amino acid.

transgenic Containing a particular recombinant DNA incorporated within the genetic material.

transition state A distorted form of a substrate that is intermediate between the starting reactant and the final product.

translation Protein synthesis; the conversion of information from an mRNA molecule into a polypeptide.

translational control Alterations in the rate at which specific mRNAs are translated.

translocation In chromosomes, a mutation in which part of one chromosome is moved to another chromosome.

transmembrane protein A protein that spans the lipid bilayer of a membrane.

transmission electron microscope (TEM) Electron microscope in which an electron beam passes through a specimen.

transmission genetics The branch of genetics that deals with patterns of inheritance.

transpiration The process by which plants pull water from the soil and release it as vapor through stomata in their leaves.

transpiration-cohesion theory The accepted explanation for the movement of water from roots to leaves; transpiration in the leaves pulls water up the stem in continuous columns.

transpiration-photosynthesis compromise The trade-off between saving water and maintaining photosynthetic productivity.

transposable element A mobile gene that can replicate only in integrated form, as part of a host's chromosome.

transposase An enzyme that catalyzes insertion of a transposable element into new sites.

tree line The upper limit where subalpine trees grow.

triceps The major muscle on the back side of the upper arm.

triglyceride (triacylglycerol) A nonpolar derivative of glycerol in which all three hydroxyl groups are attached to fatty acids.

trisomy The presence of three, rather than two, copies of a chromosome.

trophic level Each level of a food chain.

trophoblast In a mammalian embryo, a prominent outer cell layer in a blastocyst.

tropism The bending or curving toward or away from a stimulus.

true-breeding A breeding line in which offspring have the same phenotype, generation after generation.

tubal ligation An irreversible method of birth control; cutting or blocking the oviducts, or fallopian tubes, preventing eggs from reaching the uterus after ovulation.

tubulin One of two globular proteins, called α and β tubulin, from which microtubules are assembled.

tumor virus A virus that causes its host cell to lose its normal ability to regulate cell division.

turbinate Elaborately folded bones inside the noses of many vertebrates.

turgor pressure In plant cells, osmotic pressure against the cell walls resulting from the higher concentration of dissolved molecules and ions in cytoplasm than in the surrounding fluid.

typological species concept The view that each species consists of individuals that are variants of a fixed underlying plan.

unconventional gene A gene without a stable cellular address.

uniformitarianism The view, originated by Lyell, that the processes that now mold the Earth's surface—erosion, sedimentation, and upheaval—are the ones that have always molded it; geologic change is slow, gradual, and steady, not catastrophic.

universal Used by all species.

unsaturated A hydrocarbon chain (or fatty acid) that contains at least one double bond between two carbon atoms; the chain is called unsaturated because it could accept two more hydrogen atoms per double bond.

upstream In a nucleic acid, toward the 5′ end.

upwelling The upward movement of water that draws nutrients from the bottom of the ocean.

urea The compound that mammals convert into ammonia to make it less harmful.

ureter One of a pair of tubes through which urine flows from the kidneys to the bladder.

urethra The means by which the bladder drains urine to the outside of the body.

uric acid A nearly insoluble organic compound produced by insects, land snails, most reptiles, and birds as nitrogenous waste.

uterus The womb; in mammals, the portion of the female reproductive tract that is specialized to receive, nurture, and deliver a developing embryo.

vacuole Within a eukaryotic cell, a membrane-enclosed sac, without any obvious internal structure; a vacuole is larger than a vesicle.

vas deferens In the male reproductive tract, two large, thick-walled ducts that carry sperm from the testes.

vascular Having a conducting system.

vascular system In plants, the structures and cells that serve as a transport system; the vascular system forms the central core of the plant.

vasectomy An irreversible method of birth control; the cutting and tying or cauterization of the vas deferens, which prevents sperm from entering the urethra.

vasoconstriction Contraction of the smooth muscles surrounding the arterioles.

vasodilation A process causing arterioles to increase in size with consequent blood flow increase and reddening on the skin.

vasopressin (antidiuretic hormone or ADH) A peptide hormone, released by the posterior pituitary, that stimulates water reabsorption by the kidneys.

vector Any organism that transmits disease-causing microorganisms; a mobile gene such as a plasmid that can deliver genes.

vegetative reproduction The process of producing a new plant without fertilization or seed production.

veins In vertebrates, the vessels that carry blood back to the heart. In plants, the vascular tissue in the leaves.

vena cava (superior and inferior) The two largest veins, which run up through the center of the body and carry deoxygenated blood from the body to the right atrium.

ventral The part of an animal facing the Earth.

ventricle In the heart, a thick-walled pumping chamber.

ventricular fibrillation Cardiac arrest.

venules The smallest veins.

vertebrae [singular, **vertebra**] The hollow, bony segments that make up the backbone.

vertebrate An animal with a segmented spinal column and bony or cartilaginous brain case (skull), such as a fish, amphibian, reptile, bird, or mammal.

vesicle A membrane-enclosed sac; some vesicles are present inside living eukaryotic cells and others form after a cell is broken open.

vessel elements In flowering plants, structural elements of the phloem that are open at each end and connect end to end to form long open channels.

vestigial structure A part of an organism, with little or no function, that reflects evolutionary history.

vicariance The fragmentation of an already dispersed species or group of species.

villus [plural, **villi**] The fingerlike folds in the mucosa of the small intestine.

virus An assembly of nucleic acid (DNA or RNA) and proteins (and occasionally other components, such as lipids or carbohydrates) that can reproduce only within a living cell; viruses depend on cells to obtain energy and perform chemical reactions.

visceral mass The main body of a mollusk; the visceral mass contains the intestinal tract, as well as the excretory and reproductive organs.

visible light Electromagnetic radiation, with wavelengths from about 400 to about 750 nm, that can be perceived by human eyes and brains.

vitamin An organic compound that an animal cannot itself synthesize but that is required in minute amounts for normal growth and metabolism; an essential component of the diet; many vitamins are cofactors in enzymatic reactions.

viviparous Capable of bearing live young (rather than laying eggs).

voltage A measure of the potential energy that results from a charge separation.

voltage-gated Open or closed according to the voltage across the membrane.

voluntary muscles Skeletal muscles; voluntary muscles enable voluntary control over body movements.

vulva In animals, the collective name for the external genitalia of the female.

warning coloration A bright, memorable design that helps the predator remember which prey to avoid.

water potential The potential energy of water.

water-soluble signal A chemical signal that cannot pass through the plasma membrane and must act on its surface.

wavelength The distance between the crests of two successive waves.

westerlies Winds that come from the west caused by the rotation of the Earth, characteristic of latitudes between 30° and 60° in both hemispheres.

western blotting A method for detecting a particular protein after electrophoresis, by blotting the contents of an electrophoresis gel onto a paperlike support and incubating this "blot" with a specific antibody.

whorl One of four concentric circles of flower parts at the end of a specialized stem.

woody plants Trees, shrubs, and other perennial plants that have woody parts that provide support independent of turgor pressure.

work The movement of an object against a force, or the conversion of energy into electrical energy, chemical energy, or concentration energy.

x-ray crystallography The analysis of the scattering, or diffraction, of x rays by crystals.

xylem Conducting vessels that carry water and minerals from roots to the photosynthesizing leaves.

yield The amount of grain per acre.

yolk In an amniotic egg, a mixture of proteins, lipids, and carbohydrates that nourishes an embryo until it can feed itself.

yolk sac The membrane that surrounds the yolk in an amniotic egg.

Z line In muscle, the boundary of a sarcomere.

zero population growth The state of a population when its size reaches the carrying capacity and the environment cannot support any more growth; the state when a population reaches a stable size.

zygote The first cell of an embryo, formed by the union of egg and sperm.

CREDITS

This page constitutes an extension of the copyright page. We have made every effort to trace the ownership of all copyrighted material and to secure permission from copyright holders. In the event of any question arising as to the use of any material, we will be pleased to make the necessary corrections in future printings. Thanks are due to the following authors, publishers, and agents for permission to use the material indicated.

Chapter 1. xxviii: A. Dowsett/Photo Researchers, Inc. **1:** Courtesy, Dr. Barry Marshall **2:** Kenneth H. Thomas/Photo Researchers, Inc. **5:** La Seine a Herblay, by Maximilien Luce (Musee d'Orsay, Paris) **7(a):** J.H Robinson/Animals Animals **7(b):** John D. Cunningham/Visuals Unlimited **7(c):** M.A Chappel/Animals Animals **7(d):** Otto Willner/OKAPIA/Photo Researchers, Inc. **7(e):** N. Pecnik/Visuals Unlimited **7(f):** Dan Suzio/Photo Researchers, Inc. **7(g):** Jane Burton/Bruce Coleman, Inc. **7(h), left:** John Cancalosi/Peter Arnold, Inc. **7(h), right:** Alan Desbonnet/Visuals Unlimited **8(a):** G.W. Willis, M.D. **8(b):** Manfred Kage/Peter Arnold, Inc. **8(c):** Ed Reschke/Peter Arnold, Inc. **10, clockwise:** L. Grillione/Phototake **10:** R. Robinson/Visuals Unlimited **10:** Biological Photo Service **10:** F. Stuart Westmorland/Photo Researchers, Inc. **10:** R. M. Meadows/Peter Arnold, Inc. **10:** Runk/Schoenberger from Grant Heilman **11:** Larry Tackett/Tom Stack and Associates **12:** Francois Gohier/Photo Researchers, Inc. **13:** Francois Gohier/Photo Researchers, Inc. **19:** David Spears/Clouds Hill Imaging Ltd./Corbis

Chapter 2. 20: Dennis Galante/FPG/Getty Images **22:** Roy Morsch/The Stock Market/Corbis **23:** Scala/Art Resource, NY **24:** Alfred Pasleka /Peter Arnold, Inc. **24:** Paraskevas Photography **24:** Gianni Tortoli/Photo Researchers, Inc. **24:** Allan Kaye/DRK Photo **24:** AlaskaStock **24:** Runk/Schoenberger from Grant Heilman **28:** Bruce Iverson **30:** Francois Gohier/Photo Researchers, Inc. **31:** Royalty-Free/Corbis **31:** Charles D. Winters/Timeframe Photography **31:** Zefa Germany/The Stock Market/Corbis **31:** Camerique/The Picture Cube **32:** John Cancalosi/DRK Photo **33:** Larry Ulrich/DRK Photo **35:** Culver Pictures

Chapter 3. 40: John Wilkes/Octopus Photos **41:** Courtesy of Stanley Prusiner, UCSF **52:** Cabisco/Visuals Unlimited **52:** Ober/Visuals Unlimited **54:** Harcourt Photo Library **60:** K. Jakes/The Ohio State University **61:** Professor P.M. Motta and E. Vizza/Science Photo Library/Photo Researchers, Inc.

Chapter 4. 64: Scott Camazine/Photo Researchers, Inc. **65:** Science Photo Library/Photo Researchers, Inc. **66(a):** Michael Abbey/Photo Researchers, Inc. **66(b):** Eric Grave/Photo Researchers, Inc. **66(c):** Dr. Richard Kessel and Dr. C. Shih/Visuals Unlimited **68(a):** Dennis Kunkel Microscopy, Inc. **68(b):** Elizabeth Gentt/Visuals Unlimited **69:** Stan W. Elems/Visuals Unlimited **74, top:** Runk/Schoenberger from Grant Heilman **74, bottom right:** R. Kessel-G. Shih/Visuals Unlimited **74, bottom left:** Custom Medical Stock Photo **75:** Don Fawcett/Science Source/Photo Researchers, Inc. **76:** Cabisco/Visuals Unlimited **76:** Michael Abbey/Visuals Unlimited **77:** Don Fawcett/Visuals Unlimited **78(a):** Grant Heilman Photography **78(b):** Biophoto Associates/Science Source/Photo Researchers, Inc. **78(c):** Dr. Jeremy Burgess/Science Photo Library/Photo Researchers, Inc. **79 top:** David Becker/Photo Researchers, Inc. **79(a):** Richard Wade **79(b):** Roger Craig **79(c):** Roy Quinlan **80(a), left:** David M. Phillips/Science Source/Photo Researchers, Inc. **80(a), right:** Dr. Gopal Murti/Science Photo Library/Photo Researchers, Inc. **80(b):**Anne Fleury, Laboratoire de Biologie Cellulaire 4, Universite Paris Sud, CNRS **82:** Biophoto Associates/Photo Researchers, Inc. **83(a-b):** Courtesy of Amy Dunleavy **85:** Dennis Kunkel/Phototake NYC **85:** Dennis Kunkel/Phototake NYC **85:** Dennis Kunkel/Phototake NYC **85:** Michael Abbey/Photo Researchers, Inc. **85:** Michael Abbey/Photo Researchers, Inc. **85:** Michael Abbey/Photo Researchers, Inc. **87:** Kevin Collins/Visuals Unlimited **87:** Fred Hossler/Visuals Unlimited **89(a):** Gilulap, Fawcett/Visuals Unlimited **89(b):** Farquhar, Palade, Fawcett/Visuals Unlimited **89(c):** Hull, Staehelin, Fawcett/Visuals Unlimited

Chapter 5. 92: David Muench Photography **93:** Science Photo Library/Photo Researchers, Inc. **94:** Dr. Harold E. Edgerton, Harold and Esther Edgerton Foundation. Courtesy Palm Press, Inc. **96:** Superstock

Chapter 6. 108: Paraskevas Photography **110:** Dwight Kuhn **111(a):** CNRI/Phototake **111(b):** Zig Leszczynski/Animals Animals **112:** David Madison **113:** Jim Erickson/The Stock Market/Corbis **113:** David Madison **120:** D. Friend, D. Fawcett/Visuals Unlimited **122:** Lester Lefkowitz/Corbis

Chapter 7. 126: Paraskevas Photography **128, left:** Dwight Kuhn **128, right:** The Granger Collection **131, bottom:** Leonard Lessin/Peter Arnold, Inc. **131(a):** Runk/Schoenberger from Grant Heilman **131(b):** Yoav Levy/Phototake **134:** Stephen J. Krasemann/Photo Researchers, Inc. **137:** James Dennis/CNRI/Phototake **147:** Animals Animals/Earth Scenes

Chapter 8. 148: Amy Dunleavy **149:** Science Photo Library/Photo Researchers, Inc. **150, left:** Biology Media/Photo Researchers, Inc. **150, center:** Science Photo Library/Photo Researchers, Inc. **150, right:** M. Rotker/Photo Researchers, Inc. **151:** Dwight Kuhn **153:** Dr. Gopal Murti/Science Photo Library/Photo Researchers, Inc. **153:** K.G. Murti/Visuals Unlimited **154, top:** Custom Medical Stock **154, bottom:** Paraskevas Photography **156(1-5):** M. Abbey/Photo Researchers, Inc. **159(a):** John T. Hansen, Ph.D./Phototake **159(b):** Joseph Gall, Carnegie Institution **159(c):** Courtesy of Barbara Hamkalo **159(d):** Courtesy of Victoria Foe **160, top:** David M. Phillips/Visuals Unlimited **160(b):** R. Calentine/Visuals Unlimited

Chapter 9. 164: Ripsaw Inc. **165:** University of Kansas Archives **165:** Skip Moody/Dembinsky Photo Associates **169, left:** Dr. Nancy L. Segal, Entwined Lives: Twins and What They Tell Us About Human Behavior, 1999, New York, Dutton **169, right:** Dr. Nancy L. Segal, Entwined lives: Twins and What They Tell Us About Human Behavior, 1999, New York, Dutton **169(a):** Biophoto Associates/Photo Researchers, Inc. **169(b):** Kevin Schafer/Peter Arnold, Inc. **170:** Leonard Lessin/Peter Arnold, Inc. **171:** David Phillips/Visuals Unlimited **172(a-c):** Harry W. Greene/Cornell University **173:** Rob Lang **177:** Hattie Young/Science Photo Library/Photo Researchers, Inc. **177:** Dennis Kunkel/Phototake **178:** Oxford Scientific Films/Animals Animals **179:** Don Kelly/Grant Heilman Photography **179:** Photo Researchers, Inc. **181:** The Bettmann Archive **186:** J. Burgess/Science Photo Library/Photo Researchers, Inc. **186, top, bottom:** T. Kaufman

Chapter 10. 192: Paraskevas Photography **193 top:** Vittorio Luzzati **193(a):** Science VU/BMRL/Visuals Unlimited **193(b):** K.G. Murti/Visuals Unlimited **195:** Paraskevas Photography **197, top:** Paraskevas Photography **197, bottom:** Barrington Brown/Science Source/Photo Researchers, Inc. **198:** Cold Spring Harbor Laboratory Archives **202:** Ken Eward/Biografx/Science Source/Photo Researchers, Inc. **204:** Meckes/Ottawa/Photo Researchers, Inc. **207, 208:** Cold Spring Harbor Laboratory Archives **212:** D. Hogness and H. Kriegstein, Stanford University, Proceedings of the National Academy of Science, 71, 135-139, 1974 **213, right:** James D. Colandene, University of Virginia **215:** Peter Menzel/Stock, Boston

Chapter 11. 218: Paraskevas Photography **219:** UPI/Bettmann/Corbis **223(1-4):** M. Guthold, X. Zhu, and C. Bustamante, University of Oregon **232(a):** Oscar L. Miller, University of Virginia **232(b):** Dr. Barbara Hamkalo, University of California, Irvine

Chapter 12. 240, top: The Kobal Collection **240, bottom:** Runk/Schoenberger from Grant Heilman **241:** Cold Spring Harbor Laboratory Archives. Permission of Marjorie M. Bhavnani **244(a):** Jack D. Griffith, University of North Carolina **244(b):** Andrew O. Jackson, University of California, Berkeley **244(c):** Thomas Broker/Phototake **244(d):** Hans Gelderblom/Visuals Unlimited **248:** Dennis Kunkel/Phototake **252 right:** Eye of Science/Photo Researchers, Inc. **253(a):** CNRI/Phototake **253(b):** Richard Kolodner, Ludwig Institute for Cancer Research

Chapter 13. 258: Paraskevas Photography **259, top:** © 1997 Darryl Estrine Photography **259, bottom:** Paraskevas Photography **261:** Courtesy C. P. Mangelsdorf **261:** Science VU/Visuals Unlimited **267:** Jerry Cooke/Photo Researchers, Inc. **270:** Courtesy of Linda Liau, UCLA Department of Neurosurgery **271:** R. L. Binster/Peter Arnold, Inc. **272:** Gail Martin, University of California, San Francisco **272:** Charles O. Cecil/Visuals Unlimited **274:** Calgene

Chapter 14. 276: Mycoff Advertising Inc., Salem, MA **277, top:** Courtesy of Stanford University Medical Center **277(a):** Culver Pictures **277(b):** Courtesy of Alice Wexler **280(a):** Damon Biotech, Inc **280(b):** Cellmark Diagnostics **282, bottom, left:** Superstock **282, bottom, right:** Paul Grebliunas/Getty Images **282, top, left:** M. P. Kahl/Photo Researchers, Inc. **282, top, right:** Charles O. Ceci/Visuals Unlimited **283:** Superstock **285(a):** Courtesy of Ono Atsushi **285(b):** David Hall/Photo Researchers, Inc. **285, right:** L. Steinmark/Custom Medical Stock Photo **286:** © 1997 Abraham Menashe **292:** Gary Parker/Science Photo Library/Custom Medical Stock Photo **293:** Courtesy of Van DeSilva **297:** Frans Lanting/Minden Pictures

Chapter 15. 298: Daniel J.Cox/Natural Selection **300, top left:** Kevin Schafer **300, left:** Richard Gibson/Natural Selection **300, center left:** Stephen G. Maka/DRK Photo **300, center:** Daniel J. Cox/Natural Selection **300, center right:** Jack Dykinga/Bruce Coleman, Inc. **300, bottom left:** Joe McDonald/Bruce Coleman, Inc. **300, bottom center:** Mimi Forsyth/Bruce Coleman, Inc. **300, bottom right:** Frans Lanting/Minden Pictures **302(a):** Tim Flach/Getty Images **302(b):** Kevin Schafer/Peter Arnold, Inc. **304(b):** Jack Dermid/Bruce Coleman, Inc. **305, left:** Alan Desbonnet/Visuals Unlimited **305, right:** John Cancalosi/Peter Arnold, Inc. **307(a):** Bridgeman Art Library **307(b):** The Granger Collection **308, left:** Biological Photo Service **308, center:** William E. Ferguson **308, right:** James L. Amos/Photo Researchers, Inc. **309, left:** James L. Amos/Photo Researchers, Inc. **309, center:** Ted Clutter/Photo Researchers, Inc. **309, right:** J.C. Carton/Bruce Coleman, Inc. **310, left:** James L. Amos/Photo Researchers, Inc. **310, center:** Ken Lucas/Visuals Unlimited **310, right:** E.R. Degginger/Bruce Coleman, Inc. **311, left:** William E. Ferguson **311, center:** E.R. Degginger/Earth Scenes **311, right:** Tim Davis/Photo Researchers, Inc. **318(b), top:** Anthony Bannister Photo Library **318(a):** Courtesy of K. Sulik **318(b):** Professor Hideo Nishimura **321:** The Illustrated London News Picture Library

Chapter 16. 324: Frans Lanting/Minden Pictures **326, left:** Carl Purcell/Photo Researchers, Inc. **326, right:** John Gerlach/Visuals Unlimited **331, top:** Courtesy of Joiner Associates, Inc. **331, bottom left:** Klaus Payson/Peter Arnold, Inc. **331, bottom center:** Frank Siteman/Stock, Boston **331, bottom right:** F. Jackson/Bruce Coleman, Inc. **332:** © Terry Evans/Courtesy Catherine Edelman Gallery, Chicago **337:** Frank S. Balthis Photography **338:** Christopher Morris/Black Star **340:** Michael Tweedie/Photo Researchers, Inc. **342(a), top:** Kjell B. Sandved/Photo Researchers, Inc. **342(b), top:** John Cancalosi/DRK Photo **342(b), bottom:** Tom McHugh/Photo Researchers, Inc. **343(top):** Hope Entomology Collection, Oxford/Biological Photo Service **343(bottom):** Temeles, E. J. and W. J. Kress. 2003. Adaptation in a plant-hummingbird association. Science 300: 630-633 **344, left:** Fred Bruemmer/DRK Photo **344, right:** Carol Beckwith/Robert Estall Photo Library

Chapter 17. 348: Tui de Roy/Minden Pictures **349, top:** The Bridgeman Art Library **349, bottom left:** The Granger Collection, New York **349, bottom right:** Michio Hoshino/Minden Pictures **352:** Gerard Lacz/Animals Animals **357:** Tui de Roy/Minden Pictures **359(a):** N.H. [Dan] Cheatham/Photo Researchers, Inc. **359(b):** Gerald D. Carr **359(c):** C.K. Lorenz/Photo Researchers, Inc. **364:** Heather Angel/Biofotos **365:** © 1996 Dr. Nigel Smith/Earth Scenes **367(a):** Tom & Pat Leeson/Photo Researchers, Inc. **367(b):** Glenn Vanstrum/Animals Animals **368(a):** Gerard Lacz/Peter Arnold, Inc. **368(b):** M. Long/Visuals Unlimited **372(a):** Dr. Owen Lovejoy and students, Kent State University/© 1985 David L. Brill **372(b):** John Reader/Science Photo Library/Photo Researchers, Inc. **373:** Blackwell, Queens College

Chapter 18. 376: James Cotier/Getty Images **379:** Biological Photo Service **379:** Stanley Awramik/Biological Photo Service **379:** John Reader/Science Photo Library/Photo Researchers, Inc. **380(b):** Science VU-USM/Visuals Unlimited **380(c):** William E. Ferguson **380(d):** Andrew H. Knoll, Knoll and Calder, Palaeontology 26:467, 1983 **380(e):** Biological Photo Service **381:** Bill Lovejoy **384, right:** Roger Ressmeyer/Corbis **385(a,b):** D.W. Deamer, University of California, Santa Cruz **387:** Steven Brooke and Richard LeDuc **391:** Biophoto Associates **391:** Omikron/Photo Researchers, Inc. **395:** Frans Lanting/Minden Pictures

Chapter 19. 396: Middleton/Liittschwager 1994. All rights reserved **398, left:** Tom and Pat Leeson/DRK Photo **398, center:** Jane McAlonan/Visuals Unlimited **398, right:** C.K. Lorenz/Photo Researchers, Inc. **401(a):** Roger Wilmshurst/Photo Researchers, Inc. **401(b):** M. H. Sharp/Photo Researchers, Inc. **404:** Werner Forman Archive, National Gallery, Prague/Art Resource **406:** Gerard Lacz/Peter Arnold, Inc.

Chapter 20. 410: Paraskevas Photography **413:** Erich Lessing/Art Resource **415:** Bettman/UPI/Corbis **417(a):** Meckes/Ottawa/Photo Researchers, Inc. **417(b):** CDC/Science Source/Photo Researchers, Inc. **417(c):** Dr. Richard Kessel & Dr. Gene Shih/Visuals Unlimited **417, bottom:** Karen Stephens/Biological Photo Service **418:** G.W. Willis, M.D. **419:** S. Fiegler/Visuals Unlimited **420(a):** David A. Hardy/SPL/Photo Researchers, Inc. **420(b):** Wolfgang Baumeister/SPL/Photo Researchers, Inc. **422:** Courtesy of University of Illinois at Urbana-Champaign News Bureau **423, top:** Richard Thom/Visuals Unlimited **423, bottom:** F. Widdel/Visuals Unlimited **424(a):** Alfred Pasteka/Science Photo Library/Photo Researchers, Inc. **424(b):** Kevin Schafer/Corbis **425, left:** CDC/Science Source/Photo Researchers, Inc. **425(a):** Dennis Kunkel/Visuals Unlimited **425(b):** Dr. George Chapman/Visuals Unlimited **425(c):** Dennis Kunkel/Phototake **426:** C.P. Vance/Visuals Unlimited

Chapter 21. 428: Dr. Paul A. Zahl/Photo Researchers, Inc. **429:** Wadsworth Center: New York State Department of Public Health **431(a):** Dwight Kuhn **431(b):** Biological Photo Service **434, left:** Cabisco/Visuals Unlimited **434, right:** Terry Hazen/Visuals Unlimited **435, top:** Geospace/Science Photo Library/Photo Researchers, Inc. **435, bottom:** Mitsuhiko Imamori/Minden Pictures **436:** M. Abbey/Photo Researchers, Inc. **437, top:** Flip Nicklin/Minden Pictures **437, bottom:** Robert Brons/Biological Photo Service **438(a):** Dr. Dennis Kunkel/Visuals Unlimited **438(b):** Biophoto Associates **438(c):** Ric Ergenbright/Corbis **439, top:** Dr. Stanley Flegler/Visuals Unlimited **439, bottom:** Alfred Owczarzak/Biological Photo Service **441:** G. Carleton Ray/Photo Researchers, Inc. **442:** Biophoto Associates **443:** Michael Fogden/DRK Photo **444, top:** Dennis Drenner **444, left:** J. Robert Stottlemyer/Biological Photo Service **444, right:** Michael Fogden/DRK Photo **445:** Ken Barker **446:** Larry Ulrich/DRK Photo

Chapter 22. 450: @1997 The Georgia O'Keeffe Foundation/Artists Rights Society (ARS), New York, Photo courtesy Colorado Springs Fine Arts Center **456:** Pat O'Hara/DRK Photo **460, left:** Scott W. Smith/Earth Scenes **460, right:** William E. Ferguson **461:** Pat O'Hara/Corbis **462, top:** Michael Giannechinni/Photo Researchers, Inc. **462, center:** R. J. Erwin/Photo Researchers, Inc. **462, bottom :** Walter H. Hodge/Peter Arnold, Inc. **463, top:** Bohemian Nomad Picturemakers/Corbis **463, bottom:** Jonathan Blair/Corbis

Chapter 23. 468: Art Wolfe/Getty Images **469:** Brown Brothers **475:** Brian Parker/Tom Stack & Associates **477:** James R. McCullagh/Visuals Unlimited **478, left:** M. I. Walker/Photo Researchers, Inc. **478, right:** Tom Adams/Visuals Unlimited **480, left:** Robert Calentine/Visuals Unlimited **480:** Oliver Meckes/Photo Researchers, Inc. **480, right:** Raymond Mendez/Animals Animals **483, left:** Douglas Faulkner/Photo Researchers, Inc. **483, right:** Randy Morse/Tom Stack & Associates **485:** Joyce Burek/Animals Animals **489, top:** John Cancalosi/Tom Stack & Associates **489, center:** S.E. Georgia Animals Animals **489, bottom:** Tom McHugh/Photo Researchers, Inc. **490:** William E. Ferguson

Chapter 24. 494: Paraskevas Photography **498(a):** Dave B. Fleetham/Visuals Unlimited **498(b):** Cabisco/Visuals Unlimited **499:** Dave Fleetham/Tom Stack & Associates **501, top:** Courtesy of Dr. Kiyoko Uehara **501, bottom:** Berthoule-Scott/Jacana/Photo Researchers, Inc. **502(a), top:** Jeff Rotman/Getty Images **502(b):** Doug Perrine/DRK Photo **502(a), bottom:** Rondi/Tami Church/Photo Researchers, Inc. **502(b):** Sam Ogden **502(c):** Sally Bensusen/Science Photo Library/Photo Researchers, Inc. **504(a):** David M. Dennis/Tom Stack & Associates **504(b):** Joe McDonald/DRK Photo **504(c), 505, 506:** Michael Fogden/DRK Photo **509, top:** Tui de Roy/Minden Pictures **509, center:** Roger de la Harpe/Animals Animals **509(a):** Mike Bacon/Tom Stack & Associates **509(b):** Gordon Wiltsie/Peter Arnold, Inc. **509(c):** John Cancalosi/DRK Photo **510(a):** Jim Zipp/Photo Researchers, Inc. **510(b):** Russel C. Hansen/Peter Arnold, Inc. **510(c):** Barbara Cushman Rowell/DRK Photo **513, top:** Dave Watts/Tom Stack & Associates **513, bottom:** Alan Root/Okapia/Photo Researchers, Inc. **515:** Jim Brandenberg/Minden Pictures

Chapter 25. 516: NASA **517:** Kelvin Aitken/Peter Arnold, Inc. **521 left, top:** Gary Meszaros/Dembinsky Photo Associates **521, center:** PSU Entomology/Photo Researchers, Inc. **521, bottom:** Jack Rosen/Photo Researchers, Inc. **523 left bottom, top:** Richard J. Green/Photo Researchers, Inc. **523, center:** Gary Meszaros/Visuals Unlimited **523, bottom:** R. Lindholm/Visuals Unlimited **523, top:** David B. Fleetham/Tom Stack & Associates **523, center:** Nobert Wu/Peter Arnold, Inc. **523, bottom:** D.P. Wilson/Science Source/Photo Researchers, Inc. **523 right bottom, top:** Gil Lopez-Espina/Visuals Unlimited **523, center:** Jeremy Woodhouse/DRK Photo **523, bottom:** Renee Lynn/Photo Researchers, Inc. **527:** Ron Kimball Photography, Inc.

Chapter 26. 534: Matt Meadows/Peter Arnold, Inc. **535, left:** Matt Frederick **535, right:** Matt Frederick **536, top:** Superstock **536(a):** Roland Seitre/Peter Arnold, Inc. **536(b):** Tony Dawson/Words & Pictures/Getty Images **539:** Charlie Ott/Photo Researchers, Inc. **540:** John M. Roberts/The Stock Market/Corbis **541:** John Shaw/Tom Stack & Associates **542:** Tom McHugh/Photo Researchers, Inc. **543, top:** Tom McHugh/Photo Researchers, Inc. **543, bottom:** Bettmann/Reuters/Corbis **544:** © Art Wolfe **545:** G. Alan Nelson/Dembinsky Photo Associates **546:** David M. Schleser/Nature's Images, Inc./Photo Researchers, Inc. **547:** Simon Fraser/Photo Researchers, Inc. **549:** Nancy Sefton/Photo Researchers, Inc. **550(a):** David Ball/Corbis **550(b):** B. Jones/M. Shimlock/Photo Researchers, Inc. **551:** Michael P. Gadomski/Dembinsky Photo Associates

Chapter 27. 554: Krafft-Explorer/Science Source/Photo Researchers, Inc. **555, top:** Tom and Pat Leeson/Photo Researchers, Inc. **555, bottom:** John Marshall/Getty Images **558, left:** Denise Tackett/Tom Stack & Associates **558, right:** Thomas Eisner and Daniel Aneshansley/Cornell University **559, top:** W. Peckover/VIREO **559(a):** Lincoln Brower **559(b):** Thomas C. Emmel **560:** Mike Severns/Getty Images **563:** Michael and Patricia Fogden/Corbis **566:** Jorge Rey, University of Florida **567:** William J. Weber/Visuals Unlimited **569:** Dr. John Cunningham/Visuals Unlimited **570:** Barbara Gerlach/DRK Photo

Chapter 28. 574: Painting by Charles Knight. Field Museum of Chicago/Photo # Geo-CK-30Tc by Ron Testa **578:** Steve & Dave Maslowski/Photo Researchers, Inc. **580:** Kennan Ward/The Stock Market/Corbis **583:** Gianni Tortoli/Photo Researchers, Inc. **583:** Tom Stack/Tom Stack & Associates **585(b):** Leroy Simon/Visuals Unlimited **585(c):** Roger Wilmshurst/Photo Researchers, Inc. **591, left:** Peter Ginter/Material World **591, right:** Peter Ginter/Material World **592:** R. Ian Lloyd/The Stock Market/Corbis

Chapter 29. 598: Oxford Scientific Films/Animals Animals **599:** Andy Cox/Getty Images **601:** Thomas W. Martin/Photo Researchers, Inc. **603, left:** Thomas D. McAvoy/Time Life Pictures/Getty Images **603, right:** Tim Davis/Photo Researchers, Inc. **605:** Art Wolfe/Getty Images **606(a), top:** Tom and Pat Leeson/Photo Researchers, Inc. **606(b):** C.K. Lorenz/Photo Researchers, Inc. **606(a):** M.J. Ryan, University of Texas **606(b):** Juan Manuel Renijifo/Animals Animals/Earth Scenes **607(a):** Tom Brakefield/DRK Photo **607(b):** Anup & Manoj Shah/DRK Photo **608:** Gilbert Grant/Photo Researchers, Inc. **610(a,b):** George D. Lepp **611:** Michael and Patricia Fogden/Corbis **613:** Mark Moffett/Minden Pictures

Chapter 30. 614: Royalty-Free/Corbis **616:** Otto Willner/OKAPIA/Photo Researchers, Inc. **618:** Geoff Bryant/Photo Researchers, Inc. **619(a-d):** Ed Reschke **620:** Biophoto Associates **623:** Ray Evert **624(a), top left:** Barry L. Runk/Grant Heilman Photography **624(a), top right:** Runk/Schoenberger from Grant Heilman **624(c), center top left:** Dwight Kuhn **624(others):** Runk/Schoenberger from Grant Heilman **624(d), bottom left:** John Gerlach/Visuals Unlimited **624(d), bottom right:** John Gerlach/Tom Stack & Associates **625:** Darwin Dale/Photo Researchers, Inc. **626:** Holt Studios International/Photo Researchers, Inc.

Chapter 31. 628: Don Ellis **629:** University of California at Davis/Dept. of Entomology **631(a):** Nigel Cattlin/Photo Researchers, Inc. **631(b):** Jack Bostrack/Visuals Unlimited **632(a):** Lefever/Grushow from Grant Heilman **632(b):** Kjell B. Sandved/Photo Researchers, Inc. **632(c):** David Sieren/Visuals Unlimited **632(d):** W. Ormerod/visuals Unlimited **633, left:** Ed Reschke **633, right:** Dwight Kuhn **635:** Angelina Lax/Photo Researchers, Inc. **640(a):** S. Krasemann/Photo Researchers, Inc. **640(b):** M.C. Chamberlain/DRK Photo **640(c):** M. Patterson/Photo Researchers, Inc. **641:** William E. Ferguson **642:** Jerome Wexler/National Audubon Society/Photo Researchers, Inc.

Chapter 32. 646: Paraskevas Photography **649(b):** Cabisco/Visuals Unlimited **649(c):** Dennis Kunkel/Phototake NYC **651(a):** John D. Cunningham/Visuals Unlimited **651(b,c):** Jack M.Bostrack/Visuals Unlimited **653, top:** Blanche C. Haning **653, bottom:** Ed Reschke **655:** Carolina Biological Supply Company/Phototake **656:** William E. Ferguson **657:** William E. Ferguson **658:** Leslie Sieburth, McGill University, Montreal **658:** Leslie Sieburth, McGill University, Montreal **659:** A.B. Joyce/Photo Researchers, Inc.

Chapter 33. 662: Helmut Gritscher/Peter Arnold, Inc. **663:** Jerome Wexler/Photo Researchers, Inc. **664:** University of Californa Santa Cruz Photo Lab **666(a):** John D. Cunningham/Visuals Unlimited **666(b):** Ed Reschke **666, bottom:** D. Newman/Visuals Unlimited **670:** Courtesy of B.O. Phinney, University of California, Los Angeles **671:** Steve Austin, Papilio/Corbis **672:** Michelle Garrett/Corbis **673(a,b):** Yen, Lee, Klee, and Giovanoni, Plant Physiology [1995] 107:1343-1353 **675:** Paul J. Sutton/Duomo/Corbis

Chapter 34. 676: John Sibbick **677:** K. Perkins/J. Beckett, Courtesy of the Department of Library Sciences, American Museum of Natural History **679:** "Drawing Hands" by M.C. Escher © 2000 by Cordon Art-Baam-Holland. All rights reserved. **680:** Stan Osolinski/Dembinsky Photo Associates **681:** Diane R.Nelson **683:** E.R. Degginger/Dembinsky Photo Associates **686:** Joe McDonald/Corbis

Chapter 35. 688: The Field Museum/Neg #GN89677_42c Photo by John Weinstein **693, top left:** Kathleen Olson **693, center left:** LWA-Paul Chmielowiec/Corbis **693, center right:** Kathleen Olson **693, bottom right:** Photodisc/Getty Images **693, bottom left:** The Image Bank/Getty Images **693, top right:** Kathleen Olson **697:** Don W. Fawcett/Visuals Unlimited **699(a):** David M. Phillips/Visuals Unlimited **699(b):** Biophoto Associates/Science Source/Photo Researchers, Inc. **699(c):** M.I. Walker/Science Source/Photo Researchers, Inc. **700, left:** Stone/Getty Images **700, right:** Ales Fevzer/Corbis **701:** Dennis Welsh/ImageState/Picture Quest

Chapter 36. 704: Dr. Merlin Tuttle/Photo Researchers, Inc. **705:** The Granger Collection **710:** Roland Birke/Okapia/Photo Researchers, Inc. **714:** Quest/Science Photo Library/Photo Researchers, Inc. **718(a):** G. Shih-R. Kessel/Visuals Unlimited **718(b):** R.G. Kessel and R. H. Kardon, "Tissues and Organs: A Text-Atlas of Scanning Microscopy," 1979, W. H. Freeman & Co. **718(c):** David M. Phillips/Visuals Unlimited

Chapter 37. 722: © Stan Osolinski 1993/FPG/Getty Images

Chapter 38. 736: American Red Cross **737:** Time Life Pictures/Getty Images **738:** Annebicque Bernanrd/Sygma/Corbis **739(a):** David M. Phillips/Visuals Unlimited **739(b):** Meckes/Ottawa/Photo Researchers, Inc. **740:** John D. Cunningham/Visuals Unlimited **743:** Martin Rotker/Phototake NYC **744(a,b):** Biophoto Associates

Chapter 39. 754: Skull with cigarette, by Vincent van Gogh, 1885/Art Resource, New York **755:** The Granger Collection **762(b,c):** Biophoto Associates

Chapter 40. 772: Paraskevas Photography **774(a):** Robert Winslow/Tom Stack & Associates **774(b):** William E. Ferguson **774(c):** Jack Dermid/Photo Researchers, Inc. **776:** M.P. Kahl/Photo Researchers, Inc. **780(b):** F. Spinelli/Visuals Unlimited **780(c):** Fred Hossler/Visuals Unlimited **782:** SIU/Photo Researchers, Inc.

Chapter 41. 786: Don Fawcett/Visuals Unlimited **787, top:** Guido Picchio/AP/Wide World **787, bottom:** Gopal Murti/Phototake **793:** Don Fawcett/Visuals Unlimited **797:** CNRI/Science Photo Library/Photo Researchers, Inc. **800:** Ken Greer/Visuals Unlimited

Chapter 42. 804: The Granger Collection **805:** Marcus Raichle, Department of Neurology, Washington University School of Medicine **814:** G. Bredberg/Science Photo Library/Photo Researchers, Inc.

Chapter 43. 822: Tim Davis/Photo Researchers, Inc. **823:** Washington University Archives **824(a):** John D. Cunningham/Visuals Unlimited **824(b):** John D. Cunningham/Visuals Unlimited **825, top:** David Becker/Getty Images **825, center:** David Becker/Getty Images **825, bottom:** Carolina Biological/Phototake **832:** T. Reese & D. W. Fawcett/Visuals Unlimited

Chapter 44. 836: Art Resource, New York **837:** M.P. Kahl/Photo Researchers, Inc. **839:** Biophoto Associates **842:** Ed Reschke/Peter Arnold, Inc.

Chapter 45. 860: The Granger Collection **863(a-d):** David Fromson, California State University, Fullerton **864(a), all:** From Beams and Kessel, 1976 **865(b,c):** From Morrill and Santos, 1985 **867(b), 868(a):** Courtesy of K.W. Tosney **869, left to right:** Science Pictures Ltd. /SPL/Photo Researchers, Inc., Petit Format/SPL/Photo Researchers, Inc., Petit Format/SPL/Photo Researchers, Inc., James Stevenson/SPL/Photo Researchers, Inc., Petit Format/Nestle/Photo Researchers, Inc. **875(a), top:** From E.J. Kollar and C. Fisher, 1980. "Science," 207:993–995. © 1980 by the American Association for the Advancement of Science **875(b):** Custom Medical Stock Photo **875(a), bottom:** Carolina Biological Supply Company/Phototake **875(b):** Carolina Biological Supply Company/Phototake **875(c):** David Scharf/Peter Arnold, Inc. **876, left to right:** Lior Rubin/Peter Arnold, Inc., Ed Reschke/Peter Arnold, Inc., Lior Rubin/Peter Arnold, Inc., Don Riepe/Peter Arnold, Inc.

Index

Cephalochordates, 499
Cephalopods, 482–483
Cerebellum, 808
Cerebral cortex, 808
Cerebrospinal fluid, 806
Cerebrum, 808
Cervical cancer, 150
Cervical cap, 856
Cervix, 837
Cestoda, 479
CFTR, 286–287
Chamberland, Charles, 242
Chaparral, 542–543
Chaperones, 62
Character displacement, 563
Chargaff, Erwin, 199–201
Chargaff rules, 199–200
Chase, Martha, 207–208
Chelicerates, 489–490
Chemical defense, 558
Chemical reactions
    coupled, 98–99
    enzymes and, 102–104
    exergonic, 97–98
    free energy and, 96–97
    kinetics, 93
    rate of, 99–101
    rearranging atoms in, 94
    starting, 101–102
    stopping, 101
    thermodynamics, 94–96, 99
Chemical synapse, 830–831
Chemiosmosis, 121
Chemistry
    bio, 43
    definition, 21
    life, 22–23
    organic, 43
Chiasma, 173
Chiorophytes, 439–440
Chitin, 53, 441
Chloride ions, 28–29
Chlorobi, 423, 425
Chlorofluorocarbons, 595
Chlorophyll, 134, 139
Chloroplasts
    anatomy, *136*
    cyanobacteria and, 425
    DNA, 253, 255
    mitochondria, endosymbiosis, 390–391
    origins, 390
    in photosynthesis, 77–78
Cholesterol
    biochemistry, 52
    buildup, 52
    characterization, 51
    chemical structure, 51
    HDL, 52
    LDL, 52
    sources, 52
Chondrichthyes, 501

Chordata, 498–499
Chorionic villus sampling, 291
Chromalveolates, 434–436
Chromatids
    definition, 155
    distribution, 175–176
    separation, 157
Chromoplasts, 77
Chromosomes
    alignment, 157
    analysis techniques, 151
    binding, 159
    condensing, 156–157
    etymology, 70
    function, 152
    homologous, 155
    inheritance theory, 165–167, 185–190
    karyotype, *154*
    in meiosis I, 173, 175
    movement, mechanism, 157
    mutations, 214
    number, 154
    pairs, 154–155, *170*
    polyploid, 354
    sex, 179, *179*
Chrysophytes, 436
Chthamalus stellatus, 561–562
Chyme, 715, 717
Cigarette smoking
    cardiovascular damage from, 764
    death related to, 755–757
    lung damage from, 763–764
    promotion, 755–757
Cilia, 763
Cillates, 436
Circulation, 504, 738–743
Circumcision, 842, 844
Citric acid cycle
    in cellular respiration, 111
    function, 115–116
    in metabolic reactions, 117–118
Cladistis, 406–408
Classical evolutionary taxonomy, 404–405
Classification
    animals, 470–474
    cladistics, 406–408
    classical, 404–405
    divisions, 400–401
    hierarchical, *401*
    Latin use in, 401
    Linnaeus, 400–401
    phenetics, 405–406
    prokaryotes, 420–423, 425–426
    sequencing comparisons, 408
Clays, 639
Cleavage, 862, 864
Cleavage furrow, 160
Clements, Edward Frederick, 555–557
Climate, 536–539
Climax communities, 556, 570
Clitoris, 842, 844

Cloaca, 508
Cloning
    definition, 169
    dinosaurs, 273
    Dolly, 872, 874
Clostridium botulinum, 110
Cnidarians, 475–477
Cnidocytes, 476
Coacervates, 386
Cochlea, 813
Codons, 220, 231
Coelom, 473, 689
Coenzyme A, 115
Coevolution, 342–343, 560–561
Cohen, Mitchell, 410–412
Cohen, Stanley, 823
Cohesion-tension theory, 635–636
Collagen
    bone, 696–697
    function, 56
    helix, 59–61
Collenchyma cells, 619–620
Colon, 719
Coloration, warning, 558–559
Colostrum, 851
Columbus, Christopher, 646
Comb jellies, 477
Commensalism, 560
Communities
    change over time, 568–571
    characteristics, 555–556
    definition, 555
    resilience of, 571
    -species, interaction, 557–564
    stabile, 570–571
Compact bone tissue, 697
Companion cells, 643
Competition, 561–563, 604–606
Competitive exclusion principle, 561
Complement, 792
Compounds, 21–22
Compression, 695
Compstock Act, 855
Concentration, 84
Concentration gradient, 680
Conditioned reflex, 599
Condoms, 855–856
Conduction, 829
Cone cells, 816
Conformation. *See* Tertiary structure
Conifers, 461
Conjunction, 248
Connective tissue, 692, *693*
Connell, Joseph H., 561
Conservation, 209, 575
Constipation, 719
Consumers, 520, 557–558
Contact inhibition, 161
Contraception, 852–857
Contractile ring, 160
Control, definition, 5–6

Inflammation, mechanism, 792–794
Influenza virus, 246
Ingram, Vernon, 204
Inheritance
    acquired characteristics, 305, 327–329
    blending, 165, 326–327
    chromosomal, 165–167
    chromosomal theory, 185–190
    genes, 179–180
    laws, universality, 169
    Mendelian, 181–185
    particulate, 326
    transforming factor, 205–206
Inhibitors, 105–106
Inhibitory postsynaptic potential, 832
Insect societies, evolution, 607–609
Insertions, 214
Inspiration, 763
Insulin, 266, 728
Integrated community, 556
Intercourse. *See* Sexual intercourse
Interleukin-1, 794
Intermediate filaments, 78–79
Interphase, 155
Intertidal zone, 549
Intestitial cells, 841
Introns, 225–226
Inversions, 214
Invertebrates, 783–784, 806
In vitro fertilization, 856
Ions
    bonds, 28
    chloride, 28–29
    hydrogen, 34–35
    sodium, 28–29
IPSP. *See* Inhibitory postsynaptic potential
Iris, 815
Island biogeography, theory of, 565–566
Isoleucine, *58*
Isotopes
    atoms and, 312
    characterization, 25
    stability, *26*

J

Jablonski, Nina, 284
Jasmonic acid, 673
Jaws, shape, 369
Jellyfish, 475–476
Jenkin, Fleeming, 327
Jenks, Susan, 397–398
Joints, 695
Joliot, Frédéric, 35
J-shaped curve, 581
Jumping genes. *See* Transposons

K

Kamen, Martin, 140
Keller, Evelyn Fox, 241
Kendrew, John, 201
Keratin, 78–79

Kettlewell, H.B.D, 340
Keystone species, 564
Khorana, Gobind, 220
Khrushchev, Nikita, 325
Kidneys
    dialysis, 782
    failure, 782
    function, 774, 777–781
    hormones, 732, 734
    structure, 777
Kin selection, 609, 610
Kinetics, 93
King, Thomas, 870–871
Kinship, 610
Knee reflexes, *808*
Knockout genes, 272
Knockout mouse, 272
Koch, Robert, 414
Koch's postulates, 414
Korarchaeota, 421
Krebs, Hans, 115
Krebs cycle. *See* Citric acid cycle
Kyoto Protocol, 529

L

Labia minora, 842
Lacks, Henrietta, 148–149
Lac operon, 233–235
Lac repressor, 233
Lactation, 851–852
Lactic acid, 116
Lactose, 235
Lakes, 551–552
Lamarck, Jean-Baptiste, 305–306
Large intestine, 719–720
Larva, 876–877
Lateral geniculate nucleus, 817–818
Lavoisier, Antoine, 108–109, 128–129
LDL. *See* Low-density lipoproteins
Learning, 599, 601–603
Leaves, 632–634
Leclerc, George-Louis, 303
Leder, Philip, 220
Lederberg, Esther, 341
Lederberg, Joshua, 341
Leeches, 485–486
Leeuwenhoek, Antonie van, 64–65,
    68, 412
Lens, 815
Lesch-Nyhan syndrome, 603
Leucine, *58*
Leukocytes
    characterization, 738, 788
    origins, 738–739
    types, 789t
Levi-Montalcini, Rita, 822–823
Lewis, Sinclair, 243
LH. *See* Luteinizing hormone
Libia majora, 844
Lichens, 445–446
Liebig, Justus von, 109

Life
    colonizing earth, 380–381
    expectancy, 585
    extraterrestrial, 381
    formation, steps, *384*
    origins, 378
    water and, 30
Ligaments, 695
Light
    absorption spectrum, 133–134
    atom absorption, *133*
    and ATP production, 130
    climate and, 537–538
    definition, 131
    photosynthesis and, 132–133, 143
    plants response to, 132–133
    ultraviolet, 133
    visible, 131–133
    wave length, 131
Limiting resource, 561
Lingual lipase, 713
Linkage groups, 188–189
Linnaeus, Carolus
    classification system, 314,
        400, 403
    plant organization, 450–453
    role in evolution theory, 302–303
Lipids
    break down of, 117
    cellular movement, 81–82
    characterization, 46, 48–49
    classes, 40, 48, 51
    solubility, 716–717
    soluble hormones, 725
Liver, 715
Lizards, 509
Loams, 639
Lobe-finned fishes, 503
Locomotion, 682–685
Logan, Graham, 486
Loop of Henle, 778–779
Lophotrochozoa, 477–481
Lovelock, James E., 516–517
Low-density lipoproteins, 52
Lumen, 74, 711–712, 744
Lungs
    air path, 761
    artery, 742
    book, 488, 760
    cancer, smoking-related, 757
    gas exchanges, 758–760, 764–766
    oxygen uptake, 759
    semilunar valve, 742
    vein, 742
Luria, Salvador, 206–207
Luteinizing hormone, 731–732
Lycophytes, 459
Lyell, Charles, 305, 307
Lymph nodes, 745–747
Lymphocytes
    characterization, 788